MODERN COST ENGINEERING

Modern Cost Engineering: Methods and Data

Compiled and edited by
Chemical Engineering

CHEMICAL ENGINEERING
McGraw-Hill Publishing Co., New York, N.Y.

Printed in the United States of America

Library of Congress Cataloging in Publication Data
Main entry under title:

Modern cost engineering

A collection of articles which were originally published in Chemical Engineering, 1969–1978.
Includes bibliographical references.
1. Chemical engineering—Costs—Addresses, essays, lectures.
I. Chemical engineering.
TP155. Mb 658.1'55 78-23937
ISBN 0-07-099907-4 (soft cover)
0-07-010733-5 (hard cover)

Dedicated to the memory of
three *Chemical Engineering* editors who contributed much
to cost engineering and its literature.

Cecil H. Chilton
Herbert Popper
Robert B. Norden

CONTENTS

FOREWORD

By the very definition of their profession, most engineers must frequently apply principles of cost engineering. But they face problems in doing so. First, the typical engineer did not get much systematic schooling in that field. Compounding this deficiency, the methodology of cost engineering has deepened in the past few decades, becoming richer and more complex. And there is a more pragmatic difficulty—in many cases, the engineer simply does not have good cost data at hand.

This book aims to help ease those problems by offering its readers a wealth of cost-engineering information that CHEMICAL ENGINEERING has published during approximately the past ten years. It is the successor to two similar volumes, "Cost Engineering in the Process Industries" (1960) and "Modern Cost-Engineering Techniques" (1970).

There are seven sections. The first one presents the fundamentals; it consists mainly of a 20-part series on engineering economics and a 12-part series on how to apply economic principles during the development of capital projects. The second section covers a variety of more-specific topics in economic analysis and evaluation, such as risk analysis, mixed-integer programming, dynamic modeling, and technological forecasting.

In Section III, the emphasis shifts from principles to data, with actual costs given for a wide range of process equipment and other capital items. Section IV takes up cost-engineering topics related to plant construction: the cost of preparing proposals, the types of construction contracts, the economics of plant startup, and similar matters.

The articles in the next section deal with numerous specific economic alternatives, such as how much spare capacity to provide, and how to make tradeoffs between capital and operating costs.

Section VI focuses on operating costs and related topics, including the prediction of selling prices. Finally, Section VII consists of articles that are perhaps especially timely because they take up the cost-engineering aspects of inflation, energy conservation, and environmental control.

Each article includes on its title page the date of original publication. This is especially important information for the articles that offer actual cost data—not only so the engineer can update the figures with appropriate index numbers but, more fundamentally, so that he or she can decide on the kinds of estimates (differing in degree of precision) for which the data can be used with confidence. Section III, on actual cost data, begins with a May 1978 article that details the annual movement of the CE Plant Cost Index and the Marshall and Swift Equipment Cost Index since 1956, thus providing two sets of index numbers. The latest data for these appear in each issue of *Chemical Engineering.*

Nicholas P. Chopey

Section I

FUNDAMENTALS OF ENGINEERING ECONOMICS

Engineering Economics For Chemical Engineers

Systematic accounting is the basis for evaluating plant and material costs so as to provide useful predictions of product and plant profitabilities.

A New CE Refresher Series . . .

Chemical engineers are confronted with the many-faceted problems of inflation, environmental pollution, safety and health legislation, and the energy crisis. They must now examine and carefully evaluate the economic consequences of these problems.

Hence, this new CE Refresher on engineering economics for the chemical engineer. Our authors—all of whom are chemical engineers by training and experience (see biographies on p. 107)—have prepared a comprehensive economics review. This will help you understand the fundamentals of financial mathematics and show how to apply these to economic analyses of plant or product profitability and to capital-cost or manufacturing-cost estimates.

Our authors believe that training in chemical engineering, with its emphasis on flow processes and material and energy balances, will enable chemical engineers to grasp this subject far more readily than will people trained in most other disciplines, including accountancy.

To be covered in this series:

- Costs.
- Traditional methods of estimating profitability.
- Time value of money.
- Estimating profitability from cash flows and discounted-cash flows.
- Sensitivity analysis applied to profitability estimates.
- Comparative profitability of different projects.
- Statistical techniques for decision-making.
- Profitability estimates using probabilities.
- Numerical measures of company attitude to risk.
- Fixed capital-cost estimation.
- Manufacturing-cost estimation.
- Accounting methods, budgetary control, standard costing, inventory valuations.

Editor

F. A. HOLLAND, F. A. WATSON and J. K. WILKINSON
University of Salford

A company wishing to increase its profitability must carefully assess a range of investment opportunities and select the most profitable projects from those available. Further, the increasing competitiveness of modern industry requires that considerable effort be given to reducing costs of existing processes as well as minimizing the costs of future processes. In order to do this, chemical engineers should be fully aware of the relationship between technical and economic factors, in particular those factors that have the largest effect on profitability.

In the past, many engineers have concentrated on the technical side of industry and left the financial side to other specialists. Even design engineers involved in making estimates of the capital and operating costs of process plants have frequently left the overall profitability assessment and investment decision-making to others. This approach is no longer tenable for a company or for an individual desiring to be successful in modern industry.

To an increasing extent, U.S. engineers are becoming concerned with assessing projects in other countries. Therefore, it is essential that they have a firm grasp of the basic principles of engineering economics in order to deal with individual situations. It is also important that engineers know the language of accountancy to facilitate good communications between accountants and themselves.

The first article in this series will give a bird's eye view of the many kinds of costs that are a feature of manufacturing economics. These include such widely different items as fixed capital cost, annual manufacturing cost and total variable cost per unit of production. However, the detailed treatment of these costs will be left to future articles in this series.

Originally published June 1973.

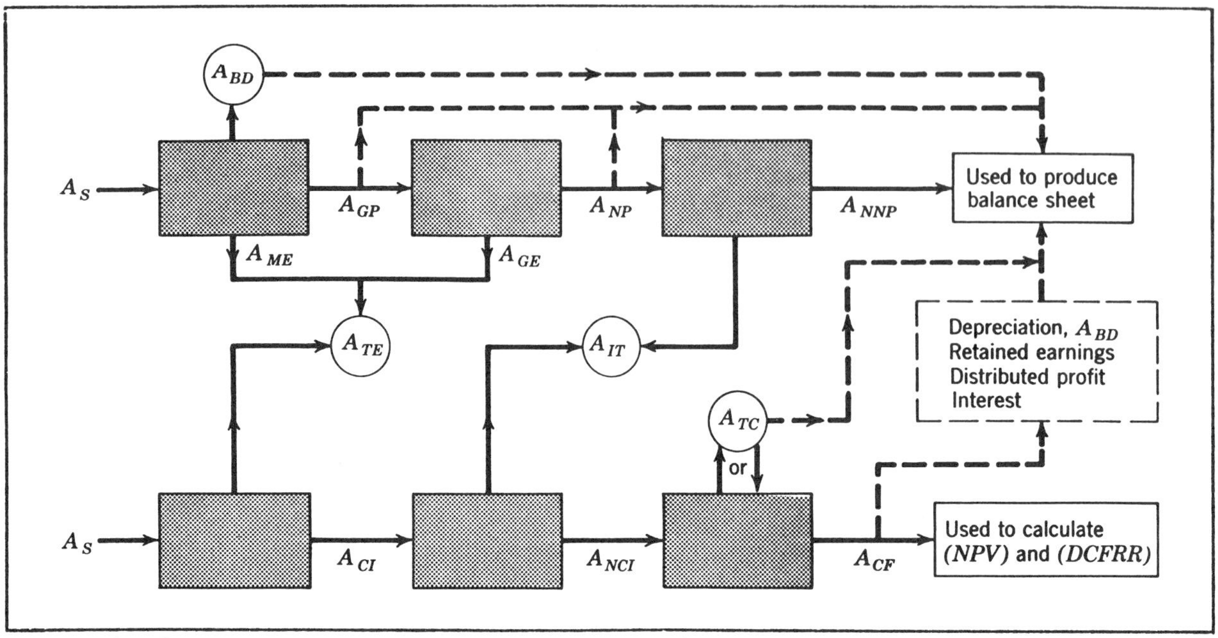

RELATIONSHIPS among annual costs, annual profits and cash flows for a project—Fig. 1

Annual Costs, Profits and Cash Flows

To a large extent, accountancy is concerned with annual costs. To avoid confusion with other costs, annual costs will be referred to by the letter A.

The revenue from the annual sales of a product, A_S, minus the total annual cost or expense required to produce and sell the product, A_{TE}, excluding any annual provision for plant depreciation, is the annual cash income, A_{CI}:

$$A_{CI} = A_S - A_{TE} \tag{1}$$

Net annual cash income, A_{NCI}, is the annual cash income, A_{CI}, minus the annual amount of tax, A_{IT}:

$$A_{NCI} = A_{CI} - A_{IT} \tag{2}$$

Taxable income is $(A_{CI} - A_D - A_A)$, where A_D is the annual depreciation charge, and A_A is the annual amount of any other allowances. Some governments give annual tax allowances, A_A, to encourage companies to build plants in otherwise unattractive areas. Neither A_D nor A_A involves any expenditure of cash, since they are merely book transactions. The annual amount of tax, A_{IT}, is given by:

$$A_{IT} = (A_{CI} - A_D - A_A)t \tag{3}$$

where t is the fractional tax rate. The value of t is determined by the appropriate tax authority and is subject to change. For most countries the value of t is about 0.5, or 50%.

The annual amount of tax, A_{IT}, included in Eq. (2) does not necessarily correspond to the annual cash income, A_{CI}, in the same year. The tax payments in Eq. (2) should be those actually paid in that year. In the U.S., companies pay about 80% of the tax on estimated current-year earnings in the same year. In the U.K., companies do not pay tax until at least nine months after the end of the accounting period which, for the most part, amounts to paying tax on the previous year's earnings. When assessing projects for different countries, engineers should acquaint themselves with the tax situation in those countries.

In modern methods of profitability assessment, cash flows are more meaningful than profits, which tend to be rather loosely defined. The net annual cash flow after tax is given by:

$$A_{CF} = A_{NCI} - A_{TC} \tag{4}$$

where A_{TC} is the annual expenditure of capital, which is not necessarily zero after the plant has been built. For example, plant additions or modifications may be required in future years. Eq. (1), (2), (3) and (4) are illustrated diagrammatically as:

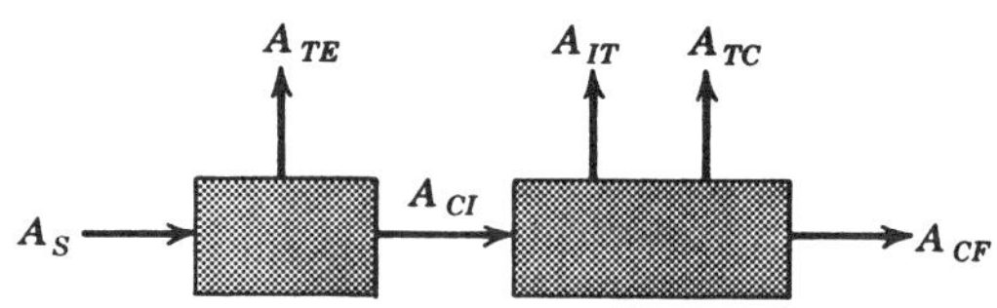

The total annual expense, A_{TE}, required to produce and sell a product can be written as the sum of the annual general expense, A_{GE}, and the annual manufacturing cost or expense, A_{ME}:

$$A_{TE} = A_{GE} + A_{ME} \tag{5}$$

Annual general expense, A_{GE}, arises from the following items: administration, sales, shipping of product, advertising and marketing, technical service, research and development, and finance.

The terms gross annual profit, A_{GP}, and net annual profit, A_{NP}, are commonly used by accountants and

Effect of Production Rate on Net Profit—Table I

Production rate, R, lb./yr.	10,000,000	20,000,000	30,000,000
Selling price, c_S, \$/lb.	0.10	0.10	0.10
Total variable expense (including materials), c_{TVE}, \$/lb.	0.09	0.09	0.09
$R(c_S - c_{TVE})$, \$	100,000	200,000	300,000
Total fixed expense, A_{TFE}, \$/yr.	50,000	50,000	50,000
Depreciation charge, A_{BD}, \$/yr.	50,000	50,000	50,000
Net profit, A_{NP}, \$/yr.	0	100,000	200,000
Net profit, \$/lb.	0	0.005	0.0067

Cost Data and Sales Prices for Problem—Table II

Product Production, units/yr.	A 2,000	B 3,000	C 10,000	D 15,000	E 20,000	F 50,000
Direct materials cost, \$/unit	5.94	7.40	7.00	6.92	7.98	8.75
Direct labor cost, \$/unit	3.20	3.20	3.20	1.60	1.60	1.00
Variable overhead cost, \$/unit	0.74	1.16	3.44	1.86	4.59	1.72
Fixed overhead cost, \$/unit	5.33	3.74	3.60	1.52	1.58	0.53
Sales price, \$/unit	14.11	14.70	17.00	11.80	15.75	12.20

misused by lots of other people. Normally, both A_{GP} and A_{NP} are calculated before tax is deducted.

Gross annual profit, A_{GP}, is given by:

$$A_{GP} = A_S - A_{ME} - A_{BD} \tag{6}$$

where A_{BD} is the balance-sheet annual depreciation charge, which is not necessarily the same as A_D used in Eq. (3) for tax purposes. For example, at the present time in the U.K., the entire fixed capital expenditure, C_{FC}, may be taken as a tax allowance in the first year of plant operation. In this case, A_D in the first year of operation may be ten times greater than the balance-sheet value, A_{BD}. Net annual profit, A_{NP}, is simply:

$$A_{NP} = A_{GP} - A_{GE} \tag{7}$$

Eq. (7) can also be written as:

$$A_{NP} = A_{CI} - A_{BD} \tag{8}$$

Net annual profit after tax A_{NNP} can be written as:

$$A_{NNP} = A_{NCI} - A_{BD} \tag{9}$$

The relationships among the various annual costs given by Eq. (1) through (9) are illustrated diagrammatically in Fig. 1. The top half of the diagram shows the tools of the accountant, the bottom half those of the engineer. The net annual cash flow, A_{CF}, which excludes any provision for depreciation is used in two of the more modern methods of profitability assessment; the net present value (*NPV*) method and the discounted-cash-flow rate of return (*DCFRR*) method. In both methods, depreciation is inherently taken care of in calculations that include capital recovery. These methods will be described in a future article.

Costs for a Manufactured Product

Annual manufacturing cost or expense, A_{ME}, can be written as the sum of the direct and indirect manufacturing costs or expenses, A_{DME} and A_{IME}:

$$A_{ME} = A_{DME} + A_{IME} \tag{10}$$

Direct manufacturing cost includes expenses incurred directly from the production operation. These expenses are: raw materials (including delivery), catalysts and solvents, operating labor, operating supervision, utilities, operating maintenance and repairs, operating supplies, royalties and patents.

Indirect manufacturing cost includes expenses incurred as a result of, but not directly from, the production operation. These expenses arise from: payroll overhead, control laboratory, general plant overhead, packaging, and storage facilities. Other indirect costs include those that are a function of the fixed capital investment. These arise from property taxes, rent and insurance.

Annual general expense, A_{GE}, can be written as the sum of the fixed and variable general expenses:

$$A_{GE} = A_{FGE} + A_{VGE} \tag{11}$$

Similarly, annual manufacturing expense, A_{ME}, can be written as the sum of the fixed and variable manufacturing expenses:

$$A_{ME} = A_{FME} + A_{VME} \tag{12}$$

Contribution and Breakeven Production Rate

Eq. (1), which gives the annual cash income, can also be written in terms of Eq. (5), (11) and (12), either as:

$$A_{CI} = [A_S - (A_{VGE} + A_{VME})] - (A_{FGE} + A_{FME}) \tag{13}$$

or as:

$$A_{CI} = A_S - A_{TVE} - A_{TFE} \tag{14}$$

where A_{TFE} and A_{TVE} are the total annual fixed and variable expenses, respectively.

If the annual sales volume can be taken as equal to the annual production rate, R, the revenue from annual sales A_S is the product of R and the sales price per unit of production, c_S:

$$A_S = Rc_S \tag{15}$$

For the most part, the total annual variable expense, A_{TVE}, can be taken as proportional to the annual production rate, R, and the total variable expense per unit of production, c_{TVE}:

$$A_{TVE} = Rc_{TVE} \tag{16}$$

Substitute Eq. (15) and (16) into Eq. (14) to give:

$$A_{CI} = R(c_S - c_{TVE}) - A_{TFE} \tag{17}$$

Eq. (17) gives the annual cash income in terms of the annual production rate. The term $(c_S - c_{TVE})$ is called

Existing Performance of Multiple-Product Plant—Table III

Product Production, units/yr.	A 2,000	B 3,000	C 10,000	D 15,000	E 20,000	F 50,000	Annual Totals 100,000
Direct materials cost, \$/unit	5.94	7.40	7.00	6.92	7.98	8.75	\$ 804,980
Direct labor cost, \$/unit	3.20	3.20	3.20	1.60	1.60	1.00	154,000
Variable overhead cost, \$/unit	0.74	1.16	3.44	1.86	4.59	1.72	245,060
Fixed overhead cost, \$/unit	5.33	3.74	3.60	1.52	1.58	0.53	138,780
Manufacturing cost, \$/unit	15.21	15.50	17.24	11.90	15.75	12.00	1,342,820
Selling price, \$/unit	14.11	14.70	17.00	11.80	15.75	12.20	1,344,320
Profit/(loss), \$/unit	(1.10)	(0.80)	(0.24)	(0.10)	0.00	0.20	\$ 1,500
Total variable cost, \$/unit	9.88	11.76	13.64	10.38	14.17	11.47	
Contribution, \$/unit	4.23	2.94	3.36	1.42	1.58	0.73	
(*CSR*)	0.30	0.20	0.20	0.12	0.10	0.06	
Profitability ranking	1	2	2	4	5	6	

the contribution to cash income per unit of production.

Separation of costs into fixed and variable components is popular with accountants. However, there are differences of opinion as to exactly which items should be included in the fixed and variable costs. Royalties and patents may be either fixed or variable costs, depending upon the way in which the agreements are written. An increasing number of companies regard labor as a fixed rather than a variable expense. Operators cannot conveniently be transferred to other duties when a plant is temporarily shut down or is operating at a reduced capacity. Similarly, there are differences of opinion as to which items should be direct costs and which should be indirect costs.

Eq. (17) can be combined with Eq. (8) to give:

$$A_{NP} = R(c_S - c_{TVE}) - A_{TFE} - A_{BD} \qquad (18)$$

Eq. (18) can also be written:

$$A_{NP} = R(c_S - c_{TVE}) - A_{FGE} - A_{FME} - A_{BD} \qquad (19)$$

Eq. (19) gives the net annual profit before tax, A_{NP}, in terms of the annual fixed general expense, A_{FGE}, the annual fixed manufacturing expense, A_{FME}, and the balance-sheet annual depreciation charge, A_{BD}.

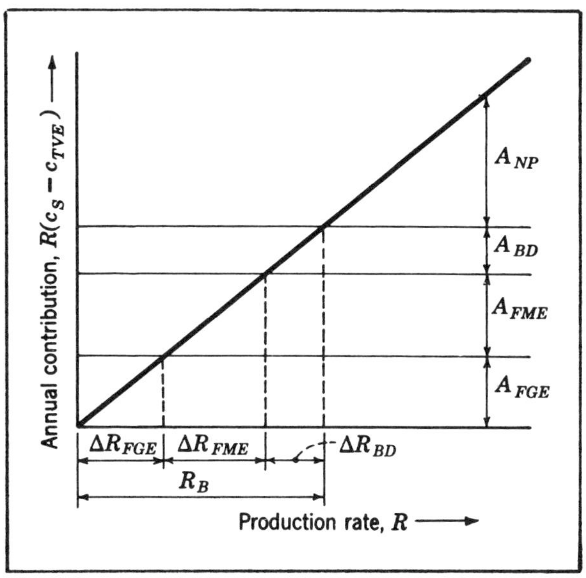

EFFECT of production rate on net profit—Fig. 2

The difference between the sales price, c_S, and the total variable cost per unit of production, c_{TVE}, represents a contribution to the fixed expenses, A_{FGE}, A_{FME}, and A_{BD}, and to the net annual profit, A_{NP}. The term $(c_S - c_{TVE})$ is the contribution to net profit per unit of production, and $R(c_S - c_{TVE})$ is the annual contribution for a production rate, R.

Eq. (19) shows that the net annual profit is positive when $R(c_S - c_{TVE})$ is more than $(A_{FGE} + A_{FME} + A_{BD})$, and negative when $R(c_S - c_{TVE})$ is less than $(A_{FGE} + A_{FME} + A_{BD})$. The net annual profit is zero at an annual production rate, R_B, given by the following equation:

$$R_B = \frac{A_{FGE} + A_{FME} + A_{BD}}{c_S - c_{TVE}} \qquad (20)$$

where R_B is the breakeven production rate. R_B can be written in terms of its component parts as:

$$R_B = \Delta R_{FGE} + \Delta R_{FME} + \Delta R_{BD} \qquad (21)$$

$$\Delta R_{FGE} = \frac{A_{FGE}}{c_S - c_{TVE}} \qquad (22)$$

$$\Delta R_{FME} = \frac{A_{FME}}{c_S - c_{TVE}} \qquad (23)$$

$$\Delta R_{BD} = \frac{A_{BD}}{c_S - c_{TVE}} \qquad (24)$$

ΔR_{FGE}, ΔR_{FME} and ΔR_{BD} are the increments of production rate required to cover the annual fixed general expense, annual fixed manufacturing expense, and the balance-sheet annual depreciation charge, respectively.

Combine Eq. (19) and (20) to give:

$$A_{NP} = (R - R_B)(c_S - c_{TVE}) \qquad (25)$$

Eq. (25) shows that above the breakeven production rate, R_B, the contribution is all net profit.

Eq. (19) is illustrated in Fig. 2. A decrease in selling price, c_S, will decrease the slope of the line in Fig. 2, and increase the required breakeven production rate, R_B. The effect of production rate on net profit is also illustrated in Table I.

Anticipated Performance of Multiple-Product Plant—Table IV

Product Production, units/yr.	F 2,000	E 3,000	D 10,000	C 15,000	B 20,000	A 50,000	Annual Totals 100,000
Direct materials cost, $/unit	8.75	7.98	6.92	7.00	7.40	5.94	$ 660,640
Direct labor cost, $/unit	1.00	1.60	1.60	3.20	3.20	3.20	294,800
Variable overhead cost, $/unit	1.72	4.59	1.86	3.44	1.16	0.74	147,610
Fixed overhead cost, $/unit	5.33	3.74	3.60	1.52	1.58	0.53	138,780
Manufacturing cost, $/unit	16.80	17.91	13.98	15.16	13.34	10.41	1,241,830
Selling price, $/unit	12.20	15.75	11.80	17.00	14.70	14.11	1,444,150
Profit/(loss), $/unit	(4.60)	(2.16)	(2.18)	1.84	1.36	3.70	$ 202,320
Total variable cost, $/unit	11.47	14.17	10.38	13.64	11.76	9.88	
Contribution, $/unit	0.73	1.58	1.42	3.36	2.94	4.23	
(*CSR*)	0.06	0.10	0.12	0.20	0.20	0.30	
Profitability ranking	6	5	4	3	2	1	

A consideration of contribution is extremely helpful in analyzing cost data. Because of the fixed components in net profit, a solution that maximizes contribution also maximizes profit. This is useful when considering whether to add or delete a particular product from a company's product range. The contribution to sales price ratio (*CSR*) is also used to analyze this kind of problem* where:

$$(CSR) = (c_S - c_{TVE})/c_S = 1 - (c_{TVE}/c_S) \qquad (26)$$

The most profitable product is the one with the highest (*CSR*). This parameter is useful for ranking products.

Example Illustrates Procedures

A company is committed to producing a range of six products, details of which are given in Table II. It is possible to adapt any of the production lines to produce the present number of units of any of the products. Compare the present cash-flow pattern with the best production schedule that could be obtained from the existing equipment, using the contribution-to-sales ratio (*CSR*) as the measure of profitability. Assume that the fixed overhead costs relate to the production line and not to the product made.

The present performance is shown in Table III. The anticipated performance if the production rates are ranked in order of (*CSR*) is shown in Table IV. Since Products B and C each have a (*CSR*) of 0.20, preference has been given to B which, although showing the smaller contribution, also has the lower total variable cost. This will minimize the working capital required to obtain the reward.

If Products B and C are interchanged (i.e., make 15,000 units/yr. of Product B and 20,000 of C), the manufacturing cost becomes $1,251,230, and the profit is $204,420.

The total manufacturing profit according to Table IV as a percentage of the annual manufacturing cost is:

$$\frac{\$202,320}{\$1,241,830} = 16.3\%$$

The marginal improvement obtained by interchanging Products B and C as a percentage of the additional manufacturing cost is calculated from the following:

$$\frac{\$204,420 - \$202,320}{\$1,251,230 - \$1,241,830} = \frac{\$2,100}{\$9,400} = 22.3\%$$

The latter strategy is the correct one to use. Therefore, the company should attempt to boost sales of Products A, C and B (at unit/yr. sales of 50,000, 20,000 and 15,000, respectively) to the maximum of production potential, while discouraging sales of Products D, E and F. It may be possible to use some of the savings in working capital that should ensue to increase production of F in the 2,000-unit line and so reduce the large unit loss as indicated in Table IV. ■

*Wright, M. G., "Financial Management," McGraw-Hill, London, 1970.

Meet the Authors

F. A. Holland

F. A. Watson

J. K. Wilkinson

F. A. Holland is professor and chairman of the department of chemical engineering at the University of Salford, Salford M5 4WT, England, and a partner in Salchem Associates, Consulting Chemical Engineers. Trained both as a physicist and a chemical engineer, Holland spent 14 years in the chemical process industries, nine of which were in the U.S. He has held management positions in production, process design, construction, and research and development. He has degrees from the universities of Oxford, London and Durham, and is a Fellow of the Institution of Chemical Engineers. He is also the author or coauthor of five books and more than fifty technical papers and articles.

F. A. Watson is a senior lecturer in the department of chemical engineering at the University of Salford and a partner in Salchem Associates. He has occupied posts in production, research and development with companies engaged in the manufacture of liquid and gaseous fuels and coal-tar byproducts. His primary interests are process design, digital computation and accountancy. He graduated from Manchester University in chemical engineering, and is a Fellow of the Institution of Chemical Engineers. He is also the coauthor of the book "Heat Transfer" (Heinemann, London, 1970—American Elsevier, New York).

J. K. Wilkinson is a senior lecturer in the department of chemical engineering at the University of Salford and a partner in Salchem Associates. He has held posts in the chemical industry in production, research and development. His main interests are in process design, financial accountancy and investment analysis. He holds a bachelor's and master's degree in chemical engineering from Manchester University, and a bachelor's degree in economics from London University, and is a Fellow of the Institution of Chemical Engineers. He is coauthor of "Heat Transfer."

Capital Costs And Depreciation

How to differentiate costs that comprise fixed capital and working capital of an enterprise, and how to understand the functions of depreciation for fixed assets.

F. A. HOLLAND, F. A. WATSON and J. K. WILKINSON, University of Salford

Capital represents the investment made to build and operate the process plant in order to manufacture a specific product. The several parameters that constitute this investment must be incorporated in an orderly manner so that plant or product profitabilities can be computed and analyzed.

Profits, however, are subject to taxes, and the amount of tax must reflect the amounts allowed for depreciation. There are several techniques used for calculating depreciation and establishing depreciation rates.

Let us now examine in detail the nature and methods for establishing and calculating capital costs and depreciation.

Capital Costs

The total capital cost, C_{TC}, of a project consists of the fixed capital cost, C_{FC}, plus the working capital cost, C_{WC}, plus the cost of land and other nondepreciable costs, C_L:

$$C_{TC} = C_{FC} + C_{WC} + C_L \tag{1}$$

The project may be a complete plant, an addition to an existing plant, or a plant modification.

The fixed capital cost includes the items listed in Table I.[1-3] Some companies make an allowance for the escalation of equipment prices, and unforeseen engineering and process-design charges.

The working capital cost of a process or business normally includes the items shown in Table II. Since working capital is completely recoverable at any time, in theory if not in practice, no tax allowance is made for depreciation. The same is also true of the value of the land occupied by the plant.

Working capital may vary from a very small fraction of the total capital cost to almost the whole of the invested capital, depending on the process and the industry. For example, in jewelry-store operations, the fixed capital is very small in comparison with the working capital. On the other hand, in the chemical process industries, land costs are normally a small fraction of the total investment while the working capital is likely to be on the order of 10 to 20% of the value of the fixed capital investment.

Bechtel[4] gave the following three methods for estimating the required working capital:

1. Compute inventory costs on the basis of one month's

Fixed Capital Costs—Table I

- ☐ Major process equipment (i.e., reactors, tanks, pumps, filters, distillation columns, etc.).
- ☐ Installation of major process equipment.
- ☐ Process piping.
- ☐ Insulation.
- ☐ Instrumentation.
- ☐ Auxiliary facilities (i.e., power substations, transformers, boiler houses, fire-control equipment, etc.).
- ☐ Outside lines (i.e., piping external to buildings, supports and posts for overhead piping, electric feeders from power substations).
- ☐ Land and site improvements.
- ☐ Buildings and structures.
- ☐ Engineering and construction (design and engineering fees plus supervision of plant erection).
- ☐ Contractors' fees.

Originally published July 1973.

supply of raw material, one month's supply of semi-finished material (based on raw material plus one-half the conversion cost), and one month's supply of finished material. Add accounts receivable, valued at one month's sales. Also, add one month's operating cash.

2. Take 30% of annual sales.
3. Take 15% of the fixed capital investment.

The three methods are listed in order of decreasing accuracy. However, where seasonal markets exist, Method 1 is subject to error. Bechtel cited a project in which the rate of return was 32.5%, excluding working capital. The rate of return with working capital included was 27.0% using Method 1, 25.8% using Method 2, and 28.5% using Method 3.

An estimated profit-and-loss statement for a new chemical, showing the rate of return on total capital used, is given in Table III.

Working Capital Costs—Table II

- ☐ Raw materials for plant startup.
- ☐ Raw-material, intermediate and finished-product inventories.
- ☐ Cost of handling and transportation of materials to and from stores.
- ☐ Cost of inventory control, warehouse, associated insurance, security arrangements, etc.
- ☐ Money to carry accounts receivable (i.e., credit extended to customers) less accounts payable (i.e., credit extended by suppliers).
- ☐ Money to meet payrolls when starting up.
- ☐ Readily available cash for emergencies.
- ☐ Any additional cash required to operate the process or business.

Depreciation

The term depreciation is used in a number of different contexts. The most common are:

1. A tax allowance.
2. A cost of operation.
3. A means of building up a fund to finance plant replacement.
4. A measure of falling value.

In the first case, the annual taxable income is reduced by an annual depreciation charge or allowance that reduces the annual amount of tax payable. The annual depreciation charge is merely a book transaction and does not involve any expenditure of cash. The method of determining the annual depreciation charge must be agreed to by the appropriate tax authority.

In the second case, depreciation is considered to be a manufacturing cost in the same way as labor cost or raw-materials cost. However, it is more difficult to estimate a depreciation cost per unit of product than it is to do so for labor or raw-materials costs. In the net present value (NPV) and discounted-cash-flow rate-of-return ($DCFRR$) methods of measuring profitability (to be discussed in future articles of this series), depreciation as a cost of operation is implicitly accounted for. (NPV) and ($DCFRR$) give measures of return after a project has generated sufficient money to recover the original investment.

In the third case, depreciation is considered as a means of providing for plant replacement. However, in the rapidly changing modern chemical process industries, it is doubtful that many plants will ever be replaced because the processes are likely to have become obsolete during operation. Furthermore, management must be free to invest its available capital in the currently most profitable projects. However, it may wish to regard a proportion of retained profits as a fund to finance new capital projects. For the most part, these are likely to differ substantially from the projects that originally generated the profits.

In the fourth case, a plant or a piece of equipment has a limited useful life. The primary reason for the decrease in value is the decrease in future life, and the consequent decrease in the number of years for which income will be earned. At the end of its life, the equipment may be worth nothing or it may have a salvage or scrap value, S. Thus, a plant of fixed capital cost, C_{FC}, depreciates in value during its useful life of s years by an amount that is equal to $(C_{FC} - S)$. The useful life is taken from the startup of the plant.

On the basis of straight-line depreciation, the average annual amount of depreciation, A_D, is given by:

$$A_D = (C_{FC} - S)/s \tag{2}$$

Book value after the first year, P_1, is given by:

$$P_1 = C_{FC} - A_D \tag{3}$$

The book value at the end of a particular year, s', is given by:

$$P_{s'} = C_{FC} - s'A_D \tag{4}$$

The principal significance of a particular depreciation rate is for tax purposes. The annual depreciation is subtracted from the annual profit before the latter is taxed. The basis for depreciation in a particular case is a matter for agreement between the taxation authority and the company, in conformity with the tax laws.

Other commonly used methods of computing depreciation are (1) the declining balance method (also known as the fixed percentage method), and (2) the sum-of-years-digits method.

On the basis of declining balance (fixed percentage) depreciation, the book value at the end of the first year is given by:

$$P_1 = C_{FC}(1 - r) \tag{5}$$

where r is a fraction to be agreed to by the taxation authority.

The book value at the end of a particular year, s', is given by:

$$P_{s'} = C_{FC}(1 - r)^{s'} \tag{6}$$

When the fraction r is chosen to be $2/s$, i.e., twice the reciprocal of the service life, s, the method is called the double-declining-balance method.

The declining balance method of depreciation allows equipment or plant to be depreciated by a greater amount

Statement of Profit and Loss for a Mature Year—Table III
(Basis: New chemical product, 10 million lb./yr.)

	$	$	Unit Values ¢/Lb.	% Sales Revenue %
Revenue from annual sales, A_S		2,000,000	20.00	100.0
Direct manufacturing expense, A_{DME}				
Raw materials	884,000		8.84	44.2
Catalysts and solvents	69,000		0.69	3.4
Operating labor	102,000		1.02	5.1
Operating supervision	20,000		0.20	1.0
Utilities	22,000		0.22	1.1
Operating maintenance	21,000		0.21	1.1
Operating supplies	4,000		0.04	0.2
Royalties and patents	10,000		0.10	0.5
Total A_{DME}	1,132,000	1,132,000	11.32	56.6
Indirect manufacturing expense, A_{IME}				
Payroll overhead	28,000		0.28	1.4
Central laboratory	10,000		0.10	0.5
General plant overhead	52,000		0.52	2.6
Packaging and storage	22,000		0.22	1.1
Property taxes	14,000		0.14	0.7
Insurance	6,000		0.06	0.3
Total A_{IME}	132,000	132,000	1.32	6.6
Total manufacturing expense (excluding depreciation), A_{ME}		1,264,000	12.64	63.2
Depreciation, A_{BD}		68,000	0.68	3.4
Other expenses				
Administration	74,000		0.74	3.7
Sales and shipping	124,000		1.24	6.2
Advertising and marketing	40,000		0.40	2.0
Technical service	10,000		0.10	0.5
Research and development	60,000		0.60	3.0
Total other expenses	308,000	308,000	3.08	15.4
Total expense, A_{TE}		1,640,000	16.40	82.0
Net annual profit, A_{NP}		360,000	3.60	18.0

Computation for rate of return on total capital

		$
Net annual profit, A_{NP}		360,000
Less taxes at 50%		180,000
Net annual profit after taxes, A_{NNP}		180,000
Total capital, C_{TC}		990,000
Fixed capital, C_{FC}	700,000	
Working capital, C_{WC}	290,000	

After-tax rate of return, $i = A_{NNP}/C_{TC} = 18.2\%$

during the early years than during the later years. This method does not allow equipment or plant to be depreciated to a zero value at the end of the service life.

On the basis of sum-of-years-digits depreciation, the annual amount of depreciation for a particular year, s', for a plant of fixed capital cost, C_{FC}, and scrap value, S, is given by:

$$A_{Ds'} = \left(\frac{s - s' + 1}{1 + 2 + 3 + \ldots + s}\right)(C_{FC} - S) \tag{7}$$

Eq. (7) can also be rewritten in the form:

$$A_{Ds'} = \left(\frac{2(s - s' + 1)}{s(s + 1)}\right)(C_{FC} - S) \tag{8}$$

It can be shown that the book value at the end of a particular year, s', is:

$$P_{s'} = \frac{2[1 + 2 + \ldots + (s - s')]}{s(s + 1)}(C_{FC} - S) + S \tag{9}$$

The sum-of-years-digits depreciation also allows equipment or plant to be depreciated by a greater amount during the early years than during the later years.

A fourth method of computing depreciation—now seldom used—is the sinking fund method. In this method, the annual depreciation, A_D, is the same for each year of the life of the equipment or plant. The series of equal amounts of depreciation, A_D, invested at a fractional interest rate, i, and made at the end of each year over the life of the equipment or plant of s years, is used to build up a future sum of money equal to $(C_{FC} - S)$. This last is the fixed capital cost of the equipment or plant minus its salvage or scrap value, and is the total amount of depreciation during its useful life. The equation relating $(C_{FC} - S)$ and A_D is simply the annual cost or payment equation, written either as:

$$C_{FC} - S = A_D\left[\frac{(1 + i)^s - 1}{i}\right] \tag{10}$$

or:

$$C_{FC} - S = A_D/f_{AF} \tag{11}$$

where f_{AF} is the annuity future-worth factor given by:

$$f_{AF} = \frac{i}{(1 + i)^s - 1}$$

Values of the future-worth factor, f_{AF}, are readily available in tables.[5-9]

In the sinking fund method of depreciation, the effect of interest is to make the annual decrease of the book value of the equipment or plant less in the early than in the later years.

However, it is preferable not to think of annual depreciation as a contribution to a fund to replace equipment at the end of its life but as that part of the difference between the revenue and the expenditure, which difference is tax free.

Some of the preceding methods of computing depreciation are not allowed by the taxation authorities in certain countries. When calculating depreciation, it is necessary to obtain details of the methods and rates permitted by the appropriate authority, and to use the information provided.

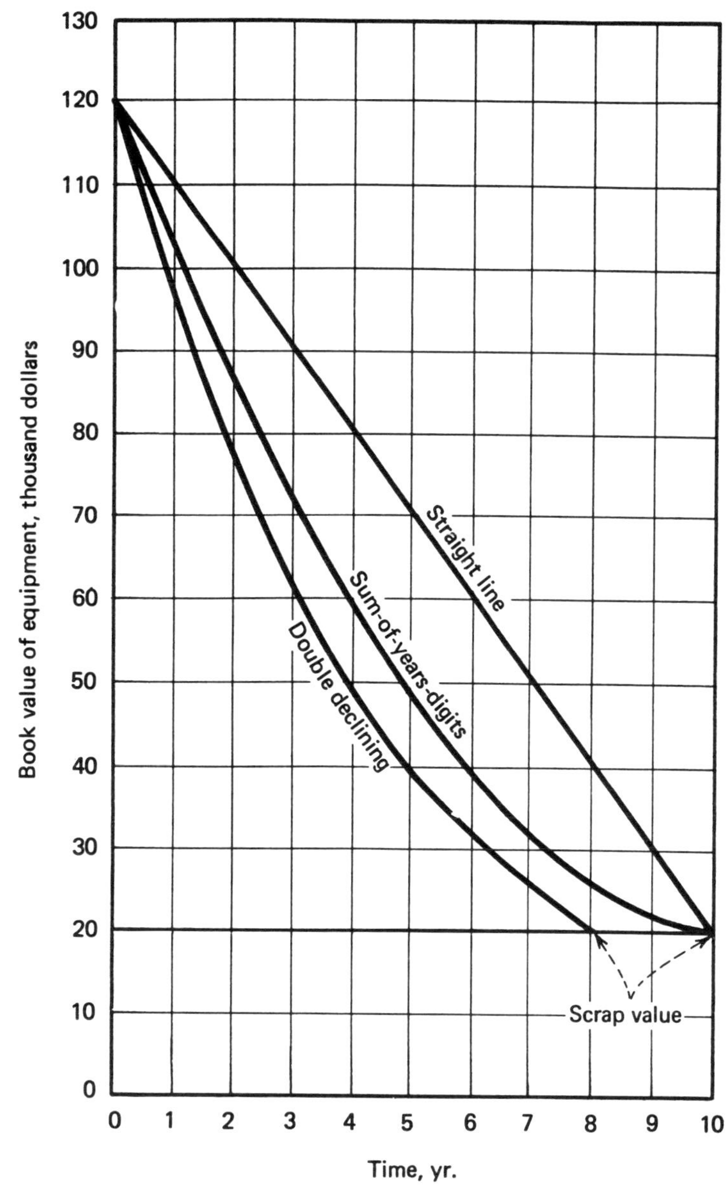

DEPRECIATION method influences the book value—Fig. 1

Fig. 1 shows the fall in book value with time for a piece of equipment having a fixed capital cost of $120,000, a useful life of 10 yr., and a scrap value of $20,000. This fall in value is calculated by using (a) straight-line depreciation, (b) double-declining depreciation, and (c) sum-of-years-digits depreciation. ■

References

1. Aries, R. S. and Newton, R. D., "Chemical Engineering Cost Estimation," p. 1, McGraw-Hill, New York, 1955.
2. Bauman, H. C., "Fundamentals of Cost Engineering in the Chemical Industry," p. 27, Reinhold, New York, 1964.
3. Peters, M. S. and Timmerhaus, K. D., "Plant Design and Economics for Chemical Engineers," 2nd ed., McGraw-Hill, New York, 1968.
4. Bechtel, L. R., *Chem. Eng.*, Feb. 22, 1960, pp. 127–128.
5. Baumeister, T. (Ed.), "Standard Handbook for Mechanical Engineers," 2nd ed., p. 1-67, McGraw-Hill, New York, 1967.
6. Perry, J. H. (Ed.), "Chemical Business Handbook," p. 20-170, McGraw-Hill, New York, 1954.
7. Perry, J. H., Chilton, C. H. and Kirkpatrick, S. D. (Ed.), "Chemical Engineers' Handbook," 4th ed., pp. 1-32 to 1-37, McGraw-Hill, New York, 1963.
8. Perry, J. H. (Ed.), "Engineering Manual," 2nd ed., pp. 1-54 to 1-59, McGraw-Hill, New York, 1967.
9. Souders, M., "The Engineers' Companion," p. 278, Wiley, New York, 1966.

Engineering Economics—Part 3

Profitability of Invested Capital

Rate-of-return methods and the payback period are traditional techniques for estimating profitability of a process, project or plant. All have the advantage of simplicity but they can yield very misleading results.

F. A. HOLLAND, F. A. WATSON and J. K. WILKINSON, University of Salford

It sometimes comes as a shock to engineers—skilled as they are in the use of modern methods for estimating profitability—to find that top-management people do not always understand the net present value (*NPV*) and discounted-cash-flow rate-of-return (*DCFRR*) methods.

Although (*NPV*) and (*DCFRR*) methods are widely used in large companies to help decide among investment opportunities, the day-to-day running of the business and budgetary control are still based on conventional accountancy practices. Thus, it is difficult to check whether past (*NPV*) estimates have proved correct or not. This may be one of the reasons why many accountants dispute the value of (*NPV*) and (*DCFRR*) methods. In fact, in some small companies, accountants still rely entirely on traditional rate-of-return methods for estimating profitability. This approach, however, can lead to both friction and confusion between engineers and accountants.

The engineer should be fully aware that he is rarely playing on his own ground. Almost invariably, he is treading on the accountant's ground. It is unusual for an accountant to talk or think in engineering terms. Since an engineer must talk to accountants if he wishes to fully appreciate the relevant business information available in his company, it follows that he must talk the accountant's language. This is important in that top management tends to be geared to accountancy thinking rather than engineering thinking.

Originally published August 1973.

Many engineers get impatient with accountants and regard their own approach as being much more flexible. Although this may well be true, it should be remembered that accountants are largely restricted to conventional and traditional practices. Many leading accountants are trying to change these conventions but since accounting practices have to be generally acceptable to be of value, the changes are slow in coming.

For these reasons, we shall take a critical look at some of the traditional methods for estimating profitability. Although these methods have the advantage of simplicity, they can yield very misleading results. This is particularly true for rate-of-return methods.

Rate-of-return methods are based on the relation:

$$PERCENT\ RATE\ OF\ RETURN = \left(\frac{ANNUAL\ PROFIT}{INVESTED\ CAPITAL}\right)100 \qquad (1)$$

Many people would be quite happy to be told that the rate of return was, say 25%, as calculated from Eq. (1), without their examining too closely what is meant either by annual profit or by invested capital. Few people would ask exactly what costs were included or excluded in the statement of annual profit. For example, the 25% return might include, say, 10% or 20% as the cost of depreciation. There is great scope in this one item alone for making the profits look high or low, depending on the aim of the estimates. This problem does not arise with modern methods for estimating profitability such as the (*NPV*) and ($DCFRR$) methods. In both of these (to be described in future articles in this series), depreciation is inherently taken care of in the calculations, which include capital recovery.

It is important to consider very carefully what is meant by annual profit and invested capital when Eq. (1) is used.

Different Kinds of Annual Profit

The quantities gross annual profit, A_{GP}, and net annual profit, A_{NP}, are normally calculated by accountants before tax is deducted. The gross annual profit is given by:

$$A_{GP} = A_S - A_{ME} - A_{BD} \tag{2}$$

where A_S is the revenue from annual sales, A_{ME} is the annual manufacturing cost or expense (excluding any annual provision for plant depreciation), and A_{BD} is the balance-sheet annual depreciation charge, which is not necessarily the same as the annual depreciation allowance used to compute the annual amount of tax, A_{IT}. Misunderstandings can arise from the fact that although A_S and A_{ME} are actual cash flows, A_{BD} is merely a book value and not an actual cash flow. However, this in no way decreases the importance of depreciation as a cost of plant operation.

Net annual profit, A_{NP}, is simply:

$$A_{NP} = A_{GP} - A_{GE} \tag{3}$$

where the annual general expense, A_{GE}, arises from the following expenses: administration, sales, shipping of product, advertising and marketing, technical service, research and development, and finance.

The total annual expense, A_{TE}, required to produce and sell a product can be written as the sum:

$$A_{TE} = A_{GE} + A_{ME} \tag{4}$$

The difference between the revenue from annual sales, A_S, and A_{TE} is the annual cash income, A_{CI}:

$$A_{CI} = A_S - A_{TE} \tag{5}$$

Net annual cash income, A_{NCI}, is:

$$A_{NCI} = A_{CI} - A_{IT} \tag{6}$$

Unlike Eq. (2) and (3), Eq. (5) and (6) contain actual cash flows only and not book values. In a sense, Eq. (2) and (3) are analogous to dimensionally inconsistent equations since they involve subtracting terms of differing types.

Eq. (2), (3), (4) and (5) can be combined to provide the net annual profit, A_{NP}, in terms of the annual cash income, A_{CI}, and the balance-sheet annual depreciation charge, A_{BD}:

$$A_{NP} = A_{CI} - A_{BD} \tag{7}$$

Eq. (6) and (7) can also be combined to give the net annual profit (after tax):

$$A_{NNP} = A_{NCI} - A_{BD} \tag{8}$$

Annual cash flows, A_{CI} and A_{NCI}, can be pinpointed much more precisely than annual profits A_{GP}, A_{NP} and A_{NNP}.

Invested Capital

The total capital cost, C_{TC}, of a project consists of the fixed capital cost, C_{FC}, plus the working capital cost, C_{WC}, plus the cost of land and other nondepreciable costs, C_L:

$$C_{TC} = C_{FC} + C_{WC} + C_L \tag{9}$$

Invested capital can be taken as the original total capital cost or value, the depreciated capital value, the average capital value, or the current capital-replacement value. Depreciated value depends on the arbitrarily chosen method of depreciation. All other methods except the first depend on the valuation of the inventory, which will be dealt with in a future article.

Rate of Return on Original Investment

There are a number of ways of assessing the rate of return on the original total capital investment. This can be done on the basis of (1) net annual profit before tax, A_{NP}, (2) net annual profit after tax, A_{NNP}, (3) annual cash income before tax, A_{CI}, or (4) annual cash income after tax, A_{NCI}.

In the first case, the fractional interest rate-of-return, i, on C_{TC} is:

$$i = A_{NP}/C_{TC} \tag{10}$$

Eq. (10) can also be written in terms of Eq. (7), as:

$$i = \frac{A_{CI}}{C_{TC}} - \frac{A_{BD}}{C_{TC}} \tag{11}$$

where A_{BD} is the balance-sheet annual depreciation.

In the second case, the rate of return is:

$$i = A_{NNP}/C_{TC} \tag{12}$$

Eq. (12) can also be written in terms of Eq. (8):

$$i = \frac{A_{NCI}}{C_{TC}} - \frac{A_{BD}}{C_{TC}} \tag{13}$$

In the third case, the rate of return is:

$$i = A_{CI}/C_{TC} \tag{14}$$

In the fourth case, the rate of return is:

$$i = A_{NCI}/C_{TC} \tag{15}$$

The main disadvantage of using Eq. (10), (11), (12) and (13) is that the fractional depreciation rate, A_{BD}/C_{TC}, is arbitrarily assessed. Its value will affect the fractional rate-of-return considerably, and may lead to erroneous conclusions when making comparisons between different companies. This is particularly true when making international comparisons.

The rate of return given by Eq. (15) is recommended in preference to the other equations. It is then quite easy to deduct any particular value of the fractional depreciation rate from the value of i given by Eq. (15) in order to get the rate of return based on the net annual profit after tax.

In Eq. (15), i can either be taken on the basis of the net annual cash income for a particular year, or on the basis of an average net annual cash income over the length of life of the project.

Rate of Return on Depreciated Investment

The rate of return on the depreciated investment can be determined on the basis of A_{NP}, A_{NNP}, A_{CI} or A_{NCI}, as previously described.

For the second case, the fractional interest rate-of-return on the depreciated total capital investment for a particular year is:

$$i = \frac{A_{NNP}}{P_{s'} + C_{WC} + C_L} \tag{16}$$

where $P_{s'}$ is the book value of the fixed capital investment at the end of a particular year, s'.

Eq. (16) can be written in terms of Eq. (8) as:

$$i = \frac{A_{NCI} - A_{BD}}{P_{s'} + C_{WC} + C_L} \tag{17}$$

For the fourth case, the rate of return on the depreciated total investment for a particular year is:

$$i = \frac{A_{NCI}}{P_{s'} + C_{WC} + C_L} \tag{18}$$

In Eq. (16), (17) and (18), the book value, $P_{s'}$, for a particular year can be calculated by the straight-line method, declining balance method, or sum-of-the-years-digits method (see Part 2 of this series, which begins on p. 8). On the basis of straight-line depreciation, the average annual amount of depreciation is given by:

$$A_{BD} = \frac{(C_{FC} - S)}{s} \tag{19}$$

where s is the service life and S is the salvage value. And the book value at the end of a particular year, s', is given by:

$$P_{s'} = C_{FC} - s'A_{BD} \tag{20}$$

Rate of Return on Average Investment

The rate of return on average investment can be computed on the basis of four cases, as previously described for A_{NP}, A_{NNP}, A_{CI} or A_{NCI}.

For the second case, the rate of return on the average total capital investment for a particular year is:

$$i = \frac{2A_{NNP}}{C_{FC} + S + 2C_{WC} + 2C_L} \tag{21}$$

Eq. (21) can be written in terms of Eq. (8), as:

$$i = \frac{2A_{NCI} - 2A_{BD}}{C_{FC} + S + 2C_{WC} + 2C_L} \tag{22}$$

If i is taken on the basis of average values for A_{NNP} and A_{NCI} over the length of the project, an average value for the working capital, C_{WC}, must be used.

For the fourth case, the rate of return on the average total capital investment for a particular year is:

$$i = \frac{2A_{NCI}}{C_{FC} + S + 2C_{WC} + 2C_L} \tag{23}$$

In Eq. (21) and (23), the computations are based on the cost of land and other nondepreciable costs, C_L, not changing. This is unrealistic, since the value of land has a tendency to rise. In such circumstances, the accountancy principle of conservatism requires that the lowest valuation be adopted.

Cash Flows for Two Projects—Table I

Year	Cash flows A_{CF}	
	Project A	Project B
0	$ − 100,000	$ − 100,000
1	50,000	0
2	30,000	10,000
3	20,000	20,000
4	10,000	30,000
5	0	40,000
6	0	50,000
7	0	60,000
ΣA_{CF}	$ 10,000	$ 110,000
Payback period, (PBP)	3 yr.	5 yr.

Payback Period

Another traditional method of measuring profitability is the payback period. Actually, this is not really a measure of profitability but of the time it takes for the cash flows to recoup the original fixed-capital expenditure.

The net annual cash flow after tax is given by:

$$A_{CF} = A_{NCI} - A_{TC} \tag{24}$$

where A_{TC} is the annual expenditure of capital, which is not necessarily zero after the plant has been built. The payback period (PBP) is the time required for the cumulative net cash flow taken from the startup of the plant to equal the depreciable fixed-capital investment, $(C_{FC} - S)$. It is the value of s' that satisfies:

$$\sum_{s'=0}^{s'=(PBP)} A_{CF} = C_{FC} - S \tag{25}$$

The payback-period (PBP) method takes no account of cash flows or profits received after the breakeven point has been reached. The method is based on the premise that the earlier the fixed capital is recovered, the better the project. However, this approach can be misleading.

Let us consider Projects A and B, having net annual cash flows as listed in Table I. Both projects have initial fixed-capital expenditures of $100,000. On the basis of payback period, Project A is the most desirable since the fixed-capital expenditure is recovered in three years compared with five years for Project B. However, Project B runs for seven years with a cumulative net cash flow of $110,000. This is obviously more profitable than Project A, which runs for only four years with a cumulative net cash flow of only $10,000.

Example Shows How To Compute Profitability

Calculate the net annual cash income rate-of-return for a project on the basis of (a) original investment, (b)

Cash Flow Data and Rates of Return for Problem—Table II

Year	Cash Flow, A_{CF} \$/Yr.	Cash Income, A_{NCI} \$/Yr.	Return on Original Capital Investment, %/Yr.	Return on Depreciated Capital Investment, %/Yr.	Return on Average Capital Investment, %/Yr.
0	− 10,000				
1	− 30,000				
2	− 60,000				
3	−750,000				
4	−150,000				
5	+100,000	100,000	11.63	12.42	17.10
6	+150,000	150,000	17.44	20.00	25.64
7	+200,000	200,000	23.26	28.78	34.19
8	+200,000	200,000	23.26	31.25	34.19
9	+180,000	180,000	20.93	30.77	30.77
10	+160,000	160,000	18.61	30.19	27.35
11	+140,000	140,000	16.28	29.48	23.93
12	+120,000	120,000	13.96	28.57	20.52
13	+120,000	120,000	13.96	32.88	20.52
14	+355,000	45,000*	5.24	14.52	7.69
Avg.			16.46	25.88	24.19

*From Eq. (24):

$A_{NCI} = A_{CF} + A_{TC}$

$= \$355{,}000/\text{yr.} + (-\$200{,}000 - \$100{,}000 - \$10{,}000)/\text{yr.}$

$= \$45{,}000/\text{yr.}$

depreciated investment, (c) average investment; also calculate: (d) the payback period for the project, and (e) on the original investment, calculate the net annual profit (after tax) rate-of-return. The following data apply:

- Cash flows as given in Table II.
- Fixed capital cost, $C_{FC} = \$750{,}000$.
- Estimated salvage value of plant items, $S = \$200{,}000$.
- Working capital, $C_{WC} = \$100{,}000$.
- Cost of land, $C_L = \$10{,}000$.
- Straight-line depreciation over 10 yr.

For calculating the several items of profitability, we must carefully analyze the problem and apply the appropriate relations. We now proceed to find the net annual cash income rate-of-return on:

a. *The original investment*—The fractional interest rate, i, given by Eq. (15) applies only for situations in which the process has a positive net cash flow. In the present example, this first occurs in the fifth year of the project, which is the first year of productive operation (see Table II). For this year, the net cash flow, after tax, is \$100,000.

From Eq. (9), we find that the total capital cost is:

$$C_{TC} = \$750{,}000 + \$100{,}000 + \$10{,}000 = \$860{,}000$$

From Eq. (24), the net annual cash income for the fifth year is:

$$A_{NCI} = A_{CF} + A_{TC}$$

$$A_{NCI} = \$100{,}000/\text{yr.} + \$0/\text{yr.} = \$100{,}000/\text{yr.}$$

The annual capital expenditure, A_{TC}, is zero in each year after startup, except for the last year of operation, at the end of which the working capital, cost of land,

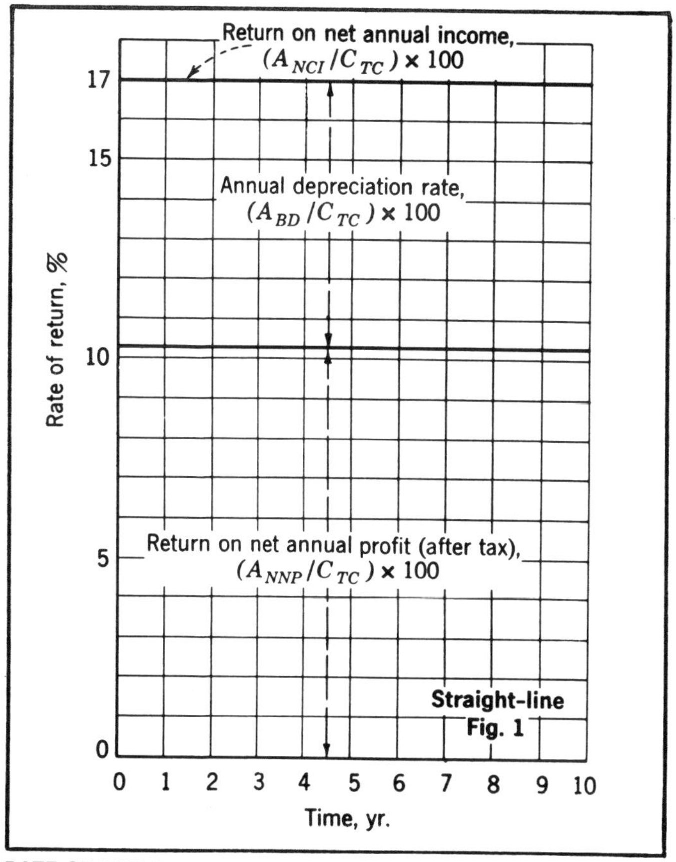

Straight-line Fig. 1

RATE OF RETURN on original capital investment shows the effects

and salvage value are recovered. Hence, in the fourteenth year, $A_{TC} = -\$100{,}000 - \$10{,}000 - \$200{,}000 = -\$310{,}000$.

From Eq. (13), the net annual cash income rate-of-return on the original capital investment for the fifth year is:

$$i = \$100{,}000/\$860{,}000 = 0.1163$$

$$i = (0.1163/\text{yr.})100 = 11.63\%/\text{yr.}$$

Values for subsequent years are calculated in the same way; these are listed in Table II.

b. *The depreciated investment*—The annual rate of straight-line depreciation of the plant items (from \$750,000 at startup to \$200,000 at the end of 10 yr.) is found from Eq. (19) to be:

$$A_{BD} = \frac{\$750{,}000 - \$200{,}000}{10 \text{ yr.}} = \$55{,}000/\text{yr.}$$

As in Part (a), $C_{FC} = \$750{,}000$, $C_{WC} = \$100{,}000$, $C_L = \$10{,}000$, and $A_{NCI} = \$100{,}000/\text{yr.}$ for Year 5. In the fifth year, the net cash flow after tax is positive for the first time, so that Eq. (18) applies from the fifth year onward. The fifth year is also the first year after plant startup, so that $s' = 1$ yr. For Year 5, the book value, $P_{s'}$, is found from Eq. (20) to be:

$$P_{s'} = \$750{,}000 - (1 \text{ yr.})(\$55{,}000/\text{yr.})$$

$$P_{s'} = \$695{,}000$$

From Eq. (18), the net annual cash income rate-of-

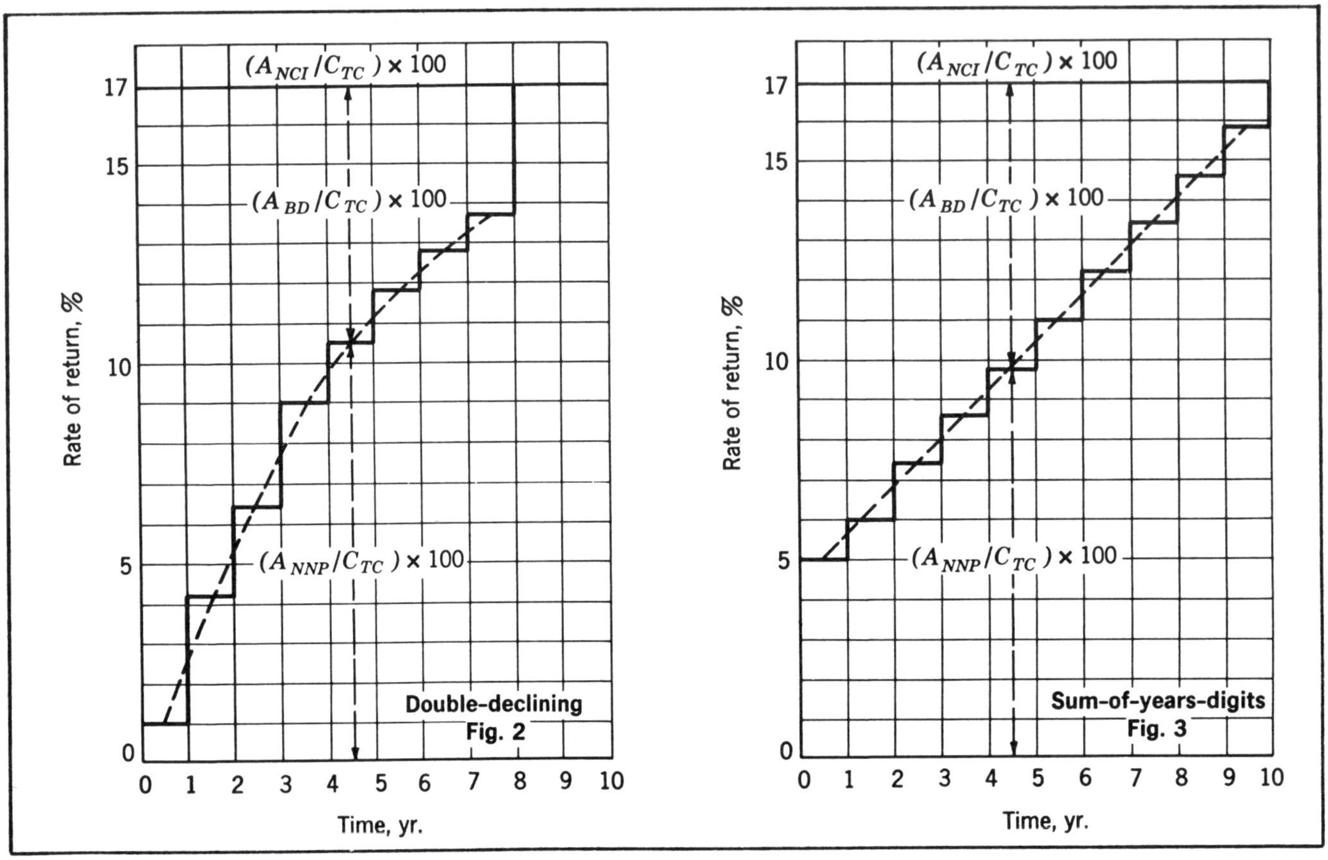

of depreciation and the depreciation method over a 10-year period for the project.

return on the depreciated capital investment is:

$$i = \frac{\$100{,}000/\text{yr.}}{\$695{,}000 + \$100{,}000 + \$10{,}000}$$

$$i = 0.1242/\text{yr., or } 12.42\%/\text{yr.}$$

Values for subsequent years are calculated in the same way; these are listed in Table II.

c. *The average capital investment*—As in Part (b), $A_{BD} = \$55{,}000/\text{yr.}$, $C_{FC} = \$750{,}000$, $C_{WC} = \$100{,}000$, $C_L = \$10{,}000$, and $A_{NCI} = \$100{,}000/\text{yr.}$ for Year 5. From Eq. (23), the net annual cash income rate-of-return on the average capital investment for the first year of profitable working (Year 5) is:

$$i = \frac{2(\$100{,}000/\text{yr.})}{\$750{,}000 + \$200{,}000 + 2(\$100{,}000) + 2(\$10{,}000)}$$

$$i = 0.1710/\text{yr., or } 17.10\%/\text{yr.}$$

Again, values for subsequent years are calculated in the same way, and also listed in Table II.

We will now use another method for measuring profitability for this example:

d. *Payback period*—The payback period (PBP) is usu-

Depreciation Method Affects Rate of Return—Table III

End of Year	Straight-Line Depreciation			Double-Declining Depreciation			Sum-of-Years-Digits Depreciation		
	A_{BD}	A_{BD}/C_{TC}	i	A_{BD}	A_{BD}/C_{TC}	i	A_{BD}	A_{BD}/C_{TC}	i
1	10,000	0.0667	0.1033	24,000	0.1600	0.0100	18,182	0.1212	0.0488
2	10,000	0.0667	0.1033	19,200	0.1280	0.0420	16,364	0.1091	0.0609
3	10,000	0.0667	0.1033	15,360	0.1024	0.0676	14,545	0.0970	0.0730
4	10,000	0.0667	0.1033	12,288	0.0819	0.0881	12,727	0.0848	0.0852
5	10,000	0.0667	0.1033	9,830	0.0655	0.1045	10,909	0.0727	0.0973
6	10,000	0.0667	0.1033	7,865	0.0524	0.1176	9,091	0.0606	0.1094
7	10,000	0.0667	0.1033	6,291	0.0419	0.1281	7,273	0.0485	0.1215
8	10,000	0.0667	0.1033	5,034	0.0336	0.1364	5,455	0.0364	0.1336
9	10,000	0.0667	0.1033	132	0.0009	0.1691	3,636	0.0242	0.1458
10	10,000	0.0667	0.1033	0	0.0000	0.1700	1,818	0.0121	0.1579

A_{BD} is the balance-sheet annual depreciation charge, \$/yr.
A_{BD}/C_{TC} is the annual, fractional depreciation rate.
i is the annual, fractional rate-of-return.

ally measured as the time from the start of production to recovery of the capital investment.

The fixed capital cost, $C_{FC} = \$750,000$, and the salvage value, $S = \$200,000$. Annual cash flows, A_{CF}, are listed in Table II. Starting from the beginning of Year 5 (the startup year), $s' = 0$. To find the payback period, we derive the following summary:

Year	s'	A_{CF}	ΣA_{CF}
5	1	+100,000	+100,000
6	2	+150,000	+250,000
7	3	+200,000	+450,000
8	4	+200,000	+650,000

The payback period, as defined by Eq. (25), is the time from startup until $\Sigma A_{CF} = (C_{FC} - S) = \$750,000 - \$200,000 = \$550,000$. Straight-line interpolation gives the time from Year 7 to be 0.5 yr. Hence: $(PBP) = 3$ yr. + 0.5 yr. = 3.5 yr. The project will break even 7.5 yr. from the start.

The final calculation requires finding the net annual profit (after tax) rate-of-return on:

e. *The original capital investment*—The rate of return, on this basis, is computed from Eq. (12), or its equivalent, Eq. (13). To use Eq. (13), we must compute the fractional depreciation rate. The annual rate of straight-line depreciation over a 10-yr. period is $A_{BD} = \$55,000$/yr. The original total capital investment, $C_{TC} = \$860,000$. Hence, the fractional depreciation rate is:

$$\frac{A_{BD}}{C_{TC}} = \frac{\$55,000/\text{yr.}}{\$860,000} = 0.0649, \text{ or } 6.49\%/\text{yr.}$$

From Eq. (13), the average rate-of-return on the original total capital investment on the basis of net annual profit (after tax), A_{NNP}, is:

$$i = (16.46 - 6.49) = 9.97\%/\text{yr.}$$

This rate of return is for straight-line depreciation over a 10-yr. period. The corresponding rate of return for straight-line depreciation over a 5-yr. period is 3.5%/yr.

Effect of Depreciation Method on Profit

Net annual profit (after tax) and the corresponding rate of return on the original investment are very much affected by the method chosen to compute the annual amount of depreciation—even if the net annual cash income, A_{NCI}, remains constant.

To show this effect, let us calculate the net annual profit (after tax), A_{NNP}, and the corresponding rate of return on the original investment on the basis of (a) straight-line depreciation, (b) double-declining depreciation, and (c) sum-of-years-digits depreciation, each over a 10-yr. period. The following data describe the project:

- Net annual cash income (after tax), $A_{NCI} = \$25,500$, in each of 10 years.
- Fixed capital cost, $C_{FC} = \$120,000$.
- Estimated salvage value of plant items, $S = \$20,000$.
- Working capital, $C_{WC} = \$10,000$.
- Cost of land, $C_L = \$20,000$.
- Values for balance-sheet annual depreciation, A_{BD}, are calculated by the methods given in Part 2 of this series. The values for this example are listed in Table III.

We must first find the total capital cost from Eq. (9). For this example,

$$C_{TC} = C_{FC} + C_{WC} + C_L$$
$$C_{TC} = \$120,000 + \$10,000 + \$20,000 = \$150,000$$

From Eq. (15), the net annual cash income rate-of-return on the original capital investment in each year is:

$$i = \$25,500/\$150,000 = 0.170, \text{ or } 17.0\%/\text{yr.}$$

From Eq. (8) and the depreciation data in Table III, we calculate the net annual profit (after tax) for each year as:

$$A_{NNP} = A_{NCI} - A_{BD}$$
$$A_{NNP} = \$25,500 - \$10,000 = \$15,500/\text{yr.}$$

From Eq. (12), the net annual profit (after tax) rate-of-return on the original capital investment in each year is found to be:

$$i = \$15,500/\$150,000 = 0.1033, \text{ or } 10.33\%/\text{yr.}$$

The rate-of-return data for straight-line depreciation are plotted in Fig. 1.

For double-declining and sum-of-years-digits depreciation, we calculate the net annual profit (after tax) and rate-of-return data for each year by using Eq. (8) and (12) and the depreciation data in Table III. The results are tabulated in Table III, and are plotted in Fig. 2 and 3.

Fig. 2 and 3 show that for both double-declining balance and sum-of-years-digits depreciation, the net annual profit (after tax) rate-of-return increases with time for a constant net annual cash income, A_{NCI}. This is in contrast to the constant net annual profit (after tax) rate-of-return for straight-line depreciation, as can be seen from Fig. 1.

It is clear that to state the net annual profit (after tax) rate-of-return without being quite specific about the basis of the inherent depreciation is misleading and uninformative.

Although various rate-of-return methods have been discussed in this article, the accountants in your company are most likely to base rates of return on the annual net profit before tax and the original historical cost of the total capital investment. You will therefore have to read your company's balance sheet and talk with your company's accountants in order to find how much depreciation has been deducted from the cash income. Although the balance sheet will give you the total annual depreciation for the company, the engineer should also know the breakdown for each project, together with the method used. This information can be obtained from the accountants. Since depreciation is assessed on an arbitrary basis, there is no consistency between various companies. Furthermore, within the same company, there may be a confusing lack of consistency. ■

Engineering Economics—Part 4

Time Value Of Money

The cost of money is an integral part of the analysis for determining profitability of an enterprise, and for establishing rates-of-return that are based on cash flows.

F. A. HOLLAND, F. A. WATSON and J. K. WILKINSON, University of Salford

A large part of business activity is based on money that can be loaned or borrowed. When money is loaned, there is always a risk that it may not be returned. A sum of money called interest is the inducement offered to make the risk acceptable. When money is borrowed, interest is paid for the use of the money over a period of time. Conversely, when money is loaned, interest is received.

The amount of a loan is known as the principal. The longer the period of time for which the principal is loaned, the greater is the total amount of interest paid. Thus, the future worth of the money, F, is greater than its present worth, P. The relationship between F and P depends on the type of interest used.

Simple Interest

When simple interest is used, F and P are related by:

$$F = P(1 + ni) \quad (1)$$

where i is the fractional interest rate per period, and n is the number of interest periods. Normally, the interest period is one year, in which case i is known as the effective interest rate.

Let us consider the situation in which $1,000 is invested in a small business enterprise that operates over a five-year period. It costs $300/yr. to run the business, and the cash receipts increase from $400/yr. in Year 1 to $800/yr. in Year 5. Table I shows that the total cash receipts of $3,000 exceed the total costs of $2,500 by $500. However, Table I presents a very simplified picture since it assumes that $1,000 is available for investment free of interest.

The calculations are reworked on the basis of $1,000 being available as a simple interest loan at 10%/yr., and are shown in Table II. It is assumed that $1,000 is borrowed at the end of Year 0 so that no interest is payable in that year. Interest payable at the end of Year 1 is $100. This increases the debt from $1,000 to $1,100. However, the net cash flow in during Year 1 is $100, which reduces the debt to $1,000 at the end of Year 1. Finally in Year 5, the net cash flow in of $500 exceeds the outstanding debt of $349.90 by $105.10. Thus, on the basis of a loan of $1,000 repayable at 10%/yr. simple interest, the total cash receipts exceed the total costs by $105.10, which is significantly less then the $500 shown in Table I on an interest-free basis.

Even with interest included, Table II presents an oversimplified picture. In practice, costs tend to rise because of inflation or the falling value of money. Therefore, it is most unlikely that the cost of running the business will remain constant at $300/yr.

Now let us consider the situation in which the costs and income are subjected to an inflation rate of 5%/yr. In this case, the net cash flow in Year 1 is ($100)(1.05) = $105, in Year 2 ($200)(1.05)^2 = $220.50, and so on. The calculations are reworked on this basis in Table III, from which it can be seen that the total cash receipts exceed the total costs by $429.89. Thus, an inflationary situation appears to increase the profitability of a project. However, this is misleading since the real value of money has fallen.

Annual Compound Interest

More commonly, compound interest is used where F and P are related by:

$$F = P(1 + i)^n \quad (2)$$

or

$$F = Pf_i \quad (3)$$

where the compound-interest factor $f_i = (1 + i)^n$. Values

Originally published September 1973.

Simplified Example of Cash Flow—Table I

End of Year	Cash In $	Cash Out $	Net Cash Flow In $
0	0	1,000	−1,000
1	400	300	100
2	500	300	200
3	600	300	300
4	700	300	400
5	800	300	500
Total	3,000	2,300	500

Interest Cost Affects Cash Flow—Table II

End of Year	Debt Excluding Interest For Current Year $	Interest Payable at 10%/Yr. $	Total Debt $	Net Cash Flow In $	Reduced Debt = Total Debt − Positive Net Cash Flow In $
0	1,000	0	1,000	−1,000	1,000
1	1,000	100	1,100	100	1,000
2	1,000	100	1,100	200	900
3	900	90	990	300	690
4	690	69	759	400	359
5	359	35.90	301.00	500	−105.10
Total	—	394.90	—	500	—

for compound-interest factors are readily available in tables similar to Tables IV and V for 5% and 10%, respectively. [The advent of computers, and especially the hand-held programmable minicalculators, enables these computations to be done without recourse to tables. See: The Engineer's Guide to Electronic Slide Rules, *Chem. Eng.,* Aug. 20, 1973, pp. 114–120. Perusal of the tables, though, is the fastest way to understand the importance of such factors in profitability assessment.]

The present value, P, of a future sum of money, F, is:

$$P = F/(1 + i)^n \quad (4)$$

or

$$P = Ff_d \quad (5)$$

where the discount factor, f_d, is:

$$f_d = \frac{1}{f_i} = \frac{1}{(1 + i)^n}$$

Values for the discount factors are readily available in tables similar to Tables IV and V.

Table V shows that it will take 7.3 yr. for money to double if compounded annually at 10%/yr. Table IV shows that it will take 14.2 yr. for money to double if compounded annually at 5%/yr.

For the case of different fractional interest rates ($i_1, i_2, ..., i_n$ in successive years), Eq. (2) can be written in the form:

$$F = P(1 + i_1)(1 + i_2)(1 + i_3) \ldots (1 + i_n) \quad (6)$$

Short-Interval Compound Interest

If interest payments become due m times per year at compound interest, mn payments are required in n years. A nominal annual interest rate, i', may be defined by:

$$F = P\left(1 + \frac{i'}{m}\right)^{mn} \quad (7)$$

In this case, the effective annual interest, i, is:

$$i = \left(1 + \frac{i'}{m}\right)^m - 1 \quad (8)$$

In the limit (as m approaches infinity), such payments may be considered to be required at infinitesimally short intervals, in which case the interest is said to be compounded continuously.

The annual interest rate equivalent to a compound interest rate of 5%/mo. (i.e., $i'/m = 0.05$) is calculated from Eq. (8) to be:

$$i = (1 + 0.05)^{12} - 1 = 0.796, \text{ or } 79.6\%/\text{yr.}$$

Most people who take out loans based on what superficially appear to be reasonable interest rates per month are not fully aware of the equivalent high interest rates on an annual basis.

Continuous Compound Interest

If we expand Eq. (7) by the binomial theorem, we get:

$$F = P\left[1 + m\left(\frac{i'}{m}\right) + \frac{m(m-1)}{1 \cdot 2}\left(\frac{i'}{m}\right)^2 + \frac{m(m-1)(m-2)}{1 \cdot 2 \cdot 3}\left(\frac{i'}{m}\right)^3 + \cdots\right]^n \quad (9)$$

As m approaches infinity, the time interval between payments becomes infinitesimally small, and in the limit Eq. (9) reduces to:

$$F = P\left(1 + i' + \frac{(i')^2}{2!} + \frac{(i')^3}{3!} + \cdots\right)^n \quad (10)$$

Eq. (10) can also be written as:

$$F = P\exp(i'n) \quad (11)$$

The same result may be obtained by writing:

$$\int_P^F \frac{dF}{F} = i' \int_0^n dn \quad (12)$$

On integrating Eq. (12), we obtain Eq. (11).

A comparison of Eq. (2) and (11) shows that the nominal interest rate, i', on a continuous basis is related to the effective interest rate, i, on an annual basis by:

$$\exp(i'n) = (1 + i)^n \quad (13)$$

From Eq. (11), we find that the continuously compounded nominal annual interest rate, i', is related to the effective annual interest rate, i, by:

$$i' = \log_e (1 + i) = 2.303 \log_{10} (1 + i) \quad (14)$$

And hence, $i' \leqq i$.

It is sometimes argued that continuous compound interest represents real-life situations more accurately

Effects of Inflation on Income and Costs—Table III

End of Year	Debt Excluding Interest For Current Year $	Interest Payable at 10%/Yr. $	Total Debt $	Net Cash Flow In $	Reduced Debt = Total Debt − Positive Net Cash Flow In $
0	1,000.00	0.00	1,000.00	−1,000.00	1,000.00
1	1,000.00	100.00	1,100.00	105.00	995.00
2	995.00	99.50	1,094.50	220.50	874.00
3	874.00	87.40	961.40	347.29	614.11
4	614.11	61.41	675.52	486.20	189.32
5	189.32	18.93	208.25	638.14	−429.89
Total	—	367.24	—	797.13	—

than annual compound interest. This is debatable since neither continuous nor annual compounding strictly represents the uneven cash flows in actual business situations. Numerically, the difference between continuous and annual compounding is small. In practice, it is probably far smaller than the errors in the estimated cash-flow data. Annual compound interest will be used in the subsequent articles in this series because this conforms more closely to current acceptable accounting practice. However, the small difference between continuous and

Factors for a 5% Compound-Interest Rate—Table IV

n Yr.	f_i	f_d	f_{AF}	f_{AP}	f_K
1	1.05000	0.95238	1.00000	1.05000	21.00000
2	1.10250	0.90703	0.48780	0.53780	10.75610
3	1.15762	0.86384	0.31721	0.36721	7.34417
4	1.21551	0.82270	0.23201	0.28201	5.64024
5	1.27628	0.78353	0.18097	0.23097	4.61950
6	1.34010	0.74622	0.14702	0.19702	3.94035
7	1.40710	0.71068	0.12282	0.17282	3.45640
8	1.47746	0.67684	0.10472	0.15472	3.09444
9	1.55133	0.64461	0.09069	0.14069	2.81380
10	1.62889	0.61391	0.07950	0.12950	2.59009
11	1.71034	0.58468	0.07039	0.12039	2.40778
12	1.79586	0.55684	0.06283	0.11283	2.25651
13	1.88565	0.53032	0.05646	0.10646	2.12912
14	1.97993	0.50507	0.05102	0.10102	2.02048
15	2.07893	0.48102	0.04634	0.09634	1.92685
16	2.18287	0.45811	0.04227	0.09227	1.84540
17	2.29202	0.43630	0.03870	0.08870	1.77398
18	2.40662	0.41552	0.03555	0.08555	1.71092
19	2.52695	0.39573	0.03275	0.08275	1.65490
20	2.65330	0.37689	0.03024	0.08024	1.60485

See Table V for explanation of factors.

Factors for a 10% Compound-Interest Rate—Table V

n Yr.	f_i	f_d	f_{AF}	f_{AP}	f_K
1	1.10000	0.90909	1.00000	1.10000	11.00000
2	1.21000	0.82645	0.47619	0.57619	5.76190
3	1.33100	0.75131	0.30211	0.40211	4.02115
4	1.46410	0.68301	0.21547	0.31547	3.15471
5	1.61051	0.62092	0.16380	0.26380	2.63797
6	1.77156	0.56447	0.12961	0.22961	2.29607
7	1.94872	0.51316	0.10541	0.20541	2.05405
8	2.14359	0.46651	0.08744	0.18744	1.87444
9	2.35795	0.42410	0.07364	0.17364	1.73641
10	2.59374	0.38554	0.06275	0.16275	1.62745
11	2.85312	0.35049	0.05396	0.15396	1.53963
12	3.13843	0.31863	0.04676	0.14676	1.46763
13	3.45227	0.28966	0.04078	0.14078	1.40779
14	3.79750	0.26333	0.03575	0.13575	1.35746
15	4.17725	0.23939	0.03147	0.13147	1.31474
16	4.59497	0.21763	0.02782	0.12782	1.27817
17	5.05447	0.19784	0.02466	0.12466	1.24664
18	5.55992	0.17986	0.02193	0.12193	1.21930
19	6.11591	0.16351	0.01955	0.11955	1.19547
20	6.72750	0.14864	0.01746	0.11746	1.17460

Factors

$f_i = (1 + i)^n$ Compound interest

$f_d = \dfrac{1}{(1 + i)^n}$ Discount

$f_{AP} = \dfrac{i(1 + i)^n}{(1 + i)^n - 1}$ Annuity present worth

$f_{AF} = \dfrac{i}{(1 + i)^n - 1}$ Annuity future worth

$f_K = \dfrac{(1 + i)^n}{(1 + i)^n - 1}$ Capitalized cost

annual compounding may be significant when applied to very large sums of money.

Let us suppose that \$100 is invested at a nominal interest rate of 5%. Let us compute the future worth of the investment after two years, and also compute the effective annual interest rate for the following kinds of interest (a) simple, (b) annual compound, (c) monthly compound, (d) daily compound, and (e) continuous compound.

The following tabulation shows the results of the calculations, along with the appropriate equation to be used:

Interest Type	Eq.	Future Worth, F	Effective Rate, i	Eq.
a	(1)	\$110.	5.%	(1)
b	(2)	\$110.250	5.%	(8)
c	(7)	\$110.495	5.117%	(8)
d	(7)	\$110.516	5.1267%	(8)
e	(11)	\$110.517	5.1271%	(8)

When computing the effective annual rate for continuous compounding (Type e in the example), the first term of Eq. (8), $[1 + (i'/m)]^m$, is equivalent to $e^{i'}$ as m approaches infinity.

Annual Cost or Payment

A series of equal annual payments, A, invested at a fractional interest rate, i, and made at the end of each year over a period of n years may be used to build up a future sum of money, F. These relations are given by:

$$F = A\left[\frac{(1 + i)^n - 1}{i}\right] \tag{15}$$

or

$$F = A/f_{AF} \tag{16}$$

where the annuity future-worth factor is:

$$f_{AF} = \frac{i}{(1 + i)^n - 1}$$

Values for f_{AF} are readily available in tables similar to Tables IV and V.

Eq. (15) can be combined with Eq. (4) to yield:

$$P = A\left[\frac{(1 + i)^n - 1}{i(1 + i)^n}\right] \tag{17}$$

$$P = A/f_{AP} \tag{18}$$

where P is the present worth of the series of future equal annual payments, A, and the annuity present-worth factor is:

$$f_{AP} = \frac{i(1 + i)^n}{(1 + i)^n - 1}$$

Values for f_{AP} are also available in tables similar to Tables IV and V.

Alternatively, the annual payment, A, required to build up a future sum of money, F, with a present value of P is given by:

$$A = Ff_{AF} \tag{19}$$

$$A = Pf_{AP} \tag{20}$$

Eq. (15) represents the future sum of a series of uniform annual payments that are invested at a stated interest rate over a period of years. This procedure defines an ordinary annuity. Other forms of annuities include: the annuity due, in which payments are made at the beginning of the year instead of at the end; and the deferred annuity, in which the first payment is deferred for a definite number of years.

Capitalized Cost

A piece of equipment of fixed capital cost, C_{FC}, will have a finite life of n years. The capitalized cost of the equipment, C_K, is defined by:

$$(C_K - C_{FC})(1 + i)^n = C_K - S \tag{21}$$

C_K is in excess of C_{FC} by an amount which, when compounded at an annual interest rate i for n years, will have a future worth of C_K less the salvage or scrap value, S. If the renewal cost of the equipment remains constant at $(C_{FC} - S)$, and the interest rate remains constant at i, then C_K is the amount of capital required to ensure that the equipment may be renewed in perpetuity.

Eq. (21) may be rewritten as:

$$C_K = \left[C_{FC} - \frac{S}{(1 + i)^n}\right]\left[\frac{(1 + i)^n}{(1 + i)^n - 1}\right] \tag{22}$$

or

$$C_K = (C_{FC} - Sf_d)f_K \tag{23}$$

where f_d is the discount factor; and f_K, the capitalized cost factor, is:

$$f_K = \frac{(1 + i)^n}{(1 + i)^n - 1}$$

Values for each factor are available in tables similar to Tables IV and V.

Example for Capitalized Cost

A piece of equipment has been installed at a cost of \$100,000, and is expected to have a working life of 10 yr. with a scrap value of \$20,000. Calculate the capitalized cost of the equipment, based on an annual compound interest rate of 5%.

Since capitalized cost is related to the fixed capital cost of the equipment, the scrap value, the fractional interest rate and the number of years, Eq. (22) applies. In this example, $C_{FC} = \$100{,}000$, $S = \$20{,}000$, $i = 0.05$, and $n = 10$ yr. Therefore:

$$C_K = \left[\$100{,}000 - \frac{\$20{,}000}{(1 + 0.05)^{10}}\right]\left[\frac{(1 + 0.05)^{10}}{(1 + 0.05)^{10} - 1}\right]$$

$$C_K = \left[\$100{,}000 - \frac{\$20{,}000}{1.62889}\right](2.59009) = \$227{,}207$$

In solving Eq. (22), the compound interest factor and the capitalized cost factor were obtained from Table IV. The answer could also have been obtained by using Eq. (23), for which the discount factor and the capitalized cost factor are found in Table IV. In this case:

$$C_K = [\$100{,}000 - (\$20{,}000)(0.61391)]\,(2.59009) = \$227{,}207 \blacksquare$$

Here are the techniques for assessing the relative investment return of potential projects. All require a projection of future cash flows that are then reduced to a present value or rate of return.

Methods of Estimating Project Profitability

F. A. HOLLAND, F. A. WATSON and J. K. WILKINSON, University of Salford

An investment in a manufacturing process must earn more than the cost of capital in order for it to be worthwhile. The larger the additional earnings, the more profitable the venture and the greater the justification for putting the capital at risk. A profitability estimate is an attempt to quantify the desirability of taking this risk.

The ways of assessing profitability to be considered in this article are: (1) discounted-cash-flow rate-of-return (*DCFRR*), (2) net present value (*NPV*) based on a particular discount rate, (3) equivalent maximum investment period (*EMIP*), (4) interest recovery period (*IRP*), and (5) discounted breakeven point (*DBEP*).

The most common modern methods for assessing profitability are the (*NPV*) and (*DCFRR*) methods. The results are dependent on the annual depreciation allowed for tax purposes. Before examining the details of these methods, let us quickly review the basic relations between sales, expenses and income of a project.

The revenue, A_s, from the annual sales of a product minus the total annual cost or expense, A_{TE}, required to produce and sell the product (excluding any annual provision for plant depreciation) is the annual cash income, A_{CI}:

$$A_{CI} = A_S - A_{TE} \tag{1}$$

Net annual cash income, A_{NCI}, is the annual cash income minus the annual amount of tax, A_{IT}:

$$A_{NCI} = A_{CI} - A_{IT} \tag{2}$$

Annual amount of tax, A_{IT}, is calculated from the following relation:

$$A_{IT} = (A_{CI} - A_D - A_A)t \tag{3}$$

where A_D is the annual depreciation charge, A_A is the annual amount of any other allowances, and t is the fractional tax rate.

Definition of Cash Flow

In this article, cash flow is the net annual cash flow, A_{CF}, after tax, and is given by:

$$A_{CF} = A_{NCI} - A_{TC} \tag{4}$$

where A_{TC} is the annual capital expenditure, which is not necessarily zero after the plant has been built.

The total capital cost, C_{TC}, of a project consists of the fixed capital cost, C_{FC}, plus the working capital cost, C_{WC}, plus the cost of land and other nondepreciable costs, C_L:

$$C_{TC} = C_{FC} + C_{WC} + C_L \tag{5}$$

Let us consider a project in which $C_{FC} = \$1{,}000{,}000$; $C_{WC} = \$90{,}000$; and $C_L = \$10{,}000$. Hence, $C_{TC} = \$1{,}100{,}000$ from Eq. (5).

If all this capital expenditure occurs in Year 0 of the project, then $A_{TC} = \$1{,}100{,}000$ in Year 0, and $-A_{TC} = -\$1{,}100{,}000$. From Eq. (4), it is seen that any capital expenditure makes a negative contribution to the net annual cash flow, A_{CF}.

Originally published October 1973.

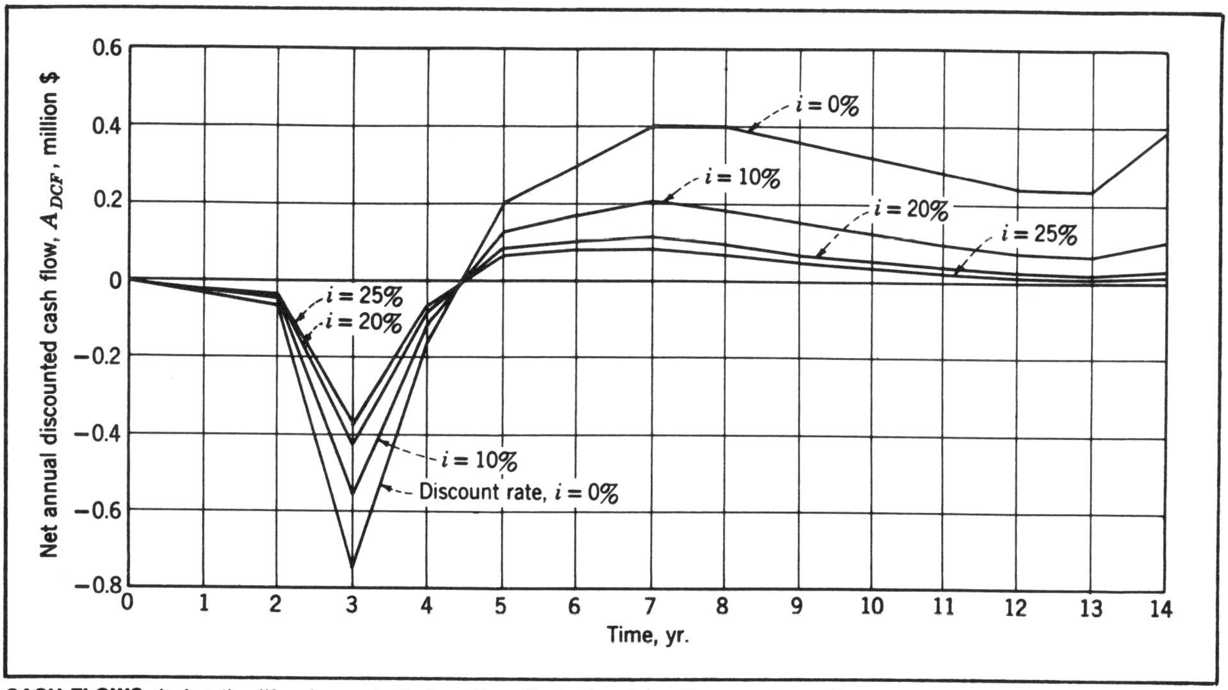

CASH FLOWS during the life of a project show the effect of various discount rates—Fig. 1

Let us consider another project in which the fixed capital expenditure is spread over two years, according to the following pattern:

$$C_{FC} = C_{FC0} + C_{FC1}$$

Year 0	Year 1
$C_{FC0} = \$400{,}000$	$C_{FC1} = \$600{,}000$
$C_L = 10{,}000$	$C_{WC} = 90{,}000$
$A_{TC} = 410{,}000$	$A_{TC} = 690{,}000$

In the final year of the project, the working capital and the land are recovered, which in this case cost a total of \$100,000. Thus, in the final year of the project, $A_{TC} = -\$100{,}000$, and $-A_{TC} = +\$100{,}000$. From Eq. (4), it is seen that any capital recovery makes a positive contribution to the net annual cash flow.

During the development and construction stages of a project, A_{CI} and A_{IT} are both zero in Eq. (2) and (4). For this period, the cash flow for the project is negative and is given by:

$$A_{CF} = -A_{TC} \tag{6}$$

Fig. 1 shows the cash-flow stages in a project. The expenditure during the research and development stage is normally relatively small. It will usually include some preliminary process design and a market survey. Once the decision to go ahead with the project has been taken, detailed process engineering design will commence, and the rate of expenditure starts to increase. The rate is increased still further when equipment is purchased and construction gets underway. There is no return on the investment until the plant is started up. Even during startup, there is some additional expenditure. Once the plant is operating smoothly, an inflow of cash is established. During the early stages of a project, there may be a tax credit due to the existence of expenses without corresponding income.

Discounted Cash Flow

The present value, P, of a future sum of money, F, is given by:

$$P = Ff_d \tag{7}$$

where $f_d = 1/(1 + i)^n$, the discount factor. Values for this factor are readily available in tables similar to Table I for 10%. Table I shows that \$90,909 invested at an annual interest rate of 10% becomes \$100,000 after one year. Similarly, \$38,554 invested at 10% becomes \$100,000 after 10 yr.

Thus, cash flow in the early years of a project has a greater value than the same amount in the later years of a project. Therefore, it pays to receive money as soon as possible and to delay paying out money for as long as possible.

Time is taken into account by using the annual discounted cash flow, A_{DCF}, which is related to the annual cash flow, A_{CF} and the discount factor, f_d, by:

$$A_{DCF} = A_{CF}f_d \tag{8}$$

Thus, at the end of any year, n:

$$A_{DCFn} = A_{CFn}/(1 + i)^n$$

The sum of the annual discounted cash flows over n years, ΣA_{DCF}, is known as the net present value (NPV) of the project:

$$(NPV) = \Sigma A_{DCF} \tag{9}$$

The value of (NPV) is directly dependent on the choice of the fractional interest rate, i. An interest rate can be selected to make $(NPV) = 0$ after a chosen number of years. This value of i is found from:

$$\Sigma A_{DCF} = \frac{A_{CF1}}{(1 + i)} + \frac{A_{CF2}}{(1 + i)^2} + \cdots + \frac{A_{CFn}}{(1 + i)^n} = 0 \tag{10}$$

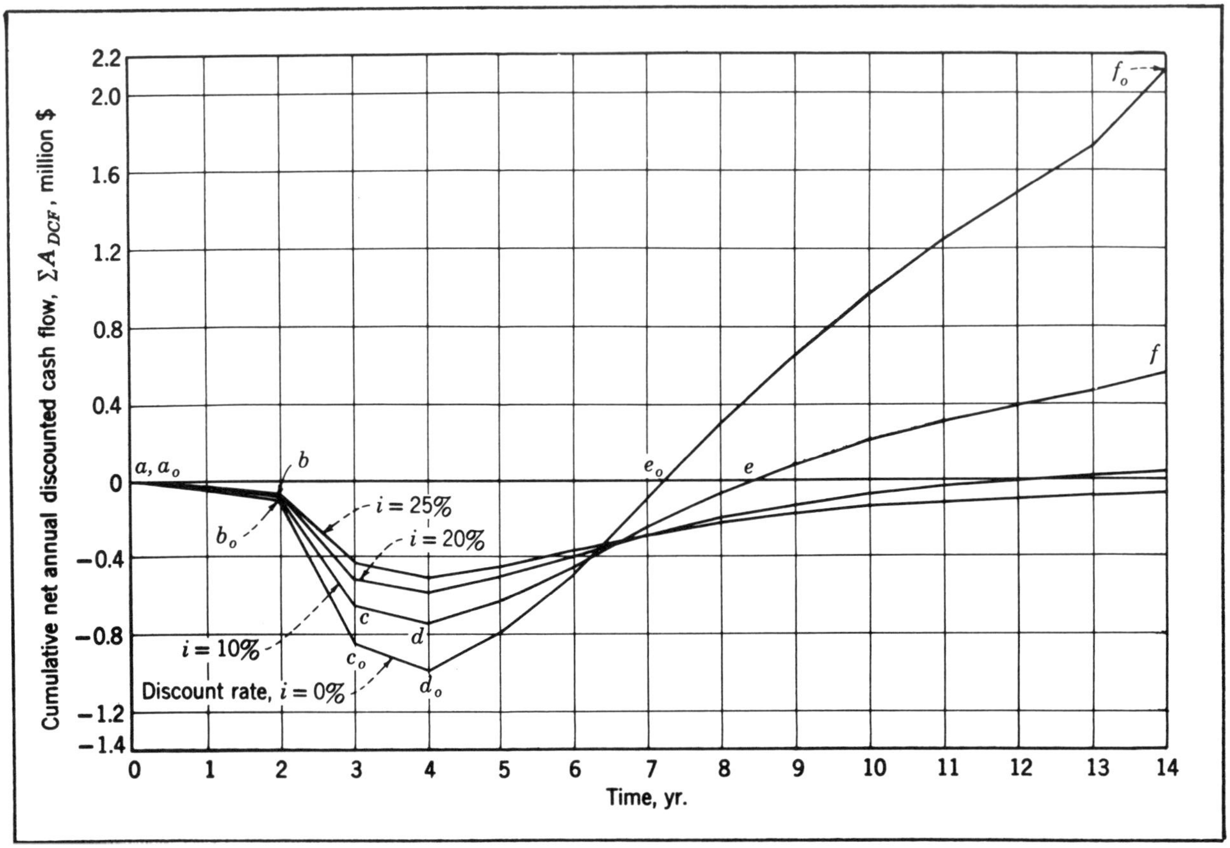

CUMULATIVE cash-flow curves establish the breakeven point and the equivalent maximum interest period—Fig. 2

Factors for 10% Interest—Table I

n Yr.	f_i	f_d
1	1.10000	0.90909
2	1.21000	0.82645
3	1.33100	0.75131
4	1.46410	0.68301
5	1.61051	0.62092
6	1.77156	0.56447
7	1.94872	0.51316
8	2.14359	0.46651
9	2.35795	0.42410
10	2.59374	0.38554
11	2.85312	0.35049
12	3.13843	0.31863
13	3.45227	0.28966
14	3.79750	0.26333
15	4.17725	0.23939
16	4.59497	0.21763
17	5.05447	0.19784
18	5.55992	0.17986
19	6.11591	0.16351
20	6.72750	0.14864

Factors

$f_i = (1 + i)^n$, Compound interest

$f_d = \frac{1}{(1 + i)^n}$, Discount

Eq. (10) may be solved for i either graphically or by an iterative trial-and-error procedure. The value of i given by Eq. (10) is known as the discounted cash-flow rate-of-return (*DCFRR*). It is also known as the: profitability index, true rate of return, investor's rate of return, and interest rate of return.

Cash-Flow Curves

Fig. 1 shows the cash-flow stages in a project together with their discounted cash-flow values for the data given in Table II. In addition to cash-flow and discounted cash-flow curves, it is also instructive to plot cumulative cash-flow and cumulative discounted cash-flow curves. These are shown in Fig. 2 for the data in Table II.

The cost of capital is also the interest rate at which money can be invested instead of putting it at risk in a manufacturing process. Consider the process data listed in Table II and plotted in Fig. 2. If the cost of capital is 10%, then the appropriate discounted cash-flow curve in Fig. 2 is abcdef. Up to Point e, or 8.49 yr., the capital is at risk. Point e is the discounted breakeven point (*DBEP*). At this point, the manufacturing process has paid back its capital and produced the same return as an equivalent amount of capital invested at a compound interest rate of 10%. Beyond the breakeven point, the capital is no longer at risk, and any cash flow above the horizontal base line, $\Sigma A_{DCF} = 0$, is in excess of the return on an equivalent amount of capital invested at a compound interest rate of 10%. Thus, the greater the area

Cash Flows and Discounted Cash Flows for a Project—Table II

			Discounted at 10%			Discounted at 20%			Discounted at 25%		
Year	A_{CF} $	ΣA_{CF} $	f_d	A_{DCF} $	ΣA_{DCF} $	f_d	A_{DCF} $	ΣA_{DCF} $	f_d	A_{DCF} $	ΣA_{DCF} $
0	−10,000	−10,000	1.00000	−10,000	−10,000	1.00000	−10,000	−10,000	1.00000	−10,000	−10,000
1	−30,000	−40,000	0.90909	−27,273	−37,273	0.83333	−25,000	−35,000	0.80000	−24,000	−34,000
2	−60,000	−100,000	0.82645	−49,587	−86,860	0.69444	−41,666	−76,666	0.64000	−38,400	−72,400
3	−750,000	−850,000	0.75131	−563,483	−650,343	0.57870	−434,025	−510,691	0.51200	−384,000	−456,400
4	−150,000	−1,000,000	0.68301	−102,452	−752,795	0.48225	−72,338	−583,029	0.40960	−61,440	−517,840
5	+200,000	−800,000	0.62092	+124,184	−628,611	0.40188	+80,376	−502,653	0.32768	+65,536	−452,304
6	+300,000	−500,000	0.56447	+169,341	−459,270	0.33490	+100,470	−402,183	0.26214	+78,642	373,662
7	+400,000	−100,000	0.51316	+205,264	−254,006	0.27908	+111,632	−290,551	0.20972	+83,888	−289,774
8	+400,000	+300,000	0.46651	+186,604	−67,402	0.23257	+93,028	−197,523	0.16777	+67,108	−222,666
9	+360,000	+660,000	0.42410	+152,676	+85,274	0.19381	+69,772	−127,751	0.13422	+48,319	−174,347
10	+320,000	+980,000	0.38554	+123,373	+208,647	0.16151	+51,683	−76,068	0.10737	+34,358	−139,989
11	+280,000	+1,260,000	0.35049	+98,137	+306,784	0.13459	+37,685	−38,383	0.08590	+24,052	−115,937
12	+240,000	+1,500,000	0.31863	+76,471	+383,255	0.11216	+26,918	−11,465	0.06872	+16,493	−99,444
13	+240,000	+1,740,000	0.28966	+69,518	+452,773	0.09346	+22,430	+10,965	0.05498	+13,195	−86,249
14	+400,000	+2,140,000	0.26333	+105,332	+558,105	0.07789	+31,156	+42,121	0.04398	+17,592	−68,657

A_{CF} is net annual cash flow. A_{DCF} is net annual discounted cash flow. f_d is discount factor at stated interest. ΣA_{CF} is cumulative cash flow. ΣA_{DCF} is cumulative discounted cash flow.

above the base line, the more profitable is the process.

When (NPV) and $(DCFRR)$ are computed, depreciation is not considered as a separate factor. It is simply used as an allowance to reduce the annual amount of tax.

A $(DCFRR)$ of, say 15%, implies that 15%/yr. will be earned on the investment in addition to which the project generates sufficient money to repay the original investment. Frequently, depreciation is considered as the means of recovering the original investment. This is done implicitly in (NPV) and $(DCFRR)$ calculations.

It is not normally possible to make a comprehensive assessment of profitability with a single number. The shape of the cumulative cash-flow and cumulative discounted cash-flow curves both before and after the breakeven point is an important factor.

Allen[1] accounted for the shape of the cumulative undiscounted cash-flow curve up to the breakeven point (Point e_o in Fig. 2) by using a parameter known as the equivalent maximum investment period $(EMIP)$, which is defined as:

$$(EMIP) = \frac{Area\ (a_o \text{ to } e_o)}{-\Sigma A_{CF\ max}} \qquad (11)$$

where the *Area* (a_o to e_o) refers to the area below the horizontal base line ($\Sigma A_{CF} = 0$) on the cumulative cash-flow curve in Fig. 2. The sum $-\Sigma A_{CF\ \max}$ is the maximum cumulative expenditure on the project, which is given by Point d_o in Fig. 2. $(EMIP)$ is a time in years. It is the "equivalent period during which the total project debt would be outstanding if it were all incurred at one instant and all repaid at one instant." Clearly, the shorter the $(EMIP)$, the more attractive the project.

Allen accounted for the shape of the cumulative cash-flow curve beyond the breakeven point by using a parameter known as the interest recovery period (IRP). This is the time period (illustrated in Fig. 3) that makes the Area (e_o to f_o) above the horizontal base line equal to the Area (a_o to e_o) below the horizontal base line on the cumulative cash-flow curve.

Sinclair[2] has considered similar parameters to the $(EMIP)$ and (IRP) based on a cumulative discounted cash-flow curve.

Consideration of the cash-flow stages in Fig. 2 shows the factors that can affect the $(EMIP)$ and (IRP). If the required capital investment is increased, it is necessary to increase the rate of income after startup for the $(EMIP)$ to remain the same. This is also the case if research and development stages were prolonged. In order to have the $(EMIP)$ small, it is necessary to keep the research and development, design and construction stages short.

Example Illustrates Procedures

The following data describe a project. Revenue from annual sales and the total annual expense over a 10-yr. period are given in the first three columns of Table III. The fixed capital investment, C_{FC}, is $1,000,000. Plant items have a zero salvage value. Working capital, C_{WC} is $90,000, and cost of land, C_L, is $10,000. There are no tax allowances other than depreciation, i.e., A_A is zero. The fractional tax rate, t, is 0.50.

Let us calculate for these data the net present value (NPV) for the following depreciation methods and discount factors:

a. Straight-line, 10%.
b. Straight-line, 20%.
c. Double-declining, 10%.
d. Sum-of-years-digits, 10%.
e. Straight-line, 10%, income tax delayed for one year.

In addition, let us calculate the discounted cash-flow rate-of-return $(DCFRR)$ with straight-line depreciation.

a. (NPV) with straight-line depreciation and 10% discount factor—We will begin the calculations for this

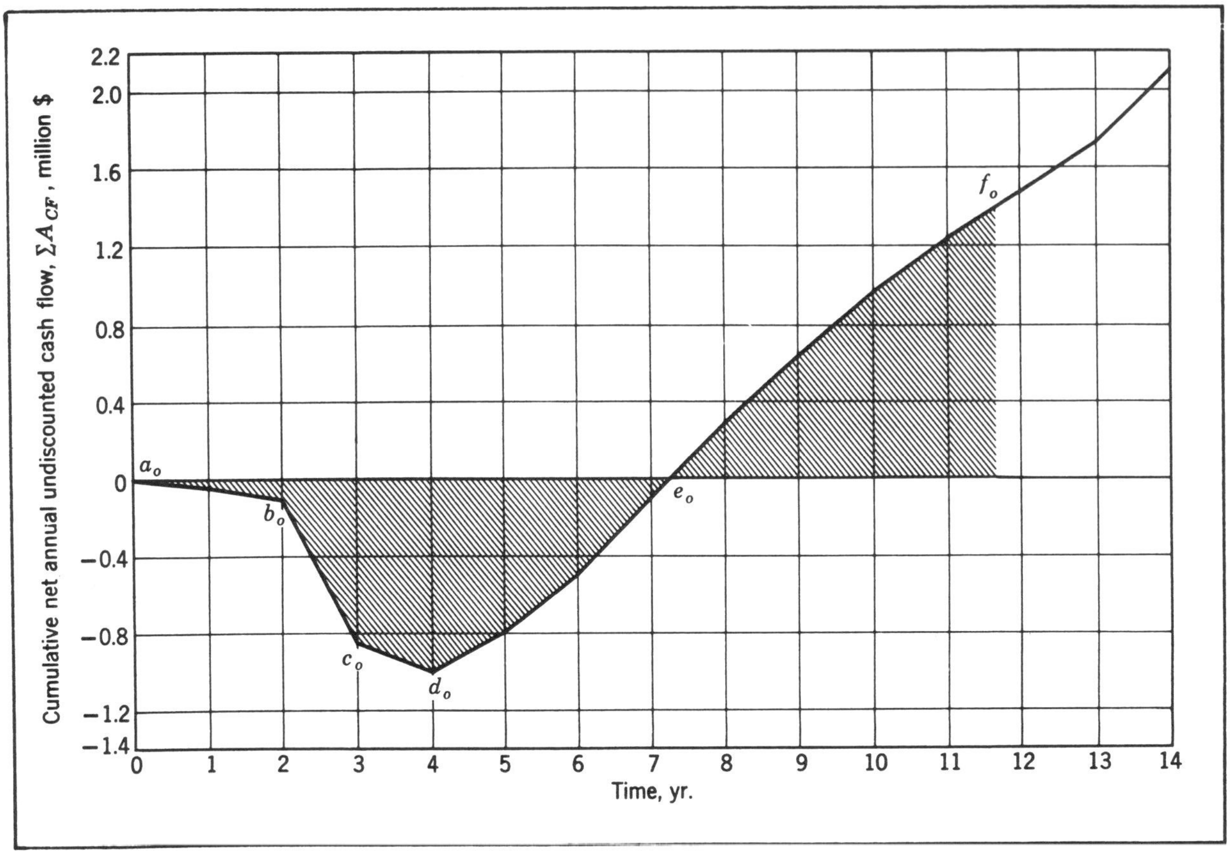

UNDISCOUNTED cumulative cash flow establishes the interest recovery period—Fig. 3

example by finding the total capital cost, C_{TC}, for the project from Eq. (5). Here, C_{TC} = \$1,100,000. In Year 0, this amount is the same as the net annual capital expenditure, A_{TC}, and is listed in Table III.

The annual rate of straight-line depreciation of the fixed capital investment, C_{FC}, from \$1,000,000 at startup to a salvage value, S, of zero at the end of a productive life, s, of 10 yr. is given by:

$$A_D = (C_{FC} - S)/s$$
$$A_D = (\$1{,}000{,}000 - \$0)/10 \text{ yr.} = \$100{,}000/\text{yr.}$$

The annual cash income, A_{CI}, for Year 1 when A_S = \$400,000/yr. and A_{TE} = \$100,000/yr. is given by Eq. (1) to be \$300,000/yr. Values for subsequent years are calculated in the same way and listed in Table III.

Annual amount of tax, A_{IT}, for Year 1 when A_{CI} = \$300,000/yr., A_D = \$100,000/yr., A_A = \$0/yr. and t = 0.5 is found from Eq. (3) to be:

$$A_{IT} = (\$300{,}000 - \$100{,}000 - \$0)/\text{yr.}\ (0.5)$$
$$A_{IT} = \$100{,}000/\text{yr.}$$

Values for subsequent years are calculated in the same way and listed in Table III.

Net annual cash flow (after tax), A_{CF}, for Year 0 when A_{CI} = \$0/yr., A_{IT} = \$0/yr., and A_{TC} = \$1,100,000/yr. is found from Eq. (4) to be:

$$A_{CF} = \$0/\text{yr.} - \$1{,}100{,}000/\text{yr.} = -\$1{,}100{,}000/\text{yr.}$$

Net annual cash flow (after tax) for Year 1 when A_{CI} = \$300,000/yr., A_{IT} = \$100,000/yr., and A_{TC} = \$0/yr. is found from Eq. (2) and (4) to be, $A_{CF} = A_{NCI} - A_{TC}$:

$$A_{CF} = \$200{,}000/\text{yr.} - \$0/\text{yr.} = \$200{,}000/\text{yr.}$$

Values for the years up to and including Year 9 are calculated in the same way and listed in Table III.

At the end of Year 10, the working capital C_{WC} = \$90,000, and the cost of land C_L = \$10,000 are recovered so that the annual expenditure of capital A_{TC} in Year 10 is −\$100,000/yr. Hence, the net annual cash flow (after tax) for Year 10 must reflect this recovery. Using Eq. (4):

$$A_{CF} = \$110{,}000/\text{yr.} - (-\$100{,}000/\text{yr.})$$
$$A_{CF} = \$210{,}000/\text{yr.}$$

The net annual discounted cash flow, A_{DCF}, for Year 1 when A_{CF} = \$200,000/yr. and f_d = 0.90909 (for i = 10%) is found from Eq. (8) to be:

$$A_{DCF} = \$200{,}000/\text{yr.}(0.90909) = \$181{,}820 \text{ yr.}$$

Values for subsequent years are calculated in the same way and listed in Table III.

The net present value, (NPV), is found by summing the values of A_{DCF} for each year, Eq. (9). In this case, (NPV) = \$276,210 for i = 10% in Year 10.

b. (NPV) *with straight-line depreciation and 20% discount factor*—The same procedure is used for i = 20%. Of course, the discount factors to be used in a table similar to Table III must be given for this rate. For this case, the (NPV) is found to be −\$151,020.

c. (NPV) *with double-declining depreciation and 10% discount factor*—The calculations are similar to those for

Annual Cash Flows, Straight-Line Depreciation and 10% Discount Factor—Table III

	Before Tax							After Tax			
Year	A_S $	A_{TE} $	A_{CI} $	$A_D + A_A$ $	$A_{CI} - A_D - A_A$ $	A_{IT} $	A_{TC} $	A_{CF} $	f_d	A_{DCF} $	(NPV) $
0	0	0	0	0	0	0	+1,100,000	−1,100,000	1.0000	−1,100,000	−1,100,000
1	400,000	100,000	300,000	100,000	200,000	100,000	0	200,000	0.90909	181,820	−918,180
2	500,000	100,000	400,000	100,000	300,000	150,000	0	250,000	0.82645	206,610	−711,570
3	500,000	110,000	390,000	100,000	290,000	145,000	0	245,000	0.75131	184,070	−527,500
4	500,000	120,000	380,000	100,000	280,000	140,000	0	240,000	0.68301	163,920	−363,580
5	520,000	130,000	390,000	100,000	290,000	145,000	0	245,000	0.62092	152,120	−211,460
6	520,000	130,000	390,000	100,000	290,000	145,000	0	245,000	0.56447	138,300	−73,160
7	520,000	140,000	380,000	100,000	280,000	140,000	0	240,000	0.51316	123,160	+50,000
8	390,000	140,000	250,000	100,000	150,000	75,000	0	175,000	0.46651	81,640	+131,640
9	350,000	150,000	200,000	100,000	100,000	50,000	0	150,000	0.42410	63,610	+195,250
10	280,000	160,000	120,000	100,000	20,000	10,000	−100,000	210,000	0.38554	80,960	+276,210

A_S Revenue from annual sales.
A_{TE} Total annual expense
A_{CI} Annual cash income
$A_D + A_A$ Annual depreciation and other tax allowances
$A_{CI} - A_D - A_A$ Taxable income
$A_{IT} = (A_{CI} - A_D - A_A)t$ Amount of tax at $t = 0.5$
A_{TC} Total annual capital expenditure
$A_{CF} = A_{CI} - A_{IT} - A_{TC}$ Net annual cash flow
f_d Discount factor at 10%
A_{DCF} Net annual discounted cash flow
$(NPV) = \Sigma A_{DCF}$ Net present value

Part a except that depreciation is computed using the double-declining method described in Part 2 of this series. The net present value for Year 10 is then calculated to be $288,530.

d. (NPV) with sum-of-years-digits depreciation and 10% discount factor—Again, the calculations are similar to those for Part a except that depreciation is computed using the sum-of-years-digits method described in Part 2 of this series. Thus the net present value for Year 10 is found to be $316,610.

e. (NPV) with straight-line depreciation, 10% discount factor, and income tax delayed for one year—The calculations follow the same procedure as for Part a although the annual amount of tax, A_{IT}, is calculated for a particular year and then deducted from the annual cash income, A_{CI}, for the following year. Thus, the net present value for Year 11 is calculated to be $341,980.

The discounted cash-flow rate-of-return, $(DCFRR)$, can readily be obtained approximately by interpolation of the (NPV) for $i = 10\%$ and $i = 20\%$.

$$(DCFRR) = 0.100 + \frac{(\$276{,}210)(0.20 - 0.10)}{\$276{,}210 - (-\$151{,}020)}$$

$$(DCFRR) = 0.164, \text{ or } 16.4\%$$

A comparison of the (NPV) values for a 10% discount factor shows clearly that double-declining depreciation is more advantageous than straight-line depreciation, and that sum-of-years-digits depreciation is more advantageous than the double-declining method. However, a significant advantage is obtained by delaying the payment of tax for one year even with straight-line depreciation.

This example is a simplified one. The cost of the working capital is assumed to be paid for in Year 0 and returned in Year 10. In practice, the working capital increases with production rate. Thus, there may be an annual expenditure, A_{TC}, on working capital in a number of years subsequent to Year 0.

Analysis of Techniques

Both (NPV) and $(DCFRR)$ methods are based on discounted cash flows and in that sense are variations of the same basic method. However, when ranking different projects on the basis of profitability, they can produce different results. This will be discussed in a future article of this series.

Since discounted cash-flow rate-of-return $(DCFRR)$ is a single percentage, it has the advantage of being readily understood. Net present value (NPV) has the disadvantage of being a sum of money. Additional information is needed before its significance can be appreciated. However, when a company is considering investment in a portfolio of projects, individual (NPV)'s have the advantage of being added. This cannot be done with $(DCFRR)$'s.

Increasing use is being made of the capital rate-of-return ratio, (CRR), which is the net present value, (NPV), divided by the maximum cumulative expenditure or maximum net outlay, $-\Sigma A_{CF\,max}$:

$$(CRR) = (NPV)/-\Sigma A_{CF\,max} \tag{12}$$

The maximum net outlay is very important since—no matter how profitable the project—the matter is academic if the company is unable to raise the money to undertake the project.

An (NPV) or $(DCFRR)$ estimation will be no better than the accuracy of the projected cash flows over the life of the project. The probability of reaching a particular (NPV) or $(DCFRR)$ will be dealt with in future articles.

Clearly, one is likely to predict cash flows more accurately for two or three years ahead than say for nine or ten years ahead. However, since the cash flows in the later years are discounted to a greater extent than for the earlier years, the latter have less effect on the overall estimation. Nevertheless, the difficulty of predicting cash flows in later years and the inherent lack of confidence in these predictions is a serious disadvantage of the $(DCFRR)$ method. In this respect (NPV)'s are more useful since they are calculated for each year of a project. Thus, a project with a favorable (NPV) in the early years is a promising one.

One way of overcoming this disadvantage of the $(DCFRR)$ method is to make estimates of the times to reach certain values of $(DCFRR)$. For example, how many years will it take to reach $(DCFRR)$'s of 10%, 15%, 20%/yr., etc.? Although $(DCFRR)$ trial-and-error calculations and (NPV) calculations are tedious if done manually, computer programs are readily available for making them easier.

It is possible for some projects to reach a stage when repairs, replacements, etc., can exceed net earnings in a particular year. In this case, the cumulative discounted-cash-flow, ΣA_{DCF} or (NPV), curve plotted against time can have a genuine maximum. The $(DCFRR)$ method has the disadvantage that no negative cash flow can appear after the first year of positive cash flow unless special evaluation techniques are used.

Other Aspects of Profitability Estimates

Inflation or the falling value of money is a current problem of modern economies. Consider a project that ran from 1960 to 1970. On the basis of the actual historical cash flows for these years, the discounted cash-flow rate-of-return is calculated to be, say, 15%/yr. However, these cash flows were subject to inflation. Let us assume that the average rate of inflation during this period was, say, 5%/yr. If there had been no inflation, the calculated $(DCFRR)$ for this project would be approximately 10%/yr., i.e., (15% − 5%).

Thus, for future projects in which the estimated $(DCFRR)$ is based on projected inflated cash flows, an approximation to the true $(DCFRR)$ on the basis of current costs can be obtained by subtracting the estimated average future rate of inflation from the estimated $(DCFRR)$.

$(DCFRR)$ estimates are liable to misuse if a company uses a particular value as a cutoff point. A company director recently told us that his company had a $(DCFRR)$ cutoff point of 17%/yr. Below this value, proposals were automatically rejected. The director was concerned since all the proposals coming forward in his company had $(DCFRR)$ values of 17%/yr. It appears that the estimators were calculating the cash flows required to ensure a $(DCFRR)$ of 17%/yr. They were probably then persuading the various people in the company that these projected cash flows were possible and obtaining estimates from them on this basis. A great deal of self-deception may have been involved, since the estimators probably came to believe quite sincerely in these numerical values.

It is important when computing (NPV) and $(DCFRR)$ not to consider the past in profitability estimations. Good money should never follow bad. Money should not continue to be put into a project if a more profitable project exists even though this may involve scrapping an expensive plant. This aspect will be dealt with in more detail in future articles.

It is our belief that no single value for a profitability estimate should be accepted without further consideration. An intelligent consideration of the cumulative cash-flow and cumulative discounted-cash-flow curves such as those shown in Fig. 2, together with experience and good judgment is the best way of assessing the financial merit of a project.

When considering future projects, top management will most likely require the discounted cash-flow rate-of-return. However, the estimators should also supply management with the:

- Cumulative discounted cash flow or (NPV) curve for a discount rate of 10%/yr.
- Maximum net outlay, $-\Sigma A_{CF\,max}$.
- Discounted breakeven point, $(DBEP)$.
- Plot of capital return ratio, (CRR), against time over the life of the project for a discount rate of 10%/yr.
- Number of years to reach discounted cash-flow rates-of-return of, say, 10 and 15%/yr., respectively. ■

References

1. Allen, D. H., *Chem. Eng.*, July 3, 1967, pp. 75–78.
2. Sinclair, C. G., *Chem. Process Eng.*, **47**, 147 (1966)

Meet the Authors

F. A. Holland

F. A. Watson

J. K. Wilkinson

F. A. Holland is professor and chairman of the department of chemical engineering at the University of Salford, Salford M5 4WT, England, and a partner in Salchem Associates, Consulting Chemical Engineers. Trained both as a physicist and a chemical engineer, Holland spent 14 years in the chemical process industries, nine of which were in the U.S. He has held management positions in production, process design, construction, and research and development. He has degrees from the universities of Oxford, London and Durham, and is a Fellow of the Institution of Chemical Engineers. He is also the author or coauthor of five books and more than fifty technical papers and articles.

F. A. Watson is a senior lecturer in the department of chemical engineering at the University of Salford and a partner in Salchem Associates. He has occupied posts in production, research and development with companies engaged in the manufacture of liquid and gaseous fuels and coal-tar byproducts. His primary interests are process design, digital computation and accountancy. He graduated from Manchester University in chemical engineering, and is a Fellow of the Institution of Chemical Engineers. He is also the coauthor of the book "Heat Transfer" (Heinemann, London, 1970—American Elsevier, New York).

J. K. Wilkinson is a senior lecturer in the department of chemical engineering at the University of Salford and a partner in Salchem Associates. He has held posts in the chemical industry in production, research and development. His main interests are in process design, financial accountancy and investment analysis. He holds a bachelor's and master's degree in chemical engineering from Manchester University, and a bachelor's degree in economics from London University, and is a Fellow of the Institution of Chemical Engineers. He is coauthor of "Heat Transfer."

Engineering Economics—Part 6

Profitability estimates are no better than the inherent accuracy of the data. Changes in sales prices, manufacturing costs and inflation affect these estimates especially when the predicted data are cast far into the future.

Sensitivity Analysis of Project Profitabilities

F. A. HOLLAND, F. A. WATSON and J. K. WILKINSON, University of Salford

Profitability estimates are made on the basis of net present value (NPV) and discounted-cash-flow rate of return ($DCFRR$). Single values of (NPV) and ($DCFRR$) are computed from a given set of cash-flow data, which are subjective estimates of sales revenue, total expense, fixed capital cost, etc.[1]

It is important to determine how sensitive a profitability estimate is to possible errors in the cash-flow predictions. For example, if the sales revenue is 10% lower than predicted, what is the reduction in the calculated (NPV)? Therefore, we shall examine the effect on profitability of variations in the forecasts of cash-flow data.

Effect of Cash-Flow Variations on Cash Income

Annual cash income, A_{CI}, is given by:

$$A_{CI} = A_S - A_{TE} \tag{1}$$

A_{CI} is the relatively small difference between two relatively large numbers, the revenue from annual sales, A_S, and the total annual cost or expense, A_{TE}, required to produce and sell the product, excluding any allowance for plant depreciation. A small error in estimating A_S or A_{TE} can make a significant difference to the cash income.

Net annual cash income, A_{NCI}, is related to A_{CI} by:

$$A_{NCI} = A_{CI} - A_{IT} \tag{2}$$

where A_{IT} is the annual amount of income tax.

An economic study should pinpoint the areas most susceptible to change. It is easier to predict expenses than either sales or profits. Fairly accurate capital-cost and processing-cost estimates can be made. However, for the most part, errors in these estimates have a correspondingly smaller effect than changes in sales price, sales volume, and raw-materials and distribution costs.

Sales and raw-materials prices may be affected by any of the following: discounts and allowances, availability of substitutes, contract pricing, government regulations, quality and form of the materials, competition.

Sales volume may be affected by any of the following: new uses for the product, new markets, advertising, quality, overcapacity, replacement by another product, competition, timing of entry into the market.

Distribution costs depend on:

- Plant location.
- Physical state of material (whether liquid, gas or solid).
- Nature of material (whether corrosive, explosive, flammable, perishable or toxic).
- Freight rates.
- Labor costs.

Originally published October 1973.

Distribution costs may be affected by any of the following: new methods of materials handling, safety regulations, productivity agreements, wage rates, transportation systems, storage systems, quality, losses, seasonal effects.

It is worthwhile to make tables or plot curves that show the effect of variations in costs and prices on profitability. This procedure is called sensitivity analysis. Its purpose is to determine which factors the profitability of a project is most sensitive to. Sensitivity analysis should always be carried out to observe the effect of departures from predicted values.

For many years, companies and countries have lived with the problem of inflation or the falling value of money. Costs—in particular, labor costs—tend to rise each year. Failure to account for this in predicting future cash flows can lead to serious errors and misleading profitability estimates.

Another important factor is the tendency of product prices to fall as the total national or international volume of production increases. Sales prices may fall by 20% for a doubling in volume of production.

No profitability estimate is better than the inherent accuracy of the data.

Effect of a Decrease in Revenue From Sales

The revenue from annual sales is related to the annual cash income and the total annual cost by Eq. (1). If the revenue from annual sales falls below the base-case value A_S by an amount ΔA_S, the annual cash income is reduced by an amount ΔA_{CI} given by:

$$\Delta A_{CI} = \Delta A_S \tag{3}$$

Annual amount of tax, A_{IT}, is given by:

$$A_{IT} = (A_{CI} - A_D - A_A)t \tag{4}$$

where A_D is the annual depreciation charge, A_A is the annual amount of any other allowances, and t is the fractional tax rate.

The effect of a reduction in A_S by an amount ΔA_S is to reduce A_{IT} by an amount ΔA_{IT} given by:

$$\Delta A_{IT} = \Delta A_{CI}t \tag{5}$$

Eq. (5) can be combined with Eq. (3), and the result written as:

$$\Delta A_{IT} = \Delta A_S t \tag{6}$$

Net annual cash flow, A_{CF}, (after tax) is given by:

$$A_{CF} = A_{NCI} - A_{TC} \tag{7}$$

Eq. (7) can also be written as:

$$A_{CF} = A_{CI} - A_{IT} - A_{TC} \tag{8}$$

where A_{NCI} is the net annual cash income, and A_{TC} is the annual expenditure of capital that is not necessarily zero after the plant has been built.

The effect of a reduction in A_S by an amount ΔA_S is to reduce A_{CF} by an amount ΔA_{CF} as given by:

$$(A_{CF} - \Delta A_{CF}) = (A_{CI} - \Delta A_{CI}) - (A_{IT} - \Delta A_{IT}) - A_{TC} \tag{9}$$

Eq. (9) can be written in terms of Eq. (8) and shown as:

$$\Delta A_{CF} = \Delta A_{CI} - \Delta A_{IT} \tag{10}$$

Substitute Eq. (3) and (6) into Eq. (10), and write:

$$\Delta A_{CF} = \Delta A_S(1 - t) \tag{11}$$

The annual discounted-cash flow, A_{DCF}, is related to A_{CF} by:

$$A_{DCF} = A_{CF}f_d \tag{12}$$

where f_d is the discount factor, $1/(1 + i)^n$.

The effect of a reduction in A_S by an amount ΔA_S is to reduce A_{DCF} by an amount ΔA_{DCF}, and is given by:

$$\Delta A_{DCF} = \Delta A_{CF}f_d \tag{13}$$

Eq. (13) can be written in terms of Eq. (11), as:

$$\Delta A_{DCF} = \Delta A_S(1 - t)f_d \tag{14}$$

The sum of the annual discounted-cash-flows over n years, ΣA_{DCF}, is the net present value, (NPV):

$$(NPV) = \Sigma A_{DCF} \tag{15}$$

The effect of a reduction in A_S by the amount ΔA_S is to reduce the (NPV) by an amount $\Delta(NPV)$, as given by:

$$\Delta(NPV) = \Sigma \Delta A_{DCF} \tag{16}$$

Eq. (16) can be written in terms of Eq. (14), as:

$$\Delta(NPV) = \Sigma \Delta A_S(1 - t)f_d \tag{17}$$

Effect of an Increase in Total Expense

The total annual cost or expense is related to the annual cash income and the revenue from annual sales by Eq. (1). If the total annual expense is increased above the base value, A_{TE}, by an amount ΔA_{TE}, the annual cash income is reduced by an amount ΔA_{CI}:

$$\Delta A_{CI} = \Delta A_{TE} \tag{18}$$

The effect of a reduction in annual cash income by an amount ΔA_{CI} is to reduce the annual amount of tax by an amount ΔA_{IT}, is given by Eq. (5). Eq. (18) can be combined with Eq. (5), and written as:

$$\Delta A_{IT} = \Delta A_{TE}\, t \tag{19}$$

The net annual cash flow (after tax) is given by Eq. (8). An increase in the annual total expense reduces the annual cash flow by an amount as given by Eq. (10). By substituting Eq. (18) and (19) into Eq. (10), we get:

$$\Delta A_{CF} = \Delta A_{TE}(1 - t) \tag{20}$$

The effect of an increase in annual total expense by an amount ΔA_{TE} is to reduce the annual discounted-cash-flow by an amount ΔA_{DCF}, as given by Eq. (13). Eq. (13) can be written in terms of Eq. (20), as:

$$\Delta A_{DCF} = \Delta A_{TE}(1 - t)f_d \tag{21}$$

The effect of an increase in the annual total expense by an amount ΔA_{TE} is to reduce net present value by an amount $\Delta(NPV)$, as given by Eq. (16). Eq. (16) can be written in terms of Eq. (21), as:

$$\Delta(NPV) = \Sigma \Delta A_{TE}(1 - t)f_d \tag{22}$$

An Increase in Fixed-Capital Investment

An increase in the fixed-capital investment, C_{FC}, will require an increase in the annual depreciation charge by an amount ΔA_D. This will cause a corresponding reduction in the annual amount of tax, as given by:

$$\Delta A_{IT} = \Delta A_D t \tag{23}$$

If the annual expenditure of capital, A_{TC}, is increased as a result of an increase in C_{FC}, the effect is to reduce the net annual cash flow, as shown by:

$$A_{CF} - \Delta A_{CF} = A_{CI} - (A_{IT} - \Delta A_{IT}) - (A_{TC} + \Delta A_{TC}) \tag{24}$$

Eq. (24) can be written in terms of Eq. (23) as:

$$\Delta A_{CF} = \Delta A_{TC} - \Delta A_D t \tag{25}$$

The effect of a reduction in annual cash flow is to reduce the annual discounted-cash-flow, as given by Eq. (13). Eq. (13) can be written in terms of Eq. (23), as:

$$\Delta A_{DCF} = (\Delta A_{TC} - \Delta A_D t)f_d \tag{26}$$

The effect of a reduction in the annual discounted-cash-flow is to reduce the net present value as shown by Eq. (16). Eq. (16) can be written in terms of Eq. (26),

$$\Delta(NPV) = \Sigma(\Delta A_{TC} - \Delta A_D t)f_d \tag{27}$$

After the fixed capital investment has been made, the value of ΔA_{TC} may be zero in subsequent years. However, the increase in annual depreciation charge, ΔA_D, will continue in subsequent years.

Example of Sensitivity Analysis

The following data describe a project. Revenue from annual sales and the total annual expense over a 10-yr. period are given in the first three columns of Table I. The fixed-capital investment, C_{FC}, is $1 million. Plant items have a zero salvage value. Working capital, C_{WC}, is $90,000, and cost of land, C_L, is $10,000. There are no tax allowances other than depreciation, i.e., A_A is zero. The fractional tax rate, t, is 0.50.

For this project, the net present value for a 10% discount factor and straight-line depreciation was shown to be $276,210, and the discounted-cash-flow rate of return was shown to be 16.4%/yr. in Part 5 of this series.

Let us use these data and the accompanying information in Table I as the base case, and calculate for straight-line depreciation the net present value (NPV) with 10% discount factor, and the discounted-cash-flow rate of return ($DCFRR$), for the project with the following situations:

Case	Modification
a.	Revenue, A_S, reduced 10%/yr.
b.	Revenue, A_S, reduced 20%/yr.
c.	Total expense, A_{TE}, increased 10%/yr.
d.	Fixed-capital investment increased by 10%.
e.	A_S reduced 10%/yr., A_{TE} increased 10%/yr., and C_{FC} increased by 10%.

Annual Cash Flows, Straight-Line Depreciation and 10% Discount Factor—Table I
(Revenue reduced by 10%/yr.)

	Base Case							
Year	A_S $	A_{TE} $	(NPV) $	ΔA_S $	f_d	ΔA_{DCF} $	$\Delta(NPV)$ $	Reduced (NPV) $
0	0	0	−1,100,000	0	1.0000	0	0	−1,100,000
1	400,000	100,000	−918,180	40,000	0.90909	18,180	18,180	−936,360
2	500,000	100,000	−711,570	50,000	0.82645	20,660	38,840	−750,410
3	500,000	110,000	−527,500	50,000	0.75131	18,780	57,620	−585,120
4	500,000	120,000	−363,580	50,000	0.68301	17,070	74,690	−438,270
5	520,000	130,000	−211,460	52,000	0.62092	16,140	90,830	−302,290
6	520,000	130,000	−73,160	52,000	0.56447	14,680	105,510	−178,670
7	520,000	140,000	+50,000	52,000	0.51316	13,340	118,850	−68,850
8	390,000	140,000	+131,640	39,000	0.46651	9,100	127,950	+3,690
9	350,000	150,000	+195,250	35,000	0.42410	7,420	135,370	+59,880
10	280,000	160,000	+276,210	28,000	0.38554	5,390	140,760	+135,450

A_S Base revenue from annual sales before tax
A_{TE} Base total annual expense before tax
(NPV) Base net present value after tax
ΔA_S Decrease in annual revenue
f_d Discount factor at 10%
ΔA_{DCF} Decrease in net discounted-cash-flow at income tax rate = 0.5
$\Delta(NPV) = \Sigma\Delta A_{DCF}$ Decrease in net present value
Reduced $(NPV) = \Sigma A_{DCF}$ Reduced net present value after tax

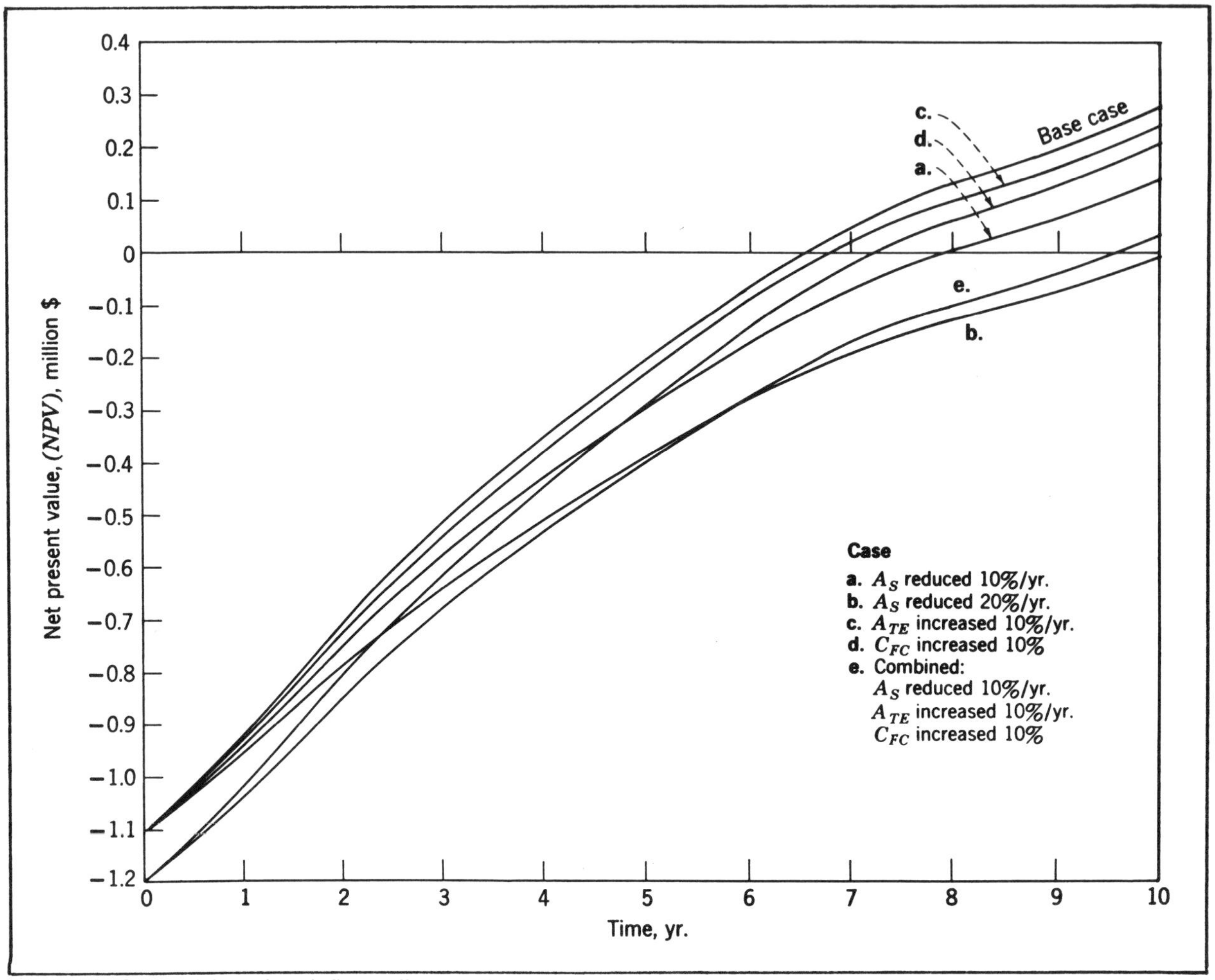

VARIATIONS in anticipated cash flows affect predictions of net present value for a project—Fig. 1

Case a—Find (NPV) and $(DCFRR)$ for A_S reduced 10%/yr. In Year 1, the revenue A_S is \$400,000/yr. (Table I). 10% of this gives $A_S = \$40{,}000/\text{yr}$. The decrease in net discounted-cash-flow for Year 1, when $t = 0.5$ and $f_d = 0.90909$, is found from Eq. (14) to be:

$$\Delta A_{DCF} = (\$40{,}000/\text{yr.})(0.5)(0.90909)$$
$$\Delta A_{DCF} = \$18{,}180/\text{yr.}$$

Values for subsequent years are calculated in the same way and listed in Table I.

The decrease in net present value for Year 10 is found by summing the values of ΔA_{DCF} in Table I, which for this case yields $\Delta(NPV) = +\$140{,}760$. The base net present value for Year 10 is +\$276,210. Therefore, the reduced net present value for Year 10 is the difference between (NPV) for the base case and the case where A_S is reduced 10%/yr.:

$$(NPV) = +\$276{,}210 - (+140{,}760) = \$135{,}450$$

The equations for ΔA_{DCF} developed in this article can be used to calculate the adjustments required in the net annual discounted-cash-flow A_{DCF} for each individual year of the project. Such a set of adjustments can be calculated for each level of variation in the factors contributing to A_{DCF}. In this way, it is a simple matter to adjust the data prior to the use of the discount factor, f_d, to allow for changes in any variable or combination of variables. The corresponding value of the $(DCFRR)$ may then be calculated by the methods shown in Part 5 of this series. For the base case, $(DCFRR)$ is 16.4%, and for the case where annual revenues decrease by 10%/yr., $(DCFRR)$ becomes 12.9%.

Case b—Find (NPV) and $(DCFRR)$ for A_S reduced 20%/yr. The procedures used here are similar to those for Case a.

The decrease in net present value in Year 10 for this case is found to be +\$281,540. Since the base net present value for Year 10 is +\$276,210, the reduced net present value becomes:

$$(NPV) = +\$276{,}210 - (+\$281{,}540) = -\$5{,}330$$

For the base case, $(DCFRR)$ is 16.4%, and for this case $(DCFRR)$ becomes 9.8%.

Case c—Find (NPV) and $(DCFRR)$ for A_{TE} increased 10%/yr. The calculations are similar to those for Case a. Decrease in net present value for Year 10 is found from Eq. (22) to be: $\Delta(NPV) = +\$37{,}780$. The reduced net present value for Year 10 becomes:

$$(NPV) = +\$276{,}210 - (+\$37{,}780) = +\$238{,}430$$

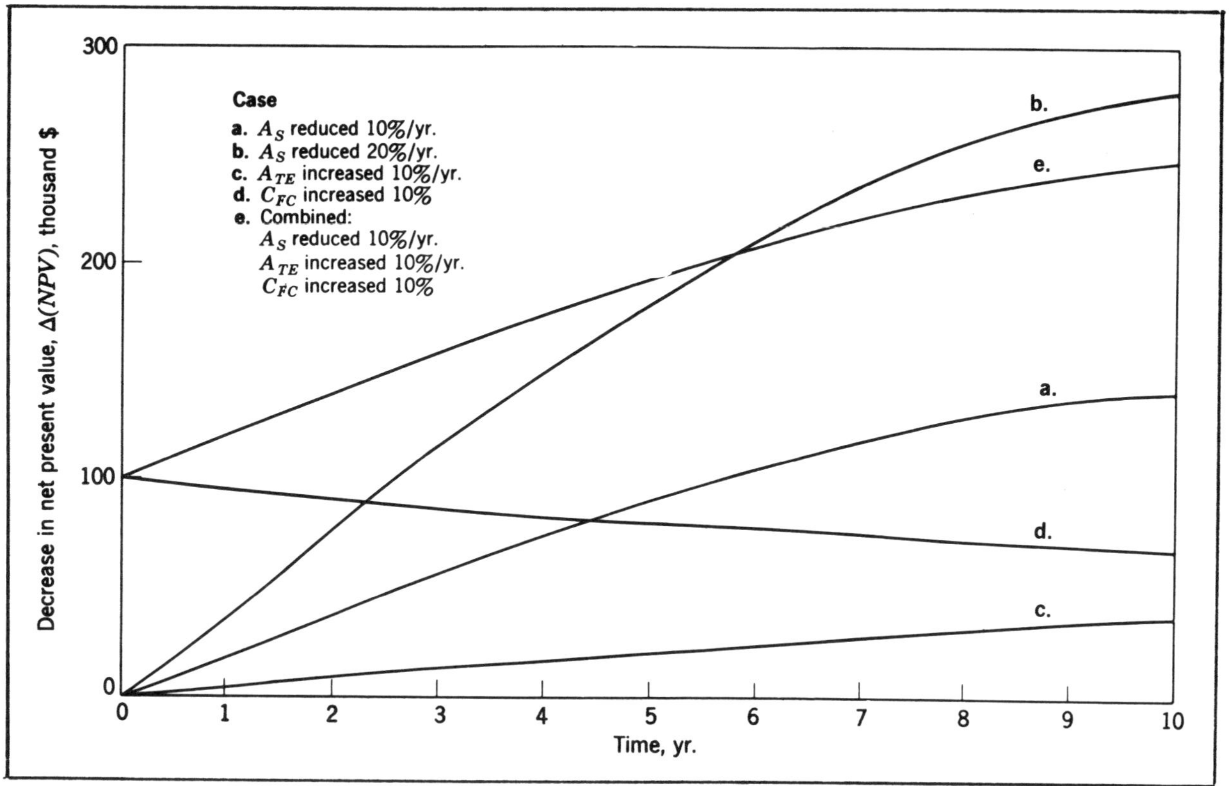

DECREASE in the net present value for a project due to changes in anticipated cash flows—Fig. 2

By a procedure similar to that in Case a, the discounted-cash-flow rate of return is found to be 15.0%.

Case d—Find (NPV) and $(DCFRR)$ where the fixed-capital investment, C_{FC}, is increased by 10%.

For the base case, the annual rate of straight-line depreciation of the fixed-capital investment from $1 million at startup to zero value at the end of 10 yr. of productive life is found from:

$$A_D = (C_{FC} - S)/s$$
$$A_D = (\$1{,}000{,}000 - \$0)/10 \text{ yr.} = \$100{,}000/\text{yr.}$$

The effect of an increase in fixed-capital cost (C_{FC} = $1,000,000) by 10% (i.e., ΔC_{FC} = $100,000) is an increase in A_D by an amount ΔA_D = $10,000/yr. In Year 0:

$$\Delta A_{TC} = \Delta C_{FC} = \$100{,}000/\text{yr.}$$

Summary of Results for Example—Table II

Case	(NPV) i = 10%	(DCFRR)
Base case	$276,210	16.4%
a. A_S reduced 10%/yr.	$135,450	12.9%
b. A_S reduced 20%/yr.	–$5,330	9.8%
c. A_{TE} increased 10%/yr.	$238,430	15.0%
d. C_{FC} increased 10%	$206,890	14.0%
e. Combined: A_S reduced 10%/yr., A_{TE} increased 10%/yr., C_{FC} increased 10%	$28,420	10.6%

The decrease in the net discounted-cash-flow, ΔA_{DCF}, for Year 0 when f_d = 1.0000, is found from Eq. (26) to be:

$$\Delta A_{DCF} = [\$100{,}000/\text{yr.} - (\$10{,}000/\text{yr.})(0.5)](1.0000)$$
$$\Delta A_{DCF} = \$95{,}000/\text{yr.}$$

For subsequent years, ΔA_{TC} = 0, and ΔA_D = $10,000/yr.

The decrease in net present value for Year 10 is found from Eq. (27) to be $\Delta(NPV)$ = +$69,320. Hence, the reduced net present value for Year 10 becomes:

$$(NPV) = +\$276{,}210 - (+\$69{,}320) = +\$206{,}890$$

By a procedure similar to that for Case a, it is found that $(DCFRR)$ is 14.0%.

Case e—Find (NPV) and $(DCFRR)$ where the following changes occur simultaneously: A_S reduced 10%/yr., A_{TE} increased 10%/yr., and C_{FC} increased by 10%.

The calculations follow the procedures previously described, and the results are summarized in Table II.

The results of this sensitivity analysis are presented graphically for net present values in Fig. 1 and 2. Of course, these are not the only factors that could have been varied. A sensitivity analysis can be carried out on the basis of any relevant factors such as delay time in startup, change in taxation rate, etc. ■

Economic factors dominate when assessing the comparative profitability of different projects or of alternate procedures in the process plant.

Time, Capital and Interest Affect Choice of Project

F. A. HOLLAND, F. A. WATSON and J. K. WILKINSON, University of Salford

Management must often decide whether, or when, to start or stop certain projects. Frequently this involves the preparation of profitability estimates for comparing an existing process with newly developed processes.

It is our view that when comparing projects of the same order of profitability, there is no single economic criterion that can take the place of good judgment. Nevertheless, economic factors tend to dominate.

In previous articles, a number of ways of assessing the desirability of a project have been considered. For the most part, the various criteria can be classified roughly under the heading of time, capital and interest.

Comparisons on the basis of time can be summarized (with reference to the cash-flow curves in Fig. 1) by the:

1. Duration of the project.
2. Breakeven point (*BEP*), shown as e_0.
3. Discounted breakeven point (*DBEP*), shown as Point e, having a discount factor of 10% in this case.
4. Equivalent maximum investment period,

$$(EMIP) = \frac{Area\ (a_0 \text{ to } e_0)}{-\Sigma A_{CFmax}} \qquad (1)$$

where $-\Sigma A_{CF\,max}$ is given by Point d_0.

5. Interest recovery period (*IRP*), which makes the Area (e_0 to f_0) equal to the Area (a_0 to e_0).
6. Payback period (*PBP*), which is the value of s' that satisfies:

$$\sum_{s'=0}^{s'=(PBP)} A_{CF} = C_{FC} - S \qquad (2)$$

where $(C_{FC} - S)$ is the depreciable fixed capital investment.

Comparisons on the basis of capital can be summarized by the:

1. Maximum cumulative expenditure on the project, $-\Sigma A_{CF\,max}$, given by Point d_0.
2. Maximum discounted cumulative expenditure on the project, $-\Sigma A_{DCF\,max}$, given by Point d, having a discount factor of 10% in this case.
3. Cumulative net annual cash flow, ΣA_{CF}.
4. Cumulative net annual discounted-cash flow, ΣA_{DCF}, or net present value (*NPV*), where:

$$(NPV) = \Sigma A_{DCF} \qquad (3)$$

5. Capitalized cost, C_K, where:

$$C_K = \left[C_{FC} - \frac{S}{(1+i)^n}\right]\left[\frac{(1+i)^n}{(1+i)^n - 1}\right] \qquad (4)$$

Comparisons on the basis of interest can be summarized as:

1. The discounted-cash-flow rate-of-return (*DCFRR*), which is the fractional interest rate, *i*, that satisfies the following relations:

$$(NPV) = \Sigma A_{DCF} = 0 \qquad (5)$$

2. Those based on:[1]

$$PERCENT\ RATE\ OF\ RETURN = \left(\frac{ANNUAL\ PROFIT}{INVESTED\ CAPITAL}\right)100 \qquad (6)$$

The rate of return can be calculated on the basis of

Originally published November 1973.

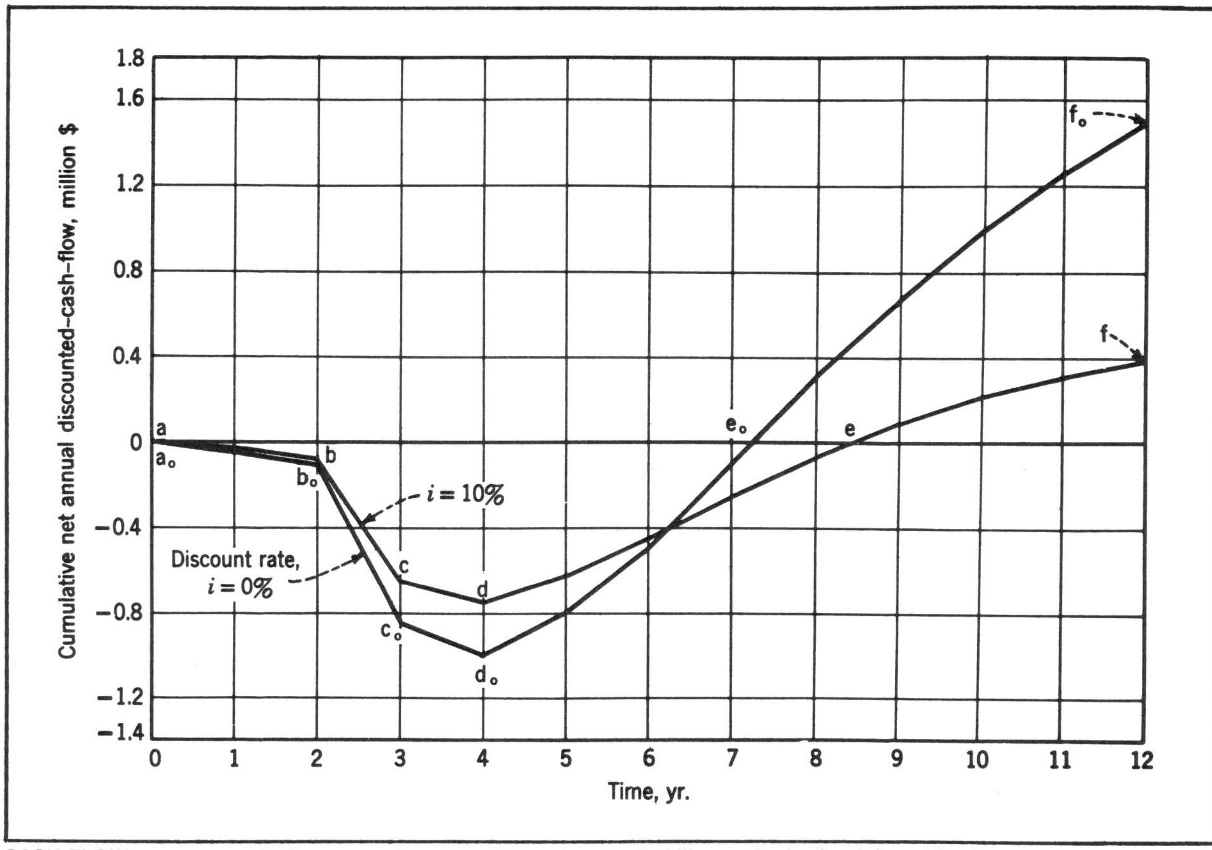

CASH-FLOW curves provide the means for assessing project profitability on the basis of time—Fig. 1

Cash Flow Data and Net Present Values for Projects A and B—Table I

(All values are thousands of dollars)

	Year	A_{CI}	$A_D + A_A$	A_{NP}	A_{IT} (t = 0.5)	A_{NCI}	A_{TC}	A_{CF}	ΣA_{CF}	A_{DCF} (i = 10%)	ΣA_{DCF} = (NPV)
Project A	0	0	0	0	0	0	200	−200	−200	−200.000	−200.000
	1	0	0	0	0	0	1,300	−1,300	−1,500	−1,181.818	−1,381.818
	2	−180	+150	−330	−165	−15	220	−235	−1,735	−194.215	−1,576.033
	3	+1,000	+150	+850	+425	+575	0	+575	−1,160	+432.000	−1,144.027
	4	+1,000	+150	+850	+425	+575	0	+575	−585	+392.733	−751.294
	5	+1,000	+150	+850	+425	+575	0	+575	−10	+357.030	−394.264
	6	+1,000	+150	+850	+425	+575	0	+575	+565	+324.573	−69.691
	7	+1,000	+150	+850	+425	+575	0	+575	+1,140	+295.066	+225.375
	8	+800	+150	+650	+325	+475	0	+475	+1,615	+221.591	+446.966
	9	+600	+150	+450	+225	+375	0	+375	+1,990	+159.037	+606.003
	10	+380	+100	+280	+140	+240	−420	+660	+2,650	+254.459	+860.462

	Year	A_{CI}	$A_D + A_A$	A_{NP}	A_{IT} (t = 0.5)	A_{NCI}	A_{TC}	A_{CF}	ΣA_{CF}	A_{DCF} (i = 10%)	ΣA_{DCF} = (NPV)
Project B	0	0	0	0	0	0	+1,520	−1,520	−1,520	−1,520.000	−1,520.000
	1	−80	+150	−330	−115	+35	+200	−165	−1,685	−150.000	−1,670.000
	2	+700	+150	+550	+275	+425	0	+425	−1,260	+351.240	−1,318.760
	3	+700	+150	+550	+275	+425	0	+425	−835	+319.309	−999.451
	4	+1,000	+150	+850	+425	+575	0	+575	−260	+392.733	−606.718
	5	+1,000	+150	+850	+425	+575	0	+575	+315	+357.030	−249.688
	6	+1,000	+150	+850	+425	+575	0	+575	+890	+324.573	+74.885
	7	+1,000	+150	+850	+425	+575	0	+575	+1,465	+295.066	+369.951
	8	+800	+150	+650	+325	+475	0	+475	+1,940	+221.591	+591.542
	9	+400	+150	+250	+125	+275	0	+275	+2,215	+116.627	+708.169
	10	+80	−50	+130	+65	+15	−420	+435	+2,650	+167.711	+875.880

Comparing Profitabilities of Projects A and B—Table II

Profitability Techniques	Project A	Project B
Time Measures		
Duration	10 yr.	10 yr.
Earning life	9 yr.	10 yr.
Breakeven point:		
Undiscounted, (*BEP*)	5.02 yr.	4.45 yr.
Discounted, (*DBEP*)	6.24 yr.	5.77 yr.
Equivalent maximum investment period, (*EMIP*)	2.93 yr.	2.81 yr.
Interest recovery period, (*IRP*)	9.37 yr.	8.57 yr.
Payback period, (*PBP*)	2.67 yr.	3.07 yr.
Capital Measures		
Maximum net outlay:		
Undiscounted, $-A_{CF\ max}$	$1,735,000	$1,685,000
Discounted at 10%, $-A_{DCF\ max}$	$1,576,033	$1,670,000
Maximum return:		
Undiscounted	$2,650,000	$2,650,000
Discounted at 10%, (*NPV*)	$ 860,462	$ 875,880
Capitalized cost, C_K	$2,457,333	$2,315,690
Interest Measures		
Average rate of return:		
On original investment	40.92%/yr.	35.72%/yr.
On depreciated investment	80.52%/yr.	64.88%/yr.
On average capital investment	65.77%/yr.	54.84%/yr.
Discounted-cash-flow rate-of-return, (*DCFRR*)	20.68%/yr.	19.68%/yr.
Capital rate-of return ratio, (*CRR*)	49.59%	51.98%

Nomenclature

A_A	Annual tax allowance exclusive of depreciation, $/yr.
A_{CF}	Net annual cash flow, $/yr.
A_{CI}	Annual cash income, $/yr.
A_D	Annual depreciation charge, $/yr.
A_{DCF}	Net annual discounted-cash-flow, $/yr.
A_{IT}	Annual amount of income tax, $/yr.
A_{NCI}	Annual net cash income, $/yr.
A_{NNP}	Net annual profit after tax, $/yr.
A_{NP}	Net annual profit, $/yr.
A_S	Revenue from annual sales, $/yr.
A_{TE}	Total annual expense, $/yr.
(BEP)	Breakeven period, yr.
C_{FC}	Fixed capital cost, $
C_K	Capitalized cost of plant or equipment, $
C_L	Cost of land and other nondepreciable items, $
C_{TC}	Total capital cost, $
C_{WC}	Working capital cost, $
$(DBEP)$	Discounted breakeven period, yr.
$(DCFRR)$	Discounted-cash-flow rate of return, %.
i	Fractional annual interest rate, dimensionless
(IRP)	Interest recovery period, yr.
(NPV)	Net present value, $
$(NPV)_W$	Weighted net present value, $
p	Fractional probability of an event, dimensionless
(PBP)	Payback period, yr.
s'	Particular year from startup, yr.
S	Scrap or salvage value, $
t	Fractional tax rate, dimensionless

net annual profit before tax, A_{NP}; net annual profit after tax, A_{NNP}; annual cash income before tax, A_{CI}; or net annual cash income after tax, A_{NCI}. In addition, the rate of return can be calculated on the basis of the original total capital investment, C_{TC}, depreciated total capital investment in any particular year, or average total capital investment over the life of the project.

Time, Capital and Interest Measures

Table I lists cash-flow data for two projects, A and B. Project A represents a new product on which a feasibility study and a market survey were carried out. The results of the studies were promising and $200,000 was spent on a design study. It is company policy to treat such expenditures as part of the fixed capital cost of any project that is completed, and as overhead if the project is terminated before production commences.

In Year 1, a site was purchased for $20,000, and $1.28 million was spent on the plant and its erection. In Year 2, the plant was commissioned and $200,000 working capital provided. The costs of completing the project and the promotional drive to launch the product exceeded the revenue from the process in Year 2 by $200,000. The company agreed with the tax authority that $20,000 of this deficit was part of the fixed capital cost of the project, while the remaining $180,000 could be treated as a negative cash income allowable for tax purposes.

It was further agreed that, for the purpose of evaluating the allowable depreciation charge, the whole of the $1,500,000 depreciable capital cost could be written off against plant earnings over the estimated working life of 10 yr., any difference to be adjusted in the last year of operation. In the tenth year of the project, the process was discontinued as part of a licensing agreement with Company B. The cost of land and the working capital were recovered, and $200,000 was also received as the salvage value of the plant.

Company B agreed to purchase a package-deal plant on license from Company A. Company B paid $1,500,000, and erected the plant on a site costing $20,000 in Year 0. In Year 1, the plant was commissioned, $200,000 working capital was provided, and a loss of $80,000 resulted. It was agreed with the tax authority that this loss would be allowable for tax purposes, and the same arrangement for depreciation was agreed as was for Company A. In the tenth year of the project, the process was discontinued due to competition from a replacement product. The cost of land and working capital were recovered, and $200,000 was also received as the salvage value of the plant.

Table I presents the calculation of the (*NPV*)s for

Projects A and B on the basis of a cost of capital of 10%, and a tax rate of 50%. The salvage value of the plant of $200,000, received in excess of the agreed value of zero, is treated as a negative capital allowance, A_A, while the agreed depreciation, A_D, of the $1,500,000 fixed capital investment, C_{FC}, is an allowable expense from the year of startup of the plant. It is assumed that the company can offset such allowable expenses against other income in the year to which the allowance applies.

Table II lists various measures of profitability for Projects A and B. These measures are derived from the data of Table I, using methods and definitions presented in earlier articles of this series. The payback period (PBP) is measured from the time of plant startup, while traditional return-on-investment methods consider only those years in which the return is positive. For this example, the traditional measure chosen is the net annual profit before tax, A_{NP}, assuming that the annual depreciation charge, A_{BD}, as shown in the balance sheet of the companies is the same as the total allowable charge for tax purposes, $(A_D + A_A)$.

Preferences From Comparative Profitabilities

In deriving Table II, it will be noted that the fixed capital cost, the cost of land, and the working capital are identical for each project—being $1,500,000, $20,000 and $200,000, respectively. In addition, the following measures are identical for the two projects: duration is 10 yr., total cash income is $6,600,000, total tax allowances received $\Sigma(A_D + A_A)$ are $1,300,000, total net profit ΣA_{NP} is $5,300,000, total tax paid ΣA_{IT} is $2,650,000, total income after tax ΣA_{NCI} is $3,950,000, total capital outlay $\Sigma A_{TC\,max}$ is $1,720,000, total after tax cash flow ΣA_{CF} is $2,650,000. Thus, the projects differ mainly in the timing of the cash flows. The various measures indicate the following preferences:

Time—Project B is favored, except for the payback period. However, if all payments prior to startup were compounded to the beginning of the startup year, then all measures would favor Project A.

Capital—The undiscounted outlay favors Project B, while discounting at 10% favors Project A. The maximum return is the same if not discounted, while discounting at 10% favors Project B. The capitalized cost favors Project B because the earning life of Project A is only nine years.

Interest—With the exception of the capital-return ratio, these measures favor Project A. Notice how erratically the traditional measures evaluate the relative worth of the projects, compared with the very small difference shown by the true rates of interest of the projects, i.e., the ($DCFRR$). Incidentally, if payments prior to startup are compounded to the beginning of the startup year at the cost of capital, an incorrect value of the ($DCFRR$) of Project A would be obtained, since such compounding should be at an interest rate equivalent to the ($DCFRR$). A plot of capital-return ratio (CRR) against time can be obtained from the (NPV) column of Table I by dividing each annual entry by $-\Sigma A_{CF\,max}$. Only the (CRR) for the final year of the project is listed in Table II.

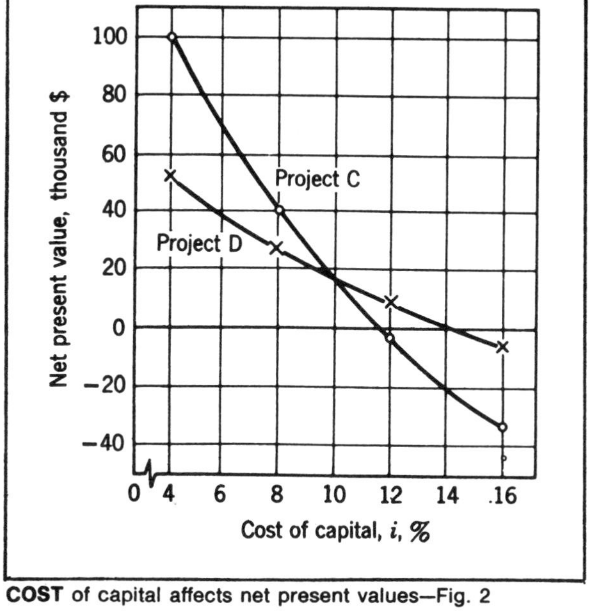

COST of capital affects net present values—Fig. 2

Comparisons on Basis of (NPV) and (DCFRR)

When comparing project profitability, the ranking on the basis of net present value (NPV) may differ from that on the basis of discounted-cash-flow rate-of-return ($DCFRR$). Let us consider the data for two projects:

Cost of Capital, i%	Project C (NPV), $	Project D (NPV), $
4	+100,000	+62,000
8	+41,000	+28,000
12	−2,000	+10,000
16	−32,000	−4,000

These (NPV) data are plotted against the cost of capital, as shown in Fig. 2. The discounted-cash-flow rate-of-return is the value of i that satisfies Eq. (5). We can estimate i from Fig. 2, and get a ($DCFRR$) of 11.8% for Project C, and 14.7% for Project D. Thus, on the basis of (DCFRR), Project D is more profitable than Project C.

The (NPV) of Project C is equal to that of Project D at a cost of capital, $i = 9.8\%$. Hence, Projects C and D are equally profitable on the basis of (NPV). If the cost of capital is greater than 9.8%, Project D has the higher (NPV) value and is, therefore, the most profitable. If the cost of capital is less than 9.8%, Project C has the higher (NPV) and is more profitable.

Benefit of Early Cash Flows

It pays to receive cash inflows as early as possible, and to delay cash outflows as long as possible.

Let us consider the net annual cash flows (after tax), A_{CF}, for Projects E, F and G, as listed in Table III. The cumulative annual cash flows, ΣA_{CF}, and cumulative discounted annual cash flows, ΣA_{DCF}, having a discount factor of 10% for these projects, are also listed in Table III. Note that the cumulative annual cash flow for each project is +$1,000.

The $(DCFRR)$ is the discount rate that satisfies Eq. (5) in the final year of the project. We can approximate the $(DCFRR)$ for each project as follows:

For Project E:

$$\Sigma A_{CF} = +\$1{,}000 \text{ in Year 3 for } i = 0\%$$
$$\Sigma A_{DCF} = +\$131 \text{ in Year 3 for } i = 10\%$$
$$\Sigma A_{DCF} = 0 \text{ in Year 3 for } i = (DCFRR)$$

Therefore:

$$\frac{1{,}000}{1{,}000 - 131} \cong \frac{(DCFRR)}{10}$$
$$(DCFRR) \cong 11.5\%$$

Similarly for Project F:

$$\frac{1{,}000}{1{,}000 + 184} \cong \frac{(DCFRR)}{10}$$
$$(DCFRR) \cong 8.4\%$$

Similarly for Project G:

$$\frac{1{,}000}{1{,}000 + 26} \cong \frac{(DCFRR)}{10}$$
$$(DCFRR) \cong 9.7\%$$

In terms of net present value (NPV), the projects in order of merit are E, G and F, with (NPV) values of +\$131, −\$26 and −\$184, respectively. In terms of $(DCFRR)$, the projects in order of merit are also E, G and F, with $(DCFRR)$ values of 11.5%, 9.7% and 8.4%, respectively.

Data for Projects E, F and G—Table III

Project E

Year	A_{CF} \$	Discounted at 10% ΣA_{CF} \$	f_d	A_{DCF} \$	ΣA_{DCF} = (NPV) \$
0	−5,000	−5,000	1.0000	−5,000	−5,000
1	+3,000	−2,000	0.90909	+2,727	−2,273
2	+2,000	0	0.82645	+1,653	−620
3	+1,000	+1,000	0.75131	+751	+131

Project F

Year	A_{CF} \$	Discounted at 10% ΣA_{CF} \$	f_d	A_{DCF} \$	ΣA_{DCF} = (NPV) \$
0	−5,000	−5,000	1.0000	−5,000	−5,000
1	+1,000	−4,000	0.90909	+909	−4,091
2	+2,000	−2,000	0.82645	+1,653	−2,438
3	+3,000	+1,000	0.75131	+2,254	−184

Project G

Year	A_{CF} \$	Discounted at 10% ΣA_{CF} \$	f_d	A_{DCF} \$	ΣA_{DCF} = (NPV) \$
0	−5,000	−5,000	1.0000	−5,000	−5,000
1	+2,000	−3,000	0.90909	+1,818	−3,182
2	+2,000	−1,000	0.82645	+1,653	−1,529
3	+2,000	+1,000	0.75131	+1,503	−26

When To Scrap an Existing Process

A company invests \$50,000 in a manufacturing process that has positive net annual cash flows (after tax), A_{CF}, of \$10,000 in each year. During the third year of operation, an alternative process becomes available. The new process would require an investment of \$40,000 that would have positive net annual cash flows (after tax) of \$20,000 in each year. Cost of capital is 10%. It is estimated that a market will exist for the product for at least six more years. Should the company continue with the existing process (Project H), or should it scrap Project H and adopt the new process (Project I)?

The net annual cash flows, A_{CF}, and cumulative discounted annual cash flow, ΣA_{DCF}, for a discount factor of 10% are listed in Table IV for the two projects. At the end of Year 9, the net present values are:

$$(NPV) = +\$35{,}390 \text{ for Project I}$$
$$(NPV) = +\$7{,}591 \text{ for Project H}$$

The difference is +\$27,799, which is numerically greater than the money lost by the end of Year 3 for Project H. Thus, Project H should be scrapped and the new Project I adopted, if only economic reasons need be considered.

End Project H and Adopt Project I—Table IV

Project H

Year	Discounted at 10% A_{CF} \$	f_d	A_{DCF} \$	ΣA_{DCF} = (NPV) \$
0	−50,000	1.0000	−50,000	−50,000
1	+10,000	0.90909	9,091	−40,909
2	+10,000	0.82645	8,265	−32,644
3	+10,000	0.75131	7,513	−25,131
4	+10,000	0.68301	6,830	−18,301
5	+10,000	0.62092	6,209	−12,092
6	+10,000	0.56447	5,645	−6,447
7	+10,000	0.51316	5,132	−1,315
8	+10,000	0.46651	4,665	+3,350
9	+10,000	0.42410	4,241	+7,591

Project I

Year	Discounted at 10% A_{CF} \$	f_d	A_{DCF} \$	ΣA_{DCF} = (NPV) \$
3	−40,000	0.75131	−30,052	−30,052
4	+20,000	0.68301	+13,660	−16,392
5	+20,000	0.62092	+12,418	−3,974
6	+20,000	0.56447	+11,289	+7,315
7	+20,000	0.51316	+10,263	+17,578
8	+20,000	0.46651	+9,330	+26,908
9	+20,000	0.42410	+8.482	+35,390

Incremental Comparisons

A company may have the choice of, say, investing \$10,000 in Project J that will give a $(DCFRR)$ of 16%, or \$7,000 in Project K that will give a $(DCFRR)$ of 18%. Should it spend \$10,000 on Project J, or spend only

\$7,000 on Project K and invest the difference of \$3,000 elsewhere?

Both projects have lives of 10 years, and constant positive net-annual-cash-flows, A_{CF}, of \$2,069 and \$1,558 for Projects J and K, respectively. The corresponding (NPV)s at a discount factor of 10% are +\$2,710 and +\$2,560, respectively. These data are summarized as:

	Project J	Project K	Project (J-K)
A_{CF}, \$ . . . . (In Year 0)	−10,000	−7,000	−3,000
A_{CF}, \$ . . . . (In each of Years 1 to 10)	+2,069	+1,558	+511
(NPV), i = 10%, \$ . .	+2,710	+2,560	+150
$(DCFRR)$, %	16	18	12.4

From the difference in cash flows between the projects, the discounted-cash-flow rate-of-return $(DCFRR)$ for Project (J-K) can be shown as 12.4%. This is significantly lower than for either Project J or Project K. Thus, if the \$3,000 can be invested to give a return greater than 12.4%, Project K should be chosen in preference to Project J.

Comparisons on the Basis of Capitalized Cost

A machine in a process generates a positive net annual cash flow of \$1,000. Two alternatives are available: Machine L costing \$2,000 requires replacement every 4 yr., and Machine M costing \$3,000 requires replacement every 6 yr. Neither machine has any scrap value. The cost of capital is 10%. Which machine is the most profitable to operate?

In this case, the lives of the machines are unequal, and the comparison is conveniently made on the basis of capitalized cost. This puts lives on the same basis, which is an infinite number of years. The net annual cash flows generated by each machine are equal.

The capitalized cost, C_K, of a piece of equipment of fixed capital cost, C_{FC}, is the amount of capital required to ensure that the equipment may be renewed in perpetuity. For a piece of equipment with no scrap value, C_K is given by:

$$C_K = C_{FC}\left[\frac{(1+i)^n}{(1+i)^n - 1}\right] \quad (7)$$

For Machine L:

$$C_K = (\$2{,}000)(3.15471) = \$6{,}309.42$$

For Machine M:

$$C_K = (\$3{,}000)(2.29607) \doteq \$6{,}888.21$$

(Values for the capitalized cost factor may be found on p. 20.) Thus, Machine L with the lower capitalized cost is the most profitable to operate.

Comparative Profitability and Uncertainty

Table V lists the results of the sensitivity analysis carried out on a specific project in Part 6 of this series.

So far nothing has been said about the probability of any of the events listed in Table V occurring. For example, what is the probability of the revenue from annual sales, A_S, falling 10% below the estimated value, or the probability of the total annual cost, A_{TE}, rising 10% above the estimate.

Results of Sensitivity Analysis—Table V

Case	(NPV) i = 10%	$(DCFRR)$
Base case	\$276,210	16.4%
a. A_S reduced 10%/yr.	\$135,450	12.9%
b. A_S reduced 20%/yr.	−\$5,330	9.8%
c. A_{TE} increased 10%/yr.	\$238,430	15.0%
d. C_{FC} increased 10%	\$206,890	14.0%
e. Combined: A_S reduced 10%/yr., A_{TE} increased 10%/yr., C_{FC} increased 10%	\$28,420	10.6%

An event or outcome that will definitely occur can be assigned a probability, $p = 1$. An event or outcome that will definitely not occur can be assigned a probability, $p = 0$. When a coin is tossed, there is an equal chance of the outcome being a head or a tail. In this case, the probability of a head is $p_1 = 0.5$, and the probability of a tail is $p_2 = 0.5$, where:

$$p_1 + p_2 = 1.0 \quad (8)$$

The net present values (NPV) resulting from the various economic circumstances listed in Table V can each, including the base case, be assigned a probability, p. A weighted average, $(NPV)_W$, known as the expected value, can be calculated from:

$$(NPV)_W = p_1(NPV)_1 + p_2(NPV)_2 + \ldots. \quad (9)$$

Where

$$p_1 + p_2 + p_3 + \cdots\cdot = 1.0 \quad (10)$$

If the base case is certain to occur, then $p_1 = 1.0$; and p_2 etc. all equal zero.

Eq. (9) gives the expected net present value $(NPV)_W$. Analogous equations may be written in terms of other quantities that measure profitability. Methods for estimating the value of the probability, p, for the occurrence of a particular outcome will be discussed in the next article.

Example Illustrates Effects of Uncertainty

The project described in Table V is for the manufacture of a product at a maximum production rate of 5.2 million lb./yr at a sales price of 10¢/lb. Although the process is protected by patents up to the end of Year 10, it is likely that competition will enter the field in Year 8 with an improved product. It is assumed that this will result in a reduction in both selling price and sales volume.

The following three courses of action are being considered in an attempt to combat competition:

a. Reducing the selling price for the base case in Year 8 by 2¢/lb., with further reductions of 1¢/lb. in each

Sales Prices, Volumes and Revenues for Optional Market Strategies—Table VI

	Option a			Option b			Option c		
Year	Sales Volume, Lb/Yr.	Sales Price, ¢/Lb.	Revenue, A_S \$/Yr.	Sales Volume, Lb/Yr.	Sales Price, ¢/Lb.	Revenue, A_S \$/Yr.	Sales Volume, Lb/Yr.	Sales Price, ¢/Lb.	Revenue, A_S \$/Yr.
1	4,000,000	10	400,000	3,600,000	12	432,000	4,000,000	8	320,000
2	5,000,000	10	500,000	4,500,000	12	540,000	5,000,000	8	400,000
3	5,000,000	10	500,000	4,500,000	12	540,000	5,000,000	8	400,000
4	5,000,000	10	500,000	4,500,000	12	540,000	5,000,000	8	400,000
5	5,200,000	10	520,000	4,680,000	10	468,000	5,200,000	8	416,000
6	5,200,000	10	520,000	5,200,000	9	468,000	5,200,000	8	416,000
7	5,200,000	10	520,000	5,200,000	8	416,000	5,200,000	8	416,000
8	4,875,000	8	390,000	4,875,000	7	341,250	5,200,000	8	416,000
9	5,000,000	7	350,000	5,000,000	6	300,000	5,200,000	8	416,000
10	4,667,000	6	280,000	4,667,000	5	233,330	5,200,000	8	416,000

subsequent year, in order to maintain the revenues from annual sales at the estimated values.

b. Making the initial selling price 12¢/lb. This should reduce sales by 10% in the first five years, and bring in competition in Year 5. The selling price should be reduced by 2¢/lb. in Year 5, with further reductions of 1¢/lb. in each subsequent year to maintain the annual sales volumes at the estimated values from Year 5 onwards.

c. Making the initial selling price 8¢/lb. This should keep competition out during the 10-yr. proposed life of the plant. The estimated sales volumes should not be affected up to Year 7. However, in subsequent years the sales volume should continue at the maximum output of 5.2 million lb./yr.

The sales data for the three options are listed in Table VI.

Determine the best pricing strategy on the basis of the following information and assumptions:

1. Sales forecasts made by the sales division for the last 10 projects were correct to ±5% in seven cases, from 5 to 15% high in two cases, and 20% high in one case.

2. In all cases, the total annual expense, A_{TE}, is 0, \$100,000, \$100,000, \$110,000, \$120,000, \$130,000, \$130,000, \$140,000, \$140,000, \$150,000 and \$160,000 in Years 0 through 10, respectively.

3. The following data are also given: straight-line depreciation, and 10% discount factor; fixed capital investment. C_{FC} = \$1,000,000; zero salvage value of plant items; working capital, C_{WC} = \$90,000; cost of land, C_L = \$10,000; no tax allowances other than depreciation, i.e., $A_A = 0$; and fractional tax rate, $t = 0.50$.

The data given above in items 2 and 3 are for the base case in the example described in Part 6 of this series.[4]

Option a—The net present value (NPV) for the base case has already been calculated and listed in Table V to bc +\$276,210. Likewise, the (NPV) for Case a in Table V is found to be +\$135,450; and for Case b is found to be −\$5,330.

Since seven out of ten previous sales forecasts were correct to ±5%, the probability for the base case, $P_1 = 0.7$. Since two out of ten previous sales forecasts were from 5 to 15% (mean 10%) too high, the probability of the revenue being reduced in each year by 10% is $p_2 = 0.2$. Since only one out of ten previous sales forecasts was 20% too high, the probability of the revenue being reduced in each year by 20% is $p_3 = 0.1$.

Weighted Net Present Values for Optional Market Strategies—Table VII

Sales Revenues	Option a (NPV) \$	Option b (NPV) \$	Option c (NPV) \$
As estimated	+276,210	+247,500	+84,760
10% down	+135,450	+109,600	−36,870
20% down	−5,330	−28,310	−158,500
$(NPV)_W$	+219,900	+192,340	+36,110

The expected value of the net present value, $(NPV)_W$, is found from Eq. (9) to be:

$$(NPV)_W = p_1(NPV)_1 + p_2(NPV)_2 + p_3(NPV)_3$$
$$(NPV)_W = 0.7(\$276,210) + 0.2(\$135,450) - 0.1(\$5,330)$$
$$(NPV)_W = \$219,900$$

Options b and c—Net present values are calculated in a similar way, and the results are as listed in Table VII. Expected values for $(NPV)_W$ are calculated as for Option a, and the results are also listed in Table VII. From these results, we see that the apparently attractive option of eliminating the competition altogether is the least profitable procedure under the circumstances proposed in this example. ■

Engineering Economics—Part 8

Uncertainty in project profitabilities requires assessing the chances of attaining forecasted cash flows, so that maximum and minimum values of profitability can be established.

Statistical Techniques Improve Decision-Making

F. A. HOLLAND, F. A. WATSON and J. K. WILKINSON, University of Salford

Discounted-cash-flow rates of return (*DCFRR*) and net present values (*NPV*) for future projects can never be predicted absolutely because the cash-flow data for such projects are subject to uncertainty. Therefore, when stating predicted values of (*DCFRR*) and (*NPV*) for projects, it is also desirable to give a measure of confidence in the predictions.

For example, for a particular project it may have been estimated that there is a 90% chance of the (*DCFRR*) being greater than 10%, 50% chance of its being greater than 16%, and only a 10% chance of being greater than 20%. Management retains the power of decision to proceed with the project or not, but the probability data provide desirable information for the decision.

The estimation of probabilities requires the use of statistics. Thus, statistical methods play an increasing role in decision-making.

Predictions From Limited Data

Predictions of future sales price, sales volume, etc., are normally based on a very limited amount of data about past events. Furthermore, it would not be convenient to use the entire population of past events even if it were available. A statistic is a measure, based on limited information from a sample, that allows the corresponding parameter of the population to be estimated.

For example, the mean value, $\bar{x}$, of a property, x, is a statistic based on a sample of n items. It is defined by:

$$\bar{x} = \frac{x_1 + x_2 + x_3 + \cdots x_n}{n} \tag{1}$$

Originally published December 1973.

The mean $\bar{x}$ is the statistic corresponding to the population parameter, μ, which is the arithmetic average of all the items in the population. The population may consist of 1,000 items, while the sample may consist of only 10 items.

In many cases, the x values will not all be different. In such circumstances, Eq. (1) can be written as:

$$\bar{x} = \frac{\Sigma x_i f(x_i)}{\Sigma f(x_i)} \tag{2}$$

where $f(x_i)$ is the frequency with which a particular value x_i occurs. If $f(x_i)$ is obtained by counting the number of occurrences of the value x_i, then $f(x_i)$ will have an integer value that may, of course, be zero; and $\Sigma f(x_i)$

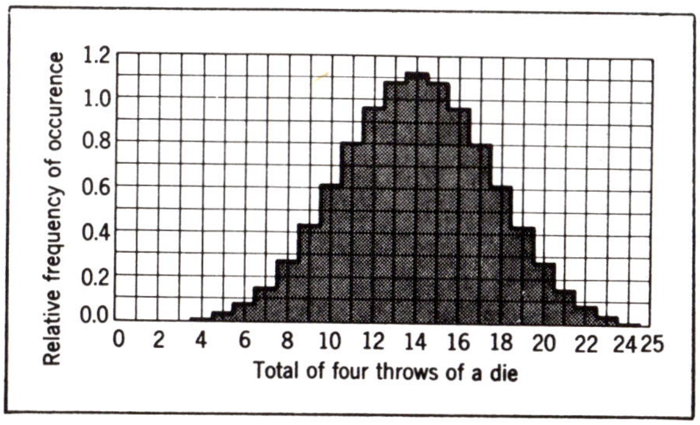

HISTOGRAM indicates frequency distribution for the total score obtained in four throws of a die—Fig. 1

will be the total number of items, n, in the sample. It is often convenient to divide the frequency of occurrence by n. In this case, $f(x_i)$ becomes the relative frequency of occurrence of the value x_i, and $\Sigma f(x_i) = 1$.

The values of x may either be discrete or continuous. The number of sales of, say, automobiles in any one day must be an integer. If a business sells four automobiles, this represents all possible values of x in the range of 3.5 to 4.5. It is normal practice to diagrammatically represent the frequency of various values of x as a bar chart, or as a histogram with the frequency constant over the range of the x-interval, as shown in Fig. 1.

When x represents a continuously variable quantity, it is sometimes convenient to take the total or relative frequency of occurrences within a given range of x-values. These frequencies can then be plotted against the mid-values of x to form a histogram. In this case, the ordinate should be the frequency per unit width of x. This makes the area under any bar proportional to the probability that the value of x will lie in the given range. If the relative frequency is plotted as ordinate, the sum of the areas under all the bars is unity.

If x is a continuous variable, and the interval ranges are made smaller and smaller, a smooth curve will eventually result. The area under such a curve between x_1 and x_2 represents the probability that a randomly selected item will have a value of x lying in the range x_1 to x_2. This is the information that is desired.

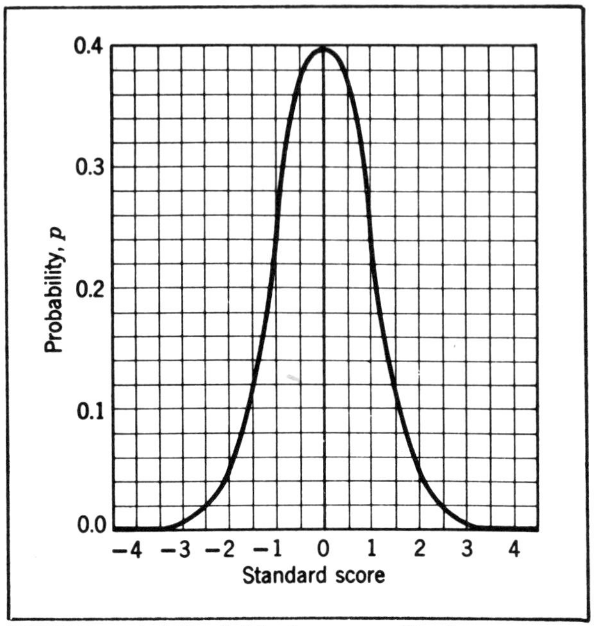

NORMAL probability distribution curve—Fig. 2

Frequency-Distribution Curves

Data available from past experience can be used to generate frequency-distribution curves. It is essential for a company to have an efficient commercial-intelligence system to assess market conditions.

Accuracy of sales forecasting can also be increased by a careful study of past sales records, price trends, etc. However, the uncertainty of an estimate increases the further into the future that the estimate is projected.

A great deal of sales and other types of forecasting is based on the opinions of experts. Experts should be able to estimate maximum, minimum, and most-likely or modal values for a quantity. The modal value is not necessarily midway between the minimum and maximum values since many distributions are skewed. An expert may be asked to estimate the probability of the occurrence of certain values on each side of the mode.

When experts are questioned separately, the procedure is known as the Delphi method. Strictly speaking, this method requires that the opinion of each expert be assessed by a coordinator, who then feeds the results back to see if the opinions of one expert are modified by those of others. The process is repeated until agreement is reached. In practice, the procedure is too tedious to be repeated more than once.

It is useful to make a comparison of the past predictions of each expert with the results obtained in practice. This information enables the opinions to be weighted by the coordinator. When the experts work in close collaboration, it is not possible to avoid some collusion. In this case, it is better to arrive at a single consensus opinion by a free and open discussion. This is the Think Tank method. Its main disadvantage is that rank or aggressiveness might unduly weight one or more opinions.

The opinions of the experts, however obtained, provide a basis for plotting a frequency- or probability-distribution curve. If the relative frequency is plotted as ordinate, the total area under the curve is unity. The area under the curve between two values of the quantity is the probability that a randomly selected value will fall in the range between the two values of the quantity. These probabilities are merely estimates and their reliability depends on the skill of the forecasters.

The estimated $(DCFRR)$ and the estimated (NPV) are both functions of the estimated cumulative revenue from annual sales, ΣA_S, the estimated cumulative total annual cost or expense, ΣA_{TE}, and the estimated fixed capital cost, C_{FC}, of the plant. The revenue from annual sales for each year is in turn the product of the sales price and sales volume. Initially it is desirable to select those values from the distribution curves of ΣA_S, ΣA_{TE} and C_{FC}, which enable the maximum and minimum $(DCFRR)$ and (NPV) values to be calculated.

If the maximum values of $(DCFRR)$ and (NPV) are not acceptable to the company, then the project should be promptly rejected. If the minimum values of $(DCFRR)$ and (NPV) are acceptable, then a detailed assessment should be made. If the maximum values of $(DCFRR)$ and (NPV) are acceptable but the minimum are not, the feasibility study should be continued.

Mathematical Models for Distribution Curves

Mathematical models have been developed to fit the various distribution curves. It is most unlikely that any frequency-distribution curve, obtained in practice, will exactly fit a curve plotted from any of these mathematical models. Nevertheless, the approximations are extremely useful—particularly in view of the inherent inaccuracies in practical data. The most common are the Binomial,

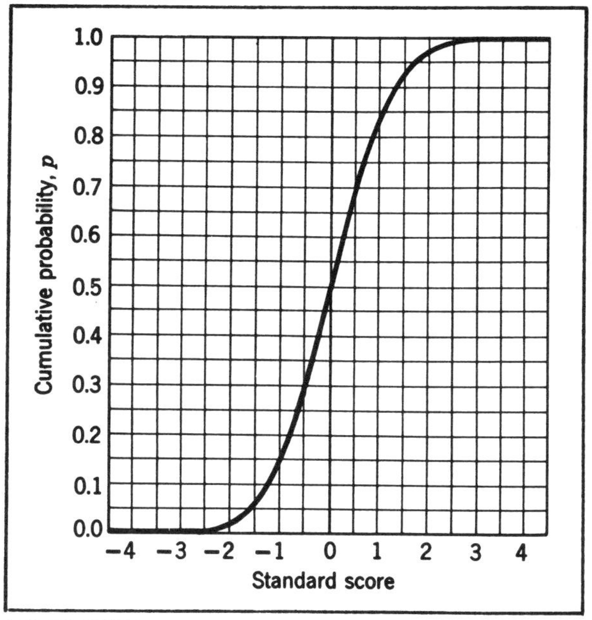

CUMULATIVE Normal probability distribution—Fig. 3

Poisson, and Normal or Gaussian distributions.[3]

A Normal distribution curve is bell-shaped, as shown in Fig. 2. The curve obeys the relationship:

$$f(x) = \frac{\exp[-(x-\mu)^2/2\sigma^2]}{\sigma(2\pi)^{0.5}} \qquad (3)$$

where σ is known as the true standard deviation. The standard deviation, s^*, from a sample is given by:

$$s^* = \left[\frac{\Sigma(x_i - \bar{x})^2 f(x_i)}{\Sigma f(x_i) - 1}\right]^{0.5} \qquad (4)$$

The standard deviation, s^*, for the sample corresponds to the true standard deviation, σ, for the whole population in the same way that the mean, $\bar{x}$, of the sample corresponds to the arithmetic average, μ, for the whole population. Eq. (3) can be written more compactly as:

$$f(z) = \frac{\exp(-z^2/2)}{(2\pi)^{0.5}} \qquad (5)$$

where the standard score, z, is:

$$z = (x - \mu)/\sigma \qquad (6)$$

The area under the curve of $f(z)$ is unity if the abscissa extends from minus infinity to plus infinity. The area under the curve between z_1 and z_2 is the probability that a randomly selected value of z will lie in the range z_1 to z_2 since this is the relative frequency with which that range of values would be represented in an infinite number of trials.

An event that will definitely occur has a probability of unity. An event that will definitely not occur has a probability of zero.

Eq. (5) can be integrated between limits to determine the probability that a random value lies between the selected limits. Extensive tables of $f(z)$ and the associated integral are available.[1] An abbreviated version is given in Table I.

Normal Distribution Curve—Table I

Standard Score, z	Ordinate of Normal Curve, $f(z)$	Area Under Right-Hand Side of Curve, $\int_0^z f(z)dz$
0.000	0.3989	0.00000
0.100	0.3970	0.03983
0.200	0.3910	0.07926
0.253	0.3864	0.10000
0.300	0.3814	0.11791
0.400	0.3683	0.15542
0.500	0.3521	0.19146
0.524	0.3478	0.20000
0.600	0.3332	0.22575
0.700	0.3123	0.25804
0.800	0.2897	0.28814
0.842	0.2798	0.30000
0.900	0.2661	0.31594
1.000	0.2420	0.34134
1.200	0.1942	0.38493
1.282	0.1953	0.40000
1.400	0.1497	0.41243
1.600	0.1109	0.44520
1.645	0.1032	0.45000
1.800	0.0790	0.46407
1.881	0.0680	0.47000
2.000	0.0540	0.47725
2.054	0.0484	0.48000
2.326	0.0266	0.49000
2.500	0.0175	0.49379
2.576	0.0145	0.49500
2.748	0.0092	0.49700
3.000	0.0044	0.49865
3.090	0.0034	0.49900
3.291	0.0018	0.49950
3.429	0.0011	0.49970
3.500	0.0009	0.49977
3.719	0.0004	0.49990
4.000	0.0001	0.49997
∞	0.0000	0.50000

Data abridged and recalculated from Fisher, R.A. and Yates, F., "Statistical Tables for Biological, Agricultural and Medical Research," 5th ed., pp. 42-43, Oliver and Boyd, London, 1957.

A frequency-distribution curve can be used to plot a cumulative-frequency curve. This is the curve of most importance in business decisions. Fig. 3 shows a cumulative-frequency curve plotted from a Normal frequency-distribution curve. The cumulative curve represents the probability of a random value, z, of having a value of, say, z_1 or less.

If a property or variable, c, is a function of several other variables, x_1, x_2, etc., it can be written in the form:

$$c = \phi(x_1, x_2, \cdots x_n) \qquad (7)$$

If each x is a Normally distributed independent variable, then:[2]

$$\sigma_c^2 = \left(\frac{\partial c}{\partial x_1}\right)^2 \sigma_1^2 + \left(\frac{\partial c}{\partial x_2}\right)^2 \sigma_2^2 + \cdots \left(\frac{\partial c}{\partial x_n}\right)^2 \sigma_n^2 \qquad (8)$$

where σ_c is the standard deviation of the variable, c, and σ_1, σ_2, etc., are the standard deviations of the variables x_1, x_2, etc.

Many distributions occurring in business situations are

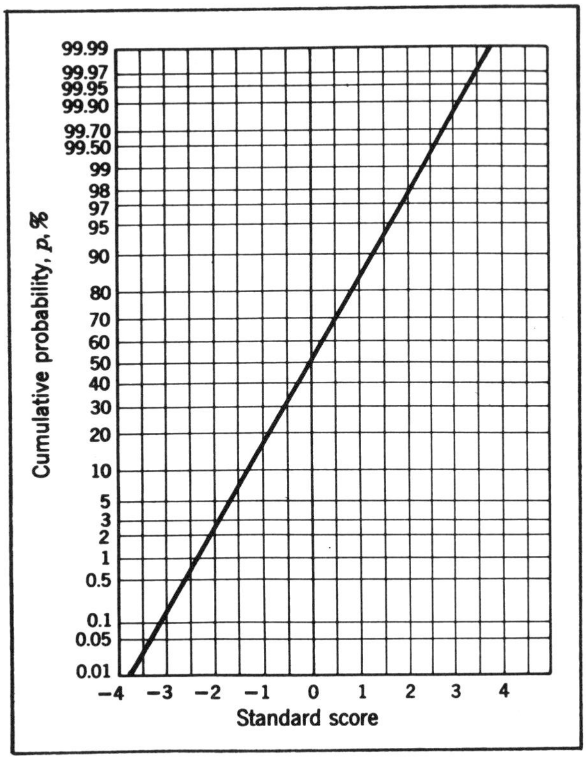

CUMULATIVE Normal distributions are straight lines on chart having probability scale—Fig. 4

not symmetrical but are skewed, and the Normal distribution curve is not a good fit. However, when data are based on estimates of future trends, the accuracy of the Normal approximation is usually acceptable. This is particularly the case as the number of component variables x_1, x_2, etc., in Eq. (7) increases. Although the distributions of the individual variables (x_1, x_2, . . .) may be skewed, the distribution of the property or variable c tends to approach the Normal distribution.

The third column in Table I gives the area under the right-hand side of the Normal frequency-distribution curve (Fig. 2) for a particular value of z. This area is also the corresponding ordinate in a cumulative Normal frequency-distribution curve such as the one in Fig. 3. Although this curve is S-shaped, it can be plotted as a straight line by rescaling the ordinate to give what is known as Normal probability graph-paper. Such paper may be constructed from the data of Table I, and is illustrated by Fig. 4.

Let us now consider an event that must have one of two outcomes. It must either occur with probability p_1, or fail to occur with probability p_2. Since these are exclusive events and the probability that something will happen is unity, it follows that:

$$p_1 + p_2 = 1 \qquad (9)$$

Provided that no learning process is involved (so that the value of p_1 is not influenced by previous results), the probability of x successes in n trials is given by the term containing p_1^x in the expansion of the binomial:

$$(p_1 + p_2)^n = p_1^n + \cdots \frac{n!}{x!(n-x)!} p_1^x p_2^{(n-x)} \cdots p_2^n \qquad (10)$$

Probability of Throwing a Certain Number A Given Number of Times on a Die—Table II

Probability per Throw	Number of Similar Throws \ Total Number Of Throws	1	2	3	4
⅙	0	0.8333	0.6944	0.5787	0.4823
	1	0.1667	0.2778	0.3472	0.3858
	2	—	0.0278	0.0695	0.1157
	3	—	—	0.0046	0.0154
	4	—	—	—	0.0008
⅕	0	0.8000	0.6400	0.5120	—
	1	0.2000	0.3200	0.3840	—
	2	—	0.0400	0.0960	—
	3	—	—	0.0080	—
¼	0	0.7500	0.5625	—	—
	1	0.2500	0.3750	—	—
	2	—	0.0625	—	—
⅓	0	0.6667	—	—	—
	1	0.3333	—	—	—

where x and n are integers, and $x!$ (read as x factorial) is the product of all integers from unity to x.

If a six-sided die marked with the numbers 1, 2, 3, 4, 5 and 6 is thrown, the probability that any given number will be uppermost is $\frac{1}{6}$. If the die is thrown twice in succession, then the probability of a given sequence of numbers occurring, say 5 followed by 6, is $(\frac{1}{6})(\frac{1}{6}) = \frac{1}{36}$. The chance of any particular number occurring zero, one, two, three or four times in four throws of the die (or in a simultaneous throw of four dice) is given by the successive terms of Eq. (10), expanded as:

$$\left(\frac{5}{6} + \frac{1}{6}\right)^4 = (1)\left(\frac{5}{6}\right)^4\left(\frac{1}{6}\right)^0 + (4)\left(\frac{5}{6}\right)^3\left(\frac{1}{6}\right)^1 + (6)\left(\frac{5}{6}\right)^2\left(\frac{1}{6}\right)^2 +$$

$$(4)\left(\frac{5}{6}\right)^1\left(\frac{1}{6}\right)^3 = (1)\left(\frac{5}{6}\right)^0\left(\frac{1}{6}\right)^4$$

$$= 0.4823 + 0.3858 + 0.1157 + 0.0154 + 0.0008 = 1$$

The distribution of the number of successes is skewed toward the low numbers. In particular, there is only a slightly better than even-money chance that any given number will occur even once in four throws. Such highly unsymmetrical distributions cannot be approximated by the Normal distribution curve.

However, let us consider the total number obtained in four throws of the die. A total of 4 can be obtained only by four throws of 1. A total of 24 can only be obtained by four throws of 6. Totals of 4 and 24 each have a relative frequency of $(\frac{1}{6})^4 = 0.0008$. The probability of a total of less than 4 or greater than 24 in four throws of the die is zero.

A total of 5 can be obtained by three throws of 1, and one throw of 2. The probability of three throws of 1 is 0.0154. The probability of throwing 2 from the possible numbers 2, 3, 4, 5 or 6 is $\frac{1}{5}$. The probability of throwing a total of 5 in four throws is, therefore, $(0.0154)(\frac{1}{5}) = 0.0031$. Similarly, the probability of throwing a total of 23 in four throws is also 0.0031.

All other totals can be obtained in more than one way. Each way involves throwing a certain number a given

Numerical Computation of the Cumulative Probability of Obtaining a Given Total By Throwing Dice—Table III

Number	4	5	6	7	8	9	10
Relative Frequency	0.0008	0.0031	0.0077	0.0154	0.0270	0.0432	0.0617
Number	11	12	13	14	15	16	17
Relative Frequency	0.0802	0.0965	0.1080	0.1127	0.1080	0.0965	0.0802
Number	18	19	20	21	22	23	24
Relative Frequency	0.0617	0.0432	0.0270	0.0154	0.0077	0.0031	0.0008

Computation Using the Normal Distribution Curve for Obtaining a Given Total by Throwing Dice—Table IV

Number	4	5	6	7	8	9	10
Relative Frequency	0.0001	0.0004	0.0005	0.0045	0.0114	0.0251	0.0479
Number	11	12	13	14	15	16	17
Relative Frequency	0.0792	0.1136	0.1413	0.1520	0.1413	0.1136	0.0792
Number	18	19	20	21	22	23	24
Relative Frequency	0.0479	0.0251	0.0114	0.0045	0.0015	0.0004	0.0001

number of times in four throws, with a probability of $\frac{1}{6}$ per throw, followed by throwing a different number a given number of times, with a probability of $\frac{1}{5}$ per throw. This is followed, if required, by throwing other numbers with probabilities of $\frac{1}{4}$ and $\frac{1}{3}$, respectively. The probability for each case is listed in Table II.

A total of 6 can be obtained by three throws of 1, and one throw of 3, or by two throws of 1 and two throws of 2. In the first case, the probability is $(0.0154)(\frac{1}{5}) = 0.0031$, and in the second case, it is $(0.1157)(0.0400) = 0.0046$. The probability of a total of 6 is, therefore, $0.0031 + 0.0046 = 0.0077$.

Probabilities for other totals are calculated in the same way, and listed in Table III. These data are used to plot the histogram in Fig. 1. Now, the distribution is symmetrical about the mean value of 14, and is almost Normal in shape. An increasing number of throws will result in totals that are even closer to the Normal distribution. This fact can be used to approximate such a distribution without the enormous labor of the calculations indicated by Tables II and III.

Possible values of the total of four throws of a die are integers from 4 to 24, and hence represent values in the range 3.5 to 24.5. The mean value, $\bar{x}$, of this range is given by Eq. (1) as: $\bar{x} = (3.5 + 24.5)/2 = 14.0$.

Table I shows that the cumulative probability of a Normally distributed variable lying within four standard deviations of the mean is 0.49997. Therefore, it is more than 99.99% (0.49997/0.50000) certain that a random value will be within $\pm 4\sigma$ from the mean. For practical purposes, σ may be taken as one-eighth of the range of certainty, and the standard deviation can be obtained:

$$s^* \cong (24.5 - 3.5)/8 = 2.625.$$

From Eq. (6), the standard score becomes:

$$z = \frac{x - \mu}{\sigma} \cong \frac{x = \bar{x}}{s^*}$$

For a total score of 4 (i.e., $x = 4$), the standard score is approximately $z = (4 - 14)/2.625 = -3.81$. Since the Normal curve is symmetrical about $z = 0$, the height of the ordinate at $z = -3.81$ is the same as that at $z = +3.81$. From Table I, the height of the ordinate is 0.0003 in units of $1/\sigma$. The relative frequency of 4 occurring is 0.0003/2.625. Values for other totals are calculated in the same way, and listed in Table IV.

This concept can be used to translate Delphi or other opinions into probability distributions, and hence into useful decision-making tools. An example will illustrate the procedure.

Predicting Size of Inventory

A store that is open 5 days/wk., is to promote a new product. The manager believes that not more than 5 units will be sold in any one day but he cannot be more precise about the probable sales pattern. Stocks are delivered once per week. What size should the first order be to give a 95% certainty of meeting demand?

Since the product is sold in units, the possible range of weekly sales is from −0.5 to +25.5 units. Therefore, the mean of the sales distribution will be:

$$\bar{x} \cong (-0.5 + 25.5)/2 = 13$$

The standard deviation for this example will be:

$$s^* \cong \frac{25.5 - (-0.5)}{8} = 3.25$$

From this, the approximate frequency distribution of daily sales can be derived by using Eq. (3), (5) or (6). The desired area to the right of $z = 0$ for the Normal probability distribution curve is 0.45. For this value, we obtain from Table I the standard score $z = 1.645$. Since the area to the left of $z = 0$ is 0.50, there is a probability of $(0.45 + 0.50) = 0.95$ that $z = 1.645$ will not be exceeded. Rearranging Eq. (6), we obtain $x = \mu + z\sigma$. When $z = 1.645$:

$$x \cong 13 + (1.645)(3.25) = 18.35$$

Hence, in order to be 95% certain of meeting demand, 19 units should be purchased.

Predicting Probability of Sales

If the value of n in Eq. (10) is large, and neither p_1 nor p_2 is too close to zero, the binomial distribution[3] can be approximated by:

$$z = \frac{x - np_1}{\sqrt{np_1p_2}} \quad (11)$$

The approximation of Eq. (11) is good enough for most purposes if np_1 and np_2 are each greater than 5.

As an example, we find that the records of a business

show that never more than one item is sold in a day, and that two sales per week can be expected. What is the probability of selling between 90 and 120 items in a 300-day year?

In a year consisting of 50 weeks of 6 days, the mean or expected value of the distribution is 100 items. The probability of a sale of an item on a given day is $p_1 = 100/300 = 1/3$, and of no sale is $p_2 = 2/3$.

By substituting the appropriate values into Eq. (11), we can write the standard score in terms of x:

$$z = \frac{x - (300)(1/3)}{\sqrt{(300)(1/3)(2/3)}} = \frac{x - 100}{8.165}$$

The integral range of 90 to 120 items contains all possible values of x from 89.5 to 120.5. For $x = 89.5$, $z = -1.286$; and for $x = 120.5$, $z = 2.511$.

From Fig. 3 or 4, the cumulative probability of a standard score of -1.286 is 0.11, while that of a standard score of 2.511 is 0.99. Therefore, the probability of annual sales in the range of 90 to 120 items is $(0.99 - 0.11) = 0.88$, or 88%.

Poisson Distribution

There are times when the frequency measurement is an integral number of events in a given segment of a continuum—for example, the number of automobiles passing a given point in one hour, or the number of leaks in a given length of hosepipe. In such cases, the correct frequency distribution is the Poisson distribution[3] in which the probability of x events per unit of continuum occurring is given by:

$$f(x) = \lambda^x e^{-\lambda}/x! \qquad (12)$$

where x is an integer, e is the base of natural logarithms, and λ is a parameter of the system, $\lambda = \mu = \sigma^2$.

As λ increases, the Poisson distribution approaches the Normal distribution, with the equation:

$$z = (x - \lambda)/\sqrt{\lambda} \qquad (13)$$

In cases where the value of p_1 is very close to zero in Eq. (10), so that the occurrence of the event is rare, the binomial distribution can be approximated by the Poisson distribution with $\lambda = np_1$ when $n \geqslant 50$ while $np_1 \leqslant 5$.

Predicting Breakdown in a Production Line

The daily chance of a breakdown in a production line operated continuously for 300 days/yr. is estimated at 1% from past performance. Estimate the probability of four or more breakdowns in the coming year.

For $n = 300$ and $p_1 = 0.01$, $\lambda \cong np_1 = 3 \leqslant 5$, the probability of no breakdown is found from Eq. (12) to be:

$$f(0) = (3)^0 e^{-3}/0! = 1/e^3 = 0.0498$$

Similarly:

Breakdowns	Probability
1	$f(1) = (3)^1 e^{-3}/1! = 0.1496$
2	$f(2) = (3)^2 e^{-3}/2! = 0.2240$
3	$f(3) = (3)^3 e^{-3}/3! = 0.2240$

Since something must happen, the probability of four or more breakdowns is:

$$1 - 0.0498 - 0.1496 - 0.2240 - 0.2240 = 0.3526$$

A simple trial will show how much more easily the preceding calculation is carried out than direct use of Eq. (10).

For example, in a production period of 100 days, 0, 1, 2, 3 and 4 machine failures occurred in a single day on 41, 37, 15, 6, 1 occasions, respectively. Fit a Poisson distribution to the data and estimate the maximum number of machine failures likely to occur in one day of a 300-day year.

The mean number of failures is found from Eq. (2) to be:

$$\bar{x} = \frac{0(41) + 1(37) + 2(15) + 3(6) + 4(1)}{41 + 37 + 15 + 6 + 1} = 0.89$$

The standard deviation is found from Eq. (4):

$$s^* = \left[\frac{\Sigma(x_i - \bar{x})^2 f(x_i)}{\Sigma f(x_i) - 1}\right]^{0.5}$$

The steps for calculating the numerator and denominator for this equation are tabulated as follows:

$(x_i - \bar{x})^2$	$f(x_i)$	$(x_i - \bar{x})^2 f(x_i)$
$(0 - 0.89)^2$	41	32.48
$(1 - 0.89)^2$	37	0.45
$(2 - 0.89)^2$	15	18.48
$(3 - 0.89)^2$	6	26.71
$(4 - 0.89)^2$	1	9.67
	$\Sigma(x_i - \bar{x})^2 f(x_i) =$	87.79
	$\Sigma f(x_i) - 1 =$	99

Therefore: $s^* = (87.79/99)^{0.5} = 0.9417$

The Poisson distribution is a good fit since:

$$\lambda = \mu \cong \bar{x} = 0.8900$$

and:

$$\lambda = \sigma^2 \cong (s^*)^2 \cong 0.8868$$

The Poisson distribution is found from Eq. (12) to be:

$$f(x) = \frac{(0.89)^x e^{-0.89}}{x!}$$

By substituting the appropriate values of x for this example into the preceding equation, we find: $f(0) = 0.4107$, $f(1) = 0.3655$, $f(2) = 0.1627$, $f(3) = 0.0483$, $f(4) = 0.0107$, $f(5) = 0.0019$ and $f(6) = 0.0003$. Hence, in 300 days the expected maximum number of breakdowns in one day is five since $(300) \times f(6) = 0.09$ occurrences. ■

References

1. Burington, R. S. and May, D. C., "Handbook of Probability and Statistics with Tables," 2nd ed., p. 367, McGraw-Hill, New York, 1969.
2. Rudd, D. F. and Watson, C. C., "Strategy of Process Engineering," p. 333, Wiley, New York, 1968.
3. Andersen, L. B., Probability Theory Provides Simple Statistical Models, *Chem. Eng.*, Nov. 26, 1962, pp. 125–132.

One way of measuring uncertainty in profitability analyses is to assess the probabilities of reaching a given value of profitability when the sample data approximate the normal distribution curve.

Probability Techniques for Estimates of Profitability

F. A. HOLLAND, F. A. WATSON and J. K. WILKINSON, University of Salford

Predicting the probability of attaining various levels of profitability requires the application of statistical techniques to the estimates of profitability obtained from a sensitivity analysis. Sensitivity analysis was covered in Part 6 of this series and statistical techniques in Part 8.

In this article we will describe, in detail, the applications of these procedures when the economic parameters conform to the normal (Gaussian) probability distributions.

Statistical Relationships

A property or variable, c, that is a function of several other variables x_1, x_2, etc., can be written:

$$c = \phi(x_1, x_2, x_3, \ldots, x_n) \tag{1}$$

If each x is an independent variable,[1] normally distributed, then:

$$\sigma_c^2 = \left(\frac{\partial c}{\partial x_1}\right)^2 \sigma_1^2 + \left(\frac{\partial c}{\partial x_2}\right)^2 \sigma_2^2 + \cdots + \left(\frac{\partial c}{\partial x_n}\right)^2 \sigma_n^2 \tag{2}$$

Originally published January 1974.

where σ_c^2 is the variance, or square of the standard deviation, of the variable c, and σ_n^2 is the variance of the variable x_n.

In many business applications, Eq. (1) can be reduced to the linear relationship:

$$c = k_1x_1 + k_2x_2 + \cdots + k_nx_n \tag{3}$$

where the k's are constants. Eq. (2) then becomes:

$$\sigma_c^2 = k_1^2\sigma_1^2 + k_2^2\sigma_2^2 + \cdots + k_n^2\sigma_n^2 \tag{4}$$

On the other hand, for a product function such as:

$$c = x_1x_2 \tag{5}$$

Eq. (2) can be written in the form:

$$\frac{\sigma_c^2}{c^2} = \frac{\sigma_1^2}{x_1^2} + \frac{\sigma_2^2}{x_2^2} \tag{6}$$

In Part 5 of this series, the discounted-cash-flow rate of return ($DCFRR$) and net present value (NPV) were both shown to be functions of the cumulative revenue from annual sales, ΣA_S, the cumulative total annual cost or expense, ΣA_{TE}, and the fixed capital cost of the plant, C_{FC}, among other factors.

Eq. (2) can be written for $(DCFRR)$ and for (NPV) as:

$$\sigma^2_{(DCFRR)} = \left[\frac{\partial(DCFRR)}{\partial \Sigma A_S}\right]^2 \sigma^2_{\Sigma A_S} + \left[\frac{\partial(DCFRR)}{\partial \Sigma A_{TE}}\right]^2 \sigma^2_{\Sigma A_{TE}} + \left[\frac{\partial(DCFRR)}{\partial C_{FC}}\right]^2 \sigma^2_{C_{FC}} \quad (7)$$

$$\sigma^2_{(NPV)} = \frac{\partial(NPV)^2}{\partial \Sigma A_S}\sigma^2_{\Sigma A_S} + \frac{\partial(NPV)^2}{\partial \Sigma A_{TE}}\sigma^2_{\Sigma A_{TE}} + \frac{\partial(NPV)^2}{\partial C_{FC}}\sigma^2_{C_{FC}} \quad (8)$$

The differentials in Eq. (7) and (8) are approximately equal to the sensitivity relationships determined in Part 6. For (NPV), these are:

$$\frac{\partial(NPV)}{\partial \Sigma A_S} \cong \frac{\Delta(NPV)}{\Delta \Sigma A_S} \quad (9)$$

$$\frac{\partial(NPV)}{\partial \Sigma A_{TE}} \cong \frac{\Delta(NPV)}{\Delta \Sigma A_{TE}} \quad (10)$$

$$\frac{\partial(NPV)}{\partial C_{FC}} \cong \frac{\Delta(NPV)}{\Delta C_{FC}} \quad (11)$$

Similar approximations can be written for $(DCFRR)$.

The revenue from annual sales, A_S, of a product at an annual production rate, R, and sales price of c_S per unit of production is:

$$A_S = Rc_S \quad (12)$$

For this case, Eq. (2) can be written as:

$$\sigma^2_{A_S} = \left(\frac{A_S}{R}\right)^2 \sigma^2_R + \left(\frac{A_S}{c_S}\right)^2 \sigma^2_{c_S} \quad (13)$$

When data are based on estimates of future trends, the accuracy of the normal probability approximation is usually acceptable. This is particularly the case as the number of component variables x_1, x_2, etc., in Eq. (6) increases. Although the distributions of the individual variables may be skewed, the property or variable, c, tends to approach the normal distribution. The use of this relationship can best be illustrated by an example.

Example: Predicting Cumulative Probabilities

Let us determine the cumulative probability of attaining various levels of (NPV) and $(DCFRR)$ for a proposed project based on the following facts, opinions and assumptions.

All factors have uncertainties that are normally distributed about their mean values. Sales volume and sales price are in the range of 13,000 to 17,000 ton/yr. and \$295 to \$305/ton, respectively. Cost of raw materials is \$135/ton of product.

A relatively simple plant can be designed, financed and completed during the current year. Fixed capital costs, obtained from five independent estimates, are \$4.2 million. \$3.85 million, \$4.5 million, \$3.7 million and \$4.25 million.

Labor requirements are 2 men on each of 3 shifts, at a labor rate, c_L, of \$5.20/hr.

The annual processing cost is given by:

$$A_P = 0.09C_{FC} + 13{,}500c_LN + A_U$$

where A_U is the annual utilities cost that is neglected for this example, i.e., $A_U = 0$. Working capital. $C_{WC} = 0.12C_{FC}$. Neglect cost of land and other nondepreciable costs. Annual general expense allocated to the product is $A_{GE} = \$450{,}000$/yr.

Assume straight-line depreciation over 5 yr., zero salvage value of plant items, and cost of capital at 10%/yr. There is no tax allowance other than depreciation, i.e., $A_A = 0$, and the fractional tax rate, t, is 0.45.

In order to resolve this problem, a number of interrelationships must be computed. These are summarized here in equation form from derivations and definitions presented in earlier articles of this series.

$$A_{CI} = A_S - A_{TE} \quad (14)$$

$$A_{CF} = A_{CI} - A_{IT} - A_{TC} \quad (15)$$

$$A_{IT} = (A_{CI} - A_D - A_A)t \quad (16)$$

$$A_{TE} = A_{GE} + A_{ME} \quad (17)$$

$$C_{TC} = C_{FC} + C_{WC} + C_L \quad (18)$$

$$A_D = (C_{FC} - S)/s \quad (19)$$

$$A_{ME} = A_P + Rc_{RM} \quad (20)$$

$$A_{TE} = A_{GE} + A_P + Rc_{RM} \quad (21)$$

$$A_P = 0.09C_{FC} + 13{,}500c_LN + A_U \quad (22)$$

$$A_{DCF} = A_{CF}f_d \quad (23)$$

$$(NPV) = \Sigma A_{DCF} \quad (24)$$

$$\Sigma A_{DCF} = 0 \text{ when } i = (DCFRR) \quad (25)$$

$$\bar{x} = \frac{x_1 + x_2 + x_3 + \cdots x_n}{n} \quad (26)$$

$$s^* = \left[\frac{\Sigma(x_i - \bar{x})^2 f(x_i)}{\Sigma f(x_i) - 1}\right]^{0.5} \quad (27)$$

$$z = (x - \mu)/\sigma \quad (28)$$

Eq. (22) gives the annual processing cost excluding depreciation. This is the manufacturing cost minus raw-materials cost. In the present case, the annual processing cost for $A_U = 0$, $c_L = \$5.20$/hr. and $N = 2$ is:

$$A_P = 0.09C_{FC} + 13{,}500(5.20)(2) + 0$$

$$A_P = 0.09C_{FC} + 140{,}400, \text{ \$/yr.}$$

The annual total expense required to produce and sell the product is computed from Eq. (21). Since A_{GE} is \$450,000/yr. and c_{RM} is \$135/ton, the annual total expense is:

$$A_{TE} = 450{,}000 + 0.09C_{FC} + 140{,}400 + 135R$$

$$A_{TE} = 590{,}400 + 0.09C_{FC} + 135R, \text{ \$/yr.}$$

where R is the sales volume of product in tons/yr.

The total capital cost is found from Eq. (18) to be:

$$C_{TC} = C_{FC} + 0.12C_{FC} + 0$$

$$C_{TC} = \$1.12C_{FC}$$

The annual rate of straight-line depreciation of the fixed-capital investment from C_{FC} at startup, to zero value at the end of five years of productive life is obtained from Eq. (19), and is:

$$A_D = (\$C_{FC} - \$0)/5 = 0.2C_{FC}, \text{ \$/yr.}$$

The annual cash income is found from Eq. (14). Here, we will substitute for A_S the value Rc_S from Eq. (12),

Nomenclature

A_A	Annual tax allowance exclusive of depreciation, \$/yr.
A_{CF}	Net annual cash flow, \$/yr.
A_{CI}	Annual cash income, \$/yr.
A_D	Annual depreciation charge, \$/yr.
A_{DCF}	Net annual discounted cash flow, \$/yr.
A_{GE}	Annual general expense, \$/yr.
A_{IT}	Annual amount of tax, \$/yr.
A_{ME}	Annual manufacturing expense, \$/yr.
A_{NP}	Net annual profit, \$/yr.
A_P	Annual processing cost, \$/yr.
A_S	Revenue from annual sales, \$/yr.
A_{TC}	Total annual capital expenditure, \$/yr.
A_{TE}	Total annual expense, \$/yr.
A_U	Annual utilities cost, \$/yr.
c_L	Hourly wage rate, \$/hr.
c_{RM}	Cost of raw materials per unit of production, \$/unit
c_S	Sales price per unit of production, \$/unit
C_{FC}	Fixed capital cost, \$
C_L	Cost of land and other nondepreciable costs, \$
C_{TC}	Total capital cost, \$
C_{WC}	Working capital cost, \$
$(DCFRR)$	Discounted-cash-flow rate of return, %/yr.
$f(x)$	Frequency or relative frequency with which a particular value x occurs, dimensionless
f_d	Discount factor, dimensionless
i	Annual fractional interest rate, dimensionless
K	Constant for a specific pattern of cash flow, dimensionless
n	Integer number of items, years, etc., dimensionless
N	Number of men per shift, dimensionless
(NPV)	Net present value of the project, \$
R	Annual production rate or sales volume, units/yr.
s	Service life, yr.
s^*	Standard deviation of sample, units of data
S	Scrap or salvage value, \$
t	Fractional tax rate, dimensionless
x	Value of the independent variable, units of the variable, or yield of product per unit of raw material, dimensionless
z	Standard score, dimensionless
μ	Mean value of population from which data came, units of data
σ	Standard deviation of population from which data came, units of data
$\bar{}$	Bar over any symbol indicates the mean value of the quantity

Normal Distribution Curve—Table I

Standard Score, z	Ordinate of Normal Curve, $f(z)$	Area Under Right-Hand Side of Curve, $\int_0^z f(z)dz$
0.000	0.3989	0.00000
0.100	0.3970	0.03983
0.200	0.3910	0.07926
0.253	0.3864	0.10000
0.300	0.3814	0.11791
0.400	0.3683	0.15542
0.500	0.3521	0.19146
0.524	0.3478	0.20000
0.600	0.3332	0.22575
0.700	0.3123	0.25804
0.800	0.2897	0.28814
0.842	0.2798	0.30000
0.900	0.2661	0.31594
1.000	0.2420	0.34134
1.200	0.1942	0.38493
1.282	0.1953	0.40000
1.400	0.1497	0.41243
1.600	0.1109	0.44520
1.645	0.1032	0.45000
1.800	0.0790	0.46407
1.881	0.0680	0.47000
2.000	0.0540	0.47725
2.054	0.0484	0.48000
2.326	0.0266	0.49000
2.500	0.0175	0.49379
2.576	0.0145	0.49500
2.748	0.0092	0.49700
3.000	0.0044	0.49865
3.090	0.0034	0.49900
3.291	0.0018	0.49950
3.429	0.0011	0.49970
3.500	0.0009	0.49977
3.719	0.0004	0.49990
4.000	0.0001	0.49997
∞	0.0000	0.50000

Data abridged and recalculated from Fisher, R.A. and Yates, F., "Statistical Tables for Biological, Agricultural and Medical Research," 5th ed., pp. 42–43, Oliver and Boyd, London, 1957.

and for A_{TE}, the value previously determined in terms of:

$$590{,}400 + 0.09C_{FC} + 135R$$

$$A_{CI} = R(c_S - 135) - 590{,}000 - 0.09C_{FC},\ \$/\text{yr.} \quad (29)$$

The annual amount of income tax is obtained from Eq. (16) for $A_A = 0$, $t = 0.45$, and the value of A_{CI} from Eq. (29), and becomes:

$$A_{IT} = [R(c_S - 135) - 590{,}000 - 0.09C_{FC}]0.45 - 0.45A_D,\ \$/\text{yr.}$$

By substituting the appropriate relations already evaluated into Eq. (15), we obtain the net annual cash flow (after tax), A_{CF}, as:

$$A_{CF} = 0.55R(c_S - 135) - 324{,}720 - 0.0495C_{FC} + 0.45A_D - A_{TC},\ \$/\text{yr.} \quad (30)$$

We now compute the net annual cash flow (after tax) for Year 0, for Years 1, 2, 3 and 4, and for Year 5 by using Eq. (15) and the appropriate values or equivalents:

For Year 0: $A_{IT} = \$0/\text{yr.}$, $A_{TC} = C_{TC} = \$1.12C_{FC}/\text{yr.}$

$$A_{CF} = -1.12C_{FC},\ \$/\text{yr.} \quad (31)$$

For Years 1, 2, 3 and 4: Use Eq. (30) for which A_D is $0.2C_{FC}$, \$/yr., and A_{TC} is 0, \$/yr. Therefore:

$$A_{CF} = 0.55R(c_S - 135) - 324{,}720 + 0.0405C_{FC},\ \$/\text{yr.} \quad (32)$$

For Year 5: Use Eq. (30) for which A_D is $0.2C_{FC}$, \$/yr., and $A_{TC} = -0.12C_{FC}$, \$/yr. from the return of working capital. We now find:

$$A_{CF} = 0.55R(c_S - 135) - 324{,}720 + 0.1605C_{FC},\ \$/\text{yr.} \quad (33)$$

Eq. (31), (32) and (33) can be used to obtain the net

annual cash flows (after tax) in \$/yr. for various values of the variables R, c_S and C_{FC}.

The net annual discounted cash flow, A_{DCF}, is found from Eq. (23) where the discount factor, $f_d = 1/(1+i)^n$. In turn, the net present value, (NPV), is found from Eq. (24).

In this example, values of A_{DCF} over the 5-yr. period are:

Year	A_{DCF}
0	$-1.12C_{FC}(1)$
1 to 4	$(Q + 0.0405C_{FC})\left[\sum_{n=1}^{n=4} \frac{1}{(1+i)^n}\right]$
5	$(Q + 0.1605C_{FC})\left[\frac{1}{(1+i)^5}\right]$

where Q in the preceding expressions is:

$$Q = 0.55R(c_S - 135) - 324{,}720$$

After summing the expressions for A_{DCF} and simplifying, we get:

$$(NPV) = Q\left\{\left[\frac{(1+i)^{n+1} - 1}{i(1+i)^n}\right] - 1\right\} + C_{FC}\sum_{k=0}^{n} \frac{K}{(1+i)^k} \quad (34)$$

where $n = 5$; $K = -1.12$ for Year 0, 0.0405 for Years 1 to 4, and 0.1605 for Year 5. R and c_S do not vary from year to year.

To find the (NPV) for this problem for a discount factor of 10%, we substitute the following values into Eq. (34):

Year	f_d at 10%
1	0.90909
2	0.82645
3	0.75131
4	0.68301
5	0.62092
Σf_d	3.79078

We also substitute the appropriate values for n and K into Eq. (34), and the final result is:

$$(NPV) = 2.0849R(c_S - 135) - 0.89196C_{FC} - 1{,}230{,}942 \quad (35)$$

Cumulative Probability of Getting a Given Net Present Value or Less—Table II

Cumulative Probability, %	Standard Score, z	(NPV), \$	(DCFRR), %/Yr.
0	−4.000	−1,076,362	0.88
2	−2.054	− 420,311	6.38
10	−1.282	− 160,048	8.56
20	−0.842	− 11,712	9.81
30	−0.524	+ 95,495	10.71
40	−0.253	+ 186,857	11.47
50	0.000	+ 272,150	12.19
60	0.253	+ 357,443	12.91
70	0.524	+ 448,805	13.67
80	0.842	+ 556,012	14.57
90	1.282	+ 704,348	15.82
98	2.054	+ 964,611	18.00
100	4.000	+1,620,662	23.50

The mean value, $(\overline{NPV})$, for Year 5 can be obtained by using the mean values for the sales volume, $\overline{R}$, sales price, $\bar{c}_S$, and fixed capital cost, $\overline{C}_{FC}$, in Eq. (35). The mean value of a quantity can be calculated from Eq. (26). Hence:

$$\overline{R} = (13{,}000 + 17{,}000)/2 = 15{,}000 \text{ tons/yr.}$$

$$c_S = (295 + 305)/2 = 300, \text{ \$/ton}$$

$$\overline{C}_{FC} = \frac{(4.2 + 3.85 + 4.5 + 3.7 + 4.25)10^6}{5}$$

$$\overline{C}_{FC} = \$4.1 \times 10^6$$

Substituting these mean values into Eq. (35) gives the $(\overline{NPV})$ for Year 5:

$$(\overline{NPV}) = 2.0849(15{,}000)(300 - 135) - 0.89196(4.1 \times 10^6) - 1{,}230{,}942 = \$272{,}150$$

The standard deviation of the fixed capital cost is obtained from Eq. (27). For this example, it becomes:

$$\sigma_{C_{FC}} \cong s^*_{C_{FC}} = \left[\frac{\Sigma(x_i - \bar{x})^2 f(x_i)}{\Sigma f(x_i) - 1}\right]^{0.5}$$

Annual Cash Flows Based on Mean Values of Factors—Table III (All values are thousands of dollars)

	Before Tax							After Tax		
Year	A_S	A_{TE}	A_{CI}	$A_D + A_A$	A_{NP}	A_{IT} ($t = 0.45$)	A_{TC}	A_{CF}	A_{DCF} $i = 12.19\%$	$\Sigma A_{DCF} = (NPV)$
0	0	0	0	0	0	0	4,592	−4,592.00	−4,592.000	−4,592.000
1	4,500	2,984.4	1,515.6	820	695.6	313.02	0	1,202.58	1,071.952	−3,520.048
2	4,500	2,984.4	1,515.6	820	695.6	313.02	0	1,202.58	955.514	−2,564.534
3	4,500	2,984.4	1,515.6	820	695.6	313.02	0	1,202.58	851.723	−1,712.811
4	4,500	2,984.4	1,515.6	820	695.6	313.02	0	1,202.58	759.206	− 953.605
5	4,500	2,984.4	1,515.6	820	695.6	313.02	−492	1,694.58	953.607	+ 0.002

Discount factors, f_d, when $i = 12.19\%$ are: 1, 0.89138, 0.79455, 0.70825, 0.63131 and 0.56274 for Years 0 through 5, respectively.

where $f(x_i)$ is the frequency of each estimate, x_i the value of each estimate, and $\bar{x}$ the mean value of the estimates. To compute the standard deviation, let us tabulate the values for the numerator and denominator of this equation as follows:

$(x_i - \bar{x})^2$	$f(x_i)$	$(x_i - \bar{x})f(x_i)$
$(4.2 - 4.1)^2$	1	0.0100
$(3.85 - 4.1)^2$	1	0.0625
$(4.5 - 4.1)^2$	1	0.1600
$(3.7 - 4.1)^2$	1	0.1600
$(4.25 - 4.1)^2$	1	0.0225

$$\Sigma(x_i - \bar{x})f(x_i) = 0.4150$$
$$\Sigma f(x_i) - 1 = 4$$

$$s^*_{\bar{C}_{FC}} = (0.415/4)^{0.5} \times 10^6 = \$322,100$$

Eq. (27) cannot be used to estimate the standard deviations of the sales volume and sales price since the data are insufficient. An alternative estimation method is required.

Table I shows that the cumulative probability of a variable that has a normal distribution lying within four standard deviations of the mean is 0.49997. Therefore, it is more than 99.99% certain that a random value will lie within $\pm 4\sigma$ from the mean. For practical purposes, σ may be taken as one-eighth of the range of certainty. Therefore, the standard deviations for sales volume and sales price are:

$$\sigma_R = (17{,}000 - 13{,}000)/8 = 500 \text{ tons/yr.}$$
$$\sigma_{c_S} = (305 - 295)/8 = \$1.25/\text{ton}$$

The variance of the revenue A_S from annual sales is given by Eq. (13), which can also be written as:

$$\sigma^2_{A_S} = \bar{c}_S^2\sigma_R^2 + \bar{R}^2\sigma^2_{c_S}$$

For this example, an analogous equation can be written in the form:

$$\sigma^2_{R(c_S-135)} = (\bar{c}_S - 135)^2\sigma_R^2 + \bar{R}^2\sigma^2_{(c_S-135)}$$

Since $(\bar{c}_S - 135) = (300 - 135) = \$165/\text{ton}$; σ_R is 500 tons/yr.; $\bar{R}$ is 15,000 tons/yr. and σ_{c_S} is \$1.25/ton, we substitute these values into the equation to get the standard deviation for the revenue from annual sales:

$$\sigma^2_{R(c_S-135)} = (165)^2(500)^2 + (15{,}000)^2(1.25)^2$$
$$\sigma_{R(c_S-135)} = \$84{,}604/\text{yr.}$$

To find the variance of the (NPV), we previously determined the mean value for $(\overline{NPV})$ in Year 5 to be:

$$(\overline{NPV}) = 2.0849\bar{R}(\bar{c}_S - 135) - 0.89196\bar{C}_{FC} - 1{,}320{,}942$$

where $(\overline{NPV})$ is in \$ for an interest rate, i, of 10%.

Since $\partial(\overline{NPV})/\partial\bar{R}(\bar{c}_S - 135)$ is 2.0849 yr., $\partial(\overline{NPV})/\partial\bar{C}_{FC}$ is $-0.89196C_{FC}$, $\sigma_{R(c_S-135)}$ is \$84,604/yr., and $\sigma_{C_{FC}}$ is \$322,100, the variance of the (NPV) is calculated from Eq. (2) to be:

$$\sigma^2_{(NPV)} = (2.0849)^2(84{,}604)^2 + (-0.89196)^2(322{,}100)^2$$
$$\sigma_{(NPV)} = \$337{,}128$$

Table I can be used to estimate the cumulative probability of a given (NPV) for Year 5 being attained on the basis of the (NPV) for Year 5 being normally distributed about a mean value of \$272,150 with a standard deviation of \$337,128. For the (NPV) in this example, Eq. (28) for the standard score can be written as:

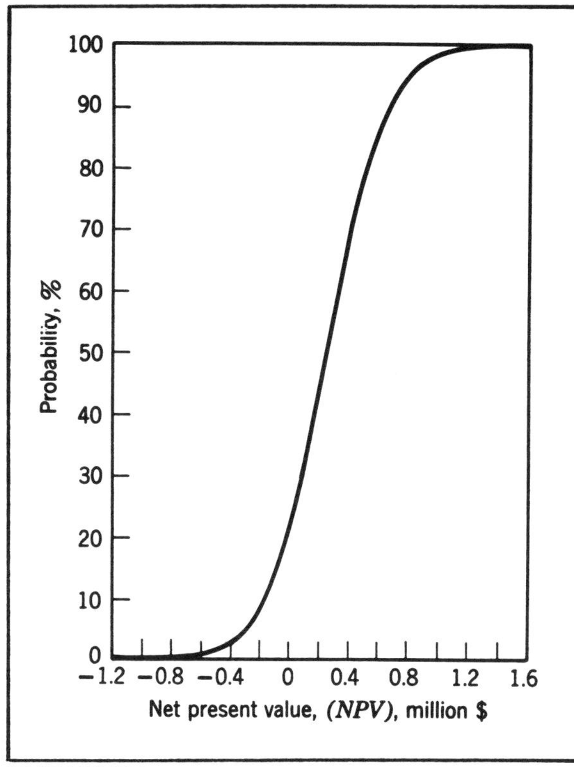

PROBABILITY for a given (*NPV*) or less—Fig. 1

$$z = [(NPV) - (\overline{NPV})]/\sigma_{(NPV)}$$

After substituting the appropriate values and rearranging, we get:

$$(NPV) = 337{,}128z + 272{,}150, \$ \qquad (36)$$

We can now compute the probability of attaining a given (NPV). For example, if the attained probability is 10%, we read from Table I that the area under the righthand side of the normal distribution curve is 0.4000 when $z = 1.282$. Since the curve is symmetrical about the mean value at $z = 0$, the area to the left of $z = -1.282$ is $(0.5000 - 0.4000)$, or 0.1000. Therefore, there is a cumulative probability of 10% that a value of $z \leqslant -1.282$ will be attained. In the example, this corresponds to an (NPV) as computed from Eq. (36) of:

$$(NPV) = 337{,}128(-1.282) + 272{,}150 = -\$160{,}048$$

Also, there is a cumulative probability of 90% that a value of $z \leqslant 1.282$ will be attained. In this case, the (NPV) as computed from Eq. (36) becomes:

$$(NPV) = 337{,}128(1.282) + 272{,}150 = \$704{,}348$$

Other values are calculated in the same way, and listed in Table II. The probability of attaining a given (NPV) or less for Year 5 is shown graphically in Fig. 1.

The net present value (NPV) for Year 5 for this example can be written in the form of Eq. (34). In Eq. (34),

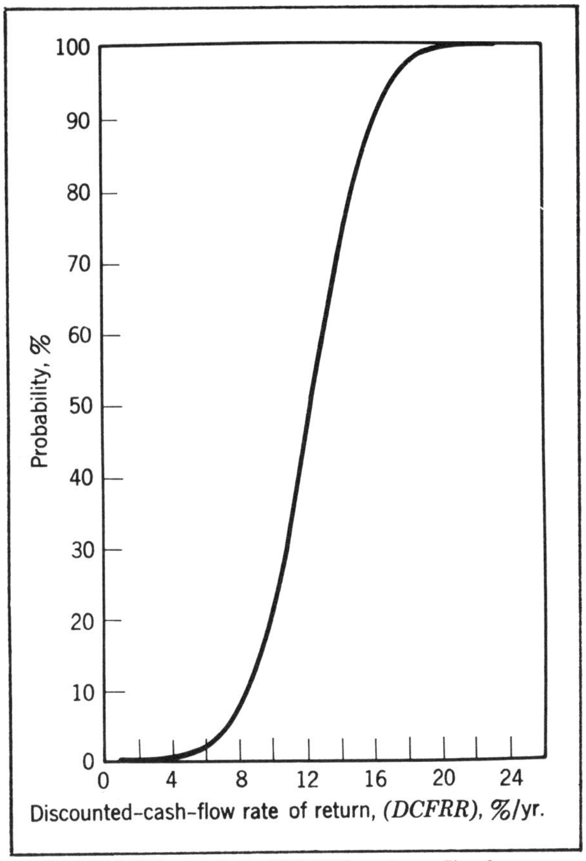

PROBABILITY for a given (*DCFRR*) or less—Fig. 2

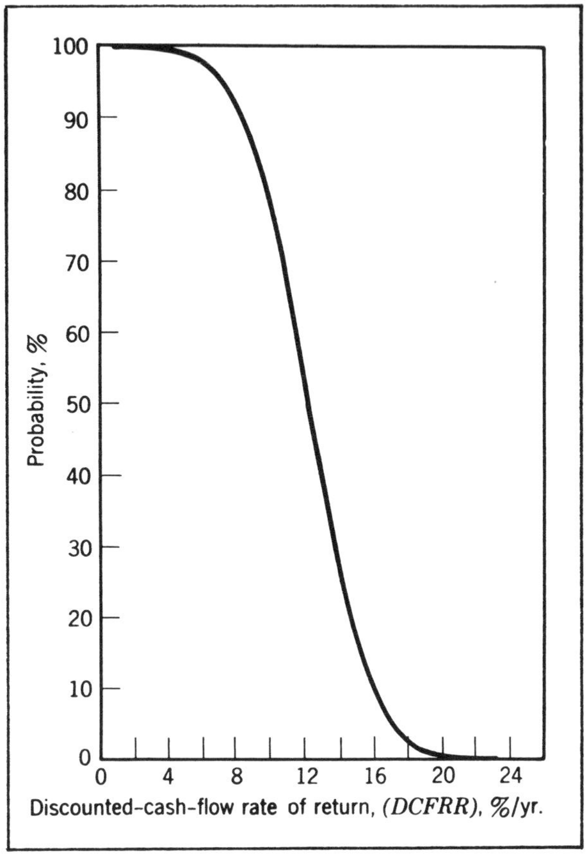

PROBABILITY for a given (*DCFRR*) or more—Fig. 3

the value of the interest rate, i, which makes $(NPV) = 0$ after 5 yr. is the $(DCFRR)$. This can be determined by an iterative procedure.

The mean value for $(\overline{DCFRR})$ of 12.19% is obtained on the basis of the mean values for the sales volume, $\bar{R}$, sales price, $\bar{c}_S$, and fixed capital cost, $\bar{C}_{FC}$. The annual cash flows are listed in Table III.

Let us put the $(NPV) = 0$ after 5 yr. into Eq. (34) and rewrite it as:

$$\frac{\sum_{k=0}^{n}\left[K/(1+i)^k\right]}{\left[\frac{(1+i)^{n+1}-1}{i(1+i)^n}\right]-1} = \frac{0.55R(c_S-135)-324{,}720}{-C_{FC}} \quad (37)$$

After differentiating both sides of Eq. (37) and substituting the following numerical values: $i = (\overline{DCFRR}) = 12.19\%$, $n = 5$ yr., $R = \bar{R} = 15{,}000$ tons/yr., $c_S = \bar{c}_S = \$300$/ton, and $C_{FC} = \bar{C}_{FC} = \$4{,}100{,}000$, we get the linear relation:

$$di = \left(\frac{-0.680296}{4{,}100{,}000}\right)d\left[R(c_S-135)\right] + \left(\frac{0.312703}{4{,}100{,}000}\right)dC_{FC} \quad (38)$$

where $i = (DCFRR)$.

For this example, the variance of $(DCFRR)$ can be related to the variance of $R(c_S - 135)$ and the variance of C_{FC} by Eq. (4), written as:

$$\sigma^2_{(DCFRR)} = \left(\frac{-0.680296}{4{,}100{,}000}\right)^2\sigma^2_{R(c_S-135)} + \left(\frac{0.312703}{4{,}100{,}000}\right)^2\sigma^2_{C_{FC}}$$

where $\sigma_{R(C_S-135)} = \$84{,}604$/yr. and $\sigma_{C_{FC}} = \$322{,}100$. Substituting these values into the equation yields the standard deviation as:

$$\sigma_{(DCFRR)} = 2.828\%$$

The probability of attaining a given $(DCFRR)$, or less, for Year 5 is determined in a manner similar to that described for the (NPV). The probabilities for various values of $(DCFRR)$, again for Year 5, are shown graphically in Fig. 2.

Decision-makers often prefer to have graphs showing the probability of attaining a value greater than a given value. Such curves are easily obtained by subtracting each probability of achieving a given value or less from 100%. Fig. 3 is obtained from the data in Fig. 2 in this way, and shows the probability of attaining a $(DCFRR)$ greater than a given one. ■

Reference

1. Rudd, D. F. and Watson, C. C., "Strategy of Process Engineering, p. 333, Wiley, N.Y., 1968.

If the distribution of factors for profitability estimates is not normal, the amount of computation can become very large. Approximations to the normal distribution can be made, and Monte Carlo methods are used to sample all possible combinations of factors.

Estimating Profitability When Uncertainties Exist

F. A. HOLLAND, F. A. WATSON, J. K. WILKINSON, University of Salford

When little is known about the true probability distribution of a factor, estimates of profitability can be made if the various factors are assumed to be normally distributed. This assumption is often used as an approximation even when it is known that the distribution is not normal, because only the normal and Poisson distributions are amenable to relatively simple mathematical manipulation.

The normal distribution is generally more useful since it contains two parameters: the mean, μ, and the standard deviation, σ. In the Poisson distribution, these two parameters are not independent. The equation for the normal distribution curve in the standardized form is:

$$f(z) = \frac{\exp(-z^2/2)}{(2\pi)^{0.5}} \tag{1}$$

where the standard score, z, corresponding to a value of the variable x is:

$$z = (x - \mu)/\sigma \tag{2}$$

The curve for Eq. (1) is single-peaked, and symmetrical about $z = 0$ (and, hence, about $x = \mu$), and is asymptotic to $f(z) = 0$ at $z = \pm\infty$. Any frequency-distribution curve that cannot be closely approximated by this formula is considered non-normal.

A single-peaked curve can differ from the normal shape being either finite in extent, asymmetrical, flatter or sharper in shape, or any combination of these.

The binomial distribution is defined by:

$$(p_1 + p_2)^n = p_1^n + \cdots \frac{n!}{x!(n-x)!} p_1^x p_2^{(n-x)} \cdots p_2^n \tag{3}$$

where the coefficient of p_1^x is the relative frequency that an event will occur exactly x times in n independent trials when the probabilities of success and failure of an individual trial are p_1 and p_2, respectively. Since the trial must result either in a success or failure, it follows that:

$$p_1 + p_2 = 1 \tag{4}$$

Unless $p_1 = p_2 = 1/2$, a binomial-frequency distribution is inherently skewed. In this case, the median (that value of the variable at which the cumulative frequency becomes 0.5) differs from the mean value, $\bar{x}$, as defined by:

$$\bar{x} = \frac{\Sigma x}{n} = \frac{\Sigma x_1 f(x_1)}{\Sigma f(x_1)} \tag{5}$$

where Σ indicates summation of all n values of x, and $f(x_1)$ indicates the number of times the value x_1 occurs.

The full description of a non-normal curve will require two extra parameters, one to define the skewness or asymmetry, and one to define the kurtosis or flatness of the curve. These parameters arise naturally by taking the mean values of successive moments about the mean. The first moment about the mean is:

$$m_1 = \frac{\Sigma(x - \bar{x})}{n} = \frac{\Sigma f(x)(x - \bar{x})}{\Sigma f(x)} \tag{6}$$

As m_1 is identically zero, it serves only as an alternative definition of the mean.

The second moment about the mean is:

$$m_2 = \frac{\Sigma(x - \bar{x})^2}{n} = \frac{\Sigma f(x)(x - \bar{x})^2}{\Sigma f(x)} \tag{7}$$

This is the classic definition of the variance of a distri-

Originally published February 1974.

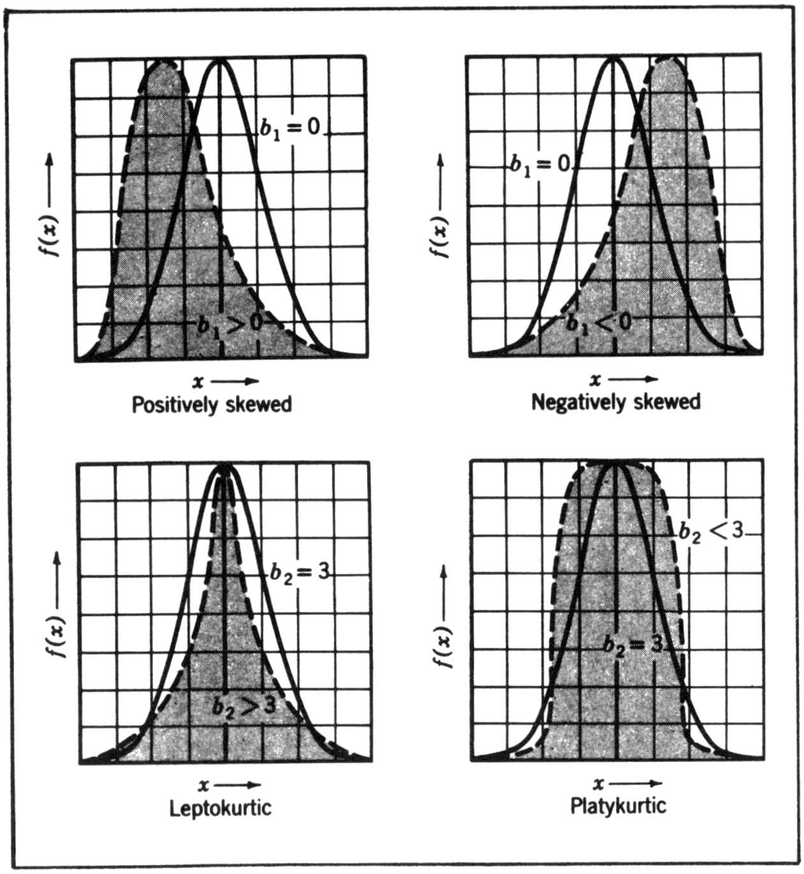

COMPARISON of curves for the normal frequency distribution with non-normal distributions illustrates the differences for the parameters that define skewness (asymmetry) of the curve and kurtosis (flatness) of the curve—Fig. 1

bution of n independent values, and is the square of the standard deviation of a large sample.

The third moment about the mean is:

$$m_3 = \frac{\Sigma(x - \bar{x})^3}{n} = \frac{\Sigma f(x)(x - \bar{x})^3}{\Sigma f(x)} \tag{8}$$

Eq. (8) is a measure of skewness, since positive and negative values will not balance exactly if the distribution is skewed. The value of m_3 can be standardized and made dimensionless by defining a parameter:

$$b_1 = m_3/(m_2)^{3/2} \tag{9}$$

This parameter will be negative if the longer tail of the distribution is toward the lower values of the variable, zero if the distribution is symmetrical, and positive if the tail is toward the higher values of the variable.

The fourth moment about the mean is:

$$m_4 = \frac{\Sigma(x - \bar{x})^4}{n} = \frac{\Sigma f(x)(x - \bar{x})^4}{\Sigma f(x)} \tag{10}$$

The fourth moment may also be standardized and made dimensionless by defining:

$$b_2 = m_4/(m_2)^2 \tag{11}$$

The parameter b_2 is influenced by the flatness of the curve. If the curve is leptokurtic, it is more pointed than the normal curve, and $b_2 > 3$. If the curve is normal, $b_2 = 3$. If the curve is platykurtic, it is flatter than normal, and $b_2 < 3$. The effect of kurtosis is illustrated in Fig. 1.

The cumulative-frequency data of the normal distribution, given in Table I, can be used to construct probability graph paper in which the abscissa is evenly divided. The cumulative normal distribution is represented by a straight line, and the probability axis is scaled in accordance with data in Table I. Cumulative-frequency data may be plotted on such paper.

If the distribution is positive skewed, the curve is concave from below at the lower end. If the distribution is negatively skewed, the curve is concave from above at the upper end. If the distribution is leptokurtic, the curve is S-shaped, flattening at the ends; if platykurtic, the curve becomes steeper at the ends. Only if the distribution is normal will the data produce a straight line.

The binomial distribution is awkward to compute, but data can often be fitted empirically to such a curve by a suitable choice of p_1 and n. For the binomial distribution, the mean is:

$$\bar{x} = np_1 \tag{12}$$

and the standard deviation is:

$$s^* = \sqrt{np_1p_2} \tag{13}$$

The Poisson distribution requires that $\bar{x} = (s^*)^2$ and hence may be used to approximate the binomial distribution if $p_2 \cong 1$, i.e., if p_1 is small and np_1 is finite for large values of n. In this case, the values of σ may be computed as for the normal distribution.

The binomial distribution may be approximated by the normal distribution if departure from normality is not too great. As a guide, $(x - np_1)/(np_1p_2)^{1/2}$ is distributed approximately normally with unit standard deviation

Normal Distribution Curve—Table I

Standard Score, z	Ordinate of Normal Curve, $f(z)$	Area Under Right-Hand Side of Curve, $\int_0^z f(z)dz$
0.000	0.3989	0.00000
0.100	0.3970	0.03983
0.200	0.3910	0.07926
0.253	0.3864	0.10000
0.300	0.3814	0.11791
0.400	0.3683	0.15542
0.500	0.3521	0.19146
0.524	0.3478	0.20000
0.600	0.3332	0.22575
0.700	0.3123	0.25804
0.800	0.2897	0.28814
0.842	0.2798	0.30000
0.900	0.2661	0.31594
1.000	0.2420	0.34134
1.200	0.1942	0.38493
1.282	0.1953	0.40000
1.400	0.1497	0.41243
1.600	0.1109	0.44520
1.645	0.1032	0.45000
1.800	0.0790	0.46407
1.881	0.0680	0.47000
2.000	0.0540	0.47725
2.054	0.0484	0.48000
2.326	0.0266	0.49000
2.500	0.0175	0.49379
2.576	0.0145	0.49500
2.748	0.0092	0.49700
3.000	0.0044	0.49865
3.090	0.0034	0.49900
3.291	0.0018	0.49950
3.429	0.0011	0.49970
3.500	0.0009	0.49977
3.719	0.0004	0.49990
4.000	0.0001	0.49997
∞	0.0000	0.50000

Data abridged and recalculated from Fisher, R.A. and Yates, F., "Statistical Tables for Biological, Agricultural and Medical Research," 5th ed., pp. 42-43, Oliver and Boyd, London, 1957.

about zero if $n \geqslant 30$, and $0.1 \leqslant p_1 \leqslant 0.9$; or if $10 \leqslant n \leqslant 30$, and $0.4 \leqslant p \leqslant 0.6$. In all other cases, the normal distribution cannot be assumed, and the methods described later in this article should be used.

Example Demonstrates Techniques

The mean and standard deviation of some positively skewed data are 4.000 and 1.530, respectively. Let us fit these data to a binomial distribution and calculate the parameters of the curve. Let us also calculate the binomial frequency distribution, and the corresponding normal approximation to that distribution.

From Eq. (12) and (13), we get:

$$\frac{np_1p_2}{np_1} = p_2 = \frac{(s^*)^2}{\bar{x}}$$

or: $$p_2 = (1.530)^2/4.000 = 0.5852$$

From Eq. (4), we compute $p_1 = 1 - 0.5852 = 0.4148$; and from Eq. (12), $n = 4.000/0.4148 = 9.6432$.

Frequency Expectation for Example—Table II

Term	Binomial Distribution		Normal Approximation	
x	$f(x)$	$F(x)$	$f(x)$	$F(x)$
0	0.0060466176	0.0060466176	0.00493	0.00493
1	0.0403107840	0.0463574016	0.02172	0.02665
2	0.1209323520	0.1672897536	0.07240	0.09905
3	0.2149908480	0.3822806016	0.16067	0.25972
4	0.2508226560	0.6331032576	0.24028	0.50000
5	0.2006581248	0.8337613824	0.24028	0.74028
6	0.1114767360	0.9452381184	0.16067	0.90095
7	0.0424673280	0.9877054464	0.07240	0.97335
8	0.0106168320	0.9983222784	0.02172	0.99507
9	0.0015728640	0.9998951424	0.00493	0.99934
10	0.0001048576	1.0000000000	0.00060	0.99994

As n must be an integer in the binomial distribution, let us take $n = 10$. Then, $p_1 = 0.4000$ and $p_2 = 0.6000$. The appropriate binomial expansion is then $(0.4 + 0.6)^{10}$. The new $s^* \cong [(10)(0.4)(0.6)]^{1/2} = 1.549$. Correctly, a test of significance should be made to see if this value of s^* is significantly different from the experimental value of 1.530. It will be assumed that the difference here is not significant.

From Eq. (3), the frequency expectation for various integer values of x is given by successive terms of the series. Therefore:

$$(0.4 + 0.6)^{10} = (0.4)^{10} + \frac{10!(0.4)^9(0.6)^1}{9!(10-9)!} + \cdots$$

The frequencies, $f(x)$, for each term have been calculated and are listed in Table II. The cumulative frequency, $F(x)$, of observing a value x or less is obtained by summing successive values of $f(x)$. Values for $F(x)$ are also listed in Table II. The parameters m_1 and m_2 for the curve have already been obtained as: zero and $(1.549)^2 = 2.400$, respectively. From Eq. (8) and the data in Table II for values of the variable x ranging from 0 to 10, we get:

$$m_3 = \frac{0.0060466176(0-4)^3 + 0.0403107840(1-4)^3 + \cdots}{1.0}$$

$$m_3 = 0.4800$$

Similarly, from Eq. (10):

$$m_4 = \frac{0.0060466176(0-4)^4 + 0.0403107840(1-4)^4 + \cdots}{1.0}$$

$$m_4 = 16.224$$

From Eq. (9) and (11), we calculate the parameters b_1 and b_2 as:

$$b_1 = 0.4800/(2.400)^{3/2} = 0.1291$$

$$b_2 = 16.224/(2.400)^2 = 2.469$$

These values for b_1 and b_2 indicate a limiting case for which the normal distribution can be applied. They are, therefore, a guide to the suitability of the normal approximation in other cases.

For the normal distribution, the standard score, z, is obtained from Eq. (2). Since the best estimates available

A Selection From a Short List of Uniformly Distributed Random Numbers—Table III

Row \ Column	A	B	C	D	E	F	G	H	I	J
1	3770	3221	4691	3352	6132	9203	0493	7174	8174	1665
2	7575	5076	5096	7807	5137	8258	0098	2829	0379	5920
3	4380	9931	8501	5262	7142	0313	2703	1484	5584	3175
4	4185	7786	4906	5717	2147	5368	8303	3139	3789	3430
5	6990	8641	4311	9172	0152	3423	6913	7794	4994	6685
46	4995	6696	2916	3827	1357	6678	2718	1649	7399	3140
47	3800	3551	8321	3282	5362	0733	7323	2304	4604	2395
48	5605	3406	6726	5737	2367	7788	4928	5959	4809	4650
49	0410	6261	8131	1192	2372	7843	5533	2614	8014	9905
50	8215	2116	2536	9647	5377	0898	9138	2269	4219	8160

for μ and σ are x and s^*, respectively, the standard score for $x = 1$ is:

$$z = (1 - 4.000)/1.549 = -1.936$$

Interpolating linearly in Table I, this value of z corresponds to a cumulative frequency for the normal approximation of:

$$F(x) = 0.5 - 0.47335 = 0.02665$$

Cumulative frequencies for other values of x have been calculated in the same way and are listed in Table II. From these values, the corresponding frequency values for $f(x)$ are obtained by subtraction, as listed in Table II. The binomial frequency distribution and the normal approximation to that distribution are shown in Fig. 2.

If the frequency distribution curve of one or more factors cannot be taken as normal, the amount of calculation can be very large. For example, if there are 10 levels of sales volume, 6 levels of sales price, 5 levels of total annual expense, and 3 levels of fixed capital cost, the total number of different ($DCFRR$) or (NPV) calculations is (10)(6)(5)(3) = 900. This is a great deal of calculation for only a modest number of levels for each factor. In most cases, it is not practical to obtain the cumulative probability curves for ($DCFRR$) and (NPV) by using all the possible combinations of factors. The problem can be greatly simplified by using Monte Carlo methods to sample the population, consisting of all possible combinations of factors.

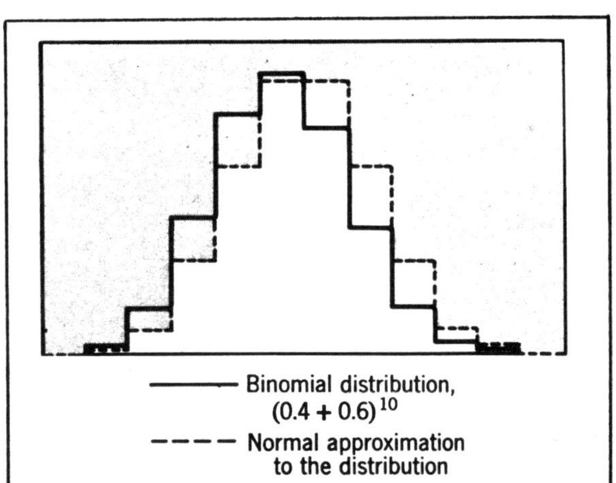

BINOMIAL distribution and normal approximation—Fig. 2

The Monte Carlo Method

The Monte Carlo method makes use of random numbers. A digital computer can be used to generate pseudo random numbers in the range 0 to 1. A partial list of such numbers, generated by a computer, is shown in Table III. To describe the use of random numbers, let us consider the frequency distribution curve of a particular factor, e.g., sales volume. Each value of the sales volume has a certain probability of occurrence. The cumulative probability of that value (or less) being realized is a number in the range 0 to 1. Thus, a random number in the same range can be used to select a random value of the sales volume.

In the same way, random values of the other factors can be obtained. These can then be combined to give random values of ($DCFRR$) and (NPV), and in turn can be used to plot cumulative probability curves for ($DCFRR$) and (NPV). The computer may be required to perform some 10,000 to 50,000 calculations.

The use of the Monte Carlo method is illustrated by the following example. [Note that cumulative probability curves of ($DCFRR$) and (NPV) can never be more accurate than the opinions on which they are based.]

Example for Monte Carlo Method

A sales manager has given the following opinions about the sales potential of a proposed new specialty product:

- Sales volume will not be less than 1,000 tons/yr. nor more than 2,000 tons/yr. All values in this range are equally likely.
- Sales price will lie in the range of \$750/ton to \$900/ton, with a most likely value of \$850/ton. The probability of intermediate values can be taken to vary linearly with sales price.

On the basis of these opinions, use the Monte Carlo technique to obtain a cumulative probability curve for attaining various levels of revenues from annual sales, A_S.

The required relation for sales revenue is:

$$A_S = Rc_S \tag{14}$$

where R is the sales volume and c_S is the unit sales price.

In considering the probability curve for the sales volume, let the total area under the curve equal unity. The area under the curve for different ranges is:

Area	Range
0	$-\infty \leqslant R \leqslant 1{,}000$ tons/yr.
$1 = 1{,}000p$	1,000 tons/yr. $\leqslant R \leqslant$ 2,000 tons/yr.
0	2,000 tons/yr. $\leqslant R \leqslant \infty$

Thus, $p = 1/1{,}000$. The probability plot for sales volume is:

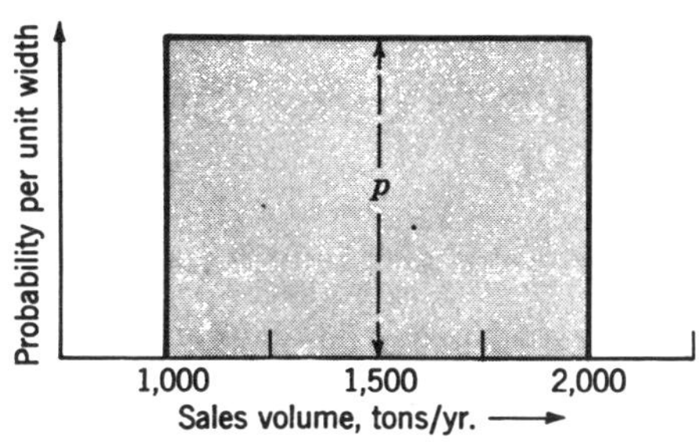

The cumulative probability of a sales volume less than, or equal to, R occurring is:

$$\int_{-\infty}^{R} f(R)dR$$

Since: $\int_{-\infty}^{1{,}000} (0)dR = 0$ and $\int_{2{,}000}^{\infty} (0)dR = 0$

the cumulative probability curve can be represented by:

$$\int_{1{,}000}^{R} (p)dR = \left(\frac{1}{1{,}000}\right)(R - 1{,}000)$$

This curve is shown in Fig. 3

Now, in considering the probability curve for sales price, let the total area under the curve equal unity. The area under the curve for different ranges is:

Area	Range
0	$-\infty \leqslant c_S \leqslant$ \$750/ton
$1 = \frac{p_m}{2}(900 - 750)$	\$750/ton $< c_S \leqslant$ \$900/ton
0	\$900/ton $< c_S \leqslant \infty$

Thus, $p_m = 1/75 = 0.01333$, which is the probability

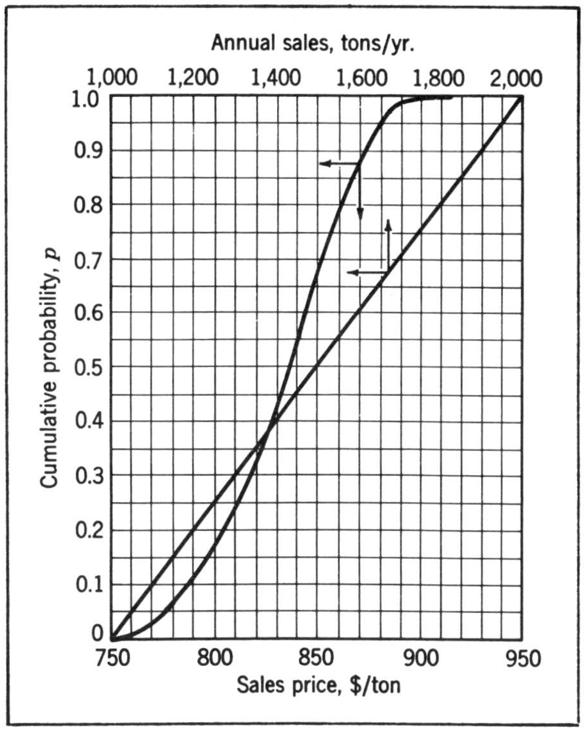

CUMULATIVE probabilities for sales and price—Fig. 3

at the most likely value of \$850/ton. The probability plot for sales price is:

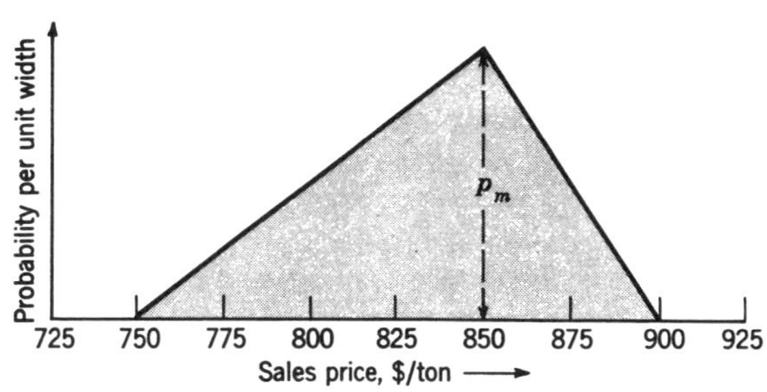

The ordinate of this probability curve can be represented by: $(0.0001333)(c_S - 750)$ for values of c_S in the range of \$750/ton $< c_S \leqslant$ \$850/ton; and $(0.0002666)(900 - c_S)$ for \$850/ton $< c_S \leqslant$ \$900/ton.

The cumulative probability of a sales price less than or equal to, c_S occurring is:

$$\int_{-\infty}^{c_S} f(c_S)dc_S$$

Since: $\int_{-\infty}^{750} (0)dc_S = 0$ and $\int_{900}^{\infty} (0)dc_S = 0$

the cumulative probability curve can be represented by:

$$\int_{750}^{c_S} (0.0001333)(c_S - 750)dc_S = (0.00006666)(c_S - 750)^2$$

where \$750/ton $< c_S \leqslant$ \$850/ton, and by:

$$(0.00006666)(850 - 750)^2 + \int_{850}^{c_S} (0.0002666)(900 - c_S)dc_S =$$

$$(0.00006666)(850 - 750)^2 + (0.000133)(900 - 850)^2 - (0.0001333)(900 - c_S)^2 = 1 - (0.0001333)(900 - c_S)^2$$

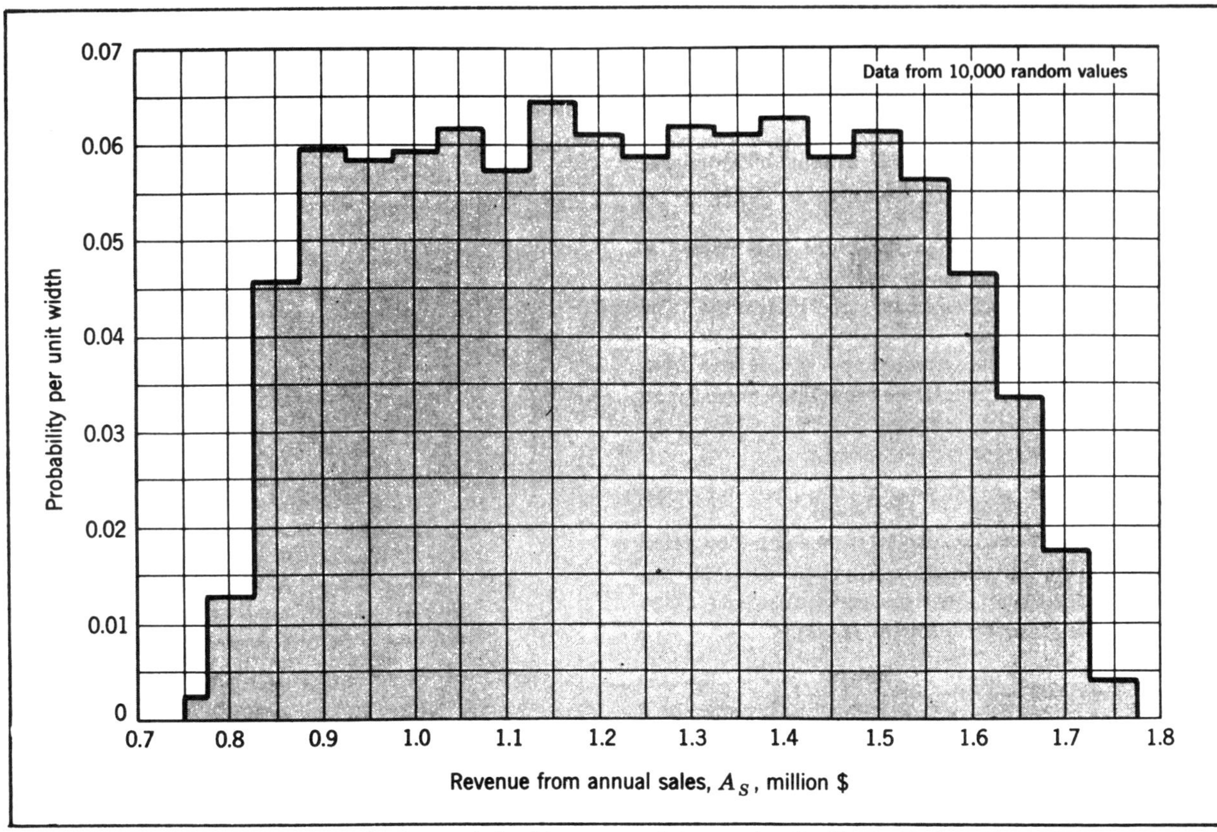

HISTOGRAM of revenue from annual sales as derived from computations using the Monte Carlo method—Fig. 4

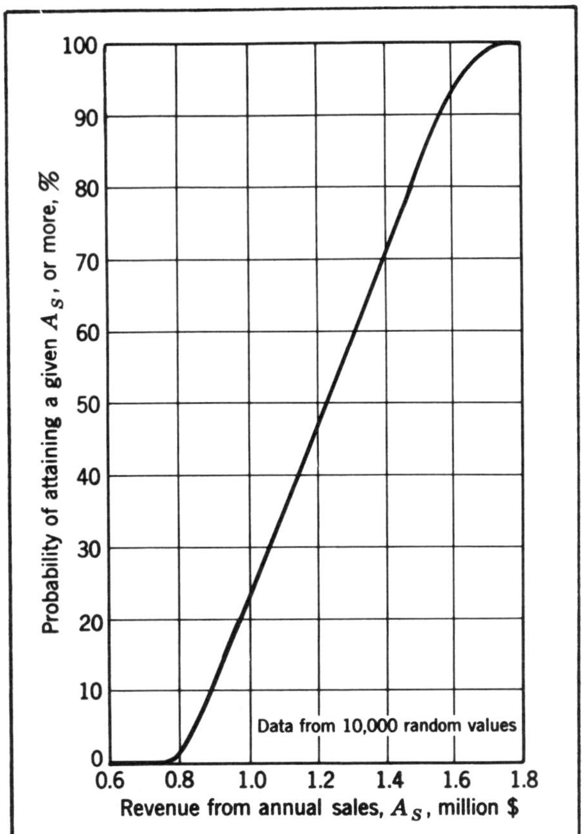

CUMULATIVE probability (for the example) of attaining a given revenue, or more, from annual sales—Fig. 5

for the range \$850/ton $< c_g \leqslant$ \$900/ton. This curve is also shown in Fig. 3.

Random Numbers and Monte Carlo Method

To generate the required cumulative probability function by the Monte Carlo method, the random numbers can be used in any sequence, provided that no number is reused. Let us read the numbers in Table III from left to right and top to bottom. Let each number of four digits represent two groups of two digits. Let the first of the two-digit figures represent a value of the sales volume, R, and the second two-digit figure represent a value of the sales price, c_g. The use of two-digit figures limits the number of possible values of R and c_g to 100 each.

The first number in Table III is 3770. Thus, the first value of R will be that which has a cumulative probability of 0.37. This is read from Fig. 3 to be 1,370 tons/yr. The first value of c_g will be that which has a cumulative probability of 0.70. This is also read from the appropriate curve in Fig. 3 and found to be \$852.5/ton. The value of the revenue from annual sales generated from the first random number is found from Eq. (14) for the appropriate amounts to be:

$$A_g = 1{,}370(852.5) = \$1{,}167{,}900/\text{yr.}$$

The minimum and maximum possible values of A_g are found from Eq. (14) to be:

$$A_g = 1{,}000 \text{ tons/yr.}(\$750/\text{ton}) = \$750{,}000/\text{yr.}$$

$$A_g = 2{,}000 \text{ tons/yr.}(\$900/\text{ton}) = \$1{,}800{,}000/\text{yr.}$$

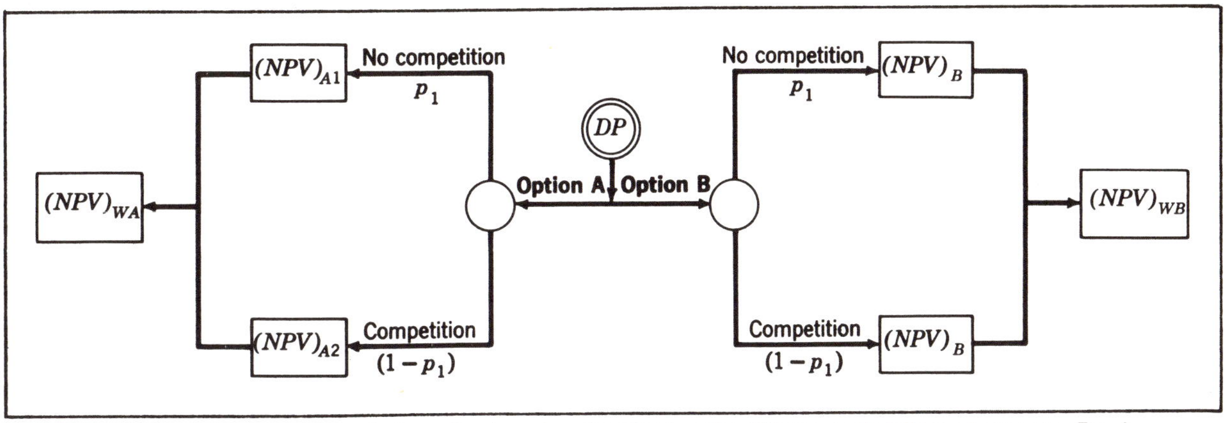

DECISION TREE enables selection of alternative investment options under differing probabilities of success—Fig. 6

Let us divide this range of values for A_S into 22 intervals, each of \$50,000/yr., and then note the frequency with which a random value of A_S falls inside a particular interval.

The first random value of A_S is \$1,167,900/yr., and falls inside the interval \$1,125,000/yr. to \$1,175,000/yr. The relative frequency with which A_S falls inside a particular interval is obtained by dividing the frequency by the total number of values calculated for A_S. Fig. 4 shows a histogram or plot of the relative frequency vs. the midpoint values of each interval. We can use the data in Fig. 4 to obtain the cumulative probability curve, as shown in Fig. 5.

With 100 possible values of R and 100 possible values of c_S, the frequency distribution of A_S can be determined exactly by 10,000 systematic calculations. However, even 10,000 random calculations do not duplicate exactly the frequencies given by systematic calculation. The frequency distribution could also be adequately assessed with about 5,000 random calculations. After making estimates of the probabilities for various options, an informed decision can be made.

Decision Trees and Investment Risk

In a typical decision tree (Fig. 6), each node represents a decision point (DP) where alternatives are available. Some quantifiable result of each alternative is chosen as a basis for comparison. In this case, it is the net present value (NPV). A value is assigned to the probability of attaining each result. Such probabilities are usually estimates made by "experts" using Delphi or other techniques. The estimates are subject to the restriction that the sum of the probabilities for branches leaving each node shall be unity.

In considering two investments, let Option B be a safe investment having a base net present value, $(NPV)_B$, that is independent of any competition. Let Option A yield a net present value of $(NPV)_{A1}$ if no competition exists, and $(NPV)_{A2}$ if competition exists. Let the probabilities of no competition and competition be p_1 and p_2, respectively.

The expected (NPV) for Option A can be written:

$$(NPV)_{WA} = p_1(NPV)_{A1} + (1 - p_1)(NPV)_{A2} \qquad (15)$$

where $(NPV)_{WA}$ is the weighted net present value for Option A, based on the probabilities of no competition, p_1, and competition $(1 - p_1) = p_2$.

The expected (NPV) for Option B can be written:

$$(NPV)_{WB} = p_1(NPV)_B + (1 - p_1)(NPV)_B$$

This relation simplifies to:

$$(NPV)_{WB} = (NPV)_B \qquad (16)$$

The gain in the expected value of Option A over Option B is:

$$\Delta(NPV)_W = (NPV)_{WA} - (NPV)_{WB} \qquad (17)$$

Eq. (17) can be written in terms of Eq. (15) and (16) as:

$$\Delta(NPV)_W = [p_1(NPV)_{A1} + (1 - p_1)(NPV)_{A2}] - (NPV)_B \qquad (18)$$

The options are represented in Fig. 6.

If $(NPV)_{A1} = 2(NPV)_B$, and $(NPV)_{A2} = 0.5(NPV)_B$, Eq. (18) can be written as:

$$\Delta(NPV)_W = [2p_1 + 0.5(1 - p_1) - 1](NPV)_B$$

In this case, if the probability of no competition is $p_1 = 0.5$, then:

$$\Delta(NPV)_W = +0.25(NPV)_B$$

Therefore, Option A should be taken.

If the probability of no competition is $p_1 = 0.3$, then:

$$\Delta(NPV)_W = -0.05(NPV)_B$$

And, Option B should be taken.

Options A and B are equally desirable when $\Delta(NPV)_W$ equals zero, i.e. when $p_1 = 0.33$. This technique has many uses that are well described in the literature (see pp. 174–181 and 190–195). ■

Engineering Economics—Part 11

Estimating the profitability of capital investments often produces a range of values. To overcome this difficulty, certain techniques are available for assessing the spread of values in conformity with a company's attitude to risk-ventures.

Numerical Measures of Risk

F. A. HOLLAND, F. A. WATSON, J. K. WILKINSON, University of Salford

Without risk, and the reward for successfully accepting risk, there would be no business activity. In a risk situation where skill plays a part, the skilled participant will prosper at the expense of his competitors provided that his resources enable him to continue in business for a sufficient length of time.

In estimating the probabilities of attaining various levels of net present value (NPV) and discounted-cash-flow rate-of-return $(DCFRR)$, we found a spread in the possible values of (NPV) and $(DCFRR)$. We will now examine a number of methods that have been suggested for assessing rewards and risks in order to become skilled participants.

Utility Functions

Let us consider a proposed project in which there is a probability, p_1, that a net present value of $(NPV)_1$ will result, a probability, p_2, that an $(NPV)_2$ will result, etc. A weighted average, $(NPV)_w$, known as the expected value, can then be calculated from:

$$(NPV)_w = p_1(NPV)_1 + p_2(NPV)_2 + \cdots \quad (1)$$

where:

$$p_1 + p_2 + \cdots = 1.0 \quad (2)$$

Analogous equations may be written for other measures of profitability such as net profit.

Let us consider a contractor who stands to make a net profit of \$100,000 on a contract. The cost of preparing the bid on the contract is \$10,000. There are four competing contractors, each with a probability $p_1 = 0.25$ of obtaining the contract. Thus, each contractor has a probability $p_2 = 0.75$ of not obtaining the contract. Therefore, the expected value of the project is:

$$0.25(\$100{,}000) + 0.75(-\$10{,}000) = \$17{,}500$$

In this case, the potential gain is ten times greater than the potential loss.

If the potential loss can bankrupt the company, then decisions are not necessarily made on the basis of expected value even though the potential gain may be very high. Also, decisions are not necessarily made on the basis of expected value if the potential loss represents a relatively small amount of money to the company. Between these two extremes, expected value can be a very useful criterion, particularly for a company with a large number of projects.

A company may be considering a project with a very high potential rate-of-return and a low risk, but it may prove impossible to raise the money to start the project. Conversely, the company may be prepared to undertake an extremely risky project if the investment is trivial. Thus, the attitude of a company to risk depends on the circumstances.

Money, Risk and Uncertainty

Money does not hold the same value for each company or each individual. A dollar may keep a pauper from starvation, while being a trivial amount to the man who gave it to him. Attempts have been made to quantify a company's attitude to money, risk and uncertainty[1] by asking businessmen a number of questions.

Your company has signed a business contract with potential after-tax proceeds of \$$P$. The probability of achieving this net gain of \$$P$ is, say, $p_1 = 0.75$; and the probability of a net loss of \$$P$ is $p_2 = 0.25$.

- If you would rather keep the contract, how much cash would you accept for your interest in it?
- If you would rather be released from the contract, how much cash would you pay to be released from it?

The same questions may then be asked for different

Originally published March 1974.

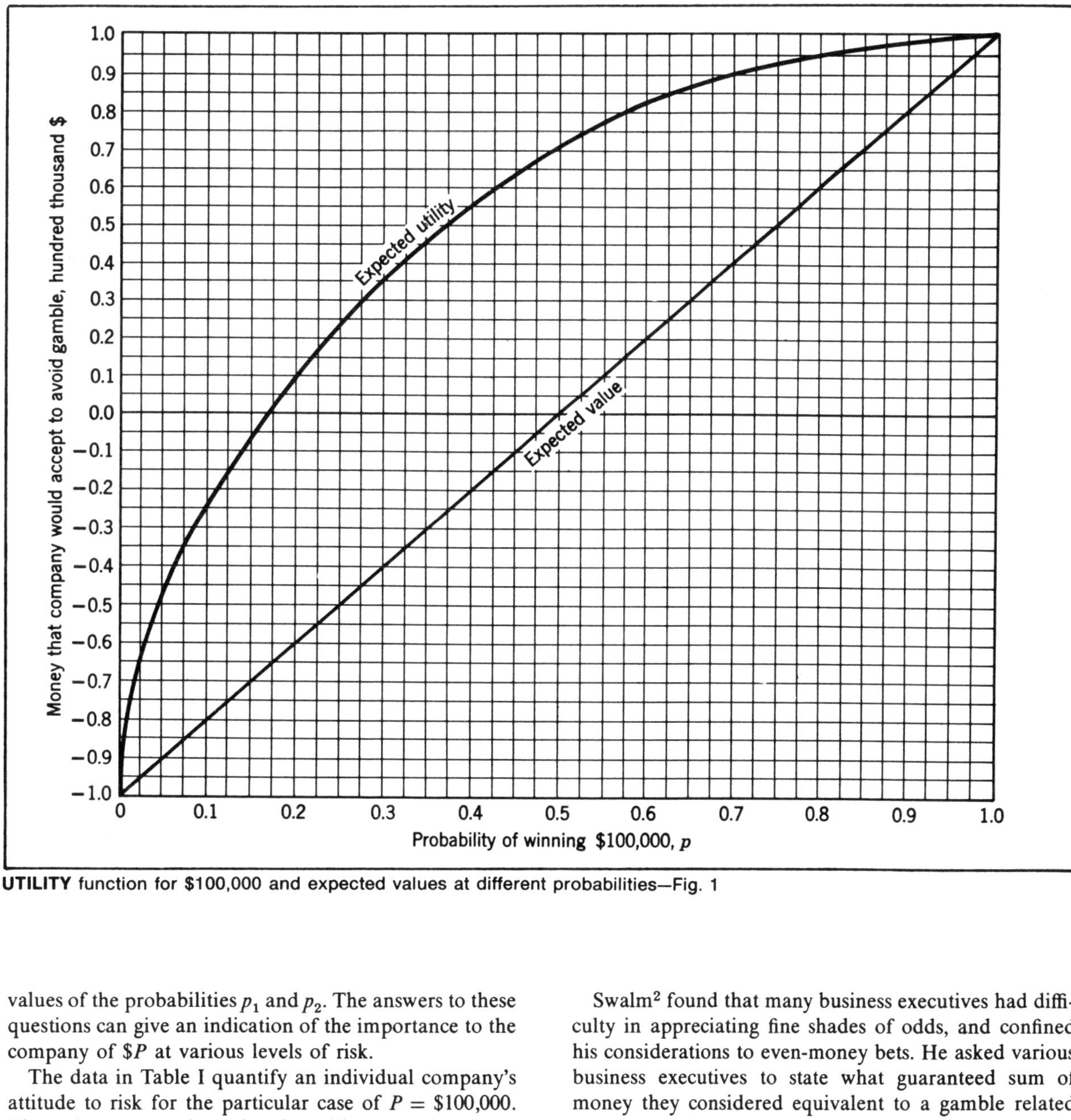

UTILITY function for $100,000 and expected values at different probabilities—Fig. 1

values of the probabilities p_1 and p_2. The answers to these questions can give an indication of the importance to the company of $\$P$ at various levels of risk.

The data in Table I quantify an individual company's attitude to risk for the particular case of $P = \$100{,}000$. These data are used to plot the utility curve in Fig. 1. Positive values are the amounts of money that the company would accept in order to forego its participation. Negative values are the amounts the company would pay in order to avoid participation.

Only when the utility value and the expected value (i.e., the straight line in Fig. 1) are the same, can net present value (*NPV*) and discounted-cash-flow rate-of-return (*DCFRR*) be justified as investment criteria.

Since the utility curve has such a subjective basis, most companies prefer the objectivity of (*NPV*) and (*DCFRR*) over the range of the normal income and expenditure budget. Subjective methods tend to be reserved for exceptionally high risk projects.

A utility curve such as Fig. 1 is specific to a certain sum of money. The curve is likely to be different for, say, $P = \$10{,}000$. Fig. 1 can only be used to consider projects that fall within the range −$100,000 to +$100,000. Other utility curves must be used to cover projects that lie outside this range.

Swalm[2] found that many business executives had difficulty in appreciating fine shades of odds, and confined his considerations to even-money bets. He asked various business executives to state what guaranteed sum of money they considered equivalent to a gamble related to the toss of a coin. If the coin fell on one side, they would win a given sum of money; if the coin fell on the other side, they would get nothing.

Swalm started by considering a sum of money equivalent to twice the maximum expenditure that the executive could authorize in one year. This was used to obtain a utility that, in turn, was used as the new prize to obtain a further utility. In this way, a utility curve could be sketched. Swalm chose an arbitrary utility scale based on a range of −120 utiles to +120 utiles. (Note: it is as incorrect to compare utiles by ratio as it is to imply that an object at 80°F. is twice as hot as an object at 40°F.)

Swalm found that most executives are conservative in their expenditure and that the patterns of utility curves are very similar if plotted with an ordinate range of ±120 utiles, and an abscissa range of ±1 unit. The unit, in this case, is the maximum authorized annual expenditure of the executive. Such curves may appear to differ quite widely when plotted in terms of absolute money values.

Utility Value of $100,000 to a Company At Various Levels of Risk—Table I

Probability of winning $100,000, p_1	Money that the company would accept to avoid the gamble
0.0	−$100,000
0.1	−$ 25,000
0.2	+$ 10,000
0.3	+$ 35,000
0.4	+$ 55,000
0.5	+$ 70,000
0.6	+$ 82,500
0.7	+$ 90,000
0.8	+$ 95,000
0.9	+$ 97,500
1.0	+$100,000

Conversely, the probability of losing is $(1 - p_1) = p_2$ because $p_1 + p_2 = 1.0$.

The curves also show that executives tend to be more conservative when considering a loss than they do when considering a reduced gain.

How To Use Probability and Utility

A company is considering investment in one or more of three projects, A, B and C. Evaluate the investment priorities on the basis of the following:

1. The probabilities of attaining various net present values (NPV) listed in the third column of Table II.
2. The company's attitude to winning or losing $100,000, represented by the data in Table I, as plotted in Fig. 1.

Eq. (1) gives the expected value for $(NPV)_w$. Hence for Project A, $(NPV)_w$ is computed from the data in Table II and found to be:

$$(NPV)_w = 0.1(\$95{,}000) + 0.8(\$45{,}000) + 0.1(-\$75{,}000)$$
$$(NPV)_w = \$9{,}500 + \$36{,}000 - \$7{,}500$$
$$(NPV)_w = \$38{,}000$$

Corresponding values for Projects B and C are calculated in the same way and are listed in Table II.

In Project A, the probability $p = 0.1$ for $(NPV) =$ $95,000. Table I and Fig. 1 show that $95,000 is the amount of money that this company would pay for an 0.8 probability of gaining $100,000. Of course, there is a 0.2 probability of losing $100,000.

In this case, a probability of $p = 0.1$ of attaining $95,000 is equivalent to a probable utility of $(0.1)(0.8) =$ 0.08 of gaining $100,000.

Eq. (1) can be used to calculate the expected utility if the probabilities p_1, p_2, etc., are replaced by the probable utilities, and the net present values $(NPV)_1$, $(NPV)_2$, etc., are each replaced by $100,000. For Project A, the expected utility, U_w, is:

$$U_w = [0.1(0.8) + 0.8(0.34) + 0.1(0.03)]\$100{,}000$$
$$U_w = \$8{,}000 + \$27{,}000 + \$300$$
$$U_w = \$35{,}500$$

Corresponding values for Projects B and C are calculated in the same way and are listed in Table II.

The straight line in Fig. 1 represents the situation in which the expected value and the expected utility are equal over the range of −$100,000 to +$100,000. In this case, decisions can be taken on the basis of the highest expected value as a routine matter. In other cases, decisions should be made on the basis of the highest expected utility. The utility curve in Fig. 1 represents the attitude of management to $100,000. This curve should be updated as the company's business position changes.

In the present case, the utility curve is above the straight line. This represents a tendency on the part of the company's decision-makers to gamble.

The investment priorities should be to implement Project A, and then, if finance is available, Project B.

It might appear that Project C should also be considered in view of the expected utility of $16,700. However, it is better to do nothing than to implement Project C. The utility of doing nothing, which is equivalent to paying $0, is read from Fig. 1 to be 0.17. This gives a

Comparison of Projects in Terms of Expected Value and Expected Utility—Table II

Project	(NPV), $	Probability, p	Equivalent Probability Of winning $100,000	Probable Utility	Expected Value, $	Expected Utility, $
	95,000	0.1	0.80	0.080	9,500	8,000
A	45,000	0.8	0.34	0.272	36,000	27,200
	−75,000	0.1	0.03	0.003	− 7,500	300
		1.0		0.355	38,000	35,500
	50,000	0.2	0.37	0.074	10,000	7,400
B	20,000	0.6	0.23	0.138	12,000	13,800
	−60,000	0.2	0.04	0.008	−12,000	800
		1.0		0.220	10,000	22,000
	45,000	0.1	0.35	0.035	4,500	3,500
C	10,000	0.6	0.20	0.120	6,000	12,000
	−60,000	0.3	0.04	0.012	−18,000	1,200
		1.0		0.167	− 7,500	16,700

corresponding probable utility of 1.0(0.17)($100,000), or $17,000. This is a better result than investing in Project C.

In this example, the order of priorities based on expected utilities is the same as that based on expected values. However, this will not necessarily be so in every case. In this example, the order of priorities is clear-cut on the basis of expected value, but much less so on the basis of expected utility.

Risk and Insurance

Capital is at risk until the breakeven point has been reached. It is common practice to consider the discounted breakeven point (*DBEP*) discounted to time zero at the cost of capital. At any time after the (*DBEP*), the project will have recovered its cost and provided a greater return on the capital than the cost of capital. It is customary for management to spread risk by diversifying the activities of a company among a portfolio of projects.

Business is an activity in which both skill and chance play a part. The theory of games[3] teaches that in the long run the following events will happen:

1. The player with the largest capital will win all the available capital.
2. Given equal capital, the most skillful player will win all the capital.
3. Any player will lose all his capital if the odds are only slightly against him and he continues to play for a long enough time.

Some of these factors can be seen to operate in the tendency of small businesses to be taken over by larger companies, and also in the bankruptcy of companies that cannot raise sufficient funds to participate in business activity for a further period. However, what happens in the short run is far less predictable.

An event or outcome that will definitely occur has a probability $p = 1$. An event or outcome that will definitely not occur has a probability $p = 0$. If the probability of success is p, then the probability of failure is $(1 - p)$. The probability of at least one success p_1 in n similar tries is given by:

$$p_1 = 1 - (1 - p)^n \tag{3}$$

If a six-sided die is thrown, the probability of a win when a given number occurs is 1/6. For this case, Eq. (3) can be written as:

$$p_1 = 1 - \left(\frac{5}{6}\right)^n$$

The probability of at least one win in four throws of the die is $p_1 = 0.9952$. If each throw costs $1, and the value of a win is $4, there is at least a 99.52% chance of breaking even in four throws.

Reul[4] has used Eq. (3) to define a parameter, which he calls the measured survival function, (*MSF*), given by:

$$(MSF) = 1 - (1 - p)^\beta \tag{4}$$

where (*MSF*) is the probability that a portfolio of bets with a similar strategy will at least break even, and β is the amount of one win divided by the amount of each bet. Reul has applied Eq. (4) to the research and development activities of a company. From the derivation of Eq. (4), it follows that $\beta \geqslant 1$, and hence, $(MSF) \geqslant p$.

Eq. (3) and (4) are based on the simplified assumption that a project either succeeds with probability p and achieves the expected reward, or fails completely with probability $(1 - p)$. Therefore, (*MSF*) is the probability of at least one success when β similar projects are undertaken, and represents a conservative measure of risk. Many projects may result in greater returns, or have an increased probability of attaining a given return, if more money is spent. Each alternative result from a given project is treated as a separate risk in the portfolio.

Research and development activities do not, in themselves, produce a salable product. Thus, they cannot directly generate a return on capital outlay. A successful research and development project is one that results in an activity that earns revenue for the company. The life cycle of the revenue from an individual product may be sketched as:

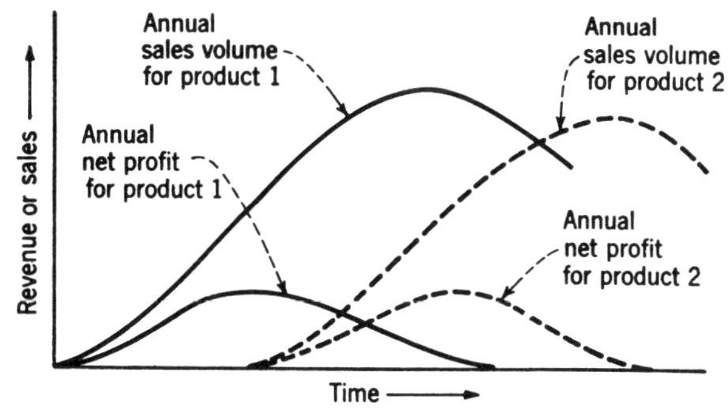

This revenue has to pay not only for the successful project but for all the unsuccessful research and development activities. It is common practice to consider all R&D as a portfolio. Disbursements for R&D are relatively flexible and can be switched from less-favorable to more-favorable projects at short notice.

When considering individual projects, β should be taken as the lesser of:

$$\beta = \frac{\text{Expected proceeds if the project is successful}}{\text{Disbursement on the project}}$$

or:

$$\beta = \frac{\text{Total expenditure on all projects over the budget period}}{\text{Expenditure on the project over the budget period}}$$

Because the projects in a portfolio will usually have different probabilities of success and different rewards for success, β and p in Eq. (4) are conservatively estimated as follows:

$$\beta = \frac{\text{Total annual proceeds if all projects are successful}}{\text{Total annual disbursements on all projects}}$$

$$p = \frac{\text{Total expected value of all projects}}{\text{Total proceeds if all projects are successful}}$$

The expected value can be calculated from Eq. (1).

The relationship between (*MSF*), p and $1/\beta$ in Eq. (4) is shown graphically in Fig. 2. It is the responsibility of management to decide on an acceptable value of the (*MSF*) for its company. The value chosen will depend on the company's attitude to risk that can be quantified

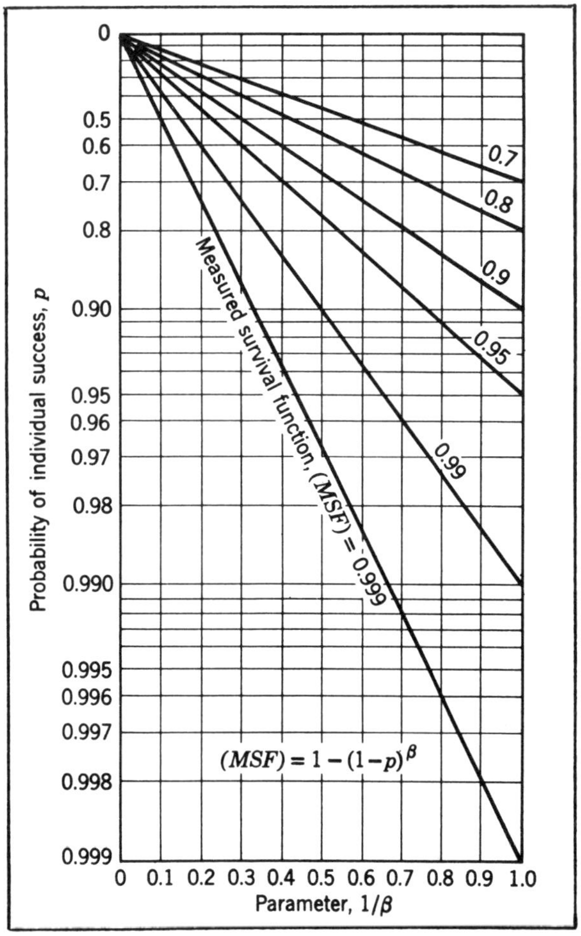

MEASURED survival function (*MSF*) vs probability—Fig. 2

in the form of a utility curve such as the one shown in Fig. 1, from which a value of equivalent (*MSF*) can be obtained.[5]

It is also the responsibility of management to estimate the probabilities for the success of individual projects after due consideration of all the data provided by the various departments. The rate of return on investment that is acceptable to management is a function of these probabilities.

Each industry has a reasonably well defined return on investment that reflects the degree of risk inherent in that industry. If management decisions are faulty, the company will either overspend or will miss opportunities.

For a disbursement of \$1,000 in year zero, the discounted breakeven point (*DBEP*) will be reached in 3 yr. for a compound interest rate of 30% if the annual net profit A_{NP} = \$550.63/yr. Thus, a discounted-cash-flow rate-of-return (*DCFRR*) of 30%/yr. corresponds to:

$$\frac{1}{\beta} = \frac{\$1{,}000}{(3 \text{ yr.})(\$550.63/\text{yr.})} = 0.61$$

For $1/\beta = 0.61$ and an $(MSF) = 0.999$, the probability of individual success is read from Fig. 2 to be $p = 0.985$. Similarly, it can be deduced that if $(MSF) = 0.999$ and $p = 0.95$, a (*DCFRR*) of 45%/yr. is required if breakeven is to be achieved in 4 yr. However, if breakeven in 20 yr. is acceptable, then a (*DCFRR*) of only 10%/yr. is needed.

Example of a Portfolio of Projects for Research and Development Program—Table III

Project	Proposed disbursement for the coming year	Annual after-tax income if successful	Probability of success
A	\$125,000	\$ 600,000	0.95
B	\$100,000	\$ 400,000	0.90
C	\$100,000	\$ 125,000	0.80
D	\$ 80,000	\$ 100,000	0.75
E	\$ 50,000	\$ 60,000	0.70
F	\$ 20,000	\$ 30,000	0.65
G	\$ 20,000	\$ 70,000	0.50
H	\$ 5,000	\$ 15,000	0.20
Totals	\$500,000	\$1,400,000	---

It is interesting to note that if an (*MSF*) of 0.95 is acceptable to management, then a (*DCFRR*) of 10%/yr. is equivalent to assuming that the process will break even in about 2 yr. [For methods of calculating (*DBEP*) and (*DCFRR*), see Part 5.]

Example: Breakeven Techniques

Details of the estimates for the current research and development program of a company are given in Table III. Estimate the probability that this portfolio will at least break even.

The total annual proposed disbursement for R&D is \$500,000. For Project A, the anticipated annual after-tax income is \$600,000. Since this amount exceeds the total annual disbursement, the correct value of β will be based on \$500,000, and is the value to be used in the computations. The effective total annual income were all projects to reach their anticipated income would be \$1,300,000. Therefore:

$$\beta = \$1{,}300{,}000/\$500{,}000 = 2.600$$

Project A has expected value of (0.95)(\$500,000/yr.), or \$475,000/yr., Project B has an expected value of (0.90)(\$400,000/yr.) = \$360,000/yr., and so on. We sum these values to obtain the total expected value of the portfolio as \$1,109,500/yr. Hence:

$$p = (\$1{,}109{,}500/\text{yr.})/(\$1{,}300{,}000/\text{yr.}) = 0.8535$$

From Fig. 2, for a probability of success, $p = 0.8535$, and a value $1/\beta = 1/2.600 = 0.3846$, we read the (*MSF*) to be 99.3%. This is the probability that this portfolio will at least break even.

Alternatively, we can substitute the values for p and β into Eq. (4) to get:

$$(MSF) = 1 - (1 - 0.8535)^{2.6} = 0.9932, \text{ or } 99.32\%$$

Example: Utility-Function Curves

Sketch a utility-function curve that is equivalent to the following pattern of measured survival functions (*MSF*) that expresses the strategy of a particular manager in spending his authorized annual budget of $1 million:

Case	Potential Proceeds Annually, $	(*MSF*), %
a.	Above 600,000	99.9
b.	300,000 to 600,000	95
c.	0 to 300,000	65
d.	Losses	75

Plot the resultant curve on a utility scale of ±120 utiles against a potential gain of ±$1 million.

The required axes range from −$1 million/yr. to +$1 million/yr., and from −120 utiles to +120 utiles. Utiles can be compared by ratio on an absolute scale only. Hence, for purposes of calculation, the axes are moved to provide a working range of $0/yr. to $2 million/yr., and 0 to 240 utiles (Fig. 3). On these axes, a potential gain of $600,000/yr. corresponds to an absolute amount of (600,000 + 1,000,000) = $1,600,000/yr., and a potential loss of $200,000/yr. to an absolute amount of (−200,000 + 1,000,000) = $800,000/yr.

a. For annual proceeds above $600,000/yr., (*MSF*) is 99.9%. If the certainty of an annual gain of $600,000 has to be abandoned in an effort to obtain an annual gain of $1 million, then on an absolute scale:

$$\beta = \frac{\$2{,}000{,}000/\text{yr.}}{\$1{,}600{,}000/\text{yr.}} = 1.2500$$

With $1/\beta = 0.8000$, and $(MSF) = 99.9\%$, we find the required probability of success by solving Eq. (4) for p:

$$p = 1 - [1 - (MSF)]^{1/\beta}$$
$$p = 1 - (1 - 0.999)^{0.800} = 0.996$$

The utility of an amount of money is its utility when it is certain to be obtained multiplied by its probability of being attained. On a scale in which an absolute annual income of $2 million/yr. has a utility of 240 utiles, the utility of $1.6 million is (0.996)(240), or 239 utiles.

b. For annual proceeds between $300,000 and $600,000, $(MSF) = 95\%$. If the certainty of an annual gain of $300,000 has to be abandoned to obtain an annual gain of $600,000, then (as before), we find:

$$1/\beta = \$1{,}300{,}000/\text{yr.}/\$1{,}600{,}000/\text{yr.} = 0.8125$$
$$p = 1 - (1 - 0.95)^{0.8125} = 0.912$$

Since, to this manager, the utility of an absolute income of $1.6 million is 239 utiles, the value of $1.3 million is (0.912)(239) = 218 utiles. On the original scales, potential annual proceeds of $300,000 have a utility of (218 − 120), or 98 utiles.

c. and d. Values of utility at other potential annual gains are calculated in the same way and listed in Table IV, and shown graphically in Fig. 3.

This strategy is extremely conservative when high gains are possible but becomes less so for smaller potential gains. Where potential losses are involved, the strategy is a fair one for which (*NPV*) would be an accurate guide for choosing alternatives.

Insurance and Risk

In the venture-premium method of assessment, risky investments are required to yield a rate of return that adds a premium to the cost of finance. Rudd and Watson[6] considered this relationship:

$$i_m = i + i_r \tag{5}$$

where i is the cost of capital; i_m is the minimum acceptable interest rate-of-return on the investment; and i_r is known as the risk rate.

Rudd and Watson[7] also suggested that each project should pay an insurance premium, i_r, to guarantee the expected profits. The magnitude of i_r is proportional to the amount of capital to be risked. It is also a function of the degree of risk involved. Working capital and capital for auxiliary facilities are assumed to be riskfree. Thus, the risk rate is only applied to the fraction of the capital investment likely to be lost if the project is unexpectedly terminated.

The main objection to the venture-premium method is that the assessment of the riskiness of a project may be too subjective. This could lead to the rejection of potentially attractive proposals, and the acceptance of projects that merely appear to be riskfree.

Insurance is protection against risk. Commercial insurance companies minimize their own risks by covering a large number of individuals against a given risk, and also by offering coverage on a wide variety of different types of risk. It is frequently quite difficult to assess the probability of success of a particular research and development project. It is much easier for an insurance company to assess its probabilities from its casualty tables.

Businesses tend to provide their own insurance cover when individual claims are likely to be a small fraction of the available capital. The cost of commercial insurance is about 30% higher than would be necessary to cover the same risk in one's own company. However, for low-probability high-cost risks, most businesses prefer to insure with a commercial insurance company. Such risks include loss of plant or buildings due to fire, losses of revenue due to delays in startup or strikes, etc.

It is also becoming necessary to insure against factors not normally considered until recently. These include possible lawsuits for polluting the environment. The cost of insurance increases the annual total expense, A_{TE}. Thus, overinsurance can lead to an unnecessary decrease in profitability. The management of a company must ultimately judge its own risks.

Example: Risk Rates

Let us calculate the required risk rate for a project that is described by the following: (a) risk strategy is equivalent to an (*MSF*) of 99%; (b) payback of risk capital is 3 yr.; (c) cost of capital i is 10%/yr.; and (d) probability of complete success of the project is estimated as 95%.

First, we must calculate the value of β in Eq. (4). For this project, with $(MSF) = 0.99$ and $p = 0.95$, we find:

$$\beta = \frac{\log[1 - (MSF)]}{\log(1 - p)} = \frac{\log(0.01)}{\log(0.05)} = 1.537$$

To recover this amount of capital and interest in 3 yr., the average net annual cash flow, A_{CF}, required is:

$$A_{CF} = 1.537/3 = 0.5124, (\$/\text{yr.})/(\$\text{ invested})$$

UTILITY-function curves express strategy of manager in spending an annual budget of $1 million for various potential annual proceeds for example—Fig. 3

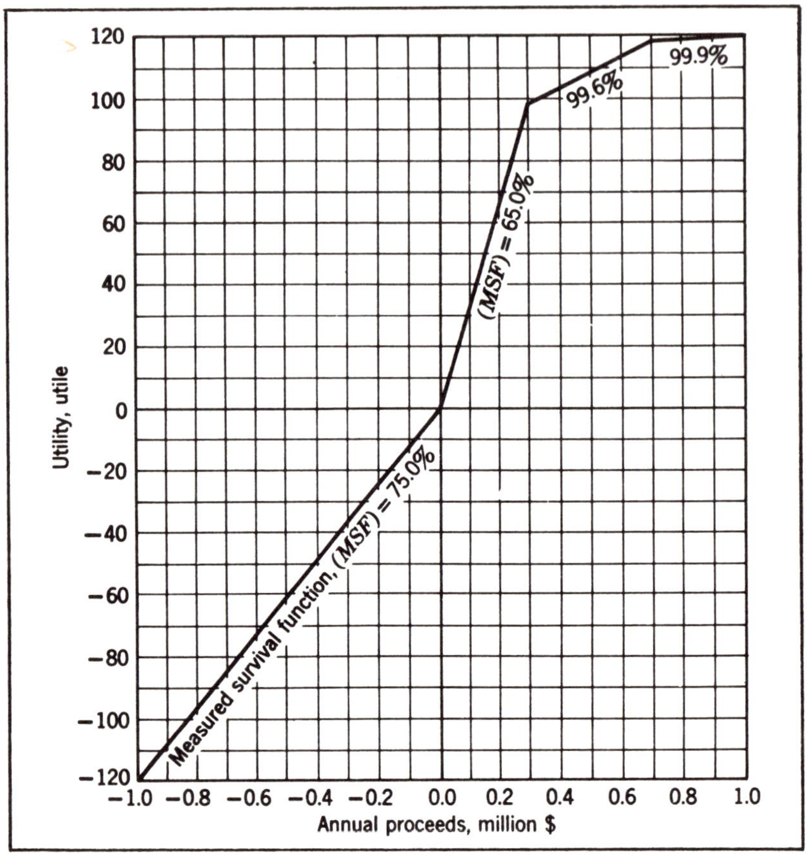

Data for Problem of Potential Gains and Utility Values—Table IV

Potential Gain, $/Yr.	Absolute Gain, $/Yr.	(*MSF*)	$1/\beta$	Probability, *p*	Absolute Utility, Utiles	Utility, Utiles
1,000,000	2,000,000	0.999	1.0000	1.000	240	120
600,000	1,600,000	0.999	0.8000	0.996	239	119
300,000	1,300,000	0.950	0.8125	0.912	218	98
0	1,000,000	0.650	0.7692	0.554	120	0
−500,000	500,000	0.750	0.5000	0.500	60	−60
−1,000,000	0	0.750	0.0000	0.000	0	−120

The relation between annual cost, A, and present worth, P, of a future sum of money is given by:

$$A = Pf_{AP} \tag{6}$$

where the annual present worth factor, f_{AP}, is:

$$f_{AP} = \frac{i(1+i)^n}{(1+i)^n - 1}$$

In effect, in computing the average net annual cash flow per dollar invested, we have obtained the value of f_{AP} in Eq. (6) for our example. From tables of the annuity present-worth factor, f_{AP}, we find the value of the interest rate to be $i = i_m = 0.25$ when $f_{AP} = 0.5124$ for $n = 3$ yr.

Hence, substituting the appropriate values into Eq. (5) and solving for the required risk rate will yield:

$$i_r = i_m - i$$
$$i_r = 25 - 10 = 15\%$$

All capital will be charged at the cost of capital $i = 10\%$. Working capital, C_{WC}, is completely recoverable without risk, and requires interest only at 10%. The unrecovered part of the risk capital, C_{FC}, attracts the additional interest rate of 15%. ■

References

1. Schlaifer, R., "Probability and Statistics for Business Decisions," p. 34, McGraw-Hill, New York, 1959.
2. Swalm, R. O., *Harvard Business Rev.*, June 1966.
3. Levinson, H. C., "The Science of Chance," p. 83, Faber and Faber, London, 1952.
4. Reul, R. I., *Trans. Inst. Chem. Eng.* (*London*), **48**, No. 238 (1970).
5. Reul, R. I., "Selection and Monitoring of Research and Development Programmes," p. A-21, Industrial and Commercial Techniques, London, 1970.
6. Rudd, D. F. and Watson, C. C., "Strategy of Process Engineering," p. 91, Wiley, New York, 1968.
7. Ibid., p. 92.

How To Estimate Capital Costs

The amount of time and money spent on any capital cost estimate depends on its intended use. When this decision is made, various procedures are available for preparing estimates that range from rapid and low-cost ones to the more formal and tedious detailed design ones.

F. A. HOLLAND, F. A. WATSON and J. K. WILKINSON, University of Salford

Capital cost estimation is a vital part of investment appraisal. For this purpose, it is necessary to estimate the total capital required for the project. The total capital cost of a project, C_{TC}, is:

$$C_{TC} = C_{FC} + C_{WC} + C_L \qquad (1)$$

where C_{FC} is the fixed capital cost, C_{WC} is the working capital cost, and C_L is the cost of land and other nondepreciable costs. The various types of capital cost are defined in Part 2 of this series, including several methods of estimating working capital cost.

The fixed capital cost is usually regarded as the capital needed to provide all the depreciable facilities. The term, C_L, in Eq. (1) includes land and items such as catalysts for which depreciation is not allowed by the Internal Revenue Service. In practice, nondepreciable items form only a small proportion of the total capital cost.

This article will be devoted mainly to the estimation of C_{FC} and C_L, for which purpose the use of a code system and checklist is essential.

For any given year, capital will generally be expended on a variety of projects that may include any of the following:

- Purchase of land.
- Construction of plant on a "green field" (i.e., grass-roots or undeveloped) site.
- Construction of plant on a developed site.
- Extensions to an existing plant to provide additional capacity.
- Replacement or modification of part of an existing plant to increase efficiency.
- Construction of nonproductive but essential facilities such as effluent treatment units, safety and welfare facilities, etc.

Often, "battery limits" are defined for a project to act as cost centers. This boundary normally includes all manufacturing equipment but excludes administrative offices, storage areas, utilities, and other essential and nonessential auxiliary facilities.

Purposes of Capital Cost Estimates

Capital cost estimates are made for a variety of reasons. Among others, to enable feasibility studies to be carried out; to enable a manufacturing company to select from alternative investments; to assist in selection from alternative designs; to provide information for planning the appropriation of capital; or to enable a contractor to tender for a new project.

The purpose of the estimate influences the amount of time and money to be spent on it. In turn, this determines the method of estimation and its probable accuracy.

The American Assn. of Cost Engineers lists five types of estimates, which are:

Type	Probable Accuracy
Order of magnitude (ratio estimate)	±40%
Study (factored estimate)	±25%
Preliminary (budget authorization estimate)	±12%
Definitive (project control estimate)	±6%
Detailed (firm estimate)	±3%

Originally published April 1974.

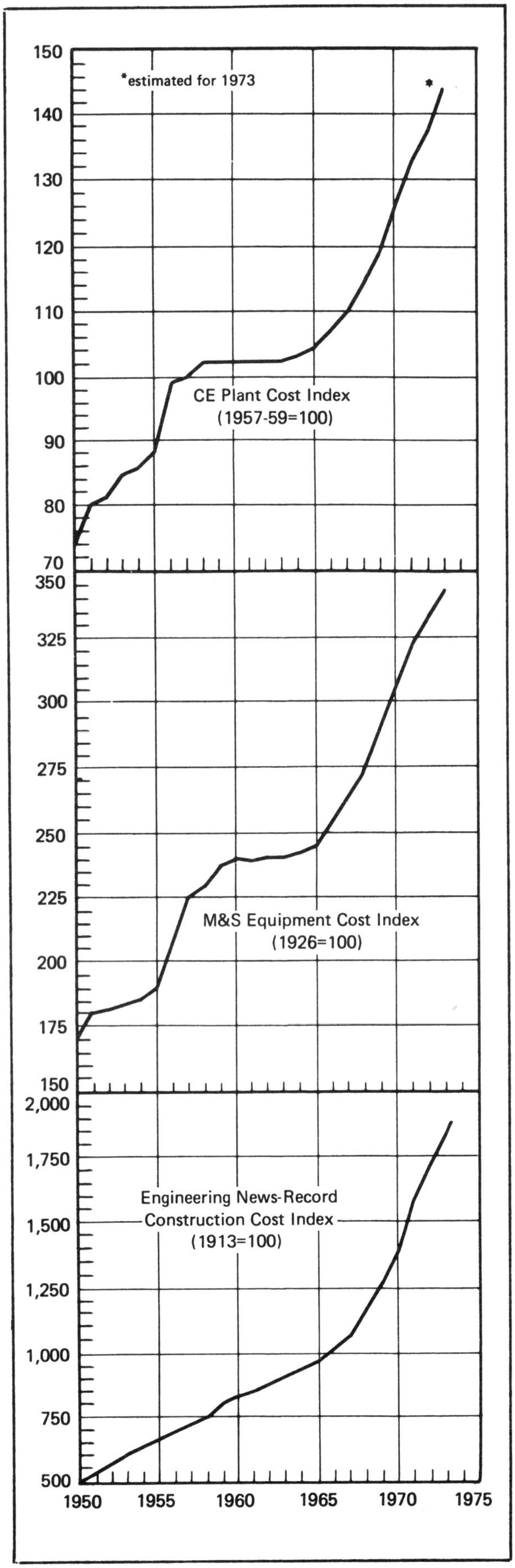

INDEXES measure material and labor costs—Fig. 1

All these methods require some knowledge of the proposed plant. It is important to realize that almost any plant can be made to work, provided enough money is spent on it, even though the basic design may be a poor one. The design, and hence, the cost of the plant depends on the skill and particular points of view of the design engineers for the project. Overdesign of control or safety factors, for example, may be used to cover up errors of judgment. Indeed, many design engineers are reluctant to practice intelligent risk-taking.

Factors that tend to increase the capital cost of a plant include: (1) overprovision for safety, (2) overprovision of standby equipment, (3) unnecessarily robust supporting structures, (4) building enclosures, (5) inclusion of nonstandard size equipment resulting from a too rigid adherence to theory, and (6) expensive materials of construction.

It may be more economic to use cheaper materials of construction, though this may lead to more frequent breakdowns and higher maintenance costs. The design engineer may feel that management will give him more credit for lower maintenance costs than for correspondingly greater savings on capital expenditure. It may well be better practice to prepare a modified design with a significantly lower capital cost. The risks involved with this second design should, of course, be clearly stated. It is quite wrong for design engineers to make implicit policy decisions in their plant design without making management aware of all the factors.

Significance of Cost Indexes

Money continuously falls in value owing to inflation. Hence, all published cost data are out of date even at the time of publication. Data from cost records of previous projects may be converted to present-day values by means of a cost index that expresses the relative cost of an item in terms of the cost at a particular base time.

There are many types of cost indexes intended to cover almost every area of interest. Some are based on national averages, others on local costs. Some are designed to cover broad areas of building, construction or equipment costs; others are highly specialized. A few of the more frequently used ones are:

- *Engineering News-Record* Construction and Building Indexes
- M&S Equipment Cost Index
- Nelson Refinery Construction Index
- *Chemical Engineering* Plant Cost Index

Some of these indexes are plotted in Fig. 1.*

The above compound indexes are normally calculated from various price indexes published by the Bureau of Labor Statistics of the U.S. Dept. of Labor, which cover changes in the price of many products and in the hourly earnings of workers in most industries. Salary trends for engineers, draftsmen, clerical and administrative staff are also included.[1]

A cost index usually reflects average national changes, and therefore, ignores local factors.

*The CE Plant Cost Index and the M&S Equipment Cost Index are compiled in *Chemical Engineering* on the page facing the inside back cover of each issue. The Nelson Refinery Construction Index appears in the *Oil & Gas Journal*, and the ENR Construction and Building Indexes appear in *Engineering News-Record*.

Cost Data for Production Units on Developed Sites—Table I

Product	Plant Capacity, Tons/Yr.	Total Capital Cost, $	Capital Cost per Annual Ton, $/Ton
Acetaldehyde[1]	50,000	3,500,000	70
Acetylene[2]	75,000	9,500,000	127
Ammonia	500,000	16,000,000	32
Ammonium phosphate	250,000	2,500,000	10
Ammonium sulfate	140,000	1,200,000	9
Butadiene[3]	100,000	50,000,000	500
Butadiene[4]	200,000	70,000,000	350
Cyclohexane	100,000	800,000	8
Ethyl alcohol[1]	75,000	3,750,000	50
Ethylene[5]	300,000	15,000,000	50
Ethylene dichloride	25,000	3,200,000	127
Ethylene oxide[1]	100,000	9,000,000	90
Glycerol	35,000	5,500,000	157
Hydrogen	60,000	6,500,000	108
Methanol[2]	210,000	9,000,000	43
Methyl chloride[6]	10,000	500,000	50
Oxygen	150,000	2,250,000	15
Phosphoric acid (as P_2O_5)	100,000	2,400,000	24
Polyisoprene	30,000	5,000,000	320
Sulfuric acid[7]	280,000	2,100,000	8

Data in this table are for 1967 costs.

Source: 1. ethylene, 2. natural gas, 3. butane, 4. butylene, 5. refinery gases, 6. methanol, 7. contact process.

Typical Values for the Capital Ratio—Table II

Industry	Capital Ratio, (CR)
Fibers, synthetic	3.44
Petroleum	3.08
Inorganics, heavy	2.24
Chemicals, general	2.02
Resins and plastics	1.90
Glass	1.46
Rubber	1.04
Pharmaceuticals	0.92
Soap and detergents	0.69
Foodstuffs, processed	0.66

Data are for the year 1958.

Equipment Estimating by Means of Capacity Ratio Exponents—Table III

Equipment	Exponent, ψ
Compressors, reciprocating	0.7
Compressors, turboblowers	0.5
Electric motors	0.8
Evaporators	0.5
Heat exchangers	0.65 to 0.95
Instruments and controls	0.0
Piping	0.7 to 0.9
Pumps	0.7
Tanks, rectangular	0.5
Tanks, spherical	0.7
Towers, constant diameter	0.7
Towers, constant height	1.0

Ratio Methods for Rapid Estimates

Haselbarth[2] published data giving the total capital investment per unit of annual production, C_{TR}/R. Table I lists data for many processes involving production units constructed on a previously developed site. Units built on a "green field" site would cost about 30 to 40% more; on the other hand, enlargements of an existing plant would cost roughly 20 to 30% less than the values given in Table I. All costs in this table apply to the first quarter of 1967 and hold for average conditions.

For plant capacities other than those given in the table, the capacity ratio exponents, ψ, (listed in Table III, and described later) may be used within a twofold ratio of sizes. Corrections for change of costs with time can be made by using the appropriate cost indexes.

Lynn and Howland[2] studied the capital ratios for 17 process industries. The capital ratio, (CR), for a "green field" site plant is defined as the ratio of the fixed capital investment, C_{FC}, to the annual sales revenue, A_S, expressed as:

$$(CR) = C_{FC}/A_S \tag{2}$$

Fixed capital was taken to include money invested in land, production facilities, warehousing, etc., R&D costs, and any auxiliary facilities necessary to support the process. In any given year, they considered C_{FC} in Eq. (2) to represent the estimated replacement value of the depreciable facilities. Typical values of capital ratios for the year 1958 are listed in Table II.

Both of these methods are useful for rough screening. They have the advantage that they do not require design work or the establishment of heat and mass balances for the process. However, both methods must be used with common sense and discretion. Obviously, they are most valid when a substantial amount of up-to-date company data exist.

Another method for making rapid capital-cost estimates uses capacity ratio exponents and is based on a company's existing cost data. This procedure also has the advantage of not requiring any design work or the establishment of heat and mass balances.

For example, let the cost of a piece of equipment or plant of size or capacity, Q_1, be C_1. The cost, C_2, of a similar piece of equipment or plant of size or capacity, Q_2, can be calculated from:

$$C_2 = C_1 \, (Q_2/Q_1)^{\psi} \tag{3}$$

where the value of the exponent ψ depends on the type of equipment or plant. Cost indexes should be used to bring the cost data to a common year. Some typical values of ψ for various items of equipment are shown in Table

III, but it is preferable to calculate such values from data in the company cost files.

From Eq. (3), it is possible to derive a unit-cost equation for equipment as:

$$\frac{C_2}{Q_2} = \frac{C_1}{Q_1}\left(\frac{Q_2}{Q_1}\right)^{\psi-1} \quad (4)$$

For complete process plants, the exponent ψ is usually taken to be 0.7. However, where the capacity increase is achieved by duplicating a number of pieces of equipment, the exponent will be higher and may even approach a value of 1.0. Continuous and highly instrumented processes will have a value of ψ lower than 0.7. Typical values of ψ for various types of complete process units are given in Table IV.[2,3] This method is most accurate where a substantial amount of company data exist for calculating base factors and ψ.

Capital Cost Estimates Based on Design

Process designs range from preliminary ones based on approximate technical data, calculations and cost estimates to final plans based on accurate technical data, calculations and firm cost quotations. Final designs require a considerable technical effort and are not justified unless sales and profitability can be forecast with reasonable accuracy. During the development of a process from the laboratory to the construction stage, the design and cost of the plant will change at frequent intervals.

Design and process economics are interrelated, and good design is not possible without adequate cost estimates. Management should be supplied with this information so that they can make the necessary decisions to avoid uneconomic commitments. In this way, many processes will not proceed beyond the preliminary design stage. Some projects may be taken much further than is economically justified when the available information is inadequate.

The main steps in preparing a capital cost estimate, based on design, are:

1. Initial idea for the process and specification of the size and type of operation.
2. Collection of physical and chemical data from the literature or from laboratory experiments.
3. Preparation of a preliminary equipment flowsheet incorporating the required unit operations, and showing the main items of equipment.
4. Preparation of heat and mass balances.
5. Specification of temperatures and pressures at various points on the equipment flowsheet.
6. Design calculations to size the main items of equipment such as pumps, heat exchangers, reactors, columns and tanks.
7. Preparation of a coded list of major process equipment such as storage tanks, reactors, columns, circulating pumps, heat exchangers, together with details such as size, capacity, materials of construction, pressures and temperatures.
8. Collection of cost data from the literature, company records and quotations.
9. Estimation of the total delivered cost of all the items of major process equipment such as storage tanks, reactors, columns, circulating pumps and heat exchangers.

In many cases, the purchased cost of equipment, free on board, is quoted by a supplier. This may be multiplied by a factor of 1.03 to give the approximate delivered cost.

For order-of-magnitude estimates, the simplest method of estimating the fixed capital cost of a plant, based on design information, is the Lang factor method. The fixed cost is given by:

$$C_{FC} = f_L \Sigma C_{EQ} \quad (5)$$

where C_{FC} is the fixed cost; f_L is the Lang factor having a value of 3.10 for solids processing, 3.63 for mixed solids-fluid processing and 4.74 for fluid processing; and C_{EQ} is the cost of a major item of process equipment. The method is most effective in companies that use Lang factors based on their own company data.

A refinement of the Lang factor method, suggested by Hand, is to use the modified equation:

$$C_{FC} = \Sigma f_L C_{EQ} \quad (6)$$

in which different Lang factors are used for different types of equipment, e.g., $f_L = 4.0$ for pumps, pressure vessels and fractionating columns; $f_L = 3.5$ for heat exchangers; and $f_L = 2.5$ for compressors.

Hand's method is also most useful when the factors are based on comprehensive company data. A list of such factors has been presented by Wroth.[3]

Several attempts have been made to break down the Lang factor into a series of other factors that more accurately represent the type of plant and process. For example, the following equation incorporates factors based on the work of Chilton:[4]

$$C_{FC} = \phi_1 \phi_2 \phi_3 \Sigma C_{EQ} \quad (7)$$

Capacity Ratio Exponents for Complete Process Plants—Table IV

Process Plant	Exponent, ψ
Acetylene	0.75
Ammonia	0.74
Ammonium nitrate	0.54
Benzene	0.61
Butadiene	0.59
Caustic soda	0.35
Chlorine	0.35
Ethanol	0.60
Ethylene	0.58
Ethylene oxide	0.79
Formaldehyde	0.55
Hydrogen cyanide	0.71
Isopropyl alcohol	0.60
Methanol	0.83
Nitric acid	0.56
Oxygen	0.64
Phosphoric acid	0.58
Polyethylene (low pressure)	0.67
Polyethylene (high pressure)	0.90
Styrene	0.68
Sulfuric acid (contact)	0.62
Urea	0.59

Eq. (5), (6) and (7) all relate the fixed capital cost of a plant to the delivered cost of all the items of major process equipment, ΣC_{EQ}. However, Chilton used installed equipment costs. The factor ϕ_1 is used in Eq. (7) to convert delivered equipment costs to installed equipment costs. In Eq. (7), the ϕ_1 factor may be taken as 1.45 for solids processing, 1.39 for mixed solids-fluid processing, and 1.47 for fluid processing. The remaining factors, ϕ_2 and ϕ_3, are evaluated from:

$$\phi_2 = 1 + f_1 + f_2 + f_3 + f_4 + f_5$$
$$\phi_3 = 1 + f_6 + f_7 + f_8$$

where the f factors have values for different items under several conditions as follows:

f_1, process-piping factors are:

Range	Condition
0.07 to 0.10	Solids processing
0.10 to 0.30	Mixed solids-fluid processing
0.30 to 0.60	Fluid processing

f_2, instrumentation factors are:

Range	Condition
0.02 to 0.05	Little automatic control
0.05 to 0.10	Some automatic control
0.10 to 0.15	Complex automatic control

f_3, building factors are:

Range	Condition
0.05 to 0.20	Outdoor units
0.20 to 0.60	Mixed indoor and outdoor units
0.60 to 1.00	Indoor units

f_4, facilities factors are:

Range	Condition
0 to 0.05	Minor additions
0.05 to 0.25	Major additions
0.25 to 1.00	New site

f_5, outside lines factors are:

Range	Condition
0 to 0.05	Existing plant
0.05 to 0.15	Separated units
0.15 to 0.25	Scattered units

f_6, engineering and construction factors are:

Range	Condition
0.20 to 0.35	Straightforward plants
0.35 to 0.50	Complex plants

f_7, size factors are:

Range	Condition
0 to 0.05	Large plants
0.05 to 0.15	Small plants
0.15 to 0.35	Experimental plants

f_8, Contingency factors are:

Range	Condition
0.10 to 0.20	Firm process
0.20 to 0.30	Process subject to change
0.30 to 0.50	Tentative process

Cost Data for Problem—Table V

Equipment Item	Quantity	Delivered Cost, C_{EQ}, $
Storage tanks	3	35,000
Centrifugal pumps	4	3,200
Reactor	1	8,000
Reboiler	1	8,800
Distillation column	1	20,000
Reflux divider	1	800
Condenser	1	7,000
Total, ΣC_{EQ}		82,800

These data are for smaller plant at 1965 prices.

Cost Factors for Use in Eq. (7)—Table VI

Description	Approximate Factor
f_1, process piping	0.55
f_2, automatic control	0.08
f_3, buildings	0.20
f_4, facilities	0.05
f_5, outside lines	0.05
Total, f_1 to f_5	0.93
f_6, engineering & construction	0.25
f_7, size factor	0.10
f_8, contingency	0.20
Total, f_6 to f_8	0.55
$\phi_2 = 1 + f_1 + f_2 + f_3 + f_4 + f_5$	1.93
$\phi_3 = 1 + f_6 + f_7 + f_8$	1.55

Unless a great deal of detailed information is available within the company, an overall factor is probably as accurate as the use of a number of component factors.

Example: Capital Cost Estimating

A company installed a small plant to produce 4 million lb./yr. of a liquid product in 1965. Let us estimate the fixed capital cost of a plant to produce 8 million lb./yr. of the same product in 1971.

For this evaluation, the following information applies: (a) same process is used in both cases, (b) delivered costs of major items of process equipment for the smaller plant are listed in Table V at 1965 prices, (c) the M&S Equipment Cost Index is 245 for 1965, and 321 for 1971, (d) $\psi = 0.63$ in Eq. (3) for scaleup of the fixed capital cost for the complete plant, (e) fixed capital cost of the plant is related to the delivered cost of the major items of process equipment by Eq. (7), and (f) factors for Eq. (7) are listed in Table VI.

For Eq. (7), we obtain $\phi_1 = 1.47$ for a fluid processing plant; and ϕ_2 and ϕ_3 as calculated in Table VI are 1.93 and 1.55, respectively. The total delivered cost of the major items of process equipment for the smaller plant at 1965 prices is shown in Table V as $82,800. If we adjust this cost to 1971 prices by using the M&S index, we find:

$$C_{EQ} = \$82{,}500(321/245) = \$108{,}500$$

Therefore, total delivered cost of the major items of process equipment for the larger plant at 1971 prices is obtained from Eq. (3) as:

$$C_2 = \$108{,}500\left(\frac{8\text{ million lb./yr.}}{4\text{ million lb./yr.}}\right)^{0.63}$$

$$C_2 = \$167{,}900$$

We now find the fixed capital cost of the larger plant at 1971 prices from Eq. (7) as:

$$C_{FC} = (1.47)(1.93)(1.55)(\$167{,}900) = \$738{,}300$$

For this example, the overall Lang factor is:

$$f_L = \phi_1\phi_2\phi_3 = 4.40$$

Computer Methods in Estimating

At present, there is an increasing use of digital computers for making capital cost estimates. For this purpose, it is then necessary to supply the following information to the computer:

1. Process design data, perhaps generated by another computer program.
2. Cost data for delivered equipment.
3. Cost indexes to update the cost data.
4. Process design equations.
5. Factors similar to those in Eq. (5), (6) and (7).

Outputs from the computer will be a process design and the capital cost estimate. This estimate can be used as a basis for cost optimization by the computer, or otherwise. The process specification should make use of standard sizes of equipment wherever feasible because these will be available and lower in cost than any nonstandard item.

A capital cost estimate, based on detailed design of a plant, may take up to several months to complete. In addition to the equipment flowsheets, piping and instrument diagrams, design specifications, equipment drawings, electrical diagrams, etc. are also required. All material is priced in detail. Labor is priced on the basis of unit cost records. The method is described in detail in the literature.[2]

Sometimes, the capital cost of only part of a plant is estimated in detail. The remainder is estimated on the basis of factors. Often, the equipment flowsheet may be divided into a number of standard systems for which a great deal of cost information is available. Then, only the nonstandards parts must be estimated in detail.

Miller developed a method of estimating that is adaptable for computer use, and usually has an accuracy of ±15%. He pointed out that in addition to the nature of the process, e.g. liquids, solids and mixed, three other factors affect estimates. These are (a) size of the equipment, (b) materials of construction and (c) operating pressures. In the main, these three factors can be taken into account by using the "average unit cost" of the process equipment. Average unit cost is defined as the total cost of main-plant items divided by the total number of main-plant items.

From the actual cost records for many plants, Miller examined the data and developed factors for all the components within battery limits. See Ref. 2 in which Miller lists a complete set of factors for estimating the cost of battery-limit plants. There are four areas involved in a plant estimate: (1) battery limit, (2) storage and handling, (3) utilities, (4) services.

The last three areas are often grouped together as auxiliaries. Their estimation has been well covered in the literature, and Miller also gives factors for their estimation.[3]

The accuracy of this method depends on the accuracy of the cost data for the main-plant items. If a good estimate has been made for the process equipment and the scope is fairly well defined, then the total estimate should be sufficiently accurate for authorization of the expenditure stage.

Advantages of Factor Estimates

In many cases, management will be satisfied with an estimating system of about ±15%. Factor estimates are usually within this range, and their other advantages are:

1. Speed in preparation.
2. Low cost compared with definitive estimates.
3. When the scope is vague, their accuracy may be better than a detailed estimate.
4. Usually the accuracy is similar to that of a budget authorization estimate.

In practice, factor estimates are often as good as or better than estimates obtained by the more tedious procedures based on detailed design.

References

1. *Eng. News-Record*, Mar. 21, 1968, pp. 77–80.
2. Popper, H. (ed.), "Modern Cost-Engineering Techniques," McGraw-Hill, New York, 1970.
3. Jelen, F. C. (ed.), "Cost and Optimization Engineering," McGraw-Hill, New York, 1970.
4. Chilton, C. H., Cost Data Correlated, *Chem. Eng.*, June 1949, pp. 97–106.

Direct, indirect and fixed costs are the principal elements in detailed analysis of manufacturing costs for one or more products. Here is how to develop and apply equations for rapid estimates of such costs.

Manufacturing Costs and How to Estimate Them

F. A. HOLLAND, F. A. WATSON and J. K. WILKINSON, University of Salford

When determining the manufacturing cost of products, an orderly procedure must be followed in order to properly allocate the several items involved. Establishing this procedure requires a decision for costing a product based on the nature of the process, which depends on whether a single product is made with associated byproducts, or two or more main products are formed.

With this procedure fixed, we can now look at the appraisal and allocation of direct, indirect and fixed manufacturing costs, and at a method for making rapid estimates of the manufacturing cost.

In this article, we will only summarize some of the myriad costs and accounting procedures used in obtaining reliable values for the manufacturing cost of products.

Manufacturing costs are best considered in the context of the manufacturing, trading and profit-and-loss accounts. Typical examples of these are shown in Tables I, II and III. These tables are based on an accountancy period of one year.

The gross annual profit, A_{GP}, in Table II depends on the balance-sheet annual depreciation charge, A_{BD}, which is not necessarily the same as the depreciation allowance used for tax purposes. Since A_{BD} is arbitrarily chosen, it can be used to make the gross annual profit high or low, according to company policy.

The gross annual profit also depends on the method of valuing the inventory. For example, raw materials may have been purchased at the beginning of the accounting year at, say, 9¢/lb. The purchase price may have risen to, say, 12¢/lb. at the end of the accounting year. If valuation of the inventory at the end of the year is made at the higher purchase price, the production cost will be lower and the gross profit higher than if the valuation were made at the lower purchase price. Although in this case the profit looks better, a higher tax is payable.

Profit also depends on the method of valuing the work in progress. Let us consider the manufacture of 100,000 lb. of product with a prime or direct manufacturing cost of 10¢/lb., and an indirect manufacturing expense of 5¢/lb. Let us further assume that 90,000 lb. of the product are sold, and 10,000 lb. stored in inventory. The value of the inventory is $1,000 on the basis of direct cost, and $1,500 when the indirect expense is included.

It is still a controversial question as to whether inventory should be valued at direct cost or whether manufacturing overheads should be absorbed in the cost. The latter is known as absorption costing, and is the traditional accounting method. Direct costing is being increasingly used and is particularly favored by engineers.

Methods for valuing inventory will be treated in detail in a future article of this series.

Estimates of Manufacturing Costs

The annual cash income, A_{CI}, before tax—in terms of the revenue from annual sales, A_S, of a product and the components of the total annual cost or expense required to produce and sell the product (excluding any allowance for plant depreciation)—is expressed by:

$$A_{CI} = [A_S - (A_{VGE} + A_{VME})] - (A_{FGE} + A_{FME}) \quad (1)$$

where A_{VGE} and A_{VME} are the annual variable general and variable manufacturing expenses, respectively; A_{FGE} and

Originally published April 1974.

A_{FME} are the annual fixed general and fixed manufacturing expenses, respectively.

Revenue from annual sales is:

$$A_S = Rc_S \tag{2}$$

where R is the production rate, and c_S the sales price per unit of production.

Similarly, A_{VGE} can be taken as proportional to the annual production rate, R, and the variable general expense per unit of production, c_{VGE}:

$$A_{VGE} = Rc_{VGE} \tag{3}$$

Likewise, for A_{VME}:

$$A_{VME} = Rc_{VME} \tag{4}$$

In practice, annual direct-variable costs such as raw materials, utilities, etc., are not always proportional to the production rate.

Substituting Eq. (3) and (4) into the first term on the right of Eq. (1) yields:

$$A_S - (A_{VGE} + A_{VME}) = R[c_S - (c_{VGE} + c_{VME})] \tag{5}$$

In Eq. (5), the term $[c_S - (c_{VGE} + c_{VME})]$ is called the contribution to cash income.

For the simultaneous production of more than one product, Eq. (5) can be written as:

$$A_S - (A_{VGE} + A_{VME}) = R_1\{(c_S)_1 - [(c_{VGE})_1 + (c_{VME})_1]\} + R_2\{(c_S)_2 - [(c_{VGE})_2 + (c_{VME})_2]\} + \cdots \tag{6}$$

where the subscripts 1, 2, etc., refer to the various co-products.

The contribution to cash income made by a particular product depends on the method of accounting. Widely different values for the contribution can be calculated by using different methods of assigning manufacturing expenses.

The cost of each item in a cost estimate should be presented in such a way that the estimate can be modified and updated at any time in the future when revised data become available.

Published data and short-cut estimating methods can be used to calculate the approximate manufacturing cost of a new product. However, most companies have extensive data on various items of cost such as overheads, property taxes, etc. These data should be used whenever possible to give the estimate that is most valid for a particular company.

For a new product, the ratio of manufacturing expense to sales price, c_{ME}/c_S, should be compared with the ratio of total manufacturing cost or expense to sales revenue for the company as a whole. If the ratio c_{ME}/c_S is less than or equal to the ratio for the company, then the proposed sales price appears to be reasonable, and the product is probably commercially viable. This comparison is, of course, only used as an approximate guide in preliminary assessments.

Manufacturing Account—Table I
(Year ended 31 Dec. 1973)

	$	$
Inventory of raw materials 1 Jan. 1973	35,000	
Add purchases	870,000	
Add carriage inwards	25,000	
	930,000	
Less inventory of raw materials 31 Dec. 1973	51,000	
Cost of raw materials consumed	879,000	
Direct wages	122,000	
Direct utilities	22,000	
Other direct expenses	104,000	
Prime cost or direct manufacturing expense, A_{DME}	1,127,000	1,127,000
Payroll overhead	28,000	
General plant overhead	52,000	
Other indirect expenses	52,000	
Indirect manufacturing expense, A_{IME}	132,000	
Depreciation, A_{BD}	68,000	
	200,000	200,000
		1,327,000
Add work in progress 1 Jan. 1973		30,000
		1,357,000
Less work in progress 31 Dec. 1973		35,000
Production cost of products		1,322,000

Trading Account—Table II
(Year ended 31 Dec. 1973)

	$
Inventory of finished products 1 Jan. 1973	200,000
Add production cost of products	1,322,000
	1,522,000
Less inventory of finished products 31 Dec. 1973	190,000
	1,332,000
Gross profit	668,000
Sales	2,000,000

Profit and Loss Account—Table III
(Year ended 31 Dec. 1973)

	$	$
Administration	74,000	
Sales and shipping	124,000	
Advertising and marketing	40,000	
Technical service	10,000	
Research and development	60,000	
	308,000	308,000
Net profit before taxes		360,000
Gross profit		668,000

One Main Product Plus Byproducts

Let one unit of raw material yield x_1, x_2, etc., weights of products 1, 2, etc., respectively. Let the variable general expense per unit of raw material be c_{VGE} and the variable manufacturing expense per unit of raw material be c_{VME}. The unit of raw material may either be a single material or a mixture of several components. Leibson and Trischman* showed the effects on manufacturing expenses of alternate feedstocks having different compositions and costs for producing the same products.

If product 1 is the main product, then this carries all the variable expenses, so that:

$$(c_{VGE})_1 = c_{VGE}/x_1 \tag{7}$$

$$(c_{VME})_1 = c_{VME}/x_1 \tag{8}$$

*See pp. 116–121.

where $(c_{VGE})_1$ and $(c_{VME})_1$ are the variable general and variable manufacturing expenses, respectively, per unit of product 1.

For byproduct 2, etc., $(c_{VGE})_2 = 0$ and $(c_{VME})_2 = 0$, etc. For this case, Eq. (6) becomes:

$$A_S - (A_{VGE} + A_{VME}) = R_1\left[(c_S)_1 - \frac{(c_{VGE} + c_{VME})}{x_1}\right] + R_2\,(c_S)_2 + \cdots \quad (9)$$

If R is the annual rate at which the raw material is consumed, then:

$$R_1 = x_1 R \quad (10)$$

$$R_2 = x_2 R \quad (11)$$

In terms of one unit of raw material, Eq. (9) can be combined with Eq. (10) and (11), and written as:

$$\frac{A_S - (A_{VGE} + A_{VME})}{R} = x_1(c_S)_1 - (c_{VGE} + c_{VME}) + x_2(c_S)_2 + \cdots \quad (12)$$

Eq. (12) gives the contribution of products 1, 2, etc., per unit of raw material. In this particular case, the contribution of the main product per unit weight of raw material is: $x_1(c_S)_1 - (c_{VGE} + c_{VME})$. The contribution of product 2 is $x_2(c_S)_2$, i.e., its selling price per unit weight of raw material.

Two Main Products by Weight

Let one unit of raw material yield x_1 and x_2 weights of products 1 and 2, respectively. Let the variable general expense per unit of raw material be c_{VGE}, and the variable manufacturing expense per unit of raw material be c_{VME}. In practice, it is rare for $x_1 + x_2$ to be exactly unity.

If the variable expenses are shared by weight, then:

$$(c_{VGE})_1 = \frac{x_1 c_{VGE}}{(x_1 + x_2)x_1} \quad (13)$$

$$(c_{VME})_1 = \frac{x_1 c_{VME}}{(x_1 + x_2)x_1} \quad (14)$$

$$(c_{VGE})_2 = \frac{x_2 c_{VGE}}{(x_1 + x_2)x_2} \quad (15)$$

$$(c_{VME})_2 = \frac{x_2 c_{VME}}{(x_1 + x_2)x_2} \quad (16)$$

where $(c_{VGE})_1$ and $(c_{VME})_1$ are the variable general and variable manufacturing expenses per unit of product 1, respectively; and $(c_{VGE})_2$ and $(c_{VME})_2$ are the comparable expenses for product 2.

For this example, Eq. (6) becomes:

$$A_S - (A_{VGE} + A_{VME}) = R_1\left[(c_S)_1 - \frac{(c_{VGE} + c_{VME})}{(x_1 + x_2)}\right] + R_2\left[(c_S)_2 - \frac{(c_{VGE} + c_{VME})}{(x_1 + x_2)}\right] \quad (17)$$

In Eq. (13) to (17), the sum of x_1 and x_2 may be equal to or less than 1. The above analysis can be extended for any number of coproducts.

In terms of one unit of raw material, Eq. (17) can be combined with Eq. (10) and (11), and written as:

$$\frac{A_S - (A_{VGE} + A_{VME})}{R} = x_1\left[(c_S)_1 - \frac{(c_{VGE} + c_{VME})}{(x_1 + x_2)}\right] + x_2\left[(c_S)_2 - \frac{(c_{VGE} + c_{VME})}{(x_1 + x_2)}\right] \quad (18)$$

Eq. (18) gives the contribution of products 1 and 2 per unit weight of raw material.

Two Main Products by Value

Let one unit of raw material yield x_1 and x_2 weights of products 1 and 2, respectively, with values of $x_1(c_S)_1$ and $x_2(c_S)_2$, respectively; $(c_S)_1$ and $(c_S)_2$ are the sales prices of products 1 and 2 per unit of production.

Let the variable general expense per unit of raw material be c_{VGE}, and the variable manufacturing expense per unit of raw material be c_{VME}. If the variable expenses are shared by value, then:

$$(c_{VGE})_1 = \frac{x_1(c_S)_1 c_{VGE}}{[x_1(c_S)_1 + x_2(c_S)_2]x_1} \quad (19)$$

$$(c_{VME})_1 = \frac{x_1(c_S)_1 c_{VME}}{[x_1(c_S)_1 + x_2(c_S)_2]x_1} \quad (20)$$

$$(c_{VGE})_2 = \frac{x_2(c_S)_2 c_{VGE}}{[x_1(c_S)_1 + x_2(c_S)_2]x_2} \quad (21)$$

$$(c_{VME})_2 = \frac{x_2(c_S)_2 c_{VME}}{[x_1(c_S)_1 + x_2(c_S)_2]x_2} \quad (22)$$

where $(c_{VGE})_1$ and $(c_{VME})_1$ are the variable general and variable manufacturing expenses per unit of product 1, respectively, and $(c_{VGE})_2$ and $(c_{VME})_2$ are the comparable expenses for product 2.

For this case, Eq. (6) becomes:

$$A_S - (A_{VGE} + A_{VME}) = R_1\left\{(c_S)_1 - \frac{(c_S)_1(c_{VGE} + c_{VME})}{[x_1(c_S)_1 + x_2(c_S)_2]}\right\} + R_2\left\{(c_S)_2 - \frac{(c_S)_2(c_{VGE} + c_{VME})}{[x_1(c_S)_1 + x_2(c_S)_2]}\right\} \quad (23)$$

The above analysis can be extended for any number of coproducts.

In terms of one unit of raw material, Eq. (23) can be combined with Eq. (10) and (11), and written as:

$$\frac{A_S - (A_{VGE} + A_{VME})}{R} = x_1\left\{(c_S)_1 - \frac{(c_S)_1(c_{VGE} + c_{VME})}{[x_1(c_S)_1 + x_2(c_S)_2]}\right\} + x_2\left\{(c_S)_2 - \frac{(c_S)_2(c_{VGE} + c_{VME})}{[x_1(c_S)_1 + x_2(c_S)_2]}\right\} \quad (24)$$

Eq. (24) gives the contribution of products 1 and 2 per unit weight of raw material.

Example for Manufacturing Costs

One pound of raw material is used to manufacture $x_1 = 0.32$ lb. of product 1, and $x_2 = 0.64$ lb. of product 2. The balance of the raw material goes to waste. Product 1 sells at $(c_S)_1 = 20¢$/lb., and product 2 sells at $(c_S)_2 = 6¢$/lb. The variable general and variable manufacturing expenses, including raw materials, $c_{VGE} + c_{VME}$, total 5¢/lb.

Let us calculate (a) the total contribution to cash income per pound of raw material and (b) the individual contributions of each product to cash income per pound of raw material for the following:

Case	Condition
1	Product 1 as the main product charged for all the variable expenses.
2	Product 2 as the main product charged for all the variable expenses.
3	Products 1 and 2 sharing the variable expenses on the basis of weight.
4	Products 1 and 2 sharing the variable expenses on the basis of value.

Case 1—The total contribution per pound of raw material is obtained by substituting the appropriate values into Eq. (12). For these conditions:

$$x_1(c_S)_1 + x_2(c_S)_2 - (c_{VGE} + c_{VME}) =$$
$$0.32(20¢/\text{lb.}) + 0.64(6¢/\text{lb.}) - 5¢/\text{lb.} =$$
$$6.4¢/\text{lb.} + 3.84¢/\text{lb.} - 5¢/\text{lb.} = 5.24¢/\text{lb.}$$

The individual contribution of product 1 as the main product is 6.4¢/lb. − 5¢/lb. = 1.4¢/lb. of raw material. The individual contribution of product 2 as the byproduct is 3.84¢/lb. of raw material.

Case 2—The total contribution where product 2 is the main product is also obtained from Eq. (12), and as in Case 1 is found to be 5.24¢/lb.

The individual contribution of product 2 as the main product is 3.84¢/lb. − 5¢/lb. = −1.16¢/lb. of raw material. The individual contribution of product 1 as the byproduct is 6.4¢/lb. of raw material.

Case 3—Where products 1 and 2 share the variable expenses on the basis of weight, the total contribution per pound of raw material is found by substituting the unit costs into Eq. (18). Values for each term are:

$$x_1(c_S)_1 = 6.4¢/\text{lb. (same as Case 1)}$$
$$x_2(c_S)_2 = 3.84¢/\text{lb. (same as Case 1)}$$
$$\frac{x_1}{(x_1 + x_2)}(c_{VGE} + c_{VME}) = \frac{0.32}{0.96}(5¢/\text{lb.}) = 1.67¢/\text{lb.}$$
$$\frac{x_2}{(x_1 + x_2)}(c_{VGE} + c_{VME}) = \frac{0.64}{0.96}(5¢/\text{lb.}) = 3.33¢/\text{lb.}$$

Therefore, the total contribution becomes:

$$6.4 + 3.84 - 1.67 - 3.33 = 5.24¢/\text{lb.}$$

The individual contribution of product 1 is 6.4¢/lb. − 1.67¢/lb. = 4.73¢/lb. of raw material. The individual contribution of product 2 is 3.84¢/lb. − 3.33¢/lb. = 0.51¢/lb. of raw material.

Case 4—Where products 1 and 2 share the variable expenses on the basis of value, the total contribution per pound of raw material is found by substituting the unit costs into Eq. (24). Values for each term are:

$$x_1(c_S)_1 = 6.4¢/\text{lb. (same as Case 1)}$$
$$x_2(c_S)_2 = 3.84¢/\text{lb. (same as Case 1)}$$
$$\frac{x_1(c_S)_1(c_{VGE} + c_{VME})}{x_1(c_S)_1 + x_2(c_S)_2} = \frac{(6.4)(5)}{6.4 + 3.84} = 3.12¢/\text{lb.}$$
$$\frac{x_2(c_S)_2(c_{VGE} + c_{VME})}{x_1(c_S)_1 + x_2(c_S)_2} = \frac{(3.84)(5)}{6.4 + 3.84} = 1.88¢/\text{lb.}$$

Therefore, the total contribution becomes:

$$6.4 + 3.84 - 3.12 - 1.88 = 5.24¢/\text{lb.}$$

The individual contribution of product 1 is found as: 6.4¢/lb. − 3.12¢/lb. = 3.28¢/lb. of raw material. The individual contribution of product 2 then becomes: 3.84¢/lb. − 1.88¢/lb. = 1.96¢/lb. of raw material.

Direct Manufacturing Costs

Direct manufacturing costs include raw materials, operating labor, utilities and some miscellaneous items. Let us quickly summarize the characteristics of each.

Raw materials—The cost of raw materials is normally the largest item of expense in the manufacturing cost of a product. The quantities of raw materials consumed can be calculated from material balances.

Material costs are conveniently presented in tables that give the following: name of material, form and grade, method of delivery, unit of measure, cost per unit, source of cost, annual consumption, annual cost, fractional consumption per unit of production, cost per unit of production.

Net consumption of materials should be used for catalysts, solvents, filter aids, etc., that may have a recovery value. Current prices of chemicals are published in various trade journals. However, quotations from suppliers should be used whenever possible.

It may be possible for a company to negotiate the purchase of a material at a cost per unit that is significantly lower than the current published price. This is particularly true if large quantities are involved. Thus, estimates should be presented for both minimum and maximum costs. Price trends, availability and quality are other factors that should be considered. A knowledge of price trends is particularly important for a product that a company may not manufacture for several years.

The yield in a chemical reaction determines the quantities of materials in the material balance. Assumed yields are used to obtain approximate exploratory estimates. In this case, possible ranges should be given. Firmer estimates require yields based on laboratory, or preferably pilot-plant, work.

Operating labor—The cost of operating labor is the second largest item of expense in the manufacturing cost. Labor requirements for a process can be estimated from an intelligent study of the equipment flowsheet, paying careful attention to the various primary process steps such as fractionation, filtration, etc. The hourly wage rate should be that currently paid in the company. Once the number of men required per shift has been estimated for a particular production rate, the annual labor cost and the labor cost per unit of production can be estimated.

Wessel[2] made a study of the operating labor requirements in the U.S. chemical industry and presented the data as a plot of man-hours per ton per processing step vs. plant capacity in tons per day. These data can be represented by:

$$\log_{10} Y = 0.783 \log_{10} X + 1.252 + B \qquad (25)$$

where Y is the operating labor, man-hours per ton per processing step; X is plant capacity, tons/day; and B is a constant having values of 0.132 where multiple units are used to increase capacity or where the process is completely batch; 0 for the average chemical processing plant; and −0.167 for large, highly automated plants or those concerned only with fluid processing.

Wessel's data for the U.S. chemical industry refer to the short ton equal to 2,000 lb. Labor requirements are higher in countries with lower productivities.

The approximate cost of supervision for operating labor is equivalent to 10% of the labor cost for simple operations, and 25% for complex operations.

Utilities—These include steam, cooling water, process water, electricity, fuel, compressed air and refrigeration. The consumption of utilities can be estimated from the material and energy balances for the process, together with the equipment flowsheet.

Let us consider a cooler in the equipment flowsheet. The required rate of heat removal is known from the balances. The rate of cooling water can be calculated once the inlet and outlet temperatures of the water have been specified. The calculation of the consumption of other utilities is also straightforward. Allowances should be made for wastage.

The current cost per unit for each utility is usually well known in a company. Thus, the annual cost for utilities and the utilities cost per unit of production can be estimated. The latter is normally much smaller than the raw-materials and labor costs. However, a great deal more work is involved in calculating the utilities cost than for any other item in the manufacturing cost.

Unfortunately, there are no satisfactory short-cut methods for doing this. When the utilities cost is relatively small, it may be possible to make an intelligent guess on the basis of known costs for similar processes in the company. Alternatively, published data for the consumption of utilities per unit of production for various processes may be used.

Miscellaneous direct costs—Estimates for the cost of maintenance and repairs, operating supplies, royalties, and patents are best based on company records for similar processes. A rough average value for the annual cost of maintenance is 6% of the capital cost of the plant. This percentage can vary from 2 to 10%, depending on the severity of plant operation. Approximately half the maintenance costs are for materials and half for labor. Royalty and patents costs are on the order of 1 to 5% of the sales price of the product.

Indirect Manufacturing Costs

Estimates for the cost of payroll overhead, control laboratory, general plant overhead, packaging, and storage facilities are best based on company records for similar processes.

Payroll overhead includes the cost of pensions, holidays, sick pay, etc., and is normally between 15 to 20% of the operating labor cost. Laboratory work is required for product quality control, and its cost is approximately 10 to 20% of the operating labor cost.

Plant overhead includes the cost of medical, safety, recreational, effluent-disposal, warehousing facilities, etc. In general, the larger the plant, the lower the overhead per unit of production. Plant-overhead costs can vary between 50 to 150% of the operating labor cost. Packaging costs depend on the physical and chemical nature of the product as well as its use and value. The cost of packaging is as high as one-third the selling price for soaps and pharmaceuticals.

Fixed Manufacturing Costs

Fixed manufacturing costs are a function of the fixed capital investment and are independent of the production rate of the plant.

Property taxes or rates depend on location. They may be taken as 2% of the fixed capital cost of the plant in the absence of specific data. Cost of insurance depends on both location and the hazardous nature of the materials handled. This cost is normally on the order of 1% of the fixed capital cost of the plant.

Rapid Manufacturing-Cost Estimates

The manufacturing cost of a product is the sum of the processing or conversion cost and the cost of raw materials. The processing cost can be roughly broken down into three parts: investment-related cost, labor-related cost and utility cost.

Companies usually include in the charge for overhead the following items: operating supplies, supervision, indirect payroll expenses, plant protection, plant office, general plant overhead, and control laboratory. This overhead charge is frequently taken as an equivalent percentage of the direct labor cost.

The percentage is best obtained from company records. Although, it can vary over a wide range, a reasonable value is 125%. In this case, the labor-related cost would be 2.25 times the direct-labor cost. For a 6,000-hr. year and N men per shift earning $\$c_L$/hr., the annual labor-related cost would be $13{,}500(c_L N)$.

Let us consider a plant of fixed capital cost, C_{FC}. If the annual property taxes are taken as $0.02C_{FC}$, insurance as $0.01C_{FC}$ and maintenance as $0.06C_{FC}$, the annual investment-related cost would be $0.09C_{FC}$. Annual utilities cost is A_U.

The annual processing cost, A_P, can be represented by:

$$A_P = \eta_1 C_{FC} + \eta_2 c_L N + A_U \qquad (26)$$

where the factors η_1 and η_2 can be obtained from the data available in a particular company, and have the dimensions of yr.$^{-1}$ and hr./yr., respectively.

Substituting the information previously given into Eq. (26) yields the relationship:

$$A_P = 0.09C_{FC} + 13{,}500c_L N + A_U \qquad (27)$$

Eq. (27) represents very closely the manufacturing costs of a particular company, and is typical of the coefficients to be expected. ■

Reference

[1]Wessel, H. E., *Chem. Eng.,* July 1952, pp 209–210.

Engineering Economics—Part 14

By comparing the deviations between actual costs and standard costs, it is possible to assess the economic performance of a product or process.

How To Budget and Control Manufacturing Costs

F. A. HOLLAND, F. A. WATSON and J. K. Wilkinson, University of Salford

The most common objective of a business organization is to make a profit. Since sales prices are normally controlled by competition, it is necessary to keep a tight control on costs in order to ensure the best profit from a particular product. The basis of cost control is forward planning, and the monitoring of all cost factors.

A budget is a plan that states the company objectives in quantitative or financial terms. Subsequent variations of actual from planned performance are then calculated and used to locate areas of substandard performance. The method is known as budgetary control and enables managerial effort to be concentrated on those areas that will affect profit the most.

The first essential for the success of a budgetary control system is full backing from management. In addition, a well-defined organization is needed. An individual manager or supervisor should be made responsible for each item of revenue or cost. In this case, a supervisor can be considered as a cost center.

Establishment of a budget requires estimates of future revenues and costs. These estimates of budget standards should be realistically made with the full cooperation of managers and supervisors at all levels of responsibility. Individuals will be less keen to operate a system that has been imposed on them than one in which they have had a part in creating and that they believe to be fair.

Every budget should be established as accurately and as fairly as possible. Unfair allocation of overheads can shift the balance of profitability from one process to another. If a budget is deemed inaccurate due to changes in the factors on which it was originally based, the budget should be revised. However, revisions should never be made in order to increase the seemingly apparent profitability of a doubtful process or to conceal inefficiency.

Standard Costs for Budgetary Control

For convenience and simplicity, we shall consider the total cost of a manufactured product to be the sum of the material, labor and overhead costs. Standard costs are those that have been predetermined and budgeted for the manufacture of a given amount of product in a given time. The deviation of the actual cost from the standard cost is called the variance. It is far easier to make comparisons between periods by using variances than by using the actual production data. The different variances for material, labor and overhead costs are listed in Fig. 1.

Standard costing is extensively used in budgetary control systems. Criteria for the establishment of standards range from the maximum possible under ideal conditions to those expected under normal conditions. Past or historical costs are not always the best basis for setting up standards because past performance may have been unnecessarily inefficient.

Static and Flexible Budgets

Overhead cost can significantly affect the profitability of a project, and is the only cost outside the control of

Originally published May 1974.

Revenue From Sales		Sales Price		Sales Volume
$Rc_S - R^*c_S^*$	$=$	$R(c_S - c_S^*)$	$+$	$c_S^*(R - R^*)$
Profit		Unit Profit		Quantity
$Rc_{NP} - R^*c_{NP}^*$	$=$	$R(c_{NP} - c_{NP}^*)$	$+$	$c_{NP}^*(R - R^*)$
Total Cost		Unit Cost		Production Rate
$Rc_{TE} - R^*c_{TE}^*$	$=$	$R(c_{TE} - c_{TE}^*)$	$+$	$c_{TE}^*(R - R^*)$
Direct Material Cost		Material Price		Material Usage
$Rc_{RM} - R^*c_{RM}^*$	$=$	$R(c_{RM} - c_{RM}^*)$	$+$	$c_{RM}^*(R - R^*)$
Direct Labor Cost		Wage Rate		Labor Efficiency
$\theta c_L - \theta^* c_L^*$	$=$	$\theta(c_L - c_L^*)$	$+$	$c_L^*(\theta - \theta^*)$

Overhead Cost = Budgeted Cost + Volume + Efficiency

$$C_{OH} - \theta^* c_{NOH}^* = (C_{OH} - \theta c_{BOH}) + \theta(c_{BOH} - c_{NOH}^*) + c_{NOH}^*(\theta - \theta^*)$$

- c_{BOH} Flexible budgeted overhead cost, \$/hr.
- c_L Actual labor cost, \$/hr.
- c_{NOH}^* Standard overhead cost based on normal production rate, \$/hr.
- c_{NP} Actual net profit before tax, \$/unit
- C_{OH} Actual overhead cost, \$/period
- c_{RM} Actual raw-material cost, \$/unit
- c_S Actual selling price, \$/unit
- c_{TE} Actual total cost, \$/unit
- R Actual quantity, units/period
- R^* Standard quantity, units/period
- θ Actual time to produce a given quantity, hr.
- θ^* Standard time to produce a given quantity, hr.
- $c_{NP} = c_S - c_{TE}$

Note: The asterisk on all items not otherwise defined indicates the standard cost for that item.

VARIANCES are the deviations between actual and standard costs for materials, labor and overhead—Fig. 1

the project manager. The project is expected to contribute a definite amount toward the expenses of the company, and will be charged this amount even if the production rate is zero. This is the fixed component of the overhead cost and will include directly allocable costs such as depreciation, and a proportion of general costs such as office salaries and heating.

Other nonproduction costs such as indirect labor may vary linearly with production rate and represent the variable component of the overhead. Costs that are neither fixed nor variable but that occur in discrete steps at various production levels (such as supervisory labor) are the semivariable component of the overhead cost. It is an easy matter to determine these various components for various production rates and list them as shown in Table I.

Two types of overhead budget are currently in use. The static, often referred to as the fixed, budgeted overhead cost is related to the standard budgeted production rate. The flexible budgeted overhead cost is that shown as the total cost in Table I. Values for intermediate production rates are often obtained by interpolation. This is only justifiable when the semivariable costs are a negligible part of the overhead costs.

Flexible budgeting is more widely used than static budgeting despite certain logical difficulties. This is because production in many cases is seasonal, and the use of a static production norm might distort the evaluation of performance. Variances are the difference between the actual costs expended and the budgeted costs expected. Variances are unfavorable if positive, and favorable if negative. Any variance should be explained and, if necessary, controlled; the largest variance should be considered first.

Let us consider the overhead cost data of Table I, with 10 million lb./mo. as the standard production rate. The static budgeted overhead is then \$150,000/mo., or 1.5¢/lb. Let us also assume that the actual overhead is \$186,000 for a month in which 12 million lb. were produced. Then, the static budgeted overhead cost would be 12,000,000(1.5), or \$180,000/mo. Therefore, the variance is \$186,000 − \$180,000 = +\$6,000, which is unfavorable because \$6,000 more was spent than was anticipated.

From Table I, we find the flexible budgeted overhead cost for a 12 million lb./mo. production rate is \$190,000. The corresponding variance is \$186,000 minus \$190,000, or −\$4,000, which is favorable because \$4,000 less was spent than was anticipated. Thus, the use of flexible budgeting makes this particular performance look better without changing either the production rate or a single cost of the planned budget.

The Standard Hour

The standard hour can be defined as the number of units of output expected to be produced in one hour.[1] It is often used as a measure of output rate by cost accountants.

Let us consider a batch processing unit that can either produce 1,000 lb. of product A in a cycle time of 5 hr., or 900 lb. of product B in a cycle time of 3 hr. Thus, for this processing unit, a standard hour is 200 lb. of product A, or 300 lb. of product B. In a budget period of, say, 1,000 hr., it is possible to produce 200,000 lb. of product

Flexible Budget For Overhead Costs—Table I

Overhead Cost, \$/Mo.	Production, Million Lb./Mo. 8	9	10	11	12
Fixed	40,000	40,000	40,000	40,000	40,000
Variable	40,000	45,000	50,000	55,000	60,000
Semivariable	40,000	40,000	60,000	60,000	90,000
Total	120,000	125,000	150,000	155,000	190,000

A, or 300,000 lb. of product B, or any appropriate combination of the two products.

For example, let us assume that the production requirements are twice as great for product A as for product B, i.e., a ratio of 600 lb., or three standard hours of product A to 300 lb., or one standard hour of product B. On this basis, for a budget period of 1,000 hr., 750 standard hours [(3/4)1,000] would be used to produce 750(200) = 150,000 lb. of product A; and 250 standard hours [(1/4)1,000] would be used to produce 250(300) = 75,000 lb. of product B.

Production efficiency, p_e, can be calculated from:

$$p_e = (p_s/p_a)100 \tag{1}$$

where p_s is the actual production rate in standard hours, and p_a is the actual hours worked.

The level of production activity, p_l, can be calculated from:

$$p_l = (p_s/p_b)100 \tag{2}$$

where p_b is the budgeted production in standard hours.

The deviation from budgeted capacity, b_c, can be calculated from:

$$b_c = (p_a/p_w)100 \tag{3}$$

where p_w is the budgeted number of working hours.

Actual Costs vs. Standard Costs

Let us consider the sales, profits and manufacturing cost data in Table II. The gross profit is \$33,129/period better than expected. Clearly, there is less incentive to investigate overall costs when the profit variance is favorable than if the profit were less than expected. However, standard costing enables an objective analysis of the data to be made—whether good or bad.

The individual variances in Table II show that the increased profit is due to reduced material costs, which affect manufacturing costs to a greater extent than increased labor and overhead costs. The individual variances also show that the increased profit is due to an even greater extent to the increase in sales revenue.

Clearly, management will wish to investigate both labor and overhead costs for any inefficiencies and to ascertain the reasons for the improved sales revenue. If necessary the standard values can be revised.

Notice that profit is obtained as the difference between two large cash sums, and that variances of some 3% in manufacturing costs and sales revenue have resulted in a variance of some 33% in gross profit.

Table II is a very simplified presentation. In a full, standard costing system, the direct material, direct labor and overhead variances are broken down into component parts to enable an even closer look at the operation. Standard costing is an invaluable aid to management for controlling a business.

Variances: Direct Material Cost

Since the variance in direct material cost, $\Delta(Rc_{RM})$, is the difference between actual cost and standard cost:

$$\Delta(Rc_{RM}) = Rc_{RM} - R^*c^*_{RM} \tag{4}$$

Sales, Profits, Manufacturing Costs—Table II

Component	Actual, \$/Period	Standard, \$/Period	Variance, \$/Period
Direct materials cost	325,080	350,000	−24,920
Direct labor cost	55,001	51,600	+ 3,401
Overhead cost	75,000	68,370	+ 6,630
Manufacturing cost	455,081	469,970	−14,889
Revenue from sales	588,240	570,000	+18,240
Gross profit	133,159	100,030	+33,129

where R is the actual quantity and R^* is the standard quantity, units/period; c_{RM} is the actual price and c^*_{RM} is the standard price, \$/unit.

Eq. (4) can be written in an expanded form:

$$\Delta(Rc_{RM}) = R(c_{RM} - c^*_{RM}) + c^*_{RM}(R - R^*) \tag{5}$$

where $R(c_{RM} - c^*_{RM})$, known as the direct material price variance, is the actual quantity multiplied by the deviation in unit price; and $c^*_{RM}(R - R^*)$, known as the direct material usage variance, is the standard unit price multiplied by the deviation in quantity.

By using the data of Table III, we can calculate the direct materials cost, and the standard direct materials cost as:

$$Rc_{RM} = 1{,}806{,}000(0.18) = \$325{,}080/\text{period}$$

$$R^*c^*_{RM} = 1{,}750{,}000(0.20) = \$350{,}000/\text{period}$$

From Eq. (4), we calculate the direct materials cost variance as −\$24,920/period. This variance is favorable. However, by using the relations of Eq. (5), we find a direct materials price variance of:

$$(1{,}806{,}000)(0.18 - 0.20) = -\$36{,}120/\text{period}$$

This variance is favorable.

Likewise, we calculate a direct materials usage variance of:

$$(0.20)(1{,}806{,}000 - 1{,}750{,}000) = \$11{,}200/\text{period}$$

This variance is unfavorable.

In this case, the favorable direct-materials-cost variance was achieved because of a lower unit price despite an inefficient materials usage, which needs to be investigated. (There is no room for complacency since the lower unit price may well be temporary.)

In the case of mixtures of raw materials, the direct material usage variance can be further subdivided into (a) a direct material mixture variance, and (b) a direct material yield variance. The former is due to the difference between the actual and standard mixture compositions, and the latter to the difference between the actual and standard yields. Here, the standard yield is the output expected from the standard input of material. The yield variance denotes the extent of loss of material.

The direct material mixture variance can be illustrated by means of an example. A standard mixture of 100 units of material contains 70% of material A at \$0.08/unit, and

Cost Data for Problems—Table III

Factor	Actual Value	Standard Value
Raw-Materials cost $/unit	$C_{RM} = 0.18$	$C^*_{RM} = 0.20$
Direct labor cost, $/unit	$C_L = 8.45$	$C^*_L = 8.00$
Production rate, units/period	$R = 1{,}806{,}000$	$R^* = 1{,}750{,}000$
Production time hr./period	$\theta = 6{,}509$	$\theta^*_N = 6{,}250$
Fixed and semivariable overhead cost, $/period	—	$C^*_{FOH} = 60{,}000$
Variable overhead cost, $/hr.	—	$C^*_{VOH} = 1.00$
Overhead cost, $/period	$C_{OH} = 75{,}000$	—

30% of material B at $0.12/unit. The standard mixture cost is:

$$70(0.08) + 30(0.12) = \$9.20$$

Let us now consider a mixture of 100 units containing 75% of material A and 25% of material B. The cost of this mixture at standard prices is:

$$75(0.08) + 25(0.12) = \$9.00$$

The direct material mixture variance is: $9.00 − $9.20 = −$0.20, and is favorable. The favorable variance has been brought about by using more of the lower priced material A and less of the higher priced material B.

The direct material yield variance is illustrated as follows. Let us assume that the standard mixture (cost $9.20 for 100 units) has a standard loss of 20%—making the cost $9.20 for 80 units, or $0.115/unit of output. Let us now consider the actual loss to be 30%—leaving 70 units of output for each 100 units of input. The direct material yield variance is: 0.115(80 − 70) = $1.15, and is unfavorable.

Variances: Direct Labor Cost

Since the variance in direct labor cost, $\Delta(\theta c_L)$, is the difference between actual cost and standard cost:

$$\Delta(\theta c_L) = \theta c_L - \theta^* c_L^* \tag{6}$$

where c_L is the actual pay or wage rate $/hr.; c_L^* is the standard pay or wage rate, $/hr.; θ is the actual time taken to produce a given quantity of product in a given period, hr.; and θ^* is the standard time taken to produce a given quantity of product in a given period, hr.

Eq. (6) can also be written in expanded form:

$$\Delta(\theta c_L) = \theta(c_L - c_L^*) + c_L^*(\theta - \theta^*) \tag{7}$$

where $\theta(c_L - c_L^*)$, known as the direct pay or wage rate variance, is the actual time taken to produce a given output multiplied by the deviation in wage rate; and $c_L^*(\theta - \theta^*)$, known as the direct labor efficiency variance, is the standard wage rate multiplied by the deviation in time taken to produce a given output.

By using the data of Table III, we find that one standard hour corresponds to:

$$\frac{1{,}750{,}000}{6{,}250} = 280 \text{ units/standard hr.}$$

Standard time for actual production is:

$$\theta^* = \frac{1{,}806{,}000}{280} = 6{,}450 \text{ standard hr./period}$$

Direct labor cost is:

$$\theta c_L = 6{,}509(8.45) = \$55{,}001/\text{period}$$

Standard direct labor cost is:

$$\theta^* c_L^* = 6{,}450(8.00) = \$51{,}600/\text{period}$$

From Eq. (6), we calculate the direct labor cost variance as $3,401/period. This variance is unfavorable. However, by using the relations of Eq. (7), we find a direct pay, or wage rate, variance of:

$$\theta(c_L - c_L^*) = 6{,}509(8.45 - 8.00) = \$2{,}929/\text{period}$$

This direct pay variance is unfavorable.

Likewise, we calculate the direct labor efficiency variance as:

$$c_L^*(\theta - \theta^*) = 8.00(6{,}509 - 6{,}450) = \$472/\text{period}$$

This variance is also unfavorable.

For this example, the adverse direct labor cost variance of $3,401 is due both to a higher wage rate per hour and a higher number of manhours.

The direct labor cost variance can, if necessary, be broken down into a direct-labor idle-time variance in addition to the direct wage rate and direct labor efficiency variances. The direct-labor idle-time variance is simply the number of idle manhours in the period multiplied by the standard wage rate. This is rarely relevant to the conditions existing in process plants other than when maintenance is involved.

Variances: Overhead Cost

Since the variance in overhead cost,[2] ΔC_{OH}, is the difference between actual overhead cost and static standard overhead cost:

$$\Delta C_{OH} = C_{OH} - \theta^* c_{NOH}^* \tag{8}$$

where C_{OH} is the actual overhead cost incurred in a given period, $/period; c_{NOH}^* is the static standard overhead cost based on normal production rate, $/hr.; θ is the actual time taken to produce a given quantity of product in a given period, and θ^* is the standard time taken to produce a given quantity of product in a given period, hr.

Eq. (8) can also be written in expanded form:

$$\Delta C_{OH} = (C_{OH} - \theta c_{BOH}) + \theta(c_{BOH} - c_{NOH}^*) + c_{NOH}^*(\theta - \theta^*) \tag{9}$$

where c_{BOH} is the flexible budgeted overhead cost at the actual production rate or operating capacity.

In Eq. (9), $(C_{OH} - \theta c_{BOH})$ is known as the budgeted overhead cost variance; $\theta(c_{BOH} - c_{NOH}^*)$ as the overhead volume variance; and $c_{NOH}^*(\theta - \theta^*)$ as the overhead efficiency variance. The last is analogous to the labor efficiency variance, and is the standard overhead rate multiplied by the deviation in time taken to produce a given output.

Also, in Eq. (9), c_{BOH} is simply the flexible budgeted overhead cost in \$/hr. for the actual production rate; and the overhead volume variance, $\theta(c_{BOH} - c^*_{NOH})$, is the actual time to produce a given output multiplied by the difference between the flexible budgeted overhead cost and the standard overhead cost in \$/hr. The budgeted overhead cost variance, $(C_{OH} - \theta c_{BOH})$, is the difference between the actual overhead cost and the actual time (in hours) to produce the given output multiplied by the flexible budgeted overhead cost (in \$/hr.).

Let us write the fixed overhead cost for the budget period as C^*_{FOH}, the semivariable overhead cost as C^*_{SVOH}, and the standard hours to produce the agreed normal production as θ^*_N.

The standard overhead cost at the agreed normal production rate can then be calculated from:

$$c^*_{NOH} = \frac{(C^*_{FOH} + C^*_{SVOH})}{\theta^*_N} + c^*_{VOH} \tag{10}$$

where c^*_{VOH} is standard variable overhead cost, \$/hr.

For production rates that differ from the agreed normal rate, the flexible budgeted overhead cost is given by:

$$c_{BOH} = \frac{(C^*_{FOH} + C^*_{SVOH})}{\theta} + c^*_{VOH} \tag{11}$$

where θ is the actual hours taken to produce a given amount of product.

For production rates lower than normal, the fixed overheads are underused, and the flexible budgeted overhead cost, c_{BOH}, is greater than the standard overhead cost, c^*_{NOH}. For production rates higher than normal, c_{BOH} is less than c^*_{NOH}.

It is common practice in cost accountancy to treat the standard semivariable cost, C^*_{SVOH}, at the normal production rate as part of the standard fixed cost. In this case, Eq. (10) and (11) can be written, respectively, as:

$$c^*_{NOH} = \frac{C^*_{FOH}}{\theta^*_N} + c^*_{VOH} \tag{12}$$

$$c_{BOH} = \frac{C^*_{FOH}}{\theta} + c^*_{VOH} \tag{13}$$

By using the data of Table III in Eq. (12), we find the standard overhead cost to be:

$$c^*_{NOH} = \frac{60{,}000}{6{,}250} + 1.00 = \$10.60/\text{hr.}$$

From Eq. (13) and the data in Table III, we find the flexible budgeted overhead cost to be:

$$c_{BOH} = \frac{60{,}000}{6{,}509} + 1.00 = \$10.22/\text{hr.}$$

By substituting into Eq. (8), we calculate the overhead cost variance:

$$C_{OH} = 75{,}000 - 6{,}450(10.60) = \$6{,}630$$

This variance is unfavorable. (Note that standard time for actual production was previously calculated to be 6,450 hr./period.)

The overhead cost variance comprises (a) budgeted overhead cost variance, (b) overhead volume variance, and (c) overhead efficiency variance. The calculations for each follow:

$$C_{OH} - \theta c_{BOH} = 75{,}000 - 6{,}509(10.22) = \$8{,}478$$

Budgeted overhead cost variance is positive and, therefore, unfavorable.

$$\theta(c_{BOH} - c^*_{NOH}) = 6{,}509(10.22 - 10.60) = -\$2{,}473$$

Overhead volume variance is negative and favorable.

$$c^*_{NOH}(\theta - \theta^*) = 10.60(6{,}509 - 6{,}450) = \$625$$

Overhead efficiency variance is positive and unfavorable.

The total variances in each category are listed in Table II.

Budgets and Costs Are Important

Chemical engineers usually make detailed evaluations of costs rather than evaluations for profits or sales. However, the latter can be analyzed in a similar manner to costs by using the equations shown in Fig. 1. For this purpose, the sign convention will be reversed because an increase in sales or profits would be considered favorable, whereas an increase in cost would be considered unfavorable. The equations can be applied to both batch and continuous processes.

Budgets can be used both for forward planning and control. Variances show a manager what his costs should have been and how near he came to meeting budgeted values. A manager will be able to assess, over a number of budget periods, the rate of improvement in performance in his area of responsibility. A good budgetary system should not only provide detailed information and an appraisal of performance but should also motivate people to improve performance.

Since operating profit is the amount by which revenue from sales exceeds operating costs, it directly depends on the variances discussed in this article. However, profit is also affected by the particular methods used for allocating overheads and valuing inventory. These will be dealt with in the next article of this series. ■

References

1. Owler, L. W. J. and Brown, J. L., "Wheldon's Cost Accounting and Costing Methods," 12th ed., p. 475, Macdonald and Evans, London, 1970.
2. Pilcher, R., "Appraisal and Control of Project Costs," p. 213, McGraw-Hill, London and New York, 1973.

Procedures for allocating overhead and appraising inventory can affect the profits of a project or product. Here is an explanation of such techniques.

How To Allocate Overhead Cost and Appraise Inventory

F. A. HOLLAND, F. A. WATSON and J. K. WILKINSON, University of Salford

How overheads are allocated can affect the total cost of a product and, hence, the estimated future cash flows for a project. Since these cash flows are used in the net present value (*NPV*) and discounted-cash-flow rate-of-return (*DCFRR*) methods for estimating profitability, erroneous allocations could result in the wrong choice of project.

The modern trend is for overhead costs to become an increasing proportion of total product costs. This results from the ever greater sophistication of process plants. Therefore, it is highly desirable that chemical engineers should have some say in the allocation of overheads and that this should not be left entirely to accountants.

Direct costs are those that can be directly charged to a single product. The most obvious direct cost is for raw materials where the quantity consumed is directly proportional to the amount of product manufactured. Direct process-labor is also considered to be a direct cost.

However, many costs cannot be directly charged to an individual product. These so-called indirect, burden or overhead costs range from the lighting and heating required for the plant and offices, to the canteen and medical facilities provided. The problem becomes increasingly difficult when several products are made in a plant. How should overheads be allocated among various products?

A number of different methods are commonly used to estimate the amount of overhead to be allocated to an individual product. These methods are necessary because the accountancy costs become prohibitive for charging all costs directly to an individual product. Unfortunately, there is always an arbitrary element inherent in the process of allocation.

Methods for Allocating Overheads

Overheads in the chemical process industries are commonly calculated as a percentage of: (a) direct materials cost, (b) direct labor cost, or (c) prime or direct costs. Other methods of allocating overheads are on the basis of: (a) plant area, (b) number of employees, (c) capital value, or (d) electrical power.[1]

These listings do not include all the methods in use. The validity of a particular method depends on the process and industry. An inappropriate method can lead to misleading and even absurd results.

Let us consider the manufacture of metal ornaments. The processing cost, exclusive of material, may vary very little for a wide range of materials. However, the direct material cost will be much greater for precious than for base metals. In this case, an overhead allocation on the basis of direct material cost could be very misleading, while one based on direct labor cost could be quite accurate.

Problems can also arise when allocating overheads on the basis of direct labor cost. Let us consider a company that evaluates overheads at 125% of direct labor cost. A process plant employs seven operators, each with a direct cost of $10,000 per budget period. As a result of a works-study exercise, it is found that the plant can

Originally published June 1974.

operate satisfactorily with six operators. The actual cost saving is likely to be far nearer to the direct labor saving of $10,000 per period than to the calculated saving of:

$$\$10{,}000 + \$10{,}000(125/100) = \$22{,}500/\text{period}$$

The $22,500 calculated saving is the direct labor cost, plus overheads taken as 125% of direct labor cost.

A thorough analysis should be made before production is stopped on a product that is losing money. Although direct costs of the discontinued product will be saved, overheads are not eliminated, as might be inferred from taking overheads as a percentage of direct material, direct labor or prime costs. The plant is still there, together with its associated costs for interest charges, insurance, painting, some maintenance, etc. Continued production can still make a useful contribution to such overheads. This contribution is lost if production is stopped, and must be borne by other products.

Problems can arise with each of the methods used for allocating overheads. Two process plants may occupy similar areas yet have vastly different material or labor costs. Problems can also arise with an individual plant that can be used to make different products, as the following example will show.

Let us consider a plant that can make either product A or product B. At normal capacity, the overhead cost is known to be $2.50/unit. Product A has a direct materials cost of $8/unit, and a direct labor cost of $2/unit. For simplicity, the prime cost is here taken as the sum of these two costs, i.e., $10/unit.

In Table I, a correct overhead cost of $2.50/unit at normal capacity is calculated by taking either 31.25% of the direct materials cost, 125% of the direct labor cost, or 25% of the prime cost. All these methods give a total cost of $12.50/unit, and a profit of $1.50/unit for a selling price of $14/unit.

The alternative, product B, has a direct materials cost of $6/unit and a direct labor cost of $4/unit. In Table I, overhead costs of $1.875/unit, $5/unit, and $2.50/unit are calculated by taking 31.25% of direct materials cost, 125% of direct labor cost, and 25% of prime cost, respectively. Total costs are $11.875/unit, $15/unit, and $12.50/unit, respectively; and profits of $2.125/unit, -$1/unit, and $1.50/unit, respectively, for a selling price of $14/unit.

An alternative to allocating overheads by using a single method is to classify the various overheads into groups and to use the most appropriate allocation for each group. For example, depreciation would be allocated on the basis of capital cost, while indirect labor might be allocated either on the basis of direct labor cost or on the number of employees. Clearly, this alternative method is more complex, increases the associated accountancy costs, and is prone to misinterpretation and possibly abuse.

Inventory Effect on Cash Income and Profit

When the annual production rate is equal to the annual sales volume, the revenue from annual sales, A_S, is:

$$A_S = Rc_S \tag{1}$$

where R is the production rate, units/yr., and c_S is the sales price per unit. In this case, the annual cash income, A_{CI}, and the net annual profit (before tax), A_{NP}, are given respectively by:

$$A_{CI} = R(c_S - c_{TVE}) - A_{TFE} \tag{2}$$

$$A_{NP} = R(c_S - c_{TVE}) - A_{TFE} - A_{BD} \tag{3}$$

where c_{TVE} is the total variable expense per unit of production, A_{TFE} is the total annual fixed expense required to produce and sell a product but excluding any annual provision for plant depreciation, and A_{BD} is the balance-sheet annual depreciation charge.

However, in a given accounting period, the sales volume may differ from the volume of production. In this case, the inventory of finished product I_1 at the beginning of the accounting period will differ from that at the end I_2, and Eq. (2) and (3) need to be written in modified forms as:

$$A_{CI} = R(c_S - c_{TVE}) - A_{TFE} + (I_1 - I_2) \tag{4}$$

$$A_{NP} = R(c_S - c_{TVE}) - (A_{TFE} + A_{BD}) + (I_1 - I_2) \tag{5}$$

If the annual sales volume exceeds the annual production rate, R, by an amount ΔR, then Eq. (4) and (5) can be written as:

$$A_{CI} = R(c_S - c_{TVE}) - A_{TFE} + \Delta R(c_S - c_{INV}) \tag{6}$$

$$A_{NP} = R(c_S - c_{TVE}) - (A_{TFE} + A_{BD}) + \Delta R(c_S - c_{INV}) \tag{7}$$

where c_{INV} is the value per unit of inventory of the finished product.

Clearly, the value of c_{INV} affects both the annual cash income and the net annual profit. Since annual cash incomes are the basic data for (NPV) and $(DCFRR)$ methods for estimating profitability, the actual value per unit of inventory is of direct importance for chemical engineers engaged in economic assessments.

Let us divide Eq. (7) by Eq. (1) to give:

$$\frac{A_{NP}}{A_S} = \left(1 - \frac{c_{TVE}}{c_S}\right) - \frac{(A_{TFE} + A_{BD})}{A_S} + \left(\frac{\Delta R}{R}\right)\left(1 - \frac{c_{INV}}{c_S}\right) \tag{8}$$

$$\frac{A_{NP}}{A_S} = (CSR) - \frac{(A_{TFE} + A_{BD})}{A_S} + \left(\frac{\Delta R}{R}\right)\left(1 - \frac{c_{INV}}{c_S}\right) \tag{9}$$

where $(CSR) = 1 - (c_{TVE}/c_S)$ is the contribution to sales price ratio. When the ratio of net annual profit to revenue from annual sales, A_{NP}/A_S, is expressed as a percentage, it is known as the profit margin.

The terms (CSR) and $(A_{TFE} + A_{BD})/A_S$ have a similar order of magnitude in chemical processing. For example, (CSR) values of 0.1 to 0.4 are typical, and these are quite close to a value of 0.2 that is common in general chemical processing for $(A_{TFE} + A_{BD})/A_S$. Thus, the profit margin is very sensitive to the value c_{INV} per unit of inventory.

For example, let us consider that both (CSR) and $(A_{TFE} + A_{BD})/A_S$ are equal to 0.3. In this case, Eq. (9) can be written as:

$$\frac{A_{NP}}{A_S} = \left(\frac{\Delta R}{R}\right)\left(1 - \frac{c_{INV}}{c_S}\right) \tag{10}$$

If the annual sales volume exceeds the annual production rate by 10%, and the inventory is valued at the sales price, then Eq. (10) shows the profit margin is

Profits of Products With Different Methods of Overhead Allocation — Table I

	Product A—Basis of Overhead Allocation			Product B—Basis of Overhead Allocation		
	31.25% of Direct Materials Cost, $/Unit	125% of Direct Labor Cost, $/Unit	25% of Prime Cost, $/Unit	31.25% of Direct Materials Cost, $/Unit	125% of Direct Labor Cost, $/Unit	25% of Prime Cost, $/Unit
Direct materials cost	8.00	8.00	8.00	6.000	6.000	6.000
Direct labor cost	2.00	2.00	2.00	4.000	4.000	4.000
Prime or direct cost	10.00	10.00	10.00	10.000	10.000	10.000
Overhead cost	2.50	2.50	2.50	1.875	5.000	2.500
Total cost	12.50	12.50	12.50	11.875	15.000	12.500
Selling price	14.00	14.00	14.00	14.000	14.000	14.000
Profit/(loss)	1.50	1.50	1.50	2.125	(1.000)	1.500

$(A_{NP}/A_S)100 = 0\%$. If the inventory is valued at the total variable cost, then the profit margin $(A_{NP}/A_S)100 = (0.1)(1 - 0.7)(100) = 3\%$. Hence, the value of the inventory is of vital importance.

Unfortunately, there is no universally accepted method for valuing inventory. The value of c_{INV} per unit can be taken on any of the following bases:

1. Direct material plus direct labor cost.
2. Direct material plus direct labor plus other direct production expenses. (This is the total variable production cost.)
3. Total variable production cost plus fixed production overhead cost.
4. Total variable cost. (This includes both production and general expenses.)
5. Total cost. (This includes the variable and fixed production and general expenses.)

Methods 1, 2 and 4 are termed direct costing, variable costing or marginal costing, respectively. Although direct costing is being increasingly used for internal accounting and control purposes, it is not acceptable to the tax authorities as a basis for calculating profit. Tax authorities and most accountants favor method 3, which is known as absorption costing. We have already seen that this method has the disadvantage that the fixed overhead cost per unit is determined for a particular normal production rate. If the production rate exceeds the normal, there is overabsorption of fixed overheads. Conversely, if production rate falls below the normal, there is underabsorption of fixed overheads. Method 5 is known as full absorption costing.

Most people agree that general expenses incurred in administration, selling, distribution, etc., should not be included in the cost of inventory. In fact, many feel that no costs should be absorbed before they have been incurred. In general, method 2 is favored by engineers, and method 3 by most accountants. However, the accountancy convention is to value on either cost or market value, whichever is the lower. In all the methods considered, either actual or standard costs can be used. Note that method 3 shows a higher profit than method 2 when sales volume exceeds production rate, and a lower profit when production rate exceeds sales volume.

The total profits (before tax) over the life of a project are independent of the method used to value inventory. Over a project life of n years, Eq. (5) can be written as:

$$\sum_0^n A_{NP} = \sum_0^n R(c_S - c_{TVE}) - \sum_0^n (A_{TFE} + A_{BD}) + \sum_0^n (I_m - I_{m+1}) \quad (11)$$

where the last term in Eq. (11) becomes:

$$\sum_0^n (I_m - I_{m+1}) = I_0 - I_{n+1}$$

Since there will be no material in inventory in the year before the project starts, or the year after it terminates, $I_0 = I_{n+1} = 0$. Hence, the total profits do not depend on individual values for I.

However, the net annual profit (before tax), A_{NP}, does depend on the value of the inventory. Since the tax payable in any individual year is based on A_{NP}, the net annual profit (after tax), A_{NNP}, is also dependent on the method chosen for valuing inventory. Frequently, a particular method for valuing inventory is chosen to delay the payment of tax as long as it is legally possible to do so.

So far, we have only considered the inventory of finished product. There are also inventories of raw materials and work in process, i.e., partially processed materials or intermediate products to be considered. It is necessary to modify Eq. (4) through (11) accordingly, to take these inventories into account.

Raw-Materials Prices Affect Profit

Raw materials for the chemical process industries are subject to relatively wide variations in price. These effects on profits will now be considered.

When the price of raw materials varies from week to week, the units in storage will not all have been purchased at the same price. Let us consider x units in storage at the start of the inventory period, purchased at a price, c_1 per unit. Additional quantities x_2, x_3, etc., are purchased at prices c_2, c_3, etc. per unit, respectively, until finally x_n units are purchased at the latest price of c_n per unit at the end of the inventory period. The total value of the inventory, C_I, at the end of the inventory period (in the absence of any withdrawal) is given by:

$$C_I = \sum_{1}^{n} x_j c_j \tag{12}$$

The value of the inventory, I, at any given time depends on the values ascribed to the units withdrawn from inventory. There are five methods for valuing inventory: (1) *FIFO* (first in, first out), (2) *LIFO* (last in, first out), (3) average cost, (4) standard cost, and (5) market value.

In the *FIFO* method, the units taken out of storage are valued at their purchase price beginning with the earliest item purchased. If a number of units, m, are removed from inventory during the period, the total cost of these items on a *FIFO* basis is given by:

$$C_{FIFO} = \sum_{1}^{p} x_j c_j + (m - \sum_{1}^{p} x_j) c_{p+1} \tag{13}$$

providing that the value of m satisfies:

$$m \leqslant \sum_{0}^{n} x_j$$

and where p is the largest integer, such that:

$$\sum_{1}^{p} x_j \leqslant m$$

Hence, the value of the inventory, I_{FIFO}, at any given time is:

$$I_{FIFO} = C_I - C_{FIFO} \tag{14}$$

In terms of Eq. (12) and (13), Eq. (14) can be written as:

$$I_{FIFO} = \sum_{1}^{n} x_j c_j - \sum_{1}^{p} x_j c_j - (m - \sum_{1}^{p} x_j) c_{p+1} \tag{15}$$

In the *LIFO* method, the m units taken out of storage are valued at their purchase price, beginning with the latest item purchased. In a similar manner, the value of the material, C_{LIFO}, taken out of inventory is given by:

$$C_{LIFO} = \sum_{p}^{n} x_j c_j + (m - \sum_{p}^{n} x_j) c_{p-1} \tag{16}$$

where p is the smallest integer, such that:

$$\sum_{p}^{n} x_j \leqslant m$$

Hence, the value of the inventory, I_{LIFO}, at any given time is:

$$I_{LIFO} = C_I - C_{LIFO} \tag{17}$$

In terms of Eq. (12) and (16), Eq. (17) can be written as:

$$I_{LIFO} = \sum_{1}^{n} x_j c_j - \sum_{p}^{n} x_j c_j - (m - \sum_{p}^{n} x_j) c_{p-1} \tag{18}$$

Example for Inventory Computations

Let us consider 10 successive batches of raw materials, each of a thousand units, purchased in a time of rising prices where c_1 = \$0.10/unit, c_2 = \$0.11/unit, etc., as listed in Table II. The total cost of the purchases in the inventory is found from Eq. (12) to be \$1,450.

Costs of Inventory With Rising Prices—Table II

p	x_j, Units	c_j, \$/Unit	$x_j c_j$, \$	$\sum_1^p x_j$, Units	$\sum_1^p x_j c_j$, \$	$\sum_p^n x_j$, Units	$\sum_p^n x_j c_j$, \$
1	1,000	0.10	100	1,000	100	10,000	1,450
2	1,000	0.11	110	2,000	210	9,000	1,350
3	1,000	0.12	120	3,000	330	8,000	1,240
4	1,000	0.13	130	4,000	460	7,000	1,120
5	1,000	0.14	140	5,000	600	6,000	990
6	1,000	0.15	150	6,000	750	5,000	850
7	1,000	0.16	160	7,000	910	4,000	700
8	1,000	0.17	170	8,000	1,080	3,000	540
9	1,000	0.18	180	9,000	1,260	2,000	370
10	1,000	0.19	190	10,000	1,450	1,000	190

Let us calculate the value of the raw-materials inventory after, say, 5,500 units have been withdrawn from inventory, first by using *FIFO,* and then by using *LIFO.*

We substitute the appropriate quantities into Eq. (15) for *FIFO,* keeping in mind that $p = 5$, $n = 10$ and $m = 5{,}500$, to get:

$$I_{FIFO} = \$1{,}450 - \$600 - (5{,}500 - 5{,}000)(\$0.15)$$
$$I_{FIFO} = \$775$$

In a similar manner, we substitute values into Eq. (18) for *LIFO* (except that p is now 6) to get:

$$I_{LIFO} = \$1{,}450 - \$850 - (5{,}500 - 5{,}000)(\$0.14)$$
$$I_{LIFO} = \$530$$

Thus, in a time of rising raw-materials prices, the *FIFO* method gives a higher value for the remaining inventory than will *LIFO*. In a time of falling prices, the *FIFO* method will give a lower value for the remaining inventory than will *LIFO.*

Average Cost Basis for Inventory

Either a simple average or a weighted average can be used to value inventory cost.

Using the simple average method, the value, C_{SAV}, of the material taken out of inventory is given by:

$$C_{SAV} = m(c_1 + c_n)/2 \tag{19}$$

The value of the inventory, I_{SAV}, at any given time is:

$$I_{SAV} = C_I - C_{SAV} \tag{20}$$

In terms of Eq. (12) and (19), Eq. (20) can be written as:

$$I_{SAV} = \sum_{1}^{n} x_j c_j - \frac{m(c_1 + c_n)}{2} \tag{21}$$

Again, let us consider the value of the inventory after 5,500 units have been withdrawn—using the data listed in Table II. On the basis of a simple average, the materials withdrawn are priced at:

$$(\$0.10 + \$0.19)/2 = \$0.145/\text{unit}$$

Since the total cost of the purchases in the raw-materials inventory is found from Eq. (12) to be $1,450, the value of the inventory after 5,500 units have been withdrawn is calculated from Eq. (21) to be:

$$I_{SAV} = \$1{,}450 - (5{,}500)(\$0.145) = \$652.50$$

For the weighted-average method, the value of the material, C_{WAV}, taken out of inventory is given by:

$$C_{WAV} = \frac{m \sum_{1}^{n} x_j c_j}{\sum_{1}^{n} x_j} \qquad (22)$$

The inventory value I_{WAV}, at any given time is:

$$I_{WAV} = C_I - C_{WAV} \qquad (23)$$

In terms of Eq. (10) and (21), Eq. (23) can be written:

$$I_{WAV} = \sum_{1}^{n} x_j c_j \left(1 - \frac{m}{\sum_{1}^{n} x_j}\right) \qquad (24)$$

Let us now use Eq. (24) to value the inventory, after 5,500 units have been withdrawn, by using the data of Table II.

$$I_{WAV} = \$1{,}450\left(1 - \frac{5{,}500}{10{,}000}\right) = \$652.50$$

For this example, the values of I_{SAV} and I_{WAV} are the same because the batches purchased are of equal size and the prices are linearly progressive. This is a combination rarely found in practice.

A more realistic example, in which the buyer seeks to purchase at the lowest price, is provided by the data of Table III. The quantities bought vary according to price, but some may have been made at high prices to maintain production.

On the basis of Table III, when 5,500 items have been removed from inventory, the value of the inventory using the *FIFO, LIFO,* simple-average and weighted-average methods, respectively, is:

$$I_{FIFO} = \$1{,}175 - \$560 - (5{,}500 - 5{,}200)(\$0.14)$$
$$I_{FIFO} = \$573.00$$

$$I_{LIFO} = \$1{,}175 - \$645 - (5{,}500 - 5{,}000)(\$0.10)$$
$$I_{LIFO} = \$480.00$$

$$I_{SAV} = \$1{,}175 - 5{,}500\left[\frac{(\$0.10 + \$0.15)}{2}\right]$$
$$I_{SAV} = \$487.50$$

$$I_{WAV} = \$1{,}175\left(1 - \frac{5{,}500}{10{,}000}\right)$$
$$I_{WAV} = \$528.75$$

Costs of Inventory With Fluctuating Prices—Table III

p	x_j, Units	c_j, \$/Unit	x_jc_j, \$	$\sum_1^p x_j$, Units	$\sum_1^p x_jc_j$, \$	$\sum_p^n x_j$, Units	$\sum_p^n x_jc_j$, \$
1	1,000	0.10	100	1,000	100	10,000	1,175
2	1,500	0.11	165	2,500	265	9,000	1,075
3	500	0.13	65	3,000	330	7,500	910
4	2,000	0.10	200	5,000	530	7,000	845
5	200	0.15	30	5,200	560	5,000	645
6	1,000	0.14	140	6,200	700	4,800	615
7	2,500	0.11	275	8,700	975	3,800	475
8	300	0.15	45	9,000	1,020	1,300	200
9	100	0.20	20	9,100	1,040	1,000	155
10	900	0.15	135	10,000	1,175	900	135

Pros and Cons of Inventory Valuation

In the standard-price method of inventory valuation, all materials are taken out at the same price. In addition to simplicity, the method has the advantage that the efficiency of raw-materials purchase is constantly checked.

In both the average-cost and standard-cost methods of valuing inventory, materials are not charged out at actual cost. Thus, the amount of profit or loss for the period may be varied by the method chosen to value the inventory. For this reason, accountants usually insist that the method of inventory valuation be consistent from period to period. This causes inertia but does not prevent a change of method when it can be justified. In such cases, it is usual to inform stockholders of the change because the influence on declared profits can be large.

Unfortunately, there is no right or wrong way to value inventory, although certain methods are not allowed in certain countries for tax-assessment purposes. For example, *LIFO* is not allowed in the U.K. As a general rule, the method used should be the one that gives the lowest tax liability. However, it is generally accepted that consistency is also a virtue in inventory valuation.

It is important to realize that the method used to value inventory for cost-accounting purposes is not necessarily the one used to draw up the balance sheet and financial accounts. In this case, inventory is valued either at the cost given by another method or at the market value, whichever is lower.

Financial Accounting

The balance sheet and the associated trading account and profit-and-loss account are prepared by the financial accountants of the company. These statements are subject to independent audit and scrutiny by the tax authority. Hence, these statements are the evidence by which the financial health and stability of the company are judged by stockholders. It is often prudent for the engineer to carry out a similar assessment before entrusting his career prospects to a particular company, or when judging the time to leave.

In order to understand balance sheets, a knowledge of the language of accountancy is needed. The next article of this series will review and explain accounting methods. ■

Reference

1. Thuesen, H. G., Fabrycky, W. J. and Thuesen, G. J., "Engineering Economy," 4th ed., p. 243, Prentice-Hall, Englewood Cliffs, N. J., 1971.

Principles of Accounting

Financial accounting involves the preparation of balance sheets and income statements. Management accounting involves standard costing, budgetary control and investment decisions. Understanding the makeup of these techniques will enable you to interpret and use accounting information.

F. A. HOLLAND, F. A. WATSON and J. K. WILKINSON, University of Salford

Accounting is the art of recording business transactions in a systematic manner. Financial statements are both the basis for and the result of management decisions. Such statements can tell a manager or an engineer a great deal about his company, provided that he can interpret the information correctly.

Since a fair allocation of costs requires considerable technical knowledge of operations in the chemical process industries, a close liaison between the senior process engineers and the accountants in a company is desirable. Indeed, the success of a company depends on a combination of financial, technical and managerial skills.

Accounting is also the language of business, and the different departments of management use it to communicate within a broad context of financial and cost terms. The engineer who does not take the trouble to learn the language of accountancy denies himself the most important means available for communicating with top management. He may be thought by them to lack business acumen. Many engineers have only themselves to blame for their relatively lowly status within the company hierarchy since they seem determined to hide themselves from business realities behind a screen of their specialized technical expertise.

Engineers involved in feasibility studies and detailed process evaluations are dependent for financial information on the company accountants, especially for information on the way the company intends to allocate its overhead costs. It is vital that the engineer should correctly interpret such information and that he can, if necessary, make the accountant understand the effect of the chosen method of allocation.

The method of allocating overheads can seriously affect the assigned costs of a project and hence the apparent cash flows for that project. Since these cash flows are used to assess profitability by the net present value (NPV) and discounted-cash-flow rate-of-return ($DCFRR$) methods, unfair allocation of overhead costs can result in a wrong choice between alternative projects.

In addition to understanding the principles of accountancy and obtaining a working knowledge of its practical techniques, the engineer should be aware of possible inaccuracies of accounting information in the same way that he allows for errors in any technical data.

At first acquaintance, the language of accountancy appears illogical to most engineers. Although the accountant normally expresses himself in tabular form, the basis of all his practice can be simply expressed by:

$$\text{Capital} = \text{Assets} - \text{Liabilities} \tag{1}$$

Eq. (1) can alternatively be written as:

$$\text{Assets} = \text{Capital} + \text{Liabilities} \tag{2}$$

Capital, often referred to as net worth, is the money value of the business, since assets are the money values of things the business owns while liabilities are the money values of the things the business owes.

Most engineers have great difficulty in thinking of capital (also known as ownership) as a liability. This is easily overcome once it is realized that a business is a legal entity in its own right, owing money to the individuals who own it. This realization is absolutely essential when considering large companies with stockholders, and

Originally published July 1974.

"The engineer who does not take the trouble to learn the language of accountancy denies himself the most important means available for communicating with top management."

is used for consistency even for sole ownerships and partnerships. If Joe Bloggs puts up $10,000 capital to start a business, then that business has a liability to repay $10,000 to Joe Bloggs.

It is even more difficult to think of profit as being a liability. Profit is the increase in money value available for distribution to the owners, and effectively represents the interest obtained on the capital. If the profit is not distributed, it represents an increase in capital by the normal concept of compound interest. Thus, if Joe Bloggs' business makes a profit of $5,000, the liability to Joe Bloggs is increased to $15,000. With this concept in mind, Eq. (2) can be expanded to:

$$\text{Assets} = \text{Capital} + \text{Liabilities} + \text{Profit} \quad (3)$$

where the capital is considered as the cash investment in the business, and is distinguished from the resultant profit in the same way that principal and interest are separated.

Profit (as referred to above) is the difference between the total cash revenue from sales and the total of all costs and other expenses incurred in making those sales. With this definition, Eq. (3) can be further expanded to:

$$\text{Assets} + \text{Expenses} = \text{Capital} + \text{Liabilities} + \text{Revenue From Sales} \quad (4)$$

Engineers usually have the greatest difficulty in regarding an expense as being equivalent to an asset, as is implied by Eq. (4). However, let us consider Joe Bloggs' earnings. During the period in which he made

ABC Co. Income Statement—Table I
(Year ending Dec. 31, 1973)

	$	$
Revenue		
Sales revenue	1,900,000	
Other revenue	100,000	
		2,000,000
Expenses		
Raw materials	953,000	
Wages	185,000	
Utilities	44,000	
Depreciation	68,000	
Other expenses	376,000	
Income taxes	194,000	
		1,820,000
Net profit (after tax)		180,000

XYZ Co. Balance Sheet (Year ending Dec. 31, 1973)—**Table II**

ASSETS
(Thousands of Dollars)

Current assets	
Cash	$ 38,893
Notes and accounts receivable	110,740
Inventories:	
Finished products	17,396
Work in process	56,690
Raw materials and supplies (at cost)	35,790
Total inventories	109,876
Total current assets	259,509
Investments and long-term receivables (at cost)	94,009
Property, plant, and equipment (at cost)	
Land	6,110
Buildings	63,848
Machinery and equipment	106,185
	176,143
Less accumulated depreciation	75,163
Net property, plant and equipment	100,980
Prepaid and deferred charges	6,094
Total assets	$460,592

LIABILITIES AND STOCKHOLDERS' EQUITY
(Thousands of Dollars)

Current liabilities	
Notes payable	$ 34,507
Accounts payable and accrued liabilities	106,433
Accrued taxes	7,264
Total current liabilities	148,204
Long-term Liabilities	67,677
Deferred income taxes	13,225
Other deferred credits	2,307
Stockholders' equity	
Common stock, $20 par value	
Shares authorized 7,750,000	
Shares issued 4,794,450	95,889
Capital in excess of par value of common stock	31,798
Retained earnings	101,492
Total stockholders' equity	229,179
Total liabilities and stockholders' equity	$460,592

a profit of $5,000, his total expenses excluding his earnings were $8,000. If he assessed the worth of his labor to the business at $12,000, then the revenue required from sales would be $25,000. Effectively, Joe Bloggs has made a personal income of $17,000 in the year but he has apportioned it to the business as $12,000 expense for his labor and $5,000 return on his capital. In larger businesses, there will also be those who receive salaries but do not hold stock and, therefore, receive no profits; and stockholders who receive profits but no salaries. Thus, the difference between expenses and profits is very practical.

The period covered by the published accounts of a company is usually one year, but the details from which these accounts are compiled are entered daily in a journal. The journal is a chronological listing of every transaction of the business, with details of the corresponding income or expenditure. For the smallest businesses, this may provide sufficient documentation but, in most cases, the unsystematic nature of the journal can lead to computational errors. Therefore, the usual practice is to keep accounts that are listings of transactions related to a specific topic such as "Purchase of Oil Account." This account would list the cost of each purchase of oil, together with the date of purchase, as extracted from the journal.

Principles of Double-Entry Accounting

Many of the accounts involve both income and expenditure. The general practice is to keep accounts by the double-entry system, which may be summarized by:

$$\text{Debits} = \text{Credits} \tag{5}$$

The principle of double entry dates from the fifteenth century, and is based on the premise that every transaction involves a giver and a receiver of value. Double entry requires that each transaction is entered into two accounts; the convention being that the account of the giver is credited and the account of the receiver is debited with the same amount of money, as noted in the journal. For convenience, each account is divided centrally, and the debit items are entered on the lefthand side of the page while the credit items are entered on the righthand side. It is also usual to provide a cross-reference to the journal entry so that errors and omissions can be checked.

Let us consider the purchase of $50,000 worth of plant equipment by Company A, paid for by check. The accounting entries are: debit the plant equipment account $50,000, and credit the bank account $50,000, as shown:

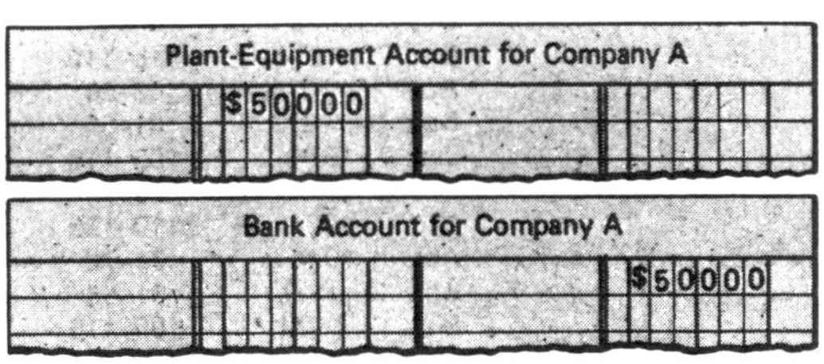

The plant-equipment account is then said to have a debit balance of $50,000, and the bank account a credit balance of $50,000, if these happen to be the only entries.

If Company A then sells $100,000 worth of product that is paid for by check, the accounting entries are: credit the sales account $100,000, and debit the bank account $100,000, as shown:

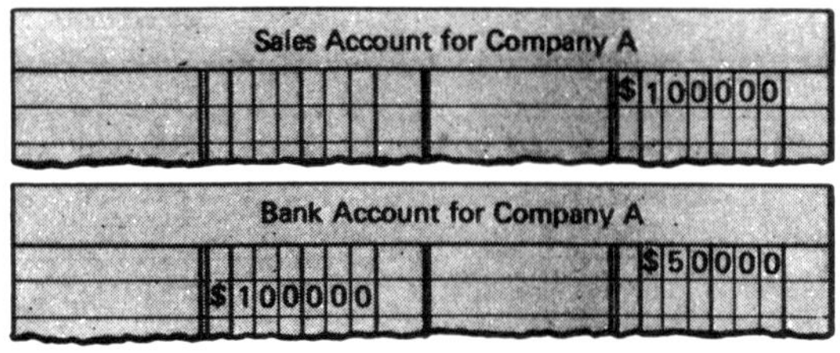

The bank account will now have a debit balance of ($100,000 − $50,000) = $50,000, and the sales account a credit balance of $100,000 if this happens to be the only sale to date in the accounting period.

In principle, the debiting and crediting of accounts is relatively straightforward. However, a great deal of practice is essential in order to achieve proficiency. Although it is not at all necessary for engineers to compete with professional accountants in this field, engineers should appreciate what accountants do and why they do it.

Of the accounts considered in the preceding illustrations, the plant-equipment and bank accounts are asset accounts, and the sales account is a liability account. To increase an asset, debit the asset account; to increase a liability, credit the liability account. Conversely, to decrease an asset, credit the asset account; to decrease a liability, debit the liability account.

Closing the Books

At the end of the accounting period, the individual accounts are closed by balancing each in accordance with Eq. (5). The balances are transferred either to the balance sheet in the case of capital expenditure, or the income statement in the case of revenue expenditure. An alternative name for the balance sheet is the position statement; the income statement is also called the trading and profit-and-loss account.

The purpose of capital expenditure, such as the purchase of a piece of plant equipment for $50,000, is to earn future revenue. In contrast, the purpose of revenue expenditure is to maintain existing business.

Revenue expenditure includes the direct material costs and direct labor costs incurred in the manufacture of a product, together with the associated overheads that include maintenance of the plant. Since these expenses are debits, the debit balance for a given accounting period is obtained by adding up the debit balances from each individual expenditure account. Similarly, since revenues from sales and other income are credits, the credit balance for a given accounting period is obtained by adding up the credit balances from each individual income or revenue account.

To ascertain profit or loss (calculated as income minus expenditure for a given accounting period), income and expenditure must be matched. For example, any rent paid in advance beyond the current accounting period should not be included in the profit or loss calculation. Similarly, goods sold but not yet paid for in a given

Provision for Depreciation of Plant-Equipment Account—Table III

1973			
		Jan. 1 Balance brought down	zero
Dec. 31 Balance carried down	$100,000	Dec. 31 Debited to income statement	$100,000
1974			
		Jan. 1 Balance brought down	$100,000
Dec. 31 Balance carried down	$200,000	Dec. 31 Debited to income statement	$100,000
	$200,000		$200,000
1975			
		Jan. 1 Balance brought down	$200,000
Dec. 31 Balance carried down	$300,000	Dec. 31 Debited to income statement	$100,000
	$300,000		$300,000

accounting period should be included in the revenue total for that period.

An income statement such as the one shown in Table I is used to obtain the profit or loss for a given period. The debit and credit balances of all the accounts that do not represent expenditure or income for a given accounting period are entered as assets and liabilities in a balance sheet, such as that shown in Table II.

There is no rigid format for either the income statement or the balance sheet. Tables I and II show common layouts for the income statement and balance sheet, respectively, but these are not the only forms. For example, vertical balance sheets, with the assets listed above the liabilities and equity, are also popular.

Some expenditures are partly capital and partly revenue. For example, repair and improvement work may be done on a plant simultaneously. In this case, the repair work should be classified as revenue expenditure, and the plant-improvement work as capital expenditure.

Accounting Concepts and Conventions

Accounting is based on the following concepts: (1) money measurement, (2) business entity, (3) going concern, (4) cost and (5) matching.

Concept 1—Money measurement means that only those facts that can be represented in monetary terms are recorded. The balance sheet and income statement for a company give no indication as to what might happen in the future. The company may be about to be successfully sued for a large sum of money, or a competitor may be launching a new product that will seriously reduce future sales of the company's products.

Concept 2—Business entity means that accounts are kept for the company quite independently of the people who may own the company. For example, if Joe Bloggs puts an additional $10,000 into his one-man business, the accounts show that the business is $10,000 richer. They do not show that Joe Bloggs' personal wealth has been depleted by $10,000.

Concept 3—Going concern means that the accounting is based on the premise that the business will continue indefinitely. It is most unlikely that the values of the assets shown in the balance sheet are what the assets would realize if sold. No attempt is made in normal accounting to measure the value of the business to a potential buyer.

Concept 4—Cost means that the assets are normally shown in the balance sheet at cost price together with their subsequent depreciation. Some assets such as land may be considerably more valuable than when originally purchased but no indication of this is given in the balance sheet. However, some governments now require a note given the current estimated value of the land.

Concept 5—Matching means that the revenue in a given accounting period should correspond to the expenses for that accounting period.

Accounting is also based on the following conventions: (1) materiality, (2) conservatism or prudence and (3) consistency.

Materiality deals with determining whether certain expenditures will have a significant effect on a company's accounting procedures. This is a matter of judgment that is to be made by each company. Obviously, the purchase of a vehicle is a material item, but what about writing-paper, or tools for maintenance? Although such items may last well beyond the current accounting period, it may not be worth the accounting effort to treat them as material items. Some companies will treat a particular item as capital, others as expenditure. Clearly, the purchase of a piece of equipment costing, say, $1,000, will be regarded as less material by a giant company than by a small one.

Conservatism or prudence means monetary values are taken that tend to understate rather than overstate the profit. In other words, chickens are not normally counted before they are hatched.

Consistency means that accounting items are normally treated in the same way over an indefinite number of years. For example, an individual item would not be treated as an expenditure during one year and as a capital item during the next, without good reason being given.

The Balance Sheet

The balance sheet, also called the position statement, presents an accounting view of the financial status of a company at a particular point in time. A typical balance sheet is shown in Table II. Although a balance sheet has two sides that balance, it is not part of the double-entry system. In fact, it is not an account but rather a statement listing all the assets of a company and the various claims against these assets on the last day of the accounting period. The assets must be equal to the claims against them at all times. Those who have claims against the assets are the owners (stockholders in a business corporation), and the people to whom the company owes money. In the case of the latter, the company is said to have liabilities to its creditors. The total claim against the assets is often labeled liabilities and owner's equity.

Assets are classified as current or fixed; and liabilities as current or long term. Fixed assets are material items that have a relatively long life, and normally include land, buildings, plant, vehicles, etc. They are held for the specific purpose of earning revenue, and are not for sale in the normal course of business. Current assets include cash, and those items that can be fairly easily converted into cash, such as raw-materials inventories, etc. In contrast to fixed assets, current assets are acquired for the specific purpose of conversion into cash in the normal course of business. However, what is regarded as a fixed asset by one type of company might be regarded as a current asset by another. For example, a chemical company would normally classify its vehicles as a fixed asset. However, a company whose primary business was to sell vehicles would classify them as a current asset.

Entries in Balance Sheets—Table IV

As of Dec. 31, 1973		
Plant equipment at cost	$1,000,000	
Less depreciation to date	100,000	
		$900,000
As of Dec. 31, 1974		
Plant equipment at cost	$1,000,000	
Less depreciation to date	200,000	
		$800,000
As of Dec. 31, 1975		
Plant equipment at cost	$1,000,000	
Less depreciation to date	300,000	
		$700,000

Similarly, the distinction between current and long-term liabilities is also not clear-cut. Current liabilities include accounts payable (money owed to creditors), taxes payable, dividends payable, etc., if due within a year. Long-term liabilities include deferred income taxes, bonds, notes, etc., that do not have to be paid within a year. The owners' equity includes the par, or face, value of the capital received from stockholders, and any retained earnings. The balance sheet shows only the nominal value and not the current or real value of this capital.

A balance sheet includes items that are not regarded as assets or liabilities in normal language, such as expenditures carried forward and accumulated profits. A balance sheet is only based on the accounting view of assets and liabilities, which does not always coincide with the commonsense view. For example, few would deny that a brilliant research team is likely to be a considerable asset to a company, yet it would not be listed as an asset on a balance sheet.

Accountants regard assets as resources that have not yet been used up. Assets are normally shown on the balance sheet at cost minus accumulated depreciation. In this sense, the depreciation charge for an accounting period is the means of converting a part of an asset into a current expenditure that is then listed as an expense in the income statement.

Let us consider plant equipment costing $1 million, and purchased on Jan. 1, 1973. Table III shows the provision for the depreciation account for 1973, 1974 and 1975 for straight-line depreciation, assuming a service life of ten years and zero scrap-value. The credit entries of $100,000 for the depreciation in each year are balanced by the depreciation charge of $100,000 debited to the income statement (or trading and profit-and-loss account) in each year. Table IV shows the corresponding entries in the balance sheets for the years 1973, 1974 and 1975. Entries for subsequent years are made in the same way.

A balance sheet is only true for one particular point in time; it tells nothing about the trends in a company. However, by comparing balance sheets for successive years, management can follow changes in the various items. If the observed trend is undesirable, management can take corrective action. Since the accounting period

Income Statement for a Mature Year—Table V
(Basis: New chemical product, 10 million lb./yr.)

	$	$	Unit Values ¢/Lb.	% Sales Revenue %
Revenue from annual sales, A_S		2,000,000	20.00	100.0
Direct manufacturing expense, A_{DME}				
Raw materials	884,000		8.84	44.2
Catalysts and solvents	69,000		0.69	3.4
Operating labor	102,000		1.02	5.1
Operating supervision	20,000		0.20	1.0
Utilities	22,000		0.22	1.1
Operating maintenance	21,000		0.21	1.1
Operating supplies	4,000		0.04	0.2
Royalties and patents	10,000		0.10	0.5
Total A_{DME}	1,132,000	1,132,000	11.32	56.6
Indirect manufacturing expense, A_{IME}				
Payroll overhead	28,000		0.28	1.4
Central laboratory	10,000		0.10	0.5
General plant overhead	52,000		0.52	2.6
Packaging and storage	22,000		0.22	1.1
Property taxes	14,000		0.14	0.7
Insurance	6,000		0.06	0.3
Total A_{IME}	132,000	132,000	1.32	6.6
Total manufacturing expense (excluding depreciation), A_{ME}		1,264,000	12.64	63.2
Depreciation, A_{BD}		68,000	0.68	3.4
Other expenses				
Administration	74,000		0.74	3.7
Sales and shipping	124,000		1.24	6.2
Advertising and marketing	40,000		0.40	2.0
Technical service	10,000		0.10	0.5
Research and development	60,000		0.60	3.0
Total other expenses	308,000	308,000	3.08	15.4
Total expense, A_{TE}		1,640,000	16.40	82.0
Net annual profit, A_{NP}		360,000	3.60	18.0

of one year is long for most businesses, it is usual to draw up balance sheets at more frequent intervals for control purposes. These may be less formal than those issued annually to the stockholders. In general, balance sheets are less useful to management than are income statements.

The Income Statement

Income statements range from the very simple presentation shown in Table I to the more informative and more complex presentation shown in Table V.

The income statement shows the revenue, and the corresponding expenses that were incurred to earn that revenue over a period of time. It is the most obvious measure of the efficiency of a business. Although published income statements are normally for one-year periods, many companies use monthly income statements for internal purposes.

Income statements are very useful tools to assist management in controlling a business and planning for the future. Since management needs to follow the trends of the normal expenses, extraordinary expenses such as those incurred as a result of a major fire or flood should be shown separately.

If revenue and expenses are not properly matched, an understatement or overstatement of profit may occur. If raw materials were previously purchased at a lower cost than their current cost, profit will be overstated. Any overstatement of profit will mean that more tax will be paid.

One of the most important items in an income statement is depreciation expense. Although depreciation should not be thought of as a means to build up a fund to replace plant, it nevertheless does enable money to be retained in the business by reducing the profit available for distribution to stockholders. It is of course a duty of both accountants and management to see that sufficient money is retained in the business to replace assets and to invest such money in other processes or outside investment.

A further duty of accountants and management is to ensure that the company always has sufficient working capital to enable it to carry on its business.

Types of Accountancy

The traditional work of accountants has been to prepare balance sheets and income statements. Nowadays, accountants are becoming increasingly concerned with forward planning. Modern accountancy can roughly be divided into two branches: financial accountancy and management or cost accountancy.

Financial accountancy is concerned with stewardship. This involves the preparation of balance sheets and income statements that represent the interest of stockholders and are consistent with the existing legal requirements. Taxation is an important element of financial accounting.

Management accounting is concerned with decision-making and control. This is the branch of accountancy closest to the interest of most process engineers. Management accounting is concerned with standard costing, budgetary control, and investment decisions.

Accounting statements only present facts that can be expressed in financial terms. They do not indicate whether a company is developing new products that will ensure a sound business future. A company may have impressive current financial statements, and yet may be heading for bankruptcy in a few years' time if provision is not being made for the introduction of sufficient new products or services.

The next article of this series will analyze and explain working capital and its effect on project and company performance. ■

Working capital is the resource that enables a company to conduct its day-to-day business. Making estimates of working capital and analyzing its effects on a business may compare a company's performance with that of its competitors.

How To Evaluate Working Capital for a Company

F. A. HOLLAND, F. A. WATSON and J. K. WILKINSON, University of Salford

There are several reasons why chemical engineers must take an interest in working capital:

First, the amount and disposition of the working capital, and the efficiency of its use, determine the immediate prospects for future growth in a company, and hence, the job prospects in that company.

Second, the bulk of managerial effort in a company is directly or indirectly concerned with the manipulation of the working capital.

Third, insufficient or misused working capital is the commonest cause of business failure.

Fourth, the universal problem of inflation, with the steady fall in the value of hoarded money, makes it essential to put the available capital to work in order to provide the highest return possible.

Let us begin by reviewing the several items of cost that comprise the total capital cost, C_{TC}, of a project, as given by:

$$C_{TC} = C_{FC} + C_{WC} + C_L \quad (1)$$

where C_{FC} is the fixed capital cost, C_{WC} is the working capital cost, and C_L is the cost of land and other nondepreciable costs.

Engineers concerned with cost estimations tend to make estimates of the fixed capital cost of a project, leaving considerations of working capital to the accountants. Although the estimation of fixed capital cost is more straightforward from an engineering point of view, the estimation of working capital is of vital importance both for an individual project and for the company as a whole. However, an elementary knowledge of accounting is an essential requirement for the engineer who wishes to understand working capital.

Rate-of-return methods for estimating profitability are based on:

$$PERCENT\ RATE\ OF\ RETURN = \left(\frac{ANNUAL\ PROFIT}{INVESTED\ CAPITAL}\right)100 \quad (2)$$

Working capital can range from about 10% to almost 100% of the invested capital, depending on the industry, and is an important factor in the profitability index of a business. For this reason, it is best to compare the performance of one's own company with others that are as similar as possible.

Gross working capital is normally defined as the total current assets, while net working capital is the current assets minus the current liabilities. Current assets normally amount to more than half the total assets of a company. When accountants wish to emphasize working capital in a company, they present the balance sheet in the vertical form, as shown in Table I. In this table, the stockholders' equity of $91,650 funds the sum of $38,650 (net working capital) and the sum of $53,000 (fixed assets less the long-term loan). The flow of working capital is diagrammatically illustrated in Fig. 1.

Originally published August 1974.

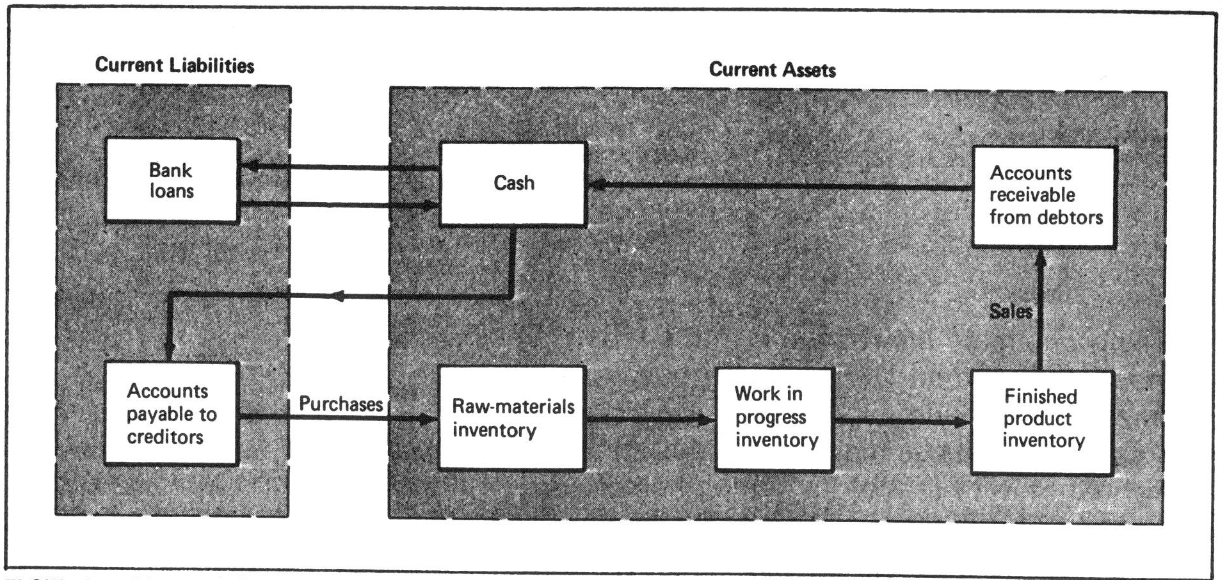

FLOW of working capital shows interrelationships between current liabilities and current assets—Fig. 1

The necessary working capital varies with sales volume or production rate. For example, sales, and hence accounts receivable, will double for a doubling in sales volume. In addition, an increase in sales volume normally requires increased inventories of raw materials, work in progress, and finished goods, all of which tie up capital. An increase in sales volume may lead to a relative shortage of working capital. In turn, this may mean that accounts payable cannot be paid in time, and valuable cash discounts may be lost or interest and penalty charges incurred. Creditors may take legal action to obtain payment, and thereby put an additional strain on the current assets of the company to provide legal fees. Often, such action leads other creditors to take similar steps, which may lead a fundamentally sound company into bankruptcy.

A shortage of cash may prevent a company from taking advantage of large discounts available for bulk purchase of raw materials. The importance of the availability of adequate cash or near cash can be seen by considering an account payable within 28 days, with a 2% discount allowed if paid within 7 days. If cash is not available to pay the account within 7 days, this is then equivalent to paying 2% interest on the money for the remaining 21-day period, or an annual compound-interest rate of more than 41%.

Adequate cash and a history of prompt payment of accounts strengthens the credit standing of a company, and makes it easier to obtain bank loans, etc. A company needs additional cash as a contingency against fires, floods, strikes, etc., as well as for additional advertising required to counteract the activities of competitors. This additional money is normally held as interest-bearing investments that can be turned into cash on short notice.

XYZ Co. Balance Sheet—Table I
(On Dec. 31, 1973)

Current assets	$	$
Cash	14,575	
Accounts receivable	35,575	
Inventories		
Raw material	6,000	
Work in process	3,750	
Finished product	10,000	
Gross Working Capital		69,900
Current liabilities		
Accounts payable	25,000	
Notes payable	3,000	
Bank loans	3,000	
Accruals payable	250	
		31,250
Net working capital		38,650
Fixed assets	75,000	
Long-term loan	22,000	
		53,000
Stockholders' Equity		$91,650

Working-Capital Ratios

Financial analysts make extensive use of ratios in assessing the economic health of a company. For evaluating the ability of a company to successfully maintain and develop its immediate business activities, analysts apply a current ratio and a quick (or acid-test) ratio, as given by:

$$Current\ Ratio = \frac{Current\ Assets}{Current\ Liabilities} \quad (3)$$

$$Quick\ Ratio = \frac{Liquid\ Assets}{Current\ Liabilities} \quad (4)$$

Liquid assets are those that can be realized almost immediately, such as cash, accounts receivable and marketable securities. Although inventories are current

Funds Statement—Table II
(Year ending Dec. 31, 1973)

	$	$
Sources of funds		
Cash income from operations		15,000
New finance		10,000
Sale of plant equipment		2,000
Total Sources of Funds		27,000
Applications of funds		
Purchase of land	3,000	
Cash dividend on stock	9,000	
Total Applications of Funds		12,000
Increase in Net Working Capital		15,000

Statement of Changes in Working Capital—Table III

	Jan. 1, 1973	Jan. 1, 1974	Change
Cash	$80,000	$94,000	+$14,000
Accounts receivable	20,000	35,000	+ 15,000
Inventories	30,000	25,000	− 5,000
Total current assets	$130,000	$154,000	+$24,000
Current liabilities	($60,000)	($69,000)	(+$ 9,000)
Net working capital	$70,000	$85,000	+$15,000
Current ratio	2.17	2.23	
Quick ratio	1.67	1.87	

assets, they must not be regarded as liquid assets because they cannot usually be converted into cash without winding up the business.

Although a high current ratio is desirable, this may be achieved by having unnecessarily high inventories that bring no profit except when commodity prices are rising rapidly. The quick ratio is less misleading in this respect.

Good management practice will hold inventories at the lowest possible levels, consistent with customer satisfaction and efficient plant operation. Excessive inventories are unproductive and are an investment having little or no rate of return. Excessive inventories should only be maintained when supplies are erratic or rising in price. Management should normally aim for a high inventory-turnover ratio, as given by:

$$\textit{Inventory-Turnover Ratio} = \frac{\textit{Revenue From Annual Sales}}{\textit{Average Value of Inventory}} \quad (5)$$

Similarly, good management practice is to hold accounts receivable at a low level, and to have a high accounts-receivable turnover ratio, as given by:

$$\textit{Accounts-Receivable Turnover Ratio} = \frac{\textit{Revenue From Annual Sales}}{\textit{Average Value of Accounts Receivable}} \quad (6)$$

Alternatively, it is good practice to have a low average-collection period, as given by:

$$\textit{Average-Collection Period} = \frac{\textit{Average Value of Accounts Receivable}}{\textit{Revenue From Sales/Day}} \quad (7)$$

Eq. (7) gives the number of days that sales are tied up in receivables.

An accounts-receivable turnover ratio of 12 is considered fairly good for a manufacturing company. This implies an average collection period of about one month. The price obtained for the goods should include an allowance for interest (otherwise obtainable) on the money tied up for such a period.

Since ratios, like balance sheets, refer to a particular point in time, they have a limited use unless compared with previous values. A study of ratio trends indicates whether or not a company is approaching a working-capital or a liquidity crisis, and may enable management to compare company performance with that of competitors.

The Funds Statement

A typical funds statement is shown in Table II. It displays the change in net working capital and can be obtained from a statement of changes in working capital, such as the one shown in Table III. A funds statement shows where the cash came from and how it was used.

Changes in working capital, ΔC_{WC}, in an annual accounting period can be represented by:

$$\Delta C_{WC} = A_S - A_{TE} - \Sigma\Delta C_{FC} - \Sigma\Delta C_L + \Sigma\Delta C_{FIN} \quad (8)$$

where A_S is the revenue from annual sales of a product, A_{TE} is the total cost or expense required to produce and sell the product, but excluding any annual provision for plant depreciation, $\Sigma\Delta C_{FC}$ is the sum of the changes in depreciable fixed assets, $\Sigma\Delta C_L$ is the sum of the changes in nondepreciable fixed assets, and $\Sigma\Delta C_{FIN}$ is the sum of changes in financial resources such as loans, bonds, preferred stock, common stock, etc.

Eq. (8) can also be written as:

$$\Delta C_{WC} = A_{CI} - \Sigma\Delta C_{FC} - \Sigma\Delta C_L + \Sigma\Delta C_{FIN} \quad (9)$$

where A_{CI} is the annual cash income, which is the main source of funds for most companies. In this case, the annual cash income excludes all noncash expenses such as the balance-sheet annual depreciation charge, A_{BD}, which is purely a book transaction.

A positive value of any term in Eq. (9) implies an increase in working capital, and a negative value a decrease. For example, the sale of fixed assets such as plant, buildings, land, etc., is a source of cash, and the purchase of fixed assets uses up cash. Similarly, an increase in financial resources in the form of loans, stock and bond issues is a source of cash, and a decrease in financial resources in the form of repayment of loans, retirement of stocks and bonds, and the payment of cash dividends uses up cash. (Note that a stock dividend as opposed to a cash dividend does not use up cash.)

In an earlier article (see Part 5), the relation between (1) net annual cash flow (after tax), A_{CF}, for individual projects, (2) the annual amount of tax, A_{IT}, and (3) the annual expenditure of capital, A_{TC}, was shown to be:

$$A_{CF} = A_{CI} - A_{IT} - A_{TC} \quad (10)$$

Eq. (10) for a single project is really analogous to Eq. (9) for a company.

An income statement or profit-and-loss account gives the net annual profit, A_{NP}, before tax. In order to assess the annual cash income, A_{CI}, as a source of funds from the value of the net annual profit, A_{NP}, given in the income statement, it is necessary to add back all noncash expenses such as the balance-sheet annual depreciation charge, A_{BD}. This practice sometimes erroneously suggests that depreciation is a source of funds, whereas cash income is the only source of funds.

Although A_{BD} does not affect working capital in any way, the annual depreciation charge, A_D, does affect the annual amount of tax, A_{IT}, given by:

$$A_{IT} = (A_{CI} - A_D - A_A)t \quad (11)$$

where t is the fractional tax rate, and A_A is the annual amount of any other allowances. Thus, the net annual cash income is affected by depreciation allowances, as follows:

$$A_{NCI} = A_{CI} - A_{IT} \quad (12)$$

In this sense, depreciation makes working capital available by reducing the cash outflow for taxes.

When a fixed asset is sold at a price that differs from its book value, an accounting gain or loss is recorded. This gain or loss does not affect working capital, which has simply been increased by the amount of cash received from the sale. For example, if an item of plant is sold for \$40,000 (whatever its book value), the increase in working capital resulting from the sale is \$40,000. If the book value of the plant had been \$50,000, the accounting loss of \$10,000 on the sale would be included as an expense when calculating the cash income. This \$10,000 must be added back to the cash income in order to get the cash income excluding noncash expenses as required for A_{CI}. Conversely, any accounting gain on the sale of a fixed asset must be subtracted if it has been included in the cash income.

Book values of fixed assets are determined by the balance-sheet annual depreciation charges, A_{BD}, that do not affect working capital. Since the accounting gain or loss on the sale of a fixed asset is based on its book value, working capital is not affected by depreciation assessments.

Transactions occur in a company that change the character of the net working capital but do not affect its value. For example, a cash payment of \$10,000 for accounts payable reduces both the current asset of cash by \$10,000 and the current liability of accounts payable by \$10,000—leaving the net working capital unchanged. However, this transaction affects both the current and quick ratios.

After the balance sheet, and income statement or profit-and-loss account, the funds statement is generally regarded as the most important financial document. However, many financial managers regard a statement showing changes in cash as being of equal importance.

As with balance sheets and income statements, there is no rigid format for funds statements. These vary in the amount of detail given and also in their layout. Funds statements help management in planning for the future—for example, in timing and making financial provision for future capital expenditures.

A typical management accountant's statement for changes in working capital, and sources and applications of funds, is shown in Table IV. This is based on the relation: an increase in applications of funds equals an increase in sources of funds. The relation can also be expressed as: an increase in assets plus a decrease in liabilities equals an increase in liabilities plus a decrease in assets.

In Table IV, the cash income for the year has been absorbed in the increase in stockholders' equity, which consists of issued stock plus retained profits or income.

Inventory Control

The optimum size of inventory depends on the type of industry and on the skills available to the individual company. Inventories are high in the tobacco industry, and low in perishable-foods businesses. The larger the inventory, the larger the warehousing and associated costs. These costs include insurance, taxes, depreciation, handling and security charges, etc., and can be taken as roughly proportional to the value of the inventory, I. The annual cost, A_{IW}, of maintaining an inventory of value I is given by:

$$A_{IW} = \alpha I \quad (13)$$

where α is the proportionality factor that is on the order of 0.25 for many industries.

In contrast, some costs can be reduced as a result of larger inventories. For example, larger discounts can be obtained on bulk purchases and deliveries. In addition, larger inventories reduce the risk of losing sales and goodwill through interruptions to production and consequently running out of stocks of finished goods.

The cost of placing an order for materials is partly fixed and partly variable. The annual cost of ordering, A_{IO}, is given by:

$$A_{IO} = FN + VR \quad (14)$$

where F is the fixed cost per order, N if the number of orders per year, V is the variable cost of ordering per unit of production, and R is the annual production rate. Although the administrative cost of placing an order will be more or less fixed, the shipping costs are proportional to the size of the order and, hence, for the year are proportional to the annual production rate. The total annual cost of inventory, A_I, is the sum of Eq. (13) and (14):

$$A_I = \alpha I + FN + VR \quad (15)$$

For a given number of orders per year, N, the value of the inventory is proportional to the magnitude of the average individual order, Q. Hence, Eq. (15) can be written as:

$$A_I = \beta_Q + (FR/Q) + VR \quad (16)$$

where β is a proportionality factor, and the number of orders per year, N, has been written as R/Q.

By differentiating Eq. (16) with respect to Q, we can

XYZ Co. Statement of Changes in Working Capital, And Sources and Applications of Funds—Table IV

	Balance Sheets		Change in Funds	
	1973 Dec. 31	1974 Dec. 31	Applications	Sources
	$	$	$	$
Cash	14,575	13,000		1,575
Accounts receivable	35,575	36,000	425	
Inventories	19,750	20,750	1,000	
Total current assets	69,900	69,750		
Accounts payable	25,000	27,000		2,000
Notes payable	3,000	2,500	500	
Bank loans	3,000	2,000	1,000	
Accruals payable	250	300		50
Total current liabilities	31,250	31,800		
Net working capital	38,650	37,950		
Fixed assets	75,000	85,000	10,000	
Net assets	113,650	122,950		
Long-term loan	22,000	22,000		
Stockholders' equity	91,650	100,950		9,300
		Totals	12,925	12,925

estimate the optimum size of order. The differential is:

$$\frac{dA_I}{dQ} = \beta - \frac{FR}{Q^2} \tag{17}$$

Setting the right-hand side of Eq. (17) equal to zero yields:

$$Q = \sqrt{FR/\beta} \tag{18}$$

where Q is now the optimum size of order.

A number of models have been developed to enable managers to handle inventories in the most profitable manner.[1] These models can be applied to other elements of working capital, such as cash.

Some General Considerations

Many people regard the management of working capital as essentially a cash-flow problem. Certainly, expansion involves increased investment in fixed assets and in the various items that comprise the current assets—all of which require cash. If this leads to a shortage of cash so that a company cannot pay its liabilities, then a situation called overtrading results and the company may ultimately be forced into liquidation. Therefore, it is essential that a company have access to readily available cash, or sources of short-term financing, preferably without having to specify a particular asset as collateral, since such collateral can only be pledged to an agreed realizable value that is much less than its true value.

Accounts payable, also called trade credit, is the major source of short-term financing. Accounts payable normally amount to about 40% of the current liabilities of a manufacturing company. Such short-term financing is relatively expensive when available discounts are lost.

The second most important source of short-term financing is notes payable from the commercial banks. Banks normally require a borrower to maintain a compensating balance. For example, if a company requires a loan of $100,000, it must borrow more than this, say $120,000 (on which it pays interest), in order to maintain a minimum checking-account balance of $20,000. Commercial banks also provide a wide variety of other services that can be of great help to companies in temporary financial difficulties.

The third most important source of short-term financing is the commercial paper, or promissory notes, of large companies. This is the cheapest form of finance. However, the amount of money available is a function of the excess liquidity of the large companies at a given time, and the money may have to be repaid at relatively short notice.

Sometimes, fixed assets are purchased via short-term loans, which can lead to liquidity problems. For the most part, fixed assets should be financed from long-term or permanent capital such as stocks or bonds. The proven ability of management to handle working capital efficiently will put a company in a better position to obtain such long-term capital when required, because the confidence of bankers and stockholders will have been obtained.

The overall progress of a company will be dealt with in the next article of this series. ■

Reference

1. Weston, J. F. and Brigham, E. F., "Managerial Finance," 3rd ed., p. 471, Holt, Rinehart & Winston, London and New York, 1970.

The pros and cons of capital accretion by floating stock and borrowing money are the means for assessing the performance of a company over a period of time and for making comparisons with its competitors.

Financing Assets by Equity and Debt

F. A. HOLLAND, F. A. WATSON, and J. K. WILKINSON, University of Salford

Most engineers who go on job interviews with companies are not equipped to assess the financial and business prospects of that company. Furthermore, many engineers already working for companies have never taken the trouble to compare the performance of their own company with that of other companies in the field.

This shortsighted attitude is surprising in a group of people who normally go to great lengths to get all the available technical data before making an assessment of a project. The attitude is even more surprising when it is realized that the data are readily available to enable an engineer to assess the prospects both of his own company and those of his particular industry. Such an evaluation should be made by any engineer interested in assessing his career prospects.

There is no single indicator of company health or potential. One cannot simply take a company's temperature and make a diagnosis. However, there are a number of financial ratios that can be used to make a rapid, although relatively superficial, appraisal.

Financial Ratios

In a previous article (see Part 17), we discussed using ratios to assess the ability of a company to handle its immediate business activities. We shall now consider ratios that attempt to represent overall company performance. Probably the most commonly mentioned ratio is the profit margin, (PM), defined as:

Originally published September 1974.

$$(PM) = \left(\frac{\textit{Net Annual Profit}}{\textit{Revenue From Annual Sales}}\right)100 \qquad (1)$$

Another common ratio is the return on investment, (ROI), defined as:

$$(ROI) = \left(\frac{\textit{Net Annual Profit}}{\textit{Investment}}\right)100 \qquad (2)$$

We also discussed in a previous article* the disadvantages of rate-of-return methods for assessing profitability. In both Eq. (1) and (2), the net annual profit can either be before or after tax. It can also include interests and dividends receivable, etc.

Obviously, the net annual profit must be clearly defined before comparisons are made with other companies. Similarly the term "investment" in Eq. (2) can have a variety of meanings. The two most common ones (used when assessing the profitability of companies as opposed to projects) are total assets, and owner's equity or capital employed.[1] In the first case, Eq. (2) can be written as:

$$(ROA) = \left(\frac{\textit{Net Annual Profit}}{\textit{Total Assets}}\right)100 \qquad (3)$$

where (ROA) is called the return on assets. In the second case, Eq. (2) can be written as:

$$(ROE) = \left(\frac{\textit{Net Annual Profit}}{\textit{Stockholders' Equity}}\right)100 \qquad (4)$$

where (ROE) is the return on equity.

Asset turnover ratio, (ATR), is a commonly used

*See Part 3.

XYZ Co. Balance Sheet—Table I
(On Dec. 31, 1973)

	$	$
Assets		
Current assets	45,000	
Investments	5,000	
Net fixed assets	50,000	
Total assets		100,000
Liabilities and Stockholders' equity		
Current liabilities	25,000	
Long-term debt and liabilities	15,000	
Stockholders' equity	60,000	
Total liabilities and stockholders' equity		100,000

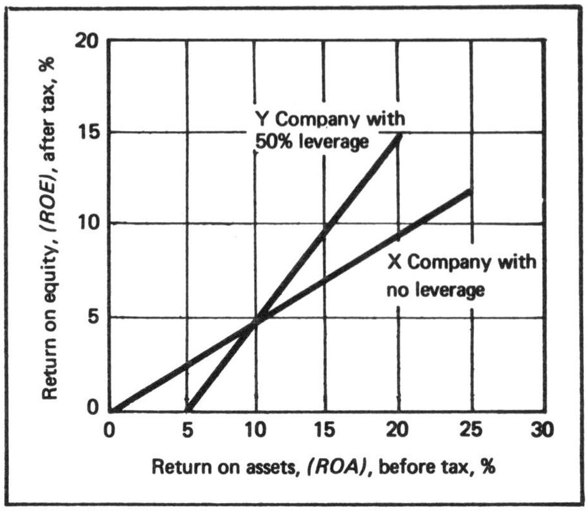

LEVERAGE affects the return on equity—Fig. 1

measure of company performance, defined as:

$$(ATR) = \left(\frac{\textit{Revenue From Annual Sales}}{\textit{Total Assets}}\right)100 \qquad (5)$$

A comparison between Eq. (1), (3) and (5) shows that:

$$(ROA) = (ATR)(PM) \qquad (6)$$

Thus, (ROA) can be improved either by increasing (ATR) or (PM). A variation of Eq. (5) is the fixed-asset turnover ratio, $(FATR)$, defined as:

$$(FATR) = \left(\frac{\textit{Revenue From Annual Sales}}{\textit{Fixed Assets}}\right)100 \qquad (7)$$

Clearly, $(FATR)$ is of less value when applied to companies that use relatively large amounts of working capital than (ATR). The $(FATR)$ is the inverse of the capital ratio, (CR), for single projects. (CR) is defined as:

$$(CR) = C_{FC}/A_S \qquad (8)$$

where C_{FC} is the fixed capital cost for a greenfields (grassroots) site, and A_S is the revenue from annual sales.

The fixed assets in Eq. (7) and those included in the total assets in Eq. (3) and (5) are usually taken at their written-down or book value, which may differ significantly from their market value. This is one disadvantage in using Eq. (3), (5) and (7).

The revenue from annual sales referred to in Eq. (1), (5) and (6) is normally taken to be the gross turnover, which includes intergroup sales. However, intergroup sales are eliminated in consolidated or group accounts. Again, revenue from annual sales must be clearly defined before comparisons are made with other companies.

Let us consider the simplified balance sheet or position statement shown in Table I. Essentially, total assets are related to liabilities and stockholders' equity by:

$$\textit{Total Assets} = \textit{Stockholders' Equity} + \textit{Total Debt} \qquad (9)$$

Eq. (9) can also be written as:

$$\textit{Stockholders' Equity} = \textit{Total Assets} - \textit{Total Debt} \qquad (10)$$

Eq. (4) and (9) can be combined to give:

$$(ROE) = \left(\frac{\textit{Net Annual Profit}}{\textit{Total Assets} - \textit{Total Debt}}\right)100 \qquad (11)$$

Eq. (11) can also be written so as to include a quantity called the debt ratio, which gives:

$$(ROE) = \left(\frac{\textit{Net Annual Profit}}{\textit{Total Assets}}\right)\left[\frac{100}{1-(DR)}\right] \qquad (12)$$

where (DR) is the debt ratio as given by:

$$(DR) = \frac{\textit{Total Debt}}{\textit{Total Assets}} \qquad (13)$$

Return on assets, (ROA), can be related to the return on equity, (ROE), by combining Eq. (3) and (12):

$$(ROE) = (ROA)/[1 - (DR)] \qquad (14)$$

(ROE) can also be related to the asset turnover ratio, (ATR), and the profit margin, (PM), by combining Eq. (6) and (14):

$$(ROE) = \frac{(ATR)(PM)}{1-(DR)} \qquad (15)$$

What are the implications for an engineer who is an employee and perhaps a stockholder of a company that has a relatively high rate of return on equity and a relatively low rate of return on assets? In order to give a meaningful answer, it is necessary to consider the nature of the financing of total assets by both debt and stockholders' equity.

Financing by Debt, or Leverage

The debt ratio, (DR), is also known as the leverage or gearing ratio. Highly levered companies have a high proportion of debt to total assets. At first glance, it may appear that the use of leverage is a simple way of increasing the return on equity, (ROE). However, interest charges have to be paid on the debt. Whether leverage is a good thing or not will depend on exactly what the interest charges are in relation to the return on assets and the return on equity.

Let us consider the simplified balance sheets of two companies, X and Y, shown in Table II. Companies X and Y have a debt (or leverage) ratio of zero and 0.5, respectively. Let us assume that the debt is of the deben-

Simplified Balance Sheets—Table II
(On Dec. 31, 1973)

X Company balance sheet

		Total debt	0
		Stockholders' equity	$100,000
Total assets	$100,000	Total liabilities and stockholders' equity	$100,000

Y Company balance sheet

		Total debt	$ 50,000
		Stockholders' equity	$ 50,000
Total assets	$100,000	Total liabilities and stockholders' equity	$100,000

Return-on-Equity After Tax—Table III

	(ROA) before tax	5%	10%	15%	20%
X Company	A_{NP}	$5,000	$10,000	$15,000	$20,000
X Company	Less tax at 50%	($2,500)	($ 5,000)	($ 7,500)	($10,000)
X Company	A_{NNP}	$2,500	$ 5,000	$ 7,500	$10,000
X Company	(ROE) after tax	2½%	5%	7½%	10%
Y Company	A_{NP} before interest	$5,000	$10,000	$15,000	$20,000
Y Company	Less interest	($5,000)	($ 5,000)	($ 5,000)	($ 5,000)
Y Company	A_{NP} after interest	0	$ 5,000	$10,000	$15,000
Y Company	Less tax at 50%	0	($ 2,500)	($ 5,000)	($ 7,500)
Y Company	A_{NNP}	0	$ 2,500	$ 5,000	$ 7,500
Y Company	(ROE) after tax	0	5%	10%	15%

ture type for tax purposes, and that the interest rate is 10% per annum. The return on equity (after tax), (ROE), is given in Table III for companies X and Y for various values of net annual profit (before tax), A_{NP}. A_{NNP} is the net annual profit (after tax). The data for Table III are plotted in Fig. 1. Fig. 1 shows that leverage has no effect on the (ROE) when the interest rate charged for the borrowed money is equal to the return on assets, (ROA), before tax. Leverage provides increased (ROE) values when the (ROA) is greater than the interest rate charged for the borrowed money, and decreased (ROE) values when it is less.

The greater the debt or leverage ratio, (DR), the more sensitive the (ROE) is to a change in (ROA), and the steeper the slope of the line in Fig. 1. Dividends to stockholders are paid out of the net annual profit (after tax), A_{NNP}, from which the (ROE) after tax in Fig. 1 is calculated. Thus, the higher the leverage, the greater the financial risk to the stockholder, and also to one's bonuses and salary increases. This risk is not the same as the business risk of the company, which is a function of its overall prospects in its particular industry. Leverage increases the return to the stockholders when the (ROA) is higher than the interest rate on debt, and decreases the return when the (ROA) is lower than the interest rate.

Whether the assets of a company are financed largely by stockholders' equity (also called net worth), or largely by debt, or some combination of the two, depends on a number of factors. If sales do not fluctuate, a company is in a good position to pay the fixed-interest charges on debt. This is also the case if the revenue from sales is steadily increasing. In this case, any new common stock issued by the company is likely to command a good price, and it also increases the attractiveness of equity financing.

The attitude of management is also an important factor in determining how much debt financing is used. In a small firm where the management owns most of the equity, management may be very reluctant to issue further amounts of common stock that would lead to a dilution of its control. Furthermore, if management has great confidence in future prospects, it will wish to ensure the maximum return for itself. In contrast, the equity in a large company is widely distributed, and the issue of further amounts of common stock has little effect on the control of the company.

So far we have not discussed exactly what constitutes equity financing and what constitutes debt financing. The difference between the two is not always clear-cut. For example, preferred stock can be classified as stockholders' equity or debt, depending on who is doing the financial analysis. We will now discuss the various components of both equity and debt financing.

Equity Financing

Typically, the company balance sheet will show the stockholders' equity and list the preferred stock, common stock and retained earnings, as in Table IV.

The issuance of common stock is the basic method of financing a company. Common stockholders take the ultimate risk in a business because they have no right to a return on their investment. However, they have the right to elect the directors of the company, who in turn are responsible for the management of the business. Stockholders are likely to vote the board of directors out if adequate dividends are not paid. Usually the liability of stockholders is limited to the nominal or par value of their stock, and hence they can lose only what they have already paid for the stock.

If the liability is not limited by law, the personal assets of the stockholders are at risk in the event of company bankruptcy, in proportion to the amount of stock held.

Preferred stock is often used as an alternative to debt when companies do not wish to issue additional common stock or to incur the fixed-interest charges required to finance debt. Preferred stockholders are not normally allowed to vote for the board of directors. They have the right to receive fixed amounts of dividends before common stockholders are paid any dividends. However, a company does not have to pay dividends. The board of directors may decide to pay little or no dividend in a particular year. Holders of cumulative preferred stock are entitled to receive compensation for the previous underpayment of dividends, when the company again pays dividends.

Common stockholders have a right to the residual assets of a company in the event of dissolution or liquidation but only after all the creditors and then any liabilities to the preferred stockholders have been paid. The larger the proportion of debt financing in a company, the smaller the amount the common stockholder is likely to receive if the company is liquidated.

Common stockholders normally have a preemptive right to the first option to purchase any additional issues of common stock. This prevents the management from using an additional issue of common stock to override the control exercised by the existing stockholders. Preemptive rights also protect the existing stockholders from having the value of their shares decreased by such dilution, since the same net earnings would be spread over more units of stock.

Let us consider the very simplified case of a company with 100,000 shares of common stock, each with a market value of $10, giving a total market value of $1 million. If a further 50,000 shares are sold at $4 each, the total market value of the 150,000 shares is $1.2 million, or $8 each. This means that the new stockholders have gained at the expense of the original ones. The preemptive right is designed to prevent this. In practice, the situation is rather more complex than is indicated here.

Both common and preferred stock normally have a par or nominal value. In the case of common stock, the market value at the time of issue usually differs from the par value. Stock can be issued either at a premium or a discount, depending on the prevailing economic conditions and the strength of the company. The difference between the actual amount paid and the par value is listed in the stockholders' equity section of the balance sheet, as shown in Table IV. The issuance of stock at a premium or a discount is done to protect existing stockholders.

In the case of preferred stock, the par value has more meaning than with common stock, since it is the amount due preferred stockholders if the company goes into liquidation—provided that this is a condition of issue.

The advantage of using common stock to finance assets is that it does not incur fixed-interest charges. Furthermore, there is no maturity date, as there is with all loans and most preference issues. Common stock can often be issued more easily than debt can be financed. However, the flotation costs for common stock can be quite high, especially when stock values are depressed, so that large discounts for the stock are needed to induce purchase.

Stockholders' equity in a company is made up of the capital contributed by the stockholders and the capital generated from retained earnings. The presence of retained earnings on a balance sheet, as shown in Table IV, does not necessarily mean that they are matched by an equal amount of cash. In fact, there may be little or no cash available.

Retained earnings can be a red herring because they may lead the unwary into thinking that the company is in a better financial position than it actually is. In practice, the retained earnings shown on a balance sheet may be largely fictitious. For example, the assets on a balance sheet may be worth less than shown, by at least the value of the retained earnings.

Stockholders' Equity as Shown in Section Of a Company's Balance Sheet—Table IV

Preferred stock, par value $100 per share, Authorized 2,000. Issued and outstanding	1,500	$150,000	
Less discount on preferred stock		($ 10,000)	
Preferred-stock equity			$ 140,000
Common stock, par value $10 per share Authorized and issued 100,000 shares	$1,000,000		
Amount paid in excess of par	$ 100,000		
		$1,100,000	
Retained earnings		$ 100,000	
Common-stock equity			$1,200,000
Total stockholders' equity			$1,340,000

Purchase and Sale of Equities

The stockholder usually requires an adequate return on his investment, and the quoted price of the stock reflects the consensus opinion of investors as to the current health of the company. Purchases or sales are normally made through stockbrokers. We will now briefly consider their function.

Most stock transactions are completed through organized security exchanges on which the stock is listed. Such exchanges have physical existence in the form of buildings located in different regions of the country. Each exchange has members who are often the nominated representatives of large brokerage firms having offices in various cities. These offices are in constant telephone and telegraph communication with the member at the exchange, passing on requests to buy or sell specified stocks. Since the broker lives by commissions and charges on transactions, he attempts to match such requests either directly or by dealings with other brokers. In the U.K., the brokers must deal through an independent 'jobber,' similar in function to a specialist broker, who quotes a low price for sales and a higher one for purchases before he knows whether the broker is buying or selling. The difference represents his margin or 'turn.' If requests to buy exceed offers for sale, the price of the stock rises until someone is tempted to sell. Conversely, if an excess of stock is offered for sale, the price is likely to fall.

It is an advantage to a company to be listed on a stock exchange since its investors can more easily sell their stock if they decide to do so. This increased liquidity makes the investor more willing to accept a lower rate of return, which effectively lowers the cost of capital to the company.

Because dealings in the stock of a listed company are published, a healthy company engenders confidence that makes it easier to obtain other forms of finance. In the absence of a regular market, stock transactions are necessarily infrequent, and prices are liable to wide fluctuation, which may make creditors wary and possibly lead to bankruptcy proceedings. Such dealings are usually referred to as 'over the counter' and are confined to the relatively few specialist brokers who hold inventories of such stock and are prepared to 'make a market' in them, or limited to private transactions.

Retained Earnings

Much confusion is caused by the practice of dividing the retained earnings under various headings such as: reserve for replacement of plant, reserve for contingencies, etc. This procedure also restricts the flexibility of management in expenditure decisions.

The amount of retained earnings, shown on a balance sheet, should not be taken as a measure of the amount of future dividends that the company is likely to pay. A contract may exist that specifies a minimum balance of retained earnings, which is then not available for dividends until bonds issued by the company have been retired.

Dividends can either be paid as cash or in the form of an additional issue of stock. A stock split is really a stock dividend, and both are used to reduce the price of stock when management considers that it is too high. A stock dividend is essentially a transfer of retained earnings to the common-stock account, and makes the amount transferred unavailable for future dividends. A stock dividend may be used in place of a cash dividend when a company is short of cash.

Debt Financing

In practice, debt financing covers a variety of fixed-income securities both long term and short term. The most common forms of long-term debt are bonds, mortgages and debentures.

A bond is simply a long-term promissory note. It is a contract established between borrower and lender in a document called an indenture. A bond indenture includes a detailed description of assets that are pledged, together with any protective clauses and provisions for redemption. A trustee is appointed to look after the interest of the bondholder. The trustee is normally a commercial bank. Bonds may be issued with a call provision that enables a company to redeem its bonds at any earlier date than scheduled. Obviously, this would be an advantage to a company in times of falling interest rates. However, a company has to pay more than the par value of the bond for this privilege. The additional amount is called the bond premium.

Sometimes a company uses a sinking fund to retire a bond issue. Sinking funds have been dealt with in an earlier article.* A series of equal annual payments, A, invested at a fractional interest rate, i, and made at the end of each year over a period of n years are equivalent to a sum of money of present value P, given by:

$$A = Pf_{AP} \tag{16}$$

where f_{AP} is the annuity present-worth factor, which is:

$$f_{AP} = \frac{i(1+i)^n}{(1+i)^n - 1}$$

Tabular values for f_{AP} are available.[2]

A company may use a sinking fund in a variety of ways but the simplest is to pay a fixed amount, A, at the end of each year to buy and retire bonds until after n years all the bonds have been retired. This annual payment may prove a significant strain on the resources of a company. Failure to make the payment could result in bankruptcy. In the case of income bonds, a company is only required to pay interest when it earns it.

A mortgage is a bond in which specific real assets are pledged as security. A senior mortgage has a prior claim on assets. A junior mortgage is normally a second mortgage on the residual value of the assets. A blanket mortgage is a pledge on all real property owned by a company.

A debenture is an unsecured bond. Strong companies are in a better position to issue debentures than weak companies since they have less need to pledge specific assets. Debenture holders are really general creditors. Subordinated debenture holders have claims on assets only after the claims of certain other claimants have been met. The issue of subordinated debentures provides a tax advantage for a company compared with the issue of preferred stock because the interest payable is a tax-deductible expense.

A financial analyst looking at a company from a potential common stockholder's point of view is likely to classify preferred stock as debt. In contrast, bondholders and general creditors are likely to regard preferred stock as additional equity. Since preferred stock is a hybrid type of security, it may be issued by a company whose management is divided over the question of whether to use equity or debt to finance additional assets. However, preferred stock does have the disadvantage that the dividends are not allowed as a tax-deductible expense.

An important factor in the financing of assets is the point at which a company is at in its life cycle. This will be dealt with in the next article of this series. ■

References

1. Abbott, J. T., Janssen, R. R. and Merz, C. M., Finance for Engineers, *Chem. Eng.*, June 25, 1973, pp. 108–114 (see this volume, p. 205).
2. Perry, J. H., Chilton, C. H. and Kirkpatrick, S. D., (Ed.), "Chemical Engineers' Handbook," 4th ed., pp. 1-32 to 1-37, McGraw-Hill, New York, 1963.

*See Part 4.

Engineering Economics—Part 19

Comparative company financial data, readily available, and your company's annual reports will enable you to predict the future prospects of your industry. Here are the necessary procedures to do so.

How To Assess Your Company's Progress

F. A. HOLLAND, F. A. WATSON and J. K. WILKINSON, University of Salford

The past and present performance of your company can give you some indications about its future. It is wrong to assume that predicting a company's future is as straightforward a matter as calculating for a single project the net present value or discounted-cash-flow rate of return that are based on fairly accurate data. Since your own prospects may depend on the future of your company, it is worth making some predictions from the available data.

Life Cycle for a Company

The life cycle of a product was referred to in Part 11 of this series. In a similar manner, a company or an industry also tends to have a life cycle,[1] which can be approximately represented by:

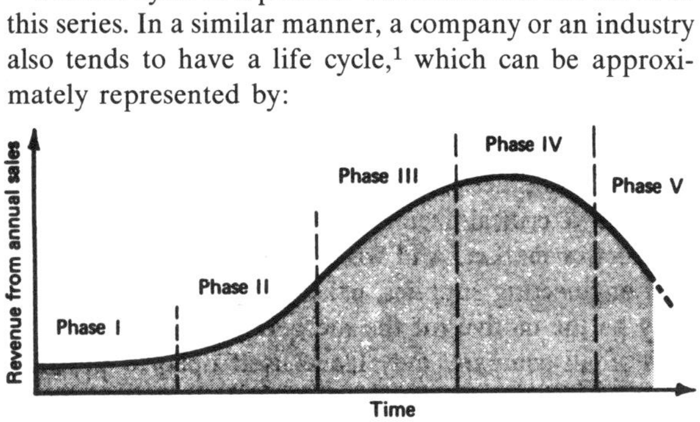

The early life of a company is represented by phase I. The growth of the company tends to accelerate in phase II, and to decline in phase III. Phases IV and V represent periods of relative maturity and decline, respectively.

This illustration, however, considerably oversimplifies the life cycles of actual companies. Furthermore, the time scales can vary widely from one industry or company to another. Some companies, particularly in the chemical industry, appear to have been very successful in prolonging phase II.

In the early phases of the life cycle of a company, assets are largely financed from stock and bond issues. As a company matures, retained earnings are increasingly used to finance assets. Finally, the annual balance-sheet depreciation charges become the main source from which to finance assets.

It is common practice to compare the performance of a company where possible with other companies in the same area of industry, e.g., agricultural chemicals, pharmaceuticals, etc. It is also preferable to compare your own company with others in the same phase of their life cycle.

Decisions as to the method of financing assets is the province of the financial manager rather than the engineer. However, engineers should make themselves as aware as possible of the implications of such financial decisions. This is not just to protect their job prospects but is also in the interests of the company. Company finance is far too important to be left entirely to financial specialists or accountants. At what point were the engineers in a blue-chip company like Rolls Royce, renowned for its brilliant engineering, aware of its impending bankruptcy? Could the existence of a significant group of financially aware engineers have averted it?

Although engineers have a professional responsibility

Originally published September 1974.

Company X Merges With Company Y on Basis of Earnings—Table I

	X Company	Y Company	Merged Company
Total assets	$100,000	$100,000	$200,000
Total debtors and liabilities	$ 30,000	$ 30,000	$ 60,000
Common-stock equity, par value $5	$ 50,000	$ 50,000	$112,500
Retained earnings and capital surplus	$ 20,000	$ 20,000	$ 27,500
Total liabilities and stockholders' equity	$100,000	$100,000	$200,000
Net worth = total assets − total debtors and liabilities	$ 70,000	$ 70,000	$140,000
Net annual profit (after tax and interest)	$ 20,000	$ 16,000	$ 36,000
Number of common stockholders' shares	10,000	10,000	22,500
Book value per share	$7	$7	$6.22
Earnings per share	$2	$1.60	$1.60
Market price per share	$42	$16	?
Price/earnings ratio	21	10	?

to give the maximum possible assistance to the management of their company, they nevertheless should keep their eyes open for any obvious risks and weaknesses in managerial decisions.

- Is your company committed to a costly development program on a new process or product that may bankrupt the company if it should fail?
- Is your company likely to be taken over for its assets, and are you and your colleagues likely to be faced with redundancy, early retirement or demotion?
- Have you and your colleagues developed expertise in an area with potential for the future? If so, your company may be taken over for its technical knowhow, and your own prospects may significantly improve.

These and other questions are worth examining in the light of available financial data.

Mergers and Consolidations

Mergers and consolidations[2] are frequently the result of efforts by companies to initiate, prolong, or return to, phase II of their life cycle. Companies of roughly equal size may merge by pooling interests, including the two managements. Alternatively, if the companies are not of equal size, the larger may purchase the smaller and take over control. One way of avoiding the disappearance of a company, as in mergers, is for the holding company to buy a controlling interest in the other company and operate it as a subsidiary. This can be done without securing the approval of the stockholders.

Mergers may be desirable for a variety of reasons. These include: (a) taking advantage of economies of scale, (b) reducing risks of various kinds, and (c) exercising greater control over the market.

Economies of scale can be effected in the areas of research, purchasing, production, advertising, sales, etc. Larger companies can afford to spend more money than smaller ones on research and development of new products and processes, essential for a company's continued existence. If one company can supply raw materials or intermediates to another in the group, the resulting vertical integration can lead to a reduction in costs without the associated risk of withdrawal of essential supplies. Sales forces can also be made stronger and more effective.

A large company that obtains its income from a relatively wide product range is usually better able than a smaller company to take risks on launching new products, and to withstand the high percentage of failures to be expected. In the event of a failure, part of the loss can be offset against income from other products, and the remainder absorbed as part of the administrative overhead. A company may well decide on a merger with another that has proven products, rather than face the risk of failure on developing its own new products.

Clearly, a larger company can exercise greater control over a market than a smaller company, provided that the former is not so big as to be regarded a monopoly and so invite government action. Market research, advertising and distribution can all be done more effectively and economically by the larger organizations, and controlled product wars can be staged to promote sales.

Mergers are sometimes used to acquire additional assets when it would be difficult to finance them from internal sources. The exchange of equity can often be done more easily than the raising of cash. Furthermore, new assets can usually be obtained more quickly and cheaply by mergers than by building them from within, especially in times of rapid inflation. Speed of acquisition may be a critical factor in the penetration of a new product or market. And with current delays in building and engineering supplies, existing plant and machinery may be the motive for the merger.

A small company may leave itself open to purchase if its net assets per share are high. It is then likely to prove attractive to a larger company seeking to improve its liquidity position. Fortunes have been made in recent years by speculators who buy a company merely to strip the easily available assets such as cash in hand or property. Such mergers usually lead to the winding up of the company, and loss of jobs.

Company X Incorporates Company Z on Basis of Market Value—Table II

	X Company	Z Company	Combined Company
Total assets	$100,000	$10,000	$187,000 (includes $77,000 goodwill)
Total debtors and liabilities	$ 30,000	$ 3,000	$ 33,000
Common-stock equity	$ 50,000	$ 7,000	$ 60,000
Retained earnings and capital surplus	$ 20,000		$ 94,000
Total liabilities and stockholders' equity	$100,000	$10,000	$187,000
Net worth = total assets − total debtors and liabilities	$ 70,000	$ 7,000	$154,000
Par value per share of common stock	$5	$2	$5
Net annual profit (after tax and interest)	$ 20,000	$ 3,500	$ 23,500
Number of common stockholders' shares	10,000	3,500	12,000
Book value per share	$7	$2	$12.83
Earnings per share	$2	$1	$1.96
Market price per share	$42	$24	?
Price/earnings ratio	21	24	?

Financial Impact of Mergers

Financial terms are ultimately the deciding factors in mergers. Let us consider the simplified, vertical, balance sheets of companies X and Y, as shown in Table I. Each company has the same breakdown of assets, stockholders' equity and liabilities, but differs in its profits and market prices. The balance sheet of a straight pooling-of-interests merger on the basis of earnings per share is shown in the last column of Table I.

Stockholders in company X receive five new shares earning $1.60 per share, for four old shares earning $2 per share, which leaves their earnings unchanged. Stockholders in company Y receive one new share earning $1.60 per share, for one old share earning $1.60 per share, which also leaves their earnings unchanged. Note that in Table I the resultant increase in common-stock equity corresponds to an equal decrease in capital surplus since this has effectively financed the issue of free stock to holders of the original shares. It is not possible to predict what the market price of the new shares will be, and this is indicated by a question mark in Table I.

Notice that a major controlling interest of $40,000 in stock of company X becomes a minority holding in the merged group. Therefore, there is often a tendency to sell some or all of the new stock after a merger. This will depress the logical market price of ($420,000 + $160,000)/22,500 = $25.78 per share, and decrease the associated price/earnings ratio below the value of $580,000/$36,000 = 16.11.

If the merger had been based on the market price per share rather than the earnings per share, the stockholders of company X would have gained in earnings at the expense of the stockholders of company Y, who would have suffered a decrease in earnings. The merger may result in a sufficiently high market price per share to more than compensate the stockholders of company Y.

Let us now consider the straight purchase of company Z by company X on the basis of market price, as shown in Table II. In contrast to company Y, company Z is much smaller than company X, having only one-tenth the amount of assets.

The current market prices of companies X and Z are $42 and $24 per share, respectively, which is a ratio of 7 to 4. Therefore, company X must give 2,000 shares of X for the 3,500 shares of Z at a market value of (2,000)($42/share), or $84,000. The purchase consideration of $84,000 exceeds the net worth or book value of company Z by ($84,000 − $7,000), or $77,000. This difference is known as goodwill. It does not mean goodwill in its normal sense but represents the additional money paid for a company that has the ability to earn higher than normal profits. Goodwill is an intangible asset and should be written off if its value is likely to decrease as a consequence of the merger.

Perhaps, company Z has a unique market for a particular product that will give company X an immediate advantage over its other competitors. However, if company X could be in the same business situation in five years without the purchase of company Z, then the goodwill should be written off over a five-year period. On the other hand, if the goodwill represents customer loyalty to company Z, which is immediately lost, the goodwill may have to be depreciated in the first year. Such a writeoff is not normally allowable for tax purposes, and will be treated as an expense.

The stockholders in company Z receive four new shares earning $1.96 per share, as shown in Table II, i.e. (4)($1.96) = $7.84, for seven old shares earning $1 per share, i.e. $7, making them better off by 12¢ per old share. In contrast, the original stockholders of company X have had their earnings decreased from $2 to $1.96 per share.

However, the data in Table II present an overly simplified picture. If the management of company X has made a shrewd purchase, then the earnings should exceed the combined earnings of the two companies, so that the stockholders of both companies X and Z will be better off. The new earnings will also have to be adjusted if the writeoff of goodwill is charged as an expense—and

Comparative Ratios for Selected U.S. Industry Groups for 1972—Table III

Row	Industry	Agricultural Chemicals	Drugs	Industrial Chemicals	Petroleum Refining	Plastic Materials and Synthetics
	Number of companies	43	67	68	51	39
Row	Ratio					
1	$\frac{\text{Current assets}}{\text{Current debt}}$, ratio	**1.92** (1.34, 3.37)	**2.78** (2.16, 3.49)	**2.59** (1.82, 3.25)	**1.68** (1.44, 2.08)	**2.28** (1.73, 3.05)
2	$\frac{\text{Net profit} \times 100}{\text{Net sales}}$, %	**2.51** (1.66, 4.21)	**6.10** (2.48, 9.31)	**3.82** (1.91, 6.29)	**5.03** (2.63, 6.89)	**4.14** (2.54, 6.40)
3	$\frac{\text{Net profit} \times 100}{\text{Net worth}}$, %	**6.77** (4.48, 11.09)	**12.76** (4.46, 20.54)	**9.71** (5.35, 14.39)	**9.86** (5.76, 12.68)	**11.30** (6.36, 18.17)
4	$\frac{\text{Net profit} \times 100}{\text{Net working capital}}$, %	**15.02** (8.39, 26.87)	**18.14** (8.97, 28.97)	**16.66** (8.00, 26.29)	**31.24** (20.49, 54.90)	**20.61** (10.85, 32.81)
5	$\frac{\text{Net sales}}{\text{Net worth}}$, ratio	**2.90** (2.14, 4.75)	**2.01** (1.59, 3.06)	**2.24** (1.70, 3.37)	**1.91** (1.45, 3.66)	**2.84** (1.99, 4.22)
6	$\frac{\text{Net sales}}{\text{Net working capital}}$, ratio	**4.50** (3.25, 7.17)	**3.19** (2.56, 4.33)	**4.34** (3.18, 6.37)	**8.16** (5.54, 11.12)	**5.21** (4.22, 7.96)
7	Collection period, days	**52** (33, 97)	**60** (71, 45)	**61** (68, 48)	**55** (81, 44)	**58** (65, 47)
8	$\frac{\text{Net sales}}{\text{Inventory}}$, ratio	**9.5** (5.7, 17.3)	**5.3** (4.2, 7.3)	**6.5** (4.7, 8.7)	**11.5** (8.9, 15.0)	**8.7** (5.9, 12.3)
9	$\frac{\text{Fixed assets} \times 100}{\text{Net worth}}$, %	**41.8** (74.3, 21.5)	**47.8** (63.8, 31.8)	**73.8** (102.3, 48.1)	**103.9** (124.4, 81.5)	**63.6** (89.9, 32.0)
10	$\frac{\text{Current debt} \times 100}{\text{Net worth}}$, %	**56.4** (124.7, 25.0)	**35.7** (61.3, 23.4)	**38.3** (66.1, 25.0)	**39.1** (55.1, 26.5)	**44.8** (67.5, 27.1)
11	$\frac{\text{Total debt} \times 100}{\text{Net worth}}$, %	**151.1** (260.8, 58.3)	**66.3** (127.0, 41.7)	**82.1** (136.7, 51.3)	**71.0** (128.7, 55.6)	**70.5** (151.2, 42.9)
12	$\frac{\text{Inventory} \times 100}{\text{Net working capital}}$, %	**49.2** (121.5, 32.1)	**62.9** (79.9, 49.8)	**73.9** (95.7, 53.0)	**73.3** (100.1, 52.6)	**65.5** (95.6, 42.8)
13	$\frac{\text{Current debt} \times 100}{\text{Inventory}}$, %	**139.4** (223.2, 109.2)	**95.8** (132.4, 66.4)	**101.0** (164.9, 77.8)	**224.7** (293.1, 156.6)	**132.9** (226.5, 84.3)
14	$\frac{\text{Funded debt} \times 100}{\text{Net working capital}}$, %	**39.2** (85.9, 11.1)	**30.9** (82.5, 9.8)	**73.2** (112.9, 47.4)	**152.2** (210.7, 88.3)	**62.6** (103.7, 24.4)

Source: *Dun's Review*, Nov. 1973. With permission of Dun & Bradstreet, Inc.
Single bold number is median value. Numbers in parentheses are lower and upper quartile values, respectively.

they cannot be offset, for example, by the sale of redundant capital assets.

Again, it is not possible to predict what the market price of the new shares will be, and this is indicated by a question mark in Table II.

Comparative Company Data

Although it is not an easy matter to predict your company's future with a high degree of accuracy, some excellent comparative data are available to help you do so. Table III gives the comparative company data that have been compiled by Dun & Bradstreet for various types of processing industries. The median value for each ratio is given—together with the lower and upper quartile values, respectively, in parentheses.

Row 1 in Table III is the:

$$\textit{Current Ratio} = \frac{\textit{Current Assets}}{\textit{Current Liabilities}} \tag{1}$$

which was discussed in Part 17 of this series.

Row 2 in Table III is the profit margin (PM):

$$(PM) = \frac{\textit{Net Annual Profit}}{\textit{Revenue From Annual Sales}} \tag{2}$$

which was discussed in Part 18. In this case, the net profit referred to is the net annual profit after tax and depreciation, A_{NNP}. The net sales is the revenue from annual

Balance Sheet for U.S. Manufacturers of Industrial Inorganic Chemicals—Table IV

Assets	%	Liabilities	%
Cash	4.3	Short-term due to banks	5.6
Marketable securities	2.2	Due to trade	14.5
Net receivables	26.2	Income taxes	3.1
Net inventory	24.6	Current maturities long-term debt	2.1
All other current assets	0.9	All other current liabilities	5.0
Total current assets	58.3	Total current debt	30.3
Fixed assets	36.3	Noncurrent debt unsubordinated	18.3
All other noncurrent assets	5.4	Total unsubordinated debt	48.6
		Subordinated debt	3.1
		Tangible net worth	48.3
Total assets	100.0	Total liabilities and stockholders' equity	100.0

Source: Abridged from "Annual Statement Studies, 1973 Edition," Copyright 1973 by Robert Morris Associates, Philadelphia, Pa. The composite figures for each industry shown in the RMA studies may not be representative of that entire industry, and they should not be automatically considered as representative norms.

Ratios for Bloggs Industrial Chemical Co.—Table V

No.	Ratio
1.	$\frac{\text{Current Assets}}{\text{Current Debt}} = 2.60$
2.	$\left(\frac{\text{Net Profit}}{\text{Net Sales}}\right)100 = 4.00$
3.	$\left(\frac{\text{Net Profit}}{\text{Net Worth}}\right)100 = 10.0$
4.	$\left(\frac{\text{Net Profit}}{\text{Net Working Capital}}\right)100 = 18.18$
5.	$\frac{\text{Net Sales}}{\text{Net Worth}} = 2.50$
6.	$\frac{\text{Net Sales}}{\text{Net Working Capital}} = 4.50$
7.	Collection Period = $\frac{\text{Accounts Receivable}}{\text{Sales per Day}} = 61$ days
8.	$\frac{\text{Net Sales}}{\text{Inventory}} = 7.14$
9.	$\left(\frac{\text{Fixed Assets}}{\text{Net Worth}}\right)100 = 74.00$
10.	$\left(\frac{\text{Current Debt}}{\text{Net Worth}}\right)100 = 35.00$
11.	$\left(\frac{\text{Total Debt}}{\text{Net Worth}}\right)100 = 65.00$
12.	$\left(\frac{\text{Inventory}}{\text{Net Working Capital}}\right)100 = 63.00$
13.	$\left(\frac{\text{Current Debt}}{\text{Inventory}}\right)100 = 100.00$
14.	$\left(\frac{\text{Funded Debt}}{\text{Net Working Capital}}\right)100 = 76.50$

Bloggs Industrial Chemical Co.—Table VI
(Balance Sheet on Dec. 31, 1973)

Assets		Liabilities and Stockholders' Equity	
Current Asssets		Liabilities	
Cash	$0.057\ A_S$	Current debt	$0.140\ A_S$
Accounts receivable	$0.167\ A_S$	Long-term debt	$0.120\ A_S$
Inventory	$0.140\ A_S$		
Total current assets	$0.364\ A_S$	Total debt	$0.260\ A_S$
Fixed assets	$0.296\ A_S$	Net worth	$0.400\ A_S$
Total assets	$0.660\ A_S$	Total liabilities and stockholders' equity	$0.660\ A_S$

sales, A_S, after deductions for returns, allowances and discounts from gross sales.

Row 3 in Table III is the return on equity (*ROE*):

$$(ROE) = \left(\frac{\text{Net Annual Profit}}{\text{Stockholders' Equity}}\right)100 \tag{3}$$

which was also discussed in Part 18. In this case, the net worth is the tangible net worth representing the sum of the preferred and common stocks, surplus and undivided profits or retained earnings, less any intangible items such as goodwill, etc.

Row 7 in Table III is the:

$$\text{Average Collection Period} = \frac{\text{Average Value of Accounts Receivable}}{\text{Revenue From Sales per Day}} \tag{4}$$

Eq. (4) was discussed in Part 17.

The funded debt (referred to in row 14) consists of mortgages, bonds, debentures, serial notes, or other obligations with maturity of more than one year from the statement data.

Robert Morris Associates[3] also compiles extensive comparative company data for various industries. In addition to ratios similar to the Dun & Bradstreet ratios, shown in Table III, Robert Morris Associates gives very useful breakdowns of assets and liabilities for various industries. Table IV shows a breakdown of assets and liabilities for U.S. manufacturers of industrial inorganic chemicals for 1972.

Application of Overall Company Ratios

One of the best ways for an engineer to develop an understanding of overall company ratios is to build up a balance sheet for a hypothetical firm, Bloggs Industrial Chemical Co. The various ratios for this firm are listed in Table V.

The balance sheet shown in Table VI has been built up from the ratios in Table V in terms of the revenue from net annual sales, A_S.

We calculate the following values for the right-hand side of the balance sheet:

From ratio 5: *Net Worth* $= A_S/2.50 = 0.4A_S$

From ratio 11: *Total Debt* $= (0.4A_S)(0.65) = 0.26A_S$

From ratio 10: *Current Debt* $= (0.4A_S)(0.35) = 0.14A_S$

$$\textit{Long-Term Debt} = \textit{Total Debt} - \textit{Current Debt}$$

$$\textit{Long-Term Debt} = 0.26A_S - 0.14A_S = 0.12A_S$$

We now calculate the following values for the left-hand side of the balance sheet:

From ratio 9: *Fixed Assets* $= (0.4A_S)(0.74) = 0.296A_S$

From ratio 1: *Current Assets* $= (0.14A_S)(2.60) = 0.364A_S$

From ratio 8: *Inventory* $= A_S/7.14 = 0.14A_S$

From ratio 7: *Accounts Receivable* $= (A_S/365)(61) = 0.167A_S$

Cash and Short-Term Investments = *Total Current Assets* − (*Inventory* + *Accounts Receivable*)

Cash and Short-Term Investments =
$0.364A_S = 0.307A_S = 0.057A_S$

In addition to the data for the balance sheet, we can calculate the net annual profit (after tax), i.e. ratio 2, to be: $A_{NNP} = 0.04A_S$.

In practice, the ratios are obtained from the information published in the balance sheet. The advantage of the above presentation for the engineer is that it relates everything to the revenue from net annual sales and, hence, underlines the importance of sales. Many engineers tend to neglect this because they consider sales to be the major unknown in process evaluation. By contrast, the marketing director will consider his sales-revenue potential to be known with reasonable accuracy as a function of selling price. He will consider production cost to be the significant unknown variable in the profit equation. This presentation simplifies the approach for the engineer, but it is dangerous to extrapolate too far from the current sales revenue.

Careful study of the ratios can produce many inferences as to the health of the company. For example, the leverage or debt ratio (DR) for this example is:

$$(DR) = \frac{\textit{Total Debt}}{\textit{Total Assets}} = \frac{0.260A_S}{0.660A_S} = 0.40$$

This value is quite low and does not present any problems of control by debtors, such as can arise when (DR) is greater than 1.

From Table VI, we can calculate the ratio for:

$$\frac{\textit{Current Debt}}{\textit{Cash} + \textit{Short-Term Investments}} = \frac{0.140A_S}{0.057A_S} = 2.45$$

Therefore, requests for early repayment by more than 40% of the debtors could be met. Hence, no liquidity problems are likely to arise, and advantage can be taken of discounts for early payment. Also, the current debt could be met by sale of the inventory, which takes $(0.140A_S/A_S)(365)$, or 51 days.

Assuming that current debtors are due for payment within 61 days, the same as the time allowed to creditors, no bankruptcy petitions are likely.

The profit of 10% (indicated by ratio 3 in Table V) will be reduced by any dividend due to preferred stockholders, because such payments are not part of fixed-debt expenses; the residue is shared among the ordinary stockholders. If all the long-term debts were in redeemable 6% preferred shares, then (from ratio 3) the net annual profit (after tax) is: $A_{NNP} = 0.10(0.40A_S)$, or $0.04A_S$. Interest due on preferred shares is $0.06(0.12A_S)$, or $0.0072A_S$. Therefore, the earnings for the ordinary shares are:

$$\frac{0.04A_S - 0.0072A_S}{0.4A_S - 0.12A_S} = 0.1171$$

This value corresponds to 11.71¢/$ of common stockholders' equity.

Assuming that available interest rates offered by banks, government, etc., for no-risk investment of capital are 10%, then the maximum economic market price of $100-stock units in Bloggs Industrial Chemical Co. is about $117. If all the debt is in bonds, etc., then the earnings on ordinary stock would be 10¢/$ of net worth, and the maximum economic price of the stock would be about $100 unless stock prices are expected to rise.

Other ratios, such as the financial ratios discussed in Part 18, can easily be deduced from those listed. For example, the return on assets, (ROA), and the asset turnover ratio, (ATR), are:

$$(ROA) = \left(\frac{\textit{Net Annual Profit}}{\textit{Total Assets}}\right)100 = \left(\frac{0.04A_S}{0.66A_S}\right)100 = 6.06$$

$$(ATR) = \left(\frac{\textit{Revenue From Annual Sales}}{\textit{Total Assets}}\right)100$$

$$(ATR) = (A_S/0.66A_S)100 = 151.5$$

The lower quartile, median, and upper quartile listed in Table III are those at which the cumulative probabilities of achieving those values are 25%, 50% and 75%, respectively. Since such curves become asymptotic to plus or minus infinity at 0% and 100% probability, the probability curves for each ratio can be sketched in by the methods of weighted assessments in a manner similar to that outlined in Part 10.

The next and final article of this series will deal with the limitations and scope of engineering economics. ■

References

1. Weston, J. F. and Brigham, E. F., "Managerial Finance," 3rd ed., p. 726, Holt, Rinehart & Winston, London and New York, 1970.
2. Bierman, H. and Drebin, A. R., "Financial Accounting," 2nd ed., p. 383, Macmillan, New York, 1972.
3. "Annual Statement Studies, 1973 Edition," Robert Morris Associates, Philadelphia National Bank Building, Philadelphia, PA 19107.

This concluding article of the series explores the limitations of discounted-cash-flow methods, reviews sources of finance and the cost of capital, and analyzes the effects of inflation.

Inflation and Its Impact On Costs and Prices

F. A. HOLLAND, F. A. WATSON and J. K. WILKINSON, University of Salford

The purpose of economic analysis is to decide how to allocate limited resources in order to obtain the maximum return. Engineers are well equipped to carry out such evaluations because they are trained to solve problems with limiting boundary conditions.

In economic analysis, money is a convenient measure of resources. However, the use of money as a measure has the disadvantage that money does not remain constant with time. First, its buying power has a tendency to decrease, which is called inflation. Second, it is capable of earning interest, so that a sum of money received today is worth more than the same sum of money received, in say, a year from now.

Until the middle of the 1950s, money was universally accepted as an invariable measure of resources. And almost all economic evaluations were done on the basis of costs and measures of profitability, as outlined in Parts 1-3 of this series. (See Table I for a diagrammatic representation of the scope of this series.)

Since then, methods of estimating profitability (based on the time value of money) have been developed, which have subsequently been extended to incorporate risk. These methods are outlined in Parts 4-11 of the series. For those interested in going beyond the statistical techniques outlined in Parts 8-10, Tocher's book[4] "The Art of Simulation," showing developments in this field, is well-worth consulting.

The basic mechanics of discounted-cash-flow methods of estimating profitability are given in Parts 5-7. Typical cumulative-cash-flow against time curves for unsuccessful, moderately successful, and highly successful projects are shown in Fig. 1. These curves are simply transients whose properties are far more familiar to the engineer than to the accountant.

In general, accountants are conservative and have been very slow to think of money as a variable measure of resources. However, a great deal of discussion is now taking place on the subject of inflation accounting.

Since company financial statements and tax systems are formulated by people trained in accountancy, top managements tend to think in that way too. It can be very disconcerting for an engineer to find that his top management does not understand his discounted-cash-flow and probability calculations. To help this matter, common accounting methods (applied to companies) have been outlined in Parts 16-18. The discussion of capital and manufacturing costs, outlined in Parts 12-15, is a useful base for the process engineer from which to tackle accounting methods.

The engineer will find a knowledge of the language of accountancy a great asset in getting his ideas over to management. However, the engineer requires perseverance and patience to appreciate what frequently appear to be arbitrary and rather illogical concepts. For example, accountants draw a distinction between capital and operating expenses. However, no such distinction is

Originally published October 1974.

drawn by engineers when they plot cumulative-cash-flow against time curves, such as is shown in Fig. 1.

Limitations of Discounted-Cash-Flow Methods

No profitability estimate is better than the inherent accuracy of the data. This is true whether the method is highly sophisticated or relatively crude. Nobody, least of all top management, is going to appreciate poor data camouflaged in a mass of sophisticated calculations.

Some years ago, a senior project manager told us about a project in his company in the early days of computers. The board of directors was not very enthusiastic about the project and questioned some of the data. When virtually the same calculations were done by the new computer, the board sanctioned the project, which subsequently turned out to be disastrous.

Experienced top managers know full well that engineers and project managers often develop a blindness to the weaknesses of a project in proportion to the work and psychological effort invested in the project. The relatively large capital expenditures nowadays required for many projects, and the evidence of some expensive failures, naturally lead top management to question the validity of the input data. An additional reason for top management to be conservatively critical is that the calculations are done by relatively junior people and the decisions are made by senior people.

The net present value (*NPV*) and the discounted-cash-flow rate-of-return (*DCFRR*) methods for estimating the profitability of future projects are based on predicted cash-flow data. In addition to the inherent technical risk in a project, there is also the risk that the chosen data are in error. Statistical and probability techniques are used in conjunction with (*NPV*) and (*DCFRR*) analyses in order to account for risk. These techniques have been outlined in Parts 8–10 and 12 of this series.

When dealing with the risk connected with the ability or inability of the analysts to forecast correctly, the probabilities used are clearly subjective. If the rate of inflation for future years could be forecast with certainty, then this could be incorporated in (*NPV*) and (*DCFRR*) estimates. However, an engineer can hardly be expected to predict that the price of a raw material such as oil will increase by 300 or 400% in the space of a few months. Clearly, such changes in the magnitude of input cost data result in serious errors in any kind of profitability estimate.

It is our belief that no single index or measurement of profitability is a substitute for a continuing and overall consideration of a project. Both (*NPV*) and (*DCFRR*) can be very helpful in ranking projects. The (*DCFRR*) is the discount rate that makes the negative and positive net cash flows equal. It is possible to have a project where you lease the plant and have only positive net cash flows. This results in a (*DCFRR*) value of infinity. Clearly, overall common sense is needed in making all assessments.

There has been criticism of (*NPV*) and (*DCFRR*) methods. Some of this has resulted from the abuse of (*NPV*) and (*DCFRR*) in validating decisions that have already been taken for other reasons. In particular, many people are questioning the value of (*DCFRR*) at the present time of rapid inflation when it is so difficult to predict future costs and prices. In the context of uncertainty about the future, we believe that estimates of the number of years to reach given discounted-cash-flow rates-of-return of say 10%, 15% and 20% are more valuable than conventional (*DCFRR*) evaluations.

The advantage of the net-present-value method is that the (*NPV*) can be calculated for each of the projects. Hence, a project that has a favorable (*NPV*) in the early years is obviously one to investigate further. Furthermore, plots of (*NPV*) against time can be done for various inflation rates. In other words, a sensitivity analysis can be carried out on the basis of inflation.

The Cost of Capital

In order to calculate the net present value, (*NPV*), of a project, net annual cash flows, A_{CF}, are normally discounted at a fractional interest rate, i, that is equivalent to the cost of capital.

In previous articles, the cost of capital has been taken as intuitively known, and indeed many companies specify what this shall be. Consequently, a project that has a positive (*NPV*) based on this value, or which has a (*DCFRR*) greater than this value, is acceptable. For the purpose of selection between alternative projects, the precise value of the cost of capital is not normally important. However, in order to accept or reject the best of the proposed projects, the value should be known with reasonable accuracy to avoid accepting loss-making activities.

There are three important categories of finance available to companies, each of which carries its own rate of interest. These are debt capital, equity capital, and retained earnings. The cost of capital is the weighted mean of these rates.

Interest on debt capital such as debentures and bank loans is usually an allowable expense for tax purposes.

New equity capital is most often raised by issue of new ordinary stock. In order to induce investment in such an issue, the return on outlay must normally be at least equivalent to that available from other sources. Although preferred stocks are strictly part of equity, they are best considered as part of the fixed interest debt.

The remaining major source of funds is retained earnings. These belong to the ordinary stockholders in the same way that distributed dividends do; but since no money has been received by the stockholder, no personal income tax has to be paid. In effect, the amount of the individual's income tax retained by the company is nominally subscribed by the tax authority. Where this is permissible, it is advantageous to finance by retained earnings rather than by raising new equity. This will also save the issue cost and any difficulties caused by increased dealing after such an issue (see Part 17 for details). For additional information on how the amounts available for investment and for distribution will grow as a function of the product of the fraction of earnings retained and the rate of return (net after corporate income tax) on these retained earnings, see Ref. 3.

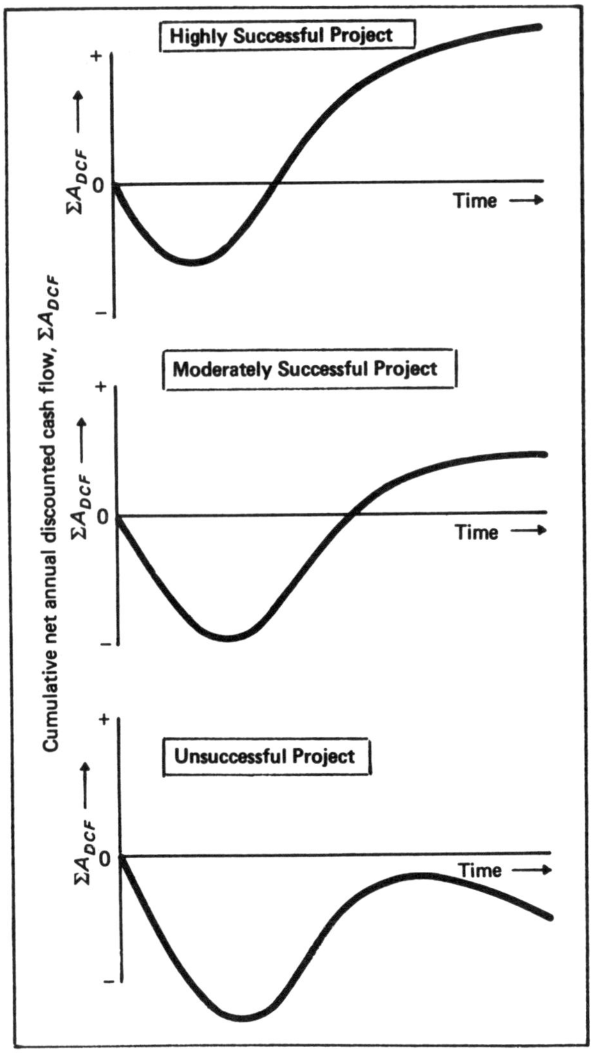

CASH-FLOW curves measure extent of payout—Fig. 1

Cost to Owner vs. Return to Investor

If earnings and dividends are both growing at a constant rate, share prices should normally grow at the same rate. The dividend and ultimate capital gain when the shares are sold will together give a (*DCFRR*) equal to the dividend (after tax) for the first year plus the rate of growth of dividends.[3] From this, it follows that the expected (*DCFRR*) is the same as the rate of return obtained (here assumed to be 15%) when all the earnings are distributed.

The (*DCFRR*) is a measure of the true interest rate before personal taxes on the capital outlay up to the time of its recovery, when both capital gains and dividends paid are taken into account. When a capital-gains tax is in force, the effective return to the investor will be reduced by an amount dependent on the rates of both capital gains and personal income taxes.

Let us suppose that the capital gains and personal income taxes are 30% and 40%, respectively, and that earnings are equivalent to 10% of the par value of the shares. If all the earnings are distributed, the investor receives (0.6)(10) = 6% net after all taxes. If all the earnings are retained and the investor keeps his stock for five years, he hopes the value of his stock will increase by the fraction $(1.10)^5 - 1$, or 0.610. Of this, 30% will be removed by the capital-gains tax, leaving 0.427 for the investor. This is equivalent to an effective annual growth rate of $(1.427)^5 - 1$, or 7.4% net after all taxes.

In the case of 50% retention of earnings, the annual growth rate will be (0.5)(10) = 5%, equivalent to 3.6% net after capital-gains tax, while the distributed earnings of (1 − 0.5)(10) = 5% represent 3.0% after income tax, for a total of 6.6% in money terms.

Such calculations establish the relationship between the cost of equity finance and the net return required by the investor after all taxes. For the above tax rates and five-year retention of stock, the linear relationships are presented in Fig. 2. For example, the cost of equity finance to provide a net return of 8.8% after all taxes is 13.1%, with 50% retention of earnings.

Inflation and Its Effect on Costs

If a particular mixture of goods that cost \$100 last year now costs \$110, then that mix has suffered a 10% rate of inflation. The purchasing power of the currency in respect to these goods has consequently fallen (\$110 − \$100)/\$110 = 0.091, or 9.1%. A nation's figures for inflation are based upon some specific mix that is chosen to represent the cost of living to the average citizen. It is this rate of inflation that is normally deducted when assessing the real cost of capital.

In countries that are largely economically self-sufficient, such as the U.S. and the U.S.S.R., inflation makes little difference to the general standard of living provided that wage rates rise at approximately the same rate as the cost-of-living index. However, during periods of rapid inflation, some groups tend to have their standard of living curtailed at the expense of improved standards of living for those powerful enough to enforce their demands for increased wage rates or prices. In a rapid inflationary situation, such as presently prevails throughout the world, such demands tend to be excessive and cause the rate of increase to accelerate.

At some stage, social conscience or an eye to electoral advantage leads to government interference on behalf of those least able to bargain for themselves. The usual approach is to attempt to redistribute wealth by changing tax rates and instituting government-financed projects. It is the effects of unpredictable changes of this kind with which we are mainly concerned here.

Another factor that affects process evaluation is differential inflation. In countries such as Switzerland, Japan and the U.K., which rely for their national wealth on upgrading imported raw materials and exporting the finished goods, sudden changes in prices of raw materials can be disastrous, especially when these increases are in no way related to the internal rate of inflation.

Modern industrial nations have achieved their standard of living over some two centuries of gradual change but some underdeveloped countries want to make that jump instantly. Those who control sources of essential foodstuffs and raw materials are currently attempting to achieve this by massive increases in the price of such goods. This adds to the inflationary spiral, and leads to corresponding increases in the cost of goods they wish to purchase.

How To Estimate the Effects of Inflation

To illustrate the relative importance of the various factors, let us consider an example:

The selling price of a particular product (reduced to a unit basis) is made up as shown:

	$	$
Manufacturing cost		
Raw materials	35	
Wages	30	
Heating and power	6	
Depreciation	2	
		73
Administrative and selling costs		
Overheads	5	
Transport	2	
		7
Production cost		80
Gross profit		20
Selling price		100

Let us assume that the current tax rates are 50% corporation tax, 40% standard-rate income tax, and 30% capital-gains tax. Also, assume that the general annual rate of inflation is 4%.

Let us assess separately the effect on sales price of a 10% increase in each of the following factors when 50% of earnings are retained: (a) manufacturing wage rates, (b) transport, heating and power costs, (c) raw-materials cost, and (d) overheads. And let us assess in a similar manner the effects of an increase in the tax rate of 10 percentage points on (e) corporation tax, (f) standard rate of income tax, and (g) capital-gains tax.

Any increase in costs results in a corresponding increase in the need for working capital. In this example, every item sold requires the investment of $80 for the period between decision to manufacture that item and receipt of cash for that item.

Factor a—A 10% increase in manufacturing wage rate increases the cost by $3. This increase must be borrowed at a cost of capital of 25% of cost (before tax). The gross profit margin is ($20/$80)(100) = 25%. Hence, the interest charge of ($3)(0.25) = $0.75. The corresponding price is, therefore, $103.75, which yields a price inflation of 3.75%.

Factor b—A 10% increase in transport, heating and power, in the same way, requires an increase of ($0.80)(1.25), or $1 in price.

Factor c—A 10% increase in raw-materials costs requires an increase in price of ($3.50)(1.25) = $4.38.

Factor d—A 10% increase in overheads requires an increase in price of ($0.50)(1.25) = $0.62.

Thus, a 10% increase in every cost that can vary after plant startup results in a price increase to:

$$\$100 + \$3.75 + \$1 + \$4.38 + \$0.62 = \$109.75$$

This is a price inflation of 9.75%.

Factor e—The increase in corporation tax from 50% to 60% will increase the required gross-profit margin. At present, this is 25% gross, or 12.5% net of the 50% corporation tax. To produce 12.5% net of a 60% corporation

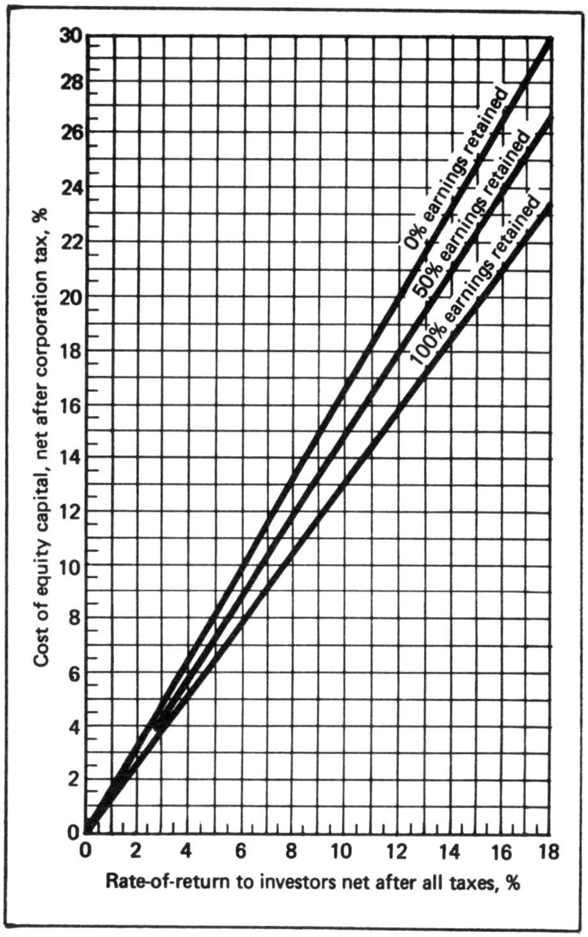

RETAINED earnings affect the cost of capital—Fig. 2

tax, a gross margin of 12.5/0.4 = 31.25% is required. The corresponding price is now ($80)(1 + 0.3125) = $105, or a price inflation of 5%.

Factor f—An increase in income tax from 40% to 50% will change the cost of both equity and retained earnings. It will be assumed, for the purpose of this example, that the capital needed will be raised by a mixture of retained earnings and 12% debt capital, with a debt ratio of 0.45.

At a debt ratio of 0.45 and debt interest of 12% (before tax), the cost of retained earnings (after corporation tax), c_r, is given by:

$$\frac{c_r + (0.45)(0.50)(12)}{1.45} = 12.5$$

$$c_r = 15.4\%$$

From Fig. 2, we find that a c_r of 15.4% for 50% retained earnings is equivalent to an investor's requirement of 10.3% net of all taxes. The change in income-tax rate will alter the slope of the lines in Fig. 2. However, by the methods used to construct Fig. 2, the cost of retained earnings to give an investor's return of 10.3% is calculated to be 16.6%. Therefore, the corresponding overall cost of capital, i_c, is 13.3% (after corporation tax), or a gross profit of 26.6%. The corresponding price required is ($80)(1.266) = $101.28—a price inflation of about 1%.

Factor g—An increase in capital-gains tax from 30% to 40% will have a similar effect on the cost of capital as an increase in income tax, since this too affects the slope of the lines in Fig. 2. To produce an investor's return of 10.3% net of all taxes, the cost of retained earnings required is 16.6%, as calculated by the method shown previously. As before, this corresponds to an overall cost of capital of 13.3%, a price of $101.28, and a price inflation of about 1%.

In this manner, the effect of any combination of inflationary increases and tax changes can readily be calculated and included into a sensitivity analysis.

For this example, a change in income tax of 10 percentage points produces a price inflation of about 1%. The same degree of inflation is produced by a 2.7% rise in wages, a 10% increase in power and allied costs, a 2.3% rise in raw-materials cost, or a 16% rise in overheads. Most of these are more likely than an increase in tax rate of 10 percentage points.

Attempts at Controlling Inflation

To restrain wage demands, governments sometimes make statutory restrictions on prices and dividends while increasing taxes such as the capital-gains tax. Every one of these decisions reduces the return to the investor. Since this does not stop all other forms of inflation, the investor must realize his capital in order to maintain his standard of living. This reduces the capital available for loan or investment, and can lead to a situation in which the government must itself invest to preserve companies from bankruptcy and to prevent increases in unemployment.

If dividends are not restricted, then every form of inflation is enhanced. If prices are controlled, then profit margins are reduced and taxation revenue is decreased. Since the cost of government is also becoming inflated, this will also require an increase of revenue, so that taxation must be disproportionately increased.

Governments are, of course, helped considerably by more people, each year, moving into higher tax brackets on the basis of higher nominal incomes. These may have no more buying power than the individuals' previous incomes that were taxed at a lower level. Governments also benefit from the greater tax on higher company profits, which may, however, be less in real terms when one considers that retained earnings are needed to replace fixed assets at inflated costs.

Sometimes, governments that employ policies of excessive deficit financing find themselves printing money without the necessary assets in gold or other surety to back the issue. Inevitably, devaluation occurs, and has the effect of making exports more competitive while making imports even more expensive. This is the classic recipe for runaway inflation as was evidenced in Germany in the 1920s and in Brazil in the 1960s. In both cases, runaway inflation led to the emergence of a government capable of enforcing control of the economy.

Brazil, where the rate of inflation was between 80 and 100% in 1963 and 1964, tackled the problems of reduced investment and a falling real growth rate of the gross national product by tying a wide variety of contracts to a cost-of-living index, so that a monetary correction was automatically made to allow for inflation.

This index is applied to things such as wages, rents, interest rates, tax rates, etc., while prices are also adjusted to offset the effects of increases in such things as raw materials. The rate of inflation has now been reduced to 15% to 20%, while permitting a real growth rate averaging about 10% over the past five years. Unfortunately, this has caused a net decrease of some 30% in the purchasing power of wages over the past decade, so that strikes are becoming commonplace in Brazil although they are forbidden by law. It is unlikely that this method of inflation control would be successful in more-advanced countries, where labor unions are more powerful.

Inflation-Accounting

Much discussion is currently taking place in the U.S. and other countries about the advisability of inflation-accounting under present conditions of rapid worldwide inflation. There is probably no simple answer to the accounting problems. It is clear that freedom to revalue assets of all kinds at will would lead to great difficulties for the tax authorities, and great uncertainty as to the real worth of a company on the part of investors.

It is unlikely that inflation-accounting could be successfully applied without governmental imposition of the rules, as in Brazil. Inflation-accounting may, however, be very useful for internal decision-making purposes, since it continually emphasizes current replacement costs of capital items.

Preparation of alternative forms of balance sheets is often too expensive unless some form of computerized accounting is used. When it is, all the necessary data are already available on tapes or cards, so that as many different balance sheets as desired can be produced in one pass. When the computer is in use, it is worth considering whether it is providing the maximum information for the time, effort and money spent.

When assessing the influence of inflation on profitability, reference should be made to the method of discounting the cash flows—first, by the cost of capital, i_c, at the base date; and then at the rate of inflation, i_i.[1] The effective discount factor, f_d, for year n is then:

$$f_d = [1 + i_c)(1 + i_i)]^{-n}$$

When this approach is valid, i_c and i_i will both be small compared to unity, so that products and powers of i_c and i_i may be neglected. In this case, inflation is taken into account by discounting at a rate $i = i_c + i_i$ when establishing (NPV) and $(DCFRR)$ values. This method is recommended for initial use, but the most promising projects should be analyzed by more-detailed methods to allow for varying inflation rates possible in each component of the selling price.

In conclusion, we believe that the implications of financial decisions are too important to be left entirely to accountants and financial analysts. It is hoped that this refresher series on engineering economics, together with our book[2] "Introduction to Process Economics," will help our fellow engineers to hold their own with the financial experts. ■

References

1. Allen, D. H., "A Guide to the Economic Evaluation of Projects," p. 13, The Institution of Chemical Engineers, London, 1972.
2. Holland, F. A., Watson, F. A. and Wilkinson, J. K., "Introduction to Process Economics," Wiley, New York and Chichester, U.K., 1974.

3. Merrett, A. J. and Sykes, A., "Capital Budgeting and Company Finance," p. 40, Longmans, Green and Co., London, 1966.
4. Tocher, K. D., "The Art of Simulation," revised impression, p. 16, The English Universities Press, London, 1967.

Meet the Authors

F. A. Holland

F. A. Watson

J. K. Wilkinson

F. A. Holland is professor and chairman of the department of chemical engineering at the University of Salford, Salford M5 4WT, England, and a partner in Salchem Associates, Consulting Chemical Engineers. Trained both as a physicist and a chemical engineer, Holland spent 14 years in the chemical process industries, nine of which were in the U.S. He has held management positions in production, process design, construction, and research and development. He has degrees from the universities of Oxford, London and Durham, and is a Fellow of the Institution of Chemical Engineers. He is also the author or coauthor of five books and more than fifty technical papers and articles.

F. A. Watson is a senior lecturer in the department of chemical engineering at the University of Salford and a partner in Salchem Associates. He has occupied posts in production, research and development with companies engaged in the manufacture of liquid and gaseous fuels and coal-tar byproducts. His primary interests are process design, digital computation and accountancy. He graduated from Manchester University in chemical engineering, and is a Fellow of the Institution of Chemical Engineers. He is also the coauthor of the book "Heat Transfer" (Heinemann, London, 1970—American Elsevier, New York).

J. K. Wilkinson is a senior lecturer in the department of chemical engineering at the University of Salford and a partner in Salchem Associates. He has held posts in the chemical industry in production, research and development. His main interests are in process design, financial accountancy and investment analysis. He holds a bachelor's and master's degree in chemical engineering from Manchester University, and a bachelor's degree in economics from London University, and is a Fellow of the Institution of Chemical Engineers. He is coauthor of "Heat Transfer."

Spotlight on Operating Cost

ECONOMIC ANALYSIS Part 1

The engineer or technical manager who knows how to evaluate operating-cost components can apply a powerful decision-making tool to such diverse problems as finding the most profitable product mix or the optimum frequency for replacing catalyst beds—as brought out in these guidelines and case studies.

I. LEIBSON and C. A. TRISCHMAN, Dart Industries Inc.

Control of day-to-day operating costs is particularly important in the manufacture of bulk materials such as chemical commodities, where cost differences of even fractions of a cent per pound can have serious effects on profitability. However, even in non-bulk areas, operating costs are coming in for close scrutiny. In fact, the present economic climate is such that the engineer and the technical manager must be able to wring every bit of information out of operating-cost data, and use it as a basis for cost-improvement decision-making.

In most plant situations, actual operating costs are reported monthly on a mill cost statement. To use this information effectively requires, first of all, an understanding of which cost elements are *variable* with production rate, as opposed to those that are *fixed* in the range concerned. Because most readers will be familiar with this basic distinction, we will just make some brief observations, and then move on to specifics.

Examples of costs that are usually variable include raw materials, certain utilities (e.g., fuel gas, water, and steam), catalysts, product additives and product packaging. The key characteristic of these cost elements when expressed as \$/mo., is that they tend to increase directly with production rate, although not necessarily in a linear relationship.

It is not uncommon to find certain of the cost elements, such as raw materials, increasing at greater than a linear rate, as the original design capacity of a unit is exceeded. This can occur due to reduced product yields at increased production rates. On the other hand, such items as steam, water, and fuel gas can increase either more or less than linearly, de-

About This New Series

The chemical engineer is continually involved in projects and decisions that have a direct impact on his firm's operating economics. To choose between alternatives, establish priorities, and sell his ideas to management, he must be able to analyze the nature and magnitude of this economic impact. To help him do this, our series will offer guidelines, case studies and other "how to . . ." information.

In the present article, we will examine the elements of a mill-cost statement, and offer specific examples on how to justify profit-improvement projects by astutely analyzing operating-cost data.

Later articles will deal with the economic justification for large and small capital projects, and will be illustrated with examples of the decision-making process as applied to plant modification, new-product commercialization, make-or-buy alternatives, and other situations that involve the economic evaluation of technical and marketing information. Various analytical tools, including decision trees, will be examined.

The case studies are designed to present a realistic picture of economic-analysis applications in today's chemical process industries. To avoid oversimplifications, some of the studies will be fairly detailed. We hope the reader will examine them carefully, because they are the best way of translating general procedures into specifics.

Originally published May 1971.

pending on the specific circumstances, as the production rate increases.

Electricity costs may contain both variable and fixed elements, depending on the cost vs. rate-of-usage structure for the particular situation involved.

Fixed costs, which can be subdivided into those that are controllable vs. those that are not, tend to be static over a fairly wide range of production rates. Examples of controllable fixed costs are operating labor, operating supplies, maintenance labor, maintenance materials, plant technical support, and quality-control laboratory.

Noncontrollable fixed costs include depreciation, real estate taxes, and insurance; also in this category is an allocation, made in most corporate accounting systems, of administrative and general expenses. Certain types of rental costs (space occupancy, etc.) are often included in this category. Noncontrollable fixed costs continue whether a plant is operating or not.

Other costs that may become part of a mill cost statement include services from a central engineering or R&D organization, as well as outside professional services (e.g., consultants). The handling of these charges is usually a function of individual corporate policy.

Accounting Systems

In the monthly reporting of manufacturing costs, one of two basic cost systems is typically used: standard costs or actual costs. In practice, there are many variants of these systems, and each company seems to have its own approach to one or the other of the two.

The *standard cost* system is generally most applicable to operations that are sufficiently standardized and reproducible from one month to the next, so that meaningful cost standards can be developed for each of the elements of the mill cost statement. In one way of using this type of system, actual costs are compared item for item with the corresponding standard costs, so as to pinpoint variances on a monthly basis. Then the significant variances are investigated to bring them into control. Another way of using standard costs involves developing a total mill cost of manufactured goods from standards, and then adjusting to actual costs via variances in a special variance report covering each cost category.

Actual cost systems are normally preferred in custom-chemical manufacturing, or in processes where the technology is changing rapidly enough to make development of meaningful cost standards very difficult. Cost variances are brought to light via comparison with earlier data or with theory-based calculations.

No matter what kind of cost system is used, accurate cost-collection systems are essential. Raw-material consumption, product-yield accounting, utilities consumption and other similar items need to be accurately monitored on meaningful time intervals so as to provide reliable input to the cost accounting system.

Operating Cost Guidelines

Let us consider guidelines on what to check in mill cost statements, and some possible pitfalls.

• *Raw Materials*—The cost of raw materials is usually a major factor in the total cost of manufacture. Improvement of product yields from raw materials can be a fertile source both for capacity-increase and cost-reduction programs. Recorded meterings of raw-materials flow, and frequent checkups on actual consumption via monitoring of inventory levels in tankage or warehouse storage, are important in providing correct usage data. Excessive consumption of individual raw materials can be pinpointed in the monthly mill-cost review if it escapes detection in day-to-day operation.

Investigation may disclose a need for: specific maintenance (such as replacement of leaking pump seals), procedural changes in starting up or shutting down equipment, technical work to provide improved material-balance closure through identification of raw-material loss-points, and improved metering.

Occasionally, a raw-material variance may be traced to a change in price or purchasing practice. Most corporate managements strive to have an approved secondary supplier to protect against uncontrollable supplier difficulties (e.g., quality, force majeure, etc.).

• *Utility Costs*—These are, of course, strongly influenced by energy costs and sources at the specific plant location. Another factor, from an accounting viewpoint, is corporate policy on how to apply overhead costs and return on investment to internal utility-producer facilities. Typical utility costs are shown below for several different regions in the U.S.; the figures are based on a 5-yr. payout for the capital investment required for corresponding internal utility-producing facilities:

Item	Gulf Coast	Mid-West	North-East
Electricity, ¢/kwh	0.6–0.8	0.7–1.2	0.8–1.2
Natural gas. ¢/1,000 std. cu. ft.	20–25	35–50	40–60
Steam, 600-psi., ¢/1,000 lb.	36	90–120	90–120
Well-water, ¢/1,000 gal.	7–12	7–12	20–30
Cooling-tower water, recirc., ¢/1,000 gal.	3–4	4–6	—

Unit costs for each of the above items may vary rather widely in the ranges cited, depending upon quantity required and plant location in a geographical sub-area. For example, in drought areas such as West Texas, water costs can be as high as 80¢/1,000 gal. As for electricity, if the power is generated from coal rather than natural gas or fuel oil, unit costs will be difficult to relate to natural-gas prices.

Steam costs are not only a function of actual plant location but also of size of boiler plant. The unit cost

* T. Weaver of Fluor Corp., and R. L. Jacks of Teller Environmental Systems Inc., provided various inputs to the table.

of steam can drop more than 30% when going from a 50,000- to a 500,000-lb./hr. generation system.

The unit cost for low-pressure steam will vary with both the method of production and evaluation; it must be calculated for each plant site and specific circumstance. Where the low-pressure steam is the result of taking backpressure steam from a turbine, the unit cost should be calculated by determining the value of the work performed by the high-pressure steam, and subtracting this from its cost.

• *Operating Labor and Supplies*—Labor requirements are a function of the particular process involved, the number of process steps and their nature (batch or continuous), the plant layout and the degree of automation. Changes in manning, or excessive use of overtime, can be pinpointed in monthly mill-cost reviews.

Operating supplies are typically about 10 to 15% of operating labor. Employee benefits range from 15 to 25% of salaries and wages, depending on company policy.

• *Maintenance Labor and Materials*—These costs typically are geared closely to investment. The range is from 3 to 7% of investment, with 5% being a good figure applicable to most situations.

Degree of plant complexity and age of the plant equipment are both important influences on total maintenance costs. Relatively simple continuous processes operate as long as two years between major scheduled maintenance turnarounds. On the other hand, batch plants handling solids, or plants operating at extremes of temperature and pressure, commonly have to undergo heavy preventive maintenance.

• *Real Estate Taxes and Insurance*—These costs are also closely related to plant investment; typically, they may total 1½% of investment, annually. Real estate valuations and tax rates strongly depend on plant location.

• *Depreciation*—In the chemical process industries, it is normal practice to depreciate capital investment on a 15-yr. base. However, in recent times, many chemical companies have adopted a 10-yr. base. Federal income tax laws are currently based on a 15-yr. depreciation allowance, tax-free; if a shorter depreciation period is used, the difference must be covered with after-tax dollars.

Depreciation is normally reported in a plant cost statement on an average or "straight line" basis. That is, the annual depreciation charge would be 1/15 of the original total investment, and would be the same year after year. However, the actual depreciation charged to the company's financial earnings statement for taxable-income reporting purposes would be on an accelerated-rate basis as prescribed by the Internal Revenue codes.

Accelerated rates do not affect the period of depreciation, but they do allow for larger depreciation charges or "writeoffs" early in the life of the project, to enable a company to more rapidly recover its capital investment dollars. The government has recently considered increasing the "early year" writeoff rates over the allowance determined from the declining balance or sum-of-the-year's digits methods* to provide added incentive for capital investment.

By using straight-line depreciation rates in routine operating-cost statements, each reporting period can be compared to others without the distorting influence of a nonuniform rate.

The depreciation account bears no real relationship to plant replacement value, due to a continuing escalation of construction costs. Furthermore, many manufacturing plants continue operating a long time after being completely written off by depreciation. However, in rapidly developing segments of technology, it is not uncommon for plants that have not yet been fully depreciated to be shut down, due to competitive pressure from larger, more-efficient ones.

It is important to note that actual cash flow to the corporate treasury is the sum of after-tax profits and depreciation.

• *Administrative and General*—This category normally includes such expenses as accounting, personnel and safety, plant security, traffic, and various supervisory salaries. Allocation costs from other company locations are sometimes included, depending on individual company accounting practices.

Now that we have defined the various cost elements in an operating plant's mill-cost statement, let us consider two typical examples of how operating costs form the basis for management decisions.

Example I: Product-Mix Optimization

A batch unit manufactures two different chemical-commodity products; it can switch from one product to the other, but cannot manufacture the two concurrently. Company sales of these products comprise less than 10% of total market size. The Marketing Group reports that equal opportunities exist for increasing the sale of either product, and the plant is currently sold out. You are a member of an economic analysis group that has been asked to recommend the optimum product mix for maximizing plant profitability.

The table below shows the pertinent factors involved in making this analysis:

	Product A	Product B
Sales price, ¢/lb.	21.0	26.0
Average freight to customers, ¢/lb.	0.8	0.5
Net sales realization (i.e., sales price less freight and allowances), ¢/lb.	20.2	25.5
Variable cost of manufacture, ¢/lb.	11.5	12.1
Onstream production rate, lb./hr.	2,200	1,600
Plant utilization, %	90.0	
Plant fixed costs, $/mo.	50,000	
Sales, R&D, misc. expenses, $/mo.	35,000	
Total nonvariable cost, $/mo.	85,000	

* For a discussion of these methods, see p. 509.

Let us calculate the profitability of Products A and B, respectively. The total nonvariable cost assignable to this operation is $85,000/mo. If the plant runs full time on Product A, 17.3 million lb./yr. would be produced (i.e., 2,200 × 24 × 365 × 0.9). The annual equivalent production rate for Product B is 12.6 million lb./yr. Distributing the total nonvariable cost* of $85,000/mo. over each individual product (as if it were the only one being manufactured in the plant), the corresponding total nonvariable cost in ¢/lb. of product is 5.9 for Product A, and 8.2 for Product B:

	Product A	Product B
Production rate, million lb./yr.	17.3	12.6
Total nonvariable cost, ¢/lb....	5.9	8.2
Variable cost, ¢/lb............	11.5	12.1
Total product cost, ¢/lb.......	17.4	20.3
Net sales realization, ¢/lb......	20.2	25.5
Profitability, ¢/lb............	2.8	5.2
Profit, pre-tax, $1,000/yr......	485	655

By subtracting total product cost from net sales realization, we can determine the profitability of each product. Thus, in the above table, A has a profitability of 2.8¢/lb., and B has a profitability of 5.2¢/lb. on a sold-out plant basis. However, A and B are produced at different rates, and this means that in order to determine the total effect on profits of substituting B for A, we must compare total profits for each product, not just profitability on a ¢/lb. basis. This comparison reveals a B-vs.-A profit advantage of $170,000/yr. before taxes (i.e., $655,000 − 485,000). Consequently, on a short-term basis, sales of Product B should be maximized.

If the plant has some idle capacity, these economics would still dictate that sales of Product B be emphasized because incremental production of B would turn out more profitable than A on an out-of-pocket cash basis. (I.e., when spare plant capacity is available, only the variable cost should be subtracted from the selling price; the product with the largest margin on this out-of-pocket basis would be the most profitable one.)

Additional Points To Study

Other important considerations in this kind of an analysis include effect of product-mix changes on quality and yield, and on maintenance and packaging expenses. Where capacity increases are involved, a careful analysis of potential plant bottlenecks is necessary, and this may reveal the need for additional facilities. The benefits of the product-mix change must be the justification for additional capital under those circumstances. Close coordination with production, marketing and various company staff groups is essential if the economic analysis is to really give the right answer.

* Including total plant fixed costs, together with the total of other operating expenses assignable to this product line.

Example II: Catalyst-Bed Utilization

For our second operating-cost example, consider a processing unit with a fixed-catalyst-bed reactor that converts a feedstock into more-valuable products.

The process engineer has an opportunity to apply operating economics in optimizing the use of the catalyst bed; let us say you are that engineer.

Your processing unit is part of an integrated refinery complex, and receives a feedstock from another unit. This feedstock is converted through a fixed catalyst-bed reactor to products A and B, both of which are consumed within the complex as blending stocks or feedstocks for other operations (see sketch). The value of the feedstock charged to your

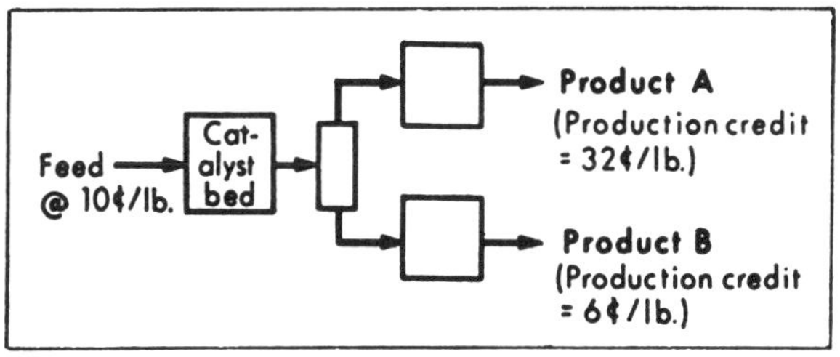

unit is set at 10¢/lb. by a refinery economics or accounting group. Your unit is credited with 32¢/lb. for producing A, and 6¢/lb. for producing B. The purpose of your unit, of course, is to maximize production of A, which has the higher value.

After one year's operation of the process unit, the reactor conversion has deteriorated from a 98% yield of Product A in the reactor effluent at the beginning of the year, to a 60% yield of A at the end of the year. The production of Product B has increased with the decline in conversion of Product A, to a point where the equipment separating B from A in the reactor effluent is loaded to capacity. Thus, a practical operating decision is made at the end of one year's operation to shut down the unit and replace the catalyst bed in the reactor.

As the process engineer on this unit, you may be asked: (1) Should the capacity of the process equipment for removing Product B be increased to allow

Catalyst-Bed Conversion History—Table I

Operating Time, Mo.	Product A Yield, %
1	98
2	97
3	95
4	93
5	91
6	90
7	88
8	85
9	80
10	75
11	70
12	60

longer reactor runs, or (2) would a higher frequency of catalyst-bed changeouts, with the resulting higher average conversion of Product A, be a more-profitable way to operate?

To answer these questions, you review the unit's operating history for the past year. The performance of the reactor catalyst bed for this operating period must be developed for an economic analysis. Table I presents the conversion data in terms of Product A's yield for the life of the first catalyst bed. You determine that the average throughput for the unit is 4,200 lb./stream-hr., with a total recovery (i.e., of A plus B) of 95% on feedstock. The unit has a 98% operating utility for the period. You learn from the Maintenance Dept. that the cost of new catalyst, labor, and equipment for changing the reactor catalyst bed is $750,000. At this price, you must determine whether changing the catalyst-bed more frequently is economically justified.

The first step is to translate the catalyst-bed conversion data in Table I into product values. You can thus determine the decline in production value commensurate with the deterioration in Product A conversion for each month of catalyst-bed operation. These values are presented in Table II, and their calculation is illustrated below:

Monthly Production (P):
Operating utility = 0.98
Overall yield = 0.95

P = 4,200 lb./stream-hr. × 730 hr./mo. × 0.98 × 0.95 = 2,854,450 lb./mo.

Product Value (Sixth Month):
Product A—2,854,450 lb./mo. × 0.90 × $0.32/lb. = $822,080
Product B—2,854,450 lb./mo. × 0.10 × $0.06/lb. = 17,130

Combined product value.............. = $839,210

The next step is to determine the profitability of the operation at a higher catalyst-bed replacement frequency than once every 12 mo., which will result in an increased average production of A. This increase must be offset by the added costs of more-frequent catalyst-bed replacements, to determine the

Product Value for the 12-mo. Catalyst-Bed—Table II

Operating Time, Mo.	Product A Yield, %	Monthly Product Value, $1,000	Cumulative Value, $1,000
1	98	898.6	898.6
2	97	891.2	1,789.8
3	95	876.3	2,666.1
4	93	861.5	3,527.6
5	91	846.6	4,374.2
6	90	839.2	5,213.4
7	88	824.4	6,037.8
8	85	802.1	6,839.9
9	80	765.0	7,604.9
10	75	727.9	8,332.8
11	70	690.8	9,023.6
12	60	616.6	9,640.2

Catalyst-Bed Changeout Costs—Table III

Catalyst replacement:	
New catalyst, labor, equipment...............	$750,000
Cost of production lost because of 48-hr. downtime:	
Product A: P* × 0.80† × $0.32/lb...........	= $48,050
Product B: P* × 0.20† × $0.06/lb...........	= 2,250
Less: Feedstock value: P × 1.00 × $0.10/lb.	= 18,770
Other var. costs: P × 1.00 × $0.006/lb.	= 1,130
Subtotal value of lost production............	$30,400
Total catalyst-bed changeout costs............	$780,400

* Where P = 4,200 lb./shift-hr. x 48 hr. x 0.98 x 0.95

† Yields for a 9-mo. catalyst-bed changeout frequency were used for this calculation. If the actual calculated frequency differs by more than a month from this pre-guess, the above values must be recalculated using the more-accurate changeout frequency.

Effect of Catalyst-Bed Changeout Frequency on Process-Unit Profits—Table IV

No. of Months Between Replacement	New Annualized Product Value, $1,000/Yr.	Less: Product Value of Base Case (12 Mo.), $1,000/Yr.	Less: Catalyst-Bed Changeout Costs, $1,000/Yr.	Added Profit, $1,000/Yr.
1	12.00 × 898.6 = 10,783.2	9,640.2	11.00 × 780.4 = 8,584.4	(7,441.4)
2	6.00 × 1,789.8 = 10,738.8	9,640.2	5.00 × 780.4 = 3,902.0	(2,803.4)
3	4.00 × 2,666.1 = 10,664.4	9,640.2	3.00 × 780.4 = 2,341.2	(1,317.0)
4	3.00 × 3,527.6 = 10,582.8	9,640.2	2.00 × 780.4 = 1,560.8	(618.2)
5	2.40 × 4,374.2 = 10,498.1	9,640.2	1.40 × 780.4 = 1,092.6	(234.7)
6	2.00 × 5,213.4 = 10,426.8	9,640.2	1.00 × 780.4 = 780.4	6.2
7	1.71 × 6,037.8 = 10,324.6	9,640.2	0.71 × 780.4 = 554.1	130.3
8	1.50 × 6,839.9 = 10,259.9	9,640.2	0.50 × 780.4 = 390.2	229.5
9	1.33 × 7,604.9 = 10,114.5	9,640.2	0.33 × 780.4 = 257.5	216.8
10	1.20 × 8,332.8 = 9,999.4	9,640.2	0.20 × 780.4 = 156.1	203.1
11	1.09 × 9,023.6 = 9,835.7	9,640.2	0.09 × 780.4 = 70.2	125.3
12	1.00 × 9,640.2 = 9,640.2	9,640.2	0.0 × 780.4 = 0.0	0.0

maximum profitable changeout frequency. From Table II, it can be determined that by replacing the catalyst bed every 6 mo., the value of production for the unit can be increased from \$9,640,200/yr. to \$10,426,800/yr. (i.e., 2 × \$5,213,400/yr.). However, the \$786,600/yr. increase in value of production must be reduced by the added costs of the additional catalyst replacement, plus the added costs associated with the loss of productivity when the process unit is shut down to replace the catalyst.

As determined previously, the Maintenance Dept. quoted you a catalyst replacement cost of \$750,000, which included new catalyst, and labor and equipment to remove the old catalyst and replace the bed. Also, you determine that the process unit downtime in terms of lost productivity must be computed, because this cost will increase directly with the catalyst replacement frequency. These costs calculations are illustrated in Table III.

When the unit is not operated for two days during the catalyst-bed changeout, the value of two days of production is lost, but feedstock charges and other variable operating expenses are not incurred. With the help of Accounting, you determine that the variable operating costs saved during this two-day outage (in terms of ¢/lb. of finished product) are:

	¢/Lb.
Operating supplies—additives, lubricants	0.2
Utilities—power, steam	0.3
Quality control—sample testing	0.1
Total	0.6

A credit for the operating manpower has not been included in the variable cost, because alternate use of this manpower is not feasible during a short, two-day outage.

Now, we are ready to determine whether an increased catalyst-bed changeout frequency will increase profitability.

The simplest and most direct approach is to calculate the added profits resulting from a catalyst-bed changeout each month, every second month, every third month, etc., for the 12-mo. operating period for which there are data. This is done in Table IV.

The annualized product value is determined by multiplying the number of catalyst-bed changeouts per year by the product value for the number of months the catalyst bed was in service (obtained from Table II). That is, for a catalyst bed changeout every 6 mo., the new annualized product value is 2 × \$5,213,400 = \$10,426,800.

To determine the added profits from a more-frequent catalyst-bed replacement mode of operation, the new annualized product value is reduced by the product value of the base operating case (catalyst bed changed every 12 mo.), and the added costs associated with the catalyst-bed changeout. Since the differences in production, feedstock cost, and operating expenses between each case have been included in the changeout cost computation, a true before-tax added profit to the current or base-case mode of operation has been determined. From Table IV, a catalyst-bed changeout every 8 mo. would be the optimum frequency to add maximum profits of \$229,500/yr. to the operation of the process unit.

Since the unit operates most profitably with an 8-mo. changeout frequency, it will not be necessary to debottleneck the Product B recovery facilities, which were loaded to capacity after a 12-mo. catalyst-bed operation.

The Examples in Perspective

Examples 1 and 2 illustrate how a chemical engineer can use economic-analysis techniques to make decisions that will result in higher profits for his plant through adjustment of operating policies, without major expenditures for capital equipment. Typically, the data affecting such decisions are available from the plant accounting staff, but the aggressive engineer must dig out and collate the information that bears specifically on his project, so that he will have all the facts needed to draw up a recommendation and justify it to management.

In the next article, we will examine one more operating-cost situation of this kind—one that will illustrate how economic-analysis techniques are used in selecting the best-suited feedstock. We will then move on to capital projects. ■

Meet the Authors

◀I. Leibson is Vice-President, Engineering and Construction Div., of Dart Industries Inc.'s Chemical Group (P.O. Box 37, Paramus, N.J. 07652). His experience includes management positions in marketing, manufacturing, R&D and commercial development both at Dart Industries and Humble Oil and Refining Co. Dr. Leibson has served as first chairman of AIChE's Petroleum and Petrochemicals Div., as well as chairman of several national committees. A Registered Professional Engineer in New Jersey and Texas, he has a B.S.Ch.E. degree from the University of Florida, and M.S.Ch.E. and D.Sc.Ch.E. degrees from Carnegie Institute of Technology.

Charles A. Trischman, Jr. is marketing manager for the ABS▶ product group at Rexene Polymers Co., a division of Dart Industries, Inc. (formerly Rexall Drug & Chemical Co.). He is located in Paramus, N.J. Since joining Rexene in 1963, he has also served as process supervisor in the firm's Odessa, Tex., polypropylene plant, and as polypropylene product-development manager for R&D. Previously, he worked for American Oil and AMOCO Chemicals in process-engineering, economic and coordinating functions. Mr. Trischman is a chemical engineering graduate of the Virginia Polytechnic Institute, a member of AIChE, and a Registered Professional Engineer.

How To Cut Operating Costs: Evaluate Your Feedstocks

Are you sure the feedstock you are now using in your process yields the highest profitability? Here is how to evaluate the effects on operating costs of alternate feedstocks, with different compositions and costs, for producing the same products.

ECONOMIC ANALYSIS Part 2

In Part 1 of this series, we pointed out that improving yields from raw materials can be a fertile area both for capacity-increase and cost-reduction programs in an existing facility.

A basic approach to such yield improvement involves making an economic evaluation of alternate feedstocks. Such an evaluation is also necessary in planning a new facility that will use hydrocarbon feedstocks, or for deciding whether changes in the feedstock supply/demand situation warrant a switch from one type of feedstock to another.

Because of the importance and prevalence of situations that require feedstock evaluation, we will present a detailed case study that concentrates on the feedstock aspects of operating costs. Although this case study involves only one type of situation, the steps would be similar for others, and the reader should thus be able to apply them to a variety of

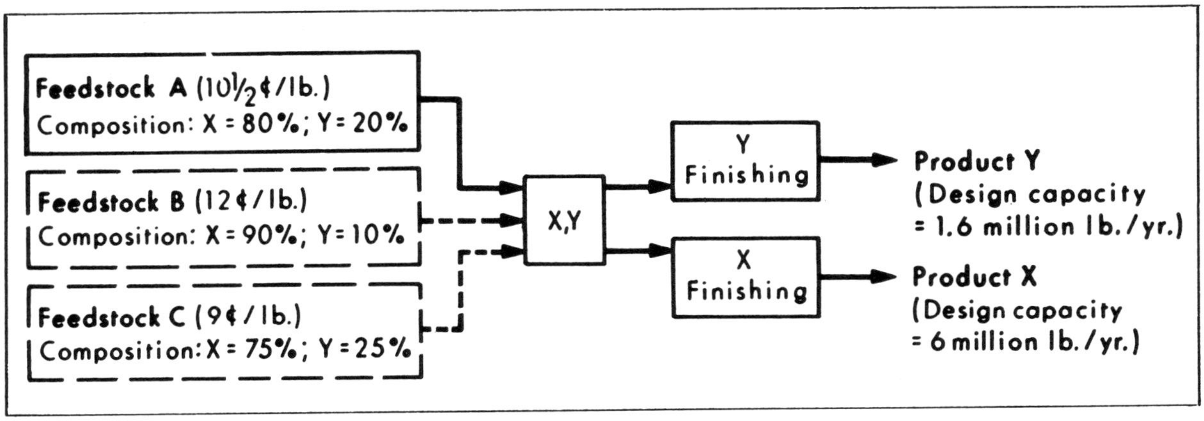

CHOOSING the "best" feedstock involves comparing profit margins for various alternatives.

Originally published June 1971.

feedstock-selection and related production-improvement problems.

Basic Data and First Steps

A process unit in your company manufactures two finished products from a single, two-component feedstock, known as Feedstock A in the diagram at left. The process unit is designed to produce 6 million lb./yr. of Product X, and 1.6 million lb./yr. of Product Y.

The Product X yield is 95% on Component X in the feedstock. The Product Y yield is 89% on Component Y in the feedstock. With this yield structure, the plant now processes 7.895 million lb./yr. of Feedstock A to obtain production of 6 million lb./yr. of Product X and 1.405 million lb./yr. of Product Y (See Table I).

The lower percentage of Component Y in the feedstock and a lower yield of Product Y than the designed basis results in the Product Y finishing section of the process unit operating at less than capacity, while the Product X finishing section operates at loaded capacity. If Product Y production could be increased from the current 1.405 million lb./yr. to the designed capacity of 1.6 million lb./yr., the additional production could be sold.

You determine that additional Product Y production is feasible if a feedstock with the proper composition of Component X and Component Y were available to balance the productions of Product X and Product Y with the respective finishing-section capacities of your process unit. The only limitation to achieving maximum production of both products is the fixed composition of Feedstock A.

You contact potential suppliers, and locate two additional feedstocks that appear to have suitable quality for your needs; Feedstock B and Feedstock C. Their composition is:

B		C	
Component X	90%	Component X	75%
Component Y	10%	Component Y	25%
Total	100%	Total	100%

The pricing structure for the current and two alternate feedstocks is:

Feedstock A (current)	10.5¢/lb.
Feedstock B	12.0¢/lb.
Feedstock C	9.0¢/lb.

Since each feedstock has a different cost, it is necessary to first determine the relative economic value of each feedstock in the process, and then the economic feasibility of achieving capacity production in the process unit with either one or a combination of two of the three feedstocks now available.

With help from your accounting department, you

Evaluating Individual Feedstocks A, B, C—Table I

	Feedstock Usage, M Lb./Yr.*		Component Content, Lb./Lb.		Product Yield Lb./Lb.		Product Yield M Lb./Yr.	Values ¢/Lb.	Values $M/Yr.
Feedstock A									
Product X	7,895	×	0.80	×	0.95	=	6,000	18.0	1,080
Product Y	7,895	×	0.20	×	0.89	=	1,405	13.0	183
Feedstock A	7,895							10.5	− 829
Variable cost							7,405	1.0	− 74
Profit margin									360
Feedstock B									
Product X	7,018	×	0.90	×	0.95	=	6,000	18.0	1,080
Product Y	7,018	×	0.10	×	0.89	=	625	13.0	81
Feedstock B	7,018							12.0	− 842
Variable cost							6,625	1.0	− 66
Profit margin									253
Feedstock C									
Product X	7,191	×	0.75	×	0.95	=	5,124	18.0	922
Product Y	7,191	×	0.25	×	0.89	=	1,600	13.0	208
Feedstock C	7,191							9.0	− 647
Variable cost							6,724	1.0	− 67
Profit margin									416

* M=1,000

Feedstock A and Feedstock C Blend To Achieve Plant Capacity—Table II

Component X: $\dfrac{6{,}000 \text{ M lb. Product X}}{0.95 \text{ lb./lb.}} = 6{,}316 \text{ M lb.}$

Component Y: $\dfrac{1{,}600 \text{ M lb. Product Y}}{0.89 \text{ lb./lb.}} = 1{,}798 \text{ M lb.}$

Total 8,114 M lb.

By simultaneous equation:
Component X: 0.80 A + 0.75 C = 6,316 M lb.
Component Y: 0.20 A + 0.25 C = 1,798 M lb.
Where A and C are quantities of Feedstocks A and C, respectively.
Component X: C = 8,421 − 1.067 A
Component Y: −(C = 7,192 − 0.800 A)

0 = 1,229 − 0.267 A

Then:
Feedstock A = 4,604 M lb., 56.7%
Feedstock C = 3,510 M lb., 43.3%

Total 8,114 M lb., 100.0%

M=1,000

determine that the value of Product X at the plant gate is 18.0¢/lb., and the value of Product Y is 13.0¢/lb. Since your ultimate goal is to increase production capacity from 7.405 million lb. to the 7.6-million-lb./yr. design capacity, the additional manufacturing cost associated with this increment of production must be determined for your economic evaluation. After consulting with your production and accounting personnel, you find that the following costs will vary directly with production rate within the range you wish to operate:

Packaging 0.5¢/lb.
Additives 0.3¢/lb.
Utilities 0.2¢/lb.

These are the variable costs (excluding the feedstock) —they will vary directly with the Product X or Product Y production rate.

Calculating Relative Profitability

The next step is to find the relative economic value of each feedstock by determining its profitability in the process unit. This is done by calculating the value of the products at the ultimate production rate of each feedstock and by subtracting from that value the cost of the feedstock and variable manufacturing costs at this production level. This resulting "profit margin" will be the potential profitability of the process unit operating on each feedstock.

The "true" operating profit for the process unit is calculated by subtracting the fixed cost from the profit margin. Fixed cost will be the same for each feedstock variation, and it is not necessary to include such cost in the calculations for relative profit value.

Such calculations for the three available feedstocks are illustrated in Table I. Composition of Feedstock A yields Product Y production of 1.405 million lb./yr. because conversion of Product X is limited to 6 million lb./yr. by capacity of the Product A finishing section.

Feedstock B yields an even lower amount of Product Y (i.e., 625,000 lb./yr.), because this feedstock contains an even lower percentage of Component Y (i.e., 10% vs. 20%) than Feedstock A. The profit margin for Feedstock B is $107,000/yr. *less* than that

Feedstock B and Feedstock C Blend To Achieve Plant Capacity—Table III

By simultaneous equation:
Component X: 0.90 B + 0.75 C = 6,316 M lb.
Component Y: 0.10 B + 0.25 C = 1,798 M lb.

8,114 M lb.

Where B and C are quantities of Feedstocks B and C, respectively.
Component X: C = 8,421 − 1.200 B
Component Y: −(C = 7,192 − 0.400 B)

0 = 1,229 − 0.800 B

Then:
Feedstock B = 1,536 M lb., 18.9%
Feedstock C = 6,578 M lb., 81.1%

Total 8,114 M lb., 100.0%

M=1,000

Profit Margins With Two Blends—Table IV

	Product Yield, M Lb./Yr.	Values, ¢/Lb.	$M/Yr.
18.9% Feedstock B Plus 81.1% Feedstock C			
Product X	6,000	18.0	1,080
Product Y	1,600	13.0	208
Feedstock B	1,536	12.0	− 184
Feedstock C	6,578	9.0	− 592
Variable cost	7,600	1.0	− 76
Profit margin			436
56.7% Feedstock A Plus 43.3% Feedstock C			
Product X	6,000	18.0	1,080
Product Y	1,600	13.0	208
Feedstock A	4,604	10.5	− 483
Feedstock C	3,510	9.0	− 316
Variable cost	7,600	1.0	− 76
Profit margin			413

M=1,000

Profitability of All the Alternatives—Table V

Feedstock	Profit Margin, \$M/Yr.	Profit Change From Base Case, \$M/Yr.
A (base case)..........	360	
B......................	253	−107
C......................	416	+56
18.9% B + 81.1% C..	436	+76
56.7% A + 43.3% C...	413	+53

M=1,000

for Feedstock A. Although this deleterious effect on the economics of the process unit by processing Feedstock B was predictable by inspection, you have now determined its exact magnitude.

Feedstock C, however, with a 25% Component Y content is capacity limited by the Product Y finishing section to 1.6 million lb./yr. production, while Product X yield falls short of the Product X finishing section capacity. The low 9.0¢/lb. cost for Feedstock C, however, results in some interesting economics. Processing Feedstock C to its ultimate capacity would result in a \$56,000/yr. increase in the profit margin of the processing unit. Although the goal of achieving capacity production in both product lines has not been achieved, you have discovered an alternate feedstock that will increase the profitability of your unit.

A Combination of Feedstocks

The next step is to find out whether a combination or blend of two available feedstocks to reach capacity production in both sections of the plant would increase the profitability of the operation over Feedstock C. Since Feedstock C contains the highest percentage of Component Y, a blend of Feedstock C with either Feedstock B or Feedstock A will give the exact Component X and Component Y content to yield capacity production in both sections of the plant.

Table II illustrates the calculations of a blend of Feedstock A and Feedstock C to obtain desired quantities of Component X and Component Y. A blend of 56.7% Feedstock A and 43.3% Feedstock C when processed at the rate of 8.114 million lb./yr. will provide 6.316 million lb./yr. of Component X and 1.798 million lb./yr. of Component Y, which will convert to 6 million lb./yr. of Product X and 1.6 million lb./yr. of Product Y, respectively.

In Table III, there are calculations for a blend of Feedstock B and Feedstock C to achieve capacity production in both products. A 18.9% blend of Feedstock B and 81.1% of Feedstock C when processed at 8.114 million lb./yr. will provide the same quantities of Component X and Component Y and the same production of Product X and Product Y as in Table II.

The profit-margin calculations for the two blends are illustrated in Table IV. A blend of Feedstock B and Feedstock C results in a profit margin of \$436,000/yr. and the blend of Feedstocks A and C results in a profit margin of \$413,000/yr. Note that variable costs are included in each calculation for the actual production rate to consistently account for the manufacturing cost change with the production rate change for each feedstock evaluation. Thus, the profit margin of \$436,000/yr. for the blend of Feedstock B and Feedstock C makes this the most profitable feedstock for the process unit.

The Answer: Blend of B and C

Comparing profit margins (Table V) for the various feedstocks and the change in profit from the base case shows that a blend of Feedstock B and Feedstock C would result in a \$76,000/yr. improvement over processing Feedstock A only, and in a \$20,000 profit improvement over the next best alternate of processing Feedstock C only. The blend of Feedstock A and Feedstock C, although more profitable than the base case, is third in the ranking of alternate feedstocks.

If, for example, the fixed cost for this unit were \$300,000/yr., the operating profit for the current mode of operation with Feedstock A only would be \$60,000/yr. By converting to the Feedstocks B and C blend, this profit would be increased by \$76,000/yr. up to \$136,000/yr. Assuming that you realistically and properly established the variable costs associated with the increased production rate, the \$76,000/yr. increase in profits is real and should be reflected in the first month's operating statement for your process unit after the change in feedstock.

Of course, the example shown here is a simplified one to highlight the calculation procedure. In an actual situation, the use of a feedstock blend may require additional equipment.

If additional handling and storage facilities are required for the feedstock blend, then the \$76,000/yr. increase in profits will serve as justification for the capital expenditures.

In other situations, the increase in production may require additional operating manpower. But this increment of labor can easily be included in the calculations. In any complete analysis, it is, of course, important to consider all the ramifications of any change in feedstock for a process.

So much for the feedstock aspects of day-to-day operating costs. In the next article, we will move on to capital projects. ■

ECONOMIC ANALYSIS

Part 3

How To Get Approval Of Capital Projects

In making modification to existing plant facilities, profit-oriented engineers must carefully analyze proposed capital expenditures so that overall profitability of the project meets management's objectives.

I. LEIBSON and C. A. TRISCHMAN, Jr., Dart Industries Inc.

More capital projects fail to be approved because of selection of an unrealistic base case than for any other single reason. What is a base case? It is what may be expected to occur if the capital expenditure under consideration is not made at all. Thus, it is a forecast of conditions expected, and involves the combined judgment of several groups. Let us consider an example.

Example 1: Cost Reduction Project

You are an engineer with responsibility for technical guidance of operations in a plant producing a chemical commodity. The management of your company has approved an expansion of a plant that came onstream 3 mo. ago, expanding its capacity from 100 to 140 million lb./yr. In a recent cost review, it was noted that recycle solvent losses have increased from a level of $20,000/mo. before the expansion to the following new levels:

Month	Equivalent Annual Production Lb./Yr.	Solvent Losses $/Mo.
Feb.	130×10^6	29,500
Mar.	135×10^6	24,000
Apr.	140×10^6	28,300

What is the base case?

You are planning a series of plant tests to determine the key sources of solvent losses. You meet with key operating and maintenance supervisors to get suggestions concerning potential solvent-loss points. Subsequently, the solvent-loss balance reveals that a reduction may be possible through equipment revisions. The bulk of the increased solvent loss appears to be occurring on the recycle solvent-recovery column. At the increased throughput, more solvent is being lost in the column bottoms as a waste stream that is incinerated.

Since the expanded plant capacity has been operative slightly more than 3 mo., a judgment is necessary regarding opportunities for improvements via changes in operating conditions (with no capital outlay). Conferences with your supervisor and appropriate operating supervisors may be useful to reach a joint decision.

You recommend a test program on the solvent-recovery column to evaluate the effect of reducing column pressure slightly, and raising reboiler heat input to drive more solvent overhead. This is accepted by all.

Your test data reveal that a 50% decrease in solvent loss is possible. However, tray revisions to permit increased vapor loading without serious instability of operation appear to be necessary in the column.

Meanwhile, two more months have elapsed. Solvent losses are: $26,800 for May at an equivalent annual production of 142 million lb./yr., and $29,200 for June at 138 million lb./yr.

What is the base case? We now have data for five months of operation at the new capacity. There is a reasonable expectation that the present production rate will be maintained through the remainder of the year. Thus, the base case consists of a continuing solvent loss of approximately $27,600/mo. at a production rate of 137 million lb./yr.

Example 1 is illustrative of the capital project that is justified primarily by cost reduction. Other cate-

Originally published June 1971.

Economic Analysis: Defining Some Terms

Companies are in business to earn a profit commensurate with the degree of risk taken in the capital investment. Let us define the key terms in this statement.

Capital is represented by anything of value having earning power that is owned by the company. A new capital investment is a sum of after-tax dollars to be spent for new equipment or plant improvements that increases total value of facilities and/or earnings. In a broad sense, a capital investment is anything not consumed during the first year of its purchase. This may include items such as supplies that become part of working capital, or even intangible assets such as patents or licenses to technology. Any cash expenditure must either be "expensed" or capitalized on the corporation's books.

Expense is represented by costs incurred in the normal course of business. Examples are: costs of production, maintenance, sales, research and other activities involved in day-to-day company operations. It is important to remember that expense dollars are normally before-tax dollars.

Profit is net earnings in a given year after all costs (including taxes) have been deducted from gross income. These dollars are available for stockholder dividends, new investments, or accumulation as cash reserves.

Degree of risk in a new capital investment may be represented by the probability of achieving the anticipated profits from the project. Investors expect profits to be commensurate with degree of risk. Speculative investments warrant higher profit-making potential.

Return on investment is the most widely used criterion for evaluating attractiveness of capital projects in the chemical process industries, at least for the early stages of evaluation.* This is the annual rate at which the net profit yields the investment expressed as a percent per year.

It is common practice from year to year to establish an internal policy for various segments of a company's business concerning the expected rate of return-on-investment for authorization of new capital projects. Often, this is a combined function of capital availability, previous record of profitability for that segment, and general business conditions for the particular market.

Typically, minimum expected after-tax rates of return-on-investment may vary from as little as 7% for low-risk, stable situations to as large as 50%, or more, for high-risk situations. Normally, a capital project with an expectation of a 10% after-tax return-on-investment will merit serious consideration by most managements.

* A subsequent article will cover the application of discounted cast flow, and other "time value of money" concepts to project evaluation.

gories of project justifications include: safety, quality improvement and capacity increase. A single capital project often may be classified in two or more categories, depending on circumstances. However, it is essential to define the base case in order to evaluate the project properly. Sufficient data must be collected to define the existing situation as well as that anticipated for the future (if no capital expenditures were made).

Scope Project Carefully

Now that we have defined the base case, the project being considered needs to be scoped so that the necessary capital investment, potential project credits, and execution schedule can be determined. Alternative solutions to the given problem need to be evaluated to determine the best project scope—not necessarily the lowest initial capital requirements.

If a modification to an existing plant is being proposed, either plant tests or pilot-plant and/or laboratory tests may be needed to provide process and product data to assess the proposed improvement. A feasibility study to guide the work as well as to pinpoint needed data may be necessary. The cases of interest are usually defined at this stage. Discussions with your supervisor, operating supervisors, and members of other groups can be helpful in providing additional ideas for the proposed project, and in critiquing cases under consideration.

There are many pitfalls to avoid in scoping projects. The most obvious is to be sure that your project scope is complete and workable. Think through startup and shutdown operations. Your design must be operable for these transient situations as well as steady-state operation. Do not make the mistake of one engineer who justified installation of two new catalyst-storage vessels, but then forgot to put money in the estimate to buy the piping and valves for hooking the tanks up to the process. Be sure that you have properly considered and specified the correct materials of construction.

If the project has primarily a capacity-increase justification, review all process steps to be certain that some other step does not become a bottleneck, and deny attainment of the proposed capacity increase. In such a situation, the justification credits must pay for the total investment actually required to achieve the capacity increase involved. It cannot be used piecemeal to apply separately to each step.

Do not fall prey to the "piggyback" situation. Occasionally, when a project shows a high return on investment, an overzealous individual may try to add other items needed for operating convenience into your scope. A careful objective review of the facts with those involved in the situation will usually suffice in confining the scope to just those items needed to secure the credits claimed.

The final scope (ready for approval) should be reviewed with operating and maintenance supervision to ensure that these groups concur in the proposal.

This is very important because new equipment or changes to existing facilities can have important effects on operating manpower, safety, operating costs, and maintenance requirements. For instance, spare parts can be minimized through equipment interchangeability. Effective plot-plan utilization in locating new equipment is important for maximizing manpower utilization and safety considerations. Operating and maintenance people are in a position to contribute heavily to the success of your project if consulted.

The scope writeup must present all of the information needed to understand what the proposal is, why it is necessary, what it is to accomplish, and how it will work vs. the base case. Project description should be accompanied by a process flow-diagram, and an equipment layout sketch. For process changes, the description should be keyed to the flow diagram so that what is being proposed and how it differs from current plant operations can be clearly understood.

Once the project scope is established, the detailed economics must be developed including capital requirements, operating costs, project credits, and return on investment.

Develop the Investment Estimate

The major equipment to be purchased should be listed for use in preparing the capital investment estimate. Many useful articles have been published concerning their preparation; we will summarize these in a future article.

At this stage, the precision of the investment estimate is a function of individual corporate policy on size of overruns and underruns, as well as budget allowance for cost of engineering for scoping capital projects. Since not every project is approved, scoping costs are a real cost element to be considered on an expense basis.

For capital projects under \$50,000, estimates are based on telephone quotations from vendors on major equipment, and factored to total installed cost. These factors are based on past experience in doing similar jobs at the same location. If time allows, written quotations may be obtained on key items based on equipment specifications submitted to vendors. However, this may not occur until after submission of the capital spending request for approval.

For larger projects, it is more usual for a definitive investment estimate to be prepared prior to final project authorization. However, a release of funds for detailed engineering is normally based on a factor-type investment estimate. The individual business situation invariably has an important influence on the approach taken.

Project Credits and Operating Cost Changes

To develop the project economics, it is necessary to determine operating costs for the situation after the project is installed vs. the base case. It is wise to establish the material balance for the base and project cases to determine the net effect on raw materials and product quantities and costs. Any change in utilities consumption or manpower requirements needs to be reflected in costs in the appropriate fashion. Furthermore, a suitable allowance must be made for depreciation, taxes and insurance. [Operating costs and their use in making management decisions were covered in the first article.] Special nonrecurring costs such as the value of downtime for installation of project equipment should be taken into account. Downtime can normally be minimized through planned prefabrication to just that needed for tie-ins to be made.

Using Example 1, let us develop the detailed project economics for installation of the necessary tray revisions to accomplish the cost reduction in solvent losses.

	Base Case	Project	Net
Production rate, 10^6 lb./yr.	137	137	—
Credits, \$1,000/yr.			
Solvent losses	332	166	166
Operating cost, \$1,000/yr.			
Fuel	274	312	38
Cooling water	88	101	13
Depreciation, 6⅔% investment (\$56,000)			3.8
Taxes and insurance, 1½% investment			0.8
Total operating cost			55.6
Net cost-reduction credits, \$1,000/yr. 166 − 55.6 =			110.4

The return on investment after federal income tax is (numbers have been rounded):

$$\frac{0.5 \times 110.4}{56} \times 100 = 100\%$$

Payout time on this investment is calculated by dividing the investment by the sum of the annual after-tax credits and the annual depreciation, or:

$$\frac{\$56,000}{\$55,200 + \$3,800} = 0.95 \text{ yr.}$$

Quality Improvement Projects

A careful forecast of market conditions is necessary to establish the base case for projects centered on quality improvement. For example, if market studies show the development of a trend toward greater product purity for a fine chemical, we dare not wait for proof in the form of business lost to competitors. In this situation, the base case becomes a forecast of potential lost business.

In some businesses, the product may be classified in quality grades for a spectrum of end-uses ranging from critical, demanding requirements to relatively broad specification uses. The sales prices often may vary as much as 50% from the most demanding to least demanding specifications.

An example of this situation may be found in the sale of low-density polyethylene for applications ranging from cable-coating grade to broad specification injection-molding compound. Price may vary as much as 5¢/lb. from one extreme to the other. In this case, quality improvement may mean an upgrading of average sales price for the product mix via improvements in the yield of the higher-quality product. Definition of the base case here depends heavily on a careful market forecast. If prices are falling preferentially for broad-specification materials due to startup of competitive new-plant capacity, then the true base case may be far less attractive than that shown by current sales statistics.

Defining the base case for a quality-improvement project can be a frustrating experience. Clear, up-to-date communications between the corporate groups (sales, engineering, production, research) are critical. Market conditions can be quite fluid, even for commodity materials. A review by all groups, prior to submittal to the approval authority, can provide improved project justifications.

Example 2: Quality Improvements

The product your company makes has an inferior color compared to a competitive product. Marketing has determined that the selling price of your product averages ½¢/lb. less than could be obtained if the product had competitive color characteristics. Research & Development has determined that a carbon-treating step in the process will remove the undesirable color bodies and produce a fully competitive product. Pilot-plant tests have established the validity of this process innovation. Management desires to proceed with the purchase and installation of a carbon-treater. Your assignment is to prepare the capital-spending-request justification.

Production capacity of the unit is 20 million lb./yr., and Marketing is selling the total capacity even with the inferior color at an average selling price of 32.0¢/lb. Marketing's premise is that the average selling price can be increased to 32.5¢/lb. Since sales are at the capacity limitation, income for the spending-request justification lies solely with the validity of the Marketing premise that production of 20 million lb./yr. can be sold at an increase in price of ½¢/lb. Annual before-tax income for the project is:

$$20 \times 10^6 \times 0.005 = \$100{,}000.$$

Expenses to be charged to the project must be determined. Engineering has estimated that the installed cost of two carbon-treating beds (to allow for continuous process operation) will be $85,000. Financial recommends the use of a 15-yr. straight-line depreciation on the new investment because the remaining economic life of the plant is estimated to exceed this. Thus, the depreciation rate will be:

$$\$85{,}000/15 = \$5{,}667/\text{yr., or } \$472/\text{mo.}$$

From the sizing of the carbon-treating vessels and from the pilot-plant carbon-treater performance data, you determine that the carbon-bed life for the production unit will be one month. The cost of new carbon to replace spent carbon will be $3,000. Maintenance predicts that about $500 in maintenance labor and equipment will be required to change the bed. Further, your calculations (based on pilot-plant input) indicate that about 2,000 lb. of product will be retained in the spent-carbon bed and cannot be recovered. This product loss must be assigned a value and included as part of the operating costs connected with the new project.

In project-justification economics, the value assigned to the lost product in the spent-carbon bed should be taken as that realized from the sale of equivalent current product, because the product could be sold at 32¢/lb. All of the costs of manufacturing the product have been incurred except the packaging and shipping costs. Thus, the company is losing the costs and profits associated with the sale of lost product. This is equivalent to net sales realization (i.e., the selling prices less the packaging and freight costs). Since average packaging cost is 0.5¢/lb. and average freight cost is 1.5¢/lb., the value of the lost product is:

Current average selling price		32.0¢/lb
Less		
Packaging cost	0.5¢/lb.	
Freight cost	1.5¢/lb.	2.0¢/lb.
Value of lost product		30.0¢/lb.

Discussions with personnel in production, laboratory and accounting indicate that other production costs such as operating labor, materials, supplies, insurance, taxes and quality control will not be significantly affected by the installation and operation of the carbon treaters.

For the carbon-treater operation requiring the change of one spent-carbon bed per month, monthly expenses to be charged against the project justification are:

	$/Mo.
Carbon cost	3,000
Maintenance cost	500
Lost product (2,000 × 0.30)	600
Depreciation	472
Total monthly expense	4,572

Hence, total annual expenses are $4,572 × 12, or $55,000.

Before-tax profit on the project is:

	$/Yr.
Income from increased market value	100,000
Operating expenses	55,000
Before-tax profit	45,000

The $45,000/yr. before-tax profit is reduced by the federal income tax at 48%. Hence, taxes are $21,600/yr. After-tax profit of $23,400/yr. divided by capital expenditure of $85,000 yields a 27.6% return on investment. Depending on your company's particular policies and financial position, this return should provide adequate incentive for management approval of capital spending.

In a market where competition is keen, products of superior quality will command a premium price. Products of inferior quality can be sold but usually at a lower price. A product with an inferior appearance compared to competitive products but performing equally well in the application can provide sufficient leverage to the purchasing agent to force price concessions from the seller. Also, superior or competitive quality products can result in lower selling costs, such as (1) fewer returns from the customer, and (2) fewer customer-service complaints.

Capacity-Increase Project

In the course of research and development activity via laboratory, pilot-plant or commercial plant-technical activity, opportunities may arise for projects leading to increase of production capacity. Definition of the base case for this type of project calls for a market forecast of price and sales—at least over the project's payout period. It is also important to pinpoint potential changes in product mix where a multiproduct grade or multiproduct production situation is involved. Often the production rates and production bottlenecks are not the same for all products. Product profitability analyses will be covered in a later article of this series.

Example 3: Incremental Capacity

One of the key objectives of a process engineer on a production unit is to determine methods for increasing capacities by removing bottlenecks. These prevent the process from operating at a production capacity that is otherwise only limited by major equipment items.

Once a new processing unit has been operated for a sufficient period of time to learn its operating characteristics, bottlenecks (e.g., undersized heat exchangers, pumps, filters, valves, etc.) to achieving higher production rates are located and defined by the process engineering group.

As a member of the process engineering group, you have determined that your production-reactor capacity is limited by heat removal in the overhead condenser. Your tests indicate that the exchanger is performing at design capacity, but reactor output could be increased by enlarging overhead-condenser capacity. Furthermore, your tests indicate that an increased reactor output could be readily handled by the other process equipment. Thus, an overall increase in plant production could be achieved by eliminating the heat-removal bottleneck. Marketing has indicated that sales can be made for the incremental production. Your task is to justify the installation of additional heat-removal capacity in the overhead condenser.

Production rate from your unit is currently 5,000 lb./hr. as limited by heat-removal capacity. The plant's operating utility is 90%. Performance tests on equipment, upstream and downstream from the reactor, indicate that the production rate could be increased to 6,000 lb./hr. if additional heat-removal capacity were installed. You investigate several methods for increasing the reactor's heat-removal capacity, and determine that the following techniques seem to be technically feasible:

Method 1. Install a new tube bundle (with finned tubes and closer tube spacing) in the existing heat-exchanger shell. This modification will provide greater heat-exchange surface at minimum cost. Installed cost of the new tube bundle would be $180,000, and the project could be completed in 8 mo. However, size of the existing shell limits the size of the modified tube bundle to a heat-removal capacity that is equivalent to a production rate of only 5,750 lb./hr.

Method 2. Purchase and install a larger heat exchanger with sufficient capacity to achieve the 6,000 lb./hr. production rate. Installed cost of the new exchanger is $375,000, and the project could be completed in 26 mo. Delivery time on the new exchanger to replace the existing one would be 18 mo. longer than the delivery time on the tube-bundle configuration.

What course of action should be recommended to management?

An examination of the economics associated with the two proposals will provide the guidance needed to decide your recommendation. Profitability of the existing operation must be developed to provide a basis for determining the incentive to increase the capacity, and to justify the capital expenditure for the new equipment.

The cost structure of the production unit must be analyzed to determine the cost elements that will increase with added production. Average operating costs for the production unit at the 5,000 lb./hr. rate are summarized in Table I.

Average Operating Cost—Table I

Cost Item	Annual $1,000/Yr.	Unit ¢/Lb.
Feedstocks	2,800	7.1
Catalyst, chemicals	315	0.8
Operating labor and benefits	550	1.4
Other supplies	80	0.2
Utilities (fuel, power, steam)	235	0.6
Maintenance	205	0.5
Quality-control laboratory	80	0.2
Administrative, technical, management	90	0.2
Depreciation, taxes, insurance	650	1.7
Total Expenses	5,005	12.7

For plant operating rate of 5,000 lb./hr. at 90% utility.

The elements of cost that will increase directly with production level must be identified. Your analysis of the operating statement, coupled with your knowledge of the operating characteristics of the unit, lead you to conclude that annual feedstock cost will increase directly with the production rate from the current 5,000 lb./hr. to the proposed 5,750 or 6,000 lb./hr. That is, unit feedstock cost will remain constant at 7.1¢/lb. Likewise, annual costs for catalyst and chemicals, and other supplies will increase directly with increased capacity.

From the consumption pattern of the utilities in your unit, and from calculations of utility consumption for major items of equipment, you determine that 50% of your current-utilities cost will vary directly with the production level over the range considered for the capacity increase. Thus, the utilities cost for the production unit will be $235,000/yr., plus 0.3¢/lb. of production over 5,000 lb./hr.

Elements of the maintenance cost will vary directly with the production rate (i.e., filter-element changes, sewer-sump cleanouts, bagging-machine repairs, reciprocating-pump rings and seal replacements, etc.). Along with maintenance personnel, you determine that about 1/5 of your maintenance costs for your unit vary directly with production rate. Hence, to base-maintenance cost of $205,000/yr., add 0.1¢/lb. for production over 5,000 lb./hr.

Analysis of operating labor and related costs, quality-control laboratory cost, and administrative, technical and management cost indicate that these expenses will not change as a result of the increase in capacity of this magnitude. Depreciation, taxes and insurance cost will increase with added investment to the production unit. These elements of cost are a function of the investment, not of the production rate, and will be determined separately.

You have now determined those elements of cost that will increase directly with the increased production rate. These variable unit costs are:

Feedstocks	7.1¢/lb.
Catalyst, chemicals	0.8¢/lb.
Other supplies	0.2¢/lb.
Utilities	0.3¢/lb.
Maintenance	0.1¢/lb.
Total	8.5¢/lb.

Thus, the cost of manufacturing additional product above the current base-cost (5,000 lb./hr.) is 8.5¢/lb.

Next, the added marketing cost for the added capacity must be determined. As some elements of the plant-operating cost remain fixed for a small increase in production, so do marketing costs. However, this requires review in each situation because some selling costs do vary directly with sales. Accounting provides you with the following unit selling costs:

Transportation, freight	1.5¢/lb.
Packaging	0.5¢/lb.
Warehousing	0.3¢/lb.
Total	2.3¢/lb.

On an annual basis, these selling costs will also increase, proportionate to the added product manufactured and sold.

Costs associated with the added investment are depreciation, taxes and insurance. For both methods of increasing the reactor's overhead-condenser capacity, accounting advises you to use 15-yr. straight-line depreciation. Since the project has been operating for 3 yr., the remaining life is 12 yr. Thus, annual depreciation for Method 1 is:

$180,000(1/12) = $15,000/yr., and for Method 2, $375,000(1/10.5) = $35,700/yr.

Taxes and insurance average ½%/yr. on investment. Note that these taxes are local property tax, not federal income tax. Added depreciation, taxes and insurance associated with proposed expansion methods are summarized in Table II.

The economics of the two proposed expansion methods can now be determined. Average selling price of the product is 14.0¢/lb. The incremental profit resulting from the capacity increase is summarized in Table III.

The incremental profit determined in Table III is additive only to the base-case operation at the current rate of 5,000 lb./hr. Therefore, it is necessary to determine only the added income and expense attributable to the capacity added in Method 1 or 2.

The justification for expenditure of $180,000 for capital equipment in Method 1 is $173,000/yr. before-tax profit, and for expenditure of $375,000 in Method 2 is a $216,000/yr. before-tax profit.

Now, these values must be converted into the standardized "payout" or "return on investment" for comparison to your company's yardstick for approving capital expenditures, and to determine which method of expansion has the best justification.

Payout time is found by dividing investment by cash flow, and is for:

Method 1: $180,000/$105,000 = 1.7 yr.
Method 2: $375,000/$148,000 = 2.5 yr.

The simple rate of return-on-investment is found by dividing the annual after-tax profit by the investment. This becomes for:

Method 1: ($90,000/$180,000)100 = 50%
Method 2: ($112,000/$375,000)100 = 30%

Payout times of 1.7 and 2.5 yr. for Method 1 and 2, respectively, are sufficiently low to justify either capacity expansion by most corporations' standards. Likewise, return on investment of 50% and 30% for Method 1 and 2, respectively, are sufficiently high to justify either one. These very favorable justification rates are not abnormal for incremental expansions of this type in the process industries.

Does it necessarily follow that Method 1 is the

Added Depreciation, Taxes And Insurance Cost—Table II

	Method 1	Method 2
Added investment, $	180,000	375,000
Depreciation, $/yr	15,000	35,700
Taxes, insurance (½% of investment), $/yr	900	1,875
Total, $/yr	15,900	37,575

obvious solution, and would provide upon completion the highest economic benefit to the company?

When there are alternate schemes to accomplish the same objectives, the time factor for implementing each becomes important. Installation of the new tube-bundle configuration (Method 1) is anticipated to be in operation 8 mo. after the decision point. The new heat exchanger (Method 2) is anticipated to be in operation 26 mo. after the decision point. Since Method 1 could be generating profits for 18 mo. longer than Method 2, this difference must be considered in the justification analysis.

The cash-flow effects for each method are presented in Table IV. The annual cash-flows generated by each method are the sums of the after-tax profit plus the depreciation. These are listed in Table III.

In Table IV, the cash outlay of $180,000 for the tube-bundle configuration is selected as time "0." This represents the completion date for installation of the tube-bundle and payment to the contractor. Installation of the new heat exchanger is complete 18 mo. later (middle of second year after time "0") with payment to the contractor of $375,000.

Method 1 accumulates cash-flow input credits at the rate of $105,000/yr., which after payback of the initial cash outlay becomes $1,080,000 at the end of project life. Starting from the same decision point, Method 2 does not start cumulating cash-flow inputs until 26 mo. later. At an input of $148,000/yr., Method 2 accumulates $1,179,000 by the end of the project life. Thus, Method 2 would earn $99,000 more than Method 1 for the total project life.

Incremental Profit From New Capacity—Table III

	Method 1	Method 2
Added production, lb./hr........	750	1,000
Annual added production at 90% operating utility, lb.......	5,913,000	7,884,000
Sales at 14¢/lb., $..............	828,000	1,104,000
Less		
Variable selling costs at 2.3¢/lb., $................	136,000	181,000
Variable operating costs at 8.5¢/lb., $................	503,000	670,000
Added depreciation, taxes, insurance, $..........	16,000	37,000
Incremental profit to base-case operation, $......	173,000	216,000
Less federal income tax at 48%, $..................	83,000	104,000
After-tax profit, $................	90,000	112,000
Depreciation, $..................	15,000	36,000
Cash flow, $.....................	105,000	148,000

Method 1 proposal is to install new tube-bundle in existing shell.
Method 2 proposal is to install new heat exchanger of larger capacity.

Cash-Flow Projections for Proposed Expansions—Table IV

	Method 1		Method 2	
Time, Yr.	Annual, $1,000/Yr.	Cumulative, $1,000	Annual, $1,000/Yr.	Cumulative, $1,000
0*	−180	−180	—	—
1	105	−75	—	—
2†	105	30	−375 +74	−301
3	105	135	148	−153
4	105	240	148	−5
5	105	345	148	143
6	105	450	148	291
7	105	555	148	439
8	105	660	148	587
9	105	765	148	735
10	105	870	148	883
11	105	975	148	1,031
12	105	1,080	148	1,179

* Time zero is 8 mo. after decision point.
† Heat exchanger for Method 2 is operational at 1½ yr.

By this cash-flow analysis, Method 2 would appear to return the higher economic benefit. Considering the very acceptable 30% return-on-investment for Method 2, this choice would seem appealing.

However, the cash-flow analysis in Table IV does not reflect the time value of money. That is, the value of future dollar earnings has not been discounted to present-value dollars at an interest rate acceptable to your company. A future article in this series will cover this concept, and will illustrate present-worth and discounted-cash-flow calculations for this example.

When the present worths of the cash flows in Table IV are determined at a 10% interest rate, Method 1 earns $49,000 more than Method 2 over the project's life. Cash earnings for Method 1 are available earlier and are subject to less discount than those from Method 2 that starts 18 mo. later.

Project Scheduling

A realistic execution schedule in the form of an arrow or block diagram should be included in the capital-spending request. This schedule should be referenced to project-approval date. The elapsed calendar time, as well as man-weeks of engineering for each phase, should be shown. The items to be covered include: (1) detailed engineering to prepare equipment specifications, engineering flowsheets, plot-plan layout and structural drawings; (2) materials procurement and delivery; (3) field construction; and (4) equipment startup.

Periodic progress reports are helpful in providing communication with other groups on job status. A smoothly executed job with a fast, easy startup is a very rewarding experience that builds management confidence in your ability, and paves the way for getting your next capital project approved. ■

Avoiding Pitfalls in Developing a Major Capital Project

ECONOMIC ANALYSIS Part 4

While emphasizing step-by-step economic analysis, the authors also offer practical guidelines on the various process-development, market-research and other technical/commercial activities that are vital to the success of new manufacturing ventures today.

IRVING LEIBSON and CHARLES A. TRISCHMAN, Jr., Dart Industries, Inc.

Having dealt with the modification of existing facilities in Part 3 of this series, we are now ready to take up a topic that brings the full scope of economic-analysis activities into play: the development of a multimillion-dollar capital project for either the manufacture of a new chemical or a major expansion of capacity for an existing product.

This type of project embraces many different research, development and design activities on the technical side, and many market-research, sales-development and financial-planning activities on the commercial side. The link that provides criteria and checkpoints for these activities is economic analysis. And we shall see that taking a market-oriented approach to these activities is a key requirement in avoiding pitfalls.

We will examine the goals and techniques of economic analysis at various stages of a project, offer checklists and other guidelines, and discuss the interplay between economic analysis and the work of the researcher, pilot-plant operator, designer and marketing specialists.

Timing, Revenue and Profit Relationships

The development of a new chemical process from the idea stage to commercialization requires from five to seven years for most companies. Any reduction in this required development time can result in large credits in terms of a more rapid buildup of product sales, a higher initial sales price (for new chemicals), and the capturing of a larger share of the total market. However, if a new venture is commercialized prematurely, without proper attention to any of the key marketing or technology factors, the penalties can be very high.

As for expansions of capacity on existing product lines, timing is extremely important in realizing the desired return on investment; delay can give the jump to competitors and boost construction costs, whereas undue haste may mean tieing up expensive capital in underutilized, unprofitable facilities—particularly if the excess capacity depresses the product's sales price.

Market research and market development must therefore start early enough in the project to provide preliminary forecasts (as a function of potential sales price), to guide project profitability studies, and to help gage the desirability of early commercialization. An error in estimating the gross sales income (i.e., either product sales price or sales volume) is usually far more damaging to profits than an error in estimating plant investment.

For example, suppose the XYZ Co. were to embark on a chemical project requiring an investment of \$10 million, with a total annual operating cost equal to 30% of investment (excluding raw materials). Let us assume that it would require an annual gross-sales income of \$10 million to earn a 20% annual rate of return after 50% federal income taxes—i.e., that

Originally published August 1971.

the total cost of sales is \$6 million/yr., and the aftertax profit for the project is \$2 million/yr. If the *investment* were underestimated by 20% (equivalent to an actual investment of \$12 million), the annual rate of return (R) after taxes on the project would be reduced only moderately—from 20% to 14.2%, with a reduction in annual aftertax profit from \$2 million to \$1.7 million, i.e.: \$4 million pretax profit, less 0.30 × \$2 million extra investment = \$3.4 million pretax profit; therefore:

$$R = [(3.4 \times 0.50)/12] \times 100 = 14.2\%$$

However, if the 20% estimating error involved *selling price* rather than investment—i.e., if the actual selling price were 20% lower than estimated, resulting in gross sales income of \$8 million/yr. rather than \$10 million/yr.—then the corresponding project return would drop to one-half of the target estimate, and the corresponding aftertax profit would decrease from \$2 million to \$1 million annually, and the annual rate of return would be a slim 10%.

Errors involving *sales volume* also tend to have a greater effect than errors involving investment (the exact effect will depend on the breakdown between fixed and variable operating costs—the higher the fixed costs, the greater the effect). Hence the need for thorough market analysis early in the development of a project.

Evaluating New Project Ideas

Ideas for new chemical projects can originate in any department of a company. Many are inspired by discoveries made in exploratory research, data revealed in literature and patent references, information made available from trade sources, market-research activities, and consultants.

The general areas for exploratory research are dictated to a large extent by the company's raw-material position, its strength in existing businesses, and by market growth patterns. Although one cannot set a time schedule for making new discoveries in exploratory research work, such discoveries have often led to ideas for a new group of chemical products or a new process for the production of an established commodity material. The development of nylon at Du Pont, and butyl rubber at Standard Oil Co. of New Jersey, are examples of this type of breakthrough.

Ideas for new projects are also inspired by product-application study groups in the course of their evaluations of potential end-uses for new chemical products. Market-research studies play an important part in highlighting customer needs and pinpointing those areas in which laboratory research may be profitable, as well as determining those commodity chemicals having the greatest growth potentials.

An economic evaluation of each new idea is a must in securing maximum results from R&D expenditures. Screening economics can be based on a comparison of long-range total raw-materials costs with potential product price. Normally, it is necessary for the total raw-material cost to be less than 40% of the potential product price for commercialization to be attractive. Raw-material position is very important for products in which raw materials represent more than 30% of the total manufacturing cost.

For example, let us say you are an engineer in a long-range planning section responsible for economic screening of new research ideas. Interest has been expressed in developing a process for manufacture of Product A. Known synthesis techniques for this material require the use of materials B and C. The current market price of material A is 30¢/lb.; the cost of B is 20¢/lb., and of C is 12¢/lb. In the synthesis, 0.8 lb. of B plus 0.2 lb. of C yield 1 lb. of A. Further, there is a nonrecoverable loss of 0.1 lb. of B and 0.05 lb. of C for each pound of A produced. Thus, the raw-material cost of producing A is:

$$(20 \times 0.8) + (12 \times 0.2) + (20 \times 0.1) + (12 \times 0.05) = 16 + 2.4 + 2 + 0.6 = 21.0¢/\text{lb.}$$

It is suggested that A may be produced via a new synthesis technique using D and E instead of B and C, and that 0.6 lb. of D plus 0.4 lb. of E will be required to produce 1 lb. of A. An allowance of a 10% loss of D and E is suggested. Both materials D and E are available on a commodity basis, and their prices are published in the *Oil, Paint and Drug Reporter* routinely. Current price for D is 7¢/lb., and current price of E is 15¢/lb. The raw-material cost for A would be:

$$1.1 [(7 \times 0.6) + (15 \times 0.4)] = 11.2¢/\text{lb.}$$

Thus—based on the potential for an almost 50% saving in raw-material costs, and the fact that such costs would account for less than 40% of the potential selling price—laboratory work to assess feasibility of this new approach for synthesizing Product A may be very much in order.

Many examples of the successful commercial development of new processes that can use lower-cost raw materials exist in the literature—e.g., the production of vinyl chloride from ethylene instead of acetylene, and the production of acrylonitrile from propylene and ammonia rather than earlier, less-economic, processes using acetylene and hydrogen cyanide.

Although many factors other than raw-material costs obviously enter into project-development decisions, screening procedures (at a very early phase of laboratory research) based on raw-material costs are quite valuable in eliminating projects that have little potential for a reasonable profit margin.

Early Interplay With Marketing Groups

Again assuming you are an engineer concerned with project development, you can expect market research activities to provide guidance with regard to long-range raw-material and product prices, product end-use consumption patterns, and market growth.

There is no shortcut to effective market research. True, business publications provide one valuable source of input information, and useful surveys of an overall field can sometimes be derived from multiple-client studies conducted by consulting companies. But at some point in time, the need will arise for a person-to-person field survey involving interviews with suppliers, customers and competitors. This survey must have clearly defined objectives so as to tie down the specific items of vital information needed.

Your group should supply samples of product to the Marketing Dept. at a very early stage, for use in market development work. If competitive products exist, get samples and make quality comparisons to evaluate comparative performance in the end-use applications. Be on the lookout for quality advantages. Many examples exist of new developments whose commercial success stemmed from important quality advantages—e.g., the development of titanium dioxide pigments via the chloride process (as opposed to the older sulfate process) so as to yield pigments with a pure white color (as opposed to the yellowish cast of sulfate-processed material).

Get any data necessary to permit material- and heat-balance calculations, and draw up a preliminary flowsheet for the process. After the feasibility of the process has been demonstrated in small-scale equipment, develop a preliminary project investment estimate, and use it to develop the cost of manufacture at various capacity levels.

Simultaneously, get the Marketing Dept. to develop a market analysis and a preliminary market forecast as a function of potential product sales-price. This involves estimation of the rapidity with which demand for the new chemical product could develop at each of several introductory prices. Include this information in the orientation economics to permit an estimate of potential product profitability.

Be sure that the market analysis takes into full account the installed capacity of each competitor now sharing in the market for the product. The U.S. Dept. of Commerce publishes statistics on sales and consumption of most commodity chemical materials. If the product involved is a new one and no statistics are available, a market-research field survey is imperative to tie down such statistics in determining your potential position in the marketplace. Use these data to make an analysis of supply vs. demand for the material involved, because this is extremely important in establishing future price trends.

Licensing, R&D

Make a thorough search as to the availability of existing processes that could be licensed for production of the chemical under consideration. Sufficient nonconfidential information can normally be obtained from a potential licensor to permit an economic appraisal. However, even after purchase of a license, some R&D work on specific aspects of the process or on product applications is usually necessary. If the knowhow purchased is complete, the way can be opened to rapid commercialization of the product.

It is normally quite difficult to justify R&D on new processes to manufacture products already commercially available, unless some breakthrough in exploratory research offers a very promising new route of manufacture at considerably lower cost, a major quality-improvement, and/or a potential unique raw-materials position.

Questions To Ask Before Piloting

Before deciding on expenditures for pilot-scale development activity, a thorough review of the project

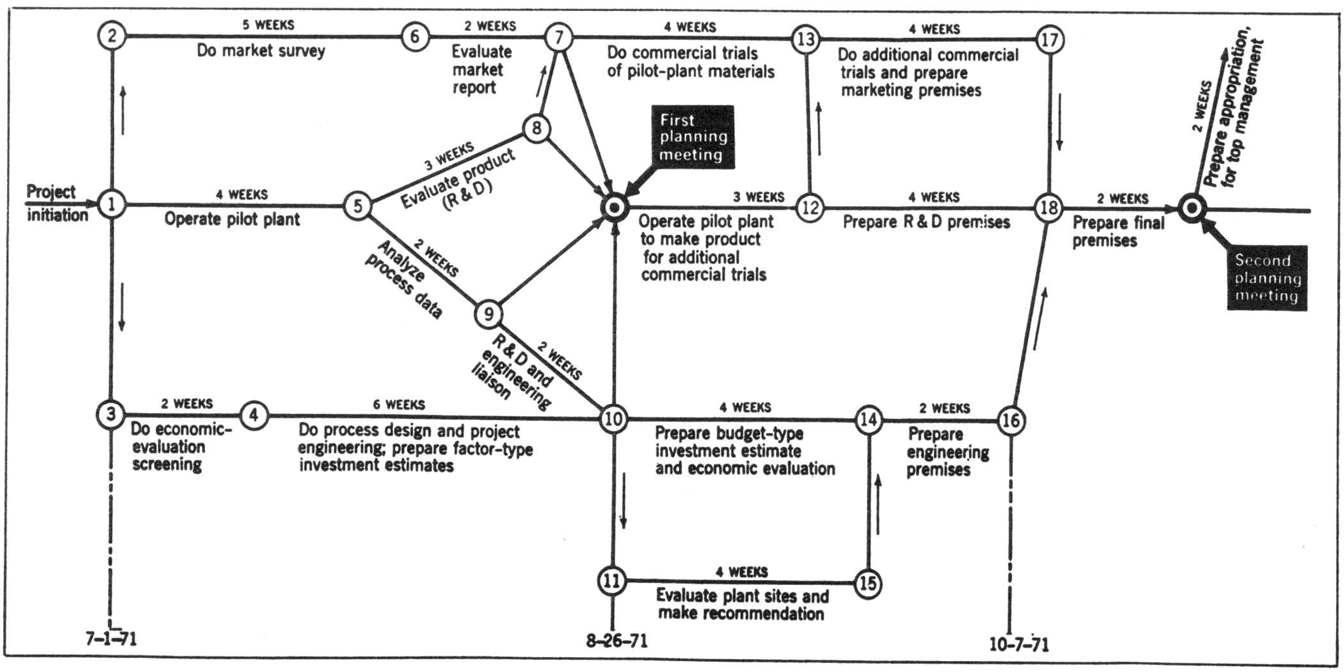

PERT DIAGRAM brings out interrelationships and time requirements of project-development activities—Fig. 1

Evaluation of Marketing Factors—Checklist I

These questions will help you assess the "marketing fit" of the new product, and provide supporting information for the economic justification and evaluation of risk:

1. What is the product's relationship to existing company markets? Any new products that can be sold through the existing marketing organization, to many of the same customers currently being served, has the advantage of lower selling cost.

2. What competition currently exists or is expected on products serving the same customer need as the new product? If other suppliers have done a good job of developing uses for the new product and have rendered effective technical service to their customers, this will have to be reflected in your company's strategy for achieving the necessary market penetration.

3. How many grades of the new product (e.g., additives, colors, purity or physical properties) will be required in order to adequately service the various needs of potential customers? This can have a very significant influence on the type of customer-service organization to be developed, the commercial-plant physical facilities needed, and the project's working-capital requirements.

4. Are there any captive uses of the new product within the company? If it seems possible to develop a captive-market position for any significant part of the new plant capacity, this can be an important risk-reducing, stability factor.

5. What technological and other changes are taking place in the markets for material such as your new product? For example, if environmental considerations are coming to the fore, and your product offers advantages in this area, its chances of exceeding its sales goals are better than if it had to rely on premium quality or better customer service alone. Or, if there is a discernible trend toward some form of process modification in your customers' plants, this may create new opportunities for your product (e.g., by permitting its substitution for a more-expensive material) or may create new risks (e.g., by lessening the need for its particular features).

6. How long will it take to obtain product acceptance by any regulatory government agencies and by the customers themselves for specific end-use applications? Testing-time requirements of acceptance by the U.S. Food and Drug Administration and other government regulatory agencies must be taken into account in determining timing for product commercialization. In some cases, customer acceptance of a new product may require extended performance testing over periods as long as one to two years.

7. What distribution facilities will be required for the new product? Must storage be dry, heated, refrigerated, corrosion proof? Will new warehouse equipment and procedures be involved? Are new distribution servicing points and facilities required? What is the pattern of critical freight rates and competitive carriers? Must inventories of product be maintained to permit overnight delivery to customers? How many different size packages must be stocked and handled? Will the market pattern include many small orders or only large bulk shipments? What will it cost to fill an order? Will an extensive advertising and promotion campaign be required? All of these factors can mean considerable additional cost if not evaluated at this stage.

8. What credit terms will customers expect? This can have a very important effect on working capital as well as allowance for bad debts.

economics should be made—one that weighs whether or not to continue development. A planning or business group consisting of representatives from marketing, manufacturing, engineering, and R&D should do this, and ask such questions as:

1. Is the indicated long-range return on investment commensurate with the degree of risk inherent in the project? (The degree of risk must be evaluated in terms of market penetration required on a share-of-market basis to achieve project profitability.)

2. Is the total manufacturing cost competitive with that of the products that might be displaced?

3. Is the breakeven production rate sufficiently low to assure at least a minimum profit during the initial, post-startup period of market development?

4. Can the initial capital investment be reduced by providing a stepwise expansion of the plant to its ultimate capacity as the market for the product is developed?

5. Is any idle equipment suitable for use in the proposed project?

6. Is the required capital investment for plant facilities, expressed as dollars per annual unit of production, comparable to that of competitive plants?

7. What specific advantages or disadvantages does the company have (compared with competitors) for entry into marketing and manufacture of the particular product?

If management renders a favorable decision as a result of this project review, the project enters its next phase of development.

The Pilot-Plant Stage

While piloting is going on, market-development activities should receive strong emphasis. Normally, the need to distribute larger-than-laboratory product samples to potential customers is as important a justification for pilot planting as is the need to collect the process-design data necessary to permit more-detailed project-investment and operating-cost studies. The semicommercial sampling activity serves to evaluate acceptance of the product in the various end-uses under consideration, and also acquaints potential customers with the possible future availability of the product in commercial quantities.

A PERT diagram should be prepared for the remaining stages of the project development.[1] Fig. 1 shows a typical diagram for a major development project. Such a diagram brings out the interrelationships and dependencies between all required activities, and puts into sharp focus the minimum time required for commercialization of the new project. Furthermore, a careful analysis of project schedule can occasionally provide attractive alternatives for speeding up development. In any event, continuous updating of this PERT diagram pinpoints the deadlines and helps coordinate the overall project effort with each of the various functions.

When sufficient process-development data have been collected to permit more-detailed process-design studies, it is time to size major commercial-plant equipment, prepare process flowsheets, and make an investment estimate for the various plant capacities under consideration. These feasibility studies provide valuable technical inputs for the R&D department's process-optimization studies carried out during this phase.

Make sure the engineer responsible for evaluating the R&D pilot-plant work and scaling it up to a commercial process takes the time to actually witness the pilot plant in operation and to discuss the development work with the operating personnel (including the technicians). Occasionally, a potentially important operating problem that has occurred intermittently in the pilot plant may be overlooked in the regular R&D pilot-plant reports. For example, if occasional plugging occurs on a catalyst feedline to a reactor, this could be regarded as merely an operating nuisance and may fail to be pinpointed as a problem that must be solved by the commercial design.

Obviously, some of these minor pilot-plant troubles may be merely problems associated with the smaller scale of equipment as compared with commercial-plant equipment. On the other hand, some minor problems can become major ones upon commercialization. For instance, if the pilot plant is designed for once-through operation on reactant materials, but a recycle of unconverted reactants will be necessary to realize satisfactory economics in the commercial plant, serious problems can result from giving inadequate attention to the effects of impurities that may build up in the recycle stream in the commercial plant.

Effect of Price and Volume

The market-development studies carried out during the pilot-plant stage should provide an indication of customer acceptance at various potential price levels, serve to define the optimum initial plant-capacity, and supply project-economics data for the commercialization decision.

As already indicated, sales volume and price are usually the most important and often the most difficult factors to tie down in a project justification.

The process design and pilot plant generally pin down estimates of key operating-cost items within reasonably accurate limits. Once the design is sufficiently well along, capital investment can be pinpointed within plus or minus 10% (assuming adequate attention is paid to the field-labor as well as equipment component of the estimate, particularly in regard to cost escalation). However, during the period between initiation of a project and startup of a full-scale production facility, price and/or sales-volume potential can fall so drastically as to wipe out project profitability completely if market conditions change. *(Text continues on p. 138)*

* Ideally, he should work a few "graveyard" shifts in the pilot plant. Technicians will have a greater tendency to share their problems and experiences with him once he has shown empathy at this level and gained their confidence.

Key Technology Factors—Checklist II

The following questions will help you assess the product's "technological fit":

1. Is the product a new composition of matter, a known product previously manufactured only on a relatively small scale, or a bulk commodity material currently manufactured by others? The risk of being copied by competitors is substantially smaller if the new product is a new composition of matter, which is protected by patents. However, all new products run the risk of early market obsolescence due to the development of cheaper or improved competitive products. The novelty of the technology involved in manufacturing the new material can be an important factor in the assessment of project risk.

2. What is the availability of raw materials, both as to quality and quantity, required by the project? Can the project use raw materials already available within the company? How many potential raw-material suppliers are available—enough to avoid dependence on a single source? If the raw material is available only in developmental quantities, will facilities to produce it in bulk have to become part of the total project justification, or will suppliers put up the necessary facilities?

3. Can existing company knowhow and/or laboratory or pilot-plant equipment be used in developing the project? If the knowhow is a logical outgrowth of current corporate technologies, it will be much easier to build the technological strengths required to ensure a successful commercial development. Furthermore, development costs can be appreciably lower if existing facilities (with some modification) can serve for pilot planting.

4. Can standardized commercial equipment be used for the new process? If custom-built equipment is required for very many of the process steps, significant equipment-development and design costs may be incurred. Furthermore, in cases where proprietary equipment must be obtained from overseas vendors, serious difficulties can develop in both schedule and costs unless careful expediting is provided.

5. Can any idle equipment or facilities be converted and put to use, so as to reduce the amount of new investment required? An example of this approach is the production of isoprene in retired butylene-dehydrogenation facilities that have been modified. Other examples involve the use of spare capacity in multipurpose batch-reaction plants to produce new chemical-specialty materials.

6. What hazardous operating conditions, maintenance problems, or waste-disposal and environmental-control problems exist in the new process? Each of these items can have a very significant effect on total investment requirements, (and or relationships with insurance carriers and regulatory agencies).

7. What is the optimum site location for the new facility? Many factors are involved in this; for example: product-distribution costs and proximity to planned markets can be very important both in project economics and in determining customer-service requirements.[2] Construction-labor productivity factors can have a significant influence on plant investment. Availability of operating labor and existing area-rate structures can also be very important in determining project economics. Utility costs and availability can be important influences on offsite investments required, as well as on operating costs. Accessibility of the site to raw-materials supply is another factor not to be overlooked.[3]

If competitors install so much new capacity for a commodity that new uses must be developed before existing facilities can be operated at full capacity, then price inevitably falls while merchant producers scramble for business. On the other hand, if the product is a unique chemical specialty with no immediate competitive equivalent in the open market, then price tends to be stable while market-development activity generates sales volume. Obviously, there are a great many intermediate situations between these two extremes.

An accurate assessment of total market potential is very important in determining the share of the market that the new project may represent. The percent market penetration required for acceptable project profitability, balanced against manufacturing economics (in regard to plant capacity), is a major factor in determining the degree of risk.

If supply is in reasonable balance with demand (i.e., 90% of the available capacity is being sold) and the market is growing, there may be room for your new venture. However, if your venture requires achievement of a share of market greater than 10% in order to attain acceptable project profitability, you had better closely review what your process and product strengths are, because you are moving into an area of higher risk.

Obviously, there are no hard and fast rules, and each situation must be judged in relation to all of the factors involved. (The case study in the next article will illustrate the assessment of such factors.) However, it can be unequivocally stated that if a mar-

Setting Forth the Premises for the Economic

Marketing

1. Sales forecast, including product mix, projected selling price for each product, distribution costs and allowances.

2. An analysis of the existing and future total market for the product(s) involved. This should pinpoint and analyze the rate of market growth by individual product applications. Furthermore, it should include a sales plan showing key accounts, and the product-quality requirements and marketing strategy planned to capture an appropriate share of the market.

3. Competitive product capacity by individual supplier, together with projected increases in such capacity. An analysis of the total projected market growth, referenced to the total projected product capacity, is very important in evaluating the supply/demand balance for the product and the corresponding effects on product pricing in future years. Furthermore, this provides another checkpoint in regard to marketing strategy in terms of share of market to be captured.

4. Planned use of any internal markets, and of any resale programs (i.e., involving products of other suppliers during the period of construction, to build scles volume so that the new capacity can be easily absorbed). Such programs should be reviewed in detail, together with applicable costs and pricing.

5. Any unique product-quality features that are needed to meet market-penetration goals. Normally, these quality requirements need to be discussed further (and specific commitments made on timing for accomplishment) in the R&D section.

6. Working capital requirements of the business in terms of finished-product inventory necessary to provide satisfactory customer service (usually developed jointly by the marketing group and the financial group).

7. A forecast of marketing and promotional expenses for the new product. This is geared to the number of new sales personnel, clerical personnel, and the costs of any promotional campaigns. In addition, the costs associated with any specific market-development programs planned (including sample costs) need to be covered.

8. A summary showing the marketing need and justification for the new plant capacity.

Manufacturing

1. A stream-rate projection by individual product grade, together with operating utility (i.e., onstream time requirement) referenced to the marketing product mix, for each year of the projected plant operation. In effect, this defines plant capacity for the product mix involved. For batch processes, it is normal to define batch size, cycle time, and operating utility where this is believed to be a significant factor for the individual product grades involved.

2. A forecast of the quality yield for each individual product, for those situations in which several quality grades are possible. This defines production that is off-specification and must be either reworked or sold at a lower sales price.

3. Plant staffing requirements, in terms of the number and types of personnel required in each individual category (e.g., operators, maintenance mechanics, laboratory technicians for quality control, clerical and accounting personnel, scheduling and shipping personnel, and supervisory and technical personnel). This is necessary in order to tie down the controllable fixed operating costs.

4. A projection of the learning curve for the startup of the new plant. This is necessary in order to estimate startup costs, and availability of product. The projection normally takes the form of a curve showing estimated actual production vs. time, and also percent prime yield of on-specification product vs. time for the startup period.

5. A detailed estimate of total product-manufacturing cost, in the form of a mill-cost sheet. This gives a breakdown of raw-materials consumption (based on estimated yield), consumption of individual utilities and the unit rate structure, operating-labor cost, production supplies, maintenance materials, maintenance labor, quality control, and noncontrollable fixed costs.[4] This must be projected for each year of plant operation, and coordinated with the market forecast and the finished-product-inventory needs for at least the first five years.

6. Working capital requirements for raw-material inventory and intermediate product storage. These data are provided for use in the financial group's economic evaluation of project profitability.

ket penetration greater than 50% is required for acceptable profitability, the degree of risk of the project may be inordinately high for prudent investment—unless important advantages in manufacturing economics, product quality or captive markets are present.

Developing the Project Justification

Checklists I and II contain key questions concerning marketing and technology factors. If the major factors appear favorable for the new project, the next step is to prepare the justification economics, and to solicit management's approval for the funds involved.

This brings us to the topic of profitability criteria. In the chemical process industries, the three most widely used criteria or methods for determining the attractiveness of capital projects are (1) payout time, (2) return on investment, and (3) discounted cash flow (which has a number of variations). Most major companies in the CPI have a standard format for justification of capital projects, and this format may require that all three of the above criteria be used.

We defined and illustrated payout time in the previous article. Although most often thought of in connection with fairly small projects (e.g., debottlenecking or cost reduction), the payout-time method also serves a purpose in large projects—particularly of the high-risk, high-return variety—because it allows management to assess the exposure time for recovering the investment.

(Text continues on p. 140)

Justification of a Major Capital Project

Engineering and Construction

1. A detailed process description referenced to a simplified process flowsheet. This not only helps tie down what is planned for the new facility, but may pinpoint any specific process problems still remaining to be resolved by the R&D group, working in close coordination with Engineering.

2. A detailed project scope, including a list of all the major items of equipment to be purchased. This provides the basis for the plant investment estimate, which should be a part of this section.

3. A detailed review of the technical status of the project, covering the specific data that have been developed to support the process design basis. Any design risks being taken should be thoroughly evaluated; specific plans for minimizing these risks should be covered, together with individual responsibility and schedules for accomplishment.

4. The detailed process-design basis. This should be pinned down in terms of: design production rates for each product, quantity yields in terms of raw materials, quality yields based on existing or anticipated specifications, operating utility, and any other specific items needed.

5. A review of offsite requirements. This should be provided in checklist form, to show that each individual item (e.g., warehouse space, offices, laboratory space, individual utility items, safety and waste-treatment facilities, maintenance shop) has been considered and coordinated with the manufacturing group.

6. The project construction schedule. This is normally in arrow-diagram form at this stage, and should show the plans for completion of each individual design and construction phase.

7. Plans for construction of the plant, in terms of using an engineering and construction contractor vs. internal personnel. If a contractor is to be used, decisions are necessary in regard to whether the job is to be handled on a lump-sum competitive-bid basis, cost-plus with a guaranteed maximum, or other means.[5] (Factors that go into this decision include timing requirements, degree of project definition, availability of internal staff to supervise contractor, etc.)

Research and Development

1. Specific requirements for R&D funds, in terms of the specific programs to be accomplished and the anticipated cost of each per year. Any new product grades to be developed in order to achieve market penetration should be reviewed in detail, and a schedule shown for the completion of each program. Also, if the technical risks described in Item 3 of the Engineering and Construction section call for R&D programs, their anticipated cost and schedule should be indicated.

2. In the case of a new process, an evaluation of competitive process technology based on the patent literature and other published sources. The strengths and weaknesses of the new process should be referenced to each of the competitive processes currently being used. This is quite important in evaluating the potential long-term competitive position of the technology involved.

3. An assessment of any specific patent problems that may be encountered, together with a recommendation for an appropriate financial reserve.

Financial

1. A calculation of project cash flow by years, for at least the first five years of project life. An accompanying calculation should show after-tax return on plant investment and also on total investment.

2. A calculation of project payout time, and discounted-cash-flow return on investment.

3. A detailed tabulation of annual cost of sales for at least the first five years of project life.

4. A detailed breakdown of other operating expenses including: administration, marketing, customer technical service, R&D, royalties, allowance for bad debts, and such other staff services as may be involved.

5. A detailed interpretive analysis of project financial risks, together with the sensitivity analysis for the various key factors such as sales price, volume, manufacturing costs, investment, and susceptibility to overall business conditions.[6]

6. A summary analysis of the present business posture of the company and the overall fit of the proposed project into the total corporate business structure.

Keep in mind that most chemical plants are highly specialized, single-use facilities that are not cheaply converted to an alternate use. The salvage value for chemical process equipment is minimal, because four to five times the purchase price of the equipment is involved in its proper installation into the process. If the process does not perform properly or a market cannot be developed for the material produced, the company is faced with a sizable unrecoverable loss. In this connection, payout time sheds light on the degree of risk involved in at least recovering the initial investment out of the project. As to what happens after that point is reached, payout time is of no help—that is one of the method's several limitations, and a reason other criteria are also necessary.

Return on investment (ROI) affords one such criterion; this method is widely used for major capital projects, and is particularly valuable in analyzing alternative proposals.* The ROI values determined for a project can be compared directly to the cost of money or alternate investment opportunities.

The usefulness of the ROI values is a direct function of the inputs used in the calculation. The inputs for the annual after-tax income can range from the (1) typical operating profit for post-startup operations, to (2) an average annualized profit for the first five years of the project, to (3) an average annualized profit for the estimated life of the project, which may be as long as fifteen to twenty years.

Input for the investment and working capital can also be varied. It can be the initial investment plus working capital, or the average depreciated investment and working capital over the life of the project. If the income is expected to fluctuate over a wide range during the life of the project, the ROI should make use of a weighted-average annual profit. For example, if pricing were expected to remain constant for the first five years and dramatically decline during the next five years as a result of new entries into the same marketing field, the ROI calculated on the profits generated during the first five years of price stability might be very favorable; but the ROI calculated on the profits generated each year for the next ten years (including the price deterioration) must be carefully considered in the final decision.

In a subsequent article of this series, we shall cover use of discounted-cash-flow return on investment, present project worth, decision-tree concepts that use computer programs to evaluate investment alternatives with widely varying annual cash flows and capital investment requirements, and some other mathematical techniques.

These techniques are all applicable, particularly for major projects—but, like ROI, their usefulness depends on the quality of the input. And this is why the present article has stressed the quality and completeness of the technical and commercial input, rather than mathematical manipulation of it.

* As discussed in Part III, the ROI is:

$$\frac{\text{Annual After-Tax Profits}}{\text{Investment} + \text{Working Capital}} \times 100$$

Completing the Input

The final step in obtaining input for the economic justification of the project is to work with each functional group (marketing, manufacturing, engineering and construction, R&D) and to cover the various "premises" tabulated on the previous page.

The format for tying down these major premises may vary from company to company; the important thing is to obtain commitments from the functional group responsible for each area. The commitments may take the form of a capital-appropriation book that is reviewed for consistency and factual content by general management prior to soliciting approval for capital spending from top management. This document later serves as a financial-control tool in measuring performance of the various functions.

In deciding how much effort to spend on the preparation of such premises, it is well to remember that there will be a "day of truth," when the actual performance of the project will be graded against the proposal. Early in the life of the project, the actual investment versus the proposed or estimated investment will be available for comparison. Later in the project, the actual operating income and operating costs will be available for comparison with the projected income and costs. Although one or more of the actual elements may be different than what was proposed, management will want an answer to the key question: "Is the project going to pay out?" or, "Is the project going to provide the return-on-investment that was promised?"

Regardless of which justification criterion is used for your project, you should be able (after a project has been implemented) to strike a comparison between actual performance and estimated performance to convince management that the profitability goal cited is in fact being realized.

In the next part of this series, we shall present a detailed case study involving a major capacity-expansion project, and show how the various inputs we have been discussing are supplied and evaluated as part of that project's commercial development. ■

References

1. For a fuller discussion of PERT diagrams and other network techniques applicable to project development, see: Fuller, D., "Organizing, Planning and Scheduling for Engineering Operations," Industrial Education Institute, New York.
2. For a more-quantitative discussion of the effect of location on distribution costs, see p. 532.
3. A review of plantsite location factors appears periodically in *Chemical Week* (the last one appeared in the Aug. 19, 1970, issue). For a good recent article on location techniques, see: Speir, W. B., Choosing and Planning Industrial Sites, *Chem. Eng.*, Nov. 30, 1970.
4. For a discussion of these operating costs, see Part 1 of the present series. To see how such costs relate to capital costs, see p. 484.
5. For advantages and disadvantages of various forms of construction contracts, and advice on negotiating with engineering contractors, see: Popper, H., ed., "Modern Cost-Engineering Techniques," McGraw-Hill, 1970, pp. 197–208 and 212–217.
6. For some simple sensitivity-analysis charts, see: Ransom, E. A., Guidelines for Evaluating New Fields and Products, *Chem. Eng.*, Nov. 6, 1967, p. 286. For a more-extended treatment, see: Strauss, R., The Sensitivity Chart—Giving Meaning to Shaky Estimates, in "Modern Technical Management Techniques," (ed. by Popper, H.), McGraw-Hill, 1971.

ECONOMIC ANALYSIS Part 5

A Realistic Project-Development Case Study

If you prefer specifics to generalities, you will be particularly happy with this case study of a major capacity-expansion project.

IRVING LEIBSON and CHARLES A. TRISCHMAN, JR., Dart Industries, Inc.

The time has come to move from principles to practice in regard to project evaluation and development.

In the previous article, we discussed a number of principles, and emphasized the continuous chain of process-development, plant-design and economic-appraisal activities that take place concurrently with market-research, market-development and sales-service activities. Let us now see how these principles and activities are actually carried out, and on what basis the final decisions are eventually made.

The medium that will allow us to do this is a hypothetical but representative case study covering the XYZ Co.'s plans to substantially expand capacity for its Product B. (Actually, most of the study would still be applicable if Product B were an entirely new product.)

The case study is structured to present a realistic picture of what normally happens during the steps leading to definition of project scope, economic justification, and final project approval. Key events will be a series of three project-planning meetings, attended by representatives from the marketing, manufacturing, R&D, engineering, and financial areas.

(No matter what your company's organizational structure is like, information collected for project evaluation must cover the above functional areas, and must usually include specific commitments by individuals representing these functions.)

To do justice to this study in terms of specific business-world practices and problems, we will present the material in some detail, and this means splitting the case study into three parts—one for each of the planning meetings. Let us proceed with the first part (the others will appear in later installments).

Background Information

The XYZ Co.—which we will refer to as "our" company—markets and manufactures a liquid resin, Product B, which is used as an additive in paints, a drying agent in inks, and a tackifying agent in adhesives. A similar product is marketed and manufactured by five competitors. As shown in Table I, our company's present sales for Product B are equivalent to about 10% of the total U.S. market for this material.

Projections of potential total market growth indicate an average rate of 11% per year for Product B. As shown in Table II, this is comprised of an average anticipated 19% annual growth rate for Product B in the ink market, 10% after 1973 in the ad-

Competitive Capacity and Sales—Table I

Company	1970 Capacity, Million Gal./Yr.	Estimated Sales Rate, Million Gal./Yr.	Share of U. S. Market, %
A	11.5	11.3	40
B	6.5	6.2	22
C	4.5	4.2	15
D	3.0	2.7	9
E	1.5	1.2	4
XYZ	3.0	2.9	10
Total	30.0	28.5	100

Percent of production capacity used for sales (1970) = [28.5/30.0] × 100 = 95.0%

Originally published September 1971.

Forecast of All-Industry Market Demand and Growth for Product B—Table II
(V = Volume, Million Gal./Yr.; G = Growth, %)

Product Application Area	1971		1972		1973		1974		1975		1976		1977		1978	
	V	G	V	G	V	G	V	G	V	G	V	G	V	G	V	G
Paint additive......	14.4	3	14.8	3	15.2	3	15.7	3	16.2	3	16.7	3	17.2	3	17.7	3
Ink-drying agent...	13.0	20	15.5	19	18.5	19	21.9	18	25.7	17	29.8	16	34.7	16	39.8	15
Adhesive..........	4.3	16	4.9	14	5.4	10	5.9	10	6.5	10	7.2	10	7.9	10	8.7	10
Total............	31.7	11	35.2	11	39.1	11	43.5	11	48.4	11	53.7	11	59.8	11	66.2	11

hesives market, and only 3% per year in the paint market. This low rate of growth is forecast for the paint market because Product B is used only in oil-based paints, and this segment of the paint market is expected to be relatively static, with the major growth slated for water-based paints.

The information contained in Table II is based on:

- A series of 35 field calls made to end-users of Product B.
- Exchanges of nonconfidential marketing information concerning growth of individual Product B applications with all five competitors.
- A combined evaluation of growth rate for final products in three markets served by Product B, with an assessment of the position of Product B in relation to other materials serving these respective markets.

In addition, statistics for total consumption of Product B in past years have been analyzed to determine growth-rate history for this material. Thus, it is believed by Marketing that the forecast of total market growth for Product B, as shown in Table II, is as reliable as can be made with the tools available.

(In actual practice, many chemical companies have invested a great deal of time and money in developing industry simulation models of the market situation for their various products. In the final analysis, such models are only as effective as the input data supplied to them. Thus, there is no substitute for detailed market surveys to establish basic industry trends regarding supply and demand in the development of long-range sales forecasts.)

As shown in Table I, total industry sales for Product B are expected to be about 95% of estimated industry production capacity in 1970. With the anticipated rate of market growth, our Marketing Dept. has proposed that the production capacity for Product B be expanded by the installation of a 4-million-gal./yr. facility to be in operation no later than July 1, 1972.

Several bottleneck-removal projects have already been completed in the existing manufacturing facility, and our Manufacturing Dept. believes that a major new grass-roots facility will be needed in order to supply the new capacity desired for Product B. The existing facility is located at a Midwest manufacturing site. Our company also has another manufacturing site at which other products are made, at a Gulf Coast location.

XYZ Co.'s Market Forecast for Product B—Table III
(V = Volume, Million Gal./Yr.; P = Sales Price, ¢/Gal.)

Product Application Area	Product Grade No.	1970		1971		1972		1973		1974		1975		1976		1977		1978	
		V	P	V	P	V	P	V	P	V	P	V	P	V	P	V	P	V	P
Paint additive........	101	1.20	69.3	1.2	62.5	1.2	66.0	1.7	61.5	1.8	65.0	1.8	64.0	1.8	63.0	1.9	62.0	1.9	61.0
Ink-drying agent.....	201	1.14	100.0	1.5	98.5	2.2	98.0	2.9	96.5	3.5	97.0	4.2	96.5	5.0	96.0	5.7	95.5	6.5	95.0
Adhesive............	301	0.56	95.0	0.8	94.0	1.1	94.2	1.4	94.0	1.7	93.5	2.0	93.0	2.2	92.5	2.4	92.0	2.6	91.5
Total..............		2.9	86.3	3.5	88.0	4.5	88.2	6.0	86.0	7.0	88.0	8.0	88.5	9.0	88.5	10.0	88.2	11.0	88.2
Share of Market:		%		%		%		%		%		%		%		%		%	
Paint additive........	101	8.6		8.3		8.1		11.2		11.5		11.1		11.0		11.0		11.0	
Ink-drying agent.....	201	10.6		11.5		14.2		15.7		16.0		16.3		16.8		16.5		16.3	
Adhesive............	301	15.4		18.6		22.4		25.9		28.8		30.7		30.0		30.0		30.0	
Total share.........		10.2		11.0		12.8		15.3		16.1		16.7		16.7		16.8		16.6	

Key- and Target-Account Sales Plan for Various Grades of Product—Table IV

Customer Account*	Grade Now Purchased	Million Gal. Sold in 1970	Sales Strategy	Million Gal. Targeted Sales for 1971	1972	1973	1974	1975	1976	1977	1978
Acme, K	101	0.30	Maintain call frequency	0.3	0.3	0.4	0.5	0.5	0.5	0.5	0.5
Alpha, K	301	0.30	Provide formulation support	0.3	0.4	0.4	0.5	0.6	0.6	0.6	0.6
Baker, K	201	0.14	Open local warehouse	0.2	0.2	0.3	0.3	0.4	0.4	0.4	0.5
Burns, K	301	0.27	Provide equip. design aid	0.3	0.4	0.4	0.5	0.5	0.5	0.5	0.5
Comet, K	101	0.50	Reduce sales effort in 1975	0.5	0.5	0.5	0.5	0.5	0.5	0.6	0.6
Crescent, K	201	0.80	Increase credit level in 1972	0.9	1.4	1.4	1.6	1.6	1.8	2.0	2.2
Capitol, K	301	—	Offer strong tech. support	0.2	0.3	0.3	0.4	0.4	0.4	0.4	0.4
Delta, K	101	0.20	Maintain call frequency	0.2	0.2	0.3	0.3	0.3	0.3	0.3	0.3
Essex, K	101	0.20	Continue special packaging	0.2	0.2	0.3	0.3	0.3	0.3	0.3	0.3
Federal, K	201	0.10	Develop special drying grades	0.2	0.3	0.4	0.5	0.6	0.8	0.9	1.0
First, K	201	0.10	Resolve processing problems	0.2	0.2	0.4	0.4	0.5	0.5	0.6	0.7
Forefront, T	301	—	Initiate sales contacts	—	—	0.2	0.2	0.3	0.4	0.5	0.6
General, T	101	—	Initiate sales contacts	—	—	0.2	0.2	0.2	0.2	0.2	0.2
Globe, T	201	—	Obtain product approvals	—	0.1	0.3	0.3	0.4	0.5	0.6	0.7
Graphic, T	201	—	Initiate sales contacts	—	—	0.1	0.3	0.4	0.5	0.6	0.7
Guardian, T	301	—	Determine product needs	—	—	0.1	0.1	0.2	0.3	0.4	0.5
Pioneer, T	201	—	Initiate sales contacts in 1972	—	—	—	0.1	0.3	0.5	0.6	0.7
Total		2.91		3.5	4.5	6.0	7.0	8.0	9.0	10.0	11.0

* K = Key account, with whom business is currently being transacted.
T = Target account, from whom XYZ Co. hopes to capture business in 1972 or later.

Competitive-Capacity Forecast—Table V

	Productive Capacity, Million Gal./Yr.								
Company	1970	1971	1972	1973	1974	1975	1976	1977	1978
A	11.5	17.0[1]	17.0	17.0	17.0	22.0[5]	22.0	26.0[5]	26.0
B	6.5	6.5	6.5	6.5	6.5	6.5	8.0[5]	8.0	8.0
C	4.5	4.5	4.5	4.5	7.0[5]	7.0	10.0[5]	10.0	11.0
D	3.0	3.0	3.0	8.0[2]	8.0	8.0	8.0	10.0	10.0
E	1.5	1.5	1.5	—[3]	—	—	—	—	—
F	—	—	—	—	—	—	—	—	4.0[6]
XYZ	3.0	3.0	3.0	7.0[4]	7.0	9.0[5]	9.0	11.0[5]	11.0
Total	30.0	35.5	35.5	43.0	45.5	52.5	57.0	65.0	70.0
Capacity utilization, %	95	89	99	91	95	92	94	92	95

Notes:
[1] Announced for startup in fourth quarter of 1970.
[2] Announced for startup in third quarter of 1972.
[3] Anticipated shutdown of plant due of uneconomic plant size and weak technology position.
[4] Recommended startup in early third quarter of 1972.
[5] Forecast capacity increase.
[6] Anticipated new entry in manufacture of Product B.

Competitive-Supplier Evaluation—Table VI

	Product Quality				
Company	Paint Additive	Ink Drying	Adhesive	Quality of Services	Pricing Strategy
A	Excellent	Excellent	Good	Excellent	Agressive
B	Excellent	Good	Limited	Good	Conservative
C	Good	Good	Good	Excellent	Competitive
D	Limited	Excellent	Good	Good	Competitive
E	Good	Limited	None	Limited	Conservative
XYZ	Excellent	Excellent	Good	Excellent	Aggressive

Meeting No. 1: Marketing Aspects

In order to start the project-evaluation studies that XYZ Co.'s top management will need, a Project Planning Meeting was held on Aug. 6, 1970. People from the marketing, manufacturing, engineering, R&D, and financial departments participated.

Table III shows the five-year forecast of sales rate and market price for the product mix proposed by Marketing. This anticipates an increase in share of market from 10% to approximately 15% over the period from 1970 to 1973. However, the adhesive grade is currently at 15.4% of market, and the ink-drying grade is at 10.6%, vs. the paint-additive grade at only 8.6% of market. The major growth of sales from 1971 through 1975 is to come from a 100% increase in share of market on adhesive grade, and a 54% increase in the ink-drying grade's share.

Table IV contains a preliminary sales plan, showing key accounts, and describing product-quality requirements and marketing strategy planned for attainment of the market forecast. Note that some product improvements are required in the adhesives area in order to preserve a strong competitive posture and to expand the high share of market currently enjoyed there. R&D product-improvement activities are currently underway to achieve this.

Furthermore, market research is underway to pinpoint additional potential customers for Product B. These alternate target accounts would form a contingency list in case the new business did not materialize at all of the primary target accounts.

Table V shows announced capacity increases planned by competitors in the near future, and estimates of further increases in later years. Note that the percent of production capacity utilized for sales on an industry basis is expected to fall to 91% in

1973 after projected new capacity comes onstream. This factor is expected to account for the price softness shown for the paint-additive grade in 1973.

The softness is projected despite anticipated shutdown of Competitor E's plant (an outgrowth of an uneconomic process for the plant size, and a weak technology position.) Company E is limited essentially to the paint-additive market. Once the new capacity scheduled to come onstream in 1973 is absorbed by market growth, prices are expected to firm even for this grade.

Table VI shows an analysis of competitive product strengths and weaknesses in each of the application areas currently being served by Product B. Field reports indicate that Competitor A is marketing a new grade of Product B that appears to be finding rapid acceptance as a viscosity-control agent in lubricating-oils for automotive use. Our company has started an R&D program to evaluate performance of existing grades of Product B for suitability in this application.

The Marketing Dept. believes that it may be possible to negotiate a contract for the purchase of Product B from Company A, to begin as soon as Company A's new plant expansion (as shown in Table V) is onstream in early 1971. If this can be negotiated on satisfactory terms, resale of this material to our company's customers could permit retention and perhaps even expansion of our share of the market during 1971 and 1972 prior to our plant expansion coming onstream. It has been agreed that the Marketing Dept. will pursue this alternative to determine its economics.

Manufacturing and Technical Aspects

At the Project Planning Meeting, the Manufacturing Dept. has presented the stream-rate projections by individual product grades, as shown in Table VII.

Table VIII shows the detailed mill cost for each of the three product grades manufactured in the existing plant. Note that raw-material costs differ somewhat from one product grade to the next because of differing costs for catalysts and additives, and varying yields of product, based on raw materials for each grade.

The first article of this series covered the various elements of mill cost in detail. Normally, mill cost is reported for the total product mix manufactured in the time period involved. However, because the individual grades can often differ greatly in regard to raw-material costs, additives, catalyst requirements and other such items, it is quite important to analyze the detailed mill cost for each individual product grade in making a study of the type covered in the XYZ Co. case study.

Furthermore, any effects of scale of manufacture on individual variable mill-cost items must be carefully considered in developing revised mill-cost statements for the proposed expansion. For example, improved recovery facilities might be installed in a new plant to permit increased yields, with more-effective raw-material utilization. Other process improvements installed in a new facility might permit a reduction in catalyst costs or additives usage. These factors must be carefully evaluated for each individual case, but for the sake of simplicity we will show variable cost of manufacture for each individual product grade

Mill Cost at Existing Midwest Site for 3-Million-Gal./Yr. Output—Table VIII

	¢/Gal. for Grade: 101	201	301	$1,000/Yr.	
Raw-material costs:					
Reactant A	11.9	5.9	7.1	258	
Reactant C	—	19.7	14.4	316	
Additives	0.2	6.7	5.2	111	
Catalyst	0.4	1.2	1.1	26	
Total raw materials	12.5	33.5	27.8		711
Packaging costs:	0.2	0.7	0.7		15
Utilities costs:					
Steam	0.7	0.7	0.7	21	
Recirc. cooling-tower water	0.1	0.1	0.1	4	
Electricity	0.2	0.2	0.2	5	
Fuel gas	0.6	0.7	0.7	20	
Total utilities	1.6	1.7	1.7		50
Controllable fixed costs:					
Operating labor				230	
Operating supplies				40	
Maintenance labor				50	
Maintenance material				48	
Technical				15	
Quality control				20	
Employee benefits				70	
Admin. and general				41	
Subtotal					514
Noncontrollable:					
Depreciation*				140	
Taxes and Insurance				29	
Subtotal					169
Total variable cost	14.3	35.9	30.2		776
Total fixed cost	21.7	23.9	22.7		683
Total costs	36.0	59.8	52.9		1,459

* Based on 6⅔% per year on $2.1-million original plant investment.

Maximum Existing Production Capacity for Any One Grade—Table VII

Product Application	Grade	Plant Production Rate,* Million Gal./Yr.
Paint additive	101	3.15
Ink-drying agent	201	2.85
Adhesive	301	3.01

* I.e., capacity at existing Midwest facility if all Product B manufacture were devoted to one grade.

as unchanged for the various expansion cases studied.

As for fixed costs, it is necessary to evaluate incremental staffing requirements for the expansion cases, and to carefully evaluate the effect of the new investment on such items as maintenance material, depreciation, and taxes and insurance. If improved materials of construction are used in the new plant, one could reasonably anticipate a corresponding reduction in maintenance costs. Thus, taking a constant percent of investment for maintenance costs for the expansion situation may not necessarily be the right answer. But, again, for the sake of simplicity, we have taken maintenance costs as a fixed percent of investment in this case study.

The prospects for achieving additional production in the existing plant are not good. Manufacturing believes that another bottleneck-removal program is not economically feasible unless a major process breakthrough is made by R&D. Although the production rate is greater for the paint-additive grade than for the other two grades, the "preferential pursuit" of paint-additive business to increase total plant production is not economically attractive compared to the adhesive and ink-drying grades, in a soldout plant condition (see Table IX).

The adhesive grade is the most profitable material, due primarily to its high sales price, lower cost, and high production rate. Marketing's strategy has been to vigorously pursue increased business for this grade so as to increase overall profitability. As the next priority, Marketing has been pursuing ink-drying-grade business because this sector of the market has been growing at an annual rate of 19% and is slightly more profitable than the paint-additive grade.

Financial Aspects

The Financial Dept. has analyzed the marketing, manufacturing, and technical information in Tables I through IX, and established the base-case situation against which to evaluate the return on investment for a capacity expansion. The base case, resale agreement, and various expansion cases as proposed by Financial are shown in Table X. Because Product B serves widely different application areas, and no individual customer buys more than a single grade, Financial believes it is possible (and Marketing concurs) to swing the product mix heavily in favor of adhesive-grade at the expense of paint-additive business in 1971 and 1972. This would retain overall profitability in the base-case situation, despite predicted price declines. Using the information in Tables III, VII, VIII and IX, Financial has generated the base-case earning statement shown in Table XI.

Let us quickly review some earning-statement relationships. The objective of an earning statement is to evaluate the profitability of the business, and this is the difference between *net sales realization* and *total cost of sales*. In turn, net sales realization is the difference between *net sales income* and *freight*. Going one step further, net sales income is the difference between *gross sales income* and *returns and allowances*.

Gross sales income is calculated by totaling up the products of sales price and sales volume for each individual grade in the product mix for that year.

Returns and allowances vary, depending on the particular kind of business. Let us say that XYZ Co.'s experience has shown that returns and allowances run at approximately ¾% of gross sales income. (This item includes price adjustments for quality complaints from customers, as well as any bulk volume discounts.)

Freight cost is calculated by summing the products of sales volume for each account (as shown in Table IV) and the applicable freight rate for the account. (We will have more to say about freight rates in the next article.)

Total cost of sales is the sum of variable and fixed costs of manufacture, and other operating expense.

Variable cost of manufacture is calculated by first multiplying the total variable cost for each grade (from the mill-cost statement shown in Table VIII) by the production volume for that grade, and then adding up these amounts so as to get a total that corresponds to the product mix (as in Table X).

The fixed cost of manufacture is the sum of controllable and non-controllable fixed costs from the mill cost (as shown in Table VIII). The sum of variable cost and fixed cost of manufacture is the total plant cost.

Profitability Analysis for the Three Product Grades, on a Soldout Plant Basis—Table IX

		¢/Gal.									
Product Application	Grade	Sales Price	Freight*	Variable Cost of Manuf.†	Fixed Cost of Manuf.†	Other Operating Expenses	Net Sales Realization	Total Product Cost	Product Profitability	Prodn. Rate, Million Gal./Yr.	Product Profitab., $/Yr.
Paint additive..........	101	69.3	7.8	14.3	21.7	2.8	61.5	38.8	22.7	3.150	713,000
Ink-drying agent.......	201	100.0	6.4	35.9	23.9	2.8	93.6	62.6	31.0	2.850	885,000
Adhesive..............	301	95.0	6.5	30.2	22.7	2.8	88.5	55.7	32.8	3.010	990,000

* Calculated from product mix and freight rates to customers.
† Taken from Table VIII.

Product Mix for Base Case and Expansion-Study Cases—Table X

Case Year	Grade	Base Case (BC) MM Gal.	Base Case (BC) ¢/Gal.	BC + Resale MM Gal.	BC + Resale ¢/Gal.	BC + 2 MM Gal. MM Gal.	BC + 2 MM Gal. ¢/Gal.	BC + 4 MM Gal. MM Gal.	BC + 4 MM Gal. ¢/Gal.	BC + 8 MM Gal. MM Gal.	BC + 8 MM Gal. ¢/Gal.
1970	101	1.2	69.3								
	201	1.1	100.0								
	301	0.6	95.0								
		2.9	**$2.504 MM**	Gross Sales Income (GSI)							
1971	101	0.7	62.5	1.2	62.5						
	201	1.5	98.5	1.5	98.5						
	301	0.8	94.0	0.8	94.0						
GSI		3.0	**$2.667 MM**	3.5	**$2.980 MM**						
1972	101	—	66.0	1.2	66.0						
	201	1.8	98.0	2.2	98.0						
	301	1.1	94.2	1.1	94.2						
GSI		2.9	**$2.800 MM**	4.5	**$3.984 MM**						
1973	101	—	61.5			0.7	61.5	1.7	61.5		
	201	1.5	96.5			2.9	96.5	2.9	96.5		
	301	1.4	94.0			1.4	94.0	1.4	94.0		
GSI		2.9	**$2.764 MM**			5.0	**$4.545 MM**	6.0	**$5.160 MM**		
1974	101	—	65.0			—	65.0	1.8	65.0		
	201	1.2	97.0			3.3	97.0	3.5	97.0		
	301	1.7	93.5			1.7	93.5	1.7	93.5		
GSI		2.9	**$2.754 MM**			5.0	**$4.791 MM**	7.0	**$6.155 MM**		
1975	101	—	64.0			—	64.0	0.8	64.0	1.8	64.0
	201	0.9	96.5			3.0	96.5	4.2	96.5	4.2	96.5
	301	2.0				2.0	93.0	2.0	93.0	2.0	93.0
GSI		2.9	**$2.729 MM**			5.0	**$4.755 MM**	7.0	**$6.425 MM**	8.0	**$7.065 MM**
1976	101	—	63.0			—	63.0	—	63.0	1.8	63.0
	201	0.7	96.0			2.8	96.0	4.8	96.0	5.0	96.0
	301	2.2	92.5			2.2	92.5	2.2	92.5	2.2	92.5
GSI		2.9	**$2.712 MM**			5.0	**$4.720 MM**	7.0	**$6.640 MM**	9.0	**$7.962 MM**
1977	101	—	62.0			—	62.0	—	62.0	1.9	62.0
	201	0.5	95.5			2.6	95.5	4.6	95.5	5.7	95.5
	301	2.4	92.0			2.4	92.0	2.4	92.0	2.4	92.0
GSI		2.9	**$2.687 MM**			5.0	**$4.690 MM**	7.0	**$6.600 MM**	10.0	**$8.710 MM**
1978	101	—	61.0			—	61.0	—	61.0	1.9	61.0
	201	0.3	95.0			2.4	95.0	4.4	95.0	6.5	95.0
	301	2.6	91.5			2.6	91.5	2.6	91.5	2.6	91.5
GSI		2.9	**$2.665 MM**			5.0	**$4.660 MM**	7.0	**$6.570 MM**	11.0	**$9.730 MM**

MM = Million

"Other operating expense" includes costs for R&D, marketing, customer technical service, royalties (if there are some that have to be paid), and other administrative and general allocations.

(The earning statements presented in this case study do not provide for escalation of labor and other overhead costs. However, in the business environment that has existed during the past 10 years, a 4% to 6% annual escalation has actually taken place in labor costs, due to inflation.)

The difference between net sales realization and total cost of sales is profit before federal income taxes. The XYZ Co. earning statements are based on a 50% federal income tax.

Annual depreciation is taken as 6⅔% of plant investment. Cash flow is the sum of aftertax profit and depreciation.

Percent return on *plant* investment is calculated by dividing aftertax profit by plant investment and multiplying by 100. Percent return on *total* investment is calculated by dividing aftertax profit by total investment (plant investment plus working capital) and multiplying by 100.

In XYZ Co.'s current Product B business, the

Earning Statement for Base Case—Table XI

	1970	1971	1972	1973	1974	1975	1976	1977	1978
Gross sales, million gal./yr.	2.9	3.0	2.9	2.9	2.9	2.9	2.9	2.9	2.9
Gross sales income, $M	2,504	2,667	2,800	2,764	2,754	2,729	2,712	2,687	2,665
Returns, allowances, $M	19	20	21	21	21	20	20	19	19
Net sales income, $M	2,485	2,647	2,779	2,743	2,733	2,709	2,692	2,668	2,646
Freight, $M	218	228	229	231	211	209	209	209	209
Net sales realization, $M	2,267	2,419	2,550	2,512	2,522	2,500	2,483	2,459	2,437
Cost of sales:									
Variable cost of mfg, $M	753	880	978	961	944	927	915	913	901
Fixed cost of mfg., $M	683	683	683	683	683	683	683	683	683
Total plant cost, $M	1,436	1,563	1,661	1,644	1,627	1,610	1,598	1,596	1,584
Other operating expenses, $M	85	85	85	85	85	85	85	85	85
Total cost of sales, $M	1,521	1,648	1,746	1,729	1,712	1,695	1,683	1,681	1,669
Profit, pretax, $M	746	771	804	783	810	805	800	778	768
Profit, aftertax, $M	373	385	402	392	405	403	400	389	384
Depreciation, $M	140	140	140	140	140	140	140	140	140
Cash flow, $M	513	525	542	532	545	543	540	529	524
Return on plant investment, aftertax, %	17.7	18.3	19.1	18.7	19.3	19.2	19.0	18.5	18.3
Return on total investment, aftertax, %	14.9	15.4	16.1	15.7	16.2	16.1	16.0	15.5	15.3

M = 1,000

aftertax percent return on plant investment is 17.7% in 1970 (on plant capital investment of $2.1 million) as shown in Table XI. The corresponding percent return on total investment is 14.9% (on a total investment of $2.50 million). Overall aftertax profitability is increased to approximately $400,000/yr. from 1970 to 1972 by preferentially capturing ink-drying and adhesive business (at the expense of paint-additive business) during this period. A corresponding improvement is shown in cash flow to approximately $540,000/yr. from 1970 to 1973.

Similarly, aftertax return on total investment is increased to approximately 16% from 1970 to 1973. Since the base case is a forecast of what will happen if no new capital is appropriated, and the business must be operated at all times for maximum profits, justification for any new capital investment to provide increased capacity must be referenced to the profits shown in Table XI.

The Action Plan

During the analysis of the material in Tables I through XI, the following action plan was established, to be completed within the next six weeks:

1. The Marketing Dept. is to develop the interest in, and terms for purchase of, Product B from competitor companies over the period from 1971 through 1975. This will permit evaluation of a "no new investment" alternative in retaining share of market. In addition, this will permit evaluation of optimum timing of new investment and plant-expansion capacity, for maximizing overall project profitability.

2. The Engineering representatives are to develop preliminary investment estimates for new plant-capacity cases of 2, 4 and 8 million gal./yr., with the product mixes shown in Table X. This study is to include the new offsite investment required for location of this expansion at either the current Midwest manufacturing site for Product B, or the company's Gulf Coast site. Mill-cost statements are to be prepared for each of these plant-expansion capacities for each site (in liaison with Manufacturing).

3. Financial is to prepare an economic evaluation of each of the expansion-capacity alternatives and of the resale program (working in liaison with Marketing, Manufacturing and Engineering). Working-capital requirements for finished-product inventory and for raw-materials inventory are to be prepared by Marketing and Manufacturing, respectively.

4. The R&D group is to define funds needed for continuing product-line technical-support functions, as well as requirements for new product-grade development, during the next five years. In addition, R&D is to report on feasibility of developing a new grade of Product B suitable for penetrating the lubricating-oil-additives market at the next meeting.

5. To provide a tool for the economic evaluation of various alternatives, Financial is to prepare a decision-tree analysis of the project. Engineering will assist in this analysis by preparing tentative project-construction schedules for each of the cases.

This brings us to Project Meeting No. 2, which will be taken up in the next article. ■

ECONOMIC ANALYSIS Part 6

Case Study Shows Project Development In Action

The capacity-expansion plans of the XYZ Co. now enter the crucial middle phase, wherein decisions relating to market strategy—as well as to the size, location and timing of the expansion—must be firmed up.

IRVING LEIBSON and CHARLES A. TRISCHMAN, JR., Dart Industries, Inc.

In the previous article, we set up a hypothetical but realistic case study involving the XYZ Co. and its Product B—a liquid resin used in paints, inks and adhesives.

The firm is considering a major capacity expansion for this product, and has already held the first of three project-evaluation meetings. At this meeting, an action plan was established involving the Marketing, Engineering, Financial and R&D departments; each of these agreed to develop certain information and to present it at the second project-evaluation meeting, to be held six weeks after the first (say, on Sept. 16, 1970).

Let us start with this meeting, which will provide a good illustration of how project development activities are carried out in today's chemical process industries.

Evaluating the Resale Agreement

You will recall that the Marketing Dept. agreed to study the purchase of Product B from competitor companies over the 1971-1975 period, with a view to using a resale agreement to optimize the timing of new investment, and to build up a market before the new capacity came onstream. (Another purpose was to shed light on the attractiveness of using a resale agreement as a substitute for new investment—i.e., as a supplement to XYZ's existing capacity, so that XYZ could retain its share of market.)

The Marketing Dept. now reports that of the competitors surveyed, only Company A is willing to negotiate an agreement with XYZ Co. The terms offered by Company A are shown in Table I. This firm is chiefly interested in boosting its sales rate for the years 1971 and 1972 so as to help sell out the new capacity it expects to bring onstream at the end of 1970. Company A has offered a 20% coproducer's discount based on published price-list figures, with a take-or-pay minimum set at 50% of the quantities shown.

Company A is only willing to provide equivalents to grades 101 and 201 (the paint-additive and ink-drying-agent grades of Product B, respectively) and is not interested in supplying an equivalent to XYZ Co.'s grade 301 (the adhesive grade). Apparently this grade runs at a considerably lower production rate in Company A's process than it does in the XYZ Co. process.

Originally published October 1971.

Table II shows what the XYZ Co. stock balance for 1971 and 1972 will be if the resale program is implemented. Note that a good match is possible between market forecast requirements for Product B and total product availability from internal production as well as from the resale program. Adoption of this program would swing the product mix in the Midwest production plant strongly toward product grades 201 and 301, with no production of grade 101 (the paint-additive grade) scheduled in 1972.

Table III shows an earning statement for the base case plus the resale agreement during the years 1970, 1971 and 1972. Note that the "other operating expense" category has been increased by $5,000 and $30,000, respectively, in 1971 and 1972; this is necessary to cover the increased technical services and sales coverage required to handle the resale volumes.

Freight, and returns and allowances, have also been increased. The net profit from the resale program, as shown in Table III, is estimated at $20,000 in 1971 and $84,000 in 1972. However, the key advantage that the resale program offers the XYZ Co. is an opportunity to retain or increase its share of market in preparation for the new production capacity, so that efficient use of this new capacity can proceed more quickly.

Other Operating Expense

A summary of "other operating expense" as a function of Product B sales rate is shown in Table IV. Increases in marketing and customer-technical-services costs stem from increases in the number of accounts to be handled as the business is expanded. For Product B, these costs run approximately 2% of gross sales income. (Costs of this type can run from as little as 1% to as much as 10% of gross sales income, depending on the type of business, the technical sophistication of the individual customers, and the number of customers to be handled.)

The R&D costs shown in Table IV represent less than 1% of gross sales income for Product B. (Again depending on the type of business and the opportunities to make technological breakthroughs, R&D costs can run from 1% to as much as 5% or more of gross sales income.) As for product-line management,

XYZ Co.'s Stock Balance With Resale Program—Table II

	1971 Million Gal.	1972 Million Gal.
Grade 101:		
Take for resale from Co. A	0.50	1.20
Internal production	0.70	—
Total available for sale	1.20	1.20
Grade 201:		
Take for resale from Co. A	—	0.40
Internal production	1.50	1.80
Total available for sale	1.50	2.20
Grade 301:		
Internal production	0.80	1.10
Total	3.5	4.5

Earnings Statement for Base Case Plus Resale Agreement—Table III

	Thousands of Dollars		
	1970	1971	1972
Gross sales income	2,504	2,980	3,984
Returns and allowances	19	23	30
Net sales income	2,485	2,957	3,954
Freight	218	263	345
Net sales realization	2,267	2,694	3,609
Variable cost of manufacture	753	880	978
Fixed cost of manufacture	683	683	683
Total plant cost	1,436	1,563	1,661
Product purchase/resale agreement	—	250	945
Other operating expense	85	90	115
Total cost of sales	1,521	1,903	2,721
Profit, pretax	746	791	888
Base-case profit, pretax	746	771	804
Net profit from resale program	—	20	84

Basis: Gross sales of 2.9, 3.5 and 4.5 million gal./yr. for 1970, 1971 and 1972, respectively.

Company A's Resale Agreement—Table I

	1971		1972	
Product Grade	Quantity,* Million Gal.	Price,† ¢/Gal.	Quantity,* Million Gal.	Price,† ¢/Gal.
101	0.50	50.0	1.20	52.8
201	—	—	0.40	78.4
301	—	—	—	—
Total	0.50	50.0	1.60	59.0

* A "take or pay" minimum is to be set at 50% of the quantities shown.

† This transfer price is based on a 20% coproducer's discount based on published price-list figures.

"Other" Operating Expense—Table IV

	Sales Rate, Million Gal./Yr.			
	3	5	7	11
Marketing expense, $1,000/yr	48	64	80	120
Customer technical services	18	27	36	50
R&D	12	24	24	35
Product-line management	7	12	16	25
Total, $1,000/yr	85	127	156	230

Investment Estimates for Various Expansion Alternatives—Table V

	Mid-West	Mid-West	Gulf Coast	Mid-West	Gulf Coast	Mid-West	Gulf Coast
Existing capacity, million gal./yr.	3	—	—	—	—	—	—
Expansion alternatives, million gal./yr.	—	2	2	4	4	8	8
Investment, million $:							
Battery-limits process unit	1.8	1.6	1.4	2.8	2.6	4.9	4.5
Utilities and offsites	0.3	0.5	0.4	0.6	0.5	0.8	0.6
Total investment, million $	2.1*	2.1	1.8	3.4	3.1	5.7	5.1

* Construction completed Apr. 1966.

this is an allocated cost from various executive-cost centers.

Samples of the Product B grades that Company A can offer have been forwarded to the XYZ Co.'s R&D Dept. for evaluation as to suitability for XYZ's current customers. A secrecy agreement is under preparation by Company A's Legal Dept. so as to permit such technical discussions as may be necessary to ensure that XYZ Co.'s product specifications can be properly met by Company A.

Capital and Operating Costs

At the project-planning meeting, the Engineering Dept. presents its estimate of the plant-battery-limits and offsite investment requirements for 2, 4, and 8 million gal./yr. of new capacity at both the current Midwest site and the possible Gulf Coast site. This investment information, shown in Table V, is preliminary in nature and based on factor-type cost estimates developed from previous construction experience at both sites, together with telephone quotations from vendors on major items of process equipment. The estimates are believed to be accurate within ±15% and contain a 10% contingency. (Several excellent articles have been written concerning development of factor-type investment estimates.* Overall factors can best be developed from your company's previous construction experience with similar units.)

Table VI compares operating costs at the Midwest site with those at the Gulf Coast site for 2- and 4-million-gal./yr. facilities. Raw-material and packaging costs are anticipated to be the same at either site. The projected difference in operating expense stems from lower utilities costs, lower operating- and maintenance-labor costs, and lower charges for depreciation, taxes and insurance (as a result of the lower Gulf Coast investment requirement). The operating-cost incentive for locating the plant at the Gulf Coast site is equivalent to $78,000/yr. for a 2-million-gal./yr. facility, and $106,000/yr. for a 4-million-gal./yr. facility.

* For instance, see: Gallagher, J. T., Rapid Estimation of Plant Costs, in "Modern Cost Engineering Techniques," (ed. by Popper, H.), McGraw-Hill, 1970.

Distribution Costs; Overall Cost Patterns

Marketing has carried out the distribution-cost study shown in Table VII, for the key-account and target-account customer list that was presented in the previous article. The freight rates were developed by the Traffic Dept. Many of the new customers to be supplied are located in the southern part of the Midwest U.S.

A distribution-cost summary that compares the Midwest site to the Gulf Coast site is shown in Table VIII. Note that it tends to favor the Midwest in 1973, but to progressively favor the Gulf Coast as the new accounts to support the expansion are captured in 1974 and 1975.

Operating-Cost Comparison: Midwest vs. Gulf Coast—Table VI

	2MM Gal./Yr.		4MM Gal./Yr.	
	Mid-West	Gulf Coast	Mid-West	Gulf Coast
Operating-cost item, $1,000				
Utilities				
Steam	14	6	28	12
Recirc. cooling water	3	2	5	4
Electricity	4	3	7	5
Fuel gas	13	7	26	14
Total utilities	34	18	66	35
Controllable fixed costs				
Operating labor	170	153	310	279
Operating supplies	30	30	55	55
Maintenance labor	40	36	80	72
Maintenance material	45	45	73	73
Technical	15	15	15	15
Quality control	15	15	25	25
Employee benefits	53	48	95	86
Admin., general	30	30	50	50
Depreciation	140	120	227	207
Taxes, insurance	29	23	47	40
Total cont. fixed costs	567	515	977	902
Total operating costs (excl. raw matls. and packaging)	601	523	1,043	937
Difference in operating costs favoring Gulf Coast Site	78		106	

MM=Million

If we combine operating and distribution costs, as in Table IX, we see that there is a cumulative cost advantage of from \$58,000 to \$110,000/yr. favoring the Gulf Coast location, depending upon size of the expansion. In addition, an investment advantage of \$300,000 exists for the Gulf Coast site whether the plant is a 2- or a 4-million-gal./yr. expansion. (This advantage stems from a reduced offsite investment requirement for the Gulf Coast site, as well as a higher construction-labor productivity factor.)

The consensus of those present at the project planning meeting is that the combination of lower operating costs and reduced capital requirements can be considered decisive in establishing the Gulf Coast as XYZ Co.'s preferred location for any new capacity expansion for Product B.

Working-Capital Requirements

Marketing and Manufacturing have estimated the working-capital requirements for the various capacities, as detailed in Table X. Working capital is based on provision of a one-month allowance for accounts receivable (taken at the average selling price), a one-month allowance of finished-product inventory (taken at the sum of variable cost of manufacture and plant fixed costs) and a half-month allowance of raw materials (taken at equivalent raw-material purchase price). This represents the funds necessary to support the going business.

The allowance on accounts receivable is based on the current XYZ Co. experience that customers normally pay invoices within 30 days. (The credit terms required in securing new accounts will have to be

Summary of Distribution Costs for Product B's Customers—Table VII

Customer Company	Prod. Grade	Customer Location	Freight Rate, ¢/Gal., From Site to Customer: Midwest	Gulf Coast
Acme	101	N. Illinois	6.0	10.0
Alpha	301	N. Illinois	6.0	10.0
Baker	201	S. Michigan	7.0	15.0
Burns	301	S. Michigan	7.0	15.0
Comet	101	S. Indiana	9.0	9.0
Crescent	201	N. Illinois	6.0	10.0
Capital	301	S. Wisconsin	7.0	10.0
Delta	101	S. Wisconsin	7.0	10.0
Essex	101	S. Illinois	8.0	8.0
Federal	201	S. Illinois	8.0	8.0
First	201	W. Ohio	8.0	11.0
Forefront	301	S. Illinois	8.0	8.0
General	101	S. Illinois	8.0	8.0
Globe	201	W. Tennessee	9.0	7.0
Graphic	201	E. Missouri	9.0	7.5
Guardian	301	E. Missouri	9.0	7.5
Pioneer	201	N. Texas	12.0	6.0

Distribution Cost Summary for All Expansion Cases—Table VIII

	Distribution Cost, \$1,000/Yr.: 1970	1971	1972	1973	1974	1975
Base Case (BC)	218	228	229	231	211	209
BC + resale	218	263	345	—	—	—
BC + resale + 2-million-gal./yr. expansion in Midwest	218	263	345	403	394	396
BC + resale + 2-million-gal./yr. expansion on Gulf Coast	218	263	345	416	389	376
BC + resale + 4-million-gal./yr. expansion in Midwest	218	263	345	465	547	564
BC + resale + 4-million-gal./yr. expansion on Gulf Coast	218	263	345	513	577	560
Advantage: Gulf Coast Over Midwest						
2 million gal./yr.	—	—	—	(13)	5	20
4 million gal./yr.	—	—	—	(48)	(30)	4

Gulf Coast's Cost Advantage for XYZ Co.'s Expansion—Table IX

	Gulf Coast Less Midwest, Credit, \$1,000: 1973	1974	1975
For 2 million gal./yr.:			
Distribution costs	(13)	5	20
Operating costs	78	78	78
Net credit for Gulf Coast	65	83	98
For 4 million gal./yr.:			
Distribution costs	(48)	(30)	4
Operating costs	106	106	106
Net credit for Gulf Coast	58	76	110

Working-Capital Requirements for Various Capacities—Table X

Working capital, \$1,000 for:	Capacity, Million Gal./Yr.: 3	5	7	11
Product inventory (one month of production)	134	239	315	492
Raw-material inventory (half month of usage)	39	69	91	143
Accounts receivable (one month of sales)	227	396	539	845
Total required (based on 1975 operating plan)	400	704	945	1,480

Earnings Projected for Base Case Plus 2-Million-Gal./Yr. Expansion at Gulf Coast Site—Table XI

	$1,000 (Except as Noted) in Year:					
	1973	1974	1975	1976	1977	1978
Gross sales volume, million gal./yr.:	5.0	5.0	5.0	5.0	5.0	5.0
Gross sales income	4,545	4,791	4,755	4,720	4,690	4,660
Returns, allowances	34	36	36	36	35	35
Net sales income	4,511	4,755	4,719	4,684	4,655	4,625
Freight	416	389	376	364	353	345
Net sales realization	4,095	4,366	4,343	4,320	4,302	4,280
Variable cost of manuf.	1,546	1,681	1,664	1,650	1,637	1,625
Fixed cost of manuf.	1,198	1,198	1,198	1,198	1,198	1,198
Total plant cost	2,744	2,879	2,862	2,848	2,835	2,823
Other operating expense	127	127	127	127	127	127
Total cost of sales	2,871	3,006	2,989	2,975	2,962	2,950
Profit, pretax	1,224	1,360	1,354	1,345	1,340	1,330
Profit, after fed. income tax (AFIT)	612	680	677	672	670	665
Base-Case profit, AFIT	392	405	403	400	389	384
For 2-million gal./yr. expansion:						
Net profit, AFIT	230	275	274	272	281	281
Return on plant invest., AFIT, %	12.8	15.3	15.3	15.1	15.6	15.6
Return on total invest., AFIT, %	11.0	13.1	13.1	12.9	13.3	13.3
Depreciation	120	120	120	120	120	120
Cash flow	350	395	394	392	401	401
Cumulative cash flow	350	745	1,139	1,531	1,932	2,333

Payout time = 4.7 years

Earnings Projected for Base Case Plus 4-Million-Gal./Yr. Expansion at Gulf Coast Site—Table XII

	$1,000 (Except as Noted) in Year:					
	1973	1974	1975	1976	1977	1978
Gross sales volume, million gal./yr.:	6.0	7.0	7.0	7.0	7.0	7.0
Gross sales income	5,160	6,155	6,425	6,640	6,600	6,570
Returns, allowances	39	46	48	50	50	49
Net sales income	5,121	6,109	6,377	6,590	6,550	6,521
Freight	513	577	535	511	507	502
Net sales realization	4,608	5,532	5,842	6,079	6,043	6,019
Variable cost of manuf.	1,684	1,996	2,194	2,183	2,172	2,161
Fixed cost of manuf.	1,585	1,585	1,585	1,585	1,585	1,585
Total plant cost	3,269	3,581	3,779	3,768	3,757	3,746
Other operating expenses	156	156	156	156	156	156
Total cost of sales	3,425	3,737	3,935	3,924	3,913	3,902
Profit, pretax	1,183	1,795	1,907	2,155	2,130	2,117
Profit, after fed. income tax (AFIT)	592	898	954	1,078	1,065	1,058
Base-case profit, AFIT	392	405	403	400	389	384
For 4 million-gal./yr. expansion:						
Net profit, AFIT	200	493	551	678	676	674
Return on plant invest., AFIT, %	6.5	15.9	17.8	21.9	21.8	21.7
Return on total invest., AFIT, %	5.0	12.2	13.7	16.8	16.8	16.7
Depreciation	207	207	207	207	207	207
Cash flow	407	700	758	885	883	881
Cumulative cash flow	407	1,107	1,865	2,750	3,633	4,514

Payout time = 4.4 years.

Earnings Projected for Base Case Plus 8-Million-Gal./Yr. Expansion at Gulf Coast Site—Table XIII

	$1,000 (Except as Noted) In Year:					
	1973	1974	1975	1976	1977	1978
Gross sales volume, million gal./yr.:	6.0	7.0	8.0	9.0	10.0	11.0
Gross sales income	5,160	6,155	7,065	7,962	8,810	9,730
Returns, allowances	39	46	53	60	65	73
Net sales income	5,121	6,109	7,012	7,902	8,745	9,657
Freight	513	577	610	663	749	834
Net sales realization	4,608	5,532	6,402	7,239	7,996	8,823
Variable cost of manuf.	1,684	1,996	2,337	2,685	3,011	3,359
Fixed cost of manuf.	2,150	2,150	2,150	2,150	2,150	2,150
Total plant cost	3,834	4,146	4,487	4,835	5,161	5,509
Other operating expenses	156	156	170	193	210	230
Total cost of sales	3,990	4,302	4,657	5,028	5,371	5,739
Profit, pretax	618	1,230	1,745	2,211	2,625	3,084
Profit, after fed. income tax (AFIT)	309	615	872	1,106	1,312	1,542
Base-Case profit, AFIT	392	405	403	400	389	384
For 8-million-gal./yr. expansion:						
Net profit, AFIT	(83)	210	469	706	923	1,158
Return on plant invest., AFIT, %	—	4.1	9.2	13.8	18.3	22.7
Return on total invest., AFIT, %	—	3.2	7.1	10.5	14.0	17.5
Depreciation	340	340	340	340	340	340
Cash flow	257	555	809	1,046	1,263	1,498
Cumulative cash flow	257	812	1,621	2,667	3,930	5,428

Payout time = 5.8 years.

carefully reviewed to ensure that proper control is maintained on working capital as the business is expanded.) Furthermore, Marketing has found that it is necèssary to have a one-month finished-product inventory on hand to provide proper customer service on orders, while simultaneously minimizing product-grade transitions in the manufacturing plant. Raw-material deliveries from suppliers are such that approximately a half-month inventory is adequate to permit purchases in bulk.

The above illustrates how working-capital requirements must reflect the needs of the business. It is important to properly consider the effect of any capacity expansion on working-capital requirements, and to evaluate return on *total* investment (not just the new-plant investment).

Summary Profitability Comparison for Expansion Cases—Table XIV

	Expansion Capacity, MM Gal./Yr			
	BC	2	4	8
Plant investment, $MM	2.1	1.8	3.1	5.1
Working capital, $MM	0.4	0.3	0.6	1.1
Total investment, $MM	2.5	2.1	3.7	6.2
6-Yr. average* return on plant investment, AFIT, %	18.8	15.0	17.6	11.1
6-Yr. average* return on total investment AFIT, %	15.8	12.8	13.5	8.5
Payout time, years	—	4.7	4.4	5.8
Breakeven operation as % of capacity based on product mix in 1978	50	49	42	41

MM=Million BC=Base Case

* 1973-1978

Analyzing the Earnings Statements

You will recall that as part of the previous project-planning meeting, the Financial Dept. prepared an earnings statement that was an estimate of what would happen if Product B capacity were merely maintained, with no further expansion and no resale agreement. This statement represented the Base Case (BC), and was shown in Table XI of the previous article.

For the present project-planning meeting, Financial has prepared statements that show payout and give the percent return on plant investment and on total investment (based on the first six years of expansion-project life) for these cases:

- BC plus a Product B resale agreement in 1971 and 1972 (Table III).
- BC plus a 2-million-gal./yr. plant expansion coming onstream in late 1972 (Table XI).
- BC plus a 4-million-gal./yr. plant expansion coming onstream in late 1972 (Table XII).
- BC plus an 8-million-gal./yr. plant expansion coming onstream in late 1972 (Table XIII).

A review of these various earning statements reveals that the most attractive case, based on return on total investment as well as payout time, is a

Breakeven Operation as Percent of Plant Capacity for 1978 Product Mix—Table XV

1. Expansion capacity, million gal./yr.	Base Case	2	4	8
2. Net sales realization, $1,000	2,437	1,843	3,582	6,386
3. Variable cost of manuf., $1,000	901	724	1,260	2,458
4. Item (2) minus (3)	1,536	1,119	2,322	3,928
5. Fixed cost of manuf., $1,000	683	515	902	1,467
6. Other operating expenses, $1,000	85	42	71	145
7. Item (5) plus (6)	768	557	973	1,612
8. Breakeven operation as % of capacity [(7) ÷ (4)] × 100	50	50	42	41

$$(\text{Breakeven operation as \% of capacity} = \frac{\text{Fixed cost of manuf.} + \text{Other operating expense}}{\text{Net sales realization} - \text{Variable cost of manuf.}} \times 100)$$

4-million-gal./yr. expansion at the Gulf Coast site. A summary of the results of these economic evaluations is shown in Table XIV.

Table XV shows what breakeven operation for the 1978 product mix would mean in terms of plant capacity. (Breakeven operation is defined as the percent of plant capacity at which the net sales realization is equivalent to the actual cost of sales for a profitability of zero, assuming no change in inventory level). The economies normally inherent in efficiently increasing scale of manufacture are apparent in the table; note that breakeven operation (as percent of capacity) decreases from 50% for the 2-million-gal./yr. expansion, to 41% for the 8-million-gal./yr. expansion, based on the 1978 product mix. This criterion is an additional useful index of degree of risk, in regard to the project's vulnerability to decline in sales volume.

A summary of total-product-line profitability is shown in Table XVI.

Reports From R&D

During the meeting, the R&D representative states that an expenditure of $35,000 will be required over a one-year period in order to upgrade quality of the adhesive grade, to preserve its competitive position.

Profitability of Total Product Line Including Expansion Cases—Table XVI

	Total Capacity, MM Gal./Yr.			
	3	5	7	11
Plant investment, $MM	2.1	3.9	5.2	7.2
Working capital, $MM	0.4	0.7	1.0	1.5
Total investment, $MM	2.5	4.6	6.2	8.7
6-Yr. avg. return on plant invest., AFIT, %	18.8	16.9	19.7	13.3
6-Yr. avg. return on total invest., AFIT, %	15.8	14.4	16.3	11.7
Breakdown operation as % of capacity based on product mix in 1978	50	50	45	44

MM=Million

R&D considers it feasible to develop a new grade of Product B suitable for viscosity control of lubricating oils in automotive applications (using the existing XYZ Co. process for manufacturing Product B). An analysis of the necessary development activities and field-evaluation trials suggests that a two-year period will be involved, at an approximate cost of $50,000 the first year and $100,000 the second. Furthermore, an additional capital investment may be required to permit manufacture of this new product grade. A very preliminary estimate of this investment is $250,000.

The R&D representative also has a report on the work his group has been doing to develop a synthesis process for internal manufacture of one of Product B's raw-material components. It is agreed that an economic evaluation of this approach (vs. the continued purchase of the component) should be made prior to spending funds for the construction of a pilot plant. Financial is to conduct this evaluation, working with R&D and Engineering. (We will have more to say on the procedure for making such "make or buy" analyses in a later article.)

Viewing the Data

At this point in the meeting, the chairman requests each representative to present his analysis of the information that has been presented in this planning meeting. The marketing representative is the first to speak, and he recommends installation of a 4-million-gal./yr. expansion on the Gulf Coast site as soon as possible, because this shows the best economics.

In the ensuing discussion, the Financial representative indicates that he believes it is too risky for the XYZ Co. to base the expansion capacity on a planned 50% increase in its share of market between 1970 and 1973. He expresses strong reservation over the ability of the marketing effort to take this amount of business away from the competition without the risk of severe price deterioration and a resultant precipitous drop in profitability.

Furthermore, the Financial representative points

Inventory Calculation and Control

The effects of changes in inventory level from one year to the next, although not shown in the case study, must be considered.

Normally, production volume is not necessarily directly equivalent to sales volume; there can be a net increase or decrease in finished-product inventory from one year to the next. In this situation, the variable cost of manufacture must account for the total production in that year.

If an increase in inventory occurs, the quantity of material involved is valued at the sum of the corresponding variable and fixed cost of manufacture (or net sales realization, whichever is less) and this is deducted from the total cost of sales for that year.

Conversely, if a net decrease in inventory occurs, this quantity of material is valued in the same way, and the dollars involved are added to the total cost of sales.

Production must be carefully controlled because very large increases in working-capital requirements can result if the plant is operated merely to produce material for inventory over any extended period of time. Most companies, as part of their financial control system, issue a monthly statement giving total inventory of finished goods carried on the books, and the value of this inventory in terms of manufacturing costs, referenced to the current average selling price.

If the inventory valuation exceeds current sales price, proper accounting practice requires that the inventory be written down, and the loss taken directly to the earnings statement for that month. This unfortunate situation can sometimes occur during a period of unstable, sharply declining sales prices. Prudent control of finished-product inventory level requires sophisticated general management that not only responds to changes, but anticipates them.

out that the profitability of the expansion depends heavily on successful completion of the R&D effort to provide further quality improvements for the adhesives grade of Product B. He expresses concern that this is necessary just to be able to retain the 15.4% share of market currently enjoyed by XYZ Co. on the adhesives grade. He recommends that a new sales forecast be prepared based on retaining existing percent share of market for each of the product grades, and on increasing the sales volume annually, commensurate with the predicted growth of the total market for the various grades of Product B.

The Marketing representative counters by expressing his faith that his department's share-of-market forecast for Product B is not only attainable but quite conservative. Furthermore, he indicates that if the new grade of Product B developed by Company A for the lubricating-oil-additives market receives widespread acceptance, then total market growth for Product B can be expected to double over the period from 1971 to 1975. For these reasons, he holds to his recommendation that XYZ Co. install a 4-million-gal./yr. facility immediately on the Gulf Coast.

After protracted discussion, the consensus of the group is that the development of a lubricating-oil-additive grade of Product B should be evaluated as a completely separate matter from the plant expansion. Financial is requested to work with Production, R&D and Marketing in developing the economic justification for this research project.*

The New Action Plan

At the conclusion of the meeting, the following action plan is set up to be completed within the next eight weeks:

1. Marketing is to prepare a new sales forecast in support of the capacity expansion. This is to be based on retention (rather than expansion) of current share of market for each individual grade, with annual increases in sales volume based on total U.S. market growth for Product B.

2. Marketing is to pursue negotiation of a purchase agreement for Product B with Company A for the years 1971 and 1972.

3. Engineering is to develop an investment estimate for a 2-million-gal./yr. plant expandable to 4-million-gal./yr. vs. the initial installation of a 4-million-gal./yr. plant, or successive installation of two 2-million-gal./yr. plants, at the Gulf Coast site. This is to include the preparation of budget-type investment estimates based on written quotations from equipment vendors on all major equipment items. Furthermore, Engineering is to develop a project construction schedule, so as to determine the long-term-delivery equipment items that should be placed on order at the earliest opportunity, as well as to define whether or not a third-quarter 1972 completion date for the new capacity is realistic.

4. R&D is to accelerate work to improve the quality of the adhesive grade of Product B to ensure that the current share of market for this grade can be retained with the expansion project.

5. Financial is to work with the other functional groups and to develop project profitability for the following three cases: (1) Base Case (BC) plus installation of a minimum-investment 2-million-gal./yr. plant, with the subsequent addition of a second 2-million-gal./yr. plant as required by the new sales forecast; (2) BC plus the initial installation of a 2-million-gal./yr. expandable plant, with the subsequent expansion of this plant to a capacity of 4-million gal./yr., as called for by the new sales forecast; (3) BC plus the initial installation of a 4-million-gal./yr. plant.

It is believed that the new sales forecast will not support any further consideration of an 8-million-gal./yr. expansion.

A date is established for Project Planning Meeting No. 3. We will examine this final meeting in the next article, and then summarize some of the most important features of this case study. ■

* A forthcoming article will cover evaluation of R&D programs of this type and will develop this as a brief case-history example.

ECONOMIC ANALYSIS

Part 7

Final Stage in the Project

After bringing XYZ Co.'s capacity-expansion study to the point where definitive decisions are made on the key marketing, technical and financial aspects, the authors supply an overall perspective on this type of project.

IRVING LEIBSON and CHARLES A. TRISCHMAN, JR., Dart Industries, Inc.

We are now ready to wind up our illustrative case study involving XYZ Co. and its expansion plans for Product B (a liquid resin, made in three grades).

You will recall that central to the development of this major capital project are three planning meetings, the first and second of which have already taken place. At the close of the second meeting, the participants from the Marketing, Engineering, R&D and Financial departments agreed to undertake certain assignments as part of the new plan of action, and to report on these assignments at the third meeting, which we will now cover.

The Marketing Dept.'s assignment was to prepare a new and more-conservative sales forecast, to be based on retention (rather than expansion) of the current market share for each individual product grade, with the annual increases in sales volume based on the total U.S. market growth predicted for Product B and its equivalents.

This revised forecast is shown in Table I. Note that the increase in sales of the paint-additive grade (i.e., Grade 101) is now forecast at 3%/yr.; of the ink-drying grade (Grade 201) at 19%/yr., and of the adhesive grade (Grade 301) at 10%/yr. Of course, these rates are consid-

Revised Market Forecast for XYZ Co.'s Product B—Table I

(V = Volume, Million Gal./Yr.; P = Sales Price, ¢/Gal.)

Product Grade No.	1970		1971		1972		1973		1974		1975		1976		1977		1978		1979	
	V	P	V.	P	V	P	V	P	V	P	V	P	V	P	V	P	V	P	V	P
101	1.20	69.3	1.2	62.5	1.3	66.0	1.3	61.5	1.4	65.0	1.4	64.0	1.5	63.0	1.5	62.0	1.6	61.0	1.7	60.0
201	1.14	100.0	1.4	98.5	1.7	98.0	2.0	96.5	2.3	97.0	2.7	96.5	3.2	96.0	3.8	95.5	4.5	95.0	5.3	94.5
301	0.56	95.0	0.6	94.0	0.7	94.2	0.8	94.0	0.9	93.5	1.0	93.0	1.1	92.5	1.2	92.0	1.3	91.5	1.4	91.0
	2.90	86.3	3.2	84.2	3.7	86.0	4.1	84.9	4.6	86.6	5.1	86.9	5.8	86.8	6.5	87.1	7.5	85.9	8.4	86.9

Originally published November 1971.

Development Case Study

erably less than in the original market forecast (which had assumed an increased share of the market as well as an increased market); note that total sales for 1978 are now projected at 7.5 million gal., vs. 11.0 million gal. in the original forecast.

Table II presents the revised product mix for the Base Case plus the 2- and 4-million-gal./yr. expansion studies, based on the revised sales forecast.

The revised earnings statement for the Base Case or existing plant is shown in Table III. As anticipated, the more-conservative revised forecast results in reduced profits for the Base Case: Aftertax annual profits for 1978 are now $373,000 as compared to $384,000 projected in the original Base Case earnings statement. Likewise, the percent return on total investment for 1978 is reduced from 15.3% to 14.9%.

The revised sales forecast for 1971 and 1972 still provides for more sales than can be produced in the existing plant—calling for the purchase and resale of 0.2 million gal. of Grade 101 in 1971, and 0.8 million gal. of Grade 101 in 1972, from Company A. The revised stock-balance for the resale program is shown in Table IV. The revised earnings-statement for the Base Case, plus resale program, appears in Table V. Net profits of $8,000 and $37,000 for 1971 and 1972, respectively, result from the resale program.

Because of the need to maintain sales momentum for the expansion program, and because of the favorable profits resulting from the resale agreement, the planning group decides that XYZ Co. should proceed immediately to consummate the resale agreement with Company A for the quantities needed. Lawyers of both companies are to review the previously discussed terms and prepare the agreement in final form, for top-management approval.

New Data From Engineering

At the previous meeting, Engineering was asked to provide Financial with capital-cost estimates for various alternatives in regard to a 4-million-gal./yr. addition to capacity. Engineering now has these investment estimates; they are of budget quality (i.e., of ±10% accuracy, with a 10% contingency included). As shown in Table VI, they cover:

- Successive installation of two minimum-investment 2-million-gal./yr. plants.
- A 2-million-gal./yr. plant expandable to 4 million gal./yr.
- Initial installation of a 4-million-gal./yr. plant.

Note that the cost of the one-step, 4-million-gal./yr. expansion is $200,000 less than the final cost for the two-step expansion program. The increased cost of the latter reflects the expected continuing escalation of construction costs, and the need for Engineering to run, in effect, two separate projects at different times. As for the successive installation of two separate, minimum-investment, 2-million-gal./yr. units, the total cost for this alternative is $700,000 greater than the initial installation of a 4-million-gal./yr. plant, and $500,000 greater than the total cost from subsequent expansion of an easily expandable 2-million-gal/yr. plant.

The Engineering Dept. has also drawn up the construction schedule shown in Fig. 1, which indicates that a Sept. 1, 1972, startup date is possible, provided that Engineering gets approval to place long-term-delivery equipment items on order no later than Mar. 1, 1971.

Evaluation Leads to Decision

The Financial Dept. has been working with the other functional groups in preparing earnings statements for the expansion alternatives. These are shown in Tables VII, VIII, and IX. After reviewing these statements, the participants at the third project-planning meeting decide that the case for the 2-million-gal./yr. plant installed in 1973, expandable to 4-million-gal./yr. capacity by 1976,

Revised Product Mix and Gross Sales Income (GSI) for Base Case and Expansion Study—Table II

Case Year	Product Grade	Base Case			Base Case + Resale			Base Case + 2 MM Gal./Yr.*			Base Case + 4 MM Gal./Yr.†		
		MM Gal.	¢/Gal.	$ MM	MM Gal.	¢/Gal.	$ MM	MM Gal.	¢/Gal.	$ MM	MM Gal.	¢/Gal.	$ MM
1971	101	1.0	62.5		1.2								
	201	1.4	98.5		1.4								
	301	0.6	94.0		0.6								
GSI		3.0		**2.568**	3.2		**2.693**						
1972	101	0.5	66.0		1.3								
	201	1.7	98.0		1.7								
	301	0.7	94.2		0.7								
GSI		2.9		**2.655**	3.7		**3.183**						
1973	101	0.1	61.5					1.3			1.3		
	201	2.0	96.5					2.0			2.0		
	301	0.8	94.0					0.8			0.8		
GSI		2.9		**2.744**				4.1		**3.482**	4.1		**3.482**
1974	101	—	65.0					1.4			1.4		
	201	2.0	97.0					2.3			2.3		
	301	0.9	93.5					0.7			0.9		
GSI		2.9		**2.782**				4.6		**3.983**	4.6		**3.983**
1975	101	—	64.0					1.3			1.4		
	201	1.9	96.5					2.7			2.7		
	301	1.0	93.0					1.0			1.0		
GSI		2.9		**2.764**				5.0		**4.368**	5.1		**4.432**
1976	101	—	63.0					0.7			1.5		
	201	1.8	96.0					3.2			3.2		
	301	1.1	92.5					1.1			1.1		
GSI		2.9		**2.746**				5.0		**4.531**	5.8		**5.035**
1977	101	—	62.0					—			1.5		
	201	1.7	95.5					3.8			3.8		
	301	1.2	92.0					1.2			1.2		
GSI		2.9		**2.728**				5.0		**4.733**	6.5		**5.663**
1978	101	—	61.0					—			1.2		
	201	1.6	95.0					3.7			4.5		
	301	1.3	91.5					1.3			1.3		
GSI		2.9		**2.710**				5.0		**4.705**	7.0		**6.197**
1979	101	—	60.0					—			0.3		
	201	1.5	94.5					3.6			5.3		
	301	1.4	91.0					1.4			1.4		
GSI		2.9		**2.692**				5.0		**4.676**	7.0		**6.463**

MM = Million

* Product mix applicable to operation with existing plant plus 2-MM-gal./yr. expansion from either (1) installation of the first of two separate 2-MM-gal./yr. plants, or (2) installation of the 2-MM-gal./yr. plant expandable to 4-MM-gal./yr. capacity.

† Product mix applicable to operation with existing plant plus 4-MM-gal./yr. expansion from either (1) installation to two separate 2-MM-gal./yr. plants, or (2) installation of the 2-MM-gal./yr. plant expandable to 4-MM-gal./yr. capacity, or (3) installation of a 4-MM-gal./yr. plant.

will be presented to the Board of Directors for approval, because:

1. This two-step expansion will achieve an acceptable return on investment of 14.1% by 1979. The cumulative cash-flow of $3,228,000 by 1979 exceeds the cumulative cash-flow for the other two proposals.

2. The case for the initial installation of a 4-million-gal./yr. expansion is not attractive because the excessive losses and low earnings in the early years. 1973-1975, do not justify the investment required. A 2-million-gal./yr. unit either expandable or nonexpandable, is the maximum new capacity that can be justified for the period 1973-1975. (Installation of at least 2 million gal./yr. of additional capacity is required by 1973 to maintain the sales momentum and market penetration required to achieve the desired revenues, earnings, and return-on-investment projected for 1978 and 1979, for either of the three expansion cases.)

Revised Earnings Statement for Base Case—Table III

	1970	1971	1972	1973	1974	1975	1976	1977	1978	1979
Gross sales, million gal./yr.	2.9	3.0	2.9	2.9	2.9	2.9	2.9	2.9	2.9	2.9
Gross sales income, $M	2,504	2,568	2,655	2,744	2,782	2,764	2,746	2,728	2,710	2,692
Returns, allowances, $M	19	19	20	21	21	21	21	20	20	20
Net sales income, $M	2,485	2,549	2,635	2,723	2,761	2,743	2,725	2,708	2,690	2,672
Freight, $M	218	228	228	238	220	211	210	209	210	211
Net sales realization, $M	2,267	2,321	2,407	2,485	2,541	2,532	2,515	2,499	2,480	2,461
Cost of sales, $M:										
Variable cost of manuf.	753	827	893	974	990	984	978	973	967	961
Fixed cost of manuf.	683	683	683	683	683	683	683	683	683	683
Total plant cost	1,436	1,510	1,576	1,657	1,673	1,667	1,661	1,656	1,650	1,644
Other operating expenses	85	85	120	85	85	85	85	85	85	85
Total cost of sales	1,521	1,595	1,696	1,742	1,758	1,752	1,746	1,741	1,735	1,729
Profit, BFIT, $M	746	726	711	743	783	780	769	758	745	732
Profit, AFIT, $M	373	363	356	372	392	390	385	379	373	366
Depreciation, $M	140	140	140	140	140	140	140	140	140	140
Cash flow, $M	513	503	496	512	532	530	525	519	513	506
Return on plant investment, AFIT, %	17.7	17.3	16.9	17.7	18.7	18.6	18.3	18.0	17.8	17.4
Return on total investment, AFIT, %	14.9	14.5	14.2	14.9	15.7	15.6	15.4	15.2	14.9	14.6

M = 1,000 BFIT = Before Federal Income Tax AFIT = After Federal Income Tax

3. The two-step expansion shows the highest Present Worth and Discounted Cash Flow.

Let us analyze the expansion decision further, particularly in regard to this last factor.

The decision really boils down to a choice between building either two separate 2-million-gal./yr. plants—one installed by 1973 and the other by mid-1976—or building a 2-million-gal./yr. plant expandable to 4 million gal./yr., with the initial phase being installed by 1973, and the second phase by mid-1976.

XYZ Co.'s Revised Stock Balance, With Resale Program—Table IV

	1971 Million Gal.	1972 Million Gal.
Grade 101		
Taken for resale from Company A	0.2	0.8
Internal production	1.0	0.5
Total available for sales	1.2	1.3
Grade 201		
Taken for resale from Company A	—	—
Internal production	1.4	1.7
Total available for sales	1.4	1.7
Grade 301		
Taken for resale from Company A	—	—
Internal production	0.6	0.7
Total available for sales	0.6	0.7
Total stock balance, all grades of product	**3.2**	**3.7**

We have noted that the investment for the expandable plant at the 4-million-gal./yr. capacity is $500,000 less than for two separate 2-million-gal./yr. plants, and that operating costs will also be lower. However, the investment for the first 2-million-gal./yr. plant with the expandable features will be $2.5 million, as compared with $1.8 million for the first 2-million-gal./yr. plant without the expandable features. Thus, $0.7 million is being invested initially to save $0.5 million (over and above recoupment) three years later.

The advantage of this approach (vs. the one involving two separate minimum-investment plants) is not obvious from the earnings statements presented in Tables VII, VIII and IX. An economic-analysis technique to clearly delineate the advantages of one project versus another is required in this example. Furthermore, the analysis must consider future earnings of the project (past the seven years that are forecast in detail) to the end of the estimated useful life of the project. The Present Worth (PW) and Discounted Cash Flow (DCF) techniques, to be examined more fully in the next article, provide the tools to guide selection of expansion methods.

Present Worth and Discounted Cash Flow

The results of the PW and DCF calculations

Revised Earnings Statement for Base Case plus Resale Agreement—Table V

	1970	1971	1972
Gross sales, million gal./yr.	2.9	3.2	3.7
Gross sales income, $M	2,504	2,693	3,183
Returns, allowances, $M	19	20	24
Net sales income, $M	2,485	2,673	3,159
Freight, $M	218	242	283
Net sales realization	2,267	2,431	2,876
Cost of sales, $M:			
Variable cost of manuf.	753	827	893
Fixed cost of manuf.	683	683	683
Total plant cost	1,436	1,510	1,576
Product resale agreement	—	100	422
Other operating expenses	85	87	130
Total cost of sales	1,521	1,697	2,128
Profit, BFIT, $M	746	734	748
Base-Case profit, BFIT, $M	746	726	711
Net profit from resale program, $M	—	8	37

M = 1,000 BFIT = Before Federal Income Tax

Investment Estimates for Expansion Alternatives at Gulf Coast Site—Table VI

	Capacity, Million Gal./Yr.	Investment, Million $
One step expansion:		
Plant for small expansion	2	1.8
Plant for large expansion	4	3.1
Two separate-plant expansions:		
First 2-million-gal. plant	2	1.8
With addition of second plant.	4	3.8
Building an expandable plant:		
First 2-million-gal. plant	2	2.5
Expanded to 4 million gal.	4	3.3

for the three project alternatives are summarized below:

Expansion Case	PW $1,000	DCF %
(A) Two separate 2-million-gal./yr. expansions	518	12.25
(B) One 2-million-gal./yr. plant expandable to 4 million gal./yr.	678	12.65
(C) One initial 4-million-gal./yr. expansion	245	10.81

Revised Base Case Plus Two Separate 2-Million-Gal./Yr. Expansions—Table VII

	1973	1974	1975	1976*	1977	1978	1979
Gross sales, million gal./yr.	4.1	4.6	5.0	5.8	6.5	7.0	7.0
Gross sales income, $M	3,482	3,983	4,368	5,035	5,663	6,197	6,463
Returns, allowances, $M	26	30	33	38	42	46	48
Net sales income, $M	3,456	3,953	4,335	4,997	5,621	6,151	6,415
Freight, $M	324	337	354	406	454	489	491
Net sales realization, $M	3,132	3,616	3,981	4,591	5,167	5,662	5,924
Cost of sales, $M:							
Variable cost of manuf.	1,137	1,286	1,442	1,674	1,914	2,148	2,336
Fixed cost of manuf.	1,198	1,198	1,198	1,400	1,673	1,673	1,673
Total plant cost	2,335	2,484	2,640	3,074	3,587	3,821	4,009
Other operating expenses	127	127	127	145	156	156	156
Total cost of sales	2,462	2,611	2,767	3,219	3,743	3,977	4,165
Profit, BFIT, $M	670	1,005	1,214	1,372	1,424	1,685	1,759
Profit, AFIT, $M	335	503	607	686	712	843	880
Base-Case profit, AFIT, $M	372	392	390	385	379	373	366
Net Profit, AFIT, $M	(37)	111	217	301	333	470	514
Return on plant invest., AFIT, %	—	6.2	12.1	7.9	8.8	12.4	13.5
Return of total invest., AFIT, %	—	5.3	10.3	6.9	7.7	10.8	11.8
Depreciation, $M	120	120	120	175	253	253	253
Cash flow, $M	83	231	337	476	586	723	767
Cumulative cash flow, $M	83	314	651	1,127	1,713	2,436	3,203

M = 1,000 BFIT = Before Federal Income Tax AFIT = After Federal Income Tax
* Assumes second 2-million gal./yr. plant is operating for last 5 months of 1976.

Revised Base Case Plus Initial Installation of a 2-Million-Gal./Yr. Plant That Is Expanded to 4-Million-Gal./Yr. Capacity in 1976—Table VIII

	1973	1974	1975	1976*	1977	1978	1979
Gross sales, million gal./yr.	4.1	4.6	5.0	5.8	6.5	7.0	7.0
Gross sales income, $M	3,482	3,983	4,368	5,035	5,663	6,197	6,463
Returns, allowances, $M	26	30	33	38	42	46	48
Net sales income, $M	3,456	3,953	4,335	4,997	5,621	6,151	6,415
Freight, $M	324	337	354	406	454	489	491
Net sales realization, $M	3,132	3,616	3,981	4,591	5,167	5,662	5,924
Cost of Sales, $M:							
Variable cost of manufacture	1,137	1,286	1,442	1,674	1,914	2,148	2,336
Fixed cost of manufacture	1,272	1,272	1,272	1,415	1,615	1,615	1,615
Total plant cost	2,409	2,558	2,714	3,088	3,529	3,763	3,951
Other operating expenses	127	127	127	145	156	156	156
Total cost of sales	2,536	2,685	2,841	3,233	3,685	3,919	4,107
Profit, BFIT, $M	596	931	1,140	1,358	1,482	1,743	1,817
Profit, AFIT, $M	298	466	570	679	741	872	909
Base-Case profit, AFIT, $M	372	392	390	385	379	373	366
Net profit, AFIT, $M	(74)	74	180	294	362	499	543
Return on plant invest., AFIT, %	—	3.0	7.2	8.9	11.0	15.1	16.5
Return on total invest., AFIT, %	—	2.6	6.4	7.6	9.4	13.0	14.1
Depreciation, $M	167	167	167	189	220	220	220
Cash flow, $M	93	241	347	483	582	719	763
Cumulative cash flow, $M	93	334	681	1,164	1,746	2,465	3,228

M = 1,000 BFIT = Before Federal Income Tax AFIT = After Federal Income Tax
* Assumes expanded 4-million gal./yr. capacity is available for last 5 months of 1976

The PW calculations assume a 10% interest rate; i.e., this rate has been chosen to represent the "opportunity value" of money. On this basis, the PW of all expenditures and earnings for the 15-yr. projected life of Case (B) —the expandable-plant proposal—is $160,000 greater than the present worth of the next alternate, and $433,000 greater than the present worth of the last alternate. Likewise, the DCF return of 12.65% for Case (B)

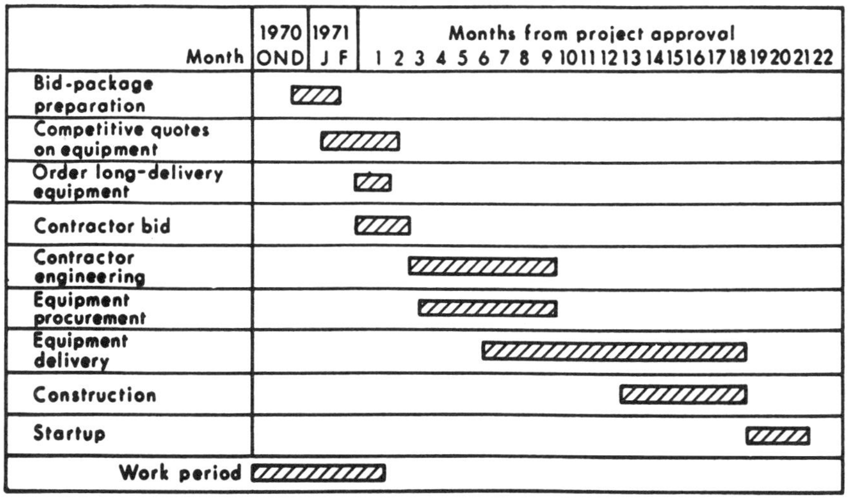

SCHEDULE for various construction phases of expandable plant—Fig. 1

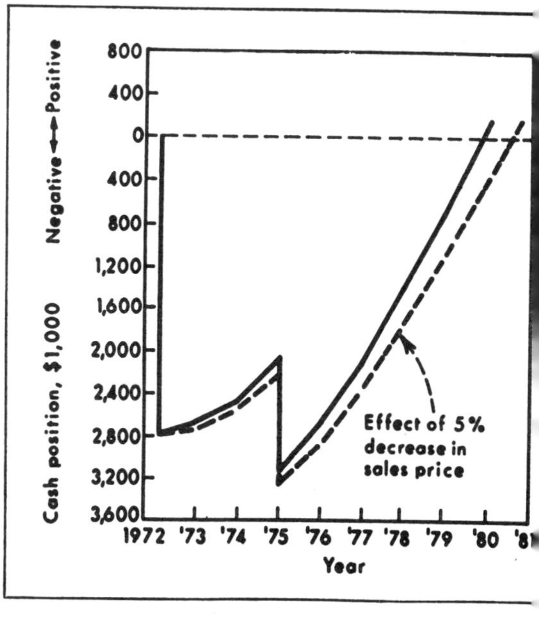

CASH FLOW for 2-million-gal./yr. expandable plant (based on the time value of money being zero)—Fig. 2

Revised Base Case Plus Initial, One-Step, 4-Million-Gal./Yr. Expansion—Table IX

	1973	1974	1975	1976	1977	1978	1979
Gross sales, million gal./yr.	4.1	4.6	5.1	5.8	6.5	7.0	7.0
Gross sales income, $M	3,482	3,983	4,432	5,035	5,663	6,197	6,463
Returns, allowances, $M	26	30	33	38	42	46	48
Net sales income, $M	3,456	3,953	4,399	4,997	5,621	6,151	6,415
Freight, $M	324	337	361	406	454	489	491
Net sales realization, $M	3,132	3,616	4,038	4,591	5,167	5,662	5,924
Cost of sales, $M:							
Variable cost of manufacture	1,137	1,286	1,455	1,674	1,914	2,148	2,336
Fixed cost of manufacture	1,585	1,585	1,585	1,585	1,585	1,585	1,585
Total plant cost	2,722	2,871	3,040	3,259	3,499	3,733	3,921
Other operating expenses	127	127	129	145	156	156	156
Total cost of sales	2,849	2,998	3,169	3,404	3,655	3,889	4,077
Profit, BFIT, $M	283	618	869	1,187	1,512	1,773	1,847
Profit, AFIT, $M	142	309	435	594	756	887	924
Base-Case profit, AFIT, $M	372	392	390	385	379	373	366
Net profit, AFIT, $M	(230)	(83)	45	209	377	514	558
Return on plant invest., AFIT, $M	—	—	1.4	6.8	12.2	16.6	18.0
Return on total invest., AFIT, $M	—	—	1.2	5.7	10.3	14.1	15.3
Depreciation, $M	207	207	207	207	207	207	207
Cash flow, $M	(23)	124	252	416	584	721	765
Cumulative cash flow, $M	(23)	101	353	769	1,353	2,074	2,839

M = 1,000 BFIT = Before Federal Income Tax AFIT = After Federal Income Tax

slightly exceeds that of Case (A), and considerably exceeds that of Case (C).

Thus, the initial installation of the expandable 2-million-gal./yr. plant is justified, subject to the realization of the market growth predicated in the revised sales forecast, and this is the case selected for presentation and approval to top management.

Breakeven Operation as Percentage of Plant Capacity for 1979 Product Mix—Table X

(Initial Installation of a 2-Million Gal./Yr. Plant Expandable to 4-Million Gal./Yr. Capacity)

	Expansion Capacity	Total Capacity
1. Capacity, million gal./yr.	4	7
2. Net sales realization, $M	3,463	5,924
3. Variable cost of manuf., $M	1,375	2,336
4. Item (2) minus (3)	2,088	3,588
5. Fixed cost of manuf., $M	932	1,615
6. Other operating expenses, $M.	71	156
7. Item (5) + (6)	1,003	1,771
8. Breakeven operation as % of capacity [(7) ÷ (4)] × 100	49.5	49.4

M = 1,000

Breakeven Point and Sensitivity

The calculations presented in Table X indicate that the selected expansion case has to be operated at 49.5% of capacity to achieve the breakeven point. Fig. 2 presents the cash-flow chart for the selected expansion case, indicating that a positive cash position will be attained in late 1979. The effect of a 5% decrease in sales price under the projected levels would delay achieving the point of positive cash position by six months.

Table XI presents the sensitivity analysis for the selected expansion case, and shows the effect of a 5% decrease in selling price on earnings and on return-on-investment, and the effect of a 10% overrun in the investment estimate on return-on-investment. The combination of a 5% decrease in selling price and a 10% increase in investment would reduce the return-on-investment in 1979 from 14.4% to 10.7%.

The Final Plan of Action

At the conclusion of the meeting, a consensus is reached on the action plan for presentation to XYZ Co.'s top management:

1. Financial is to coordinate preparation of the project premise book, using the information contained in Tables

Sensitivity Analysis for Selected Expansion Case—Table XI
(Installation of a 2-Million-Gal./Yr. Plant Whose Capacity Can Be Doubled)

	Earnings as Presented			Effect of 5% Drop in Sales Price			Effect of 10% Increase in Invest.	
Year	Profit, AFIT, $M	Cash Flow, $M	ROI,* %	Profit, AFIT, $M	Cash Flow, $M	ROI,* %	ROI,* With Sales As Presented	ROI,* With 5% Drop in Sales Price
1973	(74)	93	—	(92)	75	—	—	—
1974	74	241	2.6	44	211	1.6	2.4	1.4
1975	180	347	6.4	140	307	5.0	5.9	4.6
1976	294	483	7.6	237	426	6.2	7.0	5.7
1977	362	582	9.4	289	509	7.5	8.7	6.9
1978	499	719	13.0	412	632	10.7	11.9	9.8
1979	543	763	14.1	449	669	11.6	13.0	10.7

Breakeven 49.4% of capacity   54.2% of capacity

* ROI = Return on Total Investment   M = 1,000   AFIT = After Federal Income Tax

I through XI as refined and modified by each functional group. This work is to be ready for presentation within a month (by Feb. 1, 1971).

2. Marketing is to finalize a Product B purchase agreement with Company A for the years 1971 and 1972, as per the terms presented earlier.

3. Installation of a 2-million-gal./yr. plant expandable to 4 million gal./yr. for Product B is to be recommended on the Gulf Coast location.

4. Marketing is to actively pursue new target accounts (as shown in the previously discussed sales plan) so as to absorb the volume to be generated through the purchase agreement with Company A, as well as the anticipated Product B expansion capacity.

5. R&D is to accelerate work to improve the quality of the adhesive grade of Product B so as to justify the anticipated growth in sales for this product. The premise book is to contain a review of plans for the development of a lubricating-oil-additive grade of Product B for automotive uses.

6. A capital-project appropriation request is to be prepared for presentation to the Board of Directors' meeting, as per the project premise book.

The project was approved at the XYZ Co.'s Board of Directors' meeting held on Feb. 22, 1971.

The Case Study in Perspective

Over the course of the last three installments, we have illustrated various events in the development of a major capital project. We have seen how the basic premises were established, how various alternatives were set up and evaluated, and how this led to certain recalculations and reevaluations.

Such reevaluations are quite typical of major projects. Of course, there are two extremes to guard against:

1. Undue reluctance to consider new alternatives either because of the calculations and field-work involved, or because of emotional loyalty to the original concepts.

2. Undue reluctance to come to grips with decisions needed on options already at hand. This can result in expending a great deal of time and work on streams of aimless, foredoomed or trivial new alternatives, and may give the edge to a more decisive competitor.

With many branches of the chemical process industries still caught in a price/cost squeeze on profits, and with companies feeling the need for growth in profits through new business ventures, the capital-investment stakes today are high indeed. To succeed in this decade, companies will have to be very discriminating in how they use the capital funds available to them. To develop successful projects, they will find it extremely important to coordinate very closely the efforts of the marketing, manufacturing, R&D, engineering, construction and financial functions in making certain that good overall corporate performance is obtained.

For instance, as we have seen, carefully developed marketing inputs are vital in striking the optimum answer on initial plant capacity.

A really attractive venture will start to be profitable or at least break even in the first year of its operation. In many cases, it is prudent to tie down sufficient product sales via long-term contracts or captive uses to ensure at least break-even operation prior to the commitment of capital for plant construction. Obviously, this is not always possible; there just aren't very many projects that promise to combine a quick and high return with a minimal degree of risk. Creative risk-taking making maximum analytic use of as many facts as possible is a key challenge in today's business climate.

In the XYZ Co.'s case study, it is clear that the functional groups involved finally zeroed in on an expandable plant as the optimum answer—one that would give the firm a means of profitable growth while maintaining its share of market for an expanding product line. Assessing the degree of risk and probability of attaining an increased share of market was critical to this decision. A subsequent article will show how decision-tree analysis can be used to quantify some of these uncertainties, and to help optimize plant size.■

When and How To Apply Discounted Cash Flow and Present Worth

Economic Analysis — Part 8

Techniques based on the "time value of money" concept can be powerful tools in evaluating investment alternatives; here is how to use these tools effectively.

IRVING LEIBSON and CHARLES A. TRISCHMAN, JR., Dart Industries, Inc.

In today's world of sophisticated cash management, you cannot afford to ignore the time value of money in evaluating projects whose annual cash flows and investment requirements vary widely.

Earlier in this series, we screened investment alternatives, using the combination of rate of return on investment (after federal income taxes), and payout time. These profitability criteria, applied jointly, are often adequate for small capital projects (i.e., involving less than $100,000 in capital investment), and for some larger projects where cash flow is relatively constant from one year to the next, and only a single capital outlay at the beginning of the project is required. But we have already seen, in the final part of the XYZ Co. case study (see p. 159), that decision-making situations arise where these criteria have to be supplemented by techniques that take the time value of money into account.

In the present article, we shall examine and illustrate two of these techniques—discounted cash flow (DCF) and present worth (PW). We hope to dispel some of the confusion over the use of discount factors. Also, we will discuss proportional gain, a technique closely related to PW.

A 1963 survey[1] of 33 major chemical and oil companies showed that 64% of the respondents used return on investment and/or payout time as measures of profitability in evaluating new capital investments, whereas 33% of the respondents used DCF and PW. However, the survey also revealed that most of the respondents used more than a single method for profitability analysis on new investments—which is certainly warranted, particularly in regard to multimillion-dollar projects.

Originally published December 1971.

Discounted Cash Flow (DCF): The Basics

The DCF return on investment is the equivalent, average, continuous rate of compound interest that is earned by a project on the money invested in that project over the life assignable to it.

In the literature, various authors refer to DCF as the interest rate of return, profitability index, and investor's method.

Often, in the early years of a new project, the cash flow is not adequate to provide acceptable profitability for the initial investment (in fixed and working capital) required by the venture. An amount of money received at any future date is not equal in value to the same amount of money received today. DCF takes this into account by discounting the annual cash flows to properly reflect the time value of money on a continuous compound-interest basis.

Let us develop the DCF concept by starting with

The "F_a" Discount Factors—Table I

These continuous-interest discount factors give present worths for cash flows that occur in an instant at a point in time after the reference point.

100rn*	0	1	2	3	4	5	6	7	8	9
0	1.0000	0.9901	0.9802	0.9704	0.9608	0.9512	0.9418	0.9324	0.9231	0.9139
10	0.9048	0.8958	0.8869	0.8781	0.8694	0.8607	0.8521	0.8437	0.8353	0.8270
20	0.8187	0.8106	0.8025	0.7945	0.7866	0.7788	0.7711	0.7634	0.7558	0.7483
30	0.7408	0.7334	0.7261	0.7189	0.7118	0.7047	0.6977	0.6907	0.6839	0.6771
40	0.6703	0.6637	0.6570	0.6505	0.6440	0.6376	0.6313	0.6250	0.6188	0.6126
50	0.6065	0.6005	0.5945	0.5886	0.5827	0.5770	0.5712	0.5655	0.5599	0.5543
60	0.5488	0.5434	0.5379	0.5326	0.5273	0.5220	0.5169	0.5117	0.5066	0.5016
70	0.4966	0.4916	0.4868	0.4819	0.4771	0.4724	0.4677	0.4630	0.4584	0.4538
80	0.4493	0.4449	0.4404	0.4360	0.4317	0.4274	0.4232	0.4190	0.4148	0.4107
90	0.4066	0.4025	0.3985	0.3946	0.3906	0.3867	0.3829	0.3791	0.3753	0.3716
100	0.3679	0.3642	0.3606	0.3570	0.3535	0.3499	0.3465	0.3430	0.3396	0.3362
110	0.3329	0.3296	0.3263	0.3230	0.3198	0.3166	0.3135	0.3104	0.3073	0.3042
120	0.3012	0.2982	0.2952	0.2923	0.2894	0.2865	0.2837	0.2808	0.2780	0.2753
130	0.2725	0.2698	0.2671	0.2645	0.2618	0.2592	0.2567	0.2541	0.2516	0.2491
140	0.2466	0.2441	0.2417	0.2393	0.2369	0.2346	0.2322	0.2299	0.2276	0.2254
150	0.2231	0.2209	0.2187	0.2165	0.2144	0.2122	0.2101	0.2080	0.2060	0.2039
160	0.2019	0.1999	0.1979	0.1959	0.1940	0.1921	0.1901	0.1882	0.1864	0.1845
170	0.1827	0.1809	0.1791	0.1773	0.1755	0.1738	0.1720	0.1703	0.1686	0.1670
180	0.1653	0.1637	0.1620	0.1604	0.1588	0.1572	0.1557	0.1541	0.1526	0.1511
190	0.1496	0.1481	0.1466	0.1451	0.1437	0.1423	0.1409	0.1395	0.1381	0.1367
200	0.1353	0.1340	0.1327	0.1313	0.1300	0.1287	0.1275	0.1262	0.1249	0.1237
210	0.1225	0.1212	0.1200	0.1188	0.1177	0.1165	0.1153	0.1142	0.1130	0.1119
220	0.1108	0.1097	0.1086	0.1075	0.1065	0.1054	0.1044	0.1033	0.1023	0.1013
230	0.1003	0.0993	0.0983	0.0973	0.0963	0.0954	0.0944	0.0935	0.0926	0.0916
240	0.0907	0.0898	0.0889	0.0880	0.0872	0.0863	0.0854	0.0846	0.0837	0.0829
250	0.0821	0.0813	0.0805	0.0797	0.0789	0.0781	0.0773	0.0765	0.0758	0.0750
260	0.0743	0.0735	0.0728	0.0721	0.0714	0.0707	0.0699	0.0693	0.0686	0.0679
270	0.0672	0.0665	0.0659	0.0652	0.0646	0.0639	0.0633	0.0627	0.0620	0.0614
280	0.0608	0.0602	0.0596	0.0590	0.0584	0.0578	0.0573	0.0567	0.0561	0.0556
290	0.0550	0.0545	0.0539	0.0534	0.0529	0.0523	0.0518	0.0513	0.0508	0.0503
300	0.0498	0.0493	0.0488	0.0483	0.0478	0.0474	0.0469	0.0464	0.0460	0.0455
310	0.0450	0.0446	0.0442	0.0437	0.0433	0.0429	0.0424	0.0420	0.0416	0.0412
320	0.0408	0.0404	0.0400	0.0396	0.0392	0.0388	0.0384	0.0380	0.0376	0.0373
330	0.0369	0.0365	0.0362	0.0358	0.0354	0.0351	0.0347	0.0344	0.0340	0.0337
340	0.0334	0.0330	0.0327	0.0324	0.0321	0.0317	0.0314	0.0311	0.0308	0.0305
350	0.0302	0.0299	0.0296	0.0293	0.0290	0.0287	0.0284	0.0282	0.0279	0.0276
360	0.0273	0.0271	0.0268	0.0265	0.0263	0.0260	0.0257	0.0255	0.0252	0.0250
370	0.0247	0.0245	0.0242	0.0240	0.0238	0.0235	0.0233	0.0231	0.0228	0.0226
380	0.0224	0.0221	0.0219	0.0217	0.0215	0.0213	0.0211	0.0209	0.0207	0.0204
390	0.0202	0.0200	0.0198	0.0196	0.0194	0.0193	0.0191	0.0189	0.0187	0.0185
400	0.0183	0.0181	0.0180	0.0178	0.0176	0.0174	0.0172	0.0171	0.0169	0.0167
410	0.0166	0.0164	0.0162	0.0161	0.0159	0.0158	0.0156	0.0155	0.0153	0.0151
420	0.0150	0.0148	0.0147	0.0146	0.0144	0.0143	0.0141	0.0140	0.0138	0.0137
430	0.0136	0.0134	0.0133	0.0132	0.0130	0.0129	0.0128	0.0127	0.0125	0.0124
440	0.0123	0.0122	0.0120	0.0119	0.0118	0.0117	0.0116	0.0114	0.0113	0.0112
450	0.0111	0.0110	0.0109	0.0108	0.0107	0.0106	0.0105	0.0104	0.0103	0.0102
460	0.0101	0.0100	0.0099	0.0098	0.0097	0.0096	0.0095	0.0094	0.0093	0.0092
470	0.0091	0.0090	0.0089	0.0088	0.0087	0.0087	0.0086	0.0085	0.0084	0.0083
480	0.0082	0.0081	0.0081	0.0080	0.0079	0.0078	0.0078	0.0077	0.0076	0.0075
490	0.0074	0.0074	0.0073	0.0072	0.0072	0.0071	0.0070	0.0069	0.0069	0.0068
100rn	**0**	**10**	**20**	**30**	**40**	**50**	**60**	**70**	**80**	**90**
500	0.0067	0.0061	0.0055	0.0050	0.0045	0.0041	0.0037	0.0033	0.0030	0.0027
600	0.0025	0.0022	0.0020	0.0018	0.0017	0.0015	0.0014	0.0012	0.0011	0.0010
700	0.0009	0.0008	0.0007	0.0007	0.0006	0.0006	0.0005	0.0005	0.0004	0.0004
800	0.0003	0.0003	0.0003	0.0002	0.0002	0.0002	0.0002	0.0002	0.0002	0.0001
900	0.0001	0.0001	0.0001	0.0001	0.0001	0.0001	0.0001	0.0001	0.0001	0.0001
1000	0.0000									

*r = nominal interest compounded continuously; percent/100;
n = number of years.
Source: Ref. 7; reprinted with permission.

something very familiar. If you have a savings account in a bank, then you recognize that interest is the incremental money paid for the use of money (principal). In this vein, the capital investment for new plant equipment may be considered as principal. If interest is paid on the initial principal only, it is called simple interest. If interest is added to the principal at stated periods, forming new principal that again bears interest at stated periods, then it is called compound interest. If added each 12, 6 or 3 months, the interest is said to be compounded annually, semiannually, or quarterly, respectively.

In the typical chemical-process-industries situation, profits actually flow on a continuous basis. Most corporate accounting systems provide for monthly closings. Depreciation accounts are accrued on a monthly basis. Thus, we can regard a capital investment as principal, and the after-tax profit plus depreciation (i.e., cash flow) as a continuous, compounded interest on the investment—and this brings out the importance of the time value of money in evaluating capital investment alternatives. Furthermore, when additional subsequent investments are made on a project, one can take these into consideration in respect to corresponding changes in cash flow.

Choosing the Type of Discount Factor

Much has been published on various techniques for applying the time value of money concept to the evaluation of investment alternatives; Ref. 2 to 7 provide a sampling.

The field can be somewhat confusing to the novice and even to the experienced engineer, in that different sets of discount factors are applicable, depending on whether the discounting of cash flows is taken to occur in an instant, uniformly over individual years, uniformly over longer periods, or on some other basis.

To apply these discount factors, it is necessary, at the outset of the analysis, to define key inputs such as the date as well as value of the initial plant investment, timing and value of any subsequent investments, the working-capital requirements, debt/equity ratio, depreciation schedule, cash-flow schedule, useful life of the venture, and salvage value (if any) of the investment at liquidation.

Discount factors to convert salvage value of a plant, or working capital at the end of project life, to a present-value basis are given in Table I. These factors represent the use of continuously compounded interest to give present worths for cash flows that occur in an instant (i.e., lump sums) at a point in time after the reference point.

Discount factors appropriate for cases in which the yearly cash flow is continuous but may vary from one year to the next are shown in Table II. These factors are based on continuous interest to give present worths for cash flows that occur continuously and uniformly over one-year periods after the reference point. This is typical of the earnings cash-flow pattern for most ventures in operating companies in the chemical process industries.

Discount factors to convert cash flows that occur uniformly over a period of T years after the reference point to a present-worth basis are shown in Table III. These factors are useful in handling the cumulative

The "F_b" Discount Factors—Table II

These continuous-interest discount factors give present worths for cash flows that occur uniformly over one-year periods after the reference point.

Year*	1%	2%	3%	4%	5%	6%	7%	8%	9%	10%	11%	12%	13%	14%	15%	16%	17%	18%	19%	20%	21%
0–1	0.9950	0.9901	0.9851	0.9803	0.9754	0.9706	0.9658	0.9610	0.9563	0.9516	0.9470	0.9423	0.9377	0.9332	0.9286	0.9241	0.9196	0.9152	0.9107	0.9063	0.9020
1–2	0.9851	0.9705	0.9560	0.9418	0.9278	0.9141	0.9005	0.8872	0.8740	0.8611	0.8483	0.8358	0.8234	0.8112	0.7993	0.7875	0.7759	0.7644	0.7531	0.7421	0.7311
2–3	0.9753	0.9512	0.9278	0.9049	0.8826	0.8608	0.8396	0.8189	0.7988	0.7791	0.7600	0.7413	0.7230	0.7053	0.6879	0.6710	0.6546	0.6385	0.6228	0.6075	0.5926
3–4	0.9656	0.9324	0.9004	0.8694	0.8395	0.8107	0.7829	0.7560	0.7300	0.7050	0.6808	0.6574	0.6349	0.6131	0.5921	0.5718	0.5522	0.5333	0.5150	0.4974	0.4804
4–5	0.9560	0.9140	0.8737	0.8353	0.7986	0.7635	0.7299	0.6979	0.6672	0.6379	0.6099	0.5831	0.5575	0.5330	0.5096	0.4873	0.4659	0.4455	0.4259	0.4072	0.3894
5–6	0.9465	0.8959	0.8479	0.8026	0.7596	0.7190	0.6806	0.6442	0.6098	0.5772	0.5463	0.5172	0.4895	0.4634	0.4386	0.4152	0.3931	0.3721	0.3522	0.3334	0.3156
6–7	0.9371	0.8781	0.8229	0.7711	0.7226	0.6772	0.6346	0.5947	0.5573	0.5223	0.4894	0.4588	0.4299	0.4029	0.3775	0.3538	0.3316	0.3108	0.2913	0.2730	0.2558
7–8	0.9278	0.8607	0.7985	0.7409	0.6874	0.6377	0.5917	0.5490	0.5093	0.4726	0.4385	0.4069	0.3775	0.3502	0.3250	0.3015	0.2798	0.2596	0.2409	0.2235	0.2074
8–9	0.9185	0.8437	0.7749	0.7118	0.6538	0.6006	0.5517	0.5068	0.4655	0.4276	0.3928	0.3609	0.3314	0.3045	0.2797	0.2569	0.2360	0.2168	0.1992	0.1830	0.1681
9–10	0.9094	0.8270	0.7520	0.6839	0.6219	0.5656	0.5144	0.4678	0.4254	0.3869	0.3519	0.3201	0.2910	0.2647	0.2407	0.2189	0.1991	0.1811	0.1647	0.1498	0.1363
10–11	0.9003	0.8106	0.7298	0.6571	0.5916	0.5327	0.4796	0.4318	0.3888	0.3501	0.3152	0.2839	0.2556	0.2301	0.2072	0.1866	0.1680	0.1513	0.1362	0.1227	0.1105
11–12	0.8914	0.7946	0.7082	0.6312	0.5628	0.5016	0.4472	0.3986	0.3553	0.3168	0.2824	0.2518	0.2244	0.2000	0.1783	0.1590	0.1417	0.1264	0.1126	0.1004	0.0895
12–13	0.8825	0.7788	0.6873	0.6065	0.5353	0.4724	0.4169	0.3680	0.3248	0.2866	0.2530	0.2233	0.1970	0.1739	0.1535	0.1355	0.1196	0.1055	0.0932	0.0822	0.0726
13–14	0.8737	0.7634	0.6670	0.5827	0.5092	0.4449	0.3888	0.3397	0.2968	0.2593	0.2266	0.1981	0.1730	0.1512	0.1321	0.1154	0.1009	0.0882	0.0770	0.0673	0.0588
14–15	0.8650	0.7483	0.6473	0.5599	0.4844	0.4190	0.3625	0.3136	0.2713	0.2347	0.2030	0.1757	0.1519	0.1314	0.1137	0.0984	0.0851	0.0736	0.0637	0.0551	0.0477
15–16	0.8564	0.7335	0.6282	0.5380	0.4608	0.3946	0.3380	0.2895	0.2479	0.2123	0.1819	0.1558	0.1334	0.1143	0.0979	0.0838	0.0718	0.0615	0.0527	0.0451	0.0387
16–17	0.8479	0.7189	0.6096	0.5169	0.4383	0.3716	0.3151	0.2672	0.2266	0.1921	0.1629	0.1382	0.1172	0.0993	0.0842	0.0714	0.0606	0.0514	0.0436	0.0369	0.0313
17–18	0.8395	0.7047	0.5916	0.4966	0.4169	0.3500	0.2938	0.2467	0.2071	0.1739	0.1460	0.1225	0.1029	0.0864	0.0725	0.0609	0.0511	0.0429	0.0360	0.0303	0.0254
18–19	0.8311	0.6908	0.5741	0.4772	0.3966	0.3296	0.2740	0.2277	0.1893	0.1573	0.1308	0.1087	0.0903	0.0751	0.0624	0.0519	0.0431	0.0358	0.0298	0.0248	0.0206
19–20	0.8228	0.6771	0.5571	0.4584	0.3772	0.3104	0.2554	0.2102	0.1730	0.1423	0.1171	0.0964	0.0793	0.0653	0.0537	0.0442	0.0364	0.0299	0.0246	0.0203	0.0167
20–21	0.8147	0.6637	0.5407	0.4405	0.3588	0.2923	0.2382	0.1940	0.1581	0.1288	0.1049	0.0855	0.0697	0.0568	0.0462	0.0377	0.0307	0.0250	0.0204	0.0166	
21–22	0.8065	0.6505	0.5247	0.4232	0.3413	0.2753	0.2221	0.1791	0.1445	0.1165	0.0940	0.0758	0.0612	0.0493	0.0398	0.0321	0.0259	0.0209	0.0169	0.0136	
22–23	0.7985	0.6376	0.5092	0.4066	0.3247	0.2593	0.2071	0.1653	0.1320	0.1054	0.0842	0.0673	0.0537	0.0429	0.0343	0.0274	0.0218	0.0175	0.0139	0.0111	
23–24	0.7906	0.6250	0.4941	0.3907	0.3089	0.2442	0.1931	0.1520	0.1207	0.0954	0.0754	0.0596	0.0472	0.0373	0.0295	0.0233	0.0184	0.0146	0.0115	0.0091	
24–25	0.7827	0.6126	0.4795	0.3753	0.2938	0.2300	0.1800	0.1409	0.1103	0.0863	0.0676	0.0529	0.0414	0.0324	0.0254	0.0199	0.0156	0.0122	0.0095	0.0075	
25–30	0.7596	0.5772	0.4386	0.3334	0.2535	0.1928	0.1466	0.1115	0.0849	0.0646	0.0492	0.0374	0.0285	0.0217	0.0165	0.0126	0.0096	0.0073	0.0056	0.0043	
30–35	0.7226	0.5223	0.3775	0.2730	0.1974	0.1428	0.1033	0.0748	0.0541	0.0392	0.0284	0.0205	0.0149	0.0108	0.0078	0.0057	0.0041	0.0030	0.0022	0.0016	
35–40	0.6874	0.4726	0.3250	0.2235	0.1538	0.1058	0.0728	0.0501	0.0345	0.0238	0.0164	0.0113	0.0078	0.0054	0.0037	0.0025	0.0018	0.0012	0.0008	0.0006	
40–45	0.6538	0.4276	0.2797	0.1830	0.1197	0.0784	0.0513	0.0336	0.0220	0.0144	0.0094	0.0062	0.0041	0.0027	0.0017	0.0011	0.0008	0.0005	0.0003	0.0002	
45–50	0.6219	0.3869	0.2407	0.1498	0.0933	0.0581	0.0362	0.0225	0.0140	0.0087	0.0054	0.0034	0.0021	0.0013	0.0008	0.0005	0.0003	0.0002	0.0001	0.0001	

*Year indicates one-year period in which cash flow occurs.
Source: Ref. 7; reprinted with permission.

cash flows often occurring at a uniform continuous rate over a period of T years during the latter years of project life (as will be illustrated later).

Depreciation; Discount-Factor Mathematics

In this article, we shall consider only the straight-line method for determining depreciation; this assumes that the value of property decreases linearly with time. Other techniques for handling depreciation, such as the declining-balance method, sum-of-the-years digit method, and sinking-fund method may be used with DCF techniques in economic evaluation.[8] Use of accelerated depreciation techniques tends to increase project profitability through maximizing cash flow in the early years of the project.

A project with *acceptable* profitability can be improved via such accelerated depreciation techniques (and by borrowing funds in support of the capital requirements consistent with the corporate policy on debt-equity ratio). However, when a project of *marginal* profitability is improved to a barely acceptable level primarily via these profitability-enrichment techniques, no real service is rendered to management.

Let us now review the mathematics of discount factors that apply to lump sums.

As the original capital investment (P_0) accrues earnings at interest rate i, its value (S) increases in later time periods. Thus, after a single time period:

$$S_1 = P_0 (1 + i) \tag{1}$$

For the next time period, the previous amount is multiplied by the factor $(1 + i)$. Thus, the amount accrued after two time periods is:

$$S_2 = P_0 (1 + i) + P_0 (1 + i)\, i = P_0 (1 + i)^2 \tag{2}$$

Or, at the end of n time periods, we see that:

$$S_n = P_0 (1 + i)^n \tag{3}$$

If earnings are accrued on a continuous basis, the differential increase in the compounded amount (dS), over a differential time interval (dn), is given by:

$$dS = Sr\, dn \tag{4}$$

where r is the continuous interest rate, and n is the time.

Holding r constant with respect to time, and integrating this equation from time zero (the reference point) to time n, we obtain:

$$\int_{P_0}^{S_n} \frac{dS}{S} = \int_0^n r\, dn \tag{5}$$

$$\ln S_n/P_0 = rn \tag{6}$$

$$S_n = P_0\, e^{rn} \tag{7}$$

The "F_C" Discount Factors—Table III

These continuous-interest discount factors give present worths for cash flows that occur uniformly over a period of *T* years after the reference point

100rT*	0	1	2	3	4	5	6	7	8	9
0	1.0000	0.9950	0.9901	0.9851	0.9803	0.9754	0.9706	0.9658	0.9610	0.9563
10	0.9516	0.9470	0.9423	0.9377	0.9332	0.9286	0.9241	0.9196	0.9152	0.9107
20	0.9063	0.9020	0.8976	0.8933	0.8891	0.8848	0.8806	0.8764	0.8722	0.8681
30	0.8639	0.8598	0.8558	0.8517	0.8477	0.8438	0.8398	0.8359	0.8319	0.8281
40	0.8242	0.8204	0.8166	0.8128	0.8090	0.8053	0.8016	0.7979	0.7942	0.7906
50	0.7869	0.7833	0.7798	0.7762	0.7727	0.7692	0.7657	0.7622	0.7588	0.7554
60	0.7520	0.7486	0.7452	0.7419	0.7386	0.7353	0.7320	0.7288	0.7256	0.7224
70	0.7192	0.7160	0.7128	0.7097	0.7066	0.7035	0.7004	0.6974	0.6944	0.6913
80	0.6883	0.6854	0.6824	0.6795	0.6765	0.6736	0.6707	0.6679	0.6650	0.6622
90	0.6594	0.6566	0.6537	0.6510	0.6483	0.6455	0.6428	0.6401	0.6374	0.6348
100	0.6321	0.6295	0.6269	0.6243	0.6217	0.6191	0.6166	0.6140	0.6115	0.6090
110	0.6065	0.6040	0.6016	0.5991	0.5967	0.5942	0.5918	0.5894	0.5871	0.5847
120	0.5823	0.5800	0.5777	0.5754	0.5731	0.5708	0.5685	0.5663	0.5641	0.5618
130	0.5596	0.5574	0.5552	0.5530	0.5509	0.5487	0.5466	0.5444	0.5424	0.5402
140	0.5381	0.5361	0.5340	0.5320	0.5299	0.5279	0.5259	0.5239	0.5219	0.5199
150	0.5179	0.5160	0.5140	0.5121	0.5102	0.5082	0.5064	0.5044	0.5026	0.5007
160	0.4988	0.4970	0.4952	0.4933	0.4915	0.4897	0.4879	0.4861	0.4843	0.4825
170	0.4808	0.4790	0.4773	0.4756	0.4739	0.4721	0.4704	0.4687	0.4671	0.4654
180	0.4637	0.4621	0.4605	0.4588	0.4571	0.4555	0.4540	0.4523	0.4508	0.4491
190	0.4476	0.4460	0.4445	0.4429	0.4414	0.4399	0.4383	0.4368	0.4354	0.4338
200	0.4323	0.4308	0.4294	0.4279	0.4265	0.4250	0.4236	0.4221	0.4207	0.4193
210	0.4179	0.4165	0.4151	0.4137	0.4123	0.4109	0.4096	0.4082	0.4069	0.4055
220	0.4042	0.4029	0.4015	0.4002	0.3989	0.3976	0.3963	0.3950	0.3937	0.3925
230	0.3912	0.3899	0.3887	0.3874	0.3862	0.3849	0.3837	0.3825	0.3813	0.3801
240	0.3789	0.3777	0.3765	0.3753	0.3741	0.3729	0.3718	0.3706	0.3695	0.3683
250	0.3672	0.3660	0.3649	0.3638	0.3627	0.3615	0.3604	0.3593	0.3582	0.3571
260	0.3560	0.3550	0.3539	0.3528	0.3517	0.3507	0.3496	0.3486	0.3476	0.3465
270	0.3455	0.3445	0.3434	0.3424	0.3414	0.3404	0.3393	0.3384	0.3374	0.3364
280	0.3354	0.3344	0.3335	0.3325	0.3315	0.3306	0.3296	0.3287	0.3277	0.3268
290	0.3259	0.3249	0.3240	0.3231	0.3221	0.3212	0.3203	0.3194	0.3185	0.3176
300	0.3167	0.3158	0.3150	0.3141	0.3132	0.3123	0.3115	0.3106	0.3098	0.3089
310	0.3080	0.3072	0.3064	0.3055	0.3047	0.3039	0.3030	0.3022	0.3014	0.3006
320	0.2998	0.2990	0.2982	0.2974	0.2966	0.2958	0.2950	0.2942	0.2934	0.2926
330	0.2919	0.2911	0.2903	0.2896	0.2888	0.2880	0.2873	0.2865	0.2858	0.2850
340	0.2843	0.2836	0.2828	0.2821	0.2814	0.2807	0.2799	0.2792	0.2785	0.2778
350	0.2771	0.2764	0.2757	0.2750	0.2743	0.2736	0.2729	0.2722	0.2715	0.2709
360	0.2702	0.2695	0.2688	0.2682	0.2675	0.2669	0.2662	0.2655	0.2649	0.2642
370	0.2636	0.2629	0.2623	0.2617	0.2610	0.2604	0.2598	0.2591	0.2585	0.2579
380	0.2573	0.2567	0.2560	0.2554	0.2548	0.2542	0.2536	0.2530	0.2524	0.2518
390	0.2512	0.2506	0.2500	0.2495	0.2489	0.2483	0.2477	0.2471	0.2466	0.2460
400	0.2454	0.2449	0.2443	0.2437	0.2432	0.2426	0.2421	0.2415	0.2410	0.2404
410	0.2399	0.2393	0.2388	0.2382	0.2377	0.2372	0.2366	0.2361	0.2356	0.2350
420	0.2345	0.2340	0.2335	0.2330	0.2325	0.2319	0.2314	0.2309	0.2304	0.2299
430	0.2294	0.2289	0.2284	0.2279	0.2274	0.2269	0.2264	0.2259	0.2255	0.2250
440	0.2245	0.2240	0.2235	0.2230	0.2226	0.2221	0.2216	0.2212	0.2207	0.2202
450	0.2198	0.2193	0.2188	0.2184	0.2179	0.2175	0.2170	0.2166	0.2161	0.2157
460	0.2152	0.2148	0.2143	0.2139	0.2134	0.2130	0.2126	0.2121	0.2117	0.2113
470	0.2108	0.2104	0.2100	0.2096	0.2091	0.2087	0.2083	0.2079	0.2074	0.2070
480	0.2066	0.2062	0.2058	0.2054	0.2050	0.2046	0.2042	0.2038	0.2034	0.2030
490	0.2026	0.2022	0.2018	0.2014	0.2010	0.2006	0.2002	0.1998	0.1994	0.1990
100rT	0	10	20	30	40	50	60	70	80	90
500	0.1987	0.1949	0.1912	0.1877	0.1843	0.1811	0.1779	0.1749	0.1719	0.1690
600	0.1663	0.1636	0.1610	0.1584	0.1560	0.1536	0.1513	0.1491	0.1469	0.1448
700	0.1427	0.1407	0.1388	0.1369	0.1351	0.1333	0.1315	0.1298	0.1282	0.1265
800	0.1250	0.1234	0.1219	0.1206	0.1190	0.1176	0.1163	0.1149	0.1136	0.1123
900	0.1111	0.1099	0.1087	0.1075	0.1064	0.1053	0.1043	0.1031	0.1020	0.1010
1000	0.1000	0.0990	0.0980	0.0971	0.0962	0.0952	0.0943	0.0935	0.0926	0.0917
1100	0.0909	0.0901	0.0893	0.0885	0.0877	0.0869	0.0862	0.0855	0.0847	0.0840
1200	0.0833	0.0826	0.0820	0.0813	0.0806	0.0800	0.0794	0.0787	0.0781	0.0775
1300	0.0769	0.0763	0.0758	0.0752	0.0746	0.0741	0.0735	0.0730	0.0725	0.0719
1400	0.0714	0.0709	0.0704	0.0699	0.0694	0.0690	0.0685	0.0680	0.0676	0.0671
1500	0.0667	0.0662	0.0658	0.0654	0.0649	0.0645	0.0641	0.0637	0.0633	0.0629
1600	0.0625	0.0621	0.0617	0.0613	0.0610	0.0606	0.0602	0.0599	0.0595	0.0592
1700	0.0588	0.0585	0.0581	0.0578	0.0575	0.0571	0.0568	0.0565	0.0562	0.0559
1800	0.0556	0.0552	0.0549	0.0546	0.0543	0.0541	0.0538	0.0535	0.0532	0.0529
1900	0.0526	0.0524	0.0521	0.0518	0.0515	0.0513	0.0510	0.0508	0.0505	0.0502
2000	0.0500	0.0497	0.0495	0.0492	0.0490	0.0487	0.0485	0.0483	0.0481	0.0478

*r = nominal interest compounded continuously; percent/100
$T = n$ = number of years in time period.
Source: Ref. 7; reprinted with permission.

in which S_n is the amount accrued from the initial investment (P_0), at the continuous interest rate (r), over the time period (n). (Note that in this derivation the principal is continuously compounded along with the accrued earnings.)

Conversely, we can find the corresponding value of P_0 (i.e., the present worth) when S_n is given:

$$P_0 = S_n e^{-rn} \tag{8}$$

Or we can find the corresponding discount factor, F_A:

$$F_A = P_0/S_n = e^{-rn} \tag{9}$$

This factor may be used to discount a lump sum resulting from an instantaneous payment (such as recovery of the salvage value of a plant at the end of useful project life) to a present-worth basis. It may also be used for conversion of a lump sum gained from liquidation of working capital to a present-worth basis. In each case, the lump sum is discounted at a set continuous rate (r) over the total useful project-life (n).

Table I contains discount factors calculated by using Eq. (9). For example, if a lump sum is gained or expended in an instant five years after the reference point, and a uniform continuous-interest rate of 10% is applicable, then the corresponding discount factor F_A calculated from Eq. (9) and shown in Table I is 0.606.

Continuous Cash Flows

Let us consider the case in which a project is continuously producing a cash flow each year of project life. Even though the cash flow is uniform during the year, the flows from one year to the next are not necessarily equal. For this case, within any year (n), we can express the differential rate of appreciation of the compound amount (S) as:

$$dS/dn = C_n + rS \tag{10}$$

where C_n is the cash flow occurring in year n.

Holding r constant with respect to time (n), and integrating this equation from ($n - 1$) to n; we obtain:

$$\int_0^{S_n} \frac{dS}{C_n + rs} = \int_{n-1}^{n} dn \tag{11}$$

$$\frac{1}{r} \ln \frac{C_n + rS_n}{C_n} = 1 \tag{12}$$

$$C_n + rS_n = C_n e^r \tag{13}$$

$$S_n = C_n\left(\frac{e^r - 1}{r}\right) \tag{14}$$

In Eq. (14), we have derived an expression that yields the compounded amount for a uniform, continuous, yearly cash flow with continuous interest compounding. The corresponding present worth of cash flow C_n converted from the year n to the reference point is obtained by combining Eq. (7) with Eq. (14):

$$P_0 = S_n e^{-rn} = C_n\left(\frac{e^r - 1}{r}\right)e^{-rn} \tag{15}$$

Now, the corresponding discount factor (F_B) is:

$$F_B = \left(\frac{e^r - 1}{r}\right) e^{-rn} \qquad (16)$$

This may be used to discount cash flows that occur continuously and uniformly over one-year periods after the reference point, to give the corresponding present worth. Table II contains discount factors calculated by use of Eq. (16).

For example, if $r = 0.1$, $n = 5$ (from year number 4 to year number 5 after the reference point), the corresponding discount factor (F_B) calculated from Eq. (16) and shown in Table II is 0.638.

Handling the Initial Plant Investment

Let us now see how the initial plant investment can be treated so as to reflect the time value of money.

Most major plant-construction projects require from one to two years for completion. The shape of the cash spending curve usually varies to some extent, depending on the type of contractual commitment made with the constructor (e.g., lump sum with preset payment schedule, cost plus, or other), and the type of facility involved. The cash spending curve is typically S-shaped for most projects.

If the spending schedule is forecast or preset by contract, then plant-investment capital can be continuously compounded to reflect its value at the reference point (i.e., plant startup). This is accomplished by calculating appropriate discount factors using Eq. (9) for the corresponding monthly (or other) time periods and dividing the corresponding cash payments by these factors. However, this is usually not warranted because the cash spending pattern can seldom be rigorously defined at the time the economic justification is being prepared.

Thus, this factor is normally neglected in profitability evaluations of alternative investment opportunities, without serious effect. At most, such neglect might cause a ± 5% error in evaluating project alternatives, which is not really significant compared with the potential error in other inputs (e.g., sales price and volume).

The effect of land-value compounding from time of purchase to plant startup may be similarly neglected without any serious problem for most practical situations in the chemical process industries.

Thus, DCF return on investment is that value of r for which the following equation applies:

$$\sum_{n=0}^{n=\text{Project life}} C_n \left[\frac{e^r - 1}{r}\right] e^{-rn} + \left[\begin{array}{c}\text{Plant}\\ \text{salvage}\\ \text{value}\end{array}\right] e^{-rn}$$

$$+ \left[\begin{array}{c}\text{Working}\\ \text{capital}\end{array}\right] [e^{-rn} - 1] - \left[\begin{array}{c}\text{Total plant}\\ \text{investment}\\ \text{incl. land value}\end{array}\right] = 0 \qquad (20)$$

Thus, a trial-and-error solution is required to find the value of r that satisfies Eq. (20) using the discount factors given in Tables I, II and III.

Example of DCF Application

Let us now use discounted cash flow to analyze the investment opportunities described in the XYZ Co. case study, the concluding part of which appeared on pp. 156–163. Table IV shows the DCF return on investment for each of the three alternatives that the XYZ Co. had under consideration. Here is how this table was drawn up:

The earnings statements for the three expansion alternatives (Tables VII, VIII and IX on pp. 160–162) were the starting point; in these tables, the annual after-tax profit and cash flow has been determined for the first seven years of the project, to reflect the change in earnings from year to year as the sales volume and capacity go up. It is the cash-flow values, both input and output, that will be used in the DCF calculation. Cash flow "out" results from the initial and subsequent capital investment in land, equipment and working-capital items. Cash flow "in" results from annual earnings and from the sale of land and equipment at the end of the project.

The cash flow generated by earnings is the after-tax profits plus depreciation. (Depreciation is not an annual cash expense to the company and, therefore, must be part of the cash-flow calculation. Depreciation is included in the earnings statement to ensure that the profit value reflects an allowance for the recovery of the initial capital investment, and to expense a portion of capital each year for income tax purposes as prescribed by law.)

The projected cash-flow chronology for Case 1 (the two separate 2-million-gal./yr. expansions) is shown in Table IV-1. The first cash effect, occurring at Point 0 in 1973, is the initial investment for the first 2-million-gal./yr. plant, which includes working capital for a cash outflow (i.e., negative cash flow) of $2,100,-000. The 1973, 1974 and 1975 earnings generate positive cash flows of $83,000, $231,000, and $337,-000 each year, respectively. In mid-1976, the investment for the second 2-million-gal./yr. plant is complete, represented by a negative cash flow of $2,250,-000, which includes additional working capital.

The detailed earnings statement provides cash inputs for 1976 through 1979. Thus, cash flow by year is provided for the first seven years of the project. Earnings for the last 8 years of the project's life are assumed to be relatively constant, with a 1% decrease per year in earnings, due to price deterioration. Cash flow generated by earnings for the last 8-year period would be $5,891,000 as indicated by the calculation in the table. At the end of the 15-year life of the project, the $550,000 working capital would be recovered by sale of inventories and collection of accounts receivable. For the purpose of this project, it was assumed that the salvage value of the plant and land at this time would offset the cost of dismantling the plant.

The cash-flow chronologies for the other two expansion alternatives were constructed similarly, and appear in the second and third sections of Table IV.

Discounted Cash Flow Applied to the XYZ Co. Case Study—Table IV

1. Two Separate 2-Million-Gal./Yr. Expansions

Date, 19__	Project Year	Item	Cash Flow, $M	Table Ref.	10% DCF Discount Factor	10% DCF Present Value, $M Yr.	10% DCF Present Value, $M Cum.	15% DCF Discount Factor	15% DCF Present Value, $M Yr.	15% DCF Present Value, $M Cum.	12.5% DCF Discount Factor	12.5% DCF Present Value, $M Yr.	12.5% DCF Present Value, $M Cum.
73	0	Investment for first plant....	−2,100		1.0000	−2,100	−2,100	1.000	−2,100	−2,100	1.000	−2,100	−2,100
73	0–1	Earnings....................	83	II	0.9516	79	−2,021	0.9286	77	−2,023	0.9400	78	−2,022
74	1–2	Earnings....................	231	II	0.8611	199	−1,822	0.7993	185	−1,839	0.8296	192	−1,830
75	2–3	Earnings....................	337	II	0.7791	262	−1,560	0.6879	232	−1,607	0.7322	246	−1,584
76	3½	Investment for second plant.	−2,250	I	0.7047	−1,586	−3,146	0.5916	−1,331	−2,938	0.6440	−1,449	−3,033
76	3–4	Earnings....................	476	II	0.7050	335	−2,811	0.5921	282	−2,656	0.6462	308	−2,725
77	4–5	Earnings....................	586	II	0.6379	374	−2,437	0.5096	299	−2,357	0.5703	334	−2,391
78	5–6	Earnings....................	723	II	0.5772	417	−2,020	0.4386	317	−2,040	0.5034	364	−2,027
79	6–7	Earnings....................	767	II	0.5223	401	−1,619	0.3775	289	−1,751	0.4444	341	−1,686
80–87	7–15	Earnings[1] 8 × 767 × 0.96 =..	5,891	III & I	0.3418[2]	2,014	395	0.2037[2]	1,200	−531	0.2622[2]	1,545	−141
87	15	Working-capital recovery.....	550	I	0.2231	123	518	0.1054	58	−493	0.1526	84	−57
87	15	Net salvage value...........	0		—			—			—		
		Cumulative Total...........				+518.4			−492.5			−57.4	

[1] Assumes 1% decrease per year in earnings due to price deterioration, or 4% average for 8-year period 1980–1987.

[2] Table III × Table I
0.6883 × 0.4966 = 0.3418
0.5823 × 0.3499 = 0.2037
0.6321 × 0.4148 = 0.2622

First interpolation:

$$I = \frac{15\,(518.4) - 10\,(-492.5)}{518.4 - (-492.5)} = \frac{12{,}701.0}{1{,}010.9} = 12.56\%$$

Second interpolation:

$$I = \frac{(12.5)\,(518.4) - 10.0\,(-57.4)}{518.4 - (-57.4)} = \frac{7{,}054.0}{575.8} = \mathbf{12.25\%}$$

2. One 2-Million-Gal./Yr. Plant, Expandable to 4-Million-Gal./Yr.

Date, 19__	Project Year	Item	Cash Flow, $M	Table Ref.	13% DCF Discount Factor	13% DCF Present Value, $M Yr.	13% DCF Present Value, $M Cum.	11% DCF Discount Factor	11% DCF Present Value, $M Yr.	11% DCF Present Value, $M Cum.	12.5% DCF Discount Factor	12.5% DCF Present Value, $M Yr.	12.5% DCF Present Value, $M Cum.
73	0	Initial plant investment......	−2,800		1.0000	−2,800	−2,800	1.0000	−2,800	−2,800	1.0000	−2,800	−2,800
73	0–1	Earnings....................	93	II	0.9377	87	−2,713	0.9470	88	−2,712	0.9400	87	−2,713
74	1–2	Earnings....................	241	II	0.8234	198	−2,515	0.8483	204	−2,508	0.8296	200	−2,513
75	2–3	Earnings....................	347	II	0.7230	251	−2,264	0.7600	264	−2,244	0.7322	254	−2,259
76	3½	Investment for expansion....	−1,050	I	0.6345	−666	−2,930	0.6805	−715	−2,959	0.6440	−676	−2,935
76	3–4	Earnings....................	483	II	0.6349	307	−2,623	0.6803	329	−2,630	0.6462	312	−2,623
77	4–5	Earnings....................	582	II	0.5575	324	−2,299	0.6099	355	−2,275	0.5703	332	−2,291
78	5–6	Earnings....................	719	II	0.4895	352	−1,947	0.5463	393	−1,882	0.5034	362	−1,929
79	6–7	Earnings....................	763	II	0.4299	328	−1,619	0.4894	373	−1,509	0.4444	339	−1,590
80–87	7–15	Earnings[1] 8 × 763 × 0.96 =..	5,860	III & I	0.2502[2]	1,466	−153	0.3079[2]	1,804	295	0.2622[2]	1,536	−54
87	15	Working-capital recovery....	555	I	0.1423	79	−74	0.1921	107	402	0.1526	85	31
87	15	Net salvage value...........	0		—			—			—		
		Cumulative Total............				−73.5			+402.4			+31.5	

[1] Assumes 1% decrease per year in earnings due to price deterioration or 4% average for 8-year period 1980–1987.

[2] Table III × Table I
0.6217 × 0.4025 = 0.2502
0.6650 × 0.4630 = 0.3079
0.6321 × 0.4148 = 0.2622

First interpolation:

$$I = \frac{(13)\,(402.4) - 11\,(-73.5)}{402.4 - (-73.5)} = \frac{6{,}039.7}{475.9} = 12.69\%$$

Second interpolation:

$$I = \frac{(13)\,(31.5) - 13.5\,(-73.5)}{31.5 - (-73.5)} = \frac{1{,}328.3}{105.0} = \mathbf{12.65\%}$$

3. One Plant of 4-Million-Gal./Yr. Capacity

Date, 19__	Project Year	Item	Cash Flow, $M	Table Ref.	11% DCF Discount Factor	11% DCF Present Value, $M Yr.	11% DCF Present Value, $M Cum.	10% DCF Discount Factor	10% DCF Present Value, $M Yr.	10% DCF Present Value, $M Cum.
73	0	Investment..................	−3,650		1.0000	−3,650	−3,650	1.0000	−3,650	−3,650
73	0–1	Earnings....................	−23	II	0.9470	−22	−3,672	0.9516	−22	−3,672
74	1–2	Earnings....................	124	II	0.8483	105	−3,567	0.8611	107	−3,565
75	2–3	Earnings....................	252	II	0.7600	192	−3,375	0.7791	196	−3,369
76	3–4	Earnings....................	416	II	0.6803	283	−3,092	0.7050	293	−3,076
77	4–5	Earnings....................	584	II	0.6099	356	−2,736	0.6379	373	−2,703
78	5–6	Earnings....................	721	II	0.5463	394	−2,342	0.5772	416	−2,287
79	6–7	Earnings....................	765	II	0.4894	374	−1,968	0.5223	400	−1,887
80–87	7–15	Earnings[1] 8 × 765 × 0.96 =..	5,875	III & I	0.3079[2]	1,809	−159	0.3418[2]	2,008	121
87	15	Working-capital recovery.....	555	I	0.1921	107	−52	0.2231	124	245
87	15	Net salvage value...	0		—			—		
		Cumulative Total............				−52.1			+244.7	

[1] Assumes 1% decrease per year in earnings due to price deterioration, or 4% average or 8-year period 1980–1987.

Interpolation:

$$\frac{11\,(244.7) - 10\,(-52.1)}{244.7 - (-52.1)} = \frac{3{,}212.7}{296.8} = \mathbf{10.82\%}$$

[2] Table III × Table I: 0.6650 × 0.4630 = 0.3079; 0.6883 × 0.4966 = 0.3418

The cash flow for each period of time must now be discounted to the present value at the interest rate selected for the first trial in the DCF calculation. This is done to determine an interest rate at which the cash flow will be discounted to provide a cumulative present value of zero for the life of the project. The first trial was selected at a 10% interest rate. The initial investment occurs at time 0, which is the present. Therefore, the discount value is 1 and the present value is equal to the cash flow.

Earnings that occur one year, two years, etc., after the reference point were discounted by the factors in Table II, because these cash effects occur uniformly over a one-year period at some point in time after the reference point.

As an example, let us locate the discount factor for the 1975 earnings from Table II. First, we must determine that 1975 is the third year of the project. From Table II, for the period of time between year 2 and 3, and under the heading for 10% interest, we locate the discount factor of 0.7791. This factor, multiplied by the $337,000 cash flow for the third year, discounts this cash flow back to the present value of $262,000, which appears under the present-value column for the yearly effects.

Note that the second investment of $2,250,000 is a cash effect that occurs at a point in time 3½ years after the reference point. Therefore, the discount factor is taken from Table I. To locate the discount factor in Table I for the second investment, we multiply the interest rate by the number of years after the reference point at which the cash effect occurred; the discount factor (for $100rn = 35$) is 0.7047.

The discount factor for the cash effect resulting from the last eight years of earnings (years 7 to 15) must be determined by a special calculation. First, the total 8-year cash flow of $5,891,000 must be discounted by Table III to the beginning of the 8-year period, because these flows are cash effects that occur uniformly over the period of time after the reference point (which is the end of the first seven years of the project). From Table III, for $100rt = 8 \times 10 = 80$, we obtain the discount factor of 0.6883.

Now, we have the discount factor for the 8-year period, which will provide the present value of these earnings at the beginning of the 8-year period. This value now must be discounted back to the initial reference point, using Table I, which has a discount factor of 0.4966 for $100rn = 70$. Therefore, the discount factor for the cash effect is the product of the Table III and Table I discount factors, which is 0.3418 (see Note 2 at the bottom of Table IV-1).

The Table III discount factors are a valuable shortcut for projects whose cash flows are uniform over the life of the project or for a number of consecutive years. Present-value calculations can be greatly simplified.

At a 10% interest rate, the present worth of the cash flows in Table IV-1 adds up to +$518,000. Therefore, 10% is not the interest rate at which the cumulative present value of the cash flow equals 0. Subsequent trial at 15% interest rate results in a −$492,000 present worth. By interpolation, an interest rate of 12.5% appears to be closer to the mark (see calculation in Table IV). The third trial, at a 12.5% interest rate, leads to a small present worth of −$57,000. A second interpolation gives a 12.25% interest rate, which results in a 0 cumulative present worth for cash-flow effects from this project indicating that this is the correct rate.

DCF calculations for Cases 2 and 3 are similar to those illustrated for Case 1.

It is apparent from the three parts of Table IV that the preferred alternative—the one with the 12.65% DCF—involves the construction of a plant whose 2 million gal./yr. of capacity can be expanded to 4 million gal./yr. A 12.25% DCF is obtained from the alternative that involves constructing two separate minimum-investment facilities, each having a 2-million-gal./yr. capacity. An even lower DCF—10.81%—is shown by the alternative that involves building a 4-million-gal./yr plant right at the start.

Real Vs. Indeterminate DCF

Reversal in the direction of annual net cash flow deserves a closer look. When there is one reversal—i.e., from positive to negative, or vice-versa—the solution is valid and there is no problem. But when there is more than one or less than one, the DCF may be indeterminate.

To clarify these possibilities, let us first consider a project having a 5-yr. life and requiring a $10,000 cash outlay, whose annual cash flow will vary as shown below:

Year	Cash Flow, $
0	−10,000
1	3,000
2	6,000
3	4,000
4	3,000
5	1,000
Net:	+ 7,000

Using various DCF rates, we obtain the following discount factors (F_B) and present worths (PW):

20% DCF		25% DCF		30% DCF	
F_B	PW	F_B	PW	F_B	PW
1.000	−10,000	1.000	−10,000	1.000	−10,000
0.9063	2,718	0.8848	2,654	0.8640	2,592
0.7421	4,453	0.6891	4,135	0.6400	3,840
0.6075	2,430	0.5367	2,147	0.4241	1,896
0.4974	1,492	0.4179	1,254	0.3513	1,054
0.4072	407	0.3255	325	0.2602	260
	+1,500		+ 515		− 358

At the first trial rate of 20%, the PW column shows a net positive amount, which means that the DCF chosen is too low. Trying a 25% rate also leads to a net positive amount. But 30% leads to a negative amount, meaning that the rate is too high. The actual

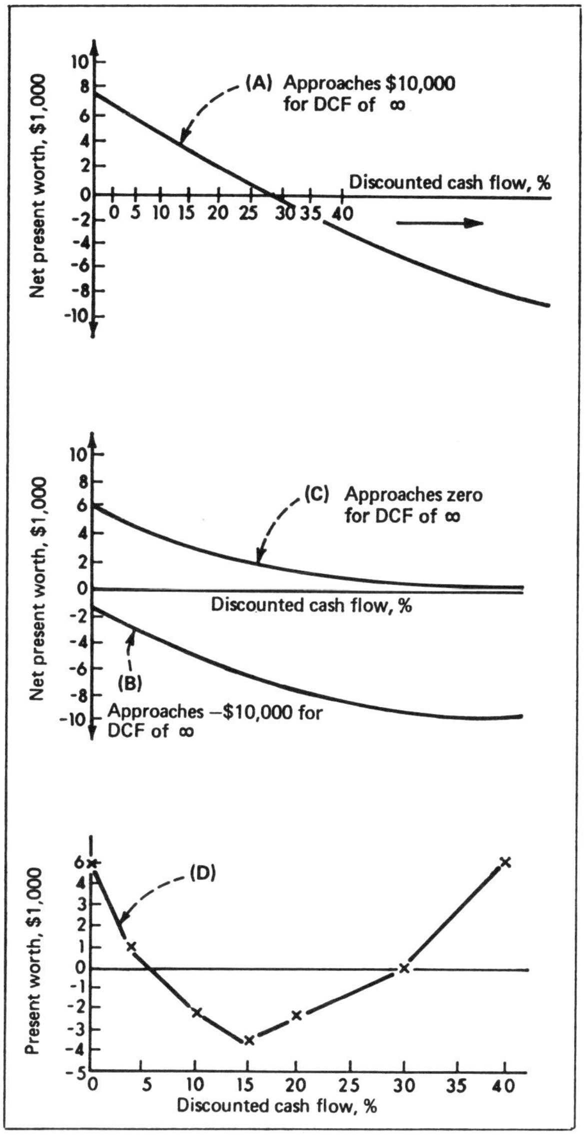

DISCOUNTED CASH FLOW: In these projects, Curve (A) represents a $7,000 cash flow over 5 years, (B) a negative cumulative cash flow, (C) an always-positive cash flow, and (D) a multiple DCF solution—Fig. 1

DCF rate appears to be less than 30% but more than 25%. The answer of 28% can be estimated by interpolation or obtained by further trials.

In Fig. 1, Curve (A) is a graphical representation of this situation. Net present worth is plotted against the DCF rate. At zero discount rate, all discount factors are unity, which means that the present worths are identical to the actual cash flows, for a net actual cash flow of $7,000 in this case. At infinite discount rate, all future cash flows are discounted to zero, leaving only the initial cash outlay of $10,000. The actual DCF return on investment of 28% is represented by the point at which the net-present-worth curve crosses the zero axis. The curve crosses the axis only once. Thus, the solution is valid.

Now let us consider a project that fails to pay off the investment. Of course, the investment might be made partly for safety or environmental-control reasons. Or, we can regard the cash-flow pattern below as a retrospective look at a project whose justification proved to be faulty because of rapidly changing market conditions:

Year	Cash Flow	Cumulative Cash Flow
0	−10,000	−10,000
1	2,000	− 8,000
2	3,000	− 5,000
3	2,000	− 3,000
4	1,000	− 2,000
5	500	− 1,500
Net	− 1,500	

This case is shown in Fig. 1 (B); note that the curve does not cross the zero axis for any positive value of the DCF rate. Thus, no valid solution via the DCF return-on-investment method exists.

In the converse of this situation, the initial capital investment may consist entirely of borrowed money. Thus, no capital investment per se may be required until the loan matures. At that time, if the cumulative cash flow is large enough to offset the loan payments, then it may be positive in every year of the project. This is illustrated below:

Year	Cash Flow	Loan Repayment	Cash Flow	Cumulative Cash Flow
0	–	–	–	–
1	+5,000	–	+5,000	+5,000
2	+7,000	–	+7,000	+12,000
3	−7,000	–	−7,000	+5,000
4	+6,000	–	+6,000	+11,000
5	+5,000	−10,000	−5,000	+6,000
Net	+16,000	−10,000	+6,000	

A graphical representation of this case is shown in Fig. 1 (C). Note that the curve approaches the zero axis at infinity. Thus, there is no valid DCF return for this example either.

Occasionally, the cumulative cash flow for a project changes sign more than once. In such a situation, the DCF return may be indeterminate. An illustration of this situation is shown in Table V. In Fig. 1 (D), which is derived from this table, note that the graph crosses the present-worth zero axis twice, giving rise to multiple solutions for DCF rates of approximately 6% and 30%, respectively. Obviously, this demonstrates an indeterminate solution to DCF return. Fortunately, situations of this sort do not occur frequently. Normally, the potential problem can be recognized by checking for the number of sign changes that occur in the cumulative cash flow, and determining whether these also occur in the cumulative present worth at various DCF levels.

Present-Worth Method of Analysis

As discussed earlier in this series, most companies have a criterion for the return on investment expected to be realized by the marginal going business in a given area. Often, this is the minimum acceptable return required in order to generate new capital for a project. The present-worth (PW) method of comparing investment opportunities takes this into account by discounting the annual cash flows calculated for a new project to their present value, using this marginal DCF rate of return as the base discount rate.

The project's PW is the difference between the present value of the annual cash flows (discounted on this basis) and the initial required investment. Thus, a direct solution is possible without any trial-and-error calculations. With this method, the project alternative having the greatest PW is the one selected, risk and other factors being judged equal.

The method is valid for comparison of projects that have identical lives or cover the same time span. The return rate used for discounting the annual cash flows of the project represents the "opportunity value" of investment capital to the company. Investments made in future years of the venture are also discounted back to their present value, using the same rate factor.

Let us consider a simple example of PW calculation. A project having a 5-year life and requiring an initial capital outlay of $10,000 has this schedule of annual cash flows:

10% DCF

Year	Cash Flow	F_B	PW
0	−10,000	1.000	−10,000
1	3,000	0.9516	2,855
2	6,000	0.8611	5,167
3	4,000	0.7791	3,116
4	3,000	0.7050	2,115
5	1,000	0.6379	638
	+ 7,000		+ 3,891

Thus, the present worth of this investment opportunity is $3,891.

Now, let us again apply PW analysis to the investment opportunities in the XYZ Co. case study. Table VI shows the calculation of PW for each of the three investment alternatives under consideration. Construction of a 2-million-gal./yr. plant expandable to a 4-million-gal./yr. capacity is the preferred alternative, with a PW of $678,000. The case involving two separate 2-million-gal./yr. expansions shows a $518,000 PW, and that for the initial 4-million-gal./yr. plant is only $245,000.

Proportional-Gain Method of Analysis

If one sums the present worth of all of the cash flows and divides that by the present worth of all of the investments made, the resultant quotient is the proportional gain for the investment alternative under consideration. This can be used as a relative measure of the efficiency with which a capital investment generates PW cash flows, and is a very valuable tool in comparing investment alternatives in a vast array of opportunities.

Table VI shows that the proportional gain for an expandable plant (from 2 to 4 million gal./yr.) is 1.198, which is greater than either the 1.145 calculated for two separate 2-million-gal./yr. expansions, or the 1.069 calculated for the initial 4-million-gal./yr. expansion. With this method, the alternative having the greatest value of proportional gain is preferred.

Choosing the Method

Now that we have discussed three techniques by which to gage the economic appeal or "ratability" of an investment opportunity, the question is which of these should be used for justifying a proposed expenditure of funds.

The most widely used of the three in the chemical process industries is the DCF method. The DCF return on investment can be compared to the cost of capital or to a minimum acceptable value set by a company for project approval.

An obvious pitfall in the DCF method crops up in a project with a sufficiently complicated cash-flow picture to produce an indeterminate answer, as previously described. Other drawbacks include complexity in re-

Example of Project Whose Cash Flow Changes Sign More Than Once—Table V

	Undiscounted		4% DCF			6% DCF			10% DCF		
n	C_n	ΣC_n	F_A	$(C_n)^{PW}$	$\Sigma(C_n)^{PW}$	F_A	$(C_n)^{PW}$	$\Sigma(C_n)^{PW}$	F_A	$(C_n)^{PW}$	$\Sigma(C_n)^{PW}$
0	−100	−100	1.0000	−100	−100	1.0000	−100	−100	1.0000	−100	−100
1	+600	+500	0.9608	+576	+476	0.9418	565	+465	0.9048	+543	+443
2	−1,000	−500	0.9231	−923	−447	0.8869	−887	−422	0.8187	−819	−376
3	+505	+5	0.8869	+448	+1	0.8353	422	0	0.7408	+374	−2

	15% DCF			20% DCF			30% DCF			40% DCF		
n	F_A	$(C_n)^{PW}$	$(C_n)^{PW}$	F_A	$(C_n)^{PW}$	$\Sigma(C_n)^{PW}$	F_A	$(C_n)^{PW}$	$\Sigma(C_n)^{PW}$	F_A	$(C_n)^{PW}$	$(C_n)^{PW}$
0	1.0000	−100	−100	1.0000	−100	−100	1.0000	−100	−100	1.0000	−100	−100
1	0.8607	+516	+416	0.8187	+491	+391	0.7408	+444	+344	0.6703	+402	+302
2	0.7408	−741	−325	0.6703	−670	−279	0.5488	−549	−205	0.4493	−449	−147
3	0.6376	+322	−3	0.5488	+277	−2	0.4066	+205	0	0.3012	+152	+5

Key: n = year; C_n = Cash flow ($1,000) in year n; $(C_n)^{PW}$ = Present worth of C_n; Σ = Cumulative total; F_A = Discount factor.

Calculating Present Worth and Proportional Gain for XYZ Co. Case Study—Table VI

				Two Separate 2-MM-Gal./Yr. Expansions			Expandable Plant: 2- to 4-MM-Gal./Yr. Capacity			4-MM-Gal./Yr. Expansion		
			Discount Factor @ 10% Int.	Cash Flow $1,000	Present-Value		Cash Flow $1,000	Present-Value		Cash Flow $1,000	Present-Value	
Date 19__	Project Year	Item			Annual	Cum.		Annual	Cum.		Annual	Cum.
73	0	Initial investment...	1.0000	−2,100	−2,100	−2,100	−2,800	−2,800	−2,800	−3,650	−3,650	−3,650
73	0–1	Earnings............	0.9516	83	79	−2,021	93	88	−2,712	−23	−22	−3,672
74	1–2	Earnings............	0.8611	231	199	−1,822	241	208	−2,504	124	107	−3,565
75	2–3	Earnings............	0.7791	337	262	−1,560	347	270	−2,234	252	196	−3,367
76	3½	Second investment.	0.7047	−2,250	−1,586	−3,146	−1,050	−740	−2,974	—	—	—
76	3–4	Earnings............	0.7050	476	336	−2,810	483	340	−2,634	416	293	−3,076
77	4–5	Earnings............	0.6379	586	374	−2,436	582	371	−2,263	584	373	−2,703
78	5–6	Earnings............	0.5772	723	417	−2,019	719	415	−1,848	721	416	−2,287
79	6–7	Earnings............	0.5223	767	401	−1,618	763	399	−1,449	765	400	−1,887
80–87	7–15	Earnings............	0.3418	5,891	2,013	395	5,860	2,003	554	5,875	2,008	121
87	15	Working-capital recovery..........	0.2231	550	123	518	555	124	678	555	124	245
87	15	Net salvage value*..	0.2231	—	—		—			—		
		PW @ 10% interest..............................				518			678			245
	(1)	PW of future earnings............................				4,081			4,094			3,771
	(2)	PW of investments plus working capital...........				3,563			3,416			3,520
	(3)	Proportional gain, (1) ÷ (2).......................				1.145			1.198			1.069

MM = Million PW = Present Worth * Salvage value equals cost of dismantling

gard to discount factors (we hope we have clarified some of the confusion), and the need to assemble quite a lot of data in regard to timing as well as amounts (although accuracy of the data becomes less crucial as you look further and further into the future, because of heavier and heavier discounting).

Also, DCF does not tell you very much about the size of the return; if used injudiciously, it can lead management to dissipate its manpower among a bevy of high-return but very small projects, rather than tackling a more-significant one that may show a slightly lower DCF.

The PW method allows profitability to be calculated at the firm's minimum acceptable interest rate for risk capital, thus alleviating the need for the trial-and-error computation required by DCF.

Also, PW reflects the size of the return and the commercial interest of the project: Obviously, a project where a $1-million investment can lead to a $2-million return over its life would stand head and shoulders in a PW analysis over a project where a $100 investment could lead to a $200 return over the same period of time (although both would show the same DCF).

Another advantage of the present-worth method is illustrated by the example in this article. The difference in the DCF values between Case 1 and Case 2 of the XYZ Co.'s alternatives is only 0.4% (12.65%–12.25%), and thus appears to be insignificant within the level of accuracy of these calculations. However, when these two projects are compared on a PW basis, the difference in present value for the life of the project at a 10% interest rate is $160,000 in favor of Case 2, which is a significant value.

The proportional-gain technique provides a ready basis for comparison with your minimum acceptable interest rate. That is, a proportional gain in excess of "one" would merit consideration. Proportional gain allows ready comparison between projects of different magnitude.

In the final analysis, your choice of methods may be governed to a great extent by the practice of your company. Certainly, calculating the values for all three methods is a relatively simple task once the techniques are understood and the earnings statements have been generated.

However, it is important to recognize that where project life is not defined by contractual or other definitive considerations, opportunities should be compared at the same lifetime level when using either present worth or discounted cash flow. Then, a meaningful analysis can be made of your project, considering the pros and cons of each method.

An additional analytical method—one involving the use of decision trees to evaluate various courses of action—will be examined in the next article. ■

References

1. Earley, W. R., *Chem. Eng.*, Mar. 30, 1964, p. 111–116.
2. Souders, M., *Chem. Eng. Prog.*, Mar. 1966, p. 79–81.
3. Weaver, J. B., *Chem. & Eng. News*, Sept. 25, 1961, pp. 94–103.
4. Winn, F. W., *Petr. Ref.*, July 1956, p. 199–210.
5. Axtell, O., *Hydrocarb. Proc. & Petr. Ref.*, Sept. 1961, pp. 221–226.
6. Weaver, J. B., Reilly, R. J., *Chem. Eng. Prog.*, Oct. 1956, p. 405–416.
7. Peters, M. S., Timmerhaus, K. D., "Plant Design and Economics for Chemical Engineers," 2nd Ed., McGraw-Hill, 1968.

ECONOMIC ANALYSIS Part 9

Decision Trees: A Rapid Evaluation Of Investment Risk

Finding the most profitable route for plant expansion requires a thorough analysis of alternative procedures. The decision tree shows at a glance the interrelationships among the various options and the time dependence of one on the others.

IRVING LEIBSON and CHARLES A. TRISCHMAN, JR., Dart Industries, Inc.

The XYZ Co. manufactures Product B for which it has considered the construction of new plant capacity. Earlier in this series, the evolution of a decision for plant expansion was outlined. [See Parts 5, 6 and 7 of this series.]

Alternatives open to the planning group may be represented by the decision tree as shown in Fig. 1. This shows all of the decisions and events that are important considerations of the XYZ Co.'s situation. In each case, the financial consequences of a given branch of the decision tree[1,2] may be shown so that the effects of making a specific decision at a given point in time can be translated through the life of the project.

By using time-value-of-money methods for financial analyses, the effects of delays in decisions and the comparison of alternative branches become clear. Furthermore, the risk factors and probabilities of success for each of the individual branches can be estimated through group effort, and these weighting factors may be applied in analyzing the financial data. Thus, a clear picture of the various risks is shown so that we can objectively evaluate various courses of action.

At the outset, the options that are open must be clearly defined. And, the interrelationship between various options and the time dependence of one on the other also must be analyzed. Each alternative is represented by a network or series of networks in the decision tree. An analysis of each network reveals valuable information concerning the specific alternative under consideration. This aids in planning and controlling the project shown by the network.

In effect, the decision tree becomes a pictorial representation of all the possible alternatives and their interrelationships, together with the economic evaluation of each and the probability of success of each.

Originally published January 1972.

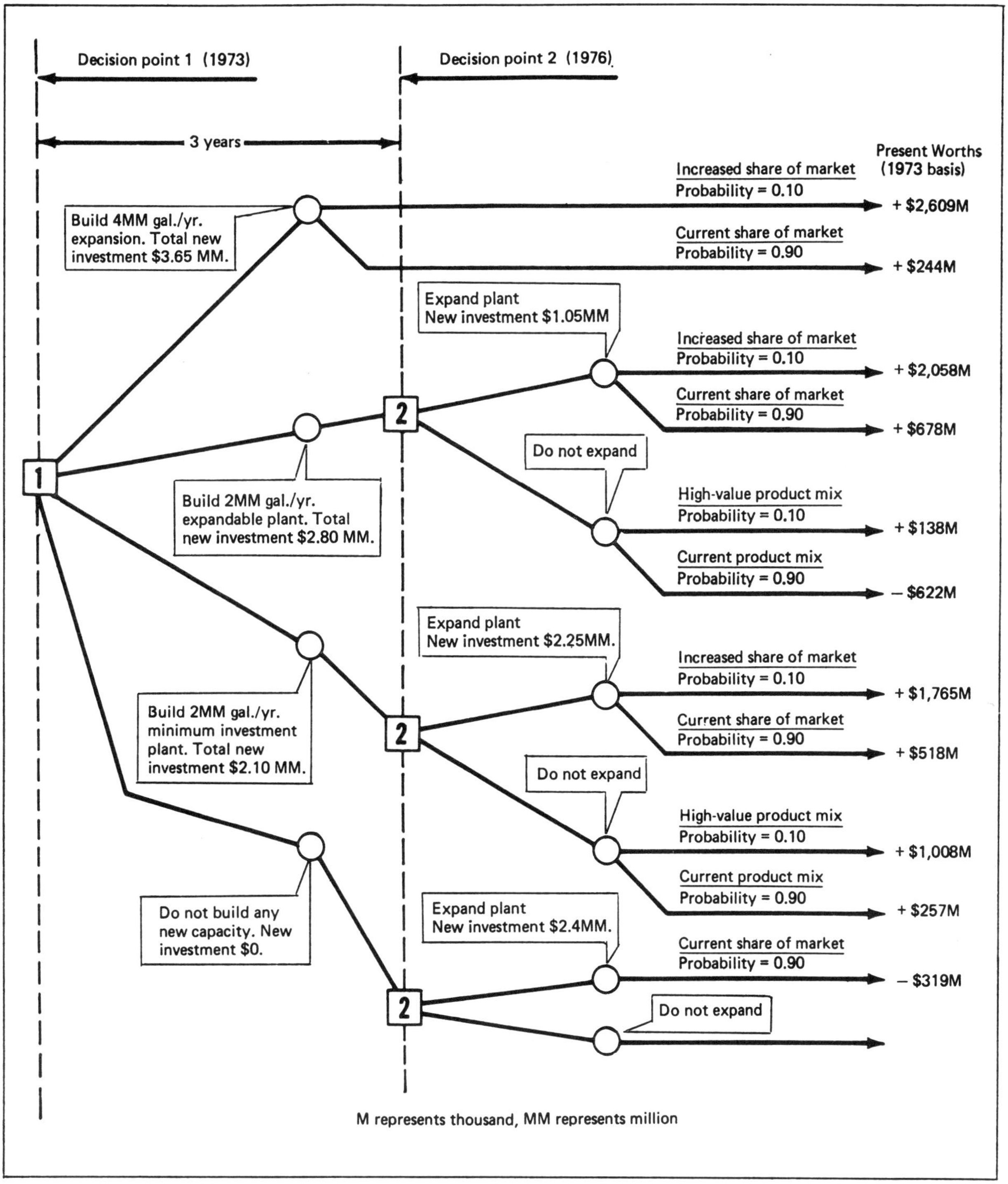

DECISION-TREE analysis for Product B expansion shows the 1973 present worths of all options—Fig. 1

This is a useful technique because it shows the key decision points from a time point-of-view, and the various responses that can be given.

Thus, the technique is particularly useful for the development of new capital projects, the strategies of marketing new products, planning research and development for results, and even the consideration of the individual personal development of employees in the organization.

Decision-tree analysis can be easily computerized by using a network-analysis approach,[3] with appropriate financial evaluation techniques such as discounted-cash-flow return, present worth, and proportional gain for each network [See Part 8 of series]. When combined with PERT diagrams for the individual networks, a very complete planning approach is possible. This may then be used to monitor progress as a program proceeds. By periodically updating the overall program, decision-tree analysis may be used as a financial control tool in business planning.

Present Worth for Two Separate 2 MM Gal./Yr. Expansions—Table I (Increased Share of Market)

				1973 Basis		1976 Basis	
Date	Year	Item	Cash Flow, $M	Discount Factor at 10%	Present Worth, $M	Discount Factor at 10%	Present Worth, $M
'73	0	Initial investment	−2,100	1.000	−2,100		
'73	0–1	Earnings	350	0.9516	333		
'74	1–2	Earnings	395	0.8611	340		
'75	2–3	Earnings	394	0.7791	307		
		Subtotal			−1,120		
'76	3.5	Second investment	−2,250	0.7047	−1,586	1.000	−2,250
'76	3–4	Earnings	700	0.7050	494	0.9516	667
'77	4–5	Earnings	758	0.6379	484	0.8611	652
'78–'87	5–15	Earnings	8,830	0.382*	3,370	0.519**	4,570
'87	15	Working capital recovery	550	0.2231	123	0.3012	165
		Total			+1,765		+3,804

* 0.6321 × 0.6065 = 0.382
Here, we find the present worth for cash flows that occur uniformly over a period of years after the reference point. Hence, we will use Table III in Part 8 of this series (*Chem. Eng.*, Dec. 13, 1971, p. 100). The discount factor is found by evaluating 100*rT* where *r* = 10%, and *T* is 10 yr. From Table III for 100*rT* = 100, we get 0.6321 as the present worth (in 1978) of uniform annual earnings that will total to $1. But, we must further discount these earnings to a 1973 basis. To do so, we use Table I in the same article on p. 98. The present worth is found by evaluating 100*rn* where *r* = 10%, and *n* = 5 yr. From Table I for 100*rn* = 50, we get the present worth (1973 basis) as 0.6065 for earnings that have a value of $1 in 1978. Hence, for earnings of 0.6321 (1978) we multiply by 0.6065 to get 0.382 (1973 basis).

** 0.6321 × 0.8187 = 0.519
The value 0.6321 is obtained as before. But, we must now further discount the earnings to a 1976 basis. Again, using Table I of Part 8 of this series, we find 100*rn* = 100(1/10)2 = 20. For 100*rn* = 20, the discount factor is 0.8187 (1976 basis). Hence, for earnings of 0.6321 (1978), we multiply by 0.8187 to get 0.519 (1976 basis).
M represents thousand, *MM* represents million.

Present Worth for Two Separate 2 MM Gal./Yr. Expansions—Table II (Current Share of Market)

				1973 Basis		1976 Basis	
Date	Year	Item	Cash Flow, $M	Discount Factor at 10%	Present Worth, $M	Discount Factor at 10%	Present Worth, $M
'73	0	Initial investment	−2,100	1.000	−2,100		
'73	0–1	Earnings	83	0.9516	79		
'74	1–2	Earnings	231	0.8611	199		
'75	2–3	Earnings	337	0.7791	262		
		Subtotal			−1,660		
'76	3.5	Second investment	−2,250	0.7047	−1,586	1.000	−2,250
'76	3–4	Earnings	476	0.7050	336	0.9516	453
'77	4–5	Earnings	586	0.6379	374	0.8611	502
'78	5–6	Earnings	723	0.5772	417	0.7791	563
'79	6–7	Earnings	767	0.5223	401	0.7050	542
'80–'87	7–15	Earnings	5,891	0.3418*	2,013	0.462**	2,720
'87	15	Working capital recovery	550	0.2231	123	0.3012	165
		Total			+518		+2,695

* 0.6883 × 0.4966 = 0.3418
** 0.6883 × 0.6703 = 0.4620
See footnotes to Table I for explanation of these discount factors.
M represents thousand, *MM* represents million.

Development of the Decision Tree

In a preceding article [Part 6], a vigorous discussion took place between the various functional representatives to the planning group in regard to the degree of market penetration that should be used as a basis for the capacity-expansion project. It was decided that the market forecast should be based on maintaining current share of market, with growth in all future sales coming from growth in total market demand for Product B.

At this stage of project development, we should attempt to quantify the project's principal uncertainties as to degree of risk. Only by doing this can we be sure that proper weight is given to each of the various factors in making the final decision.

Hence, we must assess the probability of attaining an increased share of the market relative to maintaining the current share—both early in the life of the project and in its later years. To analyze this situation, it is necessary to recognize that two different decision points exist in time. Decision Point 1 is the initial decision to build a capacity expansion. Decision Point 2 is the subsequent decision (in time) to further expand the plant built at Decision Point 1.

From the respective opinions of knowledgeable members of the group, it is determined that only a 10% probability exists for attaining a higher share of the market in either the early or late years of project life. Furthermore, a 90% probability is assigned to retention of current share of the market. The Delphi method[4] or some modification of this procedure may be used to quantify the subjective probabilities. The Delphi approach allows an individual to emphasize his views. However, he can also concede that there may be other important considerations. Application of this technique can lead to a group consensus on valuing the relative risk factors and probabilities.

In applying decision-tree analysis to this example, it is necessary to evaluate first the position value for Decision 2. By developing a monetary value for Decision 2, it becomes possible to project back to Decision 1, and compare the relative gains made by taking any one of the branches shown in Fig. 1. With this concept, we are recognizing that management does not have to make Decision 2 at the time of making Decision 1. In fact, it is not even known whether or not the opportunity will be provided to make Decision 2. The position value of Decision 2 is the expected value of the preferred branch, which is the gain anticipated (based on probability weighting of the present worths) from the respective branches leading from Decision Point 2.

In making this analysis, it is necessary to calculate the present worth of each of the branches leading from Decision Point 2, discounted to the time of Decision Point 2 (for this example, 1976). These calculations are shown in Table I through VI, and are summarized below:

Case		Present Worth (1976 Basis) \$ M	Table
A	Two separate 2-MM gal./yr. expansions		
	Increased share of market..	3,804	I
	Current share of market...	2,695	II
B	Initial 2-MM gal./yr. plant, expandable to 4-MM gal./yr.		
	Increased share of market..	5,252	III
	Current share of market...	3,882	IV
C	Initial 2-MM gal./yr. plant, no expansion		
	High-value product mix....	2,884	V
	Current product mix.......	2,440	VI

where MM represents million (10^6) and M represents thousand (10^3).

An analysis of Decision 2 to determine its decision-point value is shown in Table VII. The three investment alternatives are compared. The present values in 1976 (as shown above) are weighted by their respective percentage probabilities in regard to share of market attained. The weighted present-value of the cash flows in 1976 is calculated for each case. The respective second investment (to be made in 1976) is then deducted from the weighted cash-flow for each case in order to obtain the net value of Decision Point 2. The maximum value of Decision 2 is \$4,019 M (based on 1976 values) for an initially installed plant that is subsequently expanded.

Thus, it is worth \$4,019MM for the XYZ Co. to reach the position where it can make Decision 2. Given this value and the other data shown in Table VI, what is now the best action on Decision 1?

For the initially installed, minimum-investment plant (2 MM gal./yr.), the maximum position value

Present Worth for Expandable Plant: 2 to 4 MM Gal./Yr. Capacity—Table III (Increased Share of Market)

Date	Year	Item	Cash Flow, \$M	1973 Basis: Discount Factor at 10%	1973 Basis: Present Worth, \$M	1976 Basis: Discount Factor at 10%	1976 Basis: Present Worth, \$M
'73	0	Initial investment	−2,800	1.000	−2,800		
'73	0–1	Earnings	345	0.9516	328		
'74	1–2	Earnings	390	0.8611	336		
'75	2–3	Earnings	389	0.7791	303		
		Subtotal			−1,833		
'76	3.5	Second investment	−1,050	0.7047	−740	1.000	−1,050
'76	3–4	Earnings	732	0.7050	516	0.9516	697
'77	4–5	Earnings	790	0.6379	502	0.8611	680
'78–'87	5–15	Earnings	9,150	0.382*	3,490	0.519**	4,760
'87	15	Working capital recovery	550	0.2231	123	0.3012	165
		Total			+2,058		+5,252

* 0.6321 × 0.6065 = 0.382 } See footnotes to Table I for explanation of these discount factors.
** 0.6321 × 0.8187 = 0.519 }
M represents thousand, MM represents million.

Present Worth for Expandable Plant: 2 to 4 MM Gal./Yr. Capacity—Table IV (Current Share of Market)

				1973 Basis		1976 Basis	
Date	Year	Item	Cash Flow, $M	Discount Factor at 10%	Present Worth, $M	Discount Factor at 10%	Present Worth, $M
'73	0	Initial investment	−2,800	1.000	−2,800		
'73	0–1	Earnings	93	0.9516	88		
'74	1–2	Earnings	241	0.8611	208		
'75	2–3	Earnings	347	0.7791	270		
		Subtotal			−2,234		
'76	3.5	Second Investment	−1,050	0.7047	−740	1.000	−1,050
'76	3–4	Earnings	483	0.7050	340	0.9516	460
'77	4–5	Earnings	582	0.6379	371	0.8611	500
'78	5–6	Earnings	719	0.5772	415	0.7791	560
'79	6–7	Earnings	763	0.5223	399	0.7050	537
'80–87	7–15	Earnings	5,860	0.3418	2,003	0.462	2,710
'87	15	Working capital recovery	550	0.2231	123	0.3012	165
		Total			+678		+3,882

M represents thousand, MM represents million.

Present Worth for 2 MM Gal./Yr. Plant, No Expansion—Table V (High-Value Product Mix)

				1973 Basis		1976 Basis	
Date	Year	Item	Cash Flow, $M	Discount Factor at 10%	Present Worth, $M	Discount Factor at 10%	Present Worth, $M
'73	0	Investment	−2,100	1.000	−2,100		
'73	0–1	Earnings	350	0.9516	333		
'74	1–2	Earnings	395	0.8611	340		
'75	2–3	Earnings	394	0.7791	307		
		Subtotal			−1,120		
'76	3–4	Earnings	392	0.7050	277	0.9516	373
'77	4–5	Earnings	401	0.6379	256	0.8611	345
'78–'87	5–15	Earnings	4,000	0.382	1,528	0.519	2,076
'87	15	Working capital recovery	300	0.2231	67	0.3012	90
		Total			+1,008		+2,884

M represents thousand, MM represents million.

Present Worth for 2 MM Gal./Yr. Plant, No Expansion—Table VI (Current Product Mix)

				1973 Basis		1976 Basis	
Date	Year	Item	Cash Flow, $M	Discount Factor at 10%	Present Worth, $M	Discount Factor at 10%	Present Worth, $M
'73	0	Investment	−2,100	1.000	−2,100		
'73	0–1	Earnings	83	0.9516	79		
'74	1–2	Earnings	231	0.8611	199		
'75	2–3	Earnings	337	0.7791	262		
		Subtotal			−1,560		
'76–'87	3–15	Earnings	4,044	0.432*	1,750	0.5823	2,350
'87	15	Working capital recovery	300	0.2231	67	0.3012	90
		Total			+257		+2,440

* 0.5823 × 0.7408 = 0.432 (See footnotes to Table I for explanation of discount factors.)
M represents thousand, MM represents million.

Analysis of Decision 2 Using Probability Factors 10/90—Table VII

Case		Share of Market, %	Probability, %	Present Value in 1976, $M	Source Table	Weighted Present Value in 1976, $M
A.	Minimum Investment.	16.7	10	6,054	I	605
	Initially installed at 2 MM gal./yr.	10.2	90	4,945	II	4,451
					Total	5,056
					Second investment	−2,250
					Net	+2,806
B.	Expandable Plant.	16.7	10	6,302	III	630
	Initially installed at 2 MM gal./yr.	10.2	90	4,932	IV	4,439
					Total	5,069
					Second investment	−1,050
					Net	+4,019
C.	No expansion to be made subsequent to initial minimum investment. 2 MM gal./yr.	High-value product mix	10	2,884	V	288
		Current product mix	90	2,440	VI	2,196
					Total	2,484
					Second investment	0
					Net	+2,484

M represents thousand, MM represents million.

for Decision No. 2 is $2.484 MM for Case C, in which no expansion is made subsequent to the initial minimum investment for the 2-MM gal./yr. facility, as shown in Table VII.

Using the respective position values for Decision Point 2, it is now possible to analyze Decision 1.

The present worths (on a 1973 basis) of the project to initially install a 4-MM gal./yr. plant for (a) increased share of market and (b) retention of current share of market are shown in Table VIII and IX, respectively. Similarly, the present worths for each of the other cases are shown on a 1973 basis in Table I through IV.

Table X shows that construction of a 2-MM gal./yr. expandable plant is preferred based on an anticipated net-discounted yield of $776 M. This is deter-

Present Worth of Initial 4 MM Gal./Yr. Plant—Table VIII (Increased Share of Market)

Date	Year	Item	Cash Flow, $M	Discount Factor at 10% (1973 Basis)	Present Worth, $M (1973 Basis)
'73	0	Investment	−3,650	1.000	−3,650
'73	0–1	Earnings	407	0.9516	386
'74	1–2	Earnings	700	0.8611	602
'75	2–3	Earnings	758	0.7791	592
'76	3–4	Earnings	885	0.7050	623
'77	4–5	Earnings	883	0.6379	563
'78–'87	5–15	Earnings	8,810	0.382*	3,370
'87	15	Working capital recovery	550	0.2231	123
		Total			+2,609

* 0.6321 × 0.6065 = 0.382 (See footnotes to Table I for explanation of discount factors.)

M represents thousand, MM represents million.

Present Worth of Initial 4 MM Gal./Yr. Plant—Table IX (Current Share of Market)

Date	Year	Item	Cash Flow, $M	Discount Factor at 10% (1973 Basis)	Present Worth, $M (1973 Basis)
'73	0	Investment	−3,650	1.000	−3,650
'73	0–1	Earnings	−23	0.9516	−22
'74	1–2	Earnings	124	0.8611	107
'75	2–3	Earnings	252	0.7791	196
'76	3–4	Earnings	416	0.7050	293
'77	4–5	Earnings	584	0.6379	373
'78	5–6	Earnings	721	0.5772	416
'79	6–7	Earnings	765	0.5223	400
'80–'87	7–15	Earnings	5,875	0.3418*	2,008
'87	15	Working capital recovery	550	0.2231	123
		Total			+244

* 0.6883 × 0.4966 = 0.3418 (See footnotes to Table I for explanation of discount factors.)

M represents thousand, MM represents million.

Analysis of Decision 1 Using Probability Factors 10/90—Table X

Case	Share of Market, %	Probability, %	Present Value in 1973, $M	Source Table	Weighted Present Value in 1973, $M
1. Build 4 MM gal./yr. plant	16.7	10	6,259	VIII	626
	10.2	90	3,894	IX	3,505
				Total	4,131
				Initial investment	−3,650
				Net	+481
2. Build minimum investment 2 MM gal./yr. plant	16.7	10	980	I	98
	10.2	90	540	II	486
	Decision 2 value at end of 3 yr.*				2,085
				Total	2,669
				Initial investment	−2,100
				Net	+569
3. Build 2 MM gal./yr. expandable plant	16.7	10	967	III	97
	10.2%	90	566	IV	509
	Decision 2 value at end of 3 yr.**				2,980
				Total	3,576
				Initial investment	−2,800
				Net	+776

*Decision 2 value of $2,806 M (from Table VII) multiplied by discount factor 0.7408 = $2,085 M.
**Decision 2 value of $4,019 M (from Table VII) multiplied by discount factor 0.7408 = $2,980 M.
· M represents thousand, MM represents million.

mined (as shown in Table X) by calculating the present worth of the cash flows in the first three years of the project (after Decision Point 1), properly weighted by the percentage probability of attaining an increased or current share of the market, and adding this to the position value of Decision 2 discounted back to the begin-

Analysis of Decision 2 Using Probability Factors 90/10—Table XI

Case	Share of Market, %	Probability, %	Present Value in 1976, $M		Weighted Present Value in 1976, $M
A. Minimum Investment. Initially installed at 2 MM gal./yr.	16.7	90	6,054		5,449
	10.2	10	4,945		494
				Total	5,943
				Second investment	−2,250
				Net	+3,693
B. Expandable Plant. Initially installed at 2 MM gal./yr.	16.7	90	6,302		5,672
	10.2	10	4,932		493
				Total	6,165
				Second investment	−1,050
				Net	+5,115
C. No expansion to be made subsequent to initial minimum investment. 2 MM gal./yr.	High-value product mix	90	2,884		2,596
	Current product mix	10	2,440		244
				Total	2,840
				Second investment	0
				Net	2,840

M represents thousand, MM represents million.

Analysis of Decision 1 Using Probability Factors 90/10 —Table XII

Case	Share of Market, %	Probability, %	Present Value in 1976, $M	Source Table	Weighted Present Value in 1976, $M
1. Build 4 MM gal./yr. plant	16.7	90	6,259	VIII	5,633
	10.2	10	3,894	IX	389
				Total	6,022
				Initial investment	−3,650
				Net	+2,372
2. Build minimum investment 2 MM gal./yr. plant	16.7	90	980	I	882
	10.2	10	540	II	54
	Decision 2 value at end of 3 yr.*				2,735
				Total	3,671
				Initial investment	−2,100
				Net	+1,571
3. Build 2 MM gal./yr. expandable plant	16.7	90	967	III	870
	10.2	10	566	IV	57
	Decision 2 value at end of 3 yr.**				3,789
				Total	4,716
				Initial investment	−2,800
				Net	+ 1,916

*Decision 2 value of $3,693 M (from Table XI) multiplied by discount factor 0.7408 = $2,735 M.
**Decision 2 value of $5,115 M (from Table XI) multiplied by discount factor 0.7408 = $3,789 M.
M represents thousand, MM represents million.

Decision 1 Delayed—Table XIII (Current Share of Market)

Date	Year	Item	Cash Flow, $M	1973 Basis: Discount Factor at 10%	1973 Basis: Present Worth, $M
'73	0–1	....	....	0.9516	....
'74	1–2	....	....	0.8611	....
'75	2–3	....	....	0.7791	....
'76	3	Investment	−2,400	0.7408	−1,780
'76	3–4	Earnings	67	0.7050	47
'77	4–5	Earnings	215	0.6379	137
'78–'87	5–15	Earnings	3,210	0.382	1,220
'87	15	Working capital recovery	300	0.2231	67
		Total			−319

This table shows the effect on present worth of a decision to delay the building of the 2 MM gal./yr. minimum investment plant by 3 yr.
M represents thousand, MM represents million.

ning of the project (from 1976 to 1973), this gives a total weighted value for cash flows associated with Decision Point 1. The initial investment is then subtracted from this total weighted value of cash flows to obtain the expected net-discounted yield for Decision 1.

A Change in Probabilities

It is interesting to note from Table XI and XII, respectively, that if a 90% probability of attaining an increased share of the market were to exist (instead of 10%), then the decision-tree analysis would favor initial construction of a 4-MM gal./yr. plant.

It is also interesting to note that if Decision 1 is delayed by three years, with a 15-yr. project life, Table XIII shows that a negative present worth of $319 M is realized for construction of a 2-MM gal./yr. minimum investment plant, Thus, if the total project life is realistic, a three-year delay in decision completely nullifies any value for the expansion. Obviously, this is an exaggerated example. However, the importance of making decisions on a timely and clean basis is dramatically demonstrated by the corresponding effect on profitability.

In the next article dealing with a make or buy decision for a material, we plan to show how decision-tree analysis may be used in monitoring a commercial chemical-development project. ■

References

1. Magee, J. F., *Harvard Business Rev.*, July-Aug. 1964, pp. 126-138.
2. Chinn, J. S. and Cuddy, W. A., Project Decision and Control: A Case History, Paper 7e presented at 63rd Annual Meeting of AIChE, Chicago, Nov. 29-Dec. 3, 1970.
3. Magee, J. F., *Harvard Business Rev.*, Sept.-Oct. 1964, pp. 79-96.
4. Helmer, O., Special Report, P-3721, Rand Corp., Santa Monica, Calif. Nov. 1967.

ECONOMIC ANALYSIS

Part 10

Should You Make Or Buy Your Major Raw Materials?

Backward integration may offer a way to beat the cost-price squeeze, or it may not. Here is how to develop and evaluate the data that will help you avoid pitfalls in making individual make-or-buy decisions—as illustrated by a realistic case study.

I. LEIBSON and C.A. TRISCHMAN, JR., Dart Industries, Inc.

In today's highly competitive chemical process industries, raw-material costs are receiving more and more scrutiny. Companies are often strongly tempted to try and lower these costs by integrating backwards into raw-materials manufacture.

Corporate managements frequently ask either their internal staffs or engineering contractors/consultants for definitive studies to determine the technology and scale of operations that would be required to make the internal manufacture of certain large-volume raw materials economically attractive, and to develop the necessary cost data. These studies serve as valuable cost-information inputs during negotiation of long-term raw-material purchase agreements—thus, they often have considerable financial value even when the decision is to buy.

Du Pont recently invested a substantial amount of money in a study (made by two major engineering/construction companies) aimed at evaluating the investment requirements and corresponding economics for construction of a major petrochemical refinery. This refinery would have manufactured many of the raw materials Du Pont now purchases under long-term agreements. Plans for this refinery were subsequently shelved. However, no two cases are alike, and as a company grows to be a large producer of a given chemical, the usage of specific raw materials may become sufficiently large to justify at least serious consideration of making rather than buying.

In this article, we shall develop one more case study involving the XYZ Co. This firm has to reach a decision on whether to invest in the pilot-planting and ultimate commercialization of a new process for the manufacture of Raw Material A (RMA, for short) involved in the manufacture of Product B. The alternative which the company faces is to instead renew the current purchase agreement for RMA.

As mentioned in an earlier article, the R&D group has succeeded in producing laboratory samples of RMA that are completely interchangeable with the purchased counterpart in regard to quality and performance.

Management must now decide whether or not to appropriate funds for the modification of an existing pilot plant; this would permit scaleup of the laboratory results, and production of enough RMA for a plant-scale trial in the manufacture of Product B.

The R&D group's proposal to management calls for the expenditure of \$50,000 for the modification of the existing pilot plant to permit continued development of the RMA-manufacturing process, and \$50,000 to operate the modified pilot plant for six months. Furthermore, in Nov. 1970 (the time base for this case study), R&D and Engineering jointly developed a PERT diagram for commercialization of the project (shown in Fig. 1); this diagram calls for construction of the commercial facility to be completed by Sept. 1972 and allows two months for plant startup.

Originally published February 1972.

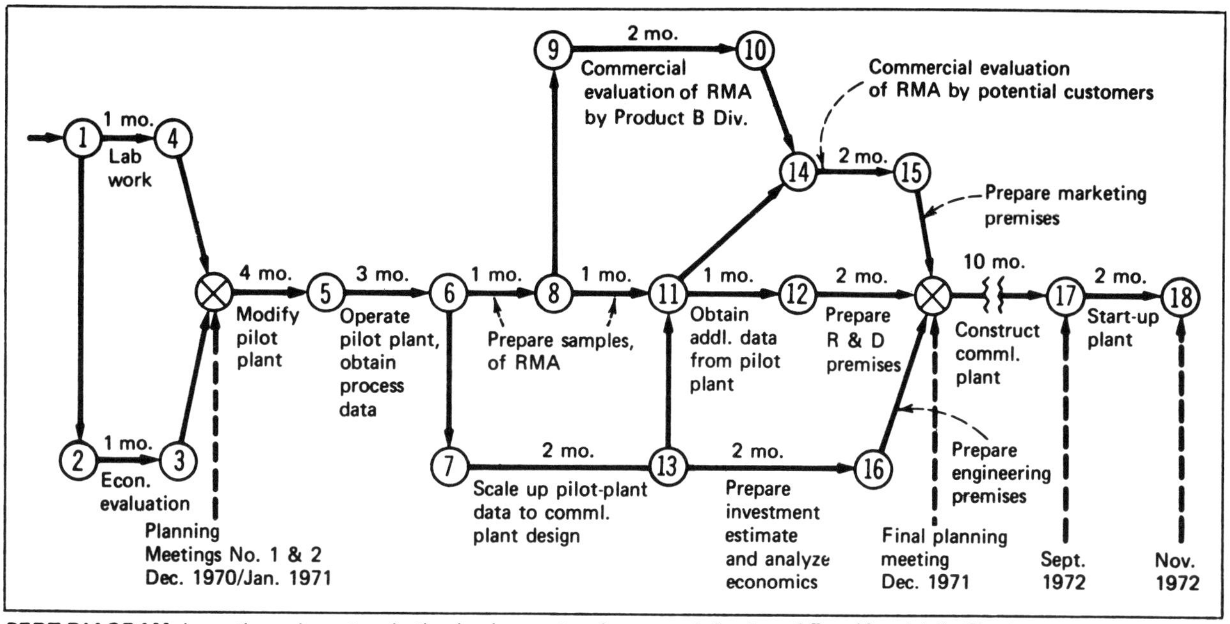

PERT DIAGRAM shows the various steps in the development and commercialization of Raw Material A—Fig. 1

The current RMA purchase commitment is due to expire in Jan. 1973. This contract calls for exercising an option to ensure continuity of supply one year prior to contract expiration. Thus, negotiations for renewal are scheduled to begin in July 1971.

As the engineer responsible for the economic evaluation of current R&D programs, you have been asked to determine the economic viability of the RMA project. Two cases will have to be evaluated:

- Case I assumes construction of a commercial facility to supply just enough RMA to satisfy the manufacturing capacity for Product B at the XYZ Co.'s Midwest and Gulf Coast sites.
- Case II assumes enough capacity to supply not only the requirements of Case I, but also merchant sales of RMA to other manufacturers of Product B.

A decision tree giving the alternatives involved in this situation is shown in Fig. 2. Decision No. 1 has the alternative of either modifying the existing pilot plant and continuing the R&D work, or dropping the work and deciding to continue to buy RMA indefinitely. Decision No. 2 has the alternatives of (1) building the commercial facilities for manufacturing just enough RMA for the company's internal requirements, or (2) building the plant large enough to supply external sales as well, or (3) dropping the entire program and continuing to purchase

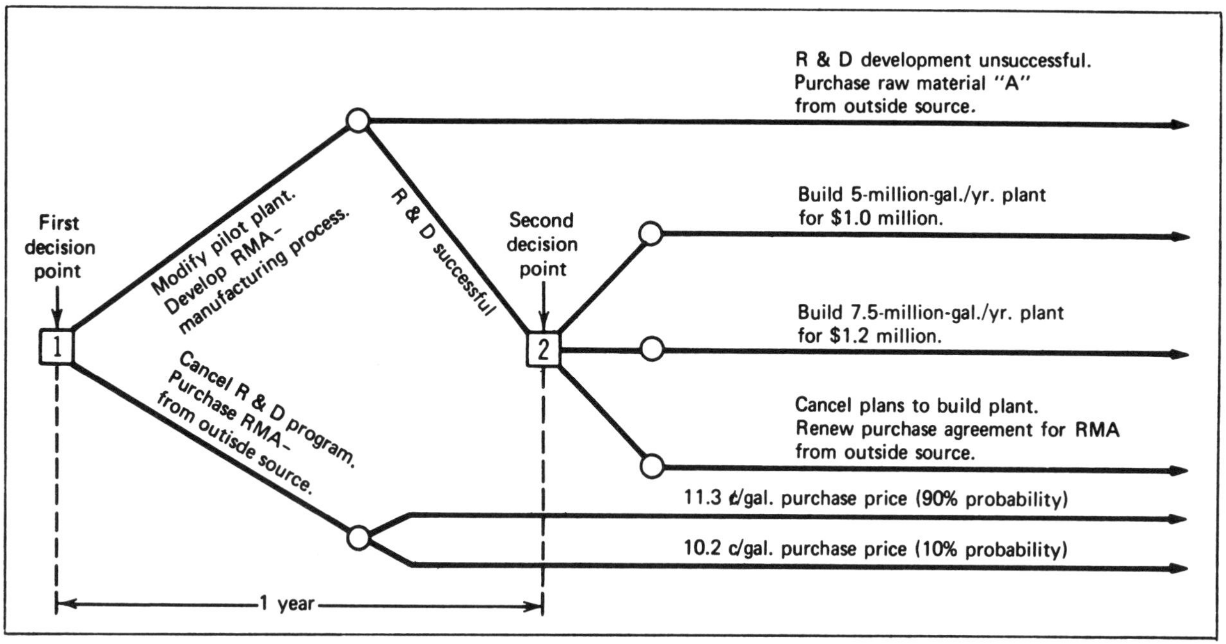

SIMPLIFIED DECISION TREE provides means of analyzing the various "make or buy" alternatives for RMA—Fig. 2

Forecast of Total Industry Demand for Raw Material A (RMA)—Table I

Year	Sales Volume, Million Gal.
1971	24.6
1972	26.7
1973	29.0
1974	31.6
1975	34.5
1976	37.6
1977	41.1
1978	44.8
1979	47.0

XYZ Co.'s Internal Usage of RMA—Table II

Year	Forecast of Internal Consumption, Million Gal.
1973	2.9
1974	3.1
1975	3.7
1976	3.9
1977	4.3
1978	4.4
1979	4.0

RMA. It is your job to develop the decision point value of Decision No. 2, and project this back to Decision No. 1, using the techniques described in the previous article of this series.

With the XYZ Co.'s captive market for RMA, the commercialization investment involves minimal business risk if R&D develops the process successfully. You have been told that the process technology being developed is unique and completely competitive with processes known to be in use by the two firms that are currently supplying the XYZ Co. with RMA.

The feedstock for manufacturing RMA via the XYZ process is relatively inexpensive, because it is a chemical-plant byproduct. The process involves chemical treatment of this feedstock, physical separation from the treatment agent, regeneration and recycle of this agent for reuse. Laboratory data suggest that the process may improve the color of RMA compared to material currently being supplied, because it will use a unique chemical treatment agent. Patent coverage is in progress.

Developing the Basic Information

Your supervisor arranges a meeting with the general manager of the Product B division to discuss the RMA project, and to arrange for development of the key marketing and technical information that you will need in order to determine the plant size for the two cases you are planning to evaluate. Subsequently, you are presented with the data shown in the following tables:

Table I is a forecast of total industry demand for RMA; it was developed from the total market forecast for manufacture of Product B, taking into account the relative quantities of RMA required to make each of the various grades of Product B.

Table II provides a forecast of XYZ Co.'s internal RMA consumption. This table (which is based on the Product B mix shown in Table II on p. 158) covers the base case of 3 million gal./yr., plus 2 million gal./yr. of added capacity in 1973, and an additional 2 million gal./yr. of added capacity in 1976. This indicates that XYZ Co.'s internal consumption of RMA is expected to reach a maximum of 4.4 million gal./yr. in 1978. Product-mix changes for Product B favor Grades 201 and 301 at the expense of 101, and these changes are expected to decrease the consumption of RMA from 4.4 to 4.0 million gal./yr. in 1979. Reason: Grade 101 consumes approximately 1.05 gal. of RMA per gal. of Product B, whereas Grade 201 consumes only 0.52 gal., and Grade 301 consumes 0.62 gal. (This illustrates the importance of reliable product-mix forecasts in assessing potential internal requirements.)

Table III is a forecast of industry manufacturing capacity for RMA. Note that Companies A and B—currently the only suppliers of RMA to all of the Product B manufacturers—are also manufacturers of Product B themselves. Furthermore, current utilization of RMA capacity is believed to be 82%. The forecast shows utilization as varying between a low of 75% (anticipated in

Forecast of Industry Capacity for RMA Manufacture—Table III

Year	Company A*	Company B*	XYZ Co.* Case I	XYZ Co.* Case II	Total Capacity* Case I	Total Capacity* Case II	Capacity Used, % Case I	Capacity Used, % Case II
1971	20	10	–	–	30		82	
1972	20	10	–	–	30		89	
1973	20	10	5†	7.5†	35	37.5	83	78
1974	20	10	5	7.5	35	37.5	90	84
1975	20	15**	5	7.5	40	42.5	86	81
1976	30**	15	5	7.5	50	52.5	75	72
1977	30	15	5	7.5	50	52.5	82	78
1978	30	15	5	7.5	50	52.5	90	86
1979	30	15	5	7.5	50	52.5	94	90

*In million gal. †Startup desired in third quarter of 1972. **Predicted capacity increase.

Forecast of XYZ Co.'s External RMA Sales—Table IV

	1973	1974	1975	1976	1977	1978	1979
Company C: Total consumption, million gal.	3.2	4.7	4.7	6.6	6.8	7.0	7.0
Sales to Co. C, million gal.	0.7	1.0	1.5	1.5	1.5	1.5	1.5
Penetration, %	22	21	32	23	22	21	21
Company D: Total consumption, million gal.	5.1	5.3	5.3	5.3	6.3	6.4	6.4
Sales to Co. D, million gal.	0.8	1.5	1.5	1.5	1.5	1.5	1.5
Penetration, %	16	28	28	28	24	23	23
Total external sales, million gal.	1.5	2.5	3.0	3.0	3.0	3.0	3.0
Internal sales, million gal.	2.9	3.1	3.7	3.9	4.3	4.4	4.0
Total sales, million gal.	4.4	5.6	6.7	6.9	7.3	7.4	7.0
Share of total market based on external sales, %	5.2	7.9	8.7	8.0	7.3	6.7	6.4
Share of available market based on external sales, %	18	25	30	25	23	22	22

1976 just after Company A brings its major new capacity onstream) to a high of 94% (resulting from continuing growth in demand for Product B) anticipated by 1979.

The above range assumes that XYZ Co. will build a 5-million-gal./yr. RMA plant. Your management suggests that a 5-million-gal./yr. RMA facility should be adequate to supply XYZ Co.'s internal requirements; startup is desired in the third quarter, 1972. (You note that this startup date is slightly inconsistent with the PERT diagram schedule prepared by R&D and Engineering shown in Fig. 1; the diagram indicates that startup cannot be achieved sooner than the fourth quarter of 1972. When you point out the discrepancy, it is agreed that this will have to be resolved in the forthcoming planning meeting.)

Table IV is a forecast of potential external sales of RMA by the XYZ Co. to two other firms that make Product B. Your management believes that if the color advantage shown by the new RMA process proves out in the pilot plant and commercial plant scaleup, then the substantial sales commitments shown to these two firms (Companies C and D) are achievable. However, management does not believe that either firm would be willing to pay a premium for the improved quality—rather, this will constitute an important "plus" in penetrating the RMA market with these two companies.

In Table IV, the 5.2 to 8.7% share-of-total-market range represents the penetration level planned for Companies C and D. However, during the ensuing discussion, it becomes apparent to all concerned that this is a grossly misleading figure. Since Companies A and B both manufacture RMA as well as Product B, their corresponding share of the RMA market is unavailable to the XYZ Co., barring any very unusual unforeseen circumstances. Thus, approximately 50% of the total market for RMA is unavailable to a newcomer. Companies C, D, E, and F constitute the only real marketing opportunity available to the XYZ Co. for external sales of RMA. Thus, when share of market is examined on this base, the targeted share of market for external sales varies from 18% to 30%, depending on the year.

Combining the maximum internal requirement for RMA with the potential external sales, a total requirement of 7.4 million gal./yr. is shown for the year 1978. Anticipating typical operation in most years at 90% of capacity (to ensure a contingency supply of RMA), a plant capacity of 7.5 million gal./yr. is set for the case involving both internal and external sales of RMA. Similarly, a plant capacity of 5 million gal./yr. is set for supply of only internal requirements.

Estimating the Capital Investment

Now that the plant capacities for the two cases have been established, you meet with the appropriate R&D personnel to draw a process flowsheet based on the laboratory data, so that you can size the major equipment items and estimate the investment needed for each of the two plant sizes.

Preliminary material and heat balances are available, based on the laboratory process data. You develop the flowsheet shown in Fig. 3, and prepare a factor-type investment estimate for each of the two commercial plant sizes, based on equipment-cost/capacity data, and on telephone quotations from equipment vendors, for all of

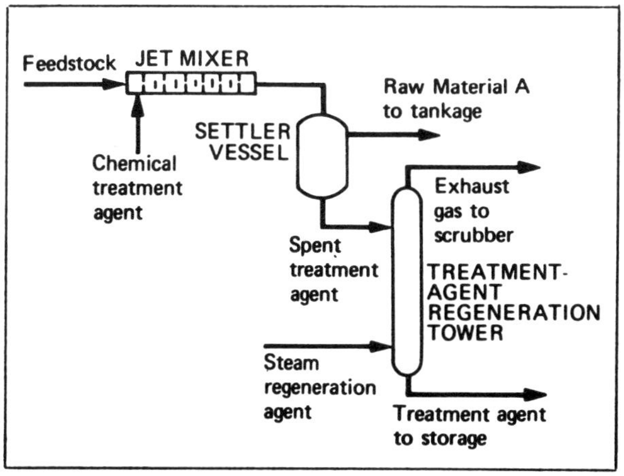

SIMPLE PROCESS FLOWSHEET for RMA is starting point for preliminary cost estimates—Fig. 3

RMA Plant-Investment Estimate—Table V

	Cost Estimate, $1,000	
	Case I*	Case II†
Major process equipment	230	275
Total plant cost (based on factor of 3.5 for similar jobs)	800	960
Contingency (for process unknowns, estimating errors, escalation)	200	240
Plant investment	1,000	1,200

*5-million-gal./yr. plant †7.5-million-gal./yr. plant

Basis: Same U.S. Gulf Coast site as new Product B facilities.

the major process equipment items. These estimates are shown in Table V. A 3.5 factor is selected based on jobs of this type executed in the past. A 25% contingency is provided to cover process unknowns, estimating errors, and escalation of equipment and construction costs.

Estimating Operating Costs and Earnings

You now prepare a mill cost estimate for the manufacture of RMA via the new process, as shown in Table VI. The cost estimated for the chemical treatment agent is based on laboratory ratios and "guestimates" of utilities requirements and losses. One operator per shift is believed to be adequate staffing for the process being developed, on the premise that the plant is to be part of a major facility on an existing site.

Based on the foregoing information, you are now ready to generate the earnings statements for a 5-million-

Forecast of RMA's Mill Cost—Table VI

	Case I*		Case II†	
	¢/Gal.	$M/Yr.	¢/Gal.	$M/Yr.
Feedstock	0.35		0.35	
Chemical-treatment agent	0.20		0.20	
Total raw materials	0.55		0.55	
Steam	0.10		0.10	
Cooling water (recirc.)	0.02		0.02	
Electricity	0.03		0.03	
Fuel gas	–		–	
Total utilities	0.15		0.15	
Operating labor		42		42
Operating supplies (@10% of operating labor)		4		4
Maintenance labor and materials (@3% of investment)		30		36
Technical and admin. services		20		20
Employee benefits		15		15
Controllable fixed costs		111		117
Depreciation (@ 6.67% invest.)		67		80
Taxes, insurance (@1.2%)		12		14
Noncontrollable fixed costs		79		94
Total variable cost	0.70	35	0.70	53
Total plant fixed cost	3.80	190	2.80	211
Total mill cost	4.50	225	3.50	264

*5-million gal./yr. plant †7.5-million-gal./yr. plant
M = 1,000

Case I: Earnings Statement for 5-Million-Gal./Yr. RMA Plant—Table VII

	Thousands of Dollars						
	1973	1974	1975	1976	1977	1978	1979
(Gross sales volume, million gal./yr.)	(2.9)	(3.1)	(3.7)	(3.9)	(4.3)	(4.4)	(4.0)
Gross sales income*	328	350	418	440	486	497	452
Returns & allowances	5	5	5	5	5	5	5
Net sales income	323	345	413	435	481	492	447
Freight†	33	33	33	33	33	33	33
Net sales realization	290	312	380	402	448	459	414
Variable cost of manufacture	20	22	26	27	30	31	28
Fixed cost of manufacture	190	190	190	190	190	190	190
Total plant manufacturing cost	210	212	216	217	220	221	218
Other operating expense	45	45	45	45	45	25	25
Total cost of sales	255	257	261	262	265	246	243
Profit, BFIT (pre-tax)	35	55	119	140	183	213	171
Profit, AFIT (after-tax)	18	28	60	70	92	106	86
Depreciation	67	67	67	67	67	67	67
Cash flow	85	95	127	137	159	173	153
Cumulative cash flow	85	180	307	444	603	776	929

Payout time: approx. 7.5 years

*At transfer price of 11.3 ¢/gal. (current contract price).
†At 2 ¢/gal. on material shipped to XYZ Co.'s Midwest Product B plant.

Case II: Earnings Statement for 7.5-Million-Gal./Yr. RMA Plant—Table VIII

	Thousands of Dollars						
	1973	1974	1975	1976	1977	1978	1979
(Gross internal sales, million gal./yr.)	(2.9)	(3.1)	(3.7)	(3.9)	(4.3)	(4.4)	(4.0)
(Gross external sales, million gal./yr.)	(1.5)	(2.5)	(3.0)	(3.0)	(3.0)	(3.0)	(3.0)
(Total volume, million gal./yr.)	(4.4)	(5.6)	(6.7)	(6.9)	(7.3)	(7.4)	(7.0)
Gross sales income	498	633	757	780	825	835	790
Returns & allowances	7	10	11	12	12	12	12
Net sales income	491	623	746	768	813	823	778
Freight	63	83	93	93	93	93	93
Net sales realization	428	540	653	675	720	730	685
Variable cost of manufacture	31	39	47	49	51	52	49
Fixed cost of manufacture	211	211	211	211	211	211	211
Total plant manufacturing cost	242	250	258	260	262	263	260
Other operating expense	65	65	65	65	65	45	45
Total cost of sales	307	315	323	325	327	308	305
Profit, BFIT	121	225	330	350	393	422	380
Profit, AFIT	60	112	165	175	196	211	190
Depreciation	80	80	80	80	80	80	80
Cash flow	140	192	245	255	276	291	270
Cumulative cash flow	140	332	577	832	1,108	1,399	1,669

Payout time: approx. 5.3 years

gal./yr. (Case I) and a 7.5-million-gal./yr. (Case II) RMA plant.

Current price of RMA sold to the XYZ Co. is 11.3¢/gal. Companies A and B are the current suppliers.

It is planned to locate the RMA manufacturing facility as part of the new Gulf Coast manufacturing facility for Product B.* In order to meet the needs of your company's Midwest manufacturing facility, about 1.6 million gal./yr. of RMA will have to be shipped via barge up the Mississippi River. Barges of 500,000 gal. capacity are available, and a total freight cost of 2¢/gal. is anticipated. Companies C and D (potential customers for Product B) are located on the Mississippi River not far from the XYZ Co.'s Product B Midwest facility, and existing barge-terminal facilities can be used.

The earning statements for Case I and II are shown in Tables VII and VIII, respectively.

The Financial Div. has specified that the R&D ex-

*Readers will recall, from earlier articles, that the XYZ Co. planned to produce Product B at a new Gulf Coast manufacturing facility, as well as continuing to produce it at the existing Midwest facility.

Case I: Discounted Cash Flow for 5-Million-Gal./Yr. RMA Plant—Table IX

Date	Year	Item	Cash Flow, $1,000	Table Ref.	Try DCF = 10%: Discount Factor	Present Value*
–	0	Plant invest. + wkg. cap.	–1,060	I	1.0000	-1,060
1973	0-1	Earnings	85	II	0.9516	81
1974	1-2	Earnings	95	II	0.8611	82
1975	2-3	Earnings	127	II	0.7791	99
1976	3-4	Earnings	137	II	0.7050	97
1977	4-5	Earnings	159	II	0.6379	101
1978	5-6	Earnings	173	II	0.5722	100
1979	6-7	Earnings	153	II	0.5223	80
1980-87	7-15	Earnings	1,200	I, III	0.342†	410
1987	15	Wkg. cap.	60	I	0.2231	13
1987	15	Net salvage**	0			0
						+3

*As of Jan. 1973; in $1,000.
†0.6883 x 0.4966 = 0.342.
**Salvage value is estimated as equal to the dismantling cost.

Case II: Discounted Cash Flow for 7.5-Million-Gal./Yr. RMA Plant—Table X

Date	Year	Item	Cash Flow, $1,000	Table Ref.	Try DCF = 16% Discount Factor	Try DCF = 16% Present Value*	Try DCF = 18% Discount Factor	Try DCF = 18% Present Value*
	0	Plant invest. + wkg. cap.	−1,320	I	1.0000	−1,320	1.0000	−1,320
1973	0-1	Earnings	140	II	0.9241	130	0.9152	128
1974	1-2	Earnings	192	II	0.7875	151	0.7644	147
1975	2-3	Earnings	245	II	0.6710	165	0.6385	156
1976	3-4	Earnings	255	II	0.5718	146	0.5333	138
1977	4-5	Earnings	276	II	0.4873	134	0.4455	123
1978	5-6	Earnings	291	II	0.4152	121	0.3721	108
1979	6-7	Earnings	270	II	0.3538	95	0.3108	84
1980-87	7-15	Earnings	2,160	I, III	0.184†	397	0.128**	276
1987	15	Working capital	120	I	0.0907	11	0.0672	8
1987	15	Net salvage value	0			0		0
								−152

*As of Jan. 1973; in $1,000.
†0.5641 × 0.3263 = 0.184.
**0.5299 × 0.2369 = 0.128.

$$\text{By interpolation, DCF} = \frac{(18)(30) - (16)(-152)}{30 - (-152)} = 16.3\%$$

penses incurred in developing the new RMA are to be amortized on a five-year base, as part of other operating expense. In Table VII (for the 5-million gal./yr. plant that would supply RMA for only internal consumption), the after-tax profit varies from $18,000 in 1973 to a high of $106,000 in 1978. In Table VIII (for the 7.5-million gal./yr. plant that would supply RMA for external as well as internal consumption), the after-tax profit varies from a low of $60,000 in 1973 to a high of $211,000 in 1978. This is based on the premise that Companies C and D will be interested in executing purchase agreements for RMA at a sales price of 11.3¢/gal. (which is the current price being paid by XYZ Co.).

Discounted Cash Flow and Present Worth

Tables IX and X show the discounted cash flow (DCF) calculations for the 5- and 7.5-million-gal./yr. RMA facilities (i.e., for Case I and II, respectively). Case I shows a DCF of 10.0%, whereas the DCF for Case II is 16.3%.

Table XI shows the present-worth calculation (based on a 10% DCF opportunity value of money). Note that the 1973 present worth is $3,000 and $597,000 for Case I and II, respectively. The corresponding 1971 present worth for Case I and II is $2,000 and $488,000, respectively. This indicates a considerable incentive for developing the external sales of RMA projected in Table IV.

Cases I and II: Present Worth Calculations (Opportunity Value of Money = 10%)—Table XI

Date	Year	Item	Discount Factor	Case I Cash Flow, $1,000	Case I Present Value*	Case II Cash Flow, $1,000	Case II Present Value*
	0	Investment + wkg. cap.	1.0000	−1,060	−1,060	−1,320	−1,320
1973	0-1	Earnings	0.9516	85	81	140	133
1974	1-2	Earnings	0.8611	95	82	192	165
1975	2-3	Earnings	0.7791	127	99	245	191
1976	3-4	Earnings	0.7050	137	97	255	180
1977	4-5	Earnings	0.6379	159	101	276	176
1978	5-6	Earnings	0.5772	173	100	291	168
1979	6-7	Earnings	0.5223	153	80	270	141
1980-87	7-15	Earnings	0.342	1,200	410	2,160	737
1987	15	Working capital	0.2231	60	13	120	26
1987	15	Net salvage value		0	0	0	0
1973 Present worth:					+3		+597
1971 Present worth:				0.8187	2		488

*As of Jan. 1973; in $1,000.
Case I = plant for 5 million gal./yr.; Case II = plant for 7.5 million gal./yr.

The First Planning Meeting

Planning Meeting No. 1 takes place on Dec. 8, 1970; your co-attendees are representatives from the marketing, manufacturing, engineering, R&D, and financial groups, as well as from Product B's general management.

When you present the information contained in Tables I through XI for discussion, the following points emerge:

• The R&D representative indicates that the probability of successfully completing the R&D pilot-plant program is very high, based on the excellent results that have been achieved in the laboratory work.

• The manufacturing representative has had periodic discussions with both of the RMA suppliers (Companies A and B), and he is convinced that at present it would be extremely difficult to negotiate a lower purchase price than the 11.3¢/gal. currently being paid.

• The marketing representative confirms the estimates of potential sales of RMA to Companies C and D shown in Table IV. However, he believes that some probability exists that either Company A or Company B, the sole current suppliers of RMA, may reduce the price by as much as 10% (from 11.3¢/gal. to 10.2¢/gal.) once these firms learn that the XYZ Co. is seriously considering a venture to manufacture RMA. The price cut's purpose would be to discourage this venture by creating a situation in which the XYZ Co.'s economics for RMA manufacture would be marginal for the scale of production being contemplated. The marketing representative believes that the probability of an RMA price reduction is substantially less if the XYZ Co. builds a facility designed to meet only internal requirements.

• The general manager of the Product B group indicates that the chances for negotiating a purchase agreement for RMA at a lower price will be substantially greater if XYZ Co.'s R&D program successfully develops the new RMA process, and Companies A and B are actually threatened with potential loss of business.

You have already recognized these various situations in the decision-tree analysis for RMA commercialization shown in Fig. 2. However, you are now requested to go a step further, as part of the action plan adopted during the meeting. Here is a summary of this action plan.

1. To complete the decision-tree analysis for commercialization of RMA, you are to develop—i.e., quantify—the probabilities to be used for each of the various alternatives in the decision tree, by applying the Delphi method.

2. The Engineering Dept. is authorized to proceed with the pilot-plant modification, subject to cancellation charges with a total allowable cost (in case of cancellation) not to exceed $10,000 by the time of Planning Meeting No. 2, which is to take place in four weeks.

3. It is agreed that Marketing should contact Companies C and D on a confidential basis to determine their interest in evaluating samples of RMA to be produced by the pilot plant, and to establish their situation with regard to potential purchase of RMA from the XYZ Co.

4. It is decided that preliminary contact should be made with the current RMA suppliers (Companies A and B) to determine the probability of obtaining a price reduction from them. (As indicated earlier, the current supply contract does not come up for renegotiation for quite some time. But XYZ Co. wants to feel out the potential for an eventual price reduction at least in a preliminary way, prior to committing the R&D funds for developing its own RMA process.)

The results of these activities are to be presented at Planning Meeting No. 2. At this meeting, the decision on whether or not to pursue the pilot-plant development of the RMA process will be finalized.

In the next article, we will examine this decision, review the activities that led up to it (with emphasis on the Delphi method), and then supply a final perspective on "make or buy" decisions in general. ■

ECONOMIC ANALYSIS Part 11

Zeroing In On "Make or Buy" Decisions

The Delphi method and other analytical techniques can help finalize your thinking on whether to purchase a major raw material or produce it internally.

I. LEIBSON and C.A. TRISCHMAN, Jr., Dart Industries, Inc.

In the previous article, we developed the XYZ Co.'s "make or buy" decision to the halfway point; we will now bring it to a conclusion, and put the procedures and strategies into perspective.

Let us again say that you are the engineer who must evaluate the feasibility of producing Raw Material A (RMA, for short), and recommend in what quantity to produce it. You will recall that XYZ Co. was considering the production of RMA not only for captive use in the manufacture of Product B, but also for external sales.

At the first planning meeting, you were given four weeks to apply the Delphi method and thus quantify the probabilities to be used for each of the alternatives in the RMA-commercialization decision-tree. Let us see what this involves.

The whole future course of the RMA-manufacturing project depends on a high degree of probability that the R&D Dept. will be able to develop a viable process, and that the RMA produced under this process will be of high enough quality to meet the needs of XYZ Co. and other manufacturers of Product B. Your task is to assign a quantitative value to R&D's probability of success of

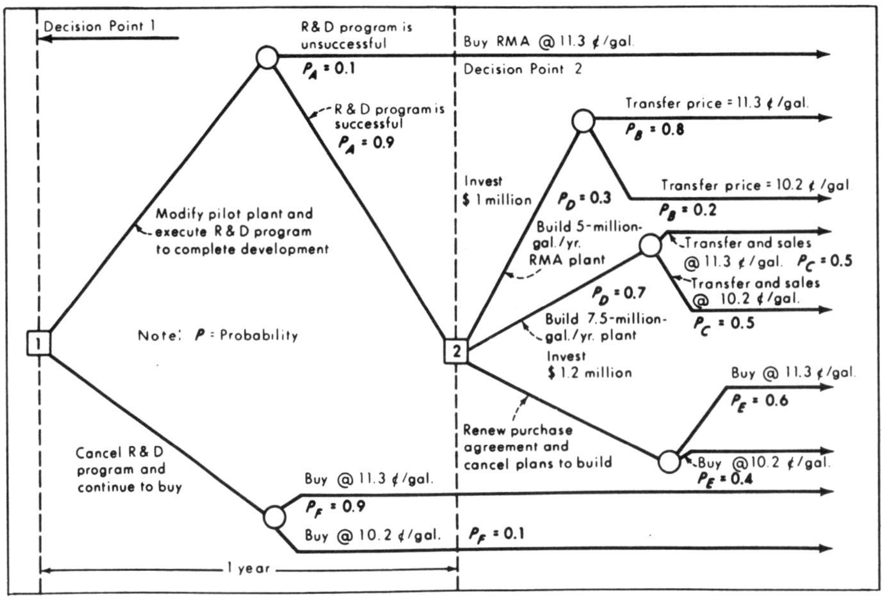

DECISION-TREE ANALYSIS for commercialization of Raw Material A shows probability (P) of various events—Fig. 1

Originally published March 1972.

Delphi Method: Scope and Results of Questionnaire for Raw Material A (RMA)—Table I

	First Round, %						Second Round, %					
	Aver-	Respondent					Aver-	Respondent				
Question	**age**	**1**	**2**	**3**	**4**	**5**	**age**	**1**	**2**	**3**	**4**	**5**
A. What probability of success do you estimate for developing the RMA process and product technology?	88	90	95	95	85	75	90	90	95	95	85	85
B. If XYZ Co. were to build just enough RMA capacity to supply internal requirements, what is your estimate of probability that the market price will hold steady at the current level (11.3¢/gal.), or decline to a 10% lower level (10.2¢/gal)?												
Hold steady @ 11.3¢/gal.	86	90	100	95	70	75	80	85	85	90	70	75
Decline to 10.2¢/gal.	14	10	0	5	30	25	20	15	15	10	30	25
C. If XYZ Co. were to build enough RMA capacity to supply internal requirements plus 2.5 MM gal./yr. for external sales, what is your estimate of probability that the market price will hold steady at current level (11.3¢/gal.) or decline to a 10% lower level, (10.2¢/gal.)?												
Hold steady @ 11.3¢/gal.	58	50	50	70	80	40	50	50	50	60	60	40
Decline to 10.2¢/gal.	42	50	50	30	20	60	50	50	50	40	40	60
D. What is your estimate of probability of success in achieving 2.5 MM gal./yr. external sales of RMA?	64	75	80	80	50	35	70	75	80	80	60	55
E. What is your estimate of probability of success in negotiating a 10% lower purchase-price agreement for RMA in December 1971, at Decision Point to construct commercial plant?												
Purchase at current level (11.3¢/gal.)	63	60	65	65	75	50	60	60	65	65	65	50
Purchase at 10% lower price (10.2¢/gal.)	37	40	35	35	25	50	40	40	35	35	35	50
F. What is your estimate of probability of success in negotiating a 10% lower purchase-price agreement for RMA in December 1970, at Decision Point to construct pilot plant?												
Purchase at current level (11.3¢/gal.)	85	65	90	95	80	95	90	85	90	95	80	95
Purchase at 10% lower price (10.2¢/gal.)	15	35	10	5	20	5	10	15	10	5	20	5

five other alternatives that are pertinent to the decision-tree—i.e., to arrive at the probability factors that are part of Fig. 1. The Delphi method will help you obtain and collate the input you need from appropriate management personnel. In order to get the best possible predictions from Delphi, you must use a segment of XYZ Co.'s managers that is broad enough to include not only those who have had previous experience in making forecasts for their own function, but also those who have had to live with the results of forecasts made by others.

Applying the Delphi Method

The first step involves preparing a questionnaire listing the six alternatives whose probabilities need to be quantified to complete the decision-tree analysis. This questionnaire, shown in the first column of Table I, is distributed to five key people whom you have selected to be your "panel of experts": (1) the national sales manager, (2) the manager of manufacturing, (3) the director of purchasing, (4) the R&D director, and (5) the manager of engineering (for whom you work).

By including representatives from the key functions within the company, the results of your questionnaire are expected to reflect a balanced judgment of the entire business activity, rather than being limited to a single department's viewpoint. Obviously, the R&D director may never predict a probability of success lower than 90% on any project his department undertakes. The sales manager may see his price projections as having similar reliability. But other executives who have lived with the sales manager's or R&D director's predictions in the past may not quite see it that way, and they must be asked to contribute judgments based on their own, more-objective (even if less knowledgeable) points of view.

As you visit each of the five panelists individually, and explain the purpose of your questionnaire, you solicit his help in predicting the future (without revealing the answers provided by the other panelists).

In averaging and reviewing the results of the first

Summary of Probabilities for Various Actions and Situations—Table II

Item	Probability, %
A. Develop process and product technology for RMA	Successful: 90 Unsuccessful: 10
B. RMA price reduction if XYZ Co. builds for internal consumption only	10% reduction: 20 No reduction: 80
C. RMA price reduction if XYZ Co. builds for external as well as internal consumption	10% reduction: 50 No reduction: 50
D. Capture of external sales to Companies C and D (requiring a 7.5 vs. a 5.0-MM-gal./yr. plant)	Positive: 70 Negative: 30
E. Renew purchase agreement for RMA at 10% price reduction at time of Decision Point 2 (Dec. 1971)	Buy @ 10.2c: 40 Buy @ 11.3c: 60
F. Renew purchase agreement at 10% price reduction at time of Decision Point 1 (Dec. 1970)	Buy @ 10.2c: 10 Buy @ 11.3c: 90

Case IA: Earnings Statement—Table III

	Thousands of Dollars						
	'73	'74	'75	'76	'77	'78	'79
(Gross sales, MM gal.)	2.9	3.1	3.7	3.9	4.3	4.4	4.0
Gross sales income	295	315	376	396	437	447	407
Returns, allowances	5	5	5	5	5	5	5
Net sales income	290	310	371	391	432	442	402
Freight	33	33	33	33	33	33	33
Net sales realization	257	277	338	358	399	409	369
Cost of sales*	255	257	261	262	265	246	243
Profit, pre-tax	2	20	77	96	134	163	126
Profit, after-tax	1	10	38	48	67	82	63
Depreciation	67	67	67	67	67	67	67
Cash flow	68	77	105	115	134	149	130

*Cost of sales is taken from previous earnings statement (Table VII of article in Feb. 21 issue). Although that earnings statement was based on a sales price of 11.3c/gal., the cost of sales is the same.

Note: Case IA is based on a 5-million-gal./yr. plant, and a sales price of 10.2¢/gal.

round of questionnaire interviews, you note that there is a 40 percentage-point spread between answers to Question C, and a 45 percentage-point spread in answers to Question D. Although not unexpected, this wide divergence of opinion casts some doubt about the validity of the "average value" for your decision-tree analysis. Therefore, you decide to proceed with a second round of questions with your panel.

This time, you reveal the single "average figure" result for each question from the first round to each respondent, and ask each of the panelists to review his prediction—i.e., his answers to the six questions—in light of the results of the first questionnaire.

The second round of answers is then collected. As shown in the last column of Table I, these answers tend to converge, with less difference between the higher and lower values predicted for any question. The percentage-point spread on Question C has been reduced from 40% in Round One to 20% in Round Two, and the percentage-point spread for Question D from 45% in Round One to 25% in Round Two. The percentage spread on the other questions has also been reduced from Round One to Round Two.

At this point, you decide that no further interviews or questioning is required of your panelists; any further converging of their answers might reflect their desire just to get done with you, rather than improved validity.

The average results of Round Two then represent the best-balanced expert opinion, and these are the values you use in the decision-tree analysis, as summarized in Table II.

Planning Meeting No. 2

Four weeks after the planning meeting described in the previous article, Planning Meeting No. 2 is convened. Again, your co-attendees are representatives from the marketing, manufacturing, engineering, R&D and financial groups, and from Product B's general management.

At this meeting, you present the group with the summary of the probability of various actions and situations, as developed by the Delphi method and shown in Table II. In response to questions, you explain that Alternative E (renewal of the purchase agreement for RMA at a 10% price reduction) is believed to have a 40% probability of success if carried out subsequent to successful completion of R&D work on the new process (i.e., in December 1971). However, an attempt to negotiate renewal of the RMA purchase agreement with a 10% price reduction now (i.e., in January 1971) is believed to have only a 10% probability of success.

Case IIA: Earnings Statement—Table IV

	Thousands of Dollars						
	'73	'74	'75	'76	'77	'78	'79
(Gross sales, MM gal.)	4.4	5.6	6.7	6.9	7.3	7.4	7.0
Gross sales income	448	573	681	702	743	751	711
Returns, allowances	7	9	10	11	11	11	11
Net sales income	441	564	671	691	732	740	700
Freight	63	83	93	93	93	93	93
Net sales realization	378	481	578	598	639	647	607
Cost of sales*	307	315	323	325	327	308	305
Profit, pre-tax	71	166	255	273	312	339	302
Profit, after-tax	36	83	128	136	156	170	151
Depreciation	80	80	80	80	80	80	80
Cash flow	116	163	208	216	236	250	231

*Taken from Table VIII of article in Feb. 21 issue.

Note: Case IIA is based on a 7.5-million-gal./yr. plant, and a sales price of 10.2¢/gal.

Cases IA and IIA: Present-Worth Calculations for RMA—Table V

Date 19_	Year	Item	Discount Factor	Case IA Cash Flow, $M	Case IA Present Value, $M	Case IIA Cash Flow, $M	Case IIA Present Value, $M
	0	Investment + wkg. cap.	1.000	–1,060	–1,060	–1,320	–1,320
73	0-1	Earnings	0.9516	68	65	116	110
74	1-2	Earnings	0.8611	77	66	163	140
75	2-3	Earnings	0.7791	105	82	208	162
76	3-4	Earnings	0.7050	115	81	216	152
77	4-5	Earnings	0.6379	134	86	236	150
78	5-6	Earnings	0.5772	149	86	250	144
79	6-7	Earnings	0.5223	130	68	231	120
80-87	7-15	Earnings	0.342	1,040	355	1,840	630
87	15	Working capital	0.2231	60	13	120	26
87	15	Salvage value less dismant. cost		0	0	0	0
1973 Present worth:					–158		+314
1971 Present worth:			0.8187		–129		+257

Basis: Opportunity value of money (discounted cash flow) = 10%. Case IA = 5-million-gal./yr. plant; Case IIA = 7.5-million gal./yr. plant.

Next, you present Tables III and IV, which are the respective earnings statements for a 5- and 7.5-million-gal./yr. plant, based on a 10.2¢/gal. sales price. The annual after-tax profit builds from $1,000 to $82,000 for the 5-million-gal./yr. facility, with corresponding annual cash flows ranging from $68,000 up to $149,000. In the case of the 7.5-million-gal./yr. facility, the annual after-tax profit is $71,000 to $339,000, with corresponding annual cash flows ranging from $116,000 to $250,000. The net effect of the lower sales price is about $20,000 and $40,000 per year, respectively, in terms of reduced after-tax profits for the 5- and 7.5-million-gal./yr. plant.

The corresponding present-worth calculation (based on a 10% opportunity value of money) is shown in Table V; the present worth is +$158,000 on a 1973 basis for the 5-million-gal./yr. facility, and +$314,000 on a similar basis for the 7.5-million-gal./yr. facility. The corresponding 1971 present-worth figures are -$129,000 and +$257,000, respectively.

Case IIIA: Present-Worth Calculations—Table VI

Date 19_	Year	Item	Discount Factor	Cash Flow, $M	Present Value, $M
73	0-1	Earnings	0.9516	33	31
74	1-2	Earnings	0.8611	35	30
75	2-3	Earnings	0.7791	42	33
76	3-4	Earnings	0.7050	44	31
77	4-5	Earnings	0.6379	49	31
78	5-6	Earnings	0.5772	50	29
79	6-7	Earnings	0.5223	45	24
80-87	7-15	Earnings	0.342	360	123
1973 Present worth					+332
1971 Present worth			0.8187		+272

Basis: Opportunity value of money = 10%. Case IIIA involves continued purchase of RMA, but at the lower price of 10.2¢/gal.

Table VI shows the present-worth calculation for continued purchase of RMA at the lower price of 10.2¢/gal., based on a 10% opportunity value of money. The annual cash flow shown in this table represents the savings that XYZ Co. can achieve by purchasing internal RMA requirements at 10.2¢ gal. (instead of the current 11.3¢/gal. price). The present worth of these savings on a 1973 basis is $332,000. The corresponding 1971 present worth of these savings is $272,000.

You then present your analysis of Decisions No. 2, as shown in Table VII. In this table, the various probabilities for each of the alternatives under consideration are combined. For example, let us consider Case II (external sales plus internal consumption from a 7.5-million-gal./yr. plant) with an 11.3¢/gal. sales price. A 50% probability exists for transfer and sales at a price of 11.3¢/gal. Furthermore, a 70% probability exists for achieving the external sales needed to support the 7.5-million-gal./yr. facility. Thus, the overall probability assignable to Case II is 35% when considered in conjunction with Cases I, IA and IIA.

Each of the probability factors arrived at in this way for these respective cases is used in weighting the corresponding present value of earnings in 1973. The results—i.e., the weighted present values in 1973—are shown in the last column of Table VII.

Now the weighted investment is subtracted from the combined, weighted, present value of earnings on a 1973 basis, to calculate the net weighted present worth of the combined alternatives, which have the same investment base as -$8,000 for the 5-million-gal./yr. facility and +$320,000 for the 7.5-million-gal./yr. facility.

Thus, the Decision-Point-No.-2 value for the combination of Cases I, IA, II and IIA is $312,000 on a 1973 basis. Similarly, the Decision-Point-No.-2 value for Cases III and IIIA is $133,000 on a 1973 basis.

Analysis of Decision No. 2—Table VII

Case		Probability	Present Value of Earnings in 1973, $M	Weighted Present Value in 1973, $M	
I	Internal sales for 5-million-gal./yr. plant @ 11.3¢/gal. sales price	0.8 x 0.3 = 0.24	1,063	255	
IA	Internal sales for 5-million-gal./yr. plant @ 10.2¢/gal. sales price	0.2 x 0.3 = 0.06	902	55	
				310	
			0.3 x 1,060 = Investment	−318	−8
II	External sales and internal consumption for 7.5-million-gal./yr. plant @ 11.3¢/gal. sales price	0.5 x 0.7 = 0.35	1,917	672	
IIA	External sales and internal consumption for 7.5-million gal./yr. plant @ 10.2¢/gal. sales price	0.5 x 0.7 = 0.35	1,634	572	
				1,244	
			0.7 x 1,320 = Investment	−924	320
			Decision point 2 value =		312
III	Negotiate renewal of purchase agreement for RMA @ 11.3¢/gal.	0.6	0	0	
IIIA	Negotiate renewal of purchase agreement for RMA @ 10.2¢/gal.	0.4	332	133	
					133

This means that it is worth as much as $312,000 on a 1973 basis for XYZ Co. to continue the pilot-plant program and complete the development of the RMA process if a positive decision is reached on commercialization at that point in time. If a negative decision on commercialization is reached, Decision Point No. 2 still has a potential value of $133,000, based on the probability assigned to successfully negotiating a lower purchase price for RMA at that time.

What would it take for the Decision-Point-No.-2 value of Cases III and IIIA (negotiation of renewal of the purchase agreement for RMA in December 1971) to be equivalent to that shown for Cases I, IA, II and IIA? This is an important question. "Equivalency" would require the probability of achieving a 10.2¢/gal. price for RMA to be 94% (i.e., 312/332 = 0.94). Thus, commercialization of the new RMA process is favored in Decision No. 2 at this time, because only a 40% probability has been

Analysis of Decision No. 1—Table VIII

Case		Probability	Present Value in 1972, $1,000		Weighted Present Value in 1973, $1,000
1A	Modify pilot plant and spend R&D funds to complete development, with successful results	0.9	Decision point 2 value + $312,000 in 1973 x 0.8187 discount factor	255	229
IB	Modify pilot plant and spend R&D funds to complete development with unsuccessful results	0.1	Cash flow for R&D program*	−75	−8
			Total		+221
IIA	Cancel program and negotiate purchase agreement for RMA in December 1970 at 10.2¢/gal.	0.1	Decision point 2 value + $133,000 in 1973 x 0.8187 discount factor	109	11
IIB	Cancel program and negotiate purchase agreement for RMA in December 1970 at 11.3¢/gal.	0.9		0	0
			Total		+11

*$50,000 capital expense + 0.5 x $50,000 operating expense = $75,000 total cash flow in 1971.

assigned to renewal of the purchase agreement for RMA at the lower cost of 10.2¢/gal. in December 1971.

The R&D Decision

The above has whetted the appetite of the group in regard to the immediate decision to be made—whether or not to continue the R&D program. You now present Table VIII, which is the analysis of Decision No. 1. Cases IA and IB cover the alternative of modifying the pilot plant and spending the R&D funds to complete the development of the process (with the outcome either being successful or unsuccessful). The probability assigned to successful completion is 90%.

The value for Decision Point No. 2 must now be projected back to the No. 1 situation. The applicable No. 2 value of $312,000 on a 1973 basis is discounted to a 1971 present value of $255,000. When weighted by the 90% probability, the weighted present value for this alternative is $229,000 on a 1971 basis.

The cash flow for the R&D program is $50,000 in capital costs, and $50,000 in operating expense, for a total cash flow in 1971 of 50 .+(0.5 X 50) = -$75,000. When weighted by the 10% probability of an unsuccessful development, the weighted present value on a 1971 basis is -$8,000. Thus, the Decision Point No. 1 value for Cases IA and IB is $221,000. Similarly, a Decision Point No. 1 value of $11,000 is calculated for Cases IIA and IIB (for cancellation of the program, and renegotiaton of the RMA purchase agreement in December 1970).

What would happen if a 100% probability existed for negotiating a 10.2¢/gal. price on an RMA purchase agreement to take effect immediately? Even for this extreme case, the prudent decision would be to continue the program, because $221,000 is the weighted present value on a 1971 basis for IA and IB (modifying the pilot plant and spending the necessary R&D funds to complete development), compared to $109,000 for a 100%-probability basis for Case IIA (cancelling the program and negotiating a new RMA purchase agreement in December 1970 at 10.2¢/gal.).

Thus, Planning Meeting No. 2 ends in a consensus of all present: to continue modification of the pilot plant, and to commit the necessary R&D funds to complete development of the RMA process.

Decision Trees and "Delphi" in Perspective

We hope that this case study has served to illustrate the advantages of the decision-tree approach in clearly delineating the "what if . . ." alternatives for a project. To use the decision-tree most effectively, the key alternatives should be brought to focus so that the dollar value of each (on a proper present-worth basis) can be weighed.

The decision needs to be made based on truly credible alternatives. As an example, in the RMA decision-tree, various capacity levels, sales volumes, and transfer prices at different probabilities could have been considered. However, the business opportunity centered on two key decisions to be made, with two alternatives at Decision Point 1, and three alternatives at Decision Point 2.

A decision-tree analysis helps to clarify the available alternatives by depicting them in a diagrammatic form. It also provides a quantitative basis for evaluation of the potential gains associated with each set of alternatives that constitutes a logical stepwide approach to a goal. Thus, it is useful in clearly establishing goals and objectives for each functional department.

For instance, in our case study, once the decision was made at Planning Meeting No. 2 to proceed with the pilot-plant development, the various functional groups were presented with a clear-cut basis for planning and executing the programs to commercialize RMA. Furthermore, if changes in basic premises were to occur subsequently—e.g., change in RMA demand, surprise announcements of new RMA capacity, or new technical processes for RMA manufacture—XYZ Co. will be able to analyze the changes to determine the effect on Decision Point No. 2 value.

In the case study, one of the key inputs to the decision-tree analysis was the assessment of probability for each alternative via the Delphi technique. XYZ Co.'s use of this technique might appear to be a way of delegating the one-man, hard-nosed business decision to a committee. Yet, the use of a formal consensus technique to focus the judgment of a group of knowledgeable people in the evaluation of credible alternatives may yield a more-balanced judgment than can be achieved by a single individual except under very unusual circumstances. Certainly, the Delphi method can prevent a particularly vocal and forceful individual opinion from overriding the less forceful but possibly more knowledgeable judgment of others. Furthermore, the method can provide probability-of-success judgments from any level of the company—e.g., field salesmen or pilot-plant engineers—and bring these judgments to bear on the managerial decision-making process.

Overall "Make or Buy" Perspective

In our case study, the XYZ Co.'s final decision was to continue developing the process for RMA, with the subsequent option of making rather than buying this material. But, in many situations, the scale of manufacture required to achieve satisfactory profitability is such that the make-or-buy decision comes out in favor of buying rather than making. Often, in this situation, the user is a large enough consumer of the raw material to command a preferred position on purchase price, but is not large enough to achieve the scale of manufacture required for satisfactory profits on proprietary facilities.

Another factor that often influences a corporate make-or-buy decision is the limited availability of R&D and capital funds for pursuit of new opportunities. In this situation, the management may decide that it is more profitable to pursue new-product developments in the company's current field of activity or in complementary fields, rather than integrating backwards.

Each individual situation requires careful evaluation; we hope that this article has shown how such an evaluation can be carried out effectively.

The next article—the final one in this economic-analysis series—will examine the marketing/R&D interface in regard to the development of new products. ■

ECONOMIC ANALYSIS Part 12

How to Profit From Product Improvement And Development

In improving an existing grade of product, as well as in developing a new grade and introducing it to the market, economic-analysis techniques help you coordinate R&D with marketing activities, and choose the most-profitable strategies.

I. LEIBSON and C. A. TRISCHMAN, Jr., Dart Industries, Inc.

In many segments of today's chemicals and plastics business, the key to profits is aggressive product-improvement activity-e.g., the development of new grades that afford important marketing pluses by improvements in end-use applications.

The life of individual product grades may be as short as two years, or even less. Keeping the product mix pruned to the key big sellers and eliminating the marginal money-makers is very important in maintaining profitable operation. Spending R&D dollars selectively on the development of tomorrow's big sellers is critical in protecting future profits.

It is well documented that the life of a typical product grade takes the shape of the curve shown in Fig. 1. Strong competitive pressures make themselves felt in providing customers with better products that do the job cheaper and more effectively. Thus, R&D must support going product lines to ensure a continuing profitable competitive position.

Individuals at the marketing/R&D interface have the challenge of properly forecasting product trends in regard to customer needs, and making prudent economic judgments on the relative merits of competing product-development programs. Close scrutiny of each profit opportunity, and tough-minded decisions on where to put the available R&D funds to protect and expand overall profits, are the keys to success here.

How many times have you heard a salesman say something along the lines of: "If I only had a product that could compete with Superchem Co.'s Z grade, I could sell 3 million lb./yr. of that material alone!" How do we decide whether it is really economically justified to spend the R&D dollars required to develop that material?

In this final article in our economic-analysis series, we shall provide some examples of the various profit opportunities that can result from healthy turbulence at the marketing/R&D interface, and examine the decision-making process in regard to (1) the improvement of an existing grade of product, and (2) development of a new grade for a new market.

Improving an Existing Product Grade

Let us once more consider a case study involving the XYZ Co. and its expansion plans for Product B. You will recall that in a previous article, XYZ Co.'s marketing group indicated that product-improvement activity would be necessary on the adhesive grade of Product B if share of market were to be maintained, and if growth projections in sales of this grade were to be realized. The adhesive grade—i.e., Grade 301—has a product profitability of 32.8¢/gal., versus a corresponding profitability for the paint-additive grade of only 22.7¢/gal. Because

Originally published April 1972.

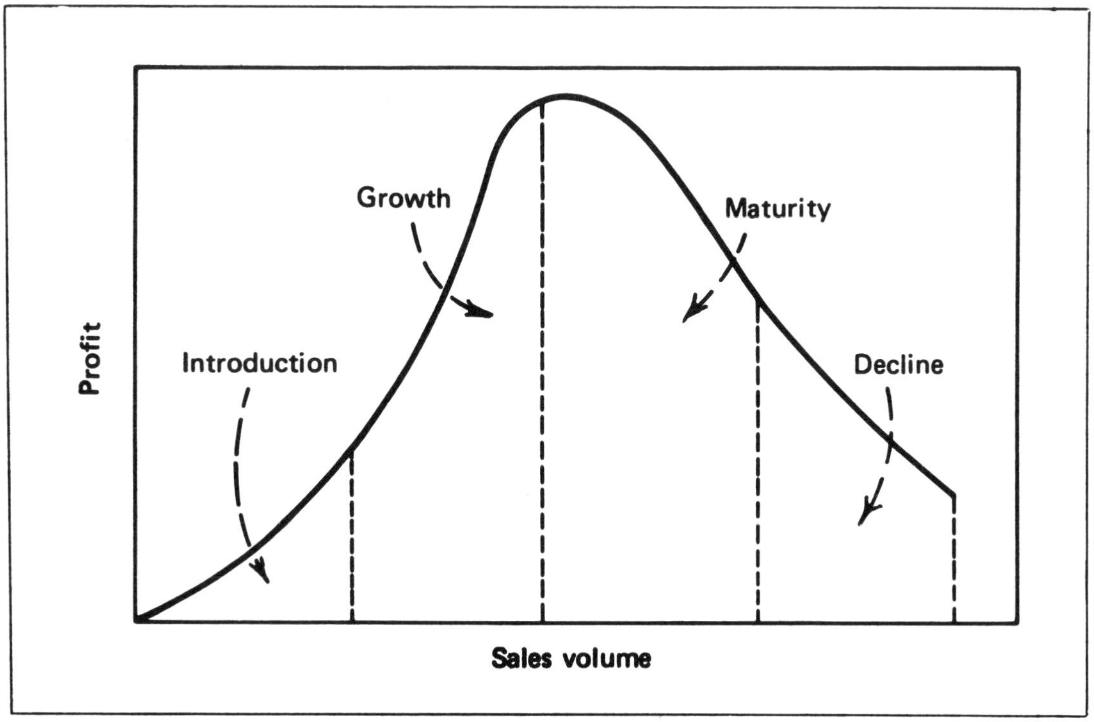

LIFE CYCLE for a non-commodity chemical shows why a company cannot just stand pat with its product line — Fig. 1

Justification of Quality Improvement for Grade 301 (an Existing Grade of Product B)—Table I

	Grade No. 101 Paint Additive	Grade No. 201 Ink Drying	Grade No. 301 Adhesive
Sales forecast for 1972 with improved adhesive grade (excl. resale program), million gal.[1] (Total = 2.90 million gal.)	0.50	1.70	0.70
Effect on sales forecast of total loss of adhesive-grade sales volume, million gal. (Total = 2.93 million gal.)	1.23	1.70	0.00
Effect on profit of the loss of adhesive-grade sales:			$1,000
Contribution to fixed costs and profit from additional 0.73 million gal. of paint-additive grade sales @ contribution of 43.9¢/gal.[2]			322
Contribution to fixed costs and profit from loss of 0.70 million gal. of adhesive-grade sales @ contribution of 57.5¢/gal.[3]			–403
Net difference in contribution			–81
Loss of potential profit from resale program[4]			–37
Effect on profit when paint-additive grade sales are substituted for adhesive-grade sales			–118

Payout time for recovery or R&D expenditures (vs. total loss of adhesive-grade business and replacement by incremental paint-additive sales, to sell out plant capacity) = $35,000/($118,000/yr.) = 0.3 yr.

Notes: [1]See Part 7, Table II (Chem. Eng., Nov. 1, 1971, p.80).

[2]The 43.9¢/gal. is the contribution to fixed costs and profit from Grade 101 sales:

	¢/Gal.
Selling price[5]	66.0
Less: Freight[6]	7.8
Variable cost[6]	14.3
Contribution	43.9

The contribution to fixed cost and profit = (0.439 $/gal.)(1.23 – 0.5 million gal.) = $322,000

[3]The 57.5¢/gal. is the contribution to fixed costs and profit from Grade 301 sales:

	¢/Gal.
Selling price[5]	94.2
Less: Freight[6]	6.5
Variable cost[6]	30.2
Contribution	57.5

The contribution to fixed costs and profit (0.575 $/gal.)(0.7 million gal.) = $403,000

[4]See Part 7, Table V

[5]See Part 5, Table III

[6]See Part 5, Table IX

the plant is expected to be essentially in a sold-out position, the penalty for not improving adhesive-grade quality is potential replacement of adhesive-grade sales by less-attractive paint-additive sales. Conversely, overall profitability can be enhanced via replacement of paint-additive sales by increased adhesive-grade sales, and this is particularly important in the later years of the project, as the total market for Product B rises in volume.

According to the R&D group, a $35,000 product-improvement expenditure will be required to develop a superior adhesive grade. On a purely defensive basis (against potential loss of profit from replacement of adhesive-grade business by less-profitable paint-grade business), the justification for this R&D expenditure is developed in Table I.

Sales of all Product B grades by the XYZ Co. in 1972 were projected to be 2.9 million gal./yr. (including 0.7 million gal./yr. of the adhesive grade). Without the R&D expenditure to improve the quality of the adhesive grade, the Marketing Dept. fears that sales of this grade will diminish rapidly.

To evaluate the dollar justification for the R&D expenditure, let us calculate fixed costs and profit, both with and without adhesive-grade sales. Sales volume of the ink-drying grade is expected to be 1.7 million gal./yr., as limited by the market, whether or not a competitive adhesive grade is available. If an improved adhesive grade is unavailable (and that segment of the sales volume is totally lost), it is anticipated that paint-grade sales will replace adhesive-grade business to achieve a sold-out plant position.

Note that the substitution of paint-grade for adhesive-grade production adds 0.03 million gal./yr. to paint-grade capacity due to the higher plant-production rate for that grade (3.15 vs. 3.01 million gal./yr. for a portion of the year).

The difference in contribution to fixed costs and profit between the two cases (i.e., with and without adhesive-grade sales) is determined by calculating the contribution to fixed costs and profit, and the additional paint-grade sales, and then subtracting the contribution of the lost adhesive-grade sales.

The contribution to fixed costs and profit for the paint-additive grade is 43.9¢/gal., as determined by subtracting the variable cost of manufacture and freight from the selling price. Multiplying this figure by the 0.73 million gal. (i.e., 1.23 − 0.5) of additional paint-grade sales, a $322,000 contribution to fixed costs and profit is obtained. Similarly, the contribution to fixed costs and profit that is lost with the total loss of adhesive-grade sales is $403,000 (corresponding to a 37.5¢/gal. contribution to fixed costs and profit, multiplied by the lost sales volume of 0.7 million gal. of adhesive grade).

If the fixed costs and other operating expenses are the same for either the 0.7 million gal. of adhesive grade or the 0.73 million gal. of paint-additive grade, then the substitution of equivalent paint-grade sales for the adhesive-grade sales results in a loss of $81,000 (i.e., 403,000 − 322,000) for 1972.

Evaluating Every Pertinent Factor

In making this type of analysis, all factors affecting the earnings statement must be considered. For instance, it was explained earlier in this series that a resale program for the paint-additive grade had been arranged to take advantage of the Marketing Dept.'s ability to sell more than 0.5 million gal./yr. of this grade. However, if adhesive-grade sales are lost, then there is enough plant capacity to supply the available paint-grade sales from internal production. Thus, the resale program is negated, and the profit contribution of this program of $37,000/yr. is also lost. Therefore, the total loss of profit if adhesive-grade sales decline to zero is $118,000 (i.e., 81,000 + 37,000) in 1972. A review with representatives from Marketing and Manufacturing indicates that no changes in

Net Substitution Loss for Adhesive-Grade Sales—Table II

	1973	1974	1975	1976	1977	1978	1979
Adhesive Grade:							
Sales price, ¢/gal.	94.0	93.5	93.0	92.5	92.0	91.5	91.0
Variable cost plus freight, ¢/gal.	36.7	36.7	36.7	36.7	36.7	36.7	36.7
Contribution to fixed costs & profit, ¢/gal.	57.3	56.8	56.3	55.8	55.3	54.8	54.3
Sales, million gal.	0.8	0.9	1.0	1.1	1.2	1.3	1.4
Contribution to fixed costs & profit, $1,000	458	512	563	614	663	712	761
Paint-Additive Grade:							
Sales price, ¢/gal.	61.5	65.0	64.0	63.0	62.0	61.0	60.0
Variable cost plus freight, ¢/gal.	22.1	22.1	22.1	22.1	22.1	22.1	22.1
Contribution to fixed costs & profit, ¢/gal.	39.4	42.9	41.9	40.9	39.9	38.9	37.9
Equivalent sales,* million gal.	0.84	0.95	1.05	1.15	1.26	1.36	1.47
Equivalent contribution to fixed costs & profits, $1,000	332	408	452	470	502	530	558
Net-substitution loss of profit, $1,000, pre-tax	126	104	111	144	161	182	203

*Equivalent sales of paint-additive grade = (Sales of adhesive grade) $\frac{3.15}{3.01}$

Present Worth for Quality Improvement of Adhesive Grade vs. Substitution of Paint-Additive Sales—Table III

Year	Time	Item	Cash Flow $1,000	Present Worth @ 10% Interest Disc. Factor	Pres. Value
1971	0-1	R&D, (0.5)($35,000)	−17.5	0.9516	−17
1972	1-2	NSP,* (0.5)(118)	59	0.8611	51
1973	2-3	NSP, (0.5)(126)	63	0.7791	49
1974	3-4	NSP, (0.5)(104)	52	0.7050	37
1975	4-5	NSP, (0.5)(111)	56	0.6379	36
1976	5-6	NSP, (0.5)(144)	72	0.5772	42
1977	6-7	NSP, (0.5)(161)	80	0.5223	42
1978	7-8	NSP, (0.5)(182)	91	0.4726	43
1979	8-9	NSP, (0.5)(203)	102	0.4276	44
80-87	9-17	NSP, (8)(203)(0.5)	812	0.278*	226
			Present worth		+553

*NSP = Net substitution profit

any other costs or programs are involved in substituting paint-additive grade for adhesive grade.

In many situations, product-mix changes can seriously affect costs for manufacturing labor or maintenance, distribution costs, marketing expenses, and packaging costs. Each cost factor must be carefully considered to be sure that no potential effect on profits is missed. Furthermore, Marketing must carefully assess whether the next available segments of substitution grade (i.e., paint-additive grade) business are truly available without serious price deterioration.

Payout Time Is Low

It is apparent from Table I that the payout time is only 0.3 yr. for the execution of this product-improvement program, if there is a 100% probability that all the adhesive-grade business will be lost without the development program.

Of course, one can say that this is a spurious payout because, at this point, no adhesive-grade business has actually been lost. But if one insists on waiting for proof of loss, the lag time in executing the product-improvement activity can make it extremely difficult to regain market position in a competitive situation.

Specifically, if XYZ Co. were to wait until complete proof were obtained, the loss in profits sustained while building replacement paint-additive business would further augment the cost of finally tackling the product-improvement job. Customers cannot be turned on and off lke a water spigot. In addition, the lost business may be extremely difficult to recover once a competitor becomes firmly entrenched with a superior product.

Effect on Profitability

Note that although the loss of adhesive-grade business (with substitution of equivalent paint-additive sales) would harm project profitability, justification for the capacity expansion of Product B cannot be undermined by this development, because the base case and the expansion cases contain essentially the same levels of adhesive-grade sales-volume in each year.

Table II shows how 1973-1979 profits would be affected by the substitution of paint-grade business for lost adhesive-grade business. The equivalency figures are based on producing paint-additive grade during the operating hours that were assigned to adhesive grade (on a sold-out-plant basis). To obtain these figures, the sales of adhesive grade are multiplied by the ratio of the paint-grade production-rate to the adhesive-grade production-rate.

Table III shows the calculation of the present worth for product improvement vs. total loss of adhesive-grade sales. The net present worth for this situation is $553,000 on a 1971 basis.

Thus the expenditure of $35,000 to prevent the potential loss of adhesive-grade sales appears to be well justified. Even if there is only a 50% probability of total loss of adhesive-grade business, the corresponding payout time of 0.3(1/0.5), or 0.6 yr., is swift enough—and the equivalent present worth of $276,000 over the life of the project is large enough—to place a high priority on this defensive product-improvement program.

Normally, most managements expect at least a two-year payout in approving defensive R&D programs of this type. In the final analysis, the potential rate of product-grade obsolescence has to be the determining factor in setting the marginal payout time required for such programs. Much shorter payout times may be required where product-grade lives of two years or less are common.

Frequent performance reviews of competitive product grades vs. your own materials are extremely important in maintaining a continuing competitive product position in the marketplace.

Developing and Marketing a New Grade

Let us now turn to the development of a new grade of chemical product. Many situations arise in which an incentive exists for marketing a new grade, to be manufactured in an existing facility; such situations often require expenditures not only for developmental activities but also to install or modify commercial process equipment to permit manufacture of the new grade.

Furthermore, if the market area is one in which the company has no current activities, additional expenditures may have to be made for facilities, and for specially qualified personnel, to provide proper customer sales and technical services.

Let us consider an example of how economic analysis can be applied to this sort of situation.

Earlier in this series, we indicated that the XYZ Co.'s R&D Div. proposed the development of a new grade of Product B—a lubricating-oil-additive grade that we shall call LOA. It was estimated that a two-year R&D effort would be required, costing $50,000 during the first year and $100,000 during the second. Furthermore, an additional investment of approximately $250,000 was believed necessary to permit commercial-scale manufacture of LOA in the XYZ Co.'s Midwest plant.

You are the individual responsible for evaluating whether or not these expenditures should be made.

A meeting has been called to review the marketing information available concerning the lube-oil-additives business, and to develop an action plan for reaching a prudent decision in regard to commercialization of LOA.

Entering the Lube-Oil-Additive Market

Representatives of XYZ Co.'s marketing, R&D and engineering groups have met to establish the guideposts for an economic evaluation. Another division which is currently active in supplying chemicals to the motor-fuels market, has also sent a representative to the meeting, to provide a high-spot picture of the requirements for successfully participating in the lubricant-additives market.

This market is characterized by strong competition, with typically two or three suppliers for every chemical additive being sold. Many of the major oil companies supply their own requirements from captive manufacturing facilities, and some of them also market the same chemical additives to competing oil companies. Suppliers market their products all over the U.S. regardless of plant location. Most materials are sold in bulk, but are also available in drums. Payment terms are typically 30 days net. Deals are often made on a competitive-quotation basis.

As is the case with many specialty materials, end-use selling is critical in getting the additive approved by the customer's evaluation group. The purchasing agent negotiates the price but can only deal with suppliers who have received approval from their firms' evaluation group. Furthermore, specifically in the case of lubricant additives, the large automobile companies establish the performance requirements necessary for their engines. This gives them a strong voice in establishing what additives get used. In addition, military specifications exist for nearly all lubricants.

Gaining product acceptance for a new lube-oil additive thus requires approval from a diverse group of entities.

The foregoing means that product-introduction costs over and above the anticipated R&D costs will be needed to gain customer acceptance for LOA. The experience of the XYZ Co. in introducing additives for motor fuel suggests that product introduction may require as long as one year after completion of the basic R&D work. Furthermore, since it is customary for the new supplier to pay the cost of getting the new material accepted, additional costs of from $100,000 to as much as $1 million may be involved.

For instance, it is quite common for the customers to insist that engine tests be run. Furthermore, in the case of an additive of new composition (such as that under consideration by the XYZ Co.), it is usual for customers to request road evaluations. Outside laboratories are available with engine-laboratory facilities for performing such tests. Testing programs can easily cost from $50,000 to $200,000 to supply the data required.

It is also necessary to conduct toxicological tests regarding any potential physiological hazards to the end-user of LOA. Again, outside laboratories are available for this purpose.

Finally, in entering this business area, it is important for marketing and technical-service personnel to maintain continuing contact with organizations active in the automotive field—in the area of setting standards and in developing testing techniques, as well as the overall development of automotive technology. Consequently, the people who are going to become involved in this project (if it proceeds) will need to become active in the American Soc. for Testing and Materials, the American Petroleum Institute, the Soc. of Automotive Engineers, and similar groups. Other employees of the XYZ Co. who are active in the fuel-additive business can serve as a point of reference and entry for this.

Getting Information From Field Contacts

With this general background of the market, it is agreed that a network of field contacts should be made by market research personnel with:

- Responsible individuals in the major automobile-manufacturing firms.
- Manufacturers of lubricant additives.
- Lubricant manufacturing companies that do their formulation with purchased additive.
- Pertinent trade-association representatives.

The objective of this network of field calls is to define the types and quantities of materials currently being used for viscosity-control in lubricants, the price of these materials, and the size and rate of growth of the market.

The plan is to identify the current producers, to seek out information concerning their relative strengths and weaknesses, and to define the specialty applications for their products. Further, it is necessary to determine whether or not a supplier with a single lubricant-additive can compete successfully for business in this market. In any business of this type, a balanced line of products is necessary in order to provide proper customer service. It is anticipated that the field survey will require approximately three months to execute, and will cost $15,000.

It is also decided that the R&D department should restudy its proposal in the light of the marketing information presented at the meeting (i.e., concerning needs for technical services and approval testing).

Engineering is to make a more careful evaluation of investment requirements, based on the process changes visualized for commercial manufacture of LOA in the Midwest plant. Marketing is to execute a field survey and develop a market forecast to be used in the economic evaluation.

Development of LOA Marketing Strategy

As part of your LOA-evaluation assignment, you meet with key personnel in the marketing group in an effort to tie down the anticipated price and sales volume for LOA. The marketing representative says that he can give you only tentative information at this time because the field survey is not yet completed.

LOA is intended as a viscosity-control agent. There are two different types of product currently being marketed by other companies to satisfy the need for viscosity-index improvement in lubricants: Agent No. 1 sells for approximately 23¢/lb., and is marketed by three suppliers. Agent No. 2 sells for 14¢/lb., and is marketed by two suppliers. The relative efficiency of the two agents is approximately the same as the ratio of the two prices. Thus, the cost of viscosity-index improvement to obtain proper viscosity control in lubricants using either of these two agents is approximately the same per unit quantity of finished lubricant.

These agents are both involved in the formulation of two types of lubricants: motor-vehicle multigrade motor oils, and automotive hydraulic-transmission fluids.

The market appears to be growing at an annual rate of about 6%. Total market size for both agents (combined) is believed to be approximately 140 million lb. in 1970.

The R&D Dept.'s laboratory work indicates that LOA appears to be approximately 20% more efficient than Agent No. 1. This means that 0.8 lb. of LOA is needed (vs. 1 lb. of Agent No. 1) to achieve the same level of viscosity-index improvement.

Establishing the Use Value

In determining pricing strategy, it is important to evaluate the *use value* of the product to the customer, and to take into account competitive materials. In this case, the sales price of Agent No. 1 is 23¢/lb. The corresponding use-value of the XYZ Co.'s LOA is: 23 ÷ 0.8 = 29¢/lb.

This means that with a price of 29¢/lb., the cost of viscosity-index improvement per unit cost of finished lubricant would be the same for LOA as for the currently used Agent No. 1. Therefore, this constitutes an upper limit on LOA's introductory price, unless other significant advantages can be shown.

Demand for viscosity-index-improvement agents in the lubricant market appears to be relatively inelastic. Normally in a situation with inelastic demand, serious consideration would be given to taking a "skimming price" approach in which the price is set high, to recover product development costs as quickly as possible before competitors enter the business with similar products.

However in this case, other producers are currently servicing the market with different products. Furthermore, it is expected that product life of LOA will be relatively short (i.e., five years). It is also known that Company A is actively introducing a similar material to the market on a current basis. Thus, XYZ Co. will not be first with the new-composition lube-oil additive. For these reasons, the Marketing Dept. believes that a "penetration price" strategy should be followed for LOA; this should enable XYZ Co. to achieve share-of-market goals more rapidly than with a "skimming price" strategy.

Because demand is inelastic, business taken at a lower-than-going market price comes at the expense of another producer. This may be expected to cause price erosion, with no compensating accelerated growth in demand at the lower price. Consequently, market entry must be very delicately executed so that the total market situation is not undermined. Quality of product and service, plus effective salesmanship, constitute the key tools for gaining market position in constructive fashion where demand is relatively inelastic.

Selling Price and Volume

The Marketing Dept. feels that under the "penetration price" strategy, LOA should be introduced to the marketplace at 23¢/lb. (i.e., $1.84/gal.). The fact that less LOA is needed per unit of viscosity-index improvement than is the case with either Agents No. 1 or No. 2 should act as a driving force to interest customers in the new material. Thus, the economic evaluation is to be conducted using this price as a basis.

Approximately one month later, in a meeting with marketing, you are given the market forecast shown in Table IV.

This forecast is based on LOA achieving a 1% market penetration in 1974 (the first full year of commercial operation for this material), increasing through an S-curve to a market penetration level of 5.5% in the fifth year, 1978. Because price has been relatively stable over a narrow range, marketing is recommending a 23¢/lb. initial price, with erosion to 22¢/lb. occurring in the third year of product life, 1976. Considering the 6% annual market growth that was found in the market research study, it is anticipated that XYZ Co. will sell 12.3 million lb. of LOA in the fifth year of product life, 1978.

Marketing believes that strong political and social pressures for reduction in air-pollution from internal combustion engines will force significant changes in the design of the automobile engine; this in turn may be expected to have drastic effects on requirements for lubricants in the automotive business. Therefore, the economic evaluation for LOA's development is to be based on only a five-year product-life projection.

XYZ Co.'s Sales Forecast for the Lube-Oil-Additive Grade (LOA)—Table IV

Year	Total Market Sales Million Lb.	XYZ Co.'s LOA Sales, Million Lb.	Percent Market Penetration	Sales Price ¢/Lb.
1970	140	...	...	...
1971	148	...	...	...
1972	157	...	...	...
1973	166	...	...	...
1974	176	1.8	1.0	23.0
1975	187	3.7	2.0	23.0
1976	198	7.9	4.0	22.0
1977	210	10.5	5.0	22.0
1978	223	12.3	5.5	22.0

Economic Evaluation

From your review of the process information concerning preparation of LOA on a laboratory scale, you develop the variable cost of manufacture to be 38.5¢/gal.,

Comparative Cost Structure for LOA vs. Paint-Additive Grade—Table V

Item	Paint Additive	Lube-Oil Additive
Variable cost of manufacture, c/gal.	14.3	38.5
Freight cost, ¢/gal.	7.8	11.5
Total, ¢/gal.	22.1	50.0
Production rate, million gal./yr.	3.15	3.00
Increase in maintenance, depreciation, taxes and insurance, $1,000/yr.*		31
Increase in other operating expense:		
Add 2 market specialists, $1,000/yr.		50
Add 1 tech. service man, $1,000/yr.		25
Add continuing R&D support, $1,000/yr.		40
Overall increase, $1,000/yr.		115

*Resulting from $250,000 incremental plant-investment to permit production of lube-oil additive grade (LOA).

as shown in Table V. Marketing has developed a freight cost of 11.5¢/gal., based on primarily bulk-transport distribution of LOA to a specific group of customers located in various sections of the U.S. Thus, the total of these two cost factors is 50¢/gal.

It is anticipated that LOA business will be developed at the expense of the paint-additive grade, using existing plant capacity, thereby substituting a more profitable product for the least profitable one in the product mix. The corresponding costs for the paint-additive grade are also shown in Table V.

From your study of the process information and its potential fit to XYZ Co.'s Midwest plant, you conclude that a production rate of 3 million gal./yr. should be obtainable for LOA once the necessary modifications are made to the Midwest plant to permit LOA manufacture. (This compares with the current production rate of 3.15 million gal./yr. for the paint-additive grade.)

In view of the comments made by representatives from XYZ Co.'s fuels-additive business, it is apparent that participation in the lube-oil-additive market will require additional personnel. Thus, the economic evaluation will assume the addition of two market specialists at a combined cost of $50,000/yr., one customer technical-service professional at a cost of $25,000/yr., and a continuing R&D support program at a cost of $40,000/yr. The total incremental cost is thus $115,000/yr. for the overall increase in "other operating expense" required to properly service the LOA business (as shown in Table V).

The economic evaluation is to be based on the difference in profits anticipated when the LOA grade is sold in place of the paint-additive grade. It is therefore necessary to calculate the annual upgrading in profits after all of the costs connected with LOA sales (on a going-business basis) have been taken into consideration. Table VI shows this calculation for each year of the five-year product-life period.

The density of LOA has been measured as approximately 8.0 lb./gal. Thus, the gal./yr. sales figures shown in Table VI were obtained by dividing each of the corresponding lb./yr. sales figures in Table IV by 8.0.

The equivalent sales of paint-additive grade have been calculated on a sold-out plant basis—i.e., by multiplying the sales forecast for LOA in million gal./yr. by the ratio of the production rate of paint-additive grade to that of the LOA.

The net profit for substituting LOA sales for paint-additive sales is calculated as the difference between the net

Net Substitution Profit for the Lube-Oil-Additive Grade of Product B—Table VI

	1974	1975	1976	1977	1978
Lube-oil additive					
Sales price, ¢/gal.	184.0	184.0	176.0	176.0	176.0
Variable cost plus freight, ¢/gal.	50.0	50.0	50.0	50.0	50.0
Contribution to fixed costs and profit, ¢/gal.	134.0	134.0	126.0	126.0	126.0
Sales, million gal.	0.225	0.462	0.99	1.31	1.54
Contribution to fixed costs and profit, $1,000	301	620	1,250	1,650	1,940
Less increase in maintenance, depreciation, taxes, insurance, and other operating expenses, $1,000	146	146	146	146	146
Net contribution to fixed costs and profit, $1,000	155	474	1,104	1,504	1,794
Paint-Additive Grade:					
Sales price, ¢/gal.	65.0	64.0	63.0	62.0	61.0
Variable costs plus freight, ¢/gal.	22.1	22.1	22.1	22.1	22.1
Contribution to fixed costs and profit, ¢/gal.	42.9	41.9	40.9	39.9	38.9
Equivalent sales,* million gal.	0.236	0.485	1.04	1.41	1.62
Equivalent contribution to fixed costs and profit, $1,000	101	203	425	562	633
Net substitution profit, pre-tax	54	271	679	942	1161

*Equivalent sales of paint-additive grade = (Sales of LOA)(3.15/3.00)

contribution to fixed costs and profit arising from LOA sales, and the equivalent contribution to fixed costs and profit from the equivalent sales of the paint-additive grade. In Table VI, this is shown to vary from $54,000 in 1974 to $1,161,000 in 1978. We need to examine whether this profit justifies the capital expenditure required in the commercial plant, as well as the R&D and product-information expenditures required to enter the lube-oil-additive business.

Since this project involves the expenditure of both capital and expense dollars, the choice of profitability criteria to be applied is an interesting one, particularly in regard to R&D. Normally, R&D costs are expensed in the year in which the work is performed, in order to take advantage of the allowable income-tax reduction for legitimate operating expenses; this maintains maximum cash flow to the company treasury. But under some circumstances, R&D expenditures associated with entering a major new business may be capitalized and then amortized over a set period of time. In our case, we expect to incur the $150,000 in R&D costs to develop LOA, and these costs will be expensed annually as the funds are spent.

This means that a simple return-on-investment calculation cannot give an accurate measure of profitability of the overall project, because the R&D expenditures made prior to the capital investment cannot readily be incorporated in the return calculation.

As for payout time, although this criterion can be employed by properly treating capital and expense, it does not give management a proper perspective as to the value of making the R&D and capital expenditures, because the project is of such limited life (i.e., only five years of active sales).

Under these circumstances, profitability criteria that properly incorporate time-value-of-money concepts must come to the fore. Thus, the present-worth and discounted-cash-flow methods are to be used.

The Cash-Flow Pattern

Marketing and R&D have determined that the economic evaluation should allow a total of $500,000 for:

- Product-testing expenses (in engine- and road-performance tests).
- Advertising and promotion.
- Other costs connected with product introduction during the third year of the development program in 1973 (at the same time that the commercial plant is being modified to permit manufacture of LOA).

Table VII shows the schedule of cash flows anticipated for the LOA project. R&D and product-introduction expenses are shown on an after-tax basis (based on a 50% Federal income tax). The cash flow from earnings gained in the sale of LOA are shown over the period from 1974 through 1978 as the sum of earnings plus depreciation (calculated on a 15-yr. base on the $250,000 investment required for facilities to permit LOA manufacture in the commercial plant). A writeoff of the remaining undepreciated investment is taken at the end of the project in 1978. This shows a positive cash flow of $82,000 (based on a 50% Federal income tax rate). Incidentally, the corresponding payout time (as calculated from the cumulative cash flow column) is 3.0 years, referenced to the start of LOA production in the commercial facility in 1974.

Present worth and DCF are calculated in Table VIII. Based on a 10% opportunity value of money, the 1971 present worth of executing the R&D program, funding

Cash-Flow Projection for Lube-Oil Additive Project—Table VII

Year	Time	Item	Cash Flow, $1,000	Cumulative Cash Flow, $1,000
1971	0-1	Development expense (0.5)(−50)*	− 25	− 25
1972	1-2	Development expense (0.5)(−100)*	− 50	− 75
1973	2-3	Product-introduction expense (0.5)(−500)*	−250	−325
1973	3	Investment in commercial plant	−250	−575
1974	3-4	Cash flow from earnings of $54,000 (0.5) plus $17,000 depreciation	44	−531
1975	4-5	Cash flow from earnings of (0.5) ($271,000) plus $17,000 depreciation	153	−378
1976	5-6	Cash flow from earnings of (0.5) ($679,000) plus $17,000 depreciation	357	− 21
1977	6-7	Cash flow from earnings of (0.5) ($942,000) plus $17,000 depreciation	488	467
1978	7-8	Cash flow from earnings of (0.5) ($1,161,000) plus $17,000 depreciation	597	1,064
1978	8	Undepreciated investment writeoff (0.5) (165.000)†	+82	1,146

Payout time = 3.0 years after initiation of commercial production and sales.

*Based on a 50% Federal income tax.

†Because the useful life of the investment is terminated at the end of the fifth year of LOA sales, the remaining undepreciated investment in commercial plant facilities is written off at this point (based on a 50% Federal income tax).

Present-Worth and Discounted-Cash-Flow (DCF) Calculation for LOA Project—Table VIII

		Cash	Present Worth, @ 10% Interest		DCF = 30%	
		Flow,	Discount	Present	Discount	Present
Year	Time	$1,000	Factor	Value	Factor	Value
1971	0-1	− 25	0.9516	− 24	0.8640	− 22
1972	1-2	− 50	0.8611	− 43	0.6400	− 32
1973	2-3	−250	0.7791	−195	0.4741	−117
1973	3	−250	0.7408	−185	0.4066	−100
1974	3-4	44	0.7050	31	0.3513	15
1975	4-5	153	0.6379	98	0.2602	40
1976	5-6	357	0.5772	206	0.1928	69
1977	6-7	488	0.5223	255	0.1428	70
1978	7-8	597	0.4726	282	0.1058	63
1978	8	82	0.4493	37	0.0907	7
				+452		− 7

the LOA-introduction expenses, and committing the capital for the modification of the commercial plant, is $452,000 for the life of the project. The corresponding DCF for this project is 30%. Thus, the project seems well worth executing.

Sensitivity Analysis

To assess the risks and sensitivities involved, you are asked to evaluate the effect of a one-year delay in realizing the first sales of LOA, with a resultant loss of the last year of earnings. For this case, you are requested to assume that R&D expenditures will continue in 1973 at the 1972 rate, with product-introduction expenditures taking place primarily in 1974.

Present Worth and DCF for One-Year Delay in LOA Project—Table IX

		Cash	Present Worth, @ 10% Interest		DCF = 16%	
		Flow,	Discount	Present	Discount	Present
Year	Time	$1,000	Factor	Value	Factor	Value
1971	0-1	− 25	0.9516	− 24	0.9241	− 23
1972	1-2	− 50	0.8611	− 43	0.7875	− 39
1973	2-3	− 50	0.7791	− 38	0.6710	− 34
1974	3-4	−250	0.7050	−178	0.5718	−143
1974	4	−250	0.6703	−168	0.5273	−132
1975	4-5	44	0.6379	28	0.4873	21
1976	5-6	153	0.5772	88	0.4152	63
1977	6-7	357	0.5223	186	0.3538	125
1978	7-8	488	0.4726	231	0.3015	145
1978	8	82	0.4493	37	0.2780	23
				+119		+6

Table IX shows that the present worth is reduced from $452,000 to $119,000 (a reduction of more than 75%). Similarly, the DCF is reduced from 30% down to 16%.

Although the project is still somewhat attractive even on this basis, the importance of executing this project on a carefully monitored schedule is apparent.

Thus, our case study—and our series—concludes with a decision by XYZ Co.'s general management to proceed with the development of LOA, and to give it a high priority.

Acknowledgement

The authors wish to express their deep appreciation to A. B. Finestone, R. M. Knight, D. L. Safley, and F. S. Valles of Dart Industries Chemical Group's top management for the opportunity to publish this twelve-part series, and to thank them for their help. The stimulating atmosphere in the Dart Chemical Group proved a valuable asset in developing many of the hypothetical case histories by which we illustrated the application of economic analysis in today's chemical process industries. ■

Meet the Authors

I. Leibson

Charles A. Trischman, Jr.,

I. Leibson is Vice-President, Engineering and Construction Div., of Dart Industries Inc.'s Chemical Group (P.O. Box 37, Paramus, N.J. 07652). His experience includes management positions in marketing, manufacturing, R&D and commercial development both at Dart Industries and Humble Oil and Refining Co. He has served as first chairman of AIChE's Petroleum and Petrochemicals Div., as well as chairman of several national committees. A Registered Professional Engineer, he has a B.S.Ch.E. degree from the University of Florida, and M.S.Ch.E. and D.Sc.Ch.E. degrees from Carnegie Institute of Technology.

Charles A. Trischman, Jr., is Marketing Manager for the ABS product group at Rexene Polymers Co., a division of Dart Industries, Inc. (formerly Rexall Drug & Chemical Co.). He is located in Paramus, N.J. Since joining Rexene in 1963, he has also served as polypropylene product-development manager for R&D. Previously, he worked for American Oil and AMOCO Chemicals in process-engineering, economic and coordinating functions. Mr. Trischman is a chemical engineering graduate of the Virginia Polytechnic Institute, a member of AIChE, and a Registered Professional Engineer.

Finance for Engineers

Your success is measured by how well you meet goals of technical performance, scheduling, and costs and profits. You can better attain cost and profit objectives if you clearly understand the underlying financial factors.

J. T. ABBOTT, R. R. JANSSEN and C. MICHAEL MERZ, Schrello Associates, Inc.

In the final analysis, an engineer's value to his company—and his own future—depends on his contribution to his company's profit. Unfortunately, most engineers fail to see this connection. So, their success often depends on intuition and chance. The way to assure success is to know what affects profit and to concentrate all your efforts on the areas that really count.

It All Starts With Sales

Whether a firm sells chemicals or coffee cups, it receives revenues from the sale of its products. But there are costs associated with producing and selling the products. Costs associated directly with producing and selling products are *operating costs*, and the resulting profit is *operating profit*. Of course, businesses have elements other than operating costs (such as nonoperating income, interest and taxes), which must be subtracted from, or added to, sales revenues to get the bottom line, or *net income*.

You get this sort of information from a company's *Income Statement*.

Measures of Profitability

Profit is, of course, a good thing. But the *degree* of goodness is worth looking at, too. One good way might be to relate the profit to what it took to generate it; that is, the *assets*. Assets, of course, are what a company owns. Obligations, or debts, are *liabilities*. The difference is what the company is worth, or *stockholders' equity*.

You get this kind of information from the *Balance Sheet*.

A measure of profitability that relates profit to the resources required to make it is often called Return on Investment, or ROI. ROI is rather an imprecise term. Any return, or profit, divided by any investment would qualify for ROI. Or, to put it another way, almost anything in the Income Statement divided by almost anything in the Balance Sheet could be called ROI. We'll discuss a couple of versions of ROI.

Fig. 1 is called the Return on Investment Financial Family Tree. It shows in a simple, streamlined way, the elements that make up the measures of profitability. When you know what these elements are, and which are the key ones, you are well on your way to improving profitability.

The owners, or stockholders, of a company are interested in the profit (net income) the company earned with the money provided by the stockholders (stockholders' equity). Net Income divided by Stockholders' Equity is called Return on Equity, or ROE. ROE is probably the most basic measure of the profitability of a company. In fact, when some 200 chief executives were asked, "What single financial parameter is most symptomatic of the success of your company?" they replied almost to a man: "Return on Equity."

Although Return on Equity is a good measure of overall company performance, it is not a good measure of operating efficiency.

To measure operating efficiency, it would be desirable to measure only those performance characteristics over which the operating manager has control. Return on Assets, or ROA, is a good measure of operating performance. ROA is the operating profit divided by the assets employed in earning that profit. The operating manager has reasonable control over both operating profit and assets. He therefore has reasonable control over Return on Assets.

The Financial Statements

In order to use the Financial Family Tree, you will need information. A company's financial reports are good sources of information for the ROI Financial Fam-

Originally published June 1973.

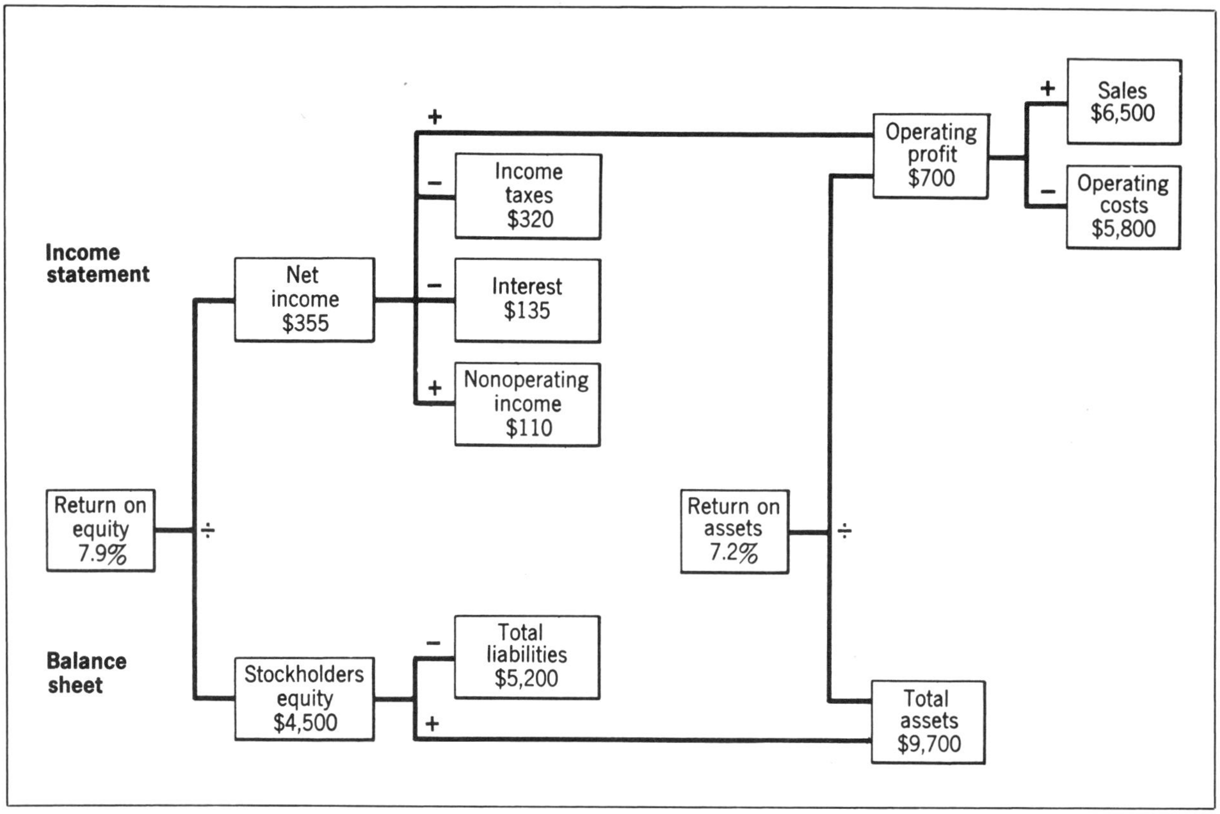

TREE NUMBERS were adapted from Merrill Lynch, Pierce, Fenner & Smith Inc.'s "How to Read a Financial Report"—Fig. 1

ily Tree. In fact, you will note that all the information in the upper portion of the Financial Family Tree comes from the Income Statement. All information on the lower portion of the tree comes from the Balance Sheet (Fig. 1).

The Balance Sheet is the oldest statement. It was originally designed for lenders and generally reflects their needs, even today. The Balance Sheet shows a company's assets and liabilities; and the difference between them, which is stockholders' equity. It is significant that the Balance Sheet is for a point in time. It reflects the net effect of the past, back to the start of the company.

The point-in-time characteristic of the Balance Sheet is contrasted with the period-of-time characteristic of the Income Statement. Remembering these characteristics can do a lot to explain certain financial-statement entries.

The Income Statement is designed for the owners. As you have seen, the Income Statement shows sales, costs, and the resulting profits.

The Retained Earnings Statement, usually issued as part of the Income Statement, shows how much of the earnings (net income) was distributed as dividends to the stockholders and how much was retained for use in the business.

The Source and Application of Funds Statement shows changes over a period of time. Its primary purpose is to show trends, especially as related to working capital. The Source and Application of Funds Statement is also known as simply the "Funds Statement" or "Changes in Financial Position."

One of the most important of the financial statements is the "Notes to the Financial Statements." You can tell they are important because they are so often printed in small type!

Seriously, the notes are important because they may have the effect of changing or modifying the numbers that appear on the statements. Much of the accounting procedure is standardized. When procedures are not standardized, or when standards are deviated from, you'll find out in the Notes. Notes also include added detail that is useful in analyzing a company.

What You Can Do

As a starter, we suggest that you:

• Fill out the Return on Investment Financial Family Tree for your company; use your company's annual report or financial statements as a source of information.

• Do this same thing for one or two of your major competitors. How do you stack up? Where do your competitors look better?

• With the help of your controller or his staff, shred out the operating cost two or three more levels. Do the same with assets.

If you do these things, you will begin to see many ways you can improve profitability—and will be on the way to good communication with your financial department. Equally as important, you'll find that you are beginning to make sense out of financial statements.

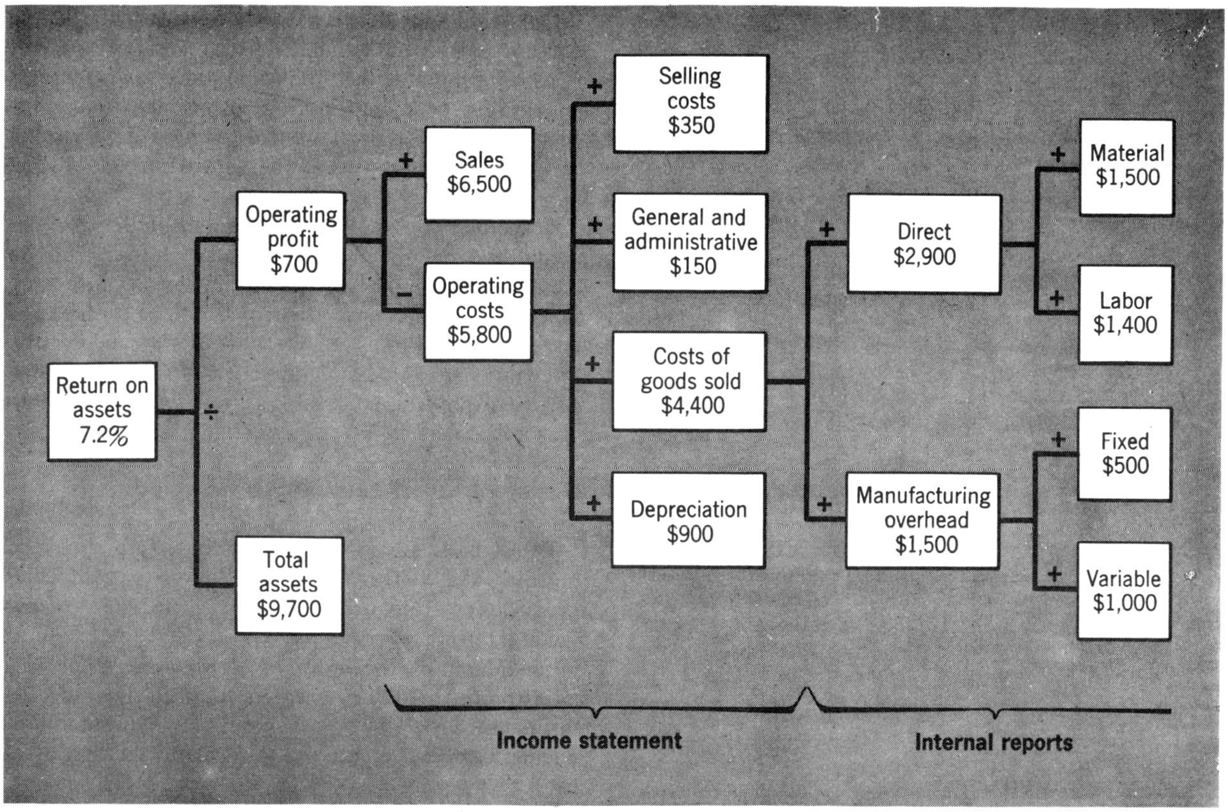

OPERATING PROFITABILITY TREE portrays financial performance of operating manager; profit he made with his assets—Fig. 2

Analyzing the Elements

The Return on Investment Financial Family Tree starts with the measure of profitability of the entire company—ROE. To determine what you can do to improve ROE, you have to know what makes it up, so we shredded-out ROE into its elements: Net Income and Stockholders' Equity. The Financial Family Tree continues the shredding through the elements that make up Return on Assets (ROA).

The scope of this article does not permit further shredding out. But, together with your financial people, you can do this for other elements of the Financial Family Tree. You could, for instance, shred out operating costs to find out what they consist of. You could do the same thing with assets. A knowledge of what the elements are will guide you to areas where you can effectively concentrate your efforts to improve profitability.

Sales Too Low or Costs Too High?

Because we are measuring profitability of operating management with the Return on Investment type criterion (ROA), the creative manager must be aware that he can improve profitability by increasing the numerator (operating profit) or decreasing the denominator (total assets).

Operating profit is the difference between sales and operating costs. If your operating profit is not high enough, either the sales from your plant are too low or your costs are too high. You can get an overall idea of how the sales and costs of your plant or division stack up against your industry by means of two simple ratios: the ratio of sales to the investment in plant and equipment (fixed assets), and the ratio of operating profit to sales. The first, sometimes called Fixed Asset Turnover ratio, gives an idea of whether your plant generates as much business as it should for its size. The second tells whether your operating profit margin is as high as it should be.

Let us see how this approach would work for the Lubrizol Corp., a manufacturer of chemicals for the petroleum industry. Lubrizol's annual report for 1971 reveals that its sales of $198 million were achieved with plant and equipment of $73 million, for a sales-to-fixed-assets ratio of 2.7, considerably lower than the industry average. Operating profit was $39 million, or 19.6% of sales, significantly higher than the industry average.

	Lubrizol	Chemical Industry Average*
Sales/fixed assets	2.7	4.1
Operating profit/sales	19.6%	12.6%

*Robert Morris Assoc., *Annual Statement Studies*

Although Lubrizol's sales per dollar of fixed assets were quite low, its operating profit per dollar of fixed assets was better than that of the rest of the chemical industry. That tells you that the poor sales-to-fixed-assets ratio was more than compensated for by Lubrizol's low operating cost (i.e., by its high operating-profit-to-sales ratio).

Such a situation would most likely result if one or more of the following conditions existed: (1) Lubrizol

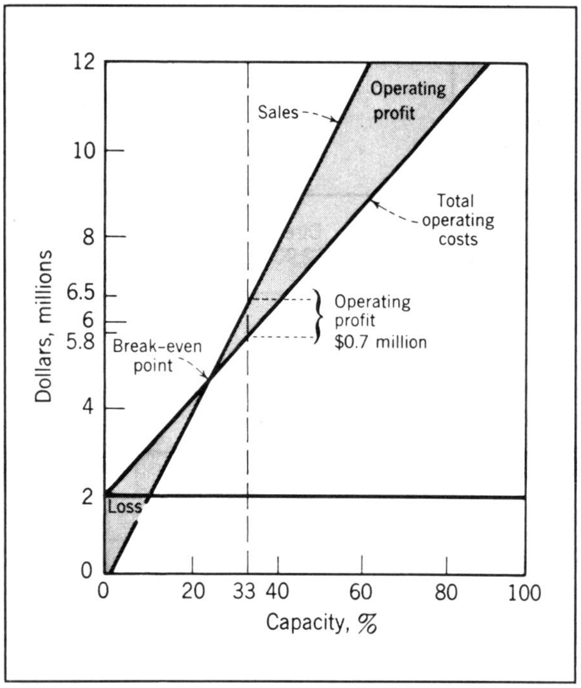

FIXED COSTS, $1.9 MM; variable costs, $2.9 MM—Fig. 3

has a more favorable price situation than the average chemical company, (2) the plant is operating at only partial capacity, and (3) the large amount spent on plant and equipment has resulted in more-efficient operations.

For a company like Lubrizol, which has been growing faster than 10% a year, it would not be surprising to have some excess capacity in anticipation of continued growth. On the other hand, a company could also find itself with excess capacity due to declining sales, perhaps because of a recession or a declining market for some of its products.

As an engineer, you should be concerned with the operating profit at your plant or operating division. Is your biggest problem not enough sales or too high costs? To answer that, compute the sales-to-fixed-assets ratio and the operating-profit-to-sales ratio for your plant or division. Compare them with the industry average tabulated in Annual Statement Studies by Robert Morris Associates, available at banks or business libraries. These two simple ratios will indicate whether you should concentrate more on getting additional business or on reducing your operating costs.

Understanding Operating Costs

As an engineer, your principal influence on the financial performance of your company is probably through your decisions on operating costs. If you completely understand the cost structure of your plant, you should be able to classify all such costs into the following categories:

- Cost-of-goods-sold, selling costs, or general and administrative costs.
- Direct or overhead costs.
- Fixed or variable costs.

Differentiating between cost-of-goods-sold, selling costs, or general and administrative costs is fairly easy. If a cost contributes to the production process, whether it is engineering or manufacturing, it belongs in cost-of-goods-sold. Otherwise, it is either selling costs or general and administrative costs, or perhaps depreciation (Fig. 2).

Cost accounting is the identification of the cost elements making up cost-of-goods-sold. Information on these cost elements is essential to intelligent profit decisions. The cost accounting system should be as simple as possible but still provide adequate visibility and control. These are improved by identifying a cost element with a particular product, i.e., calling that cost element a direct cost rather than an overhead cost.

Direct Vs. Overhead Costs

Direct costs are those that can be traced uniquely, meaningfully and economically back to a particular product. Those that cannot be so traced are called manufacturing overhead. Anything that is not a direct cost is, by definition, overhead. So, of course, general and administrative, and selling costs are forms of overhead. To distinguish between overhead costs, that involved in producing a product is referred to as manufacturing overhead.

The primary reason for breaking costs up into direct and overhead is to allow the total cost of any particular product to be determined. It is usually not too difficult to measure direct costs. In the chemical industry, however, products are produced continuously or in batches, with multiple products or byproducts resulting. Process costs, like overhead costs, must be allocated in some way. Conceptually, all the costs that are not direct costs are collected as overhead costs and then allocated meaningfully to each product. By adding the allocated manufacturing overhead costs to direct costs, you can determine the total cost of the product.

Unfortunately, the distinction between direct and overhead costs is not always clear-cut. Certain direct-material and direct-labor expenses, such as crude oil for a refinery, and operator wages, clearly fit the definition of direct cost. But certain costs may be accounted for either as direct or overhead. Material purchasing and direct-labor fringe benefits are two troublesome cost elements.

Because each product is usually the direct responsibility of a single operating manager, the direct costs of that product are usually closely scrutinized. To avoid that scrutiny, there is a temptation to put borderline costs into overhead rather than to identify the proper direct charges.

What difference does it make whether fringe benefits and purchasing are direct or overhead?

The purchasing, handling and storing of raw materials frequently costs over 20% of the purchase price. But if these costs are buried in overhead, you will not be able to accurately reflect them in a make-or-buy decision.

Should you schedule regular overtime, or hire additional employees? Fringe benefits have become quite large, ranging from 25 to 35% of direct labor costs. Because of this, many companies resort to a large propor-

Find Out Where You Stand

1. As a starting point, construct an Operating Profitability Financial Family Tree for your plant or division. Some of the data will come from your Income Statement. The objective is to gain an overall understanding of where each cost element fits into the operating cost structure. Ask your controller, or someone on his staff, to help you. It will be to his advantage, too, because after you understand the cost structure, you will be able to work with him to reduce costs and increase profits.
2. Conduct a break-even analysis of your operation. If the cost accounting records do not identify costs as fixed or variable, you will have to estimate which are which. A break-even graph will help you estimate what the costs and profits will be for any volume of production.
3. Look at the volume of sales generated at your plant and your operating profit margin. Are your sales too low or your costs too high? By comparing your company to industry averages, you will get a good idea of whether your sales are in line with the size of your plant and whether your operating costs are in the right ballpark.

tion of overtime in preference to hiring. If fringe benefits are buried in overhead, your proper decisions will be less visible.

Visibility-and-control vs. simplicity—that's the issue. Should those troublesome costs be accounted for as direct or overhead? In the long run, you as an engineering manager may want to influence the cost accounting system to get the financial data you need to make correct decisions. You will, of course, want to balance the cost of obtaining the information with the value of a more accurate decision. In the short run, you must know how each cost element is accounted for in your company if you are to adequately perform your job.

Fixed and Variable Costs

A convenient way to identify costs for managerial decisions is by fixed and variable costs. A cost is either fixed or variable, depending on how it behaves as production varies.

Variable costs change directly with small shifts in the level of production. Fixed costs remain the same irrespective of production level or of normal operating decisions made by management.

Because the chemical industry is capital intensive—that is, the cost of plant and equipment is high per dollar of sales compared to other industries—fixed costs, such as depreciation, tend to be an important part of the costs of production. This is why chemical companies often respond to cost reducing pressures by either consolidating plants or by selling or closing inefficient ones. Such action was taken by Celanese Corp. in recent years.

One of the frustrating parts of the chemical business is the wide fluctuation in raw-material costs. The importance of changes in such costs are often amplified because they represent such a large proportion of variable costs. Few industries feel the impact of raw-material prices more than the petroleum industry. One highly successful oil company reports that petroleum crude represents over ¾ of the cost of producing products for sale.

Break-Even Analysis

Having a knowledge of fixed and variable costs, you can predict operating profit as a function of production volume. This relationship between costs, profit and volume is usually called break-even analysis. Such an analysis, as presented in Fig. 3, is based on the assumption that sales dollars and production volume are closely related, which is usually the case.

Two very important aspects of the financial performance of your plant can be gleaned from a graph of cost, profit and volume. First, your break-even sales at which profit equals zero can be determined. Operating people need to know what level of production is required to be profitable. Second, you will notice that operating profit—the difference between the total cost and sales lines—increases as volume increases. Profits in the capital-intensive chemical industry are especially sensitive to volume changes.

Although most progressive companies use some form of break-even analysis, a word of caution is in order. It is difficult to isolate fixed or variable costs. Whether a cost is fixed or variable is defined by the way it varies with volume, not by the desires of the accounting department. Fixed costs will remain truly "fixed" only over small changes in volume, maybe as much as 20% but probably not more. Variable costs do not always vary directly with volume. Such costs are called semivariable. Despite its limitations, break-even analysis is a very useful tool for making decisions involving the interplay of price, volume and cost. ■

Meet the Authors

J. T. Abbott

R. R. Janssen

C. M. Merz

John T. Abbott is a Senior Associate with Schrello Associates, where he teaches "Finance and Accounting Overview for Non-Financial Managers." A Certified Public Accountant with extensive experience in auditing, management consulting, financial administration and new-product development, he holds a degree in marketing and a Master's Degree in business administration, having attended University of Notre Dame, Golden Gate College and Carnegie-Mellon University. He is chief operating officer of a distribution organization and has a CPA firm.

Robert R. Janssen, also a Senior Associate with Schrello Associates (555 E. Ocean Blvd., Long Beach, CA 90802), developed "Finance and Accounting Overview for Non-Financial Managers," which is presented nationally in both seminar and tape-cassette forms. He also assisted in the development of a "Product Evaluation and Planning" seminar. Previously, he was with North American Rockwell, holding positions of Technical Director, R&D Director and Resource Manager. He graduated from the California Institute of Technology and is a registered engineer.

C. Michael Merz, also a Senior Associate, who teaches "Finance and Accounting Overview for Non-Financial Managers," has developed analytical techniques and business models used by corporations to make investment decisions. As a marketing executive for a large company, he forecast sales for top management. He is a graduate of Villanova University, holds a Master's Degree from California State University at Long Beach and is a doctoral candidate at University of Southern California. He has been a faculty member at several universities and has published papers on economics and managerial accounting.

Section II

TECHNIQUES FOR ECONOMIC ANALYSES AND EVALUATIONS

Profitabilty, sensitivity and risk analysis for project economics

J. C. Agarwal and *I. V. Klumpar*, Kennecott Copper Corp.

By using a generalized model, one can estimate profitability, and analyze sensitivity, of processes. And one can also estimate discounted-cash-flow rate of return as a unique function of the sales-to-capital ratio.

☐ Profitability estimation, followed by sensitivity and risk analysis, should accompany experimental work and process design from an early stage of a project [*1*].

Various discounted-cash-flow methods have been widely used to calculate profitability. Sensitivity analysis determines the effect of technical and economic parameters on the profitability of a venture [*2*]. Risk analysis assigns probabilistic distributions to uncertainties in these parameters in such a way that the resulting profitability indicator is also expressed in probabilistic terms [*3*]. These techniques let management concentrate its efforts on those parameters and processes that have the controlling impact on economics and are amenable to technical improvement.

Profitability, sensitivity [*4,5*] and risk-analysis methods are very complex and require the application of computers. In order to become a routine management tool, these techniques should be simplified even if a certain degree of accuracy is sacrificed.

We investigated a wide range of possible process and economic conditions. The relationships between profitability and most of the important technical and economic parameters can be generalized on an approximate basis. Details of the analyses will be discussed later in this article.

Process classification

Profitability is a function of venture life, expenditures and reimbursements. The latter two are summarized in Table I (T/I). The greatest periodic expenditure is usually operating cost, also known as manufacturing cost in some industries. It can be conveniently divided into variable, semivariable and fixed components.

The variable component is proportional to production rate, and includes items such as raw-material cost, selling and delivery expense, as well as, in some processes, utilities, etc.

The semivariable portion varies with production rate only to a certain extent but can be approximately expressed as a quantity proportional to investment. Typically, this component comprises maintenance cost, property taxes, insurance, etc.

Finally, the fixed component of operating cost is independent of production rate within a reasonable range. It may be approximated as proportional to operating labor cost. In addition to wages and benefits of operators, it includes supervision, services and personnel-related overhead.

Originally published September 1975.

Let us now look at the operating cost of producing four typical materials:

- Polymer additives from organic intermediates.
- High-grade ethylene from naphtha.
- Sulfuric acid from sulfur.
- Copper from ore, including mining.

The variable, semivariable and fixed components are approximately equal for a typical polymer-additive process because raw materials, investment and wages of operators have a similar bearing on operating cost. On the other hand, the manufacturing cost of ethylene is controlled by the variable component, i.e., the cost of raw materials and utilities. Equipment is relatively simple and, due to a high degree of automation, there are few operators.

Operating cost of a modern sulfuric acid plant is dominated equally by raw-material cost and investment-related items since there is not much operating labor and overhead. Equipment is expensive; so is sulfur and freight for the liquid product. Finally, the operating cost of a copper mine, concentrator and smelter is dependent equally on investment and labor cost. The variable component has a minor impact because the raw material, copper ore, is free, and the product is a large-volume commodity with little selling expense and relatively low freight cost per unit volume.

This analysis for the components of operating cost provides a convenient way of classifying processes in the chemical, mining, metallurgical, and related industries (T/II). In order to normalize the process categories, 92% of the operating cost was arbitrarily assigned to the controlling component; and 4% to the other two. If two components are dominant, each gets 48%, and the remaining one, 4% again (T/III).

Profitability analysis

To apply this classification for the economic evaluation of processes, a two-step technique will be used:

Step 1—Determine the approximate profitability of the process.

Step 2—Determine how sensitive the profitability is to major technical and economic parameters on an approximate basis.

Profitability is expressed in terms of discounted-cash-flow rate of return (*DCFRR*) in this study. The first step takes advantage of the relationship between sales, capital, and the controlling components of operating cost. For each combination of the last, profitability can be approximated as a unique function of the ratio of sales to capital.

The curves, shown in Fig. 1 (F/1), can be used to directly scan process alternatives without knowing the exact value and detailed breakdown of operating cost. In essence, the chart shows how much capital can be afforded for a particular sales volume and price, or what sales are required to make a specific investment profitable for a process controlled by a particular component or combination of components.

The seven process categories, or controlling-component combinations, fall into two groups: capital-intensive processes, and other processes. While the sales-to-capital ratio varies widely between the two groups, it is almost constant for each component combination.

Classification of expenditures and reimbursements (T/I)

Expenditures

1. Depreciable capital (referred to briefly as "capital" in this article).
2. Working capital.
3. Initial nondepreciable expense (or briefly "expense") such as research and development, or land cost.
4. Periodic nondepreciable expenditures (denoted as operating cost). This is expressed on an annual basis and consists by definition of the following three components: Variable cost (directly proportional to production rate). Semivariable cost (directly proportional to capital). Fixed cost (independent of both production rate and capital cost).
5. Federal, state and local income taxes on an annual basis (or briefly "taxes").

Reimbursements

1. Annual sales, equal to sales volume times product selling price. In a multiple-product case, one product is used as a basis for expressing volume, and its price is adjusted to account for byproducts.
2. Nonperiodic reimbursement such as salvage value or final land recovery (referred to as "salvage").

Process classification by categories (T/II)

Category	Controlling components	Examples
1	Balanced: semivariable, fixed and variable costs	Polymer additives
2	Semivariable cost (investment related)	Strip mining, water and waste treatment
3	Variable cost (raw material, utilities and/or selling and delivery)	Agricultural chemicals, ethylene, smelting copper concentrates
4	Fixed cost (operating-labor related)	Underdeveloped countries
5	Semivariable and fixed costs	Sulfuric acid, oxygen
6	Variable and fixed costs	Paints, pharmaceuticals
7	Semivariable and variable costs	Copper, ammonia

Normalization of process categories (T/III)

		Contribution to operating cost		
Category	Controlling components	Variable %	Semivariable %	Fixed %
1	All three*	33.3	33.3	33.3
2	Semivariable	4	92	4
3	Variable	92	4	4
4	Fixed	4	4	92
5	Semivariable & fixed	4	48	48
6	Variable & fixed	48	4	48
7	Variable & semi-variable	48	48	4

*Operating cost consists of the variable, semivariable and fixed components.

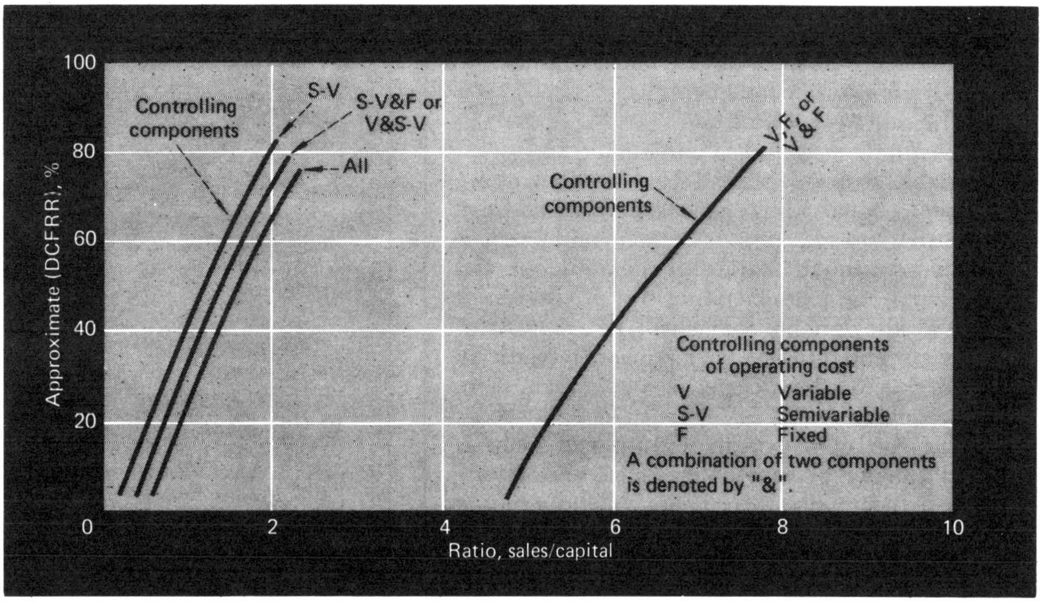

Profitability is a function of sales-to-capital ratio and the controlling components F I

The process category having all three components equally controlling will be used as an example of a capital-intensive process. Within a reasonable range of (*DCFRR*), 10 to 50%, the sales-to-capital ratio varies only in a narrow range of 0.6 to 1.6. Similar observations can be made with the other categories. The significance of this relationship is that for a practical range of profitabilities, the capital-to-sales ratio determines the controlling-component combination, and vice versa.

Let us assume the controlling components of a particular process are established, and the sales are known in terms of price and the fraction of market the company can capture. Let us further assume the company has a target (*DCFRR*) of, say, 20 to 30%. This fixes the required capital within a narrow range. Management may use this estimate for budgeting or to direct research and development.

Sensitivity analysis

Sensitivity analysis, the second step of economic evaluation, implies a set of conditions for which the profitability of a venture is known, e.g., in terms of (*DCFRR*). These conditions, or Base Case, can be expressed as a set of specific values of various technical and economic parameters, such as production rate, venture life, tax rate, or fixed component of operating cost. The Base Case is defined in T/IV.

Sensitivity analysis is conveniently performed by changing the value of one parameter at a time and estimating the corresponding change in profitability. We propose to generalize this method by normalizing the variables. The normalization includes the use of relative rather than absolute values of the sensitivity parameters and profitability, in addition to the assumptions listed in T/IV.

Relative profitability is defined as the ratio of the actual (*DCFRR*) to a reference (*DCFRR*). (We arbitrarily chose a reference value for (*DCFRR*) of 20%.) Similarly, a relative parameter is defined as a ratio of its actual value to its reference value. The set of reference parameter values corresponds to the reference (*DCFRR*).

Once the controlling components of operating cost are identified and the simplifying assumptions of T/IV made, the (*DCFRR*) becomes dependent only on volume, price, and operating cost. The latter two parameters are closely related and can be conveniently expressed as a ratio. The (*DCFRR*) is a unique function of relative volume for any ratio of the price to operating cost ranging from 1.1 to 10. This provides a method for generalizing the results of sensitivity analysis.

Let us assume that the (*DCFRR*) equals a reference value of 20%. Let us select a process equally controlled by all operating-cost components; an arbitrary reference volume, e.g., 100,000 units/yr; and an arbitrary

Assumptions for defining the base case T IV

1. Production rate equals annual sales volume. Let us denote this parameter as "volume." Any product accumulation or depletion of inventories can be expressed as variation in working capital.
2. Working capital is an empirical function of capital, operating cost, and sales.
3. Venture life, i.e., construction period plus production life, is 21 yr (arbitrary).
4. Capital factor is 10% (empirical). This is the percentage of capital that determines semivariable cost excluding depreciation.
5. Straight-line depreciation (arbitrary).
6. Tax-depreciation life of 12 yr (average value for chemical, mining, metallurgical and fuel-gas industries).
7. Tax rate is 50%.
8. No depletion allowance.
9. No expense (arbitrary).
10. No salvage (arbitrary).

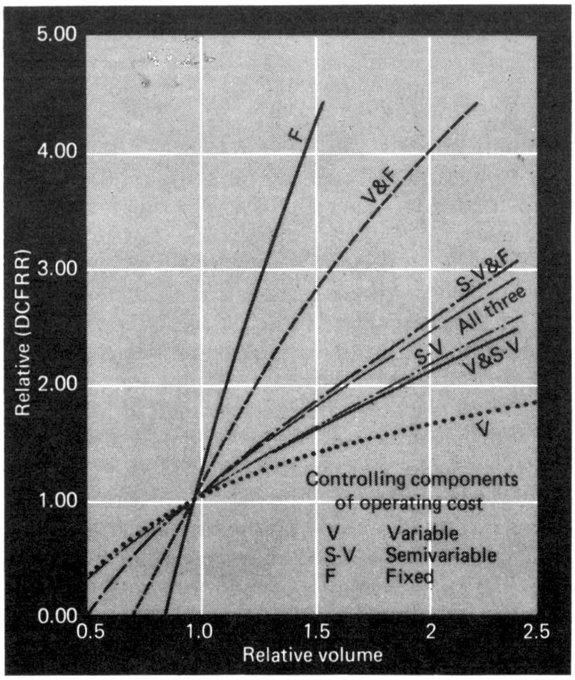

Sensitivity analysis for single parameters F/2

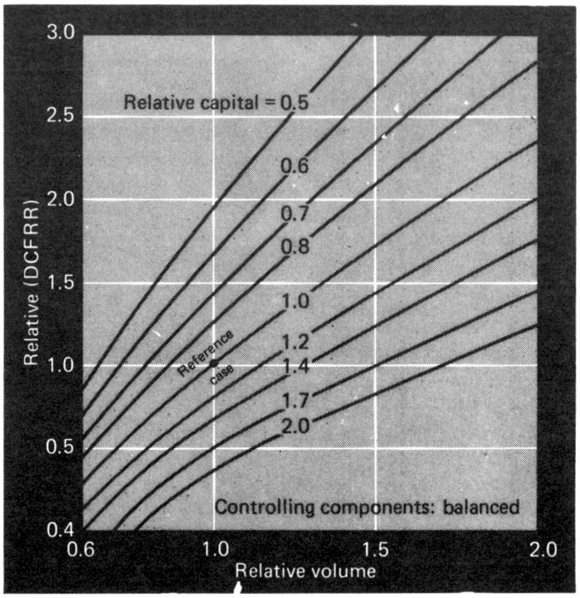

Sensitivity analysis for two parameters F/3

reference operating cost, e.g., \$100/unit for analysis.

By using the assumptions in T/IV, the corresponding price per unit volume can be calculated by trial and error. The combination of the reference (*DCFRR*), volume and operating cost, and the calculated price will define the reference case for the particular process category. Now, any parameter can be varied above and below its reference value, and the effect on (*DCFRR*) calculated, while the other parameters are kept constant. The results are conveniently plotted as relative (*DCFRR*) vs. two relative parameters.

Generalized sensitivity charts

For each process category (combination of controlling components for operating cost), relative profitability can be expressed as a function of any relative parameter. It is convenient to plot the relative (*DCFRR*) vs. one relative parameter as a family of curves, each for a particular category (F/2). This facilitates interpolation between controlling components. Alternatively, relative profitability may be plotted vs. two relative parameters, separately for each category (F/3). On such charts, the interaction of two parameters can be assessed. Let us examine the procedures for normalization of the sensitivity parameters.

We can demonstrate use of the charts by the following example. Let us assume a process that produces 100 million lb/yr of a product, priced at 75¢/lb. Required capital is \$100 million. Raw materials cost 15¢/lb; selling and distribution, 3¢/lb; utilities, 2¢/lb; and operating labor, \$12 million/yr. Cost of supervision and services can be estimated at 40% of operating labor, and capital charges at 10%. Venture life is 21 yr. and the tax-depreciation life is 12 yr.

Sensitivity to static parameters T/V

		Controlling operating-cost component			
Parameter name	Parameter change, %	All three, Relative (DCFRR)*	Variable, Relative (DCFRR)	Semi-variable, Relative (DCFRR)	Fixed, Relative (DCFRR)
Selling price	+40	1.75	3.50	1.46	3.50
Sales volume	+40	1.58	1.31	1.5	3.40
Depreciable capital	−30	1.43	1.23	1.47	1.23
Tax rate	−30	1.23	1.26	1.22	1.26
Variable cost	−30	1.14	2.92	1.01	1.09
Fixed cost	−30	1.14	1.09	1.01	1.09
Capital factor	−30	1.07	1.05	1.08	1.05
Depreciation life	−30	1.06	1.03	1.06	1.03
Working capital	−30	1.05	1.18	1.03	1.18
Venture life	+40	1.02	1.02	1.02	1.02

*Relative change in discounted-cash-flow rate of return (DCFRR) corresponding to indicated parameter change.

Sensitivity to static parameters (alternative representation) T/VI

Parameter name	All three %	Variable %	Semi-variable %	Fixed %
	Controlling operating-cost component*			
Selling price	2.7	0.80	4.4	0.80
Depletion allowance	2.7	2.3	4.2	2.3
Sales volume	3.5	6.5	4.5	0.83
Depreciable capital	3.5	6.5	3.2	6.5
Tax rate	6.7	5.8	7.0	5.8
Variable cost	11	0.79	300	18
Fixed cost	11	18	300	0.79
Expense	12	3.2	11	3.2
Capital factor	21	33	20	33
Depreciation life	27	50	27	50
Working capital	33	8.3	60	100
Venture life	100	100	100	100
Decling-balance depreciation†	100	167	91	167
Salvage**	370	635	370	635

*Increase required in given parameter to change (DCFRR) by 1%.
† e.g., double declining-balance depreciation corresponds to a 100% increment over straight-line depreciation.
**Increment based on capital.

Expressing the operating-cost components in millions of dollars, we find:

$$\text{Variable cost} = (0.15 + 0.03 + 0.02)\,100 = 20$$
$$\text{Semivariable cost} = 12\,(1.0 + 0.4) = 16.8$$
$$\text{Fixed cost} = (0.10)\,100 + (1/12)\,100 = 18.33$$

This process can be classified as being approximately equally controlled by all components. (*DCFRR*) was calculated at 14.9%.

We can estimate sensitivity to capital, productivity and capacity changes as follows: Productivity change is defined as a volume change at constant capital; capacity change is a volume change accompanied by an appropriate capital change. Let us assume that capital varies with the 0.6 power of capacity.

From F/3, we read the relative (*DCFRR*) for three combinations of relative parameter values. The relative (*DCFRR*) is multiplied by the reference (*DCFRR*) at 14.9%. The procedure is summarized and the results compared to the correct values in the following:

Changed parameter	Relative capital	Relative volume	(DCFRR) from F/3	(DCFRR) estimated, %	(DCFRR) correct, %
None	1.0	1.0	1.0	–	14.9
Capital	0.7	1.0	1.4	20.9	21.8
Productivity	1.0	1.4	1.6	23.9	24.6
Capacity	1.22	1.4	1.3	19.4	20.0

Errors in this estimate occur because the three operating-cost components are not precisely equal, and because F/3 is based on a reference (*DCFRR*) of 20%. Accuracy can be increased by interpolating between charts based on different combinations of controlling components, such as F/2, and on different reference values for (*DCFRR*). However, F/3 can be used for approximations to the (*DCFRR*) as long as it is not too far away from 20%. Profitability for the Base Case can be approximated by using F/1 if a very rough estimate is sufficient.

We can reach certain managerial conclusions from F/2. The most striking feature is the difference in the slope of the curves for fixed cost, *F*, and variable cost, *V*. Both the fixed and variable component-controlled processes fall within the group that is not investment intensive. Results of profitability analyses of variable cost-controlled processes will be valid over a wide range of sales volumes. A project director does not have to worry about marketing if he makes a decision about a process that is dominated by the raw-material cost. Increasing the sales volume raises revenues, but at the same time requires more money for raw materials.

On the other hand, a labor-intensive process needs a good market analysis first to determine what the capacity of the potential plant could be before an engineering and economic evaluation is done. Because most of the operating cost of the process is fixed and capital cost is unimportant, the profit will vary dramatically with sales volume.

Results for static parameters

In this study, about 40 technical and economic parameters were investigated. They are referred to as "static" because they can be assumed constant during venture life. The 14 parameters that have the greatest impact on profitability will be discussed. Sensitivity of the (*DCFRR*) for four selected process categories to ten parameters is shown in T/V. Combinations of two controlling components of operating cost were left out for simplicity. The parameters in T/V are arranged in descending order of sensitivity. The parameter change has been arbitrarily set at either a 40% increase or 30% decrease, whichever leads to a rise in profitability. This corresponds to multiplication or division, respectively, by a factor of 1.4.

The following conclusions can be drawn: Process categories with a dominant semivariable cost are less sensitive than those controlled by all three components to any parameter except capital. This can be explained by the stabilizing effect of investment, i.e., the higher the capital, the lesser the importance of other parameters. Volume has greater impact on fixed-cost–controlling components because production rate is increased at only slightly higher operating cost. The effect of the other parameters can be analyzed in a similar way.

The most significant conclusion is that all controlling-component combinations are highly sensitive to price. Price is the most important economic factor in any industrial process. Interestingly, this fact very often goes unnoticed.

A practical consequence for management is that more money and effort should be spent on price forecasting than on accurate engineering estimates of investment and operating cost. This is particularly true about processes that are not capital intensive, that is, those controlled by the variable and/or fixed component. Here, the effect of price on profitability is at least three times as great as that of other parameters except sales volume.

We cannot make any meaningful engineering and economic evaluation of any process of this type unless we are three times as confident in our price forecast as we are in our estimates of the costs.

Another way of expressing sensitivity is presented in T/VI. This shows the relative impact of ten selected parameters in a more dramatic way, and permits including four parameters that could not be shown in T/V. These are depletion allowance, expense, declining-balance depreciation, and salvage. Expense and salvage are incompatible with T/V because they decrease profitability. Depletion allowance and straight-line depreciation* cannot be increased by 40% because a 40% depletion and a 140% declining-balance depreciation are not legal under the tax laws. Both T/V and T/VI can be used for semiquantitative sensitivity analysis if a complete set of charts (such as F/2 and F/3) is not available.

Dynamic cases

Sensitivity to time-related, or dynamic, factors was also explored. A dynamic case is a modification of the Base Case, and represents changes in static parameters during venture life, such as plant expansion, equipment replacement, and volume fluctuations, or cost and price escalation.

Definitions of dynamic cases

Dynamic cases are defined as modifications of the static Case A, which is the same as the Base Case (T/IV). This implies a single capital outlay in year 1, constant volume in the following years, and no price or cost escalation.

The selected dynamic cases can be classified into four categories as follows:

1. *Reduced volumes* of 50% and 80%, with relative increases of variable cost by a factor of 1.2 in year 2 and 1.1 in year 3, but with a single capital outlay for both years. **Case B**

2. *Capital replacements* in year 5, 9 and 13, i.e, outlays without corresponding changes in the volume. Each replacement equals:
   a. 10% capital **Case C**
   b. 25% capital **Case D**

3. *Plant expansions,* i.e., capital outlays, each followed by a volume increase in the next year. Volume increments are uniform, and the final volume equals that of Case A. Two alternatives were considered: (a) comparison to a single train in Case A, i.e., outlay varies with the 0.6 power of the volume increment, and (b) comparison to multiple trains in Case A, i.e., each outlay is proportional to the volume increment.

In addition to variable and semivariable costs, which must change by definition, fixed cost was also adjusted for each expansion. For the single-train alternative, fixed cost was expressed as varying either with the 0.6 or 0.8 power of the volume increment. For the multiple-train alternative, it was made proportional to the volume increment. Further, starting with the second expansion, fixed cost was alternatively decreased to 75%. The following cases are defined:

Four outlays in years 1, 5, 9 and 13:
   Single train, 100% fixed cost, 0.6 exponent. **Case E**
   Single train, 75% fixed cost, 0.8 exponent. **Case F**
   Multiple train, 100% fixed cost **Case G**
   Multiple train, 75% fixed cost **Case H**

Three outlays in years 1, 7 and 13:
   Single train, 75% fixed cost, 0.8 exponent. **Case I**
   Multiple train, 75% fixed cost **Case J**

Two outlays in years 1 and 13:
   Single train, 75% fixed cost, 0.8 exponent **Case K**
   Multiple train, 75% fixed cost **Case L**

Two outlays in years 1, 5, and 1, 9, respectively, were also calculated, but the difference between these and Cases K and L were very small and have not been included in this article.

4. *Escalation* of annual prices and costs of 1%. This is to be interpreted as a relative escalation rate. For example, if only fixed-cost escalation at 1% is considered, this implies that operating labor escalates at 1% above the other costs, whose escalations are offset by price escalation. Case A is the basis in all of the escalation examples except for capital escalation, which is based on Case H. The following were calculated:
   a. Price escalation **Case M**
   b. Capital escalation **Case N**
   c. Operating-cost escalation **Case O**
   d. Fixed-cost escalation **Case P**

Let us consider plant expansion as an example. This implies an additional capital outlay, followed by an increase in operating cost and production volume. We will consider three aspects in a plant expansion: (1) number of successive expansions, (2) number of trains,

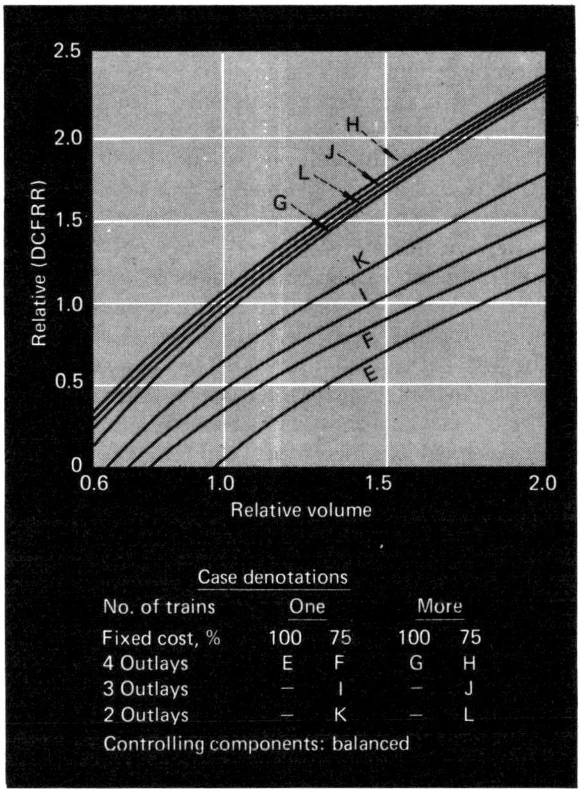

Sensitivity to plant expansions on relative discounted-cash-flow rate of return (F/4)

* Mathematically, straight-line depreciation equals 100% declining-balance depreciation.

Sensitivity to dynamic cases T/VII

Case description or parameter name	All three Relative (DCFRR)*	Variable, Relative (DCFRR)	Semi-variable, Relative (DCFRR)	Fixed, Relative (DCFRR)
	Controlling operating-cost component			
Reduced sales volume	0.73	0.01	0.87	0.01
75% equipment replaced	0.75	0.85	0.74	0.85
25% equipment replaced	0.91	0.95	0.90	0.95
Operating-cost escalation	0.95	0.53	0.99	0.53
Capital escalation	0.96	0.98	0.96	0.53
Fixed-cost escalation	0.98	0.99	1.00	0.54
Price escalation	1.08	1.25	1.05	1.25

*Relative change in discounted-cash-flow rate of return (DCFRR) corresponding to indicated parameter change.

and (3) possible variation of the fixed component of operating cost.

The importance of the number of expansions is obvious. If the expansion implies building a second, completely self-sufficient train, or more such trains, the new capital outlay will be equal or proportional to the original one. The outlay may also be expressed as proportional to a fractional power of the volume increment, a typical exponent being 0.6. This concept neglects the "learning curve" principle and cost escalation, the latter of which is treated separately. Let us call this case the multiple-train alternative.

On the other hand, let us consider an expansion that does not include any equipment duplication, but only partial replacement with large units or the addition of equipment that increases productivity. In this case, the new capital outlay will be less than proportional to the original one, and the volume exponent will probably be less than 0.6. This is denoted as the single-train alternative. Of course, there are situations that fall between the two alternatives.

Finally, the increase in the number of operators due to plant expansion may be proportional to the volume increment for the multiple-train alternative. It may also be less because of jobs that are common to all trains. For the single-train alternative, the incremental number of operators should always be less than proportional to the volume increment, and may not vary for a small expansion. The fixed component of operating cost is a linear function of the number of operators.

Results for dynamic cases

On combining various numbers of successive expansions and various adjustments of fixed cost for single-train and multiple-train alternatives, a set of plant-expansion cases was defined. The sensitivity of the (*DCFRR*) to these dynamic cases is shown in F/4 for a process that is controlled by all three components.

Fixed cost of 100% indicates that the number of operators is proportional to plant capacity. The chart shows that a phased construction of a multiple plant does not have any particular advantage over building the entire plant at the beginning. Profitability increases slightly with the number of plant expansions if operating labor does not increase at the same rate as capacity. On the other hand, phased construction of a single-train plant has a strong negative effect on profitability.

Similar charts may be plotted for other combinations of the operating-cost components and/or other relative parameters. They may be used for approximate sensitivity analysis of any process with the cautions previously discussed for static parameters.

In T/VII, we summarize a selection of results obtained by investigating other dynamic cases. Only the seven cases that have the greatest impact on profitability are shown. The combinations of two controlling components have been left out for simplicity. The table can again be used as a semiquantitative tool if charts of the types shown in F/4 are not available.

Conclusions.

The profitability and sensitivity calculations discussed in this article were performed using a special cash-flow-analysis computer program [*6*].

The sensitivity-analysis model can be incorporated into an existing computerized risk-analysis system [*7*] wherein typical sets of probability distributions are assigned to the major parameters. The process analyst may use either the resulting generalized probability distributions of (*DCFRR*) as a guide for predicting risks in similar situations or he may acquire a computer program and run his particular cases.

References

1. Agarwal, J. C., Sonstelie, W. E. and Klumpar, I. V., *AACE (Amer. Assn. Cost Engr.) Bulletin*, Vol. 14, No. 7 (1971).
2. Strauss, R., *Chem. Eng.*, Mar. 25, 1968, p. 112.
3. Holland, F. A., Watson, F. A., and Wilkinson, J. K. *Chem. Eng.*, Mar. 4, 1974, p. 119.
4. Douglas, F. R., others, *AACE Bulletin*, Vol. 6, No. 5, 63 (1963).
5. Klumpar, I. V., *Chem. Eng. Progr.*, Apr. 1971, p. 74.
6. Klumpar, I. V., *Trans. AACE 1974*, p. 263.
7. Ebeling, D. G. and Agarwal, J. C., *Trans. AACE 1973*, p. 231.

The Authors

Ivan V. Klumpar is a group leader at Ledgemont Laboratory, Kennecott Copper Corp., Lexington, MA 02173. At Kennecott, he performs preliminary engineering and economic evaluation of metallurgical and related chemical processes. He has a Ph.D. in chemical engineering from the Technical University (Prague). He is a member of the American Assn. of Cost Engineers, AIChE and ACS, and is a professional engineer in Massachusetts.

J. C. Agarwal is director of development at Ledgemont Laboratory, Kennecott Copper Corp., Lexington, MA 02173. Dr. Agarwal has written over 80 articles on minerals and metals technology, and holds some 15 patents. He has a B.Sc. from Banaras Hindu University, and an M.S. and Ph.D. from the Polytechnic Institute of Brooklyn. Among his memberships are the American Assn. of Cost Engineers, AIChE, and the American Institute of Mining, Metallurgical & Petroleum Engineers.

Risk Analysis

In theory, risk analysis is a foolproof tool for judging the reliability of capital-cost estimates. In practice, however, results are often disappointing. Success depends on understanding the reasons for failure and taking steps to avoid them.

S. S. DESHMUKH, Fluor Corp.

Uncertainty and risk are inescapable parts of planning capital projects. Those who must base important decisions on estimates and projections are made keenly aware of this. Project cost, construction time and ultimate performance depend on a variety of considerations, many of which cannot be assessed before the fact with any degree of precision. Since a wrong decision can be costly, an analysis of the risks involved is an important part of the planning process.

Risk analysis is one of the oldest management techniques. However, many of the companies that have used formal risk-analysis techniques have been disappointed with the results. The purpose of this article is to first outline the risk-analysis technique as applied to cost estimating and then discuss the reasons behind those instances that have given disappointing results.

Impact of Cost Estimates

To bring a new venture onstream—a mining and metallurgical project in the example used here—involves careful evaluation of costs, market prices and schedules.

Typically, a company will make a detailed study of profitability as it will be affected by market prices, production rates and financing methods. The company will set a price for its final product, decide upon the financing method, and select a firm for design and construction. Through contract agreements with the buyer, financier and contractor, the company will attempt to transfer part or all of the risks.

The engineering-construction firm will normally prepare a cost estimate and schedule based on a scope agreed upon by the two parties. Once approved by the owner and contractor, this becomes the basis for controlling project costs and negotiating changes that may result from additions or deletions.

It is, therefore, apparent that the cost estimate has a major impact on the project's profitability. The estimate leads to the commitment of substantial amounts of money and manpower over an extended time. From the contractor's standpoint, decisions based on cost estimates have high visibility because they are a constant reminder of performance.

After the estimate is prepared and ready for final approval, it is important that the owner and contractor have a rapid means of obtaining answers to three questions:

- How good is the estimate under consideration?
- How much contingency should be added to minimize risks?
- What is the accuracy of the estimate?

We will now show how to apply risk analysis to arrive at meaningful answers to these questions.

ABC Project Estimate Summary — Table I

Description	Estimate, $1,000
Facility – A	6,000
Facility – B	24,000
Facility – C	18,000
Intrastructure	34,000
Total Field Direct	**82,000**
Field indirect	35,000
Engineering costs	7,000
Total	**124,000**
Owner's costs: direct	42,000
Owner's cost: indirect	20,000
Escalation	14,000
Total Project Costs	**200,000**

Note: The estimate is based upon engineering design that is 30% complete.

Originally published June 1974.

Steps in the Analysis

Table I is a cost estimate for building metallurgical-processing facilities in a large mining operation. The estimate is ready for management review and contingency evaluation, and management has asked for a risk analysis, to be made in the following steps:

1. Break the estimate into those items that constitute the major risk areas. (Assume that Table I is already broken down this way.)
2. Further divide each item into components such as labor, materials, equipment, and subcontracted work.
3. Select people who are familiar with the estimate or have had experience on similar projects. Ask them to rate the components for minimum and maximum values under the best and worst circumstances. (There is a zero probability of obtaining a value smaller than the minimum or larger than the maximum.) Instruct the people to consider such factors as status of engineering drawings, labor productivity, weather conditions, market situation, pricing criteria, local political climate, etc. The resulting cost data are shown in Table II.
4. Construct a mathematical model of the cost estimate and make a computer run. Then the analyst should use triangular probability distribution and Monte Carlo simulation to obtain the final results, which are presented graphically in Fig. 1.

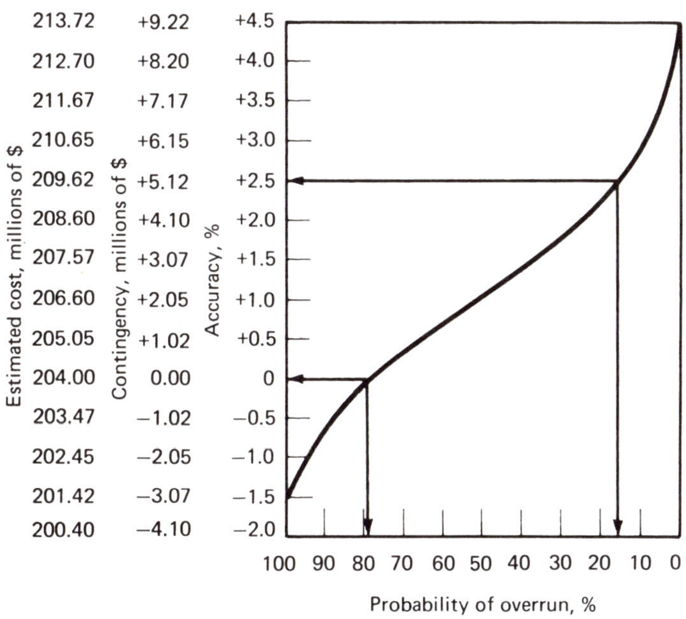

Base estimate	$200 million
Base estimate accuracy	+4.25% and −2%
Probability of overrun	15%
Contingency	$5.12 million
Estimated cost including contingency	$205.12 million
Estimate accuracy including contingency	+2.00% and −4.5%

RESULTS of risk analysis on sample project—Fig. 1

The figure indicates that there is a 78% chance of overrunning the estimated cost if no contingency is added. In other words, management should place only 22% confidence in the estimate. Assuming that management decides to take a 15% risk in a cost overrun, the corresponding contingency from the graph is $5.12 million. With this contingency added, the company can place 85% confidence in the estimate and assure itself that the estimate is +2.0% and -4.5% accurate. Thus we have answered the three questions posed earlier.

But something is wrong–intuition tells us that the answers are incorrect.

First, a contingency of $5.12 million appears too small for a $200-million project. Second, the $13.2-million range between minimum and maximum costs seems too narrow for an estimate on a project in which engineering drawings and specifications are only 30% complete. Therefore, despite a carefully prepared risk analysis, the results are disappointing, and management decides not to use the study in making its final decisions.

REASONS FOR FAILURE

Risk analysis is a process that requires thought rather than action. It does not offer a cookbook solution to estimating problems. And to get meaningful results, many factors must be considered.

To begin with, whenever a company chooses to use risk analysis as a decision-making aid, it must spell out exactly what results it wants, and make sure all parties involved have a clear understanding of the study's objectives. Moreover, company procedures for preparing estimates and controlling projects must be closely aligned with the manner in which the risk analysis is performed. Even the general climate under which the analysis is introduced is important.

In one case, for example, risk analysis was instituted as a result of a company hiring a new vice-president for engineering. He came from an organization where the technique was used successfully, and thus when the next project estimate came up for review he requested a risk analysis. The vice-president arranged a meeting with the cost estimators and asked that the estimate be broken into accounts such as concrete, piping, electrical and structural steel.

This request generated reactions that varied from enthusiasm to skepticism to confusion. Several of the people had little or no knowledge of risk analysis, and the requested cost breakdown was a new procedure requiring major effort. Traditionally, the estimators had presented their figures in terms of individual facilities such as crusher, smelter and concentrator.

After a brief explanation, the group gave its opinions of costs for each account under the best and worst circumstances. The resulting risk analysis was disappointing.

The reason for failure in the above case is obvious. Over the next two years, the company conducted classes and held discussions on risk analysis to assure that everyone had the proper background. Further, risk analysis was made a part of the company's formal estimating procedure, in a manner that closely matched the way the

Maximum and Minimum Estimates – Table II
(1,000 $)

Facility	Labor			Materials and Supplies			Permanent Equipment			Sub-Contract			Total		
	Min	Est	Max	Min	Est	Max	Min	Est	Max	Min	Est	Max	Min	Est	Max
Facility A	−15%	1,450	+10%	− 5%	1,250	+ 5%	− 1%	2,850	+ 2%	− 5%	450	+ 5%	–	6,000	–
Facility B	−15%	4,000	+10%	− 5%	8,500	+10%	− 5%	11,000	+ 8%	− 5%	500	+ 5%	–	24,000	–
Facility C	−15%	4,400	+10%	−10%	7,000	+10%	− 5%	5,500	+15%	− 7%	1,100	+ 7%	–	18,000	–
Intrastructures				− 8%	11,000	+12%	−15%	6,000	+15%	−10%	17,000	+15%	–	34,000	–
Total direct costs														82,000	
Field indirect cost													−15%	35,000	+15%
Engineering costs													− 8%	7,000	+ 8%
Total														124,000	
Owners costs: direct													− 8%	42,000	+ 8%
Owners costs: indirect													−10%	20,000	+15%
Escalation													−10%	14,000	+10%
Total project costs														200,000	

estimate was put together. Then, whenever people were asked to give their judgments under the best and worst conditions, they could relate their thinking back to the method in which they arrived at the figures.

Human Factors

Success of risk analysis depends not only on the expertise of all involved, but also on mutual cooperation. It is important that everyone have a positive attitude, which can only be achieved if the people understand the technique and its purpose. However, the meeting for assessing maximum and minimum values should not be held in an atmosphere of cross-examination. People then become defensive about their original estimates, and obtaining worthwhile input is difficult and frustrating. On the other hand, if the estimators trust the risk-analysis process they will be eager to cooperate.

Another factor is, where within the organization does the responsibility for actually carrying out the risk analysis lie? If one group is responsible for the cost estimating and another for the analysis, the chances for success are considerably reduced. This is particularly true if the two groups operate autonomously.

In one situation, risk analysis was assigned to the computer-sciences staff. This group was not directly responsible to the estimating managers, and during input interviews the estimators always felt uneasy about revealing confidential or unfavorable data. This led to a breakdown in trust and cooperation because the estimators believed they were being cross-examined by people who had little background in preparing cost estimates. The situation was eventually resolved by developing the expertise for risk analysis within the estimating department, which was then placed in charge of the procedure.

Choosing the Best Methods

Since the analysis involves selecting a probability distribution for each random variable, it is important to choose a distribution that fits the application and at the same time is easy to use. For most variables, we use triangular distribution because it requires only three input values—minimum, most likely, and maximum. This avoids mathematical terms such as mean and standard deviation. Moreover, estimators normally think in terms of three-level costs, and in our experience it is easier to specify the parameters for triangular distribution than for any other distribution.

After selecting the appropriate type of distribution, it is important to establish the relationship between statistical theory and the actual values people place on the parameters. In the case of triangular distribution, for example, theory says the minimum and maximum input values are absolute figures. However, people rarely think in absolute terms, and the mathematical model must take this into account.

Fig. 2 shows a triangular distribution, with the minimum value A, most likely value B and maximum value C. According to statistics, the probability that the actual

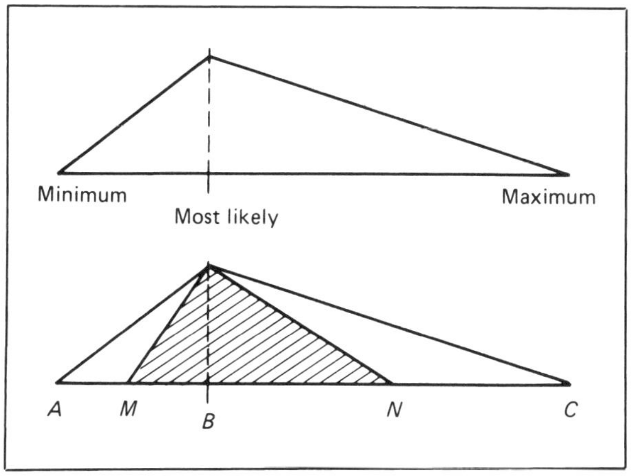

TRIANGULAR distribution applied to estimate—Fig. 2

value will be between A and C is 1.0. In other words, we can assign a 100% confidence limit to A and C.

In practice, when we asked people to give minimum and maximum values, we found that they were providing M and N in Fig. 2. We further found that the probability of the true value occurring between M and N is 68%. Thus we assign one standard-deviation confidence limit (68%) to the input values specified by the people. The mathematical model considers this factor and obtains "true" minimum and maximum values of A and C.

Likewise, values close to the "most likely" value have the highest statistical probability of occurring. However, in the original estimate, the most likely value actually may not be more likely to occur than the maximum or minimum. This is especially true if the estimate is based on limited information. The mathematical model should allow for this.

Still another consideration is that many of the items in the risk analysis respond to changes as a group. For instance, if there is an increase in electrical-labor wages, there are apt to be higher charges for piping labor. When there is an increase in the amount of structural steel involved, both labor and material costs go up. And, of course, many cost items will be affected if the project schedule is extended or contracted. The mathematical model must allow for these relationships by specifying the covariances for the dependent variables.

Overall, determining the confidence limits of the input values is a subjective process, and the limits will vary with the particular person, group, application and company. The important point is to make sure that the model attaches the same significance to the input values as the originators of them have done.

Revised Analysis

Fig. 3 shows a revised estimate based on applying one standard confidence limit to the data in Table I and II and Fig. 1. Now we find there is a 62% probability of overrunning the estimate if no contingency is added—we are 38% confident about the estimate.

The owner and contractor are now much better prepared to determine the amount of contingency needed to minimize risks. In our example (Fig. 3), the parties decided they wanted to have 85% confidence in the estimate, which requires a contingency of $12.27 million.

This contingency will not cover unusual circumstances such as sustained work stoppages, major failures in startup, severe material shortages, or unexpected design errors. The safety factor for unpredictable events depends on the contractual agreement between the owner and contractor. The risk analysis, of course, can be used to establish its magnitude. #

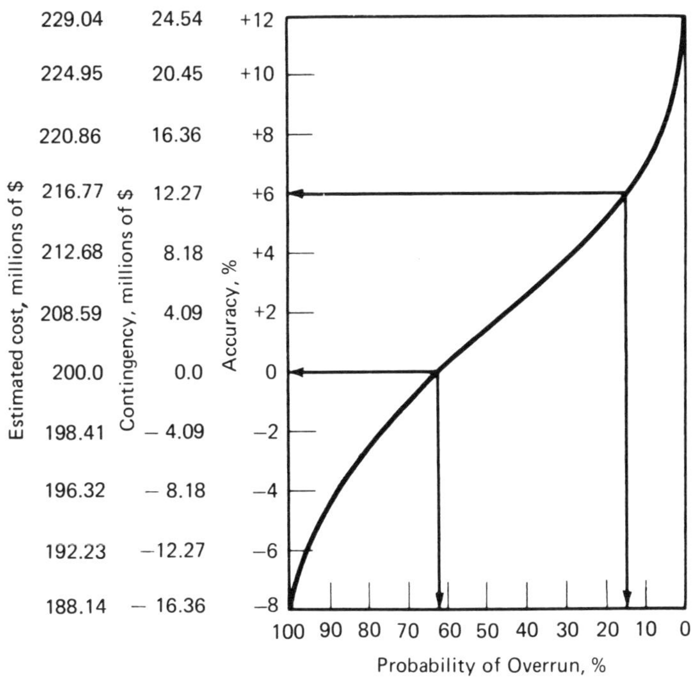

Base estimate	$200 million
Base estimate accuracy	+12% and –8%
Probability of overrun	15%
Contingency	$12.27 million
Estimated cost including contingency	$212.27 million
Estimate accuracy including contingency	+6% and –14%

FINAL ANALYSIS based on revised cost estimate—Fig. 3

Meet the Author

S. S. Deshmukh is working on implementing the concept of mobile computer centers at Fluor Corp., 2500 S. Atlantic Blvd., Los Angeles, CA 90040. Previously, as manager of computer sciences at Fluor Utah, he supervised and directed the development of computer systems for scheduling, cost control, procurement, corporate planning, finance and mining. Mr. Deshmukh has a M.S. degree in civil engineering from Stanford University, and a MBA from the University of Santa Clara.

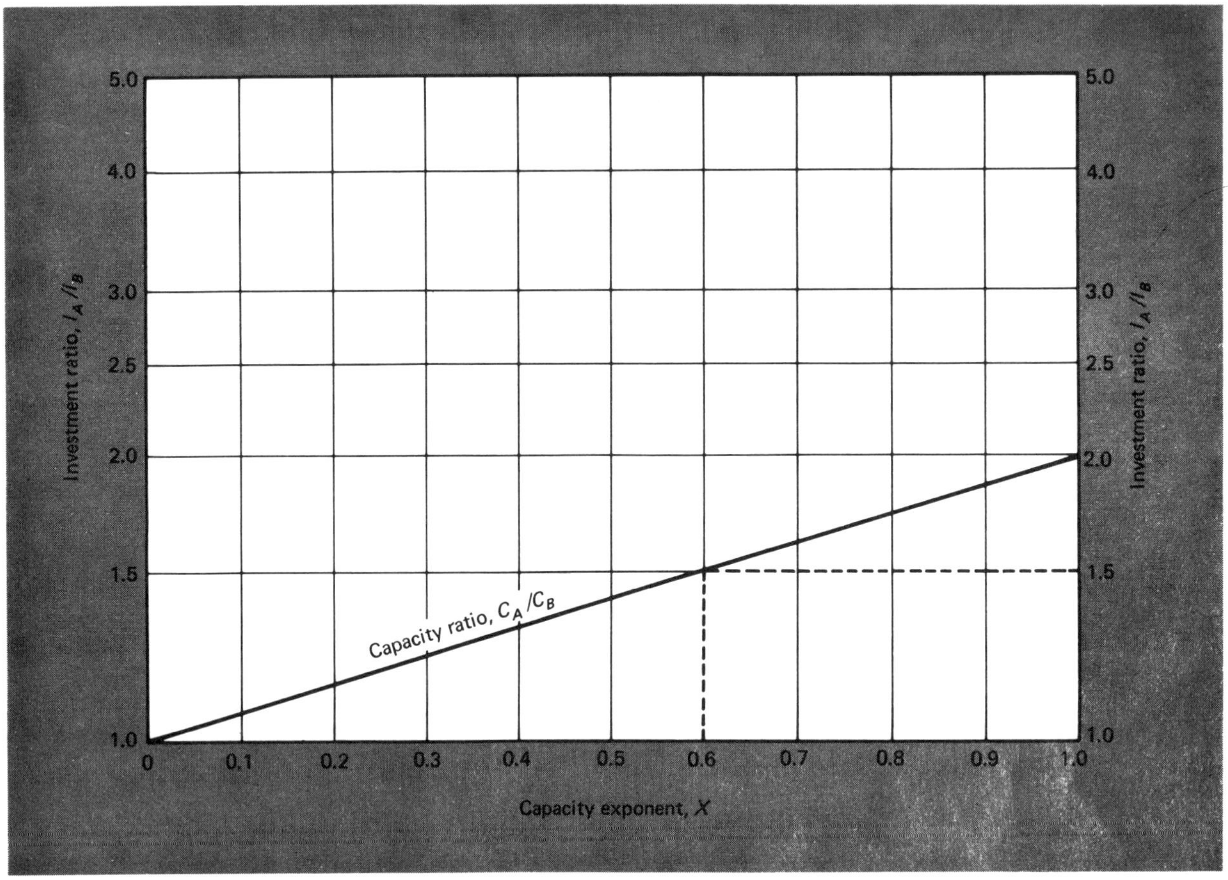

SIMPLIFYING SCALEUP COST ESTIMATION

V. F. CAPPELLO, Gulf Oil Chemicals Co.

Plant costs are frequently estimated from a known cost and a ratio of capacities. The classic exponent rule is widely accepted for estimating plant or equipment costs in this manner. The rule states that the ratio of plant investments is directly proportional to plant sizes raised to an exponential power. This may be expressed as follows:

$$I_A/I_B = (C_A/C_B)^x$$

where:

I_A = investment at capacity A (C_A)
I_B = investment at capacity B (C_B)
x = an exponential power

Since a large range of exponential powers is encountered in estimating plant and equipment costs, it would be helpful to simplify the calculation. This can be done by plotting the relationship of investments, capacities, and exponents on a semilog chart as in Fig. 1. The exponent x is plotted on the arithmetic scale, the investment ratio, I_A/I_B, on the log scale, and the capacity ratios C_A/C_B, as parameters. In this way, all the parameters are straight lines.

It is evident from the formula that, when the exponent $x = 0$, $I_A/I_B = 1.0$ for all values of C_A/C_B. Thus all the parameters will converge at the point $x = 0$, $I_A/I_B = 1.0$. Similarly, the other terminal of each parameter will occur when $x = 1.0$ and $I_A/I_B = C_A/C_B$. Accordingly, the use of this method simply requires a one-cycle sheet of semilog paper, properly labelled, and a straight edge. If desired, a whole family of parameters may be drawn for innumerable capacity ratios without resorting to a slide rule or log tables.

Example: A distillation column is to be estimated at a capacity 2.0 times the capacity for which the cost of the column is known. The exponent is 0.6. What is the ratio of investment costs? Connect a straight edge between the point, $x = 0$, $I_A/I_B = 1.0$ (on the left) and $x = 1.0$ and $I_A/I_B = 2.0$ (on the right) and read above $x = 0.60$ that $I_A/I_B = 1.52$.

A few words of caution are in order: The user should be sufficiently familiar with the process and equipment to know when all the major equipment is below maximum size. Otherwise the scaleup may result in duplicate equipment and a drastic change in the exponent. Published literature* indicates the usual range over which exponents are applicable; however, it is usually best to limit scaling to capacity ratios of 2/1 and in some extreme cases 3/1. ■

* See, for example: Dryden, Charles E., and Furlow, Richard H., "Chemical Engineering Costs," 1966 Ed., Dept. of Chemical Engineering, Ohio State University

Originally published August 1972.

Plant Cost vs. Capacity: New Way to Use Exponents

When you know the cost of a proposed plant of a certain size, you can estimate its cost at other sizes via various methods, some of which are fast but unreliable, others accurate but tedious. This exponent-based technique is fast, yet reliable enough for preliminary evaluations.

J. DAVID CHASE, Celanese Chemical Co.

An accurate scaleup exponent for a proposed process plant can be quickly developed from the scaleup exponents of the major classes of equipment that go to make up the plant. We will describe the techniques involved, illustrate their use in preliminary economic evaluations, and supply some Celanese-developed scaleup exponents for equipment items.

Exponents in Preliminary Evaluations

Preliminary economic evaluations are important at several organizational levels:

At the plant level, they help define the economic desirability of proposed plant expansions or new units. At the corporate level, evaluation groups may depend on preliminary economics to guide investment strategy. And at company technical centers, process-development research proposals may also rely on such evaluations for justification.[14, 15]

In all of the above areas, we are often interested in the cost of building and operating plants at several different capacities. To develop such data, process capital for a variety of production rates must be determined. Usually, a base case is first developed from a process design and capital estimate, from assessment data, or from the cost of an existing unit. Regardless of the source, however, the base case reflects the capital and operating costs for the process at just one production capacity.

One way the relationship between capital costs (C) and size or capacity (S) can be developed from the base case is by using the popular, but not always accurate, "six-tenths rule":[16]

$$(S_2/S_1)^{0.6}\, C_1 = C_2$$

Sometimes, we try to achieve greater accuracy by using our experience, or the literature, to "guesstimate" a better exponent—one that may be characteristic of the industry or plant type in question.

The cost of plants with various capacities can also be developed by evaluating a series of designs and cost estimates for each production rate. But this approach is timeconsuming and in many cases unwarranted. An accurate short-cut method is needed for rough screening work and rapid evaluation where time is short.

The proposed method combines the speed achieved by the single-exponent approach with the greater reliability of making several individual designs. We will consider examples of typical cost/capacity relationships: this will allow us to differentiate between straightforward and unusual situations that can occur. Once these are understood, they can be handled quite easily.

Developing the Exponent Concept

Capital estimates for preliminary economics are usually "factored" estimates—i.e., overall plant capital is a direct function of total equipment cost.[1,9,14,15] For the case where a process design has been worked up, the total equipment cost is developed from an equipment list, and company or literature equipment-cost curves. Individual equipment costs are usually subtotaled according to major equipment classes, such as heat exchangers, vessels, etc.

As a starting point, we will asume that the equipment-class subtotals described above are available.

Mathematically, the base-case total process capital may be expressed as:

$$\text{Equip. Class A} + \text{Equip. Class B} + \cdots = \text{Total Equipment Cost}$$

$$\text{Process Capital}^* = f\,(\text{Total Equipment Cost})$$

The various equipment cost curves that appear in the literature are often presented as straight lines on log-log plots of size vs. cost. The size is usually

* Represents installed cost of complete facility, including offsites, engineering and construction, and estimating contingency.

Originally published April 1970.

Scaleup Exponents for Major Types of Process Equipment

HEAT EXCHANGERS

Material: Shell/Tube	Length of Tube, Ft.	Pressure Psig.	Size Range Sq. Ft.	Exponent
SHELL and TUBE EXCHANGERS:				
CS/CS	20	150	<1,500	0.47
"	"	"	>1,500	0.78
"	"	4,800	200–5,000	0.24
"	5–12	≤600	10–3,000	0.60
All alum.	5–20	—	100–300	0.49
Adm.; SS heads or shell	5–20	—	100–4,500	0.58
CS/304 SS	20	—	20–100	0.36
"	"	—	100–3,000	0.62
CS/316 SS	20	—	10–200	0.36
"	"	—	200–4,000	0.73
"	5–20	6,000	100–700	0.80
All SS	20	—	25–500	0.39
"	"	—	500–7,000	0.73
CS/Ni	20	—	70–900	0.70
CS/Hastelloy C or Carp. 20	20	—	100–1,000	0.84
CS/graphite	7–16	—	250–4,000	0.75
CS/Ti	7–20	—	200–2,000	1.11
TANK and BAYONET HEATERS:				
CS/CS	—	≤160	20–150	0.10
"	—	"	150–1,000	0.53
SS/SS	—	"	80–1,000	0.81
" (finned tube)	—	"	50–600	0.75
DOUBLE PIPE EXCHANGERS:				
CS/CS	—	—	30–1,500	0.70
AIR-COOLED EXCHANGERS:				
CS	24	—	300–1,500	0.69
SS	"	—	300–1,500	0.93

MISCELLANEOUS EQUIPMENT

Equipment Type	Material	Size Range	Exponent
Agitators	CS or SS	1–30 hp.	0.54
Twin-cone blenders	304 SS	60–600 cu ft.	0.57
Compressors (incl. driver and gear)	—	70–5,000 hp.	0.38
Steam ejectors	—	—	0.54
Single-screw extruders	Stainless	2–6 in. dia.	1.09
"	"	6–12 in. dia.	2.60
Double-screw extruders	"	50–160 mm. dia.	1.30
Pumps	CS	0.1–25 hp.†	0.04
"	"	25–400 hp.†	0.65
"	"	5–4M hp.**	0.31
"	Stainless	0.25–5 hp.†	0.11
"	"	5–100 hp.†	0.40

VESSELS

Type	Material	Pressure Psig.	Size Range Gal.*	Exponent
Storage tank	CS	Atm.	200–1MM	0.46
Vacuum	"	—	100–10M	0.35
Pressure	"	15–35	250–15M	0.47
"	"	35–120	50–10M	0.49
"	"	35–120	10M–600M	0.65
"	"	120–160	50–10M	0.58
"	"	200–300	100–1MM	0.44
"	"	300–4,000	50–10M	0.57
Jacketed	"	50–65	100–15M	0.52
Storage	Rubber or plast. lined	Atm.	100–15M	0.57
Pressure	"	100–175	200–30M	0.50
Agitated	Glass lined	Atm.	60–6M	0.43
"	"	40–100	80–8M	0.34
Storage	Aluminum	Atm.	50–100M	0.61
Cone-bottom bin	"	"	400–10M	0.39
"	"	"	10M–25M	0.81
Storage	304 SS	Atm.	100–500M	0.50
Pressure	"	20–40	100–800	0.34
"	"	"	800–15M	0.57
"	"	40–150	250–20M	0.55
Vacuum	"	—	250–2M	0.42
"	"	—	2M–15M	0.61
Pressure	"	150–600	250–10M	0.39
Storage	316 SS	Atm.	250–50M	0.54
Pressure	"	30–260	30–800	0.41
"	"	"	800–30M	0.71
"	"	260–400	400–10M	0.92
"	"	600–800	400–10M	1.46
"	Nickel	25–100	10–100	0.30
"	"	"	100–5M	0.54
"	Hastelloy C or Carp. 20	30	400–10M	0.29
"	Ti lined	175	3–2M	0.19
"	"	"	2M–50M	0.95
"	"	500	1½M–6M	0.78

TOWERS Type	Material	No. of Trays	Diameter Range, In.	Exponent
Sieve tray	CS	10–20	60–120	1.40
"	"	21–45	30–60	0.28
"	"	21–45	60–200	0.71
"	"	46–75	24–60	0.30
"	"	46–75	60–200	0.62
"	304 SS	10–35	30–120	0.65
"	"	36–75	30–120	0.38
"	316 SS	10–80	18–60	0.34
"	"	10–80	60–200	0.72
Packed	Stainless	—	5–36	0.35
"	"	—	36–100	0.85
Bubble cap	CS	—	36–100	0.86
"	Stainless	—	36–100	0.52

Notes: *M = 1,000 †ΔP≤125 psi. **ΔP≥250 psi. **Source:** Ref. 4

related directly to the equipment's capacity. In general, the cost curve for a specific item of equipment has a slope more or less characteristic of its class. For the moment, let us also assume that the cost curve for an equipment class may be represented the same way as a specific item of equipment.

Fig. 1 shows this cost/capacity relationship for a simple unit where there are only two equipment classes, no offsites, and where process capital equals the total equipment cost. For the total plant, the assumptions are:

1. Log (capital) vs. log (capacity), if not truly linear, may be approximated by a straight line over the capacity range considered.

2. Equipment sizes are below the maximum, so that duplication of equipment is not required over this range.

Although the dashed line (Total Plant Cost) is the straight line approximation of the actual total equipment cost at each capacity, let us say that we know the total plant cost only at one point—at a capacity of 50 million lb./yr. Data for this base case are:

Equipment Class	*Cost, Million $*	*% of Total*	*Scaling Exponent*
A	2.77	60.9	0.350
B	1.78	39.1	0.760
Total Plant	4.55	100.0	Unknown

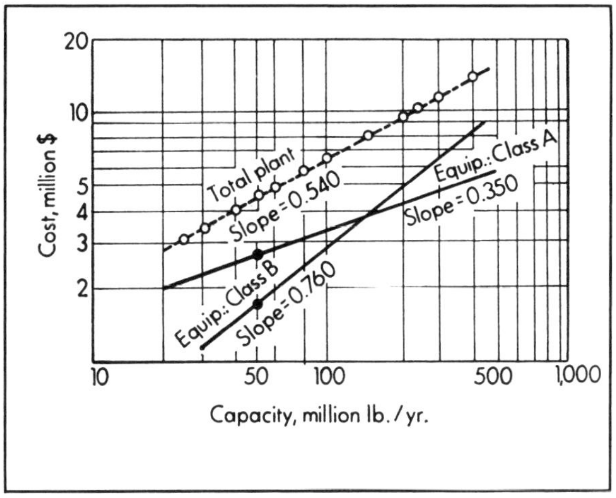

COST/CAPACITY relationship for simple plant having only two major classes of equipment—Fig. 1

slope of the dashed line in the upper part of Fig. 1. At first glance, the proposed technique may appear questionable, because you do not get a constant exponent of 0.54. To understand this point fully, note the chart below, which illustrates the differences that might exist between the straight-line approximation and individual data points (the differences are magnified for illustration):

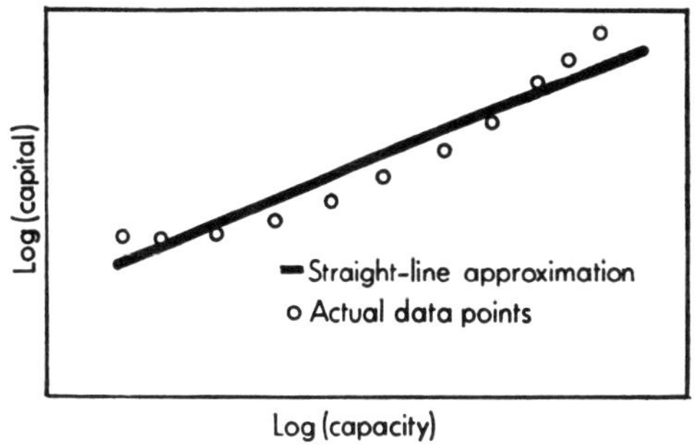

Since A + B = Total Plant, then:

$$(R)^a(A') + (R)^b(B') = (R)^t(T') = (R)^t(1.0)$$

where R is the scaling range;* a and b are the scaling exponents (slope of cost curves) for Class A and B equipment; A' and B' are the fraction of total equipment cost for each class; t is the weighted average scaling exponent for the entire plant; and T is the total plant process capital.

Therefore, we can solve for an accurate value of the plant scaling exponent, t, for this example, using:

$$(R)^{0.350}(0.609) + (R)^{0.760}(0.391) = (R)^t(1.0)$$

Calculated Scaling Exponent

The importance of both the range (R) over which the scaling exponent is estimated and the base-case capacity selected should be understood. The calculated value of the plant exponent, t, is not constant but varies slightly, depending on the extrapolation range and the base-case capacity. For the simple example above, based on Fig. 1, the following values of t were calculated for different scaling ranges, R, using two different base-case capacities:

	Calculated t	
R	*50 Million Lb./Yr. Base Case*	*100 Million Lb./Yr. Base Case*
0.5	0.50	0.52
2	0.52	0.55
3	0.53	0.56
4	0.54	0.57

To illustrate how the individual t values were calculated, consider the $t = 0.52$ figure in the 50-million-lb./yr. column. At $R = 2$:

$$(2)^{0.350}(0.609) + (2)^{0.760}(0.391) = (2)^t$$
$$1.2746(0.609) + 1.6935(0.391) = (2)^t$$
$$(2)^t = 1.4384; \text{ therefore } t = 0.52$$

By graphical measurement, the average scaling exponent for the total plant is 0.54; this represents the

* Corresponds to S_2/S_1, where S_2 is the plant size you want to scale to and S_1 is the base-case plant capacity.

Thus, if we consider the geometry of the line between any two specific points, it is apparent that the *calculated* exponent represents the actual point-to-point slope. This slope is close to, but not exactly equal to, the slope of our straight-line approximation. Fortunately, however, the inaccuracy of the straight line, compared to the data points, is usually minor.

Complex Units

Here are some pitfalls to avoid in extending the above analysis to complex systems.

A common cost/capacity relationship based on actual cost data at several capacities[12] is shown in Fig. 2. Although the log (cost) vs. log (capacity) may be approximated by a straight line over a limited range —e.g., Point *a* to Point *b*—it will appear to be curved, based on data points, over other ranges (such as Point *b* to Point *e*). Also, the slope of the curve from *c* to *e* is actually greater than 1.0; this is illogical because it indicates higher capital costs per unit with increasing capacity. It would seem at first that the concept of estimating the plant exponent from its component parts cannot be used in such a case.

Some additional design work at each capacity in the curved portion of the plot will be necessary. However, we want to reduce this work to the bare minimum consistent with reliable results, and this can be done if the cause of such behavior can be deduced. Basically, three explanations may be postulated:

Case 1: An economically significant portion of the plant equipment is paralleled at higher capacities.

Case 2: The size/cost curve for one or more major pieces of equipment in the unit makes a sharp break upward at Point *b*, due to increased fabrication costs per unit size. Cost curves of other major items break upward at Point *c*, Point *d*, etc.

Case 3: Both of the above possibilities occur.

Let us examine each case to see how we can simplify our work.

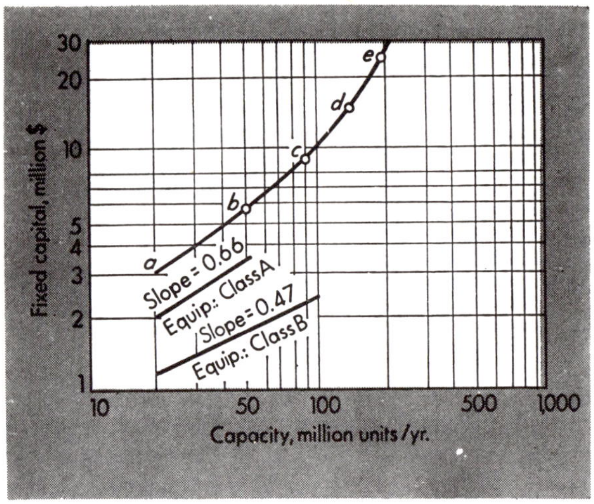

COST/CAPACITY relationship for a complex plant that involves some paralleling of equipment—Fig. 2

Parallel Equipment (Case 1)

Fig. 2 could be explained if, for example, the plant required one reaction train from Point *a* to Point *b*, two reaction trains at Point *c*, etc.

Thus, the assumed cost/capacity curve from Point *b* to Point *e*, based on the actual data points, is deceiving. When parallel equipment is involved, a complete investigation of the cost/capacity relationship for each different unit configuration would probably result in a family of straight-line approximations. Furthermore, it is also probable that the lines in the family would be parallel.

Fig. 3 illustrates this. The dashed lines through each point represent the fixed-capital/capacity relationships for the following unit configurations:

Family Point	*Number of Lines of Equal Capacity*	
	Class A Equip.	*Class B Equip.*
c	2	1
d	4	2
e	8	4

Procedure for Parallel Equipment—Once a base-case design is completed, a study of the maximum practical equipment sizes or standard sizes will tell you whether any equipment has to be paralleled over the capacity range under consideration. All other equipment can be treated in a straightforward manner. Paralleled equipment must be resized. A base-case capital is then estimated for each different unit configuration required, and the scaling exponent estimated for each configuration. The valid capacity range, particularly the upper limit, must be kept in mind.

Breaks in Equipment Cost Curves (Case 2)

Although most cost/capacity curves are single straight lines, sometimes the slope of the line will increase at some specific equipment size. This may

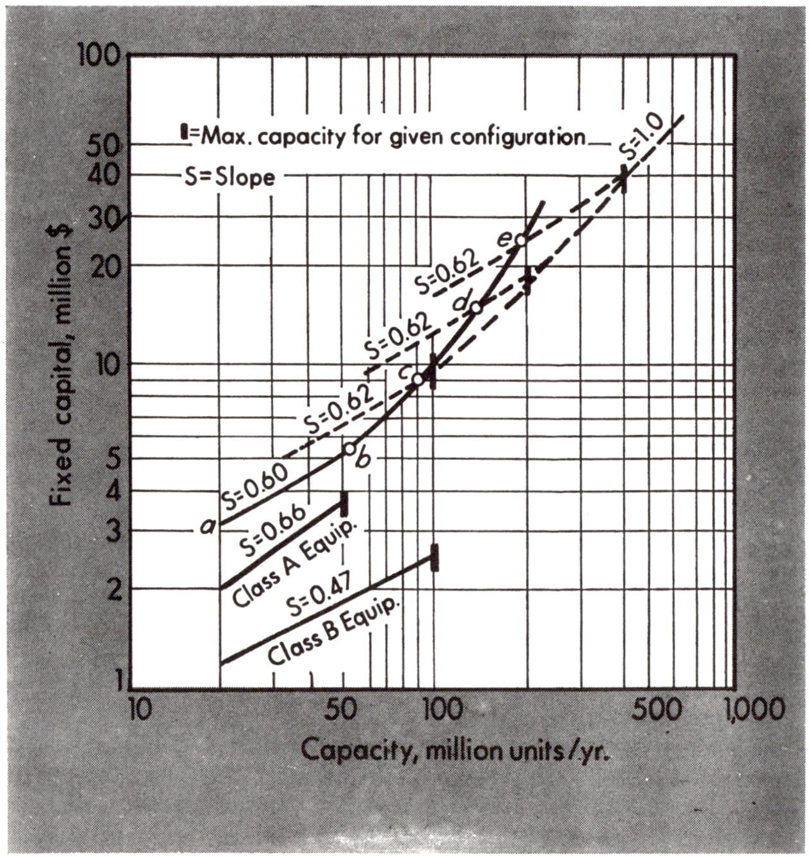

FOR PARALLEL EQUIPMENT, family of straight-line approximations (dashed lines through data points) represent the cost/capacity relationship for each unit configuration—Fig. 3

be due to increases in unit fabrication costs for larger sizes:

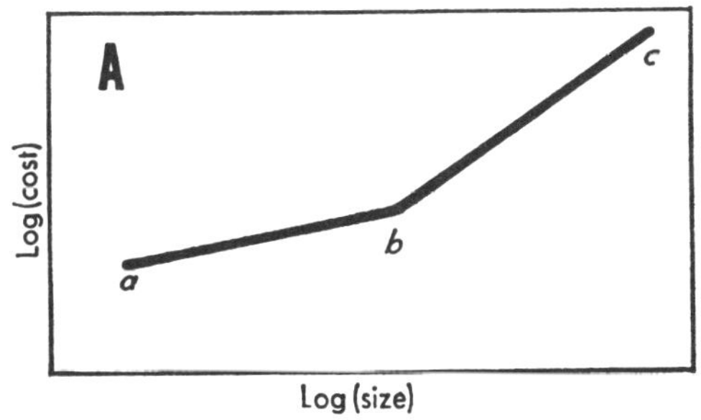

Although this behavior can produce a situation similar to the one shown in Fig. 2, it cannot fully explain the example used. The difference in the effect of a break in the equipment cost curve on capital, compared with the effect of equipment paralleling, is shown below:

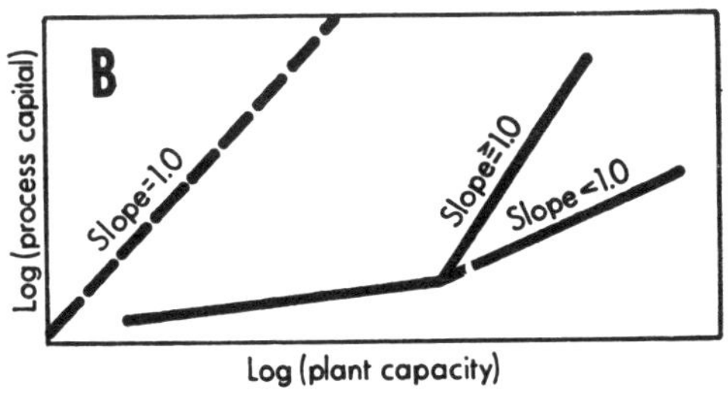

Line *b-c′* represents the parallel-equipment situation, which is usually the only situation where the apparent slope or scaling exponent can equal or exceed 1.0. (Curve *b-c′* was drawn using data points from units with different configurations or bases.) An apparent slope equal to or greater than 1 is usually fictitious upon complete analysis.

Line *b-c*, however, is typical of the result you would get if the equipment shown in Sketch (A) was an item of significant cost to the plant shown in (B). Even if all of the equipment in the unit had Type (A) cost curves, and each one changed slope at a different capacity, you still would not get a slope $\geqq$ 1.0 for the plant cost-capacity plot over the capacity range *b-c*. Hence, line *b-c* might be a result of either minor equipment paralleling or a break in the equipment cost curve.

Procedure for Equipment Cost Break—In this case, the affected equipment is resized and its cost reestimated. A new scaling exponent for the plant must be calculated. The revised exponent applies only to the capacity range in which the assumptions stated for the simple case are valid.

If Details of Capital Estimate Are Lacking

Although the proposed method is most accurate when the details of a capital estimate are known, the principles involved may also be used to advantage when less information is available.

For instance, a typical situation is a request to determine the product selling-price/capacity relationship from outside assessment data. The only information available may be a simple flowsheet, raw-materials and utilities requirements, and battery-limits capital.[13] Only rarely will the distribution of equipment costs be known.

Sometimes, it is possible to get a scaling exponent from the source of assessment data. If not, you may have to either (1) assume the process is similar to an "in house" process, for which additional details are available, or (2) rely on the similarity of the process to some other process discussed in the literature, or (3) guess at the distribution of equipment costs.

A scaling exponent for the process, or one quite similar, may have been published.[7, 10, 11, 18] If a guess at the equipment cost distribution is necessary, typical ranges for organic chemical plants are:

	% of Total Equipment	
	Liquid	*Liquid-Solid*
Heat exchangers	15–25	15–25
Machinery	10–20	40–50
Pumps and drives	8–16	6– 8
Towers (distillation and/or absorption)	20–30	8–12
Vessels	18–28	16–20

These distributions may be used to estimate scale factors in the manner discussed earlier. Of course, average exponents for each equipment class must be used. Guthrie has listed these for many different types of equipment.[9]

Some additional scaleup exponents, not previously published, are listed on p. 114.[4] Although it is usually preferable to use exponents based on your company's cost curves, you may not have these data for some equipment types.

Previously published scaling exponents for equipment can be found in Ref. 2, 3, 6, 9 and 17. Plant scaling exponents (cited earlier), or data from which exponents may be calculated, have also been published.[12, 20]

In some cases, the equipment exponents may have to be calculated from literature cost curves.[5, 19] The "Cost File" department in CHEMICAL ENGINEERING, and *Cost Engineering*, are good sources to use for this.

Step-by-Step Procedure, and Illustrative Example

To review, here is a capsule summary of the course of action for obtaining plant scaleup exponents:

First, determine the equipment cost distribution:

1. For each equipment class in the total plant.
2. Within each class (if details are known).

Next, examine the equipment sizes to see if paralleling will be required or if cost-curve breaks will occur over the prime scaling range. If either occurs, then determine the valid subranges for which:

1. No paralleling or cost breaks occur.
2. Only one unit configuration is needed.
3. Cost break begins, but no paralleling is needed, and so on.

Next, for each major or subscaling range, estimate the exponents for the various equipment classes and then for the plant.

The following example is based on actual design calculations in my files. Because I had sized and costed all the base-case equipment myself, complete details were available to begin with.

Base-case rate = 60 million lb./yr.

Desired rate = 225 million lb./yr.

Therefore, $R = 225/60 = 3.75$.

Heat Exchangers (22% of plant cost)

Subclass	X'	b'	$X'(R)^{b'}$
1	0.060	0.212	0.0794
2	0.064	0.485	0.1215
3	0.435	0.626	0.9950
4	0.166	0.644	0.3888
5	0.275	0.921	0.9290
	1.000		$\Sigma = 2.5137$

$$(R)^{t'} = \Sigma X'(R)^{b'}$$
$$t' (\log R) = \log[\Sigma X'(R)^{b'}]$$
$$t' = 0.697$$

Miscellaneous Equipment (8% of plant cost)

Subclass	X'	b'	$X'(R)^{b'}$
1	0.541	0.442	0.9703
2	0.368	0.501	0.7136
3	0.065	0.543	0.1332
4	0.026	0.829	0.0778
	1.000		$\Sigma = 1.8949$

$t' = 0.484$

Towers (26.1% of plant cost)

Subclass	X'	b'	$X'(R)^{b'}$
1	0.410	0.521	0.8163
2	0.434	0.788	1.2298
3	0.156	1.031	0.6095
	1.000		$\Sigma = 2.6556$

$t' = 0.739$

Vessels (33.9% of plant cost)

Subclass	X'	b'	$X'(R)^{b'}$
1	0.009	0.337	0.0140
2	0.073	0.293	0.1075
3	0.038	0.414	0.0657
4	0.093	0.422	0.1624
5	0.055	0.424	0.0963
6	0.378	0.434	0.6691
7	0.068	0.436	0.1210
8	0.197	0.566	0.4162
9	0.089	0.820	0.2631
	1.000		$\Sigma = 1.9153$

$t' = 0.492$

Plant Exponent

Equip. class	X'	t'	$X'(R)^{t'}$
Heat exchangers.............	0.220	0.697	0.5527
Misc. equip...................	0.080	0.484	0.1517
Towers........................	0.261	0.739	0.6932
Vessels........................	0.339	0.492	0.6496
Pumps*......................	0.010	0.489	0.1908
	1.000		$\Sigma = 2.2380$

* Estimated at 10% of total delivered-equipment cost.

In the calculations at left, X' is the fraction of total equipment cost for each class or subclass; b' is the individual scaling exponent for each equipment subclass; t' is the weighted-average scaling exponent for each major class of equipment; and t the exponent for the entire plant.

Since $(R)^t = \Sigma X'(R)^{t'} = 2.2380$, we can solve for t, and obtain a value of 0.610. This is the plant exponent.

As mentioned previously, all of the equipment sizes for the base case were known at the time this study was made. From maximum equipment sizes, it was determined that no paralleling would be necessary at four times the base-case sizes. Futhermore, no significant cost-curve breaks occurred over the scaling range. ■

References

1. Aries, R. S., and Newton, R. D., "Chemical Engineering Cost Estimation," p. 5, McGraw-Hill, 1955.
2. Bauman, H. C., Factoring Costs of Chemical Plant Installations, *Ind. Eng. Chem.*, **50**, No. 8, pp. 69A-71A (1958).
3. Bauman, H. C., Up-To-Date Equipment Costs, *Ind. Eng. Chem.*, **54**, No. 1, pp. 49-60 (1962).
4. Celanese Chemical Co., unpublished data.
5. Chilton, C. H., (Ed.), "Cost Engineering in the Process Industries," pp. 49-58, McGraw-Hill, 1960.
6. Drew, J. W., and Ginder, A. F., How to Estimate the Cost of Pilot-Plant Equipment, *Chem. Eng.*, Feb. 9, 1970, pp. 100-110.
7. Gallagher, J. T., Rapid Estimation of Plant Costs, *Chem. Eng.*, Dec. 18, 1967, p. 93.
8. Grumer, E. L., Selling Price vs. Raw Material Cost, *Chem. Eng.*, Apr. 24, 1967, pp. 190-192.
9. Guthrie, K. M., Data and Techniques for Preliminary Capital Cost Estimating, *Chem. Eng.*, Mar. 24, 1969, pp. 114-142.
10. Hackney, J. W., Capital Cost Estimates for Process Industries, *Chem. Eng.*, Mar. 7, 1960, p. 116.
11. Haselbarth, J. E., Updated Investment Costs for 60 Types of Chemical Plants, *Chem. Eng.*, Dec. 4, 1967, pp. 214-215.
12. Mapstone, G. E., Find Exponents for Cost Estimates, *Hydrocarbon Processing*, **48**, No. 5, pp. 165-167.
13. Spitz, P. H., How to Evaluate Licensed Processes, *Chem. Eng.*, Dec. 20, 1965, pp. 91-98.
14. Walton, P. R., Cost Estimating at the Research Level, *Chem. Eng.*, Aug. 15, 1966, pp. 172-176.
15. Winter, O., Preliminary Economic Evaluation of Chemical Processes at the Research Level, *Ind. Eng. Chem.*, **61**, No. 4, pp. 45-52 (1969).
16. Woodier, A. B., and Woolcock, J. W., The ABC of the 0.6 Scale-Up Factor, *European Chem. News*, Sept. 10, 1965, pp. 8-9.
17. Zimmerman, O. T., and Lavine, I., Cost-Capacity Factors—Equipment, *Cost Engineering*, **6**, No. 2, pp. 13-22 (Apr. 1961).
18. Zimmerman, O. T., and Lavine, I., Cost-Capacity Factors—Plants, *Cost Engineering*, **6**, No. 2, pp. 13-22 (July 1961).
19. Zimmerman, O. T., and Lavine, I., Equipment Cost Estimating Data, *Cost Engineering*, **7**, No. 1, pp. 4-5 (Jan. 1962).
20. Zimmerman, O. T., Unit Plant Costs, *Cost Engineering*, **12**, No. 4, pp. 12-19 (Oct. 1967).

Meet the Author

J. David Chase is a Senior Research Engineer at Celanese Chemical Co.'s Technical Center at Corpus Christi, Tex. 78403 (P.O. Box 2678). Most of the 12 years he has spent with Celanese have involved work on process design and economics. A chemical engineering graduate of the University of Arkansas, he is a member of AIChE. Among his publications is a series on sieve-tray design that appeared in CHEMICAL ENGINEERING in 1967.

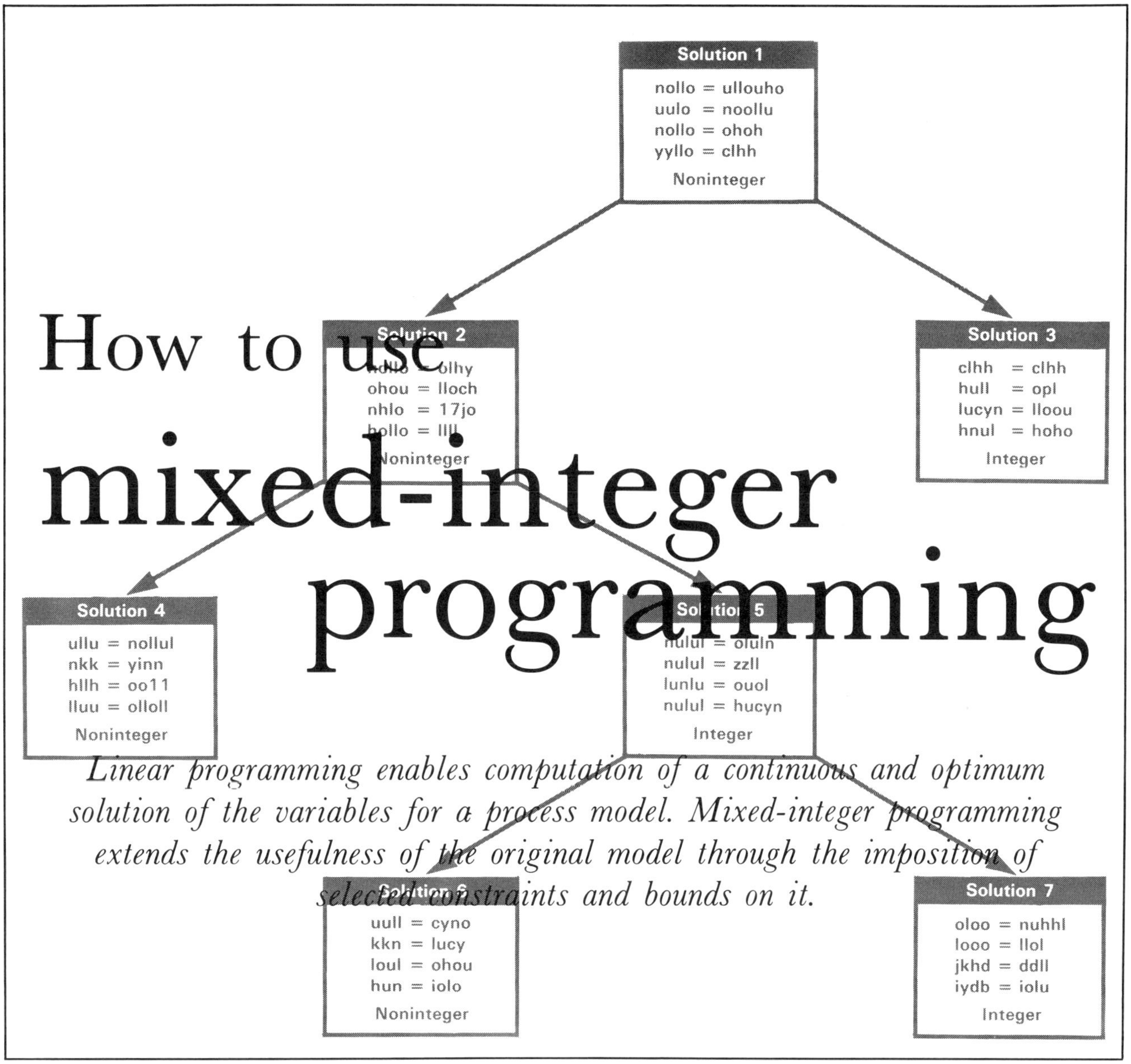

How to use mixed-integer programming

Linear programming enables computation of a continuous and optimum solution of the variables for a process model. Mixed-integer programming extends the usefulness of the original model through the imposition of selected constraints and bounds on it.

Derek H. Allen, *University of Nottingham (U.K.)*

☐ Mixed-integer programming (MIP), an extension of linear programming (LP), enables a limited number of additional special variables to be included in a model. These variables are permitted to take only integer values. Inclusion of these integer variables widens the scope of the type of problem that can be formulated and solved.

The increase in scope of MIP compared with normal LP can be seen from the following situations, whose formulation only becomes possible with the use of MIP:

- One or more variables permitted to take integer values or other discrete values.
- Decision situations and the selection between mutually exclusive variables.
- Combination of fixed and variable costs.
- Sequencing the selection of variables.

All of these situations will be discussed later; first, let us see what MIP is and how it works. (For a brief summary of the advantages and limitations of linear programming, see the box on the next page.)

A graphical example of MIP

The following example, although very simplified, will serve to illustrate MIP. A process plant makes two products, A and B, from three feedstocks, X, Y and Z, in two processing units, as shown in Fig. 1. The maximum capacities of the A and B units are 4 and 5 tons/day, respectively. The manufacture of 1 ton of A requires 0.4 ton of X and 0.6 ton of Y, and of 1 ton of B requires 0.3 ton of Y and 0.7 ton of Z. There is an upper limit on the amount of Y available, of 3.1 tons/day. The incremental profit of making A or B, i.e., selling price minus incremental manufacturing cost, is \$320 and \$400/ton, respectively. Find the optimum production that will maximize the total profit.

In this rather trivial example, the optimum production pattern of A and B that maximizes total profit can easily be found by inspection. However, let us find it by linear programming. Taking the product rates of A and B as the variables x_1 and x_2 respectively, we can

Originally published March 1976.

Linear programming: its advantages and limitations

Linear programming, a computer-based mathematical technique, is widely used in the chemical process industries for investigating process economics, for evaluating the effects of new-plant investment, and for optimizing production. It is particularly useful where complex interactions occurring between different parts of a process plant tend to obscure the effects of any changes on the performance of either the plant as a whole, or any part of it.

In a linear programming (LP) model of a process plant, the variables usually representing flowrates are related by linear equalities and inequalities. These describe the structure of the process itself, and the structure of the feedstock availability, and the product-demand pattern imposed on the plant's operation. The model has a linear objective function, in which the coefficients of the variables take account of the incremental operating costs of the plant, the cost of feedstocks, and the net from product sales. The optimal solution of an LP model provides the combination of values for the variables that maximizes the total value of the objective function (i.e., incremental profitability excluding fixed costs), within the limits imposed by the equality and inequality constraints.

The success of LP as a technique lies in the ease with which it can be applied. Comprehensive computer LP software is readily available and does not require a mathematician or a programmer. The user, typically a chemical engineer, is mainly concerned first with formulating an LP model from his technical knowledge of the process and plant, and his knowledge of the economic environment within which they operate. Second, he is concerned with the practical interpretation of the optimal solution obtained from the computer.

Literature on the mathematical aspects of linear programming is abundant. Mention will only be made of two recent books by Zionts [1] and by Cooper and Steinberg [2], which provide good introductions to linear programming.

Published material on ways to formulate LP problems and interpret their solutions is not so widespread. The initial work by Symonds [3] on refinery models is still of value. Smith's recent book [4] on business LP models is based on experience in the chemical industry, and a paper by Allen [5] describes some principles for formulating process LP models and interpreting their solution in terms of process economics. The books by Driebeek [6] and Danø [7] are concerned with more-general types of industrial LP models and contain chapters dealing with problems of their formulation.

Overcoming LP limitations

LP models must of course be linear in form, i.e., the objective function and constraints must be linear functions of the variables. Many situations in modeling process plants are inherently nonlinear, e.g., nonlinear operating costs vs. throughput, or nonlinear yields vs. operating conditions. The advantages of using linear programming compared with nonlinear optimization are such, however, that the model is usually adapted to a linear form by appropriate approximation. Since this increases the size and complexity of the LP model, a compromise is often needed between accuracy and model size (and hence, computer time and memory). Formulation of some simple types of linear approximations is discussed by Allen [5].

Although many nonlinear relationships can be approximated in these ways, others require a particular modification of LP known as mixed-integer programming (MIP). Even when conventional linear approximation is possible, the use of MIP often results in a more straightforward and efficient formulation and in a considerable saving of computer time.

By definition, LP variables are nonnegative, continuous variables. Situations can arise in the formulation of LP models where a variable should only take on integer or other discrete values. These situations cannot be accommodated in a normal LP model but can be dealt with by using MIP. An MIP facility in a computer LP package can be used to extend the type of problem and the type of formulation dealt with. In particular, it becomes possible to include logic and decision variables.

---

express the daily incremental profit by maximizing:

$$320x_1 + 400x_2 = z \qquad (1)$$

subject to the following constraints:

$$x_1 \leq 4.0 \text{ (capacity of A unit)} \qquad (2)$$

$$x_2 \leq 5.0 \text{ (capacity of B unit)} \qquad (3)$$

$$0.6x_1 + 0.3x_2 \leq 3.1 \text{ (availability of Y)} \qquad (4)$$

We obtain the graphical solution by plotting the constraint polygon for the inequalities of Eq. (2), (3) and (4), as shown in Fig. 2. The maximum permitted value of z is \$2,854/d. This occurs where the objective function, Eq. (1), just touches the constraint polygon at the point $x_1 = 2.67$ and $x_2 = 5$. So far, this is a normal LP problem and solution.

Now let us suppose that because of product instability and consequent difficulties with storage and transportation, daily production and delivery of both A and B must be in multiples of complete 1-ton lots. In other words, x_1 and x_2 may only take integer values. Thus, the normal LP solution is not acceptable, as it gives $x_1 = 2.67$ tons/d.

Therefore, the problem now is to find the best integer combination of values for x_1 and x_2 within the bounds of the constraint polygon, i.e., the permitted integer values of x_1 and x_2 that yield the highest value of the objective function z. Note that this value of z must be less than or equal to the value of z in the optimal normal LP solution (known as the "real" or continuous solution).

Graphically, the best integer solution can be found by moving the objective-function line to the left until it meets the first integer combination of x_1 and x_2 inside the constraint polygon of Fig. 2. This occurs at $x_1 = 2$, $x_2 = 5$; and therefore z = \$2,640/d. Restricting x_1 and x_2 to integer values has reduced z by \$214/d. Although x_2 has remained at its original value of 5 while x_1 has changed, this is coincidental. For example, if the slope of the objective-function line were changed, it can be

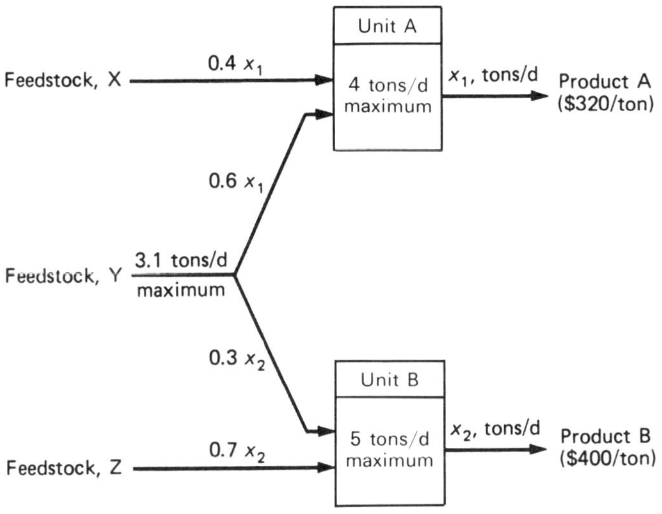

Process relations for a plant making two products illustrate the use of mixed-integer programming Fig. 1

seen that (while still giving the same continuous solution) the integer solution (Fig. 2) could become $x_1 = 3$, $x_2 = 4$ tons/d.

Branch-and-bound method of solution

In practice, a model containing integer variables is first solved as a normal LP problem, with the integer requirements relaxed to obtain the continuous solution. The optimal integer solution is then found by a search technique involving solution of a sequence of LPs, formed by adding appropriate constraints to the original model. This technique can be illustrated by again referring to the example.

Step 1—The original model is represented by the objective function, Eq. (1), subject to the constraints of Eq. (2) to Eq. (4). The optimal solution for the original model is: $x_1 = 2.67$ (noninteger), $x_2 = 5$; and $z = \$2{,}854/\text{d}$. To obtain an integer solution, x_1 must either increase to at least 3 or decrease to at least 2.

Step 2—Let us add to the original model, the constraint: $x_1 \geq 3$. The new optimal solution is: $x_1 = 3$, $x_2 = 4.33$; and $z = \$2{,}693/\text{d}$. Although x_1 is now an integer, x_2 is no longer an integer.

Step 3—Let us add to the original model, the constraint: $x_1 \leq 2$. The new optimal solution is: $x_1 = 2$, $x_2 = 5$; and $z = \$2{,}640/\text{d}$.

This solution satisfies the integer requirements of the problem, but is it the best integer solution? To find out, the search must be continued. In the Step 2 solution, z is higher than in the integer Step 3 solution. Therefore, we must see whether an integer solution can be derived from Step 2 with a value of z that is higher than that value found in the Step 3 integer solution. From Step 2, x_2 must either increase to at least 5 or decrease to at least 3.

Step 4—Let us add to the model in Step 2, the constraint: $x_2 \geq 5$. The model to be maximized is now:

$$320x_1 + 400x_2 = z$$

subject to the following constraints:

$$\begin{aligned} x_1 &\leq 4.0 \\ x_2 &\leq 5.0 \\ 0.6x_1 + 0.3x_2 &\leq 3.1 \\ x_1 &\geq 3 \\ x_2 &\geq 5 \end{aligned}$$

This model is infeasible, i.e., there is no solution satisfying all the constraints (examine Fig. 2).

Step 5—Let us now add to the model in Step 2, the constraint: $x_2 \leq 4$. The complete model is now to maximize:

$$320x_1 + 400x_2 = z$$

subject to the constraints:

$$\begin{aligned} x_1 &\leq 4.0 \\ x_2 &\leq 5.0 \\ 0.6x_1 + 0.3x_2 &\leq 3.1 \\ x_1 &\geq 3 \\ x_2 &\leq 4 \end{aligned}$$

This model's optimal solution is: $x_1 = 3$, $x_2 = 4$; and $z = \$2{,}560/\text{d}$. This solution satisfies the integer requirements, but the value of z is less than the $2,640/d of the Step 3 solution. The sequence of possible solutions to be considered has terminated. Step 3 gives the best integer solution.

The procedure for finding the optimal integer solution is summarized in Fig. 3 in the form of a "tree", which branches to new models and their solutions by the addition of constraints or bounds (hence, the name "branch and bound" given to this technique). Each branch terminates when it reaches an infeasible solution

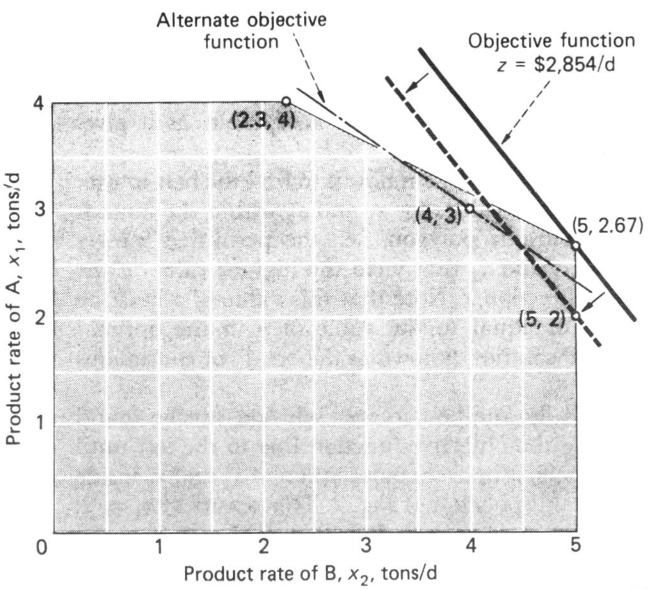

Constraint polygon for the example resolves the continuous and integer solutions graphically Fig. 2

or a solution with a value of z lower than in an integer solution already obtained. Each new branch involves adding a constraint to an earlier solution. The new, more-restrictive model cannot have a solution with a value of z greater than that in the solution for the model from which it branches.

The technique is not restricted to all-integer models as in our example. It works equally well with mixed-integer models, i.e., those containing both continuous and integer variables. Where a solution has noninteger values for more than one integer, the order in which the branches are followed is a matter of choice. More information on the mathematical aspects of MIP can be found in Zionts [1] and in Cooper and Steinberg [2].

The logical (0,1) variable

A useful integer variable is the logical (0,1) variable, i, which is only permitted to take the values of 0 or 1 by the constraint $i \leq 1$. The inclusion in a model of (0,1) integer variables (also known as decision variables) enables a variety of logical conditions to be satisfied in the model's solution.

Examples of the logical conditions that can be met by combinations of two such variables, i_1 and i_2, follow:

Condition	Comment*
$i_1 + i_2 \leq 1$	Neither or either, but not both.
$i_1 + i_2 = 1$	Either, but not neither or both.
$i_1 + i_2 \geq 1$	Either or both, but not neither.
$i_1 + i_2 = 2$	Both, but not neither or either.
$i_1 - i_2 \leq 0$	Neither, i_2 alone, or both (but not i_1 alone).
$i_1 - i_2 = 0$	Neither or both, but not either alone.

*Comments refer to the variables taking the value 1.

Examples for the application of these and other possibilities will be discussed shortly. A manual by IBM [8] helps users in formulating MIP problems.

Only integer or discrete values for variables

The earlier graphical example illustrates integer-problem variables, i.e., integer variables in the original problem as opposed to logic variables. These occur whenever a quantity is confined to integer values, e.g., the number of customers to be supplied or the number of tankers available for transportation.

For example, a model uses the total tanker capacity. This occurs in regularly spaced discrete values (not necessarily integer), and requires the addition of an integer, i, and a constraint row to the mathematical model for the problem:

$$x - Ai = 0$$

where x is a normal continuous LP variable, and A is a coefficient representing the size of the discrete jump (e.g., capacity of one tanker). Thus, as i takes values of 0, 1, 2, . . ., x must take values 0, A, $2A$, ...

If the discrete values of x are irregular, a different integer is needed, corresponding to each possible value. If the values of x are limited to 0, A_1 or A_2, we can write this as:

$$x - A_1 i_1 - A_2 i_2 = 0$$
$$i_1 + i_2 \leq 1$$

Yes/no situations

Yes/no situations involve a (0,1) variable where "0" means no and "1" means yes. An example is to allow the solution of the model to decide whether or not to

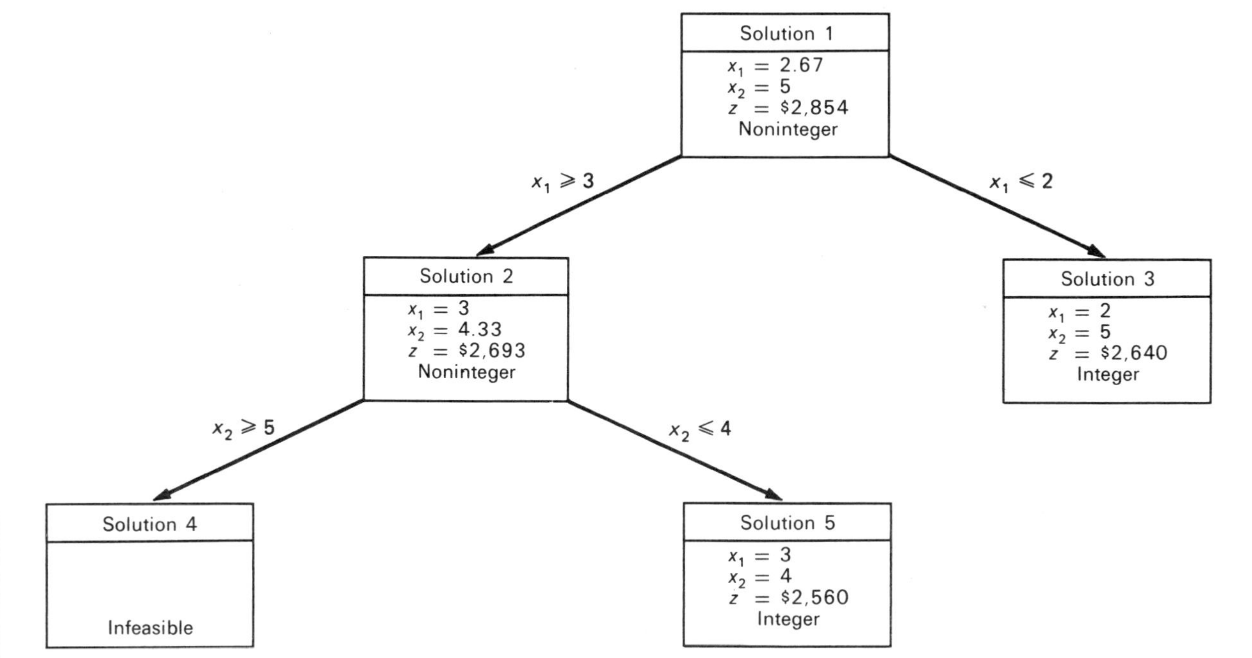

Branch-and-bound technique develops optimal integer solution in the form of a tree — Fig. 3

include a given process unit in a plant operating plan:

$$x - Ci = 0$$
$$i \leq 1$$

where x is throughput, and C is the unit's fixed capacity. If in the complete model's optimal solution $i = 0$, then $x = 0$ and the unit is not included. If $i = 1$, then $x = C$ and the unit is included at its fixed capacity. In addition, the objective coefficient of i can be a fixed cost, e.g., investment cost that is only incurred if the unit is included.

The "no/variable-yes" situation is similar. This could be when the solution of a model determines whether to include a process unit and, if so, chooses its optimum throughput, x, between specified maximum capacity, C, and minimum capacity, B:

$$x - Ci \leq 0$$
$$x - Bi \geq 0$$
$$i \leq 1$$

If $i = 0$, $x = 0$. If $i = 1$, $x \leq C$ and $x \geq B$.

The normal LP formulation for a process unit takes account of variable operating costs that are linear with throughput, and this signifies that the total operating cost reduces steadily toward zero as throughput is reduced. In many cases, however, a fixed operating cost, in addition to the variable cost, is incurred as long as the plant is not shut down. This can be taken care of by appropriate objective coefficients in the model, with the above constraints, i.e., maximize:

$$\cdots - Vx - Fi \cdots = z$$

where V and F are the fixed and variable costs. The units of V could be \$/ton and of F, \$/day. Thus, when the unit is shut down, $i = 0$ and $x = 0$, and no cost is incurred. But if throughput x is positive, i must equal 1, so that the full fixed cost and the variable cost are both incurred.

Either/or situations

Either/or situations arise where a selection is to be made by a model between mutually exclusive alternatives. Examples are where a pipeline can only be used to transfer one out of several possible products in a given period of time, or where alternative designs or capacities are being considered for a new process unit in order to select the best one.

Let us suppose that there are three alternative designs for a unit, each having different fixed and variable costs and different capacities. Using the earlier notation, we express the capacities constraints in terms of (0,1) integer variables:

$$x_1 - C_1 i_1 \leq 0$$
$$x_2 - C_2 i_2 \leq 0$$
$$x_3 - C_3 i_3 \leq 0$$

The constraint ensuring that only one design is selected is:

$$i_1 + i_2 + i_3 \leq 1$$

Thus, if $i_1 = 1$, the first unit can operate up to full capacity, and $i_2 = i_3 = 0$. Hence, the other units cannot operate at all, i.e., the first design is selected. The objective function includes the fixed and variable costs for each design:

$$-V_1x_1 - F_1i_1 - V_2x_2 - F_2i_2 - V_3x_3 - F_3i_3$$

Only the costs of the design selected are actually incurred.

This type of formulation can be extended so that a maximum of m alternatives can be selected out of n possible ones. For example, let us suppose that a multiproduct plant can make three products but that there are separate storage facilities for only two of them. Let the product rates be x_j, the maximum demand for each be D_j, and the total plant capacity be C. The required constraints are:

$$x_1 + x_2 + x_3 \leq C$$
$$x_j - D_j i_j \leq 0, j = 1, 2, 3$$
$$i_j \leq 1, j = 1, 2, 3$$
$$i_1 + i_2 + i_3 \leq 2$$

The last constraint ensures that not more than two products can be selected.

As well as the selections of alternative continuous variables, integers can also be used to select a constraint right-hand-side (RHS) appropriate to the desired variable. For example, this may occur when the maximum capacity of a process unit depends on the mode of operation, feedstock or product. Let the maximum capacity be C_1 on feed x_1, and C_2 on feed x_2, e.g. different crude oils in a refinery. If the feeds can be mixed, or if the model is considering average operations over a period of time, the capacity constraint is:

$$\frac{x_1}{C_1} + \frac{x_2}{C_2} \leq 1$$

On the other hand, if the model is only required to select a single feed, the formulation is:

$$x_1 - C_1 i_1 \leq 0$$
$$x_2 - C_2 i_2 \leq 0$$
$$i_1 + i_2 = 1$$

Thus, if x_1 is selected, the RHS becomes, effectively, C_1; and if x_2 is selected, it becomes C_2.

It is also possible to use integers to select alternative sets of constraints to be operative, depending on the particular situation. To illustrate this, suppose that a refinery can process either of two crude oils. The operating throughput-range of a downstream naphtha reformer depends on naphtha quality and, hence, on the crude oil that is run. The availabilities of crudes (or crude-unit capacities) designated by variables x_1 and x_2 are A_1 and A_2, respectively. The downstream naphtha to the reformer is designated x_3. If crude x_1 is run, the minimum and maximum permissible reformer throughputs are B_1 and C_1, respectively; for crude x_2, they are B_2 and C_2. The relevant part of the formulation is:

$$x_1 - A_1 i_1 \leq 0$$
$$x_2 - A_2 i_2 \leq 0$$
$$x_3 - B_1 i_1 - B_2 i_2 \geq 0$$
$$x_3 - C_1 i_1 - C_2 i_2 \leq 0$$
$$i_1 + i_2 = 1$$

If the first crude is run, $0 < (x_1/A_1) \leq 1$, and so $i_1 = 1$. Hence, $i_2 = 0$; and, therefore, $x_2 = 0$, i.e., the second crude is excluded. Since $i_1 = 1$ and $i_2 = 0$, $x_3 \geq B_1$ and $x_3 \leq C_1$, which are the correct reformer-throughput limitations. Similarly, if the second crude is run, $i_1 = 0$ and $i_2 = 1$, so that the first crude is excluded and the reformer throughput must now lie between B_2 and C_2. This method of formulation is simpler than that given in Ref. [8] for selecting the constraints of the problem.

Sequencing of variables

When a number of variables must be introduced in a specified order, the procedure termed sequencing of variables applies. In sequencing, the next variable is introduced (i.e., made positive) only when the one it follows has reached its maximum permitted value (its upper bound).

For example, consider an operating plan for a plant that makes a product to individual customer formulations, and exercises an order of priority so that the full order from one customer is met before starting on the next one. Further, suppose that the plant makes three formulations, x_1, x_2 and x_3, for three different customers; and the priority is to make all of x_1 before any of x_2, and all of x_2 before any of x_3. The corresponding part of the formulation is:

$$\begin{aligned} x_1 + x_2 + x_3 &\leq C \\ x_1 &\leq R_1 \\ x_1 - R_1 i_1 &\geq 0 \\ x_2 - R_2 i_1 &\leq 0 \\ x_2 - R_2 i_2 &\geq 0 \\ x_3 - R_3 i_2 &\leq 0 \end{aligned}$$

where C is the total plant capacity; and R_1, R_2 and R_3 are the sizes of the three orders. As long as $x_1 < R_1$, $i_1 = 0$; and so, $x_2 = 0$, $i_2 = 0$ and $x_3 = 0$. When $x_1 = R_1$ (i.e., the complete order), i_1 can become 1, enabling x_3 to increase to R_3. Note that the constraints $i_1 \leq 1$ and $i_2 \leq 1$ are not necessary because the existing constraints already impose these limits. In practice, however, they could be inserted to improve the efficiency of the computer program for the branch-and-bound algorithm.

Linear approximations

An important use of integers for sequencing variables is in the linear approximation of nonlinear cost curves. The incremental operating cost of a process unit may be nonlinear with throughput, as shown in Fig. 4. For example, a distillation column at low throughput may have a higher incremental operating cost than at its design throughput if it has to be operating at a correspondingly higher reflux ratio to achieve acceptable loadings.

The gradient of the curve in Fig. 4 corresponds to the objective coefficient of throughput x. The curve may be approximated by two straight lines as shown (or, by more lines, depending on the accuracy required). In the throughput range, 0 to A, the throughput variable is x_1, with an objective coefficient c_1; in the range A to B, it is x_2, with an objective coefficient c_2. The

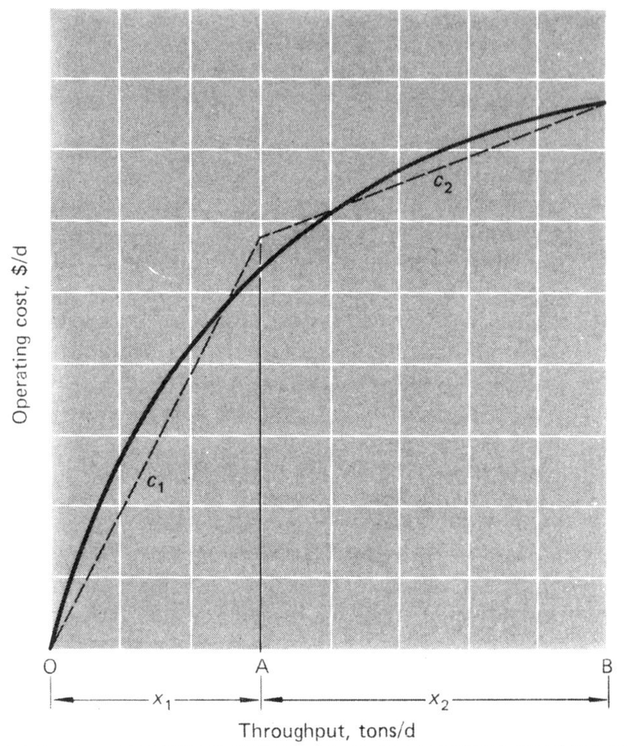

Linear approximation of nonlinear operating cost **Fig. 4**

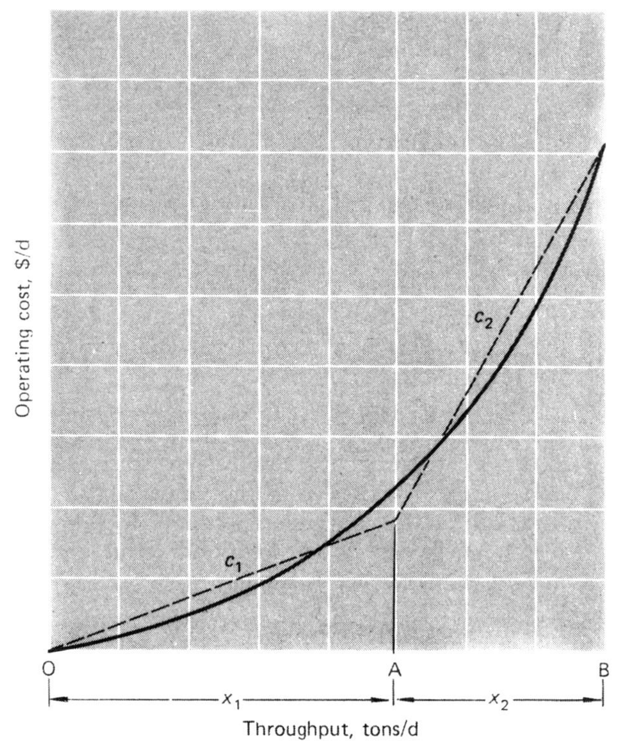

Cost curve does not require integer-based variables **Fig. 5**

formulation must ensure that x_2 remain zero as long as $x_1 < A$ (i.e., x_1 and x_2 are sequenced):

$$\begin{aligned} x_1 \quad\quad\quad\quad &\le A \\ x_1 - Ai \quad\quad &\ge 0 \\ x_2 - (B - A)i &\le 0 \\ i &\le 1 \end{aligned}$$

For the type of nonlinear cost curve shown in Fig. 5, the variables x_1 and x_2 sequence themselves naturally without the aid of an integer variable. Since $c_1 < c_2$, a permissible increase in x_1 will always cost less, and therefore be more optimal, than increasing x_2. Hence, x_2 will only be nonzero when x_1 is at its maximum. This type of curve may be shown for a pipestill where, as throughput is pushed up, the fuel-firing rate and the pumping power both rise more steeply. The integer type of formulation should only be omitted when it obviously is not needed.

Linearization of other types of nonlinear relationships frequently involves the use of "separable programming." This is a technique that in a form guaranteed to find the global (and not merely a local) optimum uses an integer type of formulation. Cooper and Steinberg [*2*] give details of this technique. It is available with most computer mathematical-programming packages that include mixed-integer programming.

Computer procedures for MIP

Most LP computer packages have been extended to provide a means for solving MIP problems. Detailed instructions are included with each package. In general, variables that are required to take integer values are "flagged" as such in the input. A constraint found in many MIP formulations is:

$$i_1 + i_2 + \cdots i_n = 1$$

This constraint implies that the i's are mutually exclusive. With many computer packages, this constraint does not have to be written explicitly but can be implied by flagging the i's as members of a special ordered set (SOS).

The primal or normal agendum is followed to obtain the real, continuous optimal solution in which the integer requirements are not enforced. The integer agendum is then called, and this explores alternative solutions with the aim of finding the best solution in which the flagged variables take integer values by means of the branch-and-bound procedure.

With some computer packages, the user can specify the search method for the best integer solution. Often, it is possible to cut the search short by specifying an acceptable reduction from the value of the objective function for the continuous solution. In this case, the search for the best integer solution is terminated as soon as an acceptable one is found.

Comments on using MIP

The complexity of the search for the best integer solution increases rapidly as the number of possible integer combinations increases. This can lead to an excessive usage of computer time. The number of variables flagged as integers should, therefore, be kept in check. Integers should only be used when they are really justified, bearing in mind acceptable limits of solution accuracy. Thus, the manual rounding of a noninteger solution may only incur small errors, e.g., 100.3 could well be rounded to 100 with little loss of precision. On the other hand, the rounding of 2.3 to 2 may not be acceptable because other variables may be significantly affected; so a search should be made for a proper integer solution.

The need for the best integer solution should also be considered. If an integer solution within, say, 5% of the real solution is acceptable, there is no point in continuing the search. The integer-solution search can also be assisted by specifying the bounds on integer variables to be as restrictive as possible—thereby reducing the number of integer alternatives to be tested.

In a normal, continuous LP problem, the computer time used is not very sensitive to the addition of variables, but it increases rapidly as the number of constraints is increased. The opposite tends to be the case for MIP problems. The addition of variables, flagged as integers, rapidly increases computing time because it extends the number of integer combinations to be tested. Addition of extra constraints involving integer variables, on the other hand, decreases the number of their possible values. The corresponding decrease in the integer-search phase usually more than outweighs any corresponding lengthening of the real continuous phase in obtaining the solution, and so tends to reduce computer time.

Many users of linear programming have facilities for MIP in their computer packages that they do not use, mainly through lack of appreciation of MIP's capabilities. It is hoped that this article may provide the stimulus for assisting you to apply MIP to LP.

References

1. Zionts, S., "Linear and Integer Programming," Prentice-Hall, Englewood Cliffs, N.J., 1974.
2. Cooper, L. and Steinberg, D., "Methods and Applications of Linear Programming," Saunders, Philadelphia, 1974.
3. Symonds, G. H., "Linear Programming: The Solution of Refinery Problems," Esso Publications, Esso Standard Oil Co., New York, 1955.
4. Smith, D., "Linear Programming Models in Business," Polytech Publishers, Stockport, U.K., 1973.
5. Allen, D. H., Linear Programming Models for Plant Operations Planning, *Brit. Chem. Eng.*, Aug. 1971, p. 685.
6. Driebeek, N. J., "Applied Linear Programming," Addison-Wesley, Reading, Mass., 1969.
7. Danø, S., "Linear Programming in Industry," Springer-Verlag, Vienna, 1974.
8. "An Introduction to Modelling Using Mixed Integer Programming," IBM Nederland N.V., Amsterdam, July 1972.

The author

Derek H. Allen is Reader in Chemical Engineering at the University of Nottingham, University Park, Nottingham NG7 2RD, England. His industrial experience includes research and development, process engineering, investment, and production planning in the gas and petroleum industries. At Nottingham, he carries out research and organizes industrial courses in this field. He has a B.Sc. and a Ph.D in chemical engineering from Birmingham University, and is a Fellow of the British Institution of Chemical Engineers.

Simple charts clearly reveal the effects of changes in either production costs or selling prices on the breakeven point, in this new technique for . . .

Breakeven Analysis

RAY STROUP, Jr., Cresap, McCormick and Paget Inc.

The traditional breakeven chart graphically portrays the short-term relationships between costs, sales and profits. Breakeven analysis is useful in:

1. Predicting costs and profits at different levels of operation.
2. Determining the impact of changes in costs and prices on the breakeven point of a firm or product.

The new technique offers advantages over traditional methods in the second of the above areas, and is also helpful in analyzing the economics of a group of similar business operations, or the economics of a group of products. Examples will illustrate these advantages.

Developing the Method

As detailed in the literature, the traditional breakeven chart is developed by first determining the sales and costs that can be attributed to a particular product. Costs are classified as fixed or variable.

The starting point for the new technique, as well as the traditional method, is the following equation:

$$\text{Sales} = \text{Variable cost} + \text{Fixed cost} + \text{Profit} \quad (1)$$

The breakeven point is defined as the sales volume where profit equals zero, so Eq. (1) becomes:

$$\text{Sales} = \text{Variable cost} + \text{Fixed cost} \quad (2)$$

Or, in other terms, sales are equal to:

$$\begin{pmatrix}\text{Price}\\ \text{per}\\ \text{unit}\end{pmatrix}\begin{pmatrix}\text{Quan-}\\ \text{tity}\\ \text{sold}\end{pmatrix} = \begin{pmatrix}\text{Variable}\\ \text{cost per}\\ \text{unit}\end{pmatrix}\begin{pmatrix}\text{Quan-}\\ \text{tity}\\ \text{sold}\end{pmatrix} + \begin{pmatrix}\text{Fixed}\\ \text{cost}\end{pmatrix} \quad (3)$$

Multiplying and dividing variable cost by sales:

$$\text{Sales} = \left(\frac{\text{Var. cost per unit}}{\text{Price per unit}}\right)\text{Sales} + \text{Fixed cost} \quad (4)$$

$$\text{Sales}\left(1 - \frac{\text{Var. cost per unit}}{\text{Price per unit}}\right) = \text{Fixed cost} \quad (5)$$

Finally, taking the log of each side of Eq. (5) and rearranging, yields an equation that can be represented by a series of straight lines on a log-log chart; i.e.: log Sales =

$$-\log\left(1 - \frac{\text{Var. cost per unit}}{\text{Price per unit}}\right) + \log(\text{Fixed cost}) \quad (6)$$

The above is shown in Fig. 1. To find the breakeven point, you must know the ratio of variable-cost-per-unit to the price, and you must know the fixed cost. Then, all you do is locate the ratio on the abscissa (R_1), next go up to the proper fixed-cost line (say, to F_3), and left to the ordinate to read the breakeven point (B). The change in breakeven point caused by any change in the ratio of fixed cost and/or variable cost to price can quickly be determined by consulting the chart.

Drawing the Fixed-Cost Lines

Two points are needed to construct the fixed-cost lines; these can be determined from Eq. (5):

$$\text{Breakeven sales}\left(1 - \frac{\text{Var. cost per unit}}{\text{Price per unit}}\right) = \text{Fixed cost}$$

When variable cost per unit = 0, then breakeven sales = fixed cost.

Therefore, in Fig. 1, the ordinates of the fixed-cost lines (i.e., at Point 1) are:

F_2 line: $y = 1.0$ where $x = 1.0$
F_3 line: $y = 2.0$ where $x = 1.0$
F_1 line: $y = 0.5$ where $x = 1.0$

When variable cost per unit equals one-half the price per unit, the breakeven sales are equal to twice the fixed costs. The ordinates of fixed-cost lines (i.e., at Point 2) are:

F_2 line: $y = 2.0$ where $x = 0.5$
F_3 line: $y = 4.0$ where $x = 0.5$
F_1 line: $y = 1.0$ where $x = 0.5$

Multicycle log-log paper can be used if a wider range of values is desired than that shown. The only limitation is that the breakeven point is undefined when the x axis is greater than 1.0 or less than zero.

Let us now look at two examples of how this new technique can be used. Here is the first.

Company A, which has been packaging an industrial oil for several years with a manual drum-filling process, has decided to automate its filling line at a substantial saving in direct labor cost. Because of this saving, the firm also expects to cut the price of its product by 50¢/gal. Using discounted-cash-flow tech-

Originally published January 1972.

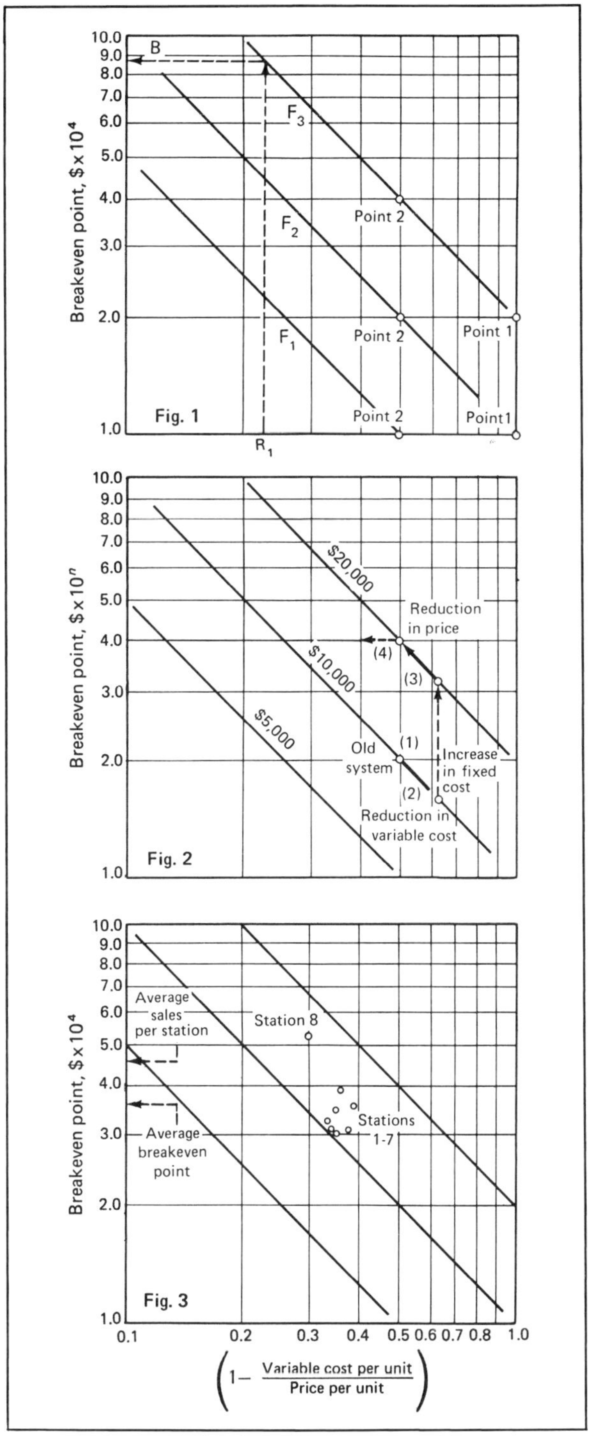

BREAKEVEN CHARTS for generalized situation (Fig. 1) and for two illustrative examples (Fig. 2 and 3) use diagonal lines to represent different fixed costs.

niques and projected volumes, this looks like a good investment. But because demand for this oil depends heavily on the general level of business activity, further analysis is necessary. Here are the cost data:

	Fixed Cost	Variable Cost	Price
Old system	$100,000	$1.00/gal.	$2.00/gal.
New system	$200,000	$0.75/gal.	$1.50/gal.

The breakeven chart, Fig. 2, reveals the impact on the product's breakeven point caused by each change in cost or price, and the total effect. Using this type of graph, it is very easy to show alternate strategies and to answer the "What if . . ." questions that decision-makers always ask.

Note that the breakeven point for this product will be doubled by the proposed change, thereby increasing the firm's risk of possible loss if sales volumes do not increase as anticipated.

Example 2: Insight Into Cost Structures

Although this example concerns the breakeven analysis of a group of gasoline service stations, the method can be applied to other revenue-producing operations, such as groups of chemical plants within a division.

Let us say that a certain marketing district of an oil company contains eight service stations, and has been profitable for several years. But seven of the station managers feel that the district could be even more successful if the eighth station would cut out some frills and emphasize profits rather than volume. The district manager is not sure whether he should go along with this thinking, because the station is profitable (although perhaps not as profitable as it should be). He suspects that some of the other station managers may resent this station merely for being "different." To see whether there really is a problem, he decides to study the eight stations individually; he graphs each of them on the modified breakeven chart (Fig. 3) and tries to find cost-structure patterns.

The chart shows a rather tight pattern for seven of the stations. The eighth station's fixed and variable cost-to-price ratios are high enough to raise its breakeven point considerably above that of any of the others.

Of course, the chart does not tell him what action to take. If the station has a significantly higher revenue potential than the other seven, perhaps effort should center on raising revenues further, and on making sure they stay well above the breakeven point. If the revenue potential is judged no better than for the others, perhaps some "defrilling" or other cost-cutting is in order, even if this means some loss of volume. What the chart does do is to single out a cost pattern that differs from the group, and to provide an insight into the effect of various courses of action on the breakeven point.

Other applications are also possible—e.g.:

1. If market influences dictate an across-the-board price cut in the district, which stations will be pushed into the red first?

2. If a ninth station is built, and if its fixed costs are higher than for the others (perhaps because real estate or construction costs have climbed), what ratio of variable expense to price is required to keep the breakeven point below a certain level? ■

Meet the Author

Ray Stroup, Jr., is Senior Associate with the management consultant firm of Cresap, McCormick and Paget Inc. (100 W. Monroe St., Chicago). Previously, he was affiliated with Sun Oil Co. as an economic analyst. Mr. Stroup has a B.S. degree in chemical engineering from Kansas University, and an M.B.A. in finance and management from the University of Texas.

Pitfalls in evaluating R&D

How do you decide whether to continue a research and development project? Indeed, should the project have been started at all? The answer is evaluation, but it can be tricky if you are not aware of the possible traps.

Anthony V. Perrella, *Mobil Research and Development Corp.*

☐ One third of measured growth in the U.S. Gross National Product has come from technological progress, progress based on R&D [*1*]. During recent years, two ominous trends have appeared to threaten continued growth [*2*]:

■ Research expenditures are declining. From 1967 to 1975, nongovernment R&D spending (in constant dollars) rose 1.8%/yr but was more than offset by a 3%/yr decline in such government spending.

■ More research effort is going into nonproductive areas. Nearly 25% of corporate research (50% in some industries) is now aimed at complying with government regulations rather than developing new processes and products.

As both trends are likely to continue, evaluation of *productive* research will become increasingly important. More companies will evaluate their R&D projects to determine which are justified, where efforts should be increased and where decreased. Most procedures for evaluating research compare the value or benefits obtained from a project with its cost; some simply involve a check list of "favorable" or "unfavorable" factors. Regardless of the method used or how detailed the analysis is, these pitfalls should be avoided:

■ Failing to quantify the potential value of a research development.

■ Failing to assess alternatives, or using the wrong alternative.

■ Omitting technical and commercial risks.

■ Overestimating the potential value.

■ Underestimating research costs.

Failing to quantify potential value

Some analysts consider only ambiguous values for research, such as compatibility with company objectives and magnitude of the potential market. Such factors may be of some use in ranking a number of research projects but do not answer the question, "Is the project justified to begin with?" Obviously, compatibility with company objectives is a consideration, but it is not a measure of research value. (It should be considered in estimating the probability that the development will be exploited.)

A research development that enables a company to enter a \$50-million/yr market is not necessarily better than one that will enable entry into a \$10-million/yr market. The latter may mean capturing 20% of a high-profit-margin, rapidly-growing market because of a superior product, and returning \$300,000/yr in profit to the company. The first may mean capturing 5% of an established commodity market, with only \$100,000/yr potential profit.

Estimating quantitative economic value for potentially new technology is necessary to properly rank research projects and determine whether they are justified. Even though considerable money has already been spent on a development, only a quantified economic analysis will determine whether further expenditures should be made.

Failing to assess alternatives

In quantifying the value of a potential research development, common practice is to assume that the alternative to using the new technology is the status quo; i.e., to do nothing. Most often, however, the firm has several alternatives—such as purchasing comparable outside technology, improving processes or products using proven technology, or even shutting down an existing facility.

For example, a potential new technology is estimated to improve an existing process by reducing costs (or increasing revenues) by \$500,000/yr, and this figure is used as the quantitative value of doing the research. However, closer examination shows the firm has an alternative, which is to purchase a competitor's technology that will accomplish the same thing at the cost of a royalty fee of \$200,000/yr. Thus, the real value of undertaking research is not \$500,000/yr, but the royalty saving of \$200,000/yr.

Even if competitive technology is not currently available, it frequently should be assumed that it will become so in the near future, especially if the existing process is in wide use. A realistic approach for the example above would be to assume the new technology will be worth \$500,000/yr for several years, and less (based on an estimated licensing fee) thereafter.

Sometimes the wrong alternative is considered. For example, evaluation of a potential development shows it will produce a new product at a cost of 22¢/lb compared

Originally published August 1977.

with an available process that costs 25¢/lb. A potential advantage of 3¢/lb is estimated and used to justify research. However, closer study reveals the available process is unattractive—i.e., if the company entered this business using the available process, it would lose 2¢/lb. Therefore, the value of the new technology is only 1¢/lb, not 3¢; in other words, it will allow the company to enter a business that will return 1¢/lb profit. The proper alternative in this case was to do nothing.

Omitting technical and commercial risks

Every research project has some risk—some chance of failure. If success were guaranteed, research would not be needed! Assigning full potential value to an R&D project is incorrect. A probable or expected value should be calculated, considering these risks:

- Probability of technical success. If research is undertaken, what is the probability that it will be successful to the degree estimated (within the time and budget allowed and the capability of the staff)?
- Probability of exploitation success. If technically successful, what is the probability the technology will be used or exploited by the firm? The company may decide not to build a plant that would use the new technology, for reasons unrelated to the success of research (change in corporate policy, declining prices, product banned by the FDA, limitation in capital availability, etc.).
- Probability of financial success. If the technology is both technically successful and exploited, what is the probability it will achieve the financial success estimated? Again, for reasons unrelated to success in developing the technology, and even after it has been exploited for several years, the estimated financial value may not be fully realized (prices decline, costs increase unexpectedly, the product is banned by the FDA, technology becomes obsolete, plant burns down, workers go on strike, etc.).

The potential value of success should be multiplied by each probability to arrive at a probable or expected value, which when related to the research costs gives a measure of the project's attractiveness, both on its own and in relation to other projects. Estimating probabilities is necessarily subjective, since they apply to the uncertain future. They should be based on the best judgment of experts in the related area: Probability of technical success should be estimated by researchers; probability of exploitation should be estimated by manufacturing or marketing people; and probability of financial success should be estimated by financial experts. Many sophisticated methods are available for evaluating risk. The degree of sophistication employed should be decided by the analyst.

Overestimating the potential value

In addition to identifying the wrong alternatives and omitting consideration of risks, other errors can be made in quantifying the potential value of an R&D project, such as overestimating the profitability of the venture itself. Product sales or price may be overestimated; capital or manufacturing costs may be underestimated. Several pitfalls may be encountered in estimating cash flow of any venture and are covered in the literature [*3–5*]. The assumption that a higher quality product will command a higher price and profit margin is particularly hazardous in a competitive market. The production cost of a new process or product using new technology is most easily underestimated, because so many factors can be overlooked. Care should be taken to compare the estimated costs with those of comparable known ventures whose costs have been established in the real world.

Underestimating research costs

Human nature impels all of us to tell management what it wants to hear. We are more likely to use the minimum cost than the maximum or most likely cost [*6*]. Furthermore, we are prone to underestimate the time and effort required for any task. The analyst should make every effort to include all costs and to use an estimate somewhere between the minimum and maximum likely.

Other pitfalls

Several other, more obvious, pitfalls could invalidate a worthwhile evaluation:

- Having researchers evaluate their own projects gives them greater insight to the business aspects of their technology but adds personal bias and inconsistency. Almost any project can be made to seem fantastically profitable if it uses a $3/h labor rate, low-cost fuel and zero maintenance. Projects should be evaluated by a separate group staffed with business-oriented technologists well versed in "real world" costs.
- Waiting to evaluate a project until complete data are available may result in a rude awakening—the realization that the project never should have been started. Evaluations should be made as early as the initial planning stage.
- Failing to reevaluate continuously creates the risk of missing changes occurring in commercial application or technical knowledge that could void an originally attractive project. This is important because R&D costs escalate rapidly as a project reaches the applied research stage and, particularly, the development stage.

Competent evaluation of research, avoiding the pitfalls discussed above, will increase the efficiency and productivity of R&D spending, which can boost U.S. Gross National Product and the profitability of a company. Also, it can avoid the embarrassment to the research manager or individual researcher of finding out the company's limited R&D funds have been misspent.

References

1. The Silent Crisis in R&D, *Bus. Wk.*, Mar. 8, 1976, p. 90.
2. Where Private Industry Puts Its Research Money, *Bus. Wk.*, June 28, 1976, p. 62.
3. Jelen, F. C., Pitfalls in Profitability Analysis, *Hydrocarbon Process.*, Jan. 1976, p. 111.
4. Buchanan, R. L., How to Avoid Common Pitfalls When Making Cost Estimates, *Chem. Eng.*, Oct. 25, 1965, p. 159.
5. Walton, P. R., Sources of Error in Operating Cost Estimates, presented at Twelfth Annual Meeting of American Assn. of Cost Engineers, Houston, Tex., June 15, 1968.
6. Souder, W. E., Pitfalls in R&D Budgeting, *Management Rev.*, Oct. 1970, p. 55.

The author

Anthony V. Perrella is an engineering associate in the Research Planning Div. of Mobile Research and Development Corp., Paulsboro, NJ 08066, where his work involves planning and economic evaluation of current and new research projects, including evaluations of individual projects and entire programs. He has a B.S. in chemical engineering from Drexel University

Analyzing Potential Research Projects

Many techniques have been proposed for evaluating the economic potential of new processes, but none of them are generally accepted. Three widely used approaches are critically reviewed here.

ARTHUR H. BOBIS and ANTHONY ATKINSON*, American Cyanamid Co.

Research managers need quantitative techniques to help them select research projects. Considerable effort has been devoted to this problem over the past ten years, and to the related problem of allocating funds to the selected projects.

Interest in the application of such strategies has coincided with a marked change in the philosophy of research.

Up to World War II, and somewhat beyond that, research was left in the hands of the scientists who could produce unique and interesting products whose utility, although uncertain, nevertheless required investigation. It was not up to the researcher to find a use for his invention, but merely to produce the noteworthy artifact (noteworthy as to scientific achievment, not necessarily with regard to anticipated profits).

During the war, when specific needs were recognized, efficiencies in research were achieved by directing efforts toward specific objectives. Thus, within a short period of time, processes for producing high-octane gasoline and synthetic rubber were developed; also, broad-spectrum antibiotics were discovered and produced with rather startling speed. Perhaps even more dramatic was the effort on the part of the British to make radar a useful defense against German aircraft. Of course, the invention, development and construction of the atomic bomb was an achievement probably unparalleled for efficiency and speed.

Not long after the war this directed-research philosophy was adopted in most industrial laboratories.

Research had always been pursued on three different levels: basic research, in which the objective was the pursuit of scientific information; applied research, where the objective was to invent products to fill needs; and process development, which was geared to the reduction-to-practice of the applied-research successes. The emphasis at this time swung dramatically toward applied research, and the products were related to some known commercial outlet.

With the orientation of research toward application, it became possible to associate each research project with a potential profit. The anticipated dollar value was used as a measure of the worth of the research program.

Industrial guidelines more or less established the total amount of money available for research by pegging it to a percentage of the sales dollar. However, the distribution of those funds among the many supposedly attractive products was subjected to intensive study. Executives tried to rank projects in order of priority, and parameters were sought that would serve as a proper measure.

Just as the potential profit was used as a measure of the worth of the projects, estimates of the cost of carrying out the project were used as a measure of difficulty. The ratio of these numbers was suggested as an appropriate ranking parameter.

These ranking techniques were just the forerunner of things to come. As competition for sales between companies became more pronounced and the cost of doing research increased, better project selection was recognized as perhaps the most direct way of improving research efficiency, and more-and-more sophisticated methods were proposed as aids in the selection process.

The proliferation of articles describing quantitative research management techniques is well illustrated by two bibliographies, one by Albert H. Rubenstein[1] and the other by N. Baker and W. Pound.[2] In their articles they review many of the proposed techniques, and describe the extent to which they have been used. The tremendous energy devoted to the development of these techniques can only be explained by

* Dr. Atkinson is now a professor at Imperial College, London.

Originally published February 1970.

need. Yet, when reading the bibliographies, one finds an obvious dissatisfaction with the methods' utility.

This article will review three of the most frequently used approaches, and point out possible reasons for their failing to achieve the desired effect. A second article, beginning on p. 248, will treat the part of the methodology that is most new—i.e., dynamic modeling techniques—and suggest modifications that could increase the general utility of these techniques.

The Problem Is Not Simple

Project selection procedures are very complicated. Of all the contributing factors, three stand out as being most confounding. First, the man who does the research is not in the best position to assess the value of his effort, nor is he totally competent to judge the direction this research must eventually take. Second, he pursues his projects in a competitive environment, and the pace at which he is able to achieve is often as important as the achievement itself. Finally, although an objective may be defined in anticipation of need, the measure of the need can only be vaguely estimated.

Because the research man is not in a position to determine the worth of his effort, he must work with those responsible for the distribution, sale and pricing of the results of his research. The research man is apt to focus his attention on the personnel and equipment requirements and concern himself only with patent literature and technical publications. The competition that he fears is technical competition. He looks through literature and observes the progress that has been made by other laboratories.

However, with the increased emphasis on application, concern must also be given to marketing the product. Consequently, the competition that must be assessed is not just technical, but commercial as well. The research man is usually not competent to identify all of the competition or to estimate its potential strength.

However, the commercial man can assess the ultimate market penetration of the invention and determine what the eventual competition might be like. It is only from his experience that he is able to determine whether an organization is constituted to handle the sale of a new product. The commercial man must also work with potential customers in establishing a product definition and then work with the researcher to help him build the product that is wanted.

Research, by its nature, is dynamic—its very purpose is to create change. The projections that are made today are being made for a product that will not be realized for several years. To interpret the market's needs, on the day the research achievement becomes available, is really the problem at hand. The estimate of the value of the invention is done with limited information as to which competitors are involved and at what rate they are progressing. A laboratory success can end in commercial failure if the result is achieved too late. Hence, in a rigorous evaluation, attention must be given to the timing of the research effort.

Another problem stems from the information on which decisions have to be based, most of which is in the form of estimates of the properties of a product that does not yet exist. Just as one was hard pressed to guess at the eventual total passenger miles that the airlines must prepare for, when the only available information concerned train travel, so it is hard to estimate the market for a hypothetical product. The estimates of research cost are also subject to this kind of difficulty.

Obviously, under these circumstances, the two measures—the worth of the project and its ultimate cost—can be described only vaguely. Specific point estimates are, of course, a valued first step, but as will be discussed later, failure to describe the uncertainty of the estimates can lead to serious misdirection of effort. In analyzing research, another consideration is whether the effort will succeed at all.

Why Use a Mathematical Treatment?

Faced with these problems, one must always consider whether mathematical treatment of the data is worthwhile. If the errors in the estimation are gross, as is often the case, what value does a rigorous mathematical treatment offer? Proponents of a quantitative assessment (and we number ourselves among them) will argue that although the accuracy of the analysis is subject to very serious interpretation, the interrelationships that can be considered through a rigorous treatment are worth establishing and these can only be recognized by some quantitative treatment. Such techniques enable the investigator to consider the consequence of alternative decisions and, although the results of such analysis are rarely definitive, the very ability to easily assess the results serves as a stimulant to the investigator's imagination.

The softness of the data can be probed by repeating the analysis at different levels of the estimation. The sensitivity of any result to variations in the estimates helps define where greater attention is required.

Decisions must be made on the basis of available data, and the weakness of the estimates is not sufficient excuse for failing to bring the best judgment to bear on the problem. For, however poor they may be, the estimates do represent our best knowledge of the situation.

Assigning a Rank

The simplest method of project evaluation is, as we have mentioned, to rank the projects in order of desirability according to some criterion. Of the many published methods we will consider one, that given by Pacifico,[3] in which the project index is defined as:

$$\frac{R_c R_t \; (P\text{-}C) \; V \; L}{T}$$

where: R_c=chance of commercial success (on a scale 0.1 to 1.0); R_t=chance of technical success (on a scale 0.1 to 1.0); P=price of product, \$/lb.; C= manufacturing and selling cost, \$/lb.; V=sales volume, lb./yr.; L=commercial life of product, yr.; T=total costs, including research, engineering, plant, market development and working capital, \$.

The main feature of this model is that the numerator contains a measure of the expected profit, while the denominator contains a measure of the estimated cost. The numerator is premultiplied by probabilities of success in the research, development and commercial stages.

In research, as was just mentioned, the chance of failure must temper any estimation of the project's vorth. Let us now deal with the problem of evaluating these probabilities.

One approach has been to ask the research man who is most knowledgeable about the project what he regards as the probability of achievement, using a scale of 0 (or 0.1) to 1. In some cases, taking the average of the estimates of several knowledgeable individuals is attempted. In other cases, the estimates are weighed on a scale that reflects the experience of the participating individuals. Another approach is to reduce the problem into categories that (collectively) describe the chance of success, so that each aspect can be carefully scrutinized.

Some criteria that have been suggested include those of Dean,[4] which focuses on such matters as patentability, search for knowledge, compatibility with other research programs, and uniqueness of approach, among others. This breakdown of the project into descriptive categories seems to be favored, in that personal bias is reduced by imposing descriptions that should be more generally obvious.

The purpose is, of course, to generate a score, and this too deserves some discussion. One approach is merely to add all the ranks and divide by the number of categories. This is the preferred approach when numerous categories are defined, but it has the weakness that categories may not be equally important. Weighting is often used to circumvent this difficulty. The larger the list of categories, supposedly the easier it is to evaluate the individual entries. Nevertheless, a point is reached where the addition of categories serves to create more problems than it solves.

Table I illustrates types of categories and the possible means of scoring. Eight categories are defined, and five possible ratings are shown. The chart is filled out by knowledgeable researchers and a score generated by averaging the values for all of the categories. A column is also included for weighting the answers to the different questions. If the weights are equal, then the answers to all of the questions bear the same contribution to the final score. However, if it is believed that the technological base is three times as important as raw materials position, then the categories would be assigned weights in that proportion.

Some of the difficulties of such a method are obvious even with this small set of questions. For example, Question 7 may be redundant relating to Questions 1 and 2. How important is Question 4, if money is available to improve the situation? What about Question 8? If this is truly important, can't this be handled by direct hiring or outside contracting without reducing the chance of success?

If the categories are few, one might determine the final score by cross-multiplication of the individual category scores, usually given from 0 to 1. A danger here is that the addition of each category serves to reduce the overall ranking. The problem can be ameliorated by taking the geometric mean of the ranks. This approach has the very distinct advantage that a single entry will substantially reduce the overall ranking, conforming nicely to the theory that the weakest link determines the strength of the chain.

Despite the variety of approaches and the weaknesses inherent in each, it is essential that any quantitative statement regarding the anticipated worth of a research project carry with it an estimation of the probability of success.

The ranking is recognized to serve as no more than a guide to the value of the research projects. The ranks pay no attention to timing, and do not differentiate between low and high risk projects, in that the compounded indexes already include a cross-multiplication of some return ratio by the probability.

Categories Suggested for Scoring—Table I

Category	Weight	Outstanding 0.9-1.0	Very Good 0.7-0.8	Good 0.5-0.6	Fair 0.3-0.4	Poor 0.1-0.2
1. Technical knowhow						
2. Experience of personnel						
3. Laboratory facilities						
4. Pilot-plant facilities						
5. Patent situation						
6. Raw material position						
7. Relationship to other projects						
8. Analytical support						

Treating the Problem as a Capital Investment

Another approach to research evaluation has been to treat it as if it were a capital investment.[5] The potential profit is estimated from anticipated sales price, volume and costs, and the investment, which includes plant development and research expenses, is approximated. With such estimates, a return on investment

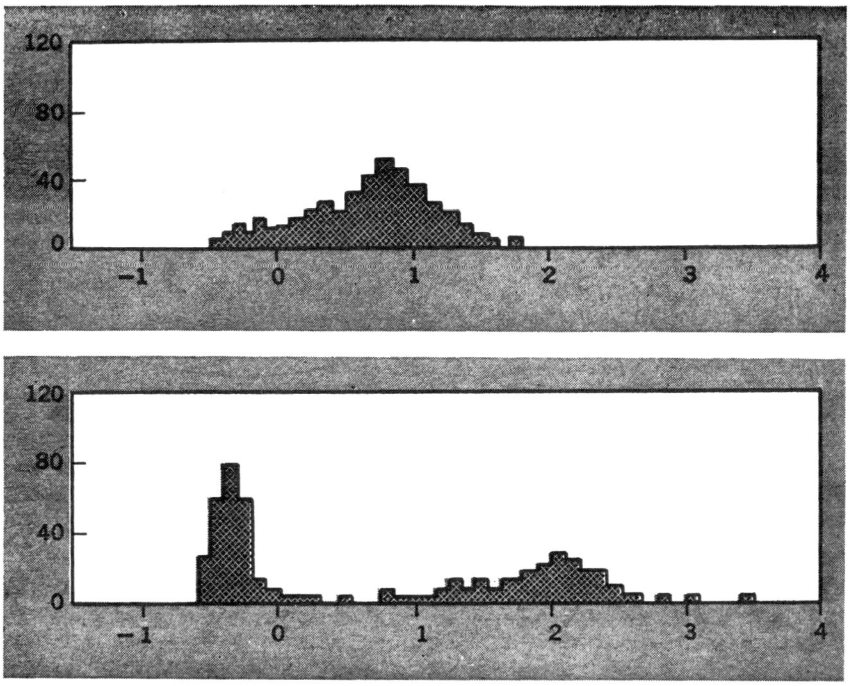

SIMULATIONS of research and capital investment decisions made on two different bases result in the two histograms shown here—Fig. 1

or payout time is then calculated, and the projects ranked on such basis.

Unfortunately, choosing an investment in research differs appreciably from that of choosing a new capital investment. Three distinct differences exist and are of such consequences that the traditional capital investment approach, if applied, sheds only modest light on the problem.

In brief, the two problems differ in the following ways:

A capital investment is a one-time expenditure which, once the commitment is made, will most assuredly be accomplished, whereas the research investment is stretched out over time, and the decision to abandon or continue must be constantly reassessed.

A capital investment is made after a technical success has been generated and the uncertainty is considerably reduced. The chance of failure is not important here, but rather the extent to which the adventure will be successful at the capital investment stage. The product has been defined and sometimes even test market data are available. Although such information is never definitive, the decision is made at a time when—compared to the time at which research is funded—a great deal is known.

Finally, the size of the investments differ markedly. The research commitment is a relatively small drain on the company's resources. Misspending will not be totally damaging, but rather reflects itself in opportunities that are lost. One can afford to take chances with research money. On the other hand, the magnitude of capital expenditures is often such that they will have a predominant effect on the company's earnings. The chances that one is willing to take under such circumstances are quite different.

Quigley and Hess[6] describe how the uncertainty in the estimates can be treated. They deal with an investment problem and show a distribution of outcomes that would result from the possible combinations of all sensible variations in the sales volume, selling price and cost. They set up distributions about each of these estimates, and then sample from them. The results are combined and the return on investment (R.O.I.) is calculated for each event according to:

$$R = \frac{V(S-C)}{I}$$

where: V=sales volume, lb./yr.; S=selling price, \$/lb.; C=manufacturing and selling cost, \$/lb.; I=investment, \$; R=return on investment, %.

This procedure is repeated a sufficient number of times until a descriptive definition of all possible events is captured and reported as either a histogram or a frequency distribution.

Hertz[7] discusses this approach as an aid in capital investment decisions. He effectively points out the danger of using point estimates for the evaluation, by giving illustrative cases where, although the most likely outcome is favorable, the probability of an unfavorable outcome is high.

Neither of these articles discusses the case of total failure, where the investment is made and then completely lost when nothing results. This possibility must be borne in mind in evaluating the research investment.

Atkinson and Bobis[8] indicate how this can be included in the analysis. Prior to sampling from the distribution of possible occurrences, sampling is conducted to determine whether the effort resulted in a

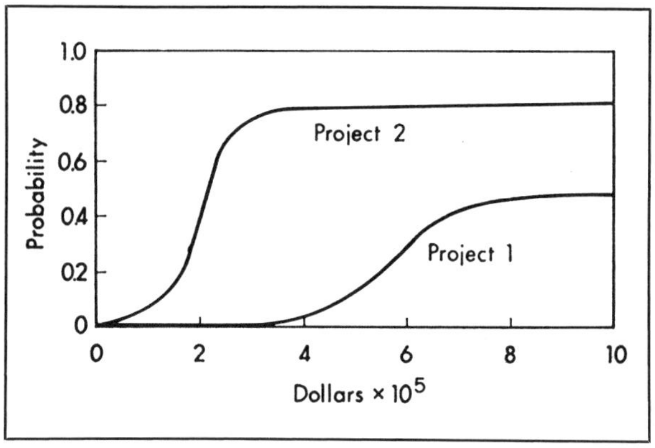

RESEARCH success vs. funding for two projects—Fig. 2

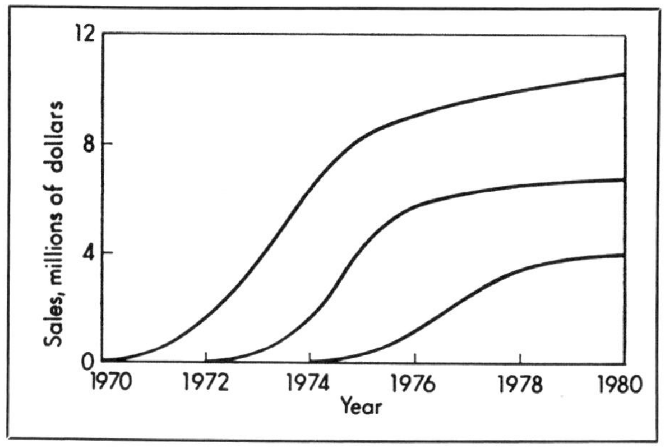

SALES growth as a function of year of entry—Fig. 3

success or a failure. The two different types of histograms that result are shown in Fig. 1. These histograms serve to illustrate the differences in the nature of the two investment decisions. For the sake of the example, 500 simulations are shown for each case.

The topmost distribution is one that could result from simulating the capital investment decision by the methods proposed by Hertz. The example shows a fairly smooth distribution, the mean of which is centered to the right of a zero return. A considerable portion of the distribution is to the left of zero, indicating an appreciable probability of loss.

Loss would result from one or all of the following causes: insufficient sales volume, lower-than-desired selling price, or higher-than-desired costs. The distribution, in addition to being relatively smooth, is also fairly concise, indicating that the individual distributions of sales volume, selling price and costs were themselves fairly well known.

The lower distribution could result from an Atkinson and Bobis simulation of a research investment situation. Here the distribution is not smooth, but bimodal; the peak to the right of the zero is the expected profit should the investment succeed. The distribution of outcomes spills over to the left of zero again, indicative of those circumstances when the less favorable combinations of outcomes result. The peak to the left of zero reflects the number of times that the investment is lost as the research fails. The frequency of such occurrences is fairly high in this particular illustration. The entire distribution is more diffuse, reflecting the greater uncertainty in the individual distributions from which these results were developed.

A Dynamic Approach to Research Budgeting

A third approach for evaluating research projects attempts to deal with the problems of uncertainty and the dynamics of the research process. A paper by Hess[9] is perhaps the cornerstone for much of this work. The details of the techniques vary from author to author, but generally the intent is the same—that of allocating funds to the research projects in an optimal fashion. Most of these efforts deal with a profit equation that is summed over the selected time, and the ultimate profits are therefore related to the date at which success is achieved. Hess' equation is shown as:

$$F_n(y_n) = \max\ G_n P_n(x_n,y_n) - x_n + \rho[1-P_n(x_n,y_n)]\ F_{n-1}(y_{n-1})\, x_n \geqq 0$$

where: G_n=total discounted gross profit, given that the project is successful in the nth period (the sum of all revenues less all costs of terminal development); ρ=discounted rate; $P_n(x_n,y_n)$=probability of technical success in nth period; $F_{n-1}(y_{n-1})$=maximum total expected discounted net profit with $n-1$ period remaining; x_n=research investment in year n; y_n=research investment to year n.

In an effort of this kind, it is necessary to relate the probability of success to some level of research spending. It is implied that spending in previous periods affects the probability of success in the period of interest. With such a relationship, it becomes possible to anticipate the expected research expenditures. The methods usually presume that if the effort is not successful in a given year, additional money will be spent in the following year. This process is repeated until all of the available money is spent, and this is usually associated with the greatest probability of success. The relationship between the probability of success and the dollars spent is usually generated through interviews with research managers.

Two different types of curves are shown in Fig. 2. One curve is for a long-range research program in which it is recognized that a considerable expenditure is required to develop the background information necessary to realize a success. The second project typifies an applied-research program in which the directions are well laid out and a minimal amount of exploratory effort is required. The construction of such curves poses serious problems in that the basis for their generation is highly subjective and, although useful information results, it is dangerous to try to use these results to compare projects.

The equation also indicates that the gain is a function of the year in which success is achieved. This is understandable, in that failure to realize the invention early not only results in a loss of sales up to the time of its introduction, but can result in serious deterioration of the market potential.

Several proposals are offered to try to associate the research expenditure to some level of profit. Once such a relationship is established, the model is then used to determine an allocation of funds that ensures a maximum return. Although this is a most desirable objective, it is so critical to the basic purpose of the effort that the method by which the profit and research cost relationship is established should be subject to considerable scrutiny.

We do not believe that a straightforward relationship exists. In fact, we feel that the only sensible relationship can be developed when the sales projections as a function of time, and the probability of research success as a function of spending, are independently developed. In this way the contribution by those people responsible for that particular aspect of the endeavor can be separately identified.

Thus, we suggest that two separate curves are useful. Fig. 3 indicates a family of sales curves, each curve based on a different starting date. The curves shown in Fig. 2, relating the probability of research success to levels of funding, are also useful. With these two relationships, one can determine the sales history and presumable profits from several rates of funding.

Although these methods cannot be faulted for lack of sophistication, their failing is that they often require information beyond that which the commercial and research departments are usually capable of providing. Furthermore, most of these methods require so much data that the enthusiasm of any of the people who try to use them soon wanes.

Summary

In recognition of their needs for developing more-objective criteria for determining the best places for their investments, researchers turned to quantitative methods for evaluating the attractiveness of potential projects. This development started somewhat after WW II and it can be directly attributed to the philosophy that research can be most effectively directed if its objectives are defined. Three methods for evaluating research projects have received considerable exposure.

One method involves the ranking of the projects based on the ratio of potential profit to potential cost. The method is unpretentious and only serves to establish priorities. Another much-discussed method is that of treating the research investment as if it were a capital investment. Such a method serves the very important purpose of focusing attention on the respective returns from the research effort, and in that way is useful in discriminating among projects. This method, nevertheless, suffers from whatever shortcomings occur when one treats an investment in research as if it were a capital investment. A third method, in which perhaps the greatest amount of activity has been concentrated but where the practice has been most limited, involves the use of sophisticated optimization procedures, which not only serve to select projects on the basis of their profitability, but also to allocate funds to these projects in some optimum way. This method, which deals realistically with the nature of the research, is somewhat unrealistic in demanding input information that is sometimes unobtainable.

All three methods have had their trial, and each continues to be used for specific cases. It is doubtful that a single procedure can be developed that totally serves research's needs, but attempts are constantly being made to find methods of greater utility than those heretofore published.

The article following describes the specific methodology used in the more sophisticated techniques and suggests modifications that render the techniques more usable. ■

References

1. Rubenstein, Albert H., Economic Evaluation of Research and Development: A Brief Survey of Theory and Practice, *J. Ind. Eng.*, Nov. 1966, pp. 615-620.
2. Baker, N. R. and Pound, W. H., R & D Project Selection: Where We Stand, *IEEE (Inst. Elec. Electron. Eng.) Trans. Eng. Mgmt.*, Dec. 1964, pp. 124-134.
3. Kiefer, D. M., Winds of Change in Industrial Chemical Research, *Chem. & Eng. News*, Mar. 23, 1964, pp. 88-104.
4. Dean, Burton V., "Evaluating, Selecting & Controlling R & D Projects," AMA Research Study 89, American Mgmt. Assn., 1968.
5. Teplitzky, Gerald, Using the Profit & Loss Statement, *Chem. Eng.*, June 19, 1967, pp. 215-220.
6. Hess, S. W., Quigley, H. A., Analysis of Risk Investments Using Monte Carlo Techniques, *CEP (Chem. Eng. Progr.) Symp. Ser.*, 42, Vol. 59, 1963, pp. 55-63.
7. Hertz, D. B., Risk Analysis in Capital Invetment, *Harvard Bus. Rev.*, Jan. 1964, pp. 95-106.
8. Atkinson, A. C., and Bobis, A. H., A Mathematical Basis for the Selection of Research Projects, *IEEE (Inst. Elec. Electron Eng.) Trans. Eng. Mgmt.*, Feb. 1969, pp. 2-8.
9. Hess, Sidney W., A Dynamic Programming Approach to R and D Budgeting and Project Selection, *IEEE (Inst. Elec. Electron. Eng.) Trans. Eng. Mgmt.*, Dec. 1962, pp. 170-179.

Meet the Authors

◀**Arthur H. Bobis** is manager of the Systems Analysis Section, Organic Chemicals Div., American Cyanamid Co., Bound Brook, N. J. 08805. It is the responsibility of his section to apply quantitative methods to the solution of manufacturing, research and commercial problems. He received a B.S. and a B.A. from Columbia University, an M.S. in Ch.E. from Pennsylvania State University and a D.Eng.Sci. from Newark College of Engineering.

Anthony Atkinson, formerly a statistician with American Cyanamid Co., Bound Brook, N. J., is now a professor at Imperial College, London, England. Educated in England, he studied chemical engineering at Cambridge University where he received an M.A. in 1960. After several years with Shell Chemical Co. in manufacturing, sales and research, he joined its mathematics group before going on to obtain a D.I.C., from Imperial College, in statistics. He then joined Cyanamid.▶

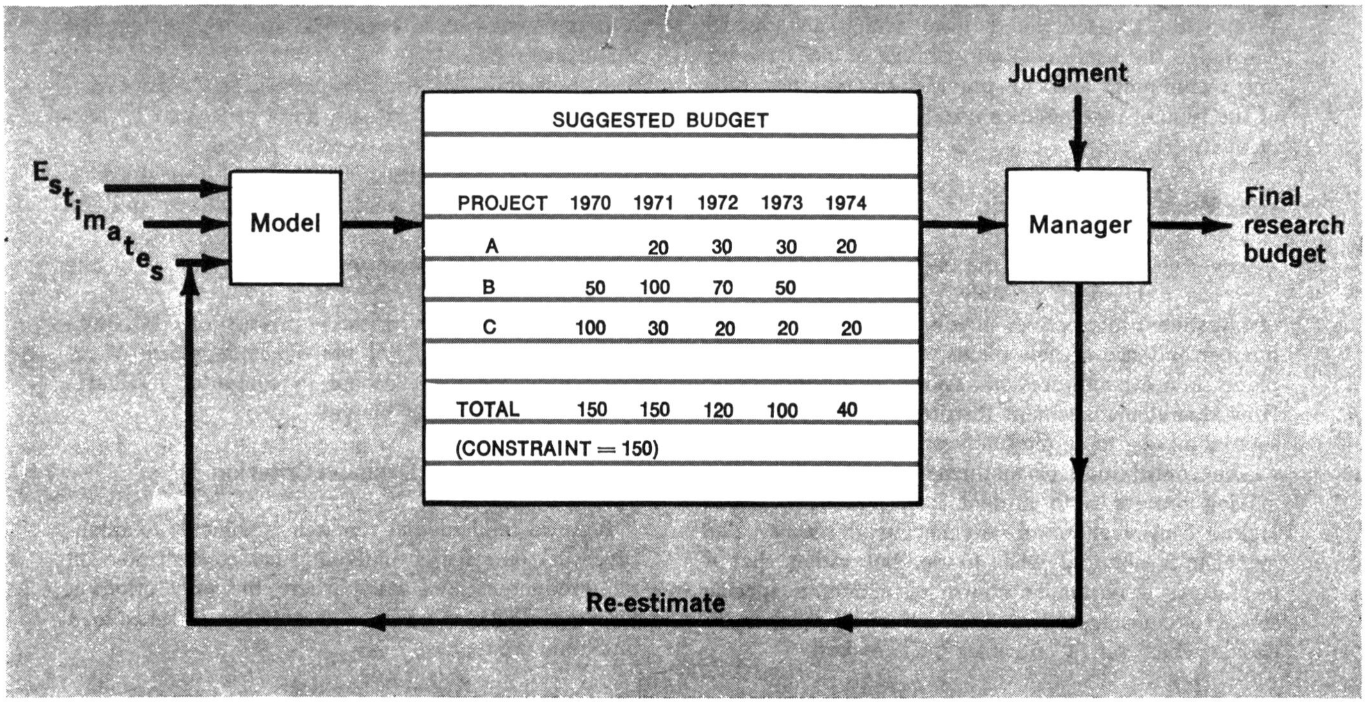

Analyzing R&D Investments Via Dynamic Modeling

To complete this two-part series on fund allocation and project evaluation in R&D work, here is what dynamic modeling is all about—including suggestions for making this new analytical tool more useful and accessible.

ARTHUR H. BOBIS and ANTHONY C. ATKINSON, American Cyanamid Co.*

In Part I, we examined some of the currently available methods for analyzing how much money should be spent on which research projects, and what profit could be anticipated. We distinguished three broad classes of approach:

1. Ranking procedures: the projects are arranged in order of attractiveness according to some criterion, and then funded in that order until all resources are exhausted.

2. Capital investment model: here the research project is treated as if it were a capital investment, and some criterion such as return on investment determines whether it should be undertaken.

3. Dynamic model: this approach recognizes that success in research depends on the timing, as well as degree, of research progress. As indicated in the table (next page), this is the most complicated of the three methods and, as such, requires the most information, while affording some important advantages.

Although most of the present article will deal with the dynamic model method, and improvements thereof, it is not our purpose to suggest that any one method of solving the problem of research budgeting is best under all conditions. Rather we hope to provide a background from which the reader

Originally published March 1970.

can judge the usefulness and pitfalls of R&D modeling techniques, and to provide an understanding of the type of mathematics usually employed in some of this work.

The Modeling Approach

Any given company in the chemical process industries has a number, possibly a very large number, of investment alternatives—involving, for instance, the number and size of new plants, extensions of existing plants, and expenditures on various forms of research. How should management distribute the available resources among these alternatives?

The contribution of mathematics to the decision-making process is to present management with the logical consequences of any set of decisions. The manager is not told what to do, but rather that if he follows a particular course of action, a specific set of circumstances will arise; it is up to him to decide which set of outcomes he likes best.

It is convenient to break the approach down into three major steps:

1. Selecting the decision criteria—e.g., sales revenues or net discounted cash flows generated by successful projects over a certain period.
2. Modeling the research process, by establishing relationships between research expenditures and the criteria.
3. Interpreting the results, so as to arrive at the final research budget.

The model of the research process can be quite complicated, especially if the dynamic nature of research is taken into account; a computer facilitates calculation of the best budget.

Step 1: Selecting a Decision Criterion

Because fundamental research is difficult to quantify, work on research budgeting has concentrated on the product-oriented area, where the only criterion for the validity of research is whether it leads to a profitable product.

Techniques for Analyzing Research Projects

Method	Advantages	Disadvantages
Ranking procedures	Calculations are easy. A single statistic (rank) is generated (people like this simple definition of priorities).	Deals only with point estimates. Does not generally recognize that economic circumstances will change before research is complete. The single statistic aspect can be a disadvantage if rankings are followed blindly, as if they reflected all factors.
Capital-investment model (e.g., discounted cash flow)	Calculations have become familiar. Required data are usually available.	It is hard to plug in the aspect of R&D failure—hence conservative projects in which the timing of R&D success can be pinpointed may become unduly favored. In research, commitment is not complete until plant construction is approved, whereas the method starts with the presumption that construction will go ahead.
Dynamic models	Method is totally comprehensive, serves to allocate as well as to select, and is concerned with rate of research budgeting as well as absolute level. Distribution of outcomes, rather than point estimates, are considered.	Confusing to use and interpret. Requires hard-to-obtain data. Some research managers do not believe the research process can be described mathematically and are reluctant to surrender prerogatives.
Modified dynamic models (as described in this article)	Same as above, but method permits use of generalized equation, and reduces necessary data base—i.e., by utilizing estimates usually made in the normal course of research budgeting. With two additional estimates, method can relate sales level to year of market entry; and the value of time gained vs. waste inherent in crash research programs.	Research managers may again consider the approach too mathematical. Easier to use, but still not easy to interpret. You get more than one result. Although method defines risk and return, it does not attempt to find a preferred mix of risky and safe projects.

Before developing a model to optimize research activity, one must be certain of what it is that he chooses to optimize. To characterize the success of research, many criteria have been suggested. All of these have their merits; no one criterion seems right for all situations. Nevertheless, the problem of optimizing a research strategy is sufficiently complicated that the optimization must be carried out with regard to only one of the criteria. It is sensible to optimize on the basis of the criterion that is most meaningful for the situation at hand, and then to calculate the values of all other statistics of interest under that optimum situation.

Although there is no firm agreement among the proponents of the more sophisticated approaches to research modeling as to the criterion of choice, almost all agree that any calculations involving the monies returned from that research effort should be discounted. The research portfolio contains projects for both long- and short-range duration. In order that these be properly compared, it is important that the monies obtained at a later date be related to those more immediately available. Discounting reflects (1) the historical evidence that inflation tends to reduce the value of future earnings, and (2) the alternative of simply letting the available money earn interest at the going rate. A discount rate of 10% is common, although the economics at any particular time might dictate a different figure.

You will recall that if the money were to be left in a bank paying an interest rate D, then after i years each dollar would be worth $(1 + D)^i$ dollars. Conversely, one dollar i years in the future is equivalent to $[1/(1 + D]^i$ dollars now.

The simplest criterion for research modeling is one that maximizes the discounted return—i.e., the present worth, which represents the net discounted cash flow over the life of the project. If there are a large number of research projects, this present worth becomes a sensible criterion, because the risks are naturally reduced by distributing the expenditures. In the work cited in Ref. 1, some of which we will discuss, this was the optimized criterion; under that set of circumstances, all of the other parameters of interest could be calculated.

Step 2: Modeling the Research Process

Having decided on a criterion for optimization, the next step is to develop a model that will enable one to calculate the quantity to be optimized. In every case, calculation of the project's profitability will be required. A cost calculation must be made, involving the product's anticipated volume and selling price, the manufacturing cost, cost of sales, and other related expenses.

Since the product becomes available only as a result of research expenditures, it is appropriate to separately calculate research expenditures, and further to develop some functionality between the expenditure of research money and the worth of the project. Two approaches have been tried in this area. In one, the research expenditure is related to some measure of product "goodness," and eventually to the level of sales; here, it is presumed that research monies are spent to provide a product with preferred characteristics, and that each additional dollar enhances the characteristics of that product. The more usual approach is to relate the research money to the probability that the research will succeed.

Another important aspect of modeling of this type is to establish a relationship between the timeliness of research expenditures and the potential sales. Here, one is concerned with trying to determine the value of a crash program as opposed to programs that proceed at a more usual pace. A balance is struck between the value of getting the product to the market early and the additional cost of research that usually occurs when work is accelerated. Finally, the sales projections must also reflect the advantages of an early market entry.

In summary, the research model should:

1. Furnish a procedure for calculating the profit.
2. Relate the research expenditures to the rate at which the research proceeds.
3. Provide for measures of the sales potential as a function of time.
4. Establish a relationship between the pace of research and the value of the time of entry.

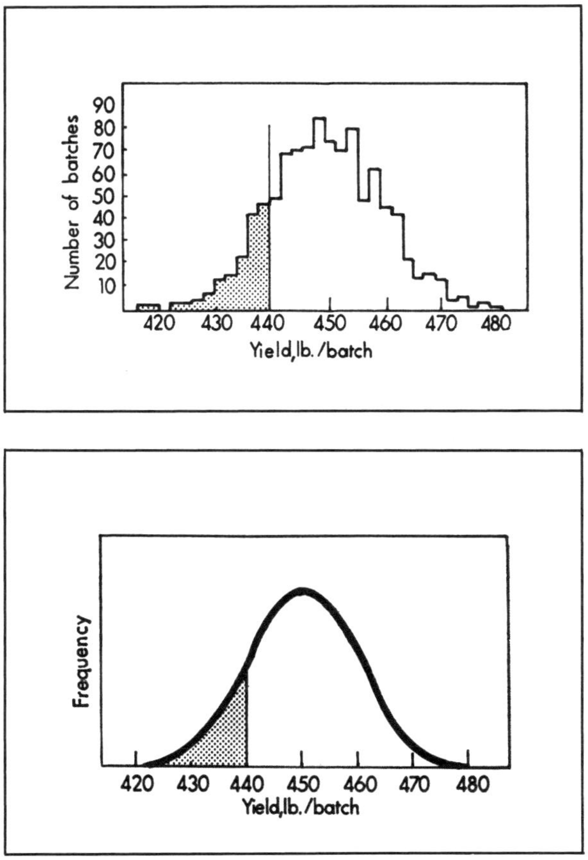

YIELD HISTOGRAM for 1,000 batches of a chemical (top); yield distribution is shown below—Fig. 1

Let us now examine the specifics of these individual four requirements.

Profit Calculations

The very simplest of the profit calculations is:

$$\text{Gross Profit} = R \sum_{i=1}^{n} [(V)(P) - (C_M + C_A + C_S) - C_R \quad (1)$$

where i represents the year of success; R the probability of success; V the sales volume in units; P the selling price in \$/unit; C_M the manufacturing costs, \$/unit; C_A the administrative expenses, \$/unit; C_S the selling costs, \$/unit; and C_R the research costs, \$.

In most cases, point estimates of each of the several numbers are made. In the previous article, we noted that reliance upon point estimates can lead to erroneous calculations; in this section, we will treat the problem of dealing with ranges rather than point estimates. For each estimate in question, it is simple to provide guesses, not only of the most likely value but also of the 95% confidence interval about that value. That is, the estimator admits to his uncertainty and then tries to express the extent of the uncertainty by providing not only his best guess but also the range for other possible results.

One set of questions that provides this information about the value of an unknown parameter is:

1. What is the most likely value of the parameter?
2. If everything went badly, what is the lowest value you would reasonably expect?
3. If all went very well, what is the greatest value you would reasonably expect?

The answers to these questions can be taken to define the mode, or most likely value, of the distribution and the 2½% and 97½% points. In the distribution of the parameter, there will be 2½% of the area of the curve to the left of the lowest value, and 97½% of the area to the left of the highest value, just as, in Fig. 1, 15.9% of the area of the curve lies to the left of the value 440. The further apart these two percentage points are, the less well defined is the situation. Conversely, the closer they are together, the more information we have. If a firm contract existed for the sales of the product, the three numbers would be identical.

In the absence of any information to the contrary, it is often assumed that these distributions have a bell shape similar to that of a normal distribution curve. There is, however, no reason why the distribution should be symmetrical. Sales could have a fairly well defined lower limit, but a much larger upper one if there were a small but non-negligible chance of capturing part of some other market.

Several distributions have been suggested that have this general shape. One of the most convenient is the logistic distribution. If $f(x)$ is the probability that the parameter of interest has a value less than x, and if a, α and β are parameters that can be estimated from the most likely value and the 2½% and 97½% points, then:

$$f(x) = 1/[1 + \exp(\alpha - \beta x^a)] \quad (2)$$

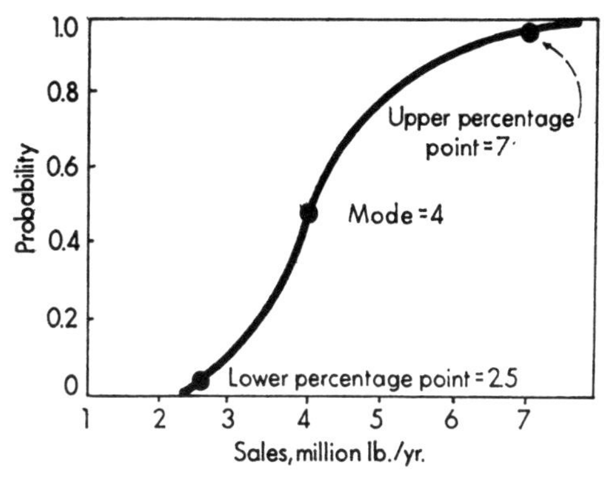

CUMULATIVE DISTRIBUTION OF SALES fitted to upper and lower percentage points and mode—Fig. 2

This so-called cumulative distribution is shown plotted in Fig. 2. The horizontal axis is the volume of sales in a particular year, the vertical axis is the probability that the sales will be less than this amount. The three defining points are the mode, which is the point of inflection of the curve, at 4 million lb./yr., and the upper and lower percentage points at 2.5 million and 7 million lb. The probability density function associated with this curve looks bell-shaped, but with a longer tail to the right.

Simulation

Uncertainty in estimates of future events is reflected in the distributions of the values of the parameters. This uncertainty shows up in the return, which is therefore also in the form of a distribution. To obtain this distribution, we resort to simulation. This is necessary because, although the distribution of the parameters in the model is known, we cannot calculate theoretically the distribution of return. The model is too complicated to allow this.

Most computers are equipped with a random-number generator. This produces numbers between zero and unity such that, at each trial, any number is equally likely. By building up a histogram of these numbers, a rectangle between zero and one is eventually obtained. If these numbers are regarded as probabilities, the cumulative distribution function of a parameter can be used to obtain a series of x values, or samples; and a histogram derived from the distribution can be generated.

Suppose sampling is taking place from the distribution of Fig. 2, which has been redrawn in Fig. 3. Two values obtained at random for the probabilities are 0.314 and 0.826. These correspond to values of sales of 3.72 and 5.30, respectively. Such a procedure is followed for each of the estimates in Eq. (1), about which a distribution was drawn. The values for each variable that is selected by simulation is then com-

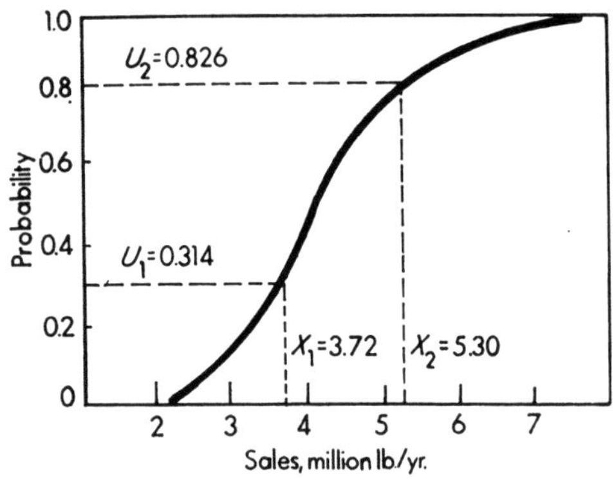

SIMULATION OF SALES; the values of U are random numbers generated between 0 and 1—Fig. 3

bined according to Eq. (1). This is repeated a sufficient number of times until a distribution of results is built up that reflects the uncertainty in the estimates.

In more-complicated examples, the relationship between the sales volume, selling price and cost of manufacturing may not be independent. This figure, or histogram, differs from the value calculated from the model values of the distributions of the parameters.

Research Expenditure and Timing

The methods discussed so far make it possible to calculate the distribution of return from a research success. We have considered what information is needed and how it can be used in some approaches. But, so far, we have said nothing about the relationship between research investment and the time when that research succeeds.

In dynamic modeling, it is recognized that research occurs over time, and that in each time period–perhaps up to five annual periods–a certain amount of money will be spent. What is therefore needed is a relationship between the expenditure and the probability that the research will succeed in a given time period.

One approach is to ask the research manager to plot a curve relating research expenditures to the probability of success. Even if the research manager is uncomfortable in developing such a curve, it is possible to construct a generalized curve based on information he is able to provide: He is accustomed to predicting the cost of the research program and, with a little effort, is able to set lower and upper bounds about that cost. Given these three estimates, we can, as before, smooth a generalized distribution through the data, where again it is assumed that the most likely value represents the mode of the distribution, and the other two bound the 95% confidence limits.

The cumulative of the logistic distribution can be smoothed through these three points, and provide curve shapes of the kind proposed in the literature by those authors who have suggested that the manager might separately draw the curves for each research project. In this way, the manager participates in the determination of the curve shape by providing the estimate, but is not encumbered by the necessity of having to structure the individual curves.

When this approach is chosen, it is necessary to distinguish between completing the project and succeeding with the project. In performing research, money is exchanged for knowledge. That knowledge may be how to produce a product with the desired properties, but it may well be that the only thing learned is that the goal is unobtainable; in this case, the project has failed. A distinction is therefore made between the probability of completing the project and the probability that the completion is successful.

The overall probability of success was discussed in the first article, and was obtained by asking specific questions about various aspects of this project. In this approach, the generalized curve just described is set equal to the cost of completion. It is this curve that is scaled by the probability of success estimates.

The curves are shown in Fig. 4. The upper curve is the cumulative of the logistic distribution, and shows the cost of completing the project after a given expenditure. The lower curve represents the probability that the completion is successful, and is obtained from the upper curve by multiplying by the overall probability of success. The difference between the two curves is the probability that the project will have failed at this total level of expenditure.

In using Fig. 4, it is possible at any point in time after a given expenditure to state the probabilities as to how the research might stand. In the example, a verticle line is struck at $250,000, and one can

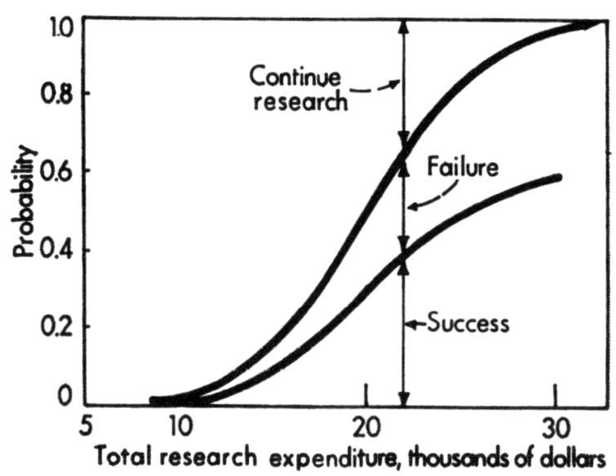

PROBABILITIES of a research project's outcome as a function of total expenditure. For expenditure of $22,-000, probability of success is 0.393; failure, 0.262; incompleteness, 0.345—Fig. 4

read that at the level of expenditure, the probability that the research has succeeded is 45%, that it has failed is 30%, and that more funds are needed is 25%.

This figure is intuitively satisfying: We can see at the beginning of research that there is a high probability that more funds will be requested, but as the research proceeds, the chances that a failure or success will have been recognized increases. When we reach that point where we have exhausted all of the monies that we would need, the chances of asking for more money have diminished to almost zero.

As a result of this kind of an approach, it is possible to calculate the probability of success and failure of the project, year by year, for a given pattern of research expenditures.

It is now possible to modify Eq. (1) in such a way as to account for the time period in which the research succeeds. This is shown in Eq. (3):

$$G_p = R_i \sum_{i=1}^{n} V_i(P_i) - (C_{Mi} + C_{Ai} + C_{Si}) - \sum_{i=1}^{n} C_{Ri} \tag{3}$$

where G_p is the gross profit, R is the probability of success, and i is the year of success.

R_i is calculated by Eq. (4), (5) and (6):

$$D = 1/[1 + \exp(\alpha + \beta x^a)] \tag{4}$$

where D is the probability that the project will be completed at an expenditure no greater than \$$x$; the other points were defined in Eq. (2). Next,:

$$d_i = D_i - D_{i-1} \tag{5}$$

where d_i is the probability that the research will succeed in year i, and:

$$R_i = Rd_i \tag{6}$$

Consideration of Sales Data

The sales data should be separately estimated, preferably by people in the sales or commercial departments. Again, the problem is greatly simplified if a generalized model that reflects the sales growth can be used and fitted to a minimum number of estimates. It turns out that the cumulative of the logistic, which was used for the description of research, also adequately defines the s-shaped growth distributions that are typical of new products.

Estimates that can be used to calculate distribution parameters are sales in three different years in the product's life. For example, estimates of sales in the first, sixth, and eleventh years yield coefficients to describe the sales growth defined by Eq. (7):

$$S_j = B/[1 + \exp(\gamma - \delta_j)] = f_{2(j)} \tag{7}$$

where S_j represents sales in a given year j; B is the asymptotic value to which sales tend; and γ, δ the parameters that can be calculated by the sales estimates.

Such a curve can represent the sales growth starting in any particular year. To determine the urgency with which the research must proceed, it is also important that the sales-growth curves for market entry in succeeding years be provided. A convenient way of doing this is to obtain the estimate for one other point in time, and then interpolate and extrapolate for entrance in other years. A useful estimate is that of the sales in the eleventh year, given a start in the sixth year. The ratio of the two fifth-year sales provides a measure of the urgency of the research. The sales curve obtained in the first year is multiplied by the coefficient K^5, which is set equal to the ratio of the sales achieved in the fifth year to the product's life, given the two different starting dates.

Sales that start in any year are then calculated via:

$$S_j = K^{i-1} f_{2(j-\delta+1)} \tag{8}$$

where i = first year of sales.

The method by which probabilities are related to the various sales curves is illustrated in Fig. 5. Here, two sales curves—one starting in 1968 and the other in 1969—are shown on the left-hand side of the figure. The probabilities of success associated with different levels of spending are shown on the right-hand side. Expenditures of \$100,000 in 1968 and again in 1969 result in the probabilities R_1 and $(R_2 - R_1)$ of following the 1968 and 1969 curves.

On the assumption that commercialization commences upon research success, the penalty for succeeding late is shown to be severe, and consideration should be given to accelerating the rate of research spending. A \$200,000 expenditure in a single year could be considered, but this may not be as effective as spending the money over a two-year period. The probability of success associated with a single large expenditure will, of course, be greater than R_1 and cannot be greater than R_2. The value of this probability is the subject of this section.

Consider the expression:

$$Z_i = Y_i\,(Y_i/X_i)^{\epsilon_i} \tag{9}$$

Here, Z_i must lie between X_i and Y_i, providing ϵ_i is constrained to lie between 0 and 1.

Such a function represents the problem we have just posed: i.e., to determine the possibility of completion at two different levels of expenditure during the same time span, given that one of the expenditures is greater than we initially thought it should be. In Eq. (9), Y_i is defined as an efficient expenditure (that is, the amount of money that can efficiently be spent in pursuit of this project); and X_i is any amount greater than this. The value ϵ_i is used to control the wastage inherent in crash programs, the wastage diminishing as ϵ_i approaches 0.

At the earliest stages of a research program, the project is usually accelerated by pursuing additional alternatives. This has the effect of increasing the probability of success; but because only one alternative (if any) will survive, money spent on other alternatives is wasted. At the latter stages of the project, research is accelerated by the in-parallel tackling of necessary and usually well-defined operations. Given this, the inherent wastage in a crash research program is greater at the beginning than at the end of the project.

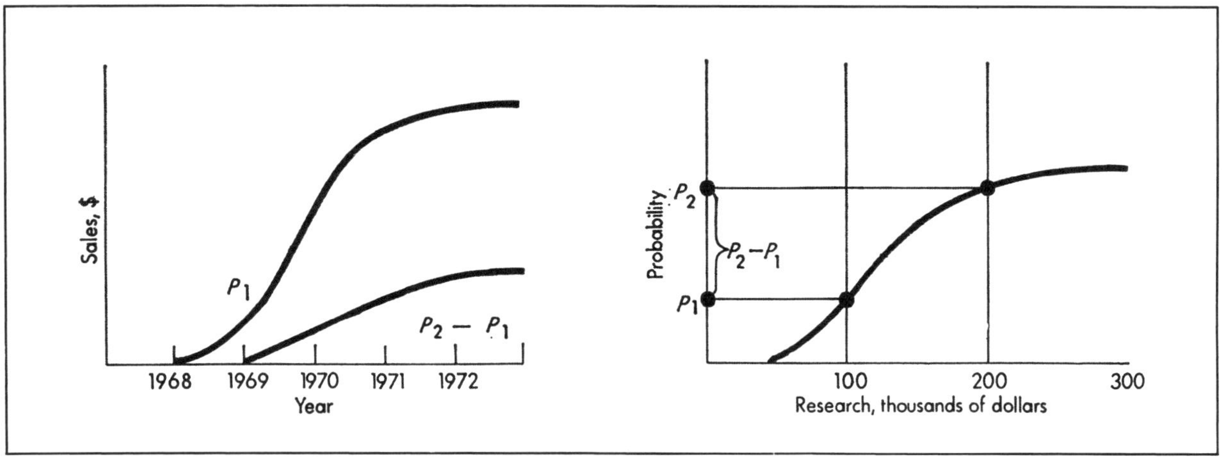

USING THE MODEL: Probabilities related to sales curves (left) and spending levels—Fig. 5

If C_3 is the estimate of the greatest cost of completing the project, ϵ_i is related to the research program's stage of development by Eq. (10) or (11):

$$\epsilon_i = (Y_i/C_3) \tag{10}$$

$$\epsilon_i = \sum_{i-n}^{n} \frac{Z_i + Y_n}{C_3} \tag{11}$$

This indicates that if as much money as would have to be spent could be spent efficiently, then the research program is well defined or near its end, and ϵ_i should be close to 1. When Y_i is small relative to C_3, which is usually true at the beginning of the program, then ϵ_i is closer to 0.

Let us consider the following example: Referring to Fig. 5, we are concerned with determining the effect of spending $200,000 all in one year when we thought we could only efficiently spend $100,000. The additional information required to make the calculation is that C_3 has been estimated at $300,000. In the example:

$$\epsilon_1 = (\$100{,}000)(\$100{,}000)/(\$300{,}000) \quad = 1/3$$

and $Z_1 = (\$100{,}000)\ (\$200{,}000)/(\$100{,}000)^{1/3} = \$123{,}000$

The inference to be drawn from this is that an expenditure of $200,000 over the same time period in which we thought we could only efficiently spend $100,000 would have the effect of increasing the probability of success up to the amount that would have been associated with an expenditure of only $123,000. We might then say that of the additional $100,000, $23,000 was in pursuit of the desired objective and $67,000 was wasted. It is the purpose of this program to try to balance the improvement in the probability of success with the additional costs of research.

Step III: Interpretation of Results

The techniques discussed so far are useful to the manager only to the extent that they help him in the selection and allocation process.

Let us now take another look at the "model and manager" sketch on the first page. Here, information in the form of estimates as to potential sales, research costs and manufacturing costs is provided to the model. The data are operated upon, using the mathematics we have discussed; they are reduced to a smaller number of parameters and presented to management.

It is most useful if the information can be reduced to the extent that a budget is suggested.

The suggested budgets rarely coincide with the manager's thoughts. Most often, attention will be focused on the initial estimates, and these will often have to be modified. (In the conventional budgeting methods, these estimates are always just as inaccurate, but there is no way of calling attention to the inadequacies.)

The procedure of comparing the computer output to the manager's judgment, and adjusting the estimate, need not be repeated until the suggested budget and the manager's budget are identical, but rather until that point in time at which the best available information has been evaluated.

Among the techniques that can help the manager arrive at the final budget are: sensitivity analysis, profitability histograms, and optimization procedures. Let us briefly discuss them in turn.

Sensitivity Analysis

In practice, there should be an iteration between management and the computer. The model can be looked upon as a means of learning about how to invest in research. By changing the inputs, new solutions are obtained. These can often shed light on the importance of various factors in determining the attractiveness of a project.

One way of obtaining further information is to make use of sensitivity analysis to discover how the variability of the return depends upon the uncertainty in the particular inputs. This uncertainty is expressed by the three estimates defining the distribution of the parameters. If these estimates cover a wide range, the input is poorly defined. The spread of the distribution of returns obtained by simulation reflects the uncertainty in the values of all the parameters. To find out how important is this lack of knowledge of the precise value of one particular parameter, a simulation is performed in which only the distribution of the parameter under study is sampled. The values of all of the other parameters are held fixed to their most likely values.

The resulting spread of the simulated distribution of return indicates the contribution that one particular parameter makes to the overall variance of the uncertainty in the estimate. If the parameter is very poorly defined and is crucial to an accurate description of the project, it will make a disproportionate contribution to the overall variation. This calls for an investigation to establish the value of the parameter more accurately. The method, known as sensitivity analysis, is of particular use where a large decision is to be made—for example, building a new plant.

Introducing a Research Model

The initiative for establishing the research-model method must come from management—i.e., from the users of the tool. Unless management believes that something useful will result, it is difficult to see how the researchers can be persuaded to undergo the inconvenience of quantifying their intuitive beliefs.

Given this initial state of enthusiasm, it is the duty of the operations-research term or whatever group is building the model to ensure that everyone involved understands as much of the model as is possible. Companies reporting successful applications have devoted a lot of effort to this communication problem. At each stage of building a model, seminars are held at which the assumptions of the model are discussed and the results of differing inputs compared.

Building the model proceeds as an interaction between those who must supply the input information and those who formulate it mathematically. In this way, the reality of the model and the significance of the inputs can be established step by step.

Another useful aid to implementation is to have the model available for experimentation by the researcher, who thus develops a feel for the significance of the values of the inputs. Especially if he is not accustomed to using a computer, it is helpful if this work can be done directly on the machine, preferably with graphical output. In this way, the researcher overcomes his distrust of the computer. Also, by having the results available almost instantaneously, an atmosphere is created in which curiosity is stimulated and the rate of learning increased. Because of the feedback it provides, this process is of equal value to those who built the model, because if the system behaves in some way that the scientist knows is intuitively wrong, then the model is at fault and must be revised.

If the project of building and using the model is treated as a process in which all groups can learn from each other, useful input information can be obtained. Only then will sensible decisions be suggested and, having been suggested, be acted upon.

Profitability Histograms

The methods suggested so far enable the calculation of the return from a research success. Several ways have been suggested to display this information.

The simplest is to calculate the expected return from a research success by averaging the results of the simulation. The expected gain from undertaking the project is then the sum of these averaged returns weighted by the probability of achieving them, less the expected research expenditure, with all quantities duly discounted. This provides a single figure for the average gain from the project for a given budget, and is a useful measure of the desirability of the project. However, due to the uncertainty inherent in the estimates of the parameters and of the year in which success will be achieved, there is a distribution of the gain about this figure, and it is important to consider this in a complete analysis.

In the capital-investment approach to describe the project, the distribution of gain is important as well as its average value. By simulation, a histogram of possible gains is built up, similar to Fig. 1. But this is not a satisfactory representation of the research process, because there is a finite probability that the research will fail. One of the more sophisticated methods proceeds by a two-stage simulation. The first stage of the sampling determines whether the project succeeded or failed, and in which year. If the project failed, the gain is negative and is the sum of the discounted research expenditures. If it turns out that the project was successful, the distributions of parameters are sampled to establish a particular case, and the return is calculated. The gain is then the return less the research expenditure.

In Fig. 6, a histogram resulting from 500 such simulations is shown. This is the distribution of gain for a particular budget for one project. Typically, this histogram is bimodal. The left-hand peak corresponds to the losses incurred if the project fails, the right-hand one to the profits from a successful project. The further to the right this peak is, the more profitable, on average, is the project. The greater the spread of the peak, the greater the range of possible outcomes.

In analyzing research projects and the related budgets, these histograms have been found to be surprisingly useful. Management must know not only that a particular project is potentially profitable, but also whether there is a significant probability of the project falling below a certain profit level. The histograms are a convenient visual means of supplying this information.

The histogram presents a description of possible outcomes for a given investment in each project. There remains the problem of how the budgets should be adjusted to determine the best overall allocation of funds.

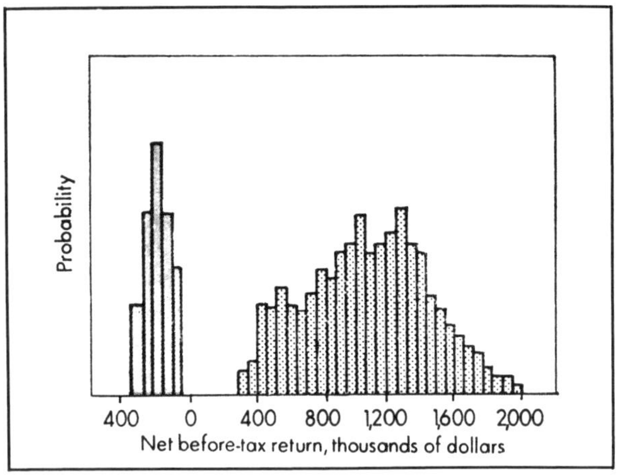

GAIN-DISTRIBUTION HISTOGRAM from 500 simulations of one research project's budget—Fig. 6

Optimization Methods

Perhaps the most useful budgeting approach is to allow the computer to suggest a preferred expenditure distribution by optimizing the criterion of choice.

In most applied-research projects, funding should be provided for a five-year period. If the number of projects, n, is large, the problem of finding the best distribution for $(5)(n)$ budgets is extremely tedious.

Several hill-climbing techniques have been described in the literature and are valuable here. The usual procedure is to take small step changes in each of the variables, and to continue making changes in the direction that tends to maximize the function. (Ref. 2, written by a chemical engineer, is a good introduction to optimization techniques; Chap. 11 of Ref. 3 is a useful source for hill-climbing techniques.)

In the problem we are considering here, the budget cannot be negative. Thus, we have the constraint $x_i \geqq 0$. There may also be a further constraint that only a certain total of money is available for research spending in any one year. The usual methods of optimization do not take account of this; they proceed automatically to find the values of the x's for which y is the greatest. It is therefore necessary to introduce transformations. Let $x_i = z_i^2$; then, whatever the value of z_i, the value of x_i cannot be negative. If the optimum budget is found in terms of the z's, on which no constraints are placed, then all the x's must satisfy the constraint that they be non-negative. In a similar way, quite-complicated constraints can be eliminated, and the power of a standard computer-function-maximization routine employed. Further examples and more information about computer programs are given in Ref. 4.

To make use of these mathematical tools, one method of optimization for a set of projects budgeted over several years proceeds sequentially, year by year. The budget for the first year is adjusted to give the maximum gain, given the budget in successive years. As there is only a limited amount of money available, the total budget for each year is subject to the constraint of not being greater than a certain amount, although it may be less than this. By using a suitable transformation, it is possible to maximize in an unconstrained space such that the budget restriction is always satisfied.

Once a better budget for the first year has been established, the probability that each project will not be completed by the end of the first year is known. The best budget for the second year, subject to a total constraint on the expected expenditure, is then found in a similar manner. This process is continued for all the years in which research funds might be spent, and then repeated until a stable solution is reached.

The optimum budget is a year-by-year allocation according to projects, such that the expected gain is maximized. The suggested budget in the sketch on the first page is typical of the kind of information available to the manager.

It has been claimed that increases in the expected gain of up to 50% have been obtained by adopting this approach. The chief cause of this increase is the concentration of research funds on fewer projects, which are therefore completed in time to make an impact on the market.

Conclusions

It is not claimed that the approach to investment problems presented here will automatically result in the best decisions. The claim is more modest: scientific decision-making can help management to a more purposeful and efficient allocation of resources.

The decisions made will only be as good as the information on which they are based. But the decisions are going to be made anyway. Anything that can be done to put decision-making on a more logical basis is a step in the right direction.

By asking for numerical answers to a comprehensive set of questions, one ensures that each person concerned has to think about what he hopes to achieve—and think about it in precise terms. By making the numbers the subject of public discussion, vain hopes can be exposed to ridicule, and more sensible estimates achieved. And, by having the process available for iteration between man and computer, management should learn what combinations of factors make a good project. In all these ways, dynamic modeling can lead to better management and a more efficient use of our limited resources. ■

References

1. Atkinson, A. C., Bobis, A. H., A Mathematical Basis for the Selection of Research Projects, *IEEE (Inst. Elec. Electron. Eng.) Trans. Eng. Mgmt.*, Feb. 1969, p. 2-8.
2. Wilde, D. J., "Optimum Seeking Methods," Prentice-Hall, 1964.
3. Davies, O. L., "The Design and Analysis of Industrial Experiments," Oliver and Boyd, London, 1956.
4. Box, M. J., A Comparison of Several Current Optimization Methods, and the Use of Transformations in Constrained Problems, *Computer Jour.*, 1966, 9, pp. 67-77.

Forecasting for Sales and Production

Industrial forecasting involves relating mathematical curves to the facts of history and the particular situation. Here is how to find and fit such curves.

GEORGE E. MAPSTONE, Swinburne College of Technology

Forecasting is as essential to civilized life as the compass is to navigation. The business man requires a forecast to plan production requirements; the Minister of Finance to plan his budget; and the university to plan its development and expansion.

Industry most commonly requires forecasts of sales and production. Since production forecasts are normally based on estimated sales, the principles of sales forecasting are generally applicable.

A sales forecast based on commercial intuition becomes difficult as the numbers of products and customers increase. In all cases, the forecast is an estimate of the value of some variable at a future date; if we know how the variable, f, changes with time, t, then we know the nature of $f(t)$, and the forecasting becomes a simple equation of mathematics, i.e. $y = f(t)$. Of course, such a mathematical procedure, which is based on the past, should also allow for future factors, such as new advertising-sales campaigns by the forecasting company or its rivals.

Kinds of Forecasts

Forecasts may be classed as either short-term or long-term. Short-term forecasts, which are used for immediate production goals, are frequently based on some form of moving average (see below).

Long-term forecasting finds its justification in regard to revised production schedules that are necessary to meet changes or growth in the market. New plants and labor-training programs both require long-range forecasts. An obsolete plant is not worth rebuilding, for example, if the product is becoming obsolete as well. With large firms,

The Moving Average

The progress of a market is seldom steady; and it is usual for the demand at some periods to differ substantially from the general trend. The shorter the time span under review, the greater this variation will be.

One method of smoothing out the random variations is to calculate a "moving average." For this purpose, a base time period is chosen that is at least as long as the periodic fluctuations. For example, if there is a seasonal variation, the period should be one year; and the average monthly sales should be computed for the periods January-to-December, February-to-January, March-to-February, April-to-March, and so forth.

When graphed, the moving average (sometimes called the running average) is plotted against the midpoint of the time period. For this reason, it is customary to choose an uneven number of time periods over which to calculate the average, unless there is a specific reason for choosing an even number, as in the foregoing example.

The moving average produces a curve that reflects the underlying progress of the market but minimizes the random fluctuation about this growth. It tends to iron out the trends, particularly nonlinear ones, when applied over longer time periods, so it is usually applied to shorter periods. It gives a measure of the random variation experienced, by giving an estimate of the average demand at the appropriate time. Extrapolation is often used for forecasting from the moving average, although this cannot be extrapolated very far without serious loss of accuracy.

One weakness of the moving average is its equal emphasis on the first and the last portions of the averaging time period. Because of this, the moving average is slow to respond to a change in sales trend.

The Exponentially Weighted Moving Average

A more realistic approach would, therefore, be to take a weighted average of the type:

$$m_t = a_1 d_t + a_2 d_{t-1} + \cdots + a_n d_{t-n+1}$$

(continued on page 258)

Originally published May 1973.

forecasts up to five years are common; and some firms prepare ten-year forecasts as guides to policy.

However, long-range forecasts are difficult. The demand for a commodity will often show a picture of steady growth over a period of years, although there are fluctuations above and below the general trend. In analyzing such cases, one is tempted to plot the demand figures against time and draw a free-hand line through the scattered points. Such a simple procedure may introduce a personal bias of the forecaster, who subconsciously would like to see the sales increase. Especially if the points are scattered, a more or less simple mathematical equation should be fitted to them, and this is done usually by the method of least squares.

When preparing for long-range forecasts, the market history should be investigated. In England, for example, the sale of motorcycles is closely tied to the minimum deposit required by law for the hire-purchase agreements; and this minimum is changed from time to time. Factors to be considered are:

- Uniformity. If there are several distinct markets for the product, the uniformity of each should be considered separately.
- Effects of supply. Is the demand a true one, or has it been limited by the available supply?
- Appearance or disappearance of a major demand (about 25% or more). Could this occur in a field of application for the product, or result from the gain or loss of a large customer?
- Demand structure. Is the demand based on the whole market or on the sales of a single company? And if it is the latter, is the share of the total market changing?
- Competition. Has any competitive product or import become a significant factor in the market?
- Politics. Are there any political factors that could affect the market?

The forecaster should be able to allow for the short-term significance of these factors. When he has done this, his long-term forecast will depend primarily on one operation:

Finding a Curve To Fit the Facts

Fitting a curve to demand-data implies choosing one that, firstly, agrees with the past figures and, secondly, represents future demand, when extrapolated.

Because of the random variation of sales figures, it is usually difficult to establish by inspection what form of mathematical curve will fit the data. However, gross misfits can be easily detected; and techniques are available for selecting the type of curve that gives the best fit.

Common trend curves can be classified into three groups: polynomials, exponentials, and modified exponentials. Each group has a recommended test for its applicability. In cases where several different equations can fit the data equally satisfactorily (or unsatisfactorily), the simplest curve is chosen, or the raw data are smoothed to a list of moving averages.

Each curve can be identified with a "slope characteristic." For example, the slope of a straight line is constant, implying a uniform growth of demand. Therefore, a plot of this slope of the demand curve against time should give a horizontal straight line. The method of applying this check can be done as illustrated in Table I and Fig. 1.

A parabolic curve implies that there is a maximum for the demand, which later falls away again. Since the slope of a parabola changes uniformly, a plot of the slope of the demand curve against time should approximate a straight line at an angle to the horizontal. A summary of such slope characteristics is given in Table II.

Polynomials

The two polynomials most used are the straight line and the parabola, but higher orders can be used if it is felt that they are justified.

These can be written:

$D = a + bt$, and $D = a + bt - ct^2$,

where:

D = demand; t = time; a, b, c = constants.

---

where: m_t = the moving average
d_t = demand (or sales) at time t
$a_1 > a_2 > a_3 \cdots$ etc.
$a_1 + a_2 + a_3 + \cdots + a_n = 1$

Holt's method of using the exponentially weighted moving average is a special case in which the coefficients a_1, a_2, a_3, etc., form a geometric progression, each coefficient after the first being a constant fraction of the one preceding it. If, for example, this fraction is taken as 0.8, it can be shown that when $a_2 = 0.2$, $a_2 = 0.16$, $a_3 = 0.128$, and so forth. The moving average can then be written:

$$m_t = 0.2(d_t + 0.8d_{t-1} + (0.8)^2d_{t-2} + \cdots + (0.8)^nd_{t-n}$$

Since m_{t-1} will have been calculated in the same way,

$$m_{t-1} = 0.2(d_{t-1} + 0.8d_{t-2} + (0.8)^2d_{t-3} + \cdots$$

Multiplying this by 0.8 and combining with the first:

$$m_t = 0.2d_t + 0.8m_{t-1}$$

Using this equation, the moving average, m_t, can readily be calculated continuously, by unskilled clerical staff, as the demand for each time period, t, becomes known.

In effect, the value of a_1 corresponds to the fraction of variants that represent real change, while $(a_2 + a_3 + a_4 + \cdots + a_n)$ corresponds to the fraction of variants that represent random fluctuations. When $a_1 = 1$, the estimate of the current demand is equal to the previous month's demand, and will therefore be subject to the full effect of monthly fluctuations in demand. On the other hand, when $a_1 = 0$, the estimate of current demand remains unchanged and is therefore unaffected by any real changes in the demand that have taken place.

When a real change in demand has taken place, it is followed by a large value of a_1; but when no real change has taken place, a large value of a_1 would give an unreliable and probably misleading estimate, particularly if the random fluctuations were large. In practice, it is necessary to strike a happy medium, and values of a_1 in the range of 0.1–0.3 have been found suitable.

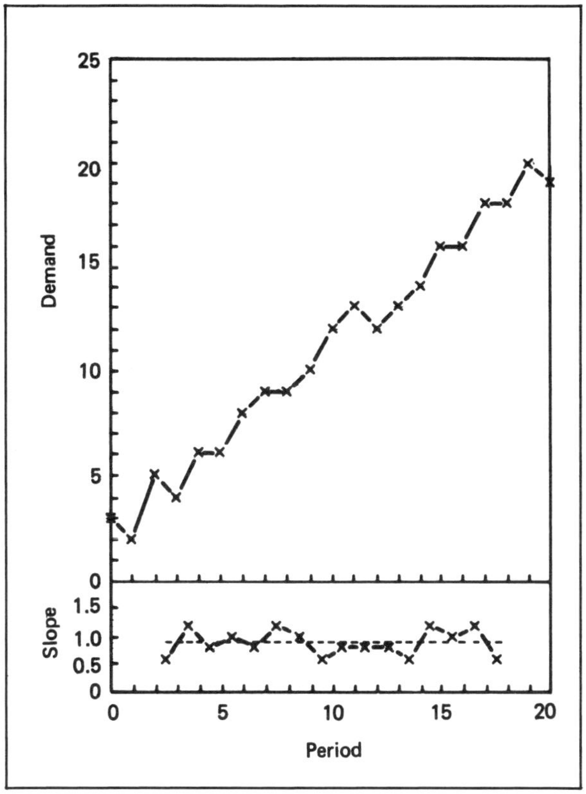

SLOPE CHARACTERISTIC by moving average—Fig. 1

Calculating the Slope Characteristic by Moving Average — Table I

t	*D*	Moving Average by 5 (*t*)	Slope (*t*-1)
0	3		
1	2		
2	5	4	
3	4	4.6	0.6
4	6	5.8	1.2
5	6	6.6	0.8
6	8	7.6	1.0
7	9	8.4	0.8
8	9	9.6	1.2
9	10	10.6	1.0
10	12	11.2	0.6
11	13	12.0	0.8
12	12	12.8	0.8
13	13	13.6	0.8
14	14	14.2	0.6
15	16	15.4	1.2
16	16	16.4	1.0
17	18	17.6	1.2
18	18	18.2	0.6
19	20		
20	19		

Exponentials

Two exponential curves are most used. The simple exponential, $\log D = a + bt$, and the logarithmic parabola, $\log D = a + bt + ct^2$.

The simple exponential is sometimes called the Compound Interest trend; this implies that the rate of growth is proportional to the demand, so that the logarithm of the demand plotted against time should approximate a straight line. The slope characteristic can be determined in either of two ways: the slope of the logarithm of the demand versus time, or the ratio versus time of that slope to the moving average (Table II). Either plot should approximate a horizontal straight line.

The logarithmic parabola implies that there is a maximum demand, after which the demand falls away again. Its slope characteristic can be determined by plotting against time either (1) the slope of the logarithm of the demand or (2) the ratio of the slope to the moving average. This should approximate a straight line at an angle to the horizontal.

The most-used modified-exponential trend curves are the simple modified exponential, $D = a - br^t$, the Gompertz curve, $\log D = a - br^t$, and the logistic curve, $D = (a + br^t)^{-1}$—where D = demand; t = time; a, b and r = positive constants; and r is less than 1.

These are all particular, or limiting, curves of the general modified exponential, $D = (a + br^t)^{\phi}$. They all assume that there will be a ceiling market, a^{ϕ}, which will be approached asymptotically. Their slope characteristics should approximate straight lines when the following slope-functions are plotted against time:

- For the simplified modified exponential, the logarithm of the slope.
- For the Gompertz, the logarithm of the ratio of the slope to the moving average.
- For the logistic, the logarithm of the ratio of the slope to the square of the moving average.

Both the Gompertz and logistic curves are S-shaped and can theoretically be applied where the data cover first the establishment, then the growth, and finally the approaching saturation of the market. Only a knowledge of the actual market can tell whether such conditions have applied or whether the inflections in the demand curve are due to other changes.

As it is difficult to make other than an arbitrary selection of the mathematical form of the trend curve without complete knowledge of the market, so it is also almost impossible to forecast future changes in trend. The problems faced by demographers in forecasting population trends, due to the reduced birthrate during the 1930–40 depression as well as the baby boom of the post World War II 1940s, are good examples.

In the vast majority of the cases, the trend will approximate either a linear or a simple-exponential growth, usually the latter. Consequently, the simpler curves are preferred.

Knowledge of the market usually suffices for ready differentiation between linear and exponential growth. Also, past demand figures can be plotted against time on arithmetic or semilogarithmic graph paper, for linear or exponential trends, respectively. If the variation is excessive, the curves can be smoothed by plotting the moving averages in the same manner.

Summary of Procedure for Use of Slope Characteristics – Table II

Plot against Time	Result Oscillates About a Straight Line Which is	Suggested Curve
Slope	Horizontal	$D = a + bt$
Slope	At an angle to the horizontal	$D = a + bt + ct^2$
Slope/moving average or Slope of log demand	Horizontal	$\log D = a + bt$
Slope/moving average or Slope of log demand	At an angle of the horizontal	$\log D = a + bt + ct^2$
Log slope	Slope down to right	$D = A - br^t$
Log [slope/moving average]	Slope down to right	$\log D = a - br^t$
Log [slope/moving average)^2]	Slope down to right	$D = (a + br^t)^{-1}$

The Method of Least Squares, With an Odd Number of Periods – Table III

Year	Sales	$y =$ log Sales	t	t^2	yt
1963	30	1.4771	–4	16	–5.9084
1964	31	1.4914	–3	9	–4.4742
1965	34	1.5315	–2	4	–3.0630
1966	33	1.5185	–1	1	–1.5185
1967	36	1.5563	0	0	0.0
1968	36	1.5563	1	1	1.5563
1969	40	1.6021	2	4	3.2042
1970	39	1.5911	3	9	4.7733
1971	45	1.6532	4	16	6.6128
	Totals:	14.0405	0	60	1.1825

$N = 9$

$$a = \frac{y}{n} = \frac{14.0405}{9} = 1.5601$$

$$b = \frac{yt}{t^2} = \frac{1.1825}{60} = 0.0197$$

$\therefore y = 1.5601 + 0.0197t$

or log sale $= 1.5601 + 0.0197t$

N.B. $t = 0$ in 1967 and progresses in years.

The Method of Least Squares, With an Even Number of Periods – Table IV

Years	Sales	$y =$ log Sales	t	t^2	yt
1962	30	1.4771	–9	81	–13.2939
1963	31	1.4914	–7	49	–10.4398
1964	34	1.5315	–5	25	– 7.6575
1965	33	1.5185	–3	9	– 4.5555
1966	36	1.5563	–1	1	– 1.5563
1967	36	1.5563	1	1	1.5563
1968	40	1.6021	3	9	4.8063
1969	39	1.5911	5	25	7.9555
1970	45	1.6532	7	49	11.5724
1971	44	1.6435	9	81	14.7915
	Totals:	15.6840	0	370	3.1790

$n = 10$

$$a = \frac{\Sigma y}{n} = \frac{15.6840}{10} = 1.5684$$

$$b = \frac{\Sigma yt}{\Sigma t^2} = \frac{3.1790}{370} = 0.00859$$

$\therefore y = 1.5684 + 0.00859\,t$

or log sales $= 1.5684 + 0.00859\,t$

N.B. $t = 0$ between 1966 and 1967, and progresses in half years.

Fitting the Curve

Linear and simple-exponential trends: The linear and simple-exponential curves can be written as $y = a + bt$, where:

$y =$ demand at year t ($= D$) for linear growth.

$y =$ log of the demand ($= \log D$) for exponential growth.

Then, for the best fit by the method of least squares:

$$a = \frac{\Sigma y \Sigma t^2 - \Sigma y \Sigma t y}{n\Sigma t^2 - (\Sigma t)^2}$$

$$b = \frac{n\Sigma t y - \Sigma y \Sigma t}{n\Sigma t^2 - (\Sigma t)^2}$$

where n is the number of periods for which the demand is known.

These formulas, and the computations, can be reduced by choosing the base year or period so that Σt becomes zero:

$$a = \Sigma y/n$$
$$b = \Sigma t y/\Sigma t^2$$

This technique is illustrated in Table III for an odd number of periods, and in Table IV for an even number.

Fitting the Gompertz Curve to Sales Data – Table V

Year	Sales	t	y = log Sales	Calculation of A	B	C
1959	1	1	0	0		
1960	1	2	0	0		
1961	2	3	0.3010	0.3010		
1962	3	4	0.4771	0.4771		
1963	3	5	0.4771	0.4771	0.4771	
1964	6	6	0.7782		0.7782	
1965	7	7	0.8451		0.8451	
1966	11	8	1.0414		1.0414	
1967	15	9	1.1761		1.1761	1.1761
1968	17	10	1.2304			1.2304
1969	20	11	1.3010			1.3010
1970	21	12	1.3222			1.3222
1971	23	13	1.3617			1.3617
Total		$n = 13$		1.2552	4.3179	6.3914
				$5A$	$5B$	$5C$

$\therefore A = 0.2510$

$B = 0.8636$

$C = 1.2783$

$$\log r = \frac{2}{13-5} \log \frac{1.2783 - 0.8636}{0.8636 - 0.2510}$$

$$= \bar{1}.95785$$

$\therefore r = 0.9075$

$$a = \frac{0.8636^2 - 0.2510 \times 1.2783}{1.7272 - 0.2510 - 1.2783}$$

$$= 2.1473$$

$$b = \frac{5.\,(1-0.9075)}{0.9075\,(1-0.9075^5)} \times \frac{(0.8636-0.2510)^2}{1.7272-0.2510-1.2783}$$

$$= 2.5135$$

$$\therefore y = 2.1473 - 2.5135\,(0.9075)^t$$

or

$$\log(\text{sales}) = 2.1473 - 2.5135\,(0.9075)^t$$

Modified exponentials: The modified-exponential trend curves can be written, $y = a - br^t$, where:

y = demand at time, t, for the simple modified exponential.

y = log demand at time, t, for the Gompertz.

y = reciprocal of demand at time, t, for the logistic.

t = 1 in year 1.

There are two standard methods of fitting a modified-exponential to the data: the "three point" method, and the Gomes method of least squares.

The three-point method is simpler, but it gives only an approximation to the best fit, rather than an exact fit. It is not recommended where there is a cyclical fluctuation about the trend, as with a trade cycle. However, it provides a starting point and check for the more precise Gomes method, and does apply in some cases.

The Gomes method makes use of tables to give an exact solution from an unbroken time series. These tables, though limited, can be found, along with a description of the method, in the references.[1,2,3]

The three-point method effectively provides three "average" values (which should satisfy the equation) by dividing the period into three equal parts. If necessary, the middle section can overlap the end section, or one or two of the data at the beginning of the period can be ignored. The averaged demands for these three periods are then calculated and used to calculate a, b and r.

If the demand data are written $y_1, y_2, \cdots y_n$, the estimate of the average, A, near the beginning is:

$$A = \frac{1}{5}(y_1 + y_2 + y_3 + y_4 + y_5)$$

and the average, C, near the end of the period is:

$$C = \frac{1}{5}(y_n + y_{n-1} + y_{n-2} + y_{n-3} + y_{n-4})$$

The estimate in the middle of the period is the average over the middle five years, when the total number of years is odd, or over the middle six years, when the total number is even. This middle-average demand is designated B. The constants a, b and r are then calculated as:

$$\log r = \frac{2}{n-5} \times \log \frac{C-B}{B-A}$$

$$a = \frac{-AC}{2B - A - C}$$

$$b = \frac{5}{(r + r^2 + r^3 + r^4 + r^5)} \times \frac{(B-A)^2}{(2B - A - C)}$$

This technique is illustrated in Table V for fitting a Gompertz curve to a set of sales data.

Confidence Limits of the Forecast

The forecaster will find that his results meet with anything from complete skepticism to blind acceptance. In the latter case, he may be criticized if sales turn out to be, say, 49,000 tons instead of his forecast 50,000 tons. Thus he is wise to include confidence limits with his forecast, although these will generally give no more than a measure of the probable variation that may occur if past trends continue.

Confidence limits are based on the extent that data points scatter from the fitted curve; the greater the scatter, the more uncertain the position of the curve and the wider the confidence limits. The derivation begins with calculation of the difference between the observed and the calculated demands. If the variation is assumed proportional to the demand (as in the simple-exponential, the logarithmic-parabola, and the Gompertz curves) the differences between the logarithms of the observed and expected demands are used.

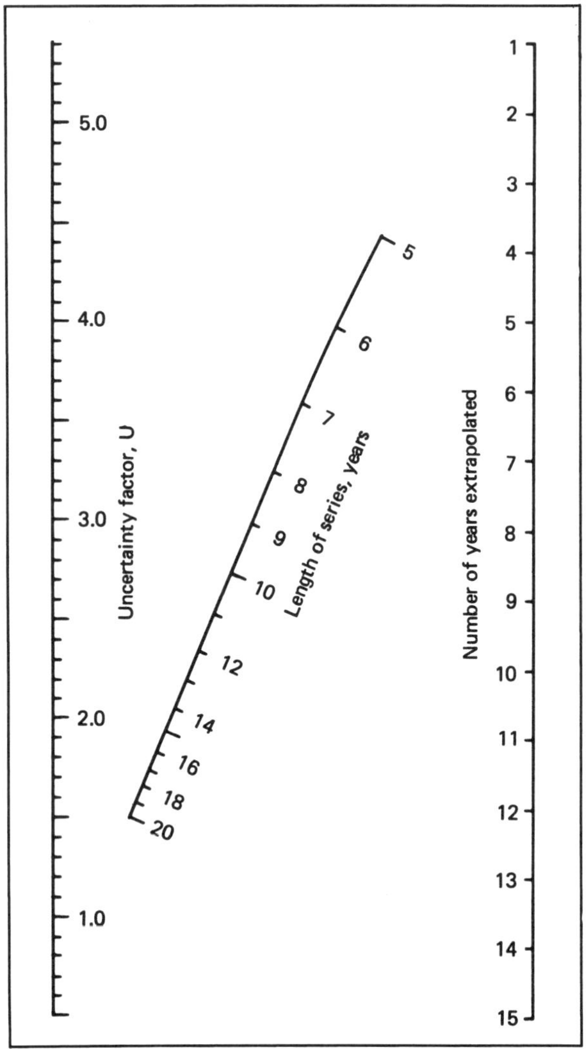

90% CONFIDENCE LIMITS to a linear trend—Fig. 2

Calculating the Standard Deviation from the Trend – Table VI

Column 1	2	3	4	5	6	7
Year	Sales	y = log Sales	t	calc.	Δ	$\Delta^2 \times 10^8$
1963	30	1.4771	–4	1.4813	–0.0042	1,764
1964	31	1.4914	–3	1.5010	–0.0096	9,216
1965	34	1.5315	–2	1.5207	+0.0108	11,664
1966	33	1.5185	–1	1.5404	+0.0219	47,961
1967	36	1.5563	0	1.5601	+0.0038	1,444
1968	36	1.5563	1	1.5798	–0.0235	55,225
1969	40	1.6021	2	1.5995	+0.0026	676
1970	39	1.5911	3	1.6192	–0.0281	78,961
1971	45	1.6532	4	1.6389	+0.0143	20,449
Total		$n = 9$			–0.0210	227,360

Columns 1, 2, 3 and 4 from Table 3

Column 5 $y = 1.5601 + 0.0197t$

Column 6 Col. 4 – Col. 5

Column 7 (Column 6)^2

$$S = \left[\frac{\Sigma\Delta^2}{n} - \left(\frac{\Sigma\Delta}{n}\right)^2\right]^{1/2} \frac{n}{n-1}$$

$$= 10^{-4}\left[\frac{227360}{9} - \left(\frac{210}{9}\right)^2\right]^{1/2} \frac{9}{7}$$

$$= 0.0178$$

N.B.: Two degrees of freedom have been used in providing the constants in the equation.

Calculation of the standard deviation for the simple-exponential trend curve of Table III is shown in Table VI. Using the procedure of Gregg, Hosell and Richardson,[1] the confidence limits (CL) are given by:

$$CL = a + b^t \pm US$$

where: S = the standard deviation of the observed data about the trend line, and
U = the uncertainty factor.

For linear trends, the value of U for 90% confidence limits can be obtained from the nomograph in Fig. 2. Thus, for example, if we use the figures from Table III:

$$\text{Log Sales} = 1.5601 + 0.0197t \quad (t = 0 \text{ in } 1967)$$

From Table VI, $S = 0.0178$.

The series covers the nine years from 1963 to 1971 inclusive, and a forecast for 1977 therefore requires extrapolation for six years. From Fig. 2, $U = 2.53$. The estimate of sales for 1977 is therefore:

$$\text{Log Sales} = 1.5601 + 0.0197 \times 10 \pm 2.53 \times 0.0178$$
$$= 1.7571 \pm 0.0450$$

Hence the sales forecast for 1977 (Log Sales = 1.7571) is 57.2, and the confidence limits (Log Sales = 1.8021 and 1.7121) are 63.4 and 51.5.

Detecting Changes in Trend

An important function of a forecast is to indicate changes in trend (particularly for the short term), in order to help the operating departments facilitate production and stock control.

One procedure for this consists of fitting a trend line by the method of least squares, either to the sales figures or to a moving average figure. (The use of the moving average smooths out much of the random variation.) Two lines are then plotted above and below the trend line at a distance of two standard variations. The trend line and these limit lines are then extended into the future, and subsequent sales figures plotted on the graph each month (Table VII and Fig. 3).

Should the sales figures fall outside the limits, a change in the trend is indicated, and a new trend line and new limits are calculated on the basis of points appearing as

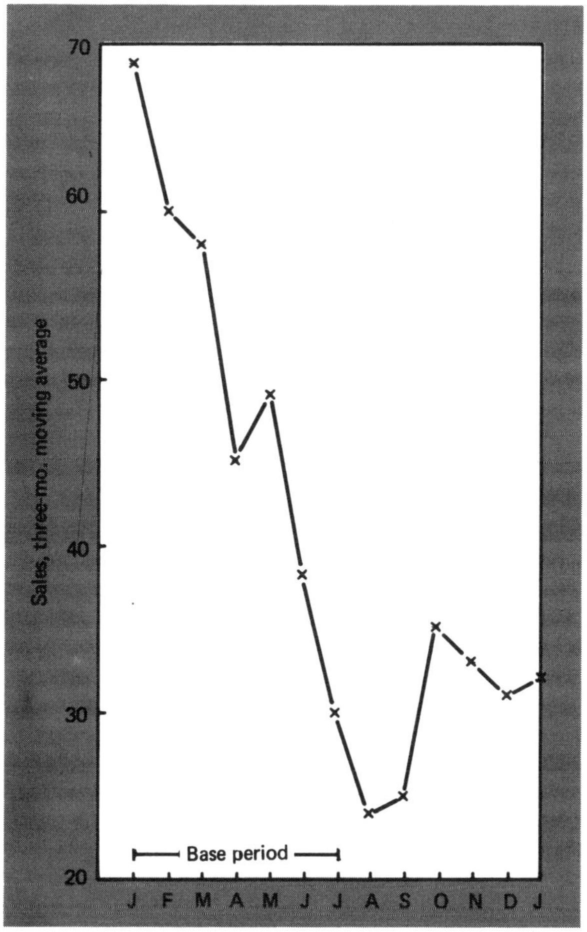

PLOTTING changes in a sales trend—Fig. 3

a new trend. If less than six points are available, the new trend should be recalculated each month, until six or seven points have been used for the basis. Reports indicate that, with this technique, a retrospective survey of seven years of sales correctly predicted a change in market trends before it otherwise would have been detected.[4] In addition, no change was predicted that did not actually occur.

A refinement of this procedure consists of plotting the limits at three standard deviations for the trend line, but with each deviation calculated as a percentage of the trend. This technique is reported to have correctly detected changes in the trend two to five months earlier than the previously mentioned procedure.

Although mathematical techniques are powerful tools when applied to forecasting, the value and accuracy of the forecaster's predictions will decrease if he becomes a slave to these techniques. Some skilled mathematicians claim that free-hand extrapolation of past-demand curves yields forecasts of the same accuracy as do the mathematical procedures—and they are often right. It is for this reason that only the simpler trend-formulas have been used in the worked examples. Application of the more sophisticated trend-curves[1,3,5] requires a better knowledge of the market, and unless the forecaster is convinced that these curves work, it is advisable to use the simpler forms. ■

Calculating the Changes in a Trend – Table VII

Month	t	y (3-month moving avg. sales)	t^2	yt	y calc.	Δ	$\Delta^2 \times 10^4$
Jan.	–3	69	9	–207	68.11	0.89	27,921
Feb.	–2	60	4	–120	62.04	–2.04	41,616
March	–1	58	1	– 58	55.97	2.03	41,209
April	0	45	0	0	49.90	–4.90	240,100
May	1	49	1	49	43.83	5.17	267,289
June	2	38	4	76	37.76	0.24	576
July	3	30	9	90	31.69	–1.69	28,561
Total	0	349	28	–170		–0.30	627,272

$$a = \frac{\Sigma y}{n} = \frac{349}{7} = 49.9$$

$$b = \frac{\Sigma ty}{\Sigma t^2} = \frac{-170}{28} = -6.07$$

$$\therefore y = 49.9 - 6.07t$$

$$\Delta = y - y \text{ calc.}$$

$$\rho = \left[\frac{\Sigma \Delta^2}{n} - \left(\frac{\Sigma \Delta}{n}\right)^2\right]^{1/2}$$

$$= \left[\frac{62.7272}{7} - \left(\frac{-0.30}{7}\right)^2\right]^{1/2}$$

$$= 2.99$$

$$\therefore 2\rho = 5.98$$

References

1. Gregg, J. V., Hosell, C. H., and Richardson, J. T., "Mathematical Trend Curves—An Aid to Forecasting," Oliver & Boyd, Edinburgh, 1965.
2. Coutie, G. A., others, "Short Term Forecasting," Oliver & Boyd, Edinburgh, 1964.
3. Woodward, R. H., and Goldsmith, P. L., "Cumulative Sum Techniques," Oliver & Boyd, Edinburgh, 1964.
4. Churchman, C. W., Ackoff, R. L., and Arnoff, B. L., "Introduction to Operations Research," John Wiley & Sons, New York, 1957.
5. Oliver, F. R., Methods of Estimating the Logistic Growth Function, *Applied Statistics,* **13,** 1965, pp. 57–66.

Meet the Author

G. E. Mapstone is Deputy Head, Dept. of Chemical Engineering, Swinburne College of Technology, P.O. Box 218, Hawthorn 3122, Victoria, Australia—a position he has held since 1967. Prior to accepting this position, he worked at shale oil refineries in Australia and in the cosmetic and chemical engineering industries of So. Africa. A member of many professional societies, he holds a B.Sc. then M.Sc. with 1st class honors in chemistry from the U. of New Zealand, a Ph.D. (applied chemistry) and a B.Com. (administration) from the U. of So. Africa, and a D.Sc. from Canterbury University in New Zealand.

Revised Technique for Predesign Cost Estimating

After investigating several methods for making predesign estimates of capital costs, the authors have developed a revised procedure that usually has an error range of less than 25%.

D. H. ALLEN, University of Nottingham, U.K., and R. C. PAGE, Towry Law Ltd.

The early stages of a project usually involve studies of alternative processes, plant configurations and types of equipment. At this so-called predesign stage, cost estimates do not need to be highly accurate because they are used to make broad comparisons rather than for preparing actual budgets.

Typical methods depend on estimating by analogy—comparing the proposed plant with similar existing plants, through approaches such as exponential adjustment for capacity, factoring, and conversion of delivered-equipment cost to total plant investment.

We have investigated four analogous estimating techniques and have developed a revised method that, while incorporating some worthwhile characteristics of other methods, attempts to avoid their shortcomings. Our method primarily aims at screening alternatives. It is only concerned with plant capital investment and does not include other initial costs such as land and working capital.

Table I lists the cost and accuracy of several kinds of capital investment estimates for chemical plants [*1*]. These data show that, irrespective of a project's size, the cost of preparing an estimate increases about fivefold as the allowable error is reduced from ±30% (study) to ±10% (definitive). Thus, one must avoid wasting time and money in preparing an estimate more accurate than is really necessary. Further, there is an incentive to reduce the cost of preparing an estimate, either by decreasing the amount of data needed or by simplifying the procedure.

Estimating by Analogy

In Hill's estimating method for low-pressure fluid processes [*2*], major equipment items identified from the flowsheet are classified as being equivalent to one or two "units." Simple items (e.g., carbon-steel towers) are single units, while more-expensive or complex pieces (e.g., furnaces, stainless-steel towers) are double units. Extra units are added as appropriate for storage facilities, and the number of units for an item that is operating at over 100 psig are multiplied by a pressure factor.

Hill concluded that for a petrochemical plant producing 10 million lb/yr of total product, each unit had an installed cost of $30,000 at a reference date corresponding to a Marshall and Swift (M&S, formerly Marshall and Stevens) equipment cost index of 185 (1954). The installed-equipment cost from this calculation is adjusted by a capacity exponent of 0.6, and then converted to grass-roots investment by using appropriate percentages for other direct and indirect costs [*3*]. Hill noted that his estimating method usually gave results that were accurate within ±25%.

In 1963, Zevnik and Buchanan [*4*] proposed a method for fluid-type chemical plants that was based on capacity, process complexity and the number of functional units. A complexity factor (CF) accounted for materials of construction and extreme operating conditions:

$$CF = 2 \times 10^{(f_t + f_p + f_m)} \qquad (1)$$

where f_t, f_p and f_m are factors for temperature, pressure, and materials of construction.

Zevnik and Buchanan defined a functional unit as all the equipment necessary to carry out a single process function. They obtained the direct investment per functional unit from a correlation relating investment to production capacity and CF. Total plant investment was estimated as:

$$I_g = 1.33\ N_f(IFU)[(ENR)/(300)] \qquad (2)$$

where I_g is grass-roots plant investment, N_f is number

This article is based on a paper presented at the International Cost Engineering Symposium, London, Oct. 6–9, 1974.

Originally published March 1975.

of functional units, IFU is direct investment per functional unit, and ENR is the *Engineering News Record* construction cost index (1939 = 100).

The factor of 1.33 allowed for utilities and general facilities, and ($ENR/300$) updated the costs. The authors obtained an accuracy of ±25% or better in their illustrations.

Modified Methods

Stallworthy's method [5] is based on Courtaulds data and was originally intended only for internal use, mainly on petrochemical plants. Similar in some respects to Zevnik and Buchanan's approach, it modifies functional units into process steps, and attempts to account for the influence of sidestreams and recycles. Stallworthy's equation is:

$$I_b = (0.0075/A)\sum_{1}^{s} (N_s F_t F_p F_m R) \quad (3)$$

where I_b is battery-limits plant investment (£); A is the cost-capacity factor; s is the number of process streams; N_s is the number of significant process steps in a stream; F_t, F_p and F_m are temperature, pressure and material factors, obtained as number-weighted averages from constituent process steps; and R is the ratio of stream flow to output stream.

In the original paper, the value of the constant (0.075) was in error. Also, data for determining F_t, F_p and F_m were not given. However, Wilson [6] provided correlations based largely on information from Stallworthy. The cost-capacity factor is correlated by Stallworthy with the production capacity. A full mass balance is necessary to calculate R for each stream, and I_b must be corrected for inflation since 1967. Stallworthy claimed his method was accurate within ±15% of a detailed estimate.

Wilson's estimating technique, published in 1971 [6], is the most recent. He combines some features of Stallworthy with the Lang factor as developed by Miller [7]. Wilson's equation is:

$$I_b = fN_m(AUC)F_t F_p F_m \quad (4)$$

where I_b is the battery-limits investment (£), f is the investment factor, N_m is the number of main-plant items, (AUC) is the average unit cost of main-plant items, and F_t, F_p and F_m are the Stallworthy factors.

Main-plant items include all major equipment except pumps. AUC is correlated with the average throughput of main-plant items. The investment factor, f, is related to AUC by the Lang factor and by applying Miller's approach, which assumes the factors are themselves functions of plant size and hence of throughput or AUC. This method also requires a cost-index correction applied to I_b.

Wilson considered data from 16 plants, most of which gave estimating accuracies within ±30%.

We have not mentioned other analogous estimating techniques here, either because they are specialized (e.g., Bridgwater's for acid extraction of metals [8]), or because they have not been published in detail outside of the originating organization (e.g., Le Page's method [9]).

How They Compare

To evaluate the estimating methods described above, we applied them to a group of four fluid-type plants, each having two different capacities. Table II presents the

Characteristics of Chemical Plant Capital-Investment Estimates—Table I

Type of Estimate	Other Names	Usual Basis	Error, %	Cost of Estimate for Various-Sized Projects, $: Less Than $1 MM	$1 to $5 MM	$5 to $50 MM
Order of magnitude	Ratio estimate	Previous similar cost information	>±30			
Study	Factored estimate	Knowledge of flowsheet and major equip.	±30	2,000–6,000	5,000–12,000	8,000–16,000
Preliminary	Initial budget, scope estimate	Sufficient data for budget preparation	±20	6,000–15,000	12,000–25,000	20,000–40,000
Definitive	Project control estimate	Detailed data, but not complete drawings	±10	10,000–25,000	25,000–50,000	45,000–100,000
Detailed	Firm estimate, contractor's estimate	Complete drawings and specifications	±5			

Comparison of Four Estimating Methods—Table II

Product	Capacity, MM lb/yr	Actual Investment, MM$	Percentage Error: Hill	Zevnik	Stallworthy	Wilson	References
Hydrogen cyanide	17.5	2.88	+98	+52	+26.5	+58	10,11,12,13
	70	8.35	+72	+13.5	+17.5	+41.5	
Acetylene	25	6.85	−11.5	+50.5	−41	−52.5	10,12,14
	100	13.5	+30	+66	−25	−37	
Ammonia	42	3.34	>+1,800	+47	−11.5	−67.5	10,12,13
	70	6.25	>+1,800	+26	−19.5	−70.5	
Phthalic anhydride	30	9.67	−59.3	−59.7	+0.5	−10.5	4,10,12,14,
	80	19.23	−63	−64.3	−3	−10	15,16
Average Equivalent (positive) error			88.2*	72.9	23.4	92.3	

*Excluding ammonia plants

Summary of Estimating Method

Information Required

1. Plant flowsheet that identifies main-plant items and process streams.
2. Total process-stream input, lb mol/yr.
3. Extreme temperature and pressure conditions, if any.
4. Materials of construction for main-plant items.
5. Operating phases of each main-plant item.
6. Whether any unusually high or low, direct or indirect, initial costs are expected.

Estimating Procedure: Find the Following

1. *CAP:* determine total process input, lb mol/yr.
2. *N:* obtain number of main-plant items from flowsheet.
3. *FF:* calculate flow factor from flowsheet and Eq. (10).
4. *PF:* obtain number of volume items from operating phase data and calculate phase factor from Eq. (11).
5. *TP:* calculate plant throughput from Eq. (9).
6. *EXP:* calculate cost-weighted plant capacity exponent from knowledge of main-plant items, Table VI and Eq. (14).
7. *BIC-TP:* construct plant log-log plot, as in Fig. 4, through the point where $TP = 2.5 \times 10^6$ and $BIC =$ \$7,000. Line has a gradient of *EXP*.
8. *BIC:* read off basic item cost from plant line, knowing plant *TP*.
9. *SF:* determine $F_{t\max}$ and $F_{p\max}$ from extreme conditions and Fig. 1 and 2. Calculate $F_{m\text{mean}}$ for main plant items, using Table III. Calculate state factor from Eq. (8).
10. *DEC:* calculate delivered-equipment cost from Eq. (7).
11. Grass-roots investment: divide *DEC* by a factor of 15 to 30%. The factor is 21% for a plant with average initial costs.
12. Updated investment: use M&S equipment cost index with June 1972 reference data value of 331.

actual grass-roots investments and the results obtained with the estimating techniques.

We were careful to be consistent in assumptions made about the plants and in the application of each method. Stallworthy's and Wilson's procedures, which estimate battery-limits investment, were converted to grass-roots costs by multiplying by 1.35, a factor developed by averaging published data. Also, the results from these methods were adjusted by 2.40 to convert from £ sterling to U.S. dollars. All results were updated to December 1971 costs, using the M&S index, and percentage errors were calculated in the usual manner (estimated minus actual, divided by actual, times 100).

The hydrogen cyanide, acetylene and phthalic anhydride units are low-pressure, fluid-type petrochemical facilities—the kinds of plants for which Hill's method was designed. Nevertheless, this method gives reasonable results only for acetylene. The meaningless figure (1,800% error) for the ammonia plant confirms that Hill's procedure is not applicable to high-pressure operations.

Hill takes no account of operating temperature, and uses a universal capacity exponent of 0.6. His method is based on the output of the main process stream, which may not be representative of throughputs inside a plant. This could be a major source of error.

The results from Zevnik and Buchanan's technique, while somewhat better and more consistent than Hill's, were nonetheless of little value. The major difficulty, also noted by others [*5,6*], was deciding the correct number of functional units. The original paper gives little guidance on this point.

The method, like Hill's, is based on the output of the main stream. We found an inconsistency in the correlations that relate *IFU* to *CF* and production capacity, which casts doubt on their validity. Referring to Eq. (1) for a normal base-case plant, f_t, f_p and f_m are all zero, so that *CF* equals 2, its minimum possible value. In the correlations, however, Zevnik and Buchanan show a line for $CF = 1$, and the lines for higher values of *CF* are parallel to it. This looks as if the correlation developed by Zevnik and Buchanan may be the result of speculation rather than calculation!

Good Results With Stallworthy

In most cases, Stallworthy's method gave very good results. It does, however, require full mass-balance information, calling for either considerable computations or assumptions. Obtaining the flowrates of all streams was the major effort in applying this method.

Stallworthy's use of flow ratios is an attempt to base the calculations on a realistic average throughput rather than solely on output. This is still not entirely satisfactory because the flow ratios are taken relative to the main output, which may not be similar to the flowrates through individual plant items.

Precisely what constitutes a process step is also not clear, although judging from the results it appears that we usually interpreted this factor correctly. A similar problem comes up in defining the main process stream.

The performance of Stallworthy's technique is influenced by the very extensive price data on which the correlations were based. These were obtained within his company, and were based mainly on large samples.

The most recent method, Wilson's, gave very disappointing results, even though in some ways it should have been an improvement on Stallworthy.

We did find Wilson's use of main-plant items more straightforward than either the functional-unit or process-step approaches. One must know the throughput in each main-plant item, calculated as the sum of all entering streams, and then find the average plant throughput. However, while better than Stallworthy's stream ratios in representing the effect of stream complexity, this still requires a full mass balance.

Wilson specifically excludes pumps from the list of main-plant items, but says that ten pumps are perhaps equivalent to one main-plant item. Because the number of pumps affects the amount and cost of piping, ignoring the pumps may lead to underestimates. The *AUC*/throughput correlation gives a straight log-log plot passing through a scatter of 16 points having a slope of 0.675, the universal capacity exponent. This also may be a source of error.

Significance of Plus and Minus Errors

All estimating methods use the conventional way of expressing accuracy as an error range of $\pm x\%$ of the actual investment. This infers that equal under- and overestimates have the same significance. However, there is no limit to an overestimate, while the limit to an underestimate is zero since a negative value is meaningless.

The actual error range is thus -100% to $+\infty\%$. If these extremes are regarded as having equal significance, it follows that an underestimate is more significant than an overestimate of the same magnitude. In fact, on this basis an underestimate of *y* times should be of equal significance to an overestimate of *y* times. For example, an error of -50% means the estimate is one-half the actual value. The equivalent positive error is therefore one having twice the actual value, i.e. $+100\%$. The relationship between equally significant errors of $+x_1\%$ and $-x_2\%$ is:

$$x_1 = 100x_2/(100 - x_2) \qquad (5)$$

$$x_2 = 100x_1/(100 + x_1) \qquad (6)$$

The average equivalent positive errors for the estimates made by the four methods are given in the bottom row of Table II.

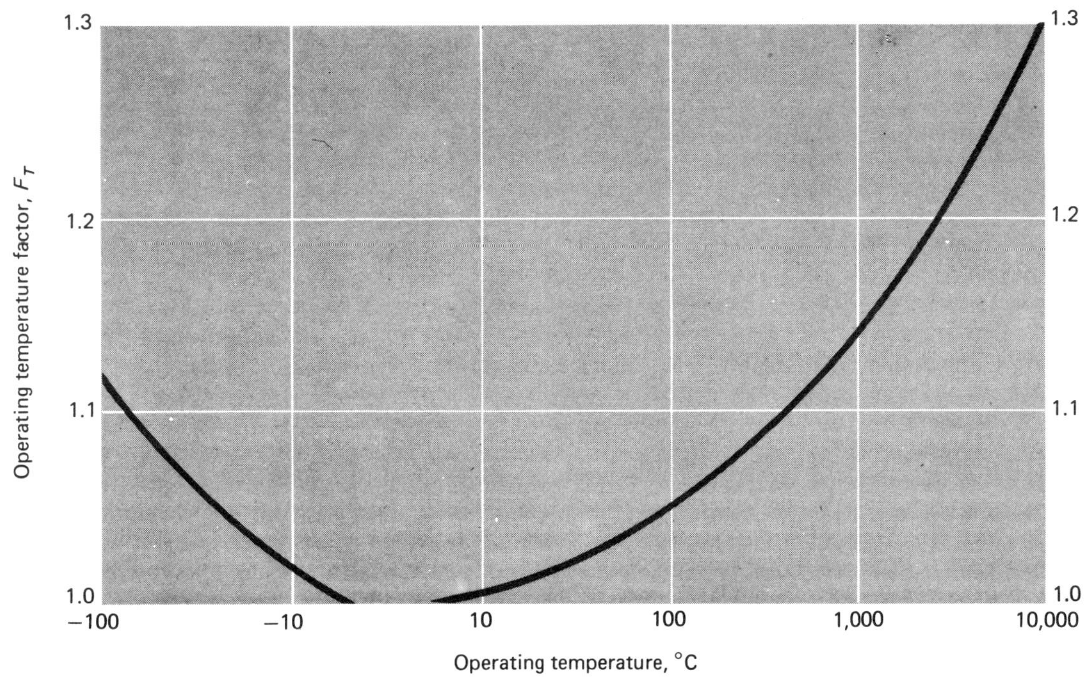

OPERATING TEMPERATURE factor for estimating main-plant items—Fig. 1

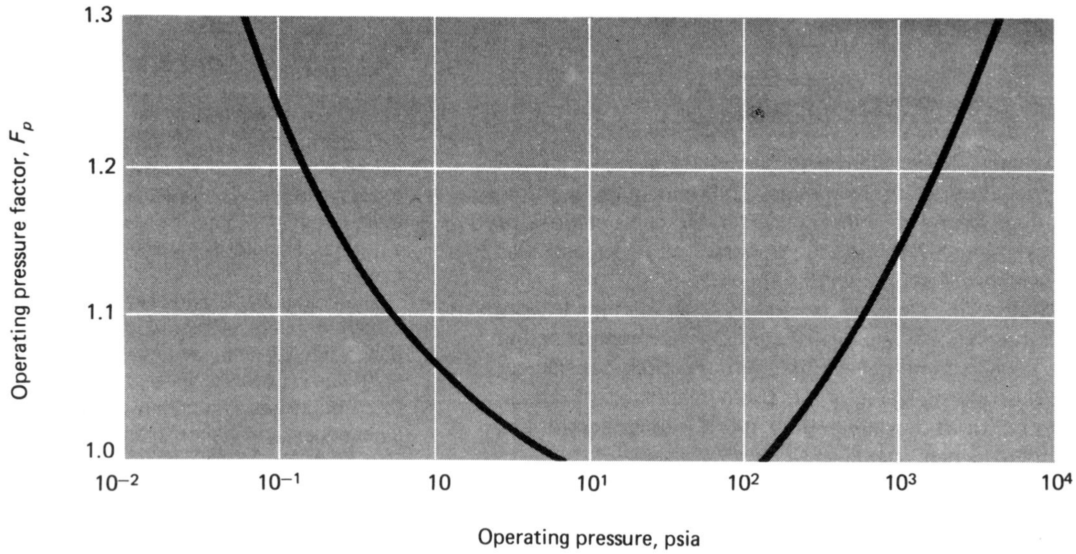

OPERATING PRESSURE factor for estimating main-plant items—Fig. 2

It is possible, of course, that the sources for the actual investment figures were significantly in error. Although all of the estimates for the hydrogen cyanide plants have positive errors, indicating that perhaps the actual figures are too low, these data were averaged from two separate sources that agreed closely. The uniformly negative errors on the phthalic anhydride estimates suggest that the actual numbers are too high. However, Stallworthy's method applied to the phthalic anhydride plants gave excellent results.

None of the estimating methods considered "learning" —that is, the real costs of comparable plants decrease with time due to the experience gained on previous projects. Even so, on an overall basis, the effect of learning would not have improved the direction of the estimates. And in any case, there are virtually no suitable data on how "learning" affects investment.

Developing a Better Method

Wilson has stated that an analogous estimating method should be versatile, accurate to within $\pm 30\%$, based on realistic throughput, require only data that are normally available early in the project, and consume less time and effort than a traditional estimate. We would go further and stipulate that for predesign purposes the estimate's accuracy should be within $+25\%$ to -20%, and that the data needed should not require comprehensive mass-balance calculations or assumptions.

From experience with the other methods, we decided that delivered equipment cost should be estimated first and converted into plant investment by applying a suitable factor. The equation for delivered equipment cost (DEC) is:

$$DEC = N(SF)(BIC) \tag{7}$$

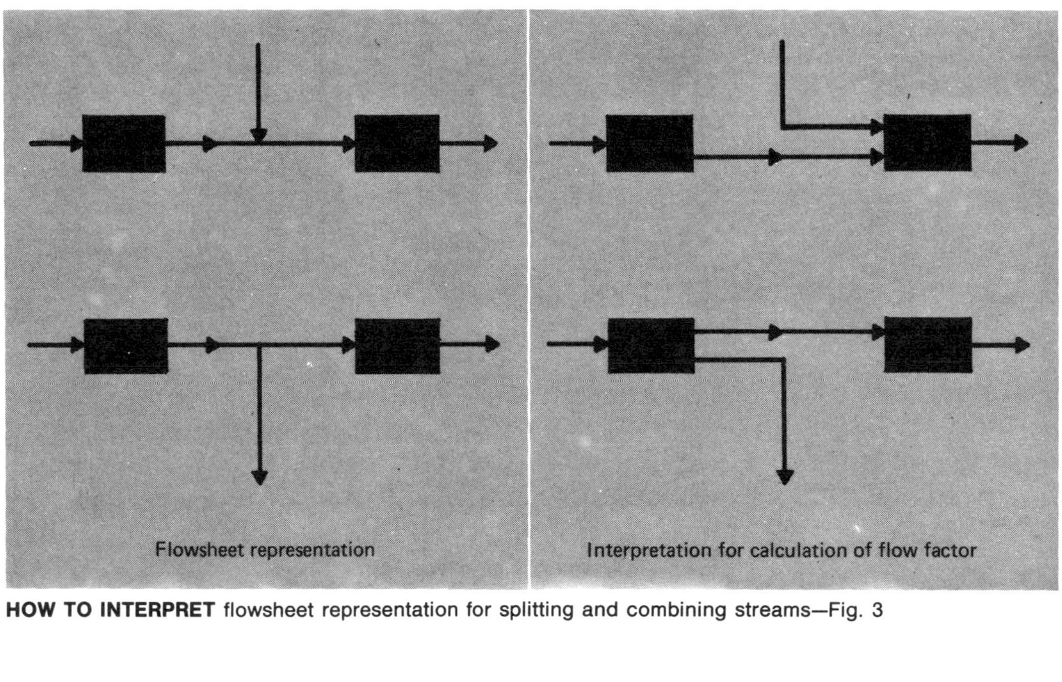

HOW TO INTERPRET flowsheet representation for splitting and combining streams—Fig. 3

where N is the number of main-plant items, SF is the state factor, and BIC is the basic item cost.

N is easily obtained from the flowsheet, and accounts for all equipment items shown, including pumps and tanks. SF takes into account operating temperatures and pressures, and materials of construction. The question arises as to whether the effects of these variables are mutually independent. Does a high process temperature, for example, also infer a special material of construction, and which factor takes account of this? This question, and its implications on whether SF should be multiplicative or additive, is discussed by at least one author [17]. The most practical approach appears to be a multiplicative function, although this is not entirely satisfactory.

We were not able to improve on Stallworthy's data as given by Wilson [6] on the effects of temperature, pressure and material of construction. These factors were included in the model, and the correlations are shown in Fig. 1 and 2 and in Table III. We found that for F_t and F_p any advantage in calculating or assuming values for all main-plant items in order to derive a number-weighted mean value did not justify the effort involved. The simpler procedure of using maximums corresponding to the most extreme conditions was satisfactory. For F_m, however, a number-weighted mean value is better. Therefore, SF is:

$$SF = F_{t\,max} F_{p\,max} F_{m\,mean} \tag{8}$$

Basic Item Cost

BIC in Eq. (7) is the effective cost of a "standard" main-plant item for the plant being estimated, i.e., before adjusting for operating conditions. It is a function of a throughput variable, TP, defined as:

$$TP = (CAP)(FF)(PF) \tag{9}$$

where CAP is the total feed to the plant in lb mol/yr, FF is a flow factor, and PF is a phase factor. The objective is to obtain an appropriate measure of plant throughput. While other methods use a mass basis for capacity or throughput, the physical size of a plant is related more to its volumetric throughput, especially when gases are involved. Hence, the use of lb mol/yr in conjunction with a phase factor.

Since we want to avoid mass-balance information, the previous approaches to process-stream complexity are not possible. Instead, a flow factor is defined as:

$$FF = \sum_{1}^{N} \left(\begin{matrix} \text{Number of input and output streams} \\ \text{of each main-plant item} \end{matrix} \right) \Big/ N \tag{10}$$

This equation accounts for all process streams when the splitting and combining streams are interpreted as illustrated in Fig. 3. The minimum value of FF, for a single-stream plant, is 2.

Main-plant items are classed either as volume items (gas or vapor phase) or mass items (liquid phase). Calculation shows that the presence of more than about 2% by mass of a gas phase in a mixed-phase plant item results in a greater volumetric gaseous throughput. Volume items are therefore extended to include those with mixed gas/liquid phases. PF in Eq. (9) is a function

Material of Construction Factor—Table III

Material	Factor, F_m
Carbon steel (mild)	1.0
Bronze	1.05
Carbon/molybdenum	1.065
Aluminium	1.075
Cast steel	1.11
Stainless steel	1.28
Worthite alloy	1.41
Stainless steel	1.5
Hastelloy C alloy	1.54
Monel alloy	1.65
Nickel/inconel alloy	1.71
Titanium	2.0

Plants Used in Developing the BIC-TP Correlation—Table IV

Plant	Stated Capacity	Actual Battery Limits Investment, June 1972, \$1,000
Isopropanol plant	30,000 ton/yr isopropanol	2,260
Alpha-olefins plant	33,600 ton/yr C_6—C_{18} alpha-olefins	3,890
Unionfining hydrogenation plant	30,000 bbl/d kerosene product	2,700
Cyclohexane plant	100,000 ton/yr cyclohexane	1,380
Arofining plant for removal of aromatics	10,000 bbl/d kerosene feed	1,900
Adip plant for removal of H_2S	925 ton/d natural gas feed (15 vol. % H_2S)	1,755
SNPA-DEA sweetening plant	220 million ft^3/d natural gas feed (26% H_2S, 6.5% CO_2	8,760
Houdriforming reforming plant	10,000 bbl/d mid-continent naphtha to pretreater	3,900

* Actual battery limits investment, June 1972.

Delivered-Equipment Cost and Other Derived Data for the Correlation—Table V

Plant	*DEC* \$1,000	*N*	$F_{t\,max}$	$F_{p\,max}$	$F_{m\,mean}$	*SF*	*BIC* \$1,000	*CAP* 10^6 Lb Mol/Yr	*FF*	*VI*	*PF*	*TP*, 10^6 Lb Mol/Yr	*EXP*
Isopropanol (1)	706.6	22	1.09	1.31	1.0	1.43	22.5	14.1	3.09	12	0.55	24.0	0.57
Alpha-olefins (2)	1,220.4	26	1.09	1.08	1.0	1.18	39.8	9.1	3.08	21	0.81	22.7	0.75
Unionfining (3)	846.2	11	1.10	1.16	1.0	1.28	60.1	18.5	2.64	8	0.73	35.7	0.79
Cyclohexane (4)	433.6	17	1.07	1.06	1.0	1.13	22.6	9.5	2.71	10	0.59	15.2	0.66
Arofining (5)	595.6	17	1.09	1.14	1.0	1.24	28.3	9.3	2.82	11	0.65	17.0	0.80
Adip (6)	533.7	10	1.06	1.06	1.05	1.18	45.2	39.6	2.80	6	0.61	67.6	0.61
SNPA-DEA (7)	2,751.8	13	1.08	1.17	1.11	1.40	151.2	222	2.85	7	0.54	341.7	0.61
Houdriforming (8)	1,222.3	14	1.11	1.11	1.0	1.23	71.0	19.8	2.79	11	0.79	43.6	0.80

of the proportion of volume items, *VI*, in a plant:

$$PF = 0.0075 + VI/N \tag{11}$$

The constant 0.0075 represents the gas- and liquid-phase density ratio of a typical hydrocarbon stream. It ensures that in the limiting case of a plant with no volume items the value of *TP* would be reasonable.

Correlating the Factors

To develop a correlation between *BIC* and *TP*, data from the investments in actual plants are needed. We made the correlation from eight fluid-type petrochemical plants, obtaining necessary data from recent literature [*18,19,20*]. Battery-limits investments were brought to a common reference data of June 1972, using the M&S index. See Table IV.

We calculated a point on the *BIC-TP* correlation for each plant after converting the battery-limits investment to a delivered-equipment cost, *DEC*, using an average percentage factor of 31.2 obtainėd from over 20 such factors reported in the literature. The *DEC* for each plant and other derived data are shown in Table V.

In each case, the number of main-plant items, *N*, was easily obtained from the flowsheet. We found the state factor, *SF*, via. Eq. (8), a knowledge of plant conditions, Fig. 1 and 2, and Table III. Then we calculated the basic item cost, *BIC*, from Eq. (7).

We developed the flow factor, *FF*, from the flowsheet and Eq. (10), determined the number of volume items, *VI*, from each flowsheet, and applied Eq. (11) to find the phase factor, *PF*. Values of *BIC* and *TP* for each plant are shown as the plot points in Fig. 4.

In comparable previous correlations, one line has in

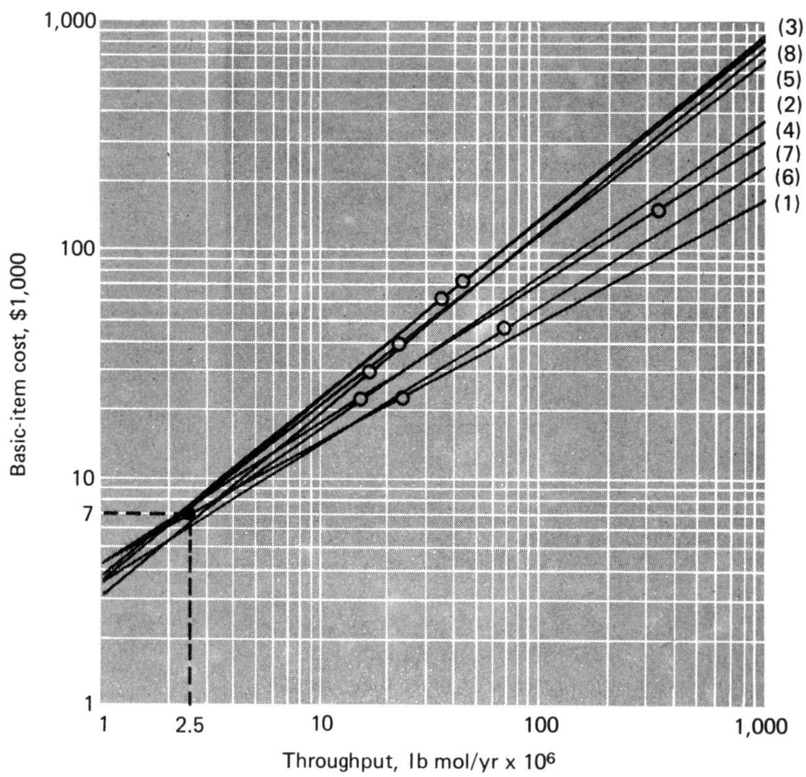

CORRELATIONS of basic-item cost with throughput—Fig. 4

effect been drawn through a scatter of points such as that in Fig. 4. Wilson's method, in particular, suffers from this. With this type of correlation, the gradient is the cost-capacity exponent. Since it is known [*14,21*] that this is not the same for all plants, we found the separate exponents for each plant (developed from the individual equipment exponents for main plant items) and drew a line having the appropriate gradient through each point. Investment in terms of main-plant items can be expressed using a plant exponent or individual-item exponents, so that:

$$\sum_{i=1}^{N} c_i R^{EXP} = \sum_{i=1}^{N} c_i R^{e_i} \tag{12}$$

where c_i is the base cost of main-plant item i, R is the ratio of new to base capacity, EXP is the plant exponent, and e_i is the exponent of main-plant item i. Rearranging:

$$EXP = \left[\log \sum_{i=1}^{N} (c_i R^{e_i}) - \log \sum_{i=1}^{N} c_i\right] \Big/ \log R \tag{13}$$

Because Eq. (13) is not a very convenient expression to include in an estimating method, we investigated the suitability of a straightforward cost-weighted exponent that is independent of R:

$$EXP = \sum_{i=1}^{N} c_i e_i \Big/ \sum_{i=1}^{N} c_i \tag{14}$$

Calculating Exponents

We used the equipment exponents and data given by Guthrie [*15*] to cost a wide range of equipment at comparable levels of performance, namely an input of about 50,000 lb/h of a "normal" feed. This rate is near the middle of the equipment capacity range. Using these data (Table VI), we evaluated, with the aid of a computer, *EXP* for several imaginary plants containing from 7 to 48 equipment items, using Eq. (13) and (14) for values of *R* from 1,000 to 0.001. The maximum deviation given by Eq. (14) compared with Eq. (13) was −11% for a case when *R* equaled 0.001. In the more realistic range from 20 to 0.05, the maximum deviation was under 5%. The corresponding values of *EXP* varied from 0.38 to 0.85, covering the range most likely to be met in practice. We therefore adopted the simpler Eq. (14).

Table V gives values of *EXP* for each of the real plants. These were developed from the flowsheets, Table VI and Eq. (14). As shown in Fig. 4, lines with these gradients converge and pass close to the point where *TP* equals 2.5×10^6 lb mol/yr and *BIC* equals $7,000.

Working back, the greatest errors in *BIC* caused by scaling up from this reference point with the appropriate exponent were +14.8% and −9.1%, for the arofining and alpha-olefins plants, respectively, well within the acceptable range. This reference point may be compared with Hill's use of $30,000 for a unit of a 10×10^6-lb/yr plant. In our method, however, plant throughput and plant items are interpreted more realistically, and scaling up from the reference point uses an exponent specific to the plant in question, instead of an invariant universal exponent.

Capacity Exponents and Relative Costs—Table VI

Equipment item	Capacity Exponent, e_i	Relative Base Cost, c_i, $1,000, 1968
Process Furnace	0.85	135.0
Direct Fired Heater	0.85	103.5
Boilers (industrial)— 15 psig.	0.5	92.0
150 psig.	0.5	101.2
300 psig.	0.5	115.0
600 psig.	0.5	138.0
Packaged Boiler Unit	0.7	60.0
Shell & Tube Exchanger	0.65	6.5
Kettle, Reboiler Exchanger	0.65	8.8
U Tube Exchanger	0.65	5.5
Cooler	0.66	6.8
Cooling Tower Facilities	0.6	9.9
Tower with Trays	0.73	33.5
Tower with Packing	0.65	35.2
Pressure Vessel—Vertical	0.65	7.6
Horizontal	0.6	5.0
Storage Tank	0.3	6.0
Horizontal Pressure Storage Vessel	0.65	4.8
Spherical Pressure Storage Vessel	0.7	8.0
Centrifugal Pump—		
Centrifugal/motor	0.52	1.5
Centrifugal/turbine	0.52	3.0
Reciprocating Pump—		
Reciprocating/motor	0.7	6.0
Reciprocating/steam	0.7	1.1
Process Gas Compressor—1000 psig.	0.82	85.0
Air Compressor—125 psig.	0.28	36.5
Crushers—Cone	0.85	12.0
Gyratory	1.2	3.0
Jaw	1.2	4.7
Pulverizers	0.35	23.4
Mills—Ball	0.65	4.4
Roller	0.65	40.0
Hammer	0.85	8.0
Evaporators—Forced Circulation	0.7	270.0
Vertical Tube	0.53	37.2
Horizontal Tube	0.53	30.4
Jacketted Vessel	0.6	32.0
Hoppers—Conical	0.68	0.1
Silos	0.9	0.4
Blowers and Fans	0.68	9.5
Crystallizers—Growth	0.65	385.0
Forced Circulation	0.55	276.5
Batch	0.7	32.5
Filters—Plate and Press	0.58	4.3
Pressure Leaf (wet)	0.58	5.3
Pressure Leaf (dry)	0.53	15.1
Rotary Drum	0.63	17.5
Rotary Disk	0.78	31.0
Dryers—Drum	0.45	30.0
Pan	0.38	12.5
Rotary Vacuum	0.45	43.4

Accuracy of Revised Estimating Method—Table VII

Plant		Percentage Error in Estimate
Product	Capacity, 10^6 Lb/Yr	
Hydrogen cyanide	17.5	+16.9
	70	− 7.2
Acetylene	25	+18.6
	100	− 2.3
Ammonia	42	−10.1
	70	−23.2
Phthalic anhydride	30	−15.5
	80	−15.1
Average equivalent error (positive)		15.3

The final step is to convert *DEC* into total plant investment, using a percentage factor to account for other direct and indirect costs. Haselbarth and Berk [22] provide a breakdown of the *DEC* percentage factor and give an indication of how *DEC* can vary as a percentage of total plant investment. For fluid-type facilities with "average" costs, *DEC* is given as 21% of the total grass-roots investment. However, it can vary between 15 and 30%, depending on whether the other costs are expected to be higher or lower than "average." The appropriate value is used to convert *DEC* to grass-roots investment, which is then updated with the M&S index.

Accuracy Is Acceptable

We applied the revised method to the eight plants previously estimated by the other techniques, assuming a normal *DEC* factor of 21%. Results are given in Table VII. As can be seen, the target error range of +25% to −20% is achieved in seven cases. The eighth, at −23.2%, is not far outside. The average equivalent error of 15.3% is well within the acceptable limit for predesign estimating purposes.

Comparing Table VII with Table II, our approach gives better results than Stallworthy's, the only other method investigated that gave acceptable results. In addition, it does not require the detailed mass balance information necessary for Stallworthy's method. It is also simpler to use.

Our method as it stands now is limited by the range of data used in its development. Thus, the *TP* range is 15 to 340 lb mol/yr; published F_t and F_p data range from −100°C to 10,000°C (the latter figure appears suspect) and from 8×10^{-2} to 5×10^3 psia; and F_m is only known for a limited range of materials.

The user may substitute data more relevant to the type or capacity of plant being estimated, or more recent cost information, using the same basic procedure. Data that may be substituted include:

1. Capacity exponents and relative costs of main-plant equipment items in Table VI.
2. Plant data used to obtain the datum point on the *BIC-TP* correlation, and the corresponding reference date for the cost index.
3. Data to derive the F_t, F_p and F_m correlations in Fig. 1 and 2 and Table III.

As mentioned, our method is intended for fluid-type plants that include some vapor processing. Further work is aimed at extending the approach to a wider range of plants. Another improvement would be to derive the factor for *DEC* as a percentage of total plant investment from information more specific to the plant in question. We feel this is the most serious potential source of error in the method at present. #

References

1. Perry, R. H. and Chilton, C. H., "Chemical Engineers' Handbook," 5th Ed., McGraw-Hill, New York, 1973.
2. Hill, R. D., *Petroleum Refiner,* Aug. 1956, p. 106.
3. Chilton, C. H., *Chem. Eng.,* June 1949, p. 97.
4. Zevnik, F. C. and Buchanan, R. L., *Chemical Engineering Progress,* Feb. 1963, p. 70.
5. Stallworthy, E. A., *The Chemical Engineer,* June 1970, p. 182.
6. Wilson, G. T., *British Chemical Engineering and Process Technology,* Oct. 1971, p. 931.
7. Miller, C. A., *Chem. Eng.,* Sept. 13, 1965, p. 226.
8. Gregory, C. A. and Bridgwater, A. V., I. Chem. E. Symposium Series, No. 35, 1972.
9. Le Page, N. J., Report MD7531, ICI Ltd., Dec. 1970.
10. *Petroleum Refiner,* Nov. 1959, p. 215.
11. Sittig, M., "Organic Chemical Process Encyclopedia," Noyes Development Corp., Park Ridge, N.J., 1967.
12. Faith, W. L., Keyes, D. B. and Clark, R. L., "Industrial Chemicals," 3rd Ed., John Wiley, New York, 1965.
13. Berk, J. M. and Haselbarth, J. E., *Chem. Eng.,* Mar. 20, 1961, p. 182.
14. Haselbarth, J. E., *Chem. Eng.,* Dec. 4, 1967, p. 214.
15. Guthrie, K. M., *Chem. Eng.,* Mar. 24, 1969, p. 114.
16. Zimmerman, O. T., *Cost Engineer,* Oct. 1967, p. 12.
17. Page, R. C., Ph.D. thesis, Nottingham University, 1973.
18. *Hydrocarbon Processing,* Apr. 1971, p. 93.
19. *Hydrocarbon Processing,* Nov. 1971, p. 113.
20. *Hydrocarbon Processing,* Sept. 1972, p. 111.
21. Zimmerman, O. T. and Lavine, I., *Cost Engineer,* July 1961, p. 16.
22. Haselbarth, J. E. and Berk, J. M., *Chem. Eng.,* May 16, 1960, p. 158.

Meet the Authors

◀ **D. H. Allen** is a senior lecturer in chemical engineering at the University of Nottingham, U.K. He holds B.Sc. and Ph.D. degrees from the University of Birmingham, and has held posts in the gas and petroleum industries. Besides cost estimation, his research interests include management of research and development, and techniques for project evaluation and operations planning. Dr. Allen is a Fellow of the Institution of Chemical Engineers.

R. C. Page is a financial adviser with Towry Law Ltd. He has a B.Sc. in chemical engineering from the University of Surrey, and a Ph.D. in chemical engineering from the University of Nottingham for research into cost estimation. ▶

Technological Forecasting for

Advances in the art of technological forecasting have produced a variety of explicit procedures for making credible forecasts. These can significantly improve the quality of decisions that must consider technological change.

JOSEPH P. MARTINO, U.S. Air Force

Technological forecasting is a subject of much current interest. There is a veritable flood of books, articles and reports that deal with it. Many, if not most, of these treat technological forecasting as though it were something brand new, previously unheard of.

What is the excitement all about? What is technological forecasting? Is it really something new and, if so, what does it replace? What significance, if any, does it have for the technical decision-maker in the chemical process industries (CPI)?

What Technological Forecasting Is

A technological forecast can be defined as a prediction of the future characteristics of a useful machine. There are three notable points to this definition.

First, the forecast describes the characteristics of some future machine. This description usually deals with the performance of the machine rather than with the details of how it will work. In other words, the forecaster is not required to invent the machine he forecasts but only to state that a machine with a certain performance will be available. Furthermore, he need not describe the means by which the performance that he is forecasting will be achieved.

Second, the restriction to useful machines is meant to exclude machines intended for luxury or amusement, which are affected more by human tastes and fads than by either technological feasibility or economic viability.

Third, the term "machine" is to be interpreted broadly, so as to include tools, techniques and procedures. Biological and behavioral technologies are as much the concern of the technological forecaster as are the more conventional hardware technologies.

When is a technological forecast needed? It is needed whenever a decision must take technological change into account. While decisions are made *in* the present, they are not made *for* the present. The right choice among the options open to a decision-maker will depend upon the conditions that will exist at the time the decision comes to fruition. These may differ from the conditions that exist at the time the decision is made. If the technological

Originally published December 1971.

the Chemical Process Industries

conditions that will exist in the future make a difference in the choice of a course of action, the decision-maker needs some estimate or assessment of what those future conditions will be. This estimate or assessment is precisely what is meant by a technological forecast.

At this point, the CPI decision-maker may well ask, "But haven't I been doing that all along?" He might well protest that of course he takes future technological conditions into account when he makes his decisions. Any assertion to the contrary is tantamount to saying he has not been doing his job right.

With this, the technological forecaster would hasten to agree. He might even recite what has come to be almost a cliche in the forecasting field, "There is no such thing as 'not forecasting.' "

Suppose the decision-maker has a choice among several options. For each option, there is some possible future in which that option would be better than all the others. Therefore, the choice he actually makes has implicit within it a forecast of future conditions.

What Is New in Forecasting

Then why the sudden interest in technological forecasting, when it actually has been going on for years? There are two new things responsible for the current excitement over it. The first is the explicit recognition of the need for forecasts; the second, the improved methods for making forecasts.

In the past, decision-makers were often not consciously aware that they were incorporating a forecast of future technological conditions into their decisions. The forecast, although inescapably a part of their decision, was completely implicit within that decision.

This has changed. Decision-makers are coming to recognize that a forecast is an essential part of a decision, and also that the forecast must be explicit, so that it can be examined and challenged, just as any other estimates or assumptions that go into a decision must be examined and challenged.

The second new feature is the use of rational methods for making forecasts. In the past, even if a decision-maker wanted an explicit forecast, he could arrive at it only by "asking an expert." The expert in the technology in question would give his subjective, intuitive estimate of what the technology might achieve by some specified date. However, this forecast could then be evaluated only by comparing it with a forecast of another expert. If the two experts disagreed, there was no way of determining which expert was more "expert," and therefore more likely to be right.

This problem has to a large extent been eliminated. A wide range of explicit forecasting procedures has been developed. And a forecast prepared by means of any one of these procedures can be reviewed by anyone familiar with the methods of forecasting. He can check to see whether each step in the procedure was carried out correctly, whether the facts upon which the forecast is based are actually correct, and whether the assumptions made are reasonable. Furthermore, these methods can be

The Delphi Procedure

The Delphi Procedure is specifically designed to gain the advantages of a committee while minimizing the disadvantages. It has the following characteristics:

Anonymity—Committee members do not know each other's identities, or who made a specific comment or suggestion.

Controlled feedback—The exchange of ideas is carried out through the Delphi Director, who filters out material not germane to the objectives of the committee, and consolidates similar arguments before presenting them to the entire committee.

Statistical response—The committee need not reach a consensus or prepare a majority opinion. After the controlled exchange of comments, each member is still free to express his opinion. The opinions of the committee are summarized statistically and all views, no matter how diverse, are included in the result. The most common form of summary is to give the median committee estimate (a number such that half the committee opinions are above it and half below) and the lower and upper quartiles (values, respectively, above one-fourth and three-fourths of the committee).

The interaction between committee members in Delphi is carried out through a sequence of questionnaires. This sequence is structured in such a way as to force the committee to concentrate on the important issues, and to elicit opinions in an efficient manner.

A typical forecast might be accomplished as follows. The first questionnaire would contain a question such as, "By what year is there a 50-50 chance that Event X will have occurred?" Each committee member is expected to make an estimate of the date, taking into account his background knowledge and specific viewpoint about how rapidly Event X can be achieved. If he feels that Event X is so unclearly defined that he is not certain what he is being asked to forecast, he may request a clarification instead of, or in addition to, making an estimate. Because of the difference in views among the committee members, the responses to this first questionnaire, or "first round," will contain differing estimates.

The Delphi Director receives the estimates and prepares the questionnaire for the second round. If necessary, he makes clarifications in the description of Event X. He also presents the median and quartiles for the first-round estimates. He then asks the members to take into account the estimates made by the other members, and make a new estimate. If a member's new estimate of the occurrence of Event X is earlier than the first-round lower quartile, or later than the upper quartile, he is asked to give reasons why he believes the majority of the committee is making too late, or too early, an estimate.

The Delphi Director again receives the estimates, and the reasons provided by the "out-liers." He prepares the third-round questionnaire by making necessary clarifications, presenting median and quartiles, and summarizing the reasons why the estimates of the date for Event X should be advanced or delayed. The members again make estimates and provide additional arguments if they are among the "outliers." In addition, any member may comment on the arguments, providing support or counter-arguments.

The Director again receives the estimates and prepares a new questionnaire. This sequence is gone through for as many rounds as necessary—i.e., until there is no longer any significant shift of opinion. This usually happens not later than the fourth round. The final-round median is then the forecast date for Event X.

The uncertainty in the forecast is indicated by the width of the Interquartile Range (upper quartile minus lower quartile). The user of the forecast then has not only a single-point estimate of when Event X will occur, but also some idea of the amount of disagreement remaining among the committee members at the end of their deliberations, which will indicate how much credence he should put in the specific date. The final summary of arguments also helps him identify those factors that have a bearing on when Event X will occur.

tested scientifically in a variety of situations to assess their validity.

What are these new methods, and how can they be used? This article does not attempt to explain forecasting methods in detail. A number of books are available that give detailed expositions of the modern, rational methods of technological forecasting.[1,2,3] Instead, some of these methods are illustrated in a variety of decision situations typical of those commonly faced by the CPI decision-maker.

SOME DECISION SITUATIONS

Virtually every decision made in the CPI is involved with considerations of technological change. Now, we will look at some fairly common decision situations, identify the information about future technology needed to make those decisions, and in each case look at one method that can help provide the information needed.

Total Market Size

How big will the total market for some chemical or class of chemicals be at some particular time in the future? The answer to this question is needed for any decision affecting plant capacity, sales goals, or the market share to be sought.

One way of forecasting total market size is known as trend extrapolation. In this method, the forecaster obtains data on market size for past years and attempts to identify a pattern of growth in the data. He usually settles for a trend rather than look for a more complex pattern. A forecast then is made by extrapolating this pattern into the future.

Fig. 1 shows a forecast of future U.S. production of ethylene. (This was prepared by the planning staff of Monsanto Co.[4]) An exponential trend was fitted to the historical data and projected into the future. This trend appears as a straight line when plotted on semi-logarithmic coordinates.

This forecasting method is straighforward and readily adaptable to visual presentation. However, it is based on the assumption that whatever forces acted to produce the trend will also extend it. As long as this assumption is valid, trend extrapolation is a successful forecasting method. If it is wrong, however, a trend forecast can be seriously in error. For instance, if ethylene were to be displaced by another chemical for some product, this would invalidate the forecast. The beginning of such replacement should be a warning to the forecaster. This technique can also be used to forecast other technological characteristics besides market size, but the same precaution applies.

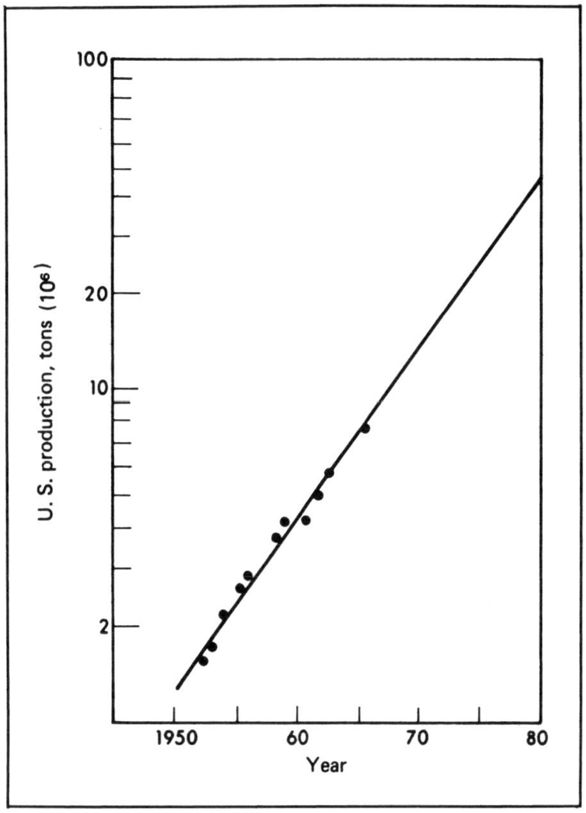

EXTRAPOLATION forecasts ethylene production—Fig. 1

Sales Growth of a New Product

When a new product is brought on the market, decisions must be made about the construction of new plants, the provision of adequate supplies of raw materials, and the setting of sales and marketing goals. These decisions will depend upon estimates of the rate at which sales will increase and new production will be needed.

Because there will be no historical data that can serve as a basis for trend extrapolation, some other method must be used for generating a forecast. In such cases, there may be no alternative but to turn to subjective methods based on expert opinion. However, even here there have been changes. Subjective methods are not as subjective as they once were. There are major differences between current and past methods for making subjective forecasts.

One is to rely on a group of experts instead of only one. This has several advantages. A group has more information available to it than does any individual. A group also can consider more factors and more points of view. Finally, an individual may have biases of which he is himself unaware, whereas if the group is properly chosen, there is a fair likelihood that the biases of its members will tend to cancel one another.

However, most decision-makers, aware of the pitfalls of committee action, are often reluctant to depend on committees. Here again, there have been some changes. Scientific studies of small-group interactions have produced methods that gain the advantages of group action while minimizing or eliminating some of the disadvantages. Many of these methods have been applied to the task of eliciting a forecast from a group of experts.

One of the most popular is known as the Delphi Procedure (see box). It was developed by the Rand Corp. in its research to find the most effective means of using groups of experts. It replaces face-to-face committee meetings with the interactions of a group whose members remain unknown to each other. These interactions are guided by a structured sequence of questionnaires presented to the group. The structure permits each expert to concentrate on only the important aspects of the issues before the group. Anonymity allows each expert to consider suggestions, comments and arguments on their merits, without regard to personalities. Experiments at Rand have shown that such group estimates and assessments are in general superior to those arrived at through the more-conventional committee.

Celanese Corp. used a Delphi panel to forecast the growth in sales of a new yarn.[5] The panel was composed of members from research, marketing, sales and top management. A member from one department was able to take into account factors that were not likely to be considered by members from other departments.

One thing estimated by the panel was the likelihood that production of the new yarn would exceed certain

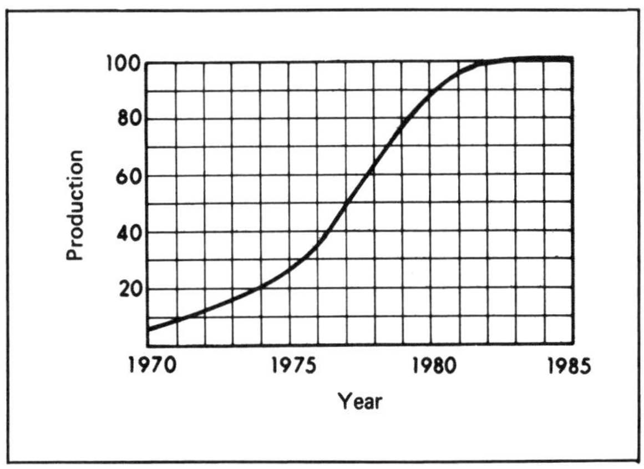

INDIVIDUAL predicts new yarn's sales growth—Fig. 2

levels by various dates in the future. Fig. 2 shows a typical estimate made by a single panel member. He felt that production of the new yarn was virtually certain to reach the specified level by 1981 (a probability of occurrence of 100%).

The estimates of all members of the panel were combined, as in Fig. 3. The circles show the median estimates for 1970, 1975, 1980 and 1985. The shaded area includes the "interquartile range" (middle half) of the panel's estimates. This composite estimate was then presented in the next questionnaire, along with the written comments of panel members as to why they believed other panel members were either too optimistic or too pessimistic. Each member then decided whether to revise his previous estimates on the basis of the opinions and arguments of the other panelists. The final result was a graph similar to Fig. 3, which indicated not only the median panel estimate but the degree of spread or uncertainty in that estimate.

Delphi is only one of a family of "scientifically subjective" methods that can be used for technological forecasting when other methods will not work. These methods can be used not only when data are lacking but also when it is known that the conditions that produced a trend are undergoing changes that would make extrapolation useless. They can also be used when external, nontechnological factors are dominant, and when ethical or moral considerations must be taken into account. Thus, even though explicit forecasting methods are desirable when these are available, there will always be a need for subjectively prepared forecasts. Delphi and similar methods help to increase the validity of these forecasts.

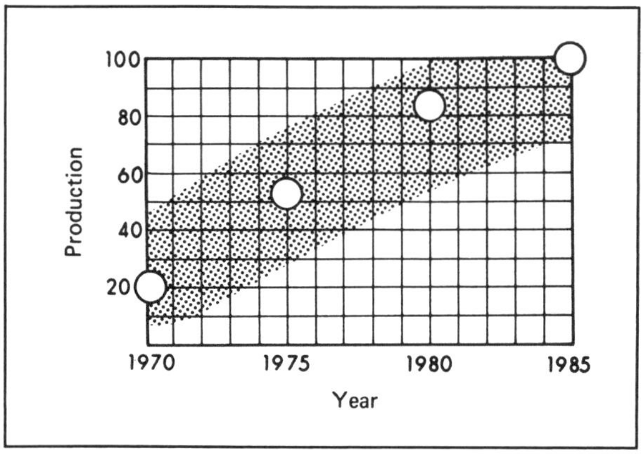

ESTIMATES of panel members are combined—Fig. 3

Product Replacement

In some cases, a new product is almost a direct substitute for some older product (e.g., margarine for butter, nylon for silk, detergents for soap). It is important to be able to estimate the rate at which a new product will take over an entire market. This estimate is needed not only to plan for the production and sales of a new product but also for phasing out the old one.

Such replacement often obeys a definite mathematical law.[6] If f denotes the fraction of the market captured by the new product, the ratio $f/(1 - f)$ appears to grow exponentially. A typical example of this behavior is seen in Fig. 4, which shows the substitution of detergents for soap in both the U.S. and Japan. Note that the slope of the trend is virtually the same in both cases, although the replacement occurred in countries with different cultures and about a decade apart.

The slope of the trend is characterized by a "takeover time," defined as the time for the new product to grow from 10% to 90% of the market (i.e., from $f = 0.1$ to $f = 0.9$). The takeover time for detergents to replace soap was 8.75 years in the U.S. marketplace and 8.25 years in Japan's.

This method is used in essentially the same way as trend extrapolation. Data on the market share of the old and new products are obtained for a period of several months or years (for example, until the new product has captured 15% to 20% of the market). The ratio of the market share of the new product to that of the old product is plotted, and the trend extrapolated. Takeover time can then be obtained from the slope of the trend.

This method makes the same assumptions as trend extrapolation. It can be used only so long as the conditions that initiated the trend continue it.

Selecting Capacity for a New Plant

When a new chemical plant is to be constructed, a decision must be made as to its size or capacity. Advances in the technology of a particular process often allow a new plant to be built with greater capacity than that of existing ones. This increase in scale usually results in a reduction in unit production cost. Once one firm has such

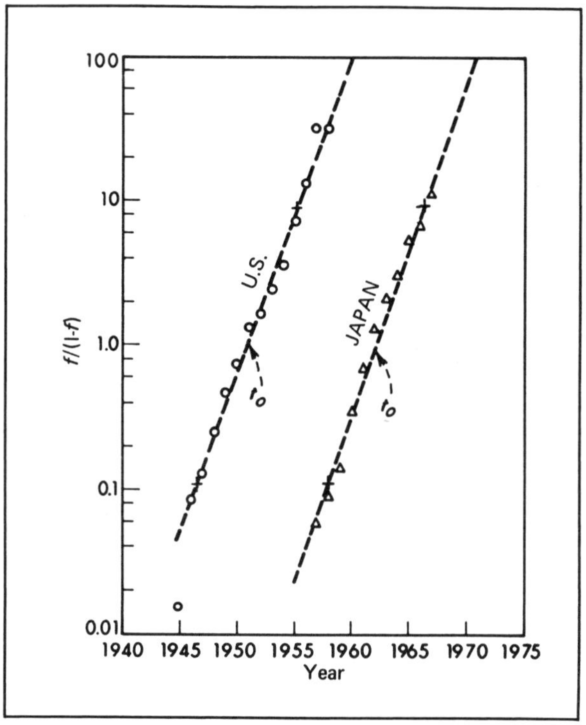

DISPLACEMENT of soap by detergent was alike—Fig. 4

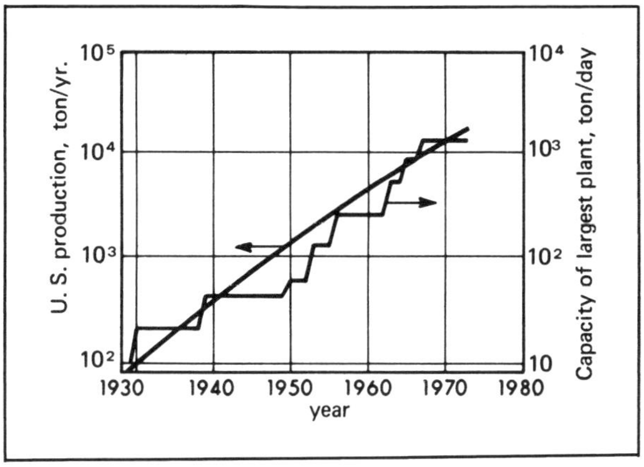

RATIO shows between production and capacity—Fig. 5

a larger and more economical plant under construction or onstream, its competitors must follow suit. Thus, it is important to be able to estimate when technology will be ready for a larger plant, and what its capacity should be.

For a number of basic chemicals (such as ethylene, benzene and synthetic ammonia), there is a fairly constant ratio between total production and the capacity of the largest single plant.[7] (This same type of relationship has also been observed in a number of other technologies, including oil tankers and electric generators, and so appears to be fairly general.) Fig. 5 illustrates it in the production of synthetic ammonia in the U.S. Between 1953 and 1967, the capacity of the largest single ammonia plant in the U.S. averaged about one-thirtieth of the total U.S. production of synthetic ammonia.[8]

This step-growth forecasting technique can be of considerable help in making decisions about the timing and capacity of new chemical plants. If, in the history of the production of a specific chemical, there has been a consistent pattern in the intervals between the introduction of larger-capacity plants, and a consistent pattern in the ratio of these plant capacities to total production, the decision-maker would be well advised to anticipate the possibility that this pattern would repeat itself.

Decisions Affected by Production Costs

Some decisions will be affected by costs of production—in particular, decisions about whether or not to enter a certain market, because these costs will have a bearing on whether the market will be profitable. In some cases, advances in technology will reduce production or unit capital costs for a particular type of installation. In these cases, a technological forecast can be helpful in making the decision of whether or not to enter a specific market.

As one instance of this type of decision, the United Kingdom Atomic Energy Authority carried out an analysis to determine whether it would be profitable for one or more British firms to attempt to export desalting plants to underdeveloped countries.

One part of the study, of course, was a market forecast of the future world need for potable water extracted from seawater, brackish water, etc. In addition, the analysis included a technological forecast of the capital costs of desalting plants. The forecast of demand for water was then converted into a forecast of the total sales of desalting plants worldwide, by year. Multiplying the number of plants required by the forecast of capital cost gave the monetary value of the total world market for desalting plants. The British share of this world market was estimated by conventional operations-research techniques, and total sales were discounted back to the present to determine whether the research and development investment in desalting processes would be recovered and a net profit realized.

In this case, the forecast of the capital costs of plants embodying future technology was based on a trend extrapolation of past plant costs, supplemented by estimates of the capital costs of plants using processes still in research and development. However, other forecasting methods such as pure trend extrapolation or the Delphi method can also be used, if necessary. In any case, when advances in technology reduce either capital or operating costs, forecasts of these reductions are an important element in decisions that will be affected by these costs.

Setting Research and Development Goals

When a project is undertaken to improve an old or develop a new product, it is necessary to set performance goals. If the product that comes out of a project has too low a performance level, it will be a commercial failure. So the goals set for a project must result in a product that will be competitive with what other firms will have on the market, or will have ready for the market at the time the project is completed. If, on the other hand, goals are set too high, the project may attain them only after a longer time and at a higher cost than what were originally anticipated; or the project may become a failure, never reaching its goals. A technological forecast can be of considerable value in helping to set research and development goals that will be high enough to be competitive but low enough to be technically feasible.

A Goodyear Tire and Rubber Co. forecast of future needs and performance of automobile tires, which was prepared to serve as a guide to company planning, provides an excellent example of the use of several techniques to predict different types of outcomes, each of significance for research and development.[10]

One part of this forecast (Fig. 6) shows the performance of automobile tires in terms of mileage before wearing out. The reduction in mileage for the bias tire, occurring after about 1953, is misleading in terms of an assessment of tire technology. The tire tread actually improved, in terms of treadwear rating, but conditions of service became so much more severe (higher automobile horsepower, greater speeds, power steering and brakes, etc.) that tire mileage dropped.

There was again an increase in actual tire mileage in the late 1960s, when the belted tire was introduced. Presumably the performance of this tire will also be described by the S-shaped growth curve that the bias tire exhibited. So, the forecaster has chosen to project the envelope of the S-curves. This forecasting method is commonly used for technologies that historically have had a

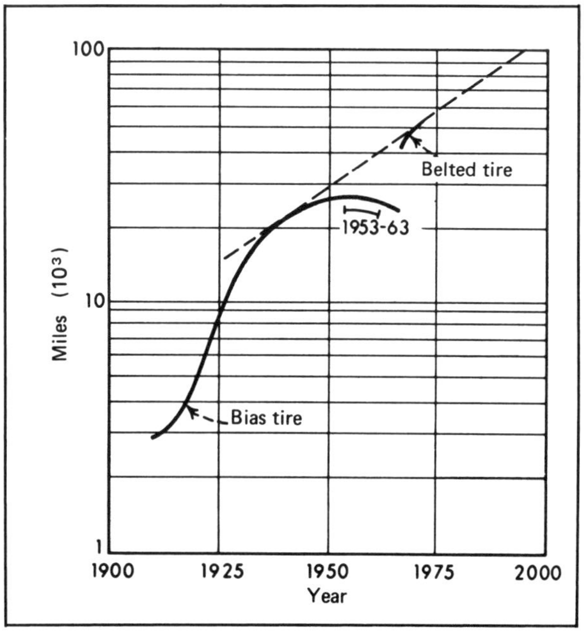

FORECASTER projects future automobile tire mileage from the envelope of the S-curves—Fig. 6

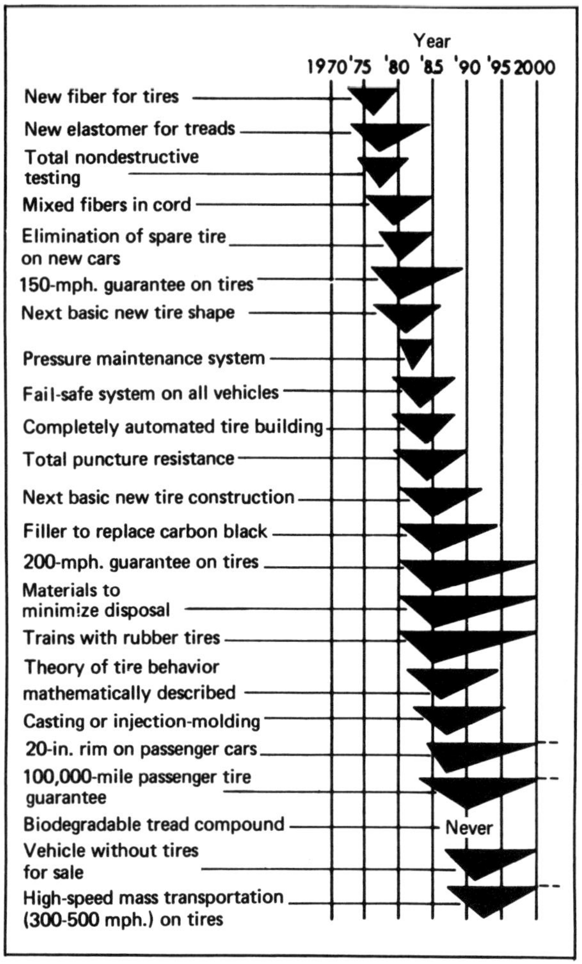

REALIZATIONS of the above events by the indicated dates are better than 50/50 probable—Fig. 7

succession of different technical approaches to performing the same function. The performance of each technical approach follows an S-curve, and overall performance is forecast by projecting the envelope of the S-curves. In this case, a tire mileage of 100,000 is forecast for some time in the 1990s.

Another part of the Goodyear forecast (Fig. 7) illustrates a set of predictions regarding tires prepared by the Delphi Procedure. In this case, the committee included experts from tire companies, companies supplying raw materials to the tire industry, consultants to the industry, and specialists with trade and technical journals.

This Delphi sequence was carried out entirely by mail. The committee members made estimates for each of the specific events shown, and these estimates were refined through several rounds of iteration. The final results are shown. The bottom points of the shaded triangles indicate the median estimates for each event, while the points at the sides indicate the lower and upper quartiles. It is worth noting the similarity between the forecast the Delphi committee made fo: the 100,000-mile tire, and that of Fig. 6, which was obtained independently. Concurrence between forecasts obtained by different methods, such as this case, increases confidence in the validity of results.

So far, none of the decision situations or forecasting methods we have discussed involved developments representing a complete break with the past. It is developments of this nature that can catch several companies, or even an entire industry, by surprise. We will now take up this type of situation.

Predicting the Unpredictable

The technological forecaster is often challenged with the question, "How can you predict a technological breakthrough?" Or, alternatively, he may be challenged with, "How can your forecasting methods, which are only based on the past, anticipate something like the DDT ban or the opposition to phosphates?"

These are legitimate questions. Their implication is that there are certain things that are inherently unpredictable, and so beyond the art of the technological forecaster. If the inherently unpredictable consists of a significant fraction—or, worse yet, most—of the factors the forecaster is asked to predict, his art would be nearly useless. It is tautologically true, of course, that no one can predict the unpredictable. So, we must ask: How much of the future is really unpredictable, and how much is merely surprising because those who should have predicted it failed to do so?

If asked to name a technological breakthrough, the average person will likely mention one of the following: the jet engine, the atomic bomb, the laser or the transistor. These are probably the most spectacular technological advances of the past three or four decades. Furthermore, in each case, the advance caught a great many people by surprise, including those in the industries most affected. Hardly anyone in the military anticipated the atomic bomb; hardly anyone in the airplane-engine business anticipated the jet engine; hardly any of the organizations that now make or use the laser anticipated it. These seem to be prime examples of breakthroughs that may be regarded as having been largely unpredictable.

Nevertheless, every one of these advances, as well as a great many more, can be traced back through three or four decades of precursory events that, seen in retrospect, should have given a fairly clear indication of where technology was going and what might have been anticipated.

It might, of course, be pointed out that hindsight is always blessed with 20-20 vision, and that when these precursors were unfolding, they really were of no help to decision-makers. This objection has some merit. During the first half of the Twentieth Century, technical decision-makers were not as aware as now of the processes of technological growth. They did not then have as much experience as we do with the succession of technical approaches that produced advances in a specific technology, neither did they have as much experience with the transition of basic research to technological application. So, one major reason they failed to recognize precursors was that they were not aware of the possibility of so doing.

There are a number of signals that can be watched. The output from basic research laboratories is one. A new reaction or chemical, a better understanding of some phenomenon—all may point to a new product, or a new way of making an old one. Another is inventions, whether patented or not, that "don't quite work" because an important element is missing or the supporting technology is inadequate. The transistor, for instance, was patented at least twice during the 1930s, but the devices described in the patents did not work because chemicals of the required "impurity" were not available. Yet-other signals are inventions or discoveries whose performances are now too low to be competitive but which might be improved.

This looking for signals has been termed "monitoring the environment" by J. R. Bright, who has been a leader in developing the technique.[2] Basic to it is that appropriate sources must be monitored continuously to identify signals.

However, no single signal ever gives a complete picture of the future. So an essential feature of the process is the search for signals that confirm or deny earlier ones. Suppose, for instance, a product has been invented whose performance is too low to be competitive (as was true of the jet engine in the early 1930s). It would then be necessary to ask what would have to be changed to make it competitive—improvements in materials, theoretical understanding, supporting technology? The answers would indicate other sources that should be monitored for signals.

Suppose this product needs to be constructed of better material. We should then watch for the appearance of such a material. If it appeared, this would be a confirming signal additionally warning of a breakthrough. Each confirming signal could indicate others to look for, so that a chain of precursors could be traced to the breakthrough.

"Monitoring the environment" is not limited to forecasting strictly technological events. It can be of even greater value in identifying events that originate outside an industry but that will have a major impact on it. Such events are often of an economic, political or social nature.

In a nation as large as the U.S., such events do not happen overnight. There are almost always precursors indicating changes. For instance, the DDT ban was not a bolt out of the blue. At least as early as the publication of Rachel Carson's "Silent Spring," there was indication that a segment of the population was unhappy with DDT. Although this one signal was not proof a change was coming, it should have been a warning to watch for confirming signals. Was this book followed by similar books and articles? Did newspapers begin to play up fish kills? Did conservation groups begin to make demands for restrictions on DDT? Did politicians begin to take notice of this shift in public opinion?

All these signals, and more, would have been caught by a forecaster or decision-maker who was "monitoring the environment." [11] He might not have been able to predict exactly when a ban would come, but should have been able to anticipate significant restrictions.

Similarly, the current concern about the overall environment did not develop overnight. A forecaster might not have predicted that phosphates specifically would be a target, but he could have predicted that the dumping of industrial wastes would become an issue.

Technological forecasters do not claim that monitoring will eliminate all surprises. There will continue to be details that will remain unpredictable, such as the opposition to phosphates rather than to some other chemical. Nevertheless, systematic monitoring for signals can go a long way towards eliminating surprises.

Using the Forecast

The CPI technical decision-maker will on some occasions prepare his own forecasts. At other times, he will depend on forecasts readied by professional technological forecasters. In either case, the fact that his information about the future can only be based on inferences from the past will affect the way in which he uses a forecast for decision-making.

Questions To Ask

When the decision-maker is presented with a forecast, he should ask questions about it, the answers to which will determine whether or not he uses the forecast at all. If he prepares his own forecast, he should ask himself the same questions:

1. What information does this forecast provide? That is, with regard to the decision to be made, what portion of the decision information is provided by the forecast? Is the forecast really relevant to the decision? What information is not provided by the forecast and must be sought elsewhere?

If the forecast is relevant to the decision, there are other questions to ask:

2. How was the forecast made? That is, what method was used? What logical structure was employed to draw inferences about the future from the data? Is the method appropriate for this particular situation? Has it been

adequately tested on similar situations and found valid?

If the forecast is acceptable to this point, the following should be asked:

3. Are the data accurate? If the data are presented explicitly, the sources should be reviewed. Are these sources the best available? Are they complete? Were significant sources overlooked? If the forecast was prepared by subjective methods, the questions should be directed at the experts who participated in it. Are they likely to be well informed about the past and present situations in their field? Do they have any known biases?

4. Are the assumptions valid? In the absence of data, every forecast will depend on some assumptions. These may involve future events, or the actions of others who will have an influence on the outcome. The assumptions should be identified and their validity assessed—are they soundly based, consistent with data available, and adequately tested?

If a forecast is satisfactory, in terms of the answers to these questions, it can be used in a decision, governed by certain considerations.

How Much Credence To Give a Forecast

If a forecast is based on an appropriate method, accurate data and valid assumptions, and the decision-maker is satisfied it cannot be improved significantly, how much credence should he give it?

Every forecast, no matter how well it is prepared, will to some extent be in error. Therefore, the decision-maker cannot give it complete credence. Even if it is the best possible forecast, he still needs to assess just how good.

Some insight into this problem can be gained from Fig. 1. Even if a decision-maker feels that a projection of a trend constitutes the best possible forecast of ethylene production, he still has a problem. The data points did not all fall on the trend line but were scattered about it. So it is unlikely that future events will fall on the projected trend line. In fact, the more scatter in the past, the more there is likely to be in the future, so the decision-maker should put even less credence in the accuracy of the trend line.

The same considerations apply to forecasts made by other methods. The more scatter or uncertainty in the data, or the greater inaccuracy a forecasting method has shown in the past, the less credence should be given to it. By thus assessing a forecast, the decision-maker can determine how much uncertainty was resolved by the forecast, and how much still remains for him to take into account in his decision.

When To Reject a Forecast

A forecast may have been acceptable when it was first prepared, but may later have to be rejected and done again. The decision-maker must be able to determine when a forecast should be rejected.

Suppose, for example, that a forecast of the sort shown in Fig. 1 had been accepted when it was prepared, but that a subsequent event falls far from the trend line (further than can be explained by scatter in the data). This may be a signal of a change requiring the revision of the forecast. If a succession of events show a consistent deviation from the trend (either all above or all below it) instead of random scattering, this also would be an indication of a change requiring revision.

This would also hold true for forecasts made by other methods. A single gross deviation, or a pattern of consistent deviations, would indicate a need for revision. In addition, a marked change in the conditions affecting the technology being forecast, such as a change in research and development expenditures or the appearance of a competing product, should call for revision.

No forecast remains valid indefinitely. For instance, a trend extrapolation should not extend further into the future than one-third to one-half the span of years covered by the data base. When this period has run out, the forecast should be done again even if it has proven highly accurate—if for no other reason than that subsequent events represent data that should be taken into account.

Much the same considerations apply to forecasts prepared by other methods. They should be reworked when their time span expires, and often even before then. In the military services, two years has been found to represent a reasonable cycle for updating a forecast. More-frequent revisions often result in little substantive change, less frequent revisions in obsolescence. Although the CPI decision-maker may find this cycle inappropriate for the technologies of concern to him, it may serve him as a satisfactory rule of thumb until his experience dictates a change. ■

References

1.Ayres, R. U., "Technological Forecasting and Long-Range Planning," McGraw-Hill, New York, 1969.

2.Bright, J. R. (ed), "Technological Forecasting for Industry and Government," Prentice-Hall, Englewood Cliffs, N.J., 1968.

3.Martino, J. P., "Technological Forecasting for Decisionmaking," American Elsevier, New York (to be published).

4.Barach, J. L., Twery, R. J., Forecasting Techniques in R&D Planning, *Chem. Eng. Prog.*, **6,** 6, June 1970, pp. 15-19.

5.Craver, J. K., Planning for Profitable Growth, paper presented at Commercial Development Assn. meeting, Chicago, Ill., Oct. 12, 1970.

6.Fisher, J. C., Pry R. H., A Simple Substitution Model of Technological Change, *Technological Forecasting and Social Change,* **3,** 1, 1971, pp. 75-88.

7.Simmonds, W. H. C., Stepwise Expansion and Profitability, *Chemistry in Canada,* Sept. 1969, pp. 16-18.

8.Simmonds, W. H. C., The Canada-U.S. Scale Problem, *Chemistry in Canada,* Oct. 1969, pp. 39-41.

9.Hunt, H., Forecasting the Need for Research and Development, *Futures,* Sept. 1969, pp. 392-390.

10.Kovac, F. J., Technological Forecasting—Tires, *Chemical Technology,* **1,** Jan. 1971, pp. 18-23.

11.Martino, J.P., Examples of Technological Trend Forecasting for Research and Development Planning, *Technological Forecasting and Social Change,* **2,** 3/4, 1971, pp. 247-260.

Meet the Author

Lt. Col. Joseph P. Martino, now at the Air War College, Maxwell Air Force Base, Alabama, has been a project engineer, statistician and operations analyst in research.

A past vice-president and member of the Board of the Military Operations Research Soc., he holds an A.B. from Miami University (Ohio), an M.S. from Purdue University and a Ph.D. from Ohio State University.

Correcting Economic Analyses

Present-worth factors based on basic compound-interest formula should be used to give accurate results in terms familiar to management.

ROBERT H. CONGELLIERE, Union Oil Co. of California

Despite the widespread acceptance in American industry of the discounted-cash-flow (DCF) method for calculating rates of return on investment, one of the significant terms used in these calculations, the present-worth factor, is still little understood or appreciated.

The Basic Compound-Interest Formula

All the types of present-worth factors used in DCF rate-of-return calculations are based on the compound-interest formula,

$$S = P(1 + i)^n$$

where S is the future worth of a present sum of money, *P*, which is invested at compound interest at a rate, *i*, per period, for *n* periods. In normal rate-of-return calculations estimates of future cash flows must be discounted to their present worths, so the formula is rearranged to:

$$P = S[(1/(1 + i)]^n$$

The factor $[(1/(1 + i)]^n$ is called the present-worth factor (or discount factor). More specifically, this particular factor is an end-of-period present-worth factor, since it is assumed that the future earnings, S, are received in one lump sum at the end of each period.

As a result of the indiscriminate use of end-of-period present-worth factors for all DCF rate-of-return calculations, those projects with earnings received uniformly throughout the year are undervalued, sometimes significantly.

For example, a sample problem (Table I) shows a rate of return of 10%, when solved with end-of-period factors, whereas the true rate of return calculated with continuous-income factors would be 14.6%, if these earnings were actually being received uniformly throughout the year.

Continuous Compounding

Many economics analysts have recognized that business income is normally spread throughout the year and recommend the use of present-worth factors that take this into account. Unfortunately, many analysts have fallaciously assumed that the adoption of continuous compounding of interest is necessary to express continuous receipts of income.[1]

However, it is apparent from a comparison of the formulas that continuous income and continuous compounding are totally different concepts. Present-worth factors based on continuous compounding are derived from a variation of the basic compound-interest formula, as:

$$S = P(1 + i/m)^{mn}$$

where *i* is the nominal rate of interest per year, and

Comparison of End-of-Period and Continuous-Income Factors—Table I

(Investment=$1,100)

Year	Annual Earnings	End-of-Period Present-Worth Factor @ 10%	Present Worth @ 10%	Continuous-Income Present-Worth Factor @14.6%	Present Worth @ 14.6%
1	$600	0.909	$545	0.935	$561
2	400	0.826	330	0.816	326
3	300	0.751	225	0.712	213
			$1,100		$1,100

Originally published November 1970.

m is the number of compounding periods per year. As m approaches infinity, it can be shown that:

$$S = Pe^{in}$$

or,

$$P = S(1/e^{in})$$

This continuous-compounding factor, $1/e^{in}$, is based on the annual income being received in one lump sum at the end of the period, the normal period being one year. There is also a continuous-compounding factor based on continuous receipts of income, $(e^{i} - 1)/ie^{in}$, which further illustrates that continuous income and continuous compounding are separate concepts.[2]

Continuous Income

Continuous compounding is not necessary to express continuous receipts of income, as shown through the following example: Suppose a loan company lends a borrower $10,000 and the borrower agrees to pay back a lump sum of $11,000 one year later. Using the basic compound-interest formula, we find that the rate, i, at which the $11,000 future income will have a present worth equal to the present investment of $10,000, is given by:

$$P = S[1/(1+i)]^{n}$$
$$10{,}000 = 11{,}000[1/(1+i)]^{1}$$
$$1 + i = 11{,}000/10{,}000$$
$$i = 0.10$$

Now, suppose the loan company asks the borrower to pay back $5,368.97 six months from the date of loaning and another $5,368.97 six months after that. Using the compound-interest formula and annual compounding of interest, we have:

$$10{,}000 = 5{,}368.97[1/(1+i)]^{1/2} + 5{,}368.97[1/(1+i)]^{1}$$
$$i = 0.10$$

To prove that the loan company is still receiving 10% on its money, assume that the $5,368.97 paid back by the borrower at the end of six months is loaned to another borrower for a six months period at the same rate of 10%. The amount paid back on this second loan is $5,368.97 $(1 + 0.10)^{\frac{1}{2}}$, or $5,631.03. Thus at the end of the year, the loan company receives one payment of $5,368.97 from the first borrower and $5,631.03 from the second, for a total of $11,000.

Following this procedure, it can be shown that the interest rate remains the same, if the time intervals are reduced to weeks, days, or even hours.

Conforming to Convention

Secondly—and this is the basic point that many analysts have apparently misunderstood—continuous compounding of interest should not be used in the evaluation of proposed capital investments because the traditional basis for expressing interest is percent per year, compounded annually.

Since a rate of return is basically the interest earned on an investment in manufacturing facilities, management should be given rates of return expressed on the traditional basis. The project's rate of return will be compared with the company's or the industry's overall rate of return, or with the cost of borrowed capital—all of which are expressed as %/yr., compounded annually.

Since management in most cases assumes that the rates of return submitted to it are on the traditional basis, the economics analyst is doing his company a disservice by presenting answers based on the more conservative continuous-compounding basis. (Rates of return based on continuous compounding are always lower than those based on annual compounding.) Injecting such conservatism into reported rates of return is usurping a prerogative that belongs to management.

The extent of such an understatement can be significant, as illustrated (Table II). This project has an actual rate of return of 35% based on end-of-year receipts and annual compounding; but it has a calculated rate of return of only 30% based on end-of-year receipts and continuous compounding.

To report the rate of return on this investment as 30% is comparable to a savings and loan company advertising 4.85%/yr. for savings without mentioning that interest is compounded on a daily basis and thus

Comparison of Annual Compounding and Continuous Compounding—Table II

(Investment=$2,220)

Year	Annual Earnings	End-of-Period Present-Worth Factor Based on Annual Compounding @ 35%	Present Worth @ 35%	End-of-Period Present-Worth Factor Based on Continuous Compounding @ 30%	Present Worth @ 30%
1	$1,000	0.741	$741	0.741	$741
2	1,000	0.549	549	0.549	549
3	1,000	0.406	406	0.407	407
4	1,000	0.301	301	0.301	301
5	1,000	0.223	223	0.223	223
			$2,220		$2,221

Continuous-Income Present-Worth Factors Based on Annual Compounding—Table III

Interest Rate of Return

Year	1%	2%	3%	4%	5%	6%	7%	8%	9%	10%	11%	12%	13%	14%	15%	16%	17%	18%	19%	20%	21%	22%	23%
1	0.995	0.990	0.985	0.981	0.976	0.971	0.967	0.962	0.958	0.954	0.950	0.945	0.941	0.937	0.933	0.929	0.925	0.922	0.918	0.914	0.910	0.907	0.903
2	0.985	0.971	0.957	0.943	0.930	0.916	0.904	0.891	0.879	0.867	0.855	0.844	0.833	0.822	0.812	0.801	0.791	0.781	0.771	0.762	0.752	0.743	0.734
3	0.975	0.952	0.929	0.907	0.885	0.865	0.845	0.825	0.806	0.788	0.771	0.754	0.737	0.721	0.706	0.691	0.676	0.662	0.648	0.635	0.622	0.609	0.597
4	0.966	0.933	0.902	0.872	0.843	0.816	0.789	0.764	0.740	0.717	0.694	0.673	0.652	0.633	0.614	0.595	0.578	0.561	0.545	0.529	0.514	0.499	0.485
5	0.956	0.915	0.875	0.838	0.803	0.769	0.738	0.707	0.679	0.651	0.626	0.601	0.577	0.555	0.534	0.513	0.494	0.475	0.458	0.441	0.425	0.409	0.395
6	0.947	0.897	0.850	0.806	0.765	0.726	0.689	0.655	0.623	0.592	0.564	0.536	0.511	0.487	0.464	0.442	0.422	0.403	0.385	0.367	0.351	0.336	0.321
7	0.937	0.879	0.825	0.775	0.728	0.685	0.644	0.607	0.571	0.538	0.508	0.479	0.452	0.427	0.403	0.381	0.361	0.341	0.323	0.306	0.290	0.275	0.261
8	0.928	0.862	0.801	0.745	0.694	0.646	0.602	0.562	0.524	0.489	0.457	0.428	0.400	0.375	0.351	0.329	0.308	0.289	0.272	0.255	0.240	0.225	0.212
9	0.919	0.845	0.778	0.717	0.661	0.609	0.563	0.520	0.481	0.445	0.412	0.382	0.354	0.329	0.305	0.283	0.264	0.245	0.228	0.213	0.198	0.185	0.172
10	0.910	0.829	0.755	0.689	0.629	0.575	0.526	0.481	0.441	0.405	0.371	0.341	0.313	0.288	0.265	0.244	0.225	0.208	0.192	0.177	0.164	0.151	0.140
11	0.901	0.812	0.733	0.662	0.599	0.542	0.492	0.446	0.405	0.368	0.334	0.304	0.277	0.253	0.231	0.211	0.193	0.176	0.161	0.148	0.135	0.124	0.114
12	0.892	0.796	0.712	0.637	0.571	0.512	0.459	0.413	0.371	0.334	0.301	0.272	0.245	0.222	0.201	0.182	0.165	0.149	0.135	0.123	0.112	0.102	0.093
13	0.883	0.781	0.691	0.613	0.543	0.483	0.429	0.382	0.341	0.304	0.271	0.243	0.217	0.195	0.174	0.157	0.141	0.126	0.114	0.103	0.092	0.083	0.075
14	0.874	0.765	0.671	0.589	0.518	0.455	0.401	0.354	0.313	0.276	0.245	0.217	0.192	0.171	0.152	0.135	0.120	0.107	0.096	0.085	0.076	0.068	0.061
15	0.866	0.750	0.651	0.566	0.493	0.430	0.375	0.328	0.287	0.251	0.220	0.193	0.170	0.150	0.132	0.116	0.103	0.091	0.080	0.071	0.063	0.056	0.050
16	0.857	0.736	0.632	0.545	0.469	0.405	0.350	0.303	0.263	0.228	0.198	0.173	0.151	0.131	0.115	0.100	0.088	0.077	0.068	0.059	0.052	0.046	0.040
17	0.849	0.721	0.614	0.524	0.447	0.382	0.328	0.281	0.241	0.208	0.179	0.154	0.133	0.115	0.100	0.086	0.075	0.065	0.057	0.049	0.043	0.038	0.033
18	0.840	0.707	0.596	0.503	0.426	0.361	0.306	0.260	0.221	0.189	0.161	0.138	0.118	0.101	0.087	0.075	0.064	0.055	0.048	0.041	0.036	0.031	0.027
19	0.832	0.693	0.579	0.484	0.406	0.340	0.286	0.241	0.203	0.172	0.145	0.123	0.104	0.089	0.075	0.064	0.055	0.047	0.040	0.034	0.029	0.025	0.022
20	0.824	0.680	0.562	0.465	0.386	0.321	0.267	0.223	0.186	0.156	0.131	0.110	0.092	0.078	0.066	0.055	0.047	0.040	0.034	0.029	0.024	0.021	0.018
21	0.815	0.666	0.546	0.448	0.368	0.303	0.250	0.206	0.171	0.142	0.118	0.098	0.082	0.068	0.057	0.048	0.040	0.034	0.028	0.024	0.020	0.017	0.014
22	0.807	0.653	0.530	0.430	0.350	0.286	0.234	0.191	0.157	0.129	0.106	0.088	0.072	0.060	0.050	0.041	0.034	0.029	0.024	0.020	0.017	0.014	0.012
23	0.799	0.640	0.514	0.414	0.334	0.270	0.218	0.177	0.144	0.117	0.096	0.078	0.064	0.052	0.043	0.035	0.029	0.024	0.020	0.017	0.014	0.011	0.010
24	0.791	0.628	0.499	0.398	0.318	0.254	0.204	0.164	0.132	0.107	0.086	0.070	0.057	0.046	0.037	0.031	0.025	0.020	0.017	0.014	0.011	0.009	0.008
25	0.784	0.616	0.485	0.383	0.303	0.240	0.191	0.152	0.121	0.097	0.078	0.062	0.050	0.040	0.033	0.026	0.021	0.017	0.014	0.011	0.009	0.008	0.006
26	0.776	0.604	0.471	0.368	0.288	0.226	0.178	0.141	0.111	0.088	0.070	0.056	0.044	0.035	0.028	0.023	0.018	0.015	0.012	0.010	0.008	0.006	0.005
27	0.768	0.592	0.457	0.354	0.274	0.214	0.166	0.130	0.102	0.080	0.063	0.050	0.039	0.031	0.025	0.020	0.016	0.012	0.010	0.008	0.006	0.005	0.004
28	0.761	0.580	0.444	0.340	0.261	0.201	0.156	0.120	0.094	0.073	0.057	0.044	0.035	0.027	0.021	0.017	0.013	0.011	0.008	0.007	0.005	0.004	0.003
29	0.753	0.569	0.431	0.327	0.249	0.190	0.145	0.112	0.086	0.066	0.051	0.040	0.031	0.024	0.019	0.015	0.011	0.009	0.007	0.006	0.004	0.003	0.003
30	0.746	0.558	0.418	0.314	0.237	0.179	0.136	0.103	0.079	0.060	0.046	0.035	0.027	0.021	0.016	0.013	0.010	0.008	0.006	0.005	0.004	0.003	0.002

Interest Rate of Return

Year	24%	25%	26%	27%	28%	29%	30%	32%	34%	35%	40%	45%	50%	55%	60%	65%	70%	75%	80%	85%	90%	95%	100%
1	0.900	0.896	0.893	0.889	0.886	0.883	0.880	0.873	0.867	0.864	0.849	0.835	0.822	0.810	0.798	0.787	0.776	0.766	0.756	0.747	0.738	0.729	0.721
2	0.726	0.717	0.709	0.700	0.692	0.684	0.677	0.662	0.647	0.640	0.607	0.576	0.548	0.522	0.499	0.477	0.456	0.438	0.420	0.404	0.388	0.374	0.361
3	0.585	0.574	0.562	0.551	0.541	0.531	0.520	0.501	0.483	0.474	0.433	0.397	0.365	0.337	0.312	0.289	0.269	0.250	0.233	0.218	0.204	0.192	0.180
4	0.472	0.459	0.446	0.434	0.423	0.411	0.400	0.380	0.360	0.351	0.309	0.274	0.244	0.217	0.195	0.175	0.158	0.143	0.130	0.118	0.108	0.098	0.090
5	0.381	0.367	0.354	0.342	0.330	0.319	0.308	0.288	0.269	0.260	0.221	0.189	0.162	0.140	0.122	0.106	0.093	0.082	0.072	0.064	0.057	0.050	0.045
6	0.307	0.294	0.281	0.269	0.258	0.247	0.237	0.218	0.201	0.193	0.158	0.130	0.108	0.090	0.076	0.064	0.055	0.047	0.040	0.034	0.030	0.026	0.023
7	0.248	0.235	0.223	0.212	0.201	0.192	0.182	0.165	0.150	0.143	0.113	0.090	0.072	0.058	0.048	0.039	0.032	0.027	0.022	0.019	0.016	0.013	0.011
8	0.200	0.188	0.177	0.167	0.157	0.149	0.140	0.125	0.112	0.106	0.081	0.062	0.048	0.038	0.030	0.024	0.019	0.015	0.012	0.010	0.008	0.007	0.006
9	0.161	0.150	0.141	0.131	0.123	0.115	0.108	0.095	0.083	0.078	0.058	0.043	0.032	0.024	0.019	0.014	0.011	0.009	0.007	0.005	0.004	0.003	0.003
10	0.130	0.120	0.112	0.103	0.096	0.089	0.083	0.072	0.062	0.058	0.041	0.029	0.021	0.016	0.012	0.009	0.007	0.005	0.004	0.003	0.002	0.002	0.001
11	0.105	0.096	0.089	0.081	0.075	0.069	0.064	0.054	0.046	0.043	0.029	0.020	0.014	0.010	0.007	0.005	0.004	0.003	0.002	0.002	0.001	0.001	0.001
12	0.084	0.077	0.070	0.064	0.059	0.054	0.049	0.041	0.035	0.032	0.021	0.014	0.010	0.007	0.005	0.003	0.002	0.002	0.001	0.001	0.001		
13	0.068	0.062	0.056	0.051	0.046	0.042	0.038	0.031	0.026	0.024	0.015	0.010	0.006	0.004	0.003	0.002	0.001	0.001	0.001				
14	0.055	0.049	0.044	0.040	0.036	0.032	0.029	0.024	0.019	0.017	0.011	0.007	0.004	0.003	0.002	0.001	0.001	0.001					
15	0.044	0.039	0.035	0.031	0.028	0.025	0.022	0.018	0.014	0.013	0.008	0.005	0.003	0.002	0.001	0.001							
16	0.036	0.032	0.028	0.025	0.022	0.019	0.017	0.014	0.011	0.010	0.005	0.003	0.002	0.001	0.001								
17	0.029	0.025	0.022	0.019	0.017	0.015	0.013	0.010	0.008	0.007	0.004	0.002	0.001	0.001									
18	0.023	0.020	0.018	0.015	0.013	0.012	0.010	0.008	0.006	0.005	0.003	0.002	0.001										
19	0.019	0.016	0.014	0.012	0.010	0.009	0.008	0.006	0.004	0.004	0.002	0.001	0.001										
20	0.015	0.013	0.011	0.009	0.008	0.007	0.006	0.004	0.003	0.003	0.001	0.001											
21	0.012	0.010	0.009	0.007	0.006	0.005	0.005	0.003	0.002	0.002	0.001												
22	0.010	0.008	0.007	0.006	0.005	0.004	0.004	0.003	0.002	0.002	0.001												
23	0.008	0.007	0.006	0.005	0.004	0.003	0.003	0.002	0.001	0.001	0.001												
24	0.006	0.005	0.004	0.004	0.003	0.003	0.002	0.001	0.001	0.001													
25	0.005	0.004	0.003	0.003	0.002	0.002	0.002	0.001	0.001	0.001													
26	0.004	0.003	0.003	0.002	0.002	0.002	0.001	0.001	0.001														
27	0.003	0.003	0.002	0.002	0.001	0.001	0.001	0.001															
28	0.003	0.002	0.002	0.001	0.001	0.001	0.001																
29	0.002	0.002	0.001	0.001	0.001	0.001	0.001																
30	0.002	0.001	0.001	0.001	0.001	0.001																	

savings actually earn 4.97%/yr. In the parlance of banking and finance, the 4.85% is called the "nominal" rate, while the 4.97% based on annual compounding is the "effective" rate. Thus in Table II, 35% is the effective rate, while 30% is the nominal rate without the compounding period specified. For proper evaluation, the effective rate must be stated in both cases.

The Correct Continuous-Income Factor

The basic compound-interest formula is capable of handling receipts of income that are truly continuous without going to continuous compounding. Present-worth factors based on continuous receipts of income (equal infinitesimal increments of money received in equal infinitesimal increments of time) and on annual compounding can be calculated with a formula derived as follows:

The present worth of \$1 at any time, n, equals

$$1/(1+i)^n$$

If a total income of \$1 for year n is received continuously during the year, each increment received would be equal to:

$$1\,dn$$

The present worth of each such increment would be

$$1\,dn/(1+i)^n$$

and the present worth of all the increments would be

$$\int_{n-1}^{n} \frac{dn}{(1+i)^n} = \int_{n-1}^{n} (1+i)^{-n} dn =$$

$$-\frac{(1+i)^{-n}}{\ln(1+i)} + \frac{(1+i)^{-(n-1)}}{\ln(1+i)} =$$

$$\frac{(1+i)^{-n}}{-\ln(1+i)} - \frac{(1+i)^{-(n-1)}}{-\ln(1+i)} =$$

$$\frac{(1+i)^{-n} - (1+i)^{-(n-1)}}{-\ln(1+i)} = \frac{\dfrac{1}{(1+i)^n} - \dfrac{1}{(1+i)^{n-1}}}{-\ln(1+i)}$$

$$= \frac{\left(\dfrac{1}{1+i}\right)^n - \left(\dfrac{1}{1+i}\right)^{n-1}}{\ln\left(\dfrac{1}{1+i}\right)}$$

Present-worth factors based on this formula are shown in Table III. Of course, infinitesimal increments of income and time are not realistic in the business world. However, continuous-income present-worth factors are identical (to 3 decimal places) to present-worth factors based on the assumption that 1/365th of the annual earnings is being received at the end of each day—an assumption that is not unrealistic.[3]

Another present-worth factor developed by analysts to express the continuous-income concept is the mid-period factor. This factor, which fairly well approximates the true continuous-income factor shown above, is based on the assumption that all income received in any year is received as one lump sum at the middle of the year, the formula being $[(1/(1+i)]^{n-\frac{1}{2}}$.

This mid-period factor does occasionally find use as a true mid-period factor. For example, the income from additional investment in higher-priced tubes for retubing a heat exchanger with material estimated to have a 3-yr. life as opposed to a 1½-yr. life is the cost of retubing the exchanger with low-cost tubes 1½ years from retubing; and the factor becomes $[(1/(1+i)]^{2-\frac{1}{2}} = [(1/(1+i)]^{1.5}$.

In sum, the DCF rate of return has been defined as the rate of interest at which funds could be borrowed by the company for investment in the project and have the project break even at the end of its life. This definition makes little sense if rate of return is calculated on a continuous-compounding basis, since funds are never borrowed at a continuously compounded rate of interest. Consequently, while present-worth factors based on continuous compounding are a mathematician's delight, they have no relevance whatsoever to the practical world of business and finance and should never be used for rate-of-return calculations.

References and Comments

1. For example, E. L. Grant and W. G. Ireson, in "Principles of Engineering Economy," 4th ed., Ronald Press (1960), state that "continuous compounding is well adapted to the assumption of a continuous flow of funds at a uniform rate throughout a stated period of time," and that "to use this . . . convention (cash flow occurring uniformly throughout the year) in trial-and-error calculations of unknown rates of return, it is necessary to assume continuous compounding of interest." Also, J. B. Weaver, and R. J. Reilly, in *Chem. Eng. Prog.*, **52**, 10, p. 405 (Oct., 1956), have said, "In most businesses, money payments occur almost continuously so that the net receipts or compounding of interest can be approximated as occurring in terms of an infinitesimal interest period." W. R. Hirschman and J. R. Brauweiler, in *Chem. Eng.*, **72**, 15, p. 210 (July 19, 1965), have stated: "Although discounting can be done on either an annual or continuous basis, we find that the continuous has several major advantages over annual. It is usually more representative of the actual flow of funds. . . . Continuous discounting more realistically reflects the actual flow of business cash than annual discounting." Other well-known writers, such as Joel Dean (*Harv. Bus. Rev.*, Jan.-Feb., 1954, p. 120) and Ray I. Reul (*Harv. Bus. Rev.*, July-Aug., 1957, p. 116) have also recommended or used present-worth factors based on continuous compounding of interest.
2. Continuous compounding is usually justified on the basis that it best reflects the continuous flow of income throughout the year. However, the continuous-compounding present-worth factors presented in the literature are frequently used for lump-sum receipts.
3. To their credit, E. L. Grant and W. G. Ireson (*op. cit.*, see reference 1 above) have recognized that rates of return based on continuous compounding are nominal rates that do not give as realistic a picture of the productivity of an investment as effective rates. Consequently, they recommend the use of present-worth tables based on continuous income and continuous compounding, but with the interest rates converted to effective rates. What they fail to point out is that when nominal interest rates based on continuous compounding are converted to effective rates, we simply end up with annual compounding. That is, the present-worth tables in the appendix of their book, which are titled "Present Worth at Zero Date of \$1 Flowing Uniformly Throughout Stated One-Year Periods, Assuming Continuous Compounding of Interest at Various Stated Effective Rates Per Annum," are identical with the present-worth factors based on continuous income and annual compounding shown in Table III. ∎

Meet the Author

Robert H. Congelliere is a Senior Engineer at the Los Angeles Refinery of the Union Oil Co. of California, Wilmington, Calif. 90744. Since joining Union Oil in 1954, he has worked as a process engineer in the Research Dept. and at Union Oil's Santa Maria and Los Angeles refineries. He holds a BS in Chemical engineering and an MS from Ohio State University.

To illustrate what you don't learn about venture-analysis work in school, here is an amusing but pointed playlet on . . .

The Human Equation in Economic Evaluations

HOMER S. SINGLETREE, Chemical Engineer

All the action takes place within an office furnished with steel desks, steel chairs and steel filing cabinets. The steel walls are bare except for a photo of a steel chemical plant. Behind the desk sits Vincent, his hair steel gray, face wrinkled as a used dishtowel. Across from him sits Norman, a man in his late twenties.

Vincent: Welcome to economic evaluation. *(He grins.)*

Norman: After doing pilot-plant work for two years, I suppose that economic-evaluation work will seem very different.

V (pontifically): It's still part of chemical engineering, Norman. Here at the R&D lab, we need good ol' economic evaluation to keep us pointed in the right direction.

N: I was thinking the other day that except for the two years I've been here with the company, most of my adult life has been spent in school learning about dimensionless numbers.

V: Didn't you take any courses in engineering economics?

N: Yes, one course. Discounted cash flow. Comparison of alternatives.

V (to himself): Damn the alternatives. Full speed ahead! *(Recovering)* I'll start from scratch, then, in explaining economic evaluation as it is done around here.

N (anxious to show that he has done some homework): It looks straightforward to me. Someone here in R&D does lab work on a new process. I take the results and design a plant. Then I estimate the capital required to build the plant, its production cost, and its profitability. Based on that, our director can tell whether to go ahead with more work on the process.

V (unimpressed): That's the way it looks, all right.

N: Let's go. What's the first thing I need to know?

V: That I'm your boss, that my boss is Rip Strong, and that Rip Strong's boss is Rex Maxim.

N (who will not really understand the significance of Vincent's last statement for some time to come): Fine. Those are the ground rules. Where now do I really begin on the evaluation of a chemical manufacturing venture?

The air shimmers in a barely perceptible way. Several times throughout this conversation there will be this peculiar agitation of the atmosphere. It is caused by the fluttering of the invisible spirits of murdered hopes, especially in the corner of the office above Vincent's desk.

V (formally): You begin with Preliminary Scouting.

N: Searching the literature?

V: Huh? Oh, yes. Of course. You have to search the literature. That wasn't exactly what I meant, though. What I really meant was, you find out first what results people *expect.*

N: People?

V: Me. Also Rip Strong. Not Maxim—he's out of the office most of the time anyway.

N: Doesn't sound very objective to me. Chemical engineers are supposed to think objectively.

V (blandly): Of course. Got to think objectively, get both arms around the barrel until they meet in back. Can't cut the mustard otherwise. But we need to cooperate, too. Find out what Rip . . . uh, the other fellow thinks.

N: I get it. Teamwork. We all need to work together.

V: Why, sure. It's everyone working together that has put the company where it *(shuddering slightly)* is today.

N (enthusiastic as a result of having heard these inspiring words): By golly, you're right!

V (pleased): Preliminary Scouting is the best way to begin any economic evaluation. Sound me out first. Then ask Rip some questions. Rip may seem a bit short-tempered at times, but he's okay when you get to know him. *(Vincent's eyes flick automatically to the door as he says the words. He makes certain that*

Originally published June 1970.

no one is standing there. He then places his elbows on the desk and cradles his head in his hands for a moment while he reflects and chooses his words.) He's not our boss for nothing, you know.

N: The plant design comes next, I suppose.

V: Not so fast, boy. Next comes Preliminary Snuffing, which means seeking out everyone else's economics.

N: Everyone else's?

V: The other companies. The ones who are already in the business.

N: What if no one else is in the business. What if it's the first time the chemical has ever been made?

V: Let me know if ever *that* happens.

N: Gosh, where do you go to look for the other fellow's economics?

V: The newspapers. The trade magazines. But best of all, Home Office Intelligence.

N (puzzled): Is the Home Office any more intelligent than we are?

V: Of course not, but it has Intelligence. Remember, a man with Intelligence is not necessarily an intelligent man.

N (alarmed): There are spies?

V: Absolutely not. Rather, there is open discussion among those in the higher levels of management. *(He clears his throat, pleased at what strikes him as having been a pleasant choice of words.)*

N: I still don't get you.

V: It's the way things are done, boy. Suppose you have a friend who works for Esso. He tells you that he had lunch last week with a Shell vice-president, and the Shell V.P. said that it took $15 million for their 100 million-lb./yr. plant.

N (depressed, because most of his friends are doing postdoctoral-fellowship work rather than lunching with vice-presidents): Aren't there other places to snuff? What about work our guys have done right here at R&D? What about studies from our own plant-engineering groups?

V: Prophets without honor, Norman. Cigarette?

There is a pause as Vincent, his hands trembling, lights up. Norman is thinking.

N (trying to reassure himself, for after all he is an engineer): It doesn't seem to me that there is much engineering involved in Preliminary Scouting or Preliminary Snuffing. Maybe the engineering comes next?

V: Not so fast. Next, there is a meeting with the chemists.

N: Why a meeting? Don't you have their reports to go on?

V (heaving a sigh): The reports tell what was *done.*

N: Isn't that what we need to know?

V: Not in the chemists' opinion. What they want to tell us is what *could* be done given six months more work in the lab.

N: So what do I do at the meeting?

V: Listen carefully to what the chemists have to say. Copy their recommendations, and also be sure to copy the flow diagram they suggest.

. . . This peculiar agitation of the atmosphere . . . is caused by the fluttering of the invisible spirits of murdered hopes, especially . . . above Vincent's desk . . .

N: And then I just plug in the data they've given me?

V (alarmed): Oh, no! Heavens, no! You subtract 10% from all yields, halve the catalyst life and add two distillation columns.

N: But why? Are you saying that chemists like Craggowitz or Thomas may not be objective?

V: Oh, they're objective all right—but they are more objective about good yields than about bad ones.

Norman does not quite understand that even a Ph.D. in organic chemistry may want to be a hero, and that heroism is more often associated with optimism than with pessimism or even with complete objectivity. He does understand, though, that further questioning at this time will do him little good. Consequently he proceeds to a new subject.

N: All right, suppose I do as you say. Now what?

V: The plant design.

N (joyous): At last!

V: At last. You make material and energy balances, size the equipment, specify materials of construction and then estimate the cost of all major items of equipment.

N (furrowing his 17-cm.-high brow): Hmm. Wait a minute. Does the company have correlations of cost with equipment size?

V: Oh, yes. We have a whole book of cost correlations. They are not right, but they are the same as the other companies use. Keeps us competitive.

N (finally getting to something he has studied in college): Now for the fixed capital. What about the factor method? Can't you arrive at the fixed capital by multiplying the cost of major equipment by a factor?

V: Correct. . . .

N: And the factor? How is it determined?

Vincent indulges in a reverie. His countenance expresses faint pain.

N: Excuse me. Are you all right?

Startled, Vincent returns to the real world, looks Norman in the eye, rises and walks around the desk. Norman rises, sensing that something important is to be divulged. Vincent seizes the younger man's arm, looks him in the eye, and speaks evenly.

V: You will learn with experience what factor to use. I can save you a lot of time, though, by giving you some good advice right now. The best way is to start with the fixed capital Rip Strong *expects* you to come up with.

N: But I meant the *factor,* the ratio of fixed capital to equipment cost . . .

V (interrupting, edging closer, intent, whispering): Exactly. You take the fixed capital Rip expects you to get, you divide it by the equipment cost you have scientifically determined, and *(here he thrusts his face close to Norman and hisses)* you then have the *best damned factor there is! (The air seems to quiver at this. The ghosts of studies past are applauding.)*

N (who dislikes such close physical contact, especially when accompanied by such evident emotion): Let's go on to operating expenses. What about those?

V: You fill in forms like this one. *(Vincent shows Norman a standard form, complete with percentages, something like a waitress's order pad.)*

N: Who publishes these forms?

V: The Home Office. Operating costs in this company are fixed by decree from the Home Office.

N: What if the plant doesn't run according to the form?

V: The company appoints a new jockey.

N (who is making less headway than he would like): And the return on investment . . .

V (breaking in): Will be low, boy, *low.*

N (smiling, since he thinks that Vincent's last statement can only have been a jest): Well, that's about the end of the calculations, isn't it?

It is Vincent's turn to smile. "Smile" is an understatement. He laughs the way a man laughs who does not laugh often, dryly, sounding like a horse trampling on a large wooden crate. After this paroxysm is over, Vincent speaks:

V: That's not quite all the calculations. Not quite. Before you lay down your sliderule you must also consider the effect on the return on investment of increasing the yields 10%, doubling the catalyst life, and cutting out two distillation columns.

N (who is no one's fool): Would those things . . . would those suggestions be made, possibly, by the chemists?

V (extending his hands widely, palms upward): Who else?

N: Then why not do it their way in the first place, and save recalculation?

V. If you did it their way in the first place, then after you finished the calculation they would want you to make it even more optimistic.

N (astounded): Oh! Barbaric. Ruthless!

V: And true. Let's move along, Norman. There is still more recalculation to do, because later on you will be asked for data on various alternatives.

N: Like what?

V: Enlarging the plant capacity. Showing sales growth at three different levels. Making all equipment of carbon steel. Eliminating in-process storage.

N: Oh, I see. Actually, those sound like pretty obvious alternatives.

V: Precisely. They are the kind that will be suggested by the Home Office after they have seen your preliminary results.

N: And how do I write up the final results?

V: Every section in your final-results letter must be constructed in the format prescribed by the company's report doctors.

N: You review the letter before it is typed, I presume.

V: That's correct, and incidentally, boy, leave a mistake or two in your work so I can find them. Makes me feel useful. I often do the same for Rip Strong.

N (anxious to please): You bet I will. And after you approve my letter and it's typed, do you send it to Rip?

V: I surely do. Then he writes a cover letter and sends it to Maxim . . .

N (gaily continuing the statement): Then Maxim writes a cover letter to the Home Office . . .

V: Right y'are.

N: Where they read all the letters.

V: Read? The Home Office is not interested in your *words.* What it really cares about are your *numbers.*

N: Then why do all that writing?

V (beginning to become frazzled): Don't knock the system, boy, until you've been around here a few months. Just remember—it's the numbers that count. And I'll tell you the best kind: Fresh numbers given to you over the phone from far away, those are the best.

A siren howls outside, signaling the arrival of lunch time. Vincent excuses himself and leaves for his bridge game. Norman, fiercely practical by now, lopes down the hall with his mind on Preliminary Scouting. ■

Meet the Author

Homer S. Singletree is a pseudonym for an experienced chemical engineer employed by a large operating company in the chemical process industries. He has spent several years of his life on the economic evaluation of new chemical ventures, and offers this play in the hope of illuminating the practical aspects of this work, especially as contrasted with the theory.

How to determine if the "best" alternative is good enough

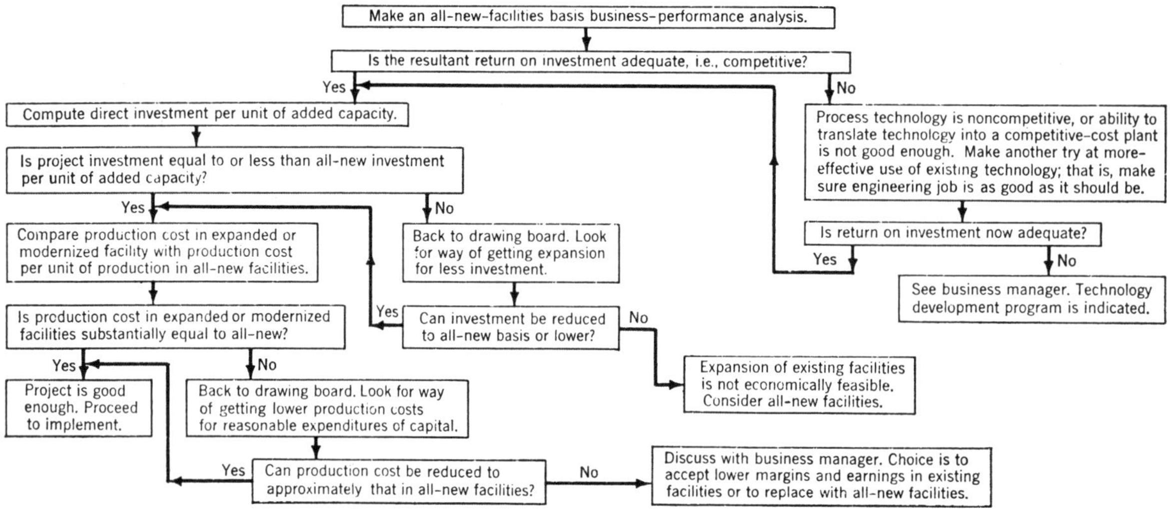

A "Business" Approach to Capital-Budgeting Decisions

You've submitted a project proposal having a high rate of return to management, but it is rejected. Why? Perhaps because the project may be in conflict with business realities. Here is a look at how business management views investment alternatives.

JAMES A. CHAMPLEY, Union Carbide Corp.

Conventional approaches to capital-investment decisions assume that the analyst is selecting a "portfolio" of projects in which to invest limited funds. Various methods are used to rank projects based on some criteria such as rate of return, present worth, profitability index, etc. It often happens, however, that a project with a high discounted-cash-flow rate of return is rejected by management. Why and how can this happen?

Basically, a seemingly promising project can be rejected because of the way management looks at investment opportunities. Management usually views this situation as one of selecting the portfolio of businesses in which to invest resources (as opposed to selecting the portfolio of projects in which to invest funds). Thus, management allocates resources to businesses rather than to projects, and management's prime concern is the overall performance of businesses as they are accounted and reported, rather than of projects as they are normally evaluated.

Originally published September 1969.

Adapt Management View

This conflict can be resolved by adapting capital budgeting to a business-oriented environment. The purpose of this article is to present the business approach to capital-budgeting decisions.

Question: When are 30% and 40% return on investment projects not good enough?

Answer: When they do not result in the ability to perform competitively in the business.

Corporations are like baseball players. They compare their batting and fielding averages (growth in earnings, return on investment, market share) against other players in the same league.

If the other players are doing better, our favorite baseball hero is dissatisfied. If other corporations are doing better, our corporation is also dissatisfied.

Our goal, whether we be baseball players or corporations, is to perform at least as well as any other player or corporation in the same league.

Let us see how this desire to perform competitively in a business influences the capital-budgeting process.

Of course, the examples that follow are not intended to be guidelines to the sometimes very complex analyses required for investment decisions. Their purpose is to show how a chosen investment alternative can be evaluated to see whether it is good enough to satisfy the goals of competitive business performance.

An Expansion Project

Sales forecasts for next year exceed production capacity. The production manager believes that his plant output can be increased by about 10% with a moderate investment.

The engineering manager is asked to evaluate methods for increasing capacity. An engineering study confirms the feasibility of an incremental expansion of the present plant.

Financial analyses of the proposed expansion project are made and the results are favorable. A capital-authorization request is drafted and submitted to the division head who has profit responsibility for the business. A summary of the financial analyses supporting the request is shown in Table I.

Let us now look at how the profit-responsible manager, i.e., the business manager, views this draft of the capital-authorization request.

Performance of the business has recently deteriorated because of sales price reductions forced by competition. This has resulted in lower margins, reduced profits, and a lower return on investment. Consequently, the business manager has come under executive pressure to improve the profit performance of his business, as measured by growth in profits before taxes and return on investment, ROI. (Return on investment is measured as the percentage ratio of the profit before tax to the total investment assigned to the business measured at original acquisition cost or gross book value.)

The business manager notes that the project promises a handsome 30% return on investment. The project will be "salable" and there should be little difficulty in obtaining a capital-budget authorization. However, his first concern is understandably his job performance, which is related to the profit performance of his business as measured on the business per-

Capital authorization request for expanding plant facilities—Table I

Annual Items	Thousands of Dollars	
Sales increase		5,000
Production cost	3,000	
Freight, warehousing, etc	500	
Total direct costs		3,500
Margin (sales minus direct costs)		1,500
Overhead costs		400
Operating income (margin minus overhead costs)		1,100
Depreciation (on fixed investment)		200
Profit before tax		900
Investment Items		
New, this project	1,000	
Share of supporting facilities (utilities, laboratory, office, etc.)	1,000	
Working capital increase	1,000	
Total additional investments		3,000
Return on investment (ROI) = (900)(100)/3,000 = 30%		

Ground rules and assumptions for Table I:

1. Evaluation criterion is return on investment (ROI), defined as the average annual increase in profit before taxes as a percentage of total added investment.
2. Official project cutoff rate is 20%.
3. Depreciation is straight-line basis, 10-yr. life.
4. Overhead costs are incremental, based on an assumed rate of overhead increase equal to ⅔ the rate of sales increase.
5. Working capital increase consists of added inventory, increased accounts receivable, and increased cash.

Business performance statement for expanding facilities—Table II

	Now		This Project		After This Project	
	Total[a]	Unit[b]	Total[a]	Unit[b]	Total[a]	Unit[b]
Sales	50,000	1.00	5,000	1.00	55,000	1.00
Production cost	30,000	0.60	3,000	0.60	33,000	0.60
Freight, warehousing, etc.	5,000	0.10	500	0.10	5,500	0.10
Total direct costs	35,000	0.70	3,500	0.70	38,500	0.70
Margin	15,000	0.30	1,500	0.30	16,500	0.30
Overhead	6,000	0.12	400	0.08	6,400	0.116
Operating income	9,000	0.18	1,100	0.22	10,100	0.184
Depreciation	5,000	0.10	200	0.04	5,200	0.095
Profit before tax	4,000	0.08	900	0.18	4,900	0.089
Investment	Total[c]	Unit[d]	Total[c]	Unit[d]	Total[c]	Unit[d]
Direct	40,000	0.80	1,000	0.20	41,000	0.75
Share of supporting	10,000	0.20	1,000	0.20	11,000	0.20
Working capital	10,000	0.20	1,000	0.20	11,000	0.20
Total	60,000	1.20	3,000	0.60	63,000	1.15
Return on investment	6.7%		30%		7.8%	

a. Thousands of dollars/yr.
b. Dollars/unit of production.
c. Thousands of dollars.
d. Dollars per unit of production/yr.

formance statement. His first reaction to the project, therefore, is: "What will this project do for my business or profit center?" The data in Table II give the answer to his question: After the project is completed, the business will show a return on investment of 7.8%.

The business manager judges a 7.8% return to be noncompetitive and, therefore, unacceptable. Overall return for the industry is about 10%, and this business should be higher than the industry average because of its specialized nature. Other businesses in his organization with similar characteristics show returns of 12 to 25%.

The business manager also is aware that other companies are building new production facilities for the same product. His company would not commit itself to major plant expansions at such a return, and likewise, neither would his competitors. Therefore, at least some of the competition must have the ability to get a higher profit margin, or use less investment, or both.

All-New Facilities

The problem becomes one of how to bring the performance of the business up to an acceptable level of return on investment. But before this can be solved, it is essential to know where the business stands today —what are the current capabilities in the business?

The business performance statement shows the historical investment to date in present facilities and today's costs of operating facilities designed and installed in the past. It does not show what today's performance would have been if today's knowhow had been used in the design of the facilities.

The business manager, therefore, asks his financial analyst, with the help of the engineering group, to reconstruct the business performance statement, using production and investment costs that could be attained in all-new facilities based on today's technology. Results are shown in Table III, together with the "after this project" column of Table II for comparison.

The result of the all-new facilities analysis is a pleasant surprise since a 20% return on investment is considered satisfactory by the business manager. It indicates that capabilities in the business are competitive.

What is the reason for the poor performance of the present facilities after the expansion project? A comparison of the two "unit" columns in Table III shows that:

- Unit production cost of the expanded facilities is higher than production cost in the all-new facilities ($0.60 vs. $0.50).
- Unit direct investment in the expanded facilities is higher than all-new-facilities investment ($0.75 vs. $0.60).

Historical investment is a sunk cost and it cannot be changed by anything we do now. Unless there is a reduction in production cost, there is nothing to be gained by scrapping old facilities and building new ones just to reduce book investment so as to increase return on investment.

However, production cost, unlike investment cost, is not a sunk cost. It is something that is continually committed anew by the continuing decision to remain in business with the current facility.

If the all-new facilities were to be operated at a production cost equal to that in the expanded facilities ($0.60/unit), profit before tax would be $0.10/unit rather than $0.20/unit, and the return on investment would be 10% rather than 20%. Thus, the earning power of the business would be halved.

A corporation, if it is to perform competitively, cannot afford to allocate its resources, particularly its scarce management resources, at 50% of the efficiency of which it is capable and at 50% of the rate required to be competitive.

Reducing Costs

Therefore, costs of production in the expanded facilities must be reduced. There are two possible ways:

- Modernize existing facilities.
- Build all-new replacement facilities.

If production costs in the existing facilities can be reduced to near the production costs in all-new facilities with the expenditure of a moderate amount of new investment, then that is the preferred path. All-new facilities and the use of new technology generally

Comparing all-new plant—Table III

	All-New-Facilities Basis		"After This Project" From Table II	
	Total[a]	Unit[b]	Total[a]	Unit[b]
Sales	55,000	1.00	55,000	1.00
Production cost	27,500	0.50	33,000	0.60
Freight, warehousing, etc.	5,500	0.10	5,500	0.10
Total direct costs	33,000	0.60	38,500	0.70
Margin	22,000	0.40	16,500	0.30
Overhead	6,400	0.116	6,400	0.116
Operating income	15,600	0.284	10,100	0.184
Depreciation	4,400	0.08	5,200	0.095
Profit before tax	11,200	0.204	4,900	0.089
Investment Items	Total[c]	Unit[d]	Total[c]	Unit[d]
Direct	33,000	0.60	41,000	0.75
Share of supporting	11,000	0.20	11,000	0.20
Working capital	11,000	0.20	11,000	0.20
Total	55,000	1.00	63,000	1.15
Return on investment	20%		7.8%	

a. Thousands of dollars/yr.
b. Dollars/unit of production.
c. Thousands of dollars.
d. Dollars per unit of production/yr.

involve greater startup delays and costs, and increased demands on skilled manpower when compared to modernization. However, if production costs in the existing facilities cannot be reduced, or if the new investment required to do so is excessive, the only viable alternative is to build replacement facilities.

If it happens that modernization is not feasible and replacement is the choice, conventional capital-budgeting practice would be to treat the replacement as a cost-saving project compared to continued operation of existing facilities. However, this conventional approach will lead to conflicting results, as will be seen below.

Normally, replacement would be compared to continued operation of existing facilities; but there is a way of expanding existing facilities. Therefore, replacement is compared to the alternative of continuing to run the existing facilities after they have been expanded. On this basis, a cost-saving project evaluation yields in thousands of dollars:

Return on cost-saving project—Table IV

Annual Items	**Thousands of Dollars**
Cost saving........................	500
Depreciation........................	100
Profit before tax....................	400
Investment Items	
New, this project....................	900
Share of supporting facilities........	100
Total additional investment......	1,000

Return on investment (ROI) = (400) (100)/1,000 = 40%

Ground rules and assumptions for Table IV:

1. Evaluation criterion is return on investment (ROI), defined as the average annual increase in profit before taxes as a percentage of the total added investment.
2. Official project cutoff rate is 20%.
3. Depreciation is straight-line basis, 10-yr. life.
4. No change in overhead costs.
5. Share of supporting investment is for added utilities usage.

Increased investment—$33,000 for replacement minus $1,000 for expansion	=	$32,000
Increased operating income—$15,600 minus $10,100 (Table III)	=	$5,500
Increased depreciation—10% of $32,000	=	$3,200
Cost saving	=	$2,300
Incremental ROI—(2,300) (100)/(32,000)	=	7.2%

The 7.2% ROI for replacement vs. continued operation of existing facilities is less than the official cutoff rate of 20% and, therefore, replacement would not seem to be the thing to do.

But there is an apparent dilemma here. Performance of the expanded facility is inadequate and unacceptable, yet the alternative of replacement vs. continued operation of existing facilities shows an inadequate return by the usual standards.

The cause of the apparent dilemma is that the assumptions underlying a differential analysis of replacement vs. continued operation of existing facilities are invalid. These assumptions are:

• Existing facilities can be operated for at least 10 years without any new commitment of funds other than the $1 million for the proposed expansion project.

This is not a realistic assumption. Continued operation of the old facilities will require a series of replacements and improvements. Because competitors are continually improving performance, a continuing upgrading program is required just to stay even. Doing this in obsolete facilities is more costly and less efficient than in modern facilities.

• There will be no difference in performance other than in production costs and investments in direct facilities.

Inadequate margins and low returns have a debilitating effect on all aspects of business performance; the market position of the business will deteriorate because competitors will have more "maneuvering room" with larger margins; organization morale will suffer, management efficiency will be lowered, and the competitive position will deteriorate further.

• Capital funds spent for replacement here at 7.2% ROI will cause other projects with returns of 20% ROI or better to be foregone.

Most companies are not fund limited but people limited. Foregoing replacement but staying in business with the existing facilities does not free people to manage other businesses and investments.

• Staying in a business at a continuing inadequate performance is a viable long-term alternative.

If a business is not to be given the resources to be competitive (which in this case means the investment for replacement), the only viable alternative is to withdraw. Otherwise resources of the organization are used up for lower profits than the competition is getting and the "batting average" becomes poor.*

Realistically, the alternative to replacement is to disengage from the business, and the return on investment for replacement vs. withdrawal from the business is 20%.

Conclusions of the Business Manager

• The business is a good one, and the company has the capabilities to perform competitively.

• The expansion project, although yielding 30% ROI, will not result in adequate performance for the business.

• The draft of the proposed capital-appropriation

* It may be necessary, however, to continue operation on a short-term basis, until new technology is developed.

request should be returned with a request to develop a new capital-appropriation request for an economically feasible way of modernizing the facilities in conjunction with the proposed expansion so as to produce lower production costs. If modernization is not feasible, then draft a request for an all-new-facilities project sized to meet market demand.

A Cost-Saving Project

Let us consider another typical problem. The production manager is under pressure to reduce costs. He asks the engineering manager to help him with a cost-saving program.

The process engineer assigned to the job soon sees that the efficiency of using a raw material can be greatly improved by adding a residue-recovery processing unit. The engineer makes design studies and then profitability analyses that show a good cost-saving opportunity. For a new investment of $1 million, the saving will be $500,000/yr.

A capital-authorization request is submitted to the division head who has profit responsibility for the business. A summary of the financial analyses supporting the capital-authorization request is shown in Table IV.

Let us again look at how the business manager views this request for capital expenditure. As before, he is primarily concerned with the profit performance of his business as shown on the business performance statement: "What will this project do for my business?" The data in Table V provide the answer to this question.

After the project is completed, the business will show a return on investment of 7.2%. The business manager judges a 7.2% return to be non-competitive and, therefore, unacceptable. The reasons for this judgment are the same as those outlined before in discussing the expansion project.

Business performance statement for cost-saving project—Table V

	Now		This Project		After This Project	
Annual Items	**Total[a]**	**Unit[b]**	**Total[a]**	**Unit[b]**	**Total[a]**	**Unit[b]**
Sales	50,000	1.00	....	.....	50,000	1.00
Production costs	30,000	0.60	−500	−0.01	29,500	0.59
Freight, warehousing, etc.	5,000	0.10	.....	.....	5,000	0.10
Total direct costs	35,000	0.70	−500	−0.01	34,500	0.69
Margin	15,000	0.30	500	0.01	15,500	0.31
Overhead	6,000	0.12	.....	.....	6,000	0.12
Operating income	9,000	0.18	500	0.01	9,500	0.19
Depreciation	5,000	0.10	100	0.002	5,100	0.102
Profit before tax	4,000	0.08	400	0.008	4,400	0.088
Investment Items	**Total[c]**	**Unit[d]**	**Total[c]**	**Unit[d]**	**Total[c]**	**Unit[d]**
Direct	40,000	0.80	900	0.018	40,900	0.818
Share of supporting	10,000	0.20	100	0.002	10,100	0.202
Working capital	10,000	0.20	.....	.....	10,000	0.20
Total	60,000	1.20	1,000	0.02	61,000	1.22
Return on investment	6.7%		40%		7.2%	

a. Thousands of dollars/yr.
b. Dollars/unit of production.
c. Thousands of dollars.
d. Dollars per unit of production/yr.

All-New Facilities

Again, the problem is how to bring the performance of the business up to an acceptable level of return on investment. But first, what are the current capabilities in the business?

The business manager asks his financial analyst, with the assistance of engineering, to reconstruct the business performance statement, using production and investment costs that could be attained in all-new facilities based on today's technology (see Table VI).

Results of the all-new-facilities analysis are disappointing. Return on investment of 11.4% is not competitive since it is below the expected 20% return. Therefore, it appears that we do not have competitive capabilities in the business.

This judgment is confirmed by the just-announced action of Competitor A in withdrawing Product X from the market—X is a byproduct in these facilities. Competitor A has been the aggressive price leader since his new plant came onstream. It appears that

Return on all-new-facilities unit—Table VI

	All-New-Facilities Basis	
Annual Items	**Total[a]**	**Unit[b]**
Sales	50,000	1.00
Production cost	28,000	0.56
Freight, warehousing, etc.	5,000	0.10
Total direct costs	33,000	0.66
Margin	17,000	0.34
Overhead	6,000	0.12
Operating income	11,000	0.22
Depreciation	4,600	0.092
Profit before tax	6,400	0.128
Investment Items	**Total[c]**	**Unit[d]**
Direct	36,000	0.72
Share of supporting	10,000	0.20
Working capital	10,000	0.20
Total	56,000	1.12
Return on investment	11.4%	

a. Thousands of dollars/yr.
b. Dollars/unit of production.
c. Thousands of dollars.
d. Dollars per unit of production/yr.

he has found a way of improving his business economics with a new process that does not involve X as a byproduct.

This indicates a need for a process-development program to improve process economics. Under such circumstances, a cost-saving project is no longer attractive because it will probably be made obsolete by the new process.

A return of 40% for the cost-saving project assumes a 10-yr. economic life. If the economic life is less than 10 yr., the return is less. Inspection of Table IV shows that the project requires 2 years or more just to break even (2 years of cost saving at $500,000/yr. is required to pay back the initial investment of $1 million). In 2 years, an improved process will probably be ready to operate.

Conclusions of the Business Manager

- The business is a good one for the company, but the company has temporarily lost its competitive capabilities. Competition appears to have developed a superior process technology.
- A research-and-development program to improve process technology should be started at once.
- The capital appropriation request should be returned with a request to stop all work on the cost-saving project.

Lessons for Functional Management

What can line people, and particularly engineers who develop capital authorization requests and make financial analyses of capital projects, learn from these examples? Engineers should remember that:

1. A business manager looks at overall business performance and not just project results. A project is a means to a business end and not an end in itself.
2. A business manager is measured by the realized profits of his business in the way the internal financial scorekeeping system reports them, and not by the promised performance of projects: (a) Project evaluation methods that involve criteria different than the ones used in business performance measurement systems, such as discounted-cash flow returns, net present worth, and profitability indexes, are not too helpful to the business manager. What he wants is the effect of the project on his business as it is accounted and as he is measured. (b) Project evaluations are based on differences between alternatives (one alternative may be to do nothing) while business performance is measured on the overall results of the chosen alternative. Project results are, therefore, not always representative of business results; a 40%-return project does not mean a 40%-return business.
3. The business manager is concerned not only that the project return is good enough to "sell," but also that it results in giving him a competitive overall business performance. This means that each project should:
   - Be the best choice among the available alternatives, such as best process, best location, best capacity, etc.
   - Promise results above the budget cutoff criterion; for example, a return better than the cutoff return.
   - Be good enough to result in competitive business performance; that is, result in a facility that has competitive costs and investments.
4. Evaluations between alternatives, if they are to be meaningful to the business manager, must be between real alternatives and not hypothetical ones.

"Do-nothing" and "maintain status quo" are not real alternatives. A decision to be in and continue in a business implies a commitment to become and to remain competitive in the business. Keeping competitive requires attention to share of the market, matching competitive customer service and product quality, and maintaining a competitive cost and investment position. In a dynamic growing market where competition is continuously improving customer service and process technology, standing still is going backward, and the only real alternative to moving ahead in step with competition is to withdraw.

Recommendations

Specifically, what can the engineer do to enhance his professional stature with the business management of his company? The following practices are recommended:

- Whenever a project proposal is submitted to a business manager for his consideration, translate the project results into business language. Show the effect of going ahead with the project on business performance.
- Determine beforehand whether current business performance is considered acceptable. If it is not, evaluate current business capabilities by making a business-performance analysis on an all-new-facilities basis. This analysis will assist the business manager in selecting the proper development and facility strategies to improve business performance.
- Do not be satisfied with simply selecting the best project alternative. Be sure the "best" alternative to be proposed is good enough to give competitive business performance. The approach outlined in the figure on p. 290 will help answer: Is the best alternative good enough? ■

Meet the Author

James A. Champley is a capital-projects consultant on the corporate staff of Union Carbide Corp., 270 Park Ave., New York, N. Y. 10017. He is responsible for corporate capital-budgeting procedures and is a consultant to operating divisions on business and project planning and analyses. Mr. Champley has also been involved in production, process design and project analysis with the Chemicals Div. of Union Carbide. He holds a B.S. (Ch.E.) from Notre Dame (1941).

Section III

COSTS OF PROCESS EQUIPMENT AND OTHER ITEMS

CE cost indexes maintain 13-year ascent

Off only slightly from the steep rates of rise that characterized 1972–74, plant and equipment costs continue their inflationary course.

Philip M. Kohn, Associate Editor

☐ When Chemical Engineering last reviewed plant and equipment cost data three years ago [*1*], we noted a record jump of 25.0% for the period 1972–74. A look at the present data shows a lower rate of climb—23.4%—for 1975–77, but one that is still 41.8% higher than the 16.5% increase for 1968–71 [*2*], and 122.8% higher than the 10.5% upturn experienced in 1964–68 [*3*]. And return to stability—last seen in 1958–64, when the overall rise was a mere 1.5% [*4*]—seems unlikely.

Plant-cost makeup

The *CE* Plant Cost Index has four major components: equipment, machinery and supports; construction labor; buildings (materials and labor); and engineering and manpower. (For details on index weighting, see Ref. 4, 6.)

Table I reveals major component trends since 1956. Equipment, machinery and supports—the most heavily weighted component—jumped 29.0% in the last three years. Interestingly, construction labor costs (corrected for productivity) were up only 9.1%, compared with its 11.7% increase in the preceding triennium.

Making up the equipment, machinery and supports component are the subcomponents listed in Table II. Pumps and compressors were the biggest gainers, leaping 36.7%, with process machinery up by 32.3% and structural supports increasing at 31.7%. Pipes, valves and fittings, though jumping by 28.8%, were off their blistering 40% rate-of-rise for 1972–74. Fabricated equipment rose by 27.3%, while electrical equipment climbed by 25.8%. Process instruments and controls was the most stable subcomponent, rising 19.0%.

Index comparisons

Reported every issue in *CE*'s Economic Indicators section is the Marshall & Swift Equipment Cost Index, which rose by 26.9% in the last three years. One should remember that while the *CE* and M&S indexes are close, M&S traces installed-equipment costs, whereas the *CE* index is designed to show trends in chemical-process-industries (CPI) plant costs. Data for the M&S index and its pertinent subcomponents since 1956 are shown in Table III.

Included on the graph is the ENR (*Engineering News-*

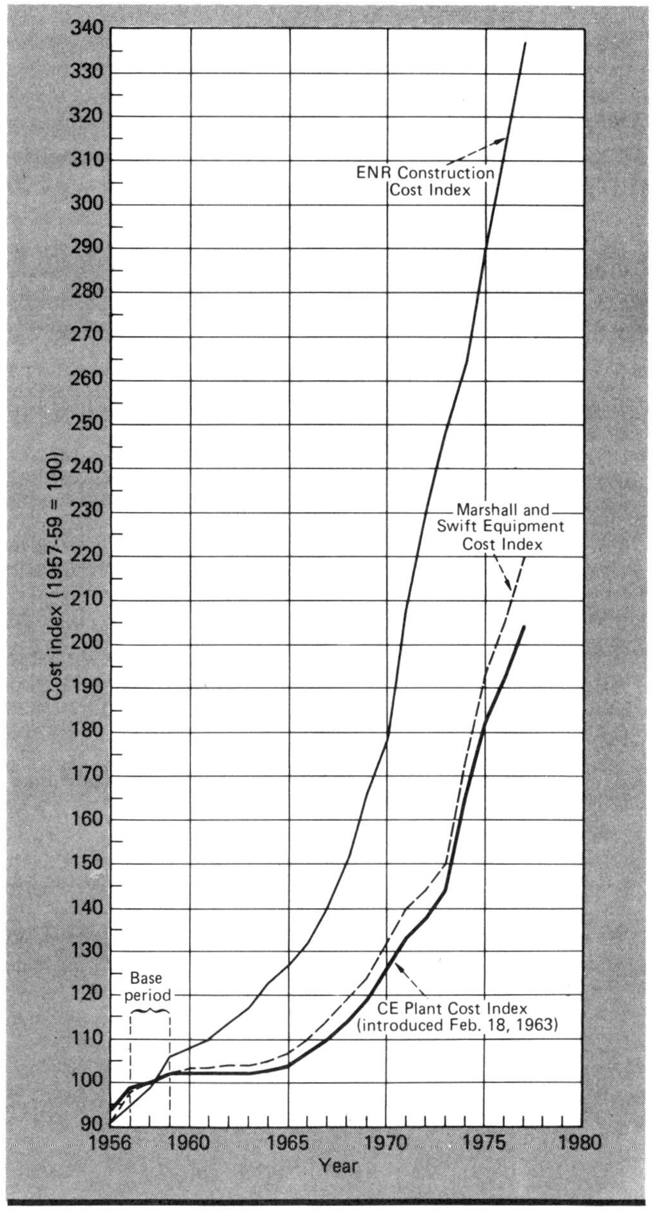

Cost indexes still on the rise

Originally published May 1978.

Trend of plant costs since 1956 — **Table I**
(Base period: 1957-1959 = 100)

	1956	1957	1958	1959	1960	1961	1962	1963	1964	1965	1966	1967	1968	1969	1970	1971	1972	1973	1974	1975	1976	1977
CHEMICAL ENGINEERING Plant Cost Index	**93.9**	**98.5**	**99.7**	**101.8**	**102.0**	**101.5**	**102.0**	**102.4**	**103.3**	**104.2**	**107.2**	**109.7**	**113.6**	**119.0**	**125.7**	**132.3**	**137.2**	**144.1**	**165.4**	**182.4**	**192.1**	**204.1**
Equipment, machinery and supports	92.7	98.5	99.6	101.9	101.7	100.2	100.6	100.5	101.2	102.1	105.3	107.7	111.5	116.6	123.8	130.4	135.4	141.9	171.2	194.7	205.8	220.9
Construction labor	95.8	98.6	100.0	101.4	103.7	105.1	105.6	107.2	108.5	109.5	112.5	115.8	120.9	128.3	137.4	146.2	152.2	157.9	163.3	168.6	174.2	178.2
Buildings	98.0	99.1	99.5	101.4	101.5	100.8	101.4	102.1	103.3	104.5	107.9	110.3	115.7	122.5	127.2	135.5	142.0	150.6	165.8	177.0	187.3	199.1
Engineering and manpower	94.2	98.2	99.3	102.5	101.3	101.7	102.6	103.4	104.2	105.6	106.9	107.9	108.6	110.9	110.6	111.4	111.9	122.8	134.4	141.8	150.8	162.1

Trend of equipment costs since 1956 — **Table II**
(History of the equipment component and major subcomponents of CE Plant Cost Index. Base period: 1957-1959 = 100)

	1956	1957	1958	1959	1960	1961	1962	1963	1964	1965	1966	1967	1968	1969	1970	1971	1972	1973	1974	1975	1976	1977
Equipment, machinery and supports	**92.7**	**98.5**	**99.6**	**101.9**	**101.7**	**100.2**	**100.6**	**100.5**	**101.2**	**102.1**	**105.3**	**107.7**	**111.5**	**116.6**	**123.8**	**130.4**	**135.4**	**141.9**	**171.2**	**194.7**	**205.8**	**220.9**
Fabricated equipment	92.5	99.5	99.6	100.9	101.2	100.1	101.0	101.7	102.7	103.4	104.8	106.2	109.9	115.1	122.7	130.3	136.2	142.5	170.1	192.2	200.8	216.6
Process machinery	92.2	98.1	100.1	101.8	101.8	101.1	101.9	102.0	102.5	103.6	106.1	108.7	112.1	116.8	122.9	127.1	132.1	137.8	160.0	184.7	197.5	211.6
Pipe, valves and fittings	94.8	97.9	98.8	103.3	104.1	101.1	100.6	100.7	101.6	103.0	109.6	113.0	117.4	123.1	132.0	137.3	142.9	151.5	192.3	217.0	232.5	247.7
Process instruments and controls	91.2	96.7	100.4	102.9	105.4	105.9	105.9	105.7	105.8	106.5	110.0	115.2	120.9	126.1	132.1	139.9	143.8	147.1	164.7	181.4	193.1	203.3
Pumps and compressors	90.0	97.5	100.0	102.5	101.7	100.8	101.1	100.1	101.0	103.4	107.7	111.3	115.2	119.6	125.6	133.2	135.9	139.8	175.7	208.3	220.9	240.2
Electrical equipment and materials	93.5	98.4	100.6	101.0	95.7	92.3	89.4	87.6	85.5	84.1	86.4	90.1	91.4	92.8	99.8	98.7	99.1	104.2	126.4	142.1	148.9	159.0
Structural supports & misc.	92.5	98.0	100.4	101.6	101.9	99.8	99.2	97.3	98.3	98.8	101.0	102.1	105.7	112.6	117.9	126.6	133.6	140.9	171.6	198.6	209.7	226.0

Marshall & Swift annual indexes of comparative equipment costs since 1956 — **Table III**
(Base period: 1926 = 100)

	1956	1957	1958	1959	1960	1961	1962	1963	1964	1965	1966	1967	1968	1969	1970	1971	1972	1973	1974	1975	1976	1977
Equipment Cost Index	**208.8**	**225.1**	**229.2**	**234.5**	**237.7**	**237.2**	**238.5**	**239.2**	**231.8**	**244.9**	**252.5**	**262.9**	**273.1**	**285.0**	**303.3**	**321.3**	**332.0**	**344.1**	**398.4**	**444.3**	**472.1**	**505.4**
Process Industries																						
Cement	199.4	216.4	222.8	228.7	232.1	231.1	231.8	232.5	235.9	239.3	249.6	258.1	268.3	280.1	298.3	317.3	328.5	340.1	398.1	454.3	482.2	518.3
Chemical	209.1	226.5	232.3	236.5	239.2	237.7	238.0	238.7	241.1	243.8	246.1	261.8	271.6	282.8	300.8	319.1	329.7	341.0	400.5	447.6	473.2	507.1
Clay products	193.8	210.2	216.8	222.2	225.7	224.6	225.5	225.8	229.2	232.6	239.0	250.7	260.7	272.2	289.9	308.4	319.3	330.3	384.5	437.2	466.9	504.2
Glass	197.5	213.8	219.3	223.2	225.3	224.4	224.7	225.4	227.6	230.1	240.6	271.1	256.3	266.9	283.9	301.2	311.3	322.0	375.4	425.3	451.3	483.5
Paint	201.2	217.6	223.2	226.9	229.5	230.0	231.5	232.1	235.0	238.1	243.9	255.7	267.4	280.3	300.1	320.7	333.3	345.9	406.0	455.0	481.2	515.3
Paper	210.5	218.2	223.8	227.8	229.9	229.0	229.3	229.9	232.3	234.8	247.5	252.1	261.6	272.4	289.7	307.5	317.7	328.4	381.7	430.1	457.6	491.2
Petroleum products	205.4	222.2	228.0	231.8	234.3	235.0	238.2	238.8	241.8	244.9	253.9	263.4	275.2	288.4	308.8	330.6	343.0	356.0	419.3	470.9	497.7	535.9
Rubber	207.9	224.9	230.8	234.6	237.3	237.9	239.2	240.0	243.0	236.2	251.2	264.7	276.5	289.9	310.3	331.6	344.7	357.8	415.2	468.5	497.3	532.8
Related Industries																						
Electrical power	211.0	229.2	235.2	239.0	241.0	236.3	235.6	234.7	236.8	239.4	246.5	257.3	264.9	274.8	290.3	307.4	316.4	327.2	391.7	438.5	465.1	494.1
Mining, milling	210.4	227.9	233.8	237.1	240.6	239.2	239.5	240.1	242.6	245.3	253.0	263.5	273.2	284.5	302.6	321.1	331.8	342.9	394.3	451.2	482.9	520.8
Refrigerating	234.3	254.2	260.8	265.1	268.2	268.8	270.4	271.2	274.6	278.2	287.1	299.1	312.5	327.5	350.6	374.7	389.5	404.6	472.4	528.1	558.4	595.9
Steam power	197.0	213.0	218.6	222.9	224.7	225.3	226.6	227.2	230.1	233.0	240.4	250.6	261.8	274.3	293.8	313.9	326.3	338.8	400.8	448.2	475.7	507.7

Record) Construction Cost Index, which went up by 27.7%. At 337, the ENR index is 133 points higher than the *CE* index, though the two were even in 1958. There are three reasons for this disparity: (1) The ENR index is not specific to the CPI; (2) it is based on the cost of construction materials, such as lumber, cement and steel, along with a labor component; and (3) ENR does not include a productivity factor in its index, as does *CE*, which applies a $2\frac{1}{2}$%/yr productivity adjustment to its labor and engineering components. (For an explanation of this factor, see p. 146 of Ref. 4, and p. 136 of Ref. 5.)

Cost escalations stable—Plant-cost gains were fairly stable in 1977, averaging about 0.5% per month; occasionally, there was an erratic excursion. Extrapolation of current data shows that the average 1978 *CE* Plant Cost Index would be about 5% higher than in 1977, barring dramatic upturns.

References

1. Ricci, L. J., CE Cost Indexes Accelerate 10-Year Climb, *Chem. Eng.*, Apr. 28, 1975, pp. 117–118.
2. Thorsen, D. R., The Seven-Year Surge in the CE Cost Indexes, *Chem. Eng.*, Nov. 13, 1972, pp. 168–170.
3. Norden, R. B., CE Cost Indexes: A Sharp Rise Since 1965, *Chem. Eng.*, May 5, 1969, pp. 134–138.
4. Arnold, T. H., Jr., and Chilton, C. H., New Index Shows Plant Cost Trends, *Chem. Eng.*, Feb. 18, 1963, pp. 143–152.
5. Popper, H., and Weismantel, G. E., Costs and Productivity in the Inflationary 1970's, *Chem. Eng.*, Jan. 12, 1970, pp. 132–136.
6. Chilton, C. H., Plant Cost Index Points Up Inflation, *Chem. Eng.*, Apr. 25, 1966, pp. 185–190.

Cost estimating for major process equipment

Step-by-step guidelines and actual cost data provide the means for estimating the fabricated cost of distillation columns and pressure vessels, storage tanks, fired heaters, pumps and drivers, compressors and drivers, and vacuum equipment as used in many plants of the chemical process industries.

Originally published October 1977.

Arkadie Pikulik and Hector E. Diaz, Scientific Design Co.

☐ The planning, engineering and construction of a chemical process plant are tied tightly into economic reality. These conditions also hold for addition of new equipment or modification of existing facilities. In determining economic soundness, one of the essential parameters is a cost-estimate of the plant, addition or modification. The cost estimate permits management to approve the work and provide the necessary monies.

At the start of a project, the estimate is of necessity imprecise, and lacking in details. However, as the project develops, the quality of the estimate is progressively improved. As engineering and construction advance, the estimate finds use in controlling project costs.

Estimates are prediction of costs and are thus only approximations. Failure to recognize this can, and often does, lead to serious misunderstandings between those responsible for completing a project at the estimated cost. It is essential that engineers understand the mechanism of economics in determining whether to approve or disapprove projects. In a competitive society, the product or services provided must be profitable. A product or service that requires a continuous inflow of monies or materials to make a product of lesser value and/or little return on invested capital can not and should not be approved. Governments often take on uneconomical projects for the good of their citizens. But even in such instances, cost estimates, controls and evaluations are necessary to plan budgets and set taxes.

How to plan for cost engineering

The principles of cost engineering are used on most investment problems, but those related to the chemical process industries (CPI) represent a special challenge because of the variety and complexity of the projects.

A typical engineering-design company has specialists centered in a division or group that is often simply called the cost or estimating department. This group performs the economic evaluations, estimates and cost control of a given project during its different phases, i.e., from evaluation and planning and preliminary estimates to detailed engineering, construction and start-up. To achieve its purpose, the estimating group is assisted by other engineering sections that update and revise the design basis and provide firm specification for process, layout, equipment, buildings, etc. The specification work is time-consuming. Depending on the type of estimate, it will provide a basis for competitive bidding, or for controlling the cost of a project once it has been approved. It is essential to understand that the more clearly a project is defined initially, the better the chance of estimating the cost.

In smaller firms, a single individual may perform all these functions, in addition to the ones related to cost estimating and cost control for the project.

The initial step in estimating may be that of collecting scattered cost data having some relation to the project. These cost data normally come from individual files, corporate files or correlations, vendor quotations and published information. The greater the accuracy desired, the greater the amount of data needed. The final result will be expressed with an indication of the probable accuracy.

Types of capital-cost estimates

Depending on the type of estimate and the inherent reliability of the cost data, the accuracy may change from a probable ±40% for an order-of-magnitude approach to a ±3% for a detailed estimate. In an inflationary economy, the positive side has the tendency to be greater. Types of estimates, useful for evaluating capital costs of CPI plants, are shown in Table I. For a summary of the information required for each type of estimate, consult Perry, R. H. and Chilton, C. H., "Chemical Engineers' Handbook," 5th ed., p. 25-15, McGraw-Hill, New York, 1973.

Typical costs for preparing three of the most common types of estimates are given in Table II. The amounts

Estimates for capital investment — Table I

Type of estimate	Usual basis	Probable accuracy, %
Order of magnitude (ratio estimate)	Previous similar cost information	±40
Study (factored estimate)	Knowledge of flowsheet and major equipment	±25
Preliminary (initial budget, scope)	Sufficient data for budget preparation	±12
Definitive (project control	Detailed data, but not complete drawings	± 6
Detailed (firm, contractor's)	Complete drawings and specifications	± 3

Source: American Assn. of Cost Engineers

Typical cost ranges for producing estimates (1976) — Table II

	Project size, $ million		
Type of estimate	Less than 1	1 to 5	5 to 50
Study estimate, $ thousand	5 to 15	12 to 30	20 to 40
Preliminary estimate, $ thousand (scope or authorization)	15 to 35	30 to 60	50 to 90
Definitive estimate, $ thousand (project control)	25 to 60	60 to 120	100 to 230

Originally published October 1977.

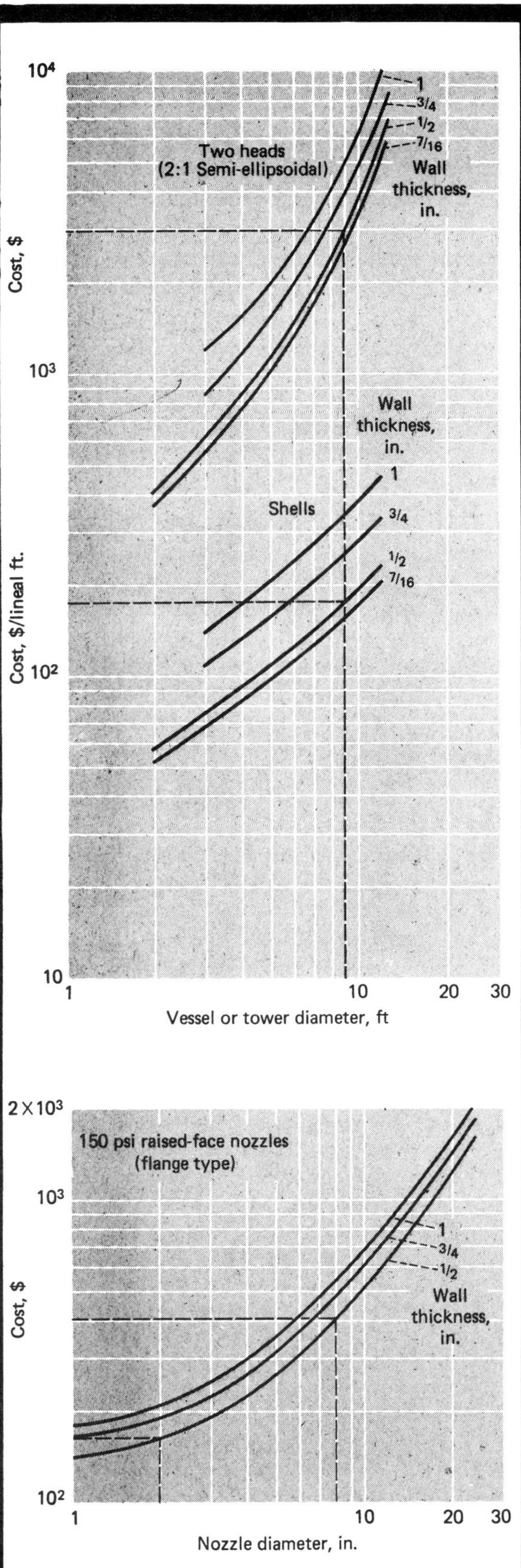

Carbon-steel pressure vessels, spot X-rayed **Fig. 1**

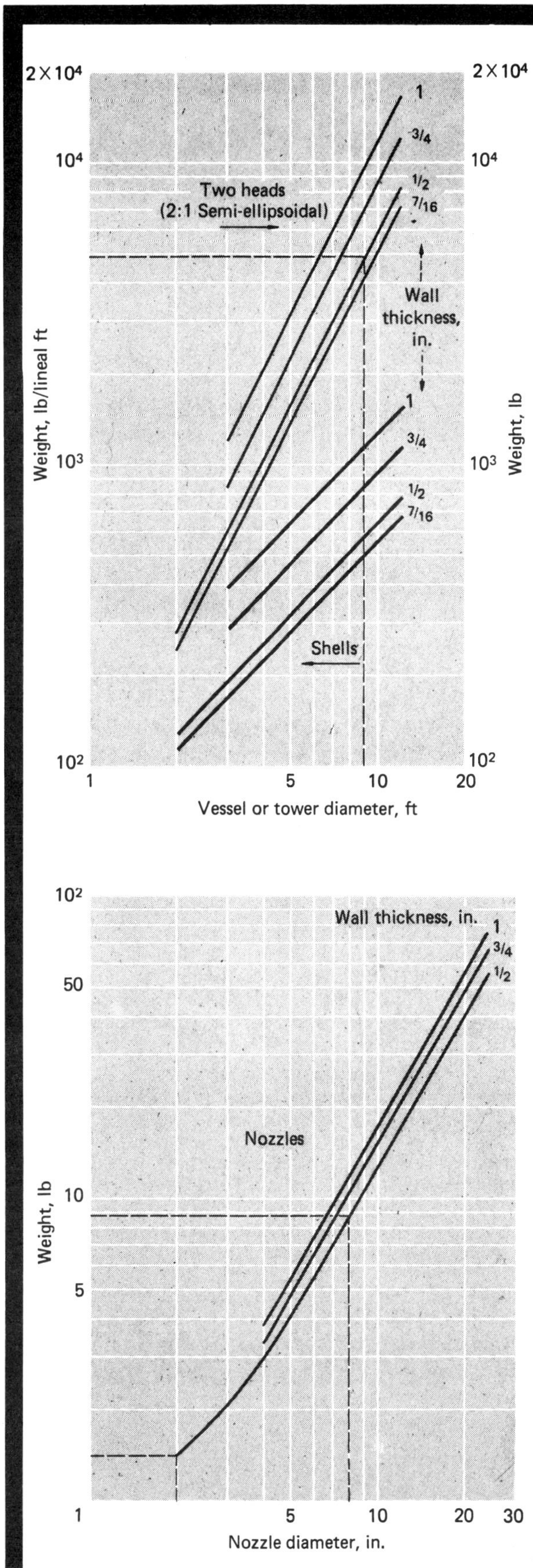

Weights of shells, heads and nozzles **Fig. 2**

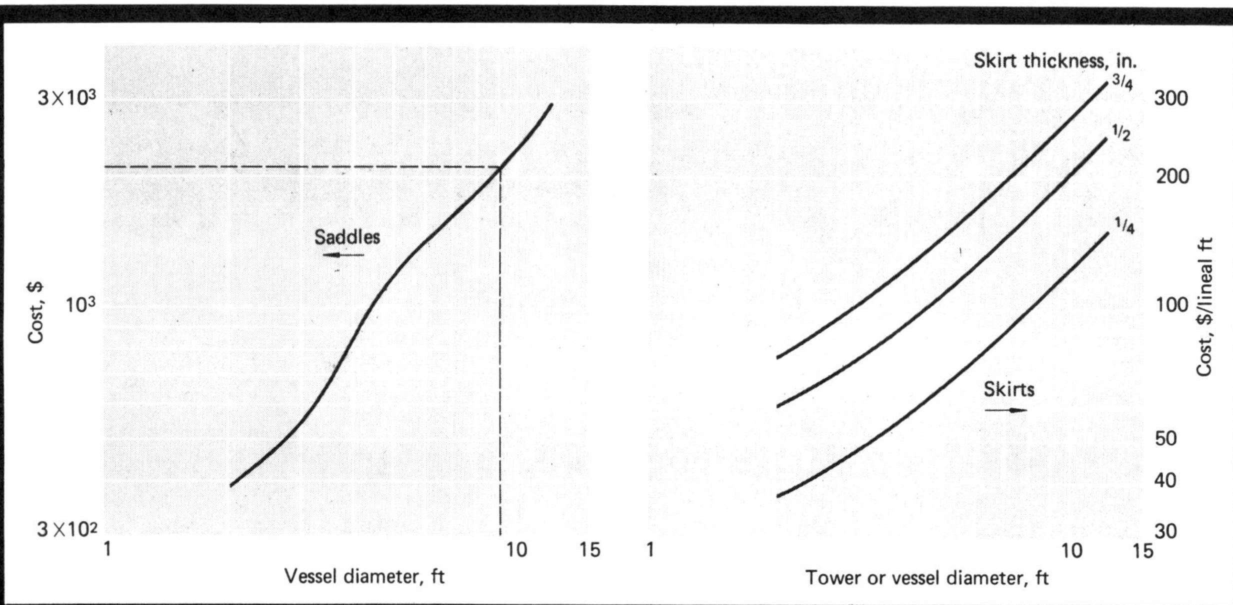

Saddle or skirt unit costs for pressure vessels — Fig. 3

represent salaries, overhead and travel for engineering, drafting and cost estimating. The cost for project-control estimates or authorization estimates may be somewhat smaller than shown because most of the engineering and drafting can be used in the final construction stages of a project. This represents an advantage and cost saving if the job is done progressively from the study estimate.

A cost estimate has a major impact on the project's profitability. The estimate leads to the commitment of substantial amounts of money and manpower over an extended time. Because of the time-span required to carry out a project, the estimate must take into account other factors such as escalation, contingency, financing, duty, taxes, etc. These vary with time, and substantially affect a project. In some cases, they may be the key for a choice among alternatives.

Our objective will be to present the steps for preparing estimates of major equipment, and to provide the tools, guidelines and instructions to do this task.

Estimating cost of major equipment

Major equipment in CPI plants consists of all those items in which process material undergoes a change of state, condition or composition. This also includes equipment for storing process materials. A list of major equipment may include: distillation columns and pressure vessels, storage tanks (built to specifications of the American Petroleum Institute), heat exchangers, fired heaters, pumps and drivers, compressors and drivers, and vacuum equipment.

Columns and pressure vessels

Pressure vessels are important elements in CPI plants and often represent the largest initial investment of a project. Failure to design them properly could lead to severe economic penalties (unexpected plant breakdowns) and could make them potential hazards to equipment and personnel.

For U.S. projects, pressure vessels are designed and constructed in accordance with Section VIII of the ASME Boiler and Pressure Vessel Code (latest edition). For foreign projects, the design should conform to the codes and requirements of the country of installation.

The data described here provide a method of estimating the fabricated cost of carbon-steel pressure vessels with a good degree of accuracy. Cost data are based as of January 1976, and include material and fabrication costs for spot X-rayed vessels without stress relief. Heads are assumed to be 2 : 1 semi-ellipsoidal.

The following information is required for estimating: vessel material, vessel diameter, tangent-to-tangent length, thickness (based on pressure-temperature rating and corrosion allowance), number of manholes, number of nozzles, and vessel position—vertical (use skirt), horizontal (use saddles).

Fig. 1 lists the cost for shells, heads and nozzles; Fig. 2 shows their weights. Fig. 3 lists cost for saddles or skirts; Fig. 4, their corresponding weights.

The same information and procedure, as required, apply to distillation towers, scrubbers and similar equipment, with the cost of trays or packing added to the value calculated for a simple vessel.

All vessels fabricated under the ASME codes are handled through the shop, and require inspection, hydrotest, cleaning, preparing for shipment and loading out. The estimated shop cost may be calculated as:

$$C_S = H(3.1D + 1.2L + 3M + N + T) \qquad (1)$$

where C_S is shop cost, \$; D is vessel or tower diameter, ft; L is the tangent-to-tangent length, ft; M is the number of manholes; N is the number of nozzles; and T is the total weight of material, tons. The sum of the quantities between the parentheses of Eq. (1) represent the total man-hours for shop fabrication. H is the wage

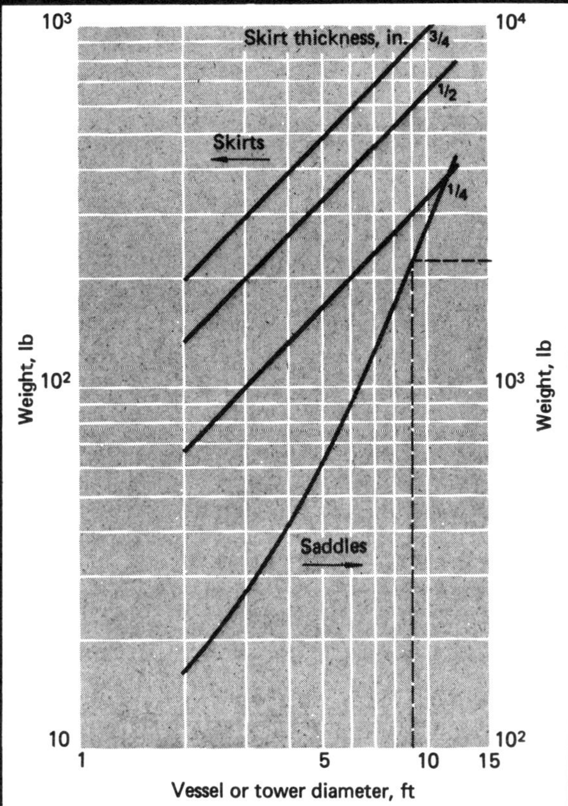

Estimated weights for vessel supports Fig. 4

rate, $16.5/manhour for vessels, and $28.5/manhour for towers.

Additional shop costs for manholes are given in Table III; and additions for engineering, purchasing, administrative and profit are shown in Table IV.

Examples illustrate procedures

Example 1—The vessel in the sketch represents an overhead condensate receiver:

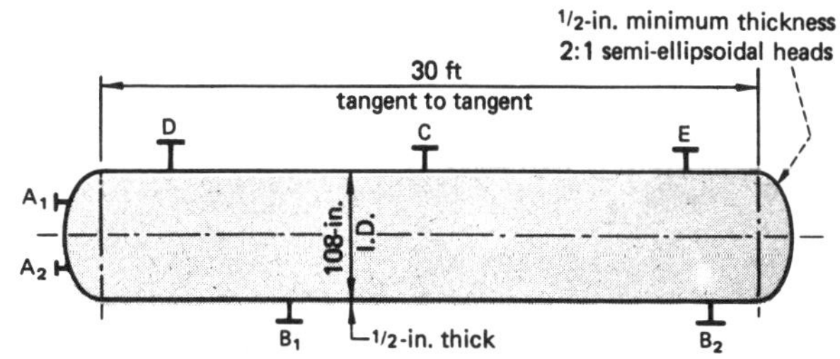

The following list shows the number, size, service and rating of each nozzle:

Mark	Size	Service
A_{1-2}	2 in.	Gage-glass connection
B_{1-2}	8 in.	Pumpout alternates
C	10 in.	Vapor outlet
D	18 in.	Vapor/liquid inlet
E	8 in.	Relief-valve connection
M	20 in.	Manhole with davit

The flange standard for these nozzles is American

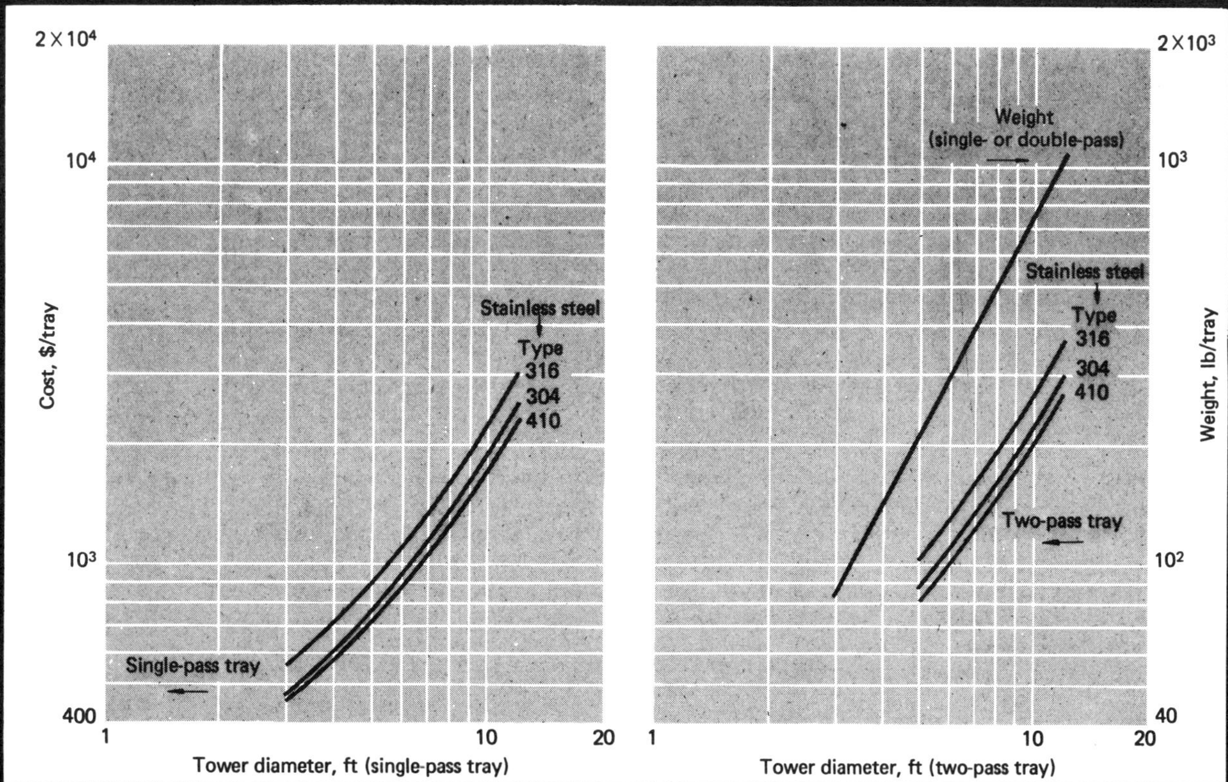

Costs and weights of column trays Fig. 5

Corrections for manholes — **Table III**

Manhole size, in.	Add to base price, $	Add to base weight, lb
16	400	270
18	500	310
20	600	390
24	800	530

National Standard Institute (ANSI), and the pressure rating is 150 psi, for raised-face flanges with manufacturer's standard facing finish.

Let us evaluate the fabricated cost of this vessel. We begin by finding the cost for the several components of the receiver:

Item	Size	Number	Cost	From	Cost
Shell	30 ft	1	\$175/ft	Fig. 1	\$5,250
Heads	108 in.	1	\$2,950	Fig. 1	2,950
Nozzle	2 in.	2	\$165 each	Fig. 1	330
Nozzle	8 in.	3	\$405 each	Fig. 1	1,215
Nozzle	10 in.	1	\$520	Fig. 1	520
Nozzle	18 in.	1	\$1,100	Fig. 1	1,100
Manhole	20 in.	1	\$1,250 + \$600	Fig. 1 & Table III	1,850
Saddles	—	—	\$2,100	Fig. 3	2,100
Total material cost					**\$15,315**

In order to compute the fabrication cost from Eq. (1), we must estimate the total weight:

Item	Size	Number	Weight	From	Weight, lb
Shell	30 ft	1	600 lb/ft	Fig. 2	18,000
Heads	108 in.	1	4,900 lb	Fig. 2	4,900
Nozzle	2 in.	2	14 lb each	Fig. 2	28
Nozzle	8 in.	3	87 lb each	Fig. 2	261
Nozzle	10 in.	1	125 lb	Fig. 2	125
Nozzle	18 in.	1	340 lb	Fig. 2	340
Manhole	20 in.	1	400 lb + 390 lb	Fig. 2 & Table III	790
Saddles	—	—	2,200 lb	Fig. 4	2,200
Total weight of material					**26,644**

We now compute the fabrication cost by substituting into Eq. (1) by converting the total weight to

Addition for engineering, purchasing, administrative costs and profit — **Table IV**

Total cost of fabricated vessel	Factor
Less than \$6,000	0.25
\$ 6,000 to \$8,000	0.24
\$ 8,000 to \$10,000	0.23
\$10,000 to \$15,000	0.22
\$15,000 to \$20,000	0.21
\$20,000 to \$25,000	0.20
\$25,000 to \$35,000	0.19
\$35,000 to \$50,000	0.18
\$50,000 to \$75,000	0.17
Over \$75,000	0.16

(26,644/2,000) = 13.3 tons and using a shop rate of \$16.5/manhour:

$$C_S = 16.5[3.1(9) + 1.2(30) + 3(1) + 7 + 13.3]$$
$$C_S = \$1,452$$

Therefore, the fabricated cost of the vessel becomes \$15,315 plus \$1,452, or \$16,767. To this must be added an amount for engineering, purchasing, etc. From Table IV, the factor for a vessel costing \$16,767 is 0.21, or \$16,767 × 0.21 = \$3,521. In summary:

Material cost	\$15,315
Fabrication	1,452
Engineering, etc.	3,521
Total FOB fabricator's shop	\$20,288
	(rounded off, \$20,300)

Example 2—A column serving as a scrubber has overall dimensions as shown in the sketch:

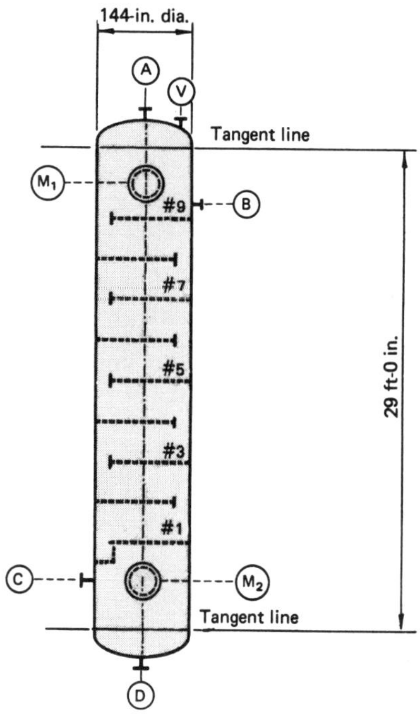

This scrubber has 9 Type 410 stainless-steel single-pass trays, equally spaced. The shell thickness calculated for the design pressure and temperature and corrosion allowance is ½ in. The heads are 2:1 semi-ellipsoidal with a minimum thickness of ½ in. The skirt required is 8 ft high and ½ in. thick.

The following list shows the connections:

Mark	Size	Service
A	24 in.	Vapor outlet
B	2 in.	Reflux inlet
C	36 in.	Vapor inlet
D	12 in.	Liquid outlet
M_{1-2}	20 in.	Manholes with davits
V	2 in.	Vent

Let us evaluate the fabricated cost of the scrubber. The procedure is very similar to that of Example 1. However, we must take into account the costs and weights of the trays as well as the different wage rate for fabrication. Trays are furnished by others and shop-installed.

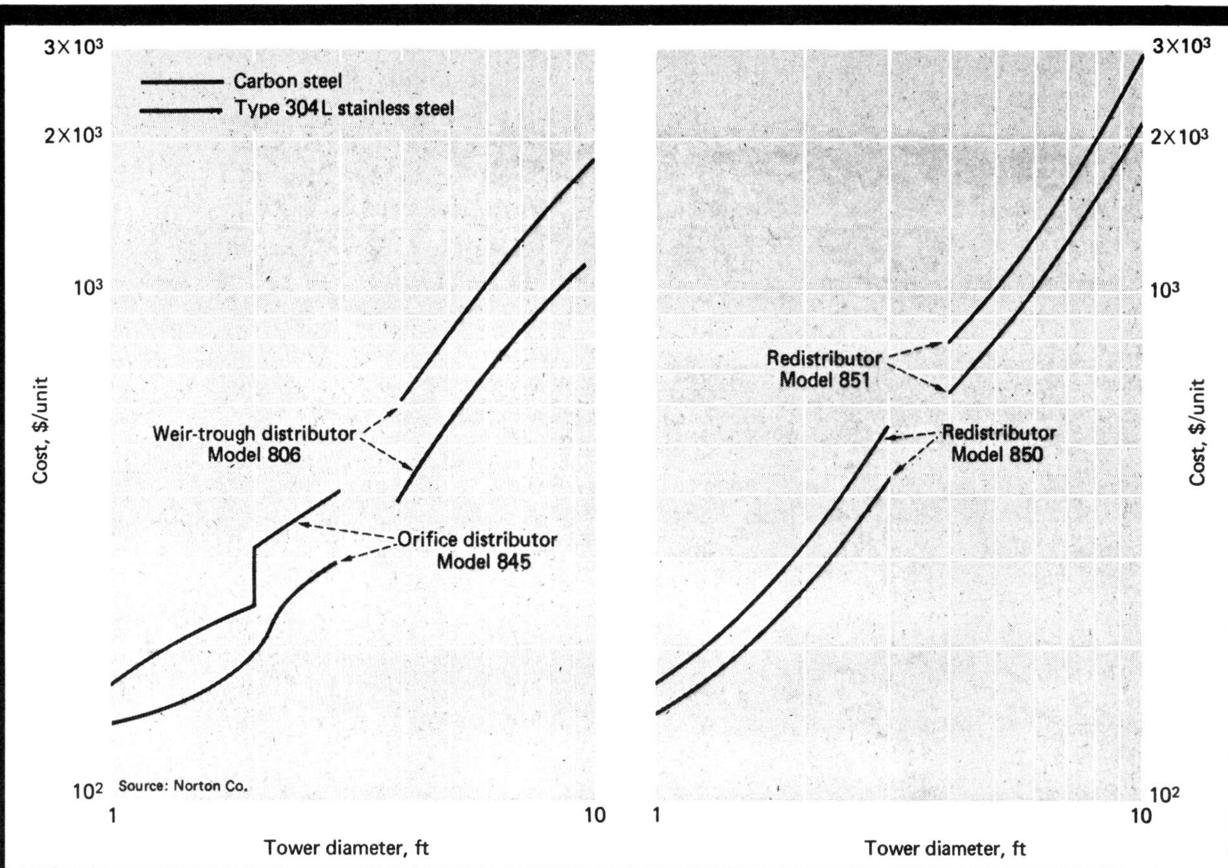

Distributors for packed columns Fig. 6

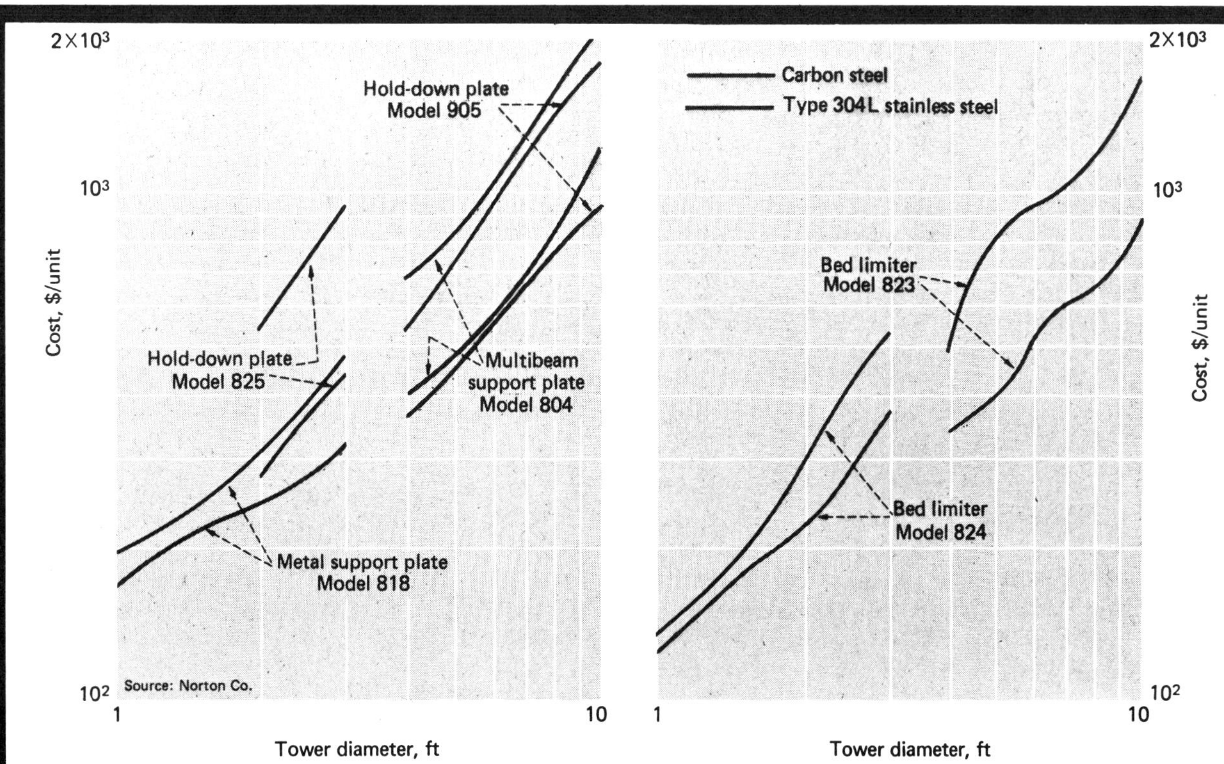

Components for packed columns Fig. 7

Costs and weights of tower packings for preparing estimates — **Table V**

Weight 1 in. lb/ft³	Weight 1½ in. lb/ft³	Weight 2 in. lb/ft³	Weight 3½ in. lb/ft³	Type and material	Estimating price 1 in. \$/ft³	Estimating price 1½ in. \$/ft³	Estimating price 2 in. \$/ft³	Estimating price 3½ in. \$/ft³
				Pall rings				
30	24	22	–	Carbon steel	19	15	13.5	–
30	24	22	–	Type 304 L stainless	54.5	43	35	–
5.5	–	4.25	4	Polypropylene	14	–	8.5	4
				Intalox saddles				
6	–	4.25	3.5	Polypropylene	14	–	8.5	4
44	–	42	37	Chemical porcelain	11.5	–	8.0	7.5
				Raschig rings				
71	48	25	–	Carbon steel	19	15	13.5	–
44	42	37	–	Chemical porcelain	11.5	9.0	8.0	–

Source: Norton Co.

We begin by finding the cost for the scrubber components as follows:

Item	Size	Number	Cost	From	Cost
Shell	29 ft	1	\$225/ft	Fig. 1	\$6,525
Heads	144 in.	1	\$6,000	Fig. 1	6,000
Nozzle	2 in.	2	\$165 each	Fig. 1	330
Nozzle	12 in.	1	\$640	Fig. 1	640
Manhole	20 in.	2	\$1,250 + \$690	Fig. 1 & Table III	3,700
Nozzle	24 in.	1	\$1,600	Fig. 1	1,600
Nozzle	36 in.	1	\$3,200	Fig. 1	3,200
Skirt	8 ft	1	\$250/ft	Fig. 3	2,000
Trays	144 in.	9	\$2,300 each	Fig. 5	20,700
Total material cost					**\$44,695**

We now estimate the total weight:

Item	Size	Number	Weight	From	Weight, lb
Shell	29 ft	1	780 lb/ft	Fig. 2	22,620
Heads	144 in.	1	8,300 lb	Fig. 2	8,300
Nozzle	2 in.	2	14 lb each	Fig. 2	28
Nozzle	12 in.	1	170 lb	Fig. 2	170
Manhole	20 in.	2	400 lb + 390 lb	Fig. 2 & Table III	1,580
Nozzle	24 in.	1	550 lb	Fig. 2	550
Nozzle	36 in.	1	1,100 lb	Fig. 2	1,100
Skirt	8 ft	1	750 lb/ft	Fig. 4	6,000
Trays	144 in.	9	1,050 lb	Fig. 5	9,450
Total weight of material (about 25 tons)					**49,798**

Using a shop rate of \$28.5/manhour, we substitute into Eq. (1) to compute C_s as \$3,078.

The fabricated cost of the vessel becomes \$44,695 + \$3,078 = \$47,773. From Table IV, we obtain the factor for a vessel costing \$47,773 as 0.18 to cover engineering, purchasing, and other costs. This amounts to (\$47,773 × 0.18), or \$8,599. In summary:

Material cost	\$44,695
Fabrication	3,078
Engineering, etc.	8,599
Total FOB fabricator's shop	\$56,372
	(rounded off \$56,400)

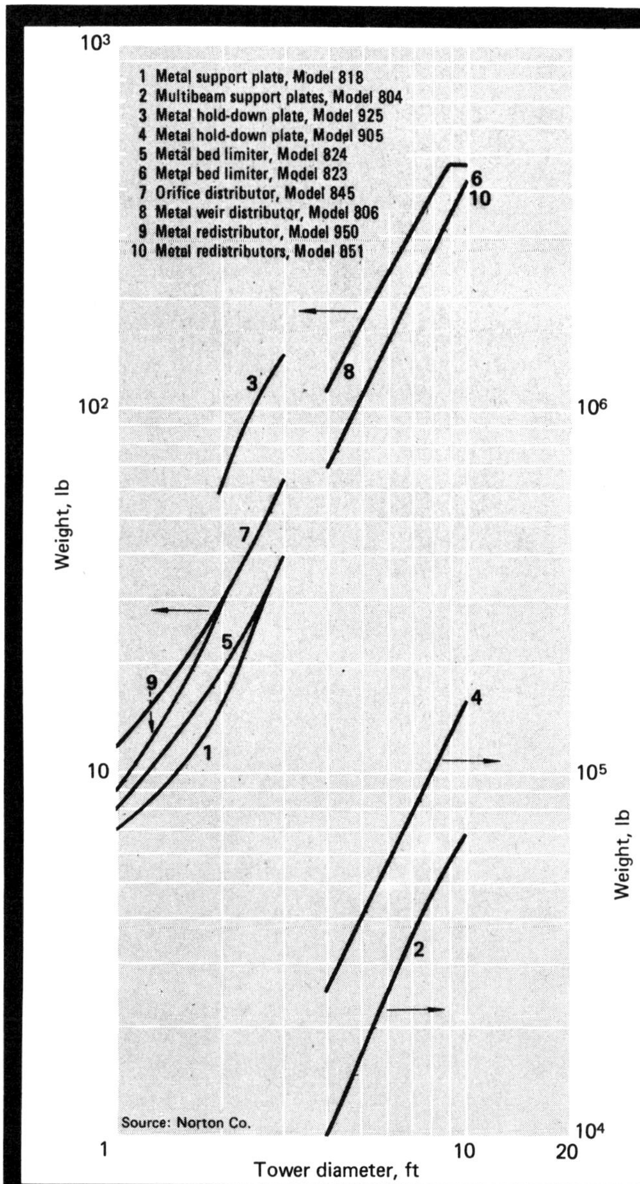

Weights of components for packed towers — **Fig. 8**

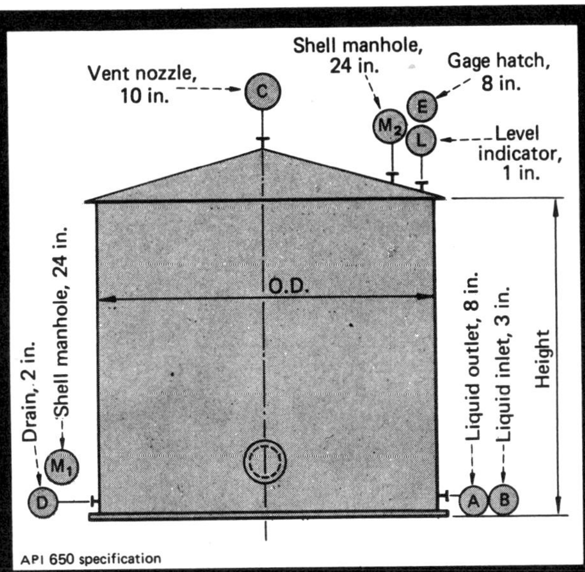

Atmospheric storage tank for liquids Fig. 9

Costs of distributors, costs of supports for packed towers, and weights of distributors and supports are given in Fig. 6, 7 and 8, respectively. Table V shows the unit costs and weights for various tower packings.

Atmospheric storage tanks

Costs for storage tanks with operating pressures close to atmospheric pressure and 2 to 3 psi above atmospheric are given for tanks designed in accordance with American Petroleum Institute standard, API 650.

Atmospheric tanks are used for liquids whose vapor pressure is atmospheric or below, at storage conditions. Local air-pollution regulations as well as physical properties of the stored liquids limit the use of atmospheric tanks. Very often, nitrogen blanketing is used to avoid contact of atmospheric air with contents of the tank.

Fig. 9 shows a typical API 650 tank whose costs are given in Fig. 10. These costs include material, fabrication and erection of a tank having the nozzle schedule shown in Fig. 9. Additional characteristics are: atmospheric working pressure at ambient temperature, corrosion allowance as required, and carbon steel for material in contact with process liquid. Flanges conform to ANSI standard for 150 psi rating, are raised face, and have manufacturer's standard facing finish. Tank is erected on foundation furnished by others. No sandblasting or painting included in costs. Additional nozzles may be added, and calculated as previously described for pressure vessels. Fig. 11 gives the weights for atmospheric storage tanks.

Example 3—Let us estimate the cost of an atmospheric tank, designed in accordance with Fig. 9, and having a diameter of 60 ft and a height of 32 ft. We will add one 8-in. connection for recirculation, one 10-in. alternate inlet, and delete the roof manhole.

Using Fig. 10, we find the fabricated cost of a tank for a 60-ft dia. and 32-ft height. We then add or deduct amounts for additional nozzles or deleted items.

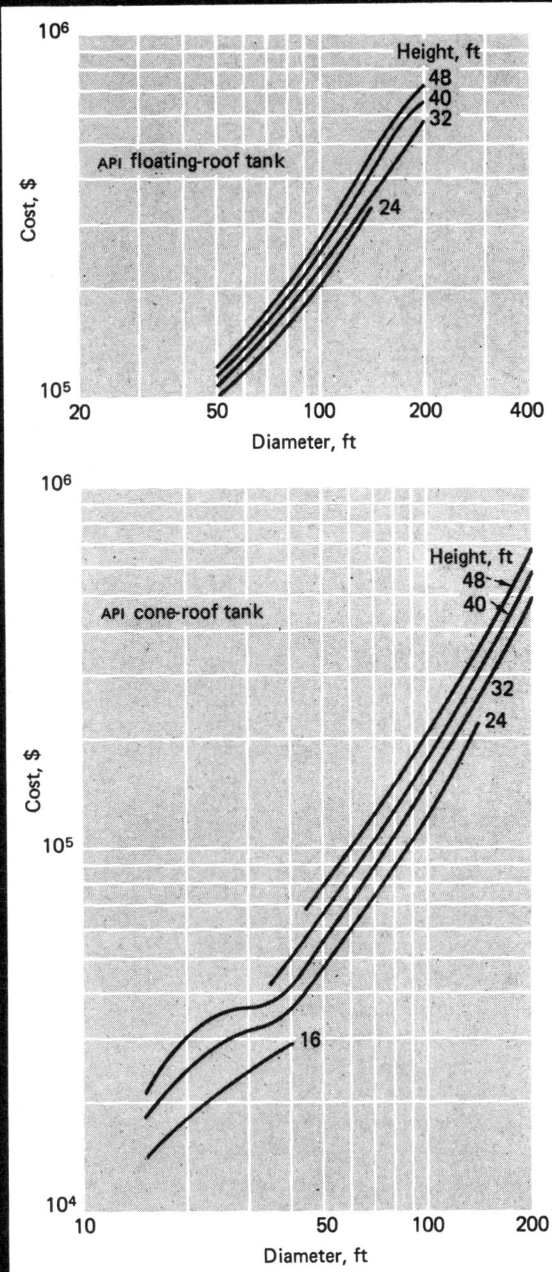

Cost of atmospheric tanks conforming to Fig. 9 Fig. 10

Vessel cost from Fig. 10	$72,000
Additions	
1 8-in. nozzle from Fig. 1	405
1 10-in. nozzle from Fig. 1	520
Fabrication (manhours)*	33
Subtotal	$72,958
Subtotal (rounded off)	$72,960
Deductions	
1 20-in. manhole (Fig. 1 & Table III)	−1,850
Fabrication (manhours)*	−50
Estimated cost of tank	$71,060

*The addition of two nozzles requires 2 man-hours at $16.50/man-hour, while 3 man-hours at $16.50/man-hour represents labor cost for one manhole.

Heat Exchangers

Heat exchangers more-often designed and/or used by us in our processes are those classified as "TEMA Class 'B' Heat Exchangers," and having designations as Type AEM, BEM, BEL, CKU, ADU, BEU, AFM, and BFM.† The first letter of the type designation refers to the front head (stationary-head types) of the exchanger; the second, to shell features; and the third, to rear heads. For the exchangers listed here, the terminology is:

A Channel and removable cover.
B Bonnet (integral cover).
C Channel integral with tubesheet and removable cover.
D Special high-pressure closure.
E One-pass shell.
F Two-pass shell with longitudinal baffle.
K Kettle-type reboiler.
L Fixed tubesheet. Like "A" stationary head.
M Fixed tubesheet. Like "B" stationary head.
U U-tube bundle.

(Consult the TEMA† standards for complete descriptions, details of construction, and component arrangements for most types of heat exchangers used by the CPI.)

To estimate the cost of the exchanger, it must be rated. The following information is required: shell diameter, tubeside temperature and pressure rating; type of exchanger; shell length, and shellside temperature and pressure rating; number and type of tubes, surface area, material, and bundle length; number and type of baffles and support plates; saddles for horizontal exchangers, or lifting lugs and brackets for vertical exchangers; and number of couplings and nozzles required on each exchanger.

From data provided by manufacturers, the base price of heat exchangers, up to 30-in. dia., follows a relationship of the type:

$$P = XD^Y \qquad (2)$$

where P is base price of exchanger, \$; D is diameter, in.; and $Y = 1.037$ as correlated from the data studies. The coefficient, X, depends on the rating and type of ends as designated by TEMA Standards. The following table lists several values of the coefficient X:

	Pressure rating		
	150 psi	**300 psi**	**400 psi**
Type of ends	***X***	***X***	***X***
A-M	170	195	217
B-M	125	145	160
B-L	170	195	217
C-M	160	180	200
C-L	187	215	240

The estimated cost of Type E and F shells can be determined from Fig. 12, and added to the base price of the exchanger, as calculated from Eq. (2). However, for shells of the kettle type, the evaulation must be made as if it were a process vessel. To do so, we follow the procedure given under pressure vessels but use 1.5 times

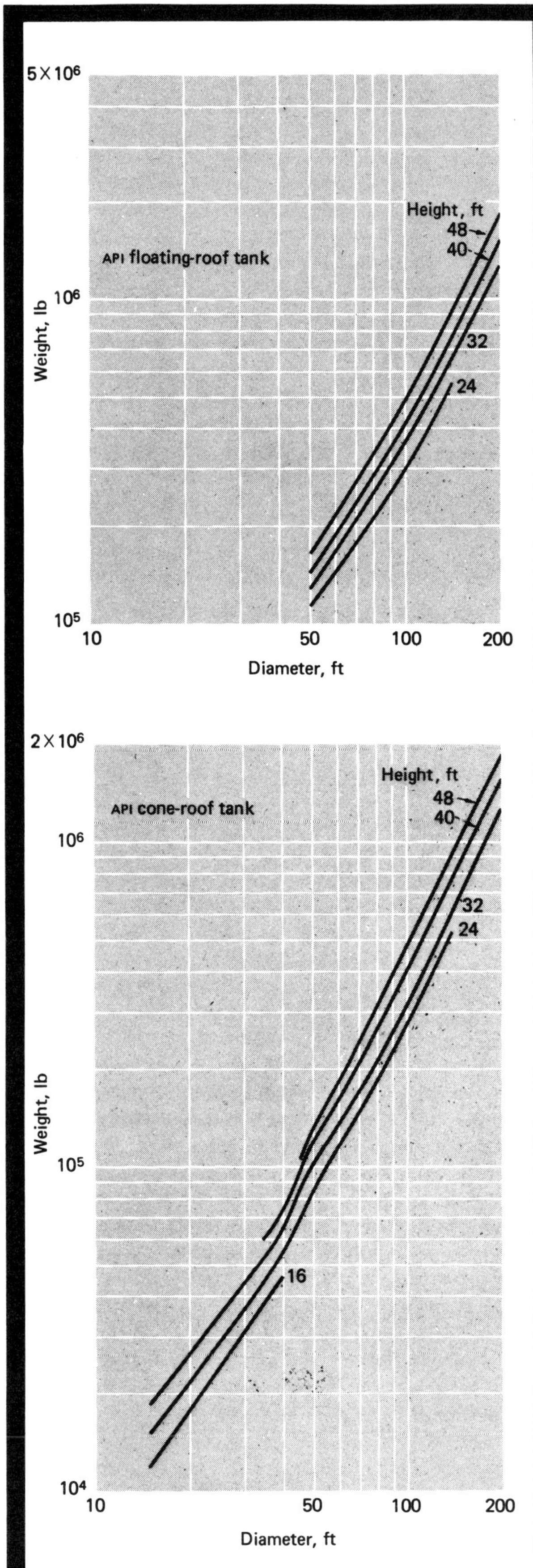

Weights of atmospheric tanks conforming to Fig. 9 — Fig. 11

†See "Standards of Tubular Exchanger Manufacturers Assn.," Tubular Exchanger Manufacturers Assn., New York, NY 10017.

Cost of carbon-steel exchanger tubes **Table VI**

Straight tubes			U-tubes		
Size, in.	Rating, psi	Cost, $/tube	Size, in.	Rating, psi	Cost, $/tube
3/4	150	1.15	3/4	150	3.85
3/4	300	1.30	3/4	300	4.15
3/4	400	1.45	3/4	400	4.50
1	150	1.30	1	150	4.14
1	300	1.45	1	300	4.50
1	400	1.75	1	400	4.60

Multiply cost/tube by total number of tubes, and add to base price of exchanger.

the cost of two semi-ellipsoidal heads. If expansion joints are required for fixed-tubesheet exchangers, prices may be estimated at the cost of 5 lineal feet of shell for each expansion joint. If ASME Code stamp and data sheets are required, add $130 to the total calculated cost of the exchanger.

Example 4—Let us estimate the cost of a 24-in. dia.,

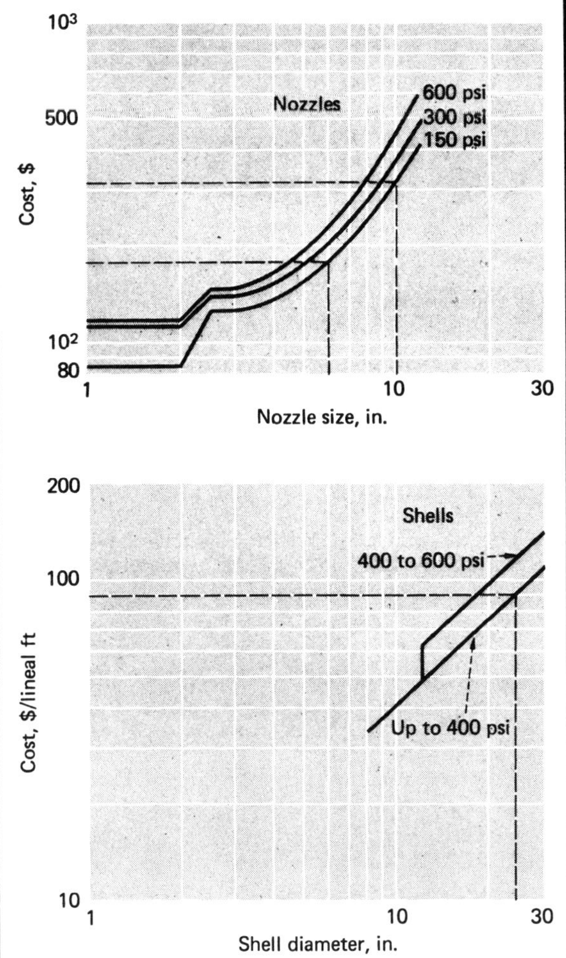

Costs of shells and nozzles for heat-exchangers Fig. 12

Cost correction for tube material **Table VII**

Material	Cost, $/ft^2
A-214 steel	3.00
Copper	6.70
Type 304 stainless steel	6.90
Type 316 stainless steel	9.35

Multiply surface area of tubes by cost/ft^2, and add to base price of exchanger.

Type AEM, TEMA Class 'R' exchanger. Tubeside and shellside rating are 150 psi. The exchanger has three 6-in. and two 10-in., 150-psi raised-face nozzles; 392 straight tubes, each $^3/_4$-in., 14 gage, Type A-214 steel. Bundle length is 20 ft, surface area 1,500 ft^2; 4 support plates, 2 support saddles, and expansion joint required.

To estimate the cost of this exchanger, we first evaluate the base price of the unit from Eq. (2), and then obtain the cost for the shell and expansion joint. To these are added the costs for nozzles, tubes, support plates and support saddles. Finally the cost for ASME code stamp and data sheets completes the procedure. The summary of estimated costs for the 24-in. dia. exchanger follows:

Base price, $P = 170(24)^{1.037}$	$4,589
Shell (24-in. dia.) from Fig. 12 ($90/ft × 20 ft)	1,800
Expansion joint: ($90/ft × 5 ft)	450
Nozzles from Fig. 12	
6-in.: 3 × $180	540
10-in.: 2 × $320	640
Tubes from Table VI (392 × $1.15)	451
Tube-material correction from Table VII A-214: (1,500 ft^2 × $3/ft^2)	4,500
Support plates from Fig. 13 (4 × $75)	300
Support saddles from Fig. 13 (2 × $97)	194
ASME code stamp and data sheets	130
Total estimated cost	$13,594
Estimated weight from Fig. 13 = 10,500 lb (1,500 ft^2 × 7 lb/ft^2)	

Air coolers for heat rejection

Air coolers are extensively used where heat removal from a process is possible with ambient air as the coolant. They are a good choice in regions of water scarcity. As costs of cooling water continue to rise and as concern for water pollution grows, applications of air coolers will increase.

Fig. 14 is based on material and fabrication costs, FOB manufacturer's shop, and includes finned tubes, steel supporting structure with plenum chambers, axial-flow fan, louvers, fan guards and positioner. (All cost curves are based on Oct. 1976 prices.)

To use Fig. 14, we need the following: bare tube area, tube material and rating, number of tube rows, and length of tubes.

If ambient-air temperatures cannot readily cool the process fluid, it is usually necessary to install a trim cooler (shell-and-tube exchanger) downstream of the air cooler to provide better temperature control of the process stream. The trim cooler must be separately estimated by using the procedure outlined under heat exchangers. Trim coolers range in cost between 15% and 20% of the air-cooler cost. Erection and installation are not included.

Example 5—An air-cooled overhead condenser has been designed for a distillation column operating at 150 psig. The calculated bare-tube surface is 7,000 ft.^{2} A trim cooler is required for control, and corresponds to 30% of the air-cooler area. Tube length is 24 ft, arranged in six tube-rows. The tubes are Type 304 stainless steel. Let us estimate the cost of the cooler system.

A summary of the steps in preparing the cost estimate for this system follows:

Bare tube area: 7,000 ft^2	
Air-cooler cost (7,000 ft^2 × \$28/ft^2)	\$196,000
Correction factor for rows from Fig. 14: 1.0 (6 tube-rows)	
Correction factor for tube length: 1.1 (24 ft)	
Adjusted cost of air cooler: (1.0 × 1.1 × \$196,000)	\$215,600
Trim cooler: (0.20 × 0.30 × \$215,600)	12,936
Estimated cost of system:	\$228,536
Estimated cost of system (rounded off):	\$229,000

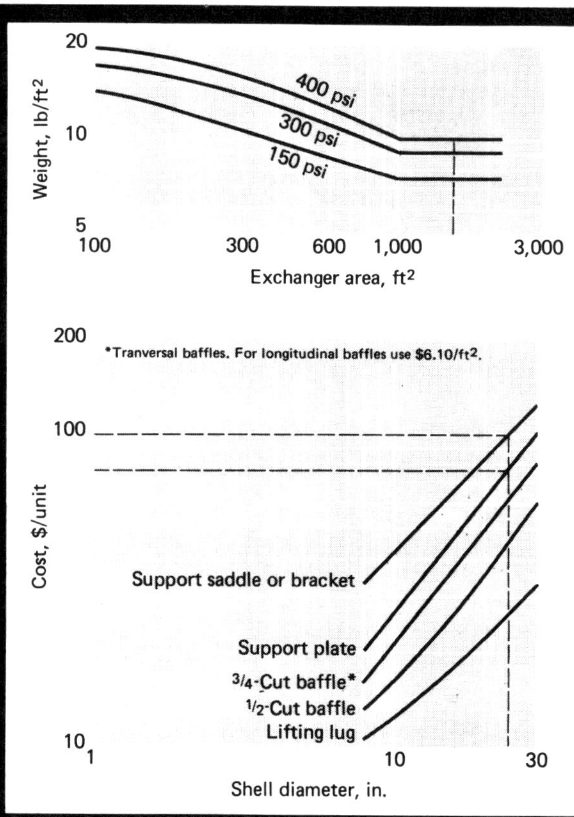

Costs for heat exchanger supports and unit weights of exchangers Fig. 13

Process furnaces

Furnaces are used for high heat-transfer rates to a processor utility stream. Indirect-fired heaters transfer the energy released by combustion of a fuel to a liquid-phase or gas-phase within tubes. The flames and products of combustion are separated from the fluid by the tube walls.

Furnaces vary in costs more than any other type of equipment because of individual design requirements. The emphasis on energy saving has increased initial investment for furnaces. For example, the use of air preheaters or other devices to recover heat otherwise lost to the stack. Also, the limitation of fuels to miminize air pollution and the addition of equipmcnt for air-pollution abatement increase the cost of the heater.

Fig. 15 shows costs for indirect-heated furnaces for different materials of construction and rating pressures as a function of the absorbed heat. Costs are based on our past projects and proposals from vendors, and updated for Dec. 1976. These curves provide a preliminary evaluation of furnace cost. Since fired heaters are fabricated on an order basis (except for a boiler package), the cost-estimating data do not have the accuracy of a direct contact with the vendor. Costs of erection, installation and subcontracting are not included.

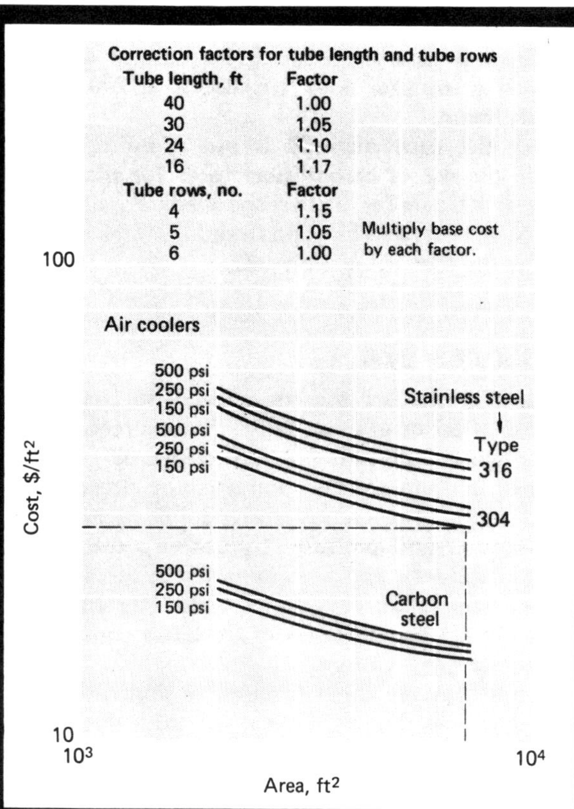

Air-cooler costs and correction factors Fig. 14

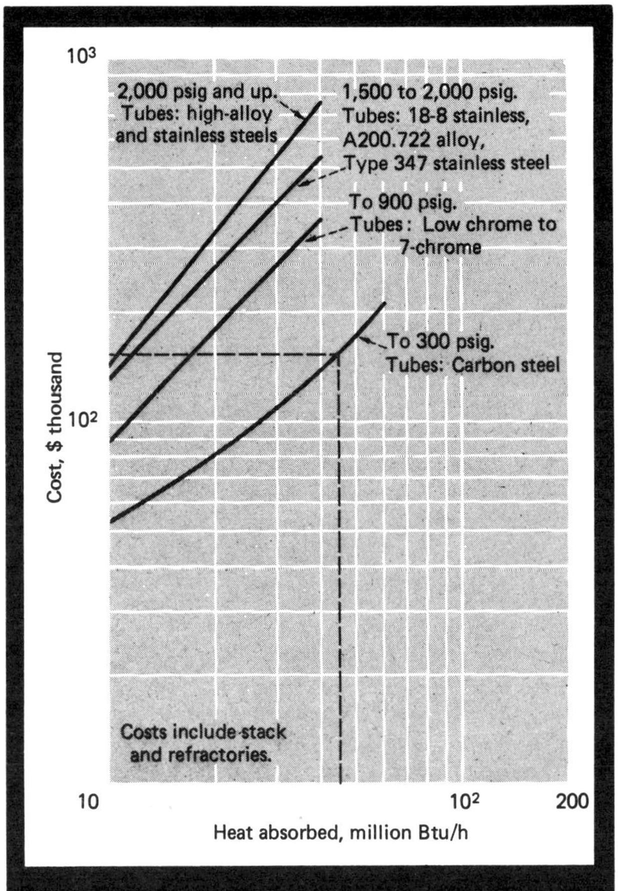

Indirect-heated furnace costs Fig. 15

Example 6—Let us estimate the cost of a furnace designed to transfer 45×10^6 Btu/h at 250 psi to a hot-oil stream.

Since the tube material is not specified, we will assume the use of carbon-steel tubes for the pressure rating of 300 psig (vs. 250 psi specified). For a heat rate of 45×10^6 Btu/h, we obtain from Fig. 15 an estimated cost for the furnace of $160,000.

Compressors

Plants in the CPI use compressors to increase the pressure of air or gases to satisfy process requirements. Process-gas compressors may be considered as custom-designed equipment. For example, the process gas is highly corrosive or carries erosive solids in suspension; temperatures and pressure requirements may lead to potential hazards; and contamination by lubricating oil (when compressing oxygen) or water (for instrument air) is not permissible whether for safety consideration or product specifications.

Compressors are either dynamic or positive-displacement machines. The dynamic types, more-often known as centrifugal, are high-capacity compressors with low discharge pressures of up to several hundred psi. Capacity range is from 400 to 13,000 ft^3/min.

In the positive-displacement types, the most-used are the reciprocal compressors with capacities much lower than centrifugals but with discharge pressures of up to 6,000 psi.

Costs for air compressors both reciprocal and centrifugal are plotted in Fig. 16 against brake horsepower. Cost includes compressor with motor drive, lubrication system, moisture separator and minimum instrumentation required. Cost data are based on 1975 prices, and corrected for Dec. 1976.

Example 7—A centrifugal air compressor has been rated to compress and deliver 26,000 SCFM (standard ft^3/min) of dry air at a pressure of 84.7 psia and 216°F. The isothermal head has been calculated as 47,700 ft-lb/lb, and power requirements are 5,000 hp. From Fig. 16, we find the estimated cost to be $460,000.

Pumps for liquid services

When specifying pumps for any service, it is necessary to know the liquid to be handled, total dynamic head, suction and discharge pressures, and the properties of the fluid, temperature at operation, viscosity, specific gravity and vapor pressure.

Pumps are specified at the early stages of a project when a plant layout is not completely defined and most of the equipment dimensions and elevations are sketchy. The easiest way to estimate the price of pumps is by using a current price-data book from one of the leading pump manufacturers, unless the pumping requirements are unique, and a special design or even a first-of-a-kind pump has to be designed. In this case, the normal procedure is to send detailed specifications to several pump manufacturers for their bids.

Fig. 17 provides costs for centrifugal pumps for AVS and API pumps. AVS pumps are chemical-process pumps designed in accordance with ANSI B 73.1-1974. They are generally supplied with open impellers. Temperatures are limited to 300°F and pressures up to 300 psig maximum, depending on the material and flange type. Maximum capacity is approximately 1,000 gpm, and discharge heads of approximately 320 ft are obtained at lesser capacities. API pumps are general refinery-service pumps designed in accordance with API Standard 610-1971. API pumps are required to have closed impellers, with case and impeller wear-rings, and are generally designed for continuous heavy-duty service under severe process conditions.

The cost curves of Fig. 17 are for pumps operating at 3,500 rpm, and are based on Oct. 1976 prices with motor and foundations included. Case material is ductile iron for AVS pumps and cast steel for API pumps.

Reciprocating or positive-displacement pumps are usually specified for low-flow and high-pressure requirements, as well as for some metering applications. Fig. 18 provide costs for various material vs. capacities for positive-displacement pumps. Costs include pump, base plate, V-belt drive and V-belt, but without motor.

Example 8—A pair of centrifugal pumps is required to deliver 1,430 gpm of a heavy organic mixture (specific gravity = 0.952) with a differential head of 77 ft. Let us evaluate the cost of this pump for (a) API horizontal type and (b) API inline type.

Since the abscissa of Fig. 17 represents a factor that is

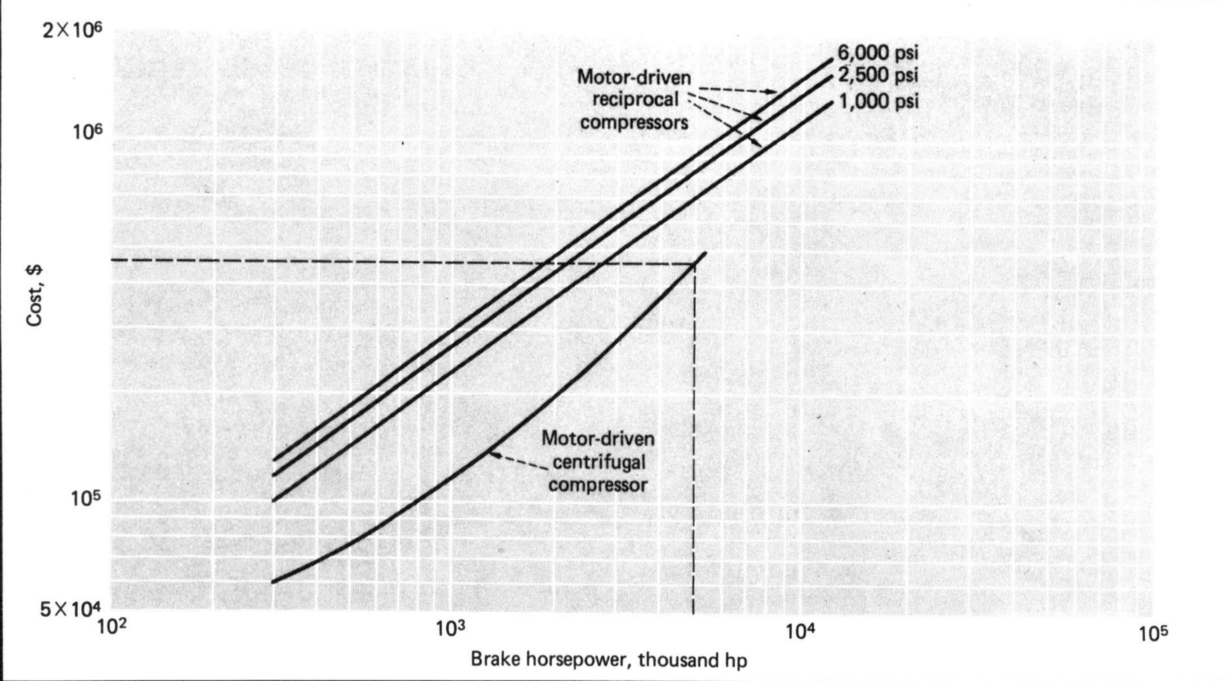

Costs for centrifugal and reciprocating air compressors — Fig. 16

the product of flow and pressure, we must first calculate the head in psi, as follows:

$$(77 \times 0.952)/2.31 = 31.7 \text{ psi}$$

where 77 ft is differential head, 0.952 is specific gravity, and 2.31 ft/psi is a conversion factor.

Hence, in this example, the size factor for Fig. 17 becomes:

$$(1{,}430)(31.7) = 4.5 \times 10^4$$

From Fig. 17, we find the cost of an:

(a) API horizontal pump as $5,100
(b) API inline pump as $3,900

Vacuum equipment

The CPI use vacuum equipment to improve some complicated separations in distillation towers and at the

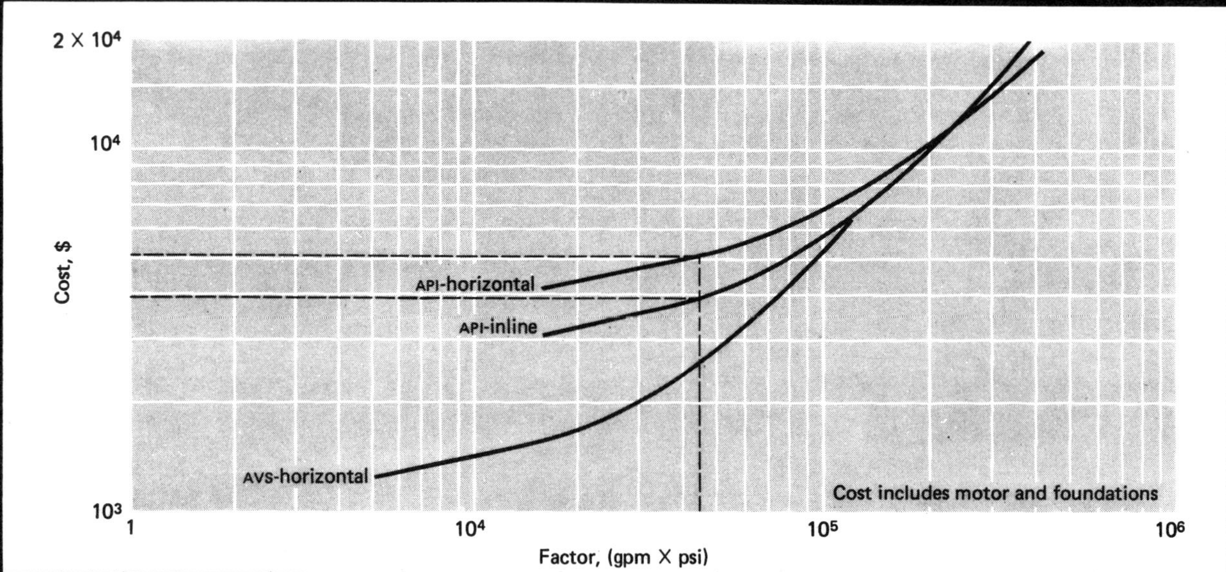

Costs for estimating standard centrifugal pumps — Fig. 17

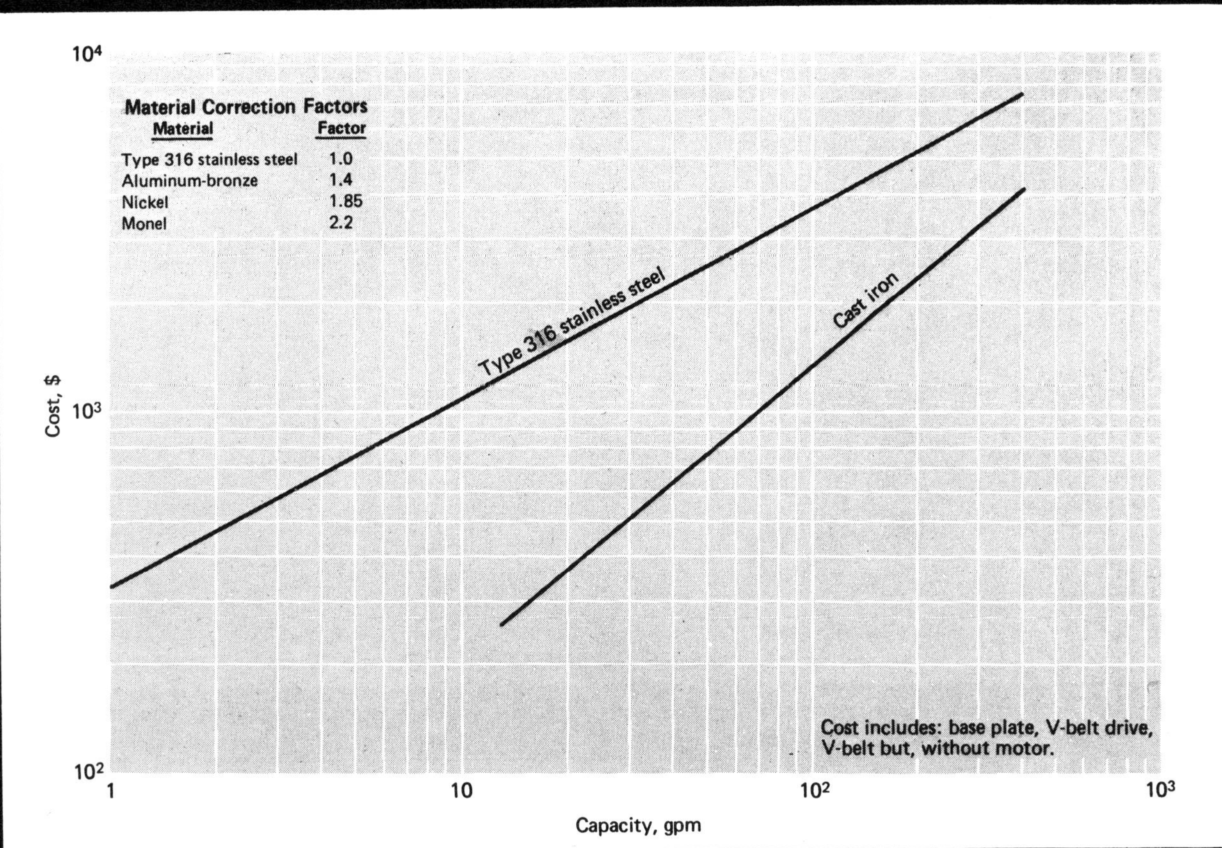

Costs for reciprocating (positive-displacement) pumps — Fig. 18

same time reduce reboiler duty, and to achieve optimum operating conditions in reactors or in extrusion processes.

The two most-common vacuum producers are steam-jet ejectors and mechanical vacuum pumps. Steam-jet ejectors have proved to be reliable and of low investment. Because ejectors have no moving parts, they are almost maintenance-free. Furthermore, they operate with motive steam that is readily available. On the other hand, mechanical vacuum pumps require higher initial investment and operating costs than the standard steam-jet but are eight to ten times more efficient users of energy.

The tendency today is toward designing with energy-saving paramount. Therefore, at the early stages of a project, a careful evaluation is necessary to determine the most-suitable equipment that can perform the job at optimum conditions.

Fig. 19 shows material and fabrication costs for mechanical vacuum pumps and steam-jet ejectors vs. a size factor. This size factor is based on studies made by R. B. Poer (*Hydrocarbon Process. Petrol. Refiner,* Mar. 1964, p. 142), and is found to correlate data for inlet pressures between 0.03 and 200 mm Hg. The size factor is defined as equivalent air load (lb/h) divided by suction pressure (mm Hg). A similar correlation has been developed for mechanical vacuum pumps from purchase order and vendor quotations for a size factor ranging from 0.3 to 15. Costs for vacuum pumps include motor, baseplate and belt drive, and these are based as of Oct. 1976.

To use Fig. 19, we need the following information:

- Steam-jet ejectors: Material, number and type of condensers, number of stages, equivalent air load, and inlet pressure.
- Vacuum pumps: Equivalent air load, and inlet pressure.

Example 9—A steam-jet ejector consisting of two stages and one surface condenser has an equivalent air load of 640 lb/h, and inlet pressure of 85 mm Hg. The material of construction is stainless steel. What will it cost? What will be the cost of a mechanical pump for the same duty? We begin by computing the size factor as: (640 lb/h)/(85 mm Hg) = 7.5.

Then from Fig. 19, we determine the cost of a single-stage ejector without a condenser as $1,430. Also from Fig. 19, we find the correction factors as:

Item	Factor
1 Surface condenser	2.3
2 Stages	1.8
Stainless steel	2.0

Then the cost of the desired ejector system becomes the product of the cost of the single-stage unit and the various factors, or:

$$(2.3)(1.8)(2.0)(\$1{,}430) = \$11{,}840$$

The cost of a mechanical vacuum pump for the same duty is found from Fig. 19 to be $29,500.

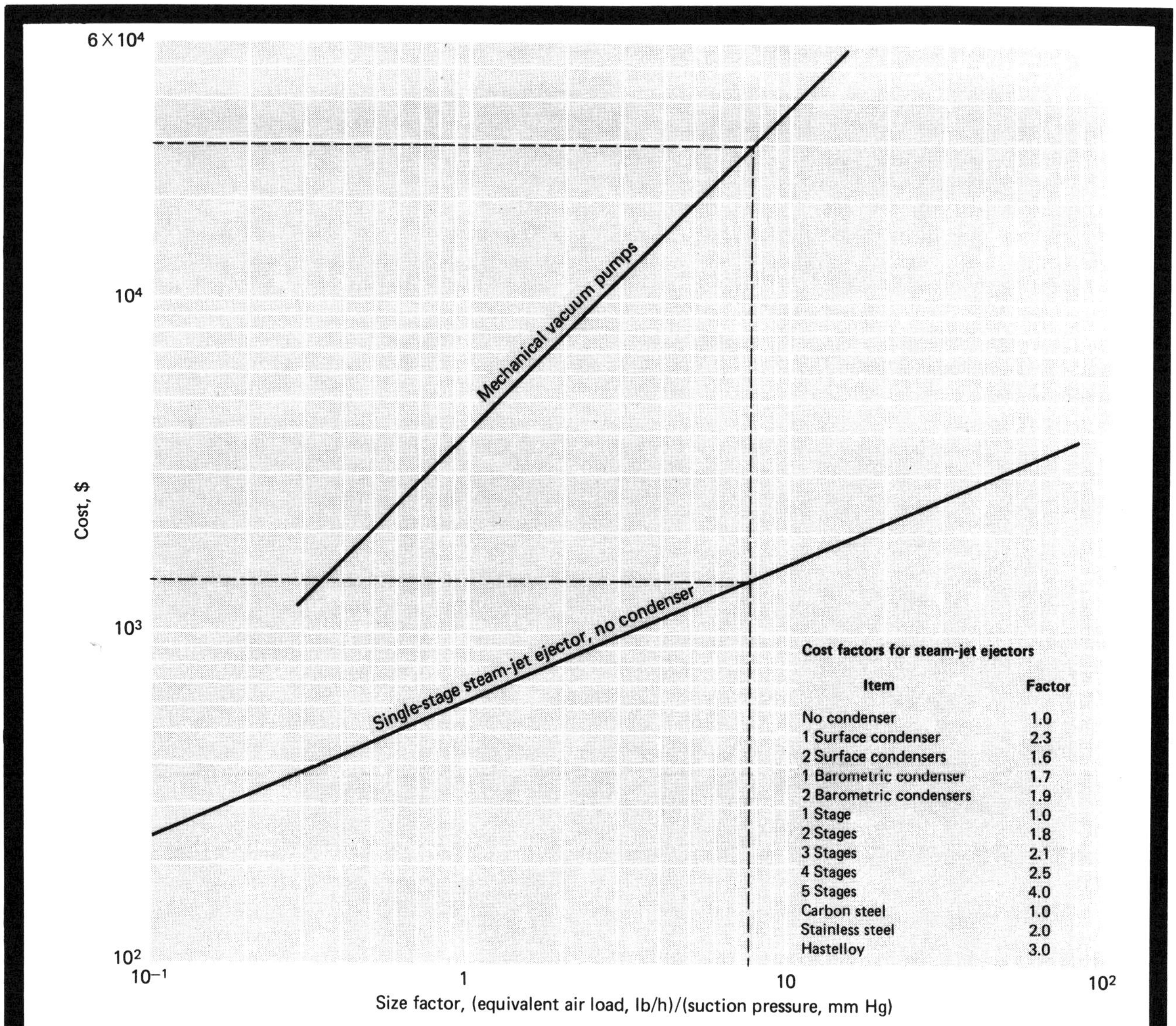

Cost factors for steam-jet ejectors

Item	Factor
No condenser	1.0
1 Surface condenser	2.3
2 Surface condensers	1.6
1 Barometric condenser	1.7
2 Barometric condensers	1.9
1 Stage	1.0
2 Stages	1.8
3 Stages	2.1
4 Stages	2.5
5 Stages	4.0
Carbon steel	1.0
Stainless steel	2.0
Hastelloy	3.0

Vacuum-equipment costs for mechanical pumps and steam-jet ejectors Fig. 19

Evaluating overall costs

The purchased cost of equipment, provided FOB the manufacturer's shop, may be increased by 3% to give the approximate delivered cost.

The fabricated cost of each item as calculated by the methods described here concentrates on the detailed cost evaluations for major equipment because these account for at least 50% of the total plant cost.

A similar method uses Lang factors, where the delivered cost of all equipment is summed, and then multiplied by an appropriate factor depending on the type of plant. For example, for a solids processing plant, the factor is 3.1; for a solids-liquid plant, 3.63; and for a fluid plant, 4.74. This type of estimate is often used in the preliminary stages of engineering, and is not very accurate. By evaluating the cost of major equipment with some degree of confidence, we will be able to improve the accuracy of the estimate, for which the Lang method is quite poor.

Providing for escalation

Escalation is the provision in actual or estimated costs for an increase in the costs of fabricated components, material, labor, etc. over those specified in an original contract. Because equipment costs are affected by price changes from the date an order is placed to the date when it is shipped, estimators use the escalation factor as a tool to incorporate these uncertainties into the cost estimate. When manufacturers of equipment submit bids, they usually include escalation formulas to determine price adjustment.

Most escalation factors are based on U.S. Bureau of Labor statistics (BLS) indexes. A typical escalation-adjustment formula is:

$$E = P_O \times F \times M \times \frac{(BLS)_A}{(BLS)_B} + P_O \times F \times (1 - M) \times \frac{(BLS)_{A'}}{(BLS)_{B'}} \quad (3)$$

where E is escalation, \$; P_O is original price, \$; F is fraction of P_O subject to escalation for which the manufacturer has no control; M is fraction of P_O related to materials cost; $(BLS)_A$ is BLS material index (selected from BLS Wholesale Price Index) at time of shipment or end of fabrication, whichever agreed to; $(BLS)_B$ is material index at time the order is placed; $(BLS)_{A'}$ is BLS Labor Index (selected from BLS Hours and Earnings) at time of shipment or end of fabrication; $(BLS)_{B'}$ is BLS Labor Index at time order is placed.

Example 10—An order is placed for an equipment item in June 1975 for delivery in Dec. 1976, at \$500,000. The manufacturer's proposal stipulates F as 54%, and the materials-to-labor ratio as 70% to 30%. Let us calculate the amount of escalation and the final cost of the equipment.

First, we find the appropriate BLS indexes for the material and labor costs as:

$$(BLS)_A = 199 \text{ at Dec. 1976}$$
$$(BLS)_B = 188.6 \text{ at June 1975}$$
$$(BLS)_{A'} = 5.62 \text{ at Dec. 1976}$$
$$(BLS)_{B'} = 5.03 \text{ at June 1975}$$

We compute the amount of escalation for the fraction of the original price subject to escalation by substituting in Eq. (3):

$$E = 500{,}000(0.54)(0.7)\left(\frac{199}{188.6}\right) + 500{,}000(0.54)(0.3)\left(\frac{5.62}{5.03}\right) = \$289{,}920.$$

The fraction of original price not subject to escalation becomes:

$$\$500{,}000(1.00 - 0.54) = \$230{,}000$$

Hence, total cost of equipment at time of shipment will be: \$289,920 + \$230,000 = \$519,920.

Escalation formulas are used when an original price is known. If it isn't known, the estimator must rely on other methods. These may be inflation indexes developed within the company and based on its experience, or published indexes.

Most published indexes are related to the construction industry. The most popular are: *Engineering News-Record* Construction and Building Indexes (ENR), *Chemical Engineering* Plant Cost Index (CE), Marshall and Swift Equipment Index (M&S), and the Nelson Refinery Construction Index.*

The ENR Index primarily measures civil construction costs, and is not adequate for chemical plant construction because civil cost is only a minor part of a chemical plant budget. Based on periodic equipment appraisals and on installation labor, the M&S Index reflects changes in installed-equipment costs, while the CE Plant Cost Index represents chemical-plant costs.

The M&S index may be used to determine the costs in Year X, C_X, for certain equipment if they are known in Year Y, C_Y:

$$C_X = (I_X/I_Y)C_Y \tag{4}$$

*The CE Plant Cost Index and the M&S Equipment Cost Index are compiled in *Chemical Engineering* on the Economic Indicators page of each issue. The Nelson Refinery Construction Index appears in the *Oil & Gas Journal*, and the ENR Construction and Building Indexes appear in *Engineering News-Record*.

where I_X and I_Y are the index values for the years, X and Y, respectively.

Example 11—The average M&S index for 1970 was 313, and for 1976 was 472. If a piece of equipment cost \$30,000 in 1970, what would be the expected cost in 1976?

$$C_{1976} = (472/313)(30{,}000) = \$45{,}240$$

The CE Plant Cost Index is used in a similar way. The CE index is not for individual equipment, but for an entire chemical plant. Very often, it is used concurrently with the well-known six-tenths capacity-correlation factor. For instance:

$$C_X = C_Y(Q_X/Q_Y)^{0.6}\,(I_X/I_Y) \tag{5}$$

where C is cost of plant, Q is capacity of plant and I is index value for the years X and Y, respectively.

Example 12—A 200-ton/d sulfuric acid plant cost \$900,000 in 1966. How much will a 400-ton/d plant cost in 1976? The CE index in 1966 was 107, and in 1976 was 192.

Substituting into Eq. (5), we find the cost of a 400-ton/d plant in 1976 as:

$$C_{1976} = \$900{,}000(400/200)^{0.6}\,(192/107)$$
$$C_{1976} = \$2{,}450{,}000$$

This procedure will give an order-of-magnitude estimate, since it does not take into account process improvement or obsolescence, productivity and the complexity of the plant.

The authors

Arkadie Pikulik is a sales executive at Scientific Design Company, Inc., Two Park Ave., New York, NY 10016 where he has been employed for twenty-three years at various positions, including pilot plant supervisor, process engineer, project engineer, project manager, project director, construction manager and start-up supervisor. He holds a B.Ch.E. from Brooklyn Polytechnic Institute.

Hector E. Diaz has been a project engineer at Scientific Design Co. since 1974. He holds a B.S. in chemical engineering from Universidad Industrial de Santander, Colombia.

The following members of Scientific Design Co. have contributed to the preparation of this report; Arthur J. Hermes (estimating techniques), Felix Murphy (cost data), and M. George Tashjian (comments and advice).

Economic pipe insulation

for cold systems

Preventing cold-pipe sweating once usually outweighed other considerations in determining insulation thickness. Nomographs in this article are geared to selecting the thicker insulation now demanded by higher energy costs.

Jules L. Abramovitz, *Coaltek Associates*

☐ Cold-insulation design presents fundamentally different problems from those of hot systems.* One basic difference is that capital costs for refrigeration equipment add substantially to the cost of generating refrigeration. This "capital equipment fuel surcharge" can tack 25% onto prevailing electric rates.

A second major difference is that cold-insulation specifications must satisfy two requirements: an economic one, and the functional one of preventing sweating, which damages insulation. A third difference is that in selection nomographs one independent variable must be stated in ¢/kWh, rather than in the 10^6 Btu of hot systems.

Occasional and moderate condensation upon the surface of cold insulation can be tolerated (it can never be completely eliminated), but frequent and excessive condensation damages the insulation. To prevent this destruction, cold insulation must be designed to limit surface wetting. In the past, this necessity has overshadowed economic considerations. But under current conditions, and those likely in the future, energy economics will prevail in many instances. Proper design now requires that thickness be determined for both criteria, and that the more demanding criterion be applied.

Cold systems vary in temperature of operation and in capacity of refrigeration machines, factors that materially affect the cost of equipment and operation. On the basis of 10^6 Btu/h capability, refrigeration equipment is far more expensive than heating equipment. The investment in refrigeration equipment cannot be ignored; annualized, it adds substantially to operating costs.

The nomographs presented are based on costs for a recent installation. Because installations differ, methods of adjusting for these costs and operations have been incorporated in the solutions.

Differences between refrigeration-system and average-ambient temperatures are generally moderate, usually in the range of 100°F or less. Nevertheless, the nomographs take into account the difference between average annual ambient temperature and that of the fluid inside the pipe, rather than only the temperature of the fluid (as do nomographs for hot insulation).

Developing a refrigeration energy audit

The basic technical data of the reference installation upon which the nomographs are based are:

Capacity, tons	1,050
Operating temperature, °F	10
Brake horsepower, full load, hp	2,300
Motor efficiency, full load, %	95
Duty, continuous, high load, h/yr	8,450

The system requires auxiliary power as follows:

Cooling-water pump, hp	145
Cooling-tower fan, hp	166
Brine-circulation pump, hp	70

The total power needed, therefore, is 2,681 hp, or 2.55 hp/ton of refrigeration.

The energy liability is therefore:

$$(2.55 \text{ hp/ton})(0.745 \text{ kW/hp})(1/0.95) = 2.0 \text{ kW/ton}$$

Or:

$$(2.0 \text{ kW/ton})(10^6/12{,}000 \text{ Btu/h/ton}) = 167 \text{ kWh}/10^6 \text{ Btu}$$

The capital equipment cost, including a part of the auxiliaries, is \$645/ton of refrigeration. Paying 9% interest on the capital financed over a 13-year economic life adds a 1.73 cost factor to the foregoing base cost. Allowing 1% per year for maintenance results in an additional cost factor of 0.13.

The total equipment cost is therefore:

$$645\ (1.73 + 0.13) = \$1{,}200/\text{ton}$$

(Text continued on p. 323)

*See J. L. Abramovitz and R. Cordero, How to Select Insulation Thickness for Hot Pipes, *Chem. Eng.*, July 21, 1975, p. 89.

Originally published October 1976.

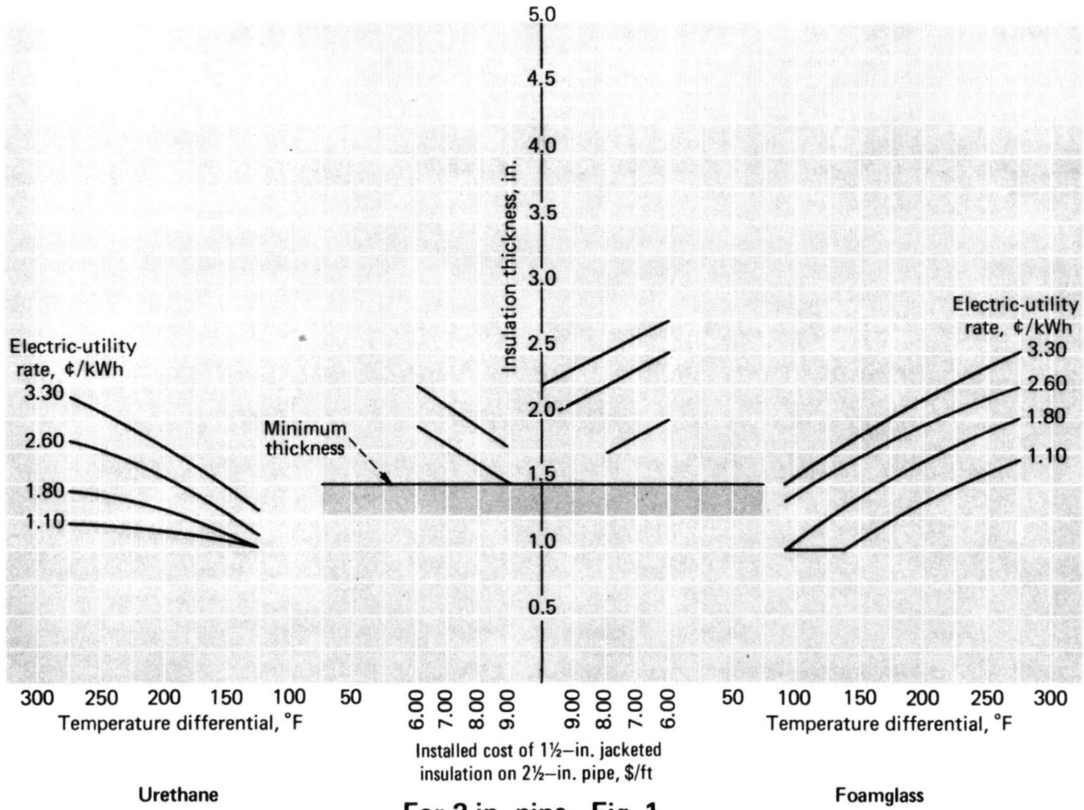

For 2-in. pipe—Fig. 1

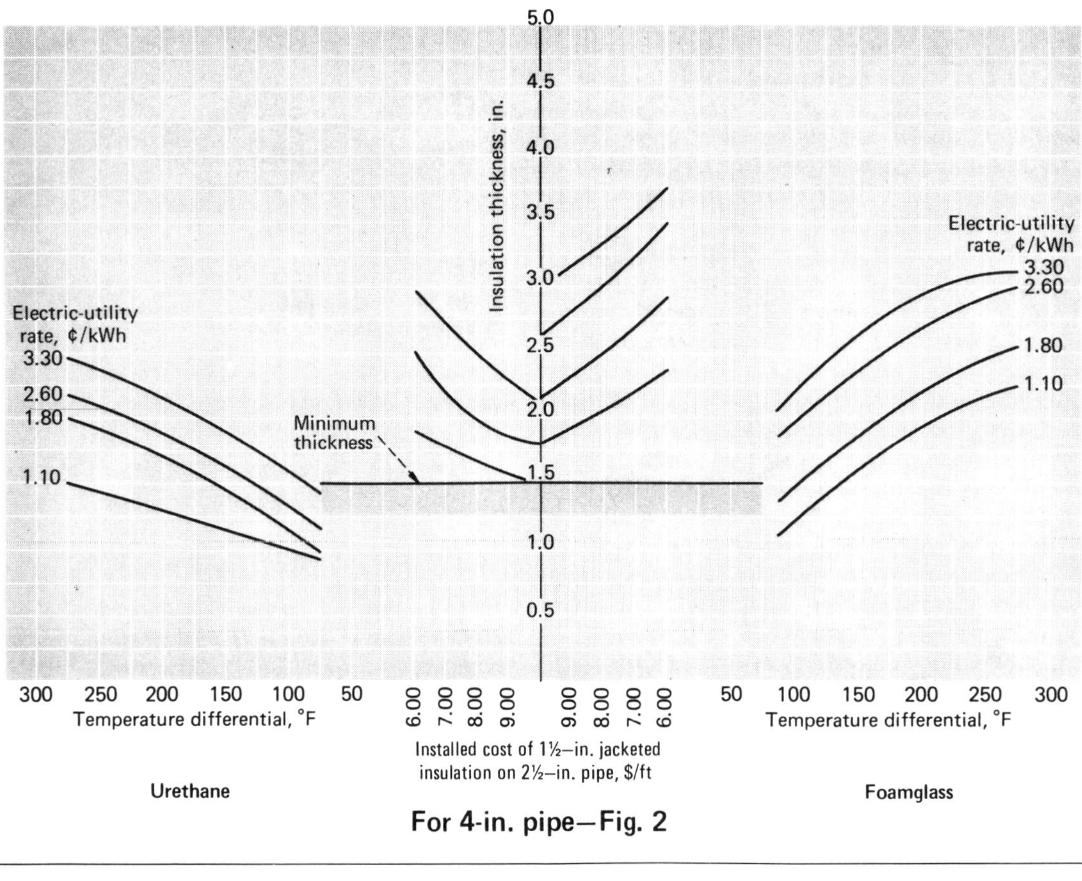

For 4-in. pipe—Fig. 2

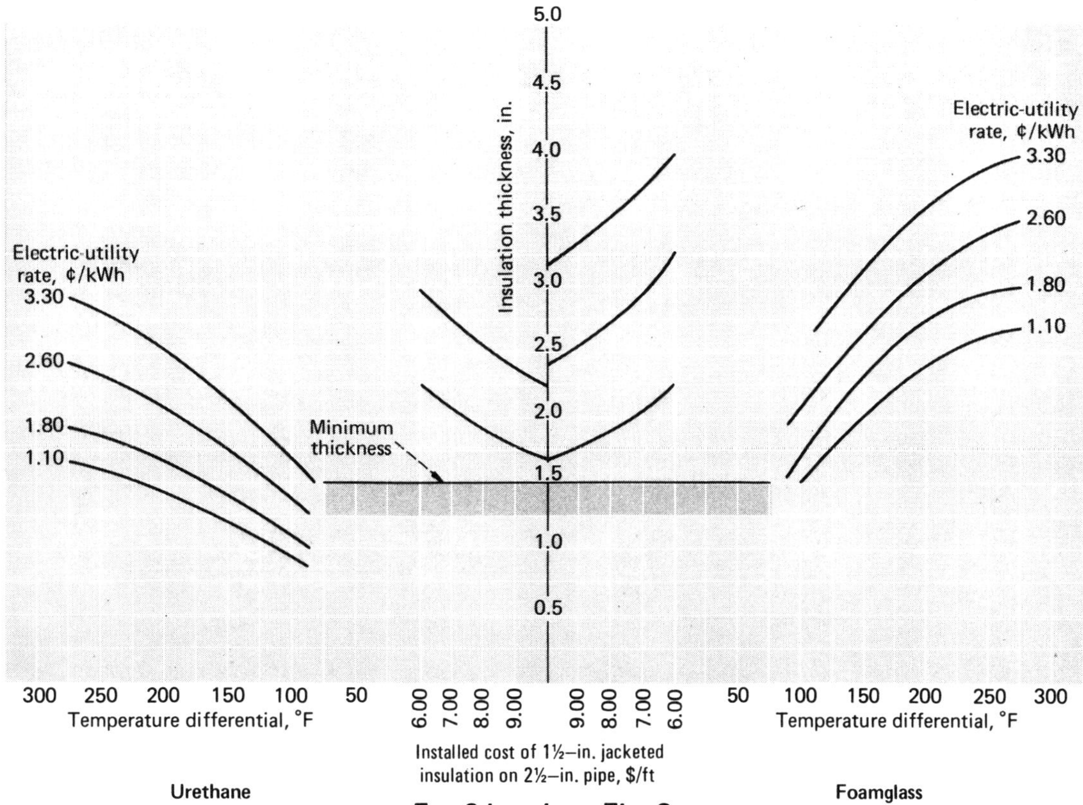

For 6-in. pipe—Fig. 3

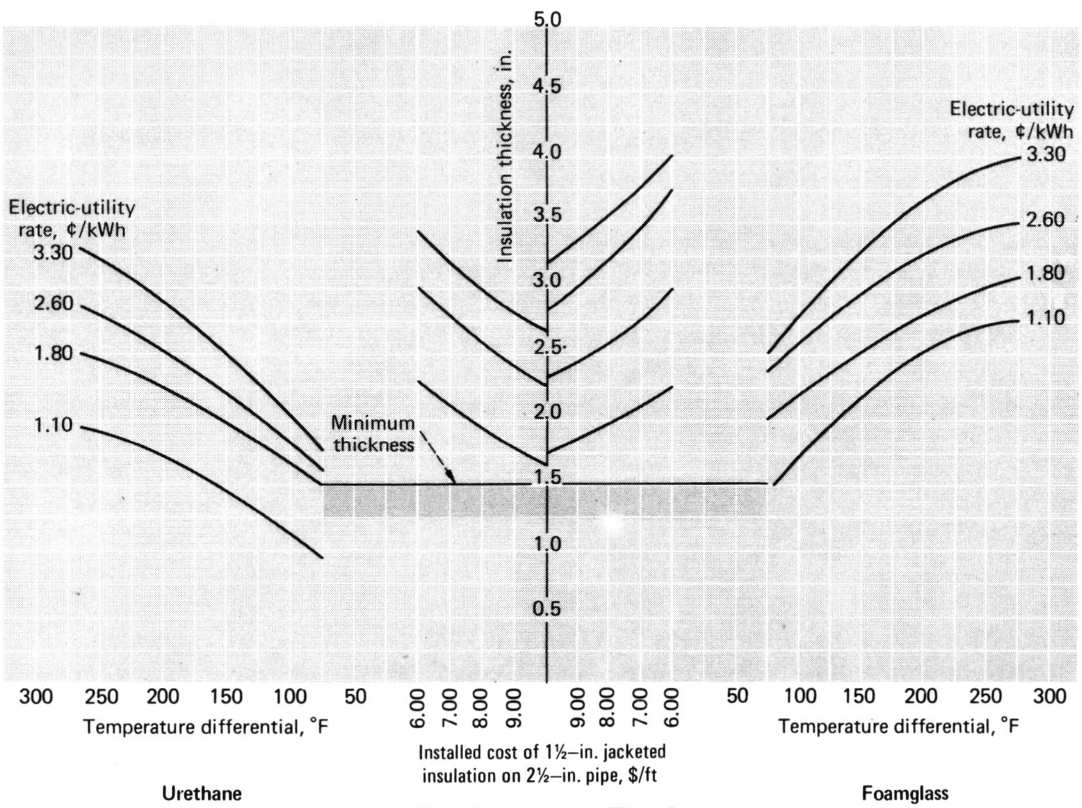

For 8-in. pipe—Fig. 4

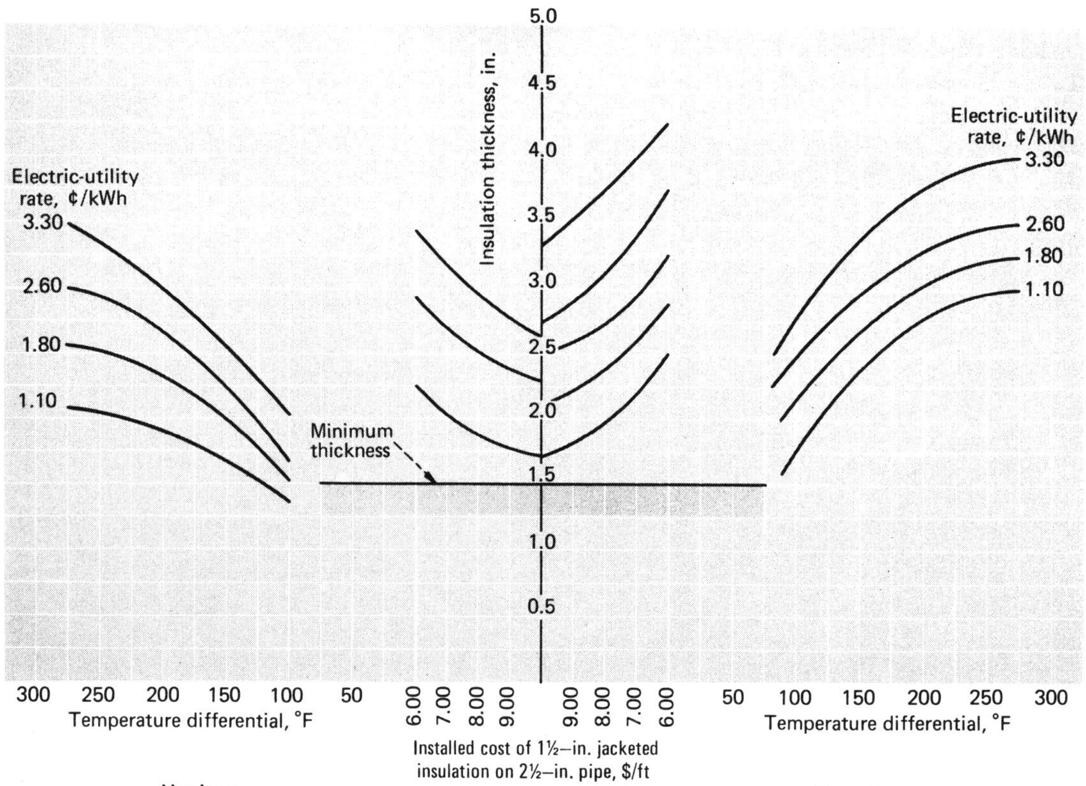

For 10-in. pipe—Fig. 5

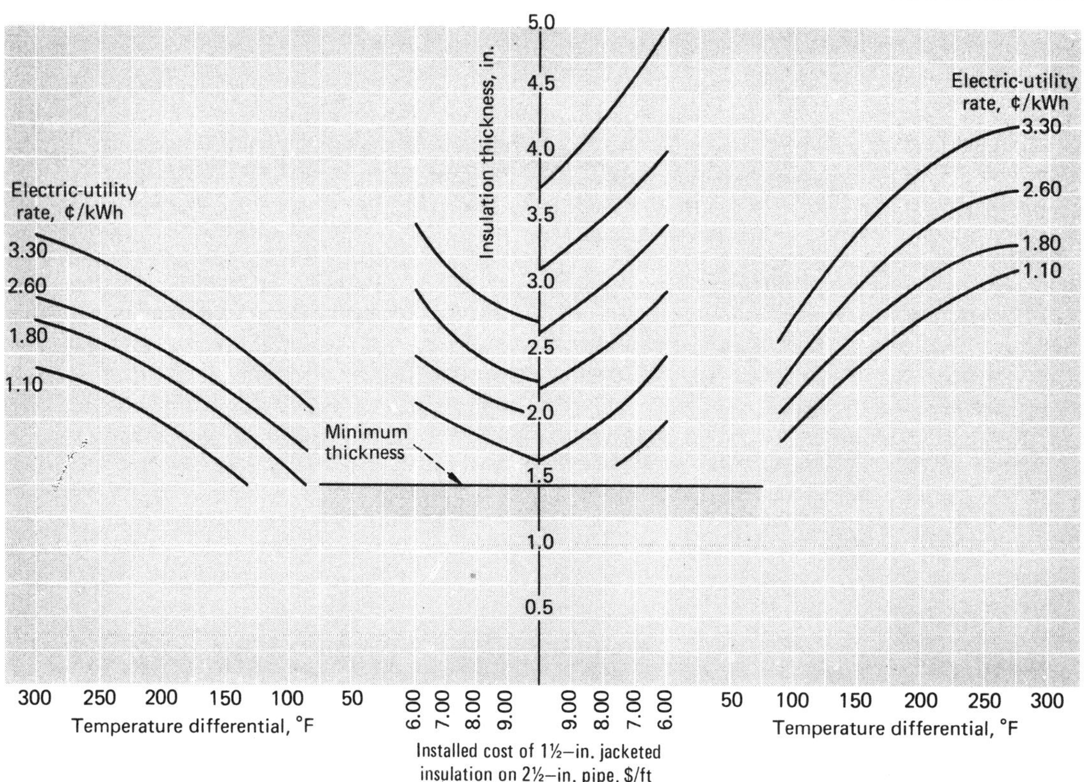

For 12-in. and 14-in. pipe—Fig. 6

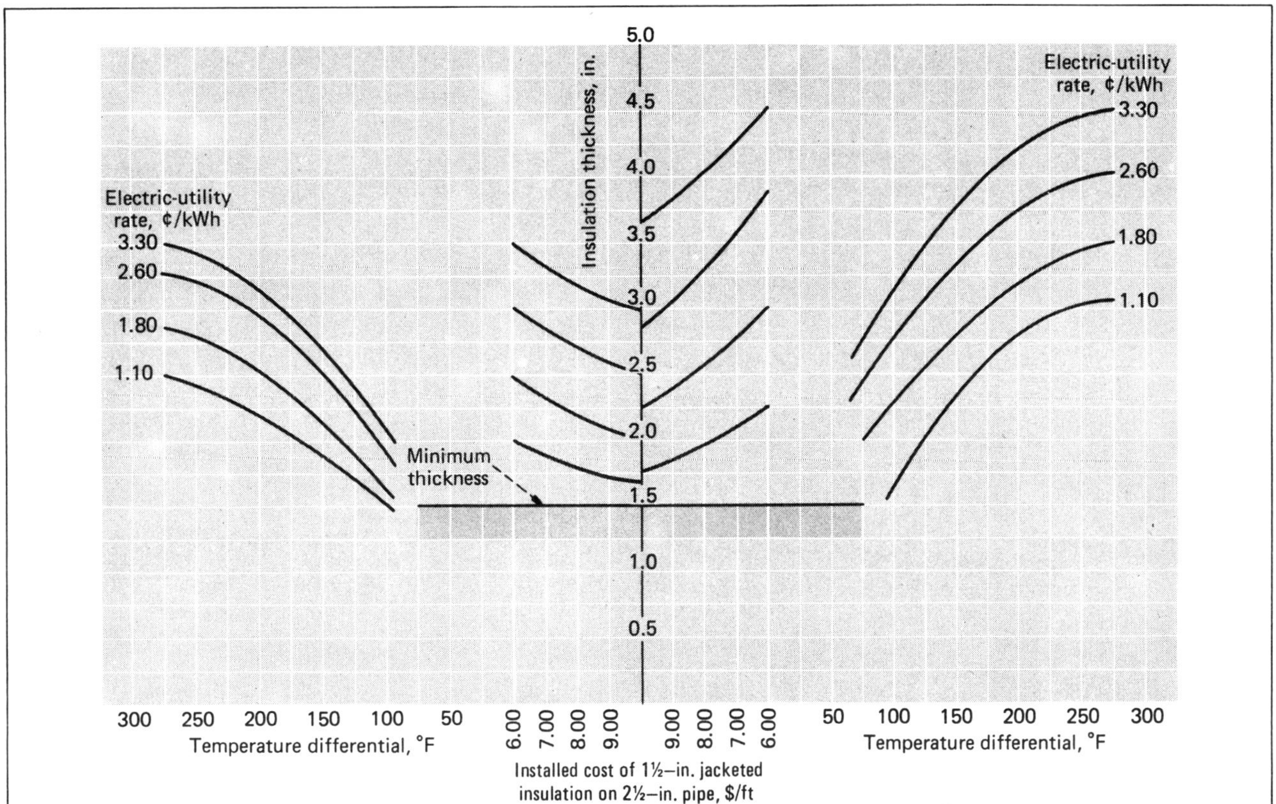

For 16-in. and 18-in. pipe—Fig. 7

Spreading this cost over 13 years (8,450 hours of operation per year) results in a capital cost ratio of:

$$\left(\frac{\$1{,}200/\text{ton}}{8{,}450\ \text{h/yr} \times 13\ \text{yr}}\right)\left(\frac{10^6}{12{,}000\ \text{Btu/h/ton}}\right) =$$

$$\$0.91/10^6\ \text{Btu, or}\ 91¢/10^6\ \text{Btu}$$

The foregoing values are subject to frequent fluctuation, and adjustment procedures are presented later.

Basis and use of the nomographs

Fig. 1 through 7 represent graphic solutions of the economic thickness of insulation plotted against temperature differential. Two arrays of curves are included in each figure, one for urethane, the other for foamglass, insulation. For each material, parameters of electrical rates are provided. Capital equipment cost does not appear separately; instead, its equivalent of 91¢/10^6 Btu has been incorporated into the stated electric rates.

The following information must be known to use the nomographs: pipe size, in.; temperature difference, °F, between the pipe fluid and the average ambient; electric utility rate, ¢/kWh, projected to the midlife of the plant; insulation material; and installed cost.

For the sake of simplicity, installed cost is based on one thickness, 1½ in., of jacketed insulation applied to a single pipe size, 2½ in. Only this installed cost is needed, regardless of the actual pipe size or insulation thickness.

Common to all the nomographs is a minimum insulation thickness of 1½ in. (Several manufacturers have indicated they are discontinuing making insulation thinner and 1½ in. because of dwindling interest.)

Fig. 8 and 9 represent graphic solutions for the insulation thickness necessary to limit surface condensation. The temperature determinants here are fluid temperature (horizontal scale) and atmospheric dewpoint at design conditions. Two dewpoints are shown (75°F and 83°F); interpolation between them is linear. (In our work, "5% values"—the atmospheric conditions that are exceeded only 5% of the cooling-season hours—are used as design points.) A 90°F dry-bulb temperature is used for setting the heat-transfer rate from the pipe outward.

The thicker insulation required by the two criteria should be installed.

Adjusting for variations from basis

Adjustments are likely when (1) a derived capital equipment cost factor other than 91¢/10^6-Btu output is necessary, all other variables remaining constant; and (2) derived capital cost and energy-consumption rate differ from the stated base.

A capital cost ratio of 91¢/10^6-Btu net output of refrigeration was calculated for an equipment service life of 13 years. At a power consumption rate of 167 kWh/10^6 Btu, this cost can be expressed as an equivalent "energy surcharge" by:

$$\left(\frac{91¢}{10^6\ \text{Btu}}\right) \div \left(\frac{167\ \text{kWh}}{10^6\ \text{Btu}}\right) = 0.55\ ¢/\text{kWh}$$

(By analogy, the capital expense for a jet costing $500,000, having a service life of 2 million miles, and consuming 2 gallons of fuel per mile would be converted into an equivalent fuel surcharge as follows:

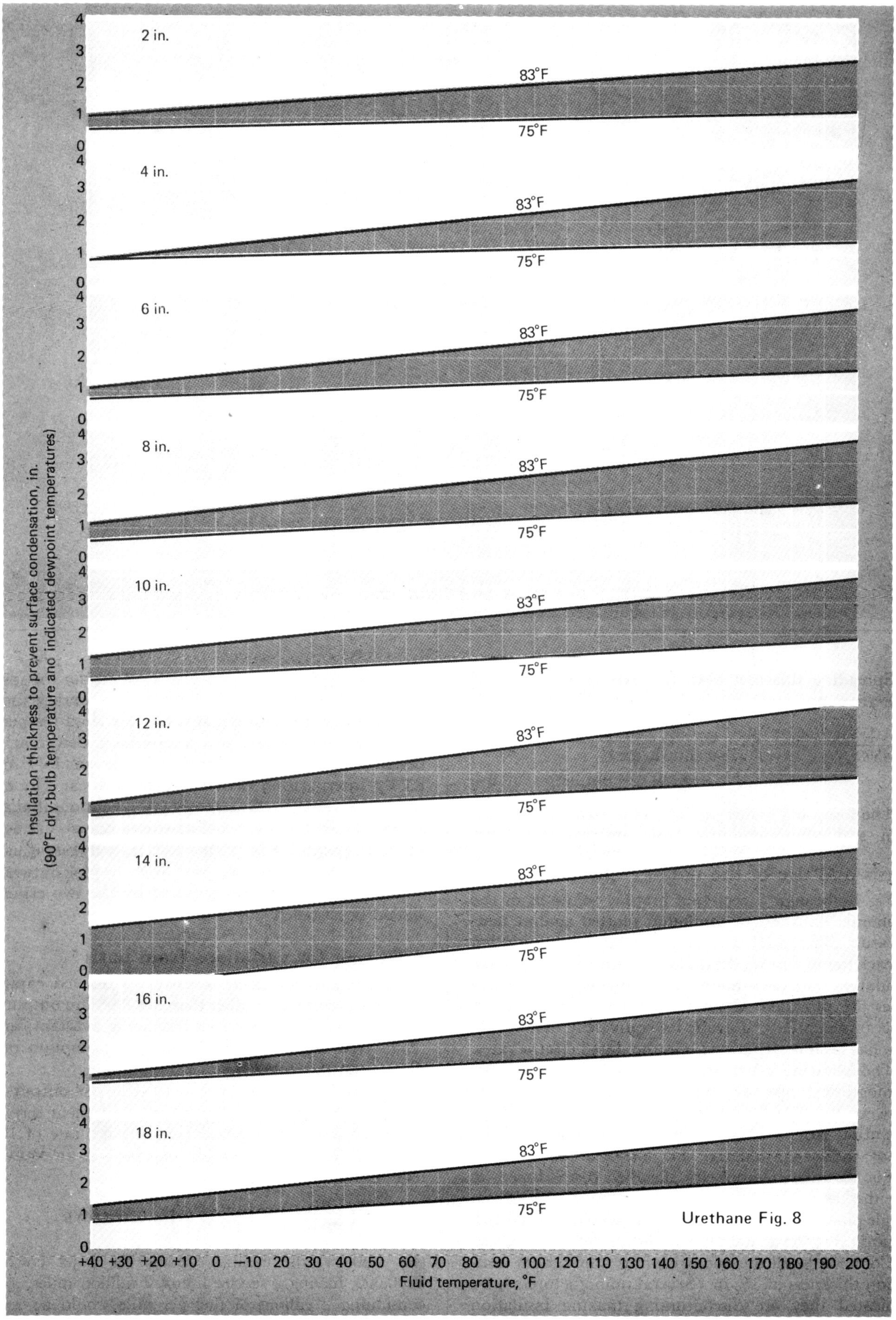

Urethane Fig. 8

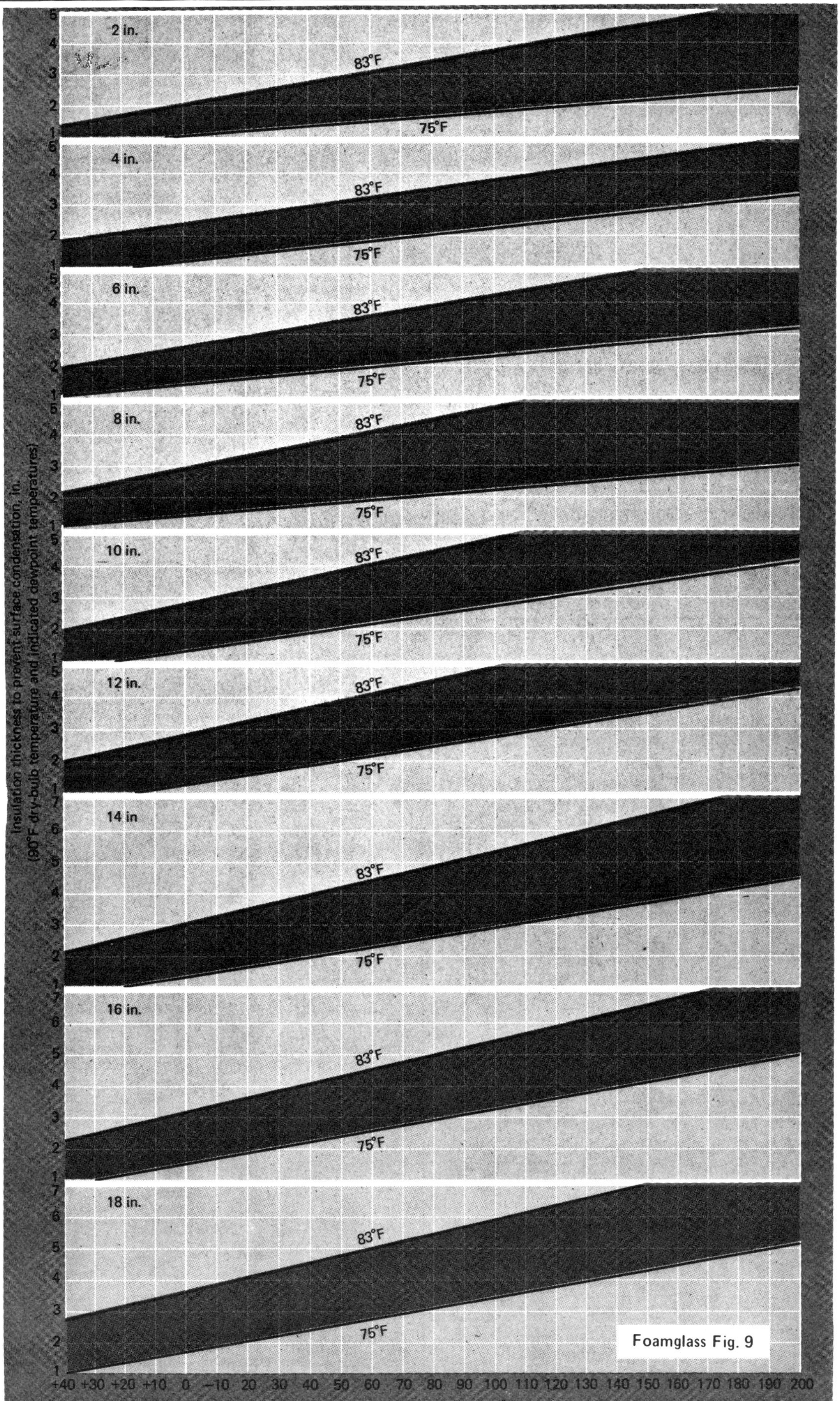

Foamglass Fig. 9

Amortization factors for Eq. (1) to solve cost of equipment

Amortization period	Interest rate, %												
	6.0	6.5	7.0	7.5	8.0	8.5	9.0	9.5	10.0	10.5	11.0	11.5	12.0
1	1.060	1.065	1.070	1.075	1.080	1.085	1.090	1.095	1.100	1.105	1.110	1.115	1.120
2	0.5454	0.5493	0.5531	0.5569	0.5608	0.5646	0.5685	0.5723	0.5762	0.5801	0.5839	0.5878	0.5917
3	0.3741	0.3776	0.3811	0.3845	0.3880	0.3915	0.3951	0.3986	0.4021	0.4057	0.4092	0.4128	0.4163
4	0.2886	0.2919	0.2952	0.2986	0.3019	0.3053	0.3087	0.3121	0.3155	0.3189	0.3223	0.3258	0.3292
5	0.2374	0.2406	0.2439	0.2472	0.2505	0.2538	0.2571	0.2604	0.2638	0.2672	0.2706	0.2740	0.2774
6	0.2034	0.2066	0.2098	0.2130	0.2163	0.2263	0.2229	0.2263	0.2296	0.2330	0.2364	0.2398	0.2432
7	0.1791	0.1823	0.1856	0.1888	0.1921	0.1954	0.1987	0.2020	0.2054	0.2088	0.2122	0.2157	0.2191
8	0.1610	0.1642	0.1675	0.1707	0.1740	0.1773	0.1807	0.1840	0.1874	0.1909	0.1943	0.1978	0.2013
9	0.1470	0.1502	0.1535	0.1568	0.1601	0.1634	0.1668	0.1702	0.1736	0.1771	0.1806	0.1841	0.1877
10	0.1359	0.1391	0.1424	0.1457	0.1490	0.1524	0.1558	0.1593	0.1627	0.1663	0.1698	0.1733	0.1770
11	0.1268	0.1301	0.1334	0.1367	0.1401	0.1435	0.1469	0.1504	0.1540	0.1575	0.1611	0.1648	0.1684
12	0.1193	0.1226	0.1259	0.1293	0.1327	0.1362	0.1397	0.1432	0.1468	0.1504	0.1540	0.1577	0.1614
13	0.1130	0.1163	0.1197	0.1231	0.1265	0.1300	0.1336	0.1372	0.1407	0.1444	0.1482	0.1519	0.1557
14	0.1076	0.1109	0.1143	0.1178	0.1213	0.1248	0.1284	0.1321	0.1357	0.1395	0.1432	0.1470	0.1509
15	0.1030	0.1064	0.1098	0.1133	0.1169	0.1204	0.1241	0.1277	0.1315	0.1352	0.1391	0.1429	0.1468
16	0.0990	0.1024	0.1059	0.1094	0.1130	0.1166	0.1203	0.1240	0.1278	0.1316	0.1355	0.1394	0.1434
17	0.0954	0.0989	0.1024	0.1060	0.1096	0.1133	0.1170	0.1208	0.1247	0.1285	0.1325	0.1364	0.1405
18	0.0924	0.0959	0.0994	0.1030	0.1067	0.1104	0.1142	0.1180	0.1219	0.1259	0.1298	0.1339	0.1379
19	0.0896	0.0932	0.0968	0.1004	0.1041	0.1079	0.1117	0.1156	0.1195	0.1235	0.1276	0.1316	0.1358
20	0.0872	0.0908	0.0944	0.0981	0.1019	0.1057	0.1095	0.1135	0.1175	0.1215	0.1256	0.1297	0.1339

$\$5 \times 10^5/2 \times 10^6$ miles ÷ 2 gal/mile = \$0.125/gal.)

For example, if the equipment capital-cost factor were \$1.50/10⁶ Btu instead of 91¢/10⁶ Btu, the equivalent "energy surcharge" would be:

$$\left(\frac{150¢}{10^6\text{ Btu}}\right) \div \left(\frac{167\text{ kWh}}{10^6\text{ Btu}}\right) = 0.90¢/\text{kWh}$$

The 0.35¢/kWh *difference* between 0.90¢/kWh and 0.55¢/kWh is *added* to the applicable electric-utility rate when using the nomographs.

In Case 2, the adjustment mechanism is more involved: an adjusted electric utility rate is derived, and the nomographs are entered with this adjusted rate.

When the derived capital cost (including the financing) and energy consumption rate differ from the reference installation, the first step is to calculate a realistic capital-cost ratio, in \$/10^6 Btu:

$$S_b = (83.3\,S/H)(A + M) = \$/10^6\text{ Btu} \qquad (1)$$

In Eq. (1), S = the cost of the equipment and a proportionate part of the cost of the auxiliaries, \$/ton; H = projected hours of operation per year; A = the amortization factor, which may be calculated with the formula $A = i/1 - (1 + i)^{-n}$ (in which i is the interest rate, and n is the number of payment periods), or may be found in the table above; and M = maintenance as an annual percentage of S.

The second step is to determine the adjusted electric utility rate, in ¢/kWh, for entering the nomographs:

$$R_{adj} = \left[R + \left(\frac{S_b - 0.91}{1.67}\right)\right]\frac{E}{167} = ¢/\text{kWh} \qquad (2)$$

In Eq. (2), R = any projected electric-utility rate, ¢/kWh; S_b = the capital-cost ratio derived with Eq. (1), \$/10^6 Btu; and E = any total-energy-consumption rate, kWh/10^6 Btu.

Example illustrates second case

Assume the following data:

Energy consumption rate, kWh/10⁶ Btu	190
Projected life of system, yr	11
Projected annual service, h	7,800
Projected electric rate, ¢/kWh	1.95
Capital equipment cost, \$/ton	750
Maintenance, %/yr	2
Cost of capital, %/yr	10.5

The adjusted capital-cost ratio, calculated with Eq (1), is \$1.42/10^6 Btu. The amortization factor, A, is 0.1575.

The adjusted electric-utility rate, determined with Eq (2), is 2.56¢/kWh. Enter this in nomographs.

Mechanically driven refrigeration units—The procedures discussed are not applicable to systems using mechanically driven units. Of course, the condensation criteria are the same whether the unit is electrically or mechanically driven, so Fig. 8 and 9 can be used for either.

The author

Jules L. Abramovitz is a senior engineer with Coaltek Associates, a partnership of Allied Chemical Coal Technology Corp. and an affiliate of Salem Corp. (P.O. Box 1013R, Morristown, NJ 07690). He holds a B.S. in mechanical engineering from Newark College of Engineering and an M.S. in mechanical engineering from Fairleigh Dickinson University.

Installed cost of a distillation column

If you are weighing the benefits of several distillation-design options, the following cost data and time-saving calculation form will streamline your evaluation.

Photo: Fluor Engineering and Construction, Inc.

J. S. Miller and W. A. Kapella, Union Carbide Corp.

☐ To estimate an installed cost for a new distillation column, the designer must carefully consider the costs of items such as piping and field labor, as well as the cost of the bare column. Given here are data that will let you make a reasonably accurate estimate (within 25%) for a completely installed column if you have fixed the dimensions; number of trays; materials of construction; pressure and temperature (or shell thickness); site location; and fabrication schedule.

The result is the cost of a column that is instrumented, piped to the next item of equipment, insulated, and ready to run. We have not included costs for the ancillary pumps, tanks and heat exchangers that, although part of the distillation system or "module," are usually optimized separately (Fig. 1).

Shell and skirt costs

The column shell is estimated in Fig. 2 as a function of diameter, height and thickness. This figure represents Union Carbide Corp.'s actual purchasing experience, rather than market price surveys. The product of diameter, height and thickness (DHT) in in.^{3}, is very nearly proportional to the volume of metal in the shell, and therefore is proportional to the column weight—the traditional cost-correlation basis for cylindrical vessels.

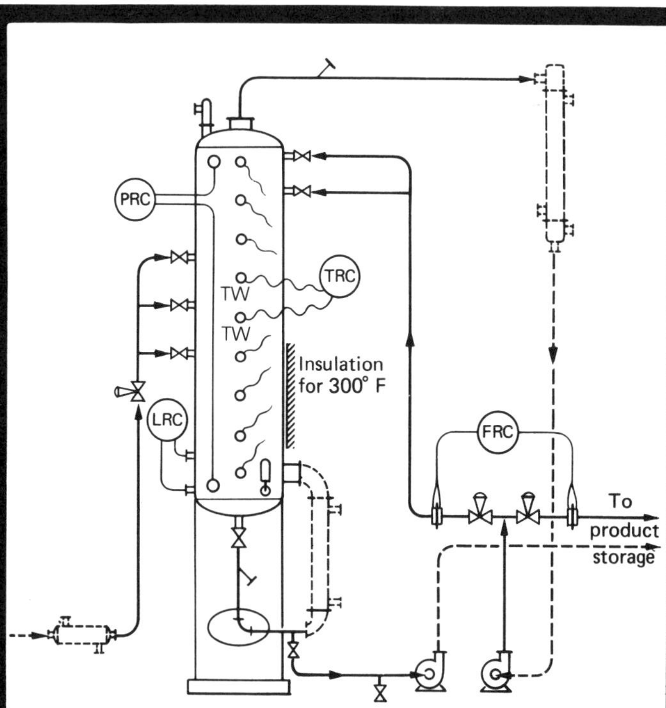

Equipment included in estimate of distillation-column installed costs. Fig. 1

Originally published April 1977.

To keep things simple, the height of the column includes the skirt, even though it may be made of a different material. Most skirts are 10–15 ft long and are thicker than the column shell. The thickness of the shell proper can be taken from Fig. 3 or Fig. 4 for either pressure or vacuum service; the latter figure, for vacuum service, should not be extrapolated. Depending on the chemical behavior of the system, an allowance for corrosion may be necessary. Also, if full radiography and stress relief are specified, these items can each add about 5% to the cost.

For the example shown in Box 1, an estimate has been made for the cost of a Type 316L stainless-steel column of the following design: 8 ft dia. x 60 ft overall height x 0.75 in. shell thickness; 30 sieve trays; full radiography required. With a *DHT* of 51,840, a Type 316L shell would cost \$64,100, and a carbon steel shell would cost \$25,700 at Index 100.

(As we will see later, it was also necessary to estimate the cost of a carbon steel column of similar dimensions (Box 2). We have assumed that both the Type 316L shell and the carbon steel shell are 0.75 in. thick, although there are circum-

Installed cost of Type 316L stainless-steel distillation column **Box 1**

(1) *DHT* = 8 ft X 60 ft X 0.75 in X 144 = 51,840
(See Fig. 3 and Fig. 4 for values of *T*)

(2) Cost of shell (Index = 100; Fig. 2):
Actual material \$ 61,000 X 1.05 (Radiography or stress relief) = \$ 64,100
Carbon steel \$ 24,500 X 1.05 (Radiography or stress relief) = \$ 25,700

(3) Cost of trays (Index = 100; Fig. 4): \$ 4,000 setup +
Actual material 30 X \$ 870 each X 1.1 Adjustment factor = \$ 32,700
Steel 30 X \$ 440 each X 1.1 Adjustment factor = \$ 18,500

(4) Tray installation (Index = 100; Fig. 5): 30 X \$ 92 each = \$ 2,800

(5) Total cost of column and trays: Actual material = \$ 99,600
Carbon steel = \$ 47,000
Ratio for use in Fig. 7 = 2.12

(6) Bulk-material cost (Index = 100; Fig. 7):
\$ 99,600 Column cost (actual material) X 0.68 = \$ 67,700

(7) Labor manhours, Fig. 7:
\$ 99,600 Column cost (actual material) X 0.066
X 1.30 Multiplier = 8,550 manhours

(8) Indexed material cost:
[\$ 99,600 (5) + \$ 67,700 (6)] X 2.50 Index = \$ 418,000

(9) Labor cost: 8,550 Manhours (7) X \$ 12 Wage rate = \$ 103,000

(10) Construction overhead: \$ 103,000 (9) X 80 % = \$ 82,000

(11) Subtotal = \$ 603,000

(12) Engineering: \$ 420,000 (11) X 20 % = \$ 84,000 *

(13) Subtotal = \$ 687,000

(14) Contingency \$ 687,000 (13) X 25 % = \$ 172,000

(15) Total Installed cost = \$ 859,000

*Same engineering as for a carbon steel column

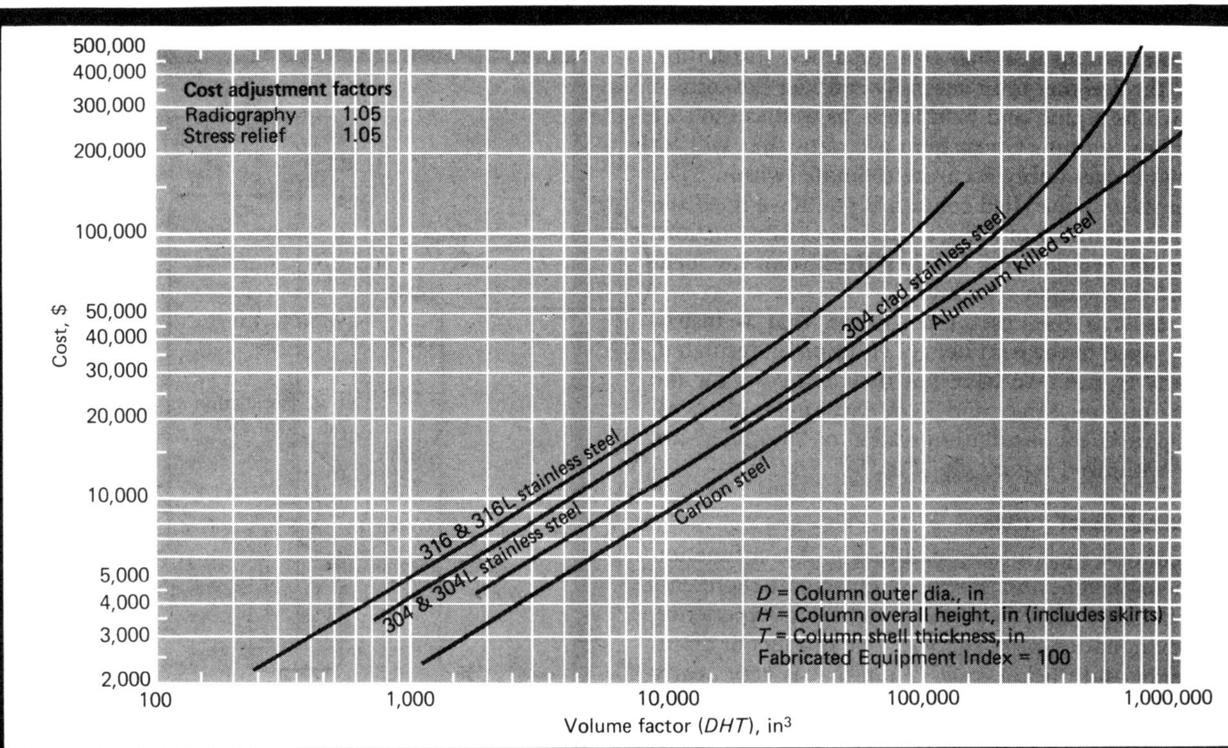

Cost of distillation-column shell as a function of dimensions. **Fig. 2**

Installed cost of carbon-steel distillation column — **Box 2**

(1) DHT = 8 ft X 60 ft X 0.75 in X 144 = 51,840
(See Fig. 3 and Fig. 4 for values of T)

(2) Cost of shell (Index = 100; Fig. 2):
Actual material $____ X ____ (Radiography or stress relief) = $____
Carbon steel $24,500 X 1.05 (Radiography or stress relief) = $25,700

(3) Cost of trays (Index = 100, Fig. 5): $4,000 setup +
Actual material ____ X $ ____ each X ____ Adjustment factor = $____
Carbon steel 30 X $ 440 each X 1.1 Adjustment factor = $18,500

(4) Tray installation (Index = 100; Fig. 6): 30 X $ 92 each = $2,800

(5) Total cost of column and trays: Actual material = $____
Carbon steel = $47,000
Ratio for use in Fig. 6 = 1.00

(6) Bulk-material cost (Index = 100; Fig. 7):
$47,000 Column cost (actual material) X 1.0 = $47,000

(7) Labor manhours, Fig. 7:
$47,000 Column cost (actual material) X 0.14
X 1.30 Multiplier = 8,550 manhours

(8) Indexed material cost:
[$47,000 (5) + $47,000 (6)] X 2.50 Index = $235,000

(9) Labor cost: 8,550 Manhours (7) X $12 Wage rate = $103,000

(10) Construction overhead: $103,000 (9) X 80 % = $82,000

(11) Subtotal = $420,000

(12) Engineering: $420,000 (11) X 20 % = $84,000*

(13) Subtotal = $504,000

(14) Contingency $504,000 (13) X 25 % = $126,000

(15) Total Installed cost = $630,000

*Use this figure for Type 316L column also

stances, especially at elevated temperatures, where they may be different.)

Index 100 refers to the Fabricated Equipment Index published in CHEMICAL ENGINEERING. Its basis is 1957–1959 = 100. You will have to forecast the value for the date that your column is to be delivered. We arbitrarily chose 250 for our example. CHEMICAL ENGINEERING reports a value of 205.9 for Oct., 1976, so 250 would probably correspond to some time in 1978.

Tray installation

The cost of trays (Index 100) comes from Fig. 5, with appropriate allowances for the manufacturer's setup charge and the penalty for having a small order of less than 40 trays. When carbon steel is used, the trays in this example run 70% of the shell cost, but only 50% when the column is constructed of Type 316L. This difference obviously reflects the greater proportion of shop-labor cost in making the carbon steel trays.

Fig. 5 applies well enough to any common type of tray: sieve, valve, or bubble cap. Our experience shows that a majority of new trays are valve trays. Very few bubble-cap trays are specified for new columns these days.

Installing the trays in a new column is nearly always done in the vendor's shop. This cost is estimated in Fig. 6. Field installation requires 25% more labor, and there is a greater chance of damage.

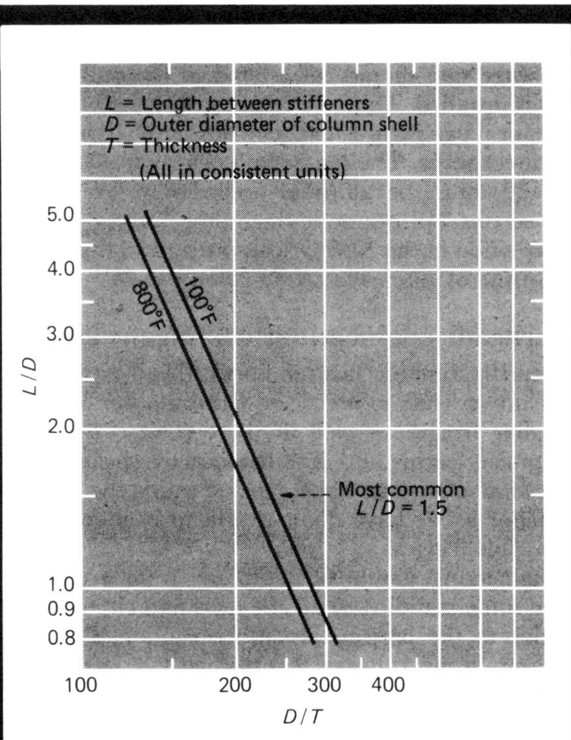

Thickness required for vacuum service: carbon and stainless steel cylindrical vessels **Fig. 3**

Bulk material and field labor

There is more than one way to estimate the cost of bulk material (Table I) and field labor. Ideally, estimates should be made from completed civil, mechanical, electrical and control-systems designs. One of the primary objectives of this article, however, is to avoid the need for those designs by providing shortcut estimating factors. These factors for bulk material and field

Breakdown of bulk material (assuming "all steel" construction) — **Table I**

Item	% of total
Piping	40.0
Instruments	24.0
Electrical	8.3
Structures	7.0
Insulation	6.7
Concrete	5.7
Fire protection	3.3
Misc. material	3.3
Painting	1.7
	100.0

Thickness correction factors

Temp., °F	Carbon steel	Type 304	Type 304 L	Type 316	Type 316 L
Ambient	1.00	0.70	0.80	0.70	0.80
100	1.00	1.00	1.00	1.00	1.00
200	1.00	1.06	1.00	1.00	1.00
400	1.00	1.14	1.07	1.04	1.00
600	1.00	1.18	1.12	1.10	1.16
700	1.04	1.18	1.16	1.15	1.22
750	1.14	1.20	1.18	1.17	1.24
800	1.35	1.24	1.20	1.18	1.27

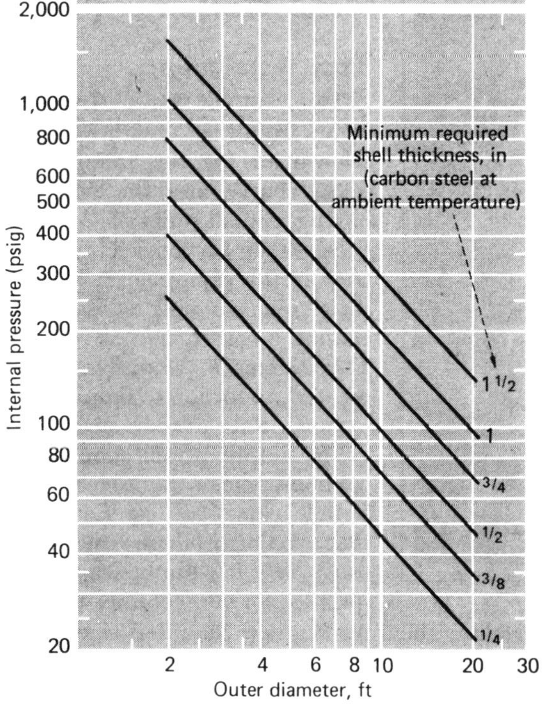

Thickness required for internal pressure and operating temperature **Fig. 4**

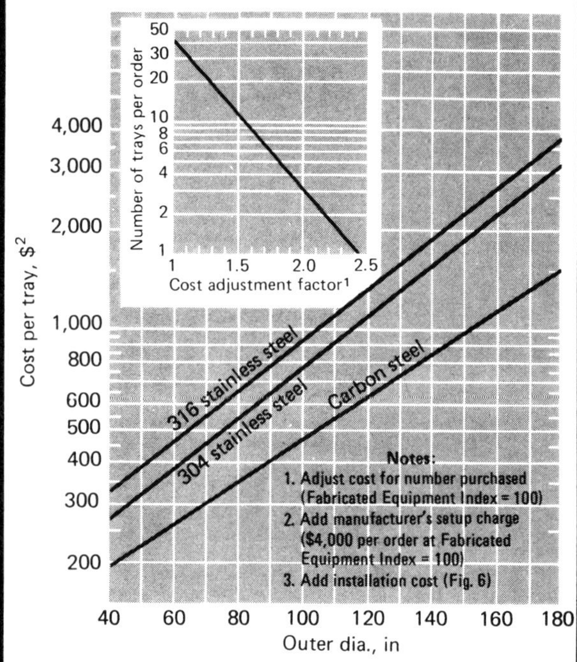

Material cost of distillation trays **Fig. 5**

labor must, of necessity, be related to the cost of the major equipment. The only real choice is whether to estimate the bulk-material components separately or as a whole. We have found that the correlation coefficient is significantly better when bulk materials are estimated as a whole. The same is true of field labor.

In Fig. 7, bulk-material cost and field-labor manhours are plotted against the ratio of total column cost (including internals) for the actual material to that for a carbon steel column. It is necessary to consider this ratio of shell material costs because bulk material and field labor are progressively smaller multiples of column cost as the column becomes increasingly "noble." This is because all bulk materials remain essentially the same, regardless of column material, except for piping, which varies proportionately with column cost. Labor does not change significantly with column material.

Labor is expressed in manhours to simplify the application of wage rates and productivity factors. Our standard productivity factor is based on open-shop construction with less than 300 workers at the construction site. The table of multipliers in Fig. 7 indicates generally the decrease in productivity (increase in manhours) that occurs when the size of the field force is increased or when closed-shop labor is involved. These multipliers are applicable along the Texas-Louisiana Gulf Coast. As a rule, productivity is not as high elsewhere. The examples use a productivity multiplier of 1.3 without reference to any particular labor situation.

In the sample calculations, Step 9 is the application of the wage rate to the manhours estimated from Fig. 7. The wage rate should correspond to the time the column is installed. This rate can be either the bare direct wage, or it can include fringes—normally about 20% of the direct wage. The factor for construction overhead, Step 10, must be adjusted accordingly. We roughly derived our $12 per hour (including fringes) for 1978 construction in the New Orleans area from the Dec. 23, 1976 issue of *Engineering News Record* (p. 74).

Overhead; engineering; contingencies

Step 10 estimates construction overhead—also known as field overhead, indirects, or distributables. Including all labor fringes, as well as the cost of construction-equipment leasing and maintenance, overhead will run somewhat more than direct labor. Putting the fringes in the labor as we have done, overhead is about 80% of labor plus fringes.

Engineering should be estimated from a clear, detailed understanding of all the facts and circumstances relative to the project size and execution plan. This, however, is not always feasible with a factored approach. For factored estimating, we find it practical, although not theoretically valid, to express engineering as a function of the sum of all other costs. For a project that includes only a distillation column (as in our example), engineering could be 20 to 25% of the sum of

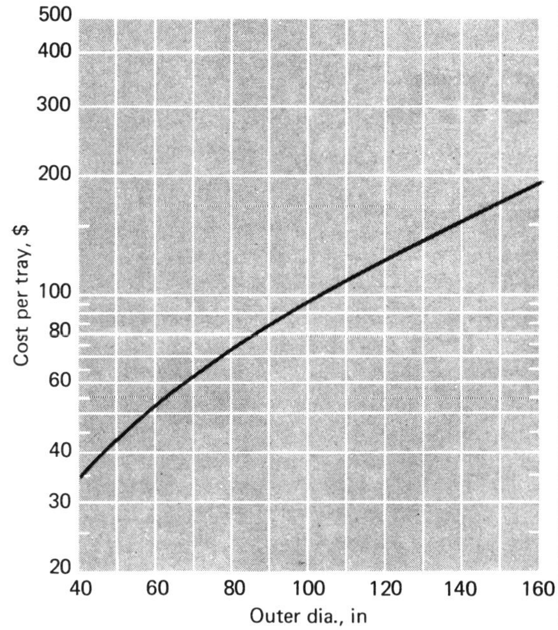

Cost of shop installation of trays (Fabricated Equipment Index = 100) Fig. 6

other costs, or 16 to 20% of the total cost excluding contingency.

Our definition of engineering includes activities in addition to process design and detailed design and drafting. We include project management, cost engineering, planning and scheduling, purchasing, and vendor expediting. There can also be a duplication of engineering supervision effort when contract engineering firms are employed.

It is probably better to base engineering on a carbon steel system and then assume that the same figure would apply to a system of any other material. For our example, therefore, we calculated the cost of a steel column system on the back of an envelope. We used engineering at 20%, which resulted in an engineering cost of $84,000. This seems like a reasonable figure for the Type 316L system as well.

Every estimate needs a contingency allowance to ensure that there is enough money in the estimate. Our standard factor is 25% for estimates made before engineering is complete. It is easy to use up that amount in design changes alone, not to mention unpredictable outside influences. Management gives no points for underestimating.

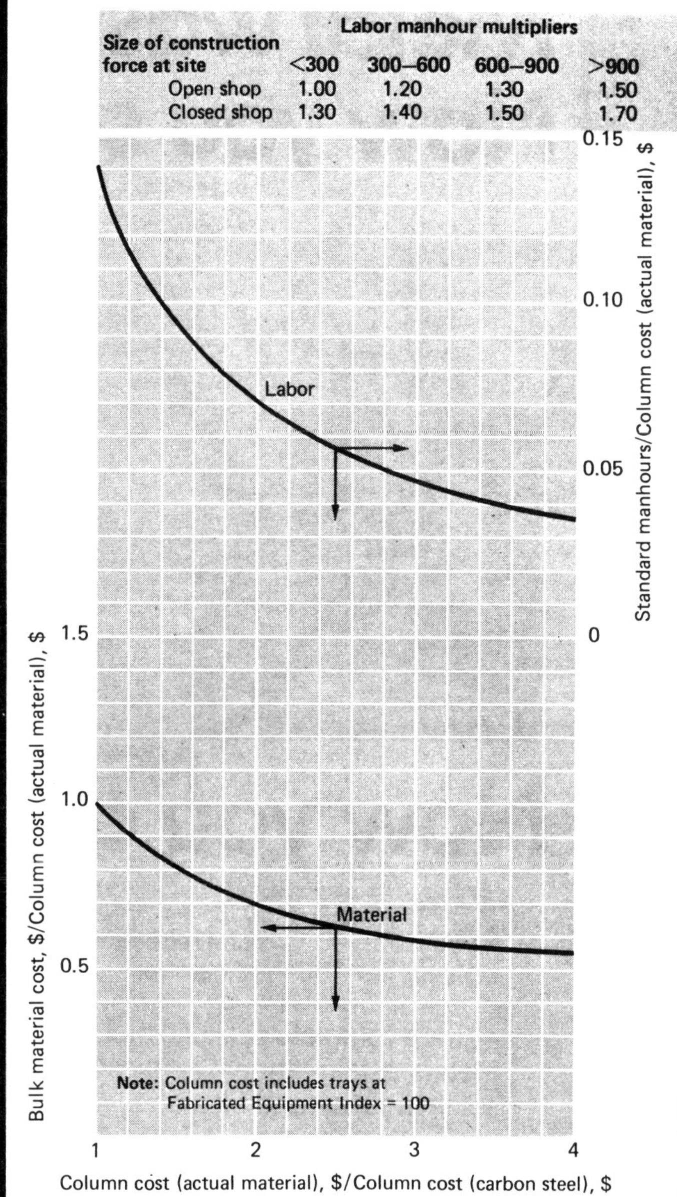

Bulk-material cost and field-labor manhours Fig. 7

The authors

J. S. Miller is a Staff Engineer, Cost Engineering Technology, Engineering Dept., Chemicals and Plastics Div., Union Carbide Corp., P.O. Box 8361, South Charleston, WV 25303, where he is concerned with developing capital-cost estimating data and methods. During his career at Union Carbide, he has been involved in production, design engineering and business analysis. He holds a B.S. in chemical engineering from West Virginia University.

W. A. Kapella is Senior Cost Engineer, Engineering Dept., Chemicals and Plastics Div., Union Carbide Corp., where he is responsible for preparing capital-cost estimates for chemical process plants. He has worked as a design engineer and as a project engineer. A member of AIChE, he holds a B.S. in chemical engineering from Purdue University and an M.S. in business administration from Northern Illinois University. He is registered as a professional engineer in West Virginia.

Cost estimation for turbine agitators

Power and speed are the primary factors affecting the cost of turbine agitators. But shaft seals, number of impellers and materials of construction also have a dominant effect on the estimated cost of complete systems.

William S. Meyer and Donald L. Kime, Chemineer, Inc.

☐ Cost estimation begins with those turbine agitators that meet acceptable process criteria and have adequate mechanical design.

Acceptable process criteria require turbine agitators to produce a scale of agitation consistent with the required process result, size, and difficulty of the agitation problem. Adequate mechanical design means that prime-mover power and shaft speed, impeller-system design, and the shaft and seal are mechanically sound and mutually compatible.

The cost-estimation logic that we will follow is shown in Fig. 1. The objective of this logic is to determine the turbine agitator's comparative present value (*CPV*). This will then be used as a measure of the agitator's total cost.

In order to obtain a realistic cost estimate for turbine agitators, we have prepared charts, shown as Fig. 2 and 3, for agitators having wetted parts made of carbon steel and Type 316 stainless steel, respectively. These cost data are useful for determining preliminary estimates for turbine agitators.

The charts for Fig. 2 and 3 are based on complete units that include motor, drive, shaft, stuffing-box shaft seal, and turbine impellers. The electric motor is a totally enclosed fan-cooled unit (TEFC) of the appropriate horsepower, a drive unit to meet the required torque and speed, and a stuffing-box shaft seal. The cost data cover a top-entering agitator configuration, as shown in Fig. 4d of Part 7. This is the most common top-entering arrangement when any type of shaft seal is required. In both Fig. 2 and Fig. 3, the data also reflect complete agitator units having a single turbine or dual turbines, respectively, of the pitched-blade type, and an appropriate shaft diameter and shaft length.

Although turbine agitators in excess of 1,000 hp have been installed, our data are restricted to horsepowers of 450 and less. These are the most frequently occurring applications of such agitators.

When either a mechanical shaft seal or no shaft seal is

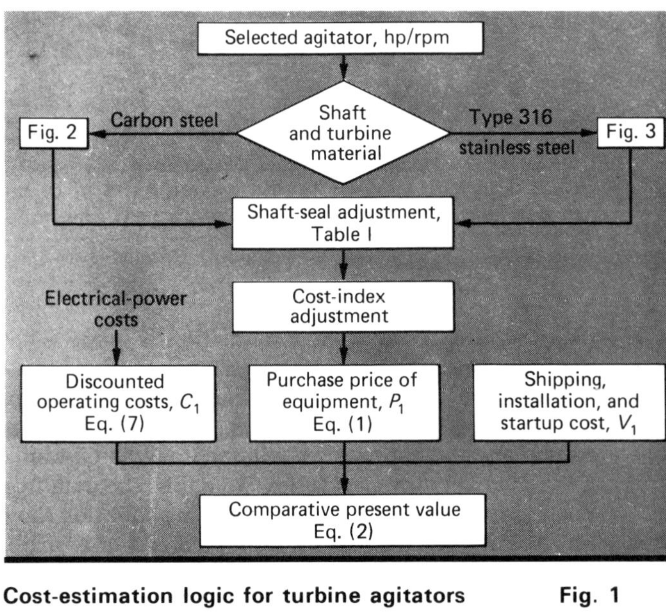

Cost-estimation logic for turbine agitators **Fig. 1**

Originally published September 1976.

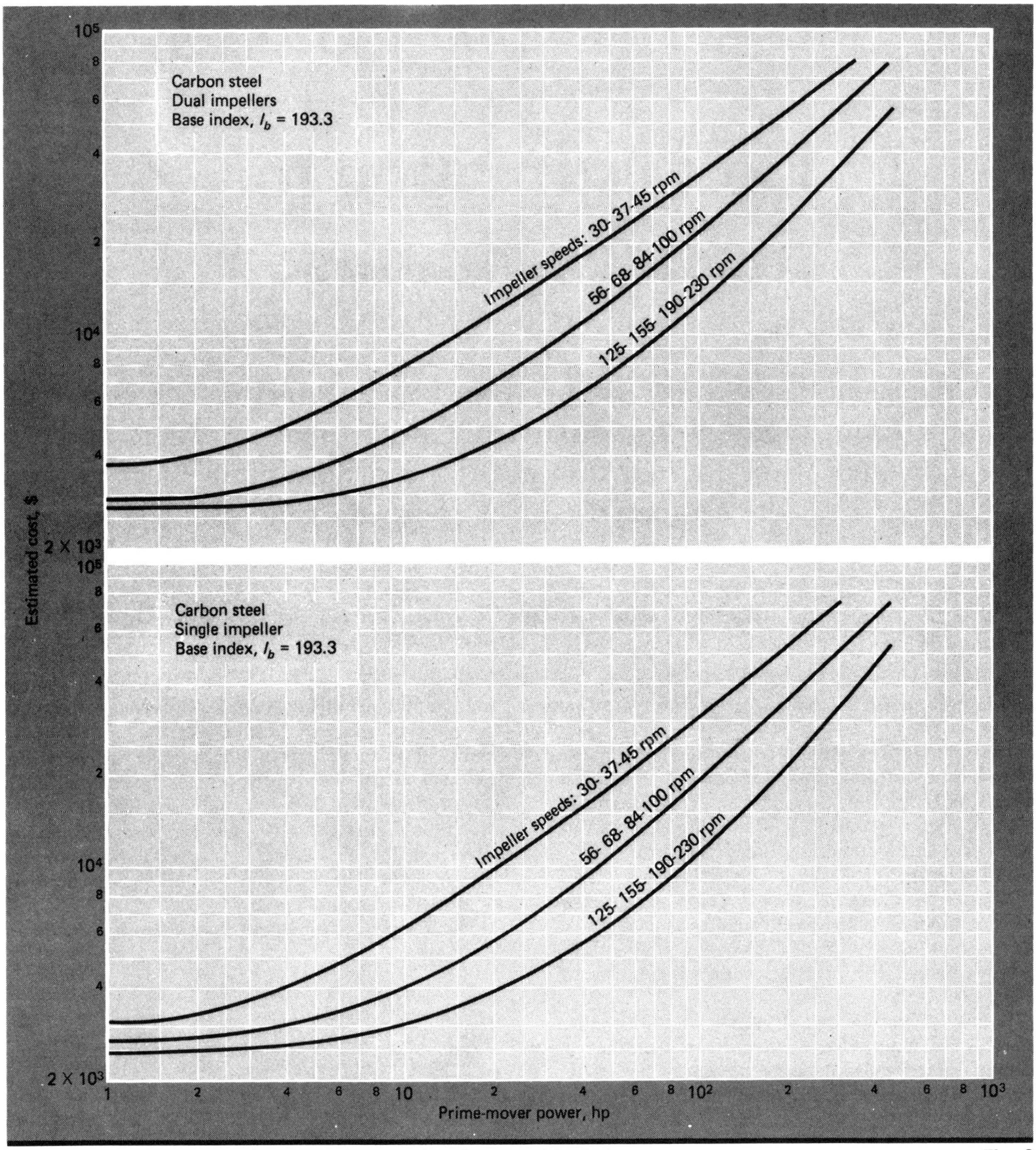

Cost data for agitators having wetted parts of carbon steel include motor, drive, shaft, stuffing-box shaft seal and impellers **Fig. 2**

required, a cost addition or deduction is made. These cost adjustments are summarized in Table I.

In order to be useful, the estimated cost of a turbine agitator must be current. We will use the "equipment, machinery and supports" component of the CHEMICAL ENGINEERING Plant Cost Index* to adjust the data from Fig. 2, Fig. 3 and Table I. The following equation shows how to compute the current estimated cost of a turbine agitator by use of the index numbers:

$$P_1 = P_b(I_1/I_b) = P_b(I_1/193.3) \quad (1)$$

where P_1 = current estimated cost of equipment, \$; P_b = equipment cost for base period (i.e., I_b = 193.3), \$; I_1 = current value of the equipment, machinery and supports component of the Plant Cost Index.

Comparative present value

The comparative present value (*CPV*) of a turbine agitator is defined as: the sum of the estimated cost of the agitator, P_1; discounted operating costs due to elec-

*This component appears in every issue of CHEMICAL ENGINEERING under the heading of Economic Indicators.

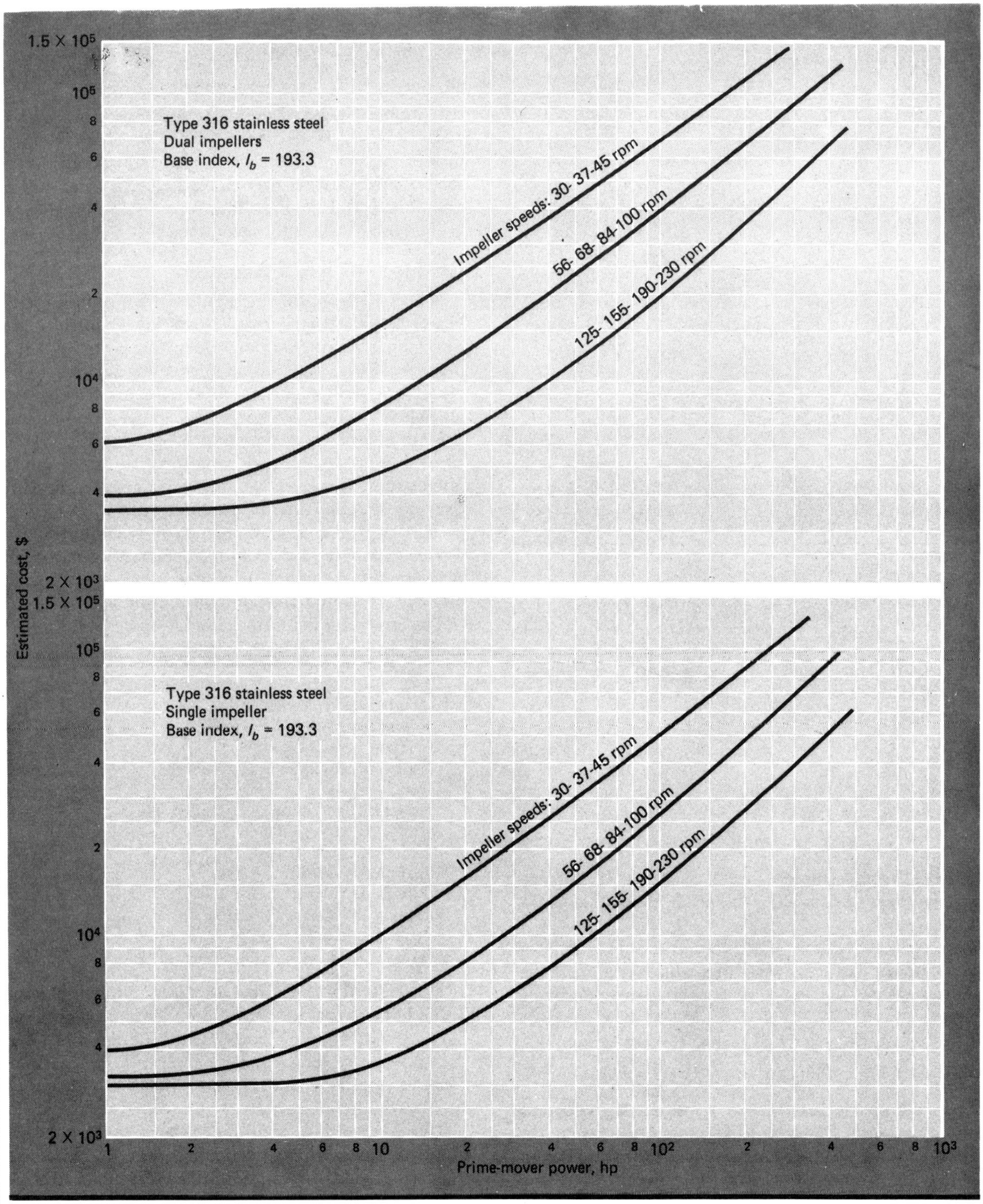

Cost data for agitators having wetted parts of stainless steel include motor, drive, shaft, stuffing-box shaft seal and impellers **Fig. 3**

tric prime-mover power, C_1; and shipping, installation and startup costs, V_1:

$$(CPV) = C_1 + P_1 + V_1 \quad (2)$$

Defined in this manner, the comparative present value is not the true total cost for the agitation equipment. However, it provides a comparative measure of cost so that final economic equipment selection can be made.

Over 95% of the prime movers used for turbine agitators are electric motors, and the principal cost of operating the agitators is the cost of electric power. We will use electric-power costs as a measure of operating costs.

To examine these electric-power costs, we will compute the present worth of the yearly power costs, C_1, that will be required to operate the agitator over a period of T years:

$$C_1 = F_c S_{cf} \quad (3)$$

where F_c is the discount factor for cash flow that will occur uniformly over a period of T years due to prime-mover power costs, and S_{cf} is the total cash flow for the T-year period due to power costs.

Adjustment of cost when stuffing box is omitted or replaced by mechanical seal — Table I

Prime-mover power, hp	Change in cost: Stuffing box omitted, Subtract, $		Stuffing box replaced by mechanical seal, Add, $	
	Carbon steel	Stainless steel	Carbon steel	Stainless steel
1 to 2	400	600	1,800	2,200
3 to 10	700	1,000	2,100	2,600
15 to 25	800	1,100	2,500	2,800
30 to 60	1,100	2,100	3,800	4,100
75 to 150	3,000	3,900	8,200	10,700
200 to 450	5,600	7,600	11,300	13,900

Basis: I_b = 193.3. This is the "equipment, machinery and supports" component of the CE plant cost index for May 1975.

The discount factor,* F_c, is expressed by:

$$F_c = (1/rT_y)(1 - e^{-rT_y}) \quad (4)$$

where T_y = total number of years, and r = nominal annual interest rate.

The total cash flow is expressed by:

$$S_{cf} = T_y Y_p \quad (5)$$

where Y_p = yearly cost of power, \$/yr.

The yearly cost of power can be calculated from:

$$Y_p = 0.745 H_p t_h C_e \quad (6)$$

where 0.745 represents kW/hp; H_p is motor horsepower, hp; t_h is hours of usage, h/yr; and C_e is cost of power, \$/kWh.

For comparative purposes, we will use the value for current power costs, knowing that as this rate increases in future years, the value of C_1 will correspondingly increase.

Combining Eq. (3), (4), (5) and (6), we obtain:

$$C_1 = 0.745(1/r)(1 - e^{-rT_y})H_p t_h C_e \quad (7)$$

For example, let us assume a 3-shift/day chemical plant, operating 7,884 h/yr in Dayton, Ohio, where the present cost of electrical power is 2.1¢/kWh. Let us calculate the present worth of the power for a 10-hp agitator over an 11-yr period at an interest rate of 12%.

The present worth of electric power over the 11-yr period is obtained by substituting into Eq. (7):

$$C_1 = 0.745(1/0.12)(1 - e^{-(0.12)(11)})(10) \times (7{,}884)(0.021)$$
$$C_1 = \$7{,}533$$

In this example, the present worth of electrical power is \$753/hp. The same calculation, done for typical electrical rates at other U.S. locations, could be as high as \$1,000/hp.

The third factor in Eq. (2) measures the influences of shipping, installation and startup costs. These costs are normally expected to be a fixed average percentage of the purchase price of agitators, and are usually about 10%. However, these costs may vary widely for different manufacturers, equipment configurations, and shaft-seal designs. A careful engineering analysis of turbine agitators for their installation and maintenance requirements is therefore recommended.

Example of cost estimation

A 5-hp/84-rpm turbine agitator with dual impellers has been found satisfactory from a mechanical design standpoint. The system on which this agitator will be used requires Type 316 stainless steel for the wetted parts (i.e., shaft and impellers), and a cartridge-type mechanical seal. It is anticipated that this agitator will operate 7,884 h/yr at present power costs of 2.1¢/kWh. The economic evaluation period will be 11 yr; annual interest rate will be 12%; and the current equipment-cost index, I_1, at time of purchase will be 210.7.

Let us estimate the cost of the complete unit for this service; calculate the discounted operating costs; and evaluate the shipping, installation and startup costs. Finally, let us find the comparative present value.

The several calculations and estimates for the 5-hp/84-rpm dual-turbine agitator are summarized as follows:

Estimated current cost	
Estimated cost (from Fig. 3) (Base period, I_b = 193.3)	\$6,500
Shaft-seal adjustment (Table I)	2,600
Total cost (base period)	9,100
Adjustment to current estimated cost, I_1/I_b = 210.7/193.3 = 1.09	
Estimated current cost, P_1, Eq. (1)	9,919
Discounted operating costs, C_1 [Use Eq. (7)]	3,765
Shipping, installation, startup, V_1 (Assumed at 10% of P_1)	992
Comparative present value from Eq. (2)	\$14,676

The cost data presented in Fig. 2 and Fig. 3 and Table I of this article are useful in determining preliminary estimates for the costs of turbine agitators. Furthermore, these data enable estimates to be made for complete agitator systems in either carbon-steel or stainless-steel construction of the wetted parts.

The charts of Fig. 2 and Fig. 3 also show how the cost of an agitator having a fixed horsepower decreases with increasing speed of rotation. The reason for this decrease in cost is that torque is lower at the higher speeds, and the machine becomes progressively smaller. In contrast, for a fixed rotative speed, the cost will increase with increasing horsepower.

The authors

William S. Meyer is administrative sales manager for the Agitator Div., Chemineer, Inc., P. O. Box 1123, Dayton, OH 45401. As manager of customer service, he has had extensive experience in the installation and field service of turbine agitators.

Donald L. Kime—For biography, see *Chem. Eng.*, Aug. 2, 1976, p. 94.

*Peters, M. S. and Timmerhaus, K. D., "Plant Design and Economics for Chemical Engineers," p. 179, McGraw-Hill, New York, 1968.

Nomograph updates process equipment costs

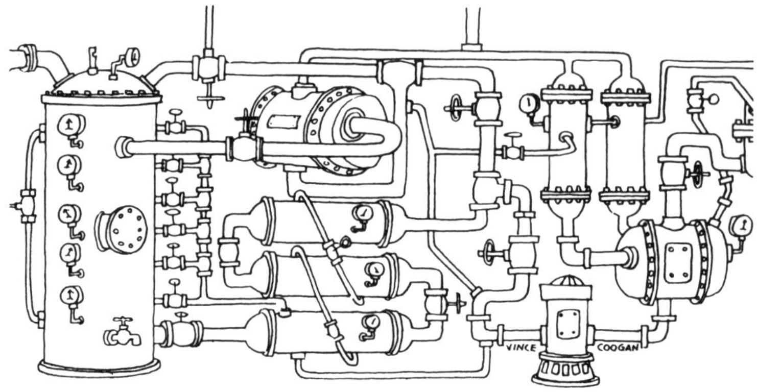

If your cost figures are out-of-date or for different size equipment, here's a nomograph that could solve either problem. It provides a quick, convenient means of estimating costs for the project you are working on now.

George M. Hoerner, Jr., Lafayette College

☐ Many chemical engineers gradually accumulate a substantial file of cost data for processes and process equipment as a result of involvement in a variety of projects.

Unfortunately, two problems generally arise regarding the subsequent use of such information in circumstances not directly related to past projects. First, the current project will frequently involve equipment of significantly different size or capacity. Second, inflation rapidly renders price data obsolete.

Generalized concepts of cost estimation

Faced with having to develop the price of a piece of process equipment or a capital-investment figure for an entire chemical plant, the project engineer has several alternatives. Depending on the time and funds available, he may choose from a spectrum of methods that extend from an order-of-magnitude approach (with a probable accuracy of ± 30%) to a detailed contractor's estimate (with an accuracy of ± 5%). The intermediate techniques, identified by a variety of names [*5*] and based on diverse assumptions, generally involve a tradeoff of expense and time against accuracy.

Given these bounds the engineer then calls upon available resources. These most frequently include individual files, corporate files or correlations, vendor quotations, and published information. As might be expected when sources are so diverse, data are frequently presented in incompatible form, which makes rapid and convenient use of the values by the engineer almost impossible.

Much work has been done in trying to overcome this problem [*2, 3, 4, 6*]. In general, these approaches attempt to compensate for size and time differences. Such methods are obviously much more useful if they also allow the engineer to incorporate private data into their general format.

Factors that scale up size

A variety of techniques have been published that correlate the cost of process equipment and components with a capacity factor. Such correlations are applied to costs of components as simple as a single valve, and to systems as complex as entire processing plants. The capacity factor may be based on geometry (pipeline diameter for a valve), energy requirement (horsepower for a motor), design parameter (transfer area for a heat exchanger), product capacity (tons per day for a facility), or other such characteristics.

The most frequent procedure of this type involves plotting, on log-log coordinates, the available cost values against the capacity factor. Interpolation will give acceptable accuracy if sufficient, reliable data are available to produce the plot. In most cases, this relationship may be adequately represented by a straight line over a reasonable range. If the straight line is satisfactory, generally more convenient is an analytical expression of the form:

$$\text{Cost, desired capacity} = (\text{Cost, known capacity}) \times (\text{Desired capacity}/\text{Known capacity})^n \quad (1)$$

In Eq. (1), n is a constant, which depends on the specific type of equipment, but is frequently assumed to be approximately 0.6 for estimation purposes.

Of course, the major difficulty with this approach is that all of the data points used to produce the plot must be based on a common time-of-purchase. If the 0.6 exponent is satisfactory, however, a single data point is obviously sufficient to establish the location of the curve.

Factors that correct for time

The practice of including escalation factors in cost-estimation calculations is well established in the chemical process industries. For short-term work, such as up-

Originally published May 1976.

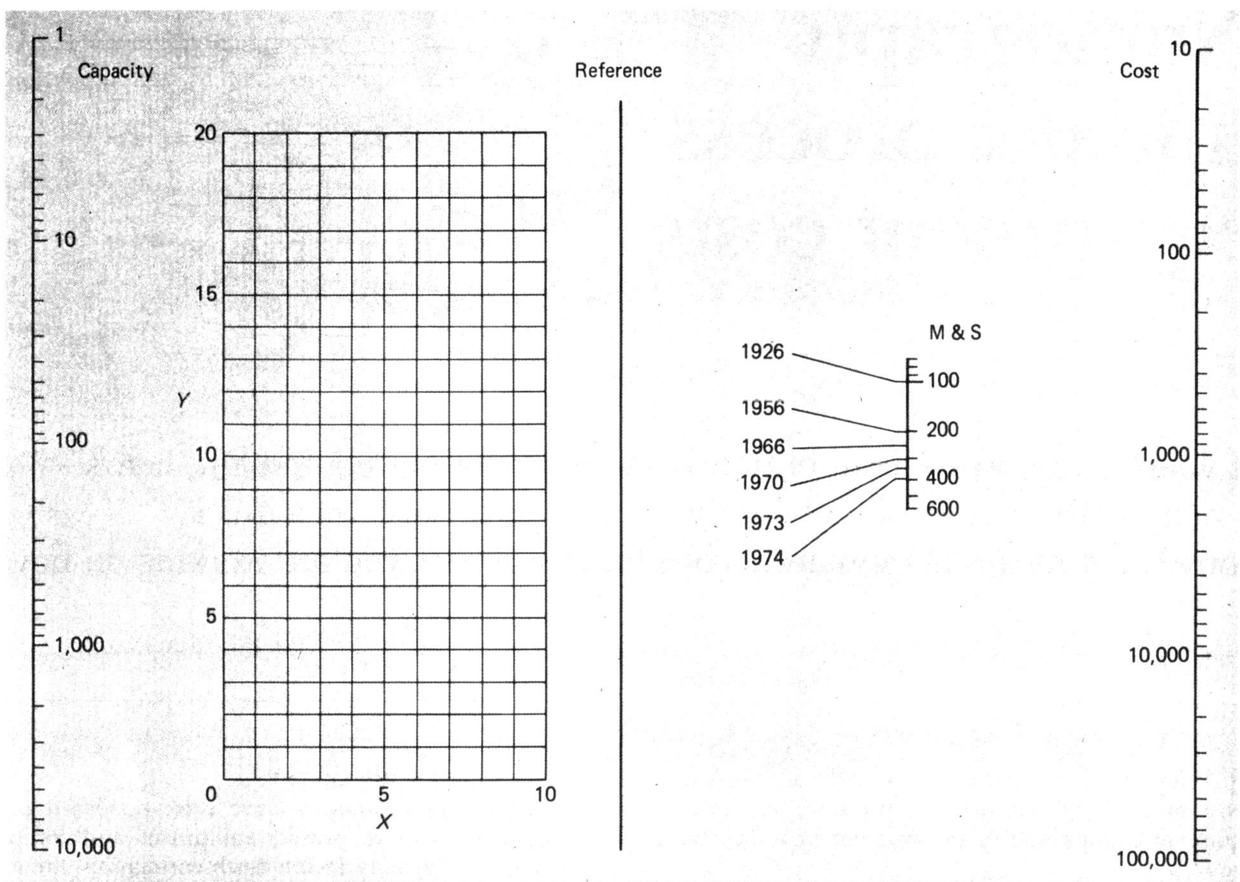

Nomograph corrects process and equipment cost data for both elapse of time and difference in size or capacity

dating a slightly obsolete quotation for a specific piece of equipment, an "add on" percentage may be adequate. For longer-range corrections, the typical method is to use one of the published price-ratio values, such as CHEMICAL ENGINEERING's Plant Cost Index or the Marshall and Swift Equipment Cost Index. Details of these systems have been well described [*1,7*], and generally involve an expression of the form:

Cost, present = (Cost, previous) × (Index, present/Index, previous) (2)

Cost information covering extensive time intervals has been published as tables and plots. The *CE* Index is based on 1957–59 costs; the M&S Index on 1926 costs.

Nomograph combines both factors

Because a design estimate will rarely involve only updating or size change, a calculational procedure that encompasses both factors offers advantages. Combining the ratio functions described previously gives the expression:

Cost, desired capacity at present = (Desired capacity/Known capacity)^n × (Index, present/Index, previous) (3)

If the necessary numbers are not readily available, using Eq. (3) can be time-consuming, particularly if n must be determined or estimated. The nomograph overcomes this problem by taking advantage of the following concepts:

1. The X–Y grid for an equipment category provides a concise method of tabulating a wide variety of systems. (Note that shifts along the X axis correspond to different values of n.)
2. The Time-Index line allows the use of whichever of the two values is known for a given application. (This avoids the need to determine the index value if only the date of a quotation is available.)
3. Updating is done by simply adding supplemental dates opposite the appropriate index numbers.
4. New equipment categories may be added by evaluating and tabulating the X–Y coordinates from two known cost values at two specified dates (or index numbers, if known).

Observe these precautions

Because of the general nature of the correlations on which the nomograph is based, several precautions should be observed when using it in the form presented, or when adding supplemental data by backplotting. The straight-line representation is only valid over a limited range, and care should be taken when extrapolating much beyond the tabulated values [3,4,6].

Improved technology may reduce the investment re-

Nomograph coordinates for equipment and processes

Equipment or process	Capacity range	Cost units	X	Y
Agitator, turbine	2-60 hp	$	6.4	14.4
Alumina from bauxite	120–330 ton/d	$1,000	6.8	13.8
Ammonium phosphate	270–750 ton/d	$1,000	6.4	10.6
Blender, ribbon	10–230 ft³	$	6.2	14.0
Blender, sigma arm	4–70 hp	$	5.4	16.4
Blower, centrifugal	10–250 ft³/s	$	5.8	15.5
Boiler, 125-psi	100–500 lb/min	$	2.7	13.0
Butadiene from butane	100–350 ton/d	$1,000	6.8	16.8
Centrifuge, solid-bowl, continuous	15-50 in dia.	$	1.8	16.2
Compressor, 125-psi, air	10–1000 ft³/min	$	7.7	15.8
Conveyor, 24-in belt	20–250-ft length	$	4.2	13.5
Conveyor, bucket (10 x 6 in)	12–140-ft length	$	6.0	13.4
Conveyor, 18-in roller	4–25-ft length	$	3.3	10.2
Crusher, ball mill	10–800 ton/d	$	4.2	15.3
Crusher, cone	40–300 ton/h	$	3.4	15.4
Crystallizer, vacuum, batch	500-7,000 gal	$	7.4	14.3
Dryer, pan, carbon steel	20–100 ft²	$	5.8	15.7
Dryer, rotary, vacuum	50–300 ft²	$	6.3	16.3
Dryer, spray	9–20 ft dia.	$1,000	2.9	10.9
Dust collector, cloth-bag	30–200 ft³/s	$	4.4	13.0
Evaporator, vertical-tube	100–10,000 ft²	$		
Fan, centrifugal	50–350 ft³/s	$	3.8	9.5
Filter, continuous-vacuum-drum	3–90 ft²	$	7.1	17.0
Filter, plate-and-frame, cast-iron	10–600 ft²	$	5.7	11.3
Gage, pressure	3–12 in, dial	$	2.5	11.0
Heat exchanger, shell-and-tube	20–1,000 ft²	$	6.2	10.8
Heat exchanger, shell-and-tube, stainless-steel type 316	20–700 ft²	$	6.5	13.1
Pump, reciprocating, stainless-steel type 316, 25 ft head	10–60 gal/h	$	5.6	12.1
Pump, rotary, bronze	10–800 gal/min	$	6.6	10.9
Reactor, glass-lined, jacketed	20–3,000 gal	$	6.7	13.6
Reactor, agitated, jacketed	70-1,000 gal	$	6.3	12.1
Refrigeration, mechanical, 20°F	4–100 tons	$	4.9	16.4
Rotameter, stainless-steel type 316	1–3 in, NPT*	$	2.8	14.9
Screen, vibrating, single-deck	3–70 ft²	$	4.8	13.4
Tank, vertical	200–10,000 gal	$	6.0	7.8
Tank, horizontal	3–40 x 10³ gal	$	5.0	14.9
Tower, cooling	20–200 gal/s	$	4.7	16.6
Tower, packed	10–150 in dia.	$	3.2	10.6
Tower, bubble cap	20–100 in dia.	$/ft	2.3	11.5
Tray, bubble cap	3–9 ft dia.	$/tray	1.3	14.7
Valve, control, stainless-steel type 316	1–5 in, NPT	$	2.5	16.2
Valve, plug, stainless-steel type 316	1–3 in, MPT	$	4.9	11.2
Valve, globe	1–3 in, NPT	$	4.9	8.8

*National Pipe Thread

quired to perform an operation, and may in fact reduce the cost more rapidly than inflation boosts it, and so reduce the validity of the nomograph. Equipment categories and processes incorporated later into the grid should be described specifically enough for accuracy in selecting and interpreting cost factors, but should not be so excessively detailed as to be relatively useless for general estimates.

Example illustrates procedure

To estimate the current cost of particular equipment:

■ Select the appropriate category for the equipment in the table. If the necessary capacity is within the specified range, note the values of X and Y.

■ Construct a line from the point on the Capacity scale corresponding to the equipment size (with units as specified in the table) through the X,Y point on the grid and to the Reference line.

■ Construct another line from this point on the Reference line through the current index value on the M&S scale (or the year point, if desired) to the Cost scale. The intersection will indicate the cost of the equipment. (Note that a few values in the table are given in thousands of dollars.)

Here is how to incorporate additional economic data into the nomograph grid:

For each two sets of Capacity index vs. Cost data available for an additional equipment category, construct a line from the Cost scale through the M&S value to the Reference line, and another line from there to the Capacity scale.

The intersection of the two Reference-Capacity construction lines will provide the X and Y coordinates for that equipment category, which may then be tabulated for later cost evaluations.

References

1. Arnold, T. H. and Chilton, C. H., New Index Shows Plant Cost Trends, *Chem. Eng.*, Feb. 18, 1963, p. 143.
2. Haselbarth, J. E., Updated Investment Costs for 60 Types of Chemical Plants, *Chem. Eng.*, Dec. 4, 1967, p. 214.
3. Lang, H. J., Cost Relationships in Preliminary Cost Estimation, *Chem. Eng.*, Oct., 1947, p. 117.
4. Mills, H. E., Costs of Process Equipment, *Chem. Eng.*, Mar. 16, 1964, p. 133.
5. Nichols, W. T., *Ind. Eng. Chem.*, Vol. 43, No. 10, 1951, p. 2295.
6. Peters, M. S. and Timmerhaus, K. D., "Plant Design and Economics for Chemical Engineers," 2nd Ed., McGraw-Hill, New York, 1968.
7. Stevens, R. W., Equipment Cost Indexes for Process Industries, *Chem. Eng.*, Nov. 1947, p. 124.

The author

George M. Hoerner, Jr., is an associate professor of chemical engineering at Lafayette College (Easton, PA 18042). He has had industrial experience with Du Pont Co. and Sun Oil Co., and is currently serving as a consultant to the J. T. Baker Chemical Co. A member of AIChE and a senior member of the Instrument Soc. of America, he is a registered engineer in Pennsylvania and a holder of a Ph.D. degree from Lehigh University.

Here are procedures and data for estimating the complete costs of . . .

Thick-Wall Pressure Vessels

F. D. CLARK and S. P. TERNI, Jr.,
Esso Research and Engineering Co.

The growing use of higher pressures and temperatures in the oil refining and chemical industries has greatly increased the need for heavy-walled pressure vessels. These vessels include the reactors used in hydrocracker, hydrofiner, and desulfurization units, and are generally designed for pressures exceeding 500 psig. and temperatures ranging from ambient to 900 F. The wall thicknesses vary from 2 to 9 in., and the weight from 150 to 750 tons.

The graphs and tables in this Cost Engineering File provide an efficient method for estimating the cost of material, fabrication, transportation and erection for shop-fabricated, heavy-walled pressure vessels.

Esso has experienced a wide price spread for these vessels under worldwide competitive bidding conditions. This variation is attributed to the problems involved in fabrication, inspection, transportation and erection, and the limited number of experienced fabricators. Since the reactors represent a large percent of the capital investment for a process unit, it is important that their cost be determined early in the life of a project. Frequently, early procurement is undertaken to ensure that the vessels arrive at the jobsite in time for startup. Esso has found fabrication times to range from 10 to 18 months, and has transported these vessels as far as 10,000 miles. Occasionally, the vessel must be field-fabricated because of access and transportation limitations. The cost of a field-fabricated, heavy-walled vessel is substantially higher than the cost of a similar shop-fabricated vessel, and is not covered by this method.

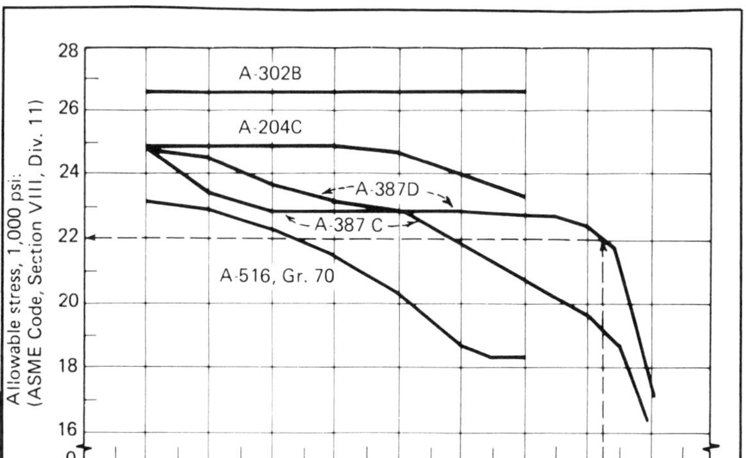

ALLOWABLE STRESS vs. design temperature for common pressure-vessel materials, ASTM specification—Fig. 1

Material, Fabrication and Inspection Costs

The cost of shop-fabricated vessels can be estimated by the following steps:

1. From process considerations, establish the vessel's design pressure and temperature, the inside diameter, overall length, and material of construction.
2. Determine the allowable stress at design temperature from Fig. 1.
3. Determine the shell thickness from Fig. 2. Enter the graph at the design pressure; move up to the inside diameter; move across to the allowable stress; then read shell thickness. Add corrosion allowance if required.
4. Calculate the total weight of the shell, heads, and nozzles from Fig. 3, 4 and 5. (Normal overweight allowances are included.)
5. Determine the unit cost from Fig. 6. Multiply the total vessel weight by the unit cost to arrive at the estimated cost of the vessel.
6. Add the cost of lining, and any additional costs. If an internal lining is required for corrosion protection, estimate the cost from Table I. Costs of the skirt, foundation, internals, etc., must be estimated separately and added on.

Transportation and Erection

The erection of each vessel or group of vessels is a unique case that must be examined independently. The costs of transporting, unloading, access-preparation, rigging of lifting equipment, vessel erection and subsequent dismantling of equipment are substantial, and require

Originally published April 1972.

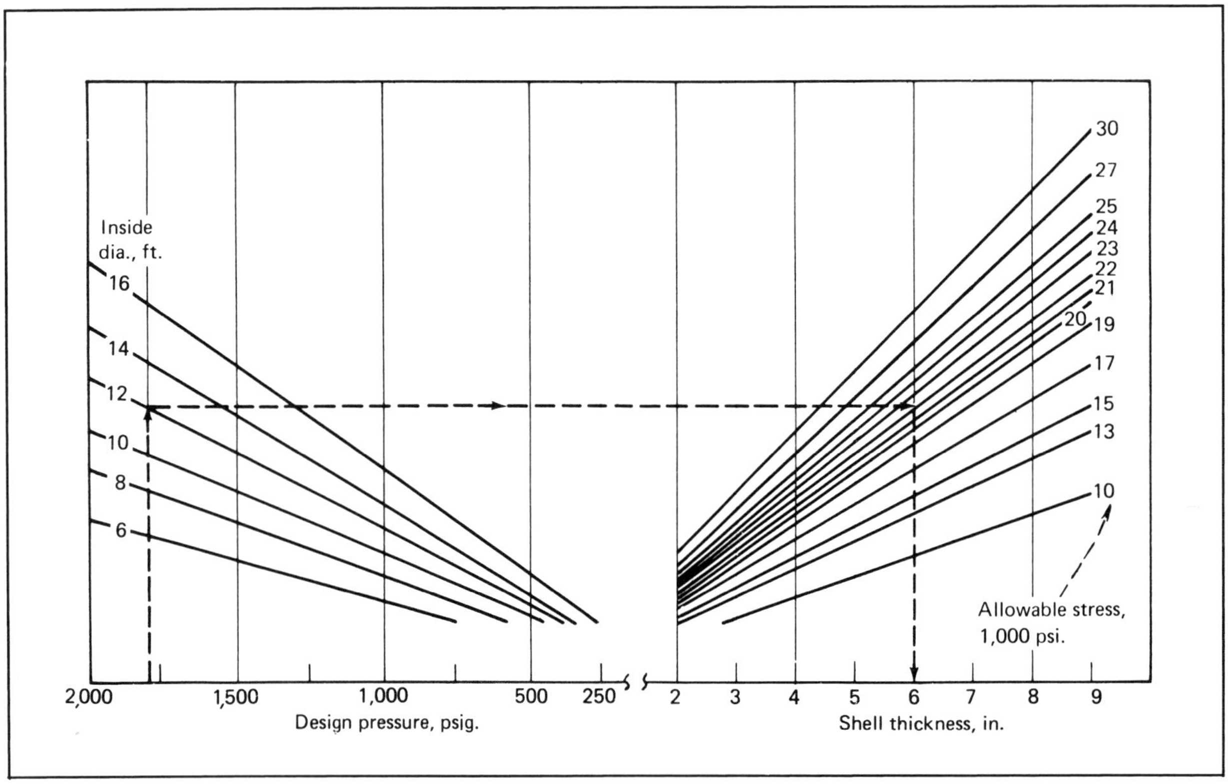

TO DETERMINE SHELL THICKNESS via this graph, (1) enter curve with design pressure, (2) move vertically up to inside diameter, (3) move horizontally across to allowable stress, and (4) move vertically down to read shell thickness—Fig. 2

detailed estimates that reflect prevailing circumstances at any particular jobsite:

Ocean Transportation—Special ocean-going ships are available for transporting large reactors and towers. These ships usually carry two or three reactors per voyage. For example, two reactors, one weighing 648 tons and the other 360 tons, were transported from France to a Caribbean refinery in one voyage. Ocean freight costs shown on Table II include special rail-mounted bogies used for transferring the vessels from the ship to the docking area, plus the actual loading and unloading operations handled by the ship's crew and officers.

Rail Transportation—The cost of U. S. rail transportation can be estimated from Fig. 7. Vessels weighing 100 tons or less, with diameters under 12 ft. 8 in., and lengths less than 90 ft., can be shipped over most railroads in the U. S. Vessels of greater weight or larger size can be shipped over select routes where bridges and clearances permit. Generally there is no unit price increase for the use of select routes. Because of weight and size limitations, it is frequently necessary to ship a large vessel in several parts. In such cases, the estimator must allow for the necessary field welding.

Ship Unloading Facilities—Most major ports in the world have loading facilities, but special temporary facilities may be needed at the destination. The special ships are shallow draft and can approach quite close to the beach. At one location, the unloading site consisted of an earth ramp enclosed by sheet piling and covered with a 2½-ft.-thick concrete slab supported on piles. Each site must be studied individually. A rough order-of-magnitude cost is given on Table II.

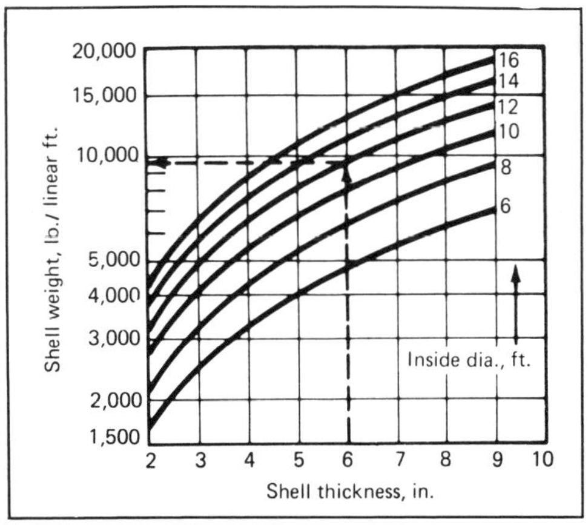

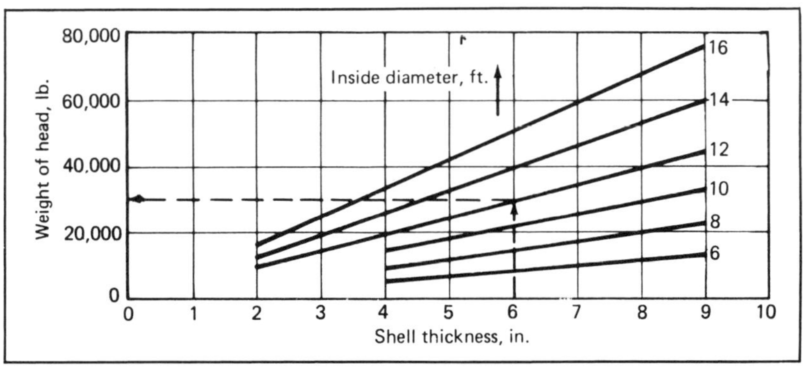

HEMISPHERICAL HEADS: weight vs. shell thickness—Fig. 4

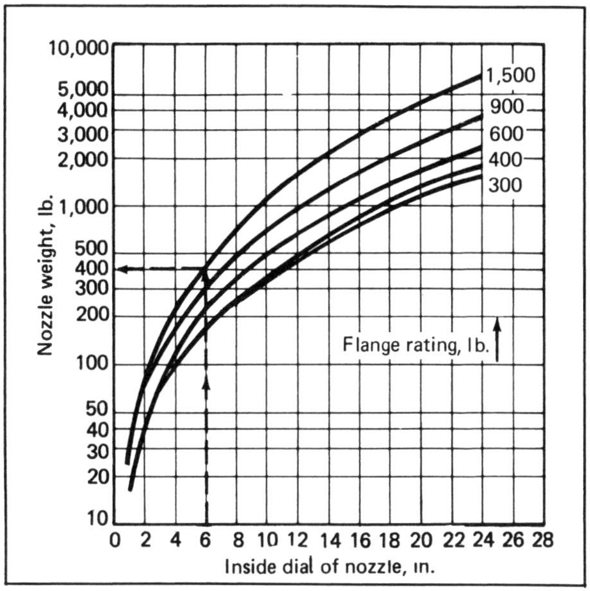

VESSEL NOZZLES: weight vs. inside diameter—Fig. 5

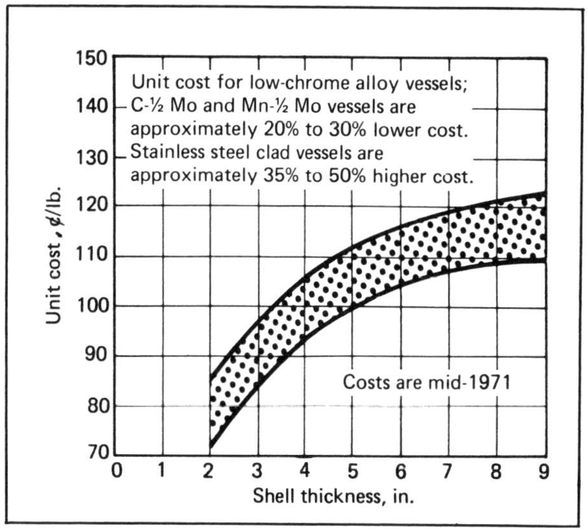

UNIT COST vs. shell thickness for heavy-walled pressure vessels constructed of low-chrome alloys, based on an analysis of bids for 50 vessels—Fig. 6

Hauling to Erection Site—If a vessel is transported by rail, a spur line may be laid directly to the site, or the vessel may be moved from the rail siding to the site by special crawlers or large low-boys.

Self-propelled, crawler type, vessel transporters can be rented. Two of these machines are necessary, and they are suitable for transportation up to several miles. At turnings, they can damage existing roads, and the estimator should allow for minor repairs.

Rental rates for a pair of vessel transporters are given in Table IV. The transporters travel 2 to 3 mph., and moving of a vessel is usually completed in one day; but to allow for preparation and coordination, at least one week's rental should be estimated per vessel. Also, the cost of strengthening existing bridges or culverts, and possible preparation of temporary roads must not be overlooked.

Temporary Bracing—Lifting lugs and temporary bracing to prevent skirt deformation are frequently required. These items are normally installed at the fabricator's shop, and should be estimated individually for each type of lift. An order-of-magnitude cost is given in Table II.

Home-Office Engineering—Planning the lifts, calculating stresses in the vessels, and designing temporary bracing require detailed engineering. This work is done by structural designers and rigging specialists working in the home engineering office. Designer and specialist manhours can be estimated from Table III.

Lifting Equipment—Gin poles are normally used for lifting heavy vessels. The major pieces of equipment consist of two gin poles, an equalizing beam, hoists, and a tensiometer. Approximate monthly rental rates are given in Table IV, and manhours can be estimated from Table III. Transportation of gin poles is expensive, and the estimator must make an allowance for preparation for shipment, in-transit rental, and freight.

Reactor Supports—Heavy reactors are frequently supported by steel space frames from 15 to 30 ft. high. If such a frame is built in place, it will interfere with lifting the vessel. This problem is frequently solved by pouring

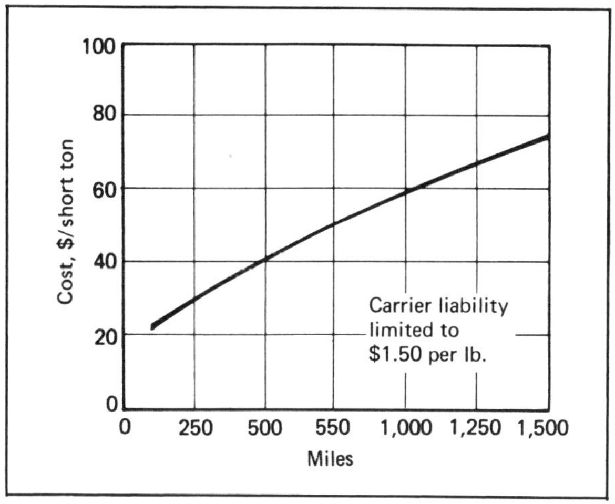

RAIL SHIPMENT: costs per ton—Fig. 7

Cost of Internal Linings—Table I

Material	Installed Cost, $/Sq. Ft.
Type 304 stainless weld overlay, 3/16 in. thick	50 to 60
Refractory (4 in. reinforced)	15 to 20
Gunite (4 in. reinforced)	5 to 10

Transportation Cost—Table II

Item		Cost, $
Ocean freight:	Atlantic crossing	35,000 to 45,000*
	Pacific crossing	75,000 to 85,000*
Rail spur line		15 to 20 per lin. ft.
Temporary bracing		5,000 to 20,000*
Unloading facility		0 to 50,000

*Per vessel

Rigging and Erection Costs—Table III

Activity	Gulf Coast Manhours
Haul from unloading to erection site (per vessel)	150
Assemble, erect and rig gin poles for 1st lift	1,700
Lift and set vessel into place	600
Relocate poles for a 2nd lift nearby	450
Relocate poles several hundred feet	750
Install dead-men for 1st lift	50
Additional dead-men for nearby lift	25
Dismantle and prepare poles for shipment	500
Manhours per vessel for home-office designers & rigging specialists (add 10% for additional vessels)	1,200

Equipment-Rental Costs—Table IV

Equipment Piece	Rental, $/Mo.*
2 Crawler-type vessel transporters†	15,000
Gin poles, hoists and accessories for lifts of 300 tons or less	10,000
Gin poles, etc., for more than 300 tons	30,000
Heavy crawler crane (165 ton)	5,000
Medium crane (40 ton)	2,200
Small crane (7½ton)	800
Miscellaneous and small tools	5,000
Dozer with winch	1,400

*Does not include transit rental or preparation for shipment.
†Rental includes lead-operator's salary.

only the concrete foundation pad in place, and constructing the steel space frame 30 to 40 ft. away. While the reactor is suspended over the foundation by the gin poles, the space frame is skidded into place. The reactor is then lowered onto the supporting frame. Manhours given in Table IV for lifting and setting a vessel are sufficient for skidding a supporting frame into position, but the estimator must remember to add the cost of a heavy crane (100 tons) and two dozers with winches.

Illustrative Example

Let us work out a sample estimate by this method. Consider the following reactor:

Inside dia.: 12 ft.	Design temperature: 830 F.
Tangent-tangent length: 100 ft.	Design pressure: 1,800 Psi.
Material: ASTM A-387 D	Nozzles: 2 x 24 in.; 6 x 6 in.
Corrosion protection: Weld overlay of 3/16 in. thick Type 304 Stainless	Source: FAS Gulf Coast
	Destination: Caribbean island, 3 miles inland

Cost of Pressure Vessel

From Fig. 1, allowable stress = 22,000 psi.		
From Fig. 2, shell thickness = 6 in.		
From Fig. 3, shell weight = 9,800 lb./ft. x 100 ft.	=	980,000 lb.
From Fig. 4, head weight = 30,000 x 2	=	60,000
From Fig. 5, nozzle weight = 6,800 x 2 + 400 x 6	=	16,000
Total weight	=	1,056,000 lb.
From Fig. 6, unit cost = $1.10/lb.		
Vessel cost = $1.10/lb. x 1,056,000 lb.	=	$1,160,000
Additional cost for weld overlay = $55/sq. ft. x 4,220 sq. ft.	=	230,000
Approximate total vessel cost, excluding internals, foundation, paint and attachments (mid-1971)	=	$1,390,000

Cost of Transportation and Erection	Manhours		$
From Table II:			
Caribbean crossing: Allow 0.67 x Atlantic crossing		=	27,000
Temporary bracing, allow			18,000
Unloading facility, allow			30,000
From Table III:	Manhours		
Haul to erecting site	150		
Assemble and erect gin poles	1,700		
Install dead-men	50		
Lift and set vessel	600		
Home office planning	1,200		
Dismantle poles	500		
	4,200		
Cost = 4,200 manhours @ $10*			42,000
Repair of existing road, allow			5,000
Strengthen and widen existing bridge, allow			10,000
From Table IV:			
Vessel transporters (1 week onsite + 3 weeks transportation and return) say 1 mo. =			15,000
Gin poles (3 weeks onsite + 3 weeks transportation and return) 1.5 mo. x $30,000			45,000
Heavy and light cranes, 0.25 mo. x $5,800		=	1,000
2 Dozers, 2 x 0.25 mo. x 1,400		=	1,000
(assume cranes and dozers available at site)			
Freight for vessel transporters and gin poles, allow			10,000
Miscellaneous and small tools			5,000
Total cost of transportation and erection		=	209,000
Grand total (mid-1971)†		=	1,599,000

Thus, for budget purposes, we can say that the installed cost of this pressure vessel is $1,600,000. ■

*Wage rate includes supervision and payroll burden.

†All costs in this Cost Engineering File reflect mid-1971 prices in U. S. dollars. The effects of the recent devaluation of the dollar on foreign purchases has not been considered.

Meet the Authors

◀**Forrest D. Clark** is a member of Esso Research and Engineering Co.'s project management department, where he is responsible for estimating methods. He is a graduate of Purdue University, and holds a Master of Engineering degree from Yale. His experience includes both project and cost engineering assignments in the Middle East, South America and the U. S. He is a professional engineer registered in the State of New Jersey, and a member of both the American Soc. of Civil Engineers and the American Assn. of Cost Engineers.

Stephen P. Terni, Jr., is a project engineer in the Cost Engineering Div. of Esso Research and Engineering Co. (P. O. Box 101, Florham Park, NJ 07932). He was graduated from Brown University in 1970 with a B.A. in Economics and a B.S. in Materials Science. He is currently working with the cost technology section, and is responsible for development of cost estimating methods. Mr. Terni is an associate member of the American Assn. of Cost Engineers. ▶

Costs of Process Instruments

This unique collection of 48 tables and graphs will enable you to quickly estimate costs of process control instruments for temperature, level and flow.

BÉLA G. LIPTÁK, Crawford & Russell, Inc.

Estimating the cost of instrumentation for a new process or a proposed plant has always been a problem, because the needed data are either widely scattered or totally unavailable. This article, and articles that will appear in the next few issues, present the required information in easy-to-use tabular or graphical form.

These data can be used for estimating the cost of control components or entire systems. But, in applying them, the reader must be aware of the following two kinds of corrections that must be made:

1. *Inflation*—Figures given in the article are based on 1969 prices; they must be updated by a correction factor. At the present rate of inflation, such can be estimated as follows:

   • Instruments with high-nickel-content alloy bodies: +20%/yr.

   • Instruments with stainless steel bodies: +10%/yr.

   • All other instruments: +6%/yr.

2. *Quantity Discounts*—Purchase quantity affects unit prices. The figures in these articles are based on the quantity of instrumentation needed for a smaller chemical plant—one in the $1- to $5-million range. The following correction factors should be used:

Plant Cost, $	Factor
Under 1 million	1.05
1 to 5 million	1.00
5 to 20 million	0.95
20 to 50 million	0.90

Instruments as a Percentage of Total Cost

The least demanding—and least accurate—technique of estimating instrument costs is to use a percentage of the total plant cost. A more-refined approach is to relate the instrument expense only to the total plant-equipment cost. This requires knowledge of total equipment (tanks, pumps, etc.) cost, but gives more reliable results because it disregards some of the cost factors not related to instruments (site development, buildings, warehouses, etc.).

The results obtained from these factor-type estimates can be improved if the nature of the control technique is also taken into account—such as pneumatic vs. electronic instrumentation, centralized vs. local control, etc.

The most time-consuming, and most reliable, results are obtained by preparing a detailed takeoff of all instruments, and pricing them individually.

Information will be given for all these approaches.

Originally published September 1970.

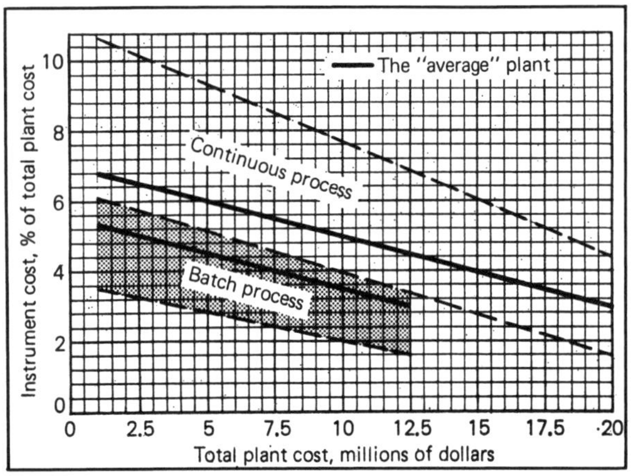

INSTRUMENT cost based on total plant cost—Fig. 1

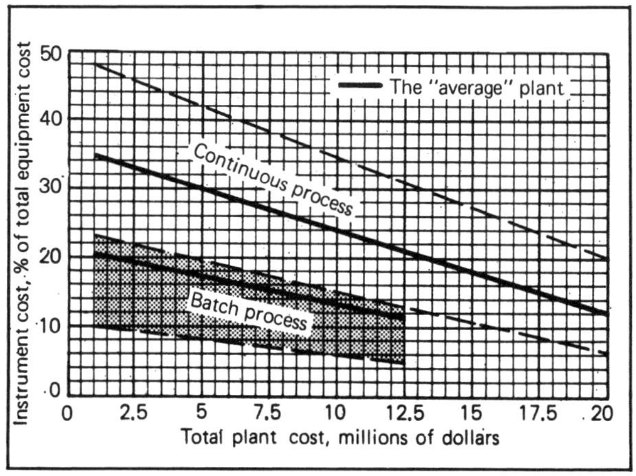

INSTRUMENT cost vs. plant equipment cost—Fig. 2

Overall Instrument-Cost Factors

Fig. 1, 2 illustrate the relationship between uninstalled-instrumentation costs and total plant costs. The plants are grouped into either continuous or batch-type ones, and in each group the "typical" or "average" plant is noted.

The distinction between batch and continuous plant designs is meaningful only if the batch operation is associated with long residence times, which allows for manual operations, and continuous operation is coupled with automated control schemes.

If one wishes to take some of the overall features of instrumentation philosophy into consideration, this can be done by using the "average plant" lines on Fig. 1, 2 and by the application of some of the adjustment factors listed in Table I. The corrected estimating factor is arrived at by the equation below:

Corrected % =
$(1 + F_1 + F_2 + F_3 + \ldots.)$ (% from Fig. 1, 2).

It should be understood that the estimating factor resulting from this procedure reflects only the instrument material cost. Further corrections are needed if one is interested in obtaining the installed-instrument cost, or if one wishes to take maintenance (over some period of years) into consideration.

As an example, let us take a continuous plant with a $20-million total cost, having a centralized control room (0.00), explosion-proof process area (+0.10), half of the samples analyzed by online analyzers (+0.20/2), and electronic control instrumentation (+0.20). By using the previous equation and Fig. 1:

Corrected % = (1 + 0.1 + 0.1 + 0.2) (3%) = 4.2%.

By applying this 4.2% factor, the instrumentation *material* cost for the particular project is estimated to be $840,000.

Installation Costs

Regardless of what method has been used to determine the instrument material cost, once that figure is known, it can be used to estimate the material and labor cost of instrument installation. The Tables II, III consider only the major factors.

Materials—The installation materials required include process- and air-side piping, valving, tubing, supports for both instruments and transmission lines, junction boxes, wiring, conduit, etc. Table II lists the percentage of the total instrument-material cost that one is likely to spend on some typical projects. If the piping that connects the instruments with the process equipment requires special considerations, such as jacketing, special corrosion-resistant materials, vacuum service, etc., the reader may adjust the factors in Table II upwards by one or two percentage points.

If, in our previous example, an intrinsically safe installation was assumed, then the total cost of installation materials can be estimated as (0.09) (840,000) = $75,600.

Labor—Estimating the cost of installation labor is probably the most difficult task because of the many variables involved. Some of the factors, such as weather, geographic location, strikes, union relations, manpower availability and the related need

Correction Factors to Reflect Instrumentation Philosophy—Table I

Features	Factor, F
Localized control	−0.20
Pneumatic instrumentation	0.00
Centralized control	0.00
Sample analysis performed in laboratory	0.00
General-purpose process area	0.00
Explosion-proof process area	+0.10
Graphic panel display	+0.10
Special alloys required for pipeline items	+0.15
Sample analysis by online analyzers	+0.20
Electronic instrumentation	+0.20
Limited-scope optimizer computer included	+0.25
All loops on computer control	+0.45

to work overtime, will not be considered here because they cannot be numerically evaluated in a generalized discussion. At the same time, few will argue that if one pays over \$10/hr. for instrument technicians, or \$9/hr. for pipe fitters and electricians, a spell of bad weather can turn out to be rather expensive. Similarly, if a construction effort is forced on an overtime schedule for any one of a dozen reasons, this can increase the labor cost (per hour of overtime) by 70% or even more. Therefore, the data given below should be used with caution, representing that hard-to-define "average" construction site, and should be checked against factual data on past performance in the geographic location being considered.

In addition to the percentage figures in Table III, one should also consider the *scope* of the task (loosely referred to as instrument installation labor) and the *party* held responsible for performing this task.

Table III assumes that the scope is limited to the physical installation of the components and to testing for leakage, polarity etc. If this scope is to be increased to include the *calibration* of all components and also a *functional checkout* of all control and interlock loops—during which all possible process conditions are simulated, and system performance is thereby checked—then another 5% can be added to the figures in Table III.

If the installation of instruments is not to be supervised by the main contractor but is given to an instrument subcontractor, this may add up to another 10% to the figures given in Table III.

If, in the previously used example, the intrinsically safe installation is coupled with a need for calibration and functional checkout (to be supervised by the main contractor), then the installation-labor cost is calculated as:

$$(0.35 + 0.05)\ 840{,}000 = \$336{,}000.$$

Therefore, in this example our estimate indicates the need for the following funds to cover instrument costs up to plant startup (assuming that the design and engineering function is performed by the user and therefore does not require the layout of capital funds):

Instrument material cost	–\$840,000
Installation material cost	–\$ 75,600
Installation labor cost	–\$336,000

Design and Engineering

When the detailed plant design is assigned to an engineering contractor, the cost of his service is also additive to the total expense. The scope of such a contract, as far as the instrumentation design is concerned, usually includes: development of measurement, control and operational concepts, preparation of hardware specifications, analysis of bids, issuance of purchase orders, review of vendors' prints, in addition to the preparation of the various design drawings for control-board layout and for the installation of both instruments and signal leads.

Fig. 3 shows the relationship between the dollar value of instrument materials and the number of engineering-design manhours required by the instrument engineers and designers. These values do not include the cost of related, nontechnical activities —such as purchasing, accounting, project, expediting, management, print room, secretarial, personnel, sales, etc.

Considering the earlier example of a plant with an instrument-material cost of \$840,000 and assum-

Installation Material Cost as a Percentage of Instrument Material Cost—Table II

		Localized Control (No Transmitters)	Centralized Control (300-ft. Average Transmission)
Pneumatic control instruments (explosion-proof process area)		10%	12%
Electronic instruments	General-purpose or intrinsically safe installation	Usually one would not consider electronics for localized control	9%
	Explosion-proof installation		11%

Installation Labor Cost as a Percentage of Instrument Material Cost—Table III

		Localized Control (No Transmitters)	Centralized Control (300-ft. Average Transmission)
Pneumatic control instruments (explosion-proof process area)		45%	52%
Electronic instruments	General-purpose or intrinsically safe installation	Usually one would not consider electronics for localized control.	35%
	Explosion-proof installation		42%

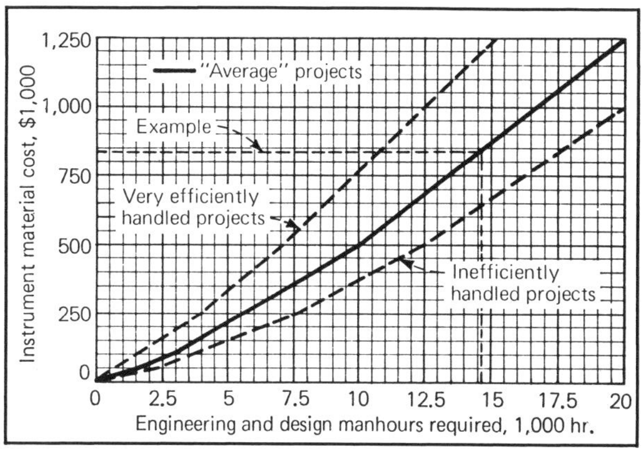

DESIGN cost vs. instrument material cost—Fig. 3

ing that this project will be handled with "average" efficiency, from Fig. 3 we see that 14,700 engineering and design manhours will be required. Considering the various overheads, this can cost the user the equivalent of 25% or more of the instrument material cost.

The overall breakdown of the various instrumentation expenses *up to the time of plant startup,* based on the example used, is as follows:

Instrument materials	60%
Installation labor	20%
Design engineering	15%
Installation materials	5%

Operating Costs

In considering the instrumentation cost of a plant, there is a tendency of counting only those expenses that occur prior to startup. This is understandable because operating expenses are usually budgeted separately from capital funds; but it is still misleading because the only meaningful evaluation of instrument expenses is one that considers both the prestartup and operating costs. In a project where false economy is applied to the point where (in order to reduce instrument material cost) the hardware quality is compromised, the result is an increase in operating costs, which is greater than the savings achieved in capital expenditure. The term "operating cost" is used here in the narrow sense, referring to that expense required to keep the instrumentation in operation, but not including the revamp or improvement-oriented costs.

In our example with $840,000 instrument material cost, one would expect a yearly maintenance cost of about 10% or $80,000 to $100,000. Most of this expenditure is for maintenance labor, and only a relatively small fraction for replacement parts. Some plants have maintenance expenses twice as great and others only half as much.

The following items tend to influence maintenance and operating costs:

• Operating expenses due to poor design or poor selection of hardware that results in maintenance tasks to open up plugged or frozen lines to eliminate leaks, etc., will not be discussed here (not that they do not occur in a well-designed plant, but because it would take too much space to cover them).

• The quality of instrument air (its oil and moisture content) can have a pronounced effect on the plant maintenance history.

• Electronic instrumentation of the solid-state type (no vacuum tubes, motors, slide wires, batteries, vibrators) compares favorably with its pneumatic counterpart from a maintenance point of view (sometimes by a factor of two), and in a few years of operation it may compensate for its higher initial cost.

• Force-balance sensors generally have a maintenance record superior to motion-balance ones. The elimination of moving parts–particularly if they are in contact with the process fluid–will usually reduce operating costs. This is why the maintenance cost of electronic temperature transducers is lower than for pneumatic ones, while the operating cost of electronic control valve actuators is higher than for pneumatic ones.

• Online analytical instruments (and particularly their sampling systems) have a very poor maintenance history; when estimating the operating cost of the plant instrumentation, their quantity should be taken into serious consideration. Some reports indicate that the yearly unit cost of keeping certain chromatographs in operation is in the thousands of dollars.

In some plants, in order to provide better job security for the permanent employees, the maintenance department is undersized and the peak workloads are handled by contract maintenance. The benefits of such an approach are usually balanced against the problems caused by the differences in salary structure between the owner's and the contractor's employees.

Computer maintenance, outside of simple module-

Level-Switch Pricing—Table IV

Type	Service: Liquids			Service: Solids
		Screwed	Flanged	
Float or displacer	With external chamber	$225	$240	—
	Tank mounted	$110	$140	
	Note: Costs based on standard materials. Add $50 for explosion-proof or dual-float design.			
Conductivity	Single-probe unit—$110 Two probes with differential control—$200			—
Diaphragm	Explosion-proof, captive air type—$70			General-purpose, standard units from $50 Explosion-proof, standard units—from $80
Rotating paddle	—			General-purpose, standard materials—$60 Explosion-proof with stainless wet parts—$110
	Liquids and Solids Service			
Capacitance	Combined probe and control unit—$110 Separate control unit with high sensitivity—$200 (Both in explosion-proof housing)			
Vibrating probe	Self-sustaining, explosion-proof design—$275			
Ultrsonic	General-purpose probe and controller—$200 Explosion-proof unit—$275			
Radiation	Source holder and on-off detector—$600 Additive source cost is function of size—$50 to $300			
Thermal	General-purpose design—$120 Explosion-proof design—$180			

Local Level-Indicator Costs—Table V

Type	Service: Liquids				Service: Solids
Gage glasses	See Fig. 4				
Tape gages	Tank Pressure	Indicator Type	Wetted Materials: Standard	Wetted Materials: Stainless	Explosion-proof indicator with transmission capability—$500 each. 18-tank remote selector and readout for $600 additional.
	Atmospheric	Board	$100	$400	
		Gage head	$200	$600	
	Up to 150 psi.	Board	$250	$600	
		Gage head	$400	$800	
	Extras: Standard oil seal—$30; explosion-proof switch—$75; stainless float—$60; steel tape block valve—$150; crank assembly—$100; transmitter—$250; 10-tank selector and readout—$1,700.				
Float gages	8 in. dial indicator with stainless-steel wetted parts—$250. Explosion-proof transmitter—$200. Remote readout and alarm for 10 tanks—$450.				

replacement types of operations are usually handled on a contractual basis by the computer manufacturer.

In general, the records seem to show a clear correlation between preventive maintenance costs and the corresponding reduction in unscheduled shutdowns and in the improved plant-safety record. This proves that if an instrument is worth installing, it is good business to keep it in good operating condition.

LEVEL INSTRUMENTS

On the following pages, estimating data are provided for individual instruments. Correct for inflation and for quantity discounts, as noted earlier.

Information concerning the cost of level sensors is grouped according to function. The main groups covered are the on-off level switches, the local level indicators and the continuous level transmitters.

Level Switches

Cost data for a group of on-off level switches are furnished in Table IV. There are some other, indirect techniques for obtaining an on-off level signal that are not included here. These are the types of installations where, for example, a pressure switch is used to detect the hydraulic head in an open tank by either mounting the switch on the tank bottom or sensing the pressure output of an air bubbler.

Pricing for such systems is obtained by adding up the component costs, which are given under the corresponding variable. Pressure-switch costs, for example, are listed under pressure detectors, and so are the differential-pressure switches that could be considered for level measurement in pressurized tanks.

It should be noted that all transmitting-type level sensors can also be employed as switches by placing an on-off detector on their output signals. Some of the level indicators, such as the flat-glass, tubular-glass, magnetic, tape or float-type level gages can

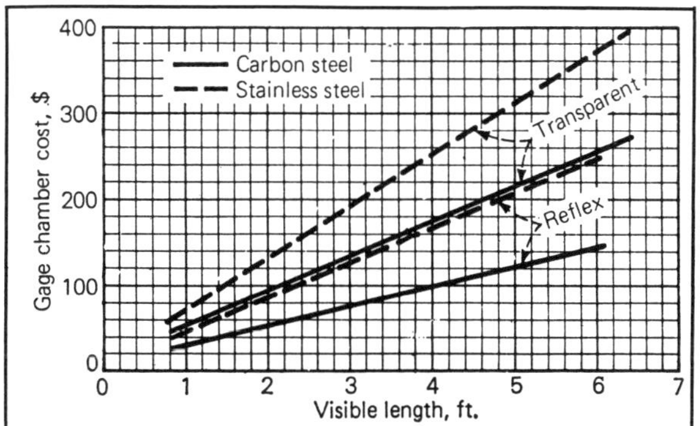

Note: For a pair of offset gage valves, with three unions, in steel materials add $50; in stainless materials add $120. If flanged vessel connections are required add $15.

COST of flat-glass type level gages—Fig. 4

also be provided with accessories for on-off level sensing. Some pricing data on these accessories will be given in the paragraphs dealing with transmitters and indicators.

Local Level Indicators

Fig. 4 and Table V provide pricing data for the more commonly used local level indicators.

Indirect level indicators such as pressure or differential-pressure gages for hydraulic-head sensing are not included here. Their costs can be obtained from the price data for pressure sensors.

All transmitting-type level sensors can also be used as local indicators by placing a local gage on the transmitted output signal. The cost of level transmitters is covered in the following paragraph.

Level Transmitters

Table VI provides pricing data for some of the common level transmitters.

In order to avoid repetition, in some cases refer-

Level Transmitter Costs—Table VI

Type	Service: Liquids — Type	Range	Cage or Flange Material — Iron	Steel	Solids
Displacer	Internal	<5 ft.	$400	$470	
		5 ft. to 10 ft.	$430	$500	
	External	<5 ft.	$500	$570	—
		5 ft. to 10 ft.	$550	$670	
	Notes: Prices above include 150 psi. flanges and pneumatic transmitter. If stainless or higher alloy construction is required, prices may double. Extra for two-mode controller is $100, and for explosion-proof electronic transmitter $200. No extra for standard control pilot.				
Resistance Tape	Standard unit with readout included—$300.				
Tape	See Table V				See Table V
Float	See Table V				—
Diaphragm repeater	See Table IX				—
Differential-pressure transmitter	See Table VIII				—
	Liquids and Solids Service				
Capacitance	Probe cost varies with material, connection, type, length—$50 to $200. Explosion-proof amplifier with mv. output—$500. Digital display for 16 tanks, in general-purpose case—$600.				
Radiation	Geiger-Mueller detector, point source, d.c. amplifier with mv. output—$1,800. Gas ionization detector, strip source, a.c. amplifier with mv. output—$3,000. Motor driven, wide-span accounting system—$8,000.				
Ultrasonic	Transducer unit cost—$250. Signal generator, scanner, power supply, readout unit for 16 tanks—$1,400.				

Vacuum-Sensor Costs—Table VII

Type	Cost
Hot cathode	Complete system with minimum-cost readout—$500.
Cold cathode	Complete system with minimum-cost readout—$260. Element only—$100.
Radiation alphatron	Sensing head and amplifier to give mv. output—$900.
Thermal vacuum sensors	Complete system with minimum-cost readout—$175. Element only—$25.
McLeod gage	Tilting type—$100. Piston type—$350.
Capacitance	Complete system depending on reference pressure, from $1,500 to $2,300. Element only—$1,200.
Quartz helix	Complete system with digital readout—$2,200.
Micromanometers	About $100.
Pneumatic and electronic transmitters	Refer to Table VIII.

ences are made in this table to some other tabulation where the pricing has already beed detailed.

Transmitters can also be used for installations that do not involve remote readouts, but the economic justification for their use in such services is questionable. Some of the devices listed are also available as local controllers. The pricing data in Table VI cover both pneumatic-displacement-type controllers and transmitters.

Pressure Detectors

Pricing data are explained below the more-frequently-used pressure detectors. These are grouped, according to their function and design, into vacuum sensors, transmitters, repeaters, manometers, gages and switches. Cost information is also furnished for accessories such as seals and for some of the special-purpose devices.

Vacuum Sensors—Table VII provides information on the cost of the more-frequently-used high-vacuum sensors.

The majority of these devices are for laboratory use, and only a few meet the ruggedness requirements for industrial installations.

Where minimum-cost readout is referred to in the tabulation, it usually means a general-purpose, portable-case unit with a d.c. mv indicator. If a remote readout or an amplified signal is required, the millivolt signal generated by the element can be fed to a converter or to a potentiometric device.

In the relative evaluation of the various designs, one should consider the maintenance aspects, realizing that the element is likely to require periodic replacement and, therefore, the less expensive it is, the more economical the installation will be.

Pressure and Differential-Pressure Transmitters—Table VIII lists the prices of some of the more common transmitters used for pressure, pressure drop, flow, level and density detection. Their motion-balance versions can be provided with direct indicators, while the force-balance designs are blind and only the output signal can be displayed on a local gage.

In evaluating the cost of the various installations, one should not overlook the expense associated with the installation materials and labor. For example, a 4-in., extended-diaphragm pneumatic transmitter at $490 is expensive, but the installation of this device does not involve isolating or manifold valving, filling or support materials (or the associated labor), and therefore a cost comparison on an installed basis is more meaningful.

Pressure Repeaters—Pressure repeaters might be employed as components in either pressure- or level-detection loops, and their prices are noted in Table IX. These devices are capable of repeating both positive and vacuum pressures and are available in the diaphragm or in the differential-pressure-cell designs.

Manometers—Simple glass-tube laboratory manometers can be obtained for $25. More-rugged or corrosion-resistant industrial glass-tube manometers will

Pressure and Differential-Pressure Transmitter Cost—Table VIII

Pressure Sensor Type	Connection	Indicating	Blind	Operation	Pneumatic: Carbon-Steel Body, $	Pneumatic: All-Stainless-Steel, $	Electronic (Explosion-Proof): Carbon-Steel Body, $	Electronic (Explosion-Proof): All-Stainless-Steel, $
Absolute	½ in. screwed		√	Force balance	420	450	570	620
Gage	½ in. screwed	√		Motion balance	170	220	490	525
	½ in. screwed		√	Force balance	250	270	440	460
Differential (for flow,	½ in. screwed		√	Force balance (above 20 in. H_2O)	280	330	500	530
level or density)	½ in. screwed		√	Force balance (below 20 in. H_2O)	325	360	540	570
	3 in. flanged		√	Force balance	330	380	520	570
	4 in. extended flanged		√	Force balance	490	575	670	750

Extra: The requirement for Hastelloy or other higher alloys can double the unit prices listed. Pneumatic output gages costs $20, explosion-proof electronic-output gages $65. Carbon-steel manifolds can be obtained for $40, stainless-steel ones for $80. The extra for filter regulators is $20. If the pressure taps require chemical-seal protection, see Table XII for the extras involved.

Pressure-Repeater Costs—Table IX

Size, In.	Flange: Steel or 304 Stainless. Wetted Parts: 316 Stainless, $	Flange: 316 Stainless. Wetted Parts: Monel, Hastelloy, Tantalum, Alloy 20, $
2	100	160
3	125	240
4	150	290

Note: Differential-pressure-transmitter type of pressure or vacuum repeaters are about the same cost as the items in Table VIII.

Cost of Pressure Gages—Table X

Element Material	2½ In., $	3½ In., $	4½ In., $	6 In., $	8½ In., $	12 In., $	16 In., $
Bronze	12	18 (27)	20 (27)	27 (36)	40 (47) ((170))	((200))	((240))
Alloy steel		22 (27)	24 (33)	31 (40)	50 (58)		
316 stainless steel			32 (43)	40 (50)	63 (72)		
K-Monel			36 (45)	43 (52)	63 (72)		

Note: Prices not n parentheses are for ±½ to 2%-accuracy industrial units. Single parentheses indicate ± 0.25%-accuracy test gages. Double parentheses indicate ± 0.1%-accuracy precision gages.

Cost of Pressure Switches—Table XI

Type of Actuation	General-Purpose: Brass, $	General-Purpose: 316 S/S, $	Weather-Resistant: Brass, $	Weather-Resistant: 316 S/S, $	Explosion-Proof: Brass, $	Explosion-Proof: 316 S/S, $
Standard, adjustable differential	22	60 (32)	50	85 (60)	100	135 (115)
Close, adjustable differential	30	70 (45)	55	95 (70)	110	145 (120)
Two-stage switch with independently set actuation	65	(85)	95	(115)	150	(165)

Note: The cost of steel bourdon-tube units is shown in parentheses.

range between $50 and $200, depending on their size and accessories.

Inverted-bell, ring-balance and float-type mercury-sealed manometers are usually applied as indicating transmitters, at a unit cost of $500 to $600.

The cost of a metallic-tube servomanometer for high-precision applications can exceed $1,000.

High-Pressure Sensors—The cost of these devices is largely a function of the pressure levels involved, the required measurement precision and the accessories needed. For pressures up to 20,000 psi., the gage pressure transmitters listed in Table VIII can be considered. These are the "C," helical and spiral bourdon-tube-type flexible-element sensors.

For higher pressures—up to 100,000 psi.—the strain-gage detectors are available. The strain-gage sensor alone can be obtained for around $300. A potentiometric amplifier with integral power supply and exciter, which will generate a mv. d.c. output signal, is priced around $450. If the same amplifier is required with a pneumatic output, this adds another $100 to the unit's cost.

For the measurement of even higher pressures, one might consider deadweight testers or the less frequently applied bulk modulus or manganin cells. A deadweight tester for 0.1% accuracy with a range of 0-2,000 psi. costs $500. Units with higher ranges and extreme precision are usually well above $1,000.

Cost of Chemical Protectors—Table XII

Wetted Housing or Flange Material	Seal Design and Connection Type: Offline* 1 In.- Sc., $	1 In.- Fl., $	2 In.- Fl., $	W.S. $	Filled Diaphragm With 2 In. Extension Flange Size 3 In., $	4 In., $	6 In., $	Flow-Through* ½ In.- Sc., $	1 In.- Fl., $	1½ In.- Fl., $	2 In.- Fl., $	3 In.- Fl., $	4 In.- Fl., $	6 In.- Fl., $
Steel	35	45	50	60	140	200	220	65	110	120	130	160	220	280
Stainless steel	55	65	75	80	150	210	235	85	150	175	210	290	340	430
Monel	65	70	100	90					185	210	235	310	390	
Hastelloy alloys	90	100	200	160										
Teflon flexible cylinder with steel flanges									140	160	195	225	280	350
Teflon flexible cylinder with stainless flanges									195	240	280	325	400	500

* **Note:** Sc. = screwed fitting; Fl. = flanged fitting; W.S. = welded saddle.

Cost of Liquid-Density Sensors—Table XIII

	Local Indicator or Recorder Only, $	Transmitter*, $
Industrial hydrometer: Brass or steel	180	($550) Impedance-bridge type, temperature-compensated transmitters.
Industrial hydrometer: Stainless steel	250	($630) Impedance-bridge type, temperature-compensated transmitters.
Industrial hydrometer: Nickel or Monel	380	($850) Impedance-bridge type, temperature-compensated transmitters.
Industrial hydrometer: Hastelloy alloys	520	($1,100) Impedance-bridge type, temperature-compensated transmitters.
Buoyant-force displacer	300 for chain-balanced	See Table VI for conventional displacer prices. ($1,500) for chain-balanced.
Hydraulic-head detector	See Table X	See Table VIII
Weight of fixed-bulb-volume	700 with recorder†	$1,350 ($1,500) with ma. d.c. output.†
Radiation gage (explosion-proof with temperature compensation.)	—	($4,500) with mv. output. Deduct $800 if d.c. amplifier is acceptable.
Weight of U-tube	—	$1,700 to $2,600, depending on compensation and other features.
Vibrating U-tube	—	($1,700) with mv. output
Angular position of displacers	—	($1,800) with mv. output
Electromagnetic suspension	—	($1,700) with mv. output

* Costs of explosion-proof electronic transmitters are distinguished from the prices of pneumatic units by parentheses. Temperature compensation is included only when so stated.
† For all-316-stainless-steel construction—$200 extra.

Pressure Gages and Switches—Pricing data for industrial pressure gages and switches are provided in Tables X, XI.

When the detection of pressure differentials is of interest, this can be accomplished by the use of duplex bourdon gages. A 4½-in.-dial gage with two bronze bourdon tubes costs about $60.

If higher precision and direct differential indication is required in the local gage, the bellows-type differential-pressure indicator can be considered. The cost of this unit with beryllium-copper bellows and brass housing is $160, and with stainless steel bellows in steel housing is $225. This differential-pressure indicator can also be furnished with explosion-proof switches at a cost of $50 extra for each.

Chemical Seals

Pricing data for the various chemical-seal designs are provided in Table XII. Where the purpose of the seal is to isolate a corrosive process from a standard instrument, "offline" designs are usually considered.

If the unit is installed as protection against plugging of the connecting piping by the viscous or slurry-type process fluid, the "flow through" design is recommended. Another method of eliminating dead-ended cavities is by the use of the extended diaphragm-type seal elements, which in larger sizes and special alloys tend to be more economical than the flow-through elements. Advantage of the flexible Teflon-sleeve design is elimination of even minor cavities.

Cost of Gas-Density Sensors—Table XIV

	Local Indicator or Recorder Only	Transmitter*
Buoyant-force displacer	$550	$1,500 ($1,600) with ma. d.c. output
Gas-column balance (temperature compensated)	$600 with recorder†	$1,200 ($1,300) with ma. d.c. output†
Centrifugal	$1,100 plus ΔP indicator†	$1,100 plus ΔP transmitter†
Viscous drag	$750	$1,200
Electromagnetically suspended dumbbell	—	$1,800 ($2,150) with mv. output
Thermal	—	($800 to $1,500) with ma. d.c. output

* Costs of explosion-proof electronic transmitters are distinguished from the prices of pneumatic units by parentheses. Temperature compensation is included only when so stated.
† Stainless-steel construction—$100 extra.

Density Detectors

The various density sensors can be estimated from the data presented in Table XIII for liquid-service units and from Table XIV for gas-service units.

The costs noted are those of industrial designs; if laboratory devices are desired, they usually cost less.

Local indication can be provided for all transmitting units by placing a receiver on their output signal, but this approach has not been considered. Where local indicators are listed as available, they are of the direct indicating type.

Thermowell Costs (304 Stainless)—Table XV

Connection	Insertion Length 6 In., $	12 In., $	18 In., $
Screwed	8	13	18
Flanged	25	30	35

Cost of Bimetallic Thermometers—Table XVI

Stem Length, In.	Every Angle 5-In. Dial, $	Fixed Angle 2-In. Dial, $	3-In. Dial, $	5-In. Dial, $
6	20	8	10	17
12	21	10	12	18
18	22	12	14	19

TEMPERATURE DETECTORS

Thermometers

Conventional temperature sensors are usually installed inside a protecting socket, which provides both for their mechanical protection and for their removal from pipelines or equipment without opening up the process. Table XV provides some cost information for Type 304 stainless steel thermowells. The prices listed are additive to all temperature-gage costs. If carbon-steel sockets are to be installed, the prices are reduced by a factor of two.

Some cost data for bimetallic thermometers are presented in Table XVI, and for filled-type temperature indicators in Table XVII. The pricing in the latter table is based on the readout dials being close-coupled, pipeline-mounted, without the use of capillary tubing. If remote dials are to be priced, the extra cost for compensated and armored capillary tubing can be estimated at $1.50/ft.

Cost of Filled Thermometers—Table XVII

	Type and Filling				
	Every Angle		Fixed Angle		
Dial Size, In.	Vapor, $	Mercury, $	Vapor, $	Gas, $	Liquid or Mercury, $
3½	27	—	22	32	—
4½	30	60	27	38	56
6	40	—	38	46	63
8½	50	—	47	57	81

Temperature Switches, Recorders, Transmitters

Estimating data for temperature switches and for filled-capillary-type temperature transmitters are provided in Tables XVIII, XIX, respectively. The cost of thermowells (Table XV) is additive.

The cost of a direct-connected, 12-in., circular-chart temperature recorder with spring-drive and 5-ft. capillary tube is usually around $225. A second pen for the same recorder costs an additional $125.

The cost of all thermal systems used in recorders, transmitters or in other temperature instruments is influenced by the following factors:

- Depending on the type of filling required, one can expect a $15-25 variation, with vapor filling being the most economical.
- The extra cost of full compensation (as contrasted with case compensation) is about $10-15 per unit.
- When the temperature range to be detected is

either extremely high or low, the thermal system cost can increase by $25-40.

Resistance Thermometers and Thermistors

The cost of resistance temperature detectors (RTD's) is a function of the intended installation and the type of certification desired. A three-wire, platinum RTD for industrial use costs about $100, to which price the cost of the protecting thermowell has to be added. If a Bureau of Standards certified RTD is desired, its cost can approach $800.

The cost of converters and readout devices will be covered later; only a brief orientation is given here. A general-purpose RTD to ma. d.c. converter costs about $250, and a differential RTD to ma. d.c. converter is about $425. A direct RTD recorder with 12-in. circular chart can be obtained for $800.

The estimating data presented for RTD's can also be used in pricing thermally sensitive resistors or thermistors.

Thermocouples

The average, industrial thermocouple assembly without a protective socket can be estimated at $20 each. Thermowell prices can be obtained from Table XV. If a dual-head unit is desired, it can be obtained for an extra $10. Noble-metal thermocouples (platinum, tungsten, gold, iridium) cost about $20 more than the more standard materials.

The cost of duplex, insulated, thermocouple extension wires can be estimated at $30/100 ft. for the typical installation in which protection against abrasion or mechanical damage is provided by installing the wires in conduit. For a more detailed cost breakdown, refer to Table XX.

The cost of converters and readout devices is covered in other parts of this article series, therefore only a brief orientation is given here. A general-purpose thermocouple to ma. d.c. converter is about $350 and an explosion-proof one is about $450. If a thermocouple-to-pneumatic conversion is desired, $100 should be added to the above prices. A direct-connected thermocouple alarm switch is about $175, and a large-case, circular-chart, single-pen potentiometric recorder is $800. Large-case, strip-chart, 24-point potentiometric recorders can be obtained at $1,500 each.

Specialized Temperature Sensors

Quartz-crystal thermometers having a singe-probe sensor and digital readout can be obtained for $2,300. If a two-probe assembly is desired for differential-temperature measurement, the complete system cost is $3,300.

Costs of the various noncontact thermometers (such as radiation, optical and infrared pyrometers) are close enough for coverage as a single group.

If a portable, manual pyrometer is desired, it will cost from $300 to $900, as a function of range, target area and design. A single-range, optical pyrometer can be obtained for $300, the minimum-cost, infrared, pistol thermometer is about $400, and a manual, radiation pyrometer with telescope and potentiometer readout is $700.

If a permanent, automatic, industrial installation is desired, the cost of the detector alone is around $400, and the cost of the desired converters or remote receivers is similar to the prices noted in connection with

Cost of Temperature Switches—Table XVIII

	Housing Type and Design					
	General-Purpose		Weather-Resistant		Explosion-Proof	
Actuation	Bi-metallic, $	Filled, $	Bi-metallic, $	Filled, $	Bi-metallic, $	Filled, $
Standard, adjustable differential	38 (25)	40 (26)	70 (40)	65	85 (50)	125
Close, adjustable differential	—	50	—	80	—	130
Two-stage switch with independently set actuation	50	85	80	120	105	180

Note: The cost of non-heavy-duty or air-conditioning-control grade units is shown in parentheses.

Cost of Temperature Transmitters—Table XIX

	Motion-Balance		Force-Balance
	Indicating, $	Blind, $	(Blind), $
Pneumatic	250	220	275
Electric (Div. 2 explosion-proof)	425	400	400

Note: The costs of capillary tubing in excess of 3 ft. and the price for thermowells (Table XV) are additive.

Thermocouple Extension Wires Cost—Table XX

Insulation	Braid	Armor	Cost,* ($/100 Ft.)
PVC (polyvinyl chloride)	—	—	15 (40)
Asbestos	Asbestos or fiber glass	—	20 (35)
Asbestos and glass	Cadmium-plated steel spiral	—	65
Asbestos	Stainless steel	—	70
Asbestos	—	Stainless steel	100

* **Note:** Prices are based on solid-wire costs. Stranded-wire prices are given in parentheses.

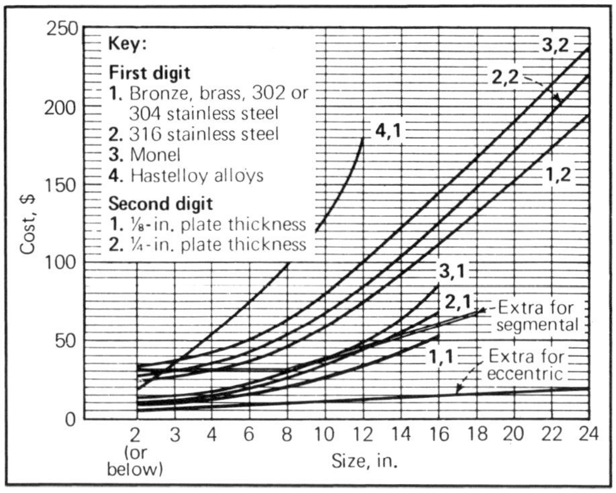

FLOW MEASUREMENT: cost of orifice plates—Fig. 5

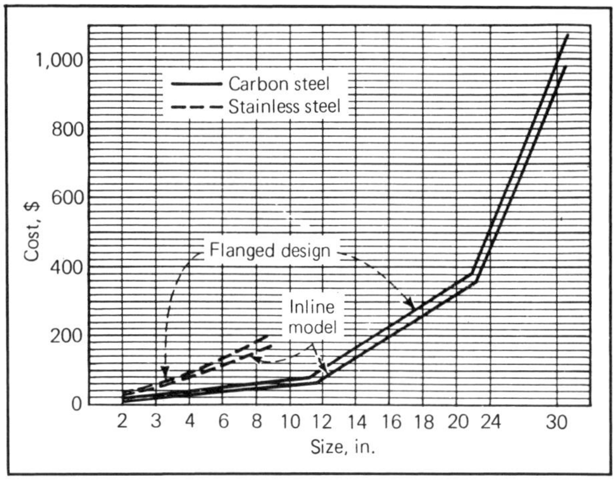

COST of straightening vanes—Fig. 8

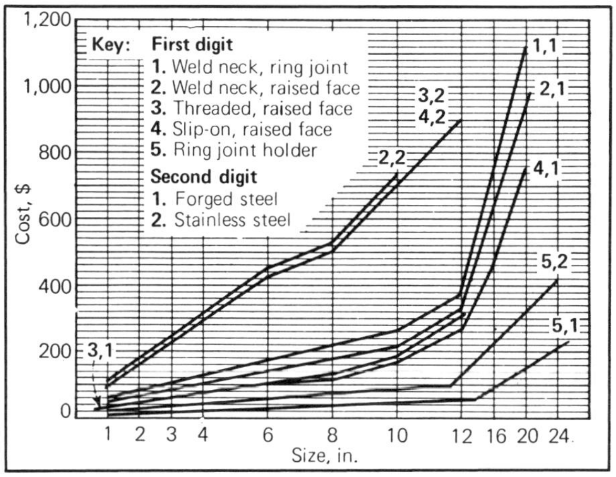

ORIFICE plate holders and 300 psi. unions—Fig. 6

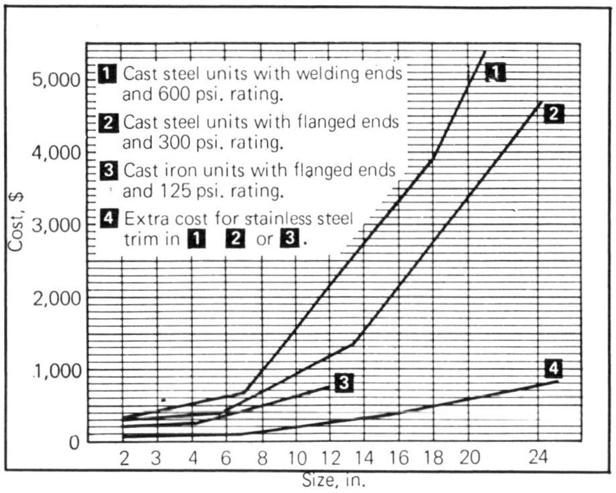

ORIFICE changer fittings cost—Fig. 9

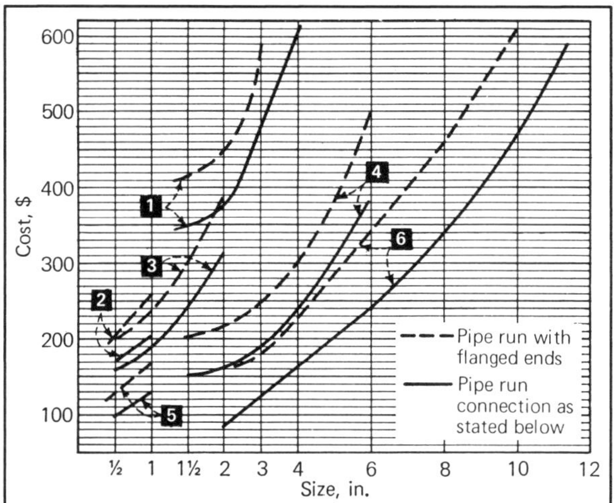

1 Orifice plate removable from welded stainless-steel holder with flange taps. Pipe run with welded ends.

2 Stainless steel unit, honed for accuracy, with corner taps in flanges and with either welded or threaded pipe run ends.

3 Stainless steel unit, calibrated, welded plate not removable, threaded or welded ends.

4 Same as **1**, but steel construction.

5 Same as **2**, but steel construction.

6 Steel construction with weld-neck flange unions, welded pipe ends. Calibration cost is extra.

COST of orifice pipe runs—Fig. 7

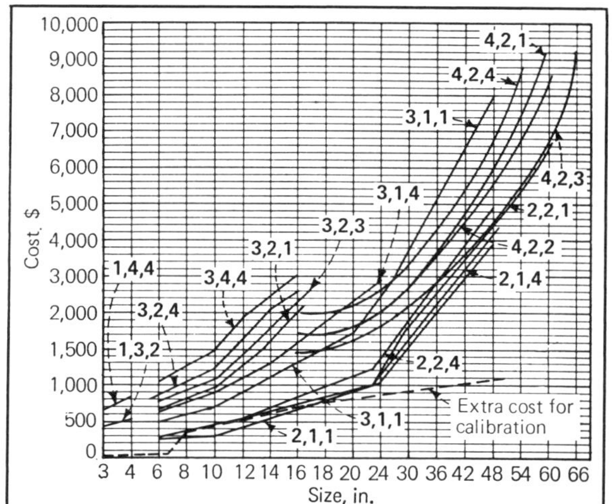

Key: First digit	Second digit	Third digit
Type of design	Housing or flange mat'l	Throat material
1. Barstock	1. Cast iron	1. Bronze
2. Insert with plastic cone	2. Steel	2. Brass
3. Cast with 125 or 150 psi. flanges	3. Brass	3. Steel
4. Fabricated with thin flanges	4. 304 stainless steel	4. 304 stainless steel

COST of flow tubes—Fig. 10

thermocouples. Depending on the type of accessories used and on the overall system complexity, an automatic pyrometer installation can cost thousands of dollars.

FLOW DETECTORS

Orifice-Type Measurement

Flow measurement using orifice plates usually requires a number of components for a complete installation. Fig. 5 provides pricing data for the orifice plates only. The cost curves, which are coded with numbers, refer to the concentric design only; if eccentric or segmental plates are to be used, the noted extras should be added.

Fig. 6 provides information on the cost of ring-joint holders and of 300-psi. steel flange unions of different designs. Pricing information for calibrated- and uncalibrated-orifice pipe runs is furnished in Fig. 7, and the cost of straightening-vanes is provided in Fig. 8.

If it is desired to change orifice plates without the need for opening up the process line, the use of an orifice-changer fitting might be considered. The cost of such a device can be found in Fig. 9.

Other components required for an orifice-type installation may include a differential-pressure transmitter and an orifice valving manifold. Cost information for these components is presented in Table VIII. If an integral-orifice differential-pressure cell is required for low flow applications, it can be priced by adding $100 to the cost figures in Table VIII. If a square-root extractor is required, it costs about $175. For pricing data on local differential-pressure indicators, refer to the paragraphs on manometers and pressure gages.

Flow Tubes and Nozzles

The costs of several flow tube designs in a variety of materials are given in Fig. 10. If an external piezometer ring is desired, it adds about $450 to the unit price for sizes up to 24 in., and $650 for sizes up to 78 in. Other accessories can be estimated by using the information provided in connection with orifice plates.

For pricing information on venturi tubes, refer to Fig. 11 and for flow-nozzle costs consult Fig. 12.

Pitot Tubes, Pipe Elbows, Weirs and Flumes

Pricing data for the various pitot tubes are provided in Table XXI, and cost information on calibrated pipe elbows is furnished in Table XXII. For estimating readout devices and accessories for these flow elements, the information presented for orifice plates can be used. The new higher accuracy "Annubar" unit is also in the noted approximate cost range.

The various weirs and flumes are usually field

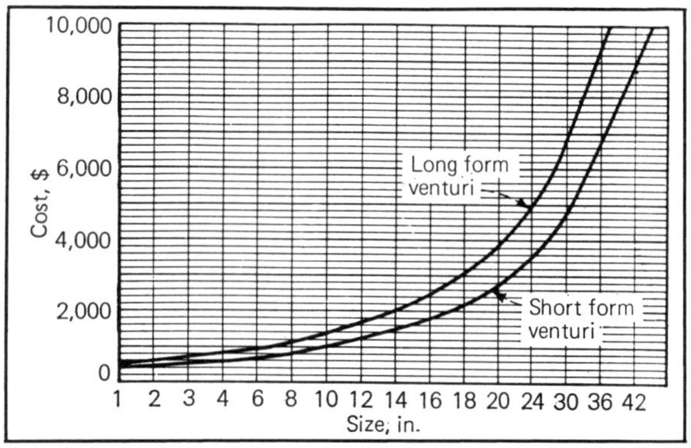

VENTURI tubes, cast iron, 125 psi.—Fig. 11

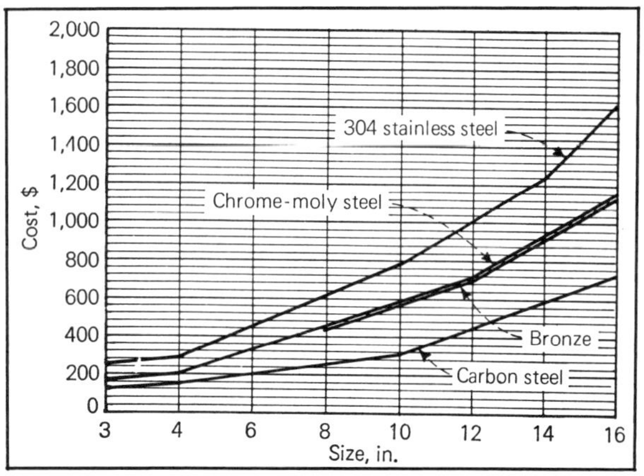

FLANGED flow nozzles cost—Fig. 12

Cost of Pitot Tubes—Table XXI

Construction Materials	Conventional Pitot Tube: Calibrated, $	Conventional Pitot Tube: Uncalibrated, $	Pitot Venturi: Calibrated, $	Pitot Venturi: Uncalibrated, $
Brass	165	100	200	130
316 stainless steel	215	140	260	180
316 stainless steel for up to 1,000 psi. service	420	350	500	430

Cost of Calibrated Pipe Elbows—Table XXII

Size and Connection Type		Carbon Steel, $	316 Stainless Steel, $
Plain or screwed ends	½ in.	160	195
	1 in.	170	205
	1½ in.	230	300
	2 in.	260	330
Flanged	½ in.	200	250
	1 in.	210	260
	1½ in.	285	360
	2 in.	315	420

fabricated and to estimate their cost would involve too many variables. For the reader's orientation, Fig. 13 provides pricing data on a plastic liner that is molded for use as a Parshall flume.

Target and Drag-Body Flowmeters

The main design variations of this flowmeter include the inline design, the wafer type that fits between standard pipe flanges, and the retractable probe version for large pipe sizes. They are available with pneumatic or electronic transmitters or with strain-gage elements. The costs of converting or receiving instruments for the millivolt signals generated by the strain gage are similar to those listed for thermocouples and RTD's. The pricing for all versions of this flowmeter type is given in Fig. 14.

Rotameters

Estimating data are furnished for glass-tube rotameters in Fig. 15 and for metallic-tube rotameters in Fig. 16. Some of the extras listed in Fig. 15 apply to both glass- and metallic-tube rotameters. Fig. 17 furnishes information on calibration charges.

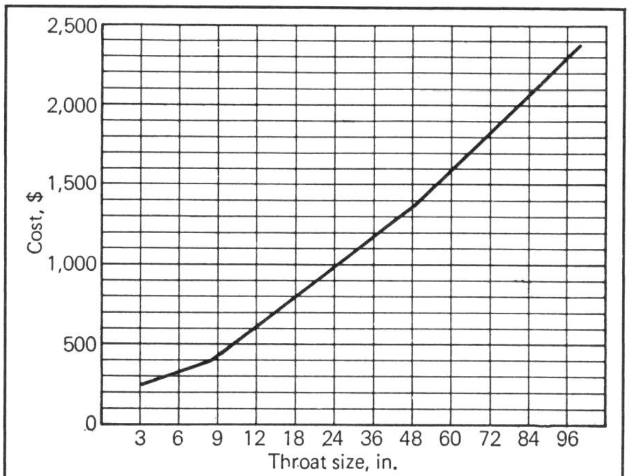

PLASTIC liner for Parshall flume, cost—Fig. 13

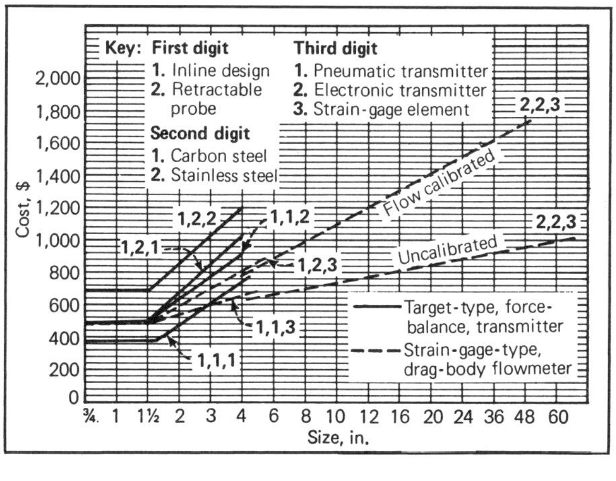

As can been seen, these charges tend to increase with calibration precision. They apply not only to rotameters but also to other standard flow detectors.

When steam or other type of jacketing is required for metallic-tube rotameters, the extra cost is a function of size and fitting materials required. To jacket a 1-in. meter in steel will cost about \$100 and in stainless steel, \$150. The same jacket for 1½-in. and 2-in. meters in steel is about \$200, and \$240 in stainless. Jacketing of a 3-in. meter in steel costs \$240, and in stainless steel \$310.

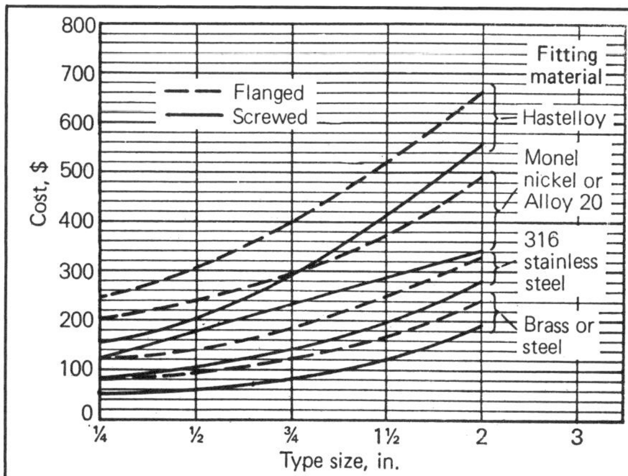

Extras:

For brass ΔP regulator and needle valve with ¼-in. purge assembly	\$ 30
Same as above but stainless steel	80
Pneumatic transmitter	280
Electronic transmitter	340
Explosion-proof electronic transmitter	480
Alarm switch	80
Explosion-proof alarm switch	150
12-in. circular chart recorder	300
12-in. recorder-controller	600
Integrator	150
Explosion-proof integrator	200

COST of glass-tube rotameters—Fig. 15

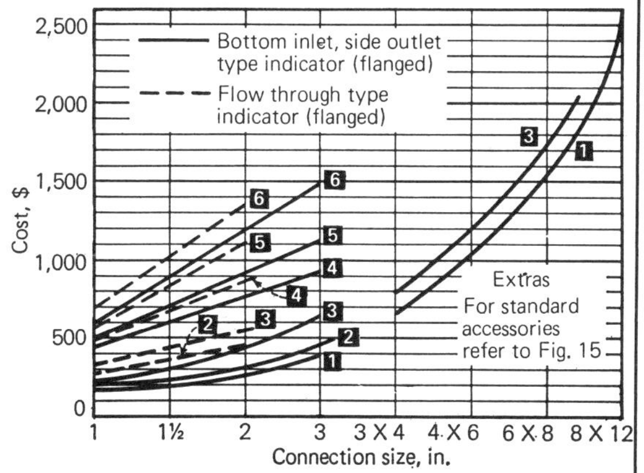

1 All steel construction.
2 Stainless tube with steel fittings.
3 All stainless construction.
4 Hastelloy or Monel tube with steel fittings.
5 All nickel, Alloy 20 or low carbon stainless construction.
6 All Hastelloy or Monel construction.

Magnetic Flowmeters

Fig. 18 provides information on the pricing of magnetic flow-elements of various sizes and materials. Other extra features of this sensor include the high-temperature (up to 300 F.) design, which can be obtained for an additional $200 in sizes up to 3 in., and for $300 in sizes between 4 and 8 in. Platinum electrodes in sizes up to 4 in. represent an extra of $100; in larger sizes, they add $200 to the price. Other special electrode materials, such as the Hastelloys, Monel, Carpenter 20, tantalum and titanium can all be obtained for about $60 extra. Explosion-proofing the flowpipe to Div. 2 requirements adds $40, and to Div. 1 requirements adds $500 for sizes up to 4 in., and $850 between 5 and 8 in. If the explosion-proof meter is to operate on low-conductivity process fluid, an additional investment of $150 is needed.

When the meter is to operate on slurry service, there are several potential methods available for keeping the electrodes clean. For about $100 extra, the meter can be electrically heated to prevent sludge or grease buildup. Brush-type cleanout electrodes add about $200, ultrasonic electrode cleaners add $300.

The low-level a.c. voltage generated by the magnetic flowmeter is carried to the receiving or converting components in cables that can be obtained at $60/100 ft. Some cost data for the receivers are listed below. Receiver costs for low-conductivity service are distinguished by parentheses.

Weatherproof direct indicator	$750 ($900)
Weatherproof indicating converter with ma. d.c. output	$825 ($960)
Weatherproof indicating converter with pneumatic output	$875 ($1,100)
General-purpose direct indicator	$500 ($850)
General-purpose direct converter with ma. d.c. output	$725 ($800)
General-purpose direct converter with pneumatic output	$550 ($950)
General-purpose batch totalizer controller	$2,100 ($2,100)
General purpose direct 12-in. dia. recorder	$870 ($870)
General-purpose 12-in. recorder-controller	$1,200 ($1,200)
General-purpose 12-in. recorder-controller with ratio	$1,350 ($1,350)
General-purpose 12-in. recorder-totalizer	$1,050 ($1,050)
General-purpose selector switch for two magnetic meters	$125
General-purpose simulator calibrator	$250

Turbine Flowmeters

Fig. 19 provides cost data for Type 316 stainless-steel turbine flowmeters in various sizes. Minor cost savings (about $100) can be accomplished by the use of threaded connections or by the use of steel flanges. The unit costs given include the price for

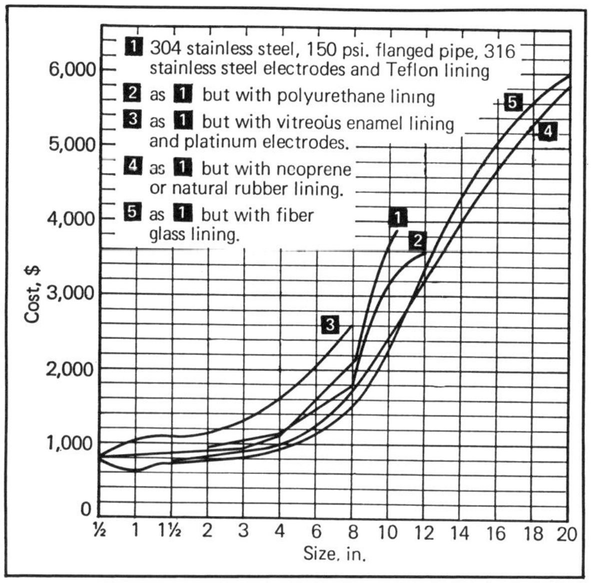

COST of magnetic flowmeters—Fig. 18

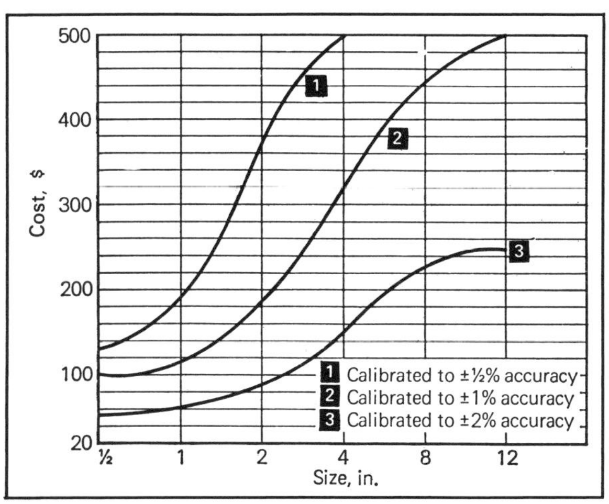

CALIBRATION costs (using water or air)—Fig. 17

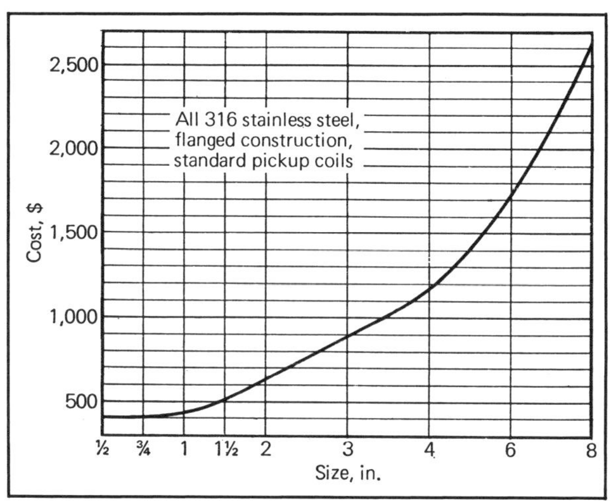

TURBINE FLOWMETERS, flanged, 150 psi.—Fig. 19

a standard magnetic pick-off coil, which is limited to under 400 ft. transmission distances and to installations where the magnetic drag can be tolerated. In this connection the following extras might be applicable:

Magnetic pickoff coil for high temperature (to 1,000 F.)	$40
Magnetic pickoff coil with explosion-proof housing	$40
Magnetic pickoff coil with preamplifier for long distances	$125
Magnetic pickoff coil with preamplifier for long distances with explosion-proof housing	$200
RF (radio frequency) pickoff coil with remote, general-purpose, oscillator-preamplifier	$125
Explosion-proof RF pickoff coil and explosion-proof oscillator-preamplifier	$250

The cost increase due to a pressure rating higher than 150 psi. usually does not exceed 5%. If carbide sleeve-bearings are to be used instead of Teflon sleeve- or stainless-ball-bearings, this might increase the unit price by $50.

The variable-amplitude millivolt pulse signal generated by the magnetic pickoff coil or the constant-amplitude square-wave pulse signal produced by the RF coil oscillator-preamplifier is carried to the receiving instruments in cables that can be obtained for $20/100 ft.

Estimating data for the various converters and receivers are listed below: (All of these items are for mounting in general-purpose areas.)

Direct-reading analog indicator with single logarithmic range	$425
Direct-reading analog indicator with dual logarithmic range	$570
Direct-reading digital indicator for single input	$1,100
Direct-reading digital indicator for two inputs	$1,250
Direct-reading digital indicator for five inputs	$2,500
Converter with analog ma. d.c. output	$500
Converter with analog 3-15 psi. air output	$625
Batch totalizer with one preset counter	$1,000
Batch totalizer with two preset counters	$1,500
Direct-reading totalizer only	$750
Digital-blending-system cost on a per stream basis	$4,000

The above price for digital blending is arrived at by estimating the flow element and preamplifier at $750, the standardizer (or sealer) at $670, the master rate unit at $550, the ratio unit at $600, the digital flow controller at $750, the per stream cost of the digital flow indicator at $600, and the cable at $80.

Other Flowmeters

The swirlmeter is a relatively new development in the flow-metering field, and it should suffice for our purposes to indicate its price range as from $1,500 to $2,000.

Laminar flowmeters contain a laminar flow element, which can be of the matrix type or a capillary tube with set dimensions. This type of pressure-drop-generating sensor can easily be field fabricated, but might also be obtained from suppliers for about $100 (uncalibrated) or $200 (calibrated). The readout devices for sensing the usually low pressure drops that are generated can also be obtained as part of a package, and their cost is the same as that of the receivers and accessories for orifice-type measurements, discussed earlier.

Thermal flowmeters are available for both gas and liquid services for installations in either laboratory or industrial environments. They are either mass-flow sensors or velocity detectors.

If the mass flowrate of a small liquid stream is to be detected in a general-purpose chemical plant with a thermal sensor that produces a zero-based mv. d.c. signal, the cost of this detector will be $1,750. Extras for the above installation might include:

For explosion-proof detector, add	$250
For dual range detector, add	$300
For sensor with three ranges, add	$750
For jacketing of the thermal element, add	$500

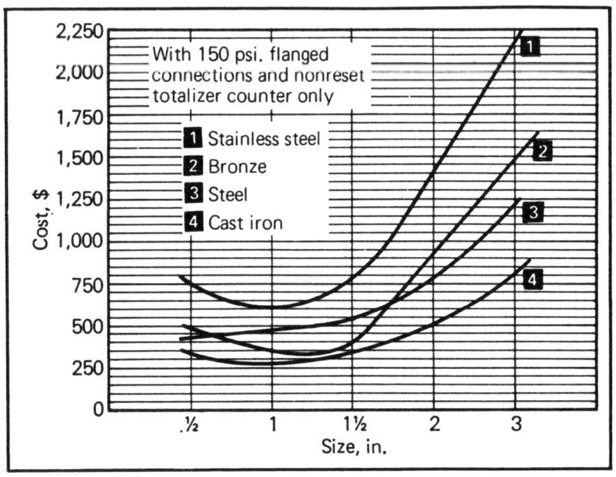

LOBED GEAR type, displacement flowmeters—Fig. 20

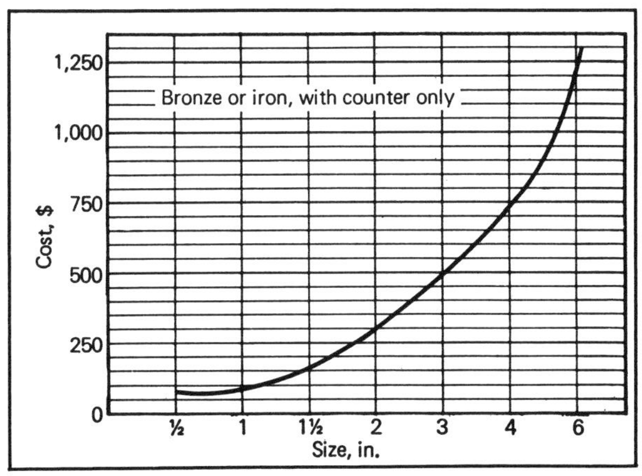

NUTATING disk displacement flowmeters—Fig. 21

For mv. d.c. receiver costs, refer to the paragraph on thermocouples.

If the mass flowrate of small gas streams is to be detected in the laboratory, a thermal flow-tube with corresponding portable readouts can be obtained for $400 to $600 as a function of flow range. (From 0.1 to 10 scfm.).

If the mass flowrate of larger gas streams is to be detected in an industrial plant, a sensor with zero-based volts d.c. output can be obtained for from $1,500 to $2,500 as a function of size. (Up to 1 in.–$1,500, 3 in.–$2,000, 6 in.–$2,500.) The extra cost for the addition of a totalizer readout is $1,000.

If it is sufficient to measure the cooling effect on a probe inserted into an industrial gas stream, such as a flare stack, in order to gain some approximate information for the total flow (basically a composition-sensitive gas-velocity detection), the following unit can be considered:

Air-velocity, calibrated, explosion-proof probe with packing gland, general-purpose power supply, and millivolt-to-air transducer with 3 to 15-psig. output for $1,700. If the millivolt signal is to be received by other instruments, the prices applicable are similar to those given in connection with the discussion of thermocouple accessories.

Angular-momentum-type mass flowmeters can be obtained for industrial installations for either liquid or gas streams. These units are available with direct-totalizer, high-pressure (up to 1,500 psi.) and explosion-proof design features for a unit cost of about $3,000. Remote (including telemetering) readout and a broad range of other accessories are available.

Positive-Displacement Flowmeters

Pricing information for the rotating-vane, displacement-type liquid flowmeter in a variety of sizes and materials is given in Table XXIII. The extras noted are approximately the same as for other meter designs. Fig. 20 provides cost information for the gear-type displacement flowmeter without any accessories; Fig. 21 gives pricing data for the nutating-disk-type water meter; and Fig. 22 contains estimating informa-

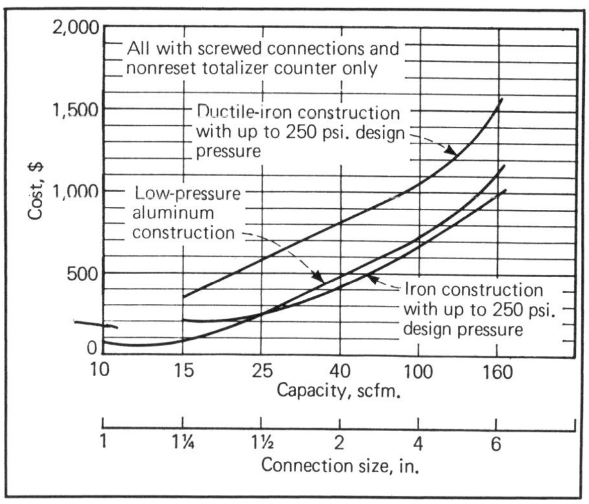

DISPLACEMENT-TYPE gas meters—Fig. 22

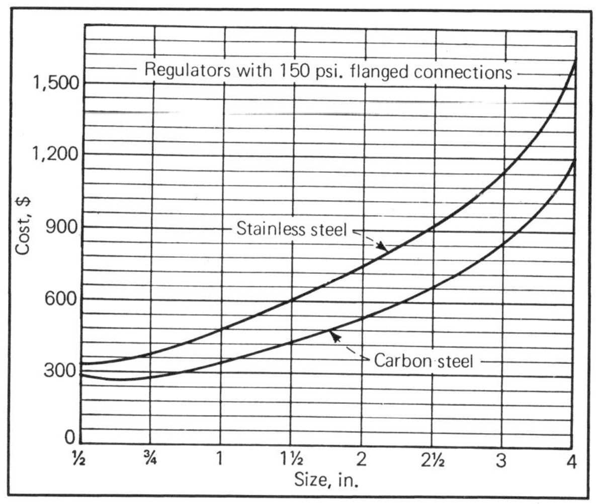

FLOW REGULATORS, self-contained—Fig. 23

Cost of Rotating-Vane Type Displacement Flowmeters—Table XXIII

Connection Size (With 150 Psi. ASA Flanges), In.	Flow Capacity, Gpm.	Construction Materials	Flowmeter and Counter Only, $	Strainer, $	Air Release, $	Preset With 2-Stage Shutoff Valve, $
1½ or 2	To 100	Anodized aluminum steel and Ni-resist 2	410	40	40	100
		Anodized aluminum and stainless steel	450	50	40	100
		Ni-resist 2 and 316 stainless steel	560	60	40	180
		All stainless steel	1,600	260	280	300
3	To 200	Anodized aluminum, Ni-resist 2, steel and brass	500	60	60	170
3 or 4	To 350		780	100	60	300
4 or 6	To 600		1,150	120	60	410

Extras: Explosion-proof micro-switch............ $70; Explosion-proof pulse generator......... $110; Local manual printer.................. $110; Flowrate indicator.................... $160; Temperature compensator.............. $210

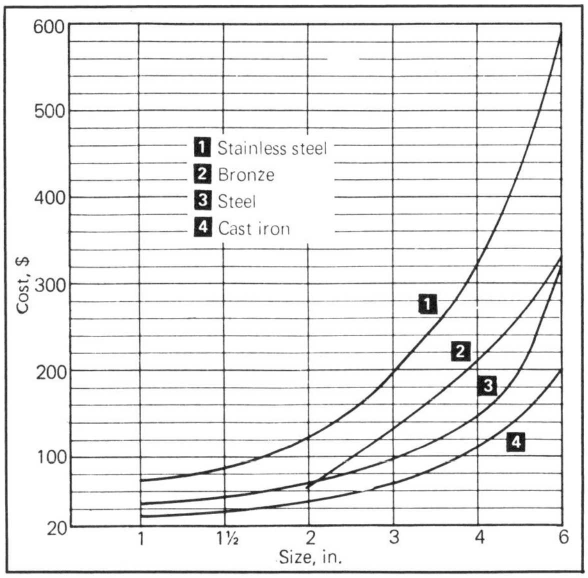

SIGHT flow indicators, flanged connections—Fig. 24

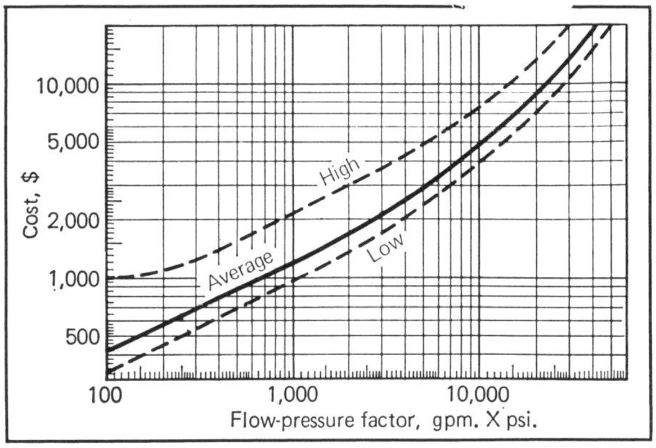

RECIPROCATING metering-pump costs—Fig. 25

tion for positive-displacement industrial gas meters.

Displacement flowmeters for miscellaneous utility services can be obtained for substantially lower charges than those given for the industrial units. For example, a totalizing meter that measures the oil consumption of mechanical equipment costs as little as $60.

Flow Regulators and Sight Flow Indicators

Estimating information for self-contained liquid-flow regulators is provided in Fig. 23. Pricing data on sight flow indicators are furnished in Fig. 47. The reader might note that only flanged designs are listed. Sight flow indicators with screwed connections are also available in the smaller (under 2 in.) sizes, and they are about 20% less expensive than the flanged ones. If special alloy construction is required, the costs shown for stainless steel in Fig. 24 should be multiplied by the following factors:

Low-carbon stainless steels	1.5
Alloy 20	2.0
Nickel, Monel, Hastelloy alloy C	2.5
Hastelloy alloy B	3.0

Metering Pumps

Peristaltic pumps are more frequently used in pilot plants and laboratories than in industrial plants. The flexible tube being the only part in contact with the process, the pump construction materials are unaffected by the nature of the process fluids. Therefore, their cost is not a function of the corrosion-resistance requirements. Constant speed, general-purpose peristaltic pumps for low flowrates can be obtained for $250. If speed adjustment capability is desired, the unit cost is about $450. Double-channel units with variable speed can be obtained for $600. The drive motors in the size range of ⅛ to ¾ hp. can be explosion-proofed for an additional $100.

Reciprocating-metering-pump prices are summarized in Fig. 25, where the product of flowrate and discharge pressure is plotted against cost. The prices shown are applicable to the flow range of 1 to 50 gpm., and for a discharge pressure range of 100 to 4,000 psi. (For flows under 1 gpm., or pressures under 100 psi., use these minimum values instead of the actual ones.)

The "high-average-low" prices shown in the figure are affected by (a) materials of construction, (b) type of flow adjustment, (c) number of cylinders.

For the purposes of this estimate, a carbon-steel pump with stainless trim is considered to be "average." Cast-iron construction would shift the cost toward "low," and stainless steel or Hastelloy alloys would shift it toward the "high" region.

If fixed-rate pumps are acceptable, their cost is likely to be "low." The cost of variable-flow pumps is considered "average" if the flow setting is manually made, and is closer to the "high" line if it is automatically set in response to a control design.

The number of cylinders affects pricing only in high-pressure applications. At discharge pressures greater than 500 psi., a simplex pump will represent the "high," a duplex the "average," and a multistage (three or more cylinders) pump the relatively "low" cost region. ■

Meet the Author

Bela G. Liptak is principal instrument engineer for Crawford & Russell, Inc., Stamford, Conn. 06904. He is responsible for the design, purchase and installation of analytical and control systems in the chemical process industries, particularly for polymer processes. He attended the Technical University of Budapest, and has an M.E. from Stevens Institute of Technology, and an M.M.E. from City University of New York. He is editor of the recently published "Instrument Engineers' Handbook" (Chilton Book Co.).

Costs of Viscosity, Weight, Analytical Instruments

Here is useful information for making a preliminary cost estimate on many process instruments and their auxiliaries that detect and measure viscosity, weight, thermal conductivity, oxygen concentration, pH, moisture and humidity. Cost data are also given for turbidity, consistency, molecular weight and conductivity instruments, as well as for chromatographs and refractometers.

BÉLA G. LIPTÁK, Crawford & Russell, Inc.

Costs for process instruments detailed here incorporate many data for the basic sensor-indicator device or the sensor coupled to a suitable transmitter. The several tables show the differences in costs for measurement, indicating, recording and/or transmitting the appropriate detailed value. Attention is called to the many factors that influence total cost, such as temperature compensation, explosionproof construction, and installation charges.

All cost data given here are based on 1969 prices. The information in the accompanying box explains how to adjust costs to reflect current prices and how to allow for quantity discounts.

How to Adjust Cost Data

Cost data given in this article are based on 1969 prices. At the present rate of inflation, these costs must be updated by the following factors:

Body Material	**Factor**
High-nickel-content alloys	20%/yr.
Stainless steels	10%/yr.
All others	6%/yr.

Purchase quantity affects unit prices. The costs in these articles are based on the quantity of instrumentation needed for a smaller chemical plant—one in the $1- to $5-million range. The following factors should be used as a measure of quantity discounts:

Plant Cost	**Factor**
Under $1 million	1.05
1 to $5 million	1.00
5 to $20 million	0.95
20 to $50 million	0.90

Viscosity Detectors

Tables I and II provide estimating data for pricing various viscometers. In the "intermittent indicator" column of Table I, the listed devices are basically automated laboratory-type units that can also be used in some industrial environments, but were not initially designed for such service. Conversely, all detectors listed under the "industrial" heading can also be used in the laboratory, although some of their features might not be required.

Because temperature substantially affects the apparent viscosity of most substances, it is desirable to maintain it constant or to compensate for its variation. Constant-temperature baths are available in a great variety of sizes and designs, but a good average estimating figure is $1,000 per bath. If temperature compensation is desired, the process temperature must be detected, and the resulting signal introduced into a computing relay. Laboratory viscometers usually depend on constant-temperature baths, while industrial ones include automatic compensation.

WEIGHT DETECTORS

Static Weighing Systems

The usual purpose of a batch-weighing system is to measure the net weight in a vessel. When the weight reaches some predetermined value, the charging or discharging operation is terminated. The main components of such a system will often include the following items:

- Weight-sensing elements or load cells. Estimating data for these is provided in Table III. Mechanical-scale costs are not included because they are not compatible with remote control.
- Weight totalizers that will combine the outputs

Originally published September 1970.

Cost of Viscosity Detectors—Table I

	Laboratory		Industrial			
			Intermittent		Continuous	
Principle	**Manual**	**Automatic**	**Indicator**	**Transmitter**	**Indicator**	**Transmitter**
Capillary timing	$100	$2,500	$2,500			
Capillary extrusion	$300	$5,000	$5,000			
Efflux cup	$100	$600	$600			
Falling ball	$300					
Bubble time	$50					
Sliding plate	$3,000[1]					
Rotational, coaxial cylinder	Portable indicator $1,200					
Rotational, cone-and-plate	$10,000	$15,000				
Falling piston				$2,500[2]		
Ultrasonic						$2,500[3]
Vibrating reed						$2,500[4]
Rotational					(see Table II)	
Two float					[5]	
Concentric viscometer					$900[6]	$1,000[6]
Capillary pressure drop					$2,500	$3,500
Plastometer, cone-and-plate		$9,000				
Plastometer, timed capillary		$5,000	$5,000			
Plastometer, fixed-flow capillary						$12,000[7]
Plastometer, fixed ΔP capillary						$20,000[8]

Notes:

1. Price includes constant-temperature bath and potentiometric receiver.
2. Cost based on explosionproof unit with general-purpose recorder. Temperature compensator not included.
3. Price includes temperature compensator and D.C. output signal in mv. Probe cost only is $300. Cost of a 12-point automatic scanning system is $800/point.
4. Price includes temperature compensator, explosionproof probe, and general-purpose converter with D.C. output signal in ma.
5. Cost of a two-float viscosity indicator can be estimated by adding $250 to rotameter price obtained from Fig. 1. This includes calibration charges
6. Price is based on steel construction. If stainless steel is required or if jacketed unit is desired, add $350.
7. Price includes constant-temperature bath, sampling unit and D.C. output signal in ma.
8. Same as 7 but also includes both MI (melt index) and CIL (Canadian Industries Ltd. flow index) capillaries, two-pen recorder, and variable-speed motor for automatic pressure control at capillary inlets.

of several load cells into a single output. Typical costs are:

Totalizer Type	Cost
Hydraulic, two-cell	$500
Hydraulic, three-cell	$600
Electronic for strain gages	$250

• Transmitters that will take the totalizer output and convert it into a pneumatic or electronic transmission signal. These transducers are specially calibrated for high-accuracy operation.

Transmitter Type	Cost
Hydraulic to pneumatic	$550
Hydraulic to electronic (D.C. in ma.)	$750
Strain gage to electronic (D.C. in ma.)	$300
Strain gage to pneumatic	$400

• Receiver, display and control components that are capable of receiving the hydraulic, strain-gage or transmitted signals, and either display them or control by them.

Display Devices	Cost
6-in.-dia. industrial pressure gage	$50 to $100
12-in.-dia. hydraulic or pneumatic indicator	$500
16-in.-dia. hydraulic or pneumatic indicator	$600
12-in.-dia. potentiometric indicator	$1,200
Digital display for D.C. electronic input (ma.)	$1,000
Single-channel D.C. electronic input (ma.) printer	$1,000
15-input, computer-compatible ticket printer	$5,000

Control Devices	Cost
Photo-electric cutoff added to any dial indicator	$400
Pressure-switch detection of pneumatic signals	$60
Electronic switch for sensing D.C. signals (ma.)	$100
16-in.-dia. potentiometric scale head to detect D.C. signals (ma.) or direct strain-gage signals, including mode selector, variable preset weight control, automatic tare adjustment, preact and dribble controls, and digital encoder for printout	$5,000
Same as above, but with digital display	$5,000

Miscellaneous expenses should not be overlooked when estimating weighing systems. These include, but are not necessarily limited to, the following:

• Engineering and design expenses to accommodate and support the weighing elements. Provisions are needed in the piping and in the electrical conduits for flexible connections; in the structural steel and tank design for lugs and jacks to accommodate the load cells; and in the design, for attaching the staybars and calibration weights.

• Installation expenses, including all modifications necessary to guarantee the free-body behavior of the weighed tank.

• Calibration expenses, including the handling of large numbers of calibration weights.

• Unit price of $100/staybar can be used for estimating staying hardware.

From the above discussion, it should be clear that the cost of weighing systems can vary greatly with

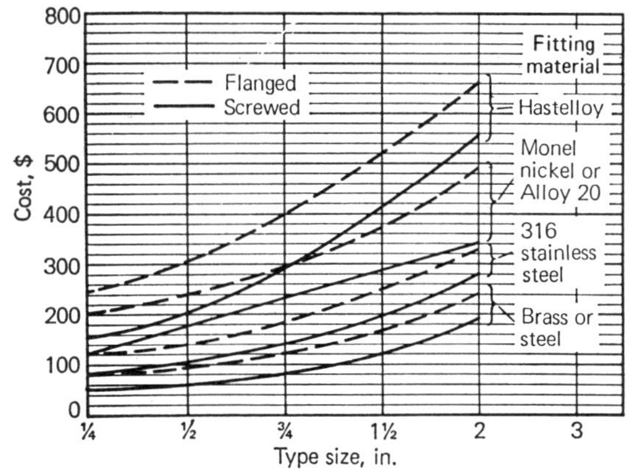

Extras: For brass ΔP regulator and needle valve with ¼-in. purge assembly ...... $ 30
Same as above but stainless steel 80
Pneumatic transmitter 280
Electronic transmitter 340
Explosion-proof electronic transmitter ... 480
Alarm switch 80
Explosion-proof alarm switch 150
12-in. circular chart recorder 300
12-in. recorder-controller 600
Integrator 150
Explosion-proof integrator 200

COST of glass-tube rotameters—Fig. 1

Cost of Rotational Viscosity Detectors—Table II

Installation	Indicator	Transmitter With D.C. Output in ma.
Mounted in atmospheric, constant-level tank	$1,300	$1,900
Mounted in pressurized, variable-level tank	$2,800	$3,400
Flowthrough design for service below 100 psig. and 150 F.	$1,600	$2,200
Flowthrough design for service up to 1,200 psig. and 600 F.	$2,600	$3,200

Extras:
If viscosity to be detected is above 1,000 cp., add $200.
If viscosity to be detected is above 50,000 cp., add $300.
If explosionproof design is required, add $300.
If thermocouple is desired in detection chamber, add $150.

Unit Costs of Load Cells and Pivots—Table III

		Load Cells	
Weight Range	Pivots	Hydraulic	Electronic Strain-Gage Type
0 to 10,000 lb.	$100	$500	$550
10,000 to 25,000 lb.	$100	$500	$900
50,000 lb.	$100	$800	$1,000
100,000 lb.	$200	$1,000	$1,750
200,000 lb., and up	$200	$1,700	$2,700

the quality and complexity of the installation. If it is sufficient to detect the weight of a tank by a single loadcell (and two pivots), and tie the cell's output directly into a local 6-in.-dia. pressure gage, the hardware cost can be as low as 10% of a potentiometric, scale-head-type, remote, automatic-weight control installation. At the same time, it is equally important to remember that no weighing system is inexpensive, because of the miscellaneous costs.

Gravimetric Feeders

Gravimetric feeders may detect the weight of moving solids and are capable of maintaining the mass feedrate constant. Several feeder designs may be considered for this purpose. The corresponding unit prices may differ by a factor of 10.

A self-powered, impact-pan mechanical feeder that is balanced by a poise costs from $400 to $1,200.

Constant-speed belt feeders with electromechanical (on-off) gate control cost from $3,000 to $4,000.

Constant-speed belt feeders with pneumatic (throttling) gate or rotary-feeder valve cost from $3,500 to $5,500.

Variable-speed-controlled belt feeder with pneumatic feedrate setpoint and control will cost from $5,000 to $7,500.

Variable-speed belt feeder with digital electronic setpoint and controls may be priced from $8,000 to $10,000.

Better control accuracies and increased system flexibility compatible with integrated control systems are obtained with the higher-cost systems.

Standard gravimetric feeder accessories include ball-and-disk-type totalizers that directly detect belt speed and pneumatically sense belt load. Their unit price is $800. Less-accurate totalizers are available for about $400. Another feeder accessory is the deaerating surge-hopper package incorporating some level and vibration controls for a price of about $1,000 (including the hopper).

ANALYTICAL DETECTORS

The success of analyzer installations depends to a great extent on the capability of delivering representative, up-to-date samples to the analyzer. This might appear self-evident, but the fact is that approximately 50% of all industrial analyzer installations fail because of problems with their sampling systems.

Most analyzers require sample handling of some sort, and their cost can vary greatly. The prices mentioned in connection with the various analyzers in the following paragraphs include what might be considered "the average" sampling system. If hard-to-handle process streams are involved, these costs can be higher. In some existing installations, more funds were invested in sampling systems than in the analyzers themselves.

Chromatographs

Composition sensing by chromatographic techniques usually means a gas chromatograph, because the gel-permeation technique is not yet widespread in the chemical process industries. Gel-permeation chromatography would usually be considered for very high-molecular-weight samples while gas chromatography is used for medium and low-molecular-weight samples.

For estimating purposes, the distinction is not important because the industrial online versions of either design fall in about the same price range.

Industrial chromatographs range from approximately $3,500 to $15,000. The lower price of $3,500 represents the cost of a single-sample installation where only a single component is analyzed, and where there are no special sampling components or special features and accessories. As the number of sample points increases, or when the purpose of analysis is for several components, the cost rises accordingly.

If a particular installation requires the analysis of 10 sample points, the application engineer has the choice of using any of the following arrangements:

No.	Chromatograph Type	Total Cost
10	Single-point units	$50,000
2	Five-point units	$20,000
1	Ten-point unit	$12,500

On this basis, it is misleading to conclude that the ten-point unit is the most economical, because these costs do not consider the expenses associated with installation, startup, maintenance, and cycle time. Cost comparisons on an installed, calibrated basis might be:

10	Single-point units	$85,000
2	Five-point units	$55,000
1	Ten-point unit	$47,500

If the substantially increased maintenance cost, possible sacrifice of product quality, and high cycle times of the multipoint installations are all considered, the result may favor single-point chromatographs.

Spectroscopy

The degree to which electromagnetic radiation is absorbed or reflected by a process sample in the ultraviolet, infrared and visible regions can be applied to composition analysis. These devices are applicable to liquid, gas or solid (colorimeters only) samples; and they provide high accuracy, and continuous measurements on single-component analysis.

Price range for these analyzers, depending on accessories and features required, is from $3,500 to $8,000. Below this is the cost of an infrared analyzer using a positive-filtering (sensitization) technique at about $1,500. Above this range, are the tristimulus colorimeters (color computers), which may cost from $15,000 to $40,000.

Thermal-Conductivity Analyzers

Thermal-conductivity detectors are applicable only to continuous, binary gas samples where moderate analysis accuracy is acceptable. They are frequently used as the detector portion of other analyzers such as chromatographs.

Standard units, depending on their features and accessories, can be obtained for a price in the range of $400 to $1,500, excluding the receiving recorder. High-sensitivity units cost $2,000 to $4,000.

Refractometers

Differential and critical-angle refractometers are used for the continuous analysis of liquid samples. The critical-angle type can be installed directly in the processing tank or pipeline. Cost range as a function of features and accessories is from $1,500 to $4,000. The inline design with accessories such as the prism wiper represents the upper cost limit.

Oxygen Detectors

Oxygen concentration of liquid and vapor samples can be measured in ppm. and percentage ranges. The estimating prices listed here for such detectors do not include the cost of special sampling systems or the price for a receiving instrument such as a recorder. In some cases, these items can require an additional $1,000.

Detector Type	Cost
Deflection type, paramagnetic (standard)	$1,800
Deflection type, paramagnetic (suppressed-range)	$2,500
Thermal, paramagnetic	$1,250
Catalytic combustion	$750
High-temperature electrochemical	$2,000
Galvanic cell for gas (portable)	$200
Galvanic cell for gas (process)	$2,000
Galvanic or polarographic cells for liquids	$1,000 to $2,000

The galvanic or polarographic cells for liquids are available in a probe-type design—eliminating the necessity for a sampling system.

pH Detectors

The average recording-pH loop can be estimated at a unit price of $1,500. Main components of a loop include the electrode, electrode chamber, amplifier-transducer, and various accessories. Unit prices are:

Electrodes:	
pH electrodes	$30
Reference electrodes	$30
Thermocompensator	$40
Electrode chambers:	
Submersible, Penton	$140
Flowthrough, Penton	$200
Submersible, stainless steel	$250
Flowthrough, stainless steel	$275
Amplifiers:	
General-purpose, D.C. output (mv.)	$600
General-purpose, D.C. output (ma.)	$675
General-purpose, pneumatic output	$800
Explosionproof, D.C. output(mv.)	$1,050
Explosionproof, D.C. output (ma.)	$1,250

In addition to the basic sensor with electrode chamber and amplifier, costs of accessories must be included:

Accessories:	
Connecting cable	$1/ft.
Selector switch, 6-point	$600
Recorder cost must be added.	

Moisture and Humidity Sensors

If the purpose of measurement is relative humidity detection, several simple units are available:

Sling psychrometer	$20
Battery-operated psychrometer	$100
Wall-mounted indicator*	$10
Hygroscopic element, indicator*	$100
Hygroscopic element, recorder*	$350

* Indicates or records humidity

If it is sufficient to indicate that the moisture level of a gas stream is above or below a certain value, a moisture indicator can be placed in the process line. This might be a sight glass with ¼-in. connections and windows, and filled with moisture-sensitive salt crystals that will respond to a change in moisture content by a change in color. Such an indicator can be obtained for $50.

Several analyzers can be considered for industrial moisture measurement in gas or liquid samples. Most of these units include a sample system, and all of them are compatible with remote recorders or controllers. Some of the measurement techniques listed, such as chromatography or infrared detection, have been already discussed, and the information given for them also applies to moisture detection. In the approximate order of increasing cost, the following designs can be considered:

Type	Cost
Impedance	$1,500
Electrolytic hygrometer	$1,500 to $2,500
Heat of absorption	$1,500 to $3,500
Capacitance	$3,500 to $4,500
Microwave absorption	$3,500 to $5,000
Infrared absorption	$3,500 to $5,000
Chromatographic	$4,000 to $5,000
Piezoelectric	$4,000 to $5,000

When the moisture content of solids needs to be detected in an industrial installation, the estimating prices are:

Type	Cost
Capacitance or resistance	$2,000 to $5,000
Infrared absorption	$4,000 to $6,000
Neutron-gage radiation sensor	$10,000 to $20,000

Turbidity Detectors

Industrial inline turbidity analyzers detect either the light-transmission or light-scattering characteristics of process fluids. For units that are suitable to produce electronic (mv.) transmission signals, the price range is from $800 to $1,200, depending on the design. Laboratory models with integral indicators are available at a unit price of $400.

Consistency Detectors

Consistency is measured by many methods. The fixed inline force-balance sensors are sensitive to flowrate variations and are subject to material buildup, but at the same time are the most economical. Their transmitter versions usually cost around $1,500. The float-type pneumatic transmitter is about $1,300, and the strain-gage pneumatic transmitter is about $1,800.

The rotating types are sensitive to variations in seal friction, and the level and flow-bridge types require purge maintenance to prevent plugging. Transmitters using these principles of operation cost between $1,000 to $2,000.

Molecular-Weight Detectors

Several techniques can be applied to molecular-weight detection. Some of the indirect methods, such as gas or gel-permeation chromatography, viscometry, or density measurement, have been discussed in previous sections. For estimating data, see the appropriate section.

Most molecular-weight sensors are laboratory instruments, and their cost is substantially affected by the method of sample handling, number of samples, etc. Some multiple-sample, indicating, vapor-pressure laboratory osmometers can be obtained for $1,400, but the majority of osmometers, light-scattering photometers, and sedimentation or diffusion ultracentrifuges are priced at above $2,000—with some at $10,000. Cost of electron microscopes is $30,000, or more.

Conductivity Detectors

Conductivity loops consist of at least two main components: (1) the conductivity cell, and (2) the conductivity detector bridge—usually the Wheatstone type. For laboratory units, the cells can be obtained for about $50. High-precision conductivity monitors are available at $275.

For industrial installations, the cell is usually of the probe type with removable construction. Therefore, it is provided with an isolating gate valve, and is frequently furnished with a temperature-compensator probe. Depending on its features, the cell may cost from $120 to $200.

The conductivity detector can be a blind Wheatstone-bridge amplifier, with relays for high- or low-alarm actuation, at a cost of about $350. It may be a manually balanced bridge priced at $275, or a servo-balanced indicating bridge, with electronic-transmitter output signals, for about $600. ■

Control-Panel Costs

Costs of control panels and their associated instruments and devices are detailed in this series of tables and graphs. Use them for estimating.

BÉLA G. LIPTÁK, Crawford & Russell, Inc.

In this article, prices are given for a medium-sized plant, in 1969. Hence, they must be corrected for inflation, and plant size. Details of these corrections are summarized in the box on this page.

How to Adjust Cost Data

Cost data given in this article are based on 1969 prices. At the present rate of inflation, these costs must be updated by the following factors:

Body Material	Factor
High-nickel-content alloys	20%/yr.
Stainless steels	10%/yr.
All others	6%/yr.

Purchase quantity affects unit prices. The costs in these articles are based on the quantity of instrumentation needed for a smaller chemical plant—one in the $1- to $5-million range. The following factors should be used as a measure of quantity discounts:

Plant Cost	Factor
Under $1 million	1.05
1 to $5 million	1.00
5 to $20 million	0.95
20 to $50 million	0.90

Control Panels

The cost of panel boards can be estimated by using Table I, where the per-foot unit cost is presented for several control-panel design variations. These prices include the following:

- All structural and support materials, together with tubing, wiring hardware, and miscellaneous minor components such as running lights, pushbuttons, relays, time delays, terminals, etc.
- All expenses associated with the preparation of design drawings to cover structural, layout, wiring, tubing, etc., aspects of the panel.
- All labor costs involved in the building, assembling, checking out and shipping of the panel.

The per-foot unit prices *do not include* the cost of the individual main instruments, such as the various controllers, recorders, converters or annunciators. The extras due to panel-material variations or to special shape requirements are reflected in this table, although their effect on the overall cost is small.

The density of instruments is referred to in terms of the number of horizontal instrument rows; it could also be expressed as the number of instruments per linear foot. Each row of large-case instruments represents a density of 0.6 instruments per foot, and each row of miniature instruments adds 1.2 units to the per-foot density of the panel.

Standby Power Supplies

Backup power supply systems consist of a number of main components, such as batteries of various types, gasoline-engine-driven alternators, battery chargers and inverters. Fig. 1-4 provide estimating data for these components.

Recorders and Indicators

Table II provides estimating data for a variety of recording instruments.

Estimating data for dial-type pressure indicators are provided on p. 349, Table X. Ribbon- or drum-type vertical indicators for pneumatic inputs can be obtained for about $110. Digital indicators have also come into use in recent years. For a 3 to 15 psi. transmitted signal, a digital readout with four significant figures can be obtained for $800. For direct process-pressure readout, the price is about the same; while a direct digital indicator for absolute or differential pressures is priced at around $1,000.

The estimating data for direct-connected temperature indicators are given on p. 351, Tables XVI, XVII. Similarly, the pricing information for turbine or

Originally published October 1970.

Control Panel Costs on a "Per Foot" Basis—Table I

Instrument Type and Electric Area Classification		Pneumatic General Purpose				Pneumatic With Explosionproof Wiring				Electronic General Purpose			
Panel material		**Steel**		**Melamine**		**Steel**		**Melamine**		**Steel**		**Melamine**	
Straight (A), or breakfront (B)		A	B	A	B	A	B	A	B	A	B	A	B
Large-case instruments................	2 rows	$550	$575	$525	$600	$800	$825	$775	$850	$525	$550	$500	$575
	3 rows	650	675	625	700	900	925	875	950	625	650	600	675
Miniature instruments................	2 rows	500	525	475	550	750	775	725	800	475	500	450	525
	3 rows	600	625	575	650	850	875	825	900	575	600	550	625
	4 rows	700	725	675	750	950	975	925	1,000	675	700	650	725
	5 rows	800	825	775	850	1,050	1,075	1,025	1,100	775	800	750	825

Extra cost per foot

Semigraphic Add $125 per linear foot of semigraphic section to the panel costs that were arrived at by using the above unit prices.

Full graphic Add $250 to the per-foot unit cost based on **equivalent** instrument density.

magnetic flowmeter output indicators is provided on p. 356, Fig. 18, 19.

In the category of electronic indicators, the following data might be of general interest:

The price of potentiometric indicators is similar to those recorders listed in Table II.

Millivolt meters with 4-in. scale length and ±2% accuracy for direct connection to thermocouples or similar elements	$75
Milliammeters with 4-in. concentric scale and ±2% accuracy	$50
Milliammeters with 4-in. vertical scale and ±2% accuracy	$70
Explosion-proof milliammeter with 4-in. ±2% accuracy scale	$75
"Quick Scan" milliammeter with 4-in. ±2% accuracy vertical scale	$120
Milliammeter with 26-in. long concentric scale and ±½% accuracy	$400

Digital displays are gaining in their popularity and are frequently used in conjunction with switching boxes to allow for the selective displaying of one of several inputs. The most popular applications involve multipoint digital indicators used as readouts on a number of thermocouple inputs. A solid-state digital indicator capable of displaying up to four significant figures and capable of accepting any one of 30 thermocouple inputs through a toggle-switch-type selector box costs about $3,000.

A similar unit for 100 inputs of either thermo-

Cost of Recording Instruments—Table II

	Large-Case Circular Chart[1]						Miniature Stripchart		
	Direct-Connected				Receivers[9]				
Number of Pens	Pressure, $	Flow or ΔP, $	Filled Temperature, $	Potentiometric, $	Pneumatic, $	Electronic, ma. D.C., $	Pneumatic, $	Electronic, ma. D.C., $	Potentiometric, mv. or RTD,* $
One pen..............	180–220[2]	330[3,4]	250[5]	760–880[6]	190	500	250	300	550
Two pens.............	260[2]	—	390[5]	—	260	680	340	500	1,000
Three pens...........	340[2]	—	540[5]	—	340	820	420	800	1,450
Four pens............	420[2]	—	680[5]	—	420	—	—	—	—
Multipoint strip-chart printer 8-in. chart				1,200[7]					
Multipoint strip-chart printer 12-in. chart				1,500[8]					

*RTD-resistance temperature detector.

Notes: 1. No extra for electric or spring chart drives. If pneumatic drive is required, add $40; for explosionproof electric drives, add $60.
2. Price range noted applies to low and medium pressures. For very high, very low, or absolute pressures, the per pen unit price is $300.
3. Add $125 if stainless steel ΔP meter body is required.
4. If a recording rotameter is being estimated, $320 should be added to the rotameter base cost.
5. Pricing is based on case-compensated thermal elements with 5 ft. of capillary tubing.
6. Price range covers several electronic input signal types, with thermocouple inputs falling on the lower, and direct magnetic or turbine flowmeter units on the upper, end of the price range.
7. Up to 12 points, printed from thermocouple or RTD inputs.
8. Up to 24 points, printed from thermocouple, RTD, strain gage, etc., inputs.
9. For a recorder-pen-actuated alarm switch, add $25 to the unit prices listed.

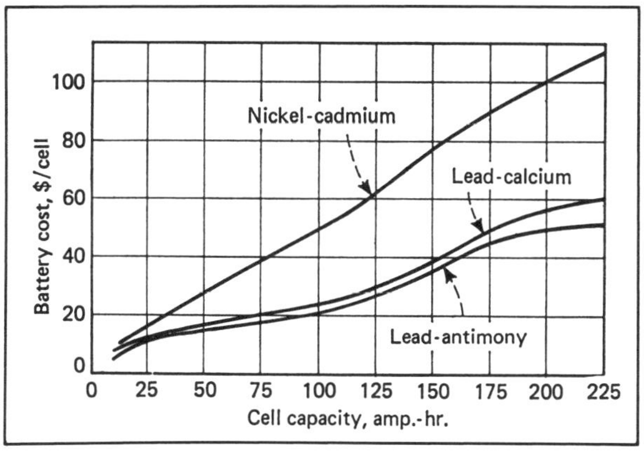

BATTERY CELL costs, for backup power—Fig. 1

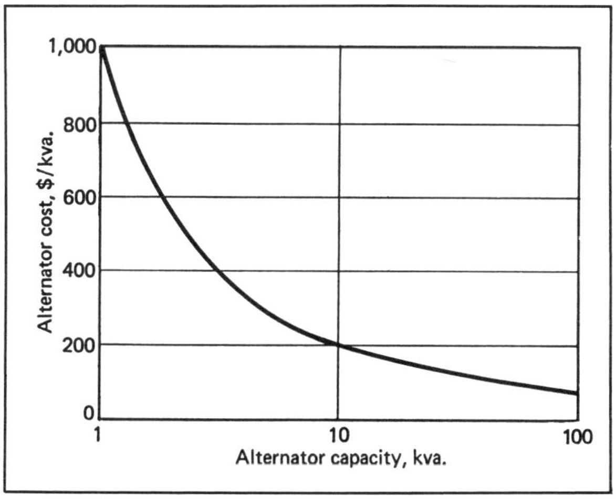

ALTERNATOR COSTS (gasoline-engine driven)—Fig. 2

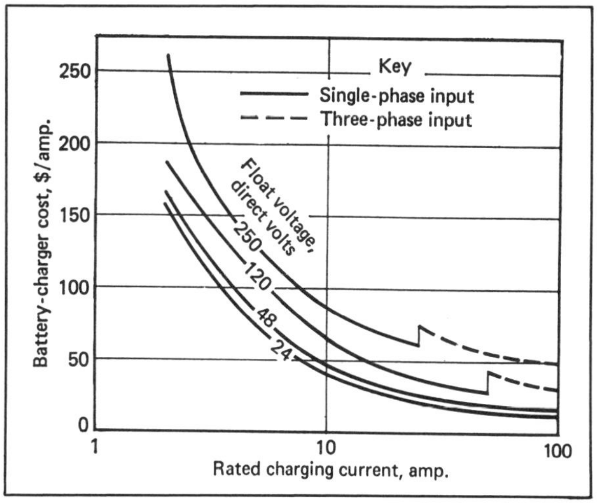

BATTERY-CHARGER costs—Fig. 3

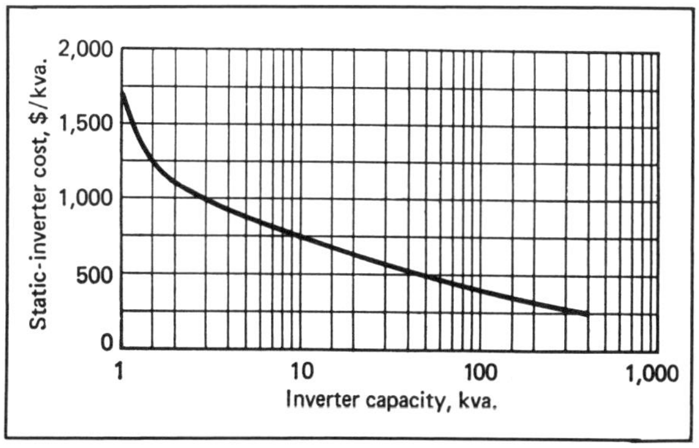

STATIC-INVERTER costs vs. capacity—Fig. 4

Cost of Various Controllers—Table III

	Direct-Connected						Receivers	
	Large Case[8]			Medium Case			Miniature	
Features	**Pressure, $**	**Flow[1] or ΔP, $**	**Filled Temperature, $**	**Pressure, $**	**Flow[1] or ΔP, $**	**Filled Temperature, $**	**Pneumatic, $**	**Electronic,[2] $**
Indicating controllers	400[3]	550[4]	470[5]	270[3]	460[4]	375[5]	330	460
Recording controllers, 1 pen[6]	430[3]	580[4]	500[5]	Not available			580	760[2]
Recording controllers, 2 pen[6,7]	500[3]	650[4]	570[5]	Not available			660	960[2]
Extra cost, 2-mode controller	60	60	60	Included in base price			35	Included
Extra cost, 3-mode controller	120	120	120	50	50	Included	90	75
Extra cost, remote setpoint[9]	60	60	60	60 (if available)			65	Included
Extra cost, auto/manual subpanel	100	100	100	Included in base prices above				

Notes: 1. For rotameter-type installations, the cost of the rotameter itself is in addition to the prices listed in this column.

2. The recorder-controller prices represent the sum of indicating controller and separately mounted recorder prices.
3. The prices given apply to the medium-pressure range. For very high, very low, or absolute pressure applications, add $100 per point.
4. Add $125 if stainless steel ΔP meter body is required.
5. Prices are based on case-compensated thermal elements with 5-ft. lengths of capillary tubing.
6. If a pen-actuated switch is required, add $25 to the prices listed. There is no extra for electric chart drives or for spring chart drives in large-case recorders. For pneumatic chart drives, add $40, and for explosionproof electric chart drives for large-case recorders, add $60.
7. First pen controls; second pen only records.
8. The base prices in this category assume proportional-only controllers without manual/automatic subpanels. For additional control modes, for A/M subpanels, refer to the extra-cost items in the lower half of this table.
9. This feature is required for cascade slaves, or for other remotely and automatically set controllers.

couple or RTD (resistance temperature detector) elements will be about $6,000.

A 200-thermocouple input package using lighted pushbuttons in the switching matrix might be priced at $10,000.

Controllers

Table III provides estimating data for a large variety of control instruments, including electronic and pneumatic, large-case and miniature, direct-connected and receiving, recording or indicating types. The table also lists the prices for some controller features that are additive as extras to the base prices.

Converters

The estimating data given below cover only the most widely used converter types. Where price ranges are given, this implies that a variety of designs and product qualities can be considered, depending on the nature and importance of the installation. For the converters used with magnetic flowmeters, the prices given in parentheses represent those for low-conductivity services.

I/P	ma. d. c. current to 3–15 psi. air converter, explosionproof	$120
E/I	Millivolt or thermocouple to ma. d. c. current converter	$350
E/I	As above, but explosionproof	$550
E/P	Millivolt or thermocouple to 3–15 psi. air converter	$450
E/P	As above, but explosionproof	$600
E/I	Magnetic flowmeter to ma. d. c. current	$725 ($800)
E/I	Magnetic flowmeter to ma. d. c. (weatherproof)	$825 ($960)
E/P	Magnetic flowmeter to 3–15 psi. air	$550–$825 ($950)
E/P	Magnetic flowmeter to air (weatherproof)	$875 ($1,100)
pH/E	pH to millivolts converter	$600
pH/E	As above, but explosionproof	$1,050
pH/I	pH to ma. d. c. current converter	$675
pH/I	As above, but explosionproof	$1,250
pH/P	pH to 3–15 psi. air converter	$800
Pu/I	Turbine flowmeter to ma. d. c. current	$500
Pu/P	Turbine flowmeter to 3–15 psi. air	$625
R/I	RTD or other resistance to ma. d. c. current	$250–$400
R/P	RTD or other resistance to 3–15 psi. air	$300–$550
dR/I	RTD differential to ma. d. c. current	$300–$600
Watts/E	Watts to millivolts converter	$200

Computing Relays

Data for pricing a variety of computing relays are given below. Where price ranges are given, it indicates that a variety of product qualities can be considered, depending on the importance of the installation. The cost of electronic devices is given in parentheses to distinguish them from the pneumatic unit prices.

Adding relay	$75–$250 ($250–$600)
Differentiating relay	($300)
Dividing relay	$240 ($275–$800)
High-limit relay	$80 ($120–$250)
High-selector relay	$20–$60 ($160–$240)
Low-limit relay	$80 ($120–$250)
Low-selector relay	$20–$60 ($160–$240)
Multiplying relay	$240 ($275–$800)
Proportioning relay	$150–$250 ($250–$600)
Ratio relay	$220–$280 ($300–$350)
Scaling relay	$150–$250 ($250–$600)
Square-root extractor	$150–$250 ($200–$300)
Subtracting relay	$75–$250 ($250–$600)

Miscellaneous Panel Instruments

Listed below are unit prices for miscellaneous panel instruments that have not been discussed earlier. Where both pneumatic and electronic designs are available, the unit price for the electronic versions is given in parentheses.

Annunciator cost per alarm point	($50)
Digital blending system (per stream blended)	($4,000)
Manual loading station	$135 ($150–$200)
Power supply—regulated (per user)	($40)
Power supply—standard (per user)	($20)
Programmer—cam type	$300–$1,000
Programmer—edge follower	$750–$1,500
Pushbutton stations (industrial)	($20)
Relays, industrial, electric, on-off	($25)
Running lights	($10)
Selector switch for low-level electronic signals	($100)
Selector switch for 110-V. 60-Hz. circuits	($20)
Selector switch for pneumatic signals	$25
Switch, high or low alarm on millivolt inputs	($175)
Switch, high or low alarm on ma. d. c. inputs	($110)
Switch, high or low alarm on pneumatic inputs	$50
Switch, high and low alarm on ma. d. c. inputs	($150)
Timer	($100)
Totalizer for pneumatic (or ma. d. c.) inputs	$250 ($400)
Totalizer and preset counter for pneumatic (or ma. d. c.)	$360 ($460)
Totalizer for turbine meter input	$750
Totalizer and two preset counters for turbine meter inputs	$2,000

■

Safety Instruments and Control-Valves Costs

This final instrumentation-costs article covers pressure relief and explosion suppression; plus control valves, including temperature, flow, pressure and level regulators.

BÉLA G. LIPTÁK, Crawford & Russell, Inc.

In this article, as in the three that precede it, prices are given for a medium-sized plant, in 1969. Adjustments for inflation and plant size are summarized in the box at the right.

SAFETY INSTRUMENTS

Pressure Relief Valves

One of the most frequent safety requirements involves protection against overpressure. Prominent members of the pressure-safety-device family are relief valves. Their basic designs include spring-loaded, weight-loaded and pilot-operated designs. Weight-loaded relief valves are seldom used anymore, and their few remaining applications are in the low-set-pressure areas (conservation vents), which are covered separately. The pricing of large, pilot-operated safety valves is slightly below that of the spring-loaded types, but for estimating purposes the use of spring-loaded unit prices should suffice.

Fig. 1 provides estimating data for spring-loaded, chemical, pressure-relief valves in standard materials, sizes and pressure ratings. If O-ring seats for tight shutoff are required, add 10%; if an open-lever accessory is needed, add 5%; and if a packed-lever is necessary, add 10% to the noted unit price.

Minimum-size relief valves are usually provided with screwed connections, and frequently are of bar-stock construction. These units in the ½-in. or ¾-in. size range and in bronze materials can be obtained for about $20, while in steel or stainless steel materials the unit price is around $50.

Rupture Disks

Three basic types of pressure-safety-disks can be distinguished. These are the metallic, the knife edge and the graphite versions. The metallic disks are the most frequently used, and their price can be estimated by summing up the costs of the disk itself (Fig. 2), the price for the insert-type base and hold-down flanges (Fig. 3) and the cost of the vacuum support (Fig. 2), if required. Special metallic disks for two-directional, low-pressure etc. service cost substantially more than the standard ones, and therefore in estimating them, a multiplier of 3 can be applied to the prices shown in Fig. 2.

For pricing the reverse-buckling knife-edge type rupture disk assembly, the reader should refer to Fig. 4. The estimating cost for the assembly is the sum of the prices of the disk and the holders.

The cost of all-graphite rupture-disk assemblies is given in Fig. 5. These units consist of an integral disk and holder assembly and can be inserted directly between pipeline flanges.

Corrections to Cost Data

Cost data given in this article are based on 1969 prices. At the present rate of inflation, these costs must be updated by the following factors:

Body Material	Factor
High-nickel-content alloys	20%/yr.
Stainless steels	10%/yr.
All others	6%/yr.

Purchase quantity affects unit prices. The costs in these articles are based on the quantity of instrumentation needed for a smaller chemical plant—one in the $1- to $5-million range. The following factors should be used as a measure of quantity discounts:

Plant Cost	Factor
Under $1 million	1.05
1 to $5 million	1.00
5 to $20 million	0.95
20 to $50 million	0.90

Originally published November 1970.

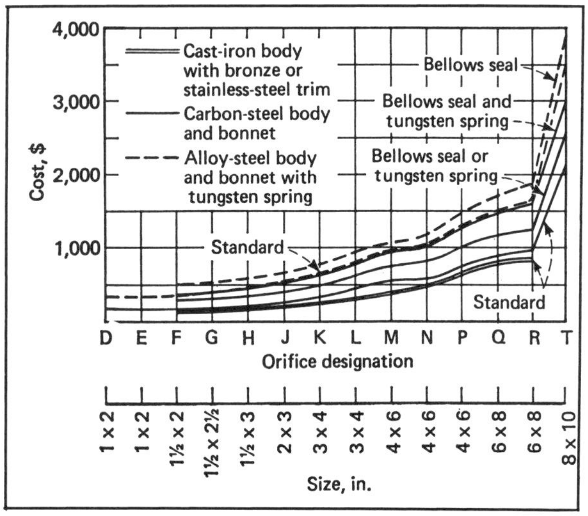

RELIEF VALVES, 125 or 150 psi.—Fig. 1

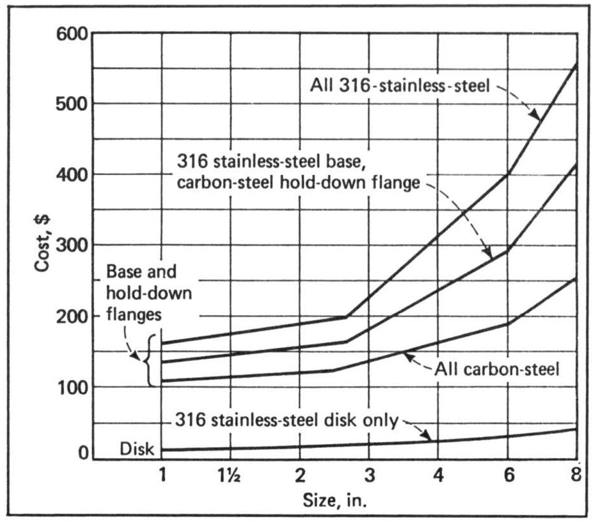

RUPTURE disks, reverse-buckling, knife-edge—Fig. 4

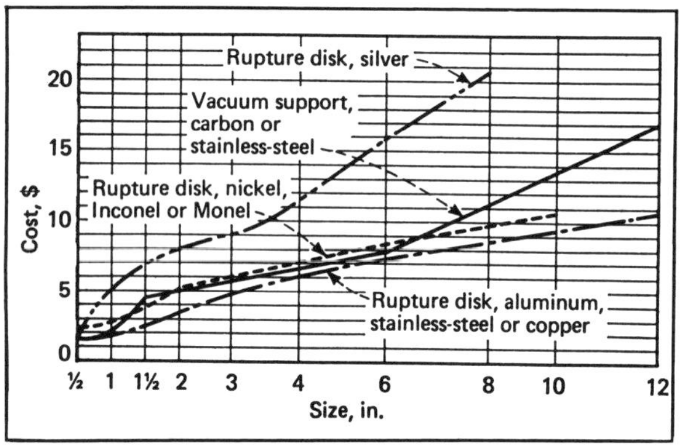

RUPTURE disks and vacuum supports—Fig. 2

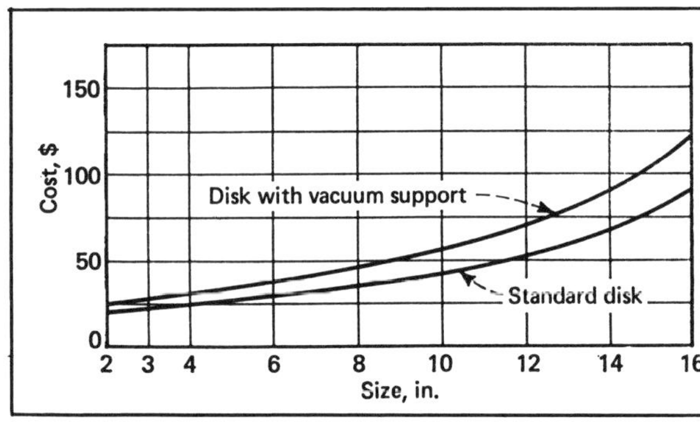

ALL-GRAPHITE rupture disk assemblies—Fig. 5

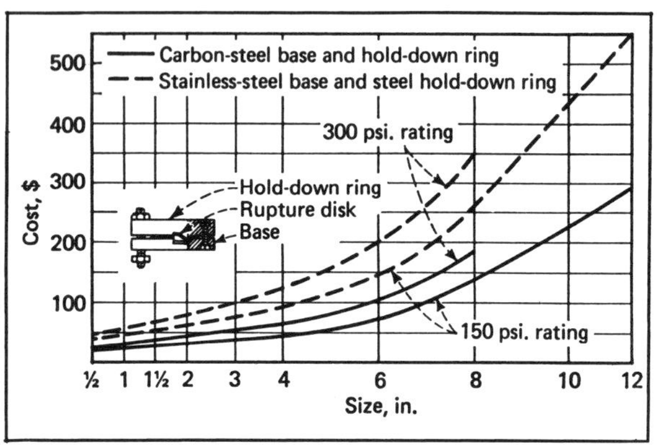

INSERT-TYPE base and hold-down flanges—Fig. 3

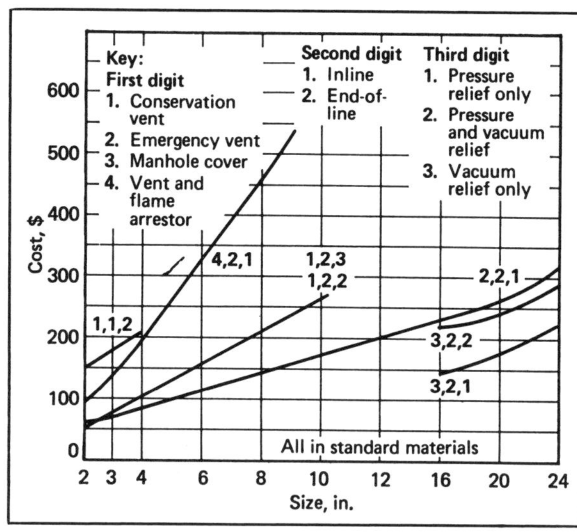

CONSERVATION and emergency vents—Fig. 6

Conservation and Emergency Vents

When the purpose of a pressure safety device is to protect against *low* positive- or negative-pressures, the spring-loaded relief valves and rupture disks are replaced by conservation and emergency vents. These devices can provide protection against high or low pressures, or both. They can be of the reclosing type (vents) or the nonreclosing type (manhole covers). In addition, they can be designed for inline or end-of-line mounting. Sometimes it is economically advantageous to combine the conservation vent and flame arrestor into a single unit.

Fig. 6 provides estimating data for these designs

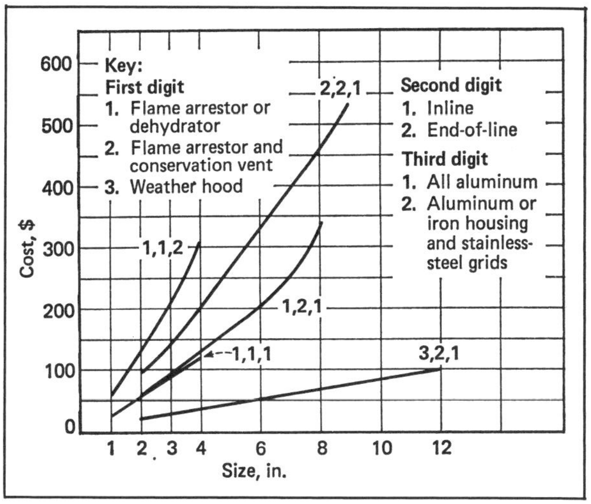

FLAME arrestors, dehydrators and hoods—Fig. 7

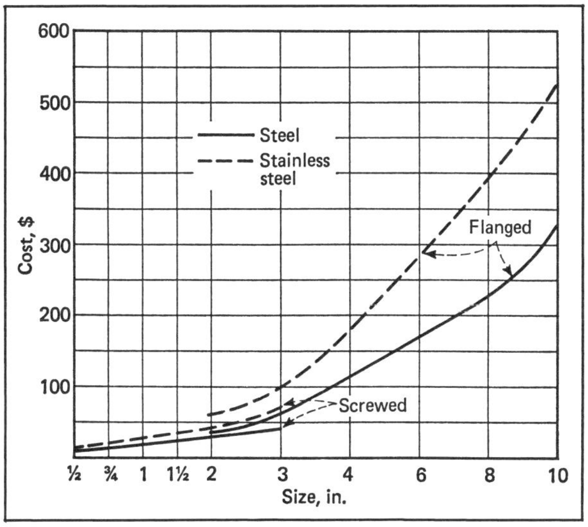

EXCESS-FLOW check valves—Fig. 8

in standard materials of construction. This usually means aluminum metallic parts and rubber gaskets. The gasket material can be changed to Teflon for a relatively small extra (under $50), but if nonstandard metallic parts are needed, the extras can be very substantial.

Therefore, for corrosive services, it is more common practice to coat all the parts that are exposed to the process vapors with Heresite.* Depending on size and design, this might add $75 to $200 to the unit prices listed.

Flame Arrestors and Dehydrators

For estimating purposes, the cost of flame arrestors and dehydrators can be assumed to be the same. Flame arrestors might be installed with weather hoods, as integral parts of conservation vents or by themselves. Cost data for these units in standard materials is given in Fig. 7. If other materials are required, the extras become rather substantial. Relative to all aluminum construction, the following multipliers can be used to estimate flame arrestors in other materials:

Material	Multiplier
Carbon-steel body, stainless-steel grids	3.0
Type-316 stainless-steel construction	4.5
Stainless-steel body, nickel grids	7.0

Excess-Flow Check Valves

Excess-flow valves protect against discharging the contents of storage tanks through broken pipe connections. They are either used as flanged inserts into tank nozzles or as threaded units screwed onto dip pipes inside the vessel. The pricing information provided in Fig. 8 covers both steel and stainless steel construction.

Explosion-Suppression Systems

As a function of system size and complexity, the cost of an explosion-suppression unit might vary from $4,000 to $20,000. If the purpose of the installation is to protect a single piece of equipment such as a dryer, the system cost will be closer to the lower end of this range. Inversely, if the unit is to protect a whole train of processing equipment, the system price will approach the upper end of the range.

As an example, the cost of an explosion-suppression system, which is to protect a single piece of equipment and which may consist of: a pressure sensor for explosion detection, a thermal sensor for fire detection, three extinguishers in the 3 to 10-liter size range and all required power and control units, is about $6,000.

Combustible-Gas Detectors

The cost of detecting the presence of flammable gases and alarming if their concentration exceeds safe limits, is affected by several factors. These include the electric area classification, the complexity of the sample-handling system required and in case of multi-sample units, the extent to which system components are shared between sample points. As one extreme, it is possible to install completely independent systems for each point being analyzed. As another extreme, one might use a sample selector and share the sample handling, detector and readout components between several points on a sequential basis.

A single-point detector with a minimum of sample-handling components, to be installed in a general purpose area might be purchased for under $1,000, but the per-point cost of more elaborate explosionproof units is likely to fall between $1,500 and $2,000.

CONTROL VALVES

Globe-Type Designs

Cost information for the various globe, control-valve designs is furnished graphically. The conventional-

* A phenol-formaldehyde thermosetting resin.

and cage-type globe valves can be priced from Fig. 9, which provides data for various sizes, construction materials and connection styles. The same type of information for three-way valves and for venturi-outlet type angle valves is provided in Fig. 10 and for split-body valves in Fig. 11.

The pricing for some of the extra features, where such is a function of valve size, is provided in Fig. 12.

In addition to the extras given in Fig. 12, the following unit prices can be used for estimating the special features and accessories listed below:

Double packing	$10
Airset	$20
Explosionproof solenoid pilot valve	$30
Explosionproof limit switch	$40
Limit stop	$40
Booster relay	$60
Pneumatic positioner	$70 to $100
Side-mounted handwheel for up to 3-in. valves	$130
Explosionproof current to air transducers	$135
Electronic positioner	$150
Side-mounted handwheel for up to 6-in. valves	$210
Side-mounted handwheel for above 6-in. valves	$400

Butterfly Valves

In Fig. 13, pricing information is furnished for butterfly control valves. The costs shown include a "standard" size diaphragm actuator and a stainless steel vane. The curves on this figure distinguish between the costs of the various body designs (standard or heavy) and body materials (iron, steel or stainless).

Accessories and special features might include:

Declutchable handwheel for up to 12-in. size	$160
Declutchable handwheel for up to 24-in. size	$250
Finned extension for up to 12-in. size in steel (stainless)	$220 ($400)
Finned extension for up to 24-in. size in steel (stainless)	$270 ($500)
Rubber liners for standard duty valves	$20

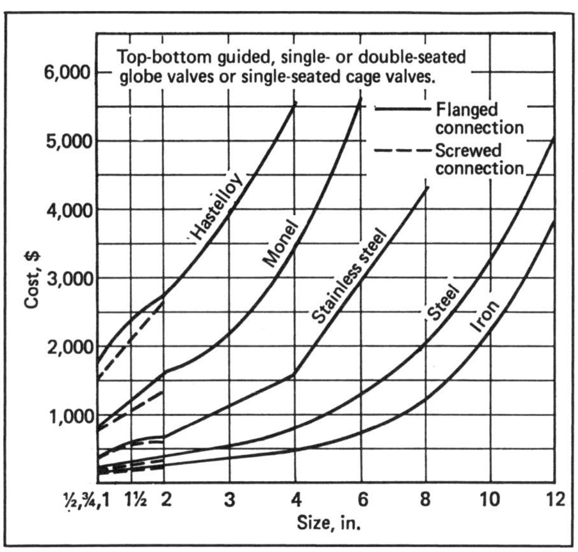

GLOBE VALVES and cage valves—Fig. 9

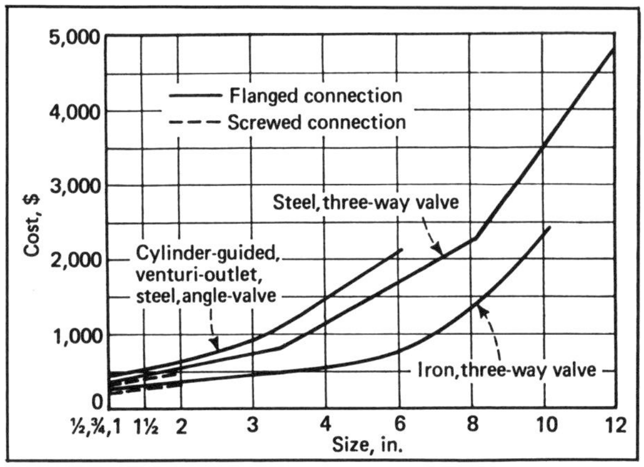

THREE-WAY valves and angle valves—Fig. 10

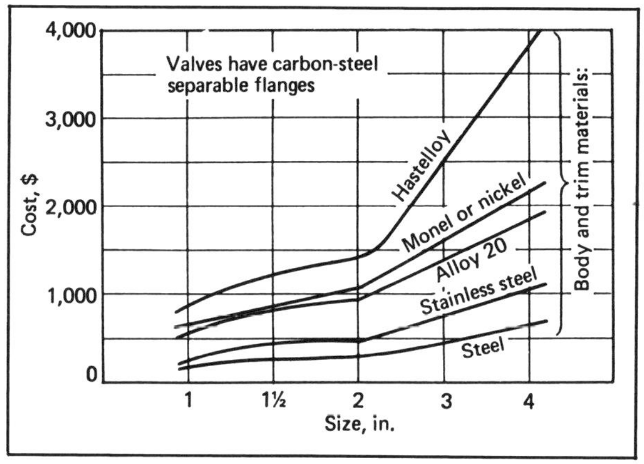

SPLIT-BODY valves with separable flanges—Fig. 11

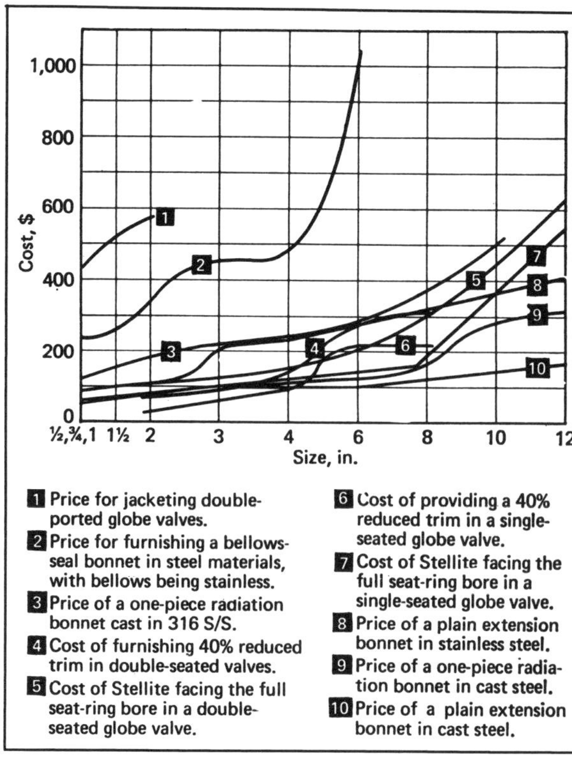

EXTRAS for globe control valves—Fig. 12

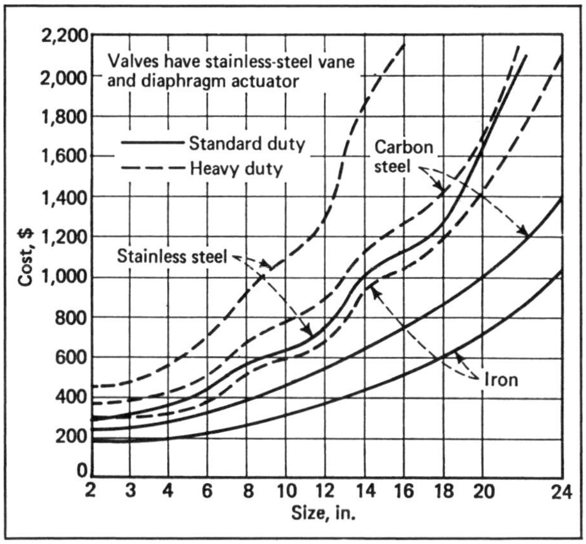

BUTTERFLY control valves—**Fig. 13**

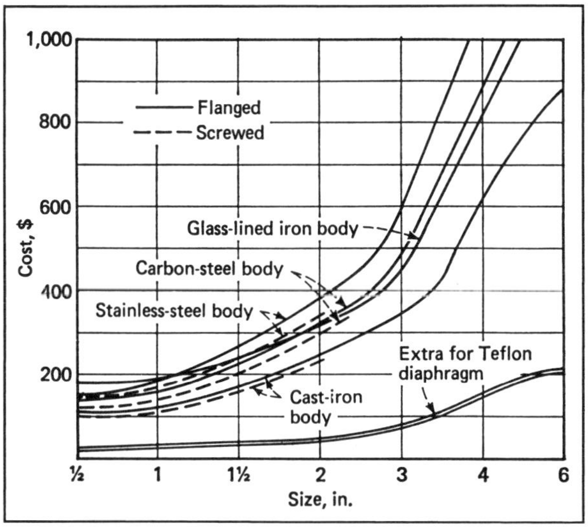

SAUNDERS valves, diaphragm actuators—**Fig. 14**

Rubber liners for heavy duty valves (for size range 2-in.–24-in.) .... $50–$100

Other accessories can be estimated by use of the unit prices given for globe valves.

Saunders Valves

Cost information for saunders valves is provided in Fig. 14. The curves on this illustration indicate the pricing structure for standard saunders valves with screwed or flanged connections and with various body materials. The unit costs shown include the price for a neophrene diaphragm. If Teflon is desired as diaphragm material, the extra cost shown in Fig. 14 should be added.

Ball and Plug Valves

Pricing information on ball valves for throttling service is given in Fig. 15. The costs are based on

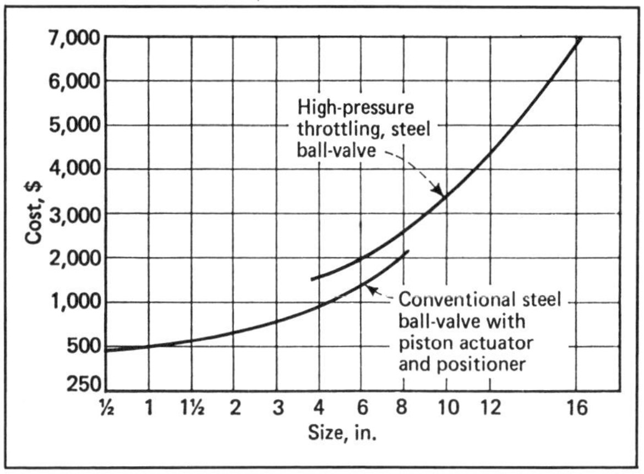

BALL VALVES, throttling type—**Fig. 15**

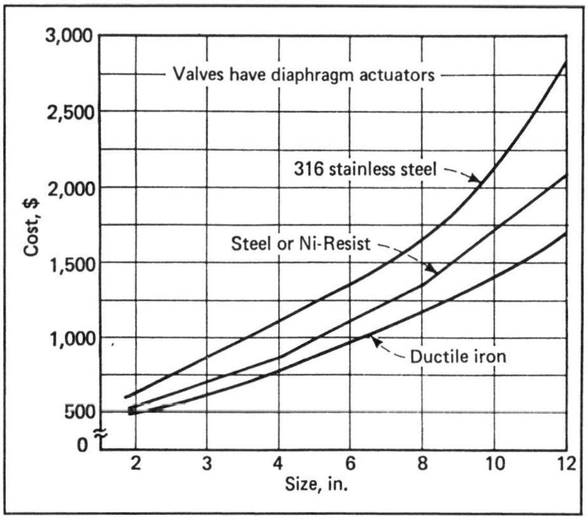

CHARACTERIZED ball valves—**Fig. 16**

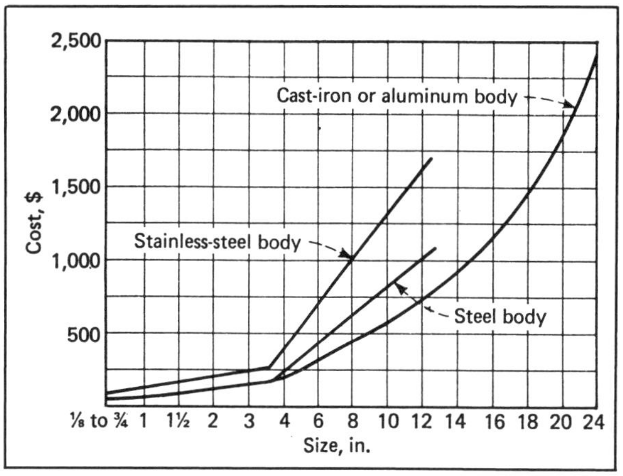

PINCH VALVES, pneumatic jacketed—**Fig. 17**

steel body materials and include the price for the pneumatic positioning actuator. One of the curves represents the conventional, the other the high-pressure, throttling ball valves. The pricing for accessories can be obtained from the information furnished under the discussion of globe valves.

Estimating data for remote-actuated on-off ball and

Cost of Plug or Ball Valves (Without Actuators)—Table I

		Body Material													
		Cast Iron		Ductile Fe or Aluminum	Bronze		Carbon Steel			Type 316 Stainless Steel			Alloy 20		Monel
Size In.	Connection	Standard, $	Full Port, $	Standard, $	Standard, $	3-Way, $	Standard, $	Full Port, $	3-Way $	Standard, $	Full Port, $	3-Way, $	Standard, $	3-Way, $	Standard, $
½	Flanged	30	...	35	26	...	35	45	...	60	70	...	80	...	80
½	Screwed	15	...	17	17	35	17	17	35	35	45	50	40	95	70
1	Flanged	30	...	35	35	...	45	55	...	80	85	...	110	...	100
1	Screwed	15	...	17	17	50	17	22	50	45	50	60	60	130	90
1½	Flanged	55	...	60	85	120	70	75	130	135	150	190	180	310	150
1½	Screwed	30	...	35	35	90	35	35	95	80	90	140	115	220	145
2	Flanged	70	100	75	100	160	85	110	180	155	180	250	220	410	190
2	Screwed	35	...	35	40	130	45	70	130	105	135	170	140	280	170
3	Flanged	125	170	145	160	370	170	200	370	250	300	520	325	825	340
4	Flanged	200	240	220	235	530	250	340	530	370	530	750	480	1,100	570
6	Flanged	270	370	300	450	...	420	540	830	670	840	1,200	885	1,730	1,010
8	Flanged	410	...	460	700	...	660	1,000	1,250	1,000	1,540	1,750	1,330	2,580	2,115

plug valves have been combined into the same tabulation. Table I gives the unit prices for ball or plug valves only, in various sizes and materials. This table also distinguishes between body styles and connection types. The costs given are based on the "fire-safe" design and on body or flange ratings not exceeding 150 psi.

The costs of the cylinder actuators for the plug and ball valves are provided in Table II; these costs are a function of the valve size and also of the body style and actuator design involved.

In order to arrive at the total price for a remote-actuated on-off valve installation, one should add up the valve cost (Table I), the actuator cost (Table II) and the price for the accessories (listed below):

Three- or four-way air selector switch	$20
Fine-adjustment type needle valve	$15
Explosionproof limit switch	$40
Explosionproof solenoid pilots, 2-way for valves up to 4 in.	$30
2-way for valves up to 8 in.	$40
3-way for valves up to 3 in.	$30
3-way for valves up to 8 in.	$65
4-way for valves up to 3 in.	$45
4-way for valves up to 4 in.	$70
4-way for valves up to 8 in.	$100

Characterized Ball Valves

Characterized ball valves can be estimated by referring to Fig. 16, which provides pricing information

Cost on On-Off Cylinder Actuators for Plug or Ball Valves—Table II

	Actuator Type			
	Spring-Loaded		Dual-Acting	
	For Valve Type			
Size, In.	Standard, $	3-Way or Full Ported, $	Standard, $	3-Way or Full Ported, $
½, ¾	85	140	55	55
1	140	140	90	90
1½	140	165	90	90
2	165	180	90	90
3	220	280	110	110
4	280	370	110	170
6	370	500	170	220
8	500	660	220	370

for various sizes and body materials. The characterized ball is stainless steel in all cases and the O-rings are of buna-N. The flanged bodies are rated for 150 psi.

Teflon packing can be obtained for an additional $50 and a declutchable handwheel is $220 extra. Other accessories can be estimated using the unit prices listed under "globe valves."

Pinch Valves

Fig. 17 shows the cost of the pneumatic- (or water-) jacketed pinch valves in various body materials and sizes. The sleeve material in all cases is either syn-

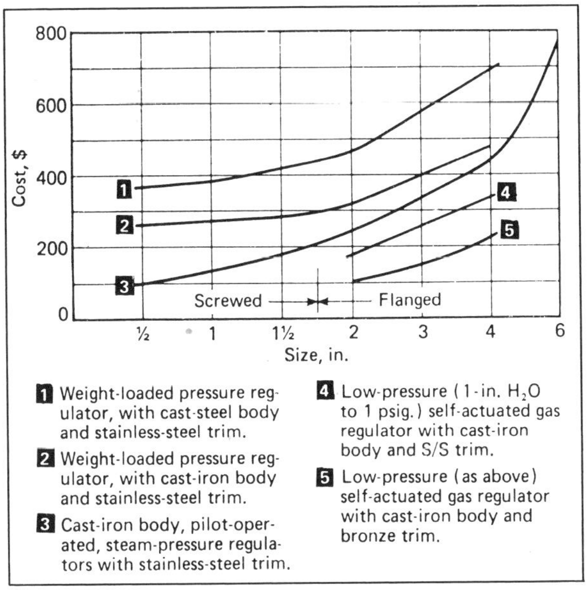

PRESSURE regulators, self-actuated—Fig. 18

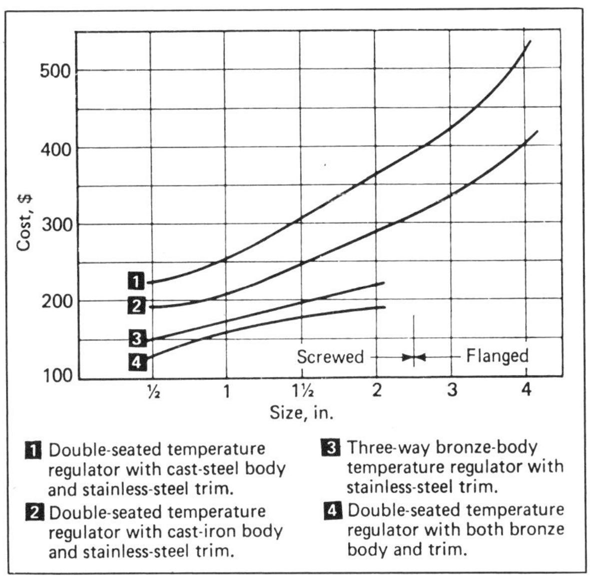

TEMPERATURE regulators, self-actuated—Fig. 19

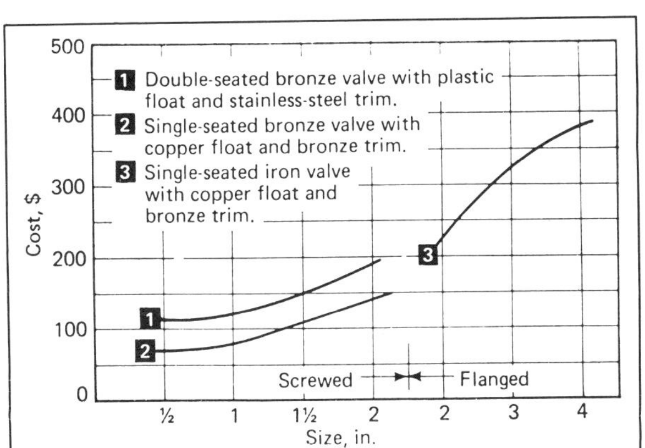

FLOAT-TYPE, level regulator valves—Fig. 20

thetic or gum rubber. In order for this valve to operate on a 3-15 psi. controller output signal, additional accessories are required to convert this signal into the corresponding pneumatic or hydraulic jacket pressure. The cost of these boosters and other components usually does not exceed $100.

REGULATORS

Pressure Regulators

Spring- and diaphragm-type pressure control valves or pump governors can be estimated by using the data provided for globe control valves.

Self-actuated, single-ported regulators for utility services in the ¼ in. to 1 in. size range, to control pressures between 2 psi. and 150 psi. can be obtained for the following prices:

With cast iron body and composition trim.....	$50
With cast iron body and stainless steel trim....	$60
With steel body and composition or stainless steel trim.....................................	$130
With bronze body and composition trim.......	$80

Estimating data for other types of pressure regulators is furnished in Fig. 18.

Temperature Regulators

Estimating information for temperature regulators is provided in Fig. 19.

Flow and Level Regulators

Pricing data for flow regulators was provided in Fig. 23 on p. 358.

Estimating information for float-operated level regulator valves is furnished in Fig. 20. ■

Meet the Author

Bela G. Lipták is principal instrument engineer for Crawford & Russell, Inc., Stamford, Conn. 06904. He is responsible for the design, purchase and installation of analytical and control systems in the chemical process industries, particularly for polymer processes. He attended the Technical University of Budapest, and has an M.E. from Stevens Institute of Technology, and an M.M.E. from City University of New York. He is editor of the recently published "Instrument Engineers' Handbook" (Chilton Book Co.).

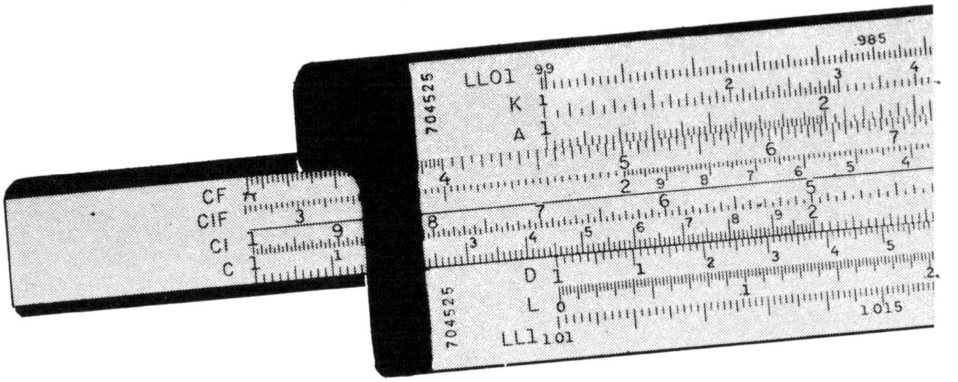

Estimating the Costs of

Here are the data for estimating equipment, installation and operating costs for virtually any type of gas-cleaning plant.

J. R. F. ALONSO, University of Melbourne, Australia

There is rarely one unique method for cleaning an effluent gas; often a number of processes are technically adequate. Thus, the final choice will frequently be guided by a preliminary cost analysis.

Even when only one method appears technically suitable, a preliminary cost estimate is necessary to give some guidelines on the economics involved, and for comparing detailed proposals from various contractors.

Comprehensive correlations for capital and operating costs for air-cleaning plants are available from the U.S. National Air Pollution Control Administration.[1] Also, extensive correlations on efficiency and costs have been published by C. J. Stairmand for dust-arresting plants[9] and for cleaning a radioactive aerosol from air.[10]

However, these are overall costs; they do not provide the flexibility necessary for assessing alternate methods and comparing proposals. The relative emphasis of various components may change in different systems, whereas a standard basis for calculation must be used for comparing cost analyses. Thus detailed analyses become essential.

For such detailed analyses, the costs of the actual cleaning units, ducting, insulation, chimneys, fans, electric motors, switch gear, erection charges, profit margins, etc., have to be added. Further, maintenance costs must be considered for the average life of critical components.

This article attempts such a broad approach. The following cost correlations are a composite of information obtained from leading manufacturers in the U.S., England, Germany and Australia. This information was made available on condition that it be used only in generalized correlations and that no specific sources be acknowledged.

To apply these data for cost comparisons, one should:

Originally published December 1971.

- Allow for cost variations with time and location.
- Follow a standardized estimating procedure in which the costs of the following components are added together stepwise: cleaning units, transportation costs of cleaning units, power and prime movers, sheet-metal elements, and erection and installation.

Cost Variations

Every effort has been made to present as many data as possible in terms, such as manhours and bills of materials, that are universal and independent of local currency. These data can be converted to local costs by applying labor costs at the construction site, and current prices of materials.

Where it has been impossible to present the data in universal terms, 1970 U.S. dollars have been used. These dollars can be converted to the local cost of the moment through multiplying by:

- The exchange rate.
- The yearly cost-index ratio.
- The National Price Factor, which reflects differences in productivity, tariffs, transportation, labor skills, etc.

Exchange rates for various currencies in 1970 are shown in Table I, along with estimates of the National Price Factor in 1970. The Marshall and Swift Equipment Cost Index, the CHEMICAL ENGINEERING Plant Cost Index, or any other suitable index can be used to prorate the data from 1970 to the date of the estimate.

Standardized Procedure

The standardized procedure is described by Eq. (2) and (3) in the box on p. 378. This procedure begins with

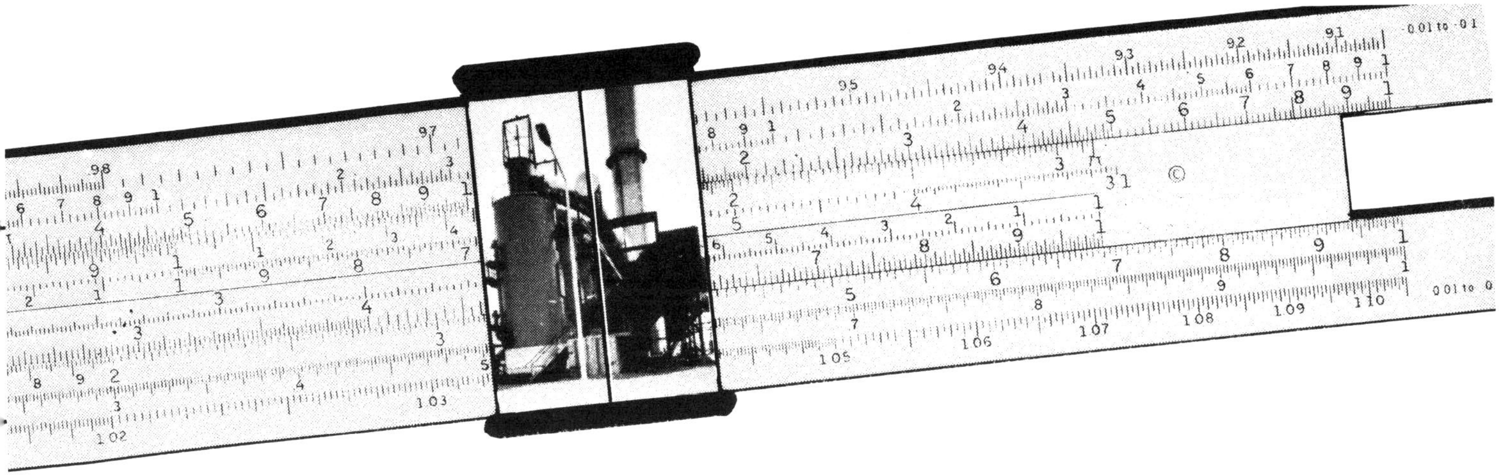

Gas-Cleaning Plants

Capital Cost Factors—Table I

	National Price Factor	Foreign Currency Equivalent to the U.S. $1 (April, 1970)		Basic Skilled-Labor Wage U.S. $1/manhr.	Relative Efficiency
Australia	1.20	$ (Aus.)	0.89	3.50-4.50	1.00
Belgium	0.94	Franc	49.6		
England	0.91	£	0.412	3.00	0.85
France	0.91	Franc	5.54	5.05	0.90
Germany	0.88	Deutsche Mark	3.65		
India	1.35	Rupee	7.50		
Italy	0.86	Lira	614		0.815
Japan	0.83	Yen	349		0.85
Netherlands	0.92	Guilder	3.52	3.50	0.85
U.S.	1.00			7.00 (Gulf Coast)	1.00
Sweden	1.09	Krona	5.18		
Spain	1.00	Peseta	69.7		

the costs for cleaning units, transportation, power and prime movers, and for erection. To these are added estimates of sheet-metal, bar-stock and labor costs for simple cyclones and ducting elements to obtain a total plant cost.

This article does not cover transportation costs since they do not lend themselves to generalized correlations. However, they should be considered in a later stage of the project when more-detailed information is available.

CLEANING UNITS

Cleaning units, which may be used singly or in various arrangements, are of seven general types:

- Dry centrifugal collectors.
- Fabric collectors.
- Electrostatic precipitators.
- Direct-flame incineration devices.
- Catalytic incineration devices.
- Wet collectors.
- Mist eliminators.

Prices for each of these units are correlated as a function of inlet-gas flow in Fig. 1 to 9. These prices apply to large-scale industrial units, supplied by most leading manufacturers.

Dry Centrifugal Collectors

Dry centrifugal collectors include: (1) gravity settling chambers with or without louvres or baffles, (2) skimming chambers where mild rotational velocity is given to the dirty gas, and (3) simple cyclones and multiple-cyclone installations where high rotational velocities and efficiencies are achieved.

The cost data in Fig. 1 refer only to the simple and multiple-cyclone systems, since designs for the settling chambers and skimmers are not standardized. Those units can only be estimated from the sheet-metal fabrication costs for the particular design.

Fabric Collectors

Fabric collectors use filters of either the bag or envelope type, which may be woven or felted fabric with ei-

Basic Equations for Estimating the Costs of Gas-Cleaning Plants

$$\text{Corrected cost} = \left(\text{U.S. cost}\right)\left(\text{Exchange rate}\right)\left(\text{Yearly cost index ratio}\right)\left(\text{National price factor}\right) \quad (1)$$

$$\text{Cleaning-plant cost} = \left(\text{Cost of cleaning units}\right) + \left(\text{Cleaning unit transportation cost}\right) + \left(\text{Cost of power and material drives}\right) + \left(\text{Cost of simple cyclones and ducting elements}\right) + \left(\text{Costs for erection of ducting and installation of fans and cleaning units}\right) \quad (2)$$

Sheet-Metal Cost

$$\text{Cost of simple cyclones and ducting elements} = \left[\left(\text{Price of sheet metal, \$/lb.}\right)\left(\text{Sheet-metal area, sq. ft.}\right)\left(\text{Thickness, ft.}\right)\left(\text{Sheet-metal density, lb./cu. ft.}\right)\right] +$$

Bar-Stock Cost, Labor Cost

$$\left[\left(\text{Bar stock used, ft.}\right)\left(\text{Cross-sectional area of bar stock, sq. ft.}\right)\left(\text{Bar-stock density, lb./cu. ft.}\right)\left(\text{Price of bar stock, \$/lb.}\right)\right] + \left[\left(\text{Manhours}\right)\left(\text{Skilled labor rate, \$/manhour}\right)\right] \quad (3)$$

ther natural or synthetic fibers. As dust deposits in the porous fabric, the pressure drop increases until the solids must be removed. This removal may be manual, automatic by rapping or shaking gear, or automatic by cyclically-reversing flow mechanisms that take individual bags out of the flow system and clean them with gas reversals.

These fabric collectors, commonly known as "bag houses," are usually supplied shop-assembled on a channel frame. The hoppers, hose boots and supports can be constructed by the user, or assembled by the contractors for an additional charge. However, for small gas-flowrate models, the unit is normally furnished completely finished and ready to use.

Cost data for fabric collectors are presented in Fig. 2. These data refer only to automatic high-efficiency fabric collectors.

Electrostatic Collectors

In an electrostatic precipitator, the ionized gas passes through oppositely charged plates, where the dust particles are deposited. The deposited solids are removed by rapping, washing or scraping. Most industrial installations use automatic-rapping vibration systems. Plate collectors are most effective for large-flowrate systems while circular-type designs are preferable for mists and low flowrates. Electrical conductivity of the dust is the most influential factor in the selection of an electrostatic precipitator for a particular application.

Electrostatic precipitators are highly efficient on small particles and allow operating temperatures that may not be reached in fabric or other collectors. Pressure drop is also very low. Despite their very high initial capital cost, electrostatic precipitators often compare favorably in a long-term cost comparison with other collectors.

Fig. 3 contains the data for high-flowrate industrial electrostatic collectors.

Direct-Flame Incineration Devices

Direct-flame incinerators find applications in which combustible dust, mists or soluble gases can be burned to produce innocuous combustion products. The combustion usually requires addition of extra fuel to maintain an adequate chamber temperature.

Increasing the thermal efficiency of the direct-flame incinerators is possible by recycling the flue gas leaving the combustion chamber and by adding a feed preheater. Cost data for direct-flame incineration devices are presented in Fig. 4.

Catalytic Incineration Devices

Fig. 5 contains the cost data for catalytic incineration devices.

Catalytic burners are used instead of direct-flame incineration devices under conditions of very low percentages of combustible material in the gas stream, or prohibitively high fuel costs, or where it is not advisable to have very high temperatures in the process stream. As with direct-flame incinerators, the use of recycle and heat exchangers greatly improves the system's thermal efficiency. This reduces operating costs while increasing the initial capital costs.

Catalytic incinerators require more-sophisticated control systems than their direct-flame counterparts. Also, catalyst replacement costs are high when needed. It is absolutely necessary to choose the proper catalyst for the process to reduce this expense. The experience of the

Cost Correlations for Cleaning Units and Electric Motors

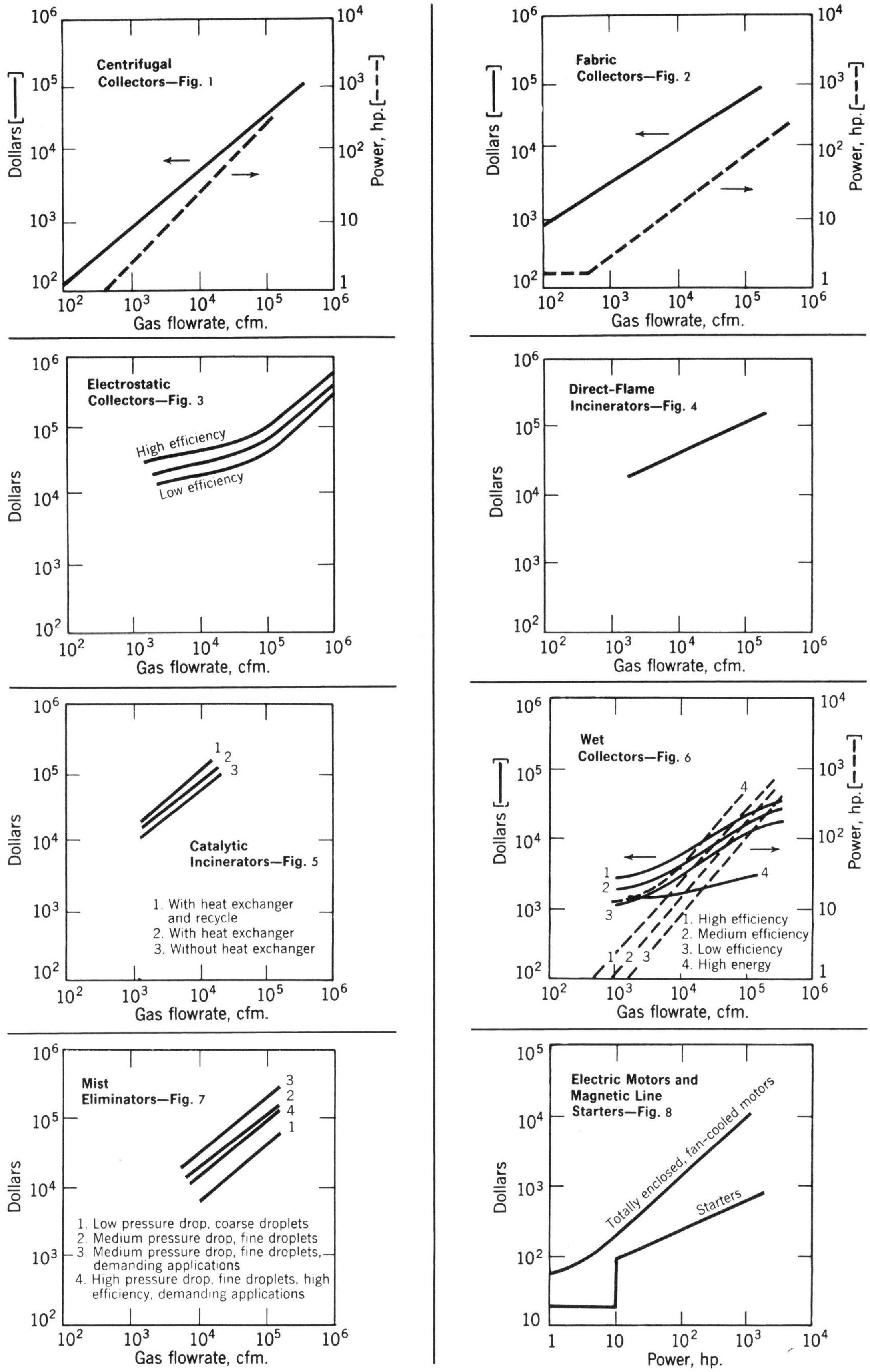

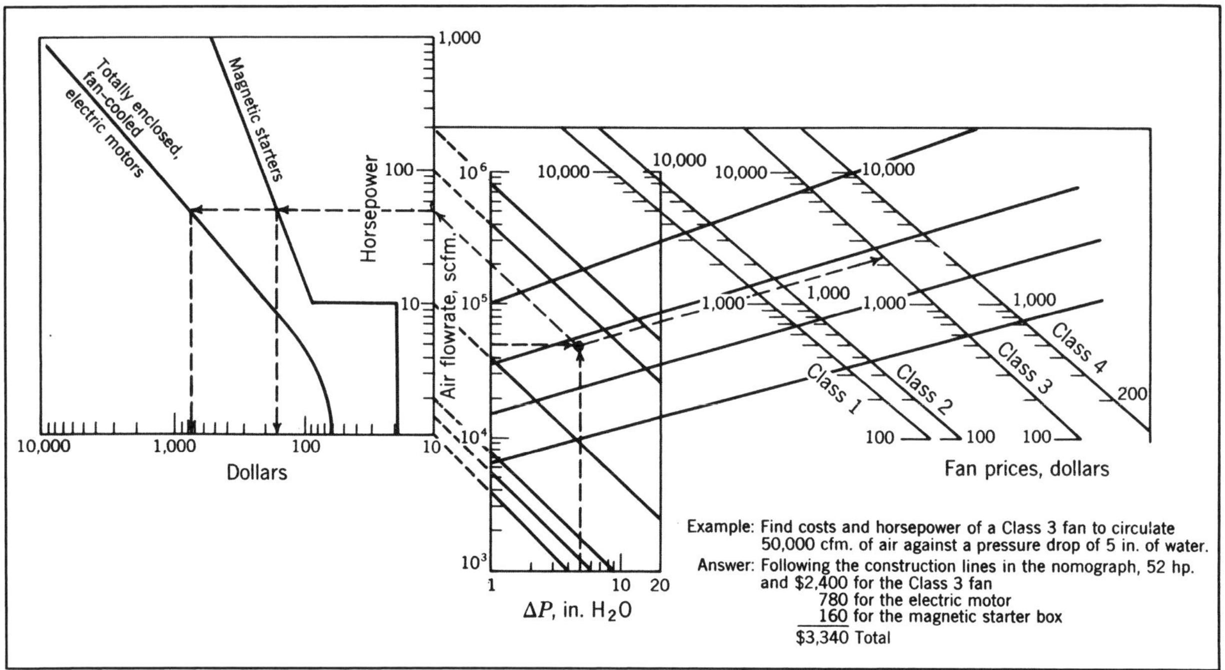

COST correlations for the four standard classes of fans, associated motors and switch gear—Fig. 9.

manufacturer is invaluable in the choice of the catalyst and the selection of the optimum operating conditions.

Not all process gases can be burned without damage to catalyst. As a result, the system may not prove as flexible as other methods. However, where ideally suited for the application, catalytic oxidation does provide a very economical (on a long-term basis) and efficient gas-cleaning method.

Wet Collectors

Cost data for wet collectors are presented in Fig. 6.

Wet collectors, more widely known as scrubbers, are essentially of two types. In the first, water or some other suitable scrubbing liquid is sprayed into the gas stream. In the second, a stream of liquid is injected at or near a venturi throat through which the gas stream flows at a high velocity. The first type is simple, offers low pressure drop in the gas stream, and presents no construction problems. On the other hand, the venturi is more complex, consumes more power due to its higher pressure drop, and provides better agglomeration of the particles.

In the spray collector, the accumulated water will contain a considerable quantity of the dirt that was originally in the gas stream. The venturi scrubber requires a collecting unit downstream from the nozzle to remove the dirty water.

Both types of wet collectors need mist eliminators. In addition, temperatures should be carefully controlled to avoid unsightly water-vapor plumes at the outlet.

The number of commercially available designs is bewildering, as may be seen in Fig. 6. Close attention should be given to the power-cost curves: they may be the determining factor in selecting one system over another.

For homemade units, the spray nozzles and the sheet metal are the main cost items. Spray-nozzle costs vary over a large range due to the many different types of spray patterns available. Each pattern requires careful machining of the individual nozzles. The material of construction also has a major effect on nozzle costs.

No estimates are given here for these costs because the number needed in a particular installation may not be known at the time of the preliminary cost estimate. In addition, homemade units are not recommended when the present state of the art is considered. An experienced contractor should be consulted to ensure good design and efficient operation of the scrubbing device.

Mist Eliminators

Mist eliminators do not fall into any general class of gas-cleaning equipment, and a separate chart for them has been provided in Fig. 7.

Prices for the standard glass-fiber and Type 316 stainless-steel construction are quoted in U.S. dollars fob, their place of manufacture. The prices include all the internals: cleaning elements, element gaskets, cap screws, washer, and polygon framework. However, cost of the tank in which the elements are installed has not been accounted for.

POWER AND MATERIAL DRIVES

Power and material drives consist of the elements in a gas-cleaning installation that provide either the power for or the means of conveying the process gases. Included are:

- Fans for the gas.
- Electric motors.
- Pumps for scrubbing liquids.
- Conveyors for the collected dusts.

Fans and electric motors are the only elements that will be discussed here. The prices of pumps and their motors are included in the scrubber price charts, and therefore a separate chart is not necessary. Prices for conveyor belts, dust valves and other dust-handling equipment are not included in this report because too many items would have to be listed. Furthermore, this equipment only accounts for a very small fraction of the total cost of a well-designed installation.

Electric Motors and Fans

Prices for electric motors and their associated switchgear are included in Fig. 8 for 1970, in U.S. dollars.

Cost charts for fans are presented in Fig. 9. These correlations include prices for the four standard classes of fans, as well as those for associated motors and switchgear.

Fan selection is a very difficult job that relies on long experience and on extensive testing by the fan manufacturers. The information in Fig. 9 will not necessarily provide the optimum selection for the particular application. It should be used in combination with details furnished by experienced fan suppliers.

Common Gages of Sheet Metal for Uncoated Iron and Steel—Table II

Gage No.	Thickness Decimal, In.	Thickness Fractional, In.
7	0.1878	3/16
8	0.1719	11/64
9	0.1563	5/32
10	0.1406	9/64
11	0.1250	1/8
12	0.1094	7/64
13	0.0938	3/32
14	0.0781	5/64
15	0.0703	–
16	0.0625	1/16
17	0.0563	–
18	0.0500	–
19	0.0438	–
20	0.0375	–
21	0.0344	–
22	0.0313	1/32

SHEET METAL AND BAR STOCK

Sheet metal is the general term applied to large thin sheets of metal. Common materials are iron, steel, copper, aluminum and brass. The thicknesses available are identified by standard gage numbers. Sheets of thickness greater than 3/16 in. are usually referred to as metal plate. Table II shows the most common gage numbers and thicknesses in decimals and fractions of inches for uncoated iron and steel sheet metal. Tables for other metals, coated and clad composite sheets, and special-alloy sheets, are readily available in most engineering and trade handbooks.

Ducting circuits, settling chambers and simple cyclones are the elements of a gas-cleaning plant manufactured from sheet metal. Fig. 10 provides correlations to determine the sheet-metal area, bar-stock length, fabrication manhours and, where applicable, bar-stock cross-sections and sheet-metal gages as a function of the nominal inlet diameter of the duct. Combining the sheet-metal, bar-stock and labor costs obtained from the graphs according to Eq. 3 gives the total cost for ducting circuits and simple cyclones.

The cost data obtained from Fig. 10 include a material-waste factor. This explains why the values for the sheet-metal area are somewhat larger than those that would be calculated from purely geometric considerations.

Ducting Circuits

The ducting systems used in gas-cleaning plants are manufactured from sheet metal following standardized techniques.

Isometric design drawings and specifications are first made for each item in the ducting circuit. Next, developments showing the true size and shape of each surface are drawn by parallel line, radial line or triangulation procedures. Detailed explanations of these drafting procedures can be found in drawing textbooks. After developing the pattern of the item from orthographic projections, allowances are made for the material required for the hems, seams and joints that will be necessary to hold the pattern together and to attach it to the supporting structure. These detailed calculations and drawings exactly define the sheet metal required for the particular application.

Next, the sheet metal is cut out to the desired shape, and formed by bending, rolling or folding the surfaces. The pieces obtained are then joined by crimping, seaming, riveting, screwing or spot welding, or a combination of these. The selection of the joining method depends upon structural and other practical considerations. Once again, the choice relies largely on the experience of the assemblers.

Soldered seams are usually the lap, lock or butt type. They are used when liquid or gas tightness is essential. The seams may be riveted or screwed for less-demanding

Steel Prices and Densities—Table III

Steel Type	Australia, ¢/Lb. (1970)	U.S.* ¢/Lb. (1970)	Density Lb./Cu. In.
301	48.0	53.8	0.29
304	54.0	60.5	0.29
316	72.0	80.6	0.29
321	55.0	61.6	–
430	41.5	46.5	0.28

* 1.12 conversion factor was used to change from Australian to U.S. currency. Prices may vary depending on conditions.

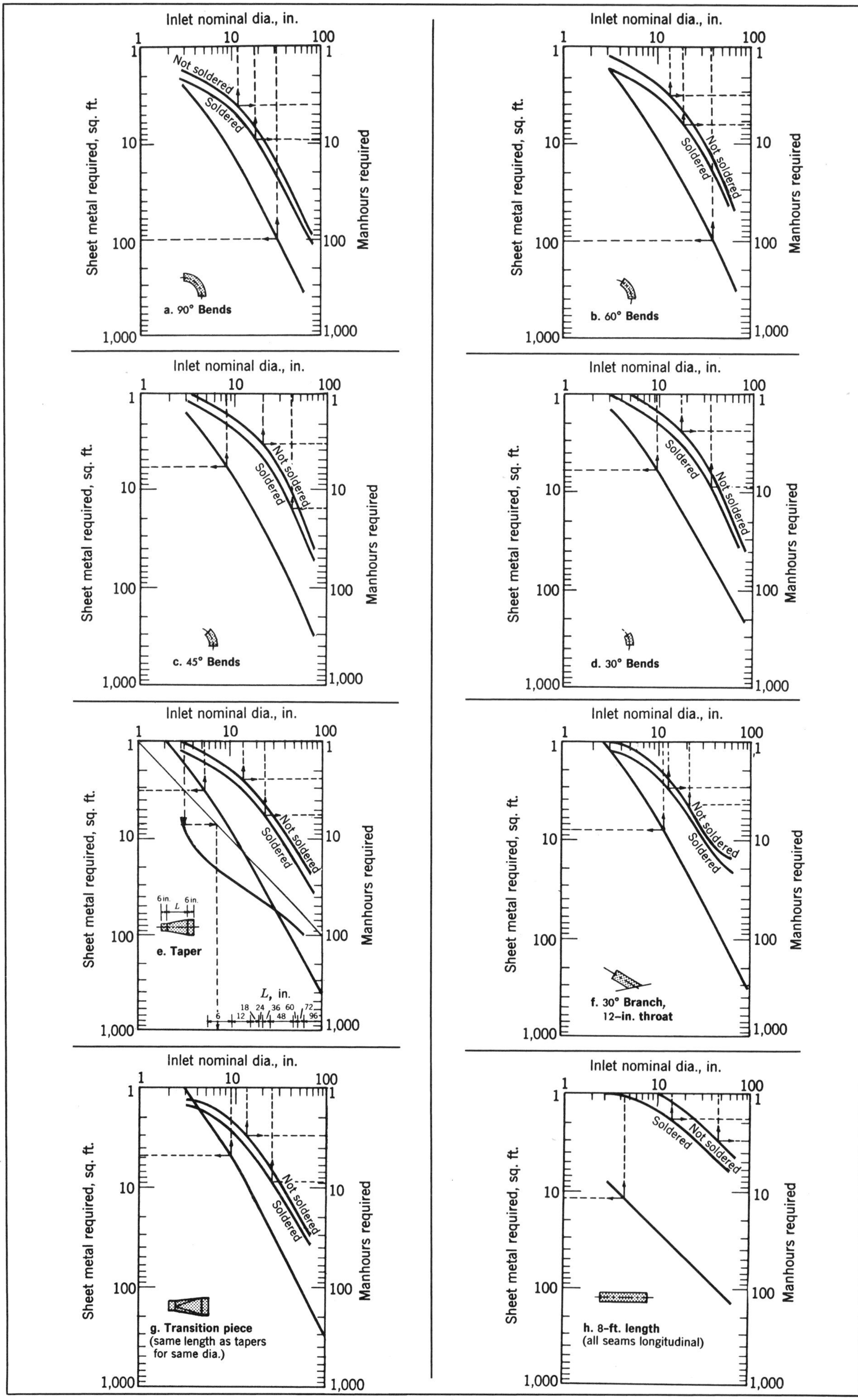
Inlet nominal dia., in.
Sheet metal required, sq. ft.
Manhours required
Not soldered
Soldered
a. 90° Bends
b. 60° Bends
c. 45° Bends
d. 30° Bends
6 in.
L
6 in.
e. Taper
L, in.
f. 30° Branch, 12-in. throat
g. Transition piece (same length as tapers for same dia.)
h. 8-ft. length (all seams longitudinal)

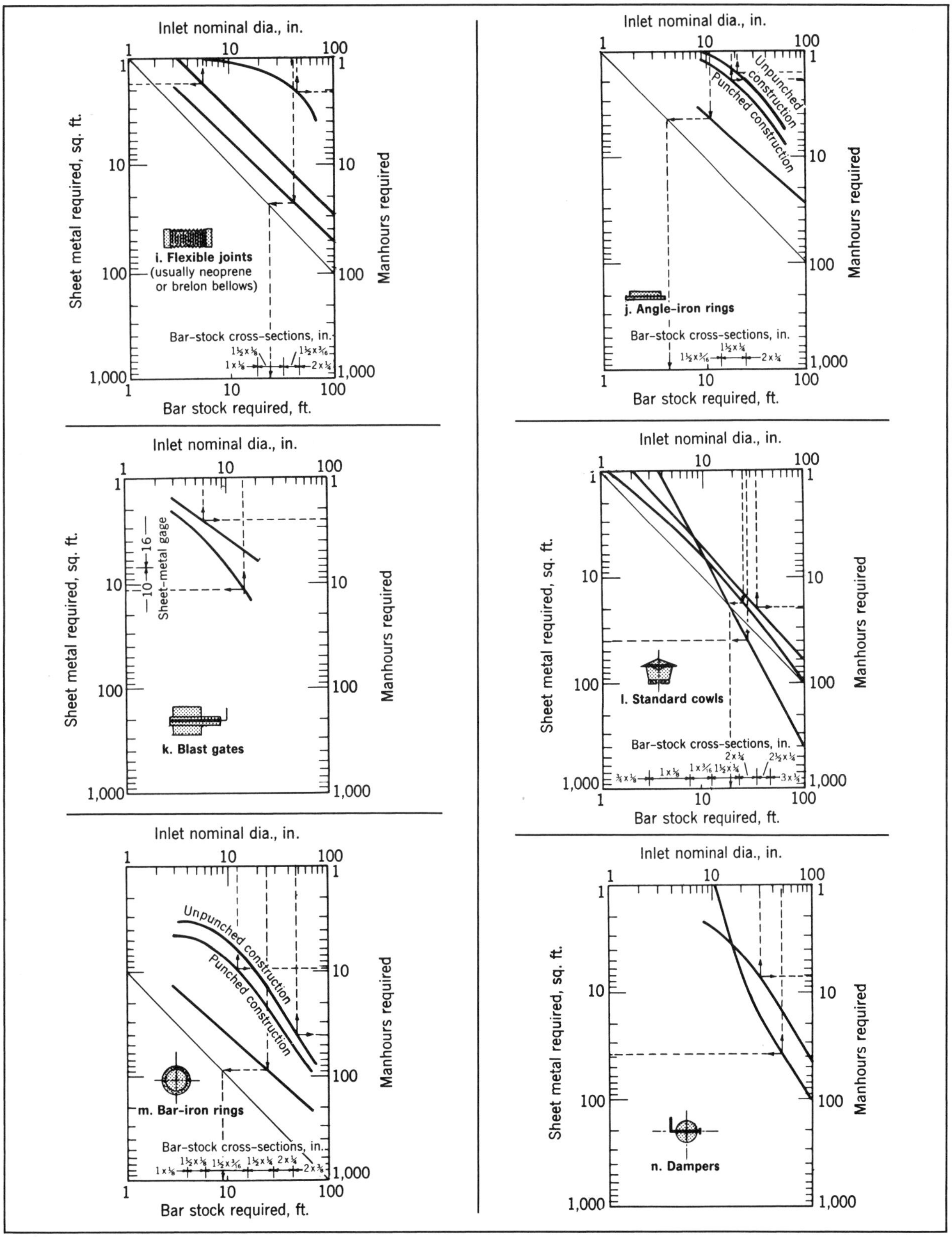

COSTS for sheet metal, bar stock and labor for ducting-circuit components—Fig. 10

applications. However, heavy plates are normally soldered. Hems, essentially stiffeners, also eliminate sharp edges. Wire edges may be considered as particularly strong hems.

The welding of galvanized materials releases noxious fumes. To avoid this hazard, special galvanized sheet metal, like Galvabond, is used. This material is capable of withstanding folding and locking without deterioration of its protective galvanized film.

The most common cutting sizes of sheet metal are 6 x 3 ft., 6 x 4 ft., and 8 x 4 ft. In the U.S., a greater variety of standard sizes is available than in other countries. An experienced assembler who has a large workshop at his disposal can use this variety of stock sizes to provide significant savings in material cost and construction time.

The standard thickness of sheet metal used in ducting circuits is 16 gage. Light-wear systems and air conditioning equipment may be manufactured from 20-gage sheet metal. Systems subjected to the heaviest wear conditions use 10-gage material. Wear plates should always be specified on critical points in the ducting system that handle abrasive material.

The cost of the ducting circuits is the sum of the sheet-metal, bar-stock and labor costs determined from Fig. 10. A good rule of thumb to remember is that the sheet-metal cost for common iron is roughly one fifth of the labor cost. The ducting correlations are calculated for 16-gage soldered ducting unless otherwise specified. For lower gage numbers, labor costs should be doubled to account for the increased weight of the material handled.

There are many factors that can increase the cost estimates obtained from the correlations. For example, the cost for bends may be reduced by 20% in large orders. Rings and other elements made from bar or angle stock may receive large quantity discounts of as much as 50%. Galvanization adds 20% to the material cost. Type 304 stainless steel multiplies the material cost five times, while Type 316 stainless multiplies it by seven. Table III shows the 1970 stainless steel prices in the U.S. and Australia.

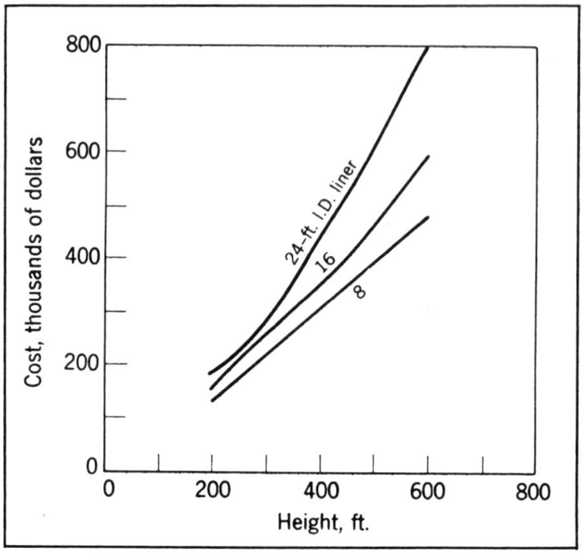

CHIMNEY cost correlations—Fig. 12

Simple Cyclones

Simple cyclones are manufactured from sheet metal, and their cost may be evaluated using correlations similar to those presented for standard ducting elements. Fig. 11 shows the manhour and sheet-metal requirements for simple cyclones as a function of their nominal inlet diameter.

Settling Chambers

Settling chambers, manufactured from sheet metal, can have almost an infinite variety of shapes and sizes. In general, they are not advisable except for removing the coarsest and heaviest dusts from low-flowrate streams. Although internal baffles may increase the efficiency, they also increase the cost of an already bulky and inefficient device. The costs of settling chambers have not been correlated since they can be evaluated from sheet-metal costs and from an estimate of the labor involved.

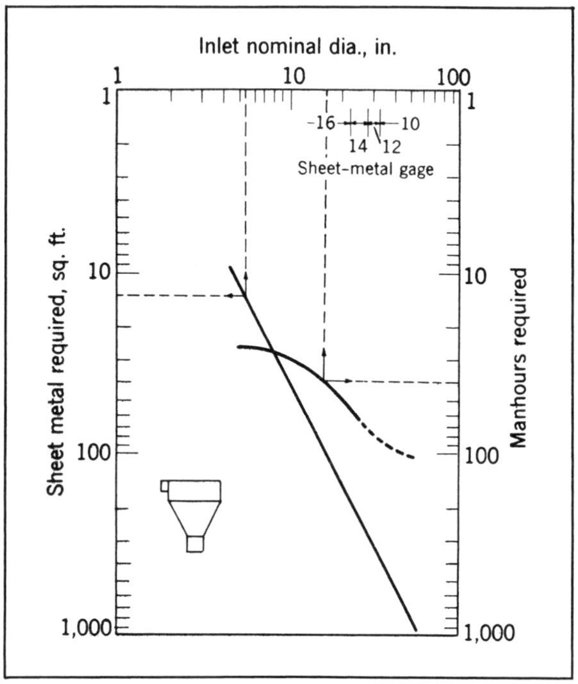

COSTS for simple cyclones—Fig. 11

CHIMNEY COSTS

Fig. 12 contains the cost data for chimneys of standard construction in the U.S. in 1968. The correlations are based on a normal windloading, a concrete column, a masonry liner and a potassium-silicate mortar. Freight and labor costs vary widely both in the U.S. and overseas because construction methods such as continuous casting may not be practicable in some locations.

It should be remembered that a chimney requires maintenance just as any other operating unit of the plant. Since maintenance costs are small items, no estimates for them are given here, although they should not be neglected in the more detailed studies.

For further information on chimney construction and costs, the reader should consult the literature.[14]

ERECTION AND INSTALLATION COSTS

Erection and installation costs can be divided into two broad categories: structural work; fabrication and installation work.

Structural work is usually priced on a weight basis. Structural mild-steel costs range from \$155 to \$176 per ton. Stainless-steel costs about \$900 per ton.

Fabrication costs vary with the complexity of the job. A good rule of thumb is that 20 to 60 sq. ft. may be manufactured by one man in an average 8-hr. day. Smaller jobs, of course, cost more per ducting-unit installation. The cheapest gas-cleaning components that can be fabricated from 20-gage sheet metal, installed and spot welded on the site, correspond to the highest value of 60 sq.ft. per 8 manhours. Heavier gage, fully soldered complex systems correspond to the lower value of 20 sq.ft.

Running Costs And Engineering Factors—Table IV

	Electrostatic High Voltage	Electrostatic Low Voltage	Scrubbing	Fabric	Centrifugal	Direct Afterburner	Catalytic Afterburner
Fan Efficiency	0.60	0.60	0.60	0.60	0.60	0.60	0.60
Pump efficiency	0.50	0.50	0.50	0.50	0.50	0.50	0.50
Annual operating time, hr.	8,760	8,760	8,760	8,760	8,760	8,760	8,760
Contact power, hp./actual cfm.							
Low eff.	0.00019	0.000015	0.0013				
Med. eff.	0.00026	0.000030	0.0035				
High eff.	0.00034	0.000040	0.015				
Power cost, $/kwh.							
Low	0.005	0.005	0.005	0.005	0.005	0.005	0.005
Typical	0.011	0.011	0.011	0.011	0.011	0.011	0.011
High	0.020	0.020	0.020	0.020	0.020	0.020	0.020
Maintenance cost, $/actual cfm.							
Low	0.01	0.005	0.02	0.02	0.005	0.03	0.07
Typical	0.02	0.014	0.04	0.05	0.015	0.06	0.20
High	0.03	0.02	0.06	0.08	0.025	0.10	0.35
Head for liquid, ft.							
Low			1				
Typical			30				
High			60				
Liquid circulation, gal./actual cfm.							
Low			0.001				
Typical			0.008				
High			0.20				
Liquid consumption, $/(1,000 gal.)(hr.)							
Low			0.35				
Typical			0.50				
High			1.00				
Pressure drop, in. H_2O							
Low	0.1	0.1		2.0	2.0	0.5	0.5
Typical	0.5	0.5		5.0	3.0	1.0	1.0
High	1.0	1.0		8.0	4.0	2.0	2.0
Fuel costs, $/(actual cfm.)(hr.) (50% excess air)							
With heat exchanger						0.00023	0.00014
Without heat exchanger						0.00057	0.00028

per 8 manhours. Common high-quality 16 gage, welded and flanged ductwork corresponds to 24 to 32 sq. ft. per 8 manhours.

Erection and installation costs have been found to approximate the cost of labor for ducting and simple cyclones fabrication. Since these costs have been comprehensively correlated earlier in this article, it is advisable to use those data for preliminary estimates.

ANNUAL COSTS

Annual costs are made up of three elements:

- Depreciation of the total capital investment (on the basis of the installed capital cost).
- Capital charges: taxes, interests and insurance.
- Maintenance and operating costs.

The first two of the above items will differ from installation to installation, depending on the methods used to calculate depreciation, taxes, insurance rates and many other variables. Consequently, this report will not attempt to cover them. Estimates have been made for a specific set of assumptions,[1] but no generally applicable set of values can be developed. However, maintenance and operating costs are easily expressed as functions of gas throughput and a few other equipment-dependent parameters.

Maintenance and Operating Costs

Maintenance costs involve the labor and replacement expenses required to keep the cleaning units in top operating condition. Maintenance costs representative for the U.S. in 1968 are included in Table IV.

Operating costs include the expenses involved in running a cleaning unit at its specified collection efficiency. They depend on:

- Gas volume handled.
- Pressure drop across the system.
- Operating time.
- Electrical consumption and cost of electricity.

• Mechanical efficiency of the motors, fans and pumps.

• Scrubbing-liquid consumption, where applicable.

Table IV presents values for the above factors for the U.S. in 1968.

Operating and maintenance costs are directly proportional to the flowrate of the gas handled and to another parameter that depends on the type of unit under consideration. Eq. (4) through (8), obtained from the literature, [1] are presented below for the five most common types of cleaning devices.

Gravitational and Centrifugal:

$$G = S\left[\left(\frac{0.7457}{6356\,E}\right)\left(PHK\right) + M\right] \tag{4}$$

Fabric Filters:

$$G = S\left[\left(\frac{0.7457}{6356\,E}\right)\left(PHK\right) + M\right] \tag{5}$$

Afterburners:

$$G = S\left[\left(\frac{0.7457}{6356\,E}\right)\left(PHK\right) + M + Hf\right] \tag{6}$$

Electrostatic:

$$G = S\left[JHK + M\right] \tag{7}$$

Scrubbers:

$$G = S\left[\left(\frac{0.7457}{6356\,E}\right)\left(PHK\right) + M + WHL + \left(\frac{0.7457}{1722\,F}\right)\left(HKQg\right) + \left(\frac{0.7457}{3960\,F}\right)\left(HKQh\right)\right] \tag{8}$$

$$= S\left[\left(0.7457\,HK\right)\left(Z + \frac{Qh}{1980}\right) + WHL + M\right] \tag{8a}$$

Note: Eq. (8a) is a simplified version of Eq. (8). For details, see Ref. 1.

Where:

f Fuel cost, \$/ (actual cfm.)(hr.)
g Liquid pressure at collector, psig.
h Physical height of liquid in circulation system, ft.
E Fan efficiency, assumed 60%
F Pump efficiency, assumed 50%
G Annual cost, \$
H Annual operating time, assumed 8,760 hr.
J Power requirement, kw./actual cfm.
K Power cost, \$/kwh.
L Liquid cost, \$/(gal.)(hr.)
M Annual maintenance cost, \$/(gal.)(hr.)
P Pressure drop, in. H_2O
Q Liquid circulation, gal./actual cfm.
S Design capacity of the collector, actual cfm.
W Makeup liquid consumption, gal./actual cfm.
Z Total power input required for a specified scrubbing efficiency, hp./actual cfm.

Representative values of all the parameters above may be obtained from Table IV for the U.S. in 1968, or they may be determined independently for different locations or conditions. No analysis is given here for the variations in the cost for different applications. Operating and maintenance costs vary widely with the different uses of the equipment.

It should be noted that these costs can be a very small fraction of the annual costs in simple, low-efficiency units such as gravitational settling chambers; or they can comprise a substantial fraction, as in the case of high-pressure-drop scrubbers.■

Acknowledgements

The author is most grateful to the many firms that made available confidential data, and to the individuals in those firms who helped collate, so that these cost-estimation procedures could be developed. The author is also most appreciative to Dr. W. Strauss, who made available some of his personal cost-files.

References

1. "Control Techniques for Particulate Air Pollutants," National Air Pollution Control Admin. Pub. No. AP-51, U.S. Govt. Printing Office, 1969.
2. Sargent, Gordon D., Dust Collection Equipment, *Chem. Eng.*, Jan. 27, 1970, p. 130.
3. Environmental Engineering, Deskbook Issue, *Chem. Eng.*, Oct. 14, 1968.
4. Guthrie, K. M., Capital Cost Estimating, *Chem. Eng.*, Mar. 24, 1969, p. 114.
5. Guthrie, K. M., Liquid Handling Costs, *Chem. Eng.*, Apr. 14, 1969, p. 201.
6. Gallagher, John T., Efficient Estimating of Worldwide Plant Costs, *Chem. Eng.*, June 2, 1969, p. 196.
7. Johnson, Robert J., Cost of Overseas Plants, *Chem. Eng.*, Mar. 10, 1969, p. 146.
8. CE Cost Indexes, *Chem. Eng.*, May 5, 1969, p. 134.
9. Stairmand, C. J., The Design and Performance of Modern Gas Cleaning Equipment, *J. Inst. Fuels*, February 1956.
10. Stairmand, C. J., Removal of Grit, Dust and Fume from Exhaust Gases from Chemical Engineering Processes, *The Chemical Engineer*, December 1965.
11. Forbes, M. C., Cost Accounting for Pollution Control, *Hydrocarbon Process.*, October, 1969, p. 145.
12. Johnstone, R. Edgeworth, Pre-design Estimation of the Capital Cost of Chemical Plants, *Trans. Instn. Chem. Engrs.*, **32**, p. 151 (1954).
13. Werner, Klaus-Dieter, Catalytic Oxidation of Industrial Waste Gases, *Chem. Eng.*, Nov. 4, 1968, p. 179.
14. Carlton-Jones, Dennis, and Schneider, H. B., Tall Chimneys, *Chem. Eng.*, Oct. 14, 1968, p. 166.
15. Nowak, F., Preise, Gewichte und Raumbedarf von Elektrofiltern für hohe Abscheidegrade, *Mitt. Ver. Grosskesselbesitzer*, **83**, p. 84 (1963).

Meet the Author

J. R. F. Alonso is presently completing his Ph.D. requirements at the University of Melbourne, Victoria 3052, Australia. Mr. Alonso received his B.Sc. in chemical engineering from Columbia University and his M.Sc. from Worcester Polytechnic Institute. Mr. Alonso's professional experience includes two years in the Computer Dept. of The M. W. Kellogg Co., and one year as an IBM research assistant at the MIT Computer Center.

Cost of Standard-Sized Reactors and Storage Tanks

In addition to new cost data, here is an examination of the savings that stem from using standard-dimensioned process equipment.

LAWRENCE D. EPSTEIN, Perry Products Co.

The charts on the next page provide cost/capacity data for two types of corrosion-resistant, standard-designed process equipment:

1. Storage tanks for chemicals. Construction: Type 304 stainless steel with #2B finish. Joints are butt-welded; seams are pickled and wire-brushed, and all surfaces are passivated.

2. Jacketed reactors. Construction is in accordance with ASME code; the internal shell and dimple jacket are both of Type 304 stainless steel; the shell has a #1 mill finish. All welds are pickled and brushed, and the entire vessel passivated.

Material Correction Factors for Reactors

The above type of reactor, constructed from Type 304 stainless steel, is suitable for heating and cooling chemical reactions that are not severely corrosive, and is used in plastic-resin processing, foam manufacturing, acrylic makeups, etc. More-corrosive applications would require Type 316 stainless steel, which contains molybdenum and which would increase the cost of the reactor by about 12% (i.e., costs taken from Fig. 2 would have to be multiplied by a 1.12 correction factor).

For very high temperatures, either Type 321 or 347 stainless steel would have to be used, resulting in a reactor that is about 25% more expensive than one of Type 304.

For nitric acid service, reactors must use the L grades (such as Type 304 ELC and 316 ELC); these reactors would cost about 8% more than ones of regular Type 304 stainless.

The Case for Standardization

Although the charts are based on my own company's products, readers familiar with current costs will note that the price structure is quite competitive. There is no secret to this situation. Rather, it is primarily the result of dimensional standardization, which produces marked economies for the fabricator of specialized equipment—and these economies can be shared with the user.

My own company fabricates a great many non-standard as well as standard tanks and pressure vessels, and we recognize that the purchaser may have space requirements or special process conditions that may make a nonstandard, custom-built unit either desirable or essential. However, the purchaser who can make use of standard-dimensioned equipment can get such equipment at considerably lower cost, for these reasons:

The manufacturer of standard sizes can produce for inventory, without fear of being stuck with his products. (And, of course, the buyer who can purchase from stock effects the economies of immediate delivery.) By adhering to standard sizes, the producer orders his sheets, tops, and bottoms, so that every bit of it is utilized; there is no waste.

Standard units take standard-size hardware and fittings. Piping, valves, etc., all can be purchased as

Originally published October 1971.

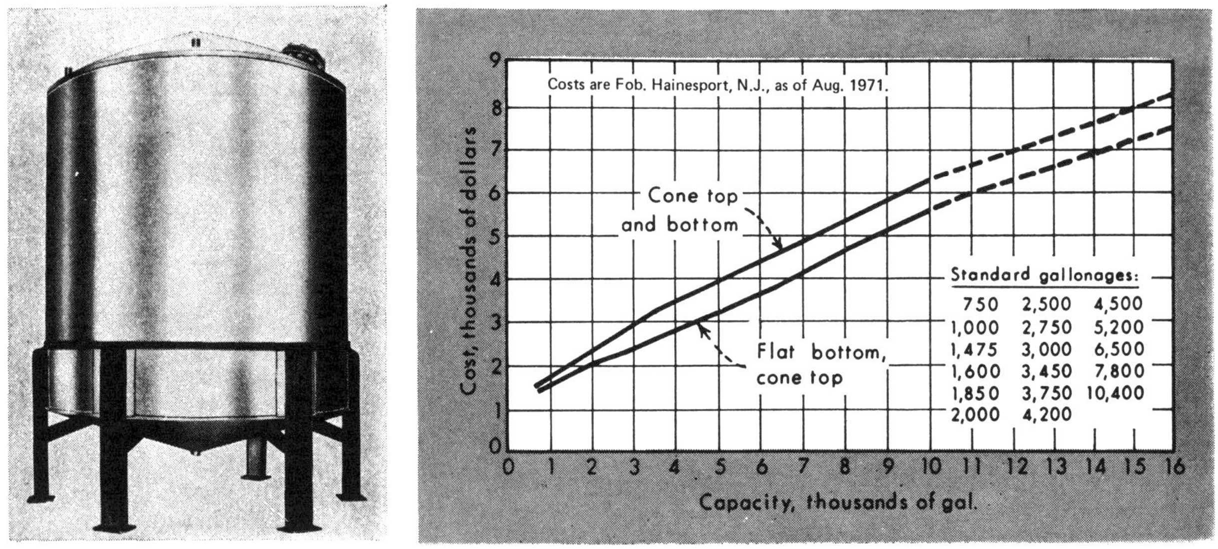

STAINLESS-STEEL TANKS for atmospheric storage: costs vs. capacity—Fig. 1

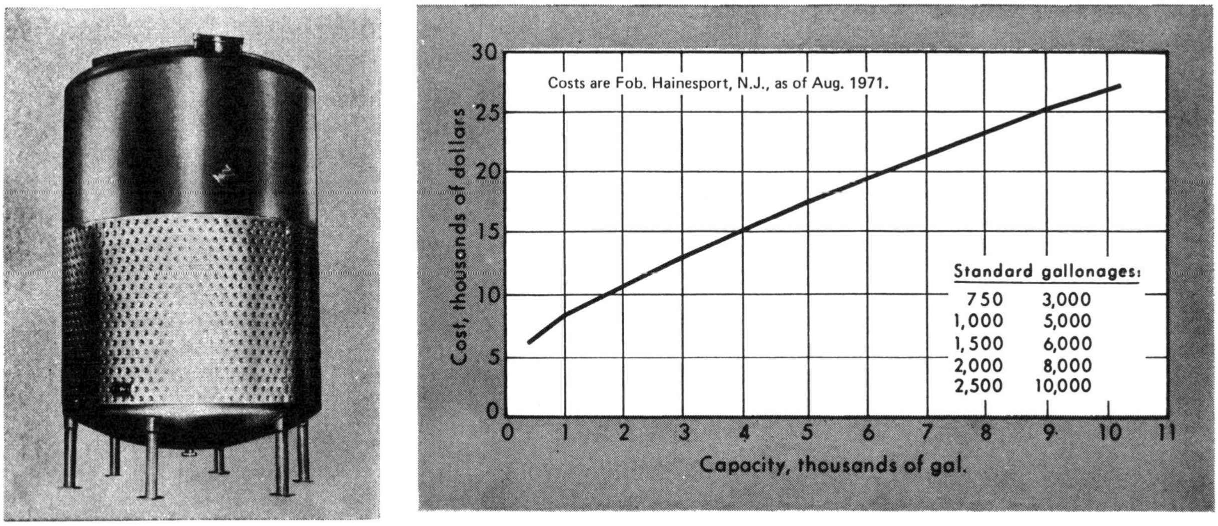

STAINLESS-STEEL, DIMPLE-JACKETED REACTORS: costs vs. capacity—Fig. 2

standards and, consequently, at a favorable cost.

In a fabrication shop geared to produce in standard sizes, the machinery and equipment have been specifically purchased, resulting in a very efficient manufacturing process.

To illustrate these factors, let us compare a 3,000-gal., light-gage tank fabricated to standard dimensions—i.e., 8 ft. dia. x 8 ft. 2 in. high—with another 3,000-gal. light-gage tank of 10 ft. dia. x 5 ft. 9 in. high. The latter would require trimming a standard 6-ft. sheet, leaving 3 in. of waste. In addition, because of the larger diameter, the bottom will have to be made of a heavier-gage material. Then, you need two additional legs to support the added weight. The larger bottom will require more lineal feet of welding, with acccmpanying cleaning, testing and passivation costs. Result: the standard tank will be at least 10% cheaper.

Because of such savings, as well as the desire to conserve design manhours, some of the largest names in the chemical process industries have already programmed standard-dimensioned tanks and reactors into their computers, making the selection of available units an easy operation. We look for this trend to continue.

Of course, savings also occur when allied equipment often found in installations with tanks and reactors can be purchased in standard sizes and capacities—e.g., stainless-steel heat exchangers, columns and absorption sections. ■

Meet the Author

Lawrence D. Epstein is president of Perry Products Co., Hainesport, N.J., a leading fabricator of stainless-steel process equipment. He has an M.B.A. degree from Temple University, and a B.S. in industrial management from Pennsylvania State University. He joined Perry in 1961 as a sales engineer, and assumed his present position in 1969.

Capital and Operating Costs For 54 Chemical Processes

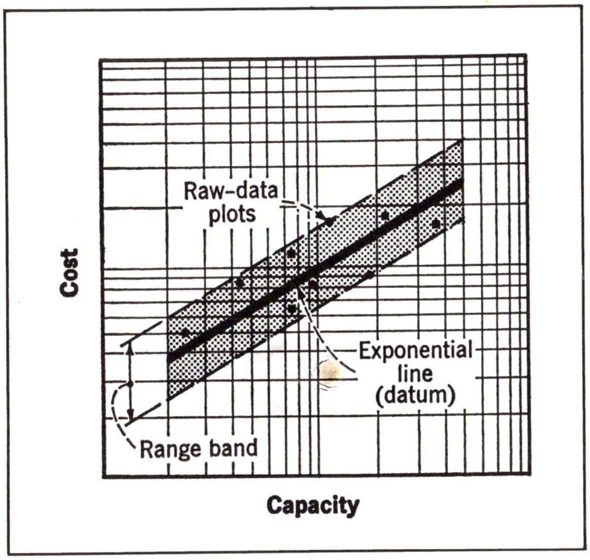

With these new cost/capacity charts (based on scatter diagrams such as this one), you can not only estimate the required investment for a wide range of chemical and oil-refining processes but also the cost of operating the facilities per unit of output.

KENNETH M. GUTHRIE, Fluor Corp.

Charts that relate the capital and operating cost of a process plant to its capacity occupy a prominent place in preliminary venture-analysis work. However, not many comprehensive, consistently-plotted sets of such charts can be found in the literature.

To fill this gap, I have tried to pull together a great deal of operating- and capital-cost data from personal files as well as secondary sources, and to present the data so that every chart not only includes the same cost elements but covers the same time base and location base. This makes it easy to project costs into the future, to adjust them for regional variations, and to use them in preliminary venture-analysis work with little fear of cost omission or cost duplication.

Plotting the Data

Observed data can be plotted on a scatter diagram from which the statistical "line of best fit" can be established (see sketch above). The slope of this line is known as the exponent, and is usually less than 1.0. This line provides the datum from which to generate anticipated costs from given capacities.

The range of the data can be indicated by dotted lines on either side of the datum, so that the distribution of the plotted points is contained within the limits of the range band.

Each of the 54 capital-investment charts in this report is based on a series of scatter plots maintained over a number of years. Constant analysis has gradually refined these data into a set of significant observations for each process characteristic, and has reduced the range band to practical limitations of accuracy.

The data points remaining within the bands have been classified as "significant" in these categories:

- Actual observed costs (designated as *A*).
- Estimated cost experience (*E*).
- Published cost sources (*P*).

The resulting *A/E/P* mix of these data points is included alongside each capital-cost chart.

What Is Included in Capital Cost?

Capital costs forming the basis for the charts include the following components:

- Major equipment items.
- Field materials and labor.
- Project management and engineering.

Originally published June 1970.

Typical Operating-Cost Relationships for Chemical Plants and Refinery Units

	Capacity, Thousand Ton/Yr.	Average Capital Investment, Million $	Operating Labor and Supervision Manhours/Ton	Maintenance Labor and Supervision Manhours/Ton	Power and Utilities, per Ton or Bbl.			
					Fuel MM Btu./Hr.	Steam Lb./Hr.	Power Kwh.	Water Gpm.
Chemical Plants:								
Acetone	100	7.5	0.518	0.315	—	1.73	310	5.18
Acetic acid*	10	2.1	1.483	0.984	—	—	180	0.58
Butadiene	100	7.5	0.345	0.285	—	0.012	130	0.73
Ethylene oxide	100	29.0	0.232	0.104	—	4.88	140	0.148
Formaldehyde	10	4.5	0.259	0.328	—	34.6	200	0.029
Hydrogen peroxide	100	7.8	0.288	0.352	—	2.62	160	0.186
Isoprene	100	16.5	0.230	0.325	—	0.81	710	0.001
Phosphoric acid	10	1.2	1.85	0.442	—	0.18	40	0.03
Polyethylene	10	8.5	0.259	0.295	—	0.23	450	0.0004
Urea	100	3.8	0.238	0.215	—	0.33	135	0.0002
Vinyl acetate	100	20.0	0.432	0.528	—	1.34	275	0.27
Refinery Units:	**(Thousand Bbl./Day)**		**(Manhours/ Bbl.)**	**(Manhours/ Bbl.)**				
Alkylation	10	6.2	0.007	0.0895	—	10.83	0.07	1.48
Coking (delayed)	10	7.0	0.011**	0.0096	0.007	1.85	0.07	—
Coking (fluid)	10	4.1	0.0096	0.0058	0.012	2.55	0.06	0.64
Cracking (fluid)	10	8.2	0.0122	0.0115	—	(4.73)*	0.02	0.33
Cracking (thermal)	10	1.8	0.0096	0.0025	0.012	(2.55)*	0.06	0.64
Distillation (atm.)	100	10.1†	0.0048	0.0042	0.004	0.25	0.03	0.16
Distillation (vac.)	100	6.5†	0.0024	0.0154	0.003	0.95	0.04	0.18
Hydrotreating	10	1.3	0.0048	0.0028	0.006	0.92	0.01	0.14
Reforming, catalyt.	10	8.6	0.0048	0.0078	0.002	1.38	0.23	0.28
Polymerization	10	10.5	0.0024	0.0158	—	4.85	0.07	0.43

* Net steam generated. † 65% vaporized. ** Includes two coke cutters (1 shift/day).

- Indirect costs (freight, taxes, contractor overhead, contractor profit, etc.).

Major variables such as site development, buildings and offsite facilities (other than storage) are not included, in order to present a consistent cost module,* and to avoid possible distortion of the data. As indicated alongside each chart, the cost of storage facilities has been included in some cases—i.e., when the plant is not normally installed within an integrated plot arrangement.

The total of included components cited above is equivalent to the "total fixed capital cost" of the plant. An amount representing 15 to 20% of this total cost has been added to it, to represent the cost of startup, working capital, capitalized interest and other capitalized costs. (The exact percentage depends on the process.) This new total represents the total capital investment, and this figure—stabilized for a construction site located in the U.S. Gulf Coast area in mid-1968—is the one used in the charts.

What Is Included in Operating Cost?

An evaluation of operating costs is essential to the early stages of venture analysis; these costs have been collated for each process, related to plant capacity, and presented on the secondary charts.

These items are included under operating costs:

- Materials (operating and maintenance supplies as well as raw materials).
- Labor and supervision maintenance.
- Utilities (fuel, steam, power, water).
- Indirect costs (fringe benefits, statutory payroll burdens).
- Plant overhead expenses (plant administration, technical services, accounting, etc.).

These costs have been plotted on a unit basis for each process, and again represent approximate Gulf Coast conditions in mid-1968. The chart band includes the range of significant data where possible. The data are again based on actual cost records, estimated cost experience and published material, but it is not practical to cite the number of data points in each category because published data have sometimes had to be combined with estimated data, etc.

The table on this page provides typical values for some of the operating-cost components mentioned above; it presents the data on the same basis as for the cost/capacity charts that appear later on—i.e., in terms of ton/yr. rate values for chemical processes, and bbl./day for petroleum processes.

* For a full discussion of cost modules as a systematic way of collecting and assembling cost of data, see: Guthrie, K. M., Data and Techniques for Preliminary Capital Cost Estimating, *Chem Eng.*, Mar. 24, 1969.

Adjusting for Time and Location Factors

Returning to capital costs, the estimates needed for preliminary venture studies usually involve a forecast of what will happen to plant costs several years from now. The upward trend of the past, as reflected in CHEMICAL ENGINEERING's Plant Cost Index is likely to continue well into the future.

By considering the trends in the cost of direct materials, labor and engineering, the following index multipliers have been generated for this report; they can be applied to the capital costs taken from the charts to convert the costs from mid-1968 to some other year:

Year	1968	1969	1970	1971	1972	1973	1974	1975 (Mid)
Index	1.00	1.05	1.12	1.15	1.20	1.23	1.29	1.32

Of course, the years after 1970 reflect long-term approximations, and the reader may want to modify these figures based on his own cost-trend appraisal.

As for location factors, these have been developed on a similar basis: they represent a numerical evaluation of relative equipment and material costs, freight charges, labor rates, burdens and productivity, weather and site conditions, etc. Domestic and overseas locations have both been analyzed; some approximate multipliers for the U.S. are tabulated below:

Gulf	West		Central		East			Puerto
Coast	No.	So.	No.	So.	No.	So.	Hawaii	Rico
1.00	1.08	1.10	1.12	1.04	1.04	1.06	1.17	1.07

Approximate multipliers for some overseas locations are:

U.S.	England	France	Germany	Brazil	Peru	Australia	Japan
1.00	1.03	0.98	0.95	1.10	1.04	1.05	0.95

The above discussion of time and location factors applies to capital costs only. It is left up to the reader to decide whether any adjustments need be made to operating costs.

Example: How to Use the Data

A company is considering the feasibility of a new sulfuric acid facility to be constructed on land already owned, in the vicinity of Lima, Peru. Adequate shipping installations exist in the area.

A market deficiency of some 60,000 tons/yr. of sulfuric acid is expected to develop by the time the new plant is ready for startup. The selling price of the product is expected to be 1.25¢/lb. ($25/ton) at that time.

Cost of sales is based on historical data and collated as follows:

Capacity (tons/yr.)	20,000	50,000	100,000
Cost of sales (% of mfg. cost)	15.0	12.0	9.0

Using the appropriate charts in this report, prepare a preliminary sensitivity study for management review. Assume that $1 million of capital can be made available for the project, and that the minimum acceptable return on this capital is 25% after taxes. A maximum payout of five years would be expected.

Here is the step-by-step procedure:

1. Select the capacities for study, and establish the total capital investment from the "capital cost" chart for sulfuric acid:

	Case A	Case B	Case C	
Selected capacities, tons/yr.	20,000	50,000	100,000	
Total investment, $*	330,000	620,000	1,000,000	(1)

2. For first approximations, establish a selling price slightly below the anticipated maximum market value (say, 1.0¢/lb., or $20/ton), and calculate annual sales income by multiplying the selected capacities by $20/ton:

Annual net sales, $	400,000	1,000,000	2,000,000	(2)

3. Establish annual operating costs from the sulfuric acid operating-cost chart (i.e., by multiplying unit costs shown by the selected capacities); then add depreciation, property taxes and insurance to obtain total manufacturing cost:

Annual operating costs, $	260,000	575,000	1,050,000	
Depreciation,† 10% of (1)	33,000	62,000	100,000	
Taxes, ins., etc., 2½% of (1)	8,300	12,700	25,000	
Manufacturing costs, $	301,300	649,700	1,175,000	(3)

4. Apply appropriate percentages to the manufacturing costs to establish the cost of sales (which includes freight, packaging, advertising, general expenses, corporate overhead, etc.) For our three selected capacities, assume that these percentages are 15%, 12% and 9%, respectively:

Cost of sales, $	45,200	78,000	106,000	(4)

5. Establish total annual costs by adding the cost of sales (4) to the manufacturing costs (3):

Total annual costs, $	346,500	727,700	1,281,000

6. Establish profit potentials:

Gross profit, (2) − (5), $	53,500	272,300	719,000
Less federal taxes @ 50%	26,750	136,150	359,500
Annual net profit, $	26,750	136,150	359,500

7. Calculate profitability and payout:

Annual return on cap. invest., (6)/(1) x 100	8%	22%	33%
Payout, (1)/(6), yrs.	12.3	4.5	2.7

Analysis—The above indicates that an attractive economic structure for a sulfuric acid plant can be assembled within the constraints of the stated problem, and will lie somewhere between Case B and Case C. The low rate of return and excessive payout eliminate Case A entirely.

It is also clear that the economic capacity of the

* For Lima, Peru, the chart readings are multiplied by 1.04.
† Straight-line basis.

(*Text continues on p. 405*)

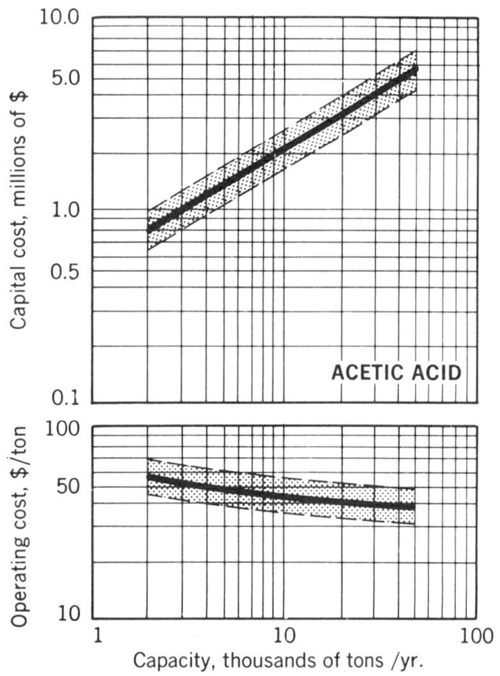

Acetic acid: Via exothermic conversion of methanol and carbon monoxide in presence of catalytic agents.

Size Exponent: 0.68 **Data:** A = 2; E = 1; P = 2

Included: Process unit and storage facilities

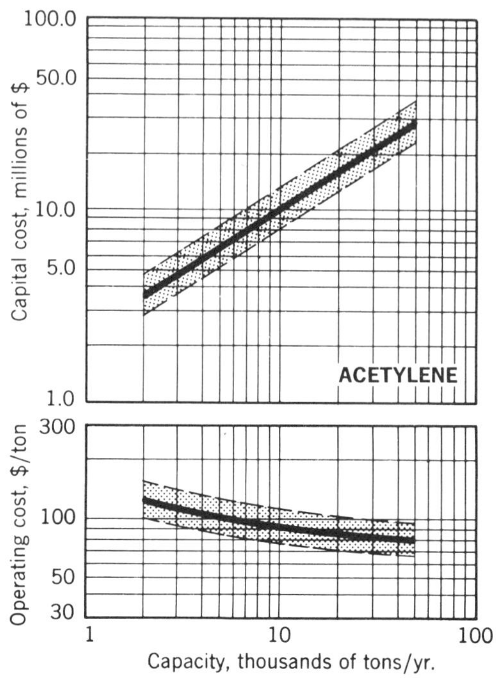

Acetylene: Via pyrolysis or regenerative conversion of hydrocarbon feedstock. Byproduct: Ethylene.
Both product and byproduct are 99.9% pure

Size Exponent: 0.65 **Data:** A = 1; E = 3; P = 2

Included: Process unit and storage facilities

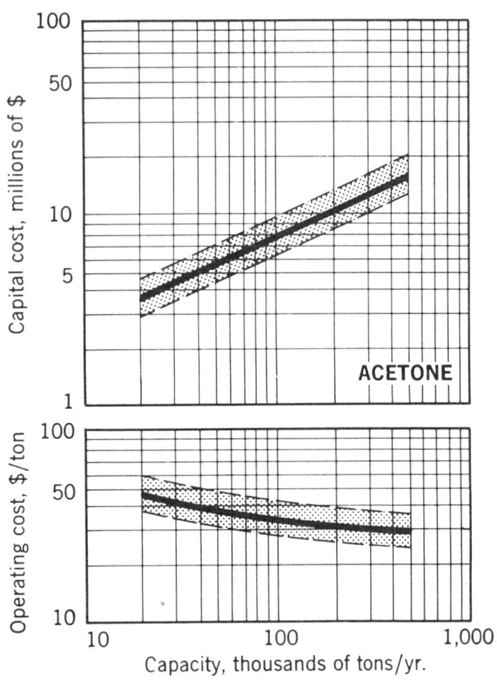

Acetone: Via exothermic conversion of propylene, using copper chloride catalyst and compressed air.

Size Exponent: 0.45 **Data:** A = 0; E = 2; P = 2

Included: Process unit and storage facilities

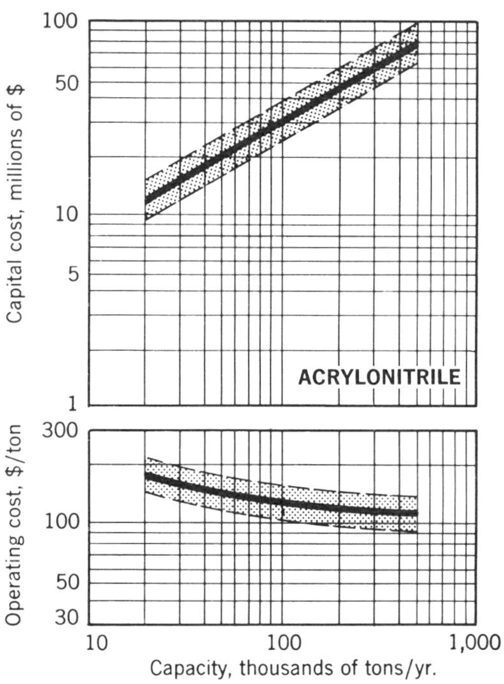

Acrylonitrile: Via catalytic reaction involving acetylene and hydrogen cyanide feedstock.

Size Exponent: 0.60 **Data:** A = 0; E = 1; P = 2

Included: Process unit and storage facilities

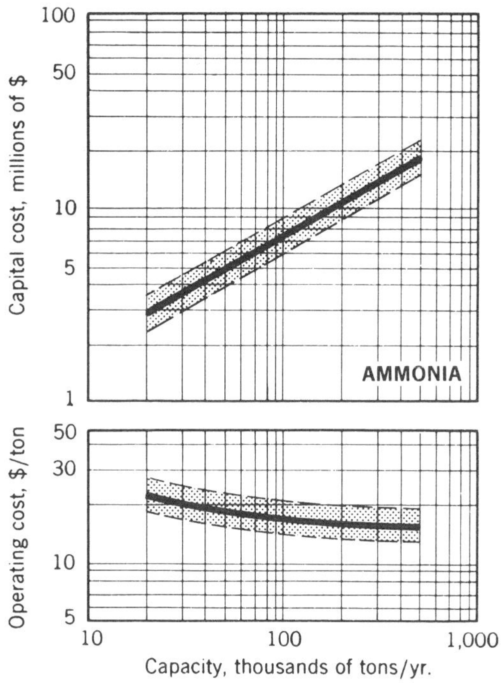

Ammonia: Via steam reforming to convert light-hydrocarbon gases or naphtha into 99.9% pure anhydrous ammonia.

Size Exponent: 0.58 **Data:** A = 4; E = 5; P = 2

Included: Process unit and storage facilities

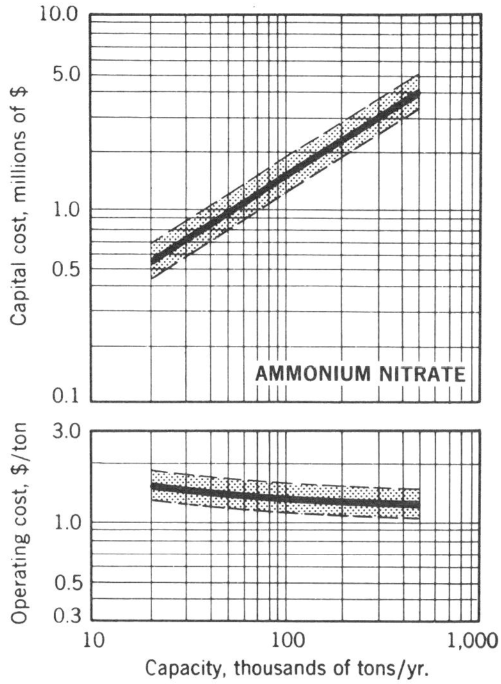

Ammonium nitrate: From ammonia and nitric acid, via a neutralizing and prilling process, to produce fertilizer-grade product.

Size Exponent: 0.65 **Data:** A = 2; E = 4; P = 2

Included: Process unit and storage facilities

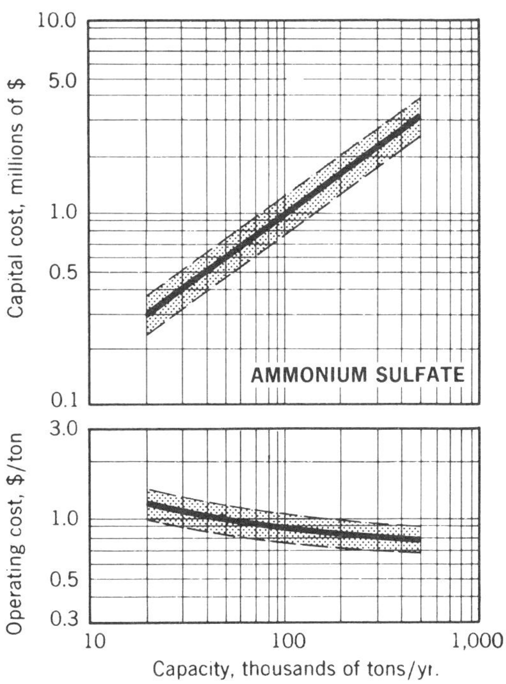

Ammonium sulfate: Reaction of ammonium hydroxide and sulfuric acid, with evaporative process resulting in solid nitrogenous fertilizer.

Size Exponent: 0.73 **Data:** A = 1; E = 3; P = 4

Included: Process unit only

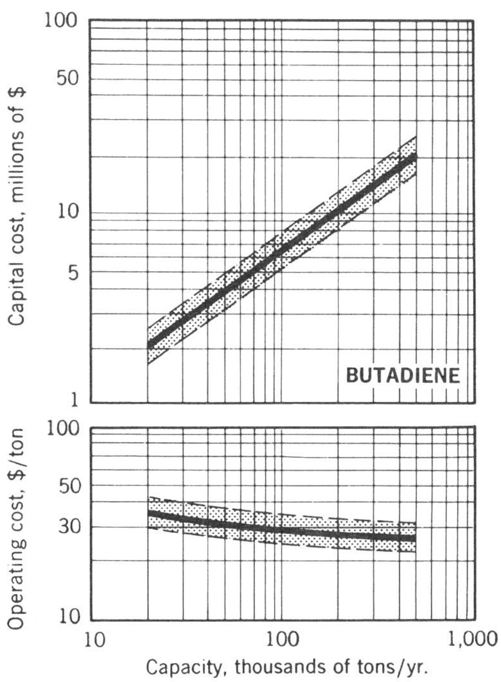

Butadiene: From butane, via dehydrogenation and catalytic reaction. Butylenes are byproducts.

Size Exponent: 0.68 **Data:** A = 3; E = 2; P = 3

Included: Process unit and storage facilities

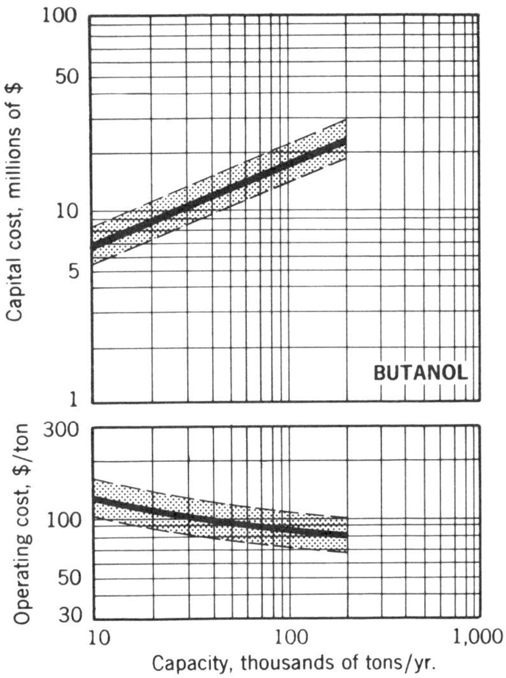

Butanol: From propylene, carbon monoxide and water, via catalytic reaction to produce high-purity *n*-butanol and *iso*-butanol, with CO_2 as byproduct.

Size Exponent: 0.40 **Data:** A = 0; E = 2; P = 4

Included: Process unit and storage facilities

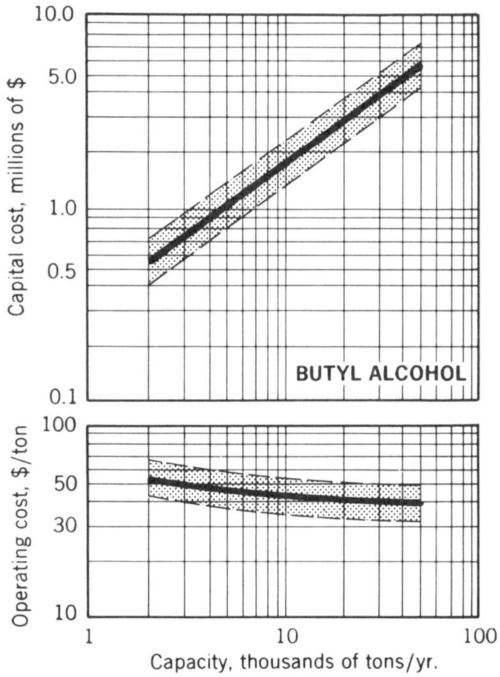

Butyl alcohol: Via catalytic hydration of butylene.

Size Exponent: 0.78 **Data:** A = 0; E = 1; P = 2

Included: Process unit only

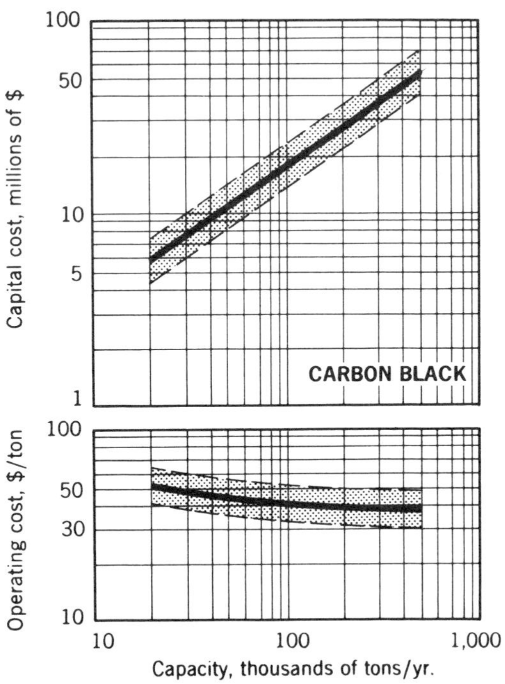

Carbon black: From aromatic oils, fuel gas and air via an atomizing and reaction process.

Size Exponent: 0.70 **Data:** A = 0; E = 2; P = 2

Included: Process unit and storage facilities

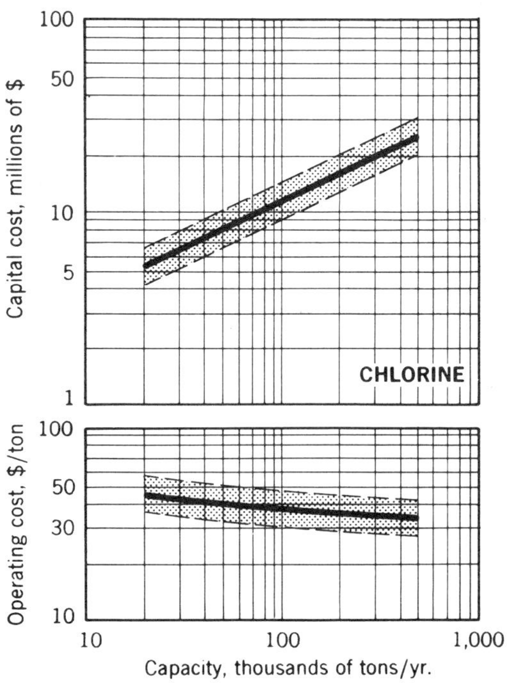

Chlorine: Via electrolysis of sodium chloride. Byproduct: Caustic soda

Size Exponent: 0.45 **Data:** A = 0; E = 2; P = 3

Included: Process unit only

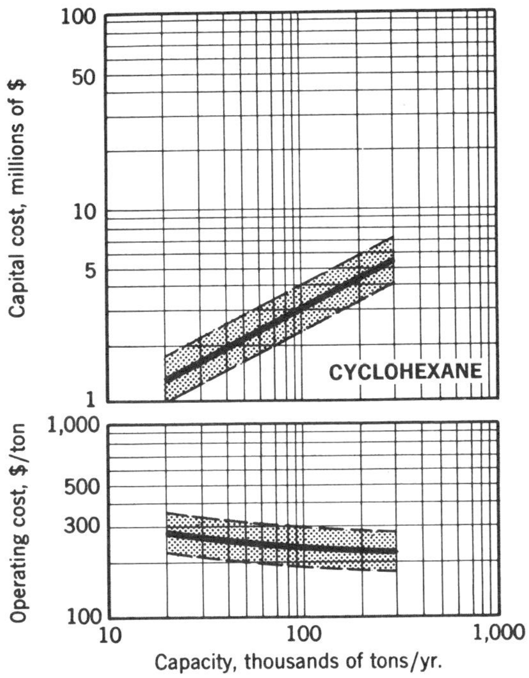

Cyclohexane: From benzene and hydrogen-rich gas, via catalytic hydrogenation to produce 99.9% pure product.

Size Exponent: 0.50 **Data:** A = 1; E = 0; P = 4

Included: Process unit, hydrogen plant and storage facilities

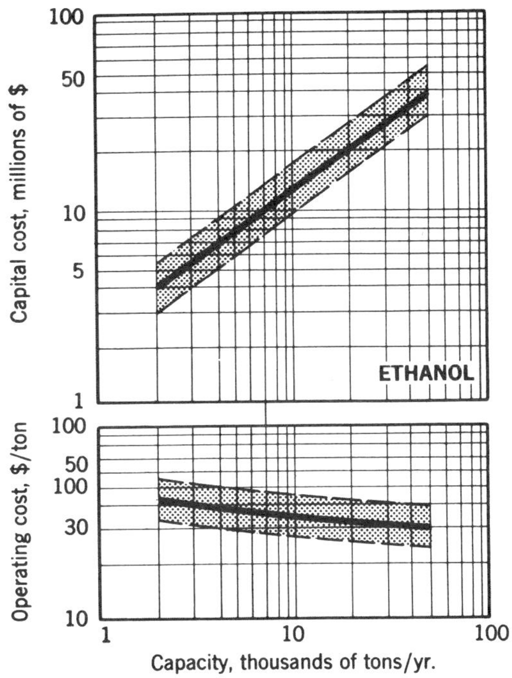

Ethanol: From ethylene gas and condensate, via catalytic reaction and purification.

Size Exponent: 0.73 **Data:** A = 0; E = 1; P = 3

Included: Process unit and storage facilities

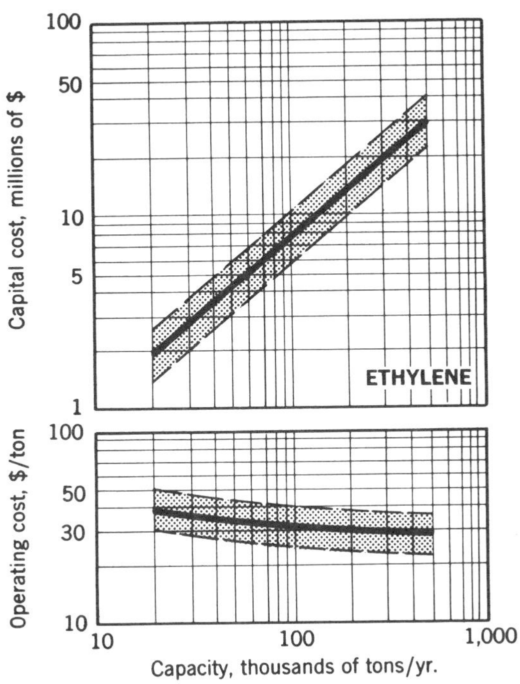

Ethylene (high-purity): From refinery gases, naphthas and gas oils via steam pyrolysis and low-temperature distillation. Byproducts: Propylene, methane, H_2.

Size Exponent: 0.83 **Data:** A = 1; E = 3; P = 4

Included: Process unit and storage facilities

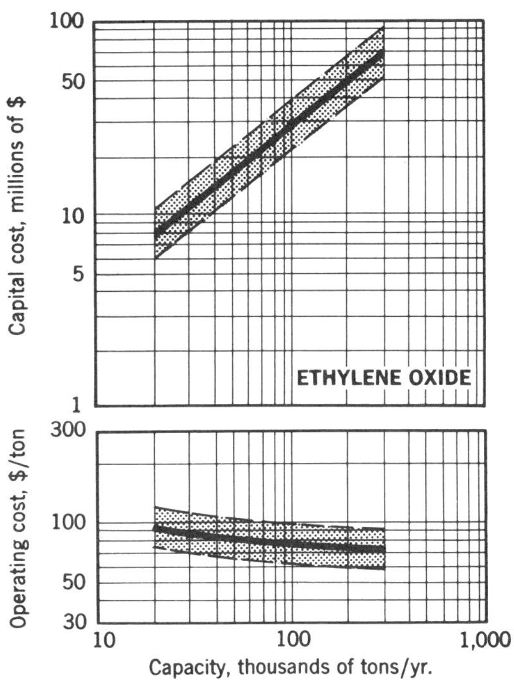

Ethylene oxide (high-purity): From commercial ethylene, air or oxygen, via catalytic oxidation and isothermal process. Byproducts: High and low grades of ethylene glycols.

Size Exponent: 0.78 **Data:** A = 0; E = 2; P = 2

Included: Process unit and storage facilities

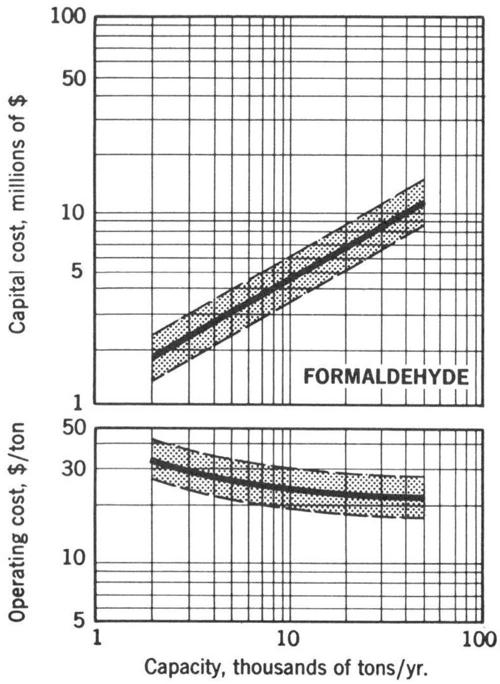

Formaldehyde (aqueous): Via catalytic oxidation and absorption.

Size Exponent: 0.55 **Data:** A = 1; E = 1; P = 3

Included: Process unit and storage facilities

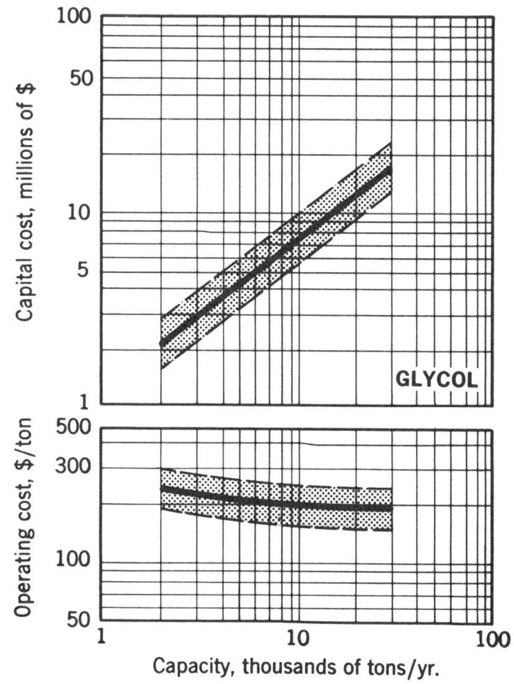

Glycol (high-purity): From ethylene and chlorine, via reaction and chemical treating.

Size Exponent: 0.75 **Data:** A = 1; E = 1; P = 3

Included: Process unit only

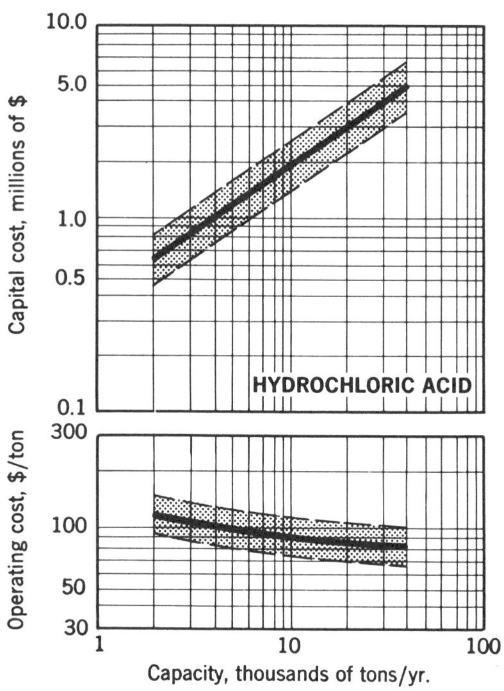

Hydrochloric acid: From hydrogen chloride and water via dissolution process.

Size Exponent: 0.68 **Data:** A = 1; E = 2; P = 2

Included: Process unit and storage facilities

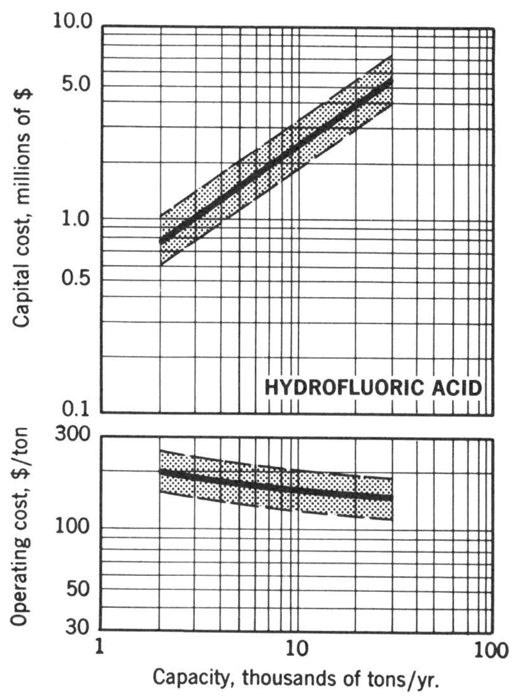

Hydrofluoric acid: From hydrogen fluoride and water via dissolution process.

Size Exponent: 0.70 **Data:** A = 2; E = 0; P = 2

Included: Process unit and storage facilities

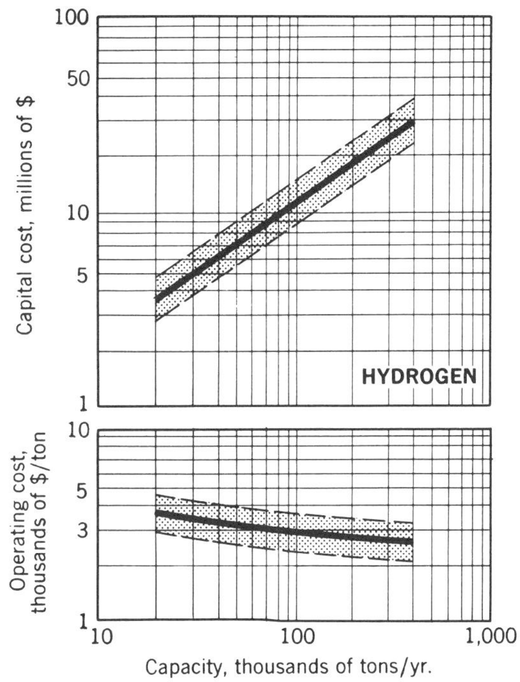

Hydrogen (high-purity)**:** Via separation from petroleum and petrochemical off-gas. Byproducts: Argon, nitrogen, methane.

Size Exponent: 0.70 **Data:** A = 2; E = 1; P = 4

Included: Process unit only

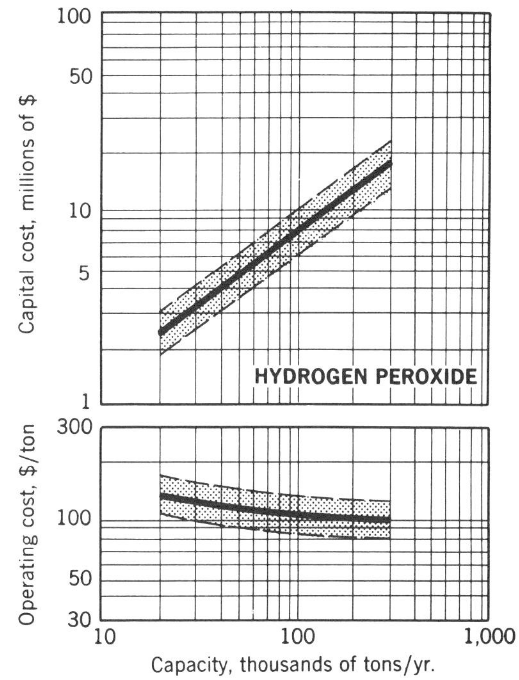

Hydrogen peroxide: Via oxidation and concentration, using isopropylene alcohol and oxygen as feedstock.

Size Exponent: 0.75 **Data:** A = 0; E = 1; P = 2

Included: Process unit and storage facilities

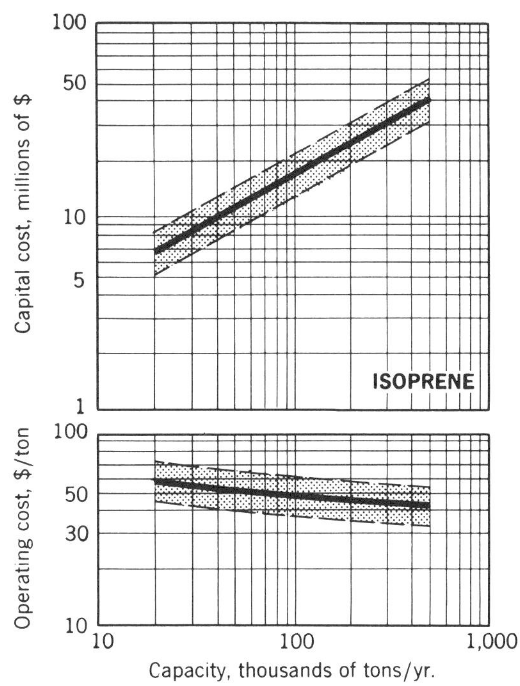

Isoprene: Via a catalytic reaction and distillation, using propylene, methanol and oxygen as feedstock. Byproducts: Isobutylene and off-gas.

Size Exponent: 0.55 **Data:** A = 1; E = 2; P = 1

Included: Process unit and storage facilities

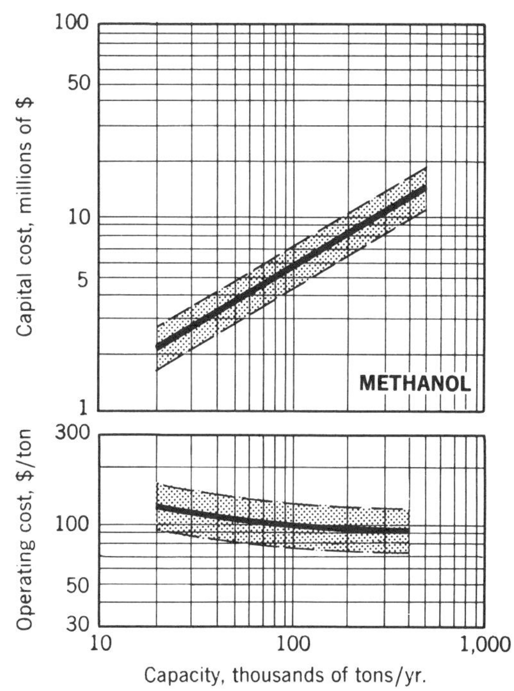

Methanol: From carbon dioxide, natural gas and steam, via a reforming and synthesis process. Byproduct: Fuel gas.

Size Exponent: 0.60 **Data:** A = 2; E = 2; P = 4

Included: Process unit and storage facilities

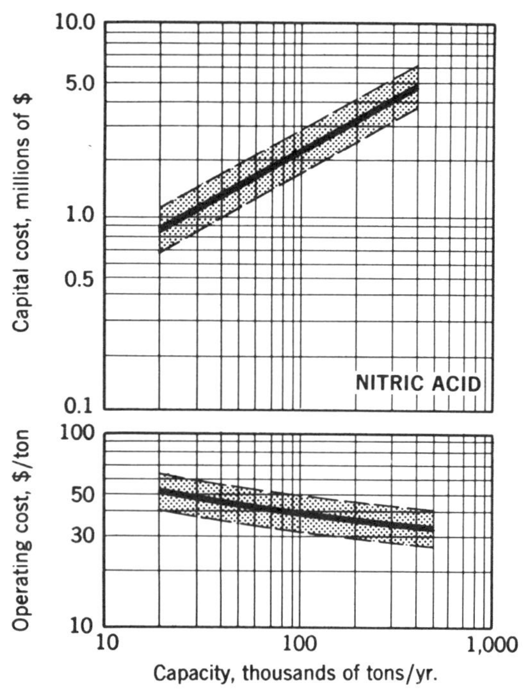

Nitric acid (high-strength, high-purity): From anhydrous or gaseous ammonia, or urea off-gas, via catalysis and absorption.

Size Exponent: 0.60 **Data:** A = 2; E = 3; P = 2

Included: Process unit and storage facilities

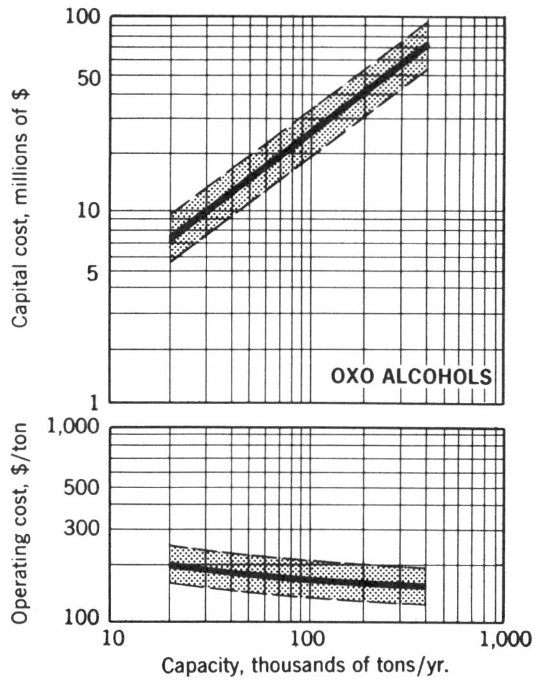

Oxo alcohols: From olefin streams, CO and H_2, via a reactive fractionation or hydrogenation process to yield aldehydes (butyraldehyde) and alcohols. Byproducts: Butanol, isobutanol, high alcohols, light hydrocarbons.

Size Exponent: 0.75 **Data:** A = 1; E = 2; P = 2

Included: Process unit and storage facilities

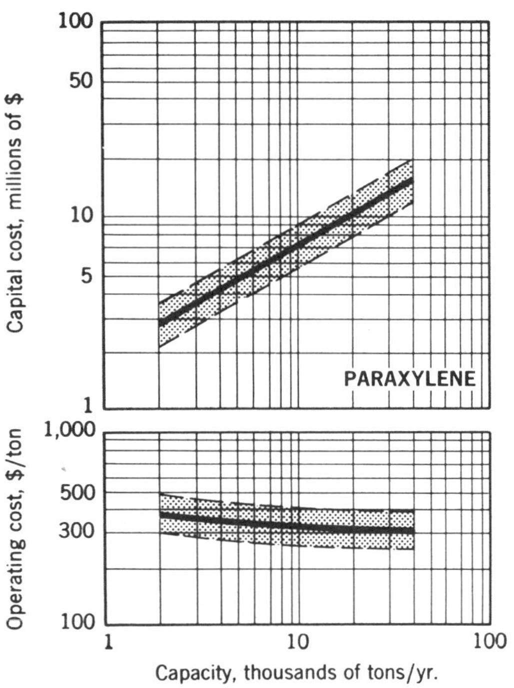

Paraxylene: From xylene-rich hydrocarbon streams and H_2, via a crystallizing and catalytic recycle process. Byproducts: Light ends, aromatics

Size Exponent: 0.55 **Data:** A = 1; E = 2; P = 2

Included: Process unit and storage facilities

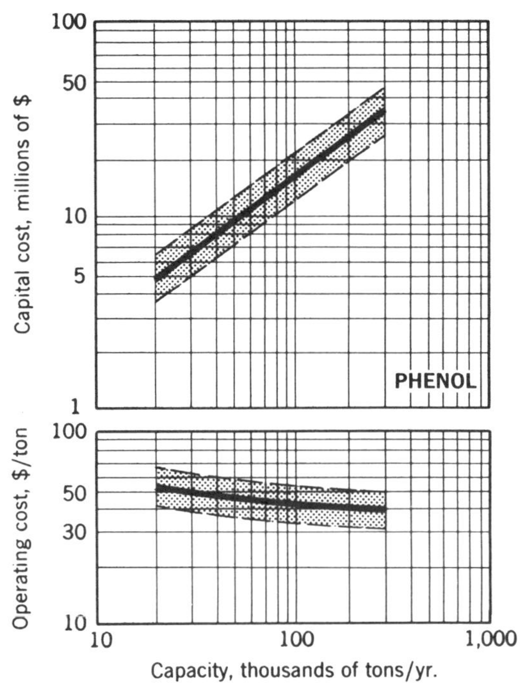

Phenol (high-purity): From toluene and oxygen or air, via oxidation and catalytic conversion.

Size Exponent: 0.75 **Data:** A = 1; E = 2; P = 3

Included: Process unit and storage facilities

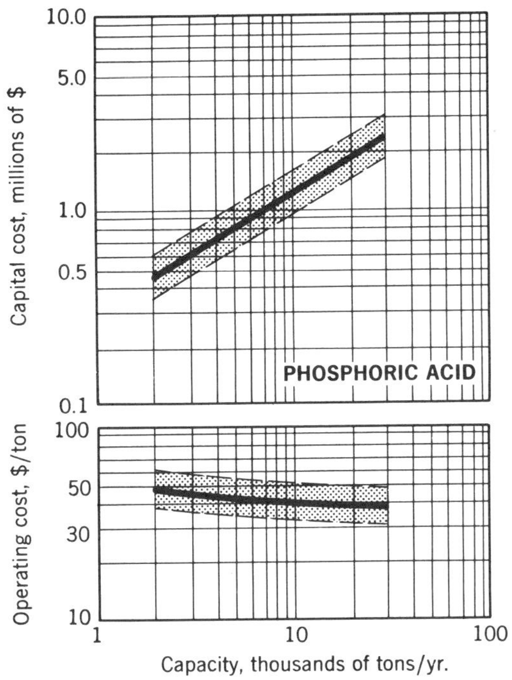

Phosphoric acid: From calcium phosphate and sulfuric acid, via a reaction process.

Size Exponent: 0.60 **Data:** A = 3; E = 2; P = 4

Included: Process unit and storage facilities

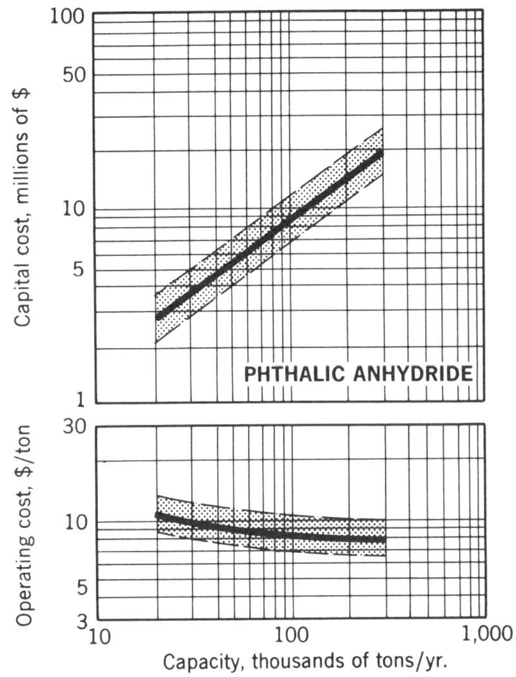

Phthalic anhydride (high-purity): From naphthalene and air, via an exothermic reaction and distillation process.

Size Exponent: 0.70 **Data:** A = 2; E = 3; P = 4

Included: Process unit and storage facilities

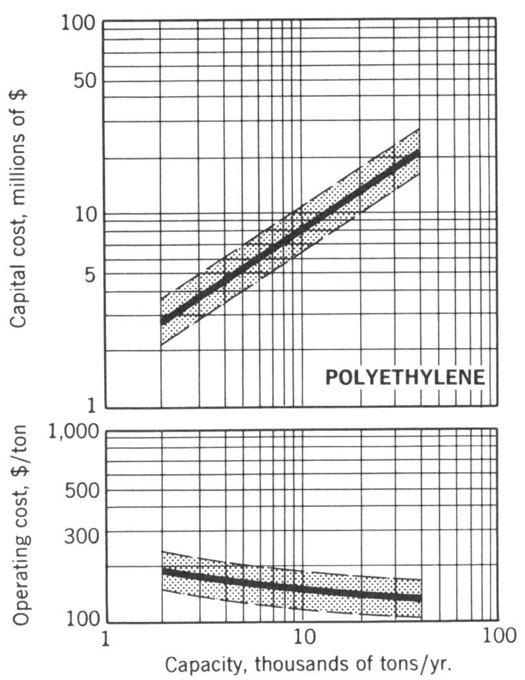

Polyethylene (high-density): From pure ethylene, via a catalytic reaction and high-pressure separation.

Size Exponent: 0.65 **Data:** A = 2; E = 2; P = 4

Included: Process unit and storage facilities

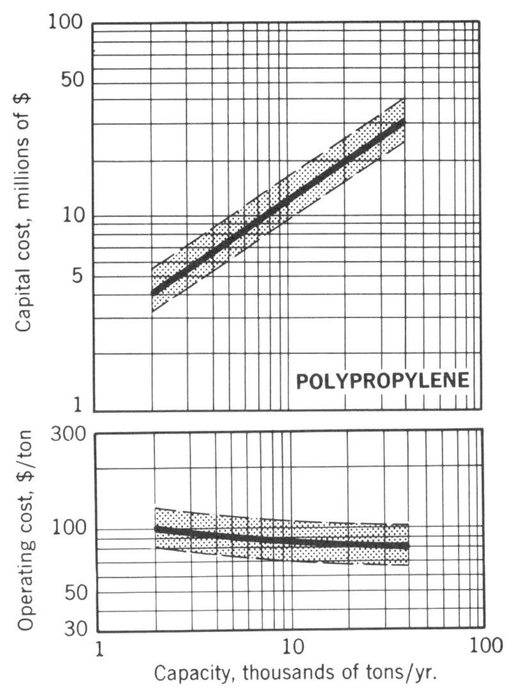

Polypropylene: From propylene-rich streams, via a low-pressure catalytic reaction and polymerization.

Size Exponent: 0.70 **Data:** A = 0; E = 2; P = 3

Included: Process unit only

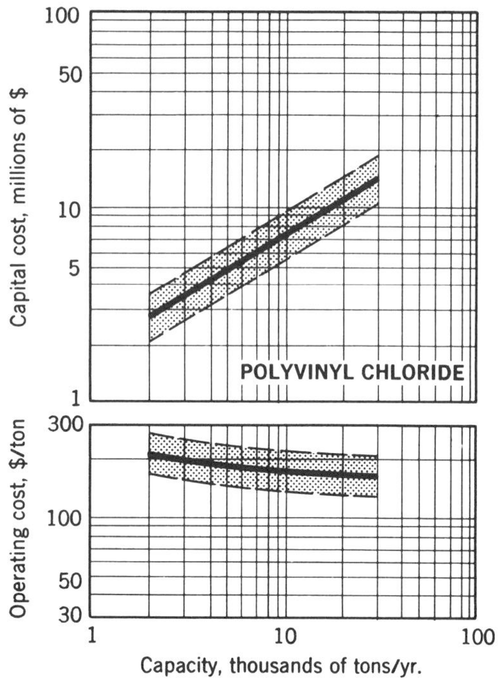

Polyvinyl chloride: From vinyl-chloride-rich monomer streams, via polymerization and filtration.

Size Exponent: 0.60 **Data:** A = 0; E = 2; P = 4

Included: Process unit only

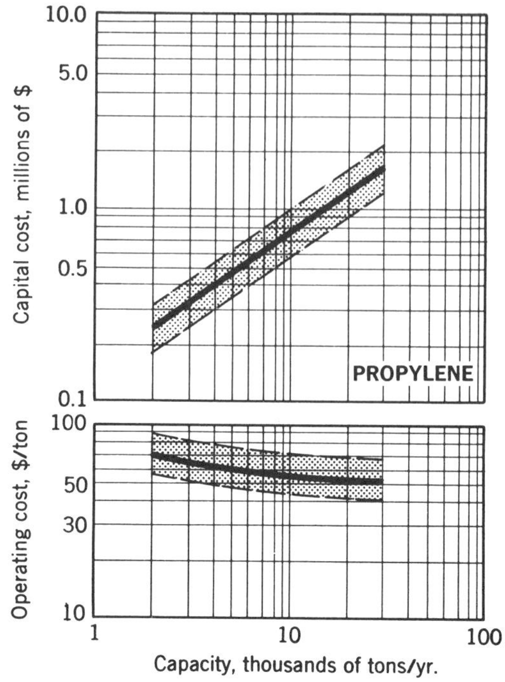

Propylene: From refinery gases, naphtha or gas oil, via thermal cracking, gas compression and fractionization. Byproduct: Ethylene.

Size Exponent: 0.70 **Data:** A = 1; E = 3; P = 3

Included: Process unit and storage facilities

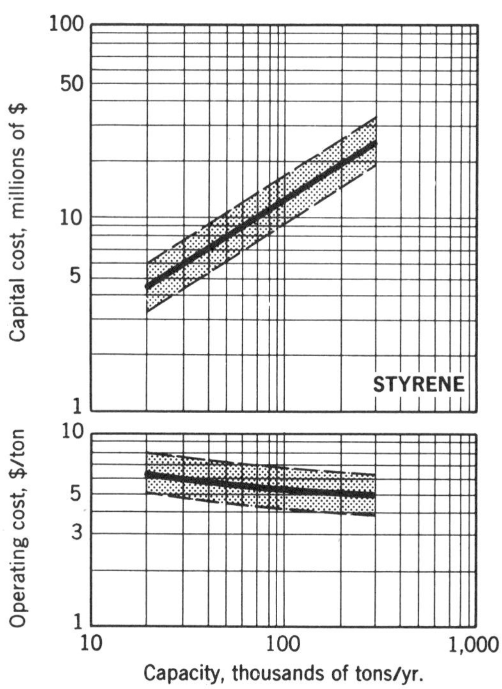

Styrene: From benzene, ethylene, steam, via a catalytic reaction, with alkylation and distillation to recover styrene monomer.

Size Exponent: 0.60 **Data:** A = 0; E = 2; P = 4

Included: Process unit only

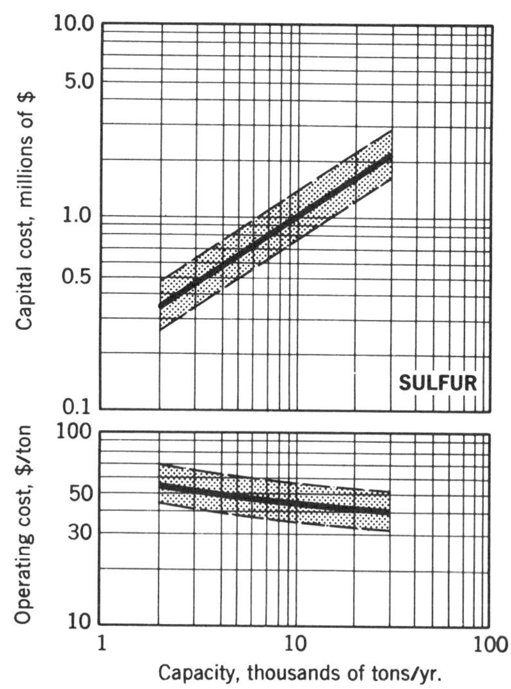

Sulfur: Via an oxidation reaction to recover basic sulfur from hydrogen sulfide gas.

Size Exponent: 0.65 **Data:** A = 1; E = 2; P = 4

Included: Process unit only

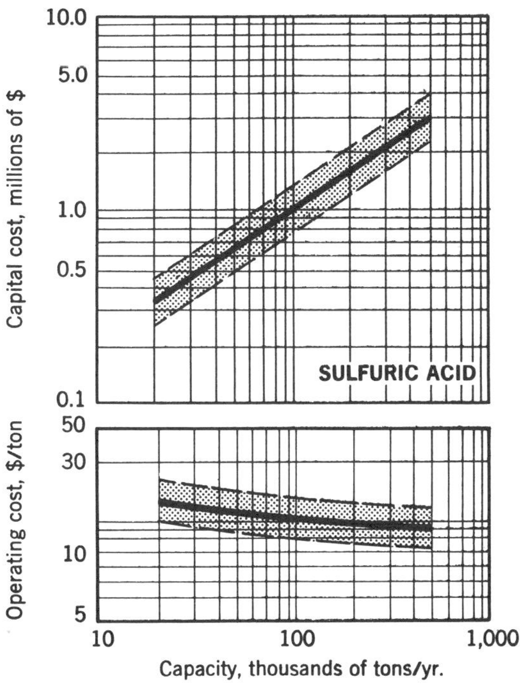

Sulfuric acid: From sulfur, via a catalytic reaction.

Size Exponent: 0.65 **Data:** A = 1; E = 1; P = 3

Included: Process unit and storage facilities

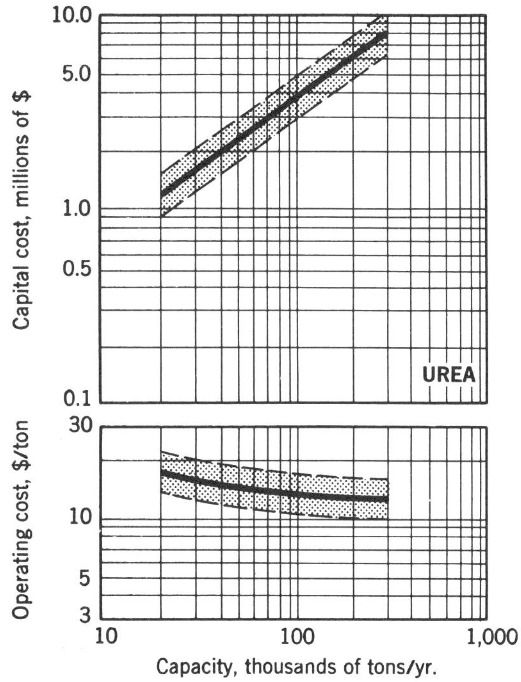

Urea (prills or crystals): From ammonia and carbon dioxide, via exothermic reaction, decomposition and prilling.

Size Exponent: 0.70 **Data:** A = 2; E = 3; P = 3

Included: Process unit and storage facilities

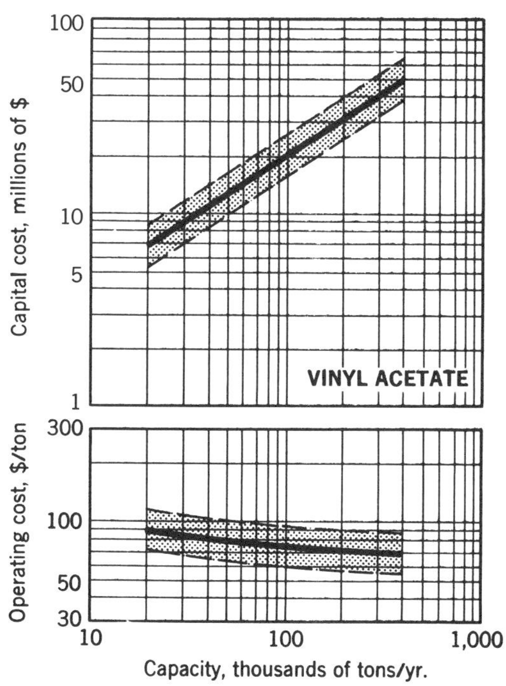

Vinyl acetate: From ethylene and oxygen, via a catalytic reaction and distillation.

Size Exponent: 0.65 **Data:** A = 2; E = 3; P = 2

Included: Process unit only

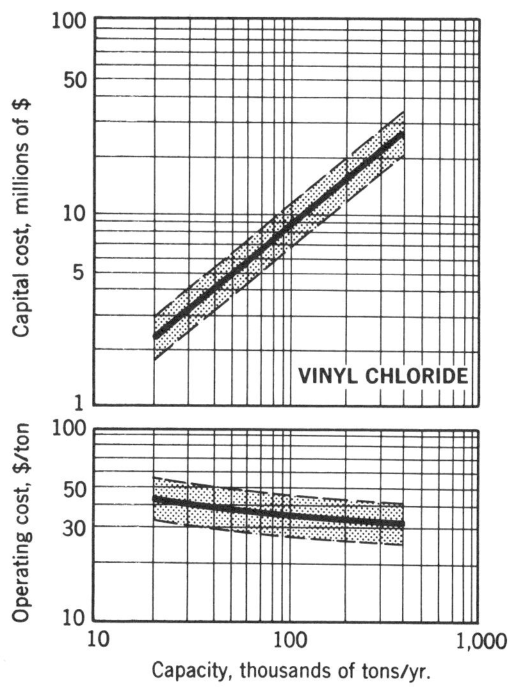

Vinyl chloride: From ethylene, chlorine and air, via thermal cracking, high-pressure compression and distillation.

Size Exponent: 0.80 **Data:** A = 0; E = 3; P = 2

Included: Process unit only

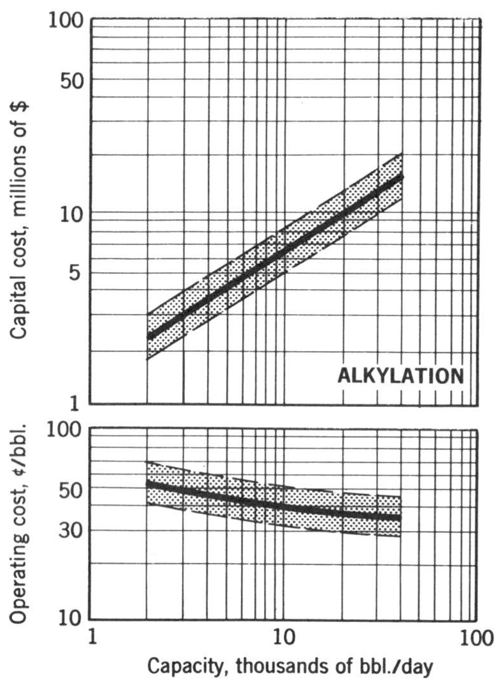

Alkylation: A catalytic process to upgrade light gaseous hydrocarbons (propylene, butylene or amylene) into high-octane gasoline and aviation fuels, in the presence of sulfuric acid.

Size Exponent: 0.60 **Data:** A = 2; E = 4; P = 5

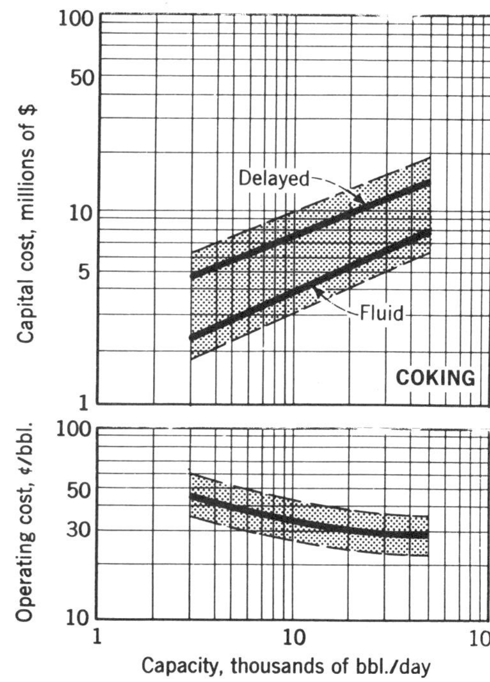

Coking: A thermal cracking process to upgrade reduced crudes, cracked tars, asphalts and heavy vacuum residues into gas oil, gasoline, naphtha and light gases.

Size Exponents: 0.38 for delayed; 0.42 for fluid

Data: A = 1; E = 3; P = 3

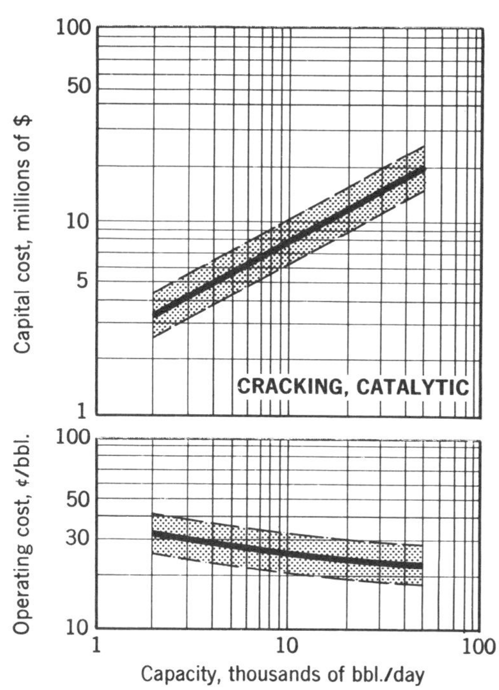

Cracking, catalytic: Catalytic conversion and regeneration to upgrade deasphalted stocks (reduced crudes, gas oils, naphthas) to high-octane gasolines. Byproducts: Butane, olefins, heavy gasoline, diesel stock.

Size Exponent: 0.55 **Data:** A = 4; E = 4; P = 2

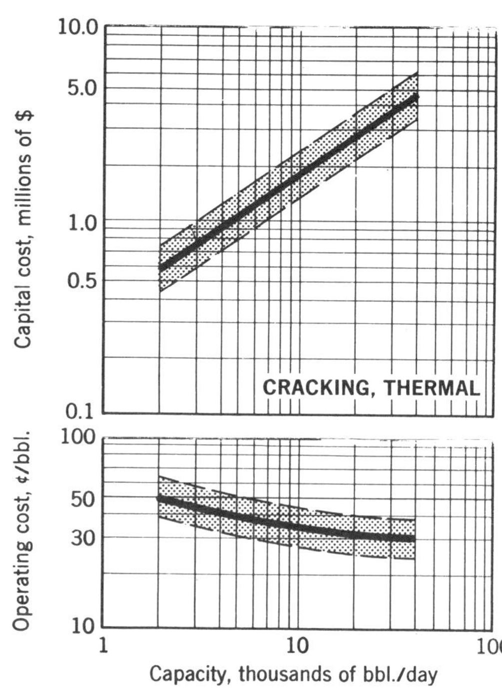

Cracking, thermal: Reduction of heavy hydrocarbon streams (reduced crude oil, residual oil and other distillates) to produce stabilized gasoline. Byproducts: Butylene, propylene, light and heavy fuel oils.

Size Exponent: 0.70 **Data:** A = 3; E = 4; P = 2

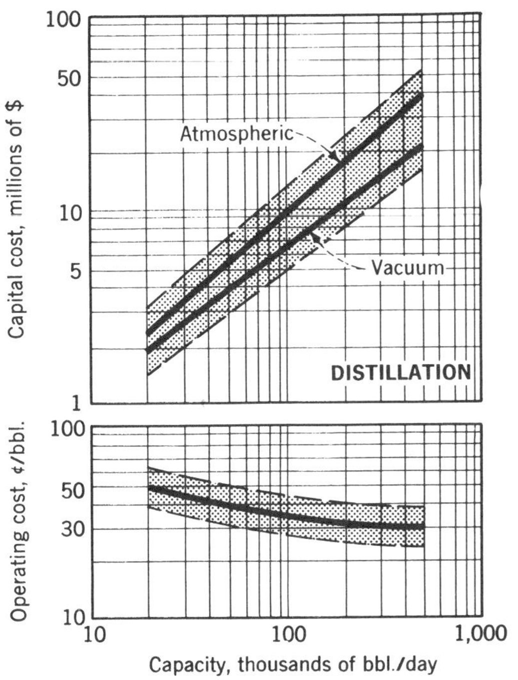

Distillation: Separation of crude oil into specific fractions and intermediates—naphtha, kerosene, diesel oils, lube oils, gas oils, asphalts.

Size Exponent: 0.70 for vacuum; 0.90 for atmospheric

Data: A = 3; E = 5; P = 4

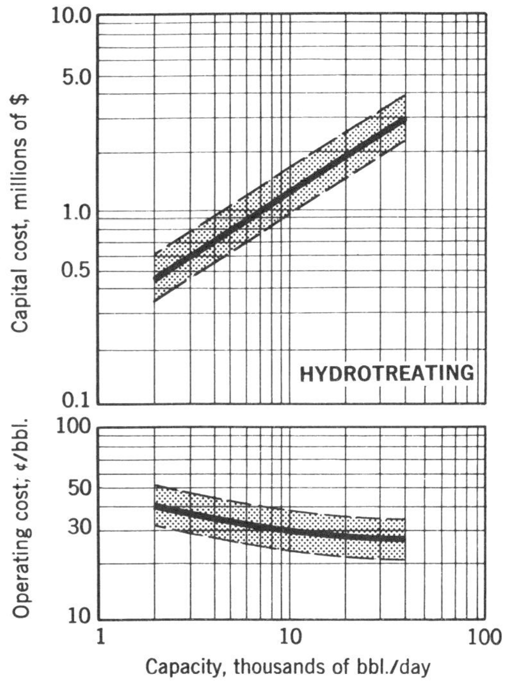

Hydrotreating: Hydrocatalytic desulfurization to convert high-sulfur residuals and hydrogen-rich gas into low-sulfur lube oils and purified naphtha feeds. Byproducts: Fuel gas, desulfurized fuel oils.

Size Exponent: 0.65 **Data:** A = 3; E = 4; P = 2

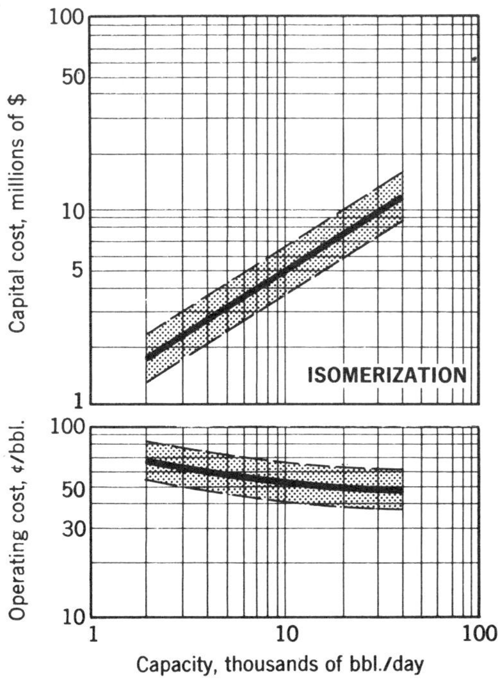

Isomerization: Desulfurization and hydrogenation to upgrade low-octane feedstocks (pentane, hexane, reformer off-gas) into high-octane gasoline blending components.

Size Exponent: 0.65 **Data:** A = 2; E = 3; P = 2

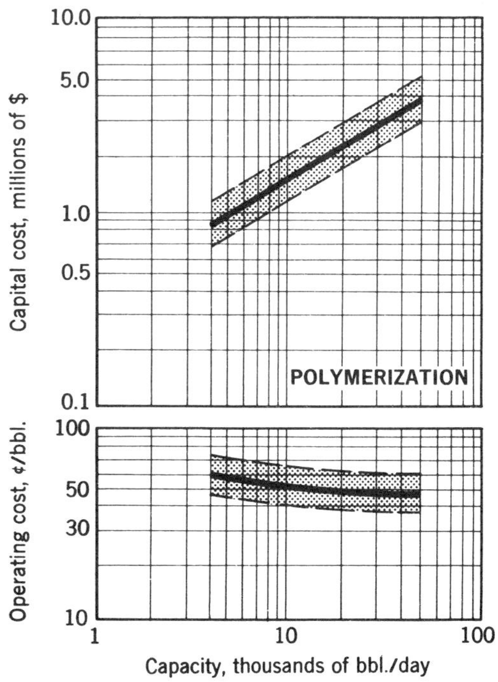

Polymerization: Catalytic reaction to convert hydrocarbon olefinic streams into high-octane additives for gasolines.

Size Exponent: 0.58 **Data:** A = 1; E = 3; P = 2

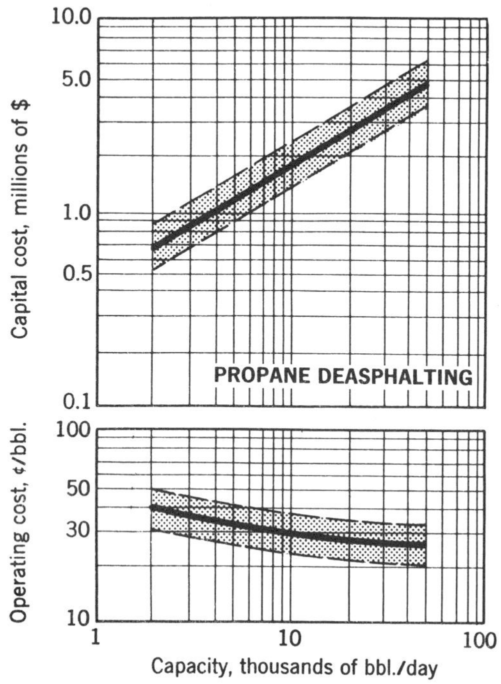

Propane deasphalting: Extraction, separation and stripping to produce deasphalted oils, cylinder stock and asphalt from propane and vacuum reduced crudes.

Size Exponent: 0.60 **Data:** A = 2; E = 3; P = 1

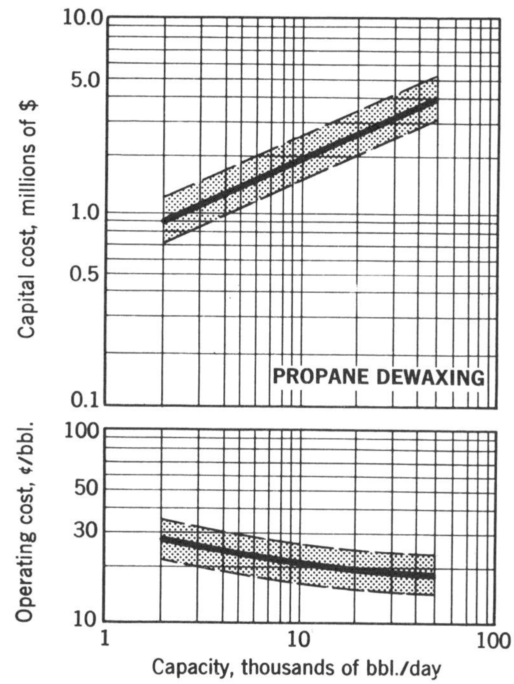

Propane dewaxing: A filtration and stripping process to extract waxy materials from diesel oils, cylinder stock and other waxy distillates.

Size Exponent: 0.45 **Data:** A = 2; E = 3; P = 2

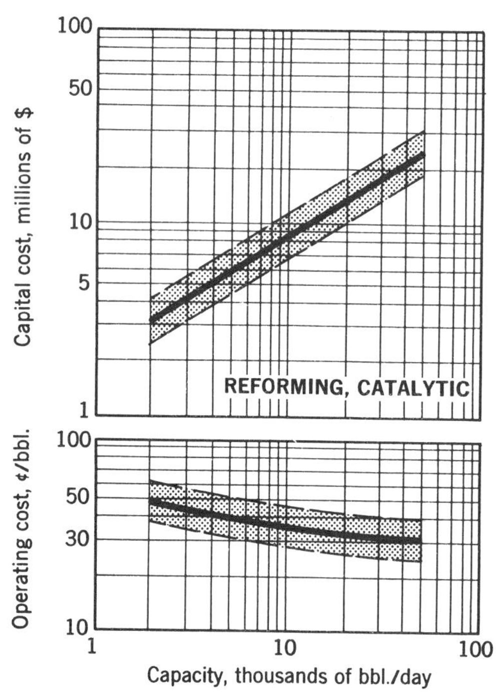

Reforming, catalytic: A process to improve the octane ratings of low-octane naphthas to produce high-octane blending stocks and aromatics. Byproduct: Hydrogen

Size Exponent: 0.61 **Data:** A = 2; E = 4; P = 2

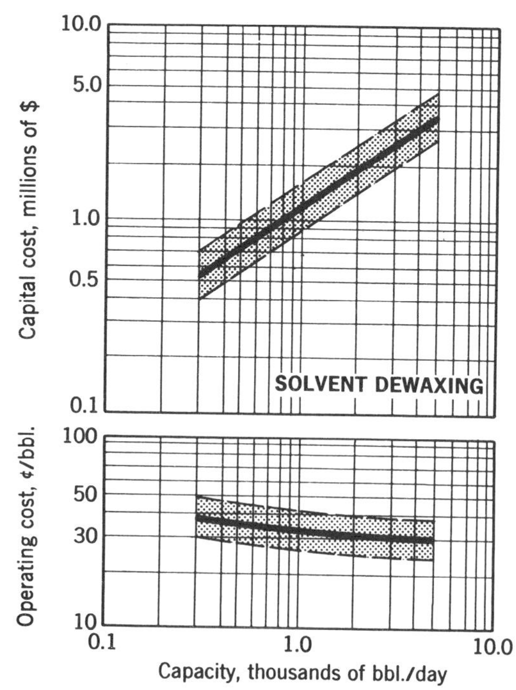

Solvent dewaxing: A solvent extraction process to remove waxes from lube oil stocks (distillate or residual).

Size Exponent: 0.68 **Data:** A = 2; E = 3; P = 2

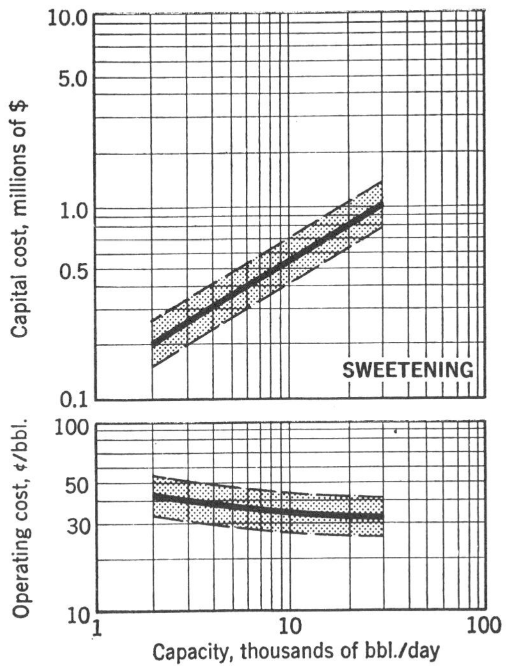

Sweetening: Treatment of straight-run and cracked gasolines to convert mercaptans to disulfides and to remove hydrogen.

Size Exponent: 0.65 **Data:** A = 2; E = 3; P = 2

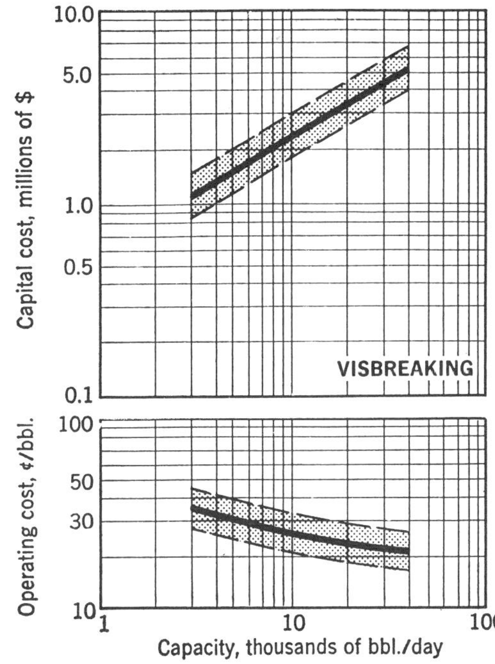

Visbreaking (viscosity breaking): A thermal cracking and vacuum flash process to break down topped crudes into gasoline, light gas oils and heavy distillates.

Size Exponent: 0.60 **Data:** A = 3; E = 3; P = 2

new plant will have to at least match, and perhaps exceed, the anticipated market deficiency in order to achieve the 25% rate of return on the investment. In a rough way, we have anticipated the weakening in selling price that this may bring, by assuming a selling price of 1¢/lb. rather than 1.25¢/lb. (We may now want to do an additional study at, say, 1.2¢/lb., based on part of the plant's capacity remaining idle at the start.)

Capital requirements will not exceed the limitation of $1 million, and the payout requirement can be easily met.

Further conceptual studies can now focus on capacities and other variables between Case B and Case C, using the charted data and similar calculation formats, until a satisfactory basis for cash-flow analysis is established.

Note—The example problem is intended to illustrate how the data in this report can help you rapidly assemble a series of preliminary project-evaluation studies for review, before the application of more-sophisticated cost-estimating and cash-flow-analyses techniques. It is for such preliminary "rough estimate" studies that these cost/capacity charts are intended.

Petroleum-Refining Units

To provide flexibility, petroleum refinery costs have been isolated into a series of separate unit processes, and these are charted on pp. 402–405.

The unit process costs are based on integrated refinery feedback, and include a proportionate allocation of the refinery distribution-system costs, such as fuel-oil/fuel-gas, steam, cooling water, plant and instrument air, overpressure to flare stack, sewers and drainage, interconnecting piping, etc.

Offsite costs such as general site preparation, buildings, yard piping, major tankage, loading and unloading facilities are not included and have to be added as required. Use 35 to 40% of process battery limit costs if no other data are available. Again, all data are as of mid-1968, U.S. Gulf Coast. ■

Meet the Author

Kenneth M. Guthrie is with Fluor Corp. (2500 S. Atlantic Blvd., Los Angeles, Calif. 90022), where his responsibilities include capital cost estimating. Before joining Fluor, he was chief estimator and cost analyst in the Engineering Div. of W. R. Grace & Co. His dozen years of experience in cost engineering also include work with Mobil Oil Corp. and M. W. Kellogg Co. Mr. Guthrie's last article dealt with the cost of liquid handling systems, and appeared in the Deskbook issue of Apr. 14, 1969.

How Much Working Capital Will the New Project Need?

More funds may be needed to supply a new venture with working capital than to supply it with pumps or pressure vessels—hence the need to estimate working-capital requirements with insight and consistency.

THOMAS B. LYDA, Gulf Oil Chemicals Co.

Working capital—the money required for the day-to-day operation of a business over and above the fixed investment for land, buildings and equipment—may amount to more than 20% of the total investment for a project. In some cases, working-capital requirements may even exceed the fob. cost of the process equipment. Failure to make adequate provision for working capital has been a classic cause for bankruptcy among companies whose expansion policies ran too far ahead of their cash flow.

Some engineers may tend to think of working capital as "money in the bank" that can easily be recovered and reassigned should the project fail—but a substantial part of the working capital may be tied up in inventories that can only be liquidated at a substantial loss. This is particularly true if a new product of high unit value has to be withdrawn.

For all these reasons, a realistic estimate of working-capital requirements is extremely important in venture analysis. It must be remembered that there is a danger in over-estimation as well as under-estimation—i.e., danger that inflated working-capital estimates will reduce the returns on investment to a point where good projects are passed up, and business opportunities are missed.

Averaging the Requirements

In its most obvious form, working capital appears as money that must be supplied from a company's treasury to buy inventories for raw materials and finished products, and to provide credit to customers who have not yet paid for delivered products.

Any funds coming into the treasury that can be retained, or any payments that can be delayed until further funds are received, reduce a company's working-capital requirement. One example would be an account payable—an unpaid bill—for a raw material: The goods for which the bill is payable have been used and thus provided a revenue that can be retained in the company's treasury until the bill's due date.

A less obvious form of a reduction in working-capital requirements is an accruing liability. A good example is the company's payroll. Suppose a company has a monthly payday: Halfway through the month the firm owes half of the payroll for services rendered by its employees. It has accrued a liability—in effect an unpaid bill—payable to the employees. And because the efforts of the employees presumably produce continuous income, the cash reserved in the treasury for the payment of these wages temporarily reduces the amount of outside money required to operate the business.

It is obvious from the above example that the requirement for working capital does not have a constant value, but is a quantity that varies from day to day and even from hour to hour. This variation is an inconvenience in venture analysis where it is desirable to have a single, time-independent quantity that can readily be used in return-on-investment or discounted-cash-flow calculations. Fortunately, two reasons make it possible to justify using an average working capital requirement for such calculations:

First, consider a large company that operates many units at many locations. Here, a peak demand for working capital at one location may be offset by a simultaneous low demand at another location. The company's total working-capital requirements under such conditions thus approach the average total demand of all the units or all the various ventures.

The second averaging effect is possible even though a peak demand at one location may not coincide simultaneously with a low demand at another. Should a fiscally sound company find itself temporarily short of working funds, it may borrow the money for a month, a week, or even a few days. Similarly, it may "loan" any temporary excess of funds (e.g., by buying U.S. Treasury Bills) and realize a return on its short-term investment. Income

Originally published September 1972.

earned from temporary investments can be used to pay at least some of the interest on temporary loans. The company's total working-capital requirement thus approaches an average value, and the net balance of temporary investment and borrowing approaches zero.

Of course, in some cases, borrowing to alleviate temporary shortages of funds cannot be counted on (e.g., if the firm's credit rating is low, if restrictions have been imposed by the terms of previous loans, or if a "tight money" period is anticipated). In these situations, it will be necessary to estimate the *maximum* working-capital requirement for use in venture analysis, rather than the *average* requirement. (Also, some companies prefer to work with *maximum* requirements in economic analysis for reasons of fiscal conservatism—i.e., to take care of temporary surges in the need for funds without recourse to borrowing even if such borrowing is likely to be readily available.)

Working Capital for a Single Venture

A second inconvenience in estimating a value for working capital for venture analysis is deciding just what part of a company's total working capital should be included in the investment for the particular venture under evaluation. Often (as in the example that will be developed) the distinction between included and excluded components of a venture's working capital is obvious. Sometimes, however, it is necessary to apply the general rule that the working-capital investment chargeable to the venture is the incremental increase of total working capital caused by the venture—i.e., the difference between the company's total working capital after the project's startup and what the total working capital would be had the venture never been undertaken.

For instance, let us say that, to save on transportation costs, a petrochemical plant is considering buying a feedstock from a neighboring refinery rather than continuing to ship the feedstock from a distant refinery owned by its parent company. Consider the accounts-payable aspect of this situation: Any "payables" that the petrochemical plant may have for feedstock under the present practice is offset by "receivables" at the refinery, because both are part of the same company and presumably share the same treasury. However, when the petrochemical plant buys the feedstock from outside the company on, say, a 30-day account, the company's working capital requirement is reduced by the amount of the payables, for there is no longer an offsetting receivable in the same company. Were there no other considerations, this venture would have a negative incremental working-capital requirement.

Actually, it is quite possible for an individual venture's working-capital requirements to be zero or even negative. The "negative" situation occurs if, on an incremental basis, the firm's working-capital requirements are reduced by initiation of the venture.

The Components of Working Capital

Although working capital is an important part of economic evaluations, relatively little has been published on its estimation.[1,3] It can be shown, however, through a rather detailed study of the day-to-day cash requirements of a venture, that the working capital required at any moment of the venture's life is the algebraic sum of five items:

1. The value of the raw-material and supplies inventories at cost.
2. Plus the value of the product inventory at cost (excluding non-cash depreciation costs).
3. Plus the accounts receivable at selling price.
4. Less the accounts payable.
5. Less the sum of all accrued liabilities (e.g., wages, taxes, dividends).

The detailed study also reveals that an accurate estimate of a project's working capital can be made from the preliminary estimate of its economics (as will be done in the later example) and that a time-independent estimate useful in discounted-cash-flow and return-on-investment calculations can be made by properly averaging each of these five component items individually.

Item-by-Item Averaging

There are two methods of averaging the five component items listed above. The first four items are best considered as being set by operating and sales policy. For instance, sales considerations may dictate that, say, 60 days of production be maintained in inventory so that a "dependable supply in the event of plant breakdown" pitch can be made to prospective customers. Similarly, shipping and distribution considerations—storage capacities, the size of barge liftings, etc.—may fix the average inventories of products and raw materials. Accounts payable and accounts receivable are, of course, set by business policies.

The last of the five items, made up of all of the accruing liabilities, best lends itself to a different type of averaging. These accruing values vary between two predictable values: a minimum, usually zero, and a maximum reached after some fixed period of time. Wages are again a good example: immediately after payday, no wages are owed to the operators of the unit. However, just before the next payday (say, a month later), the full payroll has accrued. Because preliminary economics are usually tabulated on an annual basis, it is convenient to take as an average amount a fraction of the estimated annual amount determined by the duration of the accrual. Since, in the wage example, the accrual period was a month or 1/12 year, the average amount is reached in half that time or 1/24 year. Therefore, the average of the amount of accrued wages is one twenty-fourth of the annual wages.

Payables Associated With the Profit

The liabilities created by a profit in the operation deserve special mention. A profit made by a unit is divided into three parts: part is paid to the government as income taxes, part is paid to stockholders of the company as dividends, and the remainder is retained for the expansion of the business. The first two of these parts are not paid immediately, but are normally paid quarterly. Thus, when profits are made, funds are reserved for these purposes, and these cash balances held in the company's treasury decrease a profitmaking venture's working-capital requirements.

The tax rate is set by law, so determining an average liability, or account payable, for taxes is not difficult. The division of the remainder of the profit,however, is a matter of management's discretion. Although the division may vary from year to year as profits vary or management policies change, a particular company will usually maintain the same approximate ratio of dividends to retained earnings over the years. The ratio can usually be determined by examining a company's annual reports or other published data.[2] It will be necessary, of course, for the engineer to establish the ratio for his company before making estimates of working capital. Often the ratio is about 50:50, and we will assume this ratio in a later example.

The portion of the profit not paid out in taxes or dividends is used for expansion of the business. Corporate expansion in its various forms (including normal growth of existing businesses, as well as new ventures) creates an ever-present demand for cash, and it can be assumed that these profit-derived expansion funds are not allowed to accrue for later payment, but are paid immediately. Thus this portion of the profit does not make a negative contribution to working-capital requirements.

Two Paradoxes

The effects of taxes and dividends create two paradoxes in venture analysis:

1. Although a venture's profits do not in themselves reduce its working-capital requirements, a profitmaking venture nevertheless requires less working capital than does a breakeven venture. Assuming that the revenue (and thus the accounts-receivable part of working capital) are the same, this paradox can be explained by considering the taxes payable. A breakeven venture creates no net revenue nor does it create a tax liability, while a profitmaking venture produces cash income that may be retained in the treasury until the taxes are due. The part of profit that is dedicated to future tax payments temporarily reduces the treasury's need to borrow outside working funds, and thus *incrementally* reduces the firm's working-capital requirements.

2. The average working capital required by a venture undertaken by a company that pays its stockholders a dividend is lower than that required by the same venture if undertaken by a company that pays no dividends. This second paradox is again explained by the venture's incremental effect on the firm's total working capital. During a quarter, the treasury of a dividend-paying firm is required to accumulate the money that will be paid to the stockholders; this reserve can temporarily reduce the outside financing required, and thus incrementally reduce the firm's working-capital requiremets. Indeed, this reserve can constitute a major portion of a firm's total working capital.

A company that pays its stockholders no dividends need not accumulate this reserve, and may well have a lower "total company working-capital" requirement.

Estimated Economics of Hypothetical New Plant Venture—Table I

Investment for 33-Million Lb./Yr. Plant:	**$M**	
Process	4,500	
Utility, storage, offsite facilities	1,100	
Total plant investment for product P	5,600	
Raw-Material and Operating Costs:	**$M/Yr.**	**¢/Lb. of P**
Raw material, R	1,332	4.04
Operating labor (incl. maint. labor)	161	0.49
Labor burdens	184	0.56
Utilities	100	0.30
Catalysts	62	0.19
Chemicals	82	0.25
Supplies (incl. packaging, maint.)	140	0.42
Other operating costs (except deprec.)	90	0.27
Depreciation	419	1.27
Total operating cost	1,238	3.75
Total cost (breakeven)	2,570	7.79
Revenue @ 12 ¢/lb.	3,960	12.00
Profit margin	1,390	4.21

Working-Capital Estimate—Table II

Item	Description, Basis and Calculation	Working-Capital Contrib., $M
1	Value of raw-material and supplies inventories at cost:	
	Inventory requirement for R is set at 30 days (1/12 yr.) (1/12) ($M 1,332)........................	111
	Inventory of chemicals, catalysts and supplies is set at 30 days (1/12) ($M 284)..........................	24
2	Plus value of product inventory at cost (Excluding depreciation):	
	Inventory of P set at 45 days of production (1/8) ($M 2,570 - $M 419	269
3	Plus accounts receivable:	
	Sales basis is 30 days cash (1/12) ($M 3,960)..........................	330
4	Less accounts payable:	
	Payable for 30 days of raw material (1/12) ($M 1,332)..........................	-111
5	Less the sum of accrued liabilities:	
	Wages and other operating expenses accrue for 30 days; average amount is 15 days, or 1/24 year's amount (1/24) ($M 1,238 - $M 419)..........	- 34
	Taxes accrue for 90 days; average amount is 45 days, or 1/8 year's amount (1/8) (0.48) ($M 1,390).................	- 83
	Half of the after-tax cash flow is paid in quarterly dividends (1/8) (1/2) [(0.52) ($M 1,390) + $M 419]	- 71
Total average working-capital requirement		435

However, in evaluating a single venture, the absolute value of the company's total working capital is immaterial—only the incremental increases or decreases caused by the venture are pertinent. (The incremental-economics concept is well described in Ref. 4.)

An Example of Working-Capital Estimation

Let us now estimate the average working capital required for the hypothetical project whose economics are shown in Table I. A plant costing $5.6 million is to make 33 million lb./yr. of product *P* from raw material *R*; the entire production is to be sold for 12¢/lb.

We will assume that a study of company policy and the business situation reveals that:

1. Inventories of raw material, chemicals and supplies will be maintained at an average of 30 days' requirements. To assure customers of continuity of supply, finished-product inventories will be maintained at an average of 45 days of production. There are no byproducts to inventory.
2. When sales are made, payments are due from the customers 30 days from the shipping date. Past experience has shown that almost all of the customers take the full 30 days to pay their bills.
3. The vendors of raw material and supplies require payment 30 days from delivery.
4. There is a monthly payroll—i.e., wages will accrue for 30 days.
5. As a reasonable approximation, it can be assumed that the operating expenses other than labor also accrue for 30 days before payment.
6. Taxes are paid quarterly; the tax rate is 48%.
7. As a rule, management retains half the after-tax cash flow for expansion of the business, and pays out the rest in quarterly dividends.

Table II summarizes the makeup of the working capital for the hypothetical project, and shows how the total requirement of $M 435* is calculated from the five items listed earlier. Note that in several places, an amount of $M 419 for depreciation is used in the working-capital calculation. Although depreciation is considered an expense, it is not a cash expense. Working capital involves hard cash.

Rounding off the calculated working capital requirement to $M 400, the total investment required by the proposed venture is $M 5,600 for total plant, plus the $M 400 for working capital, or $M 6,000. The pre-tax return on investment is therefore (100) ($M 1,390)/($M 6,000) or 23.2%.

Some observations are pertinent:

1. In our example, work-in-process inventories are insignificant, but had the venture required several discrete stages of processing with, say, a 30-day work-in-process inventory, this would equal the cost of the 30-day raw-material inventory plus (approximately) half the conversion (operating) cost on a per-lb. basis.
2. Although, in the example, the requirement for a 30-day raw-material inventory is exactly offset by the 30-day accounts-payable credit, this would not always be the case. For instance, a different venture might require only a 15-day raw-material supply on average, or perhaps a 60-day supply.
3. The example considered raw material to be a 30-day account payable (i.e., payment is due 30 days after the date on the bill), whereas the other operating costs were considered as accrued liabilities (i.e., payment for services rendered during the month are due around the first day of the next month). If chemicals, catalysts and supplies had been treated as accounts payable instead of as accrued liabilities—i.e., if they had been bought on 30-day-after-date-of-bill basis—this would not have made a very significant difference in the working-capital requirement. (In effect, it would have reduced the requirement by about $12,000 or 3%.)
4. Often, significant byproduct inventories must be considered. These inventories are taken at selling price, because it is usually impossible to assign a cost to them. (The revenue derived from byproducts then reduces the cost of the primary product, in the profit-margin calculations.)

*M = 1,000 throughout the example.

Working-Capital Buildup

As is usually done in return-on-investment calculations, capacity production rates were assumed in the example. Often, it is more realistic to assume initial production at a rate somewhat less than capacity with a gradual increase to capacity rates after a few years. In such a situation, the requirements for working capital also build over a few years, a fact that is very important in making discounted-cash-flow calculations.

The technique demonstrated is no less applicable to a *pro forma* buildup situation than it is to capacity production. It is only required that the averaging be applied to the economics of each of the first several years—economics that are already required in the calculation of the annual cash flows for the DCF. Similarly, changes in working-capital requirements caused by anticipated changes in raw-material or product price, and anticipated changes in inventory and other policies, can be handled. ■

References

1. Weaver, J. B., and Lyndall, F. S., *Ind. & Eng. Chem.*, 51, No. 6, p. 67A (1959); and *Ind. & Eng. Chem.*, 51, No. 8, p. 49A.
2. *Chem. & Eng. News*, Facts and Figures, 48, No. 38, p. 15A-25A (1970).
3. Dickens, S. P. and Douglas, F. R., *Chem. Eng. Prog.*, 56, No. 12, p. 44 (1960).
4. Hegarty, W. P., *Chem. Eng.*, Aug. 12 and Sept. 9, 1968.

Meet the Author

Thomas B. Lyda is a Staff Analyst in the Petrochemicals Div. of Gulf Oil Chemicals Co., P.O. Box 2100, Houston, TX 77001. Prevously, Mr. Lyda served in process engineering capacities with Ethyl Corp. and Elcor Chemical Corp. A graduate of the University of Texas (with a B.A., a B.S. in Ch.E., and an M.S. in Ch.E.), he is a member of AIChE and a Registered Professional Engineer in Texas.

Section IV
COST ENGINEERING AND PLANT CONSTRUCTION

Cost Plus vs. Fixed Fee Construction Contracts

Need to choose between a fixed-fee and a cost-plus contract? Would the best solution be a contract that combines features of both? Here are guidelines on which to base a decision, and for selecting the right engineer-contractor.

JOSEPH J. JACOBS, Jacobs Engineering Co.

Competitive pressures and the rising cost of construction are forcing managers of operating companies to become more sophisticated in choosing from the several options available in the design and construction of new facilities.

The possibilities available range from a company doing both the design and construction with its own staff, to having a single outside firm do both. From among a host of architects, architect-engineers, engineering consultants, engineering-contractors, general contractors and subcontracting specialists any combination in between these two extremes can be assembled. Each of these service organizations and the combinations of them, has a place, so that each building plan calls for specific analysis. Unfortunately, many managers lack information on hidden consequences of the alternatives. So, they often tend to follow habit and convenience rather than develop the flexibility for fitting the proper approach to the specific plan.

For most manufacturing facilities, the single-responsibility—or, turnkey—approach is the most economical because of the large cost benefit of schedule compression. With a reputable and professional engineering contractor, the cost-plus-fixed-fee contract is undoubtedly most economical.

If, however, the owner prefers to pay a premium for the assurance of a fixed ceiling price, he can retain the basic advantages of single-responsibility performance by contracting the design on a cost basis, then negotiating a guaranteed maximum price for the construction.

Regardless of the type of contract, however, the selection of an engineering-contractor still requires an intensive investigation of reputations, integrities, technical qualifications and size appropriateness.

Internal Vs. Outside Staffs

An operating company must first decide whether to do the design and construction itself, or do the design itself but contract out the construction, or hire out both the design and construction. If the company chooses to rely at least partly on outside help, it must then decide on the types of firms it wishes to employ, the best type of contractual arrangement, and finally, the specific firm or firms.

Companies whose annual capital expenditures are substantial are more inclined to establish their own engineering design group. There is, however, an accelerating trend away from this approach, despite some notable holdouts.

Some companies maintain a moderately sized design staff to accommodate base load requirements and to protect proprietary information. Although this policy has merit, it is sometimes used by company engineering managers to create overly large staffs.

The advantages of relying on outside staffs have convinced an increasing number of companies to limit the size of their own groups. They have found value in keeping a small engineering group on hand to serve as a cadre of experienced project managers who can represent the company with outside design groups, and control their activities.

Originally published January 1973.

Separate Vs. Single Responsibility

It has long been assumed that professionalism and ethics require that the architectural and engineering design functions be separated from construction. Indeed, architects and engineers often take upon themselves the role of owner's representative.

Contractors and construction companies have, on the other hand, gotten the image of venality, opportunism and irresponsibility, probably because of the ease with which craftsmen in the past became general contractors, "out of their hats," so to speak. It is largely these contrasting images that are behind the strongly held opinions of many managers that design and construction should be separated. This view is, of course, archaic, as even architects, who are the most vocal holdouts for separatism, are beginning to admit. Because licensing has become increasingly stringent, and construction companies have become larger, better financed and more ably managed, professionalism in construction companies has become the rule rather than the exception, particularly in that part of the industry that serves manufacturing companies.

When design and construction functions *are* separate, the selection of a construction company should follow a pattern. Financial stability within the limits of the size of the contract is important, but it is also vital to recognize specialization as to function and as to the project size wherein each contractor operates most efficiently.

A company whose experience has been mostly with office buildings would be lost trying to build a chemical plant. It is surprising how often the simple requirement of technical qualification is ignored, mainly because of the false assumption that a fixed-price bid automatically ensures a good job and protection to the owner.

The matter of matching contractor size to job size is often handled perfunctorily. Financial requirements usually prevent a contractor accustomed to $100,000 jobs from being considered for $10,000,000 ones. But having a large company handling the smaller job can be equally bad. It is not that the large contractor cannot do a small job well; it is rather that it may not be important enough to assure his best efforts. Except under very special circumstances, his best people will be assigned to the large jobs.

Conversely, the small contractor will give the small job his best effort, and the man assigned to it may be every bit as capable as the men assigned by the larger company to the larger job. Because the man in the small company does everything himself, he would be hopelessly lost on the $10,000,000 job. On the other hand, the larger company has the organization, methods and staff to control and organize it properly.

Turnkey Design and Construction

The chemical process industries pioneered the turnkey, or single-responsibility, design-and-construction concept. Its success and economic advantages are likely to lead to its adoption in most industries. Even architects, who have traditionally resisted the integration of design and construction, are coming to accept the trend.

This integrated approach to building new facilities is often resisted by the unsophisticated and by those who have always worked under the concept that separating design from construction divides ethical professionals from venal contractors. Under the assumption that the designer (or architect) represents the owner, the constructor is regarded as an adversary who will cheat or skimp on materials if not closely watched. Resistance to the turnkey approach also often results from fear of the unknown and worry over becoming committed to a company that will perform poorly or dishonestly.

Such concern can be relieved by simply adhering to a selection procedure geared to technical qualifications, size appropriateness to the job, financial stability and references. Final selection can be made on the basis of proposals describing each company's project approach, overhead and fee structure, wage rates, proposed schedule and other pertinent information.

A Stepwise Approach

Client concern about too early a commitment to a project or contractor can also be allayed by having the engineer-contractor accept a limited, stepwise commitment (which almost any reputable one will do), with continuance dependent upon satisfactory performance, or even upon arriving at a binding cost at the appropriate time, that assures a viable project.

So the selection of an engineer-contractor need not entail any more risk than that involved in choosing an architect or engineer. With the stepwise approach, the client can maintain full control and flexibility while enjoying the schedule benefits and good client relationship of the turnkey approach. However, the foregoing is not meant to suggest that the choice of a properly qualified engineer-contractor is not of prime importance even with the stepwise approach. Even though it is possible to change engineer-contractors if satisfactory negotiations for continuance of the project cannot be concluded, doing so will incur additional costs and schedule delays because of the break in continuity.

For the companies that spend vast sums of money each year, the selection is easy, because they thoroughly know most of the engineer-contractors, and can cautiously test new ones as they come along by trying them on smaller, less critical jobs. Therefore, the likelihood of their not continuing with a firm is small. Smaller companies, however, must depend more on evaluating the reputation and experience of contractors they may not have used before.

The advocates of separatism claim that the adversary system saves money. To the owner—caught between the engineer blaming the contractor for not meeting specifications and the contractor blaming the engineer for specifications that were unclear, missing, wrong or impossible to meet—the savings may not be worthwhile. Such conflicts usually cause inefficiencies and overreactions that increase costs, which the owner, who must act as referee, pays anyway.

In contrast, the single-responsibility method permits no buck-passing if the design, construction, or both, are faulty. This does not imply that a watchdog is unneces-

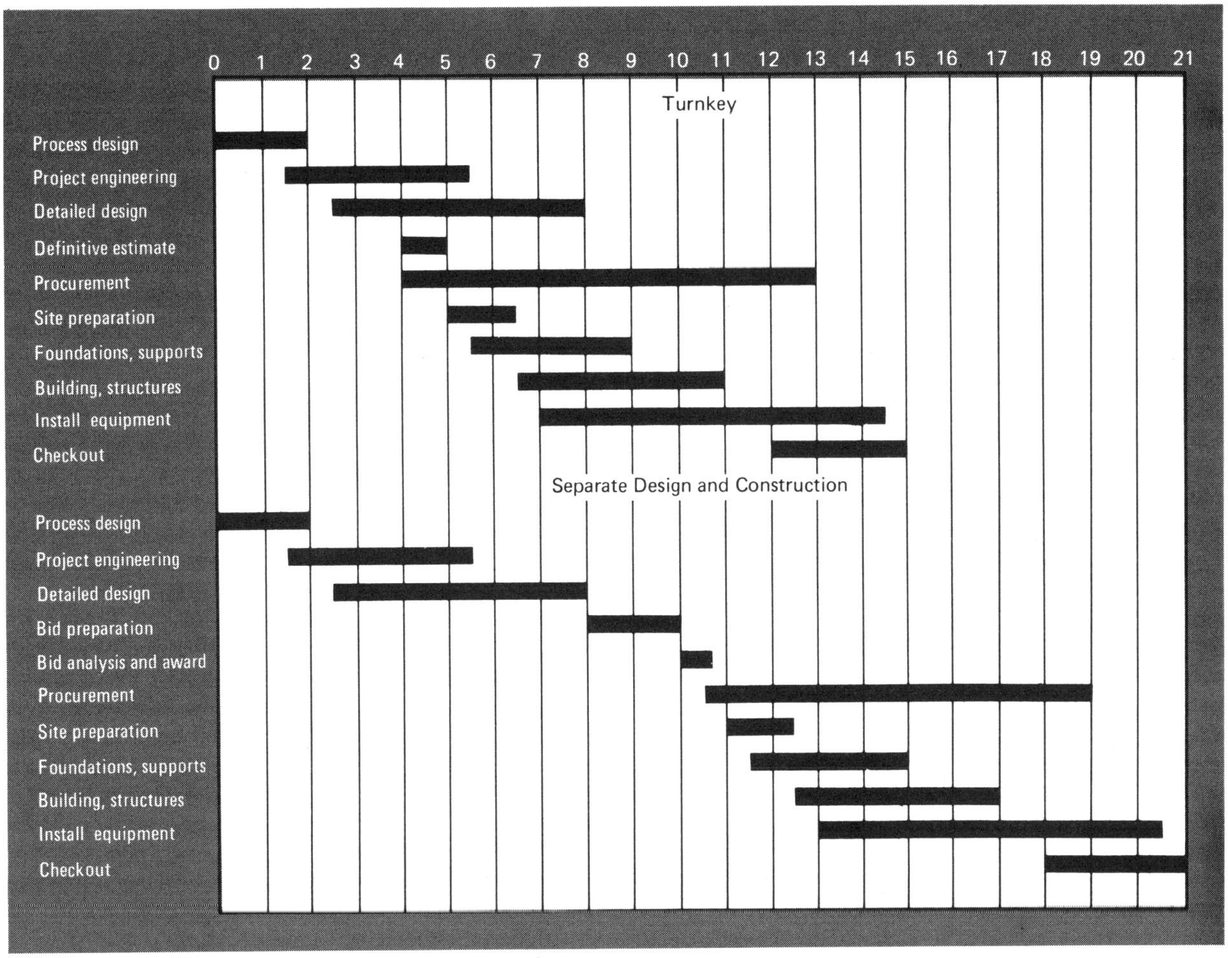

SCHEDULE COMPRESSION is the chief advantage of the turnkey approach over separate design and construction.

sary. He should be, and usually is, a part of the owner's staff because of two advantages: he has no axe to grind, and he has control over the money spent, a powerful tool. Of course, the owner's representatives (or project managers) should be senior men with extensive engineering and construction experience. Because they monitor the job from the initial design to final construction, they are as familiar with the requirements as anyone, and can make quick and fair judgments on variations in specifications and drawings with their only object being the owner's best interest.

The Advantage of Schedule

Although some advantages of single-responsibility contracts, such as better communications and a sense of common purpose, are difficult to evaluate quantitatively, one of the most important that can be so evaluated is that of schedule.

As has been pointed out, the traditional separation of design and construction assumes that contractors will be asked to submit lump-sum bids on the basis of complete drawings and specifications, and after the bids are evaluated a contract will be awarded. This system does not permit any time overlap between design, ordering of equipment or construction, unless the client is willing to place orders before receiving bids.

Under a single-responsibility contract, however, reliable capital estimates will be available earlier, which permits the client to release orders sooner for equipment requiring long delivery. Time is not wasted preparing formal bid books, waiting for contractors to prepare their bids, or converting the various bids to a common basis for accurate analysis.

To illustrate the differences, a $5-million process plant to be built by either the split-responsibility or the turnkey approach is postulated.

The bar chart presents two schedules for this hypothetical plant, one based on the turnkey method and the other on separate design and construction. The overall schedule savings of the turnkey approach is six months, made up of the following factors: (1) under split responsibility with competitive bidding, the continuity break between design and construction adds a total of three months, including eight weeks for the contractors to prepare and submit bids, three weeks for the client to analyze the bids and award the contract, and two weeks for the bidder selected to get equipment and manpower into the field; and (2) split responsibility also requires that design be completed before construction can begin, whereas the turnkey method permits going into the field at the 60% design point, which saves another three months. The cost estimate made at the 60% design point (called a definitive estimate) is usually very accurate and can be used for budget control during the construction stages of the plant.

The schedules assume no equipment requires delivery times longer than eight months. When they are longer, the turnkey contractor can, with the owner's permission, place orders for such equipment before a firm capital estimate is available. This can save an additional two to four months, depending on the design work necessary to specify the equipment.

(It should be noted that the accelerated schedule results from the single-responsibility approach and not from whether construction is performed on a cost-plus-fixed-fee basis or some negotiated variation that guarantees or limits project cost.)

Economics of Schedule

A common way of expressing the economic value of schedule acceleration is in terms of "profit lost" for each day the plant is not operative. Although this measuring stick has value, it is somewhat oversimplified because of built-in assumptions regarding product markets and because it ignores the fact that the delays in income are to some extent compensated for by a slower rate of investment in the project.

More accurate and informative is an evaluation using discounted-cash-flow analysis and present-value concepts, which take the time value of money into account. Using reasonable assumptions* as to investment policy and project profitability, such a technique is applied to the hypothetical \$5-million project in order to convert cash flow schedules first to present cash values and then to equivalent capital investment. This permits expressing the benefits of schedule acceleration in terms of the capital that would be willingly invested to get it. The results indicate that the owner should be willing to increase capital expenditure by up to \$500,000, or 10%, to move startup ahead by six months. It is difficult to demonstrate savings of this magnitude by competitive fixed price construction bids and separate designs.

Contractual Arrangements

Comparisons between single-responsibility and separated design and construction strongly depend on the type of contract and the distribution of financial risk between client and contractor. When the design and construction are separated, construction is often contracted on a competitively bid, fixed-price basis. For a complicated processing plant, complete drawings and specifications are developed by the design firm at a cost of about 8 to 10% of the total project. Several construction firms are then invited to study the information and to submit bids specifying the price for which they will build the facilities.

With single responsibility, there is not sufficient information at the signing of the original contract to fully define the scope of the project and develop a firm cost estimate. In such cases, the basic contract has been the cost-plus type, wherein the client reimburses the engineer-contractor for all costs plus a negotiated fixed fee.

Between the extremes of lump-sum bidding and cost-plus-fixed-fee are a number of popular compromises, such as the guaranteed maximum price, the guaranteed maximum with sharing of cost under-running, and the "target price" with sharing of both under- and of over-running. These represent attempts at correcting, or at least alleviating, the drawbacks supposedly inherent in the cost-plus type of contract, while retaining some of its advantages of flexibility, control and schedule.

In the typical arrangement involving the stepwise variation of the turnkey approach to design and construction mentioned earlier, the engineer-contractor would first perform sufficient engineering on a cost basis to define the scope and prepare a firm capital estimate. This so-called "definitive estimate" entails expenditures of 3-5% of the total cost of the project. At this point, the contract can be converted to a fixed-price or guaranteed-maximum one. This gives the client the risk protection of the competitive-bid type of contract, and it does so with less financial exposure, because competitive bids are based on full detailed engineering costing 8-10% of that of the total project. Another advantage of having the earlier estimate provided by the turnkey method is that design changes can still be introduced at moderate cost if the first estimate is higher than the budget.

Cost-Plus Vs. Fixed-Price Contracts

The general indictment of the cost-plus fixed fee contract is that the engineering contractor will not exert full effort and operate efficiently, or use his most capable people, on such a project. It is fairly common for the financial personnel (accountants, controllers, etc.) of a company, to maintain that a fixed price arrived at through competitive bidding is the most desirable from the company's viewpoint. Only under the stress of time, or if the project's definition and specifications are incomplete, will they agree reluctantly to a negotiated contract. The charge of inefficiency is, except in isolated cases, baseless. The exceptions, which are rare, occur when the facilities are highly duplicative and relatively simple; that is, when the requirements and limitations are unequivocal.

A mistake often made by companies is to assume that lump-sum bidding eliminates, or at least minimizes, the need to establish the contractor's qualifications and competence. Because it is often the incompetent, poorly qualified and thinly financed contractor who will be the low bidder, more investigation, rather than less, is required than for a negotiated cost contract, which permits the owner to exercise substantial control.

This anomaly arises from the rules imposed by the type of contract. Essentially, a negotiated cost contract makes the contractor temporarily a part of the owner's staff. Although it is true that the owner inherits both the weaknesses and the strengths of the organization he retains, he also gets something of utmost importance—con-

*The assumptions are: the project will generate \$2 million before taxes, after a two-year buildup, to yield an internal rate of return of 17% after taxes on \$5 million of fixed capital and \$1 million of working capital. Owner investment policy is based on having a reservoir of investments available at an "opportunity rate" of 8% after taxes. Specific incremental opportunities not affecting overall project risk are valued on their ability to exceed this rate. The value of schedule acceleration will, of course, increase as the gap between the opportunity rate and the projected rate increases. Obviously, the more profitable the project, the more value there will be in speeding up the schedule.

trol. He usually has, in a final sense, control similar to that which he has over his own employees. It is not unusual for an owner to pick and choose individual technical people from the contractor's staff or to request removal of somebody from his job who is not performing well. This is, after all, his protection against the criticism that inferior people will be assigned to his job.

The competitive, fixed price contractor operates under sharply different rules. He is, in a sense, the enemy. And the owner has little control over his weaknesses.

Financial officers usually make a misjudgment when they assume that the threat of monetary loss will ensure that a contractor will do a good job and will automatically assign his best people. In reality, it may only forcc him to find a way not to lose money, which is not the same as doing a good job. The owner's chances are better with a contractor who values his reputation above monetary loss.

The Incentive Argument

Some claim that a contractor's employees work harder if they know the contract is fixed-price, and lackadaisically if it is a cost contract. Superficially, this argument has appeal. But the facts are that most engineer-contractors do not use different efficiency factors for labor, or different prices for equipment, when bidding a fixed-price job as compared to a cost-plus-fee one. Obviously, if employees were observed to work harder on fixed-price jobs, contractors would take that into account to win more jobs. But the fact is that under equal conditions (with a contractor of high integrity), the cost-plus job will more often cost less than a fixed-price one.

The mathematics are quite simple. If the contractor is required to ensure the final price, he must charge what amounts to an insurance premium, which can easily add 5% to the total price. If costs run below the estimate, the owner does not share in the savings. Also, it costs money to prepare a fixed-price bid and ultimately the owner pays for this as well.

Bidding costs vary with the type of facility and the amount of project definition. Even when bids are based on a complete design, this cost will run at least 0.5% of the job and may go as high as 1.5% on less completely defined projects. With six bidders (and some owners ask for many more), there is a waste of at least 3%. No one benefits from this waste, and each contractor will, in one way or another, add about 3% (assuming a one-in-six success ratio) to his bids to recover it. Add it to the 5% premium, the total cost is up by 8%. It is doubtful that anyone can show that increased diligence resulting from competitive pressure and fear of losing money will save this much. To see why, consider the following breakdown of construction costs incurred by the contractor for our hypothetical $5-million plant:

	Typical Distribution	Maximum Controlled By Contractor,	
	$1,000	%	$1,000
Purchased plant equipment	1,400	0	0
Other purchased materials	1,000	2	260
Field labor	1,300	20	60
Other field costs (supervision, tools, etc.)	600	10	60
Project management and coordination (scheduling, purchasing, expending, cost control, accounting, etc.)	150	10	20
	$4,450		$360

The table excludes $400,000 in design costs and $150,000 for contractor fees, because it is assumed that the design would be done at a cost basis in any event, and the fee is of course not a cost item to which diligence can be applied.

Even granting that the fixed-price contractor could achieve the maximum potential savings of $360,000, this would amount to only 7% of the total $5-million cost of our hypothetical project. So, if both the cost-plus and lump-sum contractors are equally diligent, the cost to the owner will be approximately 8% less with the cost-plus contract because of the bidding costs and "price insurance."

Even with the unlikely assumption of a maximum "diligence gap," the lump-sum contract would just fail to pull even with the cost-plus contract. Furthermore, if to this 8% differential is added the 10% "cost" of schedule extension when design and construction are separated (as discussed under Economics of Schedule), then the total added cost of separation, combined with fixed-price bidding, is far beyond any mystical incentive saving.

Bids Vary More Than Performance

Advocates of fixed-price bidding often cite variations of 20% or more in competitive bids to combat this argument without recognizing that these wide variations are in estimates, not in costs. A low bidder may in fact lose 10% or a high bidder, if he got the job, may actually make a 30% profit. But these variations merely represent the lack of precision in estimating, not differences in performance. Only the very experienced are aware how imprecise is the art of construction estimating. Differences in actual cost between reputable contractors cannot be expected to vary nearly so widely.

Of course, on any one contract, the owner may get a bargain—a dubious one—if the contractor loses money or, at best, breaks even. In such a case, the owner can only benefit at the contractor's expense because (as has been shown) there is no way a contractor has of significantly cutting the costs under his control without sacrificing quality. It is either that or lose money. When a reputable contractor loses money, it is almost always because of inexact estimating and not because of poor execution. ■

Meet the Author

Joseph J. Jacobs is founder, president and chairman of Jacobs Engineering Co. (837 So. Fair Oaks Ave., Pasadena, CA 91105), an engineering and construction firm serving the chemical process industries. He received B.S., M.S. and Ph.D. degrees from Polytechnic Institute of Brooklyn, and has contributed to the "Kirk-Othmer Encyclopedia of Chemical Technology" and authored numerous articles and patents. A registered engineer in three states, he is a member of AAAS, ACS, AIChE and API, and is listed in American Men of Science, Who's Who in the West, Who's Who in Engineering and Chemical Who's Who.

Economics of Plant Startups

Startup cost and length can be estimated with a fair degree of accuracy by means of correlations presented in this article. Better estimates can improve your control of cost and strengthen your business decisions.

RICHARD P. FELDMAN, Dow Badische Co.

You may ask, "Why worry about startup cost?" There are basically two reasons.

First, startup costs can be large; in fact, they can often make the difference between a profitable and an unprofitable project. Second, an understanding of startup costs can help you make better business decisions.

As an example of what can happen when sufficient attention is not paid to the measurement and prediction of startup cost, take Union Carbide's experience at Taft, La. Startup problems at Taft reduced Carbide's after-tax earnings $30 million in 18 months, lowering earnings per share $.34 in the first six months of 1968, and putting the complex 18 to 24 months behind schedule.

No procedure could have predicted all the problems Carbide experienced at Taft. However, accurate measurement and prediction can give a better idea of what to expect.

An accurate estimate of startup cost and time can help management by providing a more realistic appraisal of when products from new plants will be available, the means by which to plan market entry, and a better understanding of overall product profitability, because cash flow and net-present-value calculations will be more accurate.

Originally published November 1969.

Defining Startup Cost

The lack of an adequate definition of startup cost and time has frustrated their measurement, prediction and control. Because engineers have slightly different concepts of startup cost and time, it is often meaningless to compare costs of one plant with that of another.

To develop a procedure for measurement and prediction, it is necessary first to define startup cost and time.

Startup cost is the total of only those costs directly related to getting a new production facility into operation; it should not include costs associated with entering or expanding a business. This definition distinguishes this method of measurement from others currently followed.

If the cost of starting up a plant in six months is expected to be $100,000, and the business manager estimates that the company will lose $20,000 of profit for each month beyond the third month of scheduled startup, he can, with this information, tell engineering how much of a premium he is willing to pay to have the plant running in three months.

If, however, business costs were included in the cost of getting the plant operating, the manager would not be able to decide which policy would be in the company's best interest. The following are examples of which costs should be included in the definition of startup cost.

(1) All expenses for changes in process and equipment after completion of construction, if their purpose is to help meet the original plant production goals, but not if the project scope has been redefined.

(2) All cost for labor beyond that for construction, including that for technicians and engineers to check out equipment.

(3) All costs normally incurred during operation. These costs will be corrected later when net startup costs are calculated.* Including normal operating expenses as startup cost makes it easier for field personnel to collect consistent data, because they will not have to decide whether a charge belongs to the startup or to normal operations. Because all costs incurred during the startup are included, bona fide startup costs cannot be "buried" or hidden.

(4) Expenses for training plant personnel, even if done before the startup has officially begun.

(5) All research and development costs actually incurred during the startup.

The following should not be included as startup cost:

(1) Marketing costs.

(2) Expense of training salesmen.

(3) Penalties for shipping outside optimum freight areas.

(4) Costs associated with starting a new company. Include only costs associated with starting the production facility.

(5) Lost sales, unless there is a contract with a penalty.

(6) Profit lost due to timing.

Defining Startup Time

Startup time is defined as the span between the end of construction and the beginning of normal operation.

Starting when the contractor finishes the whole plant or a section of it offers a clear demarcation line from which to begin measuring startup time and collecting startup costs. When the same point is defined as the beginning of all startups, the results of one startup can be compared with that of another. For field personnel gathering startup cost data, this definition takes the guesswork out of when to begin collecting data.

The beginning of "normal" operation may be defined differently for each project: as operations at certain percent of design capacity, as a specific number of days of continuous operation, or as the capability for making a specified product purity. Project and production people should agree beforehand on a definition.

Obtaining agreement on what constitutes normal operation is very important for several reasons: (1) it sets a goal for the field personnel; (2) it makes certain that everyone is striving for the same goal; (3) it makes it possible to compare the results from one plant with another; and (4) it sets a cutoff point by defining when the job is complete.

"Normal operating expenses" for a new plant can be constructed from the estimate of production cost in the original authorization. If, however, the plant required major modification to get it started, then the original estimate of the cost of production may not be valid. If this is the case, it will be necessary to wait until the plant is running well so as to use actual cost data.

Controlling Startup Cost and Time

To keep within your startup cost and time estimate, you must start with a good estimate.

How well you stay within your estimate will be an important factor, perhaps the most important, in evaluating your performance. Because of this, you will most likely spend a great deal of time making actual startup costs conform to the estimate. But how much time should you have spent on the estimate itself? Many engineers are overwhelmed by the idea of making a detailed estimate of startup cost.

Although the chemical industry includes a seemingly infinite variety of plants and processes, certain types of startup problems are common to it. The following is an attempt to analyze and estimate these problems.

Although you now have a basis for measuring startup cost, and a consistent procedure for collecting data, you may still need some way of estimating startup cost, because it will take a long time, perhaps 2 or 3 years, to gather factual startup data under the procedure outlined.

General Parameters

The parameters that affect startup cost and time are general enough so as to apply to virtually all types of chemical plants. Not a perfect mathematical representation but a description of the important parameters is all that is important. Even if the final result is not useful to your particular project, thinking about these parameters and trying to describe them will get you on the road to better estimation and control of startup cost and time.

This procedure is only the first step in a continuing process to achieve better control of startup cost and time. These parameters can always be adjusted later. After a few startups, you will be able to build a model

* Net startup cost is equal to gross startup costs minus normal operating expenses.

that describes the parameters important to your segment of the chemical industry.

The first parameter is direct fixed capital for the plant.* Although different types of plants and processes have different startup costs, direct fixed capital permits estimates based on order of magnitude, e.g., startup costs may vary anywhere from 2% to 20% of direct fixed capital.

The second parameter is the newness of the process and technology, not newness to the industry necessarily, but to the company and the personnel involved in the startup. Obviously, there should be some difference in startup cost and time between a new and an established process. This also applies to different sections of one plant.

The third general parameter is the newness of type of equipment, newness not only to plant personnel but also to the equipment vendor. Even if the process is familiar, quite often the equipment is not. Many companies have had trouble starting up giant versions of familiar processes, primarily because the equipment's size and characteristics were unfamiliar.

The fourth general parameter is labor quality and quantity. This one is difficult to assess beforehand without some familiarity with the workers. It is important because even the most established process with standard equipment can be run badly by an inexperienced or poorly trained staff.

The fifth parameter is "interplant dependency." This parameter can most significantly affect all the other startup parameters. "Interplant dependency" is defined as how much the startup of one plant depends on that of another. This dependency applies to major sections of a large plant as well as to separate plants.

If a plant depends on another for its feedstock, it cannot be started up until the other one (or several others) is running.

Obviously, the longer the chain of plants, the greater the risk of a long, costly delay. Highly automated plants grouped in complexes offer tremendous efficiency, but they lose a great deal of the flexibility of operation that is needed during startups.

The Mathematical Expression

Formulating the parameters into a mathematical expression results in the following:

Startup cost $= f\ (A, B, C, D, E)$

The term A represents the direct fixed capital cost, B the newness of the process and technology, C the newness of the type of equipment, D the quality and quantity of labor available, and E the "interplant dependency" factor.

The equation for estimating startup time is similar to that for estimating startup cost, the only difference being the substitution of the parameter of construction time for that of direct fixed capital.

These parameters are general, applying to virtually all chemical plants. More-specific parameters (such as that for a winter startup in a cold climate) can also be included.

The following examples show how the general expressions can be applied. The values shown apply to large (1,000-1,400 ton/day) air-separation and ammonia plants:

Startup cost $= A\ [0.10 + B + C + D + N(E)]$

In this equation, A again is direct fixed capital cost, 0.10 is an empirical constant, while B, C, D and E vary as follows: B is 0.05 if the process is radically new, 0.02 if only relatively new, and −0.02 if old; C is 0.07 if the equipment is radically new, 0.04 if very new, 0.02 if only relatively new, and −0.03 if old; D is 0.04 if labor is in very short supply, 0.02 if in short supply, and −0.01 if there is surplus labor; E is 0.04 if the plant is very dependent on another, 0.02 if moderately dependent, and −0.02 if independent. The term N represents the number of plants, or sections of a large plant, that make up the process chain.

To calculate startup time, A becomes construction time instead of direct fixed capital, the constant becomes 0.15, while the numerical values for the various categories of B, C, D and E become (following the sequence above): B, 0.15, 0.05 and −0.01; C, 0.15, 0.08, 0.05 and −0.01; D, 0.15, 0.05 and −0.01; and E, 0.25, 0.10 and −0.02. The term N remains unchanged.

Examples will illustrate the simplicity of using these correlations. Startup cost for a plant built around an old process but with very new equipment, where labor is in short supply and dependent on only one other plant (two in a chain), would be calculated as follows:

Startup cost $= A\ [0.10 - 0.02 + 0.04 + 0.02 + (2)(0.02)] = 0.18$ (direct fixed capital cost).

Startup time for the same plant would be estimated as:

Startup time $= A\ [0.15 + (-0.1) + 0.08 + 0.05 + 2(0.10)] = 0.47$ (construction time).

The values chosen for these calculations were based on particular experiences. Those you would choose would depend on the type of plant for which you were estimating startup cost and time (e.g., whether inorganic, petrochemical, etc.). Your correlations might also be modified by variables that apply to your location, such as winter vs. summer startup.

* Direct fixed capital is that for only one plant, not the capital shared by any other plant. Some call it battery-limits capital.

Meet the Author

Richard P. Feldman is an economic evaluation analyst in the Dept. of Economics and Pricing of Dow Badische Co., Williamsburg, Va. 23185. He earned a BS degree in chemical engineering at Newark College of Engineering in 1963 and an MBA from State University of New York at Buffalo in 1968.

He has had experience in engineering plant design, plant instrumentation and economic evaluation, and is a member of ACS.

Exactly what does a contractor do when you ask him to bid on a capital project? To prevent misunderstandings and wasted effort, here's what you should know about . . .

The Proposal Manager's Work

ROY J. LORING, Allied Chemical Corp.*

The proposal manager plays a key role in owner/contractor relationships. But many owner-company people with whom he must deal, and even some people within his own contracting company, seem to lack a clear view of his functions, and of how he must proceed in order to carry them out efficiently. Hopefully, this Cost File will clear up some misconceptions, pave the way to better communications, and lead to greater efficiency in defining key aspects of a project. Let us start by taking a look at the proposal management system.

After an operating company decides to build a new plant or to modify an existing one, it may choose to negotiate with only one contractor or, more often, will solicit bids from several contracting companies. In evaluating these bids, the lowest price is not always the only consideration. The proposal selected must contain the proper engineering to meet the owner's requirements, and sufficient information to inspire confidence that the contractor can indeed build the plant in the time and way he has indicated. To do this, the contracting company needs someone who is close to the bid development, and who can be the liaison between the owner and the contractor—someone who can represent the owner and inject his thinking into the proposal.

Just such a key person does exist. And although he works for the contractor, he can provide a very valuable service to the owner company.

Implementing the System

To both the owner and contractor, the proposal is vitally important; its quality will directly affect the execution of the project. The proposal document, upon award of a contract, will become the scope of work or guiding instructions for the project-execution phase. A badly prepared proposal can seriously hurt the owner or contractor, and destroy any chances of future agreements between them. And this is one reason why a capable individual must manage the preparation of a proposal.

Contractors use many different approaches or organizational setups in preparing a proposal. Some assign a project manager to supervise development of one or more proposals. At award of a contract, he would follow through in his normal position as project manager. Other companies have a proposal department with several proposal managers who have complete supervision and control over the development of one or more proposals. Still other companies separate the technical and commercial portion of the proposal; responsibility for each under a separate department or individual.

Common to all process-plant construction contractors is the concept of executing proposals under a proposal-management system, whereby one person will represent the contractor to the client and the client to the contractor, and will be responsible for all management aspects of the proposal to the satisfaction of both parties. For the sake of simplicity we will call this person the proposal manager (PM). He will have various responsibilities, depending on the particular company, but in most cases he will be directly responsible for the development of the proposal—from the initial inquiry, through the planning and actual preparation, to the submission of the finished document to the client.

The PM's function is much like that of a project manager, only instead of managing a project, he manages a proposal. In doing so, he has an organization or team to direct, and a budget to follow.

If the scope of the project is large, one or more project engineers will be assigned to the PM, with responsibility for directing and approving the technical portion of the proposal. The PM will also direct the efforts of the process, design engineering, estimating, purchasing, construction, scheduling, and legal departments toward the common goal of preparing the proposal, and will maintain a balance between the scope, budget and time available. All information from the owner will be screened through him and passed on to the proper recipients, and he will likewise pass all questions from the proposal team to the owner.

* Mr. Loring has recently joined The Lummus Co.

Originally published August 1970.

Proposal Bid Package

The bid package basically consists of these items, which can be divided into three main headings:

1. Technical Proposal

- Project schedule
- Process description
- Operating requirements
- Plot plans and elevations
- Process flow diagrams
- Engineering flow diagrams
- Utilities flow diagrams
- Heat and material balance
- Equipment list
- Equipment data sheets
- General facilities, such as piping, instrumentation, electrical, civil, construction, etc.
- Contractor's or client's specifications or standards
- Services provided by contractor
- Services provided by client
- Model and/or rendering of proposed plant

2. Commercial Proposal

- Introduction and background to contractor's company
- Organization of contractor's company
- Schedule of professional personnel
- Resumé of key personnel or a resume summary
- Project-management policy or philosophy
- Description of contractor's Engineering Department
- Description of contractor's Construction Department
- Description of contractor's Procurement Department
- Description of contractor's Financial Controls Department
- Experience List of similar plants built
- Experience List of large chemical complexes built
- Experience List of all plants built by contractor
- Experience List of using a client's process
- Photographs of plants built by contractor
- Draft contract

3. Cost Proposal

- Price for services offered in proposal
- Breakdown of price (materials, labor, etc.)
- Escalations (lump sum contract)
- Amount for subcontract work
- Amount for offsite facilities
- Taxes
- Royalty payment
- Alternative systems
- Optional equipment
- Price adjustments (labor efficiency, etc.)
- Schedule of payments

In general, a proposal must convey the ideas, data, and knowhow that qualify the contractor to perform the work specified by the owner. The proposal must be prepared in a way to adequately reflect the owner's specification, scope of work and, where possible, alternative approaches that will demonstrate to the owner that this proposal will save him time and money.

The bid document or proposal is generally made up of a technical section supported by a commercial and financial section; we will refer to the total package as the proposal. The box on this page outlines the contents of each section. Because each proposal is unique, not all the items listed will be used each time; on occasion, more information than that shown will be incorporated in the proposal.

Types of Bids and Proposals

Process-plant projects can range anywhere from a feasibility study, through modifications of existing facilities, to the complete design, procurement and construction of a multiplant complex. Many times, the process and engineering has been done by the owner, and the project is primarily a procurement and construction effort—sometimes, just a pure construction job. Whatever the project may be, the bid request and type of information desired may differ radically.

Bids are solicited for a variety of reasons, such as for plant-feasibility evaluation, as an aid in capital expenditure planning, or for a firm commitment to build. For the purpose of defining the work to be performed by their staff, contractors categorize the incoming request according to the amount of information and the accuracy desired:

An *order-of-magnitude* bid or estimate is made without detailed engineering data, and has an accuracy of ±25% within the scope of the project. This type of estimate may make use of cost-capacity curves, approximations of past estimates (not necessarily similar), or scale factors.

Next, there is the *approximate* estimate, which is also made without detailed engineering data, but is pro-rated from previous projects that are similar in scope and capacity. This type of estimate has an accuracy of ±15%.

A *definitive* estimate is prepared from defined engineering data including (as a minimum) fairly complete plot plans and elevations, piping and instrument diagrams, one-line electrical diagrams, equipment data sheets and quotations, structural sketches, soil data and sketches of major foundations, building sketches and a complete set of specifications. This type of estimate would be accurate within ±5%, depending on basic information.

Proposal inquiries usually begin with either a request brought in by a sales engineer, or through the formal soliciting of bids by the owner. The first step taken for any proposal request is to get authorization to go ahead. The procedure for obtaining authorization can be formal or informal, and in most cases it is done through a committee made up of upper

PROPOSAL AUTHORIZATION REQUEST

To: J. J. Smith, Executive Vice President — Date

Client : Typical Chemical Co.
Location: Midland, Tex.
Proposal Due Date: Feb. 13, 1970
Approximate Value of Facility: $2,500,000
Purpose of Proposal:
[ ] Budget [ ] Preliminary [ ] Formal Offering

Approved by : ______ Exec VP
Authorized by: ______ VP Sales
Charge Number: E-1249

Description of Facility:

Lump-sum estimate for complete, battery-limits, 1,100-ton/day sulfuric acid plant, tied in to owner's existing facilities.

A site visit will be required to examine plot location relative to adjacent facilities.

Estimate will be for engineering, procurement, and construction. Owner's specifications will be used, supplemented by our specifications where necessary.

Job Scope:

[ ] Preliminary Spec	[X] Plot Plans	[X] Utility Requirements
[ ] Spec Review	[X] Process Flow Dia	[X] Engineering Manhours
[X] Final Spec	[X] Eng'ing Flow Dia	[X] Cost Estimate
[ ] General Arrgt	[ ] Material Bal	[X] Schedule

The manhours and cost for the preparation of a proposal for the facility described above are estimated as follows:

Item	Department	Manhours Estimated	Manhours Actual			COST Estimated	COST Actual
1)	Production Engineering	900					
2)	Process Engineering	200					
3)	Construction	100					
4)	Financial Control	500					
5)	Sales	300					
6)	Purchasing	100					
	Total Manhours	2,100		@ $ 9.00	per hour =	$ 18,900	$
7)	Est Trips Expense	1		@ $ 300	per trip =	300	
8)	Est Days Lvg Expense	2		@ $ 50	per day =	100	
9)	Computer Hours	____		@ $ ____	per hour =		
						$ 19,300	$

COSTS are estimated on authorization form—Fig. 1

management. An authorization form is prepared; it contains the type of information shown in Fig. 1, including the manhours and cost required to prepare the proposal.

On obtaining the approved authorization form, the proposal manager will start the preparation machinery moving.

The first phase will begin with a preliminary review of the owner's bid specification to determine what the owner is looking for, what type of information to provide him and in how much detail. While reviewing the owner's documents, the proposal manager must consider his own company's intermediate and long-range objectives, as well as the prime concerns of the moment.

If his company is anxious to move into a new process area or to gain a larger share of an existing area, then a reduction in profit may be called for in order to get the job. Another reason for reducing profit might involve a "showpiece" plant, such as a government-agency facility that will be visited by many foreign business men. If the profit picture is already tight, then possibly an all-out effort to do a superior job may be the answer.

The opposite side of the coin would be where management, for economic reasons, has decided to reduce the overhead costs of proposal, thereby imposing only a minimum manpower effort, and using off-the-shelf designs as much as possible. Also to be considered are possible political implications, such as U.S. requirements when dealing with Communist-bloc countries, dealing with countries that have unstable governments, and even state restrictions that give a competitor from that area a competitive edge.

More usually, the prime considerations will be in the areas of manpower scheduling and the competitive situation. If a contractor is exceptionally busy, with many projects being done simultaneously, there may not be enough manpower available to complete the proposal-development work in time to meet the bid submission date. Only a foolish contractor will try to bid on every job. Not finishing in time to submit a bid, or submitting a hastily written incomplete proposal, may preclude the contractor from being on future bid lists.

An analysis is therefore made of every incoming request to bid to determine the likelihood of getting the job as compared to other contractors who are bidding. If a choice must be made as to which of two projects to bid on, the contractor is less likely to choose one where he feels some other contractor is the owner's favorite.

Defining, Staffing, Scheduling

To ensure that the proposal contains the required information and in the proper format, the PM must take these steps, once the decision to bid has been made:

1. Thoroughly acquaint himself with the owner's bid specification.
2. Distribute all pertinent client documents to proper persons.
3. Organize a planning meeting to determine the basis of proposal.
4. Detail all responsibilities and assignments, and state overall schedule dates.
5. Publish a detailed proposal schedule.
6. Coordinate engineering, estimating, purchasing and sales in selecting vendors for major items.
7. Distribute preliminary proposal to member of top management and key people for review.
8. Distribute draft sales letter, guarantees and contract terms for review.
9. Conduct final review meeting before submission to client.
10. Transmit all proposal documents to client in quantity specified.

After the initial conceptualization of the proposal's scope and the types of manpower needed to prepare it, the PM will distribute copies of the client's inquiry documents to the supervisors who will be involved with the proposal, and who will make the actual manpower assignment.

After allowing a few days for review, the PM will schedule a meeting that will include the managers of process and project engineering, estimating and sales, as well as any specialists required to cover unusual requirements the owner may have included in his bid request. At this in-depth proposal-review meeting, the PM will provide background information, define the scope and requirements, indicate any special considerations, and recommend the approach to be taken in preparing the proposal. He will lead the participants in covering such areas as:

- Amount of bidding time allowed.
- Process scheme and number of options.
- Amount of detailed engineering information required for the proposal specifications, and for the estimate.
- Amount of time required to prepare the engineering information.
- Special requirements of the client—e.g., standard specifications, required vendors.
- Major equipment to be sent out for quotation.
- Schedule of work.
- Special legal, guarantee, royalty, etc., problems to be investigated.
- Additional information required from client.
- Necessity of a site or installation survey.

Next, the PM will prepare a scope memorandum recording the results of the meeting, establishing the scope of the work, and containing the proposal work-schedule on a group-by-group basis, as agreed to by the participants.

The schedule for the proposal can be submitted on a form similar to Fig. 2, which indicates when various information is required, as well as the overall completion dates. In establishing the schedule, the PM considers two critical requirements that are often conflicting: (1) to expedite the proposal so that it is completed on time, and (2) to stay within the proposal budget.

This requires the careful budgeting of time (e.g., via a network-diagram approach) so that work done by one group that is crucial to a second group, is completed and given to the second group with enough time left for that group to complete its portion of the work, and possibly pass it on to still another group.

Keeping the Proposal Work on Course

Once the scope and schedule has been established, Phase II will begin for the PM, and this involves the monitoring, appraisal and control of the entire proposal effort. Unless the PM keeps a close watch on the progress of the proposal and evaluates the performance of the personnel assigned to work on it, the various groups can lose their direction and either overlap each other's efforts, or engage in a "let's do the easy part first" philosophy that can lead to a work-load crisis at deadline time.

One method of control is to have all proposal information, drawings, instructions, data, schedule revisions, etc., flow through the PM (or his technical

Proposal No. .... Issue No. .... Date ....

Proposal Schedule

ITEM	START DATE	COMPL. DATE
All Departments		
Written questions regarding client's specifications and requirements		
Financial and Legal		
Comments on legal and financial terms		
Process Engineering		
Equipment list		
Major equipment specifications for use in obtaining price quotations		
All remaining equipment specifications to be estimated		
Process flow diagrams		
Motor list		
Engineering flow diagrams		
Assist plot plan preparation		
Process description		
Operating requirements (raw material and utility)		
Steam-condensate power balance		
Guarantees		
Review proposal volume		
Electrical Engineering		
One-line diagram		
Description of electrical system for use in proposal		
Furnish other data to Estimating Department		
Water Engineering		
Provide technical assistance as required		
Design process-waste and storm-drainage system		
Purchasing		
Data sheets for quotes		
Furnish firm price quotation on selected items of major equipment		
Contract Administration		
Review technical proposal volume		
Review contract		
Planning and Scheduling		
Prepare project schedule including erection and startup		
Estimating		
Equipment estimate		
Piping, insulating and painting estimate		
Civil estimate		
Buildings estimate		
Engineering and drafting manhours		
Other costs		
Civil Engineering		
Description of civil work for proposal		
Paving and draining plan		
Structural design information for use in estimating		
Construction		
Plantsite survey		
Report on survey		
Field-labor estimate		
Material take-off completed and priced		
Piping Engineering		
Size lines and provide necessary valving		
Furnish underground piping plan		
Instrument Engineering		
Instrument list		
Instrument flow diagram		
Description of instrumentation for use in proposal		
Instrument cost estimate		

SCHEDULE for typical process-plant proposal—Fig. 2 ▶

representative), who examines the work performed by one group for completeness and accuracy before forwarding it to the next group.

The soliciting of vendors' quotations is a critical input to a good proposal, especially if it is for a lump-sum job. Depending on the company's estimating setup and on the amount of effort going into the proposal (i.e., order of magnitude vs. all-out effort), bids are usually obtained on 50 to 90% of the process equipment. One method of ensuring proper use of vendors' quotations is for the PM to indicate on an equipment list released by the process group the dates each group must complete a certain item, and the deadline for receiving a quotation.

The actual items to be sent for quotes are decided upon by the PM, the project estimator, the process and project engineers, the purchasing manager, and other key individuals for specialty items. The PM, as part of his progress monitoring, will make sure that verbal quotations are followed up with a written confirmation, and that the quotations are complete to the point of including prices, expiration dates, escalation requirements, and delivery terms.

A form of overall control is to have a checklist of areas of primary consideration that can be checked against a proposal so as to identify any missing area. Some people use mental checklists, but experience has shown that written checklists are more valuable in preventing oversights.

Final Reviews: Putting It All Together

After the proposal preparation has reached its endpoint and a draft copy has been put together (but before the proposal is released for management's approval), a final-review meeting is held. There, the various aspects of the proposal are checked for their validity, and the design is completely reviewed.

Along with the interdepartmental final review, there may be separate departmental ones, such as a cost or estimating review, a proposal-management or commercial review, an engineering review, etc.

As part of the PM's final review, he will check to make sure that if the owner is receiving any proprietary information he has first signed a secrecy agreement.

At the completion of the final review meeting, a draft copy is sent to key management people in the organization for their comments and approval. This is the final hurdle to overcome and, hopefully, if the PM has done his job well, will only involve minor changes—such as deletion or rewording of a paragraph, addition of a table or figure, or possibly a rearranging of the proposal contents.

Once the PM has obtained final approval from management to complete the proposal and to print copies, he must make sure that the proposal is technically, aesthetically and commercially complete. Phase III in the development of a proposal is the actual putting together of the information. Writing a complete and reliable proposal of the type shown earlier in Fig. 1 requires many inputs, all of them important. A well-organized proposal, free of unclear or sloppy data sheets and flow diagrams, and with an orderly arrangement of information may indicate to a prospective owner that information transmittal will be equally well organized during the job phase.

Help the Proposal Manager Help You

Now for the all-important question: What can you, the owner's representative, do to help the PM prepare a good proposal—one that will avoid confusion, and conserve your time, as well as his. Here are a few simple rules:

- Prepare the bid request carefully, clearly stating what is required. A hastily written or sketchy inquiry will inevitably lead to difficulties and erroneous conclusions.
- Decide on a realistic due date for the bid, remembering that the bid request will first be passed around through management before it reaches the people who will work on it. A tight schedule helps no one.
- Select only competent contractors to bid. Substandard contractors submitting inferior bids will cloud true engineering proposals, waste your evaluation time, and cause unnecessary delay in making an award.
- Keep a channel of communication open to the PM, work through him, and provide him with all the data possible. This minimizes the number of assumptions he has to make, and helps him lend direction to his staff.
- If the contractor requests an extension of the bid due date, try to grant it. An added few weeks may well result in a better designed plant, for the same price.
- Understand the contractor's viewpoint when he recommends some change. In many cases, the change offered will result in a better or cheaper plant. If you cannot accept the change, explain your reasons in such a way as not to discourage other possibly money-saving suggestions.
- In general, put yourself in the contractor's position. Just as an owner selects a contractor to work on a project, the contractor has to select the bids on which he will submit a proposal. Because bid preparation can cost anywhere from a few hundred dollars to $300,000 or even more, a contractor working within a budget must be selective. An owner who develops a reputation for flouting the above rules can discourage a busy, competent contractor from submitting a detailed proposal. ■

Meet the Author

Roy J. Loring wrote this Cost File while Proposal Manager for Allied Chemical Corp.'s Process Plants Construction Div., at Morristown, N.J. Very recently, he has joined The Lummus Co. (1515 Broad St., Bloomfield, N.J. 07003) as a Proposal Manager. His experience includes project engineering as well as proposal work in the chemical and petrochemical fields. Mr. Loring has a B.S. degree in mechanical engineering from New York University, and is completing his course work for an M.B.A. from the City University of New York.

For plant construction projects, here are costs, causes and possible solutions for . . .

Overtime-Pay Problems

L. R. JONES, Standard Oil Co. of Calif.

Although work schedules that embody heavy overtime seldom give satisfactory results, such overtime is still a factor in a considerable amount of construction work in the chemical process industries.

There are a number of reasons for this—we will discuss their validity, examine the cost impact of overtime (in part, by charting available data), compare this cost impact with that of various alternatives to overtime, and discuss what both the owner and contractor can do to improve and control the situation.

The focus will be on "scheduled" overtime—i.e., work that is scheduled in excess of the straight-time work week for several consecutive weeks, or even the duration of the project—rather than on "incidental" overtime that may be required for a few days to meet some specific need or emergency.

Why Overtime?

The frequency with which overtime programs fail to produce the hoped-for results suggests a misunderstanding of what overtime can and cannot do. The users of construction labor usually hope that the scheduling of overtime will achieve one or more of these objectives:

1. Attract additional manpower. When the demand for construction labor exceeds the available work force, any undermanned projects must either attract additional manpower or accept a later completion date. An undermanned project makes inefficient use of supervision and equipment; furthermore, product commitments or economic factors may make a delayed completion unacceptable. The obvious solution is to go on an overtime schedule in the hope that the incentive or premium pay will attract competent manpower from other projects in the area or from other areas. This is by far the most common reason for scheduling overtime.

2. Maintain or improve completion schedule. Here the prime objective is not to attract men, but to increase productivity through longer hours of work. Perhaps half as many extended work schedules are initiated to improve completion as to draw manpower.

3. Take advantage of weather. The objective of completing a specific part or all of a project prior to the advent of inclement weather is often given as a prime reason for overtime.

4. Attract manpower to remote locations. Obviously, some incentive is required to attract craftsmen to undeveloped locations having primitive living conditions and adverse climates. Only by offering an overtime schedule—usually at least a 60-hr. week—can such projects be manned. Plants involved in the recovery and primary processing of raw materials often fall in this category.

5. Make maximum use of equipment. In types of work requiring high investments in construction equipment, it is often economically justifiable for a contractor to work an extended schedule. Earthmoving contractors frequently justify lengthened work schedules on this basis.

6. Maintain schedules in limited work space. Overtime is sometimes used to avoid delaying specific types of construction when space, sequence or safety considerations limit the number of workers. Rolling tubes in a boiler drum is an example. Such overtime can normally be limited to a specific craft.

7. Avoid contract penalties, or win incentive awards. (The cost of the overtime program must be weighed against the potential penalty or award.)

8. Complete emergency construction as soon as possible. (Normally, only natural disasters involve construction so extensive that emergency work on a protracted basis is required.)

9. Avoid strike deadlines. No contractor likes the prospects of having his supervisory personnel and construction equipment idled by a strike, and no client relishes the possibility of an idle, almost-complete plant. In such cases, using an overtime schedule to complete a project before the deadline seems like a bargain.

10. Avoid higher wage rates. Paying premium rates to avoid higher wages seems like an anomaly, but some recent wage agreements in the construction trades add more to labor costs than would a modest overtime program before rate increases take effect.

Note the interrelationships. Using an overtime program to take advantage of weather is a justification for maintaining or improving completion. Peripheral

Originally published June 1971.

objectives could include winning an incentive award, avoiding a strike deadline, improving equipment usage and attracting additional manpower. Attracting skilled craftsmen is nearly always a factor.

Why Not Overtime?

First, everyone recognizes the most obvious disadvantage of an overtime schedule to the contractor or owner—premium wages. Assuming an overtime rate of twice the regular rate (this is usual in the building trades), and assuming a 40-hr. base week, a craftsman would receive 60-hr. pay for 50-hr. work, or 80-hr. pay for 60. The average wage for *each hour* on the job, then, would be 117% and 133% of straight-time rates, respectively. The contractor's evaluation of overtime would be simple if this were the only variable, but it is only the most obvious.

Second, overtime work has an adverse effect on labor efficiency. Many studies have delved into the physiological factors that guide the energy expenditures of a worker under varying conditions, and all concur that the human body cannot sustain the same output over prolonged periods. The maximum work schedule that can be maintained for a number of weeks without a measurable drop in efficiency varies with the physical and mental output required, working conditions, attitude, and various off-the-job factors. It follows that any schedule exceeding that maximum will reduce labor efficiency. Quantitative data will be presented later.

Third, overtime increases absenteeism. Most workers are able to justify taking time off to attend to seemingly urgent personal matters when working an extended schedule. Too, they are better able to absorb the loss in pay when working a six- or seven-day week. A single day's pay represents a smaller percentage of their income. (The opposite could be expected for a work week shorter than five days: time off for personal matters would be difficult to justify, and a day's pay would be a more significant part of income.) The loss in productivity incurred while reorganizing work crews to make up for no-shows is still another costly aspect of absenteeism.*

Fourth, accidents increase. Safety records reveal a disproportionate increase in accidents for schedules exceeding 8 hr./day and 40 hr./wk.

Fifth, and frequently overlooked, is the effect of overtime on workmanship. Although no quantitative statistics are available, the incidence of faulty workmanship and the resultant cost of rework increases with longer work schedules.

Sixth, an overtime schedule creates labor unrest. Area contractors and union business agents usually must man straight-time as well as overtime projects. The disappointment of workers not drawing premium pay affects their productivity. The reaction is similar when only a part of the crafts on a project are put on an overtime schedule: Other crafts tend to reduce productivity in the hope that they, too, may be placed on premium schedules. Most contractors now recognize this problem, and do not attempt to mix straight-time and overtime schedules on the same project. Small specialty contractors are the exception.

Seventh, competition for labor soon offsets the attraction of premium pay. The craft labor supply within a given area is relatively inflexible. When one project initiates an overtime program, others must follow suit by equaling or exceeding the first schedule to hold or regain their labor force. Such a tug-of-war for the available labor supply can drastically increase costs without expanding the work force.

Eighth, overtime draws less proficient workers. Because of the relative stability of a local work force, any increase in construction employment must be supplied from other areas. But the most proficient craftsmen are usually employed on projects in their own labor market, and the premium pay may not be sufficient to offset the cost and inconvenience of living away from home. Thus it the less desirable workers who may be most attracted by overtime projects in another area, and particularly so if the base wage rates are low. (This is not intended as a blanket condemnation of "boomers" or "travelers." Many highly proficient construction workers, particularly in specialized fields, are attracted to the nomadic existence of job-followers and wouldn't live any other way. Too, highly skilled workers may be available from areas of depressed construction activity.)

Finally, there is the "overtime syndrome." It is characterized by the workers' desire, consciously or unconsciously, to force a project to an overtime schedule and then perpetuate that schedule. Few contractors will deny its existence, although it cannot be identified or measured. One can only conjecture on the potential savings if a contractor precluded overtime at the beginning of a project.

Measuring Overtime Productivity

Quantitative data on the effect of work schedules on productivity in the construction industry are scarce. Perhaps the most exhaustive study of overtime efficiency is the "U.S. Dept. of Labor Bulletin No. 917," published in 1947.[1] It presents 79 case studies covering 2,445 men and 1,060 women in 34 plants during and immediately after World War II.

Despite its almost universal use as a reference, the applicability of the bulletin to industrial construction in 1971 must be viewed with caution for a number of reasons: E.g., the bulletin is based on mostly indoor work, some of it machine paced and most of it highly repetitive. Construction projects were not included. Many of the workers were women. Patriotism during the war years may have affected productivity. And so on.

If we ignore isolated data for single projects or

* Work schedules and wage agreements have a definite effect on absenteeism patterns. Workers are rarely absent for Saturday of a six-day week if weekend work draws premium pay regardless of the weekly total of hours worked, but more workers fail to show up on straight-time days. The cost for such workers is thus increased with absolutely no increase in hours of work. This practice is discouraged by both unions and employers.

broad-brush approaches that uniformly condemn overtime productivity, we find our data sources seriously restricted. The following three sources were used in preparing Fig. 1:

• *Mechanical Contractors' Assn. of America Bulletin No. 18A.*[2] The avowed intent of this bulletin is to strengthen the mechanical contractor's case for increased compensation for overtime work. Information on efficiency was compiled from the abovementioned Bulletin No. 917.

• *Construction Users Anti-Inflation Round Table Report* on the effect of scheduled overtime on construction projects.[3] This excellent report, compiled under the guidance of Weldon McGlaun of Procter & Gamble and issued in early 1970, includes a compilation of efficiencies for seven different overtime schedules presumably based on P&G experience. All crafts were included. That McGlaun made reference to a 1918 publication to cite studies made in Europe in 1900 only emphasizes the dearth of applicable modern-day investigations.

• *National Electrical Contractors' Assn.* Survey of overtime work-efficiency. This survey, prepared by the Southeast Michigan Chapter in 1962, summarizes survey forms returned by 289 NECA members. Plotted points are for electricians only. A 14-day overtime period is assumed for the plotted points .

That's all we have to work with. Why such limited data? Actually, there are two interrelated obstacles to collecting meaningful data on overtime productivity in the construction industry: measuring productivity, and isolating the overtime factors from all others affecting productivity. The construction of chemical, refining and other process facilities differs from most manufacturing work in that each day's work is different. There are, of course, standards of productivity for specific tasks—so many manhours for making a weld, for laying a number of bricks, or for installing a specified length of conduit. But the time required for even these relatively well-defined tasks may vary from day to day because of job conditions and other factors, and only relatively long-term averages are meaningful. The box on the next page, listing but a few of the factors affecting productivity, could be expanded almost indefinitely with detailed examples.

Note that the "reporter's viewpoint" is included as a factor there. Obviously it does not have a bearing on productivity per se, but it most certainly does affect the measurement of it. A contractor interested in attracting a project to an area sees productivity differently than another in the same area who is attempting to justify additional reimbursement for labor overruns. Similarly, the other factors affecting productivity may be emphasized or overlooked to support a specific viewpoint. If, for example, a contractor initiates an overtime program to complete a project before winter, how much of any drop in productivity should be attributed to worsening weather, to the normal decline as any project nears completion, and to the effect of overtime?

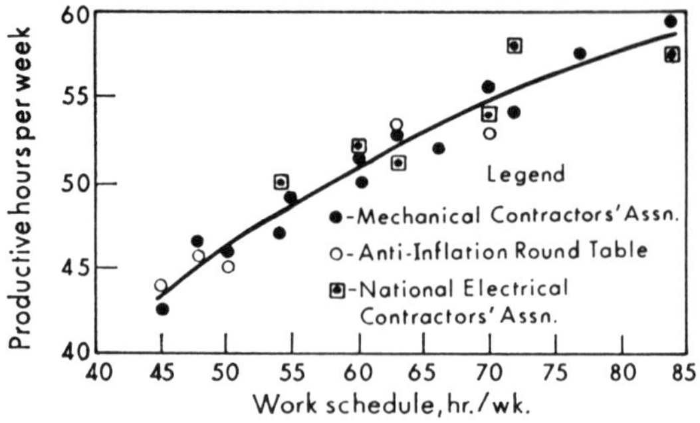

PRODUCTIVE HOURS PER WEEK tend to go up at a much slower rate than the work schedule—Fig. 1

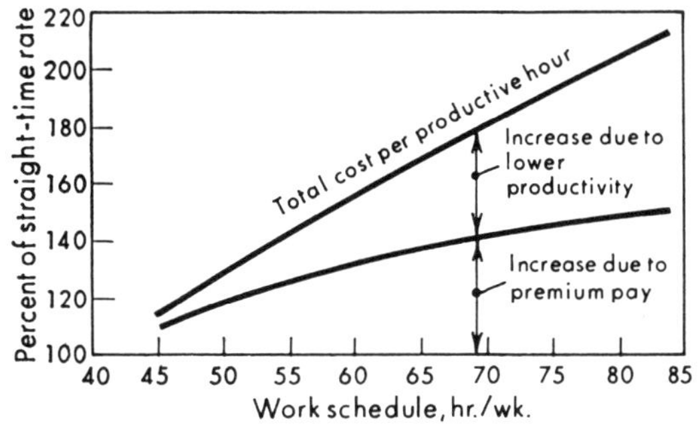

PRODUCTIVITY LOSS as well as premium pay influence the cost per productive hour of work—Fig. 2

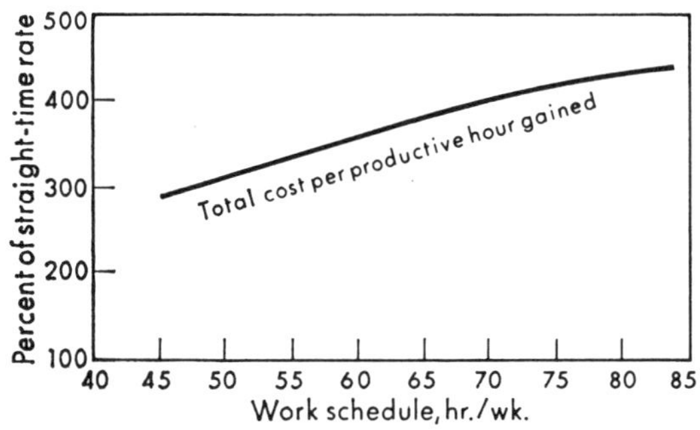

EACH PRODUCTIVE HOUR gained from overtime tends to cost more and more as the schedule lengthens—Fig. 3

The Story in the Charts

Fig. 1 shows that our three sources correlate rather well—amazingly well considering the different origins. It tells us that, compared to a 40-hr. week, we can expect 46 productive hours from a 50-hr. schedule, 51 from a 60, and 55 from a 70.

But that's only part of the story. Using the same basis and assuming a double-time rate for all premium hours, the total cost for all productive hours increases markedly. From Fig. 2, we see that each of the 46 productive hours obtained on a 50-hr. schedule costs 131% of a normal work-week hour. For a 60-hr. week, the cost increases to 157%, and for 70 hr., 182%. The lower curve shows the effect of premium rates; the difference between the two curves therefore measures the cost attributed to inefficiency.

Fig. 3, derived by dividing the total added cost by the total added productive hours, shows that the cost of each productive hour gained exceeds 300% of a normal straight-time hour for even a relatively modest overtime schedule!

Avoiding Overtime

Aside from supporting apprenticeship and special training programs, and getting behind other moves to reduce regional skilled-craftsmen shortages (many of which seem to persist despite the general rise in unemployment of the last year), what can be done?

Both the contractor and the client can take positive action against being placed in a position where overtime seems to become a must. One big step is to schedule realistically. The need for a new plant, coupled with the high cost of money, encourages design and construction schedules based on the unrealistic assumption that everything will proceed in the most desirable manner. Many project managers will assert that the exact opposite is nearer the truth. Schedules must be reasonable.

Competitive overtime schedules should be avoided. When a contractor uses overtime to attract workers in a restricted labor market, the resultant competition will increase costs without accomplishing its objective. A survey of potential sources is in order before embarking on any overtime program intended to attract additional labor.

Too many contractors overlook the value of quality supervision in avoiding overtime. They are quick to agree to the importance of good supervision, but fail to compare the relative incremental cost of the best available supervisory personnel with the less obvious cost of reduced productivity and a potential overtime program. Quality supervision is but one of many controllable factors that increase productivity and hence, decrease the potential need for overtime. The box on this page includes many more that can be regulated to some extent.

Designing for minimum field labor has been another common approach to combatting craft labor shortages that have often led to overtime. New materials and larger components are a natural product of any expanding technical-based society, and high field costs give impetus to their use, despite the less-than-enthusiastic attitude towards prefabrication by some construction unions.

Shift work is seldom an alternative to overtime work, because an extended work week would rarely be considered if enough manpower to staff an additional shift were available. One evaluation of manpower concentration[4] indicates that doubling an optimum density of 10 men per 1,000 sq. ft. will reduce efficiency less than 5%—an obvious bargain when compared to either overtime or shift work. Disadvantages of shift work include the cost of the shift differential (up to 15% or so) added lighting and supervision, lost productivity due to turnover between shifts, and the loss of identity between the craftsman and his work.

In the final analysis, scheduled overtime in the construction industry can best be combatted by education. It is unlikely that overtime programs would be initiated in the construction of process facilities if the probable effects on cost and productivity were better understood. Unfortunately, available data are limited in scope and application, and pertinent up-to-date information is badly needed.

A proposal for conducting a nationwide survey has been submitted to the national board of the American Assn. of Cost Engineers for their consideration. We may soon be able to better identify and evaluate a factor that added an estimated quarter of a billion dollars to the cost of industrial and commercial construction in 1970. ■

Productivity Is Affected By . . .

- Worker attitude, ability and health
- Quality of supervision
- Foreman ratio
- Comfort factor
- Lighting
- Time of day and week
- Density of workers
- Labor availability
- Subsistence pay and wage agreement
- Size of project
- Quality and availability of materials and tools
- Work rules
- Phase of project
- Measurement standards
- Reporter's viewpoint

References

1. U.S. Dept. of Labor, Bull. 917: "Hours of Work and Output," Washington, D.C., 1947.
2. Mechanical Contractors' Assn. of America, Bull. 18-A, "How Much Does Overtime Really Cost?" 1969.
3. Construction Users Anti-Inflation Round Table, "Effect of Scheduled Overtime on Construction Projects," Cincinnati, 1970.
4. Hackney, J. W., "Control and Management of Capital Projects," p. 64, Wiley, 1965.

Meet the Author

L. R. (Les) Jones is group supervisor, cost engineering methods, for Standard Oil Co. of California (555 Market St., San Francisco, Calif. 94120). His current assignment follows nearly 20 years of supervisory experience in the design and construction of oil refining and petrochemical facilities. A registered professional engineer, he has an M.S. (M.E.) degree from Oregon State University, and is a member of the American Assn. of Cost Engineers.

Contracting for new construction

Many types of contracts are used in construction of new plants or plant additions. This article examines three of the most common, with special emphasis on the design/build (or turnkey) contract.

Eric M. Bergtraun, *National Semiconductor Corp.*

☐ Even though the cost of buildings and leasehold improvements is very important to the corporate balance sheet, very little is done to ensure that firms get their money's worth for each dollar spent on building projects. Companies that owe their existence to selling products on low bids do not hesitate to award building contracts either to high bidders or to builders who cannot show that their proposals will provide real financial advantage.

Even more startling, some of our most progressive companies never bother to evaluate a building project after its completion. It would be wise for these companies to ask the questions listed in Table I.

Questions to ask after new construction is completed — Table I

1. Did we get the type of building we wanted?
2. In order to get what we wanted, did we have to make costly last-minute changes, or even tear things out and redo them?
3. Did we get the building in the time promised?
4. If we got it on time, did we have to spend premium money to meet the time schedule?
5. Have provisions been made for future expansion?
6. Has the building been designed for the greatest flexibility and ease of maintenance?
7. Has the building been overdesigned?
8. Have some consultants been allowed to build monuments for themselves, with little regard for costs?
9. Has there been a lot of experimenting with as-yet-unproved building materials and concepts?
10. In order to get the most for the money, has there been enough coordination among: consultants, contractors, code-enforcing agencies, and insurance carriers?

Choosing a contract

One thing management seems to spend considerable time on is selecting the type of contract to be used. Purchasing agents write a great variety of building contracts. A few of the common types are shown in Table II.

Convertible contracts can start out as any one of the other types in Table II and then change at certain phases to other types. Thus, a master contract can be drawn up encompassing several of the contract types for different phases and different subcontracts.

Without going into depth on all (or any one) of the mentioned types of contracts, let us discuss the three most generally used.

Standard government-type contract

In the past, the most popular building contract has been one where the owner picks an architect who designs the project. For the design job, the architect employs the services of many consultants and specialized design experts, from surveyors and structural engineers to heating, ventilating and air-conditioning (HVAC) engineers and materials-handling consultants. Some of these are listed in Table III.

While some architectural firms employ such experts full-time, many go outside for this specialized help.

Unfortunately, owners often cannot clearly define their needs. They do not have a person or staff that can devote the necessary time to such an important project. Architects are therefore often left to their own devices to fully design the building or project. However, in doing this, architects have over the years been blamed for anything that goes wrong on a project, especially in disputes with the owner when things have not been clearly defined. In order to protect themselves, they quite commonly overdesign. How often has it happened that a plant engineer takes over a new building and says: "I wish they had spent less money on this fringe area and instead given us a little better HVAC system, or paid more attention to the maintainability of the mechanical and electrical system!"

In this first type of building contract, after the design is completed the plans and specifications are bid upon by a group of general contractors. The successful bidder gets the job, takes bids from his subcontractors, and builds the project under the guidance of the architect.

If the owner wants to make changes during the building period, he has to go through a very complicated approval chain: architect - general contractor - subcontractor consulting engineer - project manager - etc.

For many owners, this type of contract has become too

Originally published June 1977.

Some of the many types of construction contracts	Table II
1. Lump sum, based on plans and specifications.	
2. Lump sum, based on preliminary plans and specifications.	
3. Straight bid.	
4. Cost plus.	
5. Cost plus, fixed fee.	
6. Cost plus, guaranteed maximum.	
7. Cost plus, guaranteed maximum with incentive.	
8. Cost plus, guaranteed maximum with provision for escalation.	
9. Bonus/penalty time and completion.	
10. Bonus/penalty operation and performance.	
11. Unit price, flat rate.	
12. Unit price, sliding scale.	
13. Time and material.	
14. Time and material, guaranteed maximum.	
15. Time and material, guaranteed maximum with incentive.	
16 Time and material, guaranteed maximum with provision for escalation.	
17. Convertible.	

Some of the architect's consultants and design experts	Table III
1. Facility planner.	
2. Surveyor.	
3. Structural engineer.	
4. Civil engineer.	
5. Soil engineer.	
6. Mechanical engineer.	
7. Heating, ventilating and air-conditioning (HVAC) engineer.	
8. Electrical engineer.	
9. Controls engineer.	
10. Public-address-system engineer.	
11. Communications engineer.	
12. Environmental consultant.	
13. Acoustical consultant.	
14. Fire-protection engineer.	
15. Landscape architect.	
16. Irrigation designer.	
17. Materials-handling consultant.	
18. Interior designer and color consultant.	
19. Cafeteria designer.	
20. Traffic engineer.	

difficult and time-consuming. A new type of contract has become very popular.

The team contract

This contract involves a team of owner, architect/engineer and general contractor. Here, the owner defines a performance specification and picks both the architect/engineer firm and the general contractor, based on a set of proposals and, usually, a predetermined budget. From the beginning, the team works closely together and comes up with a design that is then put out for bids by the general contractor to a list of mutually approved subcontractors. An advantage of this contract is that the owner has the services of the general contractor early in the job. The latter is usually able to influence the designer and come up with practical solutions to many problems.

One problem is that the architect/engineer firm has a much greater influence on design decisions than does the contractor. As much as 60% of the project is in the hands of the main subcontractors, but they are usually left out of the picture until the design is complete and ready to be bid upon.

Lately, a third system of contract has become even more popular.

The design/build concept

This type is also often called the turnkey contract. The owner selects a firm that has the capability to architecturally design, engineer, manage and build the new facility. The owner can either find a large contracting organization that has the entire capability in-house, or a contracting firm that will hire an architect/engineer firm to do the design work.

This type of contract is often used (the contractor being picked competitively) for the design of a shell building. The owner spells out what he needs, and the contracting firm designs the best possible building. A budget is established, and the building shell is erected in accordance with the owner's needs. By dealing directly with the builder, the owner has the benefit of an easily available estimating service. This will provide him with correctly costed proposals throughout the job. The owner works closest with the contractor who delivers the goods but still has the benefit of the team of consultants working for the contractor.

Company needs competent engineer

A company that has a competent plant engineer, who can devote full time to a building project, can best use the design/build concept. At first, a general contractor is picked to design/build a complete shell building. Then, the owner can act as his own general contractor and issue separate design/build contracts to the subcontractors outfitting the building for him.

If the owner's project manager can devote full time to the project, he can easily coordinate the subcontractors. Some of this work can get started with the help of the general contractor while the shell building is erected. Each subcontractor should be signed up to also work via the design build/concept.

Typical specialties for which subcontracts are issued are shown in Table IV.

In each case, a capable plant engineer with construction experience can pick the subcontractors who have in-house design capability. The subcontracts can then be competitively bid. Through these turnkey contracts, the engineer will be able to get a firm, fixed price for all individual contracts on the project. This, of course, should be extremely important since he will then know how much each subcontract will cost.

Bear in mind that in a turnkey project there is no open end left on either the engineering cost details or the

Typical subcontracts in plant construction — Table IV

1. Electrical installation.
2. Sanitary and process piping.
3. HVAC
4. Fire-protection sprinkler system.
5. Landscaping and irrigation.
6. Partition and drywall installation for the building.
7. Hardware and lock installation.
8. Acoustical ceiling.
9. Floor covering.
10. Public address system.
11. Interior decoration.
12. Cafeteria installation.
13. Furniture.
14. Signs.

Some advantages of the design/build concept — Table V

1. It usually provides the lowest cost per square foot of space.
2. It cuts communication time, because the design team and construction team work for the same firm.
3. It eliminates a lot of the buck-passing between the architect, engineers, consultants, the general contractor, and subcontractors.
4. It gets the owner, through his representative, directly involved with the people who build the facility.
5. It makes changes and additions easier.
6. It provides the owner with a continuous, competent estimating service.
7. It saves time, since work can start before the design is completed.
8. It can be set up with easier and better cost controls.

construction. Another advantage is the performance guarantee that can be obtained from each contractor, based on specifications provided by the contractor and approved by the owner at the start of the job. The value to the owner of having a performance guarantee is that the contractor-engineer very clearly knows what is required of him. There should be no argument at the completion of the project.

The plant engineer deals with one person from each contractor at all times; this will eliminate dividing time between architect/engineer firms and contractors. In this way, the owner can be sure that he gets his money's worth out of each design/build concept contract. The beauty of this concept is that the owner, through his project manager, gets directly involved in the most important subcontracts that will make the new building project a success or a failure.

Some advantages

Let us take, for example, the HVAC contract for a new building, especially when items such as clean-rooms, humidity control, heavy heat removal, hood exhausts, etc., are involved. Under a straight contract of the first two types, the owner has to go through the architect/engineer firm, the general contractor and the HVAC design engineer before he can reach the subcontractor whose job it will be to build an acceptable system and maintain it for the owner after the job is completed. An HVAC installation might make the difference between a profitable or unprofitable product line, so it is important that the owner directly benefit from the contracting organization's practical experience.

How often have you, as a project manager, plant engineer or maintenance manager, seen that a specific system in your new building did not perform properly? Who usually ends up fixing it, at either his own or the owner's expense? It is almost never the architect, consulting engineer, or general contractor. Usually, the subcontractor fixes it and keeps it running for the owner. Since the subcontractor is around longest to keep the system operational, train the maintenance personnel, and guarantee the system, he is also best suited to:

- Discuss with the owner what is really required.
- Provide the owner with a budget proposal.
- Direct the engineers to design the system.
- Provide job management to assure that the system is built in accordance with plans and specifications.
- Supervise the project to assure that it remains within budget.
- Complete the project to the owner's satisfaction, guarantee its continuous operation, set up a maintenance program, and train the owner's maintenance personnel to take care of it.

Disadvantages are minimal

The design/build (or turnkey) concept has many advantages. A few are listed in Table V.

The only disadvantage is that the owner can lose the control and inspection provided by an independent architectural and engineering firm. That is why the owner should have his own well-qualified representative. This representative will be supported by mandatory city, county, state and insurance inspections. He also can have the help of supplier representatives and other independent experts.

During the past ten years, I have worked with different building contracts and concepts and have been very pleased with the good results obtained from the "design/build approach to contracting for new construction; my employer is sold on it as well.

The author

Eric M. Bergtraun is manager of plant engineering and maintenance at National Semiconductor Corp., 2900 Semiconductor Drive, Santa Clara, CA 95051, where his responsibilities include both headquarters and multidivision plant operations totaling a half-million ft². He holds a B.S. in electrical engineering from Heald Engineering College in San Francisco. He is a member of the Institute of Electrical and Electronics Engineers and the Electrical Maintenance Engineering Assn., and has very recently been elected a Fellow of the American Institute of Plant Engineers.

With construction costs refusing to "hold still," here is a timely, relatively simple way of combining Cost-Index components with regional cost-trend data and other information in estimating . . .

Building-Cost Escalation

L. R. JONES, Standard Oil Co. of California

As everyone knows by now, construction costs are going up rapidly in some regions, and very rapidly in others. This has reinforced the need for a simple, effective means of predicting the amount of escalation that must be added to an estimate that is based on today's costs if the project will not be completed until some future time.

The five-step procedure that we will examine and illustrate was developed to meet this need, and applies to oil refineries, petrochemical plants and other process facilities. It is not intended to replace the more-sophisticated methods available for projects in the advanced design stage, but many engineers and estimators may find it acceptable for general use. Primary applications include order-of-magnitude estimates, feasibility studies, comparative evaluations, and appropriations or budget estimates.

Although devised for the owner, the approach is valid for contractors' applications not requiring more detailed and time-consuming analyses.

Evaluating Other Methods

For maximum accuracy, it may seem that a separate rate of escalation for each item contributing to the overall cost of a project would be desirable. But this approach has several shortcomings:

1. The cost of designing a plant to the extent necessary is normally unacceptable until the project has been authorized.
2. The time required for such an approach is seldom available.
3. Re-estimates changing the design, scope or schedule are considerably more complex.
4. Separate escalation rates are rarely available for all items, and the assumed advantage of accuracy is lost if average rates are used.

The informal approach of estimating everything on the high side and hoping that the "fat" will cover increasing costs has been used by many without serious difficulties in times of relatively stable costs. But at present, such seat-of-the-pants estimating is highly questionable even for small jobs of short duration. It is certainly unthinkable for major projects.

Estimating future escalation by extrapolating one of the published cost indexes is fast, simple and inexpensive, but what about accuracy? The *Engineering News-Record* (ENR) and Nelson Inflation indexes have been increasing much faster—twice as fast in some recent years—than the CHEMICAL ENGINEERING and Marshall and Stevens indexes. Which index should you use? Each index has its champions, but all cannot be equally suitable for a specific project. What are the characteristics of each? Which is best for a specific project? The estimator should certainly become familiar with the makeup, intended applications and relative merits of each index used.

There's no denying the usefulness of published indexes in determining relative price levels for the base conditions assumed in preparing the index. Each is also flexible to some degree. Components are published periodically, and the user is able to make at least some adjustments for type of facility and/or location.*

Our experience with an index compiled from the components of published indexes to suit our specific needs has been good, and particularly so in updating costs. Carefully selected components of published indexes are frequently the best data available for resolving specific cost and escalation questions. Unfortunately, the as-published indexes are not always satisfactory for estimating escalation for a specific project. No single index is sufficiently detailed to meet all the following criteria:

1. Suitable for a specific location, U.S. or foreign.
2. Adjustable for various processes, sizes and types of installation.
3. Applicable to projects for which engineering,

Based on the author's presentation at the 14th National Meeting of the American Assn. of Cost Engineers, San Francisco, June 22-24, 1970.

* See, for instance, Chilton, C. H., Plant Cost Index Points Up Inflation, Chem. Eng., Apr. 25, 1966, p. 190.

Originally published July 1970.

procurement and construction are each at a different location.

4. Detailed sufficiently to permit comparison of estimated and actual escalation wherever practical.

5. Usable for determining the effects of a range of escalation rates for one, two or all three of the major cost categories.

6. Flexible enough to allow for unusual schedules.

7. Simple enough to be easily understood.

Step 1 of Our Method: Basic Estimate Only

The first step in estimating the degree of escalation is to prepare the estimate without any escalation at all. Any method consistent with accuracy requirements may be used, but it is important that all costs have the same base date.

Even a contracted labor-rate increase should not be included if it becomes effective after the date of the estimate. If a common date is not used, any overall application of an escalation rate loses accuracy. Further, subsequent revision to an estimate becomes unnecessarily complicated if the time base is not fixed.

Step 2: Distribute All Costs

The next step is to determine the percentage cost breakdown for engineering, materials and equipment, and construction labor. These are broad classifications, and the breakdown cannot be precise. The following criteria may serve as a guide:

1. *Engineering.* Include all costs and overhead items normally associated with an engineering design contract. Markups or margins are normally based on the engineering payroll, and the overall cost of engineering closely parallels engineering salaries. Include field engineering and supervision.

2. *Material and equipment procurement.* Use delivered cost including freight and taxes. Include tools, equipment and consumable supplies.

3. *Labor.* Include all construction labor and labor-related charges. Include overhead and fee for construction contracts.

This breakdown is normally available from the basic estimate. Our experience with a wide range of refinery and petrochemical units shows the following ranges: Engineering, 6 to 28%; materials, 40 to 72%; labor, 12 to 37%.

Occasionally a distribution is not readily available. (An order-of-magnitude estimate prepared from cost-capacity data is one example.) For refinery and petrochemical units an engineering/material/labor breakdown of 12%/63%/25% may be used.

Step 3: Assign Escalation Rates

In determining the rate of escalation for each of the cost components, factors to be considered include:

1. *Engineering.* The lack of standard categories or classifications for engineers make it difficult to evaluate salary escalation. A merit increase may be offset by increased output or responsibility, and thus not be inflationary, but to assume all salary increases fall into this category is naive. Engineering costs *are* increasing.

Some engineering firms have fixed classifications with rates for each. Changes in rates thus represent escalation. The percentage markup over payroll is a common method of billing for engineering services, and only a change in the markup is identifiable as cost escalation. For in-house engineering, an average hourly rate for various classifications of employees may be used, and increases in the average rate represent escalation.

If sufficient experience data are not available, a published index may be used. The "Engineering and Supervision" index published by *Chemical Engineering* is one. Recently this index has been increasing by less than 1% annually—an unexpectedly low figure when compared to other known increases. This is attributed at least in part to the assumed 2.5% annual productivity improvement used (in compounded form) in preparing this index. Such advances as the use of computers and models tend to reduce engineering costs, but it is questionable whether higher engineering productivity can continue to offset higher salaries to this extent. The user may wish to adjust this index for local or other specific conditions.

2. *Material and equipment procurement.* Past company experience is of questionable value in determining material cost escalation for a complex installation. Much of the equipment is fabricated to order or purchased with special features, and identifying a portion of the cost as escalation is impractical if not impossible. Certainly, many basic materials such as concrete, reinforcing rod, lumber, and aggregates have easily traceable cost histories that can be projected, but such materials represent a relatively small portion of the overall cost.

The material and equipment components of one of the published indexes are usually the best source of cost escalation data for complete plants. *Oil and Gas Journal* (Nelson) and *Chemical Engineering* (*CE*) publish the materials and equipment components of their indexes. Each component is derived from weighted cost data for the many types of materials and equipment used in plant construction. Nelson's materials and weights are based on refinery construction, *CE*'s from chemical plant experience. Each uses average U.S. prices, ignoring the effect of location. The majority of materials and equipment follow a nationwide cost pattern, and the effect of location on the escalation of process plant materials is usually insignificant. There has been close agreement between the material components of Nelson and *CE* since 1960.

3. *Labor.* Escalation of labor costs varies considerably in different parts of the country. The composition of the construction force is also significant. Refinery facilities and liquid petrochemical plants require a work force composed of nearly 50% pipefitters and pipefitter welders. Any sizable change in the costs

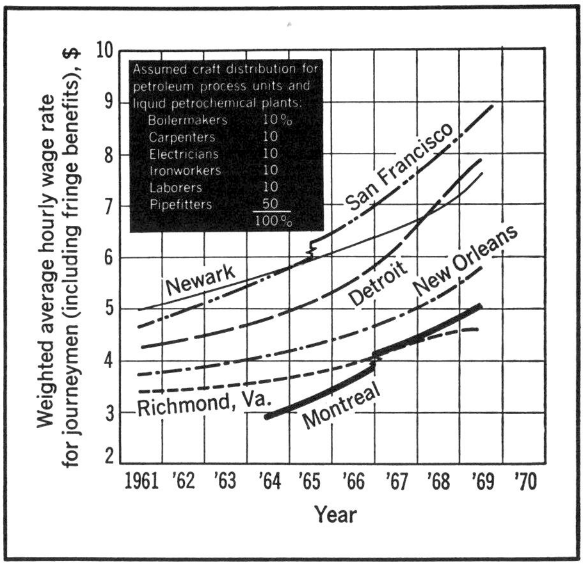

CONSTRUCTION PAY in selected cities—Fig. 1

for this craft obviously has a severe effect on total labor costs, yet the Nelson and ENR Construction Indexes do not include pipefitters in their labor component.

ENR does publish a skilled-labor index derived from the average rates for bricklayers, carpenters and structural ironworkers. But trends based on only three crafts can be very misleading, and this source is mentioned only as a rough indicator of the sizable differences in labor escalation within the U.S. Of the 22 cities reported, the increase between May 1969 and May 1970 ranged from zero for Cleveland to a high of 34.8% for Denver. During the same period, ENR's common labor indexes varied from zero for four cities to +42.3% for Kansas City.

Fig. 1 shows average hourly rates for a simplified labor distribution for six selected cities. Compounded annual rates for the 1966-1969 period vary from a high of 13.0% for Detroit down to 5.3% for Richmond, Va. An abrupt change in the escalation rate is illustrated by the jump caused by San Francisco pipefitters changing from 40 to 35 hours per week with a compensating increase in rates in 1965. Note that this significant increase is not evident from published indexes.

Among the best sources for wage rate information are the bulletins titled "Union Wages and Hours: Building Trades," published annually by the U.S. Dept. of Labor. Detailed data for essentially all construction crafts are given for 68 cities. From these, (and using craft-distribution patterns based on experience) a composite wage pattern for a specific location and type of construction can easily be determined.

Step 4: Determine the Schedule

The weighted rates are meaningless unless we make a reasonable judgment of the time interval applicable to each rate. Fig. 2 is plotted from our experiences on over 40 refinery and petrochemical projects. Everything from single-plant additions to grass-root projects and major additions totaling over $100 million were used. Each of the lower curves is a plot of the midpoint of firm commitments for the category indicated. In this context, we define "firm commitment" as one no longer subject to escalation. Any fixed-price purchase order or contract is firm, whereas an agreement having an escalation clause is not firm until the costs can no longer be escalated.

In these times of rapidly increasing costs, it is becoming more and more difficult to obtain firm prices, particularly for projects of long duration. Further, a fixed price may include an abnormally large contingency for escalation, and the purchaser frequently saves money by accepting an escalation clause.

As might be expected, the commitment and expenditure patterns vary considerably. Type of process certainly affects the patterns, but the more severe differences seem to be due to unpredictable causes such as labor disturbances, major design changes and abnormal equipment deliveries. For refining and petrochemical processes, the curves shown are reasonable averages of our experiences. The user may want to make adjustments for some unusual aspect of a project. For example, if the major components of a plant are sufficiently standard to be placed on firm-priced orders very early in the project, the average duration of materials escalation is lessened. Usually such corrections are minor.

We assume that the commitment curve for a cost category is symmetrical about the midpoint with respect to time, and that the escalation rate is uniform during the project. These are generalizations, and are seldom true for an actual project, but the errors

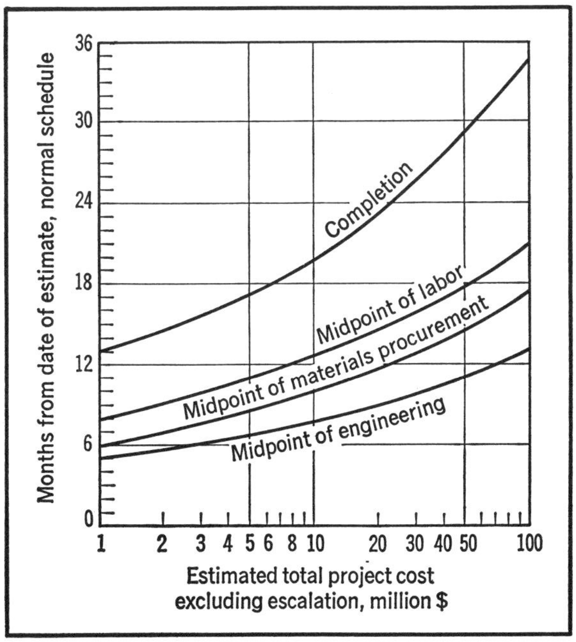

DURATION OF ESCALATION for cost components—Fig. 2

introduced by these assumptions are negligible on all but the most unusual situations.

One final point. Note that the time scale originates at the date of the estimate. The elapsed time to the various curves is based on a normal design and construction schedule—i.e., completing the project as rapidly as possible without incurring excessive premium costs. Sometimes, however, we are involved in a crash construction program, and the durations indicated are compressed. Obviously a part of the added cost of the "crash" is offset by the savings in escalation.

Common questions are, "What will it cost if we start the project now? Three years from now?" We might therefore divide escalation into two components:

1. The unavoidable portion, which occurs even while a project is proceeding on a normal schedule.
2. The avoidable portion, which occurs during delays in starting the project, whether by plan or by lack of a decision to proceed.

To determine the latter, we need only ignore Fig. 2, and apply an annual rate instead.

Step 5: Calculate the Composite Escalation

To illustrate the final computations, consider this example: The estimated cost of a liquid petrochemical facility to be built in San Francisco is \$10 million without escalation, as of Jan. 1, 1970. The cost breakdown is 12% engineering, 63% material and 25% labor. It is desired to find the unavoidable escalation assuming normal project duration, and also the escalation if start of the project is delayed three years.

Escalation Rates: The engineering rate is assumed to be 6.0%, based on recent salary experience in the San Francisco area. For materials, use 6.5%, based on recent Nelson and *CE* Index component rates (6.2% and 5.8%, respectively) and allowing for a continuation of the uptrend of early 1970. San Francisco labor rates have risen at a compounded rate of nearly 9% since 1967, and 10% in the last year. Recent wage settlements indicate a continued rise. Use 11%, which allows for a slight gain in productivity.

Escalation Durations: From Fig. 2, we see that engineering materials procurement and labor medians are 8, 10 and 13 months after the date of the estimate if the project proceeds normally. Completion will be about 20 months after the estimate.

The minimum normal (unavoidable) escalation is: Eng., $\$10,000,000 \times 0.12 \times 0.06 \times 8/12 = \$48,000$; Mat., $\$10,000,000 \times 0.63 \times 0.065 \times 10/12 = \$341,000$; Labor, $\$10,000,000 \times 0.25 \times 0.11 \times 13/12 = \$298,000$; Total, \$687,000 (6.87%).

Similarly, the cost of delay (avoidable escalation) is determined using the annual rates. In this example, it is \$757,000 for the first year's delay (7.57% annual rate). Assuming a constant escalation rate, the total escalated cost if the start of the project is delayed for three years is:

$$\$10,000,000\ (1.0757)^3 \times 1.0687 = \$13,300,000$$

Note that the rates have not been compounded in calculating the minimum normal escalation. The effect of compounding is relatively insignificant for the periods involved, and errors in the assumed rates and breakdown probably introduce greater differences. For delayed projects, compounding can be more significant, particularly if several years are involved. A constant rate of inflation over several years is unlikely, and the more intrepid estimator may wish to predict different rates for each year. In the above example of a three-year delay, the differences between compounded and straight rates is \$340,000, or over 11% of the uncompounded escalation.

Using the same approach, a \$100-million project in Detroit would have a minimum escalation cost of over \$12,500,000, or 12½% of the cost of the project! On the same basis, delay would cost \$8,100,000 the first year, or 8.1% annually.

Verification of Results; Computerization

Unfortunately, the verification of material and equipment escalation is impractical if not impossible even after a project is completed. As previously mentioned, usually only the standard materials have traceable cost histories. Unless escalation clauses are included in the purchase agreement, it is normally impossible to determine escalation on fabricated and specialized equipment. The only available check for the total material and equipment escalation, then, is a comparison between published and assumed rates.

For engineering, the actual escalation can frequently be determined from contract markups, from standard rate schedules, or from average rates.

Labor, usually the most volatile component of escalation, is the most easily checked. It is relatively simple to compare the labor ratio in effect at the time of the estimate with actual rates paid.

Obviously our method is easily computerized, although the inherent simplicity may not make this attractive. For more sophisticated approaches (i.e., the ones applicable in the advanced design stages of a project), the computer makes practical the use of historical distributions and schedules for each craft, and the known raises in pay. This increases the accuracy of the labor escalation cost, particularly on projects that can be completed under existing labor agreements. ■

Meet the Author

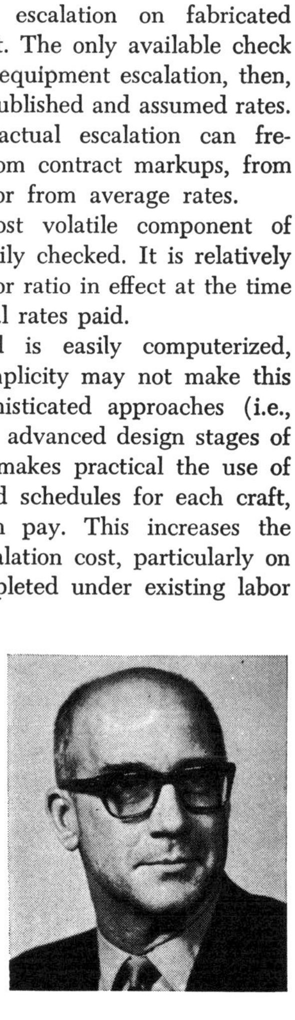

L. R. (Les) Jones is group supervisor, cost engineering methods, for Standard Oil Co. of California (555 Market St., San Francisco, Calif. 94120). His current assignment follows nearly 20 years of supervisory experience in the design and construction of oil refining and petrochemical facilities. A registered professional engineer, he has an M.S. (M.E.) degree from Oregon State University, and is a member of the American Assn. of Cost Engineers.

Cost of Preparing Proposals

ROY J. LORING, Allied Chemical Corp.*

When an operating company asks a contractor to bid on a $10-million process-plant project, what does it cost the contractor to prepare such a bid? As shown in Table I, the answer may be $1,000, or $10,000, or $100,000, depending on the type of plant, scope of the work, and type of estimate.

In other words, the budgets established for the proposals in Table I (proposals that were prepared over the past two years) depend on what type of information was required, and in what detail. For example, a qualifications brochure required a much smaller effort and budget than a definitive estimate.

* Mr. Loring has recently joined The Lummus Co.

Typical Allocated Budgets for Various Recent Process-Plant Proposals—Table I

Type of Plant	Scope of Work*	Type of Estimate†	Plant Cost, Million $ (Approx.)	Proposal Budget, Thousand $
Ammonia+urea	E, P, C	D, LS	30.0	30.0
Ammonia	E, P, C	D, LS	20.0	50.0
Ammonia	E, P, C	D, LS	5.0	20.0
Chemical	E, P, C	Q	60.0	5.0
Chemical	E, P, C	Q	10.0	1.0
Crude oil	Study	P, LS	100.0	5.0
Dewaxing	E, P, C	P, CP	3.0	4.0
Ethylene	Study	B	70.0	5.0
Ethylene	Study	D, CP	15.0	5.0
Ethylene	E, P, C	D, LS	25.0	150.0
Ethylene	Basic E	D, LS	45.0	145.0
Ethylene	E, P, C	Q	25.0	5.0
Ethylene	E, P, C	Q	50.0	6.0
Fertilizer	E, P, C	Q	50.0	3.0
Fertilizer	E, P, C	Q	20.0	1.0
Fertilizer	E, P, C	Q	20.0	3.0
Hydrogen	E, P, C	D, LS	30.0	60.0
Hydrogen recovery	E, P, C	P, LS	2.5	8.0
Lube oil	E, P, C	D, LS	15.0	50.0
Lube oil	E	D, LS	5.0	20.0
Lube oil	E, P, C	D, LS	20.0	50.0
Lube oil	Study	Q, LS	5.0	2.0
Olefins	Study	Q, CP	30.0	1.0
Petrochemical	Study	Q	100.0	50.0
Petrochemical	E, P, C	Q	60.0	7.0
Petrochemical	E, P	P, LS	100.0	60.0
Petrochemical	Basic E	P, LS	200.0	20.0
Petrochemical	E, P, C	Q, LS	150.0	15.0
Petrochemical	E, P, C	Q, CP	120.0	10.0
Pharmaceutical	E, P, C	Q	5.0	1.0
Pharmaceutical	E, P, C	D, LS	45.0	200.0
Polyethylene	E, P, C	Q, LS	80.0	5.0
Refinery (oil)	E, P, C	D, LS**	100.0	70.0
Refinery (oil)	E, P, C	D, LS	70.0	90.0
Refinery (oil)	E, P, C	D, GM or LS	40.0	3.0
Refinery (oil)	E, M, C	D, LS	30.0	135.0
Refinery (oil)	E, M, C	D, LS	30.0	130.0
Sulfuric acid	E, P, C	D, LS	3.0	10.0
Sulfuric acid	E, P, C	D, CP	3.5	5.0
Sulfuric acid	E, P, C, D	D, LS+CP	4.0	25.0
Sulfuric acid	E, P, C	D, LS	2.5	3.0
Sulfuric acid	E, P, C	D, LS	3.5	20.0
Sulfuric acid	E. P, C	D, LS	3.5	10.0
Urea	E, P	D, LS	6.0	15.0
Urea	E, M, C	D, LS	10.0	100.0
Urea	E, P, C	D, LS	2.0	10.0
Vinyl chloride	E, P, C	P, CP	8.0	10.0
Vinyl chloride	E, P	B	10.0	10.0
Waste disposal	E	Q	300.0	8.0
Waste-water control	E, P, C	D, LS	2.0	5.0

*** Key for Scope of Work**
E—Engineering
P—Procurement
M—Material specification
C—Construction or supervision of construction
D—Demolition

† Key for Type of Estimate
B—Budgetary
D—Detailed
P—Preliminary
Q—Qualifications
CP—Cost plus
GM—Guaranteed maximum
LS—Lump sum
**—Plus reimbursable

For a discussion of the differences between various types of construction contracts, and how they may affect proposal preparation, see: Gallagher, J. T., A Fresh Look at Engineering Construction Contracts, Chem. Eng., Sept. 11, 1967 (included on pp. 121–217 of "Modern Cost Engineering Techniques," ed. by H. Popper, McGraw-Hill, 1970).

Originally published November 1970.

Departmental Cost Components of Proposal-Preparation Charges—Table II

Type of Plant	Eng'g., Drafting	Proposal, Estim'g.	Other Depts.*	Other Charges**
Ammonia	$22,000	$7,000	$100	$12,000
Ammonia	6,000	2,000	1,500	500
Ammonia	1,500	700	100	200
Chemical	43,000	12,000	5,000	8,000
Chemical	2,500	1,000	——	50
Chemical	1,200	1,000	——	400
Chemical	2,000	500	3,000	700
Chemical	800	1,500	4,500	3,000
Fertilizer	600	700	100	500
Fertilizer	2,000	1,500	100	300
Fertilizer	4,500	2,000	——	1,000
Refinery (oil)	120,000	24,000	17,000	45,000
Refinery (oil)	15,000	8,000	4,000	12,000
Refinery (oil)	4,000	4,000	600	1,000
Sulfuric acid	4,500	5,600	2,000	2,000
Sulfuric acid	1,500	1,500	500	300
Sulfuric acid	1,800	2,000	2,000	1,500
Sulfuric acid	10,500	9,000	1,000	2,000
Sulfuric acid	11,000	10,000	3,000	5,000
Urea	8,000	4,500	100	9,000
Urea	3,000	200	——	200

* Includes departments such as Purchasing, Construction, Scheduling, Legal, etc.

** Includes such costs as sales charges, reproduction services, computer time, communication charges, order of special materials, overhead, etc.

Manhours Allotted to Proposal Preparation by Various Departments—Table III

	Departmental Manhour Allocations				
Type of Plant	Process	Eng'g.	Proposal	Estim'g.	Others
Ammonia	700	1,200	1,200	500	300
Ammonia+urea	150	200	500	300	100
Ammonia+urea	600	340	200	160	200
Chemical	300	1,000	800	350	150
Chemical	350	200	300	250	100
Chemical	400	400	400	500	200
Ethylene	1,000	100	200	100	100
Ethylene	700	350	1,000	500	100
Nitric acid	250	200	250	350	150
Nitric acid	280	180	230	460	200
Petrochemical	——	100	150	250	50
Petrochemical	800	600	850	300	250
Petrochemical	800	3,000	800	1,200	1,000
Refinery (oil)	820	2,100	2,000	1,100	400
Refinery (oil)	1,100	300	800	300	200
Sulfuric acid	200	50	100	100	100
Sulfuric acid	200	500	500	600	375
Sulfuric acid	300	450	450	900	350
Sulfuric acid	200	150	300	200	250
Urea	250	400	400	150	100
Urea	1,100	6,800	2,000	2,400	2,000
Urea	1,000	3,600	1,000	3,500	1,200
Waste treat.	120	175	200	200	100

* Includes departments such as Purchasing and Construction.

(Proposals in general range from information contained in a letter, to a bound qualifications proposal that contains various technical and nontechnical information but makes no formal commitment, to a detailed proposal that contains a definitive estimate and makes a firm commitment to perform certain services for a specified amount of compensation.)

It is interesting to probe further and examine the amount of money spent by each department in preparing a given proposal. Such a breakdown of expenditures is shown in Table II. It can be clearly seen that the bulk of the expenditures fall in the engineering (both process and general) and drafting area, and rightly so. However, note that the sum of the "other" charges that can develop during proposal preparation can be significant, and can at times even exceed the expenditures incurred by the proposal and estimating departments.

Table III shows the manhour allocations for the process, engineering (including drafting), proposal and estimating departments, as well as for the miscellaneous departments (such as purchasing and construction) that are combined together as "others." This breakdown indicates that for the typical proposal lasting four to five weeks and requiring flowsheets and other common technical data, the proposal department itself will use approximately the same number of manhours as the engineering department. This relationship will hold as a general observation, although we must repeat that the amount of manhours will be related to the amount of detail required of that department: if the operating company issues a multivolume bid package that the contractor must evaluate, this will require more manhours than if the bid request is only a twenty-page letter.

The necessity of using the operating company's standards instead of the contractor's standards will also require more manhours.

Staying Within the Budget

The cost of preparing proposals is a necessary—and very significant—part of a contractor's operating expense. This cost has become so critical that a lot of attention is being paid to ways of getting the most good from it, both by improving the percentage of successful bids submitted, and by applying budgetary controls on manhours expended and on other cost items (similar to the controls and monitoring used on an actual engineering and construction job).

Experience has shown that out of a sample of 50 proposals, 18 underran the allocated budget and 21 overran the budget, with the remaining 11 approximately meeting the established budget.

In addition to the department-by-department budgets that are established for each individual proposal, a total or overall budget is set for the proposal department, and the cumulative cost of preparing proposals is continuously checked against it. For a fair-size contractor, a departmental budget would range anywhere from $400,000 to $700,000 annually. ■

Section V

CHOOSING THE RIGHT ECONOMIC ALTERNATIVE

How to make the correct economic decision on spare equipment

Probability of failure, loss of production, cost of repairs, and time value of money are some of the major factors affecting evaluation of alternatives for maintaining plant capacity when unscheduled shutdowns of major equipment occur.

Gerald O. Davis, *Allied Chemical Corp.*

☐ There is a recurring need to decide when, or whether, to purchase a spare for major equipment in the typical process plant. And such decisions are both difficult and important. Most plants have several items such as pumps or heat exchangers that encounter variable service conditions, and these items have certain reliabilities inherent in their design. This makes a realistic standard policy on buying spares near impossible. Also, the dollar impact of an incorrect spare decision is very significant.

Obviously, the relative initial cost of a spare is small when compared with the value of lost production experienced when an entire plant is shut down due to an equipment failure, and no spare is on hand. This is particularly true when the replacement item has the typical long delivery time. Not so easy to evaluate is the equally real cost of committing funds for several expensive equipment spares, and then going for years before a need arises.

How can we estimate the value of a spare when a failure of the inservice unit can at worst only cause a 20% cutback in production? How do we factor in the possibility of making an onsite repair on a failed item? The engineer who is asked for a recommendation on the important matter of spares that are essential in maintaining normal plant production for various pieces of equipment faces a complex task.

We will outline a procedure for logically analyzing and reaching conclusions on spares. This procedure is simple to describe, and is intended for items costing more than $1,000.

For the first alternative, we establish the important costs and time variables, then adjust each cost to the same economic present-value time basis, and finally

Originally published November 1977.

Costs and other variables for spares decisions — **Table I**

Variable	Definition
C_s	Current cost of the spare item.
t_d	Time required for delivery of the spare.
C_{ss}	Cost per chosen time period to carry the item in the plant's store inventory. This cost varies. As an approximation, let us use C_{ss} as 5%/mo of C_s for items $10,000 and under, and 2%/mo of C_s for items over $10,000.
C_i	Cost of removing the old item and installing the new. Applies to actual manhours and equipment costs needed.
C_{Lp}	Cost per chosen time period of lost production caused by not having the item in full service.
C_e	Above-normal cost per chosen time period for increased energy and utility usage caused by the item not operating, or operating below normal.
C_{mh}	Above-normal cost per chosen time period for increased manhours in production or maintenance departments caused by the item not operating, or operating below normal.
C_r	Total cost for labor and materials involved in making a repair on the item.
F_P	Point in time at which failure of the normally operating item occurs. A "failure" can be: a 100% shutdown of the item; a decrease in operating level of the item so as to pose a safety hazard, or abnormal energy usage, or release of a pollutant; or any situation judged to be a failure.
t_{MAX}	Most probable maximum time between the time the analysis is made and the next F_P for the item.
t_{MIN}	Most probable minimum time between the time the analysis is made and the next F_P for the item. The value used must be basis in accord with the 66% Special Provision.
t_F	Most probable pseudo-mean time between the time the analysis is made and the next F_P for the item. $t_F = (t_{MAX} + t_{MIN})/2$

66% Special Provision—Because of the costs associated with the unscheduled shutdown or failure of a piece of equipment, there is a tendency to buy a spare irrespective of the facts in a particular situation. To make this analysis agree with such a decision, many users will give a very low value to t_{MIN} and a very high value to t_{MAR}. To avoid this, there must be at least a 66% probability that the item could actually fail at t_{MIN}, or before; and there must be at least a 66% probability that the necessary repair will actually take t_{MAR} to complete.

Variable	Definition
t_{MAR}	Most probable maximum time, after F_P has occurred (or a regularly scheduled shutdown period has elapsed), that is required to make the needed repair to the item. Just what is a needed repair will depend on an honest judgment of what will be the most likely failure mode of the item. The value used must be basis in accord with the 66% Special Provision.
t_{MIR}	Most probable minimum time, after F_P has occurred (or a regularly scheduled shutdown period has elapsed), that is required to make the needed repair on the item.
t_r	Most probable pseudo-mean time required to make the needed repair on the item. $t_r = (t_{MAR} + t_{MIR})/2$
$(PVF)_{ONE}$	Economic present-value factor used to bring one future cost back to its present-time value. This factor multiplied by the future cost converts that cost to its equivalent present-time value. The factor to be used in a particular situation depends on the number of chosen time periods (i.e., weeks, months, years) between the present and that point in the future where the cost is expected to occur. The factor also depends on the interest rate.
$(PVF)_{SER}$	Economic present-value factor used to convert a series of regularly occurring equal future costs into a single equivalent future cost that occurs at the time of the first cost in the series. This factor is multiplied by one of the equal future costs to make the conversion. The factor to be used in a particular situation depends on the number of chosen time periods over which the equal future costs occur, and the interest rate used.

total the adjusted costs. We follow this method for other alternatives, and compare the resulting equivalent time-basis total costs. All other factors being equal, we then pick the least-cost alternative.

Of course, the saying is a lot easier than the doing. The engineer will have much information to search out for a particular problem, and the application of present-value factors may not be familiar. This is difficult to do at first, but the exercise of trying to apply the correct factor is useful. Here, the engineer must think through the problem and identify both the important and the insignificant costs and times. To help with this, we will show a list of the costs and time variables that are often important in making a spares decision. Examples will illustrate and go into the practical details of applying the technique.

Definitions of cost and other variables

The important elements for this analysis are listed in Table I. In some decision situations, one or more of these elements may be zero, or too small to be considered relevant. The present-value factors, $(PVF)_{ONE}$ and $(PVF)_{SER}$, in Table I adjust for the fact that a cost incurred today cannot be compared to a cost occurring in the future. This is due to the interest-bearing or revenue-producing potential of invested money. This time-value-of-money concept is a primary characteristic of this procedure, as will be seen in the solved examples.

Uncertainty controls the decision

In most situations, the results will greatly depend on the value assigned for the time between the present and failure points of the item, or for the time necessary to make a repair. One assumed value may result in Alternative A being the favorite, and a somewhat different time value may show Alternative B as best. Uncertainty is inherent in decisions on spares, and this procedure will not change that fact.

Hopefully, the risks are lessened by the way in which the time value in the final calculations is obtained. This value is a pseudo-mean of the maximum and minimum expected-time intervals between the present and failure points of the item. This is not a true mean because usually there will not be a linear increase in the probability of failure over the service life of an item. The idea of using this pseudo-mean time to failure is an admitted compromise. The approach should give sufficiently accurate results for practical purposes when the "66% Special Provision" is used. This provision requires that there be at least a 66% probability that the item will actually fail at the assumed minimum value.

Besides avoiding the tendency to be too conservative and using a minimum time to failure that is too small, the 66% rule has another role. It narrows the gap between the minimum and maximum times to failure, and thus lessens the error-producing effects of nonlinearity. Realistically, the user of this procedure will not be able to pick minimum and maximum time-to-failure values that have exact 66% and 100% probabilities. However, to approach a given situation with this general method and definitions will tend to improve the accuracy of the results.

A characteristic of this or any other prediction model is that the results can only be as good as the input data. For example, an error in the values used for the time of delivery or for the point of failure of the item will definitely be felt to varying extents. Sometimes, seemingly different alternatives will produce final total costs that are very similar. This may cause the user to rethink his assumptions and to check the input data for possible errors. However, the reevaluation may also show that what appeared to be dissimilar economic alternatives are really not, when all the costs are considered in the analysis. The user can also easily pick the input data to make the procedure fit his prejudices about a given spare decision. This is easy after a quick look at the method. The procedure's credibility is largely dependent on the engineer's conscientiously searching out the best available information, and then applying it with honest engineering judgment.

Let us now review two examples that reflect typical operating situations in a chemical plant, and show how to apply these techniques. The first example considers a case where repair is impractical; and the second, a case where a repair is valid.

Examples for major spare procedures

Example A—A special design pump, CP-219, handles a corrosive liquid in Area R of a chemical plant. The auxiliary process stream necessarily varies in its corrosiveness. Thus, the pump's effective life is variable and unpredictable. Because of the nature of the pump's design, it is not possible to make inplant repairs, and sending it back to the vendor's shop for repairs is impractical for other reasons.

When CP-219 is fully operational and all other equipment is normal, Area R and the entire plant operate at 100% of design capacity, and the plant uses no more than its design level of energy. However, when CP-219 fails (i.e., CP-219 is not able to provide its full design flow and head), Area R must be operated differently, so that the production of the entire plant must be cut back to 99.5% of design and the plant's energy usage is 101% of design.

Due to the secondary importance of CP-219 in its role on the auxiliary process stream, no installed spare or stored spare unit was originally provided. Now, an engineer has been assigned to reevaluate the spare CP-219 situation from a total cost standpoint, and make a recommendation.

The following alternatives will be evaluated:

1. Buy a spare CP-219 now, and install when a failure of the present unit occurs.
2. Buy a spare CP-219 only when a failure of the present unit occurs.

(Possibly better alternatives exist but for purposes of these examples only simplified cases are considered to better illustrate the technique.)

In order to perform the necessary computations, additional information must be obtained from various sources. For this example, the data so developed are:

1. A new CP-219 costs \$4,300; its current delivery is 2 mo; and the cost to keep the spare in plant stores is 0.05(4,300) = \$215/mo.
2. The present unit has been fully operative for 10 mo. Based on past history for such pumps, and applying the 66% Special Provision, the minimum expected full-service life is 12 mo, and a maximum expected life is 20 mo.
3. Value of the product output from the normally operating entire plant is \$70,000/d, or \$2.1 million/mo. Lost production is equivalent to 0.5% of design.
4. Cost for energy for the normally operating entire plant is \$8,000/d, or \$240,000/mo. Increase in energy is 1% of design.
5. Cost for additional production and maintenance manhours to operate Area R in the different manner when CP-219 is down is \$30/d, or \$900/mo.
6. Cost required to remove a failed CP-219 and install a new one is \$300. Time required is ½ day.

In summary, the resulting values to be used in the computations are:

C_s = \$4,300
t_d = 2 mo
C_{ss} = \$215/mo
t_{MIN} = 2 mo
t_{MAX} = 10 mo
t_F = 6 mo
C_{Lp} = 0.005(2,100,000) = \$10,500/mo, or \$350/d
C_e = 0.01(240,000) = \$2,400/mo, or \$80/d
C_{mh} = \$900/mo = \$30/d
C_i = \$300

Factors for 10%/yr interest **Table II**

n mo	$(PVF)_{ONE}$ for $i = 0.10/12$	$(PVF)_{SER}$ for $i = 0.10/12$
1	0.9917	0.9917
2	0.9836	1.9753
3	0.9754	2.9507
4	0.9673	3.9181
5	0.9594	4.8774
6	0.9514	5.8289
7	0.9436	6.7725
8	0.9358	7.7083
9	0.9280	8.6363
10	0.9204	9.5567
11	0.9128	10.4695
12	0.9052	11.3747
13	0.8977	12.2725
14	0.8903	13.1629
15	0.8830	14.0459
16	0.8757	14.9216
17	0.8684	15.7900
18	0.8612	16.6513
19	0.8541	17.5055
20	0.8471	18.3526
21	0.8401	19.1928
22	0.8331	20.0259
23	0.8262	20.8522
24	0.8194	21.6717
25	0.8126	22.4835

Formulas for present-value factors

$$(PVF)_{ONE} = \frac{1}{(1 + i_M)^n}$$

$$(PVF)_{SER} = \left[\frac{(1 + i_M)^n - 1}{i_M(1 + i_M)^n}\right] = \left[\frac{1 - (1 + i_M)^{-n}}{i_M}\right]$$

where $i_M = i/12$

We begin by calculating the present-value total cost for Alternative 1, $(PV)_1$:

$$(PV)_1 = C_s + C_{ss}(PVF)_{SER(4)}(PVF)_{ONE(2)} + (C_{Lp}/2)(PVF)_{ONE(6)} + (C_e/2)(PVF)_{ONE(6)} + (C_{mh}/2)(PVF)_{ONE(6)} + C_i(PVF)_{ONE(6)}$$

where the first term, C_s, is the cost of the spare; and the second term converts the four future monthly storage costs, C_{ss}, into a single cost when multiplied by $(PVF)_{SER(4)}$, which equals 3.9181 from Table II. This will occur two months from now when the new CP-219 is delivered. Multiplication by $(PVF)_{ONE(2)}$, which equals 0.9836 from Table II, converts the one equivalent storage cost that will occur in two months to its present value.

The remaining terms are: the lost production, $C_{Lp}/2 = \$175$; additional energy, $C_e/2 = \$30$; manhour costs, $C_{mh}/2 = \$15$ experienced six months from now when the present pump fails; and a half day required to remove the pump and install the stored spare. We multiply each of these costs and C_i by $(PVF)_{ONE(6)}$ to get their present values. We find $(PVF)_{ONE(6)} = 0.9514$ from Table II. Therefore:

$$(PV)_1 = 4{,}300 + 215(3.9181)(0.9836) + 175(0.9514) + 40(0.9514) + 15(0.9514) + 300(0.9514)$$

$$(PV)_1 = \$5{,}632$$

Now, we determine the present-value total cost for Alternative 2, $(PV)_2$:

$$(PV)_2 = C_s(PVF)_{ONE(6)} + C_{Lp}(PVF)_{SER(2)}(PVF)_{ONE(6)} + C_e(PVF)_{SER(2)}(PVF)_{ONE(6)} + C_{mh}(PVF)_{SER(2)}(PVF)_{ONE(6)} + C_i(PVF)_{ONE(8)}$$

$(PVF)_{ONE(6)} = 0.9514$ from Table II converts the cost of the new pump from the cost-commitment point, six months from now, to its present value. This factor also converts the equivalent single lost-production cost, additional energy cost and additional manhour costs that will occur six months from now to their present values.

$(PVF)_{SER(2)} = 1.9753$ from Table II converts the lost-production, additional-energy and manhour costs that occur during the two-month delivery period for a new pump into a single equivalent cost for each.

$(PVF)_{ONE(8)} = 0.9358$ from Table II converts the one cost of removing the old CP-219 and installing the just delivered pump from its occurrence point eight months from now to its present value.

Therefore:

$$(PV)_2 = 4{,}300(0.9514) + 10{,}500(1.9753)(0.9514) + 2{,}400(1.9753)(0.9514) + 900(1.9753)(0.9514) + 300(0.9358)$$

$$(PV)_2 = \$30{,}306$$

Alternative 1, buying the spare CP-219 now and installing it when the present unit fails, is the least-cost option. This is primarily true because of the value of the lost production and additional energy used during the two-month delivery period required for the new pump. If this pump were a shelf item that the vendor could deliver in a matter of days, the predicted costs of the two alternatives might be very close. In such a case, the engineer might recommend that management base its spare-equipment decision on other criteria besides those evaluated here.

Example B—A large shell-and-tube heat exchanger, HE-12, operates in Area 10 of a chemical plant. The unit must operate under severe process-upset conditions so that its service life is variable. When the unit fails, its failure mode is such that inplant repairs are possible. Due to this characteristic, the past policy has been to repair the unit when it fails, and not to stock a spare in inventory.

There are drawbacks to this repair-only policy. First, the cost of the average repair is relatively high, and there is also lost production during an HE-12 shutdown. Moreover, the time necessary for a suitable repair is variable and dependent on the degree of damage to the unit. This is not known until the unit is disassembled and inspected.

An engineer has been assigned to evaluate the repair-only policy, and make a recommendation about its continued use, or provide a better solution. The following three alternatives are to be evaluated in order to determine total comparative costs:

1. Continue the repair-only policy—i.e., repair when a failure occurs.
2. Buy a spare HE-12 now, and store until a failure of the present unit occurs.
3. Buy a new HE-12 when the present unit fails.

The following additional information must be obtained in order to make the necessary analyses:

1. A new HE-12 costs $72,000 and has a 4-mo delivery. Storage cost is 2%/mo of $72,000.
2. Cost to remove a failed unit and install a repaired or new unit is $2,500, and takes one day.
3. Value of one day's production from the entire plant is $160,000. When HE-12 is not in service, the plant rate must be cut back to 95% of design level.
4. There is little detectable change in total plant-energy usage whether or not HE-12 is in service.
5. At the time of this analysis, a newly repaired HE-12 has just been installed. The minimum and maximum times between past failures have been 12 and 24 mo.
6. Average cost for a repair in the past has been $28,000 (for repair work itself).
7. Minimum past time required to make sufficient repair to a failed unit has been three days, and maximum required time has been eight days. These times do not include time for removing the failed unit and installing the repaired unit.

In summary, the resulting values to be used in the computations are:

C_s = \$72,000
t_d = 4 mo
C_{ss} = 0.02 (72,000) = \$1,440/mo
C_i = \$2,500, and takes 1 day
C_{Lp} = 0.05(160,000) = \$8,000/d, or \$240,000/mo
C_e = 0
t_{MIN} = 12 mo
t_{MAX} = 24 mo
t_f = 18 mo
C_r = \$28,000
t_{MIR} = 3 d
t_{MAR} = 8 d
t_r = 5.5 d

Alternative 1 (repair when a failure occurs) has a present-value total cost, $(PV)_1$, of:

$$(PV)_1 = C_i(PVF)_{ONE(18)} + C_r(PVF)_{ONE(18)} + C_{Lp}(6.5)(PVF)_{ONE(18)}$$

$$(PV)_1 = 2{,}500(0.8612) + 28{,}000(0.8612) + 8{,}000(6.5)(0.8612)$$

$$(PV)_1 = \$71{,}049$$

Note that 6.5 in the last term represents t_r = 5.5 d plus one day to remove and install the unit.

Alternative 2 (buy a spare now, and store until a failure occurs) has a present-value total cost, $(PV)_2$, of:

$$(PV)_2 = C_s + C_{ss}(PVF)_{SER(14)}(PVF)_{ONE(4)} + C_i(PVF)_{ONE(18)} + C_{Lp}(PVF)_{ONE(18)}$$

$$(PV)_2 = 72{,}000 + 1{,}440(13.1629)(0.9673) + 2{,}500(0.8612) + 8{,}000(0.8612)$$

$$(PV)_2 = \$99{,}378$$

Alternative 3 (buy a new unit when the present unit fails) has a present-value total cost, $(PV)_3$, of:

$$(PV)_3 = C_s(PVF)_{ONE(18)} + C_i(PVF)_{ONE(18)} + C_{Lp}(PVF)_{SER(4)}(PVF)_{ONE(18)}$$

$$(PV)_3 = 72{,}000(0.8612) + 2{,}500(0.8612) + 240{,}000(3.9181)(0.8612)$$

$$(PV)_3 = \$873{,}983$$

On the basis of the information supplied for this analysis, Alternative 1 (the present, repair-only policy) has the least total cost and should be continued.

One value of this type of analysis is that the relative importance of the values of the different variables is clearly identified. For example, in Alternative 3, the 4-mo delivery of a new HE-12, coupled with the value of the plant's total production and HE-12's effect on it, far outweigh the cost of a spare unit.

It is also clear that the estimated values used for variables are very important. For example, if the times to make a repair had been t_{MIR} = 6 d, t_{MAR} = 12 d, and t_r = 9 d, then the lost-production term in Alternative 1 would have been \$68,896, and the total cost for Alternative 1 would have been \$95,163 instead of \$71,049 as shown. Due to this increased value for t_r, Alternatives 1 and 2 would have been within 5%. In that case, the engineer's recommendation might be the safer, buy-a-spare route.

Present value factors

The value of $(PVF)_{ONE}$ depends on the number of time periods, n, between the present and the point in the future at which the cost will occur, and on the annual interest rate, i. The value of $(PVF)_{SER}$ depends on the number of months, n, over which the series of equal monthly costs occur, and the annual interest rate, i. The interest rate should be equal to the company's cost of capital. This is what the company must pay for the use of the money to fund its projects. The cost of capital continually varies, but a typical annual rate of 10% if consistently applied in these calculations will give satisfactory results. Formulas are included at the bottom of Table II for computing the present-value factors for any positive values of n and i.

References

1. Perry, R. H., ed., "Engineering Manual," 2nd ed., McGraw-Hill, New York, 1967.
2. Weston, J. F. and Brigham, E. F., "Essentials of Managerial Finance," Holt, Rinehart & Winston, New York, 1968.

The author

Gerald O. Davis is a project engineer in the Plant Engineering Dept., Allied Chemical Corp., P.O. Box 761, Hopewell, VA 23860. He is responsible for a variety of project-engineering duties. Previously, he did engineering work in the U.S. Air Force, at Shell Oil Co. and Ingersoll-Rand Corp. He holds a B.S. in mechanical engineering from Virginia Polytechnic Institute and State University, an M.B.A. from Golden Gate University, and is a member of ASME.

Economic evaluation of future equipment needs

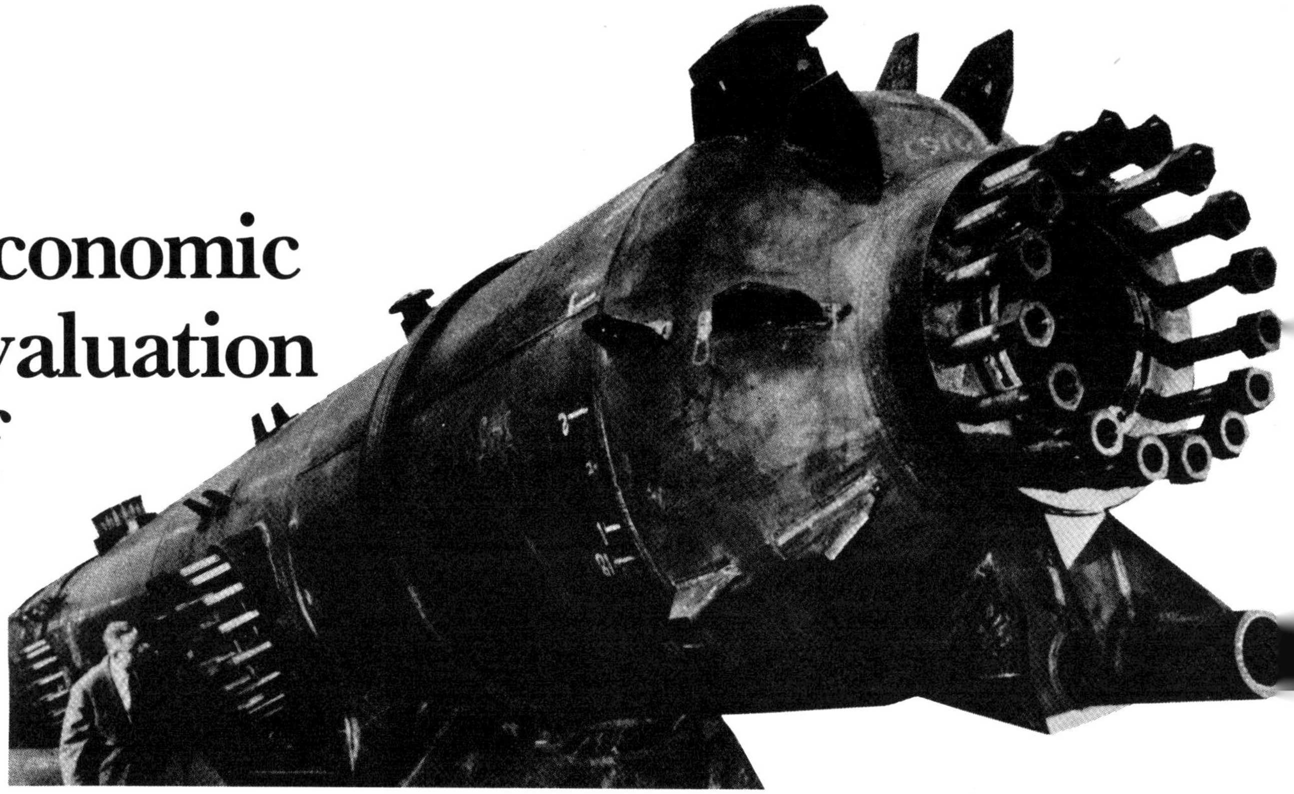

This article presents an analytical way to decide whether to buy a piece of equipment for today's needs or for some future capacity. Significant amounts of money depend on the decision.

Thane R. Brown, *Procter & Gamble Co.*

☐ An easy-to-use technique helps determine whether to oversize new equipment for future expansion needs. The method described herein is based upon comparisons of the time value of money, and takes into account rates of escalation. It is adjustable to the different economic policies of companies.

Since the engineer has few analytical tools to help him, he normally uses his judgment to decide whether or not to oversize. In doing so, there is a high risk of improperly spending capital. With today's economic situation, this can be serious.

The method presented here compares the time value of money for two options:

1. Buy the higher-capacity equipment, even though all the capacity will not be used for several years.
2. Buy the lower-capacity equipment, and invest the differential cost between the larger and the smaller equipment.

The advantages and disadvantages of the options are presented in the accompanying table. An analysis of the future worth versus time curves, shown in the figure, provides the key to deciding which option should be selected.

To arrive at a decision, consider the following situation: (1) the small unit has been purchased, (2) extra capacity will be needed before the breakeven point, (3) the cost difference between the large and small units ($P_L - P_S$) will be invested, and the future worth of this money will increase. When the extra capacity is needed, the larger unit will be installed, and the money earned

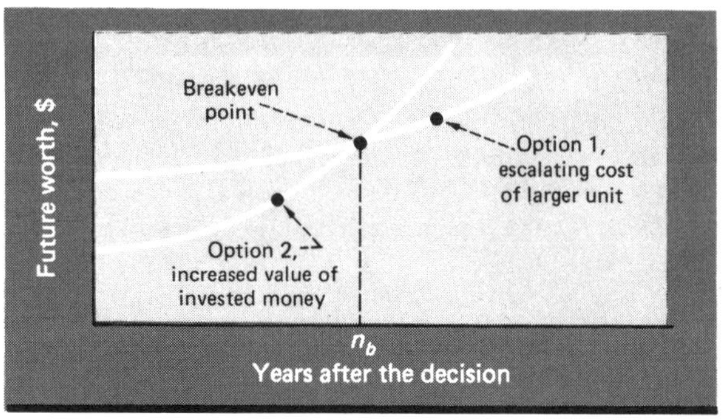

Present worth of equipment versus time

Originally published January 1977.

Nomenclature

e	Annual rate of escalation
L_S	Salvage value of smaller unit, when replaced by the larger unit, \$
n	Years after the decision
n_b	Breakeven point, yr
P_L	Installed cost of the larger unit, \$
P_S	Installed cost of the smaller unit, \$
r_{AT}	Rate of return after taxes, $r_{AT} = \approx 0.52 r_{BT}$
r_{BT}	Rate of return realized on invested capital before taxes (determined by the company's policy, which sets a minimum acceptable rate of return)
S	Future worth, \$
S_1	Future worth of larger unit due to escalation, \$
S_2	Future worth of the cost difference of the larger and smaller units, when invested at r_{BT}, \$

from the investment will be used to help pay for the new equipment.

An examination of the figure shows that additional money will be required because the escalated cost of the large unit is greater than the future worth of the invested money. Therefore, less capital would have been required had the larger unit been installed originally.

From this situation, the decision-making guidelines are as follows: (a) if the extra capacity will be needed *before* the breakeven point, buy the larger equipment (Option 1); (b) if the extra capacity will be needed *after* the breakeven point, buy the smaller piece of equipment (Option 2).

Formula derivation

Rather than plot the future worth of a piece of equipment, an expression can be derived to find the breakeven point. To do this, consider the general formula* for future worth:

$$S = P(1 + i)^n \tag{1}$$

For Option 1:

$$S_1 = P_L(1 + e)^n$$

For Option 2:

$$S_2 = (P_L - P_S)(1 - r_{BT})^n$$

Since, at the breakeven point, $S_1 = S_2$, and $n = n_b$:

$$P_L(1 + e)^{n_b} = (P_L - P_S)(1 + r_{BT})^{n_b}$$

$$\text{or} \quad \left(\frac{1 + e}{1 + r_{BT}}\right)^{n_b} = \frac{P_L - P_S}{P_L}$$

Taking logarithms and solving for n_b:

$$n_b\left[\ln\left(\frac{1 + e}{1 + r_{BT}}\right)\right] = \ln\left(\frac{P_L - P_s}{P_L}\right)$$

$$n_b = \frac{\ln\left(\dfrac{P_L - P_s}{P_L}\right)}{\ln\left(\dfrac{1 + e}{1 + r_{BT}}\right)} \tag{2}$$

If an after-taxes rate of return is used (Eq. 2), then:

$$n_b = \frac{\ln\left(\dfrac{P_L - P_S}{P_L}\right)}{\ln\left(\dfrac{1 + e}{1 + \dfrac{r_{AT}}{0.52}}\right)} \tag{3}$$

If the smaller unit has a salvage value, Eq. (2) and (3) become:

$$n_b = \frac{\ln\left(\dfrac{P_L - P_S - L_S}{P_L}\right)}{\ln\left(\dfrac{1 + e}{1 + r_{BT}}\right)} \tag{4}$$

and

$$n_b = \frac{\ln\left(\dfrac{P_L - P_S - L_S}{P_L}\right)}{\ln\left(\dfrac{1 + e}{1 + \dfrac{r_{AT}}{0.52}}\right)} \tag{5}$$

Examples

The three examples that follow show how these equations can be used to improve the quality of equipment-sizing decisions.

Example 1—An engineer is sizing a stripping column with a reboiler. There is a high probability that 50%

*Grant, E. L., and Ireson, W. G., "Principles of Engineering Economy," p. 43, Roland Press Co., New York (1960).

Advantages and disadvantages of options 1 and 2

Options	Advantages	Disadvantages
1. Buy the higher-capacity equipment	No money is lost by having to scrap the smaller equipment when more capacity is needed. The larger equipment is installed for a lower actual cost.	Capital that could be earning profits is tied up in equipment whose capacity cannot be used.
2. Buy the lower-capacity equipment	The differential cost between the higher and lower-capacity equipment can be invested to earn profits.	The lower-capacity equipment must be scrapped before it is worn out., whereas the higher-capacity equipment will cost more, due to escalation.

more capacity will be needed in five years. In today's dollars, the installed cost of the lower-capacity equipment is \$75,000 (\$75M), and the installed cost of the larger unit is \$100M. The small unit has a salvage value of \$5M. If the larger unit is not purchased, the unused money can be invested to return 20% before taxes. The annual rate of escalation is 7%.

By using Eq. (4) to find the breakeven point:

$$n_b = \frac{\ln\left(\dfrac{P_L - P_S - L_S}{P_L}\right)}{\ln\left(\dfrac{1+e}{1+r_{BT}}\right)} = \frac{\ln\left(\dfrac{100\text{M} - 75\text{M} - 5\text{M}}{100\text{M}}\right)}{\ln\left(\dfrac{1+0.05}{1+0.20}\right)} = 12.1 \text{ yr}$$

Since the extra capacity is needed in 5 yr, or prior to the breakeven point, the extra capacity should be purchased.

Example 2—An engineer is sizing a reactor system. He has been told that more capacity will be needed in 6–8 yr, due to the introduction of several new brands. It is expected the added volume will be 60%. The installed cost of the system that will meet today's needs is \$2.6 million (\$2.6MM). The cost of the system having 60% extra capacity is \$3.5MM. Since the reactor is of such a specialized design, it has no salvage value. The company's investment policy states that capital investments should pay out within 2½ yr on a before-tax basis. The annual rate of escalation is anticipated to average 5% over the next eight years. To return an investment in 2½ yr, the rate of return must average 40%/yr.

Using Eq. (2):

$$n_b = \frac{\ln\left(\dfrac{P_L - P_S}{P_L}\right)}{\ln\left(\dfrac{1+e}{1+r_{BT}}\right)} = \frac{\ln\left(\dfrac{3.5\text{MM} - 2.6\text{MM}}{3.5\text{MM}}\right)}{\ln\left(\dfrac{1+0.05}{1+0.40}\right)} = 4.7 \text{ yr}$$

Since the added capacity will not be needed for 6–8 yr, or after the breakeven point, the smaller reactor should be purchased and the unneeded capital invested.

Example 3—This example deals with the sizing of a heating, ventilating and air-conditioning system for a new building. Because there is potential for a major expansion of this building in 6–7 yr, the engineer is considering sizing the refrigeration equipment 40% larger than is needed today. If the extra capacity is not installed now, additional parallel refrigeration equipment will be installed in 6–7 yr. Thus, the engineer has two options: (a) buying the larger equipment, whose installed cost is \$7.4MM, or (b) buying the smaller equipment now, and in 6–7 yr, buying the additional parallel equipment. The cost of the smaller equipment is \$6.0MM in today's dollars, whereas the cost of the additional equipment is \$3.0MM. This cost will escalate at 5%/yr.

The estimated annual rate of escalation is 5% and the company's capital-investment guidelines state that incremental capital investments should be justified against a potential after-taxes return of 25%/yr.

The first thing that must be done is to find the equivalent present worth for Option (b), which is composed of two parts: (1) the cost of the smaller equipment, and (2) the capital that must be invested to provide money for the equipment that will be installed in 6–7 yr.

To find the investment capital, first determine the capital required in the 6–7th yr. Use Eq. (1) to do this: $S = P(1+i)^n$. By letting $n = 6.5$ yr; and $i =$ the rate of escalation $= 0.05$; then:

$$S = \$3.0\text{MM}(1+0.05)^{6.5} = \$4.12\text{MM}$$

Again, using Eq. (1), determine how much capital must be invested now, so that \$4.12MM will be available in 6.5 yr. If $S = P(1+i)^n$, the term i is determined by the company's investment policy. Thus:

$$i = r_{BT} \approx \frac{r_{AT}}{0.52} = \frac{0.25}{0.52}$$

Rearranging Eq. (1):

$$P = \frac{S}{(1+i)^n} = P = \frac{\$4.12\text{MM}}{\left(1 + \dfrac{0.25}{0.52}\right)^{6.5}} = \$0.32\text{MM}$$

Now P_L can be calculated:

$$P_L = \$6.0\text{MM} + \$0.32\text{MM} = \$6.32\text{MM}$$

By using Eq. (3) to find the breakeven point:

$$n_b = \frac{\ln\left(\dfrac{P_L - P_S}{P_L}\right)}{\ln\left(\dfrac{1+e}{1+\dfrac{r_{AT}}{0.52}}\right)} = \frac{\ln\left(\dfrac{7.40\text{MM} - 6.32\text{MM}}{7.40\text{MM}}\right)}{\ln\left(\dfrac{1+0.05}{1+\dfrac{0.25}{0.52}}\right)} = 5.6 \text{ yr}$$

Since the additional capacity is not needed for 6–7 yr, or after the breakeven point, the smaller unit should be installed now.

The author

Thane R. Brown is presently the project manager for a major capacity expansion at Procter & Gamble Co., P.O. Box 599, Cincinnati, OH 45201. He has experience in food and paper engineering, doing feasibility studies, selecting and designing equipment and managing projects. He holds a B.S. degree in chemical engineering from Oregon State University and is a registered professional engineer in Ohio. A member of AIChE, he has written a number of articles in technical journals.

Saving energy and costs in pumping systems

For lowest cost in pumping systems, looking at first costs is not enough. You must evaluate the overall system, including flowrates and variable-capacity and material requirements.

John A. Reynolds, *Union Carbide Corp.*

□ There are many ways to unknowingly squander energy in pumping systems. Some of the more prominent ones will be discussed in this article. But first, we should look at what energy is costing us, and find out how we can save operating horsepower by investing in a more efficient pump.

Energy costs

Let us compare two pumps of different manufacture for water-booster service. Brand A requires 10 hp at the specified performance conditions; Brand B requires 9 hp. The electric utility rate is 2¢/kWh. The direct operating cost for a ten-year-life project would be (2¢/kWh)(8,760 h/yr)(0.746 kW/hp) (10 yr) ÷ 0.85 motor efficiency = $1,538/hp.

However, a dollar spent in the tenth year of the project is worth less than a dollar now, due to the time value of money. (Using a discounted-cash-flow method of analysis, one can enter the various elements of consideration, such as investment, direct costs, interest rates, project life, etc., and arrive at a precise number.) For simplicity in later comparisons, assume that $1,000 is the amount we can afford to spend in first cost to save one operating horsepower, which we will call the *investment equivalent* of the lifetime operating costs. Thus, if Brand A pump costs $1,500 and Brand B costs $1,700, Brand B would have the best overall economics, as shown by the comparison presented for Brand A and Brand B in Table I.

Once we have established an investment equivalent of utility usage, it is advisable to pass this value on to the supplier (even to having an entry on the equipment data sheet). This will help him to offer the pump with the best overall economics; it will encourage him to avoid limiting his bid to one that is lower in first cost only.

Originally published January 1976.

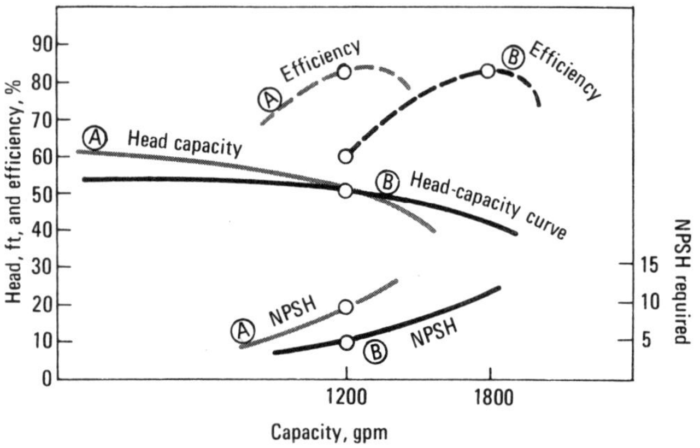

NPSH requirements boost pumping costs — Fig. 1

Calandria circulating pumps

Quite often, the design engineer decides upon the elevation above grade of his distillation column before he selects the calandria circulating pump. In the interest of economy, he makes the height as small as possible, not realizing that he will be forever paying an energy penalty for a smaller first-cost saving.

For example, he may need a 1,200-gpm, 50-ft-head pump. But with a small elevation, perhaps only 5 ft of NPSH (net positive suction head) will be available. Pump A, shown in Fig. 1, has the best efficiency for the application, but requires 9 ft of NPSH at the rated capacity. Consequently, the vendor must offer an oversized and more expensive pump, operating at a lower efficiency to meet the 5 ft of NPSH available. (There must always be more NPSH available than is required by the pump, or it will cavitate.)

If an expensive alloy is required for the application, the first cost of the larger pump is even greater. The higher first cost of this pump may of itself outweigh the cost of additional height for a steel column. Table II presents a comparison of typical economies of pumps A and B. We can realize a $6,800 savings by adding height and buying the more efficient pump. It is important to check the NPSH requirements for a pumping application before establishing elevations above grade.

Total evaluated costs — Table I

Pump	First cost	Investment equivalent	Total cost
Brand A	$1,500	$1,000	$2,500
Brand B	1,700	0	1,700
		Difference in favor of Brand B	$ 800

Larger pump vs. added skirt height — Table II

Pump	First cost	Other costs	Total cost
A (18.3-hp rating)	$3,500	$1,200*	$4,700
B (25.3-hp rating)	4,500	7,000†	11,500
		Net savings of pump A is	$6,800

*for extra column skirt and piping
†cost of 7 hp at $1,000 hp (investment equivalent)

Variable capacity requirements

Some processes have capacity requirements that vary over a wide range. A common way of handling this is to buy a pump with a fixed-speed electric motor drive and install a control valve to reduce the flow during lower demand periods. In systems where the pressure drop is a significant part of the total head requirement, a variable-speed drive merits consideration. Take the example shown in Fig. 2: the normal capacity is 500 gpm, but at certain times up to 1,000 gpm is required.

Obviously, a 1,000-gpm pump with an electric motor and a control valve can handle the 500-gpm demand. But notice that at that rating, such a pump's efficiency is only 65%, as compared with 83% at the 1,000-gpm level. Also note that the actual head requirement of the system is only 30 ft, not the 100 ft that the pump is developing at the high rpm.

The head developed by a centrifugal pump varies as the square of the rpm ratio. Its capacity varies directly with the rpm ratio. Thus, it can "track" the system pressure-drop curve very nicely, while keeping at or near its best efficiency point, if the rpm can be varied. Looking at Table III, it is apparent that a variable-speed drive is the best choice.

Before leaving this example, let us look at the system's pressure-drop curve, keeping in mind the affinity laws concerning pump head and capacity versus rpm. If demand is equally divided between 500 and 1,000 gpm, with no in-between requirements, then a two-speed motor having a high rpm twice the lower rpm and costing about $1,000 would be the most economical choice. If there were a few in-between demands, a control valve could be used to handle them.

One last word about variable-speed drives for capacity control of centrifugal pumps: They can be used to vary the flow *exclusive* of a control valve, and can reduce horsepower requirements as much as 50% or more.

Proper pump selection

Often, the design engineer will specify a standard centrifugal pump for very small flowrates. Consider an example where there is a 3-gpm requirement for deionized water at a 324-ft differential head, and Type 316 stainless steel materials of construction are necessary. Table IV illustrates that a non-standard, small, high-speed centrifugal pump has the most attractive overall economics (when compared with a standard

Fixed-speed drive vs. variable-speed drive **Table III**

Drive	Investment equivalent	Motor	Total cost
Fixed rpm* (19.4 hp)	$19,400	$ 500†	$19,900
Variable rpm* (4.6 hp)	4,600	4,500‡	9,100
Net savings for variable-rpm driver			$10,800

*At 500 gpm.
†Cost of standard 30-hp motor.
‡Cost of 30-hp variable speed motor and controls.

Types of pumps—very small flowrates **Table IV**

Pump*	First cost	Investment equivalent	Total cost
Standard centrifugal (10.9 hp at design)	$1,200	$10,900	$12,100
Metering (0.58 hp at design)	3,500	580	4,080
Small high-rpm centrifugal (1.5 hp at design)	1,300	1,500	2,800

*All pumps handle 3 gpm at 324 ft (140 psi)head.

Types of pumps – small flowrates **Table V**

Pump*	First cost	Investment equivalent	Total cost
Standard centrifugal (17.2 hp at design)	$1,500	$17,200	$18,700
Small high-rpm centrifugal (4.7 hp at design)	1,500	4,700	6,200
Triplex plunger (1.7 hp at design)	5,500	1,700	7,200

*All pumps handle 12 gpm of deionized water at 400 ft (173 psi)

Pumping high-viscosity fluids **Table VI**

Pump*	First cost	Investment equivalent	Total cost
Standard centrifugal (34 hp at design)	$1,600	$34,000	$35,600
Rotary-gear (4.5 hp at design)	1,600	4,500	6,100
Difference in favor of gear pump			$29,500

*All pumps handle 50 gpm, at 230 ft (100 psi) head of fluids with 500-cp viscosity.

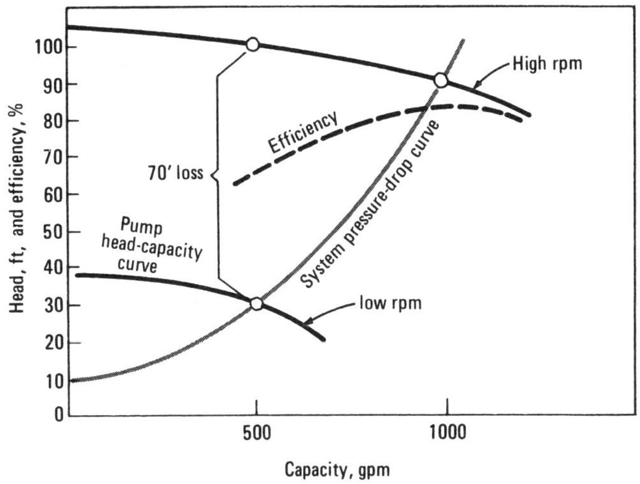

Variable-speed drive saves pumping costs **Fig. 2**

centrifugal pump or a metering pump), although it is not low in first cost, nor is it the most efficient. (A gear pump would also have good economics, but has not been considered because of the poor lubricity of the water, which could cause galling of the 316 SS gears.)

If exact capacity regulation is required for the flow of the deionized water, then the costs of both centrifugal pumps must include that of a measuring and regulating system. This system could be omitted from the metering-pump cost since that pump has an inherent accuracy of flow control, and since a pneumatic stroke-control is included in the cost shown. Thus, in this example, the metering pump may turn out to have the best economics.

If we take another example, where we want to pump 12 gpm of deionized water at 400 ft head, the capacity needed is greater than that of a metering pump, and we might consider a triplex plunger pump, a standard centrifugal, or a small high-speed centrifugal. Table V shows how the small high-speed centrifugal pump has the best economics.

However, if carbon-steel were an acceptable material, the triplex pump would cost much less, e.g. $1,800, which would give it the lowest overall cost. On the other hand, if capacity control were again a requirement, the added cost of a variable-speed drive or of bypass provisions for the triplex might cause the small centrifugal pump to be the best choice.

Our final example will illustrate the economics of selecting pumps for high-viscosity fluids. Let us say we want to pump 50 gpm of a clean product having a 500-cp viscosity at 230 ft of differential head. In Table VI, the first costs of a centrifugal pump and a gear pump put them at a standoff. But the required horsepower for the centrifugal at that rating makes a clear case for the rotary gear pump, even if a variable-speed drive or a bypass to suction is needed for flow variations.

These various examples show that the entire system

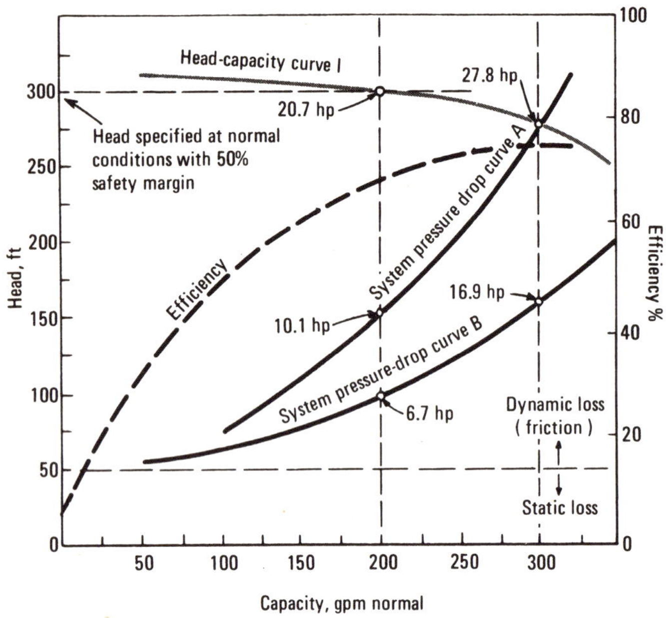

Safety margin makes valve a power consumer — Fig. 3

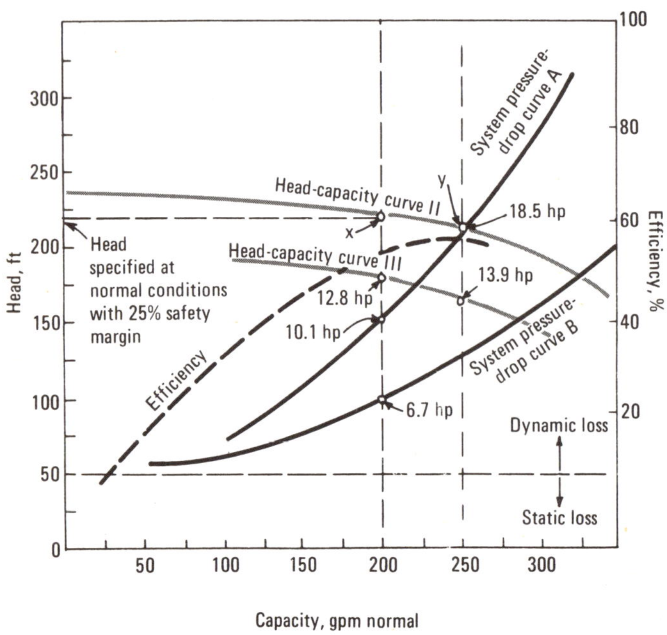

25% overdesign reduces power requirements — Fig. 4

must be considered, in order to make a sound decision—looking at the type of pump alone is not enough.

Rated vs. normal flows—valves and lines

The flow for which a centrifugal pump is rated is the calculated normal flow, plus an added margin of safety. This margin of safety can cause a waste of power in two ways, if it is overly generous. First, the selection is usually based on a pump that would be near its highest efficiency at the rated flow; and subsequent operation at normal flow would be away from this highest efficiency. Second, the large margin of safety in flow results in a much higher calculated head, since the pressure drop of the system increases exponentially with increase in flow.

In the system of Fig. 3, for example, curve B is made up of 50 ft of static loss (difference in elevation between suction and discharge vessels) plus a variable dynamic loss (friction in lines, heat exchangers, etc.) which varies with flowrate. Curve A consists of curve B plus the pressure drop through the control valve in the pump discharge. One accepted method for sizing such valves is to make the valve-pressure drop equal to the sum of all the other dynamic pressure drops in the system. Thus, a 50% margin of safety and very conservative control-valve sizing criteria would lead one to specify a pump having a 30-hp motor and using 20.7 hp at the normal flowrate. The dynamic loss over the control valve becomes more than five times all other dynamic losses at that flowrate.

Curve II of Fig. 4, by contrast, shows that if a 25% safety margin is used, a 20-hp pump with a lower head will do the job, even when the same conservative control-valve sizing criteria are used. Furthermore, by making the pressure drop through the control valve only half all other dynamic losses, we can use a pump with a head-capacity curve like curve III. This pump requires a 15-hp motor and uses 12.8 hp at normal flow.

In many pumping systems, the pressure drop of the piping is the major dynamic loss. We have seen that the control-valve pressure drop is predicted on such dynamic losses. Therefore, one of the first analyses to be made in pump selection is the balance between a low first cost of small piping versus a higher operating cost caused by high dynamic losses. After that, a choice of reasonable margins of safety and control-valve sizing criteria should bring some significant power savings.

The author

John A. Reynolds is a staff engineer in the Chemicals and Plastics Engineering Dept., Union Carbide Corp., South Charleston, WV 25303. He is responsible for specification, procurement and consultation to plants on pumps, seals and related equipment. Mr. Reynolds holds a B.S. degree in mechanical engineering from West Virginia University and has been active in mechanical evaluation of machinery since 1951. He has specialized in pumping equipment since 1962.

The cost of missing pipe insulation

Here's a graphical method for finding the annual expense of heat loss from uninsulated piping—at any efficiency of energy conversion, and for any cost of fuel and wind condition.

Rene Cordero, *Allied Chemical Corp.*

☐ For piping whose surface has been left uninsulated or whose insulation has been so severely damaged as to be useless, the yearly (8,760 h) cost of the heat loss per square foot can be found from Fig. 1. Costs so determined will generally be within the accuracy of data that can be obtained in the field, and are suitable for most applications without further refinement.

Fig. 1 costs are based on a net efficiency of energy conversion of 100%, a fuel price of $1.00/million Btu, and heat transfer to still air. Corrections are necessary to arrive at annual costs for other conditions:

- *Net efficiency of less than 100%*—To correct for a different net efficiency, divide the cost derived from Fig. 1 by the actual net efficiency of energy conversion. (For intermediate-size boilers most common in Allied plants, the net efficiency is typically 72%.)
- *Fuel cost other than $1.00/million Btu*—To correct for a different cost, multiply the cost derived from Fig. 1 by the actual cost of the fuel.
- *Outdoor installation exposed to moving air*—To correct for variant wind conditions, multiply the cost obtained from Fig. 1 by a correction factor from Fig. 2 for the actual wind velocity. For most of the continental U.S., 10 mi/h represents a reasonable annual average.

Assumptions made in developing the graphs

The unit costs provided by Fig. 1 are average values for a range of pipe sizes in still air. Although there is some variation for different pipe sizes and ambient temperatures, the variations are small compared to the cost of heat loss caused by low air velocities.

The heat loss due to missing insulation is the difference between the heat loss from bare pipe and that through normal insulation. Insulation standards have changed over recent years because of varying fuel and insulation costs and other economic factors. And although the loss through insulation will vary with the standard adopted for a project, such variations will be negligible compared to the heat loss from bare pipe.

In general, corrections for precise ambient conditions, pipe size and insulation thickness will not be jus-

Originally published February 1977.

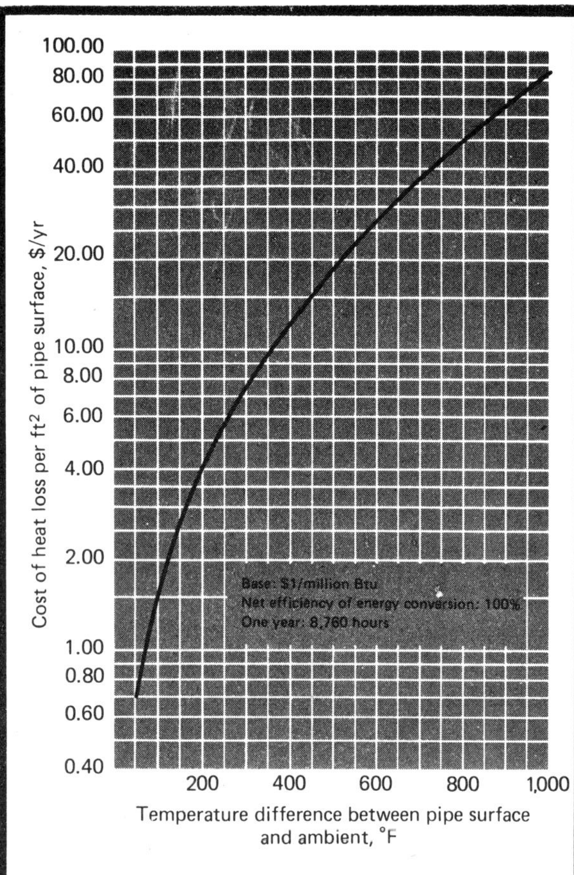

Cost of heat loss when insulation is missing Fig. 1

tified because they will go beyond the reasonable accuracy of field data.

Sample problems illustrate the procedure

The first problem requires estimating the annual cost for the extra energy lost from a pipe in still air without insulation:

Pipe dia., in.	6
Pipe temperature, °F	375
Average ambient temperature, °F	75
Length of uninsulated pipe, ft	10
Fuel cost, $/million Btu	1.44
Net energy conversion efficiency, %	72

For the ΔT of 300°F, Fig. 1 gives a unit economic loss of $7.60/yr/ft^2 of pipe surface. This figure is converted into the cost per linear foot by multiplying it by 1.734 ft^2/ft, the surface area per foot of a nominal 6-in. pipe. The answer is $13.20/ft of pipe.

The annual cost for the entire 10 ft of pipe is, of course, $132.00. The corrections for net efficiency and fuel cost (132.00 × 1.44/0.72) give an annual cost for missing insulation of $264.00.

The second problem is similar to the first, except that the pipe is exposed to an average wind velocity of 10 mi/h during the year.

From Fig. 2, the wind correction factor is 2.14. Correcting the previously calculated annual cost for the uninsulated pipe in still air ($264.00 × 2.14), gives an annual cost for missing insulation of $565.00.

References

1. Crocker, S., and King, R. C., "Piping Handbook," 5th ed., McGraw-Hill, New York, 1967.
2. "Engineering Manual on Building Materials and Power Products," Johns-Manville.

The author

Rene Cordero is a piping and process mechanical equipment design engineer for Allied Chemical Corp. (Corporate Engineering Dept., P.O. Box 2105R, Morristown, NJ 07960). Holder of a B.M.E. degree from Catholic University and an M.M.A.E degree from the University of Delaware, he is a registered engineer in the State of New Jersey and a member of ASME. He is a co-author of "How to Select Insulation Thickness for Hot Pipes" (*Chem. Eng.*, July 21, 1975).

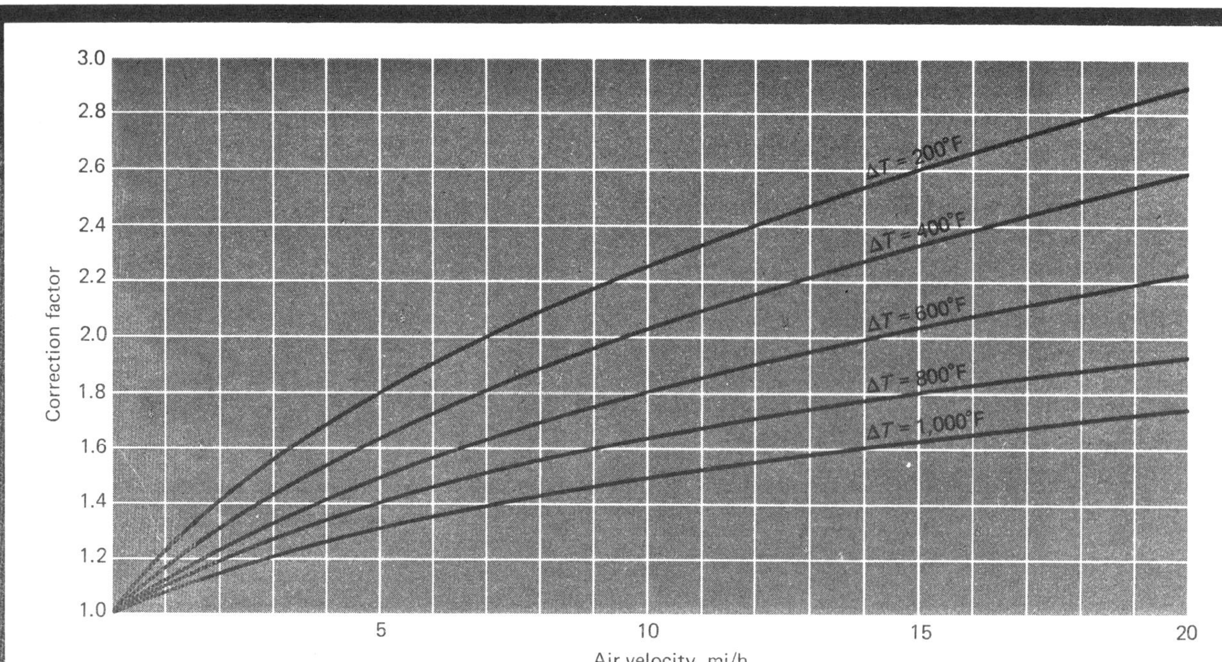

Correction for wind velocity; ΔT is difference between pipe and ambient temperature Fig. 2

Economic penalties of operating a process at reduced capacity

A method that can be made specific to a project enables calculation of reduced profitability and of price rises necessary to offset losses.

*F. A. **Holland** and F. A. **Watson**, University of Salford (U.K.)*

☐ Projects are normally started when they have the potential to be profitable. However, after the capital outlay has been committed, something may happen to disturb the projected cash flows and thereby alter the expected profitability.

Cash flows can be affected by delays in plant startup, delays in reaching design capacity, reductions in output in the interest of conservation, or for a variety of other reasons. In these cases, it is important to determine required cash flows after the disturbing events, in order to retain the original profitability set for the project.

Development of the model

A proposed project will normally be approved only if it meets the company criterion of minimum profitability. This may be the (a) payback period, (*PBP*), within which the income must be sufficient to equal the non-recoverable capital outlay; (b) the discounted cash-flow rate-of-return (*DCFRR*)*; (c) the net present value (*NPV*). The payback period is related to the life of the project and to the (*DCFRR*) [1,2]. (*DCFRR*) may be considered to be the cost of capital, i, with an extra interest charge to allow for entrepreneurial risk.

Any of these criteria, and others that are equivalent, may be chosen. For the present purpose, we will use the net present value, (*NPV*), as defined by:

$$(NPV) = (A_{CF})_O + \sum_{n=1}^{n=n} \frac{(A_{CF})_n}{(1+i)^n} \tag{1}$$

where $(A_{CF})_n$ is the annual cash flow, after tax but before provision for capital repayment in year n. The fractional interest rate chargeable on the invested capital is i/yr, based on annual compounding at the end of each year. This is used here because major payments such as taxes due are often annual, and the necessary data can be obtained from conventional balance sheets. There is a mathematical simplification if payments are assumed to be uniformly distributed throughout the year. In this case, the summation of Eq. (1) is replaced by an integration.

*An alternative and unique measure of the profitability of a project is obtained by determining the value of i that makes the (*NPV*) zero. This value is called the (*DCFRR*). It indicates that the project will make a profit over the postulated working life, provided (*DCFRR*) is greater than the cost of capital, i.

Originally published January 1977.

Possible disturbances to the original assumptions on which approval for the project was obtained fall into several categories:

- Startup may be delayed due to difficulties with labor, constructional and other materials, planning permission, and other legal interventions.
- After startup, the project may not achieve its anticipated after-tax disposable income, due to: technical or labor problems, reducing production below the design level; consumer resistance, reducing sales level and/or sales price while increasing promotional expenses; changes in the expected level of government-support grant and/or tax levels; increased cost of labor or materials without a corresponding increase in sales income.
- After startup, production may be reduced for a limited period, due to strikes, damage, restricted availability of raw materials, governmental policy, etc.
- Maximum working life of the process might be limited by availability of raw materials, technological obsolescence, physical durability of the equipment, etc. Such limits may be affected by the manner of working. For example, the working life of a mineral deposit will depend on the rate of extraction, while idle plant may rot faster than well-maintained working equipment.

Any or all of these factors can be built into a model. As an example of reasonable generality, the following assumptions will be made:

1. All expenditures prior to startup have been discounted to the end of year zero at the prevailing cost of capital, i.
2. Maximum working life for the original project is n_m years.
3. The original project provided for an immediate production rate of R_m units/yr, from which the disposable cash flow (after-tax) but before provision for plant depreciation would be $(A_{CF})_m$, \$/yr.
4. All cash flows are in "real" terms, corrected for general inflation of the currency.
5. Project operates at a production rate of R_1 units/yr for n_1 yr.
6. Project operates at R_2 units/yr for n_2 yr, and then continues at a production rate of R_3 units/yr until termination.

Necessary after-tax cash-flow ratio **Table I**

Period of reduced production, n_2, yr \ Delay in start of restricted period, n_1, yr	0	2	4	6	8	10
0	1.000	1.000	1.000	1.000	1.000	1.000
2	1.193	1.195	1.198	1.202	1.207	1.214
4	1.431	1.437	1.444	1.454	1.468	1.486
6	1.723	1.735	1.751	1.771	1.789	1.836
8	2.084	2.106	2.133	2.170	2.219	2.290
10	2.531	2.566	2.613	2.675	2.761	2.886
12	3.086	3.142	3.217	3.320	3.464	3.681
14	3.777	3.865	3.984	4.149	4.388	4.759
16	4.639	4.776	4.962	5.226	5.620	6.256
18	5.722	5.931	6.221	6.641	7.291	8.401
20	7.087	7.405	7.856	8.528	9.614	11.610
22	8.819	9.303	10.005	11.091	12.943	16.717
24	11.029	11.768	12.871	14.652	17.925	25.677
26	13.874	15.009	16.763	19.757	25.854	44.368
28	17.571	19.331	22.172	27.396	39.770	102.202
30	22.431	25.201	29.930	39.560	68.803	∞

Data for the computations leading to the cash-flow ratios in Table I and Fig. 1, and to the price ratios in Table II and Fig. 2.

Number of years for original project, n_m, yr	25
Production rate for original project, R_m, 10^6 bbl/d	4
Production rate for first period, R_1, 10^6 bbl/d	4
Production rate for second period, R_2, 10^6 bbl/d	2
Production rate for third period, R_3, 10^6 bbl/d	4
Net annual cash flow for original project, $(A_{CF})_m$, 10^9 \$/yr	2.92
Net annual cash flow for first period, $(A_{CF})_1$, 10^9 \$/yr	2.92
Net annual cash flow for second peirod, $(A_{CF})_2$, 10^9 \$/yr	0.365

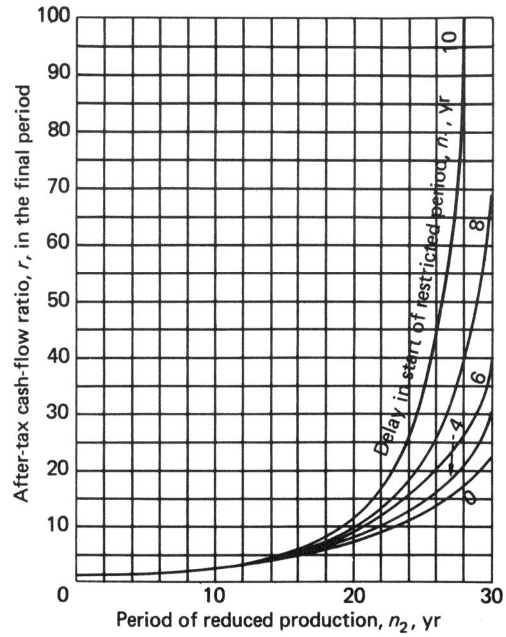

After-tax cash flow required in final period **Fig. 1**

7. Limitation on working life is due to available mineral deposits. Any other restriction may be used to determine the working life of the modified project in a specific case.

8. During the period of restriction, the annual cash flow will be $(A_{CF})_2$ and will afterwards change to $(A_{CF})_3$.

9. Effect of underusage of the installed-equipment capacity will be measured by the value of $(A_{CF})_3$, required to maintain the original profitability level as measured by the (NPV).

For the above conditions, the relationship between the net annual cash-flow ratio, $r = (A_{CF})_2/(A_{CF})_m$, and other variables may be developed as follows:

	Units
Anticipated capacity of working reserves	$= R_m n_m$
Reserves used in initial period	$= R_1 n_1$
Reserves used in restricted period	$= R_2 n_2$
Residual reserves available	$= R_m n_m - R_1 n_1 - R_2 n_2$

Working time, n_3, after restricted period becomes:

$$n_3 = \left(\frac{R_m}{R_3}\right) n_m - \left(\frac{R_1}{R_3}\right) n_1 - \left(\frac{R_2}{R_3}\right) n_2 \qquad (2)$$

From Eq. (1), we obtain the $(NPV)_O$ of the original project as:

$$(NPV)_O = (A_{CF})_O + \sum_{n=1}^{n=n_m} \frac{(A_{CF})_m}{(1+i)^n} \qquad (3)$$

For the revised project:

$$(NPV)_{Rev} = (A_{CF})_O + \sum_{n=1}^{n=n_1} \frac{(A_{CF})_1}{(1+i)^n} + \sum_{n=n_1+1}^{n=n_1+n_2} \frac{(A_{CF})_2}{(1+i)^n} + \sum_{n=n_1+n_2+1}^{n=n_1+n_2+n_3} \frac{(A_{CF})_3}{(1+i)^n} \qquad (4)$$

If $(NPV)_O$ and $(NPV)_{Rev}$ are equal, then Eq. (3) and (4) are equal, and we can write:

$$\sum_{n=1}^{n=n_1} \frac{(A_{CF})_1}{(1+i)^n} + \sum_{n=n_1+1}^{n=n_1+n_2} \frac{(A_{CF})_2}{(1+i)^n} + \sum_{n=n_1+n_2+1}^{n=n_1+n_2+n_3} \frac{(A_{CF})_3}{(1+i)^n} = \sum_{n=1}^{n=n_m} \frac{(A_{CF})_m}{(1+i)^n} \qquad (5)$$

Eq. (5) can readily be extended to cope with periods of varying A_{CF} for whatever reason such variation occurs and is, therefore, general in application. If the preferred criterion is $(DCFRR)$, then we simply replace the cost of capital, i, in Eq. (5) by $(DCFRR)$. The A_{CF} values occurring under the summation signs are mean values for the period in question but would not be expected to vary widely since major changes signal the beginning of a new period. The value of $(A_{CF})_m$ would normally have to be calculated by amortizing the income over the n_m years of the project.

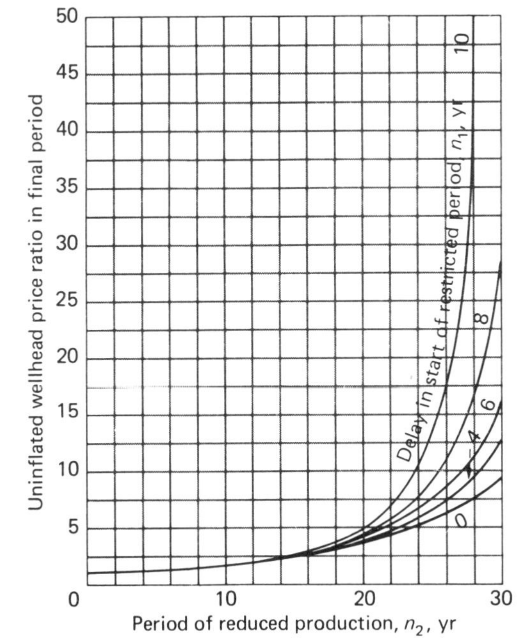

Uninflated wellhead price of oil in final period

Fig. 2

Uninflated wellhead price ratio

Table II

Period of reduced production, n_2, yr. \ Delay in start of restricted period, n_1, yr	0	2	4	6	8	10
0	1.000	1.000	1.000	1.000	1.000	1.000
2	1.077	1.078	1.079	1.081	1.083	1.085
4	1.172	1.175	1.178	1.182	1.187	1.194
6	1.289	1.294	1.300	1.308	1.319	1.334
8	1.434	1.442	1.453	1.468	1.448	1.516
10	1.613	1.627	1.645	1.670	1.704	1.755
12	1.834	1.857	1.889	1.928	1.986	2.072
14	2.111	2.146	2.193	2.260	2.355	2.503
16	2.456	2.510	2.585	2.690	2.848	3.102
18	2.889	2.972	3.088	3.257	3.517	3.960
20	3.435	3.562	3.742	4.011	4.446	5.244
22	4.127	4.321	4.602	5.036	5.777	7.287
24	5.012	5.307	5.748	6.461	7.770	10.871
26	6.150	6.604	7.305	8.503	10.942	18.347
28	7.628	8.332	9.469	11.559	16.508	41.481
30	9.572	10.680	12.572	16.242	28.121	∞

Example illustrates the use of Eq. (5)

Let us assume that the installed capacity for oil extraction from the North Sea is 4 million bbl/d while the current demand in the U.K. is 2 million bbl/d. For simplicity, let us also assume that at an extraction rate of 4 million bbl/d, the oil reserves will last 25 yr (in practice, the extraction rate will dwindle toward the end of the field's life).

It has been suggested in the interests of conservation that oil should only be extracted to meet the current needs of the U.K., which for this example may be assumed static. It has further been suggested that such conservation would be repaid by the extra revenue obtainable when world shortages further increase the oil price. Let us assume for simplicity that the current price of oil is \$10/bbl, and that the marginal tax rate is 50%. What is the required after-tax income necessary to maintain the original profitability of the project in inflation-corrected currency?

In order to solve this problem, certain other factors must be specified—particularly the profit margins and taxation practice on which the original profitability study was based. For our present purposes, let us assume that at a production rate of 4 million bbl/d the disposable after-tax cash flow represents 20% of the sales income, while at the lower rate of 2 million bbl/d, this figure is reduced to 5%, based on a wellhead price of \$10/bbl. Cost of capital is 10%.

The problem now resolves itself into determining the price to be charged at the wellhead if production at the rate of 4 million bbl/d for n_1 years is then restricted to 2 million bbl/d for n_2 years, after which the wellhead price is increased to \$$P$/bbl for 4-million-bbl/d production until the wells run dry.

To make the calculation specific, let us consider the initial period, n_1, at full production to be 5 yr, and the restricted period, n_2, to be 10 yr. Then, the estimated annual income and after-tax cash flow from the original project are:

$$(A_S)_m = (4 \times 10^6 \text{ bbl/d})(365 \text{ d/yr})(\$10/\text{bbl})$$

$$(A_S)_m = \$1.46 \times 10^{10}/\text{yr}$$

$$(A_{CF})_m = 0.20(\$1.46 \times 10^{10}) = \$2.92 \times 10^9/\text{yr}$$

Sales income and after-tax annual cash flow with

Nomenclature

A_{CF}	Net annual cash flow, \$/yr
A_{FE}	Annual fixed expense, \$/yr
A_{IT}	Annual amount of tax, \$/yr
c_S	Sales price per unit of production, \$/unit
c_{TVE}	Total variable expense, \$/unit
$(DCFRR)$	Discounted cash flow rate-of-return, yr^{-1}
i	Rate of return at the cost of capital, yr^{-1}
n	Number of years, dimensionless
(NPV)	Net present value, \$
P	Required oil price, \$/bbl
(PBP)	Payback period, yr
r	Net annual cash-flow ratio, dimensionless
R	Annual production rate, units/yr
t	Fractional tax rate, dimensionless
Subscripts	
m	Referring to original project
n	Referring to year n
O	Original project
Rev	Revised project
1,2,3	Referring to first, second, third periods

production restricted to 2×10^6 bbl/d are:

$$(A_S)_2 = (2 \times 10^6 \text{ bbl/d})(365 \text{ d/yr})(\$10/\text{bbl})$$
$$(A_S)_2 = \$7.30 \times 10^9/\text{yr}$$
$$(A_{CF})_2 = 0.05(\$7.30 \times 10^9) = \$3.65 \times 10^8/\text{yr}$$

The present example satisfies the conditions on which Eq. (2) was based. Substituting the above values into Eq. (2), we find:

$$n_3 = (1.0)25 - (1.0)5 - (0.5)10 = 15 \text{ yr}$$

We now substitute these values into Eq. (5):

$$\sum_{n=1}^{n=5} \frac{\$2.92 \times 10^9}{(1 + 0.1)^n} + \sum_{n=6}^{n=15} \frac{\$3.65 \times 10^8}{(1 + 0.1)^n} + \sum_{n=16}^{n=30} \frac{(A_{CF})_3}{(1 + 0.1)^n} = \sum_{n=1}^{n=25} \frac{\$2.92 \times 10^9}{(1 + 0.1)^n}$$

To simplify our computations, we make note of:

$$\sum_{a+1}^{b} \frac{1}{(1 + i)^n} = (1/i)[(1 + i)^{-a} - (1 + i)^{-b}] \quad (6)$$

By using Eq. (6), we reduce the numerical summations of the problem to:

$$(3.791)(\$2.92 \times 10^9) + (3.815)(\$3.65 \times 10^8) + (1.821)(A_{CF})_3 = (9.077)(\$2.92 \times 10^9)$$
$$(A_{CF})_3 = \$7.712 \times 10^9/\text{yr}$$

We can now calculate the net annual cash-flow ratio:

$$r = \frac{(A_{CF})_3}{(A_{CF})_m} = \frac{\$7.712 \times 10^9}{\$2.92 \times 10^9} = 2.641$$

Other values of $(A_{CF})_3$ or r can be calculated in the same way for other values of n_1 and n_2, and the results presented as in Fig. 1 and Table I.

$(A_{CF})_3$ represents the minimum after-tax income that will maintain the profitability of the project at the original level. The wellhead price will be dependent on tax policy. If it is assumed that all income greater than 2-million bbl/d production is taxed at 50% after expenses, an estimate of the selling price can be obtained.

In years where no capital expenditure is made, the annual cash flow may be obtained from:

$$A_{CF} = R(c_S - c_{TVE}) - A_{FE} - A_{IT} \quad (7)$$

where R is the production rate, units/yr; $(c_S - c_{TVE})$ is the contribution, \$/unit; A_{FE} is the annual fixed expense, \$/yr; and A_{IT} is the annual amount of tax, \$/yr.

If the fractional tax rate, t, is chargeable on income after allowable variable and fixed expenses, then:

$$A_{CF} = [R(c_S - c_{TVE}) - A_{FE}](1 - t) \quad (8)$$

Since the after-tax cash flows have been given for two production rates, and the tax rate is 50%, Eq. (8) may be solved to give:

$$(c_S - c_{TVE}) = \$7.00/\text{bbl}$$
$$A_{FE} = \$4.38 \times 10^9/\text{yr}$$

In order to produce a cash flow of $\$7.712 \times 10^9$/yr from a production rate of 4 million bbl/d, we substitute into Eq. (8) and find that the contribution must be increased to \$13.56/bbl at the wellhead. Thus, the wellhead price ratio is 1.656 in stabilized currency. If currency inflation is assumed constant at a modest 5%/yr (which appears acceptable in developed nations), then the asking price would be $(\$16.56)(1.05)^{16} =$ \$36.15 at the start of the final period.

If $(A_{CF})_3$ is set at some other value than that calculated in this manner, then the difference in value of the two sides of Eq. (5) reflects the gain or loss in profitability of the project. If, in the present example, the value of $(A_{CF})_3$ had been set at $\$2.92 \times 10^9$/yr by keeping the wellhead price at \$10/bbl, (NPV) would have been:

$$(3.791)(\$2.92 \times 10^9) + (3.815)(\$3.65 \times 10^8) + (1.821)(\$2.92 \times 10^9) - (9.077)(\$2.92 \times 10^9) = -\$8.725 \times 10^9$$

For the conditions of the example, it is clear from Fig. 1 that if the decision to conserve North Sea oil stocks is to be made then it should be made at once unless a maximum period of 12-yr conservation is envisaged. Such a period would require an after-tax distributable cash flow in the final period some 3½ times the present cash flow. Assuming that labor and energy charges do not rise disproportionately in the intervening period, this would require (as shown by Fig. 2) an uninflated wellhead price of about twice the current price—say \$20/bbl. If the 2-million-bbl/d rate were reached in 1980 and an immediate conservation period of 10 yr were imposed then, a wellhead price of \$16.13/bbl (from Table II or Fig. 2) would have to be agreed on from1990 forward to justify such conservation. On any other basis, the loss in (NPV), implied by a lower wellhead price, would be a disguised subsidy by the oil companies to the conservationists.

References

1. Bates, A. G., Managing Capital Expenditures, Part 3, *Hydrocarbon Process.*, Mar. 1966, p. 181.
2. Estrup, C., Evaluation of the Payback Methods by Means of the DCF Technique, *Brit. Chem. Eng.*, Feb./Mar. 1971, p. 171.

The authors

F. A. Holland is Chairman of the chemical engineering department at the University of Salford, Salford, M5 4WT, U.K. He was originally trained as a physicist and later as a chemical engineer. He holds the Certified Diploma in Accountancy and Finance of the Assn. of Certified Accountants, London, and holds degrees (including the Ph.D.) from the Universities of Oxford, Durham and London. He has held a wide variety of technical and managerial positions in his 14 years in industry (including 9 in U.S. industry).

F. A. Watson is a senior lecturer in the department of chemical engineering at the University of Salford. His primary interests are process design, digital computation and accountancy. He graduated from Manchester University in chemical engineering and has a masters degree in process economics from Salford. He is a Fellow of the Institution of Chemical Engineers, and coauthor of "Heat Transfer" and "An Introduction to Process Economics."

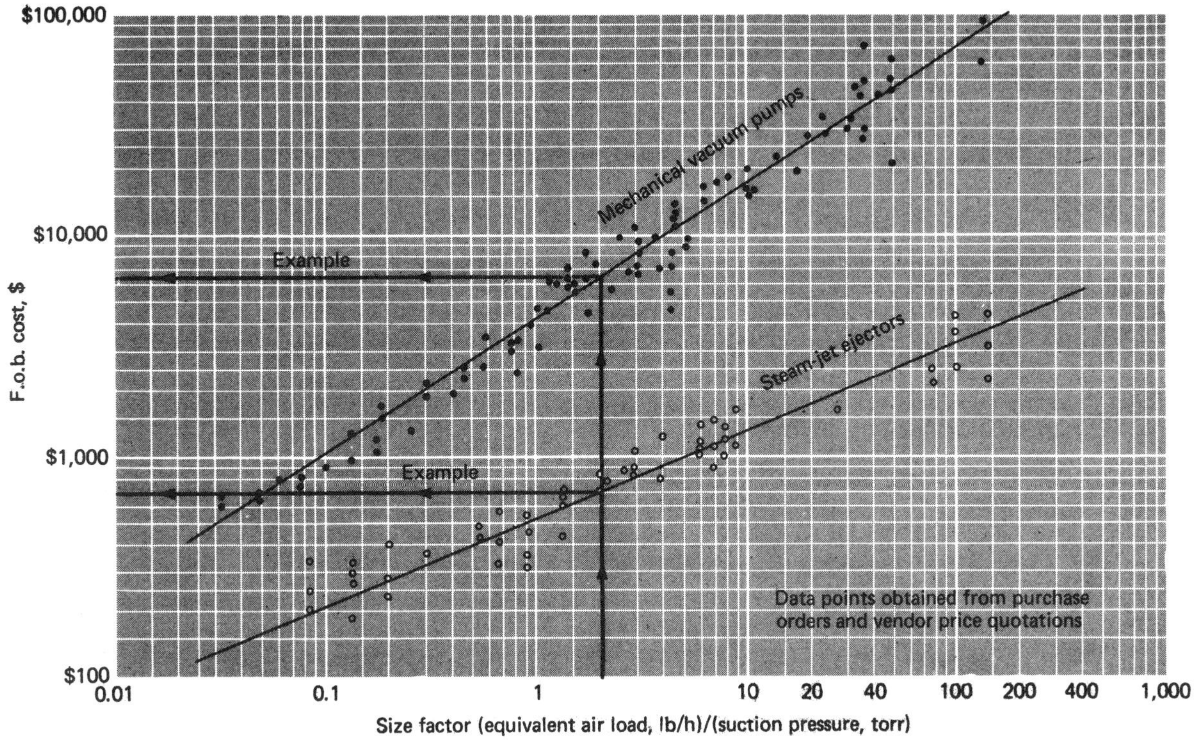

Example: How much will a stainless-steel, 3-stage ejector with 2 surface condensers for 50 lb/h air load at 25 torr cost?

From Table I, obtain the factor 2.0 for stainless steel, the factor 1.6 for 2 surface condensers, and the factor 2.1 for 3 stages. Then divide 50 lb/h by 25 torr to obtain the size factor 2. Now draw a line from size factor 2 to the steam-jet ejector line to obtain a cost of $700. Thus, the solution becomes: 2.0 × 1.6 × 2.1 × $700 = $4,700. On the other hand, a mechanical liquid-ring pump rated for 50 lb/h at 25 torr would cost $6,500. This cost is obtained by drawing a vertical line from size factor 2 to the mechanical-vacuum-pump line, and then, from this point, horizontally extending a line to the left-side vertical cost line.

Base costs for mechanical vacuum pumps and steam-jet ejectors (first-quarter 1975) **Fig. 1**

Selecting a vacuum producer

Rising fuel costs have altered the economics of choosing a vacuum producer. You can use this article and its charts to evaluate the costs of an inexpensive steam jet versus energy-efficient mechanical vacuum pumps for your application.

George A. Huff, Jr., Monsanto Co.

☐ Many chemical processes require vacuum services on distillation columns, on reactors and polymerizers, or on extruders. Until recently, engineers have tended to favor steam jets over mechanical pumps in vacuum systems.

Past experience with steam ejectors shows that they are reliable, versatile, inexpensive to purchase, and that they operate with motive steam that is readily available. However, mechanical vacuum pumps are eight to ten times more efficient users of energy than steam-jet ejectors [*1*]. Therefore, with the rising costs of fuel, it becomes necessary to reevaluate the tendency to favor steam-jet ejectors over vacuum pumps.

Although mechanical vacuum pumps require higher investment and maintenance costs than steam ejectors, the pumps' energy savings justify their economic applicability for suction pressures greater than approximately 25 torr.

If the process vapors present a waste-disposal problem, a separate economic study should be undertaken, depending on the nature of the pollutant. But even though mechanical vacuum pumps reduce the amount

Originally published March 1976.

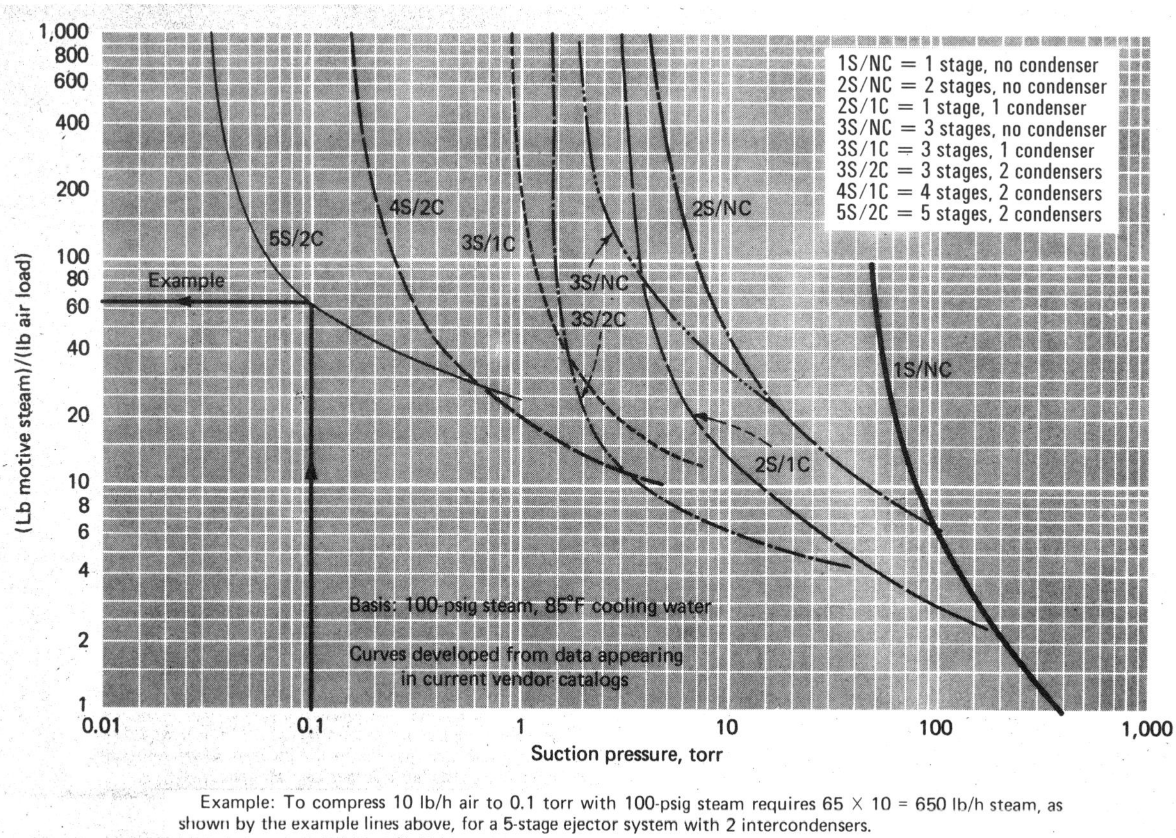

Motive-steam requirements for steam-ejector systems with and without intercondensers **Fig. 2**

of waste for treatment, they may not always provide the most economical vacuum for a system.

Cost estimates, energy requirements

A detailed study [2] of cost estimates for single and multistage steam-jet ejectors shows that purchase costs are based on a size factor, equivalent dry-air load (lb/h at 70°F), divided by the suction pressure (measured in torr). Typical data from vendor estimates and Monsanto purchase orders are found to yield a linear relationship (Fig. 1), if correction factors are applied to the material of construction, type of intercondenser, and the number of ejector stages (Table I). The data incorporated in this linear correlation have been limited to steam-jet ejector systems, operating with the suction pressures shown in Table II [3].

An analogous correlation between purchased cost and size factor has also been developed for mechanical vacuum pumps from purchase orders and vendor quotations (Fig. 1). The purchased cost of a cast iron, mechanical system includes the price of a vacuum pump, the totally enclosed fan-cooled motor, the baseplate and the belt drive. Operating ranges for the various mechanical vacuum systems considered here are shown in Table III.

Installed costs for vacuum systems are estimated from a correlation developed by W. A. Scholle of Monsanto Co. The following equation fits best for vacuum systems whose costs range from \$1,000 to \$100,000:

$$\text{Installed cost} = 55.2 \times (\text{f.o.b. costs})^{0.678} \tag{1}$$

Fig. 2 may be used to estimate the steam requirements for ejector systems, if the load consists of noncondensable vapors. For multistage steam-ejector systems

Price multipliers for steam-ejector systems **Table I**

Item	Multiplier
No condenser	1.0
1 surface condenser	2.3
2 surface condensers	1.6
1 barometric condenser	1.7
2 barometric condensers	1.3
1 stage	1.0
2 stages	1.8
3 stages	2.1
4 stages	2.5
5 stages	4.0
Cast iron	1.0
Carbon steel	1.3
Stainless steel	2.0
Hastelloy	3.0

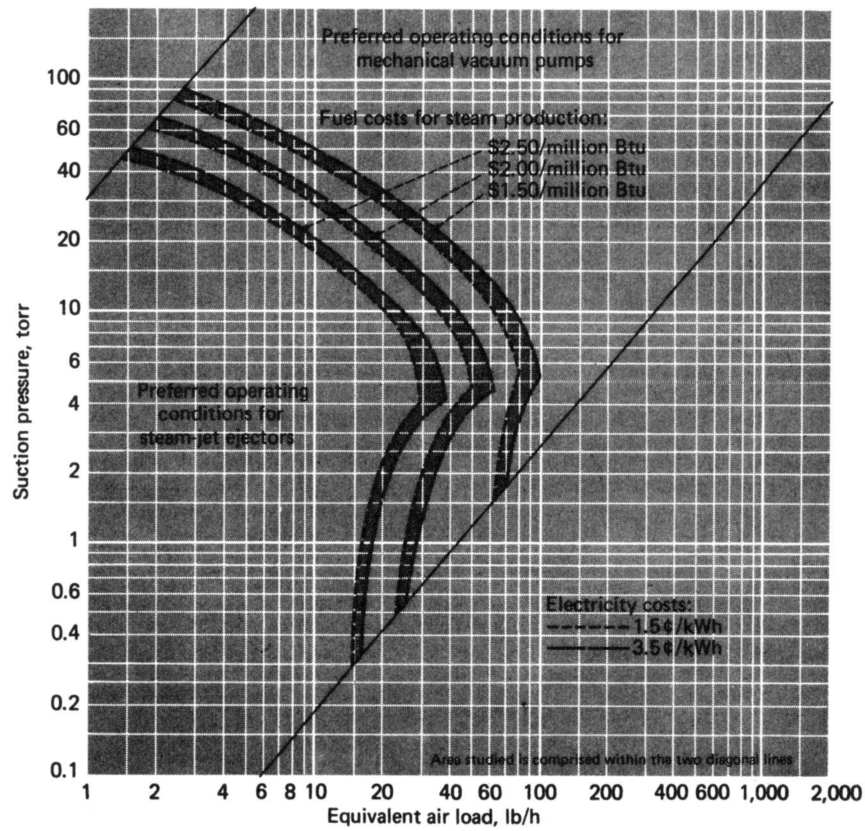

Economic comparison of mechanical and steam-jet vacuum systems at two electricity costs **Fig. 3**

Pressure ranges for standard-size ejectors Table II

Steam-ejector system	Suction pressure, torr
1 stage, no condenser	45-200
2 stages, no condenser	4-100
2 stages, 1 condenser	3-200
3 stages, no condenser	2-20
3 stages, 1 condenser	0.9-7.5
3 stages, 2 condensers	1.5-40
4 stages, 2 condensers	0.15-5
5 stages, 2 condensers	0.03-1

Operating range for mechanical vacuum pumps Table III

Mechanical vacuum system	Equivalent dry-air load at 70°F, lb/h	Suction pressure, torr	Size factor, lb/(h) (torr)
Liquid-ring	1.0-20,000	25-200	0.05-35
Air-ejector, liquid-ring	0.2-400	5-40	0.05-10
Rotary piston	0.1-2,000	0.01-760	0.03-8
Lobular, rotary piston	0.1-40	0.01-10	9.50-45
Reciprocating	1.0-7,000	15-200	1.00-25

handling condensable vapors, the steam usage will be less than indicated, because the last stages handle only the noncondensable vapors.

The following equation provides the 85°F cooling-water requirements for multistage steam-ejector systems with direct-contact condensers:

$$\text{Cooling water, gpm} = 0.611 \times (\text{lb/h steam})^{0.957} \quad (2)$$

Correlations have also been developed whereby the electrical-power requirements for mechanical systems—based on operating brake-horsepower (BHP)—may be approximated. The following equations represent such relations:

Reciprocating vacuum pumps:

$$\text{BHP} = 7.14 \times (\text{size factor})^{0.963} \quad (3)$$

Rotary piston pumps:

$$\text{BHP} = 7.63 \times (\text{size factor})^{1.088} \quad (4)$$

Lobular/rotary piston pumps:

$$\text{BHP} = 0.621 \times (\text{size factor})^{1.043} \quad (5)$$

Liquid-ring pumps:

$$\text{BHP} = 13.8 \times (\text{size factor})^{0.924} \quad (6)$$

Air-ejector/liquid-ring pumps:

$$\text{BHP} = 22.5 \times (\text{size factor})^{0.810} \quad (7)$$

Mechanical systems that operate with a liquid-ring pump also require a sealing fluid. Although various

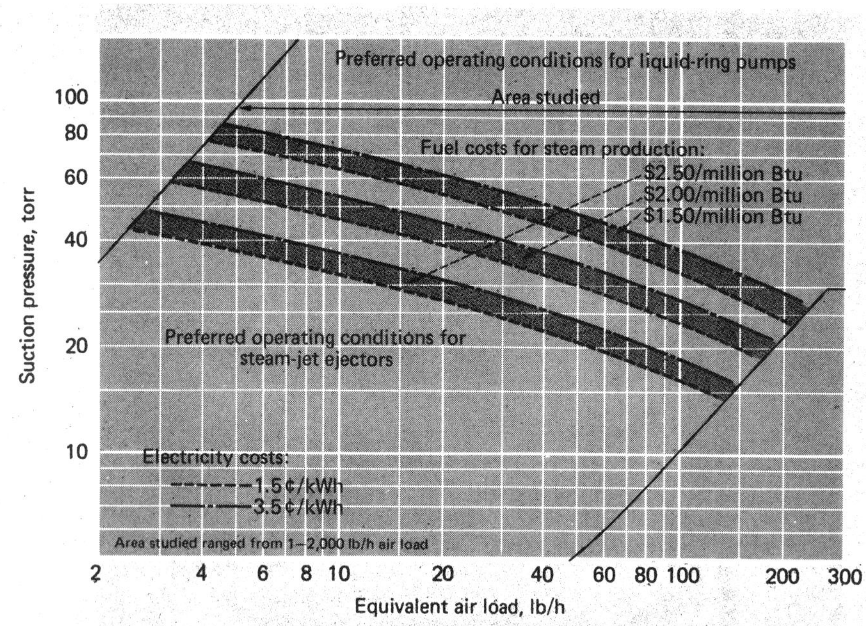

Economic comparison between liquid-ring-pump and steam-jet vacuum systems **Fig. 4**

such fluids are available, the following equation provides the pumping fluid requirements when 60°F water is used:

$$\text{Sealing water, gpm} = 7.815 \times (\text{size factor})^{0.889} \quad (8)$$

Total economic cost

The most economical steam-jet ejectors and mechanical pumps have been compared (Fig. 3) over suction pressures ranging from 0.1–200 torr, and equivalent air loads of 1–2,000 lb/h, based on a discounted-cash-flow analysis. The assumptions built into this figure are:

1—As many vacuum pumps as required are considered.

2—Yearly operating time = 8,0C0 h.

—Initial investment is made in the first year (1976) of analysis.

4—Utility costs are based on the estimated prices of purchased energy for nine years of consumption (energy costs indicated in Fig. 3 are the 9-yr arithmetic mean values over this period).

5—Standard material of construction for either steam jets or mechanical pumps is cast iron.

6—Barometric intercondensers are used on steam-ejector systems.

7—Income-tax rate = 50%.

8—Annual maintenance for mechanical vacuum systems = 9% of fixed investment.

9—Annual maintenance for steam-ejector systems = 2% of fixed investment.

10—Process vapors contain neither contaminants nor condensable vapors other than motive steam.

The figure is divided into two sections, one of them being where steam ejectors are economically advantageous; the other, where mechanical systems are more economical. As indicated on the figure, the size of these areas is fairly sensitive to energy costs.

At nearly all operating conditions, reciprocating and rotary piston pumps were found to be the most economical mechanical systems in their range of application. The irregularity in the curve dividing the two regions may be attributed to a small range where combinations of lobular and rotary piston pumps were found economical.

Liquid-ring pumps do not appear in Fig. 3 because they were never economically attractive when compared to all other mechanical vacuum pumps. On the other hand, liquid-ring pumps are reportedly more simple and reliable than other mechanical systems. Therefore, the most economical steam-jet ejectors and liquid-ring pumps are compared over suction pressures of 5–200 torr, and equivalent air loads of 1–2,000 lb/h, based on a discounted-cash-flow analysis (Fig. 4 only shows range 5-100 torr, and 1-300 lb/h).

References

1. Monroe, E. S., Vacuum Pumps Can Conserve Energy, *Oil & Gas J.*, Vol. 73, No. 5, p. 127, 1975.
2. Power, R. B., Steam-Jet Air Ejectors—Evaluation: Bids, Design, Costs, *Hydrocarbon Proc. Petrol. Refiner, Vol. 43, No. 3, p. 142, 1964.*
3. Graham Mfg. Co., "Ejector Manual" (1955).

The author

George A. Huff, Jr., held a 1975 summer staff position in the Corporate Engineering Dept., Monsanto Co., 800 N. Lindbergh Blvd., St Louis, MO 63166. He is currently a Du Pont graduate fellow at the Massachusetts Institute of Technology, working towards an Sc.D. in chemical engineering. He received his B.S. degree in chemical engineering from the University of Utah. His interests include coal gasification, catalysis, and combustion.

Designing for usable capacity, usable quality

Plant and equipment overdesign may be either an unnecessary waste of money or a wise investment. An engineer's proper assessment of a particular project will make the difference.

Martin Kladko, *Jacobs Engineering Co.* *

☐ The title of this article raises an interesting question—Are there such things as unusable capacity or unneeded quality? Yes, there are—they are called useless overdesign. Well, then, is there such a thing as *useful* overdesign? Yes, there is such a thing, too. The ability to discern between the two is often a function of the experience, judgment, and courage of the responsible project manager/engineer. In order to make the design decisions, the engineer must ask himself questions such as:

1. If the equipment were overdesigned to provide more future capacity, could the capacity be utilized at relatively little additional cost? If the answer is "yes," the overdesign may be justified. If the answer is "no," we should save our money for investment elsewhere either in this facility or some other.

2. Will a future plant expansion result in a prolonged plant shutdown, or scrapping of an existing unit? If the answer is "yes," we should identify the piece of equipment that will cause the problem and consider overdesigning it.

3. Will increased quality result in substantially reduced maintenance costs? The key here is "substantially." Obviously, it would be great to work in a plant requiring zero maintenance, but the money implications preclude any such wishful thinking. The engineer must figure his tradeoffs.

4. Will more-extreme processing conditions (higher high temperatures, lower low temperatures, higher pressures, lower vacuums, greater flowrates, etc.,) really improve the operations with respect to cycle time, yield, etc.? The engineer should perform a careful analysis of the effect of these variables on the process.

Materials of construction

One area where useless quality is built into process equipment is in the materials of construction. Particularly in the fine-chemicals and pharmaceutical industries, stainless steel is used far too often when ordinary carbon steel will do. I can accept the need for stainless steel reactors and pipelines where the product must be protected from discoloration or iron contamination. (In fact, for multipurpose reactors, I would recommend 316L stainless steel over 304, because it gives added insurance for 10–15% additional cost.) But what is the rationale for using stainless steel equipment in a solvent recovery system when the distilled solvent is stored in mild-steel tanks (and the purchased solvent is shipped in mild-steel tank cars or trucks)?

Judgment should be used in choosing which parts of the system should be stainless. There is justification for constructing the internals of the column of stainless steel, as they may be constructed of thin sections; and boiling water (in the lower section of the column) when combined with small amounts of inorganic salts can be corrosive to steel over a period of years. However, con-

* Since writing this article, the author has left Jacobs Engineering Co. See "The author" at the end of this article.

Originally published February 1976.

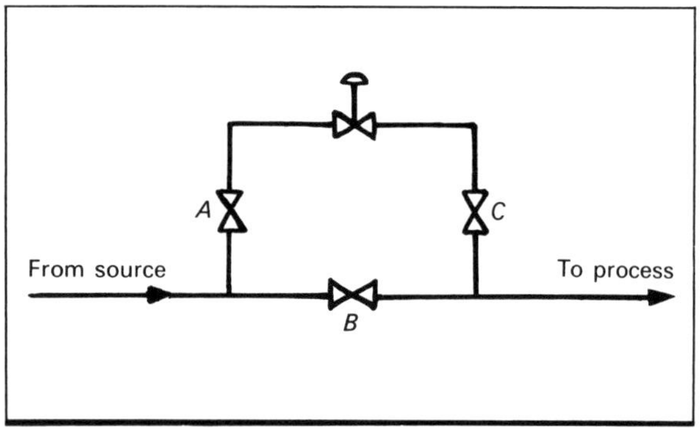

Control valves are generally piped in this way Fig. 1

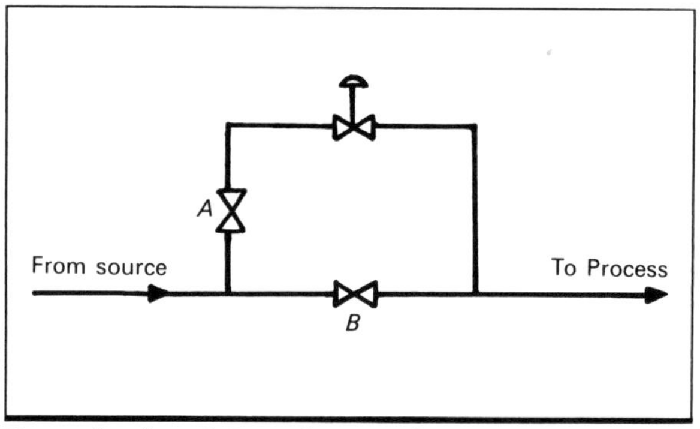

Control valve piping often used on batch reactors Fig. 2

sideration should be given to constructing the column walls out of mild steel since the walls must have considerable thickness. Definitely, the vapor riser should be steel. The condenser may have stainless tubes to minimize fouling and cleaning but the distillate lines and reflux drum can certainly be steel. The distillate pump can be stainless (or steel with a stainless impeller), and the line going to the steel storage tank should certainly be steel. If a horizontal still-pot is used as a reboiler, it would be advisable to use stainless steel for the steam tubes, but mild steel would suffice as the material for the still-pot itself.

You may disagree with some of my choices of materials. Fine. But at least I started you thinking, so that the next distillation system you design will not be all stainless steel right up to the product storage tank.

Equipment capacity

The question of where to install increased production capacity in a new plant is one that must be resolved by the project manager, requiring his experience and a feel for the future direction of the business. For example, if he is installing a system requiring three reactors that then feed to a single, large, rotary filter, it would be prudent to oversize the filter by perhaps 50% or 100%. Future increases in production requirements can usually be met relatively easily by the addition of more reactors, but the addition of a new rotary filter is a major endeavor that might necessitate a plant shutdown, unless the original design took this need specifically into account.

In a similar vein, the project manager who increases the diameter of a distillation column (to provide future capacity) without increasing the size of the condensers, reboiler, pumps, etc., is shortchanging his company because future increases in the capacity of the overall system will require substantial expenditures and production shutdown.

Pumps

In the installation of pumps, first-cost considerations will dictate the purchase of 3,500-rpm units, while the maintenance manager and production manager will opt for 1,750-rpm pumps, hoping to minimize seal wear and impeller erosion. Both sides have valid points. However, let us look a little deeper. Are there installed spares? If so, then 1,750-rpm pumps are a luxury. Does the process require the purchase of several pumps having somewhat similar characteristics? If so, it might be wise to specify all the pumps with the same characteristics, meeting the most rigorous requirement. Then you can purchase one spare for all of these pumps, go with a 3,500-rpm design, and install the spare when one pump goes down—assuming the process can be shut down (and is not equipped with installed spares). Again, to make an intelligent decision in the 1,750 vs. 3,500-rpm controversy, one must consider:

- Does the system have installed spares?
- Is it high head or low head? The cost differential increases with higher head requirements.
- Is there a spare pump in inventory that can be rapidly installed?
- Can the process be readily shut down?
- How many hours per year does the pump normally run? A transfer pump used three times a day for 1-h intervals does not deserve 1,750-rpm construction.

Whether the pumps will have usable or unusable capacity or quality will depend on how carefully the project manager weighs the above considerations.

Pipelines

The sizing of pipelines is not a simple matter that is dependent only upon design flows. The engineer who specifies utility lines (steam, water, air, chilled water) for the exact size as determined by the plant's design requirements is doing his client a disservice. New utility lines should be oversized by at least a factor of 2×, and in many cases even higher. Any new plant is bound to grow, and will add additional reactors, cooling towers, chillers, etc. To attempt to add utility lines after the plant is in operation becomes a horrendous, expensive task. Therefore, design the additional capacity into the system right from the start. This is usable capacity.

What is unusable capacity? Take as an example a steam line from a header to the jacket of a glass-lined reactor. The engineer might arbitrarily make this a 2-in line. However, if he assumed a reasonable value for both the overall heat-transfer coefficient for the vessel and the largest temperature differential that might be obtained, and thereby determined the heat-transfer

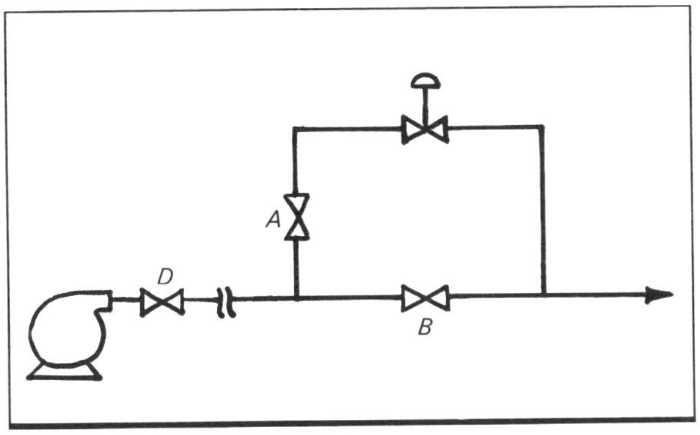

Piping may have valve at source (Valve *D*) Fig. 3

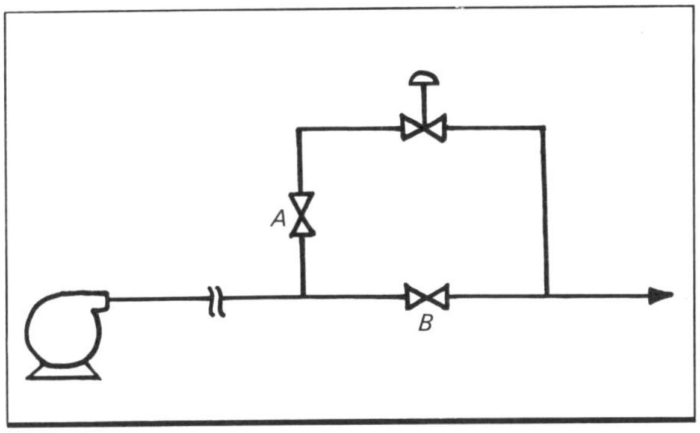

Pump itself may be locked out to isolate control valve Fig. 4

area, he could calculate the maximum heat transfer possible. If he translated this into pounds of steam, he might find out that a steam line less than 1 in would suffice. So a safe size might be 1½ in, but not 2 in. This is unusable capacity.

And speaking of pipelines, the next time you have to cut a trench across a busy plant road to bury a new line, consider putting in another line and capping it off for future use. Your boss may scream at the extra cost, but think of the hero you'll be three years later when an expansion requires that there be a new line across the road—assuming you are still there to accept the plaudits.

Control valves

One of the most obvious instances of useless overdesign is the simple control-valve station. Many engineers, when preparing a Piping & Instrumentation Diagram for a control valve, will draw it as shown in Fig. 1, without giving adequate thought to the function the various parts will play.

When questioned, any engineer can tell you the functions of the various valves. Valves *A* and *C* are used to isolate the control valve, so that it can be removed from service for repair without shutting down the process, while flow and control can be maintained through Valve *B* (a throttling-type valve). However, the next question the engineer should answer is, "How long will it be until the process shuts down?" If the process is truly continuous, such as a refinery that shuts down just once a year, then an arrangement as shown in Fig. 1 is proper. However, how many times have we seen this arrangement on a batch reactor, which turns around every 8, 12 or 24 h? For such a system, consider Fig. 2.

In the event of malfunction of the control valve, the valve should "fail safe," and the batch process can be completed by manual operation of Valve *B*. After completion of the batch, the maintenance mechanic can close Valve *A*, remove the control valve for shop servicing, plug or blank off the open ends left by removal of the valve, and turn the system back to the operator in a matter of 1–2 h. The operator can then run the next few batches under manual control with Valve *B*, until the control valve is repaired and installed at the next convenient batch turnaround. Savings: one installed valve, times the number of control stations, with no loss of operating capability. But do not stop here. Remember the left-hand side of the line—the source? Does it look like Fig. 3? Or like Fig. 4?

If it looks like either, what is the function of Valve *A*? Merely to shut off the flow so that the pipefitter can break the control-valve connection. But in Fig. 3, he can isolate the control valve with Valve *D*, and in Fig. 4, he can isolate the control valve by locking out the pump (assuming that the pumped material is not higher than the control valve so that it can run out through the pump via gravity). The point is that Valves *D* and *A* are redundant, and only one is needed (we assume that there are no branches coming off the line between these points). Again, some thought is necessary by the engineer; if the situation calls for it, the control station can look like Fig. 5.

Savings? Two installed valves, times the number of control stations.

Distillation column control

Even more refinements are possible. Take a case frequently seen in the chemical process industries, that of a continuous distillation column. In much of the CPI, particularly where solvent recovery is concerned, the distillation columns are not truly continuous. That is,

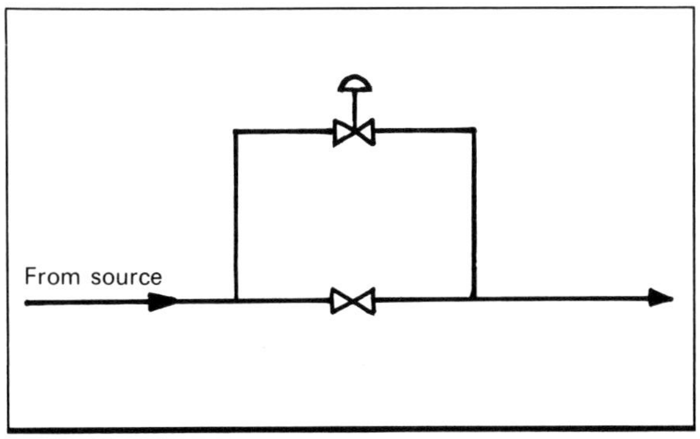

Only two valves are needed in many situations **Fig. 5**

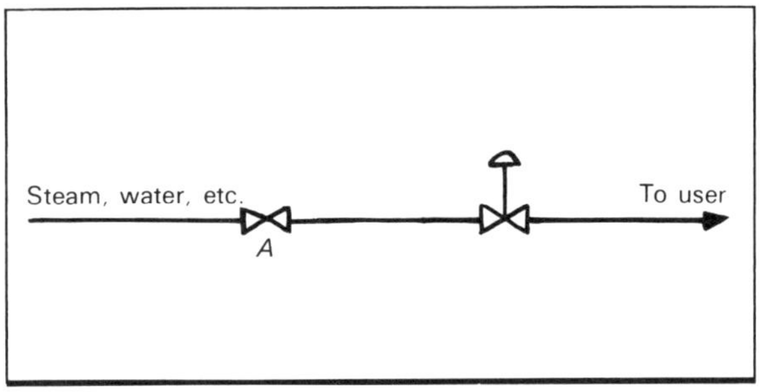

Another arrangement adequate for many lines

Fig. 6

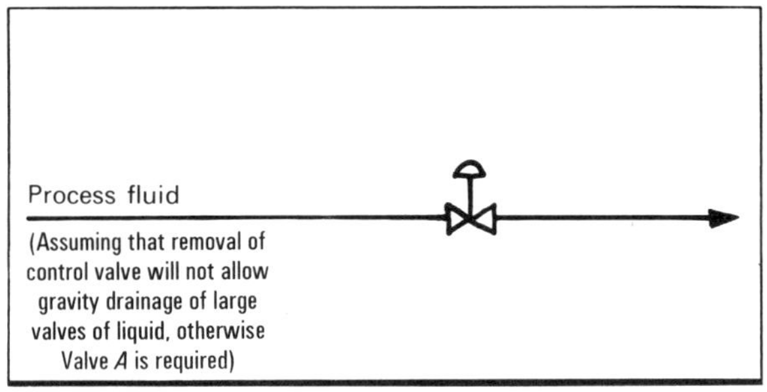

Often, the control valve itself will suffice

Fig. 7

they are not run all the time until the annual turnaround. Rather, they are run in a continuous mode for several days until the feed inventory is depleted, shut down until the feed inventory builds up, then started up again. Their operation can be described as dis-continuous.

Control valves for dis-continuous operation

Now, back to our control-valve stations. Go out to your dis-continuous distillation column and inspect the steam-control-valve station, which probably looks like Fig. 2. The control valve is probably either on steam-flow control or bottom-temperature control. If the steam valve malfunctions, do you expect an operator to sit out on the platform and operate the bypass valve manually? How will he know whether to open or close the valve? Will there be a communications link between him and the operator at the control panel watching the temperature or steam flow and advising him which action to take? Probably not. More likely, you will shut the column down, repair the valve, and then start up again.

Other automatic valves

What about all those other automatic valves, which control reflux split, reflux-drum level control, bottoms level control, etc.? Is it reasonable to assume that if the control valve malfunctioned, the operator would be able to operate the bypass valve and still perform his other duties? If the answer is "no," then Valve *B* represents useless overdesign, and the control station can be designed like Fig. 6 or 7.

Elimination of the bypass Valve *B* can save two to three installed valves per station, in addition to a substantial amount of process real estate.

The foregoing is an example of useless capacity (or capability). What about some useful capability? Well, the next time you design a utility distribution system, give consideration to putting in a few extra tees and valves (or blanks, if money is tight) for future expansion. Utilities always seem to have a way of getting themselves extended, and having the tee already there is a real bonus.

Ejectors

Design of vacuum ejectors is another area where engineers tend to specify useless capacity. Take the example of a vacuum-drying operation in which volatile solvent is removed from a filter cake, followed by the solvent being condensed in a heat exchanger by a cold-brine system, and then collected in a receiver. The design suction-pressure at the ejector inlet should be somewhat lower than the vapor pressure of the condensed solvent (the vapor pressure of the condensed solvent is a function of the temperature of the brine system in the condenser—and perhaps in the jacket of the receiver). Yet I have had Ph.D.'s try to convince me that an operating three-stage jet rated at 1.2 mm Hg shutoff, was inadequate and should be replaced by a four-stage jet. Discussions with the manufacturer revealed that the four-stage jet was rated for 0.7 mm Hg shutoff. An increase of 0.5 mm Hg vacuum will not do much for you when the condensed solvent has a vapor pressure of 10 mm Hg at the temperature it is being condensed and maintained. The urge to add another jet to increase drying rate is tempting until one actually analyzes the situation.

(Incidentally, the problem turned out to be a filter, in the vapor line, that rapidly plugged up during operation, and showed itself when we drilled and tapped a nipple in the dryer and measured the pressure drop across the filter.)

Summary

The foregoing gives an idea of how a project manager/engineer should avoid designing-in useless capacity and quality, and yet should be farsighted enough to include useful capacity and quality. In order to differentiate, the project manager should be an experienced individual, have a thorough understanding of the process and the future direction of the business, and have the courage to defend his actions against those who think he is spending too much, as well as those who think he is cutting corners.

The author

Martin Kladko has recently been appointed Manager of Engineering for E. R. Squibb & Sons, Georges Rd., New Brunswick, NJ 08903, where he has responsibility for capital expenditures for Squibb's U.S. facilities. He received his B.Ch.E. from The Cooper Union and his M.Ch.E. from Rensselaer Polytechnic Institute. He is a member of AIChE (and Secretary of the New Jersey Chapter) and the National Soc. of Professional Engineers. He is a registered professional engineer in New York and New Jersey, and has been a frequent contributor to the *Plant Notebook* department of CHEMICAL ENGINEERING.

Shall We Buy That New Equipment?

You are considering modernizing your operation by purchasing a new, but expensive, piece of equipment to replace the old—but still good—unit. What factors should you consider?

ERNEST W. FAIR, Technical Writer

When is the appropriate time to replace present equipment? Although the conventional one is when depreciation has been absorbed and useful life is nearing an end, drastic changes within an industry may make the replacement necessary much earlier.

Today, for example, the problem comes up because of the many improvements being made in every kind of chemical-plant equipment. It is not easy to answer the question "Shall we replace this old unit even though it is not fully depreciated, is still doing a good job, and still has a reasonable useful life?"

No ready formula exists to provide an answer, but here are a number of check points—obtained from the practical experience of men in the chemical process industries—that will make the decision easier.

Economic Considerations

Production Costs—Is the proposed equipment change of sufficient scope to assure a reasonable reduction in production costs? If this is so, there is a strong probability that the move will be a good one. If such savings are sufficient to permit other company improvements that are now being delayed because of budgetary reasons, any delay should be out of the question.

Resale Value—Is the value of present equipment sufficiently high on the resale market to provide a reasonable amount above the depreciated and useful life remaining? A favorable figure should be weighed as an asset in favor of change. Investigate also if the present market for used equipment is a good one, with high-level prices that may not be sustained; this could seriously affect the cost of replacement in the future.

Another factor to consider is whether industry experience with the new equipment—within one or two years—will cause a drastic decrease in the resale value of present equipment. If this is likely to happen, selling the old unit now might be the thing to do.

Availability of Funds—Are funds available to make the replacement without cutting deep into capital, reserve, or operating budget, which might affect other programs of the firm? If the funds are available, the new machine will be an asset; if they are not, the unit can be a liability.

Forced Additional Changes—Can the replacement be done without extra modifications or remodeling of other units now being used? The true cost of any replacement cannot be known without consideration of this factor.

True Cost of New Unit—What are the actual dollar-and-cents costs involved in making the replacement, other than the original purchase price? This should include freight, local moving costs (if any), hiring of outside skills to make the installation, additional or new costs involved in power or other hookups, site-preparation costs, etc.

Reflect also on whether the price of the new equipment has been highly inflated over its actual value, merely because it is new. In some instances, this factor may not be of too much importance, but if the change is not absolutely necessary, it may well wait until the price has been stabilized.

Ask yourself what will be the overall effect of a replacement on other equipment in the plant. Will the new investment be only a relatively small one leading to almost immediate major capital outlay for replacement of other units? Too often this condition is not foreseen but becomes apparent after the initial step has been taken.

Personnel Changes—Would new features of the unit require changes in personnel because of lack of background knowledge on the part of the present staff? How would this affect payroll costs?

Remodeling Old Equipment—Is it possible to remodel present equipment to add the new features at low enough cost? This may be the best step to take.

Employees' Morale—Will the change make possible very decided improvements in morale of workers involved at the particular station? When this is decisive, it

Originally published October 1973.

can mean a great deal of difference in the payroll costs.

Product Changes—Will the replacement lead to changes in product or container that could be expensive and require considerable changes all down the line? There are occasions, though, when this is a positive, rather than a negative, condition.

Weight of Equipment—When a new unit is much heavier than the old one, the costs of rebuilding plant floors and substructures to carry the additional weight needs consideration from a money standpoint, as well as others. Normally, if the unit being replaced is a very old one, the reverse is true, for all equipment has moved toward lightness in weight. However, replacement equipment often involves the consolidation of two or more old units into a single machine, so that a significant weight-increase factor can exist.

Safety Accessories—One should also investigate if accessory safety guards and other protection existing around and attached to the present unit can be transferred successfully to the new one. On a major piece of equipment, this can be a very sizable expense. On small ones, it is seldom a factor worth considering.

Performance, Competition Factors

Competitive Position—Will failure to make the replacement be a detriment to the firm's competitive position in the immediate future? Or the reverse: Can the new equipment provide system improvements that will enhance the firm's competitive position in the industry? Either one of these factors could be of major importance.

Production Capability—Will the new machine permit production that at present must be turned down or jobbed out elsewhere?

Technical Soundness—Are the improvements and changes in the new unit based on sound engineering principles and practical needs, or do they contain an element of fad or novelty?

It might also be well to find out if the efforts being made to secure the replacement are based on such sound facts, or whether they are due solely to the determined efforts of one individual. If the latter is true, it is advisable to investigate what, if any, are the personal reasons involved.

Accessories—Will present accessories be usable on the new unit, or must they also be replaced? Even if this is necessary, the change may still be worthwhile; the answer depends on the extent involved.

Versatility of Equipment—Does the new unit have the same versatility as the present equipment, or have the changes involved limited its use? Generally, added versatility is part of the improvements, but this should never be taken for granted.

You must be certain whether the unit under consideration will be adaptable to possible product changes in the future. Even when no such move is on the company's program, it is always a possibility. Should the new equipment be less versatile in this area than the old one, it may be advisable to postpone replacement.

Also consider if changes coming up in the immediate future (within the company's program) may make the new equipment obsolete at that time, and also whether the present unit can be used economically until then. It could be that purchasing the new unit would accelerate such planned changes. This, in many instances, could dictate an immediate change.

Plant-Layout Considerations—There may exist planned changes in plant layout, in which case it will have to be determined whether the new unit would be an asset or a liability to this program.

Delivery of Equipment—Is the new equipment available for immediate delivery, or will there be a long delay? The effect of a delay on operating costs, with reference to competitive conditions, could be serious.

Product Improvement—Will using the new equipment permit improvements that would be too costly, or even impossible, to adopt with the present unit? Find out also whether the improvement offered by the new machine would be of a permanent nature or only a first step leading toward later and even more important—and possibly more expensive—changes.

Obsolescence of Old Unit—Determine how much danger there is of obsolescence of the present unit. Could the sudden appearance of such a situation place the company's production department in an unprofitable position? When such possibilities exist, it is often advisable to make changes ahead of their actual need.

Reliability and Maintenance Considerations

Equipment Testing—Has the new equipment been properly field tested to make certain the claims being made for it are justified? When possible, such tests should go well beyond the laboratory stage.

Recent, thorough checkups on the present unit are also important. Are there indications of a major overhaul need that will be serious enough to justify replacement ahead of schedule?

Repair Costs—Have maintenance and repair costs been above normal on the present equipment during the last year? In most cases, this is a warning of continued high-level maintenance expenditures.

Equipment Parts—Have the parts composing the proposed new equipment been checked? Are they easily available, and are the materials they are made of as durable and maintenance-free as those on the present unit? If less durable, will the extra costs involved be justified by the improvements obtainable in making the switch?

It is never safe to assume that a breakdown will not occur, even during the first week of operation. Should replacement parts be very remote, a breakdown could very well run into major costs through downtime. It may be best to put up with the old unit until this problem is solved satisfactorily.

Maintenance Tools—Are there adequate repair and maintenance tools in stock to handle the new equipment, or will it be necessary to acquire entirely new ones at considerable expense? ■

Meet the Author

Ernest W. Fair (13900 SE Highway 212, #112, Clackamas, OR 97015) has been a business and technical magazine writer for over 35 years. During his long career, he has traveled extensively throughout the whole country. To gather material for his articles, he has interviewed countless people—having all kinds of positions—in almost every business and industrial field.

Cost comparisons of analog-control and computer-control systems

Here is an analysis for advanced control schemes in which economic justification includes factors such as fail-safe logic, onstream reliability, and supervisory control.

A. Eli Nisenfeld, Olin Corp.*

☐ The question of backup frequently occurs in cost comparisons between digital-computer and analog-control systems for processes. Including backup will increase the cost of the digital system and therefore place it at a disadvantage, because no backup is normally suggested for the analog-control scheme.

Backup considerations for digital computer systems have continued unabated for more than 12 years. The environment in which these concepts were originally developed was one of low reliability of the computer hardware. As hardware reliability grew, software complexities became the limiting factor. Today, with excellent hardware reliability and straightforward software systems having been developed for process control, the question of backup must be reexamined.

This discussion can have several orientations: direct-digital control (DDC), supervisory computer control, data-logging and alarming systems. In this article, we will discuss primarily the criteria for backing up a computer control system for supervisory control. The DDC system requires special considerations and is a subject unto itself. The question of backing up a data-logging system is so seldom raised that discussing it here is not justified.

Hundreds of advanced control systems based on analog hardware are now in operation. These systems perform functions that a supervisory computer might do. The thought of backing up these analog systems is seldom considered. Even though the risks are now no greater in a digital system than in an analog system, a proposal for a digital system generally includes funds for a measure of backup, either analog or digital. This extra cost, and it can be large even for minimum backup, can make justification of the digital system marginal. Furthermore, the increased complexity of the system, especially in the area

* The author, who is no longer with Olin Corp., now works as an independent consultant.

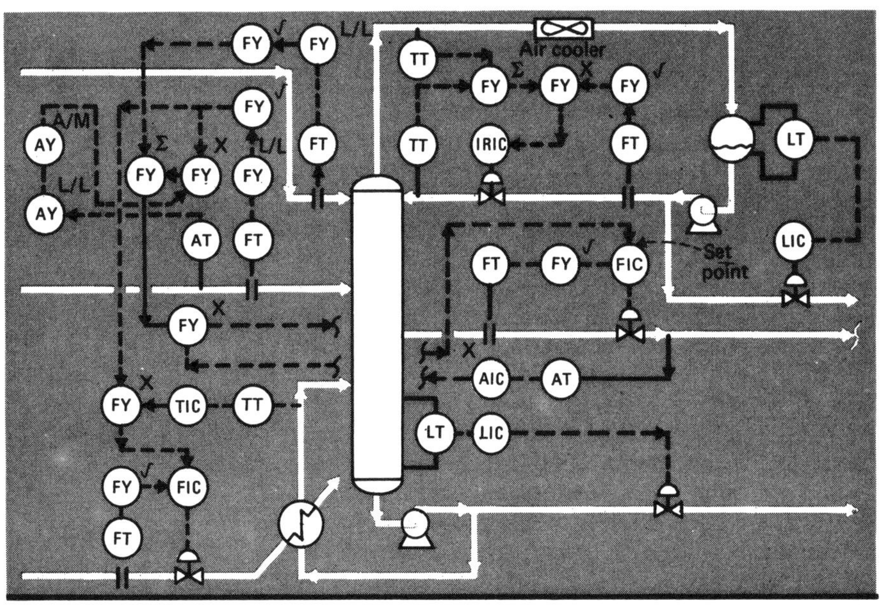

Feed-forward system coupled with analog components controls distillation column

of programming, may actually decrease the overall system reliability.

Control schemes: analog vs. digital

We can see what happens when the same standards of reliability and safety are applied, by examining an advanced control system for a train of distillation columns. In Fig. 1 (F/1), we find a distillation column with an analog feedforward system; and in F/2, the same column with the same feedforward system implemented on a digital computer. The column shown is a deisobutanizer (one of six columns in the distillation train).

Justification for the advanced control system is dramatic. The original backup philosophy for digital-computer control (in F/2) was a fail-safe system. If the computer were to fail, its outputs would be frozen, and the column transferred to manual control. Similar backup was not provided for the analog system.

On this basis, it is less expensive to implement the advanced control system by using analog hardware. In fact, it seems less expensive to implement advanced control for all six columns with analog hardware, as shown by the cost data in Table I (T/I). The backup philosophy, shown in F/2, is the one commonly used, and is probably an expensive method.

Reliability of analog-control systems

To make a fair comparison, we must question the reliability of the analog systems and the results of failure. It is apparent from inspection of F/1 that the analog system does not have the protective features of the digital system should any one of the computing components fail. There are 12 computing relays in this control system; the possibility of failure is real. Therefore, fail-safe intelligence is required for the analog system to provide the same protection as the digital ones.

Comparative costs of basic analog- vs. digital-control systems for a six-column distillation train (T/I)

Distillation column	Analog Hardware	Analog Other*	Digital Hardware	Digital Other*
1	$9,300	$3,500	$51,000	$8,000
2	3,950	1,750		1,800
3	4,300	2,200	3,100	1,250
4	4,300	2,200		300
5	3,950	1,750		300
6	4,300	2,200	2,300	450
Subtotal	30,100	13,600	56,300	12,100
Total	$43,700		$68,400	

*Other costs for analog systems result from special calibration and documentation, and for digital systems, from programming and documentation.

Note: Installation costs between the two types of systems are about equal and are not included.

F/3 shows part of one of the analog loops with fail-safe logic added. This system is much more complex and expensive than the original analog system of F/1. The cost of adding this fail-safe function to the advanced control loops on all six columns is $26,504, including engineering, and is based on the hardware additions shown in F/3. The total cost of the analog system is now $70,204, or $1,804 more than digital. Besides the basic computing relays, the analog system with backup components now has 52 separate logic elements and 8 special controllers.

Control schemes

The safety of the analog and digital control schemes of F/2 and F/3 are now almost equivalent if the logic elements are considered perfect. These equivalent systems would be a reasonable basis for a cost comparison. However, the digital system still performs functions that the analog system cannot. The digital system increases the overall reliability of the control scheme by checking the validity of analyzer signals, detects impending failures, and actually initiates calibrations.

When these features are included in an evaluation, a properly designed digital system is competitive with an analog system—maybe even for the control of a single distillation column. T/II shows a comparison of the costs and justifications for the two systems on the single column. The basic justification uses the same operating improvements, that is, increased recovery of

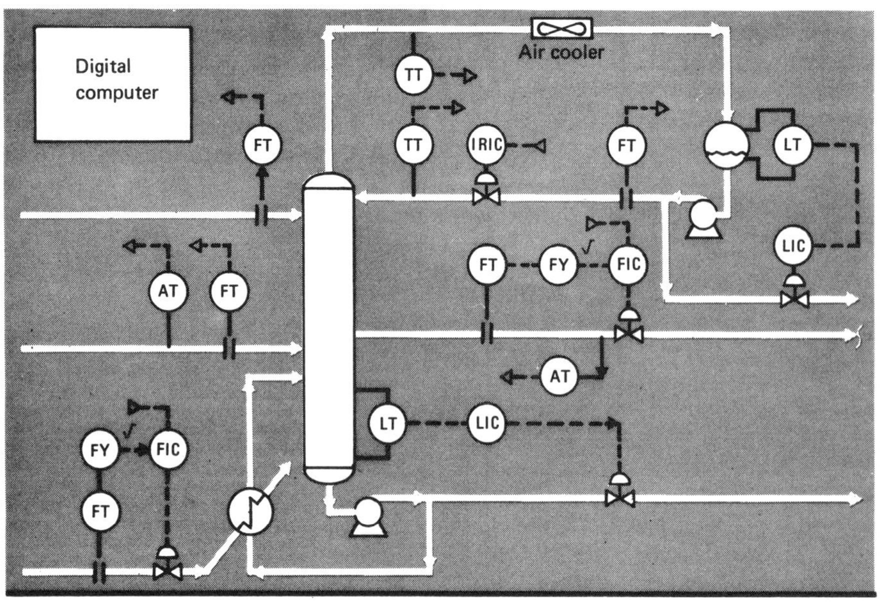

Digital-computer simulates feed-forward system and controls distillation column (F/2)

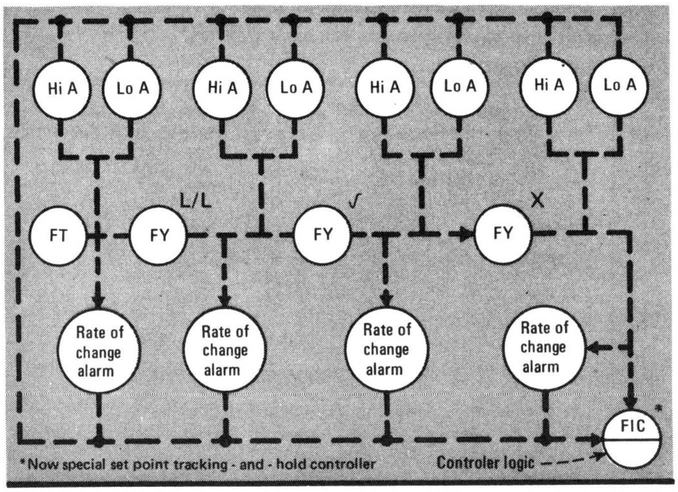

Fail-safe logic necessary to back up each analog loop (F3)

Nomenclature

AIC	Analysis, indicator/controller
AT	Analysis, transmitter
AY	Analysis, relay or computer
IRIC	Internal reflux, indicator/controller
FIC	Flowrate, indicator/controller
FT	Flowrate, transmitter
FY	Flowrate, relay or computer
HiA	High alarm
LoA	Low alarm
LIC	Level, indicator/controller
LT	Level, transmitter
TIC	Temperature, indicator/controller
TT	Temperature, transmitter

Function designations

A/M	Automatic/manual
L/L	Lead/lag
X	Multiply
Σ	Sum
√	Square root

more-valuable products and energy savings.

Justification for the analyzer validating and calibrating functions of the computer can be taken in either of two ways: manpower savings, or improvement in the onstream factor of the control system. It is difficult to estimate manpower savings, so an estimate of the increase in onstream time will be used. By monitoring analyzer performance, the computer provides for scheduling of preventive maintenance and for online calibration to improve the analyzers' onstream factor.

The onstream factor of the analog system with fail-safe logic is 0.68 (analyzer downtime is the major problem). The onstream factor for the digital system is 0.92. Here, too, analyzer downtime is the major problem—even with the improvements effected by the digital computer. As T/II shows, the return on investment of the digital system is about the same as the analog system for control of a single distillation column. If a multiple-train distillation system were being evaluated, the justification for the digital system over the analog system would be even greater, as shown in T/III.

In this analysis, justification of the computer system was based on its control functions only. Although the computer will be doing a reasonable amount of monitoring of process variables, no attempt was made to include the value of this monitoring in the cost evaluations. By removing the intangible value for data monitoring and logging from the justification, the control engineer is able to guarantee the integrity of the control functions against the computer's time-and-memory demands for logging and monitoring.

The logic presented in this article and the cost comparisons of T/II and T/III show that the cost of digital computer control is competitive with an equivalent-quality analog control.

Performance comparison of digital vs. analog control on one distillation column (T/II)

	Digital	Analog
Hardware	$51,000	$16,005
Programming and documentation	8,000	–
Special engineering, calibration and documentation	–	5,176
Total cost	**$59,000**	**$21,181**
Potential increase in annual profit	$150,000	$150,000
Onstream factor	0.92	0.68
Net improvement in profit (before taxes)	$138,000	$102,000

Performance comparison of digital vs. analog control on six-column distillation train (T/III)

	Digital	Analog
Hardware	$56,300	$51,332
Programming and documentation	12,100	–
Special engineering, calibration and documentation	–	18,908
Total cost	**$68,400**	**$70,240**
Potential increase in annual profit	$475,000	$475,000
Onstream factor	0.89	0.63
Net profit before taxes	$422,750	$299,250
Return on investment: $\frac{\text{(50\% of net return)}}{\text{(Total cost)}} \times 100$	309%	213%

The author

A. Eli Nisenfeld is an engineering consultant in process control, 355 Purdy Hill Road, Monroe, CT 06468. Previously, he was manager of the control and statistics group at Olin Corp. He also worked for a large engineering contractor for five years, and spent an additional three years with a major instrumentation firm. Mr. Nisenfeld has a B.A. and the B.S. in chemical engineering from Drexel University, and is a member of AIChE, Instrument Soc. of America, and the International Federation of Automatic Control.

New Method Gives Quick, Accurate Estimate of Distillation Costs

With these new correlations, requiring a minimum of input data, you can develop a quick, order-of-magnitude estimate of investment and operating costs for distilling organic chemicals.

ROBERT F. SOMMERVILLE, Sun Oil Co.

A number of rigorous methods are available for estimating the investment necessary for distillation processes.[1,2,3] But in some instances, such lengthy and often costly approaches may not be required.

There is a need for a quick, but accurate, estimating technique for making rough checks on cost estimates prepared by others, or for preparing preliminary economic evaluations for new ventures where a number of potential projects must be screened quickly and with reasonable accuracy.

This article will describe a new technique that I believe meets these requirements.

Two-Variable Method

As many variables as possible have been eliminated consistent with obtaining order-of-magnitude accuracy: my method is based on only two inputs: (1) number of actual plates, and (2) charge rate. The more accurate the estimate of the number of plates, the more valid the correlation becomes. It is, then, very desirable to have a rigorous design method for calculating the number of distillation trays.

If this is not possible, a simplified technique for estimating the number of actual trays can be used. This method is based on the minimum number of theoretical plates as calculated using the Fenske[4] equation and an empirically derived multiplication factor for converting this figure to the number of actual plates.

The Fenske equation for minimum number of theoretical plates (occurring at total reflux) is:

$$N_{TM} = \frac{\log [(X_{LK}/X_{HK})_D (X_{HK}/X_{LK})_B]}{\log (\alpha_{LK/HK})}$$

Before using this correlation, a binary system must first be identified. For more than two materials, (usually the case), the two key components have to be established. For our purposes, the key components are the principal compounds between which the "split" is made.

Once the appropriate binary system is established (together with the mole fractions of the keys in the distillate and bottom products) it is necessary to calculate relative volatility, α. Assuming ideality:*

$$\alpha = p_{LK}/p_{HK}{}^{**}$$

For correlating purposes, relative volatility has been estimated by calculating p_{LK}/p_{HK} at the average of the reboiler outlet and overhead vapor temperatures.

Now it is possible to calculate the minimum number of theoretical plates using the Fenske equation. But, of course, what is needed is the number of actual plates for a specific separation. In equation form, the problem is:

$$N_A = F(N_{TM})/E = Y(N_{TM})$$

where N_A = number of actual plates.
F = Multiplication factor
E = Overall plate efficiency
N_{TM} = Minimum number of theoretical plates
$Y = F/E$

If a sufficient number of actual engineering designs ex-

*Many multicomponent systems of industrial importance may be considered nearly ideal for practical purposes. This is particularly true for hydrocarbon mixtures of the same homologous series, such as the paraffin series or the low-boiling aromatic hydrocarbons. Activity coefficients must be used to correct for nonideality, but use of activity coefficients is beyond the scope of this article.

**Personnel involved in engineering design work generally calculate α by using equilibrium K values. However, K values, which yield more accurate α's, are not as readily available as vapor-pressure data.

Originally published May 1972.

ist where number of actual plates, overall plate efficiency, and minimum number of theoretical plates are known or calculated, it is possible to get an average value for F. Unfortunately, the overall plate efficiency is generally not specified in published estimates, although approximate methods such as by O'Connell[5] are available. One of the inputs to O'Connell's method is the viscosity of the column feed; and in many cases, this is difficult to determine. Because of the problems in defining the overall plate efficiency and, hence, deriving empirical values for F, I have concentrated on determining empirical values for the ratio Y.

I was able to obtain what appears to be reliable design data from several internal and external sources for 18 distillations carried out at ambient temperatures or above and at moderate pressures. The separations include several organic compounds such as paraffins, olefins, aromatics, alcohols, ketones, acids, and miscellaneous categories. Because these estimates are purported to be final designs, I have assumed they are optimum. The average value of Y for the 18 cases was 3.0 (standard deviation = 0.6). Fourteen cases fell within ±25% of the mean.

Nomenclature

B	Bottoms product
D	Distillate product
E	Overall plate efficiency
F	Multiplication factor used in number-of-plates correlation
HK	Heavy key component
K	Equilibrium value
LK	Light key component
N_A	Number of actual plates
N_T	Number of theoretical plates
N_{TM}	Minimum number of theoretical plates
RR	Reflux ratio
S_T	Number of theoretical stages
S_{TM}	Minimum number of theoretical stages
p	Vapor pressure
X	Mole fraction in liquid
Y	$Y = F/E$; Used in number-of-plates correlation
α	Relative volatility

Literature Support for Multiplication Factor

According to an analysis by Happel,[6] the economically optimum reflux ratio often is only a few percent greater than the minimum, although total cost is not increased greatly by a moderate increase in reflux ratio (up to 1.5 times the minimum). Happel further notes that the optimum number of trays for a wide range of conditions varies from 2.0 to 3.0 times the minimum number. He suggests that a convenient, rapid approximation is to multiply the minimum number of theoretical trays derived from Fenske's equation by 2.5 to obtain optimum number of theoretical trays.

As a further comparison, the Gilliland[7] correlation may be used. This equation presents a relationship between the number of theoretical stages, S_T, and reflux ratio. Most contract-engineering firms consider the optimum reflux ratio as about 1.3 times the minimum reflux ratio. At this reflux ratio, the abscissa of Gilliland's correlation nearly always lies between 0.04 and 0.22. The ordinate $(S_T - S_{TM})/(S_T + 1)$ would thus fall between 0.45 and 0.60. Within this range, the number of theoretical plates, N_T, varies between 1.8 and 2.5 times the minimum number of theoretical plates, N_{TM}.

In a recent article, Van Winkle and Todd,[8] using proprietary cost information, determined the optimum fractionation design for 12 feed mixtures comprised of paraffins and olefins in the C_4–C_6 range. The average ratio for N_T/N_{TM} for the 12 cases was 1.75, with individual values ranging from 1.5 to 2.2. The corresponding average optimum reflux ratio was 1.5 times the minimum reflux ratio (individual values varied from 1.25 to 1.7).

According to my correlation, the average number of actual plates is 3.0 times the minimum number of theoretical plates.

For those 15 cases where I could calculate overall plate efficiencies using O'Connell's method, I obtained an average value of about 0.65. This suggests that the number of actual theoretical plates would be about 2.0 times the minimum number of theoretical plates. This agrees with the number derived from the Gilliland correlation at 1.3 times the minimum reflux ratio, is somewhat less than that suggested by Happel and is slightly higher than that suggested by Van Winkle and Todd.

Vacuum- and Low-Temperature Distillations

As an interesting sidelight, 11 additional cases involving vacuum distillations and 3 involving subzero temperatures were examined. In these situations, average values of Y of 2.5 and 3.5, respectively, were obtained. Such results are logical because the normal approach when operating a fractionation unit under vacuum is to reduce the number of plates (hence, increasing reflux ratio) to minimize column pressure drop.

Conversely, in distillations at subzero temperatures, one normally maintains very close tray spacing and reduces the reflux ratio to keep refrigeration costs down. By so doing, the number of plates to achieve a desired degree of separation is increased appreciably. Typical examples of subzero distillations would be methane/ethane and methane/carbon monoxide.

Investment Estimate

Even though we now have an easy-to-use correlation useful for estimating the number of plates required to achieve a separation of two key components, it is of little value unless it can be translated into investment costs. To solve this problem, I have prepared a second correlation (see figure, p. 474), based on investment estimates from several internal and external sources. The only inputs required are charge volume per unit time, and number of actual plates.

The data presented represent the mean of 20 investment estimates for the fractionation of various organic mixtures. It should be noted that at a charge rate of 100 million lb./yr., 14 of the 20 estimates fall within ±25% of the mean (standard deviation = 26%). This set of data is independent of that used to estimate number of plates (there are no data points in common). Therefore, its use is not dependent on the previous correlation for estimating the number of plates. If the number of trays is derived

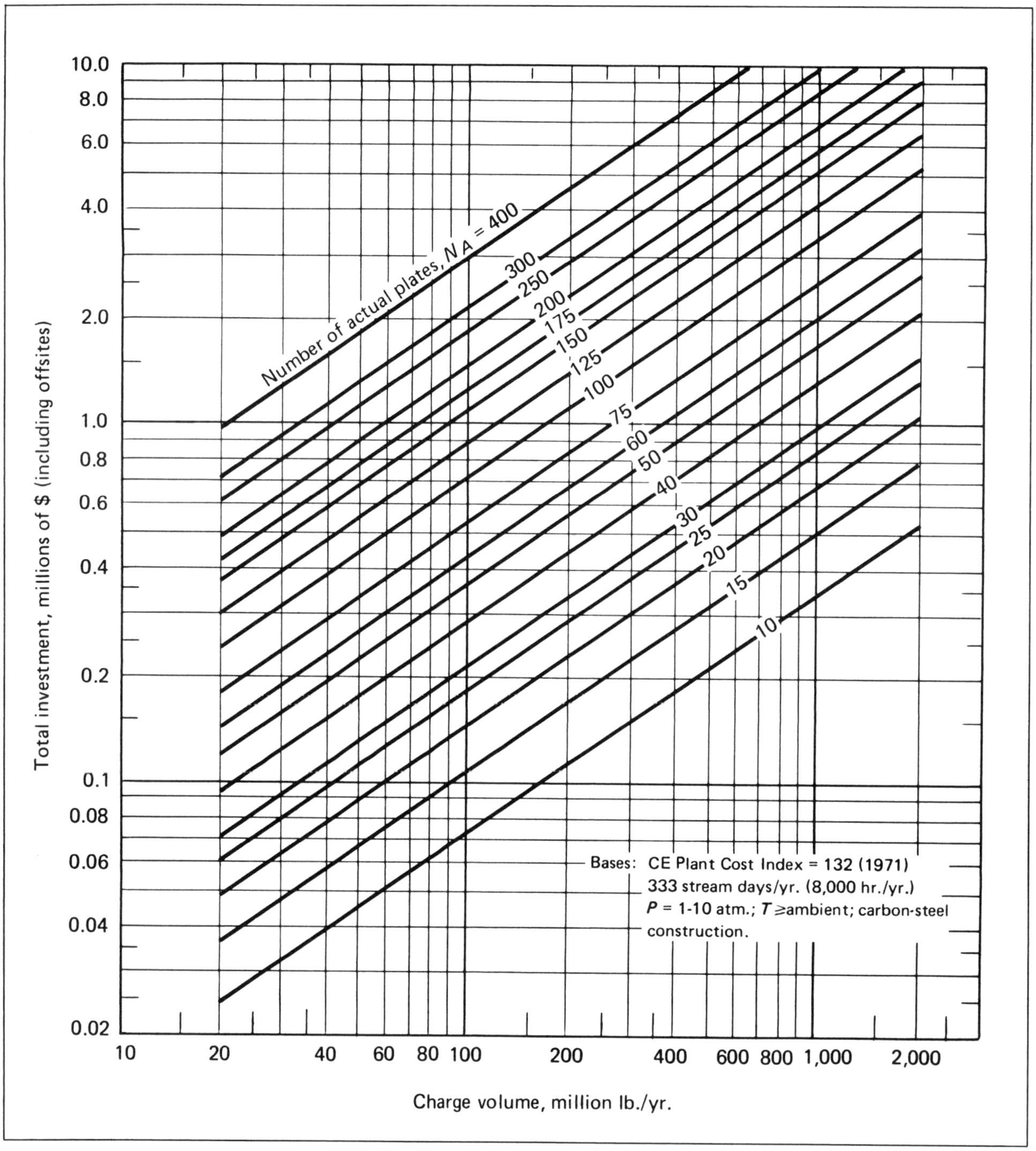

TOTAL INVESTMENT is a function of charge rate in this correlation for organic-chemical distillations.

from another source (such as from a rigorous design), the cost chart will still be applicable.

The investments reported in the chart above include (1) the tower, its installation, foundations, structures, piping, and instrumentation, (2) auxiliary equipment such as pumps, condensers, reboilers, preheaters and accumulators, (3) indirect engineering and construction costs, (4) offsite facilities such as raw-material and product storage, shipping facilities, and essential services (utilities, sanitary, etc.). Table I presents a more detailed listing of the cost factors included.

Correction Factors

In the correlation, an offsite investment equal to 35% of the onsite investment was assumed in all cases. This is a typical value for additions to existing plants. The factor for a well-integrated facility could be substantially less, whereas for grass-root facilities, it would probably be about 100%. If your situation differs from the 35% figure, the correlation, of course, must be adjusted accordingly to reflect the difference.

The investment correlation is based on an operating

Cost Items Included in Investment Estimates
Table I

Onsite Investment	
Major equipment:	
Towers	Accumulators
Reboilers	Pumps
Condensers	Miscellaneous equipment
Preheaters	
Labor to install major equipment	
Other items:	
Piping	Stacks
Electrical	Grade, pave, fence
Instruments	Start-up
Foundations	Painting
Buildings	Scaffolding
Structures	Dismantling
Fireproofing	Relocations
Insulation & jacket	Temporary & incidental
Sewers	
Field distributables:	
Supervision & overhead	Other (tools, temporary facilities, etc.)
Construction equipment	
Owner's inspection	
Home office costs:	
Design & drafting	
Other (eng., etc.)	
Offsite Investment	
Utility resources (steam, electricity, water, etc.)	
Storage tanks (raw material, intermediate, & products)	
Loading & shipping facilities (warehouse, spur lines, etc.)	
Change houses, cafeteria, etc. (an allocation may be charged).	

pressure of 1 to 10 atm. and carbon-steel construction. If either of these variables differs significantly in your case, a correction factor must be incorporated.

As a "rule of thumb," I would suggest adding 10–15% to the total investment cost (including offsites) of a comparable all-carbon-steel facility if 410 stainless steel were used: (1) to line the column and pumps and seals, and (2) as the only material to construct the trays as well as the tubes in the bottoms heat exchanger and fired heaters. Likewise, for 304 or 316 stainless steel, I would add 20–25% to the total investment cost given in the chart. It should be noted that these correction factors assume that no alloy is used in the instrumentation, piping (other than in close proximity to the column or other alloyed major equipment), or offsite facilities.

At a column pressure of about 350–400 mm. Hg, I would suggest adding about 20–25% to the total investment (including offsites) for a comparable atmospheric unit. For full vacuum operations (50–100 mm. Hg), the total investment (including offsites) for a given number of plates could easily run three times the cost obtained from the chart. Thus, a 20-plate distillation column operating at 75 mm. Hg with a charge rate of 165 million lb./yr. might cost about 3 × 200,000, or $600,000. Conversely, for a fractionation carried out at about 200 psig., total investment would probably be about 5–10% higher than that at atmospheric pressure. Usually, this pressure is not exceeded unless the column is operated at less than ambient temperature. (A word of caution at this point. The correction factors for vacuum operations and alloys recommended above are based on limited data. If a vacuum or alloy estimate is critical in your own particular economic study, I would suggest that you contact your company's experts for guidance.)

The effect of temperature is not considered because most distillations are performed at moderate temperatures. The exception is the subzero temperature operation where costly refrigeration facilities are required. I was unable to find sufficient empirical data to correlate the effect of subzero temperatures on fixed capital. However, Zevnik and Buchanan[9] presented such a relationship.

Limitations of Method

As noted earlier, this approach is not intended for detailed engineering design work. Rather it provides data for preliminary economic evaluations of proposed new ventures or for checking estimates prepared by yourself or others. The advantages of the correlations are in their simplicity coupled with order-of-magnitude accuracy. To achieve this objective, certain assumptions and omissions had to be made.

Errors in estimating actual number of plates might result from an inaccurate estimate of the relative volatility (possibly because of a large deviation from ideality), or if the overall plate efficiency differed considerably from the "average" value.

The applicability, nonapplicability, and possible sources of error associated with the investment correlation are:

- The chart may be used for essentially all organic chemical fractionations (including petrochemicals) in which the number of actual plates is known or has been estimated (from the first correlation or otherwise). It is based on ambient temperatures or above, moderate pressures, and carbon-steel construction. As discussed previously, variations in any of these items must be handled via the use of correction factors.
- The chart may not be used for other separation columns such as extraction or stripping, or for distillations carried out at temperatures below ambient (because of the absence of empirical data).
- Some errors might result because effects of (a) reflux ratio, (b) type of tray, tray spacing and tray configuration, (c) wall thickness and (d) column diameter (possibly different in enriching and exhausting sections) have not been considered in developing the correlation. Further, if the first correlation is used to estimate the number of actual plates, any error in this determination could carry over. This particular error could either partially offset or add to any error inherent in the chart.

Even with these limitations (the consideration of which would cost time and money), I feel that the correlations presented can be very useful.

Accuracy of Investment Correlation

To obtain an unbiased check on the accuracy of the method, I consulted with Sun Oil's Engineering Dept. after I had plotted and analyzed the data in my posses-

Check the Method Against Actual Data

Let us examine the economics of recovering Aromatic D[a] (98% purity). Pertinent details are:

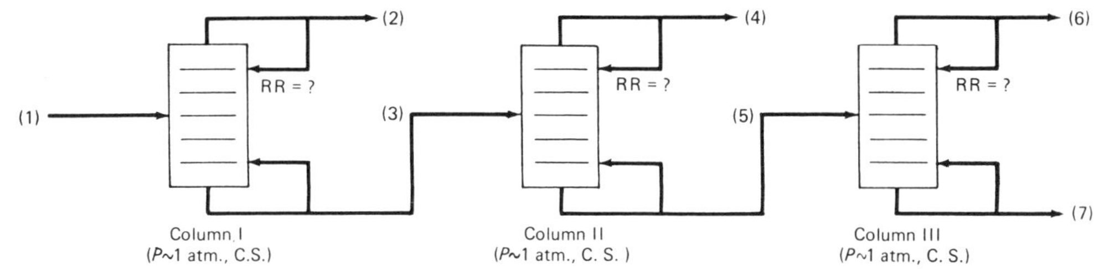

Streams; Flowrates, Million Lb./Yr.

Component	(1)	(2)	(3)	(4)	(5)	(6)	(7)
A (Low boil. fract.)	2.4	2.4					
B	1.9	1.8	0.1	0.1			
C	0.7	0.4	0.3	0.3			
D	67.0	1.0	66.0	49.8	16.2	16.1	0.1
E	15.6		15.6	0.4	15.2	15.0	0.2
F (Polymer)	0.8		0.8		0.8	0.1	0.7
Total	88.4	5.6	82.8	50.6	32.2	31.2	1.0

Relative Volatility Calculations

	Vapor Pressure,[b] p, MM.Hg	$\alpha = p_{LK}/p_{HK}$
Column I		
Key component, B[c]	1,050	1.6
Key component, D	640	
Column II		
Key component, D	860	1.3
Key component, E	660	
Column III		
Key component, E	?	4 (guess)
Key component, F	?	

Calculation of Number of Actual Plates

Fenske's equation:

$$N_{TM} = \frac{\log\left[(X_{LK}/X_{HK})_D\,(X_{HK}/X_{LK})_B\right]}{\log(\alpha_{LK/HK})}$$

Column I

$$N_{TM} = \frac{\log\left[\frac{1.8(66.0)}{1.0\,(0.1)}\right]}{\log 1.6} = \frac{\log 1{,}190}{\log 1.6} = \frac{3.08}{0.20} = 15.4$$

No. of actual plates ~3.0 (15.4) ~46[d]

Column II

$$N_{TM} = \frac{\log\left[\frac{49.8(15.2)}{0.4(16.2)}\right]}{\log 1.3} = \frac{\log 117}{\log 1.3} = \frac{2.07}{0.11} = 18.8$$

No. of actual plates ~3.0 (18.8) ~56

Column III

$$N_{TM} = \frac{\log\left[\frac{15.0(0.7)}{0.1(0.2)}\right]}{\log 4.0} = \frac{\log 525}{\log 4.0} = \frac{2.72}{0.60} = 4.5$$

No. of actual plates ~3.0 (4.5) ~14

Capital Investment Estimate

The estimated actual plates required and annual charge rates were used as inputs to the chart to yield capital-investment estimates. The results are:

	Column I	Column II	Column III	Total Process
No. of actual plates (N_A)	46	56	14	116
Fixed capital, thousands of $ (1971-CE Plant Cost Index = 132)	300	350	42	690

Venture Economics

The economics associated with this Aromatic D venture may now be estimated using the available data:

	¢/Lb. of Aromatic D	
Feed cost = 1.75 lb./lb. of Aromatic D @ 2¢/lb.[e]	3.50	
By product credit:		
Lower boilers = 0.11 lb./lb. of Aromatic D @ 1¢/lb.[e]	−0.11	
Higher boilers = 0.64 lb./lb. of Aromatic D @ 2¢/lb.[e]	−1.28	
Net charge cost	2.11	
Utilities = 0.043¢/lb. of charge to each column	0.17	
Chemicals & catalysts	0.00	
Total variable costs		2.28
Labor-related cost = $40,000/(yr.)(column)	0.24	
Maintenance = 4% of investment/yr.	0.05	
Taxes & insurance = 2% of investment/yr.	0.03	
Depreciation = 10% of investment/yr.	0.14	
Total fixed costs		0.46
Total mill cost		2.74
Sales cost ($250,000/yr.)[f]		0.49
Net income required for 25% ROI (before taxes); includes working capital[g]		0.55
Required Aromatic D selling price, Fob. works in bulk		3.8[h]

[a] Actual compounds cannot be identified because of Sun Oil security considerations.
[b] Vapor pressure at average tower temperature.
[c] Desired split is between (B) and (D) rather than (C) and (D).
[d] Use of (C) and (D) as key components would have yielded an N_A of 53.
[e] Internal transfer price.
[f] Estimate.
[g] Working capital (2.5 months of cash costs excluding sales expense, depreciation, and by-products credits) = $420,000.
[h] A change in investment of ±50% would change the required selling price by ±0.3¢/lb.

Check of Estimate

In 1964, Sun Oil's Engineering Department prepared a process design and onsite investment estimate for the case just cited, i.e., the recovery of 50 million lb./yr. of 98% purity Aromatic D. In the following table the Engineering Department's estimated (updated to 1971 [CEI = 132] with a 35% offsites allocation added) are compared to the result obtained using the correlations given in this article.

	Column I		Column II		Column III		Total Process	
	No. of Actual Plates	Fixed Capital Thousand $	No. of Actual Plates	Fixed Capital Thousand $	No. of Actual Plates	Fixed Capital Thousand $	No. of Actual Plates	Fixed Capital Thousand $
Eng. Dept. estimate, E_1	60	N.A.	60	N.A.	16	N.A.	136	680
Estimate from article, E_2	46	300	56	350	14	42	116	690
Difference, $E_1 - E_2$	+14	N.A.	+4	N.A.	+2	N.A.	+20	−10
% Diff., $E_1 - E_2/E_2$	+30	N.A.	+7	N.A.	+14	N.A.	+17	−1

sion. They provided nine independent cost estimates prepared in the course of their regular design and estimating activities. A comparison of the results from the Engineering Dept. studies and those derived from the chart are given in Table II.

It should be pointed out that three cases involved column diameters of 19 to 20 ft. These diameters are considerably higher than are normally encountered in organic distillations. In at least two of the cases, reflux ratios of 33 to 1 were used.

Operating Costs

To provide additional guidance in preparing preliminary economics for distillation operations, I collected published operating cost data for several organic distillations. This information has produced the following costs bases:

Labor-related costs

Operating labor—$\frac{1}{3}$ man/column for charge rates in the 25- to 100-million-lb./yr. range; $\frac{1}{2}$ man/column in the 100–500-million-lb./yr. range. (There is little variation with throughput).

Supervision—20% of operating labor.

Direct employee benefits—40% of operating labor + 20% of supervision.

Laboratory—30% of operating labor.

General overhead—85% of operating labor and supervision.

Thus, assuming an average operator wage rate of \$4.50/man-hr., the total labor-related cost for a single column would range from about \$40,000 to \$60,000/yr. in most instances. This labor deployment is based on an integrated facility where the distillation section represents only a part of the total operation. The labor requirement for a segregated distillation unit would probably be higher.

Cost of utilities

Expressed in terms of ¢ per weight of column charge, the average cost of utilities for 11 "typical" organic chemical distillations examined was 0.043¢/lb. (standard deviation = 0.015). Individual values ranged from 0.026 to 0.065¢/lb. Of course, this value is dependent on reflux ratio, thermal condition of feed, etc.; but for preliminary estimates the average cost can probably be used.

Two special cases (not included in the above average determination) were considered. These involved columns with reflux ratios of approximately 100 to 1, with 300 to 400 trays each. These probably represent a maximum utility cost for an organic distillation. In both instances, the total cost of utilities came out to about 0.15¢/lb. of column charge.

Comparing Accuracy of New Estimating Method—Table II

Case No.	No. of Columns	Total No. of Actual Trays	Eng. Dept. Investment Estimate Including Offsites (1971), E_1, Million \$*	Investment Estimate from Chart (1971), E_2, Million \$	% Diff. $(E_1 - E_2)/E_2$
1	1	132	\$4.9	\$4.0	+23
2	3	360	7.2	5.7	+26
3	1	90	2.1	2.3	−9
4	1	60	1.7	2.0	−15
5	1	64	0.97	1.0	−3
6	1	50	1.3	1.9	−32
7	1	60	3.0	4.6	−35
8	2	240	5.1	5.2	−2
9	3	280	7.2	6.2	+16
				Average =	18

*Essentially atmospheric pressure and carbon-steel construction. These estimates have a reported accuracy of −10% to +20%; that is, a plant estimated to cost \$3.0 million might actually cost between \$2.7 and \$3.6 million. A 35% offsites allocation was assumed in each case.

The utility costs cited in this article are based on the following rates:

Fuel gas . . . . \$2.40/F.o.e. bbl. (40¢/10^6 Btu.)
Electricity . . . . 0.8¢/Kwh.
Steam . . . . 75¢/10^3 lb.
Cooling-tower water . . 2¢/10^3 gal. of circulation

Average cost must be adjusted if other rates are applicable in your area. This is especially true of the reboiler fuel, i.e. either fuel gas or steam, which normally represents 80 to 90% of the total utilities cost for an organic distillation.

Investment-related costs are as follows:

Maintenance . . . . 4% of investment/yr.
Taxes and insurance . . . . 2% of investment/yr.
Depreciation . . . . 10% of investment/yr.

■

Meet the Author

Robert F. Sommerville is in the Systems Planning Dept. of Sun Oil Co.'s Systems and Computers Div., 1616 Walnut St., Phila., Pa. 19103. He joined this group in 1971 after 10 years with Sun Oil's R&D Div. in economic evaluation. Mr. Sommerville also worked for Monsanto Co.'s Research Dept. at Texas City, Tex. A chemical engineering graduate of the University of Louisville, Mr. Sommerville is a registered professional engineer in Pennsylvania and Kentucky, and a member of AIChE and ACS.

References

1. Zimmerman, O. T., and Lavine, I., "Distillation Towers," *Cost Eng.*, April 1963, pp. 14–19.
2. Prater, K., and Antonacci, P., "How to Estimate Fractionating Column Costs," *Pet. Refiner*, July 1960, pp. 119–126.
3. Nelson, W. L., "How to Estimate Fractionator Costs," *Oil and Gas Journ.*, Feb. 20, 1956, pp. 155–156.
4. Fenske, M. R., *Ind. Eng. Chem.*, **24,** 482 (1932).
5. O'Connell, H. E., *Trans. A.I.Ch.E.*, **42,** 741 (1946).
6. Happel, J., "For Mass Transfer—Quick Method Finds Optimum Trays and Reflux Ratio," *Chem. Eng.*, July 14, 1958, pp. 144–146.
7. Gilliland, E., *Ind. Eng. Chem.*, **32,** 1220 (1940).
8. Van Winkle, M., and Todd, W., *Chem. Eng.*, Sept. 20, 1971, pp. 136–148.
9. Zevnik, F., and Buchanan, R., "Generalized Correlation of Process Investment," *Chem. Eng. Prog.*, February 1963, pp. 70–77.

You can analyze and compare the attractiveness of easy payment terms, minimum-maintenance features and other expense-related variables by applying cost engineering techniques to . . .

Equipment Buying Decisions

SERGIO MARTINEZ, Parsons & Whittemore, Inc.

In today's economic climate, engineers are particularly anxious to cut operating and maintenance costs, and this often involves buying new, more efficient, easier-to-maintain equipment. Vendors are anxious to sell him such equipment, and sometimes offer better terms, higher quality or faster delivery than was true a few years ago. But because we are still in a period of tight money and pinched profits, the engineer must know how to really prove to management that the purchase is justified after considering all factors—including the time value of money, especially when interest rates are high. (And they *are* still quite high by historical standards, although down somewhat from their 1970 peak.)

Because payment terms have become an important factor in the purchase of individual items of equipment, the engineer must know how to evaluate them properly in presenting his proposal or in comparing quotations. Also, he must know how to quantify maintenance savings in deciding between the purchase of an initially expensive but easy-to-maintain piece of process equipment, and one that is initially cheap but expensive to maintain.

Total vs. First Costs

In order to really reduce costs, rather than merely postpone them or shift them from one account to another, it is necessary to think in terms of "total" and not just "first" costs. In some pieces of equipment, the first cost may be a large percentage of the total cost but in others it may be a very small percentage.

For instance, in the case of an oil storage tank that costs $66,000, the only other cost will be that of painting every five years at $1,200, and inside cleaning every ten years at $2,000. The magnitude of the costs, assuming a 30-year life span, is illustrated in Fig. 1(A).

If we take now an air compressor that costs $11,000, we have a number of additional expenses. The oil must be changed every six months at a cost of $50, a minor overhaul is required every year at a cost of $400, and a major overhaul is required every 5 years at a cost of $3,200. Furthermore, the estimated life is only 20 years. An equivalent graph for total costs will look like Fig. 1(B).

The above examples are quite simplified to show extremes—note that original cost is 83.3% of total cost for the oil tank and only 32.5% for the air compressor—but they serve to illustrate the pitfalls of applying the same cost-reduction thinking to nonsimilar types of equipment.

In actual practice, many more factors must be considered. One of the most important is the truism that a dollar spent or saved today is not the same as a dollar spent or saved next year, or five years from now. Although this "time value" concept is old hat to cost engineers evaluating major capital projects, it is not used as well as it should be by chemical engineers involved in buying individual pieces of equipment, or in operating or maintaining them. Before seeing how these concepts apply to the total cost of hardware, let's take a quick look at how this hardware can be paid for.

Evaluating the Payment Terms

We will confine our discussion to relatively small pieces of equipment—i.e., somewhere in the range from $1,000 to $100,000. (Below that range, the purchase is usually a cash transaction; above it, the purchase usually involves complicated financing.

The suppliers of such "small" items of equipment are often willing to accept delayed payments that amount to equipment financing. In some cases, a 10

Originally published April 1971.

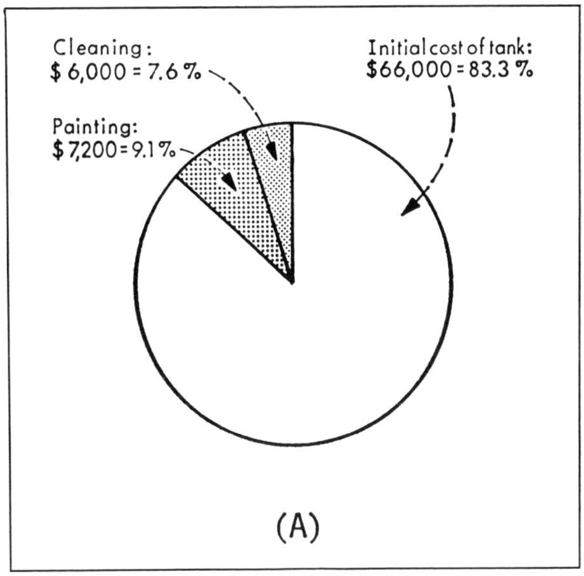

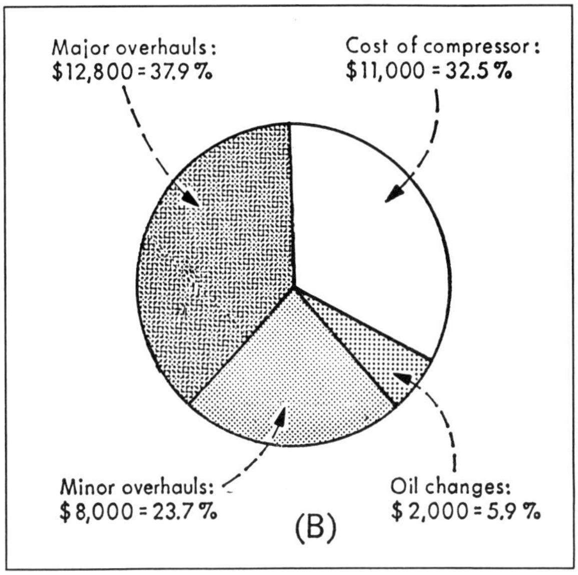

INITIAL COST accounts for a much larger part of total cost for tank (A) than for compressor (B)—**Fig. 1**

to 20% down payment will be required when the order is placed. However, if the purchaser has a good credit rating and a record of dependability with the different suppliers, it is sometimes possible to get the supplier to accept a purchase order without any down payment, unless the order is for a custom-made or very specialized piece of equipment.

Depending again on the supplier, balance of payments can be a percentage against delivery, sometimes as low as 10 to 20%, and the balance spread over a number of years, sometimes as high as five years. Besides the financial advantage of the delayed payments, this plan has the added benefit that if any trouble develops with the piece of equipment, and payment has not been completed, service from the supplier may be better than if payment had been completed.

Let's consider one piece of equipment with two possible alternatives for payment:

- Alternative A will be 20% down payment with order and the balance within 30 days after delivery. Assuming an 11-month delivery period, the balance will be due one year from now.
- Alternative B will be 20% against delivery, or one year from now; 30% six months after delivery, or 1½ years from now; 25% one year after delivery, or 2 years from now; 25% three years after delivery, or 4 years from now.

In order to have a realistic comparison between any two alternatives, it is necessary to evaluate what is called the "present-worth value." This includes the actual expenses to be met at the present time, plus all future expenses converted to their equivalent "present worth" as of today, so that we can quantify the difference between the money paid now and that paid sometime in the future. (Of course, any money that is not paid now can be deposited in a bank at compound interest.)

Present-worth factors (F_{PW}) for various interest rates are tabulated in the literature; we will use a 10% rate in finding the present-worth value of our two payment-term alternatives. For Alternative A, this value is:

$$0.20 + 0.80\ (F_{PW} \text{ for 1 yr. @ } 10\%) = 0.20 + 0.80(0.91) = 0.93$$

For Alternative B, the present worth value is:

$$0.20(F_{PW} \text{ for 1 yr.}) + 0.30(F_{PW} \text{ for } 1\tfrac{1}{2} \text{ yr.}) + 0.25(F_{PW} \text{ for 2 yr.}) + 0.25(F_{PW} \text{ for 4 yr.}) = 0.20(0.91) + 0.30(0.86) + 0.25(0.82) + 0.25(0.68) = 0.182 + 0.258 + 0.205 + 0.170 = 0.815$$

In other words, the equivalent present-worth cost of Alternative A is 93% of the total value, vs. 81.5% for Alternative B. For a $50,000 piece of equipment, the difference in present worths would be $5,750.

The above type of calculation is useful not only in choosing the most favorable payment terms (which may be a rather obvious decision, as in this case) but also in:

- Choosing between two quotations that differ in amount as well as in payment terms—e.g., how much are generous terms worth in regard to price?
- Recalculating marginal investment situations, to see if favorable payment terms can change the expenditure to "clearly justified."

Comparison of Purchase Alternatives

The evalution of alternatives is one of the most frequent problems encountered in engineering practice, and can involve a simple case—like whether to buy a diesel motor or a gasoline motor—or a complex one—like what fuel should be used on a new process plant.

In making such an evaluation, both capital and operating costs are usually considered. Capital costs include not only the basic outlay (whether in the form of cash, promissory note, bonds, etc.) but also any freight charges, sales taxes, foundation costs and actual installation costs; all of these expenses are grouped to form the "capital expenditure" that must be depreciated over a number of years in accordance with Bulletin F of the Treasury Department. As for operating costs, these include all those expenses needed to keep the equipment running, such as labor, operating supplies (fuel and lubricants), maintenance, etc.

In general, in evaluating alternatives it is assumed that:

- No inflation occurs—that is, replacement after a number of years would be at the same cost as the initial capital expenditure. (However, we will have something to say on how to allow for inflation.)
- The different alternatives will all do the job—that is, no one is superior to the other in such a basic way that the choice is preordained.
- Necessary data are available, possibly from previous experience.

To illustrate, we will evaluate two high-pressure pumps. Alternative X will be the conventional, relatively low-speed multistage type with a steam turbine drive through a reduction gear. Alternative Y will be a newly developed type consisting of a high-speed, single, centrifugal impeller mounted on the same shaft as a high-speed turbine. The data:

	Alternate X	Alternate Y
First cost	$18,500	$29,000
Yearly maintenance cost	$2,600	$800
Estimated life	20 years	15 years

Let's assume again 10% interest, that there is no salvage value for either unit at the end of the estimated life and that the efficiency is the same for both alternatives.

Since the life of the two units being considered is different, the present-worth values will be misleading unless properly corrected. It is easier and quicker to use "total annual costs," including the cost of capital. For this type of problem, the original investment is spread over a life of the equipment via a constant annual charge called "Capital Recovery," which takes into account the compound interest values on all remaining amounts each year.

The calculations for capital recovery are quite simple, using the factors that can be found in the tables of any economic-analysis book. The tables are arranged for various interest rates, and show the factors for different number of years. In the case of Alternative X, the factor at 10% interest and 20 years is 0.117, which (when multiplied by the initial investment of $18,500) gives a capital recovery cost of $2,160 for each of the 20 years. Similarly for Alternative Y, the factor is 0.131 for 15 years, which gives a capital recovery cost of $3,800 for an initial investment of $29,000.

Our comparison will be as follows, taking into consideration only the items that differ between both alternatives:

	Alternative X	Alternative Y
Capital recovery cost = (original investment) × (factor for years of life)	$2,160	$3,800
Yearly maintenance costs	$2,600	$800
	------	------
Total annual costs	$4,760	$4,600

Thus, if our basic assumptions are correct, Alternative Y is better despite the higher initial cost and shorter estimated life.

Obtaining and Applying the Data

Let's analyze the preceding comparison in more detail. Original investment (which includes freight, taxes and erection) is easy to determine: The cost that will be paid for each pump is available from the supplier's quotation, freight and taxes are easily estimated based on present rates, and erection cost can be "guessed at" fairly accurately based on previous experience. If no previous experience is available, we can use a percentage of the total equipment cost based on similar equipment.

The factor for capital recovery cost depends on the rate of interest to be used. Management must determine what this rate should be based on expected return of the capital if invested in some other project, actual bank interest rates, cash and debt position of the company, etc.

Yearly maintenance costs are sometimes difficult to estimate accurately. However, for equipment purchases, enough internal data are usually available to act as a guide for at least one of the alternatives, and this can be the base case. For the other alteratives it may be possible to get data from someone (i.e., a noncompetitor) who has used the equipment. If the equipment has new features, an analysis by a person with maintenance-engineering experience should be able to put a value on these features in terms of projected savings in labor manhours, labor skill requirements, and maintenance materials. The manufacturer may be willing to make some maintenance guarantees (e.g., that lubricant consumption will not exceed a certain figure).

If in doubt about some completely new equipment, you may want to adopt a conservative stance, and put the burden of proof on the supplier in regard to whether you are being *too* conservative for the intended application.

Keep in mind that the intended location for a piece of equipment may affect maintenance savings—e.g., a 25% reduction in overhaul-frequency requirements may lead to greater savings for equipment in a cramped, inaccessible location than for equipment in a wide-open area next to the maintenance shop.

Evaluating an Additional Variable

Now for a comparison that is slightly more complicated because it involves differences in operating as well as maintenance costs.

A refinery can provide water storage either via a steel tank on a tower adjacent to the plant, or via a concrete standpipe on a hill some distance away.

The elevated tank will cost (erected) \$82,000, while the standpipe will cost \$54,000 plus an additional \$6,000 for pumps and control. Operating and maintenance costs of the pumping station will be \$3,000 per year, whereas such costs will be only \$270 per year for the elevated tank (i.e., because these costs will be confined to maintenance).

Using a 10% rate of return and an estimated life of 30 years, we obtain the following comparison.

	Elev. Tank	Conc. Standpipe
Capital recovery	\$8,690	\$6,360
Annual expenses	270	3,000
Total	\$8,960	\$9,360

(The capital-recovery factor for 10% interest and 30 years is 0.106. For the elevated tank, \$82,000 × 0.106 = \$8,680; and for the concrete standpipe, \$60,000 × 0.106 = \$6,360.)

If the interest rate is 15%, the values will be:

	Elev. Tank	Conc. Standpipe
Capital recovery	\$12,460	\$9,120
Annual expenses	270	3,000
Total	\$12,730	\$12,120

(The capital-recovery factor for 15% interest and 30 years is 0.152.)

In other words, based on a 10% return, the elevated tank is cheaper; but based on a 15% return, the concrete standpipe is cheaper.

The graph for this situation will look like this:

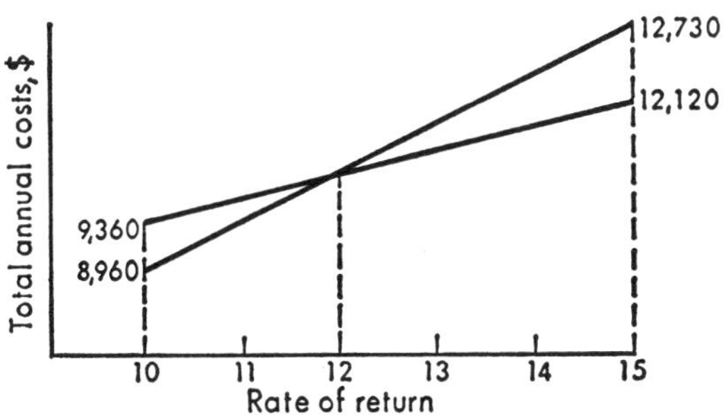

It can be seen that at 12%, both alternatives will cost exactly the same; consequently, the decision must be based on the actual rate of return specified by the company. Similar comparisons can be made using different lengths of life.

The Effect of Inflation

So far, we have assumed no inflation. One possible way to consider it would be as follows:

Let's say that an inflation cost of approximately 5% per year should be added to the annual expenses. To simplify the calculations, we can assume the actual increase is the same every year: \$15 per year on the annual expenses of the elevated tank, and \$150 per year on the annual expenses of the concrete standpipe. Using the same 15% rate as before, but taking into consideration the yearly increase in annual expenses (converted to an equivalent uniform annual expense), we obtain:

	Elev. Tank	Conc. Standpipe
Capital recovery	\$12,460	\$9,120
Annual expenses	363	3,932
Total	\$12,823	\$13,052

As can be seen, the yearly increase in maintenance has changed the picture, and now the concrete standpipe is actually more expensive than the elevated tank.

For a more detailed treatment of inflation effects, particularly in regard to major capital projects, the reader is referred to the article on p. 484.

Meet the Author

Sergio E. Martinez is chief project engineer with Parsons & Whittemore, Inc. (200 Park Ave., New York City), a firm that has built or is building, pulp and paper mills in 23 countries on 4 continents. Mr. Martinez, who has 30 years of experience in project management and engineering, holds B.S. degrees in electrical and mechanical engineering from M.I.T., and an M.S. in industrial engineering from New York University (where he is a part-time graduate-school instructor).

Relative Metal Economy of Pressure-Vessel Steels

Here is a simple numerical procedure for comparing the relative economy of the wide variety of ASTM plate steels approved in Section VIII of the ASME Code for Unfired Pressure Vessels.

LUDWIG ADAMS, Pittsburgh-Des Moines Steel Co., Neville Island, Pa.

Although chemical engineers seldom get involved with the detailed design of pressure vessels, they do often participate in general-design aspects such as choosing the material of construction, and hence should be interested in the metal-cost comparison shown in Table I. The comparison is in terms of $/sq. ft. of plate surface (stressed, under design pressure, to identical loading in lb./linear in. of cross section).

Before illustrating the use of this table, let us quickly review the metal selection process.

If you had to design a pressure vessel in accordance with Section VIII "Pressure Vessels," Div. 1, of the ASME Boiler and Pressure Vessel Code, as you turned to p. 102 of the 1968 Edition thereof, you would be faced with many choices of approved plate steels.

Under "carbon steels" are listed A-7; A-36; A113 Grade C; A283 Grades A, B, C, D; A285 Grades A, B, C; A299; A442 Grades 55, 60; A455 Grades A, B; A515 Grades 55, 60, 65, 70; and A516 Grades 55, 60, 65, 70.

Under "low-alloy steels" are listed A202 Grades A, B; A203 Grades A, B, D, E; A204 Grades A, B, C; A225 Grades A, B; A302 Grades A, B; A357; A387 Grades A, B, C, D, E; A410; and A533, Class 1, Grades A, B, C. These present a choice of at least 46 approved steels.

On p. 156 you would have a choice of 17 grades of high-alloy A240, the stainless steels. On p. 186 you have A353; A517 Grades A, B, D, E, F; and A553.

These tabulations give the "maximum allowable stress values in tension" for various service temperatures.

Based upon past experience and current pricebook awareness, your choice would quickly narrow down to perhaps a half dozen, or less. The final choice could simply be the one that resulted in the lowest cost in $/sq. ft. of vessel surface for the chosen dimensions of height and/or diameter. I say *"could* simply be" because the relative metal cost does not necessarily indicate the relative total cost of the vessel in place on foundations. The total cost, of course, involves variables in fabrication, shop assembly or field erection, freight and foundations.

Perhaps as a simplification, our past experience indicates that the relative metal cost is a fairly reliable index of relative total cost. Where alternate steels are included in the buyer's specifications, the final proof is in the bid prices for the alternates.

Example of Cost Calculations

Consider the specific example of a sphere of 23-ft. dia. or 138-in. radius, to store gaseous nitrogen at a design pressure of 100 psig., at ambient weather conditions, at a minimum of −20 F.

The tensile loading by the internal gas pressure is 0.5 pressure radius, or 6,900 lb./linear in.

The minimum plate thickness would depend upon the maximum allowable stress value. For A285 Grade C Firebox Quality, this stress value is 13,750 psi. for −20 F. to +650 F. The minimum thickness would then be 0.501 in. (at 100% welded joint efficiency, such as for complete radiography butt-weld inspection). This thickness means a weight of 20.44 lb./sq. ft. of plate surface.

The price of this steel, at the mill, would include the "plate base" of $6.10 per hundred lb. (cwt.), an "Association Specification" extra of $0.85 per cwt. (for Grade C, Firebox Quality, 1.5 in. thick and under) and a "width-thickness" extra of $0.55 per cwt. (plate widths of 90 to 100 in., inclusive, for 0.5 to 1.0-in. thick, inclusive). The 0.501-in. thick-

Based on a paper presented at the Thirteenth National Meeting of the American Assn. of Cost Engineers, Pittsburgh, Pa., July 1, 1969.

Originally published December 1969.

Cost Comparison for Pressure-Vessel Plate, 90 In. to 100 In. Wide, Loaded at 6,900 Lb./Linear In.—Table I

ASTM Spec. & Year	Grade	Yield, Psi., Min.	Ultimate Psi., Min.	Stress, Psi., Max.	Thick., In., Min.	Plate Weight Lb./Sq.Ft.	Pricing, ¢/Lb. Base + Extras (1969) Base	Spec. Grade	Qual. Chem.	Width Thick.	Tests	Heat Treat	Total Price ¢/Lb.	Cost $/Sq.Ft.
Carbon Steels														
A36 -67		36,000	58,000	12,650	0.545	22.24	6.10	0.15	–	0.55	–	–	6.80	1.51
A113-66	C	26,000	48,000	11,050	0.625	25.50	6.10	–	0.35	0.55	0.10	–	7.10	1.81
A283-67	C	30,000	55,000	12,650	0.545	22.24	6.10	–	–	0.55	0.10	–	6.75	1.50
A285-66	C-FB	30,000	55,000	13,750	0.501	20.44	6.10	0.85	–	0.55	–	–	7.50	1.53
A299-67		42,000	75,000	18,750	0.368	15.01	6.10	–	1.35	0.95	0.15	–	8.55	1.28
A442-67	60	32,000	60,000	15,000	0.460	18.77	6.10	–	0.95	0.75	0.15	–	7.95	1.49
A455-66	II B	36,500	63,000	18,250	0.378	15.42	6.10	1.80	–	0.75	–	–	8.65	1.33
A515	70	38,000	70,000	17,500	0.394	16.08	6.10	1.45	–	0.75	–	–	8.30	1.33
A516	70	38,000	70,000	17,500	0.394	16.08	6.10	1.85	–	0.75	–	–	8.70	1.40
Low-Alloy Steels														
A202-67	B	47,000	85,000	21,250	0.325	13.26	8.65	1.80	0.55	1.55	0.10	–	12.65	1.68
A203-67	B	40,000	70,000	17,500	0.394	16.08	8.65	4.40	0.55	1.30	0.10	–	15.00	2.41
A203-67	E	40,000	70,000	17,500	0.394	16.08	8.65	6.45	0.55	1.30	0.10	–	17.05	2.74
A204-67	C	43,000	75,000	18,750	0.368	15.01	8.65	2.35	0.55	1.55	0.10	–	13.20	1.98
A225-67	B	43,000	75,000	18,750	0.368	15.01	8.65	2.25	0.55	1.55	0.10	–	13.10	1.97
A302-67	B	50,000	80,000	20,000	0.345	14.08	8.65	3.00	0.55	1.55	0.10	–	13.85	1.95
A357-67	Anld.	30,000	60,000	15,000	0.460	18.77	–	25.50	–	3.75	–	–	29.25	5.49
A387-68	A	33,000	65,000	16,250	0.425	17.34	8.65	3.25	0.55	1.30	0.10	2.25	16.10	2.79
	B	33,000	60,000	15,000	0.460	18.77	8.65	3.35	0.55	1.30	0.10	2.25	16.20	3.04
	C	35,000	60,000	15,000	0.460	18.77	8.65	4.30	0.55	1.30	0.10	2.25	17.15	3.21
	D	30,000	60,000	15,000	0.460	18.77	8.65	8.40	0.55	1.30	0.10	2.25	21.25	3.98
	E	30,000	60,000	15,000	0.460	18.77	8.65	9.15	0.55	1.30	0.10	2.25	22.00	4.13
A410-67		30,000	60,000	15,000	0.460	18.77	8.65	4.40	0.55	1.30	0.10	2.25	17.25	3.23
A533-67	1A	50,000	80,000	20,000	0.345	14.08	8.65	2.65	0.55	1.55	0.10	3.50	17.00	2.39
	1B	50,000	80,000	20,000	0.345	14.08	8.65	3.60	0.55	1.55	0.10	3.50	17.95	2.53
	1C	50,000	80,000	20,000	0.345	14.08	8.65	4.00	0.55	1.55	0.10	3.50	18.35	2.58
Heat-Treated Steels														
A353-67		75,000	100,000	23,750	0.290	11.83	8.65	22.05	0.55	1.75	0.25	5.00	38.25	4.52
A517-67	B	100,000	115,000	28,750	0.240	9.79	8.65	2.35	0.55	2.10	0.50	3.50	17.65	1.73
A517-67	F	100,000	115,000	28,750	0.240	9.79	8.65	5.25	0.55	2.10	0.50	3.50	20.55	2.01
A553-67		85,000	100,000	23,750	0.290	11.83	8.65	22.05	0.55	1.75	0.25	3.50	36.75	4.35
Stainless Steels														
A240-63	304	30,000	75,000	18,750	0.368	15.21	–	43.00	–	3.75	–	–	46.75	7.11
	304L	25,000	70,000	17,500	0.394	16.29	–	51.00	–	3.75	–	–	54.75	8.92
	310	30,000	75,000	18,750	0.368	15.21	–	92.50	–	3.75	–	–	96.25	14.64
	410	30,000	65,000	16,250	0.425	17.57	–	31.75	–	3.75	–	–	35.50	6.24

ness is an "odd gage," for which there is an extra of $0.15 per cwt., unless the thickness is increased to the next "even gage," which would be 0.5625 in. (with a plate weight of 22.95 lb./sq. ft).

Other extras could be applicable, such as quantity, marketing, stamping, loading, etc., but these can be avoided—for example, by placing a minimum-item order of 20,000 lb., with no "special" marking, stamping or testing.

For the 0.50-in. plate, the base price plus the above-discussed extras would total $7.65 per cwt. The 0.5625-in. plate price would total $7.50 per cwt. The relative metal costs would be $7.65/lb. × 20.44 lb./sq. ft. or $1.57 per sq. ft. versus $7.50/lb. × 22.95 lb./sq. ft. or $1.72 per sq. ft.—indicating that it would be cheaper, in this example, to use the thinner plate and pay the odd-gage extra. Other benefits include less dead weight, less weld wire, less weld time (albeit insignificantly so).

Other possible choices of steels, besides A285 Grade C, could then be evaluated in the same manner and results tabulated for comparison and final selection.

Table I compares several steels, loaded at 6,900 lb./linear in., at recent base prices and extras. The steels are grouped in ASTM numerical order.

It may be interesting to note that A517B at about 17¢/lb. is more economical than A113C at about 7¢/lb. On the other hand, A517 at a minimum yield of 100,000 psi. is less economical than A285C at a minimum yield of 30,000 psi.

As you may suspect, the plate loading of 6,900 lb./linear in. (rather than some more convenient loading such as 10,000) was chosen to be able to include A113C in the tabulation. The ASME code limits its recommendation of this steel to a maximum of 0.625-in. thickness.

The most economical steel would seem to be A299, because of its relatively high allowable design stress. It is frequently bypassed, however, in favor of A515 or A516, depending upon operating temperatures; these newer steels are easier to weld.

You don't always have as wide a choice of steels as in the simple example given above—but if there is any choice at all, the selection should be based upon metal cost in $/sq. ft. of vessel surface. ■

Operating vs. Capital Costs: Evaluating Tradeoff Benefits

The current economic climate gives emphasis to this flexible, easily used method for expressing the value of anticipated operating-cost savings in terms of the capital investment justified by those savings.

JOHN D. BUEHLER and GEORGE J. FIGGE
Jacobs Engineering Co.

In today's environment of tight money and inflation, there is even more incentive than usual to locate the right balance between operating costs and capital investment. To do this, we need both a well-defined policy and an evaluation method through which that policy can be implemented. The method should be easy-to-use, "programmable" by management, and flexible enough to recognize all of the important factors affecting a proper decision. The present-value tradeoff method, which we will present here, meets these needs and spells out the important effects of construction schedules, inflation, and length of savings.

Tradeoff Opportunities: Questions to Ask

The opportunities to trade operating costs for capital investment are always with us. We can use more automation to decrease labor costs, or higher-priced alloys to hold down maintenance costs. More-expensive furnaces will improve fuel efficiency, and costlier motors can mean lower power bills. Additional recovery facilities can be installed to cut losses and improve yields. Or we can put in a whole new unit to make rather than buy raw materials.

The list goes on, and for each opportunity we must ask ourselves whether the savings are worth the price. We usually pose this question in one of three ways: What is the investment's level of profitability; or does the investment meet some pre-established benchmark of profitability; or, alternatively, how much could be invested while still meeting such a benchmark?

The latter question, which does not require a capital estimate until later, is the one employed in our present-value tradeoff method.

Choosing the Profitability Yardstick

The ways in which profitability is measured range from payout time to the various forms of discounted-cash-flow analysis. Simple payout time, calculated on a before-tax basis, is the easiest and quickest to calculate. However, by ignoring the time value of money and the post-payout period, it relegates too many factors to seat-of-the-pants evaluation.

Average return on original investment brings in such things as administrative overheads, depreciation, and working capital, and it casts the results in a familiar accounting format. However, it is not greatly superior to payout time as a decision tool, and fails to recognize the importance of construction lag, the length of the savings period, and the time pattern of these savings. In short, although it yields a percentage figure that looks misleadingly like an interest rate, it completely ignores the time values that are the essence of a true interest rate.

In order to rationally quantify the important effects of construction time, length of savings period and rate of inflation, we must turn to one of the variations of discounted-cash-flow analysis: either the internal rate-of-return variation (which treats savings as a combination of principal repayment and interest at whatever rate the project can support), or the net-present-value variation (which treats as a bonus any savings over those required to repay principal plus interest at a pre-established rate).

The advantages of the discounted-cash-flow methods easily outweigh the often quoted and usually exaggerated disadvantages—i.e., of increased complexity and need for more detailed information. Perhaps a more important deterrent has been that these methods share in many minds the place reserved for such abstractions as Einstein's time slowdown and Heisenberg's uncertainty principle: interesting, but not needed to navigate familiar waters.

Originally published February 1971.

Properties of a Successful Method

Having outlined our problem and chosen our measuring stick, we can now move more directly toward our goal: a comprehensive, flexible, easily used method for expressing the value of operating-cost savings in terms of the capital investment justified by those savings. Such a tool should be responsive to many factors, some of which are unique features of the particular project and should be established by the project evaluator. These include anticipated savings schedule, construction schedule, estimated downtime during tie-ins, and the time and expense required for development, pre-operating preparations, and startup.

Another group of important factors involves management's policy and judgment, and can either be built into the method or supplied to the evaluator. Some of these, such as acceptable profitability level, are policy factors to be set by management. Others, such as anticipated income tax rates and depreciation schedules, require a projection of future government policies. Closely related to these are judgments on future inflationary trends and on the relative uncertainty of the projected savings.

Last, but not least, the method should lend itself to the implementation of company growth policy and long-range plans. For instance, does management wish to emphasize new products or consolidation of existing lines? Specialty products or commodities? Modernization or new facilities? Present locations or new complexes? Such considerations are beyond the scope of the individual project and can best be handled by adjusting parameters of the evaluation method (such as acceptable rate of return).

The Present-Value Approach to the Problem

We have chosen a present-value approach, because we feel it best meets all of the above requirements. It takes note of all potential timing variations. The pre-selection of interest rates provides a powerful tool for reflecting management policies and judgments. Even within a given case, multiple interest rates can be used to recognize the differing risk levels attached to various elements of total cash flow. No trial-and-error solutions are required, and the method is amenable to fast but flexible graphical solution. It permits various judgments to be made at whatever organizational level is desired.

The form of our problem is as follows: Certain savings can be realized. How much capital investment would be justified? To solve it, we first establish the present value of the projected savings schedule. Then we determine the negative present value associated with each dollar of capital expenditure. By dividing the first by the second, we come up with the number of capital dollars that can be justified. Because present values are additive, we can break down both the savings-related and the investment-related cash flows into several component parts for separate study and analysis.

The primary *savings*-related items are tabulated below; this list should be scanned on any project, to make sure that we consider all tradeoff opportunities:

- Raw materials
- Catalysts and chemicals
- Utilities
- Operating labor and supervision
- Maintenance labor and materials*
- Local taxes and insurance*
- General plant administration and services
- Packaging, warehousing and freight

The primary *investment*-related items can be grouped as follows:

Plant Investment—Initial capital expenditure for plant; tax benefit of depreciation writeoffs; future salvage value.

One-Time Expenses Associated With Investment—Development costs; Pre-operating and startup expenses.

Operating Costs Dependent on Plant Investment—Local taxes and insurance;† maintenance.†

Working Capital—Initial investment; Future recovery.

Note that maintenance, local taxes and insurance have been shown on both the savings- and the investment-related list. Obviously, these items should be considered only once on any given project, and the particular circumstances will dictate where they can best be handled.

Example Calculation

Now let's look at a simple case. The operating department of a process plant estimates that certain plant modifications will result in before-tax operating cost savings of $60,000/yr. The capital will be required six months before savings begin. The Division Manager knows that general corporate policy calls for a 10% rate-of-return after taxes for low-risk projects. However, this particular plant's future is somewhat uncertain because of pollution problems, so word has been passed to employ a 15% discount rate for low-risk cost-cutting projects and to limit savings projections to 10 years. Government policies permit double-declining-balance depreciation based on a 15-yr. tax life. What is the maximum investment that can be justified?

Present Value of Cash Savings Schedule—First, we find the value of the proposed cash savings, based on a 50% income tax rate. For now, we will ignore the depreciation deduction, bringing it in later as an offset to the capital investment. Using midyear discount factors and a 15% discount rate, we set up the following table for the 10-yr. savings schedule.

* Include either here or with investment-related items.
† Include either here or with savings-related items.

Year	Savings, Thousand $ Before Tax	After Tax	Midyear 15% Discount Factor	Present Value, Thousand $
1	60	30	0.933	28.0
2	60	30	0.811	24.3
3	60	30	0.705	21.2
4	60	30	0.613	18.4
5	60	30	0.533	16.0
6	60	30	0.464	13.9
7	60	30	0.403	12.1
8	60	30	0.351	10.5
9	60	30	0.305	9.2
10	60	30	0.265	8.0
				161.6

The present value of the after-tax cash savings is approximately $162,000.

Present Value at Initial Capital Outlay—Next, let's look at the present value of all cash flows directly arising from each dollar of investment. First, the initial investment outlay six months before savings begin.

Year	$	Compounding Factor	Present Value, $
−½	1.00	1.075	1.075

Depreciation effects are next. Each dollar of depreciation saves 50¢ in taxes, so we have the following schedule, which establishes 24.9¢ as the present value of the cash flow resulting from depreciation writeoffs associated with each dollar invested.

Year	Depreciation, $	Tax Saving, $	Discount Factor	Present Value, $
1	0.133	0.067	0.933	0.063
2	0.116	0.058	0.811	0.047
3	0.100	0.050	0.705	0.035
4	0.087	0.043	0.613	0.026
5	0.075	0.037	0.533	0.020
6	0.065	0.033	0.464	0.015
7	0.057	0.028	0.403	0.011
8	0.049	0.025	0.351	0.009
9	0.042	0.021	0.305	0.006
10	0.037	0.019	0.265	0.005
11	0.032	0.016	0.231	0.004
12	0.038	0.014	0.200	0.003
13	0.024	0.012	0.174	0.002
14	0.021	0.010	0.152	0.002
15	0.018	0.009	0.132	0.001
	0.884	0.442		0.249

Salvage Value—Next, salvage value is taken as the undepreciated 11.6% of the original investment and recovered at the end of the fifteenth year. Present value is 1.4¢ per dollar of plant investment.

Year	Salvage	Discount Factor	Present Value, $
15	0.116	0.123	0.014

Maintenance, Local Taxes and Insurance—The projected $60,000/yr. savings did not include any allowance for maintenance or for local taxes and insurance on the new equipment. We estimate that these will total 7% annually of the new investment, or $0.07 annually per dollar invested. The after-tax effect will be half of this, or $0.035 annually. Using the same schedule of discount factors as for the depreciation calculation, we come up with a present-value cost of 21.8¢ per dollar of plant investment.

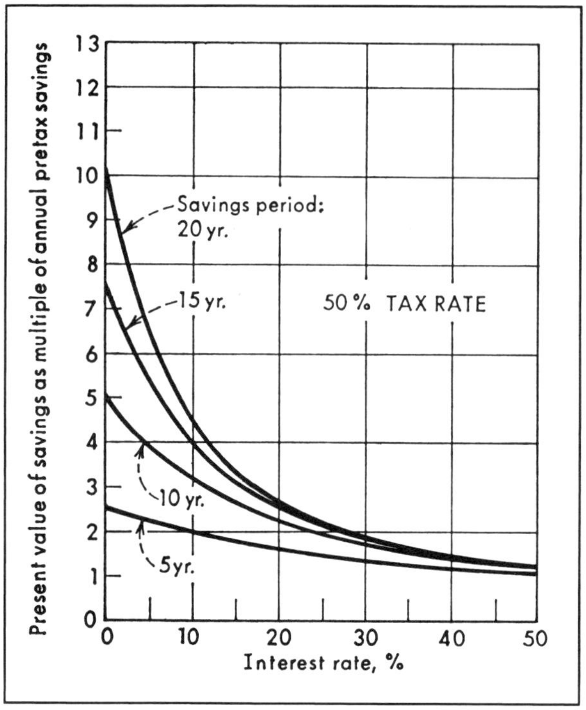

EFFECT OF SAVINGS-PERIOD DURATION on present value of savings-related items—Fig. 1

Total Value of Investment-Related Items—Adding together the present values of our investment-related items gives us the total present value of all of the cash flows directly resulting from installation of the facilities:

	Present Value Per Capital Dollar
Initial outlay of $1......................	−1.075
Depreciation tax credits.................	0.249
Salvage credit..........................	0.014
Maintenance, taxes, insurance...........	−0.218
Net......................	−1.030

The result for this particular case, including all maintenance, local taxes and insurance, turns out to be $1.03 for each dollar of plant investment. Without the maintenance, taxes and insurance, it would have been 81¢ per dollar.

Solution and Variations—Now, we merely divide the present value of our savings-related costs ($162,000) by the present value of our investment-related costs ($1.03 per $1) and come up with an allowable expenditure of $157,000. This corresponds to a 2.6-yr. simple payout before taxes.

Before you protest that you already knew a 2.6-yr. simple payout was acceptable, let's look at some pos-

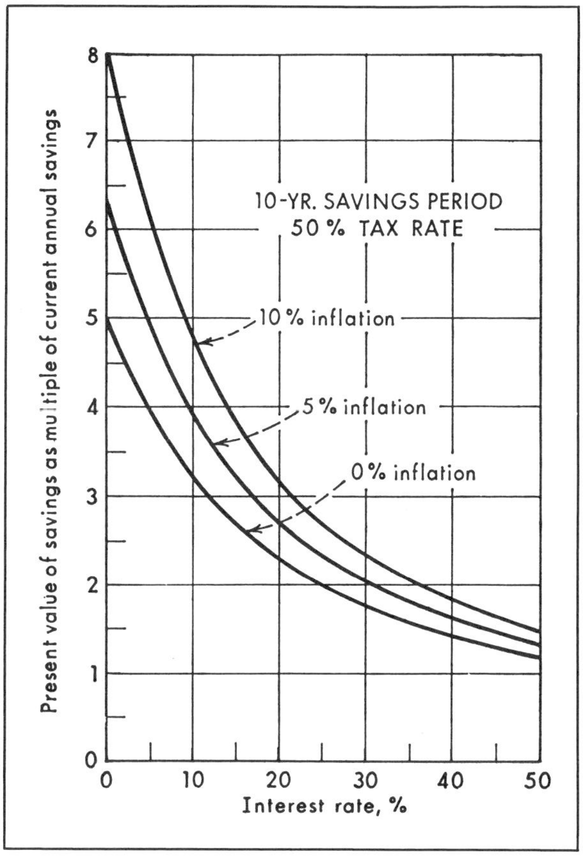

EFFECT OF VARIOUS INFLATION RATES on present value of savings-related items—Fig. 2

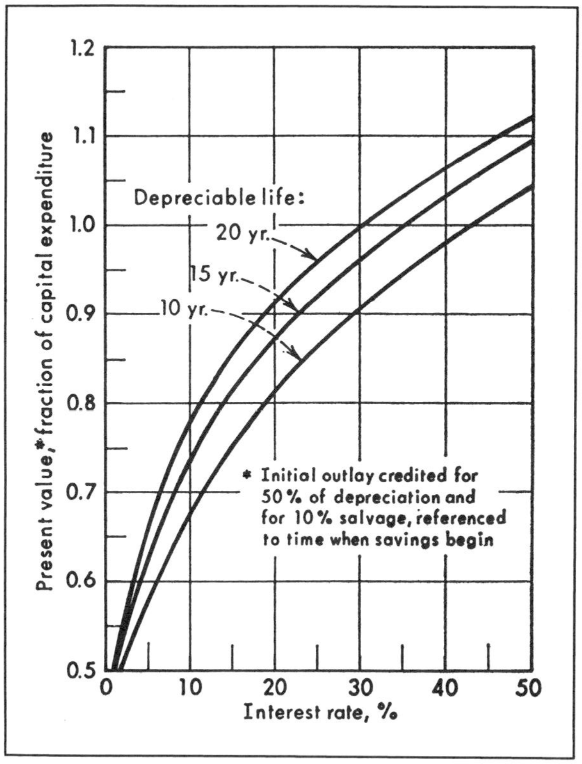

EFFECT OF DEPRECIABLE LIFE on present value of investment-related items—Fig. 3

sible variations. Suppose the $60,000 savings disappeared after only 5 years rather than 10. The adjustment can be made easily. The value of the savings schedule decreases to $108,000, the sum of the present values for the first 5 years. Everything else stays the same, so we now divide $108,000 by $1.03 to get an allowable expenditure of $105,000. But now the acceptable payout time is 1.75 years.

What if we're quite sure of half of the savings, but feel the other half is risky? We can arbitrarily decide to discount half of the savings at a higher interest rate—say, 30% rather than 15%. This cuts the value of the savings schedule from $162,000 to $134,000 which, divided by $1.03, gives a permissible expenditure of $134,000 rather than $157,000. Simple payout is now 1.9 years.

The really important feature of the method is not its apparent precision (because obviously the setting of the discount rate is primarily an act of judgment, and many of our forecasts are inexact) but rather, its ability to respond to various policy judgments, timing patterns, and risk assessments in a rational, consistent and quantifiable way.

Further, as we shall see in the next paragraphs, the method can be made very simple and fast to use, without losing its responsiveness to changes in construction schedule, risk, inflation, project life, and delay.

Graphical Methods for Evaluating Tradeoffs

The present-value method lends itself to flexible graphical solution. Sometimes, graphs can be used exclusively; in other cases, they may be only partially applied. Two useful graphical variations depicting savings-related items are shown in Fig. 1 and 2.

In Fig. 1, the present value of a string of identical annual savings, expressed as a multiple of annual savings before taxes and depreciation, is plotted against interest rates. This is done for savings periods ranging from 5 to 20 years and is nothing more than a graphical annuity table with midyear discounting and built-in tax adjustment.

Inflationary Effects—Fig. 2 illustrates the effect of inflation, which should receive serious consideration in a cost-saving project. For instance, suppose we can eliminate one around-the-clock operating job for a 10-yr. period, by installing certain equipment that will not increase maintenance costs. We estimate the cost of manning such a job at $50,000/yr., including all relief and benefits. Using a 10% interest rate and neglecting inflation, we consult Fig. 2 to place a value of 3.2 times $50,000, or $160,000, on this savings. But suppose we anticipate that labor rates will inflate 5% annually during this period, so that 10 years hence we expect the cost of manning the job to be $81,000/yr. In this situation, the present value of the string of savings increases to 3.9 times $50,000, or $195,000. This would support 22% more capital expenditure than if we expected no inflation.

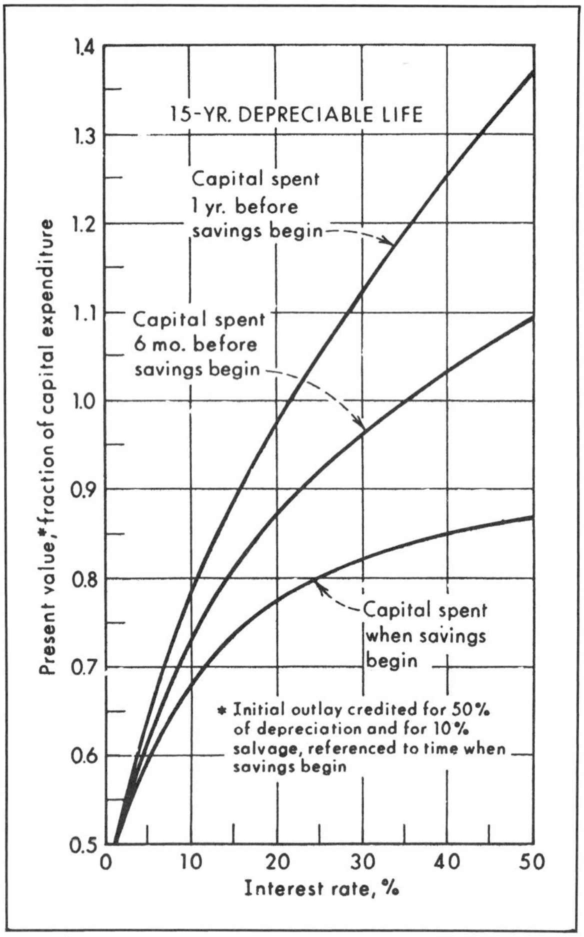

EFFECT OF LAGS IN CONSTRUCTION on present value of investment-related item—Fig. 4

Graphical Valuation of Investment-Related Costs—Investment-related costs are shown graphically in Fig. 3 and 4. The factors included are the initial capital expenditure, depreciation tax credits, and final recovery of salvage value. Not included are maintenance, local taxes and insurance, or working capital. The present value of the investment is referred to at the time when savings begin and is expressed as a fraction of initial investment. Fig. 3 illustrates the effect of depreciable life, and is based on an 18-mo. design-construction period, which is approximated by a lump sum expenditure six months before savings begin.

Fig. 4 fixes depreciable life at 15 years, and looks at the effect of the lag between investment and the beginning of savings. It can be seen that this factor is very important, especially at the higher interest rates. At a 15% rate, the net present value of a $1-million expenditure with a six-month lag before savings begin is $810,000, as compared to $735,000 without the lag. This means that, other things being equal, we could justify spending 10% more if there were no construction lag and if the costs of maintaining the new equipment remained unchanged.

Reworking the First Example via the Graphs—First, we use Fig. 1 to establish that the value of a 10-yr. savings schedule discounted at 15% is 2.7 times annual savings before taxes. Applying this multiple, we obtain:

$$2.7\ (\$60{,}000) = \$162{,}000$$

Then we use Fig. 3 to establish the value of investment-related items at 0.81 times initial plant investment for a discount rate of 15%, a 15-yr. depreciable life and a six-month delay.

Now, how do we handle maintenance, local taxes and insurance as a function of capital? We put Fig. 1 back in service, noting that a 15-yr. uniform cash flow is valued at 3.1 times the annual flow. Since the uniform flow for local taxes, insurance and maintenance has been estimated at 7% of initial plant investment, this becomes:

3.1 (0.07) = 0.22 times initial plant investment

Adding this to the 0.81 that we determined from Fig. 3 for other investment related costs, we obtain:

0.81 + 0.22 = 1.03 times initial investment

Now, setting 1.03 times the capital investment equal to $163,000:

$$\text{Permissible capital investment} = \$163{,}000/1.03 = \$157{,}000$$

We have read only three graphical points and performed only four operations of basic arithmetic. Yet we have retained the flexibility to consider variations in inflation rate, depreciable life, savings life, construction lag, and interest rate.

For a nonuniform savings schedule, we depart from the graphs only to perform manual discounting of the savings schedule, remembering to put it on an after-tax basis. The rest is done graphically.

To illustrate the sensitivity of permissible investment to various parameters, we have used the graphs to develop the data shown in the table on this page.

The most striking effect, aside from interest rate, is that of chopping the savings period in half. This lops one-third off the justified level of investment; yet it would be ignored by the simple payout method and easily overlooked by the return-on-original-investment method. Construction lag and inflation rate are of approximately equal importance, with inflation being more important at low discount rates and construction lag being more important at higher rates.

In conclusion, it should be noted that the present-value tradeoff method does not in any sense serve as a substitute for sound management judgment. On the contrary, it both permits and requires the manager to incorporate more of his judgment directly and quantifiably into the evaluation method, rather than having him try to apply it later via nonquantified observations or reservations.

Sensitivity of Justified Investment to Various Parameters (via Graphical Approach)

Interest, %	Savings Period, Yr.	Time Before Savings, Mo.	Inflation, %/Yr.*	Depreciable Life, Yr.	Justified Investment, $	Simple Payout, Yr.†
15	10	6	0	15	157,000	2.6
15	5	6	0	15	105,000	1.8
15	10	0	0	15	170,000	2.8
15	10	6	5	15	176,000	2.9
15	10	6	0	10	167,000	2.8
25	10	6	0	15	113,000	1.9
25	5	6	0	15	85,000	1.4
25	10	0	0	15	128,000	2.1
25	10	6	5	15	125,000	2.1
25	10	6	0	10	119,000	2.0

Basis: Cash savings of $60,000/yr. in operating costs, before income taxes, depreciation, maintenance, local taxes and insurance. (Maintenance, local taxes and insurance are estimated separately at 7% of investment.)
* Affects savings, maintenance, local taxes and insurance.
† Calculated without including maintenance, local taxes, insurance.

Meet the Authors

◀ **John D. Buehler** is manager of the Process and Economics Dept. for Jacobs Engineering Co., headquartered at 837 So. Fair Oaks Ave., Pasadena, Calif. 91105. His responsibilities include all technical and economic aspects of project feasibility and economic studies, as well as overall direction of engineering, design and construction projects. Past experience with the firm includes process engineering, project engineering and project management. He holds a B.S. in chemical engineering from the University of California, and is a Registered Professional Engineer.

George J. Figge is chief economist and assistant manager of the ▶ Process and Economics Dept. for Jacobs Engineering Co. Included in his responsibilities are preparation of economic evaluations, management of company computer activities and administrative assistance to the Process and Economics Department. Previously, he served as senior project evaluator and principal planning engineer for Dart Industries (then, Rexall Chemical Co.). He has a B.S.Ch.E. degree from the University of Idaho.

Section VI

OPERATING COSTS AND RELATED TOPICS

When making operating or marketing decisions about a particular product in a multiproduct plant, you can avoid pitfalls by making sure you are using . . .

Meaningful Production Costs

CARROLL R. KEIM, CPC International, Inc.

"How much does it cost us to make Product X?" Decision-makers keep asking this question in perfectly good faith. Cost people keep answering, in equally good faith. And their answers, reported to several decimal places, give the impression of tremendous precision.

Unfortunately, this impression is misleading–often very significantly so. Another cost department, using exactly the same figures but other accepted accounting conventions, might report the cost of Product X as being 50% lower or higher.

In a great many chemical processes, two or more products are manufactured simultaneously in the same unit. There may be a combination of one or more primary products coupled with one or more byproducts from the process. The main products may have different market values, and the byproducts may have yet other market values, or they may represent a cost (i.e., of disposal).

Recent articles on economic analysis,* have contained examples of multiproduct profitability calculations. In order to handle a wide range of such calculations, and to avoid pitfalls in manufacturing and marketing decisions, it is vital to know what the figures supplied by the accounting department mean and don't mean. This is the aspect the present article will examine and illustrate.

Most companies report and analyze profitability by each product and/or product line. The form of the report may vary, but in the chemical process industries it often looks like Table I (in $, $/unit, or %).

This type of information is used by different departments for different purposes–i.e., by:

• Accounting–to calculate, at the end of an accounting period, the value of the inventory and the cost of the goods sold during the period.

• Manufacturing–to compare actual results with forecast or standard costs as a means of controlling plant operations.

• Production Scheduling–for assigning orders to the least-cost plant or production line.

• Marketing–for (1) comparing profitability of different products to determine optimum allocation of money and effort; (2) making selling-price decisions such as tradeoffs between price and volume (e.g., "Can we afford to reduce prices 10% for a 40% volume increase?"); (3) deciding manufacturing priorities when orders exceed plant capacity.

• Top Management–to assist in decisions regarding allocation of resources, and to help measure the performance of various corporate subdivisions.

The trouble with this widespread dissemination and varied use of these figures is that they really are completely valid only for the accounting use.

Typical Product-Profit Report—Table I

	Product X	Product Y
Sales		
Gross sales	110	115
Sales deductions	–10	–15
Net sales	100	100
Factory Cost		
Raw material	30	10
Direct mfg. expenses	20	20
Factory overheads	10	10
Product factory costs	60	40
Gross Profit	40	60
Selling and Administrative		
Selling	5	15
Advertising	10	20
Administrative & general	5	5
	20	40
Total Profit for Product	20	20

Product Costs for Profit Reporting

In calculating a company's gross profit, the accountant proceeds on this stepwise basis: First, goods are produced and put into inventory–not necessarily physically, but their value or "cost of production" is added to the inventory; the inventory value is reduced by the cost of those goods that were sold (cost of sales). Then, the dif-

Originally published November 1971. *See pp. 116–125.

What Does It Cost To Produce Gluten and Starch? Well, It All Depends!

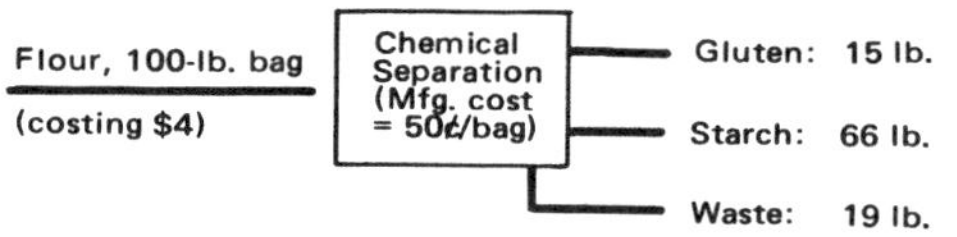

By a relatively simple process, ordinary wheat flour is separated into its main components, starch and gluten (used in the paper industry and in bakeries, respectively). It is assumed that a 100-lb. bag of flour, which costs $4, yields 15 lb. of gluten selling at 25¢/lb., and 66 lb. of starch selling at 5¢/lb. (The shrinkage is due to loss of moisture and of soluble and insoluble materials.) Total manufacturing expenses are 50¢ per bag of flour consumed, and for purposes of this example are considered to be not directly associated with either end product.

Let us see how raw-material cost and manufacturing expenses would be assigned to starch and gluten inventory costs, under these typical procedures:

Byproduct Accounting; Starch as Main Product

1. Raw-material costs: Gluten sales per bag of flour processed are $3.75 (15 lb. @ 25¢). This is credited to raw material, leaving a net flour cost of 25¢ per bag ($4.00 – $3.75). The raw-material cost for starch is then 25¢/66 lb., or 0.38¢/lb.
2. Manufacturing expense: Since starch is the main product, starch bears all of the common manufacturing expenses of 50¢ per bag of flour. This is equivalent to 50¢/66 lb., or 0.76¢/lb.
3. Total product costs, in ¢/lb., are thus:

	Gluten	Starch
Raw-material cost	25.0	0.38
Mfg. expense	--	0.76
	25.0	**1.14**

Byproduct Accounting; Gluten as Main Product

1. Raw-material cost: Starch sales per bag of flour processed are $3.30 (66 lb. @ 5¢). This is credited to raw material, leaving a net flour cost of 70¢ per bag ($4.00 – $3.30). The raw material cost for gluten is then 70¢/15 lb., or 4.67¢/lb.
2. Manufacturing expense: Being the main product, gluten bears all of the common manufacturing expenses of 50¢ per bag of flour. This is equivalent to 50¢/15 lb. = 3.33¢/lb.
3. Total product costs, in ¢/lb., are thus:

	Gluten	Starch
Raw-material cost	4.67	5.0
Mfg. expense	3.33	--
	8.0	**5.0**

Coproduct Accounting: Case 1

For this case, costs will be allocated according to product weight, i.e.:

Starch	66 lb.	81.5%
Gluten	15 lb.	18.5%
	81 lb.	100.0%

1. Raw-material cost: The $4 bag of flour is split 81.5% ($3.26) to starch and 18.5% ($0.74) to gluten. Raw-material costs per pound will be equal at 4.93¢/lb. ($3.26/66 and 0.74/15).
2. Manufacturing expense: Following the same procedure, the manufacturing expense allocated to starch and gluten are equal at 0.62¢/lb.
3. Total product costs, in ¢/lb., are thus:

	Gluten	Starch
Raw-material cost	4.93	4.93
Mfg. expense	0.62	0.62
	5.55	**5.55**

Coproduct Accounting: Case 2

For this case, costs will be allocated according to product value; i.e.:

Starch 66 lbs. @ 5¢ =	$3.30	46.8%
Gluten 15 lbs. @ 25¢ =	3.75	53.2%
	$7.05	100.0%

1. Raw-material cost: For starch, 46.8% x $4.00 = $1.87 per bag of flour; 1.87/66 = 2.84¢/lb. For gluten, 53.2% x $4.00 = $2.13 per bag of flour; 2.13/15 = 14.19¢/lb.
2. Manufacturing expense: For starch, 46.8% x $4.00 = $0.23 per bag of flour; 0.23/66=0.36¢/lb. For gluten, 53.2% x $4.00 = $0.27 per bag of flour; 0.27/15 = 1.77¢/lb.
3. Total product costs, in ¢/lb., are thus:

	Gluten	Starch
Raw-material cost	14.19	2.84
Mfg. expense	1.77	0.36
	15.96	**3.20**

Reviewing these pairs of costs, we see that depending upon the system used, gluten may seem to cost 25¢, 8¢, 5.6¢, or 3.2¢ per lb.

Each pair of product costs properly accounts for the total raw material and manufacturing costs, but there is an enormous difference in the product profit and loss reports. And this can lead to widely different corporate strategies and responses.

---

ference in dollars between the revenue from the sales and the cost of sales is the gross profit.

In a multiproduct operation, the total cost of production is distributed among the different products, and this permits the valuing of inventory and the costing of sales, even though the product-mix keeps changing.

It is the cost accountant's responsibility to assign costs to individual products in the most careful and logical way he can, but there are an infinite number of ways to choose from. Necessarily they are all highly arbitrary, and although under certain circumstances some may seem more appropriate than others, there is no such thing as one "right" way.

Some of the alternatives faced by the accountant in creating costs for individual products are described below. For purposes of discussion, they will be grouped according to their closeness of identity with a specific product.

Direct Costs—As we all know, costs whose totals rise and fall in direct proportion to the volume of an item that is produced are usually called "direct" or "variable" costs. Raw materials, ingredients, packages and supplies are closely identified with products in a physical sense and are best classified as direct. However, even with these, there are often problems in assigning their costs to individual products. For instance, if a package is purchased, its cost is pretty well established, but if it is made in the plant, its "cost" is subject to all the vagaries of product costing that will be observed later. Furthermore, if a number of products are made from one raw material, there can be a great deal of question as to how to assign raw-material cost to products. For instance, there are various ways in which to charge the cost of crude oil to gasoline, heating oil, petrochemical feedstocks, wax and coke.

Fixed (or Period) Costs—These are much more closely related to the passage of time than to the volume of production. Some, such as plant taxes and insurance, and usually depreciation, will continue even if the plant is closed down. Others, which are usually budgeted, continue with time, no matter whether production is high or low, as long as the plant is operating at all, or even if the plant is temporarily shut down. These costs would include such things as the salaries, fringes and expenses of plant-management and plant-service personnel, as well as some maintenance expenses.

Other Costs—Even though these costs are related to production volume, the relation is not necessarily directly proportional. Examples would be the so-called indirect labor and its fringes, first-line supervision, and the utilities consumed in the process.

When product costs are built up for inventory- and profit-reporting purposes, they usually must account for every cent of factory expense. Thus all the direct, fixed and other costs must be allocated to one product or another. Any logical system that absorbs all the costs into inventory or into cost of sales will meet accounting requirements for profit determination.

In practice, each company develops its own assumptions and rules for the allocation of each type of expense to products. Since there is a wide range of methods that can be logically used, different accountants and different companies can, and do, assign them differently to products.

Byproduct and Coproduct Accounting

When two or more products are derived from a single raw material, there are a number of ways of allocating the raw-material cost among the products. The two basic approaches are referred to as "byproduct" and "coproduct" accounting.

Byproduct accounting is most appropriate where a process that is carried out solely to obtain one product (called the "main" or "head" product) unavoidably produces others of relatively little value ("byproducts"). In this situation, whatever money is recovered from the sale of the byproducts is credited to the raw-material cost of the main product. All the manufacturing expenses—or at least all those that do not specifically apply to improving the value of the byproducts—are charged to the main product. This results in the total cost (inventory value) of a byproduct equalling its selling price, and its showing zero profit when it is sold.

When the processing of a raw material results in a mix of products of relatively equal volume and/or value, "coproduct" accounting usually seems more appropriate. With this approach, raw-material cost and the manufacturing expenses that are incurred before the point of product separation are divided between the products. Thus the inventory costs of coproducts are not specifically their selling price, and each coproduct shows a profit or loss when it is sold.

Very often, there is a question as to which is the better method to use. The selection of one or the other is highly subjective, and drastically affects product costs. A vivid example, taken from real life, is shown on the opposite page; it illustrates the great disparity in the reported cost of producing starch and gluten under various coproduct or byproduct accounting procedures.

Each company must choose which of these (or possibly other) accounting methods to employ. If coproduct, should the raw material be allocated by weight or value? Should the allocation be made before or after separation from other products? If byproduct, should the value at first or later stages be credited? How should intermediate products be valued? And so on. Once these choices have been made, reported product costs will depend entirely on this set of assumptions.

Allocating Other Than Direct Costs

It is hard enough to logically assign direct costs to products, but the assignment becomes even more arbitrary with other than direct costs.

For example, where there is no guaranteed annual wage, operating labor is usually considered a direct cost, but even here there are problems: If a shift is called in, the cost is fixed, irrespective of the production rate.

Many utility charges act similarly, in that demand or stand-by costs are essentially fixed, while costs vary with production above this level, but not directly so.

Fixed costs are often allocated to products by relating them to other costs that are more nearly direct, such as to machine hours or to "direct" labor, or in proportion to profit indicators such as selling price or value added.

In sum, it is quite evident that the figures used as product costs for inventory valuation are anything but clear-cut facts, since they stem from a great number of arbitrary decisions. Even so, these figures are completely valid for profit reporting. They are the ones that nearly all companies use for external reporting, and that most use for internal reporting as well.

Standard Costs for Each Product

Many companies control their manufacturing costs by the use of standard costs; these are the anticipated costs that will result from attainable factory performance under a set of conditions that is planned for the period. The subsequent performance of each cost center may be evaluated by comparing its actual costs with those standards.[1]

Standard costs are usually developed each year as part of the profit planning cycle. Sales forecasts are converted into production schedules; the Manufacturing Dept. provides details on material yields, supply-consumption rates, product-throughput rates, station-manning requirements, etc., as well as budgets for overheads and maintenance; the Purchasing Dept. provides expected costs for raw materials and supplies; and the Industrial Relations Dept. provides anticipated rates for labor and fringes. By assembling the resulting costs for each cost center, and allocating all costs according to the predetermined procedures, standard costs are calculated for each product.

During the year, these are the costs that are used to

Example: Effect of Volume on Cost Variance

A company's standards are based on a monthly production of 10,000 units, with direct costs of $60 per unit and nondirect costs of $400,000 per month:

	$/Mo.	$/Unit
Direct costs	600,000	60
Nondirect costs	400,000	40
Cost of product	1,000,000	100

In January, production was the standard 10,000 units but this was reduced to 8,000 in February because of a drop in sales. Then in March, production was increased to 11,000. In each of the three months, the plant performed exactly as originally planned: direct costs of $60 per unit; nondirect costs of $400,000 per month. Total plant costs were thus:

	Jan.	Feb.	Mar.
Units produced	10,000	8,000	11,000
Direct costs, $, @ $60/unit	600,000	480,000	660,000
Nondirect costs, $	400,000	400,000	400,000
Cost of product	1,000,000	880,000	1,060,000

But at the standard cost of $100 per unit, the value added to inventory was:

	Units	Value, $
Jan.	10,000	1,000,000
Feb.	8,000	800,000
Mar.	11,000	1,100,000

The cost variance would thus be unfavorable in February and favorable in March:

	Actual Cost, $	Inventory Value @ Standard, $	Variance, $
Jan.	1,000,000	1,000,000	
Feb.	880,000	800,000	+ 80,000
Mar.	1,060,000	1,100,000	- 40,000

value the production and inventory. Differences between the value of production at standard and the total plant costs for a period are a gross profit variance and are applied directly to profit and loss. Sometimes they are mistakenly called manufacturing variance, and it is left to the plants alone to explain them. But such variances are composed of a number of elements besides those for which the plant is responsible, such as the effects of variance from standard total production (responsibility of marketing and inventory control), different product mix than planned (marketing), differences in prices of purchased materials (purchasing), differences in labor rates (industrial relations) etc.

In fact, variances due to elements outside the plant's control are often much larger than the ones under their jurisdiction. The example on this page shows how one of these—production volume—can cause very large gross profit variances. It will be noted that in the example, while the plant performed exactly as planned, substantial gross-profit variances were caused by volume differences that should be laid at the door of the company's marketing department.

Where there are numerous departments and many different products, it is extremely cumbersome to analyze such variances by working backwards from the total variance in production cost. This is because standard costs necessarily include great numbers of allocations, even allocations of allocations, and it is extremely difficult to "unallocate" in searching for the causes of variance.

On the other hand, it is quite simple and highly useful to determine variances and their causes in detail at the very lowest reporting levels, and combine them in the condensed reports at succeedingly higher levels. This approach is especially practical because the lower-level reports can also be expressed in the physical terms that are most pertinent to first-line control.

The monetary control of a production cost center begins with information on such things as manning and overtime, consumption of supplies and utilities, production rates and downtime; product yields and reprocessing. This information is processed by assembling it in physical units and then placing dollar values on it. Thus it is simple to provide first-level operating-center reports in physical terms, with dollar values applied to the variances. Cost-control reports at any level should normally cover only those costs that can in fact be controlled at that level. Thus, any variances due to volume, mix, prices, wage rates, etc., must be ascribed to the responsible department, with informational copies to the plant if desired.

Nonproduction expenses should be controlled in each area of responsibility by comparing actual vs. budget in dollars and also in physical units where this is pertinent.

No useful purpose can be served by providing plant people with figures of allocations or the product costs that are used for inventory valuation, as they are confusing to operators and do not help them do a better job.

Product Costs and Decision-Making

Decision-making is the choosing between alternate possible courses of action after having compared the probable future results to be expected from each of the alternatives. Product costs are developed for inventory and profit reporting for only *one* course of action and using only one specific set of costing procedures. Thus, inventory costs are rarely useful in decision-making situations.

However, it is possible to arrange the cost information so that it is useful in comparing the results to be expected from various conditions. To answer many "what if . . ." questions, the use of product direct costs is very helpful.

To show how this may operate, let us consider a packing station whose capacity is 2,500 bags per day, but whose output is limited by slow product feedrates to only 2,000 bags. Daily fixed costs are $150, and the direct costs of supplies are 20¢ per bag. What would happen if the feed were increased to the 2,500 capacity by the packer?

	At Standard Rate:		At Increased Rate:	
	$/Day	¢/Unit	$/Day	¢/Unit
Fixed costs	150	7½	150	6
Direct costs	400	20	500	20
	550	27½	650	26

The cost of packing the *additional* bags is neither the old standard at 2,000 bags or the standard that would be calculated at the new rate of 2,500. Rather, the cost of packing the additional 500 bags would be $650 - $550, or

$100, or 20¢ per bag—which, of course, is the direct cost. (The cost of increasing the feedrate itself is a separate study of the labor and investment involved.)

By using direct costs instead of inventory product costs in evaluating alternative courses of action, many erroneous decisions can be avoided. This approach uses the concept of product "profit contribution." This is the change in revenues less costs resulting from changes in sales (and production) volumes of a product. The costs involved are essentially the direct costs, and the concept is valid within very broad, but not extreme, limits.

There is a significant difference between this approach and the inventory product cost method. Direct costs are lower than inventory costs, so unit profit contributions are larger than classical gross profits. Furthermore, profit contribution is applied to pay for fixed expenses first, and only after these are covered does it apply to profit. Classical gross-profit accounting assumes a profit from every unit sold—beginning with the very first—but this is correct only if the final sales for the period have been exactly equal to the planned volume.

Where decisions are to be made involving noticeable changes from normal conditions, with current profit being the deciding factor, improved decisions will result from using the profit contribution approach. The examples on this page demonstrate how this comes about. The examples have been deliberately simplified (e.g., by ignoring changes in the costs of carrying inventories and receivables) in order to illustrate the main point as clearly as possible; some more-detailed and broader examples can be found in the articles on pp. 116–125.

Of course, there are situations where tools even more sophisticated than direct costs are required to evaluate the effect of alternate choices. These generally occur when plants are very complex with a broad product line, when there is an interrelation between variable costs, or where there are several plants making the same products.

The cost system really shouldn't be expected to handle such complex questions, as complete recalculations are required to answer them. Since this can be an extremely lengthy process, it is usually desirable to have all, or at least some of the relationships set up in computer models. ■

Use of Inventory Costs Vs. Direct Costs

Example I—Product A, which has sales of 400,000 units per year, shows the following cost picture:

Direct costs, $	3.50
Fixed costs, $	1.00
Total costs, $	4.50
Selling price, $	4.25
Product profit (loss), $	(0.25) = (100,000/yr.)

With no shortage of plant capacity, should this product be discontinued?

Based on product costs, the product could be dropped for an apparent $100,000 increase in profit. However, dropping the product would save only its direct costs, and result in a loss to the company of $300,000:

Selling price, $	4.25
Direct costs, $	3.50
Profit contribution, $	0.75 = $300,000/yr.

Example II—Product B, which has annual sales of 200,000 units, shows this cost picture:

Direct costs, $	3.50
Fixed costs, $	1.00
Total cost, $	4.50
Selling price, $	5.50
Product profit, $	1.00 = $200,000/yr.

There is no shortage of plant capacity, and it would be possible to double the sales if the price were reduced by 75¢. Should this be done?

This would look like a losing proposition:

	Present	Suggested	Difference
Volume	200,000	400,000	
Product profit margin, $	1.00	.25	
Profit per year, $	200,000	100,000	−100,000

But with the direct costs being only $3.50, the real result is:

	Present	Suggested	Difference
Volume	200,000	400,000	
Selling price, $	5.50	4.75	
Direct cost, $	3.50	3.50	
Profit contribution, $	2.00	1.25	
Annual contribution, $	400,000	500,000	+ $100,000

Example III—Plant capacity is limited; it can produce either 6,000 units per day of Product C, or 7,800 units of Product D. Which product should be given preference?

	C	D
Selling price, $	5.25	4.50
Direct costs, $	2.50	2.25
Profit contribution, $	2.75	2.25
Contribution per day, $	16,500	17,550

Thus, it would appear that C would produce $750 per day more profit and should be preferred. However, the real results obtained by the direct costing approach would give the opposite decision:

	C	D
Direct costs, $	2.50	2.25
Fixed costs, $	1.00	1.00
Total costs, $	3.50	3.25
Selling price, $	5.25	4.50
Product profit, $	1.75	1.25
Profit per day, $	10,500	9,750

Meet the Author

Carroll R. Kelm joined CPC International, Inc. (then, Corn Products Co.) in 1947. After assignments in product development, marketing and manufacturing, he spent 10 years as general manager of CPC's Colombian operations. He is presently staff assistant to the president, at the corporation's world headquarters at International Plaza, Englewood Cliffs, N.J., where his assignments include cost-system planning. Mr. Keim is a chemical engineering graduate of Case Institute of Technology.

Cost-Improvement Techniques

Control of costs by monitoring budgets and standards is not enough; to survive strong competition, cost improvements must be managed. An intensive program that reaches every employee can have a major effect on costs.

N. S. EASTMAN, Union Carbide Corp., Chemicals and Plastics Div.

For the past decade, the petrochemical industry in the U.S. suffered—along with others—from two enormous pressures: cost inflation and a downward trend in product sales prices. There was very little that management could do about sales-price attrition in a free-enterprise system. But management did have control over one remedy for inflation, and that was cost improvement.

In Union Carbide's Chemicals and Plastics Div., where an operations inprovement program has been highly developed, cost-improvement rates have approximately doubled from 3.7 to 7.0% in a six-year period (Fig. 1). This increase appears to be significant because the improvements, less all capital and expense costs to implement the changes, rose from 10% of pretax income in 1966 to an average of 45% for 1968 through 1972. Indeed, during the 1971-1972 recession, net cost improvements were more than half of pretax income.

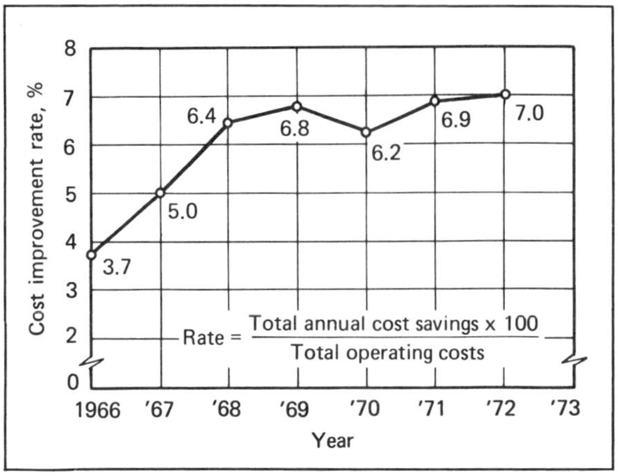

COST-IMPROVEMENT RATES for C & P Div. – **Fig. 1**

How was all this achieved? The key was to concentrate on making improvements in existing procedures, systems and processes, and to require high employee participation in the program. The details of our experience that follow will show how any organization can set up similar cost-improvement programs.

Operations Improvement Program

The purpose of this Union Carbide management system is to improve profit performance by accelerating the

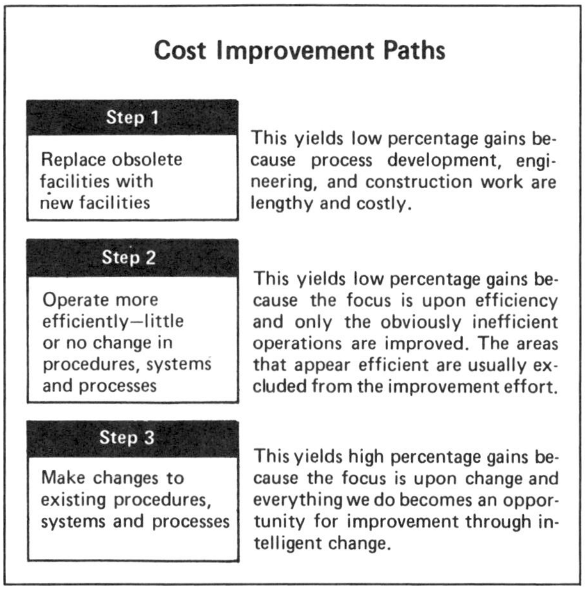

THREE BASIC WAYS to improve costs; Step 3 is best because it focuses on every operation – **Fig. 2**

Originally published December 1973.

Spur Widespread Savings

rate at which changes are made to improve costs. Called the Operations Improvement Program (OIP), the system worked reasonably well in some of Carbide's small divisions before it was more fully developed by Chemicals and Plastics (C&P).

There are basically three ways to improve costs, as shown in Fig. 2. How an organization orients itself toward cost improvement will have a marked effect upon the results. Carbide attributes much of its success to focusing on Step 3 of the figure for two reasons. First, it tries to obtain the broadest possible field for finding improvement projects. Every operation can be challenged from the view of "Why do we do it this way–is there a better way?" Second, since everything that is currently done will eventually be improved through intelligent change, it is critical that all employees participate, because high cost-improvement rates require high employee participation.

The OIP program that we use is designed to meet the following criteria. I believe that your organization must meet them also to achieve outstanding results in cost-improvement performance:

1. Every employee must understand the need for cost improvement, and how he can contribute.
2. The system must motivate employees to produce a continuous flow of new projects throughout the year.
3. Each part of the organization must track its cost-improvement rate against goals.
4. Cost-improvement rates must be clearly visible to all managerial personnel.
5. The overall cost-improvement rate should offset the effect of inflation or rising costs plus any product-price attrition.

In the early states of a pilot program at one of Carbide's large plants, several facts stood out:

- ☐ Cost-improvement rates could be readily doubled by using a management system.
- ☐ There was little difficulty in finding cost-improvement projects.
- ☐ More than half the projects or changes could be implemented without cost.
- ☐ For projects requiring capital investment, the return on investment averaged over 100%.
- ☐ Cost improvement was one of our most profitable businesses.

Seeking a High Level

In 1966, when I became manager of OIP, the C&P div. sought a cost-improvement rate that would offset inflation and product-price attrition. This necessitated increasing the then-current rate of 3.7% to approximately 7.0%. Obviously, such a large increase cannot be achieved by a few people; it requires the entire organization working toward a common objective. Therefore, the program was spread to all functions–manufacturing, distribution, engineering, R&D, accounting and sales.

Our cost improvement experience at the beginning of the program seemed to be typical of that of many outside organizations I visited: an unorganized or spontaneous approach to the problem yields 3.0 to 4.0%. This is probably your current rate if you are not managing cost improvement effectively.

When launching an OIP, anticipate resistance and problems, such as those summarized in Fig. 3. Item 1 will probably be the biggest headache; how we overcame this and other problems will be detailed below, as well as the five basic elements of our OIP, which are:

- ☐ Program mechanics (procedures and guidelines).
- ☐ Motivating employees (to gain high participation).
- ☐ Measuring performance (achievement vs. goal).
- ☐ Cost-improvement techniques (cost teams, etc.).
- ☐ Resource allocation (manpower and funding).

It is essential to recognize that all parts of the organization had previously and spontaneously worked on cost

Problems

1. The attitude of "We have always improved, why have a program?"
2. The attitude of "I manage cost improvement by closely monitoring budgets and product costs."
3. Disbelief of reported savings
4. Complaints that program required too much paperwork.
5. Complaints that program cost too much to operate.
6. Can't relate savings to costs and profits.

PROBLEMS in program implementation – Fig. 3

improvement to some degree, but the new program developed a highly sophisticated system to organize these efforts. Cost-improvement efforts were put on a managed basis like cost control vs. budget, production volume vs. production plan, sales volume vs. sales forecast, etc. Once this was done, many more improvement ideas were generated, per employee and in total:

Year	Proposals Accepted
1967	4027
1968	4849
1969	5249
1970	5764
1971	5653
1972	7387

PROGRAM MECHANICS

To introduce the Operations Improvement Program to a location, the location manager or director appoints an OIP coordinator to assist him in establishing and promoting the use of the system. The coordinator should be at assistant-manager or assistant-director level.

The manager also appoints an Advisory Board, usually composed of the manager/director, his immediate subordinates, and the coordinator. The board is formed for two reasons. First, it provides a mechanism to help the managers screen, review, approve, or assign for study the improvement proposals or ideas that do not have accompanying solutions. Second, it provides a means for top management to participate in and vigorously support the program in a manner visible to all employees.

A guide manual is distributed to employees. It contains a set of standard procedures and forms for submitting projects or ideas, maintaining project inventories, and reporting project savings.

Establishing Goals

As revealed in Fig. 4, the cost-improvement rate that is needed to offset inflation and product-price attrition, plus an "x" amount, is estimated annually. This is done by converting the increase in costs for labor, materials and services into a composite cost inflator. Available data permit this to be done quite readily; the main effort involves the use of invoice payment data to determine the percentage increase in unit costs for materials and contract services. A product price index is available from a computer program that is run monthly.

These two factors, the composite cost inflator and price index, are equated to a cost-improvement rate that will offset them. This rate is subsequently adjusted for the "x" amount, the previous year's performance, and the availability of funds. The final cost-improvement rate with the supporting data is then communicated to all functional managers who oversee the setting of a goal for each location. Location goals are determined democratically.

Each location department submits its own goal based on project inventories and past experience, and these goals are consolidated to comprise the overall location goal. A large plant location may be composed of 25 to 50 departments. Departmental goals are expected to be challenging and to contain some "stretch" beyond existing project inventories. Low department goals are frequently recycled.

The final local goal is approved by the functional officer. Goals may vary from location to location but should average out so that the cost improvement level within each function improves at, or exceeds, the rate projected for the division. If there is a substantial deviation, some location goals are recycled.

Reportable Projects

All reportable projects must be deliberate, describable, intelligent changes in the way something is done, and must reduce costs or avoid expenditures that would otherwise occur. Changes that would reduce the desired level of product quality, customer service, employee safety, etc., are rejected.

Every product is classified as an operating cost improvement or a construction (capital) cost improvement. Projects in the former category include all formal budgets for operation of manufacturing plants and auxiliary functions such as distribution, R&D, etc. Projects in the latter category save capital (investment money) and usually originate in the engineering function.

A further classification is made according to whether projects are recurring or nonrecurring in nature. In C&P, recurring projects are those with continuous or repetitive savings that will be in effect two years or longer. All others are nonrecurring.

Projects either reduce or avoid costs. Cost reduction refers to a reduction in costs that are already occurring. This is the most common type of project. Cost avoidance denotes a cost that would otherwise certainly have occurred, and that is being avoided by an OIP project.

The cost savings must be identifiable and measurable with high accuracy; the credibility of reported savings cannot be overstressed. This excludes reporting any activities that are clearly desirable and that have a long-term beneficial economic impact but to which a measurable cost savings cannot be attached.

Projects affecting a normal cost base must result in an improvement; returning to a previously normal base from a temporary abnormally high level, while desirable, is not reportable in the system.

The project savings must be claimed and reported by the segment of the organization on whose budget the savings occur, even though the idea originated in another part of the organization. This is necessary to avoid duplication of claims and to permit auditing of savings.

Cost avoidance projects must have the original cost base approved by that level of management with final authority. Cost avoidance savings are, from a practical standpoint, very difficult to audit. However, in C&P, only recurring savings are goaled and only 15% of this category are cost avoidance.

Introducing Proposals or Ideas

The process is very simple. There are two forms readily available (Fig. 5 and 6) for the employee to use. Technical and supervisory personnel use the more detailed form in Fig. 5. Most of the other employees use the other. It should be noted that the latter form can be completed by such simple action as checking the box for "I have a cost improvement idea and would like to discuss it." One copy of the form is always mailed to the location OIP coordinator. This permits him to keep close watch over all ideas. He frequently inquires about the action being taken by line management on the other copy.

All ideas require evaluation and a response to the submitter. The evaluation may vary from a judgment by a supervisor to an analysis by technical personnel. All approved ideas are given a priority and placed on the location's project inventory (Fig. 7). All rejected ideas, or ideas requiring substantial study, are reviewed with the location advisory board. The decision of the board is final; it is responsible for assigning projects to study teams or to the technical sections. At least monthly, the progress of all approved projects is reviewed by the advisory board.

Most department heads post on a bulletin board each month an inventory of all projects with their status, to permit employees to follow the progress of their projects.

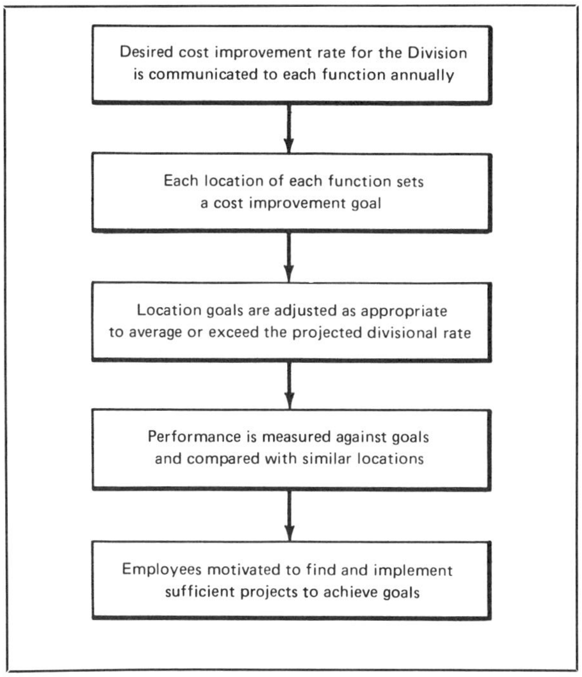

COST-IMPROVEMENT PROCESS—how it works – Fig. 4

OPERATIONS IMPROVEMENT PROPOSAL

DEPARTMENT	OPERATIONS TEAM	DATE	OIP PROJECT NO.

TO: 1. Department Head 2. OIP Coordinator | Submitted by:

TITLE:

Description: (Describe present method and proposed method. Use supplementary sheet if necessary.)

Equipment Requirements:

TO BE FILLED IN BY DEPARTMENT HEAD Capital $ ______ Non Capital $ ______

Estimated Cost to Implement: Estimated Completion Date:
Estimated By:

Estimated Savings: $ ______ ______ Period Recurring ☐
______ Direct Non-recurring ☐
Cost Avoidance ☐

Basis for Estimated Savings:

TO BE FILLED OUT BY COORDINATOR | Date Acknowledged:
Reviewed by:
(Advisory Board Members) | Date Reviewed:

Action: Responsibility Assigned to ______ For Implementation ☐
For Study ☐
Comments: Rejected ☐

Completed ______

FORM for operations-improvement proposal submitted by technical personnel asks for details – Fig. 5

OIP "IDEA"

Send Copy to each:
1. Supervisor
2. OIP Coordinator

Department: ____________ Submitted By: ____________ Date: ______

Check Box

☐ I have a cost improvement idea and would like to discuss it.

☐ Here is a brief description of my idea.

____

____

____

____

SHORT FORM simplifies submitting of OIP ideas; it is used mostly by nontechnical personnel – **Fig. 6**

MOTIVATION

Motivation is by far the most important element in a successful cost-improvement effort. Experience has taught us that it comes from the top down in an organization. You cannot maintain enthusiasm or commitment to any program for long unless the officers, managers and directors are solidly behind it. Therefore, a great deal of time was spent in earning their support.

Since motivation is so important, and since many of the measures taken are interrelated, let us examine these measures chronologically.

First we published a program guide-manual; then, the program manager and staff visited officers, managers and directors to explain the program and answer questions. During the first year, our staff also held information sessions concerning the program with all supervisory personnel in the division. An effective step was initiation of next year's goals, tracking performance against goals, and comparison of performance of similar locations.

STATUS OF IMPROVEMENT PROJECTS

____________ LOCATION

OIP PROJECT NO.	DATE REC'D.	TITLE	ESTIMATED SAVINGS ANNUALIZED	SUGGESTER	ACTION DATE	ASSIGNED TO FOR STUDY	STUDY COMPLETE DATE ESTIMATED	STUDY COMPLETE DATE ACTUAL	ASSIGNED TO – FOR IMPLEMENTATION	ESTIMATED COST OF CHANGES: CAPITAL	ESTIMATED COST OF CHANGES: NON-CAPITAL	IMPLEMENTATION COMPLETED	SAVINGS REPORTED ($)	REMARKS

PROJECT INVENTORY form for given location helps keep track of ideas through every step – **Fig. 7**

Startling Findings

After the program was operational for about 1½ years, the attitudes of C&P personnel were measured by a questionnaire. It was startling to find that only 40% of the questionnaire recipients supported the program. The main complaints centered around credibility of the reported savings and the attitude that "we have always improved—why have a program?"

Immediately, a number of changes were made. First, complete implementation of the improvement project and verification of savings were required before reporting. We issued clearer and more-detailed standards for defining a reportable project, and monitored accepted projects very closely. This was followed with a series of analyses and reports to the officers, managers and directors concerning the importance of cost improvement and the credibility of reported savings, and dealing with:

- ☐ Effect of cost-improvement rates upon profits.
- ☐ Return on investment being realized from cost-improvement projects.
- ☐ Cost of operating the program.
- ☐ Balance between cost savings and cost inflators plus product-price attrition.
- ☐ Random audits of reported location savings compared to budgets.
- ☐ Reconciliation of reported location savings to budgets.

In between these analyses, other companies were visited to exchange expertise about the management of improvement. Information sessions were then held at all locations to discuss the analyses and the programs of other companies. Such sessions were very well received by our organization.

In addition, a statement of policy firmly signifying the need for managing cost improvement and that all employees were expected to support the OIP was issued by the C&P operating committee, usually composed of four top officers of the division.

Turnaround, and More Motivation

About 1½ years after the first questionnaire, a second questionnaire was processed. Support had risen to 70% and most of the previous complaints had disappeared.

Our division was on the way now, but additional steps were taken to further improve employee motivation. First, almost all of the information obtained in the analyses listed above was placed on slides and shown to all employees. Then, personal recognition of each employee by his superior for improvement ideas was highly stressed; and this is unquestionably the most effective motivator found to date. Every employee receives a response to his idea whether or not it is acceptable.

Many locations implemented token award systems—which amount to modest merchandise awards, participant drawings for savings bonds, free dinners, and favorite parking spots—to give further recognition to participating employees. Token awards appear to be beneficial, but their value is not clear at this time. The program is further promoted in all location news-media, usually by showing pictures of the employee and his project.

Participation of employees is heavily promoted, tracked, and compared between similar locations. With minor exceptions, all location employees know throughout the year the cumulative percent of salary and hourly employees that have participated in the program. This develops, to some extent, a competitive atmosphere; some locations have reached participation levels as high as 100% of salary personnel and slightly over 30% of hourly personnel. A participant is defined as an individual who has submitted at least one acceptable cost-improvement idea in the current year.

In the past few months, the program staff has held discussions with panels made up of about 20 members that represent a cross-sectional slice of the personnel at that location. The format of these meetings has been as follows:

- ☐ A 20-min. information session about cost improvement is given.
- ☐ The panel is requested to tell the program manager how to improve OIP.
- ☐ All inputs from the panel are logged on a blackboard for all to see; later they are classified and evaluated. There are no debates or refutations permitted concerning any input.
- ☐ The panel findings are then reported to the panel members and to management.

The panels have been enlightening, and in many cases have brought about beneficial procedural changes in the program for their location. But most important of all, the employees have expressed an appreciation for being able to improve the program from their point of view.

MEASUREMENT

To successfully manage improvement, it is essential to measure the effect of changes made. One of the unique qualities of our OIP is that the reporting system gives excellent visibility of achievement versus goals. This permits each department, location, and function, as well as the division, to follow performance.

Furthermore, the reporting system permits comparison of performance between locations (Fig. 8). The plants, for example, are lined up quarterly, in descending order of cost-improvement level reached. This tactic of comparing results from similar locations was acquired from another company where it worked very well; it has also worked very well for us. While the comparison may not be quite fair, there is a very competitive desire among plants to "be on top of the list."

In general, the reportable savings are the difference between the costs associated with certain activity before and after the implementation of the project. All operating cost savings must be verified before being reported. This requires each change to be completely implemented and operational.

The savings value for recurring improvements is annualized at the "business plan" level. Nonrecurring improvements are calculated on the basis of the level of activity at the time a change is implemented. Capital savings are treated differently in that they are reported as potential savings. Potential refers to a change that may or will produce savings in the future. This is necessary be-

Study Teams are the Most Effective Cost-Improvement Technique

Types, Purpose, Organization

Type	Purpose	Organization: Composition	Organization: Leadership
Interfunctional	Find and identify cost and productivity improvements in process facilities.	Representatives from functions outside location (i.e. engineering, R&D, distribution, staff specialists) and location representatives.	Usually a manager or supervisor from location of study.
Interdisciplinary	Find and identify cost and productivity improvements in process facilities.	Representatives from several departments within the location (i.e., production, utilities, shipping, maintenance).	Usually a manager or supervisor from location of study.
Technology	Solve a known technology problem (i.e. product quality-efficiency-capacity, conservation of utilities).	Representatives from more than one function, inside or outside of location (i.e., engineering, R&D, distribution, staff specialists).	Usually changes from one member to another, depending upon aspect of project.
Survey	Review how well an activity is being performed, identify improvement opportunities and recommend changes (i.e. materials management, maintenance, machinery reliability).	Representatives from inside and outside of location (i.e. maintenance, materials management, etc., and staff specialists).	Usually an outsider from a specialty group or from another location.

Advantages and Disavantages

Advantages

- Uncovers improvement opportunities not seen by normal line management due to familiarity.
- Compensates for individual bias.
- Overcomes parochialism of department or location.
- Provides a way to define vague projects that people mean to do when time permits.

Disadvantages

- Ties up too much manpower for large studies.
- Requires expenditures not budgeted.
- Making up team is often difficult, especially for remote locations.
- Disturbs existing working relationships if improperly superimposed on present organizations.

A Team Should Be Formed:

1. When the technology problem, process facility or service area is so large or complex that a better solution will result if viewpoints, ideas and experiences of outside personnel are utilized.

2. When the skills required for the study reside in more than one function, department or individual.

3. When interplay of various departments (such as process, maintenance, utilities) within a location is needed to obtain the best results.

Procedure for Study Teams:

1. The team should usually contain only three or four members to conserve manpower.

2. The team should have at least one or two "outside" members (an "outsider" is an individual from another function or department). The outsider is essential in order to have the "whys" asked.

3. The team should be formed jointly by the plant, business and specialty* managers. Great care should be taken to select team members who will work together effectively. In order to be assured that the team receives the necessary management support, an appropriate manager should be designated as sponsor. The sponsor assists the team in establishing the required relationships for a successful study. The sponsor should also participate with the team sufficiently to stay abreast of its progress.

4. One or more of the team members should have the following qualifications: A thorough knowledge of the department's operation (people, procedures, facilities, processes, etc.); complete knowledge of all business and administrative objectives, as well as all departmental costs and accounting practices involved; a broad experience in the activity being studied (production experience if a unit is being studied, materials-management experience for a materials-management-department study, maintenance experience for a maintenance-area study, etc.).

*In C&P, specialty managers, usually with small staffs, provide staff support for such areas as materials management, cost improvement, maintenance, machinery reliability, utilities, etc.

After the Team is Formed:

It should go to the plant and visit the appropriate managers to inform them of the objectives of the study and how it will be conducted. Plant management should guide and advise the team at this point.

The team's outsiders should study the detailed cost analysis, and process or work flowsheets provided by an "inside" member of the team. During the review of costs and flowsheets, the outsiders should formulate a number of questions to be pursued later. This is done prior to the on-site study and takes two to five days, depending upon the complexity of the department.

The team should now methodically interview all supervision in the department under study and in associated departments (e.g., process, materials, maintenance, shipping, laboratory and accounting). The interview should focus on departmental problems, high-cost items, waste, and "Why do we do it this way?" The cost-improvement projects found should be defined as accurately as possible.

When the on-site study is completed (estimated time required is three to ten days, depending upon the complexity and size of the department), the team should verbally inform plant management (and business team management if the latter is involved) about its preliminary findings. As soon as possible a report should be issued by the study team to the appropriate members of management.

Plant management should follow through with further screening and defining of the improvement projects to enable entering them into the location's Operations Improvement Program.

cause of the long time-lag between the design and construction phases of these projects.

Quarterly, achievement is tracked against the goals. At that time, each department and location manager submits a progress report to his superior, using the same quarterly report format division-wide.

TECHNIQUES

In our program, *the most effective cost-improvement technique is the study team.*

The study team is a management-instituted, formal team made up of members from two or more disciplines. It is assigned to a specific problem or area, usually to improve costs, productivity, product quality, or service; it is disbanded when the assigned study is completed.

In C&P, about 80 team studies were made each year during 1971 and 1972, working primarily on costs and productivity projects spread over 14 plants. The teams found cost-improvement projects equal to about $10 million/year or one-fourth of the production function's total cost and productivity improvements. Savings averaged about $2,000 per man-day for the study team.

One of our larger operations areas used the study-team technique in all five of its processing units. When the studies were started, the process units had on inventory improvement projects equivalent to $680,000 of annualized recurring savings. The study teams found new cost-improvement opportunities equivalent to an additional $1 million of annualized recurring savings.

A recent survey showed 64 study teams active, with cost improvement projects ranging in value from $15,000 to $2 million of annualized recurring savings.

The study team has been used regularly in such areas as materials management, maintenance, and machinery reliability. It is termed a survey team in these areas, and the focus is slightly different. The main objectives are to ascertain how efficiently an activity is being conducted, make recommendations for improvement, and spread the best procedures and systems found to all other locations. Many of the improvements in the service activities also improve the process unit being serviced. For example, improvements in machinery reliability not only reduce maintenance costs but invariably result in increasing "onstream time" of a processing unit.

The purpose and organization of the four basic types of study teams most frequently used by C&P are shown in the box, p. 504. Each type is usually sponsored by the location manager and is formed by the location, business and specialty-management groups. Further description of advantages and disadvantages, suitability, and procedures for teams, are also detailed.

Training for Systematic Problem Solving

A presentation by N. R. Baker of Purdue University at the 1968 International Symposium on the Management of Improvement, sponsored by Georgia Institute of Technology, convinced us of the value of training. In numerous industrial studies of the improvement function, Baker found that two pieces of information are usually necessary for generating an improvement idea:

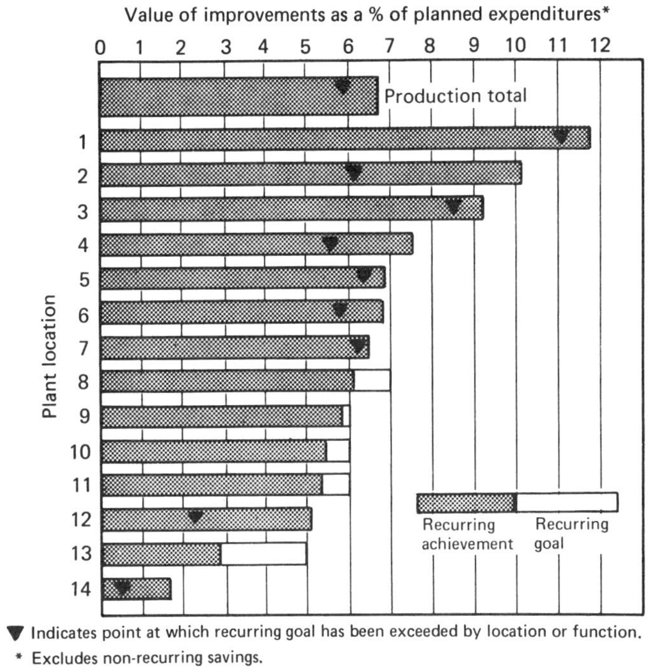

GOALS, ACHIEVEMENT record for operating plants – Fig. 8

1. Recognition of an organizational need, a problem, or an opportunity that is perceived to be relevant to organizational objectives.

2. Recognition of a means or technique by which to satisfy the need, solve the problem, or capitalize on the opportunity.

There is considerable evidence that about 80% of the time an individual will not submit an improvement idea unless he can accompany it with a solution, even if the solution is incorrect.

Consequently, in order to assist employees in finding solutions to their ideas, a member of the Operations Improvement manager's staff developed and conducts a training course in cost-improvement techniques and systematic problem-solving.

The course has been gradually adjusted to focus entirely upon the latter. The first day's study deals with techniques to find problems—cost and deviation analysis, etc.; then there is a four-week break so participants can identify and select actual problems. A two-day session covers techniques to solve problems, and a similar meeting informs on methods of implementing solutions. On-the-job applications are followed up by instructors.

Problems have not only offered an opportunity for improvement and are in endless supply, but the team mode can be used in their solution. About 600 employees from the technical- and middle-management groups have taken the course. We are now putting the material on video tape.

RESOURCE ALLOCATION

Resource allocation is divided into two parts—funding and manpower. The more-critical resource has shifted

Quarterly Report Form

	Full Year		
	Current Year Actual	Current Year Plan	Previous Year Actual
Cost Improvement Projects			
Annualized Recurring Savings Rate, %			
Annualized Recurring Savings, $MM			
From No-Cost Projects			
From Expense-Only Projects			
From Projects Requiring Capital			
Implementation Cost of Projects, $MM			
Annualized Gross Margin, All Projects, $MM			
ROI on Projects Requiring Cap. Invest., %			
Annualized ROII, All Projects, $MM			

OIP QUARTERLY REPORT for return on investment – Fig. 9

back and forth; in 1970, it was funding and currently it is manpower.

Funding

The cost for implementing improvement projects, expense and capital, is reported quarterly, at the same time savings are reported. The return on investment (ROI) is calculated for projects requiring capital and also displayed in the divisional quarterly report (see Fig. 9). Because historical data have been accumulated, estimating fund requirements has been relatively easy.

Expenditures for project implementation are controlled by varying the guideline for ROI of approved projects. Over the past five years, the required ROI has ranged from 67 to 150%. The ROI guidelines used by

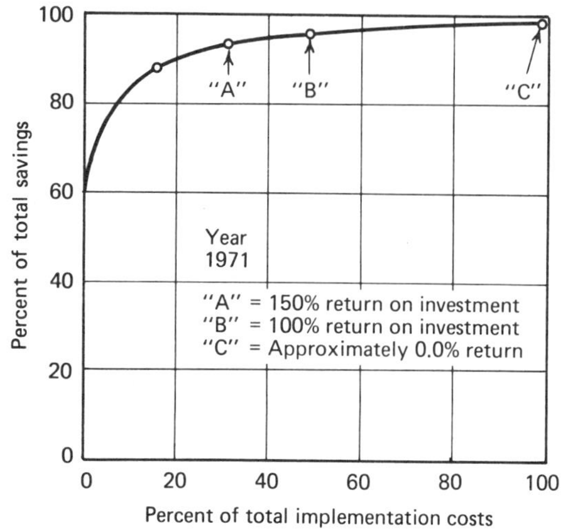

SAVING-TO-COST relationship for plant OIP projects – Fig. 10

C&P would not necessarily be applicable to other organizations. Each organization would need to establish an ROI guideline to suit its business situation.

Analyses of projects to establish savings-to-cost relationships are very helpful in determining the required funding support and ROI guidelines. Fig. 10 represents plant OIP projects implemented in 1971, in which cumulative savings expressed as a percent of total savings are plotted against cumulative implementation costs expressed as a percent of total costs. Projects are listed in order of decreasing savings-to-cost ratio. The savings include both recurring and nonrecurring projects; costs include both expense and capital. This type of analysis is useful in determining:

1. Capital and expense funds required to support the cost-improvement program.
2. Percent of savings achieved for no-cost; this is evidenced by the zero cost intercept, or 63%, in Fig. 10.
3. Relationship of project ROI to program savings and costs. Points "A" and "B" represent the 150 to 100% ROI points in the project listing. It is interesting to note that point "A" shows that 93% of the total savings were achieved for 33% of the total costs. Point "B" represents 96% of the total savings and 50% of the costs. In other words, half of the costs were required to achieve the 4% of total savings from points "B" to "C".

The average ROI for this group of projects was about 40%, which is still very good. There are two causes for the lower-return projects between points "B" and "C". First, many of them were justified for some other purpose than cost improvement, but were reported under OIP because some incidental cost savings were obtained. Second, and more significant, they represent cost-improvement projects that did not work as well as planned.

Manpower

The distribution and manufacturing functions contribute about 90% of the C&P division's operating cost savings. In these two functions, the manpower to find a sufficient number of projects to meet high goals is no problem. However, manpower to evaluate, engineer and implement projects is usually limited.

In all other functions (engineering, R&D, accounting, sales), this resource is not a constraint to the improvement effort. The finding of improvement projects is usually the limiting factor.

SOLVING THE DIFFICULT PROBLEMS

The excellent results obtained through the use of the system were very difficult to achieve. This is quite expectable because acceleration of the rate of change was being attempted, and most people tend to resist change in the first place. The crucial problems concerned attitudes, so there was difficulty in defining these problems and uncertainty about when they were solved. The major problems:

Chemicals and Plastics developed the program as a part of the production function, and the other functions (distribution, accounting, engineering, R&D, sales) felt, to a noticeable degree, that a production-function program was being forced upon them. This problem has been largely overcome by maintaining the same general structure of the system while permitting each function to modify some of the program detail to suit its business.

This problem could have probably been avoided if the development of the program had been assigned to a team having representatives from each function. This multifunctional team should have also managed the pilot program, regardless of the location selected for the test.

The C&P pilot program was very successful. During the three years 1962-1964, there was a rudimentary cost-improvement system in operation; projects were submitted annually, progress was reviewed at irregular intervals, and records of achievement were kept.

However, there were no goals, no continuous input of ideas, and only minimal emphasis on cost improvement. The average saving for these three years was $962,000 per year.

In 1965 and 1966, following introduction of the pilot program, the average cost-improvement saving was $2,564,000 per year, a 167% increase. These results were very impressive to us because we were part of the test, but many of the other plants and most of the other functions were quite dubious.

The attitude among management personnel of "We have always improved; why have a program?" was a very substantial problem during the implementation phase of the program. We realized that improvements had always been made, but argued that they had not been done on a managed basis. Most people agreed with us that if cost improvement were managed, more improvement ideas would be generated and carried out than if improvements were left to individual spontaneity.

There are still a few people today who claim the improvement level would have risen from 3.7 to 7.0% of costs without the improvement system. However, the overwhelming majority of our management view this position as illogical today, since we did not know and were not tracking our cost-improvement rate prior to implementing the system. Furthermore, the concept that improvement could offset the rising costs of labor and materials had not occurred to a large portion of the organization prior to that time. Higher product sales prices and volumes were looked to as the solution. Patience, education and results were required to gradually change this attitude.

The attitude of "I manage cost improvement by closely monitoring budgets and product cost per unit" was a persistent problem. Some of our most cost-conscious people believed that controlling costs within budget or standard was sufficient. This can be readily understood because of the formal systems that had been in existence. It was argued that this was only part of good cost management, and that cost-improvement goals and tracking of achievement against these goals would also be necessary to reach and maintain the lower costs desired.

Many analyses and reports—covering subjects such as the effect of cost improvement upon profits, cost improvement versus cost inflation plus product–price attrition, the results of cost-improvement study teams, and the participation of employees in the improvement effort—slowly eroded this attitude.

Disbelief of reported savings was used as an excuse for not supporting the program. The C&P cost-improvement program is based upon defining, implementing and reporting the verified savings from individual projects. An add-up of all projects equals the total savings achieved.

Initially, many of the C&P people thought that the reported cost savings were badly inflated. In fact, some people used this as an excuse for not participating in the program. The first approach to solving this credibility problem was an accounting reconciliation of the reported savings to plant costs. Using the reconciliation formula, the preceding year's cost, plus or minus changes from the preceding to the current year, minus Operations Improvement savings, should equal the current year's actual cost. Any major error in cost-improvement claims would result in an obvious variance from the current year's costs.

This reconciliation process was used on several plants, and the deviations were 0.4 of 1% or less of the actual current year's cost. If this deviation were assigned entirely to the Operations Improvement Program, it would have amounted to less than 10% of the savings claimed. (See Fig. 11 for one of the reconciliation analyses.) The reconciliation analyses greatly strenghened the credibility of the reported savings, but the process was too lengthy.

Currently, an auditing process is used. This calls for reviewing at year-end all projects reported throughout the year on an individual basis with the person reporting the project savings. Since projects are submitted, implemented, and savings reported, throughout the year, the audit will result in some adjustments, up and down, but locations have averaged verifying 95% or more of their reported savings. The combination of reconciliation analysis plus random audits has essentially solved the problem.

Complaints that the program required too much paper-

Reconciliation Statement

		Millions of Dollars
1. Previous Year's Actual Plant Costs		111,469
2. Net Changes - Previous Year to Current Year		
Period Costs (Ex OIP)	+ 2,065	
Direct Costs (Ex OIP)	+ 1,310	
Previous Year's Nonrecurring OIP	+ 227	
Savings in Current Year Only from OIP	- 4,737	
3. Previous Year's Actual Plant Costs + All Above Adjustments (Should = Current Year's Actual Plant Costs)		110,334
4. Current Year's Actual Plant Costs		109,869
5. Unaccounted For		465

TOTAL PLANT COST reconciliation form – Fig. 11

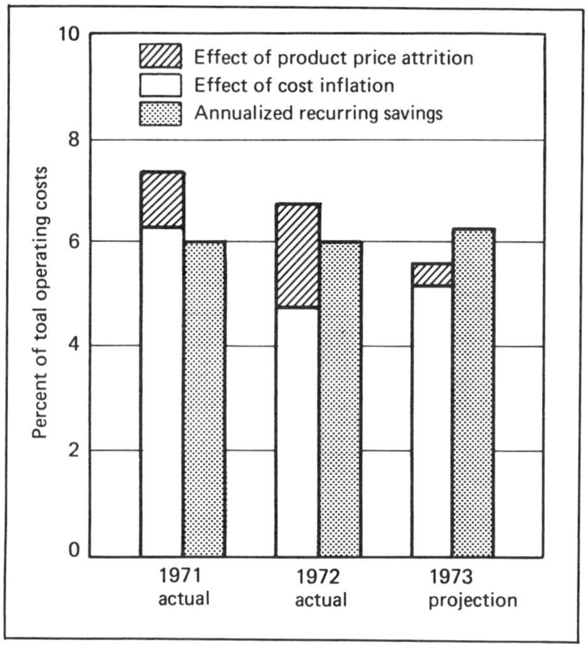

PRICE ATTRITION plus inflation vs. savings—Fig. 12

work. Early in the program's development, excessive paperwork was clearly a problem according to each of the attitude surveys of 1968 and 1969. Each part of the system has been studied repeatedly and the paper flow has been reduced about 50%. The top officers receive only a one-page report quarterly. Each management level is given only enough report detail for the position. All projects saving less than $5,000 per year are lumped together in quarterly reports; details for these projects are held at the locations concerned. All projects to be circulated to the other locations are coded; this results in the circulation of about 10% of the approximately 7,000 projects implemented per year.

The attitude surveys also indicated almost unanimously that a report was needed. The agreement was equally strong, 87%, that the report should not be published by the accounting and financial function because it could not effectively evaluate the projects against reporting standards.

Complaints that the program cost too much to operate. There was very little substance to this problem but until the cost of the program was determined and publicized, a number of people resisted on this basis.

The average administrative cost per project was determined to be 10 man-hours, or about $66. This is equivalent to administrative costs for all C&P locations in 1972 of about $500,000, which was less than 1% of the annual savings. Not only was the cost very small, but most of it was not "out of pocket" dollars. Once the facts were known, the problem vanished.

Difficulty in relating savings to costs and profits. Initially, cost-improvement savings were reported as recurring or nonrecurring, which made it difficult to relate them to profits. This has been largely solved by converting the savings to annualized gross margin, which deducts all expense costs, and to return-on-investment income (ROII), which deducts both expense and depreciated capital costs. (See Fig. 9.) Gross margin and ROII are familiar and well-understood terms by all C&P management.

An additional step taken recently to respond to some remnants of this problem is to show the balance between recurring savings and cost inflation plus product-price attrition (Fig. 12). This type of chart has received a great deal of praise from a large portion of the organization because it displays in an easily understood manner how cost savings are being offset by rising costs plus decreasing product prices. The chart is very educational in that it has shown unfavorable but improving balances in 1971 and 1972 and a projected favorable balance for 1973. Today there is no apparent problem in relating cost savings to profits.

Need for assistance of location management in explaining the need for cost improvement. This deficiency was consistently found when discussions were held with cross-sectional panels about how to improve the program. The cause for this deficiency appears to be the lack of cost data to develop inflators for labor, materials and services. Employees were consistently enlightened and impressed by the visual aids that related cost-improvement savings to rising costs and decreasing product prices. This type of visual aid is now in general use by all location management, and the problem is essentially solved.

Realization that acceptance of a formal cost-improvement program will come slowly. The point here is that the causes for not accepting a formal program are very numerous—but none are really substantial. In C&P, questionnaires were used to determine the personnel attitudes toward the Operations Improvement Program. All levels of personnel were covered. In the first questionnaire, early 1968, only 40% of the respondees supported the program. In late 1968, this level had risen to 70%, and today support is essentially unanimous. This problem was overcome primarily through an extensive educational process conducted by the locations in cooperation with the program manager and staff, coupled with the impressive results that locations were achieving. ■

Meet the Author

N. S. Eastman is manager of the Operations Improvement Program and also manager of materials-management activities for the Chemicals and Plastics Div. of Union Carbide Corp., South Charleston Technical Center, P.O. Box 8361, South Charleston, WV 25303. Previously an assistant plant manager before assuming current staff-management assignments, he holds a B.S. in chemical engineering from Newark College of Engineering; he is a member of AIChE and the American Production and Inventory Control Society.

How To Estimate Operating Costs and Depreciation

Meaningful estimates of operating costs are important not only in evaluating the overall economic potential of a new process but also in optimizing the design and scope of any capital project. Here is a practical guide to modern practice in this key area.

LESLIE C. JENCKES, Sun Oil Co.

Even if he is not a cost specialist, the engineer in the chemical process industries is constantly exposed to situations where operating costs must be estimated, evaluated or optimized. Hopefully, the techniques and guidelines in this article will prove helpful in preventing errors of omission or commission; Part II, which begins on p. 513, will show how the techniques are applied.

Types of Estimates, and Their Purpose

In the early phases of developing a new process, when it is necessary to evaluate economic possibilities (e.g., compared to other processes that are competing for time and equipment in the laboratory), it is usually not yet possible to optimize either capital or operating costs. Sensitivity analyses must therefore play a major role in evaluations, and in guiding future development work. Estimation of operating cost must come from laboratory data, or literature articles on similar processes, or a mixture of the two. Often, this results in a simple "cents per unit of charge or product" for total operating cost.

When a project advances into the preliminary engineering stage, the detail involved in estimating operating cost should expand into several major areas, as shown in the "component" box on this page.

At this stage, the estimating of operating costs is a most important part of the evaluation and optimization of process and mechanical design alternates. For instance, unit-cost comparisons for steam and electric power will go a long way in deciding what type of drivers to use on rotating equipment. The cost of fuel plus the efficiency of fired heaters—versus the cost of steam, insulation or heat exchange—might have a major effect on process design. Exotic instrumentation or computer control to save operating manpower, and the use of special materials to lower maintenance costs, are further examples of how the estimation of operating costs can help optimize the mechanical design of a plant.

Setting up the basis for costing can be the most difficult part of this type of operating-cost estimating. For instance, should steam be valued on a heat, pressure, or some other basis? Can you realistically predict manpower savings with today's labor-union involvement?

It is also possible that a comparison between actual and predicted operating costs can reveal non-optimum procedures after the plant is operating.

Components of Operating Costs

Fixed charges:	Operating labor Maintenance labor Maintenance materials
Consumables:	Fuel, steam, power, water Chemicals and catalysts*
Departmental charges:	Maintenance overhead Instrument department Laboratory Technical service
Distributed charges:	Taxes and insurance Distribution and shipping Corporate overhead
Non-cash costs:	Depreciation

* Raw materials are not considered part of operating costs, but are part of manufacturing costs. (The illustrative example in Part II will include development of raw-material costs and byproduct credits.)

This article is partly based on a presentation at the 14th National Meeting of the American Assn. of Cost Engineers (San Francisco, June 22, 1970), and partly on additional material supplied by the author.

Originally published December 1970.

Departmental Manning Estimate for a Typical Grass-Roots Plant

Department	Job Title or Function	Salaried People	Hourly People
Administration	Manager	1	
	Superintendent	1	
	Personnel Director	1	
	Safety Director	1	
	Medical Director	1	
	Secretaries		3
Accounting	Chief Accountant	1	
	Accountants	12	
	Clerks		8
	Secretaries		2
Maintenance	Superintendent	1	
	Craftsmen		30
	Clerks		6
	Secretary		1
Engineering	Director	1	
	Engineers	6	
	Clerks		2
	Secretary		1
Laboratory	Chief Chemist	1	
	Chemists	3	
	Clerks		2
	Secretary		1
Operations	General Foremen	3	
	Shift Foremen	13	
	Operators		26
	Helpers		35
	Clerks		3
	Totals	**46**	**120**

When the process and mechanical design bases for a plant are reasonably well finalized, it is time to reevaluate operating costs on a more firm basis and to analyze these costs to determine needs for further optimization. At this stage of a project, funds for final design and construction will normally be approved if the economic evaluation still looks satisfactory.

Estimating the Fixed Charges

Operating Labor—Firming up the estimate of operating labor usually requires working with plant management personnel for number of men needed, and with accounting people for proper unit costs. If the capital project is closely associated with a plant that is already operating, perhaps the estimator need only think in terms of adding some people to an existing labor setup. But for a large grass-roots installation, or for major modernization steps involving shutting down of existing plants, he may have to set up a complete manning table, similar to the one shown in the table above.

Setting the basis for these labor estimates usually requires the advice of a top plant-management man who is familiar with long-range company staffing goals, and with the intricacies of union agreements. It should be remembered that only those labor changes associated directly with addition of the new process should be considered. General optimization of labor or any other operating cost should be considered accomplished in the Base Case.

Modernization steps that involve overall reductions in operating manpower should include a study of average labor age, pension possibilities, early retirement, etc., to evaluate the true amount of money to be saved.

Maintenance—Getting a legitimate basis for accurately forecasting maintenance costs is very difficult. The most used method of determining these costs is a yearly cost based on a percent of investment. This method requires either previous experience with the same or similar type plants, or a complete reliance on the literature.

If maintenance work is going to be contracted out, it might be possible to get cost estimates from the outfit that will do the work. The contractor, however, would probably not have many more tools for making the estimate than the owner.

On grass-roots plants or major modernization jobs, it might be desirable to set up a maintenance department, complete with manning table and supplies, whose costs could be estimated from scratch.

One big drawback to using the percent-of-investment method is that it does not take into account investments that are made to produce maintenance savings (e.g., via alloys or other premium materials, designs with easy accessibility, and other labor-saving features) unless your experience or literature references also include these items. Using a standard percent of investment will increase the maintenance-cost estimate as more-expensive alloy materials are used, rather than decreasing it. If the plant is likely to put particular emphasis on ease of maintenance, a lower percentage should be used than for a plant where minimizing the initial investment will be the primary goal.

One method of estimating maintenance costs that appears to work reasonably well (especially when applied to classes of processes) was developed by a major pulp and paper manufacturer and published in the Feb. 9, 1959, issue of CHEMICAL ENGINEERING.* This method relates maintenance dollars to plant investment, times plant age. Recent and predicted future inflationary trends would have to be taken into account when using this method today.

Estimating the Consumables

Requirements for the various consumable items can be estimated from preliminary process design, experi-

* See: "Maintenance Cost Estimation . . . Is This a Better Tool for You?," *Chem. Eng.*, Feb. 9, 1959, p. 140.

ence with similar processes, or from literature unit-cost estimates (adjusted for the correct location).

Care must be taken to see that the correct unit prices are applied: The cost of chemicals, catalyst, power and water must include all expenses necessary to make them available at the consuming site—such as delivery, handling, storage, treating, etc.

One pitfall in the application of unit costs could be power used to pump water. This cost must be included with either power or water but not with both.

Fuel costs may be based either on the purchase of externally obtained fuel or on a portion of internally produced fuel that has value as a salable product.

All consumable volumes should be firmed up, using the results of cost and process-design optimization studies.

If steam is purchased from an outside source, this cost is relatively easy to determine. If, however, steam-producing facilities must be installed as an offsite adjunct to the process unit, then the cost of producing this steam must be calculated, and must include, at the very least, boiler-feedwater and fuel needs. Labor and depreciation charges are sometimes included. If the process being evaluated also produces some steam for export, then the value of this exported steam should be set at the cost of an alternate source of steam to the user, and credited to the operating cost of the producing process.

Estimation of catalyst costs is limited to the replacing of worn-out catalyst, because the original loading is capitalized with plant construction costs. This operation could take the form of physical replacement or simple regeneration, and will involve both labor and material charges.

Evaluation Types: What To Include

Before discussing departmental and distributed charges, let us quickly compare the various kinds of evaluations, which include (1) Full Cost, (2) Out-of-Pocket or Net Additional, and (3) Incremental. Full Cost may be defined as the total of all costs *including distributed overheads* associated with providing a service or producing a product. Out-of-Pocket or Net Additional costs are those costs actually incurred because of the addition of the new productive facilities being evaluated. Incremental Costs are the costs associated with producing an additional increment of product or service. These are limited to cost components that vary in dollars per unit of time as plant prodution volume changes—e.g., the additional cost of producing at full capacity vs. at some partial capacity.

Normally, Out-of-Pocket (net additional) costs are the main costs used in evaluating the profitability of a project. Incremental Costs could be used in determining the most economic size for the plant. The estimation of Full Costs in evaluating a new process is valuable for post-audit purposes since accounting records will include distributed overheads.

For simplicity, this Cost File will assume that the process (plant) being evaluated is being added to a rather large complex of associated plants so that its effect on existing labor and overhead costs will be minimal.

Departmental Charges

Net Additional costs for departmental charges are estimated on the basis of new manpower hires or additional equipment needed. Departmental charges for Incremental production are usually zero. Full Costs are based on adding new costs to total existing costs for the complex and distributing the new total over all units by some agreed-upon method, e.g.:

Maintenance Overhead....Percent of maint. labor
Instrument Dept.........Number of instruments
Laboratory...........Number of tests required
Technical Service................Value of plant

Estimating the Distributed Charges

Distributed Charges to entities other than "Departments" include taxes, insurance, distribution, and corporate overhead.

Accurate estimation of the cost of taxes and insurance could be a very involved process. For instance, some States base assessments on total value of the plant, while others base it on only a part of this value, such as land and buildings. One reasonably efficient way of estimating this cost is to determine the average historical cost of taxes and insurance as a percentage of the replacement value of plants covered (including offsites). No savings should be taken for plants that will be shut down unless they are demolished. Both Net Additional and Full Cost should be identical, while no cost is applied for Incremental production.

The type of operation determines costs for distribution, shipping, etc. For instance, the Incremental Cost of pumping additional liquid through a pipe would be zero in the "distribution and shipping" category, because the power or steam providing the motive force would be charged under utilities, whereas the Incremental Cost of packaging additional production would include at least the materials involved, and may include labor (i.e., if an idle line will be started up).

Full Cost for distribution and shipping is normally charged to plants on the basis of units of charge or product. For instance, the total cost of all loading and shipping operations in a petroleum refinery might be distributed over a group of plants, based on their production of certain products common to all.

Corporate overhead costs include support activities such as engineering, accounting, corporate management, research and development, etc. In the normal estimation of Net Additional or Incremental Costs for adding equipment to an existing installation of reasonable size, these corporate overhead costs would be nil. For Full Cost, however, overhead costs may be assigned on the basis of labor required or plant value, or on any other basis that management feels will give the most realistic picture. This is the category that is often used for reflecting the cost of direct employee benefits such as pensions, vacations, company-sponsored activities, etc. This particular portion must be included in Net Additional evaluations as well as Full Cost.

Depreciation: Three Alternatives

In most cases, the recovery of funds invested in depreciable assets—i.e., depreciation—probably would not be considered as an operating cost. However, an estimate of such depreciation is just as necessary to a complete economic evaluation as is an estimate of labor and utility costs, so that depreciation should preferably be handled as a "non-cash" operating cost. In this way, an additional operating-cost item is subtracted from pretax profits, which reduces tax liability.

Investments in land and working capital are not subject to depreciation; this money can presumably be recovered in full.

Yearly depreciation charges are a function of the type of depreciation, and the depreciable life of the project. The type of depreciation is usually determined by historical company preference and/or special considerations associated with the project, providing agreement can be obtained from the Internal Revenue Service. Government publications are available to assist in setting the allowable depreciable life of assets.

The term depletion is applied to the recovery of the value of mined mineral deposits, and amortization to recovering items such as the paid-up cost of royalties.

Three different types of depreciation are available for use in negotiations with the Internal Revenue Service: (1) Straight Line, (2) Sum of the Digits, and (3) Double Declining Balance. The Straight Line method allows recovery of depreciable value in equal increments over the depreciable life of the asset. The other two methods allow greater recovery of value early in the project life because of lower tax liability during this period; the higher early cash flows can result in greater profitability.

The Sum of the Digits method applies a decreasing fractional part of the total depreciable value for each succeeding year of depreciable life. The denominator of this fraction is constant, and is equal to the sum of the digits from 1 to (and including) the last year of depreciable life. For example, if a project has a 10-yr. depreciable life, the denominator would be $1 + 2 + 3 \ldots\ldots + 10$, or 55. The numerator of this fraction for the first year is set equal to the depreciable life, or 10 in the case of our example. The numerators then decrease by one for each succeeding year, so that the fraction for the tenth (last) year of depreciable life will be 1 over 55.

The Double Declining Balance method doubles the yearly percentage allowable under the Straight Line Method, but applies this percentage only to the remaining depreciable value. Since the depreciation rate is applied to the undepreciated value, there is always some remaining investment at the end of the depreciation period. For purposes of economic evaluation, this balance is usually charged off in the last year of depreciable life.

The following comparison of depreciation methods applies to a project with $1 million of depreciable value and a 10-yr. depreciable life:

Year	Straight Line, $	Sum of Digits, $	Double Decl. Balance, $
1	100,000	181,818	200,000
2	100,000	163,636	160,000
3	100,000	145,455	128,000
4	100,000	127,273	102,400
5	100,000	109,091	81,920
6	100,000	90,909	65,536
7	100,000	72,727	52,429
8	100,000	54,545	41,943
9	100,000	36,364	33,554
10	100,000	18,182	26,844
Total	**1,000,000**	**1,000,000**	**892,626**

In making studies, the evaluator must select the type of depreciation to be used and make an educated guess as to the number of years the IRS will allow for investment recovery. It is usually advisable to set different depreciation-recovery periods for onsites and offsites. Additions being made to existing plants that are not fully depreciated usually take special handling. It would be advisable to check with the company tax department for advice in such cases.

So much for the main operating-cost categories. In the next article, we will illustrate the development of these costs via an economic-evaluation example. ■

Meet the Author

Leslie C. Jenckes is a Senior Special Studies Analyst in Sun Oil Co.'s Manufacturing and Engineering Dept. (1608 Walnut St., Philadelphia, Pa. 19103). A graduate of Temple University with a B.A. in economics, he has worked in the field of cost engineering for the past 18 years. He is an active member of the American Assn. of Cost Engineers, and a past chairman of its Operating Cost Committee.

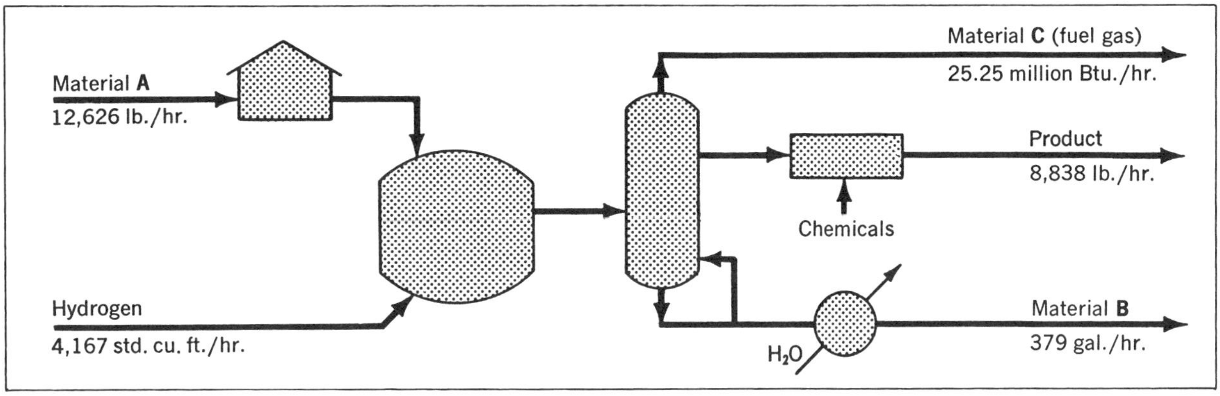

This flowsheet is a basis for a concise case study that illustrates efficient ways of . . .

Developing and Evaluating a Manufacturing-Cost Estimate

LESLIE C. JENCKES, Sun Oil Co.

Having reviewed the various components of operating costs in the previous article, let us now develop the data needed to perform a simple but realistic economic evaluation. To make this illustrative case study more meaningful, we will expand the area of consideration from operating costs to manufacturing costs, which allows us to include the cost of raw materials and the value of byproducts.

The purpose of this evaluation is to determine the cost of producing a chemical under varying circumstances. (The method can also be used for nonchemical products.) Unit costs and volumes have been set to give reasonable results and are not meant to apply to any specific processes or geographic areas.

The starting point for making an evaluation is often a flow diagram such as that shown on this page. The amount of detail contained in such a diagram depends on the developmental stage of the process at the time the evaluation is being made. Note that the quantity/time measurements may be expressed in different units for different streams.

Originally published January 1971.

Table I shows developments of the net cost (cost of charge material minus byproduct credit), and the material balance around the process. Checking the material balance is especially important if a flowsheet is not available. Often, the engineer supplying the basic information does not realize just what is needed for an economic evaluation.

Development of operating costs is the next step. Table II shows details of these costs on a Net Additional (Out-of-Pocket) basis.† I am assuming that the project is far enough along so that most of the items can be estimated in a reasonably definitive manner. The calculation shows the manufacturing cost of the main product to be 5.14¢/lb. at full-capacity operation on a Net Additional basis. (Remember that this cost does not include sales expense, distributable

†As discussed in the previous article, Net Additional Costs represent only the out-of-pocket operating costs incurred because of the addition of the new production facilities, whereas Full Costs include overhead distributions. The capital investment for the new production facilities is assumed to be $8 million; this figure is used as basis for maintenance expense in Table II.

Material Balance and Net Cost in Illustrative Case Study—Table I

	Std. Cu.-Ft./Hr.	Million Btu./Hr.	Gal./Hr.	Lb./Hr.*	Thousand Units/Yr.†	$/Unit	Thousand $/Yr.
Charges: Material A	......	......	....	12,626	100,000	0.025	2,500
Hydrogen	4,167	......	....	22	33,000	0.0005	17
Subtotal				12,648	....	....	2,517
Byproducts: Material B	......	......	379	2,577	3,000	0.10	300
Material C	......	25.25	....	1,233	2×10^8	0.40**	80
Subtotal				3,810	....	....	380
Main product	......	......	....	8,838	70,000	?	?
Net cost (charges minus byproduct credit)							2,137

* Basis for conversion to lb./hr.: Material B, 6.8 lb./gal.; Material C, 20,500 Btu./lb.; Hydrogen, 5.32 lb./mcf.
† A year made up of 330 days, with 24-hr. operation.
** I.e., 200,000 million Btu. @ 40¢/million Btu.

Development of "Net Additional" Operating Costs—Table II

Cost	Basis	Thousand $/Yr.
Hourly labor	10 men @ $5/hr.	104
Maintenance labor	$0.02 \times \$8,000,000$	160
Maintenance material	$0.01 \times \$8,000,000$	80
Fuel, 3×10^{11} Btu	@ 40¢/10^6 Btu.	120
Power, 20,000,000 kwh.	@ $7/$10^3$ kwh.	140
Water, 600,000,000 gal.	@ 5¢/10^3 gal.	30
Chemicals and catalysts.		150
Instrumentation	1 man @ $12,000	12
Technical service	1 man @ $15,000	15
Taxes, insurance	$0.005 \times \$8,000,000$	40
Dir. employee benefits	$0.40 \times \$264,000$	106
Operating-cost subtotal		**957**
Depreciation, onsite, $5 million over 14 yr.		357
Depreciation, offsite, $3 million over 20 yr.		150
Total		**1,464**
Net cost (charges minus byproduct value)		2,137
Operating cost (Net Additional basis)		1,464
Cost of producing main product		3,601
Cost per lb. @ 70 million lb./yr.		5.14¢/lb.

Development of Operating Costs on a "Full Cost" Basis—Table III

Additions	Thousand $/Yr.
Operating supervision	16
Fuel	6
Power	2
Water	6
Maintenance overhead (100% of maint. labor).	160
Laboratory (8 tests/day @ $2/test)	6
Dist. and shipping (0.2¢/lb. of product)	140
Corporate overhead (70% of labor + 1% of investment	275
	611
Summary	
Net additional cost	1,464
Less offsite depreciation	− 150
Less direct employee benefits	− 106
Plus above additions	+ 611
Total	1,819
Net cost (charges minus byproduct value)	2,137
Operating cost (Full Cost basis)	1,819
Cost of producing main product	3,956
Cost per lb. @ 70 million lb./yr.	5.65 ¢/lb.

overheads, or funds for capital recovery over and above depreciation.)

The next step is to convert to Full Cost by adding distributable overheads, as shown in Table III. The "additions" cover the following costs, of which some will cause a corresponding decrease in distributed costs charged against existing operating facilities:

- *Supervision*—Partial cost of supervisors already on the payroll who will now assist in the new operation.
- *Fuel, Power, and Water*—Mainly depreciation charges for offsite items associated with treating and distribution of utilities.
- *Maintenance Overhead*—Distributable costs of maintenance supervision, planning, office space, heat, lighting, etc.
- *Laboratory*—Distributable costs of labor and material for running laboratory tests.
- *Distribution and Shipping*—Mainly depreciation

Cost of Operation at 75% Capacity—Table IV

Fixed Costs at 75% Capacity	Thousand $/Yr.
Hourly labor	104
Maintenance labor	160
Maintenance materials	80
Fuel (25,000 × 10^6 Btu. @ 40¢/10^6 Btu.)	10
Water (400,000,000 gal. @ 5¢/10^3 gal.)	20
Instrumentation (1 man @ $12,000)	12
Technical service (1 man @ $15,000)	15
Taxes and insurance (0.005 × $8,000,000)	40
Direct employee benefits	106
Subtotal	547
Depreciation (onsite and offsite, as before)	507
Variable Costs	
Fuel (2,750 Btu./lb. of charge @ 40¢/10^6 Btu.)	83
Power (0.2 kwh./lb. of charge @ $7/$10^3$ kwh.)	105
Water (2 gal./lb. of charge @ 5¢/1,000 gal.)	8
Chemicals and catalysts ($150,000 × 0.75)	113
Subtotal	309
Summary	
Net cost ($2,137,000 × 0.75)	1,603
Operating cost (see above)	1,363
Cost of producing main product	2,966
Cost per lb. @ 52.5 million lb./yr.	5.65 ¢/lb.*

* Compared with 5.14¢/lb @ 70 million lb./yr.

charges associated with distribution, storage and shipping of raw materials and products.

• *Corporate Overhead*—Distributable costs of top company management, accounting, engineering, R&D employee fringe benefits, etc.

Certain items must be deducted from the Net Additional Costs since even though they are Out-of-Pocket Costs, they are handled in a different manner on a Full Cost basis. Net Additional offsite depreciation charges are deducted and distributed to "utilities" and "distribution and shipping," as shown above. Also, direct employee benefits (fringes) are now handled as part of "corporate overhead."

As can be seen in Table III, the manufacturing cost of the main product on a Full Cost basis and at full-capacity operation is 5.65¢/lb. or 0.5¢/lb. higher than Net Additional. This cost breakdown should compare with the accounting books that are set up for the project, so that it can be used as a basis for later post-audit work.

More often than not, it is important to know how costs change with varying production rates. Table IV shows this calculation for manufacturing costs at 75% of capacity. Total costs must be divided into fixed costs (those that do not vary with production rate) and variable costs. Note that some elements of cost can have a basic quantity that is fixed and not subject to production rate, and also an incremental amount that is variable.

I have assumed that the raw-material and byproduct cost and credit are not subject to supply and demand changes, and vary directly with production rate. This is, however, not always true and should be investigated before assumptions are made. Usually, this calculation is made at several production levels, and a curve is included in the economic evaluation.

Payout and Sensitivity

Calculating the gross income needed to show a given investment payout merely involves dividing the funds to be recovered by the payout years desired, and back-calculating through depreciation, taxes, operating cost and net cost.

For instance, if we desire a 5-yr. cash-flow payout for our $8-million investment, this represents $1,600,000 per yr., of which depreciation will supply $507,000 per yr. The balance is $1,093,000, and we must add a like amount to it because of taxes (assuming a 50% tax rate). To the subtotal of $2,186,000, we then add $1,464,000 for operating costs, and $2,137,000 for net cost (i.e., as per Table II). The total is $5,787,000 per year, and in order to produce this gross income from 70 million lb. of annual production, we need an income 8.27¢/lb. This is on a Net Additional basis and does not take the time value of money into account.

A valuable type of cost breakdown (in ¢/lb. of product) for analysis might be:

	Net Additional	Full Cost
Charges	3.60	3.60
Byproducts	(0.54)	(0.54)
Operating costs	2.09	2.60
5-Yr. cash recovery	3.12	3.12
Total	**8.27**	**8.77**

A more detailed analysis to show the comparative effects of such things as a 10% variation in investment cost, a 1¢/lb. change in raw-material costs, and variations of operating-cost items most subject to inaccurate estimation will supply information on sensitivity. ■

Before You Buy, What About Maintenance Costs?

Here are some guidelines to apply—with respect to maintenance and repair costs—when selecting new equipment to replace old units.

ERNEST W. FAIR, Technical Writer

"Since this equipment has a new design, its maintenance and repair needs will be less than on the present unit. We will be able to save enough on maintenance costs alone in just a few years, to more than pay for the new unit."

Statements such as the preceding one—which I am sure you have heard before—occur far too often. They quickly dismiss other cost factors involved in replacing old equipment in chemical plants.

Yet, improvements in basic design or construction of a new unit do not, in themselves, guarantee that maintenance costs will be below those involved with existing equipment.

One should never take for granted that the replacement unit will be less expensive to operate than the present equipment. It is equally unsound to assume that, because the new unit may be a small one, the actual maintenance and repair costs will be insignificant enough to make this expense unimportant in selecting a specific unit from among several that may be available.

The guidelines that follow represent practices of several companies that are well aware—through costly experience—of the value of applying such practices.

Compare Areas of Maintenance

1. Compare the new equipment's areas of required preventive maintenance with those of the present unit. Similar areas can have marked differences in how much maintenance attention and repair may be needed. Cost estimates can be made, based on existing records.

2. Make a careful search for new areas of maintenance on the new machinery that are not covered by present experience within your company. It may be worthwhile to inquire outside, among other concerns in the field to secure such information. As accurate a picture as possible should be obtained before any new purchase is confirmed.

3. Metals and alloys of which the new unit is made can affect future maintenance and repair costs. In most cases, improvements in materials during the past few years have lowered maintenance and repair costs. It is never safe, however, to take this for granted. Make a study to determine whether such changes will definitely result in lower costs.

4. Examine each specific part in the new unit for possible maintenance cost. This procedure may be tedious, but will produce a true picture of future maintenance and repair costs. Doing this does not require taking apart for visual or technical inspection each part. Studying a diagram of the machine, or even an actual overall inspection, can reveal much information. This can even be done by focusing attention primarily on the new and different parts involved in the new equipment.

5. Automated operation of the new unit may or may not affect maintenance and repair costs. If automation

Originally published March 1974.

does away with high repair costs caused by damage done by operators, a decisive gain is made. The cost of keeping a complex, automated unit in operation, however, may be higher than for the existing equipment. Ordinarily, such an increase is more than offset by the gain in productivity.

6. Compare lubrication costs for the new unit with those of the present equipment. When lubrication is a steady maintenance requirement, this can be of dollars-and-cents importance. A good look at not only the time and amount of lubricants needed in each specific case should be made, but consideration should also be given to the type and kind of lubricants required. Great differences in quality can add to costs.

Moving Parts, Speeds

7. The operating speed of the new unit may be a major maintenance factor. In general, the higher the speed, the more maintenance attention will be needed. If the new equipment is fast enough to increase production, the maintenance and repair costs are likely to be higher. Check records on present equipment for maintenance and repairs directly attributable to operating speed; then project such probable costs as they may apply to the new unit being considered.

8. Consider next the difference in the number of moving parts involved in performing each specific task within the unit's complex operation. More moving parts may create added maintenance and repair costs, fewer may reduce both. Wherever moving parts exist, maintenance will be needed.

9. Have changes been made in the number of moving elements on the equipment under consideration (as compared to the existing unit) to the extent that maintenance will, in all probability, become insignificant? This has been an area of great advance during the past few years; and it may well be even greater in the immediate future.

10. Has the amount of vibration in the new equipment reached a considerably lower level than that existing in the old unit? If so, this is certainly a major plus in favor of the new unit.

11. Compare closely the prime points of conventional wear in the new and existing equipment. Most maintenance involves moving parts, bearings, etc., which will be greatly similar to the ones on the existing equipment. When improvements in the new unit are extensive in such parts, the reduced maintenance costs may be of major importance.

Maintenance, When and by Whom?

12. Is it possible that maintenance costs on the new unit can be materially reduced by contracting outside the company for handling such repairs? If outside maintenance service is readily available, it is highly probable that the same services can be applied to other new equipment recently purchased.

13. Will the new equipment lend itself to maintenance and repair functions at any time when they are needed, or must these be done at periods when overtime may be involved, or when shutdown of other units in line with it occurs? The more complex modern equipment becomes, the more important this factor becomes in maintenance systems and costs.

14. Wage factors with respect to equipment maintenance always require close study. A glance at maintenance payroll records will highlight how much these costs have increased over the past five years; they could take a similar jump in the next five. Wages are also of particular importance with respect to the skill required in maintenance and repair work on the new equipment as compared with the old. Higher workmen skills will call for higher wages.

The presence of elements in the new equipment that create considerable added efficiency in its performance is never a guarantee that they will provide as favorable an aspect in maintenance. The very existence of such elements may contribute to marked increases in maintenance costs. This is certainly worth weighing at the time a selection is being made.

15. Do not neglect to consider possible forced downtime for maintenance or repair work. If downtime is needed, both production and profit may suffer. Whether or not this downtime affects the continued flow of production in other areas of the plant is also worth studying. If only the one unit is involved, it is of much less importance than stopping an entire flow of production through that area of the plant.

Accessibility of Parts

16. How much elbow room will be required for maintenance and repair on the equipment being considered? If a great deal more dismantling is required than what the present unit calls for, this could increase maintenance costs. It could, under some circumstances, even require removal of the unit to another area for major repairs, whereas these repairs may be done on the spot with the existing unit.

17. Poor accessibility of all parts for preventive maintenance, adjustment and repair, can raise maintenance and repair costs. This applies not only to payroll costs, but also to the human weakness of avoiding anything that requires too much trouble to accomplish.

This is particularly applicable when preventive maintenance is an assigned duty of the employee operating a particular piece of equipment. His training is never as thorough on maintenance as that given to or acquired by a maintenance man himself. Every element of machine construction must be in favor of making this as easy as possible; otherwise the employee will neglect attending the machine as he should.

Summarizing, make certain that any analysis of maintenance and repair costs is complete in all details. A decision made on the basis of only one or two factors may likely be wrong; those ignored can be the very ones that would alter the whole assessment.#

Meet the Author

Ernest W. Fair (13900 SE Highway 212, #112, Clackamas, OR 97015) has been a business and technical magazine writer for over 35 years. During his long career, he has traveled extensively throughout the whole country. To gather material for his articles, he has interviewed countless people—having all kinds of positions—in almost every business and industrial field.

Earned value: tool for measuring R&D?

New rules of the Financial Accounting Standards Board call for R&D expenditures to be charged against current earnings. A company may no longer afford the luxury of a poorly measured return on R&D costs.

Michael D. Pollock, Professional Engineer

☐ Engineers and scientists in R&D have not yet agreed on how their performance should be measured, even on whether it can be, and should be, measured at all. This attitude is fostered by the prevailing reliance of R&D management on funds control—that is, on allowing R&D progress to be controlled by the rate at which budgeted funds are spent.

The task of getting the corporate dollar's worth from R&D can be improved by a performance-measurement method whose mainstays are visibility and early warning, and whose aim is cost control rather than funds control. The difference between cost control and funds control is the application of earned value to results-oriented accountability.

Planning is the key

In an earned-value measurement system, an R&D project is organized by an operational plan that breaks up the project into bits of work by means of a work-breakdown structure. In a typical breakdown, the project work is divided into tasks, which are successively subdivided into lower-level ones, until the level at which the work is actually done is reached. At this last level, the work is broken into manageable, measurable parts called work packages, which are handled by specific functional groups.

The earned-value concept demands a realistic plan for each work package. The plan constitutes the yardstick by which performance is measured. If, during the life of the project, new factors upset the plan, it must immediately be replaced with another one. When a plan is scrapped, a determination must be made as to whether any earned value can be assigned to the work already done. It may be that a large part, or even all, of the money spent to date will have been spent without value being received. Only with the consent of the project manager can new funds be infused into the work package from a reserve.

Responsibility for the work package

A work package describes the scope of the work, specifies starting and stopping dates, budgets a monthly fund for the scheduled work, and assigns responsibility to an individual.

This individual devises the work-package plan, and it is against this plan that he will be measured. He is the work-package manager. Before the plan is implemented, however, he and his supervisor agree on it, in management-by-objectives style.

Each month, the work-package manager evaluates his accomplishment, problems and outlook. He calculates, as objectively as he can, his "budgeted cost of work performed"—i.e., the earned value of the scheduled work actually performed up to that point. If he is too generous to himself in claiming earned value during the earlier stages of the project, his generosity will be automatically disclosed later.

For a work package, the sum of all the monthly budgeted costs of work scheduled should add up to the budget at the completion of the package. If the man-

Originally published June 1976.

ager has encountered larger problems than anticipated, he announces the fact by increasing his completion estimate. This completion estimate equals the amount of money spent so far plus how much more will be needed to complete the package.

These monthly tasks serve to discipline the manager to think out his technical approach, to consider tradeoffs, or even determine whether or not to abandon his original plan before he becomes too committed.

Tracking variances

Accounting records show the actual cost of the work performed and the variances:

- A negative difference between the "budgeted cost of work performed" and the "budgeted cost of work scheduled" indicates a schedule problem.
- A negative difference between the "budgeted cost of work performed" and the "actual cost of work performed" indicates a cost problem.
- A negative difference between the "budget at completion" and the "completion estimate" indicates the possibility of a project overrun.

Significant negative variances signal that the work-package plan has gone awry. The manager must explain these variances by means of a short, written analysis of the problems. Not only must he report why the variances occurred but, more importantly, what he plans to do to correct them. The variances and his report focus the attention of the higher levels of supervision upon the problem project. In other words, the project and its problems have been rendered highly visible. At least one benefit of the manager's written analysis is that it virtually forces his supervisors to respond with questions, advice and direction. If at no other time, now is when he will find that those above him are interested in his progress.

What makes a variance significant?

An often-asked question is: "What is a significant variance?"

The answer must be qualified. There is no "range of significance" that can fit all cases. Significance may depend on the type of work being measured, which functional group is performing the work, and when the work is performed in relation to the overall schedule.

A common technique for establishing significance is to set percentage tolerance-limits at the work-package level. Often, simple "eyeballing" can be relied upon without much risk that a significant variance will be overlooked.

Here is how an unfavorable variance can be used as an early warning signal:

A work-package manager's report concluded: "No problem." However, the month's earned-value cost performance read:

Budgeted cost of work scheduled: 12,790
Budgeted cost of work performed: 10,615
Actual cost of work performed: 13,672

These, of course, result in a schedule variance of -2,175, and a cost variance of -3,057.

The work-package manager's supervisor called him and said, "We show an unfavorable schedule variance of 2,175. This indicates some sort of schedule problem. Do we have one?"

The manager replied, "As my report says, we have no problem. I am right on my manhour expenditure curve. My costs are in line."

"Well," answered his supervisor, "suppose we tick off where we stand. You finished phases 1, 2 and 3 of your project and took an earned value of 10,615. But we were supposed to have completed phase 4 last month. I see we are still working on it."

"We will complete it next week."

"What seems to be the problem?"

"Well, we were not able to assign as many men to it as I had planned."

"Why?"

"We were busy troubleshooting problems of the third phase."

Discussion about the -3,057 indicator also established that, despite the work-package manager's soothing report of "no problem," he had in fact worked his men overtime to catch up. This overtime had forced his costs to exceed his plan for work to be accomplished to that point.

His reliance solely upon his manhour expenditure curve for the analysis of the status of his project had caused the work-package manager to disregard two danger signals: (1) the time spent troubleshooting completed work hindered further progress, and (2) overtime was necessary to catch up.

Manhours planned for future work but spent ahead of schedule on current work will only delay the inevitable cost overrun. Corrective action must always be taken immediately. Furthermore, manhour expenditure curves, when relied on exclusively, act as a palliative and are misleading because they simply measure the passage of time and not the work accomplished.

Treat a work package as a contract

R&D characteristically involves uncertainty and risk, so it can be no surprise that cost and schedule plans can be fragile and transient. The issue should not be that more money has been spent than planned, but rather whether the money has been spent wisely toward attaining the objective.

Earned-value measurement of performance applied imprudently is wasteful. Its application can be futile unless goals are stated as specific objectives of a work package, which should be treated like a contract.

The imposition by management of rigid budget limits upon an ill-defined work scope is self-defeating. On the other hand, management's lack of interest in significant variances can create apathy overall.

Finally, too mechanical an interpretation of variances can also be stifling. Variances are only indicators.

The author

Michael D. Pollock (13561 Old Tree Way, Saratoga, CA 95070) is Manager of Planning and Control for a U.S. Dept. of Defense contractor. He has had considerable experience with the application of earned value to R&D, and has taught a course on "Performance Measurement for R&D Programs" at the University of California. Besides a background as both an electrical and mechanical engineer, he is a registered engineer in the states of California and New York, and has had controllership business experience. His experience also includes program-management responsibility for the world's first two research nuclear reactors to be built commercially overseas.

Value Engineering—A Money-Saving Tool

Here is what this important technique is all about and how it is applied in industry to effect considerable savings.

PAUL V. DOBROW, Army Corps of Engineers

Value engineering (VE) is an organized approach to analyzing operations, systems, methods, constructions, equipment and supplies to achieve required functions at the lowest total cost, but at no sacrifice in performance, reliability, quality or maintainability.*

There are no tradeoffs just for the sake of reducing cost. The term cost, as used here, means the total, life-cycle cost that includes not only the initial price of an item, but also the costs related to operating and maintaining it throughout its life. Quite often, in pricing an item, significant factors are overlooked only to discover later that the cost of operating and maintaining a facility goes far beyond reasonable limits. It is for this reason that early application of VE to operation and maintenance is so important.

Value engineering had its formal beginning in industry immediately following World War II, at a time when business was reverting to production of commercial products. At first, VE was applied to items produced in large quantities, where incremental savings multiplied as production continued to increase. Because this resulted in appreciable savings, the method was tried in other areas with equal success.

The following discussion explains what the method consists of. Some experiences that the Corps of Engineers has had will show how VE can also be effectively applied in industry.

A Functional Approach

The function or specific purpose of an item is determined by considering the user's needs, as well as such traits as justify the item's existence. This usually leads to the question "What does the item do?" rather than "What is it?"

The functional approach is very useful in isolating the basic functions—which have value and required performance features—from secondary functions that have no value but contribute to the price of an item. In the VE system, an attempt is made to eliminate all secondary functions, or at least as many as possible, to obtain the greatest value for the user.

*The term *value analysis* is used interchangeably with VE.

What the Job Plan Consists of

Once the basic functions are determined, the next step is to commit the required resources in terms of qualified manpower and time to perform the task in an organized manner. This plan of action—which is called the *VE job plan*—consists of five distinct phases: (1) gathering of information; (2) speculation, or generation of alternatives; (3) analysis, or evaluation of alternatives; (4) development of firm proposals; and (5) presentation, or selling the proposals to management.

The first phase of the job plan has two basic objectives. One is to obtain a thorough understanding of the system, structure or item under study by a rigorous review of all factual data. The other one is to define the value of the problem by means of function descriptions, which should be accompanied by estimates of the worth of accomplishing such basic functions.

Key Questions To Answer

All of these points concerning the subject under study should be defined:

- What is it?
- What does it do?
- What *must* it do?
- What does it cost?
- What is the value of the basic functions?

The objective of the second phase of the VE job plan is to generate numerous alternative means for accomplishing the basic functions. Consideration of alternative solutions should not formally begin until the problem is thoroughly understood. But as soon as the *required* function has been identified, speculation upon alternatives can begin. All members of a VE study team should actively participate, for the greater the number of ideas conceived, the better the chances of developing an outstanding alternative.

Originally published August 1972.

A proper frame of mind is important at this stage of the study; creative thinking should replace the conventional. Unique flights of the imagination should be undertaken to generate numerous alternate methods of accomplishing the necessary functions. Judicial thinking does not belong in this phase.

Every attempt should be made to depart from ordinary patterns, typical solutions and habitual methods. Experience indicates that it is often the new, fresh and radically different approach that uncovers the best-value solution. An individual may supplement his ideas with those of others, but everyone should make a contribution.

It may happen that the best solution turns out to be the complete elimination of the present function or item. This possibility should not be overlooked. Or perhaps some aspect of the subject under consideration can be modified that will permit eliminating the function under study. Only after determining that the function must remain, should the study group look for alternate ways to perform the same function at the lowest conceivable cost.

The speculation phase answers the question "What else will do the job (perform the basic functions)?" The completeness of the answer determines to a very high degree the effectiveness of the work done.

The objective of the third (analysis) phase is to evaluate, criticize and test the alternatives generated to determine which ones offer the greatest potential for savings. While during the speculation period there is a conscious effort to inhibit judicial thinking, the reverse is now true: the alternatives must be critically evaluated.

Questions to be asked are "What does each alternative cost?" and "Will each perform the basic functions?" Thus, the first step is to assign a gross dollar cost to each alternative, and to rank these according to their savings potential.

Next comes the judging of the advantages and disadvantages of each alternative, which reduces the list, as only those alternatives that have passed the standards set by the team are allowed to remain. Areas that require further investigation should be determined at this phase of the study.

Refining the Alternatives

To refine further the alternatives that survived, more detailed cost estimates are prepared from additional data. This estimating for each alternative proceeds only if the preceding step indicates that the alternative is still a good candidate. Although this evaluation is the responsibility of the VE study team, the assistance of other knowledgeable persons should be sought to fully estimate the potential of the various possible choices.

Cost estimates must be as complete, accurate, and consistent as possible to minimize the possibility of errors. Specifically, the method used to cost the original subject under study should also be used to cost the alternatives that have been proposed.

The key question now becomes "Will this alternative meet all necessary requirements?" To answer this satisfactorily, the few remaining alternatives must undergo (1) a careful analysis to ensure that the user's needs will be satisfied; (2) a determination of technical accuracy;

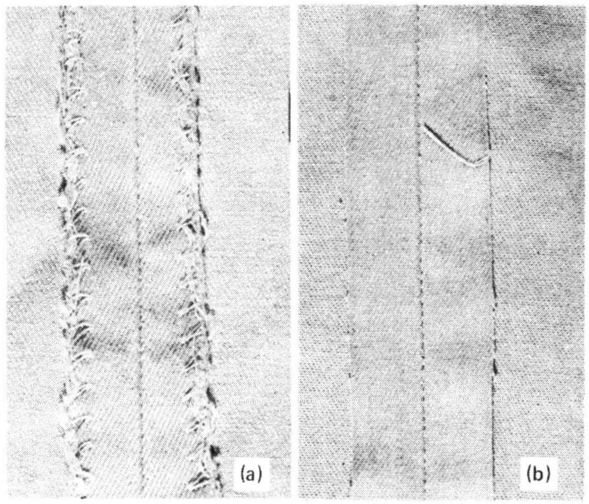

APPEARANCE of (a) stitched military-trouser seam and (b) sealant-treated seam—Fig. 1

and (3) the development of more-accurate cost estimates, and of implementation expenses and schedules (including schedules and costs of conducting all necessary tests and evaluations).

Technical problems, for example, must be approached with the assistance of personnel experienced in what the item must do so as to determine with certainty such things as within what constraints the item must perform, how dependable it must be, and under what environmental conditions it must operate. In addition, any problems related to design, fabrication, construction, implementation, or operation and maintenance must be determined.

Concerning costs, conferences with specialists are most helpful in such areas as engineering, law, materials, cost estimating, and construction/operation/supply/safety.

Finally, the alternative with the greatest cost-reduction potential is selected for implementation. One or more others are also selected for presentation, in the event the first choice is rejected by management. The implementation schedule that will yield the greatest cost reduction should also be specified.

A formal report is then prepared, which summarizes the results of the investigation, recommends specific action, and requests implementation approval.

The Real Challenge

The objective of the fifth and most important phase of the VE job plan includes the presentation of the VE report to management, and the followup necessary to ensure implementation. The best proposal ever prepared can be a wasted effort if it is rejected. Herein lies the real challenge. The value engineer must be able to see the problem through the eyes of management; he must be a salesman, a psychologist, an engineer, an opportunist, and a student of human nature.

There may be a number of well-founded reasons for not implementing a proposal, or the reasons given may be of a purely personal nature. Generally, lack-of-cooperation problems are caused by insecurity, habitual or

Before

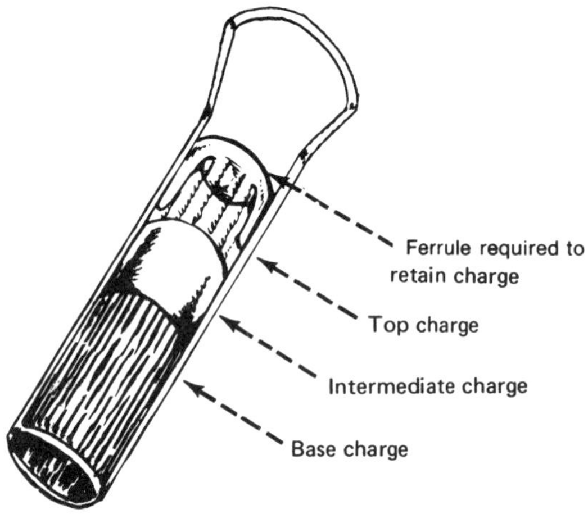

After

Lower charges now retained by top charge

Ferrule eliminated

Loading procedures simplified

Improved safety of assembly operation

Unit cost reduced by 8¢

Validated savings: $346,900

CHARACTERISTICS of blasting cap before and after modification of top charge—Fig. 2

negative thinking, fear, wrong beliefs, ignorance, laziness, self-defense, pressure of time schedules, reluctance to seek outside advice, and undesirable habits. It is the job of the value engineer to identify the problem and gain cooperation through tactful advice, a diplomatic approach, and logical reasoning and understanding.

Even if the proposals are accepted, they must yet be converted into action. Therefore, those who performed the study must maintain an active interest in the project until it becomes operational. Continued followup is essential to provide assistance, clear up misconceptions, and resolve problems that may arise, as well as to minimize delays and to compare actual results with what was originally anticipated.

When unexplained delays are encountered, a polite followup note may serve as a reminder to the responsible authority that those who made the study are available for assistance, and savings are being lost because of the delay. A VE proposal should not die because of inaction in the implementation process. The project must be monitored until complete.

Maximum effectiveness through early application of VE can be achieved only if management becomes involved and supports VE programs with positive actions. A basic requirement, therefore, is the appointment of a qualified VE manager who reports to top management and who identifies areas of good VE potential. VE is, by far, the most effective means of reducing costs and of improving operation and maintenance in a plant.

Value Engineering Applied

Here are some recent examples of how VE has proven to be a money-saving tool in plant operation and maintenance:

1. After reviewing the large usage of softwood pallets, a VE study team recommended four-way hardwood pallets that could be used to ship raw materials in and finished products out. This saved $10,000 in the first year.

2. A similar VE study on incoming shipping containers brought about a selection process whereby all salvageable ones were retained, instead of being discarded. The containers were then used to ship goods at a savings of $12,000/yr.

3. A study was made of the high failure rate—through pressure pulsations and machinery vibration—of 2½-in. pressure gages used for oil, steam and water systems aboard ships. By substituting lubricated stainless steel for brass in the geared movement, the life of the gages increased from 20 to 6,000 hr., and resulted in savings of $38,000 during the first year.

4. A chemical sealant was evaluated by a VE team at the U.S. Army Natick Laboratories for application in the manufacture of military trousers, in lieu of overedge stitching inside the raw edges of the trousers. The sealant performed as well as the stitching (Fig. 1), and annual savings of $286,000 were effected.

5. Another VE team at Picatinny Arsenal recommended modifying the top charge of a nonelectric blasting cap to permit the elimination of ferruling. The loading procedure was thus simplified, and safety features improved in the assembly operation. Although unit savings were only 8¢, the total savings in the first year amounted to $346,000 (Fig. 2).

Other VE proposals that resulted in savings were (a) conversion of forklift trucks to use of liquefied propane gas fuel; this reduced operating costs, maintenance costs and air pollution, and improved working conditions; (b) deletion of light diffusers from fluorescent-light fixtures that were about 10 ft. from the floor (it was determined that the diffusers were ineffective at that distance); this resulted in initial and maintenance cost reductions; (c) replacement of air-powered equipment having wet-type paint shields by electrostatic paint-spray equipment; from 50 to 60% savings in paint were effected. ■

Meet the Author

Paul V. Dobrow is senior value engineer in the Army Corps of Engineers, where he is responsible for value engineering training. A former staff advisor, consultant and coordinator for Dept. of the Army on Value Engineering, he also monitored value engineering in the Army Cost Reduction Program. He holds degrees in engineering and law, is a registered professional engineer, has completed the DOD Defense Weapon Systems management course, and is a graduate of the Industrial College of the Armed Forces, where he lectured and authored the thesis "Aspects of Value Engineering in the Dept. of Defense."

Estimating Marketing Costs

In selecting research projects, and in making some types of marketing decisions, it is necessary to estimate a product's selling, technical-service and general-overhead costs. Here is a new and logical way of doing this.

ERNEST O. OHSOL, American Cyanamid Co.

A very important factor in determining the potential profitability of a new product is its selling cost—and this is a cost item often misjudged by technical men.

Usually the engineer, and very often the business executive as well, takes a percent of selling price based on the experience of his own company and applies this to the estimated selling price of the new product. Of course, if the product isn't really *new* to the company, but is quite like a product in its existing line, this procedure may be quite valid and reliable.

However, what happens when a really new product is discovered and developed—one that must be sold in different markets and by different sales techniques from those applying to products in any existing line that the company has?

Sometimes, management will simply refuse to consider such a product, saying it is unfamiliar with the business, cannot assay the commercial risks and doesn't want to set up a whole new organizational arrangement just to handle one item. This attitude is undoubtedly correct if the new product has relatively little long-range potential or will lead the company in directions it does not want to pursue. However, the new development is often quite valuable, has good potential and does fit in with desirable future trends. How can the company assure itself of a reasonable evaluation of the commercial risks?

An examination of selling costs in various industries produced the interesting tabulation shown in Tables I and II. The data were obtained from company annual reports for the year 1969 (or 1970, where available).

In Table I, relating to typical chemical companies, we see two trends—(1) a decrease in "selling and administrative" (S&A) expenditures as percent of sales as the size of the company gets larger (e.g., comparing Du Pont with any of its competitors), and (2) an *increase* in relative S&A expense as one proceeds from bulk chemicals toward specialty products.

The trend of S&A costs as a function of the type of industry is even more visible in Table II. Those industries that sell large tonnages of materials in bulk, such

Selling and Administrative (S&A) Costs In the Chemical Industry—Table I

Company	Type of Business	Total Sales Million $ (1969)	S&A Costs (Excl. R&D) % Sales	S&A Costs (Incl. R&D) % Sales
Stauffer	Chemicals (mainly industrial)	500	8.8	11.3
Monsanto	Chemicals (some specialties)	1,900	11.4	16.6
Hercules	Chemicals (some specialties)	800*	11.2	13.9
Union Carbide	Chemicals (some specialties)	2,900	10.6	13.1
American Cyanamid	Chemicals (considerable specialties)	1,200*	16.5	20.2
DuPont	Chemicals (some specialties)	3,600*	9.9	12.0
Nalco	Specialty chemicals	165	18.3	22.5
Eastman Kodak	Specialties and chemicals	2,750	11.5	16.9

* 1970 figure

Selling and Administrative Costs in Industries Other Than Chemical—Table II

Company	Type of Business	Total Sales Million $ 1969	S&A (Incl. R&D) % Sales
Alcoa	Metals	1,580	9.2
Texaco	Petroleum	6,270	9.1
U. S. Gypsum	Building materials	470	15.9
General Foods	Consumer products—food	2,060	28.0
Procter & Gamble	Consumer products—soap	2,730	24.5
Celanese	Textiles and chemicals	1,250	15.1
General Electric	General equipment	8,550	18.9
Leeds & Northrop	Specialty Equipment (instruments)	80	26.9

Originally published May 1971.

as the metals and petroleum industries, have a very low percentage of S&A costs. The textile-fibers industry, selling to a larger number of customers, has a higher ratio. The strictly consumer-oriented industries such as foods, cosmetics, etc., have a still higher ratio, including large amounts spent for advertising. Finally, those industries selling products that include a great deal of technical services for specific applications—such as specialty instruments—have a very high S&A cost.

Recognizing the qualitative nature of the above comments, is there any quantitative approach to estimating S&A and related costs that compares in reliability to techniques for estimating direct manufacturing costs? To date, no such approach has been published, to my knowledge, so we offer the following as a basis for synthesizing a rational estimation procedure.

Breakdown of "Overhead" Costs

Let us examine the makeup of typical overheads:

1. Salaries of "Overhead" People
   - Direct salesmen.
   - Sales supervision.
   - Other marketing personnel—market researchers, planners, sales development people, supervisors.
   - Accountants and clerks (sometimes a major portion of these people are charged to direct manufacturing expense).
   - Technical-service personnel
   - Product and process development personnel.
   - Packaging, handling and traffic personnel.
   - Purchasing personnel.
   - Personnel-department people.
   - General-service personnel — for buildings, grounds, library, cafeteria, etc.
   - Top management.
2. Employee benefits associated with the above.
3. Cost of maintaining buildings and equipment for the above people.
4. Supplies and services purchased by the above people—e.g., photographic and reproduction services, computer, telephone and telegraph services, advertising, travel, entertainment, and the like.

In general, estimation of costs involved in running an overhead organization can be simplified by establishing cost centers for various functions and developing an average $/yr. figure per man for each function. Such functions might be:

1. Top management (a minimum number of people essential to the business as a whole).
2. Salesmen: outside (includes expense account); inside (only minor expense account); advertising assessment, usually based on number of salesmen.
3. Accounting and clerical staff.
4. Purchasing, warehousing and traffic.
5. Technical service and R&D.

Other supporting activities—such as maintenance, services, buildings—can be prorated among the above functions, along with employee benefits, on a percentage-per-person basis.

The benchmarks given in Tables I and II provide a starting point for estimating the overall total costs of these functions in various types of business. The question now comes up, how can these costs be estimated for some new product line, the nature of which is *different* than the existing business of the company?

For this purpose, we have developed a concept called proliferation coefficient, N, which in essence may be defined as the average number of individual transactions per thousand dollars of sales.

For example, if a producer makes low-quality dioctyl phthalate in large volume, and sells it all to three customers on 3-mo. contracts, he might have 12 transactions for \$10-million worth of sales, giving a coefficient, N, of 12/10,000, or 0.0012. He could have a very low overhead operation, with almost no clerical or technical staff, and no salesmen. A 1¢ or 2¢ profit margin over manufacturing cost could represent a very profitable business.

On the other hand, a producer manufacturing specialty industrial finishes might sell to 100 customers, each of whom had a dozen or two special formulations that he needed at different times. The producer might thus have 15,000 individual orders totaling \$10-million worth of sales. In this case, N would be 15,000/10,000, or 1.5. An average accounting or production clerk can handle perhaps 25 transactions per day, and there may be 3 or 4 paper transactions for each order (order, invoice, manufacturing sheet, quality-control sheet). The 15,000 orders would thus require some 12 clerical people plus perhaps 6 in production control, 5 in quality control, 10 technical people and 10 salesmen, with other supporting staff—a total of perhaps 80 people, representing an annual cost of some \$1.8 million over and above direct manufacturing cost. This corresponds to 18¢ on each sales dollar.

Interestingly, it is often the smallest orders that require the most technical service and quality control. An additional cost that might be significant and that could be allowed for is the cost of downtime and cleanout in changing from one product to another in manufacturing facilities used for short runs of many products.

In summary, these overheads might be represented by this formula:

$$O = \sum_{n=1}^{n=N} (C + P + Q + T + S)(1 + F)$$

where O represents dollars of overhead per \$1,000 of sales, and N represents the proliferation coefficient or number of individual orders per \$1,000 of sales. The items parenthesized in the above equation are as follows:

C = Clerical cost required per individual order, obtained as outlined in the second example above.
P = Production-control cost per individual order.
Q = Quality-control cost per individual order.
T = Technical-service cost per individual order.
S = Selling cost per individual order.
F = Factor representing the overhead on the functions listed above, as described previously (support-

ing staff, general services, employee benefits, etc., and including top management).

Applying the Formula

An example of the calculation for a simple bulk chemical such as a plasticizer ester—we will call it Chemical X—is shown below:

Overhead Function	Annual Cost Per Person In Function	Transactions Per Person Per Year	Cost Per Trans.
Clerical (*C*)................	$9,000	4,000*	$2.25
Purchasing			
Cost accounting			
Billing, collection			
Production overhead (*P*)...	10,000	1,000*	10.00
Scheduling, prod. cont.			
Equipment setup			
Shipping, receiving			
Quality control (*Q*).........	10,000	4,000	2.25
Technical service to sales (*T*)	10,800	2,000	5.40
Selling cost per order (*S*)...	15,000	1,000	15.00
			$34.90

F, the "overhead on overhead" factor, generally runs at least 50%. If we use the 50% figure, our cost per transaction is $34.90(1 + 0.50), or $52.30.

If it takes about 200 individual orders to sell 5 million lb./yr. of Chemical X, the cost per pound for these overheads will then be (52.30 × 200)/5,000,000, or 0.209¢/lb. This might be realistic for selling a product like a plasticizer ester.

Note that in the above example the average-size order, 25,000 lb., is a *number* average rather than a weight average. On the other hand, if the average order were only 2,500 lb., the overhead cost contribution would be 2.09¢/lb.

The figures in the above case are indicative of only one type of production; estimates for other cases could yield widely different costs per transaction in the individual categories. For example, if a long clean-out time between batches is required, the production overhead could be $1,000 instead of $10 (e.g., because far fewer transactions could then be handled per person per year). Or, in making high-priced products, the quality-control overhead could be greatly increased.

With the aid of this type of correlation, we can predict selling, technical and administrative costs—if not accurately on an absolute basis, at least on a comparative basis to select projects among alternates.

Research costs as opposed to technical service are of course more a matter of company policy, but can be assumed proportional to technical service as a reflection of technical sophistication of the type of business. They may, therefore, as a first approximation be added on to straight technical service in a pro-rata fashion.

Another use of this type of analysis is as an aid in product line planning, to help decide on how many types, grades and styles of product to offer in any given line. If too many individual items are offered, overhead costs can balloon.

* Reflecting the time needed for subfunctions listed under C and P.

Let us consider a final example, involving a choice between two new chemical products, A and B. Product A has a material cost of 40¢/lb., a labor cost of 20¢/lb. and a selling price that will yield a 40% gross margin, using a conventional "factory overhead" figure of 100% on direct labor. The total factory cost thus comes out to be 80¢/lb., the selling price 0.80/0.60 = $1.33/lb., and the gross margin is thus 0.53¢/lb.

Product B has a material cost of 25¢/lb., a labor cost of 15¢/lb. and a selling price (computed on a 30% gross-margin basis) of [25 + (2x15)]/70 = 79¢/lb. The gross margin is thus 24¢/lb.

Ordinarily, management would overwhelmingly choose A over B—but let us see what our analysis shows. Let us suppose that A, being a specialty chemical, is sold in single drums or, at most, four drums at a time, while B is sold in lots of at least 10,000 lb. Furthermore, A will reach a sales volume of 2 million lb./yr., while B will probably reach a volume of 10 million lb./yr. The number of transactions (*N*) per year for A, assuming a "number average" order size of 1,000 lb., will be 2,000, whereas *N* for B will be on the order of 500, assuming a "number average" order size of 20,000 lb. Estimating the cost per transaction to be $150 for A and $75 for B, we obtain these transaction-related overheads:

For A, (2,000 x $150)/2 million lb. = $0.15/lb.
For B, (500 x $75)/10 million lb. = 0.004/lb.

In other words, there is an additional cost differential of about 14½¢/lb. favoring B. Assuming the plant investment is roughly proportional to the 0.6 power of plant capacity, the plant for B will cost 2.62 times more than A. Comparing the returns on investment for A and B, we obtain:

	Product A	Product B
Sales volume, lb..........	2 million	10 million
Sales income, $...........	2,667,000	7,860,000
Margin, gross, $..........	1,067,000	2,400,000
Margin, net, $............	767,000	2,360,000
Investment, $.............	*I*	$(5)^{0.6}I = 2.62I$
Ratio, net margin to investment	$\frac{767{,}000}{I}$	$\frac{2{,}360{,}000}{2.62I} = \frac{900{,}000}{I}$

Thus, the return on investment as well as net profit in $/yr. will be greater for the lower-priced, moderate-volume chemical than for the higher-priced, low-volume one. ■

Meet the Author

Ernest O. Ohsol is director of process development at American Cyanamid Co.'s Stamford Research Laboratories (1937 W. Main St., Stamford, Conn. 06904). He has had 30 years of experience in R&D, manufacturing engineering, economic evaluation and technical management. A chemical engineering graduate of the College of the City of New York, he has a Sc.D. degree from M.I.T. Dr. Ohsol has been very active in AIChE, both locally and nationally.

Estimating Mill Costs At Low Production Rates

With these correlations and a minimum amount of data, you can quickly make order-of-magnitude estimates for the manufacturing costs of chemicals produced in volumes less than 2 million lb./yr.

ROBERT F. SOMMERVILLE, Sun Oil Co.

Manufacturing-cost determinations associated with low-volume processing have had very limited coverage in the literature. Normally, low-volume operations involve either market-development quantities of a precommercial product for customer evaluation, or small amounts of material for inhouse testing.

It is hoped that the correlations presented here will prove a useful tool in preparing project budgets for such undertakings, for making a rough check of estimates prepared by others, and for preliminary economic evaluations of proposed new ventures.

Components of Mill Cost

In general, manufacturing costs (also known as mill costs) can be broken down into two major components—i.e., variable and fixed costs.

Variable costs normally remain constant on a cost per unit volume basis regardless of plant output, and include such items as raw materials, utilities, chemicals and containers.* Although sales and freight costs are usually variable in nature, they are omitted from the mill cost calculation.

On the other hand, fixed costs such as labor, maintenance and depreciation tend to be constant on a total dollar basis regardless of whether a unit is operating at 50% or 90% of capacity.

Mill Cost for Low-Volume Operations

A distinct characteristic of the *low* production rate operation is that the variable costs tend to become insignificant in comparison to labor and investment-related costs (fixed costs) and thus can usually be omitted in the mill cost calculation. This fact is important, because the total unit-variable-cost differs widely for individual processes and cannot be incorporated into a general correlation.

As for fixed costs, small-volume operations seem to be amenable to a generalized correlation of fixed cost vs. production rate *essentially independent of the actual process or product involved,* because:

1. In most accounting procedures, all of the fixed costs (including depreciation and maintenance charges) associated with small-volume units are normally allocated as a percentage of the operating labor requirement. Thus, the fixed cost becomes a function of labor cost only, and is normally independent of capital investment.
2. As a manufacturing operation becomes small, not only does the number of men per shift diminish, but there is less variation in labor force for different processes.
3. The percentage accuracy is not as critical as it is for large commercial units, because the cost in total dollars is relatively small.

These combined factors make it possible to establish the desired correlation.

We will consider two types of low-volume operation: (1) a high-labor process, where several steps are carried out concurrently, and (2) a low-labor process involving either only a few steps, or else many individual steps that are carried out in sequence.

For these typical operations, reasonably accurate cost estimates at any specific output rate under 2 million lb./yr. can be readily determined without detailed background information.

A New Correlation

From various company and outside sources, I have collected fixed-cost data for materials manufactured at rates ranging from 30 lb./yr. to 2 million

* Actually, variable costs per unit are not entirely independent of volume because of the need to buy raw materials in small-lot sizes at low volumes, etc.

Originally published April 1970.

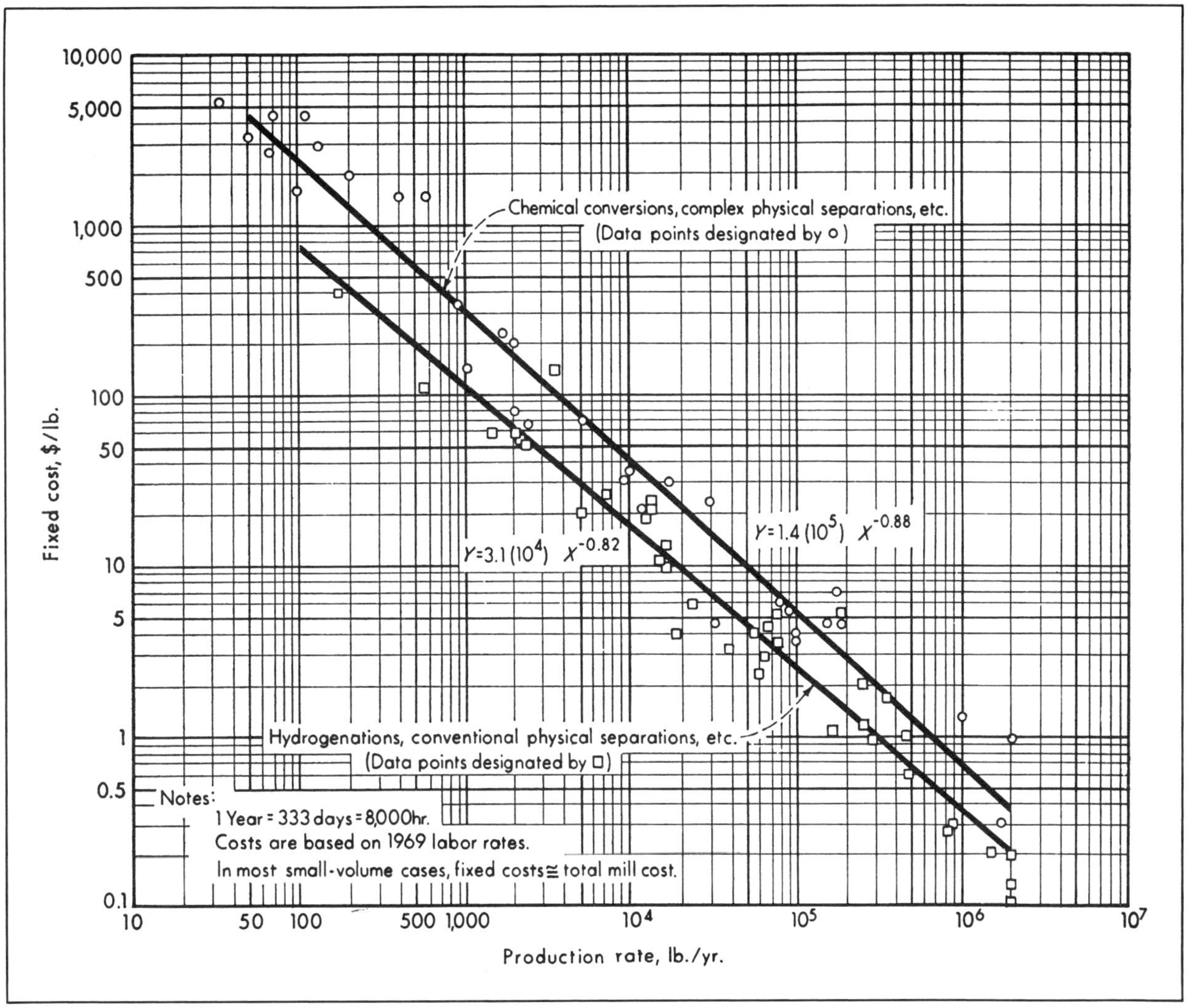

FIXED COST (the major cost component at low outputs) shows this correlation with annual production rate—Fig. 1

lb./yr. A total of 75 data points have been obtained, 36 of which are believed to represent Type (1) operations, while the other 39 are apparently associated with Type (2) operations.

In gathering the data for the correlation, the specific labor deployment was unknown in most instances. Here, I assumed (perhaps incorrectly in some cases) that a chemical conversion or complex physical separation was a high-labor operation, while a hydrogenation or conventional physical separation represented a low-labor process. Errors made in classifying the labor level of various processes could explain some of the data scattering found in Fig. 1. About one-half of the data have been derived from paper studies, while the remainder represent the actual manufacture of various products.

The fixed-cost estimates are based on a 1969 operator wage rate of about $4-$4.50/hr., and are calculated on a full-cost basis to also include direct employee benefits, supervision, analytical and shop services, depreciation and overhead. The total full cost used in the correlation is about $20-$25/manhour of operation. If the full cost in your company varies significantly from this figure, the correlation must be adjusted accordingly.

The data are given in Fig. 1, which presents the relationship of fixed cost ($/lb.) to *annual* production rate. The use of the annual production rate (versus monthly, weekly, etc.) was selected arbitrarily. Note that the production capability must be expressed in lb./yr. For instance, a unit manufacturing 10 lb. of chemical in an 8-hr. shift is capable of producing about 10,000 lb. in one year (assuming 8,000 hr./yr. of available operating time, allowing for planned shutdowns, etc.).

Don't forget that the correlation is based on output rate, not output per se. Let's assume that someone contracts the toll manufacture of 1,000 lb. of a development chemical for the calendar year 1970. He might incorrectly interpret the output rate to be 1,000 lb./yr.; but in reality, if the material were made in batches requiring only six months of actual running time, the equivalent annual rate would be 2,000 lb.

Straight lines were plotted through the data points

on logarithmic graph paper using the method of least squares. The resultant equations are:

$$\text{Chemical conversions, etc.: } Y = 1.4(10^5)X^{-0.88}$$
$$\text{Hydrogenations, etc.: } Y = 3.1(10^4)X^{-0.82}$$

where Y represents fixed costs, \$/lb.; and X is the production rate, lb./yr.

As mentioned previously, at small production volumes, the total fixed cost generally approximates the total mill cost. However, if the raw material cost and/or other variable costs are excessive, they can be added to the fixed costs to yield the total mill cost.

Under most situations involving small volumes, I believe that this particular correlation offers a distinct advantage over the larger-volume methods of earlier studies[1] in that the user does not have to be concerned with the exact process sequence, the hourly labor requirements, equipment investment estimates, maintenance needs, etc. In such rare cases as require a detailed analysis, the correlation can be used as an order-of-magnitude check.

Example Calculation

Let us say that 100 lb. of Chemical C is to be manufactured via the reaction of 75 lb. each of Chemicals A and B in the presence of 4 lb. of Catalyst Y. Several concurrent operations are involved. Materials A, B, and Y cost 25¢/lb., 40¢/lb. and \$2/lb., respectively. The available processing unit is capable of producing 10 lb./hr. of Chemical C.

To find the approximate mill cost for this operation:

1. Calculate the annual output rate of Chemical C. 10 lb./hr. (8,000 hr./yr.) = 80,000 lb./yr.

2. Determine the estimated fixed cost from the top line in Fig. 1, i.e., \$6.3/lb. of Chemical C.

3. Estimate the raw material and catalyst costs (assuming no byproduct credit).

Chemical A	= \$0.25/lb. × 75/100	= \$0.19/lb. of C
Chemical B	= \$0.4/lb. × 75/100	= \$0.30/lb. of C
Catalyst Y	= \$2/lb. × 4/100	= \$0.08/lb. of C
		\$0.57/lb. of C

4. Calculate the approximate mill-cost by summing the fixed, raw material, and catalyst costs; i.e.: \$6.3 + 0.57 = \$6.9/lb. of Chemical C.

5. The total absolute mill cost will be about \$6.9/lb. (100 lb.) = \$690. (For most purposes, it would have been sufficiently accurate to assume that the fixed cost was equal to the total mill cost.)

Literature Support for Correlation

Chilton[2] has reported that for several *large*-volume chemical and petroleum processing units, plant investment tends to vary with capacity to approximately the 0.6 power. That is, if the plant size is doubled, the investment should be $2^{0.6}$ times as much. Further, based on Wessel's[3] plot of operating-labor requirements vs. plant capacity for a number of multimillion-lb./yr. organic and inorganic chemical facilities, the annual labor cost (or man-hrs.) appears to vary with capacity to a power of roughly 0.2. Substantiation of the latter value is provided by Isard and Schooler,[4] who listed exponents for use in scaling labor costs from one capacity to another for each of 31 commercial chemical plants. The average exponent for the units reported is 0.23, with individual values ranging from 0.15 to 0.4.

Because the total fixed cost is the summation of those costs related to investment and labor, it follows that the annual fixed cost for large plants should vary with capacity to an exponent between 0.6 and 0.2, depending on the specific process. However, this relationship does not apply to a small-volume operation, because the fixed cost is normally determined on the basis of operating labor only. Assuming that the 0.2 labor factor holds for small units (not previously established), it follows that my correlation should also obey this rule. Based on the results of the analysis just discussed, this actually appears to be the case.

For production rates under 2 million lb./yr., fixed costs (in \$/yr.) apparently vary with annual output to a power of roughly 0.2 (actually 0.15). This can be derived from the correlation:

$$Y \simeq f(X^{-0.85})$$
$$YX \simeq f[X(X)^{-0.85}] \simeq f(X^{0.15})$$

where Y = fixed costs, \$/lb.; X = production rate, lb/yr.; YX = fixed costs, \$/yr.

The 0.15 factor could be quite useful if one has fixed-cost information based on a cost accounting procedure or cost basis widely different from that used in the correlation. With just a single datum point, one could very easily prepare cost estimates at other production rates via the relationship:

$$\frac{Y_1X_1}{Y_2X_2} = \left(\frac{X_1}{X_2}\right)^{0.15}$$

by using the known values for X_1, X_2, and Y_1, and solving for either $Y_2 X_2$ or Y_2. ■

References

1. Berry, R. M., Estimate Manufacturing Costs, *Chem. Eng.*, June 27, 1960, pp. 123-126.
2. Chilton, C. H., Six-Tenths Factor Applied to Complete Plant Costs, *Chem. Eng.*, Apr. 1950, pp. 112-114.
3. Wessel, H. E., New Graph Correlates Operating Labor Data for Chemical Processes, *Chem. Eng.*, July 1952, pp. 209-210.
4. Isard, W., and Schooler, E. W., "Location Factors in the Petrochemical Industry With Special Reference to Future Expansion in the Arkansas-White-Red River Basins," U.S. Dept. of Commerce, July 1955, Appendix B.

Meet the Author

Robert F. Sommerville is in the Economics Evaluation Group of Sun Oil Co.'s R&D Div., Marcus Hook, Pa. 19061. He joined that group in 1961, following service with Monsanto Co.'s Research Dept. at Texas City, Tex. A chemical engineering graduate of the University of Louisville, Mr. Sommerville is a registered professional engineer in Pennsylvania and Kentucky, and a member of AIChE and ACS.

Thermodynamics Helps You Predict Selling Price

With this new correlation, a bulk chemical's free energy of formation can help you estimate what that chemical will sell for if produced at any particular volume.

J. T. SOMMERFELD and C. T. LENK, Wyandotte Chemicals Corp.

Long-range price patterns of many chemical products can be predicted from free energies of formation and estimates of future volumes. By combining thermodynamics with economics, we can make better estimates than by relying on economic factors alone, subject to these limitations:

1. Supply and demand are assumed to be in reasonable balance, and hence short-range fluctuations in price based on distress sales or distress purchases (or on sudden scarcities or gluts of raw materials) cannot be predicted.
2. The method is not accurate for chemicals produced in low volumes (less than 100 million lb./yr.), or that have extremely hazardous properties, or are subject to stringent government regulations, or are produced as cheap byproducts.

These limitations still leave a fairly large area of application for the new correlation. But before we introduce and justify it, it is necessary to take a brief look at estimating methods that depend on volume/price relationships.

One of the obvious factors influencing the selling price of a product is its annual volume—i.e., the total amount of that material produced by a company and its competitors.[1] There is considerable value in being able to predict the price of a given product as a function of this annual volume; many recent articles on economic analysis have dealt with various facets of this relationship, including profitability analysis as related to supply and demand,[2] evaluation of long-range economic price,[3] and the determination of optimum plant sizes.[4, 5]

One of the most well-known tools for relating selling price to volume involves the Zabel Exclusion Charts,[6, 7] whereby an individual chart is required for each class of chemical products—e.g., pesticides, medicinals, plasticizers, etc. A recent article[8] related the selling price of a chemical to the cost of the raw materials going into its manufacture. In an even more recent article,[9] cumulative market values vs. time were correlated for 53 individual chemical products. Each one of these tools, however, requires *a priori* knowledge of the price-volume history of a given chemical.

Why Introduce Thermodynamics?

The theory behind our approach is that the relative ease or difficulty associated with the manufacture of a chemical product should be related to its free energy of formation. To account for this factor in a quantitative fashion, a ground free-energy state (G) is defined: This quantity, in units of thousand Btu./lb., corresponds to that energy state below which no chemical compound has a free energy of formation from its elements. A free energy consumption (C) associated with the manufacture of a chemical compound may thus be calculated from its standard free energy of formation ($\Delta F_f°$, thousand Btu./lb.) as follows:

$$C = \Delta F_f° - G \qquad (1)$$

This quantity may then be used to transform the selling price (P_S ¢/lb.) of this chemical to a unit price for the free energy consumed (P_E, ¢/thousand Btu.) relative to the hypothetical ground state, in its manufacture:

$$P_E = P_S/C \qquad (2)$$

This quantity, rather than the selling price, is then correlated with the annual volume of the product in order to recognize the ease or difficulty of product manufacture.

Results: Less Estimating Error

Price-volume data for 61 chemical products—inorganic and organic—were analyzed; these 61 chemicals are listed in the table. The analysis consisted of correlating the logarithm of the selling price (P_S) or free energy price (P_E) with the logarithm of the

Originally published May 1970.

Basis for Correlating Sales Price, Annual Volume, and Free Energy of Formation

Chemicals Studied: Inorganic

Alumina trihydrate	Magnesium oxide
Aluminum sulfate	Manganese dioxide
Ammonia	Nitric acid
Ammonium sulfate	Phosphorus oxychloride
Barium hydroxide	Phosphorus pentoxide
Boric acid	Phosphorus trichloride
Calcium carbide	Potassium chloride
Calcium carbonate	Potassium nitrate
Calcium chloride	Sodium carbonate
Calcium oxide	Sodium chlorate
Calcium phosphate, dibasic	Sodium fluoride
Carbon	Sodium hydroxide
Carbon dioxide	Sodium sulfate
Chlorine	Sulfur
Hydrochloric acid	Sulfuric acid
Hydrogen cyanide	Zinc chloride
Magnesium chloride	Zinc sulfate heptahydrate
Magnesium hydroxide	

Chemicals Studied: Organic

Acetaldehyde	Heptenes
Acetone	Isobutylene
Benzene	Isobutyl alcohol
Benzoic acid	Methanol
Butenes	Methyl chloride
Carbon disulfide	Monochlorobenzene
Carbon tetrachloride	Nitrobenzene
Chloroform	Nonene (tripropylene)
Cyclohexane	Propane
Ethylbenzene	Propionic acid
Ethyl chloride	Propylene
Ethylene	Toluene
Ethylene dichloride	Urea

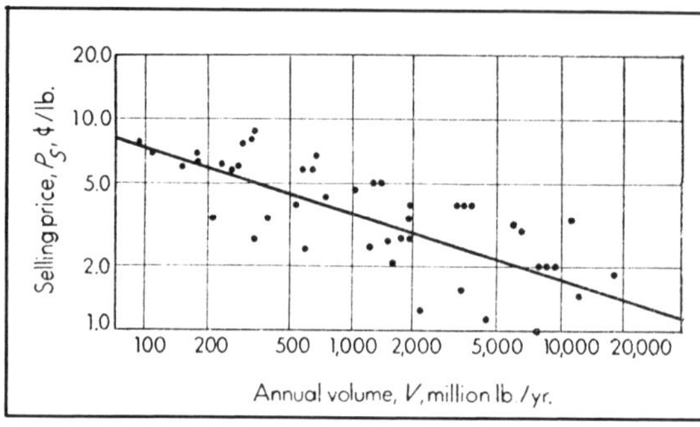

ANNUAL VOLUME VS. SELLING PRICE: Data and correlation for 61 chemicals listed in table—Fig. 1

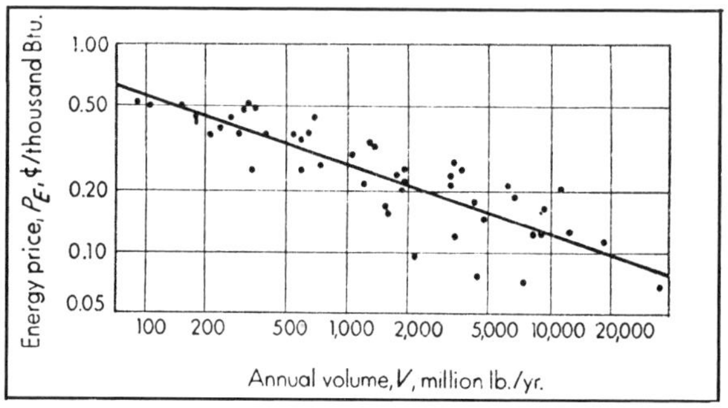

FREE ENERGY VS. ANNUAL VOLUME for 61 chemicals; ground free-energy state = −16,000 Btu./lb.—Fig. 2

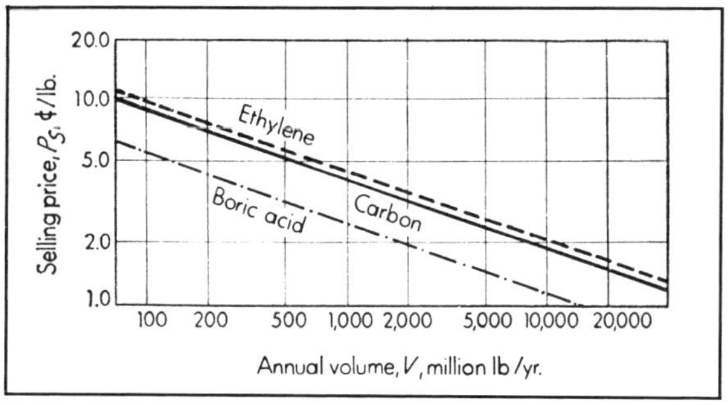

SELLING-PRICE/VOLUME CURVES for ethylene, carbon and boric acid as calculated from Eq. (5)—Fig. 3

annual volume (V, million lb./yr.) via regression methods. It was found that a first-order (or linear) model represented the best fit for correlating the logarithm of either the selling price or the free energy price with the logarithm of the annual volume. For the selling price, the resulting correlation is:

$$\log P_S = 1.517 - 0.3132 \log V \quad (3)$$

or:

$$P_S = 32.91\, V^{-0.3132} \quad (4)$$

A plot of Eq. (3), as well as the raw data on the 61 chemicals used, is given in Fig. 1. The residual standard deviation for this selling price correlation is 1.335 ¢/lb.

The correlation for the energy price is as follows:

$$\log P_E = 0.4122 - 0.3285 \log V \quad (5)$$

or:

$$P_E = 2.583\, V^{-0.3285} \quad (6)$$

This correlation and the computed energy price data, resulting from Eq. (1) and (2), are plotted in Fig. 2. The residual standard deviation of *the selling price* for this energy price correlation is 0.958 ¢/lb. Thus, the latter correlation results in a 28% reduction in the error of estimating the selling price of a given chemical product. This reduction indicates the value of incorporating the free energy of formation into the price-volume correlation.

Eq. (5) was obtained only after a non-linear search for the optimum value of the ground free-energy state (G); i.e., for the value that minimizes the residual standard deviation of the selling price. The optimum value derived this way was $G = -16{,}000$ Btu./lb. It is interesting to note that this is significantly less than the standard free energy of formation for any known chemical compound.

Applying the Correlation

To illustrate the use of the energy price-volume correlation in developing price-volume curves for chemical products, let us consider three chemicals with widely different free energies of formation:

1. Boric acid, $\Delta F_f{}^\circ = -230.2$ kcal./mol
2. Carbon, $\Delta F_f{}^\circ = 0$
3. Ethylene, $\Delta F_f{}^\circ = +16.28$ kcal./mol

From the previously developed correlation between free energy price and volume, and with a ground free-energy state of −16,000 Btu./lb., individual selling price curves can be developed for each of these

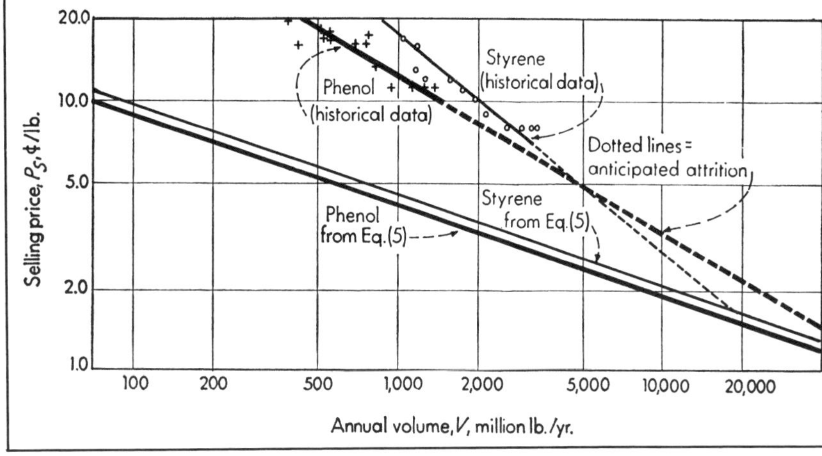

PRICE ATTRITION of phenol and styrene based on extrapolated historical data, and on the correlation in Eq. (5)—Fig. 4

three chemicals. Such curves are plotted in Fig. 3. It is seen that at any given volume, the selling price for ethylene should be higher than that for carbon, which in turn is higher than that for boric acid.

Similar curves for any chemical can be readily constructed from this correlation, given a free energy of formation. There are reliable thermodynamic methods[10] for estimating the latter if handbook or literature values are not available.

A recent article[11] presents an interesting method for predicting future annual volumes of chemical products; it is based on a log-log correlation of *cumulative* production versus time. This method, together with the energy-price correlation presented herein, could be used to predict normal price attrition with time resulting from increased production. One thus has a tool for representing what may be called the "learning curve" phenomenon.[1]

Predicting Process Improvements

At the beginning of this Cost File, we indicated the type of chemicals whose price-volume behavior cannot be accurately predicted by either of the correlations.

Another group of chemical products that deviate from both correlations are those for which significant process improvements should be possible. Two such examples—styrene and phenol—are depicted in Fig. 4. Selling-price/volume curves, as computed from the energy price correlation, are plotted for each of these two materials. Historical as well as anticipated price attrition for each of these chemicals is also shown. Styrene appears to be experiencing more rapid price attrition than phenol. This has also been reported in other work.[12]

It should be noted that our correlation does not reflect inflation. Most of the raw data were taken from a relatively narrow time interval. Inflation could easily be represented, however, merely by changing the intercept, but not the slope, of the correlation. Also, the prices for styrene and phenol in Fig. 4 were not adjusted for inflation. Such an adjustment would move the low-volume prices upward and would accentuate the price attrition trends. The general conclusions regarding the price behavior for these two chemicals should remain unchanged, however. ■

References

1. "Perspectives on Experience," Boston Consulting Group, 1968.
2. Moore, C. S., Supply and Demand Curves in Profitability Analysis, *Chem. Eng.*, Oct. 7, 1968, p. 198.
3. Cheslow, R. T. and Bates, A. G., Long-Range Economic Price, *Chem. Eng.*, Nov. 22, 1965, p. 161.
4. Urban, W. J. and Holland, F. A., How to Determine Optimum Plant Size, *Chem. Eng.*, Mar. 28, 1966, p. 103.
5. Corrigan, T. E. and Dean, M. J., Determining Optimum Plant Size, *Chem. Eng.*, Aug. 14, 1967, p. 152.
6. Zabel, H. W., The Exclusion Chart, *Chem. Eng. Prog.*, **52**, No. 5, p. 183 (1956).
7. Zabel, H. W. and Marchitto, M., What Price Can I Get for My Chemical?, *Chem. Eng.*, Oct. 19, 1959, p. 112.
8. Grumer, E. L., Selling Price vs. Raw-Material Cost, *Chem. Eng.*, Apr. 24, 1967, p. 190.
9. Massey, D. J. and Black, J. H., Predicting Chemical Prices, *Chem. Eng.*, Oct. 20, 1969, p. 150.
10. Reid, R. C. and Sherwood, T. K., "Properties of Gases and Liquids," McGraw-Hill, New York, 1958.
11. Jackson, J. W. and Black, J. H., Predict Petrochemical Production, *Hydrocarbon Processing*, **48**, No. 7, p. 143, (1969).
12. Fedor, W. S., Commodity Forecasting by Computer Time Sharing, *Chem. & Engrg. News*, Sept. 12, 1966, p. 80.

Meet the Authors

Jude T. Sommerfeld is director of process engineering for Wyandotte Chemicals Corp. (Wyandotte, Mich. 48192), where he is primarily concerned with process improvement and computer applications. Previous employment was with Monsanto Co., Parke, Davis & Co., and Ethyl Corp. Dr. Sommerfeld received a B.Ch.E. from the University of Detroit, and M.S.E. and Ph.D. degrees in chemical engineering from the University of Michigan. He is a member of AIChE and ACS.

Carl T. Lenk is assistant to the Director of Organic Research and Development, Wyandotte Chemicals Corp. He was previously with Syntex, S.A., Mexico City. His experience with Wyandotte has been in organic and inorganic research management, technical-economic analysis, and research planning. A graduate of Wayne State University with a B.Sc. and Ph.D. in chemistry, he is a member of ACS and the Electrochemical Soc.

Your company need not be an industrial giant to get maximum benefits from a well-engineered distribution system. Here is how the smaller company can find offsets to the rising cost of transportation (see chart), save money on warehousing and packaging, and thus . . .

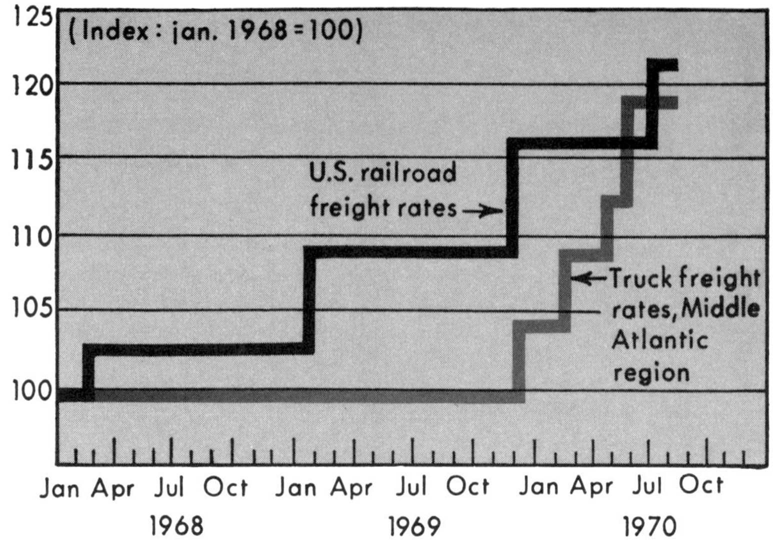

Stretch Distribution Dollars

W. F. FRIEDMAN, Walter Frederick Friedman and Co.

A small Midwest chemical process industries firm is providing 24-hr. delivery service on small orders to the New York metropolitan area, without a regional distribution center and with minimum field inventories. The company's 1,000-lb. pallet-boxes share space in a pubic warehouse specializing in chemicals storage. In the delivery truck, the order rides next to a small order by another chemical company, frequently going together to the same customer, with a substantial saving in transportation cost.

Another small company in the CPI, facing a decision to either build a new plant somewhere in mid-America or expand its Eastern plant, used a computer-based plant location model that was originally developed for one of the giant oil firms.

Other small companies are sharing computer services for sales analysis, forecasting and inventory management, and are applying analytical techniques to decision-making in these areas—yet these companies do not have a computer, nor do they have systems analysts or programmers on their staff.

Chemical Engineering's 1970 Forecast contained a section entitled "Distribution Is on the Move,*" which described Allied Chemical Corp.'s organizational approach to distribution, Union Carbide's huge new automated warehouse, and other large-company trends. But, as illustrated in the initial examples, there is also a lot that *smaller* companies are doing, and can do, to fight the high cost of distribution. In such companies, nonspecialists such as chemical engineers and production managers often become more intimately involved in distribution decisions than is the case in larger companies, so the topic is of considerable interest—particularly in today's cost and profit climate.

* Pp. 88-91 of the Jan. 12, 1970, issue.

The First Step

Do you really know what the true costs of distribution are in your company? Most organizations' accounting systems do not separately record all the costs involved in a product. At best, we can obtain 50 to 65% (usually in-and-out transportation and inventory costs) by referring to financial documents. Other costs are "hidden" as marketing or production costs; these include such important items as packaging materials, production-control and inventory-management personnel, and manufacturing warehouse space. Table I can guide you in collecting such costs.

Working with a company having $10 million in annual sales, we found that primary emphasis had always been given to the transportation cost of finished goods, which had increased from 3% to 4% of sales (i.e., from $300,000 to $400,000) during the last three years. But when all costs were compiled in our study, they showed that the company was really spending $1.5 million yearly for distribution. For instance, we found that the costing system had treated transportation of raw materials as a product-cost element. These elements increased total transportation costs to $600,000 per year, and uncovered some real opportunities for consolidated shipments and backhauling for a limited company-owned truck operation.

Proper identification of costs serves two purposes:

1. It gives management a true picture of total costs, and what the potential savings are if distribution responsibilities are efficiently organized and controlled.
2. It puts the spotlight on specific problem areas that might be immediately attacked.

Let us take a look at some possible problem areas.

Originally published September 1970.

Guide to Uncovering Total Distribution Costs—Table I

Distribution Cost Elements	Where to Look for Cost Data; and What Data to Look for
Transportation: inbound	Financial ledgers re freight-bill payments; raw-material cost standards.
Transportation: outbound	Financial ledgers re freight-bill payments; invoice summaries.
Inventory-carrying cost for raw materials and finished goods	Book value of inventories (converted to cost, for finished goods) from accounting records.
Warehouse-space cost: manufacturing	Allocated space cost (including all facility-occupancy cost items suitably apportioned), from accounting records.
Warehouse-space cost: field	Total occupancy cost, from accounting records.
Materials handling: raw materials to finished goods	Allocated indirect-labor costs, from manufacturing and warehousing cost records, plus fringe costs.
Receiving and shipping	As above.
Packaging materials	Purchase costs, from purchasing records or manufacturing standards.
Administrative and clerical costs (order processing, customer service, production planning, inventory management, billing, traffic)	Labor cost and fringe-benefit cost for supervisory and administrative personnel (suitably apportioned where overlap between marketing and manufacturing occurs), from general administration, marketing and production cost records.
Data-processing costs, telecommunications, office supplies	From general administration cost records, suitably apportioned between other functions such as finance and marketing.

Consider New Expendable Pallets for Reducing Storage, Shipping Costs

A flurry of new developments in both materials and manufacturing techniques offers the user a greater choice of expendable pallets. These nonwooden pallets are generally lighter in weight, less costly, save space, and reduce damage to the product. But before you change to such units, testing in long-term storage and in shipping will be necessary.

Some of the new expendable pallets are made of:

1. Die-cast hardwood (and designed to nest so as to reduce cube).
2. All-plastic, such as polystyrene foam (which permits custom fabrication).
3. Combination of foam and corrugated (which permits a more versatile pallet than does all-plastic construction).
4. All-kraft honeycomb construction, with high-impact absorption characteristics (see photos below). The honeycomb pallet-runners can be directly attached to bulk boxes by simply rolling on the adhesive, and can also be used for circular containers. In carload quantities, unit cost is less than 19¢ for a 36-in. runner, 6 in. wide and 2 in. thick.

Union Camp Corp.

Checklist of Public-Warehouse Facilities and Services

TYPE OF STORAGE

- ☐ Dry heated
- ☐ Dry unheated
- ☐ Freezer space: 10 F.
- ☐ Humidity-controlled space
- ☐ Refrigerated space 32 F. to 50 F.
- ☐ Silo storage: dry
- ☐ Tank storage: liquids
- ☐ Bonded storage

BUILDING CONSTRAINTS

- ☐ Ceiling height, ft.: —— in.: ——
- ☐ Floor loads, lb./sq. ft.: ———
- ☐ Single-story
- ☐ Multi-story
- ☐ Sprinklers
- ☐ Security alarms

SPECIALIZED STORAGE

- ☐ Red label*
- ☐ White label*
- ☐ Yellow label*
- ☐ Other

ACCESSORY SERVICES

- ☐ Inspection of incoming shipments
- ☐ Fumigation
- ☐ Break bulk and repack
- ☐ Inventory management
- ☐ Consolidate shipments
- ☐ Bagging equipment

TYPES OF HANDLING AND EQUIPMENT

- ☐ Clamp handling
- ☐ Palletized handling
- ☐ Slip-sheet handling†
- ☐ Pneumatic conveying
- ☐ Conveyor movement
- ☐ Tow line
- ☐ Electric fork trucks
- ☐ Gasoline fork trucks
- ☐ Propane fork trucks
- ☐ Fork truck, capacity ——— lb.
- ☐ Automatic dock levelers

TRANSPORTATION FACILITIES

- ☐ Private siding
- ☐ Inside rail docks
- ☐ Inside truck docks
- ☐ Sheltered truck docks
- ☐ TOFC/COFC (Piggyback)
- ☐ Barge facilities
- ☐ Deepwater facilities

* These designations refer to the storage and shipment of explosives and other dangerous articles, in accordance with applicable regulations by the U.S. Dept. of Transportation.
† This method of unit-load handling involves a sheet of fibreboard (or plywood).

PUBLIC WAREHOUSES can help you reduce distribution costs; they offer a great many services, although not every warehouse can provide every service. This checklist can help you evaluate the services and facilities that are offered, so that you can compare them with your needs.

Transportation and Packaging

As shown in the chart on the previous page, surface transportation costs have been climbing steeply, and the end of this trend is not in sight. The recent Teamster settlement provides for annual wage increases of 13% for each of the next three years. Because labor is such an important cost component for the trucking industry, these increases virtually assure major boosts in trucking rates through 1972.

The railroad picture is not much different. There have been five freight-rate increases in the last three years, and the railroads are readying a new request before the Interstate Commerce Commission.

With per-unit transportation costs likely to rise faster than the prices obtained for CPI products, the pressure is on to increase the "productivity" of each transportation dollar.

For the small company—say, your company—this can take many forms. For instance, you may benefit by a new service provided by the railroads to take advantage of carload rates to a central point in your marketing area. It works like this:

You consign a full carload of your product (either dry or liquid chemicals) to one of the railroad's bulk terminals. After the car arrives, the product is stored in the car until you issue instructions, at which time it is pumped into trucks for delivery to your customers. This service lets you obtain low carload rates for the long haul and low truckload rates for the short haul.

As for packaging: improved ways of filling, stacking, shipping and protecting various types of containers can often provide offsets to the rising cost of containers themselves (and of the handling labor).* Two examples—expendable pallets and shrink-wrapped pallet loads—are described on pp. 533 and 535.

Getting the Most From Warehousing

The public warehouse industry offers a method of expanding your market without any capital expenditure for the brick and mortar of branch warehouses or regional distribution centers. Use of such services can

* For data on the costs and applications of various types of containers, see: Uncles, R. F., Containers and Packaging, *Chem. Eng.*, Oct. 13, 1969, pp 87-97 (Deskbook Issue).

be increased or decreased in line with seasonal demand, or for market testing. Payment is on a pay-as-you-go basis for space used and for handling charges, for movement in and out of the warehouse, and for other services that you may elect to use (such as generation of inventory reports, invoicing, repacking, etc.). You can use the checklist on the previous page to summarize what you need, so that you can readily compare it with what any particular public warehouse has to offer.

Generally speaking, public warehousing (in lieu of a captive branch operation) will be economical if replenishment is in full-car or truckloads. Local shipments in less than truckloads are made by common carrier, or a local cartage firm that may be part of the public warehousing organization. It is particularly desirable to select a local cartage company that services your customers with other products, which may allow you to negotiate for a favorable delivery rate. Fig. 1 shows the overall logistics of such a network.

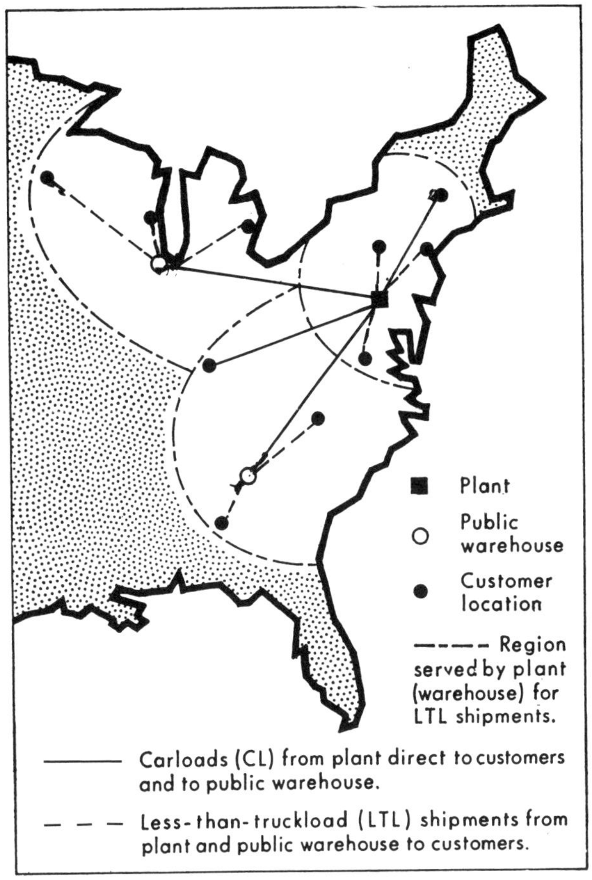

PUBLIC WAREHOUSES are the basis for this efficient, one-plant distribution network—Fig. 1

What You Should Know About Shrink-Wrapped Pallet Loads

The advent of shrink-wrapping has narrowed the gap between industrial packaging and materials handling because this technique minimizes or eliminates the need to protect the individual item or unit, yet ties the product into a solid pallet load for efficient mechanical handling, storage and transportation.

Advantages of shrink-wrapped loads include lower shipping and packaging costs (i.e., compared to other, heavier, forms of protective packaging), better protection from moisture and dirt (e.g., permitting outdoor storage of bagged and boxed materials), improved security, stabler loads, and easy product identification. Major differences in pallet shrink-wrapping techniques and equipment are summarized below.

Methods of Overwrapping

- Shrink-bag method: Polyethylene bags are simply slipped over the pallet load, then exposed to a controlled flow of heated air.
- Wraparound method: Plastic-sheet film is wrapped around the load then exposed to heated air.

Methods of Heating

- Shrink chamber: This is a single-door unit; the pallets enter and exit through the same door.
- Shrink tunnel: This two-door system is equipped with a through conveyor; the pallets enter one door and exit through a second door.
- Heating frame: Here, a metal frame containing heating elements moves up and down a pallet-load of merchandise that is covered with a plastic bag.

Degree of Mechanization and Production Rates

- Low volume: Handwrapping with sheeting can accommodate one pallet every 8 min.
- Medium volume, semi-automatic: The in-and-out tunnel can wrap one pallet every 2 min.
- High-volume, automatic operation: As above, but can wrap one pallet every 45 sec.

Exploiting Available Services

So far, our discussion has centered on individual components of the distribution picture, and on the potential for individual savings. But the most significant savings are brought about when distribution is considered as a whole, and when management is fully aware of the interactions involving important tradeoffs in the design of a real supply-and-demand system.

To obtain some of these benefits requires a skilled team of specialists experienced in data processing, transportation, warehousing and operations research techniques. By going outside its internal organization—i.e., to consulting groups—the small company can share these specialists with other companies, so that the cost is spread over a much broader base.

Consider computer services. These may be very crude, taking the form of periodic sales analysis and inventory reports; or they may provide your company with a complete management information system whereby you will be tied to a central computer with a remote terminal.

As an example of a very simple system used by one small company, prepunched cards for each item in the product line are maintained in a tub file. Orders by salesman or over the telephone are handwritten and, for fast customer service, are used for picking and shipping. A customer master card and the prepunched

Alternatives Considered in Plant and Warehouse Location Study—Table II

Year	Code	Plants (Facilities Configuration)	Warehouses (Facilities Configuration)
1969		Baltimore (present capacity)	—
1972	I	Baltimore (expanded)	Cincinnati, Ohio; Cleveland, Ohio; Chicago, Ill.; Birmingham, Ala.; Dallas, Tex.
	II	Gary, Ind.	
	III	St. Louis, Mo.	
	IV	Memphis, Tenn.	
1975	I	Baltimore (expanded)	Same as above
	II	Gary, Ind.	
	III	St. Louis, Mo.	
	IV	Memphis, Tenn.	

product cards are then pulled, the quantity shipped is entered, and the card deck is daily dropped off at a service bureau.* The next day, the completed invoices are picked up, checked and mailed. From this basic order information, inventory status reports are run on a weekly basis, sales analysis report and sales commissions at monthly intervals. Such systems can be designed for any degree of sophistication to meet a company's needs, with minimum investment or rental in peripheral equipment and in data processing staffs.

* A company that provides computer information services for a fee, on a continuing basis.

Computer-Based Location Study

For another example of how the smaller company can exploit external resources, here is how a simulation model provided an inexpensive, quantitative basis for decision-making. This study answered the following questions for an $8-million CPI firm presently located in one single plant facility in the East—a company that expected to triple its volume over the next five years:

1. How many warehouses, plants or combinations of each should be provided?
2. Where should the various facilities be located?
3. What volume should each facility handle?

Based on present and predicted locations of customers and sales volumes, feasible locations were heuristically selected—i.e., we used judgment to come up with alternatives that appeared promising and realistic enough to warrant evaluation.

Because a substantial investment would be required for a new or expanded plant, it became extremely important to evaluate both the short- and long-term im-

Location Study: Fixed and Variable Manufacturing and Distribution Costs for 1975—Table III

Run No.	Parameter	Plant: Baltimore (Expanded)	Plant: Gary	Plant: St. Louis	Plant: Memphis	Whse.: Chicago	Whse.: Dallas	Whse.: Cincinnati	Whse.: Birmingham	Whse.: Cleveland
	Cost to carry inventory, ¢/lb	0.277	...	...	...	1.225	1.333	1.333	1.333	1.333
	Warehouse cost, ¢/lb	0.063	...	...	...	0.267	0.267	0.267	0.267	0.267
	Bulk transp. "in," ¢/lb	—	...	...	...	1.100	0.820	0.980	1.070	1.690
	Var. mfg. cost, ¢/lb	0.599	...	...	...	0.599	0.599	0.599	0.599	0.599
I	Var. mfg. cost + dist. cost, ¢/lb	0.939	...	...	...	3.191	3.019	3.179	3.269	3.889
	Fixed distribution cost, $/yr	0*	...	...	...	2,620	2,080	2,350	2,350	2,350
	Cost to carry inventory, ¢/lb	0.250	0.500	...	...	1.225	1.333	1.333	1.333	1.333
	Warehouse cost, ¢/lb	0.063	0.063	...	...	0.267	0.267	0.267	0.267	0.267
	Bulk transp. "in," ¢/lb	—	—	...	...	0.240	0.590	0.580	0.980	1.190
	Var. mfg. cost, ¢/lb	0.578	0.627	...	...	0.627	0.627	0.627	0.627	0.627
II	Var. mfg. cost + dist. cost, ¢/lb	0.901	1.190	...	...	2.359	2.817	2.807	3.207	3.417
	Fixed distribution cost, $/yr	0*	0*	...	...	2,620	2,080	2,350	2,350	2,350
	Cost to carry inventory, ¢/lb	...	...	0.500	...	1.225	1.333	1.333	1.333	1.333
	Warehouse cost, ¢/lb	...	...	0.063	...	0.267	0.267	0.267	0.267	0.267
	Bulk transp. "in," ¢/lb	...	...	—	...	0.480	0.770	0.590	0.910	1.070
	Var. mfg. cost, ¢/lb	...	...	0.625	...	0.625	0.625	0.625	0.625	0.625
III	Var. mfg. cost + dist. cost, ¢/lb	0.901	...	1.188	...	2.597	2.995	2.815	3.135	3.295
	Fixed distribution cost, $/yr	0*	...	0*	...	2,620	2,080	2,350	2,350	2,350
	Cost to carry inventory, ¢/lb	...	...	...	0.500	1.225	1.333	1.333	1.333	1.333
	Warehouse cost, ¢/lb	...	...	...	0.063	0.267	0.267	0.267	0.267	0.267
	Bulk transp. "in," ¢/lb	...	...	...	—	1.100	0.830	0.590	0.710	0.980
	Var. mfg. cost, ¢/lb	...	...	...	0.301	0.301	0.301	0.301	0.301	0.301
IV	Var. mfg. cost + dist. cost, ¢/lb	0.901	...	...	0.864	2.893	2.731	2.491	2.611	2.881
	Fixed distribution, $/yr	0*	...	...	0*	2,620	2,080	2,350	2,350	2,350

* The fixed costs for plant warehouse are set to zero to ensure their staying in program; they are added at the end so that analysis is complete.

Location Study: Determining the Total Cost of Manufacturing and Distribution for 1972 via a Computer Run—Table IV

Facility	Customers	Total Volume, Thousand Lb.	Total Var. Mfg. Cost, $	Total Var. Dist. Cost, $	Fixed Cost, $ Dist.	Fixed Cost, $ Mfg.	Total Cost of Dist. and Mfg.
Baltimore	145	16,900	98,000	302,319	8,075	802,641	1,211,035
St. Louis	92	12,100	80,710	268,355	8,075	751,230	1,108,370
Chicago warehouse	16	920	4,880	20,021	Included	—	24,901
Birmingham warehouse	11	1,600	21,660	44,952	Included	—	66,612
Dallas warehouse	5	750	3,883	25,587	Included	—	29,470
Totals	269	32,270	209,133	661,234	16,150	1,553,871	2,440,388

plications of predicted customer location-pattern on the sites selected for plants and warehouses. Table II identifies the three time-periods that were used, and the plant and warehouse configurations that were analyzed for each period. For each computer run, the Baltimore plant was considered—either expanded to meet total demand, or adjusted proportionately with the capacity required at another plant location.

Table III illustrates the basic cost factors that had to be developed in order for the computer simulation runs to properly compare costs for inbound transportation, warehousing, inventory carrying, manufacturing, and outbound transportation. The accuracy of these cost factors is not crucial, but it is important to use the same basis for the various locations being considered. In other words, the cost *differentials* are the most sensitive to the final result of the simulation runs.

In completing an analysis of this type, many companies fall into a data-collection trap. We found that this could be eluded by averaging or estimating the cost for factors that have a minimal effect on the outcome.

Referring back to Table II, note that there were nine computer simulation runs: one for the present, four each for 1972 and 1975. The result of the 1972-III run, shown in Table IV, illustrates how the simulation establishes the total manufacturing and distribution cost, identifies the most economical warehouse locations to be included, and establishes the sales volume and number of customers to be serviced from each facility location. The results of each of the nine runs were evaluated, and it was determined that the 1975-IV run, a two-plant configuration with five warehouses, would provide the overall lowest manufacturing and distribution cost.

Although this type of study could also be accomplished without the aid of a computer, it would be a horrendous job, and would entail greater risk regarding the accuracy of the result. With the computer, many simulations can be completed with minimum effort once the initial cost factors and sales volumes are established.

The Big Picture for the Small Company

In summary, there are a number of levels at which the war against distribution costs must be fought.

1. The internal, technical level—whereby people in production, design and industrial engineering stay alert to improvement in materials handling, packaging, shipping, receiving and warehousing needs.

2. The external (consulting) level—whereby specialized personnel and other resources can provide the small CPI company with the economy of scale without creating a major burden.

3. The internal, managerial level—whereby an organizational framework is established to effectively use both internal and external resources in streamlining the company-wide distribution system, and keeping it streamlined.

We have not said much about this sort of organizational framework because this may be a rather specialized topic for the technical reader, but basically we feel that responsibility for the physical distribution function should be placed as high in the organization as possible. This is important because of unavoidable conflicts between the marketing and production groups: Marketing wants the best customer service, and thus high inventories close to the market; whereas Production wants to keep unit costs low with long production runs, which may often cause inventory imbalances and inadequate service to customers.

One effective approach for the smaller company is to make the chief financial officer responsible for the measurement and control of distribution cost tradeoffs, and to let people such as traffic and warehouse managers report to him. For the larger company whose annual distribution costs exceed \$2 million, we recommend a professional manager reporting directly to the president or an executive vice-president. ■

Meet the Author

Walter F. Friedman is president of Walter Frederick Friedman and Co., management consultants, 280 Madison Ave., New York, N.Y. 10016. A specialist in physical distribution, he has many years of experience in this area, and is coauthor of the book "Industrial Packaging," published by Wiley. Mr. Friedman has a B.S. in mechanical engineering from the University of Wisconsin, and an M.S. in industrial engineering from Columbia University.

Predicting Chemical Prices

By correlating the cumulative market value as a power function of time, the authors have developed a useful method for predicting the selling prices of chemical products—one that resembles the "six-tenths" rule for cost-capacity correlations.

DIANNE J. MASSEY and JAMES H. BLACK, University of Alabama

As a result of studying the relationship between time and cumulative market value (i.e., volume multiplied by price) for 53 chemical products, we have developed an empirical equation of the six-tenths rule form, and have computed the constants via a "least-squares fit" computer program. Based on the years 1946-1965, the correlation appears strong enough to be useful in predicting future market values not only for these 53 chemicals, but also for others for which historical market-value data are available. Selling price can be estimated from the market value, by making an independent prediction of volume.

If historical data are meager or nonexistent, the average values for the constants found from the 53-product study can be used in estimating future market values, and hence future market prices.

The results show surprising accuracies for this correlation. Before discussing them in detail, let us establish a framework by taking a look at the problem of selling-price prediction, and at some of the previous work done in this field.

Over the past few years, there have been a number of attempts to improve estimating procedures for market studies. As Miller points out,[1] capital- and operating-cost estimates are commonly made with accuracies of ±10%. Market estimates, on the other hand, have sometimes been in error by over 50%.

Thorne[2] stresses that the marketing variables are often the key variables in profitability studies. They are not only the most important ones in project economics, but also they are the hardest to predict.

The Profitability Committee of the N.Y. Section of the American Assn. of Cost Engineers, in an extensive study of the interrelationships of project variables,[3] states that the most significant marketing variable affecting profitability is product pricing. The next most important factor is market volume. Below, we see that the other variables have a much smaller effect on profitability; the table shows the effect of a 10% adverse change in these variables on the Profitability Index (P.I.):

Variable Being Changed	P.I. (Base = 15.3)	Δ P.I.
Investment	14.0	1.3
Raw materials	12.5	2.8
Operating costs	14.0	1.3
Market volume	6.0 to 14.5*	0.8 to 9.3
Market price	−1.0 to 13.2*	2.1 to 16.3

(* Depends on pattern)

Note that a 10% error in market price could change a project from a comfortably profitable one to a complete financial failure; no other factor in an estimate has such a dramatic effect.

A similar conclusion was reached earlier by Twaddle,[4] who showed the dramatic effect of price and volume on profitability. Of these two, price has the greater impact, because any decrease in it comes straight out of profits.

Previous Work

About 12 years ago, articles began to appear in which attempts were made to correlate market variables. The first of these correlations[5] were the so-called "exclusion charts," which are plots of price vs. production any one year for the various members of a class of chemical products, such as heavy chemicals, cyclic intermediates, or drugs. The result of such a plot is an L-shaped area within which the points represented by the price and production of any prospective member of that group or class should fall. Thus, if the point for any prospective new product should fall outside of the L-shaped area, that product is "excluded" from being competitive.[6] This L-shaped area is roughly equivalent to the usual hyperbolic demand curve.

These exclusion charts are useful in determining whether a new product would be priced competitively

Based on a paper presented at the 13th Annual Meeting of the American Assn. of Cost Engineers, Pittsburgh, June 30, 1969, this article contains subsequent revisions by the authors.

Originally published October 1969.

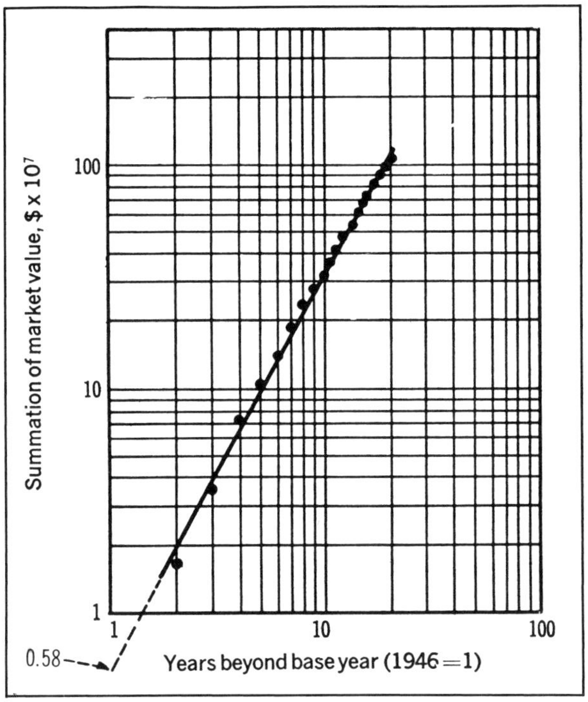

SYNTHETIC METHANOL: Market-value summation vs. number of years beyond 1946 base year—Fig. 1

today, but they are of little value in predicting price variations as a function of time.

More recently, Twaddle and Malloy[7] proposed an equation for projecting market price. The market price is made up of a price floor and a price margin, both of which decay with time, or they follow a "learning" curve that is exponential in form. Because big margins encourage new capacity, the margin curve also decreases as the price floor decreases. One problem with this correlation is the extrapolation difficulty attending curvilinear relationships.

In another study, Perkins and Enyedy[8] proposed a learning curve for correlating and predicting chemical prices. In this method, cumulative production is plotted against selling price on log-log coordinates. There are two difficulties: First, the slopes of the learning curves vary from chemical to chemical. Second, these relationships have no element of time in them. The time element is supplied by a separate semilog plot of production vs. time, which is curvilinear and thus involves the concomitant extrapolation problems.

Another approach to pricing has been proposed by Seglin,[9] who emphasizes the need to base prices on demand rather than on costs. The optimum price should be independent of fixed operating cost, how rapidly the market potential is achieved, or whether or not future profits or losses are discounted. The price, he feels, should be dependent upon variable operating costs and on the price elasticity of the market potential. The elasticity is defined as the percentage change of potential demand with percentage change of price. Again, this method requires some kind of assumption about the growth pattern for the product. This usually takes the form of a time-production curve, which again is a curvilinear relationship.

A more recent study by Moore[10] analyzes demand curves, or price-volume curves, to determine the price-volume combinations that maximize the discounted cash flow rate of return for a venture. Again, external sources were recommended for the production-vs.-year data.

It is apparent that there is a need for a simple procedure for predicting price trends. Particularly desirable would be a linear relationship between price and time, so that price trends can be extrapolated.

Proposed Correlation

A study of the price and production trends for 53 chemical products shows that market value can be correlated with time, in an equation of the "six-tenths rule" form. This relationship is surprisingly accurate and yields a straight line on log-log paper. The recommended equation is

$$\Sigma V = kt^m \tag{1}$$

in which V is the market value, t is the number of years since the base year, and k and m are constants.

The market value is the market price or selling price multiplied by the annual production. The determination of the value of t, the number of years since the base year, requires some explanation. In short, the year for which the first information is collected becomes year one. However, because the total annual market value is not realized until the end of year one, the first t value is, for all practical purposes, two. Each succeeding cumulative market value is likewise plotted at the end of the year in which it was accumulated.*

Therefore, Eq. (1) can be utilized by plotting cumulative market values vs. t on log-log paper. The slope of the resulting line is m, and the intercept at $t = 1$ is k. This plot is analogous to the cost-capacity plots that have appeared in the literature for many years.[12]

Fig. 1 is a plot of Eq. (1) for the market value of methanol, beginning with the year 1946. The value of m is 1.725, and k is 0.5825×10^7. The plot shows that the relationship between the log of the cumulative market value and the log of the years since the base year is linear. This linear relationship so closely represents the data that correlation coefficients are very near 1.0. Fig. 1 shows that the practice of plotting the cumulative production at the end of the year in which it was accumulated gives excellent results.

Graphs similar to Fig. 1 can be prepared for other chemical products. The results are uniformly close correlations. In each case, a k and a slope m can be

* The situation here is analogous to the situation with dates, such as B.C. and A.D. Thus, the day after Dec. 31, 1 B.C. was Jan. 1, 1 A.D.; there was no year 0. Similarly, there is no 0 year since the base year. We then begin counting years since the base year with the year 1.[11]

calculated. The data were obtained from U.S. Tariff Commission reports.[13] (For a few of the chemical products, there were some missing data; these were estimated from other sources, such as Faith, Keyes, and Clark.[14])

Results of the Calculations

The data from the literature were correlated, according to Eq. (1), by means of a least-squares computer program, run with the data for all 53

Slopes and constants for use in Eq. 1, for 53 chemicals—Table I

Class	Number	Product Name	Years	m	$k\,(10^3)$	$k\,(10^3)/V_1$	% Error
Acids	1	Acetic acid	20	1.508	9,096	0.4122	7.3
(organic)	2	Formic acid	17	1.323	952	0.4826	−5.7
	3	Lactic acid	18	1.346	664	0.4673	−2.7
	4	Oxalic acid	20	1.401	953	0.5009	−7.7
	5	Propionic acid	14	1.678	630	0.3433	−11.4
Alcohols	6	Methanol	20	1.725	5,825	0.3569	−3.8
	7	Ethanol	20	1.585	9,451	0.2532	14.2
	8	Isopropanol	20	1.549	10,263	0.4325	1.0
	9	n-Butanol	20	1.444	6,915	0.4565	1.0
	10	Other butanols	20	1.621	3,761	0.3885	−3.6
	11	Iso-octanol	11	1.487	2,715	0.4017	−1.0
	12	1, 2 Octanol	11	1.035	1,455	0.4905	−0.6
Aldehydes	13	Amylcinnamaldehyde	8	1.625	84	0.3289	2.9
	14	Anisaldehyde	16	1.439	144	0.3512	16.5
	15	Benzaldehyde	15	1.308	356	0.3729	8.1
	16	Cinnamaldehyde	7	1.809	89	0.2902	−7.8
	17	Formaldehyde	20	1.922	3,315	0.2338	27.3
	18	Methylcinnamaldehyde	7	1.673	5	0.3455	−10.7
	19	Pentylcinnamaldehyde	8	1.453	196	0.3779	−2.7
Chlorinated	20	Dichloromethane	19	2.056	374	0.2390	−2.9
hydrocarbons	21	Chloromethane	20	1.420	1,446	0.3624	14.8
	22	Chloroethane	18	1.321	15,815	0.5051	−6.5
	23	1, 2 Dichloroethane	18	1.752	4,417	0.2638	12.2
	24	Chloroform	20	1.793	445	0.2565	8.6
	25	Carbon tetrachloride	20	1.765	2,501	0.3466	3.4
	26	p-Dichlorobenzene	20	1.576	1,169	0.4093	−14.5
	27	Vinyl chloride monomer	15	1.468	18,835	0.3263	5.5
Esters	28	Ethyl butyrate	19	1.330	44	0.4504	3.1
	29	Isoamyl butyrate	20	1.164	15	0.4916	8.8
	30	Glyceryl monostearate	20	1.778	378	0.4173	−0.6
	31	Glyceryl mono-oleate	20	2.149	25	0.2259	−46.5
	32	Diethylene glycol mono-oleate	19	0.891	188	0.6815	−23.4
	33	Diethyl glycol mono-laurate	14	1.365	85	0.5555	−15.4
	34	Oleic acid esters	20	1.500	573	0.4125	−17.8
	35	Acyclic phosphoric acid esters	17	1.796	433	0.5449	−17.6
	36	Stearic acid esters	20	1.221	1,841	0.8328	−25.3
	37	n-Butyl stearate	20	1.415	226	0.4816	−8.2
Hydrocarbons	38	C_2 hydrocarbons	19	2.261	3,905	0.2355	6.5
	39	C_3 hydrocarbons	19	2.132	1,607	0.2193	9.3
	40	C_4 hydrocarbons	20	1.577	37,562	0.2530	4.2
Inorganic	41	Total acetic acid salts	19	1.432	850	0.4105	2.3
acetates	42	Ammonium acetate	16	1.496	37	0.4014	16.6
	43	Cobalt acetate	9	1.504	44	0.3602	−4.4
	44	Copper acetate	14	1.323	31	0.4689	−0.4
	45	Potassium acetate	19	1.555	32	0.2471	19.7
	46	Zinc acetate	18	2.019	6	0.2229	9.7
Mixed	47	Phenol	20	1.830	5,862	0.3129	5.2
organics	48	Phthalic anhydride	20	1.750	5,623	0.3836	−17.0
	49	Acetone	20	1.458	10,232	0.4840	−6.2
Organic	50	Total butyl acetates	20	1.032	9,921	0.6608	1.4
acetates	51	n-Butyl acetate	11	1.328	3,747	0.5534	−7.1
	52	Ethyl acetate	20	1.200	5,228	0.5352	0.5
	53	Vinyl acetate	15	2.304	889	0.2779	−13.4

Summary of mean values and standard deviations for 53 chemicals—Table II

Product Class	Mean Value, $\bar{m}$	Standard Deviation, δm	Mean Value of k/V_1	Standard Deviation $\delta k/V_1$
Acids	1.4512	0.1183	0.4412	0.0565
Alcohols	1.4922	0.2046	0.3975	0.0663
Aldehydes	1.6041	0.2019	0.3288	0.0469
Chlorinated hydrocarbons	1.6438	0.2256	0.3386	0.0830
Esters	1.4609	0.3443	0.5094	0.1545
Hydrocarbons	1.9900	0.2966	0.2359	0.0100
Inorganic acetates	1.5548	0.2197	0.3518	0.0883
Mixed organics	1.6793	0.1596	0.3935	0.0692
Organic acetates	1.4660	0.4949	0.4793	0.1392
Overall for 53 products	1.5589	0.2891	0.3963	0.1236

chemical products. In addition to the least-squares fit, the program also computed the slope of the line, the constant k, and the ratio of k to the initial market value, V_1. (This was done because there was a noticeable constancy in k/V_1 indicating that this ratio might be useful for estimating purposes if data are meager.) The calculated value of the cumulative market value, using the least-squares line developed in the program, was also printed out for each year. Table I is a summary of the results. The last column shows the percent error in estimating cumulative market value.

Table II presents the mean values of the slope m and the mean values of the ratio k/V_1 for each of the nine classes of chemical compounds studied (and also for the 53 chemicals on an overall basis). The standard deviations of the slopes and ratios about their means are also listed.

Table III shows how the slope m changes, for the same chemical product, as the number of years changes. Four different compounds were selected at random, and the m values were calculated for several arbitrarily chosen spans of years.

Inspection of the results in Table I shows a surprising uniformity among the m and k/V_1 values for all the chemicals, regardless of type. One might reasonably expect that each type of chemical product would have a characteristic m or k/V_1 value, much as there are exclusion charts characteristic of different types of chemicals. The fact that different types of chemical equipment have characteristic cost-capacity slopes lends strength to this assumption. Such a grouping of m values or of k/V_1 values was not found, however.

The m values in Table I vary from a low of 0.891 to a high of 2.304. The mean m value, $\bar{m}$, was found to be 1.5589 for all 53 chemicals, with a standard deviation of 0.2891. Although the spread in m values is large, about 70% of the m values lie between 1.3 and 1.7.

The results in Table III show again that the m values can vary rather widely as the number of data points changes. In general, as one might expect, the m value approaches the overall mean $\bar{m}$ value more closely as the amount of data increases. In only one instance did the m value vary from the overall mean by more than ±15 to 16%.

Effect of differing spans of years—Table III

Product Name	Number of Data Points (Years)	m	Average m	% Dev. from Overall Average m
Methanol	15	1.371		−12.1
Methanol	20	1.725		10.8
			1.548	−0.7
Ethanol	17	1.781		14.2
Ethanol	20	1.585		1.7
			1.683	8.0
Phthalic anh.	20	1.750		12.2
Phthalic anh.	5 (1946–50)	2.049		31.5
Phthalic anh.	15 (1951–65)	1.328		−14.8
			1.709	9.6
Acetone	20	1.458		−6.5
Acetone	5 (1946–50)	1.802		15.6
Acetone	15 (1951–65)	1.303		−16.4
			1.521	−2.4
Average m for 53 chemicals			1.559	

The errors in predicting cumulative market values from the derived equation are shown below. Note that the error was within ±10% for 66% of the 53 products, and within ±20 for 92%.

Error Range, %	Frequency
50+	0
40–50	1
30–40	0
20–30	3
10–20	14
0–10	35

As for the relationship between the constant k and the initial market value V_1, although neither the consistency nor the closeness of the extreme values were as good as they were for the values of the slopes (see Tables I and II), the k/V_1 ratio can still find application in predicting market values when insufficient data are available.

Illustrative Example

Let us say we want to estimate the selling price of polyvinyl chloride in 1973 from the historical production and market-value data given in the U.S. Tariff reports for the 12 years from 1954-1965.[13]

The production for 1973 is estimated to be 2,140 million lb., by a method described earlier.[15] (Or, a production estimate can be made by some other method.) By plotting the market value vs. t for the 12 data points, one can find the best line through the data via the least-squares method, and then derive the slope m. The constant k can be found from any point on this line and the value of m. Using Eq. 1 and the computer program, m is found to be 1.451 and k is 6.44535×10^7.

Then, $\log \Sigma V = \log k + m \log t$

For $t = 20$ (for 1972):

$\log \Sigma V = 7.809255 + 1.451 \times 1.30103$

or $\Sigma V = 4.978$ billion for 1972

For $t = 21$ (for 1973):

$\log \Sigma V = 7.809255 + 1.451 \times 1.32222$

or $\Sigma V = 5.343$ billion for 1973

V in 1973 = \$5.343 billion − \$4.978 billion = \$365,000,000

Selling price, 1973 = (\$365,000,000)/(2,140,000,000 lb.) = \$0.17/lb.

Effect of Inflation; Limitations of Method

One might wonder whether inflation would affect adversely the correlation represented by Eq. (1). Although the effect of inflation on selling prices can be taken into account by adjusting the selling prices for the decrease in purchasing power of the dollar, the resulting correlation between market value and time is no closer than between uncorrected market value and time. This correction was therefore not deemed necessary. The same conclusion was reached by Perkins and Enyedy.[8]

Naturally, correcting market value for inflation gives different values for the slopes and intercepts in Eq. (1).

The accuracy of the proposed method appears to correspond to the accuracies attending the "six-tenths" rule or cost-capacity correlations, which this method closely resembles. It is best suited to intermediate term (i.e., three- to ten-year) trend prediction.

Violent economic upsets, such as World War II, interrupt the continuity of the correlation. Probably other types of upsets in the economy, such as the 1929-1939 depression, would also have a profound effect on the continuity of the correlation. Minor recessions would seem to have little effect.

As for sudden technological advances, this question has not been studied thoroughly yet. The chemicals used in the study were chosen because they fell conveniently into groups of similar compounds, and not because of their technological histories. Additional work is needed to find out how this correlation holds up with chemicals that have had very erratic price and volume histories because of breakthroughs.

Care should therefore be exercised in relying on this method if a market study reveals that technology or uses for the product are about to go through revolutionary rather than evolutionary changes. ■

References

1. Miller, C.A., "New Cost Factors Give Quick, Accurate Estimates," *Chem. Eng.*, **72**, No. 19, 226-36 (1965).
2. Thorne, H.C., "Analyze Ventures by Market Research," *Hydrocarbon Processing*, **45**, No. 7, 145 (1966).
3. Douglas, F.R., et al. "Effect of Key Marketing Variables on Profitability—I," *AACE (Amer. Assn. of Cost Engineers) Bulletin*, **5**, No. 3, 63-76 (Sept. 1963).
4. Twaddle, W.W., "The Power of Certain Important Factors in Economic Evaluation," 6th Annual Mtg., Amer. Assn. of Cost Engrs., Chicago, Ill., June 1962.
5. Zabel, H.W., "The Exclusion Chart," *Chem. Eng. Prog.*, **52**, No. 5, 183-6 (1956).
6. Zabel, H.W., Marchitto, M., "What Price Can I Get for My Chemical?," *Chem. Eng.*, **66**, No. 22, 112-117 (1959).
7. Twaddle, W.W., Malloy, J.B., "Evaluating and Sizing New Chemical Plants In a Dynamic Economy," *AACE Bulletin*, **8**, No. 3, 135-9 and 142-5 (1966).
8. Perkins, J.H., Enyedy, G., "Use of the Learning Curve to Forecast Trends of Chemical Prices," Paper No. 66-A. 12, 10th Annual Mtg., Amer. Assn. of Cost Engrs., Phila., Pa., June 1966.
9. Seglin, L., "An Approach to Pricing of New Products," *AACE Bulletin*, **8**, No. 2, 52-56 and 85-91 (1966).
10. Moore, C.S., "Supply and Demand Curves in Profitability Analysis," *Chem. Eng.*, **75**, No. 21, 198-205 (1968).
11. Jackson, J.W., Black, J.H., "Predicting Markets for Chemical Products," *Trans. Amer. Assn. of Cost Engrs.*, 13th Annual Mtg., Pittsburgh, Pa. (1969).
12. Chilton, C.H., "Cost Data Correlated," *Chem. Eng.*, **56**, No. 6, 97-106 (1949).
13. United States Tariff Commission, Synthetic Organic Chemicals, Report No. 159 (1946); 162 (1947); 164 (1948); 169 (1949); 173 (1950); 175 (1951); 190 (1952); 194 (1953); 196 (1954); 198 (1955); 200 (1956); 203 (1957); 205 (1958); 206 (1959). TC Publication 34 (1960); 72 (1961); 114 (1962); 143 (1963); 167 (1964); 206 (1965).
14. Faith, W.L., Keyes, D.B., Clark, R.L., "Industrial Chemicals," 3rd Ed., Wiley, New York (1965).
15. Jackson, J.W., Black, J.H., "Predict Petrochemical Production," *Hydrocarbon Processing*, **48**, No. 7, 143-6 (1969).

Meet the Authors

◀ **Dianne J. Massey** is a senior in the Chemical Engineering Dept. at the University of Alabama. (University, Ala., 35486). She will receive a B.S. in Chemical Engineering in Jan., 1970. She is a student member of both AIChE and the American Assn. of Cost Engineers (AACE), and is one of the first women to be initiated into Tau Beta Pi. She has been a consistent scholarship student, and has received the Outstanding Junior Award of the Central Alabama AIChE Section.

James H. Black is Professor of Chemical Engineering at the University of Alabama. He received an A.B. degree in chemistry from Cornell University in 1943, and B.S. (1948), M.S. (1949), and Ph.D. (1954) degrees in Chemical Engineering from the University of Pittsburgh. His work experience includes assignments with Mellon Institute of Industrial Research, Standard Oil Co. (Indiana), and U.S. Steel Corp. Dr. Black is a registered professional engineer in Pennsylvania and Alabama, a member of AIChE and AACE, and serves as Executive Secretary of AACE. ▶

Section VII

TOPICAL COST ENGINEERING: ... INFLATION ... ENERGY CONSERVATION ... ENVIRONMENTAL CONTROL

Machinery Costs vs. Energy Conservation

The increased cost of energy and the critical need to conserve fuels have upset the chemical process industries' traditional approach to specifying equipment. The author shows how the additional expense of more-efficient machinery can be justified in today's economy.

R. W. NEWKIRK, Fluor Corp.

INCREASING COST of fuel oil makes heat-recovery equipment on fired heaters an attractive capital investment.

In the days of low-cost energy, the most economical plant design often involved equipment that was relatively wasteful as far as power consumption was concerned. This was true even for energy-intensive devices such as pumps, compressors, and air coolers. Now, with the higher price of electric power and the emphasis on energy conservation, cost estimators and management will need to reexamine their approach to process plant design.

For example, 434 pumps were recently installed in a new petrochemical plant. These were excellent U.S.-supplied centrifugals, mostly API-type specifications or better. The total operating power required by the drivers was 32,675 hp.

The average efficiency for the entire set of pumps was only 49.1%. If this could have been boosted by just 10%, the connected hp would have dropped by about 2,925 and the annual electric power bill would have been cut by some $291,000 at 1¢/kWh.

Obviously, both management and engineering need to change their design and economic philosophies so that future installations will have higher pump efficiencies. The data in Table I indicate that an overall efficiency increase of 5 to 10% is attainable in many cases.

Oversizing Can Waste Energy

If one analyzes the installed cost of pumps relative to total plant cost, this may account for only 3 to 5% of the

Originally published June 1974.

Typical Efficiency Ranges of API Pumps – Table I

Gal/Min	Vendor	Range, Gal/Min	Single-Stage Head Range, Ft	Efficiency Range, %
100	A	80-110	80-170	49-54
	B	70-100	75-180	46-60
	C	90-160	60-160	54-64
200	A	160-235	125-260	50-63
	B	150-250	120-270	57-66
	C	185-250	140-250	57-63
500	A	360-525	160-390	62-71
	B	360-490	220-420	68-72
	C	425-575	105-225	65-73

total. However, the power required to run these pumps is a substantial portion of the owner's operating expense. It is poor planning for the 1980s when 1970 guidelines force engineers to design a unit with less than the best possible hydraulic efficiency.

To allow for future increases in flow and head, almost all of the present design standards of major oil and petrochemical companies dictate the selection of pumps and drivers that are oversized. Unfortunately, the original reasons for oversizing are rarely checked after plant startup or questioned during the life of the equipment. This approach wastes energy in two ways:

- The pump cannot be designed so that its maximum efficiency is at the point where it will actually operate. Therefore, it is less efficient than a pump that is sized for normal service.
- The extra head or flow from an oversized unit is wasted across a control valve until such time as the additional capacity is needed.

A review of 225 pumps in another recently constructed grass-roots refinery showed that eliminating the future head or flow capacity would have meant a 10 to 20% efficiency increase for many of the smaller pumps. Table II illustrates how the higher first cost of a more efficient pump can be offset by lowering operating expense.

Some Specs Too Conservative

As energy costs rise, one should remember that the efficiency of many pumps can be improved by reducing internal clearances or taking more risk of parts rubbing together and failing. It was not unusual in the late 1940s for European companies to bid—at low prices—on pumps with close clearances and thin metal sections. This approach has not gained favor here, even though many U.S. standards are too conservative and unnecessarily sacrifice efficiency. Granted, conservative design is desirable in critical, nonspared pumps. But such specifications on a total plant basis will result in a lifetime waste of energy.

Pump standards—API, ASME, DIN, American Water Works, etc. usually take years to develop and are updated slowly, even though they may be out-of-date by the time they are published. To reduce power bills and obtain maximum machinery efficiency in the future, standards for new projects must be given a thorough modernization.

As one example of the effect of pump specifications, some old standards limit the maximum inside impeller diameter to 90% of what the case can accommodate. The reason is to provide the option of installing a larger impeller, if required to account for design mistakes or changed operations. Management should consider that undersized impellers waste energy—a typical situation is presented in Table III.

Motor and Turbine Drives

In the past, the chemical process industries generally bought small motors (under 75 hp) on the basis of price rather than efficiency. As in the case of pumps, this may not be the path to follow in the future—we will be able to justify higher-efficiency motors because energy is becoming more expensive. Changes are also likely in the noisy, integral fans that remove motor heat. It is very doubtful that this type of unit will meet noise specifications for air-cooler and pump drives.

Although U.S. suppliers have been developing quieter motors with better insulation and lower energy losses, there will be only limited quantities available for some years.

Small steam turbines are also due for reevaluation because of the energy situation. It is apparent that many steam-driven units are wasting energy. Replacement of small turbines with multistage or other more efficient types can be easily justified in view of the increased cost of fuel oil for generating steam.

Large motors in the 10,000 to 20,000-hp range have been used primarily for driving centrifugal compressors or blowers. In the future, I believe steam and gas turbines, instead of electric motors, will be the first choice of engineers. Supersize motors (50,000 to 100,000 hp) will be installed only in plants that generate their own power or have special arrangements with a utility for obtaining low-cost electricity.

In any event, there will be a strong trend to specify large drivers that have maximum efficiency. To utilize 30,000 to 100,000 hp from one driver requires a central station with multiple compressors—say, three or four units driven by a large steam or gas turbine. This is now practical because with new technology and high-strength materials manufacturers can build single compressor cases that absorb 10,000 to 30,000 hp.

The increased efficiency of a single, large driver could justify the investment for centralized compressors. For example, if you improve the average efficiency of 50,000 compressor hp by 10%, the savings of 5,000 hp represents about $400,000 in fuel and capital over a period of four or five years.

With major compressor plants driven by very large turbines, there will be many modifications to plot-plans so that these units comply with community noise specifications and electrical area-classifications. Also, there probably will be many more large units with gear boxes,

Pump Efficiency Versus Cost – Table II

	First Cost	Efficiency, %	10-Yr Evaluated Cost Including Operating Cost and First Cost*
Lowest-possible cost unit, non-API.	2,500	62	$55,000
Pump with end suction, single stage, API standard, lowest cost.	3,500	69	45,900
Double-bearing pump with side suction, slow speed, and longer life.	4,500	76	44,800

*10% interest and 1¢/kWh power.

Undersized Impellers Waste Energy – Table III

	Pump Efficiency, %	Total Hp	Annual Savings, Power at 1¢/kWh
100 refinery pumps with 90% impeller.	62.5	5,260	
Same refinery pumps with 100% impeller.	70	4,640	$33,000

as designers try to match peak driver efficiency with compressor operations.

Air Coolers

During the last 20 years, the CPI have greatly expanded their use of air coolers because these units do not present the environmental problems associated with shell-and-tube exchangers. However, air coolers waste substantial heat to the atmosphere, and in many plants there are opportunities for transfering excess heat from one process unit to another.

An additional energy disadvantage of air coolers is the amount of electricity required to operate fans. This may equal or exceed any other electrical load in some plants. One recently constructed petrochemical facility has over 700 motors (25 hp each) driving fans on air coolers. Obviously, we need new approaches so we can exchange more heat in the process and throw less away to the air and water.

Fired Heaters Are Wasteful

We have purchased 225 major fired heaters over the last 10 years. These cost approximately $49 million and have a total fired duty of about 18 billion Btu/h. The heaters exhaust one-quarter of this heat to the sky—equal to as much oil as is consumed daily by many medium-size cities. It is now critically important that those people responsible for new projects review their company's position on energy losses from heaters and boilers.

In one refinery, fired heaters for 17 process units cost $4 million and have an average efficiency of 70%. Just 10% better performance from these units would save 260,000 bbl of oil annually. The problem here, as in many recently constructed plants, is that management decisions on furnace efficiency were made over five years ago for units that are only now being installed and operated. During that time, the relationship between capital costs and fuel costs has changed greatly.

Another pitfall is that design engineers might not revise equipment specifications in line with improvements in the state of the art. For example, some specifications call for massive convection sections with bare tubes on wide spacing. This is based on a 20-year-old theory that says smooth, bare tubes are needed to facilitate soot removal. Experience has shown, however, that extended-convection surfaces will boost efficiency 5 to 15% with a minor amount of risk.

It is no news to most refinery and petrochemical operators that many of their "1960 era" fired heaters should have been equipped with heat-recovery devices or had such units added soon after initial startup. A recent review of 30 heaters installed during the 1962-1965 period shows that management and engineering were short of the mark in providing the best furnaces for later modification with heat-saving devices.

Consider a fired heater installed in 1962 with a first cost of $470,000. It would have required about $200,000 to add heat recovery. With fuel oil at $1/bbl, the payout would have been 14.7 yr, too long to justify the investment. But with fuel at $3/bbl, savings over the last four years would have been $335,500. Moreover, in the next three years, assuming fuel at $4/bbl, the heat-recovery equipment would return some $557,000.

Overall, it is obvious that the new era of high-cost energy changes the traditional approach to specifying equipment. First-cost will remain a consideration, but purchasing policies heavily weighted toward maximizing efficiency will usually be the best economic path to follow. #

Meet the Author

Robert W. Newkirk is manager of projects at Fluor Corp., 2500 S. Atlantic Blvd., Los Angeles, CA 90040. After nine years with Standard Oil of California, Mr. Newkirk joined Fluor in 1956. He has had a wide range of project experience, and is currently manager of 140 engineers and technical consultants who are involved in $100 million annually of refinery and petrochemical plant construction. He is a mechanical engineering graduate of the University of California, and a registered professional engineer in Arizona, California and Nevada.

How to Estimate and Escalate Costs of . . .

Wastewater Equipment

Here are charts and tables of cost data that will give you a rough, preliminary cost estimate for a new facility, and a technique for escalating the estimate to a future time.

HERBERT G. BLECKER, HARVEY S. EPSTEIN, and THOMAS M. NICHOLS, Icarus Corp.

These are difficult times for preparing cost estimates on new facilities. With material costs going up at a rapid rate, many equipment manufacturers will not give a firm price on an equipment item. The trend now is to bid on an "as delivered" basis.

Nevertheless, cost estimates often must be prepared for evaluating the economics of a new project before any equipment is delivered.

The cost data presented here are intended for that purpose. This information should be helpful in preparing conceptual or rough estimates of facilities for treating wastewaters, or comparing alternative processing schemes. And field labor manhours and labor-to-material (*L/M*) ratios as a function of size, as presented here, should be helpful in updating cost information to a different time or productivity base.

Basis of Cost Data

Cost data were generated by a computer-aided design and estimating system developed by Icarus Corp. The system arrives at equipment costs in the same manner as the equipment vendor and assembles the complete installed costs the same way a contractor does at the definitive stage.

The graphical data (Figs. 2 to 13) include purchased-equipment costs (or field-erected cost), equipment-setting manhours, installed-equipment cost, total field-labor manhours, and where appropriate, the labor-to-material (*L/M*) ratio.

Equipment Costs

Data are based on equipment-cost models involving a series of cost engineering equations that perform a mechanical- and civil-engineering design; generate required material quantities (e.g. number of plates, nozzles, flanges, weld rod, etc.) for fabricating the item in the vendor's shop; develop required vendor's fabrication labor by craft; calculate labor cost, machine burdens as a function of machine-tool operating time, shop overhead, office overhead, G & A, profit; and summarize these costs to obtain the estimated purchased-equipment cost.

Each model has been verified by an in-depth analysis of current vendor "bid tabs" and vendor design and pricing data.

Equipment Setting

Manhours for equipment setting shown on the graphs include transporting the equipment item from a local field storage area to its final site, and setting and aligning the item on its foundation.

Not included with the equipment-setting manhours are any field-labor economies in handling multiple equipment items, unallocable equipment-setting manhours, double handling, inefficiencies caused by improper field scheduling, and bad-weather conditions.

Installed Equipment Cost

Installed-equipment costs shown on the graphs include all direct-material (bulk items) and field-labor costs required for installing the particular equipment item under "typical" or "average" conditions.

Field-material (bulk) quantities as a function of the size, weight, and other characteristics of the equipment item are determined by a computerized quantity-material takeoff. This takeoff relates the material required to install the item for each bulk account (piping, concrete, steel, instrumentation, electrical, insulation, and paint) to the appropriate sizes, dimensions, weight, material and other characteristics of the item. For example, concrete material is developed by considering the equipment (centrifugal pump, vertical tank, heat exchanger, etc.), weight, required depth of footings, and item dimensions to determine: (a) foundation type (small slab, octagon, ring wall, etc.), (b) base area, (c) footings size (if required), (d) volume of foundation and footing, (e) area of form work, (f) required back-up lumber, (g) accessories. The direct material includes cost of all piping, concrete, steel, instrumentation, electrical, insulation, and paint associated with installing the equipment item.

Not included with these "equipment item" material costs are direct materials that are "operational unit dependent"—a term used in the computer-aided estimating system to define a geographically discrete area of a

Originally published October 1974.

facility that is usually a unit operation, unit process, or offsite facility.

Operational-unit-dependent direct-material costs are those items derived from the area (piping for the fire-water loop, electrical for area lighting), or building structure (concrete for area paving, foundations for open-steel structures and pipe racks, and structural steel for open-steel structures and pipe racks).

Field-Labor Manhours

Total field-labor manhours shown on the graphs represent the sum of all crew manhours for all work items required to install the equipment item in a "norm," "typical," or "average" manner.

A typical work item in the computer system's data base is shown in Fig. 1. The work item is a series of cost-engineering equations that relate field-labor manhours to material quantities for a specific labor activity. Data for the relationships come from field labor on recently completed projects. Production rate used by the system corresponds to a man working in the U.S. Gulf Coast area 42 min./h.

How to Determine Escalation

The following procedure shows how to escalate cost to a future point in time and/or "move" the data to a location other than the U.S. Gulf Coast.

Let us assume we want to calculate the installed cost of an equipment item in our South Philadelphia plant. Our plans call for procurement to peak during the 2nd quarter of 1975 and construction in the field to be at its peak during the 2nd quarter of 1976. Table I illustrates a table of cost elements and their value at the different time periods. The first period is the data base (4th quarter 1973); the second, the procurement peak (2nd quarter 1975); and the third and last period, the field-labor peak (2nd quarter 1976).

For simplicity, the procurement peak will be used both for equipment and other material. This would not be true in the case of large construction projects where other material will lag equipment procurement by at least one quarter. However, in considering construction of small industrial pollution-control facilities (i.e. under $2 million, 4th quarter 1973), all material procurement will probably occur during the same quarter.

Items 1, 2, 3, 4, 6, 7, 8 and 11 in the data-base time period are values that we would pick off the graphs presented with this article.

Item 2 in the procurement peak-time period is determined by extrapolating the Marshall & Swift equipment cost index that appears at the back of each issue of CHEMICAL ENGINEERING.

Items 6, 7, and 9 in the field-labor peak time period depend on the location of the facility and expected labor conditions during construction.

To determine the total installed cost for the equipment item in South Philadelphia in the 2nd quarter 1976 we must calculate:

- Estimated purchased-equipment cost for the 2nd quarter 1975.

ICARUS CORPORATION — ACCOUNT: Concrete

WORK ITEM: Erecting Forms and Shoring for Solid-Slab Type Concrete Construction

BY:

DATE:

DESCRIPTION: Work item includes all labor required to erect forms and shoring for solid-slab type concrete construction.

CREW MIX:	EQUIPMENT:	MATERIAL:
0.5 Foreman 4 Rough Carpenters 4 Carpenter Helper	N. A.	N. A.

PRODUCTION RATE (100 % = 42 WORKING MINUTES PER HOUR)

Form Height (feet)	Slab Thickness (inches)	Production Rate (square ft/hr): Forms With 4x4 inch Wood Shores	Forms With Adjustable Shores
10-12	4-8	70	78
12-14	4-8	65	74
14-16	4-8	61	70
16-18	4-8	56	65
18-20	4-8	52	61
10-12	8-12	61	70
12-14	8-12	56	65
14-16	8-12	52	61
16-18	8-12	48	56
18-20	8-12	43	52

TYPICAL CONCRETE account work item—Fig. 1

- Material cost (less equipment) for 2nd quarter 1975.
- Field-labor cost for 2nd quarter 1976.

Estimated purchased-equipment cost for 2nd quarter 1975 is calculated as follows:

$$\left(\begin{matrix}\text{Equipment cost,}\\ \text{4th Q 1973}\end{matrix}\right)\left(\frac{\text{M\&S index, 2nd Q 1975}}{\text{M\&S index, 4th Q 1973}}\right) = \begin{matrix}\text{Est. equip. cost,}\\ \text{2nd Q 1975}\end{matrix}$$

$$(10{,}000)(440/350) = \$12{,}600$$

Estimated material cost (less equipment) for 2nd quarter 1975 is calculated as follows:

$$\left(\begin{matrix}\text{Total installed}\\ \text{cost, 4th Q 1973}\end{matrix}\right) - \left(\begin{matrix}\text{Purchased equip.}\\ \text{cost, 4th Q 1973}\end{matrix}\right) - \left[\left(\begin{matrix}\text{Field labor,}\\ \text{manhours}\end{matrix}\right) \times \left(\begin{matrix}\text{U.S. Gulf Coast composite}\\ \text{wage rate, 4th Q 1973}\end{matrix}\right)\right] \times \left(\frac{\text{M\&S index, 2nd Q 1975}}{\text{M\&S index, 4th Q 1973}}\right) = \begin{matrix}\text{Est. material cost,}\\ \text{2nd Q 1975}\end{matrix}$$

$$[(15{,}000 - 10{,}000) - (220 \times 7)](440/350) = \$4{,}350$$

Then, field-labor costs for 2nd quarter 1976 are calculated:

$$(220\ \text{h})(100/74)(\$14/\text{h}) = \$4{,}160$$

Adding the three costs ($12,600 + $4,350 + $4,160 = $21,110), we obtain the estimated installed cost of the item at a different point in time and at a different location.

Such a procedure can be used for equipment items not covered in the graphs in this article by using the L/M ratio (Table II) that most closely approximates the type of equipment being considered. Remember that all data are based on U.S. Gulf Coast labor (approx. $7.00/manhour in 4th quarter 1973), U.S. Gulf Coast productivity (100%), and material costs as of 4th quarter 1973 (M&S = 350.0).

(*Text continued page* 554)

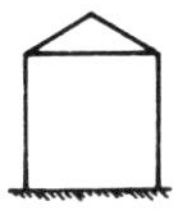

Storage Vessels
Material = Carbon Steel

Fig. 2

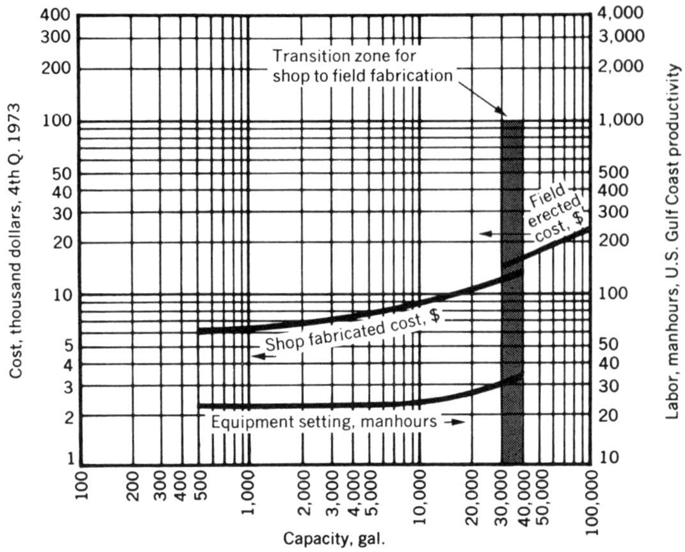

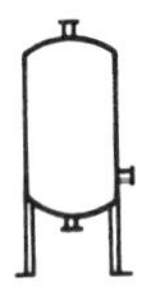

Vertical Pressure Vessels
15 psig design pressure
Material = Carbon Steel
L/D ratio = 3:1 to 6:1
L/M ratio = 0.380 to 0.302

Fig. 3

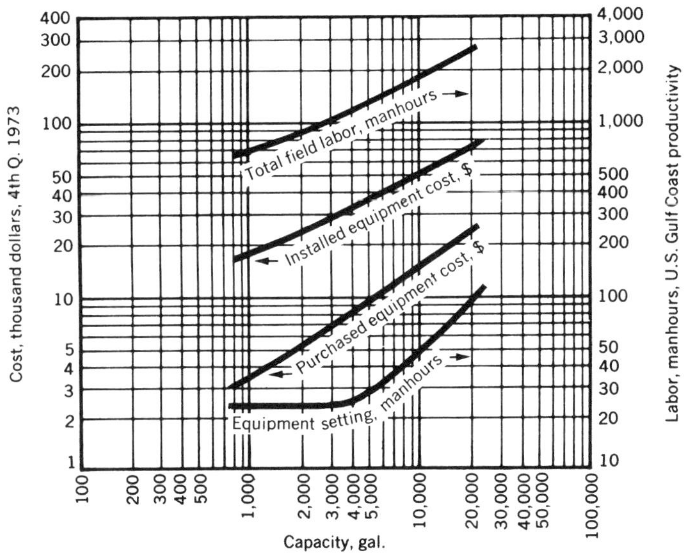

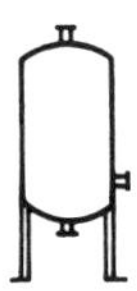

Costs of
Vertical Pressure Vessels
250 psig design pressure
Material = Carbon Steel
L/M = 0.380 to 0.302

Fig. 4

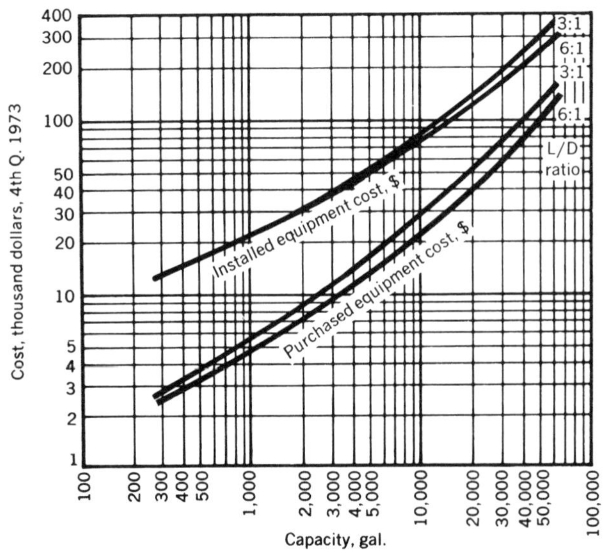

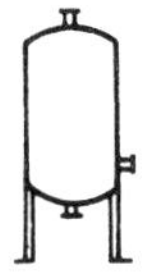

Field Labor for
Vertical Pressure Vessels
250 psig design pressure
Material = Carbon Steel
L/M = 0.380 to 0.302

Fig. 5

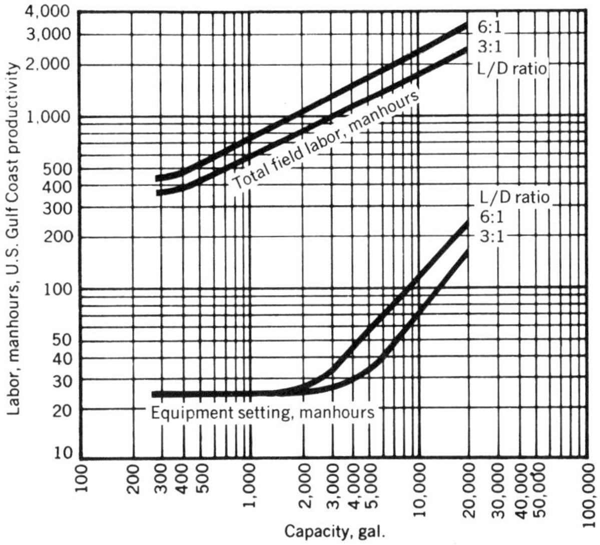

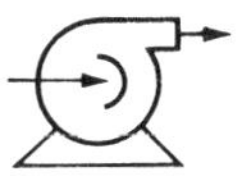

Costs of
General-Service Centrifugal Pumps
Material = Carbon Steel
L/M = 0.529 to 0.452

Fig. 6

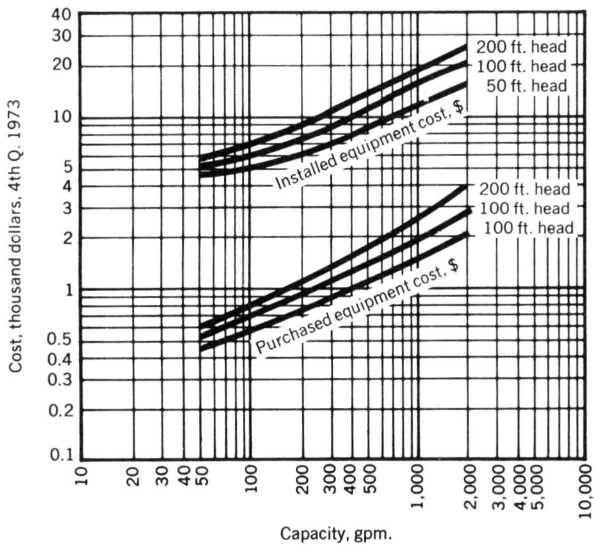

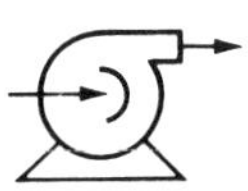

Field Labor for
General-Service Centrifugal Pumps
Material = Carbon Steel
L/M = 0.529 to 0.452

Fig. 7

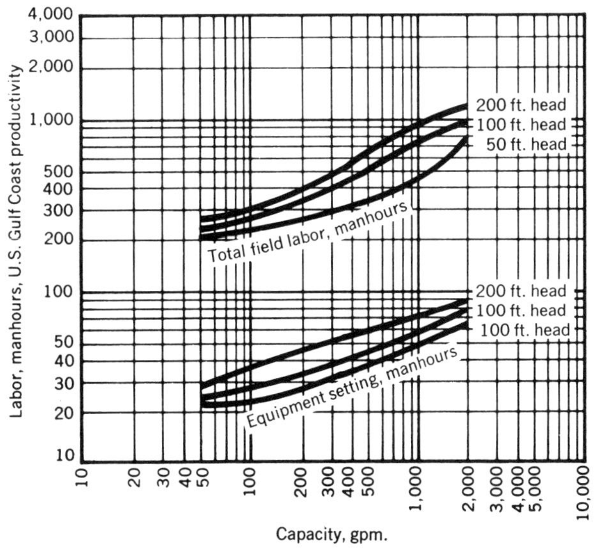

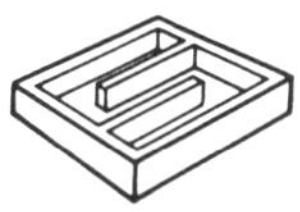

Reinforced Concrete Aeration Basins
Tank Depth = 17 ft.

Fig. 8

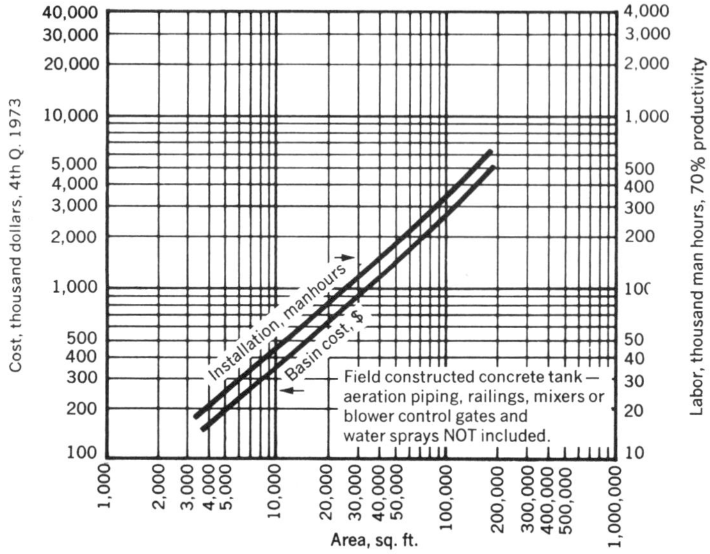

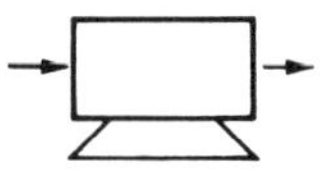

Motor-Driven Reciprocating Pumps
Material = Carbon Steel
Discharge Pressure = 1,000 ft. head
L/M = 0.299 to 0.249

Fig. 9

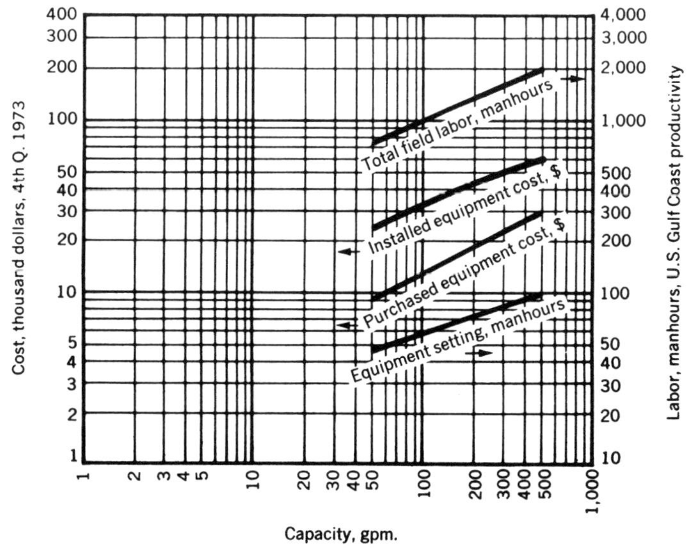

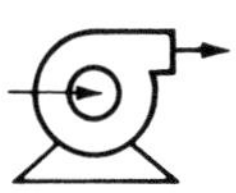

Motor-Driven Rotary Blowers
Material = Carbon Steel
Discharge Pressure = 10 psig.
L/M = 0.530 to 0.402

Fig. 10

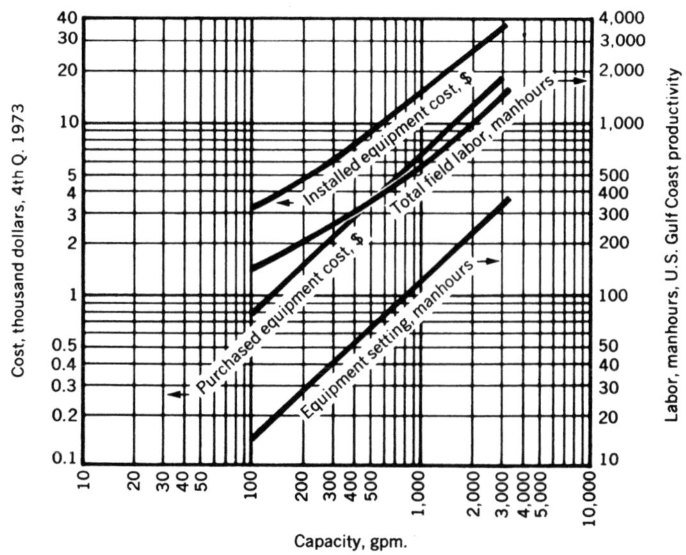

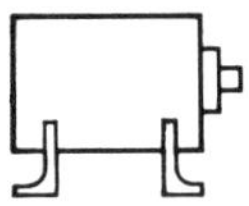

Variable Speed Motors
1200-1800 rpm.
L/M = 0.603 to 0.348

Fig. 11

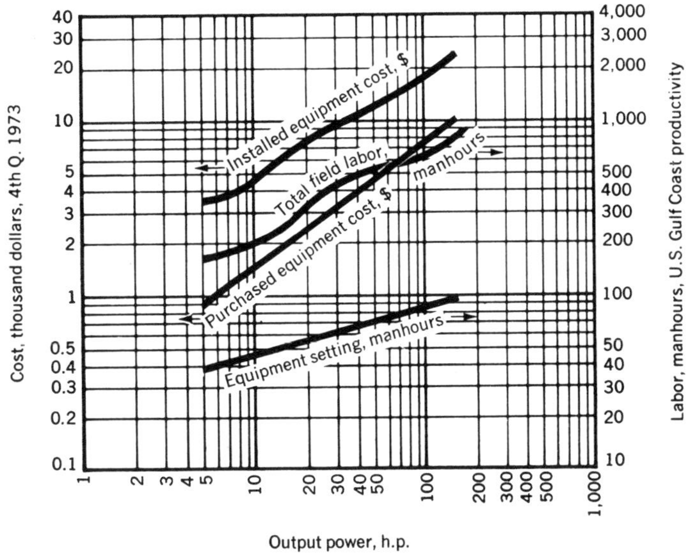

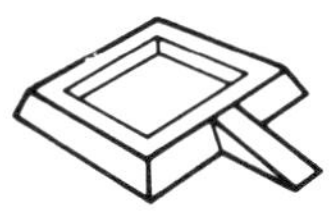

Rectangular Excavations

Fig. 12

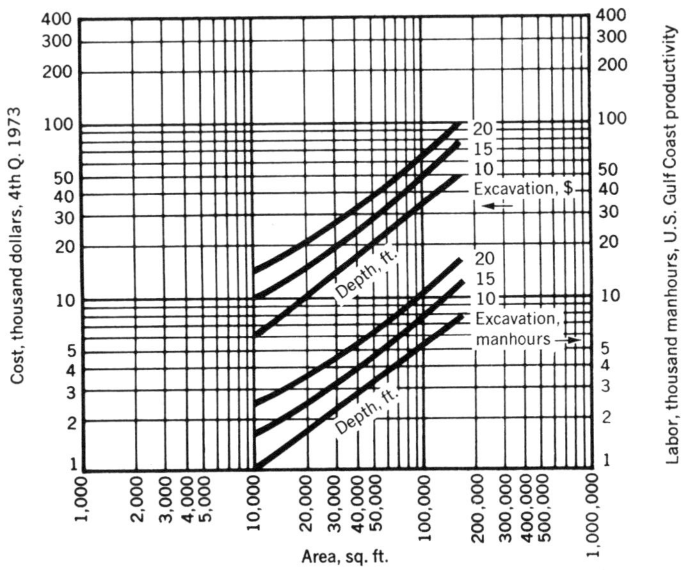

Clarifier Mechanisms – Fig. 13

Clarifier Dia., Ft.	Clarifer Depth, Ft.	Installed Cost, $*	Field Erection, Manhours**
30	13	65,000	4,700
40	14	80,000	6,100
55	14	99,000	8,300
70	14	120,000	10,700
75	15	132,000	11,900
85	15	148,000	13,700
100	15	174,000	16,800
105	16	211,000	20,000
125	16	254,000	25,000
140	16	287,000	29,000
150	17	317,000	33,000

*Includes concrete tank and clarifier mechanism.
**Manhours are for field constructing the concrete tank only—does not include installation manhours for the clarifier mechanism.

To separate the labor and material elements from Table II, use the following procedure:

For an ion exchanger, L/M range from 0.579 to 0.652 (Table II, large to small units). Assume we want to determine the installed cost in South Philadelphia of a large ion exchanger with an equipment cost of $27,000 as of 4th quarter 1973, and estimated material costs for installation of $18,000 in 4th quarter 1973. Installation is to be made during the 2nd quarter 1975. Using an L/M ratio of 0.579, we calculate labor cost:

$$(L/M)(\text{Equip. cost} + \text{Material cost}) \times \left(\frac{\text{U.S. Gulf Coast productivity}}{\text{Productivity at location}}\right)\left(\frac{\text{Location composite wage rate}}{\text{U.S. Gulf Coast composite wage rate}}\right) = \text{Labor cost}$$

$$(0.579)(27{,}000 + 18{,}000)(100/74)(12/7) = \$60{,}400$$

The total installed cost is then the sum of equipment cost, material and labor—27,000 + 18,000 + 60,400 equals $105,400.

Other Costs

In using these cost data, the user should be aware that L/M ratios are for "average" installations and "typical" processing schemes. Also, construction overhead, home-office engineering and design costs are excluded from the data, and no allowance has been made for contingencies, contractor's fees, or subcontracting part or all of the construction.

Acknowledgment

The authors gratefully acknowledge those contractors and vendors who have assisted in our efforts, especially Ted Zaferatos of Envirotech Corp.

Meet the Authors

Herbert G. Blecker is president and founder of Icarus Corp., 8630 Fenton St., Silver Spring, MD 20910. Under his direction, the company has developed several computer systems for process simulation, design and cost estimating. Mr. Blecker has had extensive consulting experience, and holds a B.S. degree in chemical engineering from the City University of New York.

Thomas M. Nichols is vice president of Icarus Corp. He is a founder of the company, and has experience in developing computer-aided design and estimating systems for contractors, operating companies and the federal government. Mr. Nichols has a B.S. degree in chemistry from the University of Maryland.

Harvey S. Epstein is manager of environmental systems at Icarus. He has a background in design and cost estimating of wastewater treatment facilities, and is the developer of the Environmental Protection Agency's BICEP (Brief Input Cost Estimating Program). Mr. Epstein holds B.S. and M.S. degrees in chemical engineering from the University of Maryland.

How to Escalate Equipment and Labor Costs—Table I

Material Cost	4th Q 1973	2nd Q 1975	2nd Q 1976
1. Equipment	$10,000	$12,600	—
2. Marshall & Swift index	350	440	—
3. L/M ratio	0.114	—	—
4. Material	$13,460	—	—
5. Material less equipment	$3,460	$4,350	—
Labor Cost			
6. Labor market	U.S. Gulf	—	So. Phila.
7. Productivity	100%	—	74%
8. Field-labor manhours	220	—	297
9. Composite wage rate	$7.00/h	—	$14.00/h
10. Labor cost	$1,540		$4,160
11. Total cost	$15,000		$21,110

Labor-To-Material Ratio For Selected Installed Equipment Modules—Table II

Installed Equipment Module	Range of L/M, Large Unit–Small Unit
Air compressors	0.330–0.872
Agitated open tanks	0.410–0.592
Agitated pressure tanks	0.299–0.571
Condensers	0.437–0.643
Conveyors	0.966–1.147
Crushers	0.988–1.113
Crystallizers	0.952–1.031
Centrifuges	0.968–1.021
Cutters	0.859–0.965
Dryers	0.825–0.937
Dust collectors	0.755–0.838
Drum dryers	0.837–0.846
Evaporators	0.336–0.695
Electrical generators	1.224–1.389
Ejectors	0.868–0.898
Extractors	0.962–0.975
Filters	0.992–1.121
Feeders	0.997–1.083
Fans	0.987–1.049
Flotation machines	1.021–1.084
Furnaces	0.616–0.674
Gas compressors	0.448–0.870
Gear pumps	0.394–0.410
Heat exchangers	0.227–0.766
Heating units	0.385–0.580
Hoists	0.706–0.729
Ion exchangers	0.579–0.652
Mills	0.990–1.151
Mixers	1.040–1.196
Packed columns	0.292–0.740
Presses	1.040–1.062
Pumps	0.318–0.720
Reactors	0.611–0.735
Rotary dryers	0.898–0.956
Stacks	0.937–0.995
Tray towers	0.435–0.581
Vacuum pumps	0.437–0.705
Vibrating screens	1.189–1.270

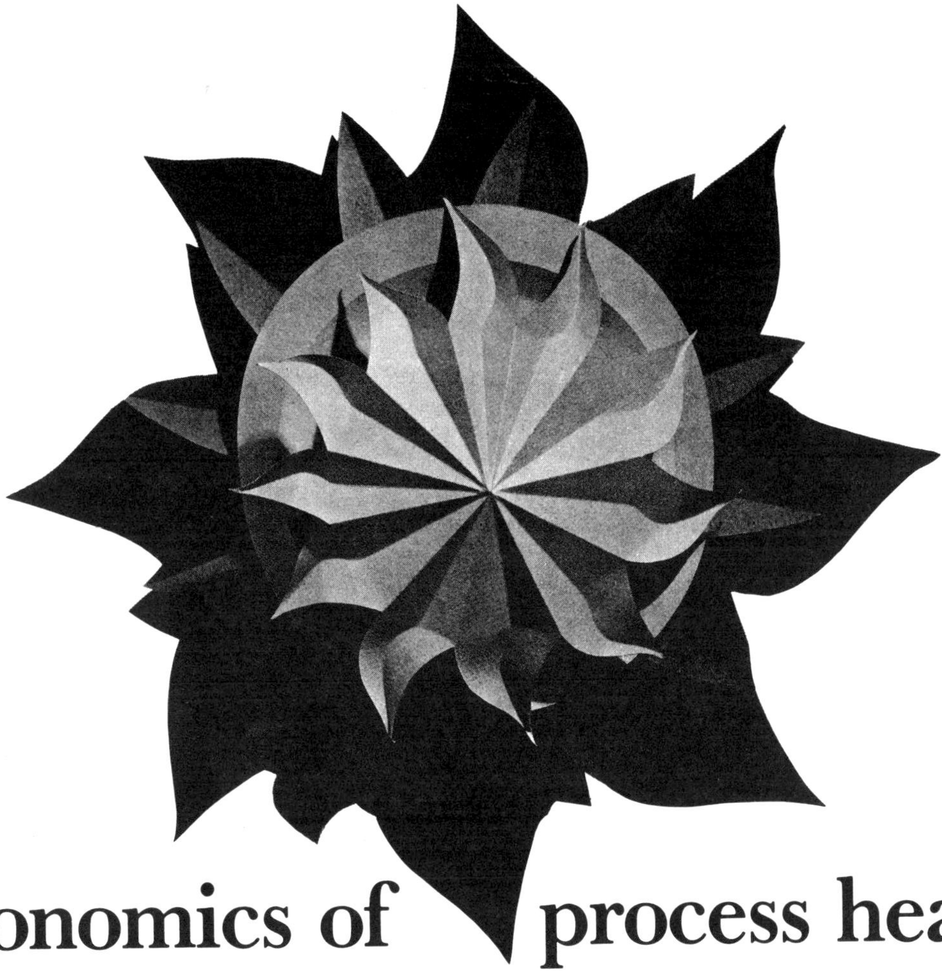

Economics of process heat from solar energy

Steep rises in fossil-fuel prices could in time make solar energy economically attractive for producing process heat. Federal tax incentives, however, could do this right now.

W. C. Dickinson, *Lawrence Livermore Laboratory*

☐ About one fourth of the total energy consumed in the U.S. goes into producing heat for industrial processes. About two thirds of this process heat (17% of total U.S. energy consumption) is in the form of hot water or steam [*1*]. This is equivalent to the consumption of 2.5 billion barrels of oil each year. If solar heating systems could provide even 5% of this energy for making hot water and steam, an equivalent of 125 million bbl of oil would be saved annually.

Because of the high cost of fossil fuels and their possible scarcity, industries that use large quantities of process heat can be expected to have an interest in solar energy. A recent report concluded that at least one of three conditions must exist for a company to switch from oil and natural gas [*2*]:

1. The economics of using the new energy will provide an acceptable return on investment.
2. The company is expanding and will install an alternate energy facility on the basis of long-term expectations.
3. Scarcity forces a switch.

A solar heating system to produce process hot water will generally supplement rather than replace a fossil-fuel heating system. Such a solar heating system will require a large area of solar thermal collectors. For example, the total annual solar radiation falling on 1 ft^2 of horizontal surface in many areas of the southwestern U.S. averages only about 0.65 million Btu.

(Compared to that for a horizontal surface, the total annual radiation incident on a surface tilted to the south at a fixed angle equal to the latitude is about 20% greater, and for a surface always held normal to the sun's rays, about 50% greater.)

Annual collector efficiency for converting radiation into hot water averages about 40% to 45%. If a facility's annual requirement for process hot water totaled 10^5

Originally published January 1977.

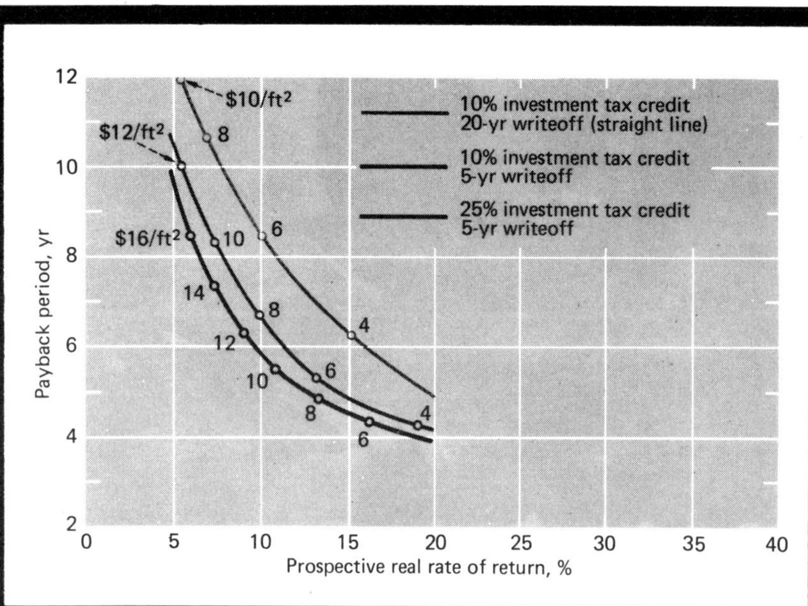

Numbers on the curves represent total allowed solar system cost **Fig. 1**

million Btu, and half of this were supplied by a solar system, a collector area of about 170,000 ft^2 (about 4 acres) would be needed.

Reliable flat-plate collectors are commerically available for about \$10 to \$15/ft^2. Adding site preparation, plumbing, storage and the control system would bring the total cost for a solar system to about \$20/ft^2. Therefore, the initial capital investment for a solar system that would supply half the process heat required annually would be about \$3.5 million.

Competitive solar system

What cost for a solar heating system would be economically attractive to a company? Rule-of-thumb says that 1 ft^2 of horizontal collector in the Southwest can provide about 0.30 million Btu/yr. This is equivalent to the energy supplied by \$1.10 worth of \$15/bbl fuel oil (converted into delivered heat at 70% efficiency). To be economical, therefore, the installed cost of the collector system obviously can be only a few dollars per square foot.

Answering the question more definitively calls for a life-cycle cost analysis that takes into account the cost of fossil fuel, the necessary rate of return on the solar system investment, extra expenses for operation and maintenance, additional income tax due because of fossil-fuel savings, and depreciation of the solar system. The standard method of making such a comparative cost analysis is the present worth, or discounted-cash-flow, method.

The following represents a typical set of economic conditions that a company considering the installation of a solar heat system might encounter:

Heat for process hot water	10^5 million Btu/yr
Heat from solar system	5 x 10^4 million Btu (50% of total)
Price of fuel oil	\$15/bbl
Fuel-oil conversion efficiency	70%
Cost of oil without solar system	\$370,000/yr
Cost of oil with solar system	\$185,000/yr
Collector area (Southwest U.S.)	170,000 ft^2
Boiler amortization and operation and maintenance	\$0.20/million Btu (based on total process heat required)
Solar system operation, maintenance and insurance	\$15,000/yr
Income tax rate	50%
Solar system lifetime	20 yr
Salvage value at end of 20 yr	\$0

For a particular real rate of return* on investment (after taxes) in a solar system under the foregoing conditions, a present-worth calculation will determine the

* Real rate of return is based on dollars of uniform purchasing power. Market rate includes an inflation factor; that is, market rate is about equal to the real rate plus the projected inflation rate.

ERDA solar process heat program

In addition to the Sohio project, ERDA's Div. of Solar Energy initiated in 1976 several new government-funded experimental demonstrations of solar energy systems to provide process heat. Four projects have recently been funded for the construction of process hot-water systems; six other projects, now in the design stage, are for industrial-scale dehydration and drying.

The hot-water projects consist of: (1) a 7,000-ft^2 system of flat-plate and parabolic-trough collectors for washing cans in a large soup company; (2) a 9,000-ft^2 parabolic-trough collector array for a commercial laundry; (3) a 6,000-ft^2 evacuated-tube collector system for textile dyeing; and (4) a 9,000-ft^2 concentrating collector system for concrete-block curing. Construction of these projects will get underway early in 1977.

The dehydration collector systems include: (1) air heating to dehydrate grapes for raisins; (2) water heating with heat exchanger for dehydrating onions with hot air; (3) air heating for drying alfalfa; (4) water heating with heat exchanger for kiln drying of lumber; (5) air heating for the drying of soybeans; and (6) steam-generating-concentrating with heat exchanger for drying textiles. The best of the six designs will be chosen by ERDA for construction in 1977.

The objectives of the projects are: first, to establish the performance of commercially available solar systems in providing process heat for industry and, second, to determine the current economic feasibility of such systems. In addition, ERDA plans to sponsor R&D programs aimed specifically at developing better and less-costly solar process heat systems for industry.

allowed total solar-system cost for which the annual cost of process heat to the company will be the same for either the system without solar heating or a 50% fuel oil / 50% solar energy arrangement.

For the economic conditions listed, Fig. 1 shows the relationship between prospective real rate of return and payback period† for three different sets of tax conditions. The numbers attached to the curves are the total allowed solar-system cost in dollars per square foot of collector area. The curves would apply with reasonable accuracy to the heat requirement of any size industrial process, when a solar system supplies an average of half the heat, and fossil fuel is priced at the equivalent of \$15/bbl of fuel oil.

Generally, a company not only desires a certain minimum rate of return but also a relatively short payback period (perhaps 5 years or less). With the 10% investment tax credit and a 20-year straight-line writeoff (top curve of Fig. 1) currently permitted, a 10% real rate of return gives a payback period of 8.5 years and a total allowed solar-system cost of \$6/ft^2. However, if the federal government were to allow a 25% investment tax credit and 5-year writeoff (bottom curve of Fig. 1), the payback period would be reduced to 5.8 years and the total allowed solar-system cost would become a more realistic \$11/ft^2. This illustrates the leverage that federal tax incentives would have in promoting the industrial use of solar energy.**

Fig. 1 would, of course, not be affected by general inflation if the solar system were purchased with equity, rather than debt, financing. General inflation merely increases the overall cost of process heat (solar plus fossil fuel) in current dollars. However, a steeper rise in fossil-fuel prices would, of course, make solar heating more attractive by decreasing the payback period and increasing the allowed cost for a solar system. Fig. 1 does not take into account this possibility.

The ERDA-Sohio process heat project

The shallow solar pond—a large-area, low-cost solar collector being developed by the Solar Energy Group at Lawrence Livermore Laboratory—performs very much like a conventional metal/glass flat-plate collector, but its unit area cost is considerably less. In fact, it appears that a shallow solar-pond system will be able to provide process hot water in the Southwest at a cost competitive with \$15/bbl fuel oil.

Under a no-funds-exchanged contract between the U.S. Energy Research and Development Administration (ERDA) and the Sohio Petroleum Co., Lawrence Livermore is designing an experimental solar water-heating facility for Sohio that is to be built at the latter's new uranium mining and milling plant, located about 50 miles west of Albuquerque, New Mexico [3]. The plant went into operation in the fall of 1976.

† Payback period is defined as the number of years required to recover the net initial solar investment (total investment minus the investment tax credit) out of annual incremental cash flow.

** Tax laws permit a 5-year writeoff if the asset is deemed important to national interests. Examples are pollution control facilities and mine safety equipment. In such cases, system lives need not be estimated when computing depreciation rates.

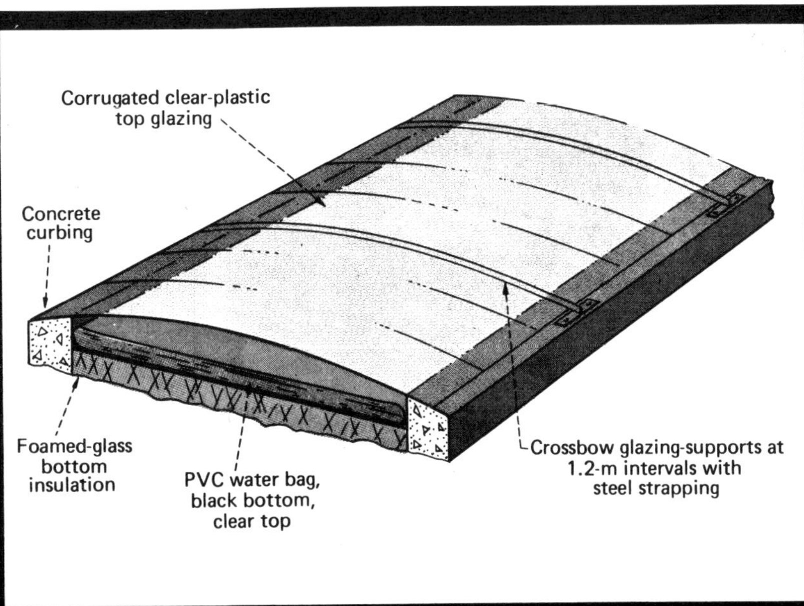

Full-size module measures 12 ft x 200 ft, and holds 25 tons of water Fig. 2

Uranium ore is being concentrated to U_3O_8 in a chemical leaching process using a dilute hot acid solution. This process requires about 300 gal/min of 140°F water, 24 h/d—a total volume of about 400,000 gal/d. The plant has oil-burning boilers to heat water but, because of the high cost of fuel oil and the possibility of occasional shortages, Sohio plans to supplement oil heating with solar heating.

A system of shallow solar ponds that are simple, reliable and low cost is being designed. Each pond module consists of a long, narrow plastic bag, 11.5 ft x 200 ft, containing 4 in. of water. The bag's transparent top allows transmission of sunlight and prevents evaporation, and a black bottom absorbs sunlight. Under the bag, a layer of insulation limits heat loss to the ground. One layer of clear, plastic, corrugated panel is arched over the water bag to suppress upward convective and radiative heat losses. At night, the hot water is stored in an underground insulated reservoir, ready for use. (Fig. 2 shows a view of the currently preferred pond design.)

Each module heats 25 tons of water per day in a single-batch process. In summer, the average water temperature at the end of the day lies between 130°F and 140°F; in winter, from 85°F to 90°F. Oil burners will bring the water up to the required 140°F.

The onsite project work is being conducted in two phases: a prototype and final system. The first phase, now completed, consisted of the design, construction, operation and testing of three full-size shallow-pond modules. Fig. 3 shows the prototype facility.

Performance measurements on the New Mexico ponds as well as similar ponds at Lawrence Livermore have been in progress for the past year. On an annual

Prototype shallow solar ponds near Grants, New Mexico Fig. 3

average basis, the heat collected amounts to about 825 Btu/(ft^2)(d), in good agreement with calculated values.

The total cost to Sohio for the prototype—including site preparation, cold- and hot-water reservoirs, piping, pumps, and controls—came to \$6.90/ft^2 of pond area. Detailed working drawings for a 36-module system have been prepared by an Albuquerque-based consulting engineering firm. The projected full-size system consists of 72 modules.

The lowest bid for building the 36-module system, from an Albuquerque-based general contractor, totals \$463,315 and breaks down as follows (in \$/ft^2 of shallow solar pond):

Site preparation	0.82
Collectors,	
Materials	2.56
Installation	0.46
Hot-water reservoir	0.56
Pumping stations	0.22
Piping	0.43
Hot-water pipe to mill	0.25
Controls and electrical	0.30
Total system cost	\$5.60/ft^2

The economic conditions listed earlier apply to the Sohio plant [3]. For a prospective real rate of return on the solar investment of 10%, the payback period is 8 ½ yr and the allowed system cost is \$6/ft^2 (Fig. 1).

The PVC water bags and top glazings, which make up about one fourth of the total system cost, are expected to have an average life of at least 5 yr. Including this in the economic analysis would reduce the allowed system cost by about 10%. Therefore, the \$5.60/ft^2 low-bid cost is consistent with a 10% rate of return.

It can be seen from Fig. 1 that if the tax laws permitted a 10% investment tax credit and a 5-yr writeoff of an industrial solar system such as this, the real rate of return would be increased to about 14% and the payback period reduced to about 5 yr. This would undoubtedly make this type of solar process-heat system more attractive for wide industrial use.

Prospects for steam from solar heat

The shallow pond collectors described will heat water efficiently and competitively. However, in many industrial applications, process steam rather than hot water is required. In fact, industrial steam (not including that for generating electricity) accounts for about 12% of total U.S. energy consumption.

Heating water first with solar energy and then converting the hot water into steam with fossil fuels makes little use of solar energy. The energy required to bring 50°F water to 200°F is 150 Btu/lb, but the energy of steam at 212°F is 1,130 Btu/lb. The energy of 100 psig, 340°F saturated steam, which is widely used in industry, is 1,190 Btu/lb. Therefore, solar-heated water would make only a 10% to 15% contribution to the total energy required for steam production.

When steam is generated with fossil fuels, the cost per Btu is virtually the same for both 350°F and 220°F steam. For solar-generated steam, however, the cost per Btu will be less for 220°F steam than 350°F steam, because the collectors will be less costly and more efficient at the lower temperature.

In the food canning industry, for example, the practice is to distribute 100-psig, 340°F saturated steam throughout the plant, even though considerably lower temperatures are often needed at points of use. If solar energy is to become competitive for making industrial steam, designs and temperature and pressure requirements will most likely have to be changed.

Also, a low-cost, low-concentration solar collector needs to be developed specifically for this application. Heliostat-tower boiler systems, such as are now under development for converting solar energy into electricity, will produce steam at 1,000°F, a considerably higher temperature than is required for most industrial processes. However, cost will probably put these systems out of range for supplying industrial heat.

References

1. Study of Effectiveness of Industrial Fuel Utilization, Thermo Electron Corp., Waltham, Mass., Report TE 5357-71-74, 1974.
2. Decker, G. L., Barnes, R. W., Sampson, R. E. and Prentice, V. L., "Evaluation of New Energy Sources for Process Heat," Dow Chemical Co., Midland, Mich., and Environmental Research Institute of Michigan, Sept. 1975.
3. Dickinson, W. C., Clark, A. F. and Iantuono, A., "Shallow Solar Ponds for Industrial Process Heat: The ERDA-Sohio project," UCRL—78288, June 17, 1976.

The author

William C. Dickinson is Solar Project Leader for Lawrence Livermore Laboratory (University of California, P.O. Box 808, Livermore, CA 94550), whose solar energy program is directed toward developing cost-competitive solar heating systems for process and space heat. Previously, he was engaged in nuclear research at the Laboratory, was Head of the Dept. of Physics at the University of Indonesia, and worked in nuclear research at Los Alamos. He holds a Ph.D. in physics from Massachusetts Institute of Technology and an A.B. in physics from the University of California.

Putting inflation into profitability studies

A sound economic evaluation of a proposed project cannot ignore the effects that inflation will have on cash flows from the project. This article shows the impact of across-the-board inflation on two measures of profitability.

F. A. ***Holland*** *and F. A.* ***Watson,*** *University of Salford (U.K.)**

☐ Chemical engineers regularly must evaluate the profitability of proposed investments whose future earnings are virtually certain to be eroded by inflation. Unfortunately, it is common practice to ignore inflation. Perhaps this is done on the not-unreasonable grounds that predicting the market rate of interest (and thus the appropriate discount rate for future cash flows) is hard enough, without having to worry about inflation as well. But failure to at least try to predict inflation rates and take them into account can greatly distort a project's economics—especially at the double-digit rates that now can be found widely throughout the world.

This article shows the effects of inflation on profitability, and how these effects can be taken into account.

Inflation—what it means

Although the word "inflation" comes before the public every day, most people would find it difficult to define. The average person is only too aware that a given amount of money buys less and less of goods and services as time goes by, but would probably be unable to express this knowledge quantitatively.

A nation's figures for inflation are based on some particular mixture of goods and services that is chosen to represent the material wants of the average citizen. If a given quantity of this specific mixture cost $100 last year and now costs $120, then the mix has suffered a 20% rate of inflation. The purchasing power of the currency (i.e., of the $120) in respect to these goods and services has consequently fallen by a factor of ($120 − 100)/$120, or 16.7%.

*For biographies of the authors, see p. 7.

Two kinds of inflation can be considered: general or open inflation, and repressed or differential inflation.

In the first case, all costs and prices increase at a uniform rate. Thus, the same rate of inflation will be calculated regardless of the particular mixture of goods and services chosen.

In the second case, the rate of inflation will depend on the spending spectrum of the individual or company. For instance, a given company's labor costs and material costs may inflate at different rates. To quite a large extent, inflation becomes repressed or differential rather than open or general due to government policy in such fields as taxation, import control, and price restriction.

This article deals only with the effects of general or open inflation on project profitability. The article in an upcoming issue will discuss differential inflation, as well as an approach to analysis of risk.

The effect of inflation on the real value of future earnings from a project should not be confused with the "effect" of the market rate(s) of interest on those earnings. Because of each effect, a dollar of project income next year has a smaller true value than does a dollar in hand today. But the interest-rate effect comes about because the dollar in hand could be financially invested today at the prevailing interest rate and the resulting dollar-plus-interest-earnings recouped next year. By contrast, the inflation effect comes about simply because $1 can buy more today than next year, due to the rise in prices. This distinction emerges more clearly in the following sections.†

†Strictly speaking, the market interest rate and the inflation rate are not fully independent, at least according to some economic theorists. However, we treat them as separate in this article.

Originally published February 1977.

If there is no inflation . . . **Table I**

Year, n	Net capital expenditure, A_{TC}	Revenue from sales, A_S	Total expenses, A_{TE}	Cash income, A_{CI} $(=A_S-A_{TE})$	Depreciation charge, A_D	Taxable income, $(A_{CI}-A_D)$	Amount of tax at $t = 0.5$, A_{IT} $(=[A_{CI}-A_D]t)$	Net cash flow after tax, A_{CF} $(=A_{CI}-A_{IT}-A_{TC})$	Discount factor at i=10%, f_d $\left(=\frac{1}{[1+0.1]^n}\right)$	Discounted net cash flow, A_{DCF} $(=A_{CF}\,[f_d])$	Net present value, NPV $\left(=\sum_0^n A_{DCF}\right)$
0	1,100,000	0	0	0	0	0	0	−1,100,000	1.00000	−1,100,000	−1,100,000
1	0	500,000	100,000	400,000	200,000	200,000	100,000	300,000	0.90909	272,727	− 827,273
2	0	500,000	100,000	400,000	200,000	200,000	100,000	300,000	0.82645	247,935	− 579,338
3	0	500,000	100,000	400,000	200,000	200,000	100,000	300,000	0.75131	225,393	− 353,945
4	0	500,000	100,000	400,000	200,000	200,000	100,000	300,000	0.68301	204,903	− 149,042
5	−100,000	500,000	100,000	400,000	200,000	200,000	100,000	400,000	0.62092	248,368	+ 99,326

Two common modern yardsticks for assessing the profitability of a proposed project are its (1) net present value (NPV), and (2) discounted-cash-flow rate of return ($DCFRR$). For a definition and discussion of each, see the article on p. 22. The rest of this present article deals with the effect that inflation has on each.

Effect of inflation on *NPV*

Consider a plant designed to manufacture 1 million widgets per year, which generates a positive net annual cash flow, A_{CF}, of \$1 million in 1977. Assume that the plant will generate a cash flow of \$2 million in 1987 for the same production rate. When computing the NPV for this proposed project, what happens if the actual cash flows are simply added together, instead of adjusting all the values to their purchasing power in a particular year? To answer this question, let us look at the basis for NPV calculations.

The NPV of a project is the sum of the net annual discounted cash flows over n years:

$$NPV = \sum_0^n A_{DCF} \tag{1}$$

In the case of discrete annual compound interest,

$$\sum A_{DCF} = A_{CF_0} + \frac{A_{CF_1}}{(1+i)} + \frac{A_{CF_2}}{(1+i)^2} + \cdots + \frac{A_{CF_n}}{(1+i)^n} \tag{2}$$

where $A_{CF_0}, A_{CF_1}, A_{CF_2}, \ldots A_{CF_n}$ are the net annual cash flows in years 0, 1, 2, . . . n, respectively. The fractional market-interest rate, i, is normally taken as the cost of capital funds, and all interest payments occur annually at the end of the year. (Eq. 1 and 2 could alternatively be based on continuous compounding; however, we shall use discrete annual compounding, because it more closely conforms to accounting practice.) Eq. 1 and 2 can be combined to give:

$$NPV = A_{CF_0} + \sum_1^n \frac{A_{CF_n}}{(1+i)^n} \tag{3}$$

Eq. 3 is valid for the case of no inflation. In the case of general inflation at a fractional rate i_i, this equation can be written in the modified form:

$$NPV = A_{CF_0} + \sum_1^n \frac{A_{CF_n}}{(1+i)^n(1+i_i)^n} \tag{4}$$

Eq. 4 enables all the net annual cash flows to be corrected to their purchasing power in year 0. If the inflation rate is zero, Eq. 4 becomes identical with Eq. 3.

The following example underscores this effect of inflation on NPV, as well as on the taxes the company pays.

NPV example

Consider a simplified project in which \$1.1 million of capital is spent in year 0, consisting of \$1 million of fixed capital and \$100,000 in working capital. The fixed

If inflation is present but is not allowed for . . . **Table II**

Year, n	Net capital expenditure, A_{TC}	Revenue from sales, A_S	Total expenses, A_{TE}	Cash income, A_{CI}	Depreciation charge, A_D	Taxable income, $(A_{CI}-A_D)$	Amount of tax at t =0.5, A_{IT}	Net cash flow, A_{CF}	Discount factor at i=10%, f_d	Discounted net cash flow, A_{DCF}	Net present value, NPV
0	1,100,000	0	0	0	0	0	0	−1,100,000	1.00000	−1,100,000	−1,100,000
1	0	500,000	100,000	400,000	200,000	200,000	100,000	300,000	0.90909	272,727	− 827,273
2	0	600,000	120,000	480,000	200,000	280,000	140,000	340,000	0.82645	280,993	− 546,280
3	0	720,000	144,000	576,000	200,000	376,000	188,000	388,000	0.75131	291,508	− 254,772
4	0	864,000	172,800	691,200	200,000	491,200	245,600	445,600	0.68301	304,349	+ 49,577
5	−100,000	1,036,800	207,360	829,440	200,000	629,440	314,720	614,720	0.62092	381,692	+ 431,269

Nomenclature

A_{CF}	Net annual cash flow after tax, \$/yr
A_{CI}	Annual cash income, \$/yr
A_D	Annual depreciation charge, \$/yr
A_{DCF}	Net annual discounted cash flow, \$/yr
A_{IT}	Annual amount of tax, \$/yr
A_S	Revenue from annual sales, \$/yr
A_{TC}	Net annual capital expenditure, \$/yr
A_{TE}	Total annual expense, \$/yr
$DCFRR$	Discounted-cash-flow rate of return, yr^{-1}
i	Interest charged on capital, yr^{-1}
i_e	Effective interest rate, yr^{-1}
i_i	Inflation rate, yr^{-1}
f_d	Discount factor $(1+i)^{-n}$, dimensionless
n	Number of years from start of project, dimensionless
NPV	Net present value of project, \$
PBP	Payback period, yr
t	Fractional tax rate, dimensionless

capital is depreciated on a straight-line basis to a book value of zero at the end of year 5. The annual sales revenue in years 1 through 5 is \$500,000. There is no inflation. The \$100,000 of working capital is recovered at the end of year 5. The taxation rate is 50%, and the market interest rate 10%. Table I spells out the cash-flow data for this project, showing that the *NPV* at the end of year 5 is \$99,326, per Eq. 3.

Now, modify this example by assuming that there is a general inflation rate of 20% per year, but further assuming that the project analyst ignores the inflation and (inappropriately) applies Eq. 3. The revenue and expense data for this case are shown in Table II on p. 88, yielding an *NPV* of \$431,269. When Eq. 3 is (inappropriately) used on the same example but with various other rates of inflation, the resulting *NPV*'s can be plotted as the upper line in Fig. 1.

If the inflation is instead correctly taken into account, by applying Eq. 4, the results are strikingly different. The table below shows the appropriate inflation discount factors, $1/(1 + i_i)^n$ for the example cited above, and reveals that the project actually incurs a negative

If the inflation is allowed for . . . Table II

Year, n	Inflation discount factor at i_i = 20%, $\frac{1}{(1+i_i)^n}$	A_{DCF} corrected for inflation, when i = 10% and i_i = 20%	*NPV* corrected for inflation
0	1.00000	−1,100,000	−1,100,000
1	0.83333	227,272	− 872,728
2	0.69444	195,133	− 677,595
3	0.57870	168,696	− 508,899
4	0.48225	146,772	− 362,127
5	0.40188	153,394	− 208,733

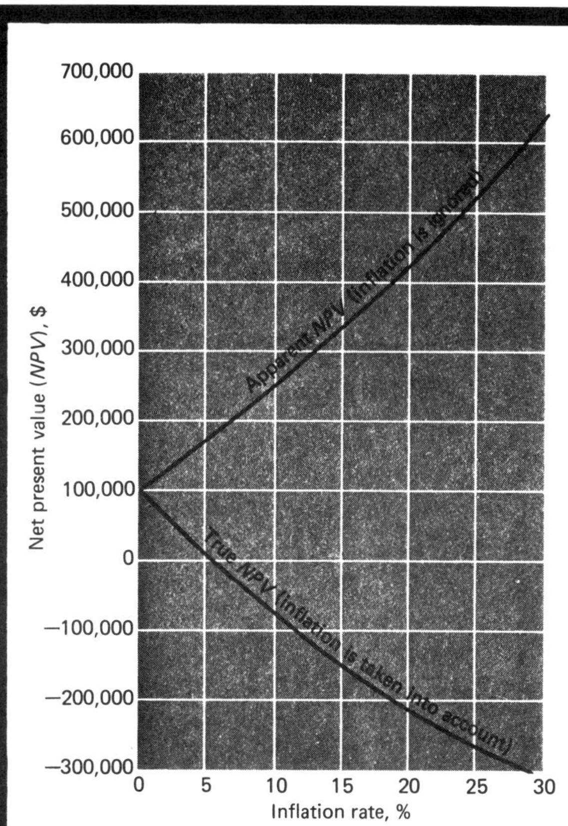

Inflation makes a project look deceptively attractive Fig. 1

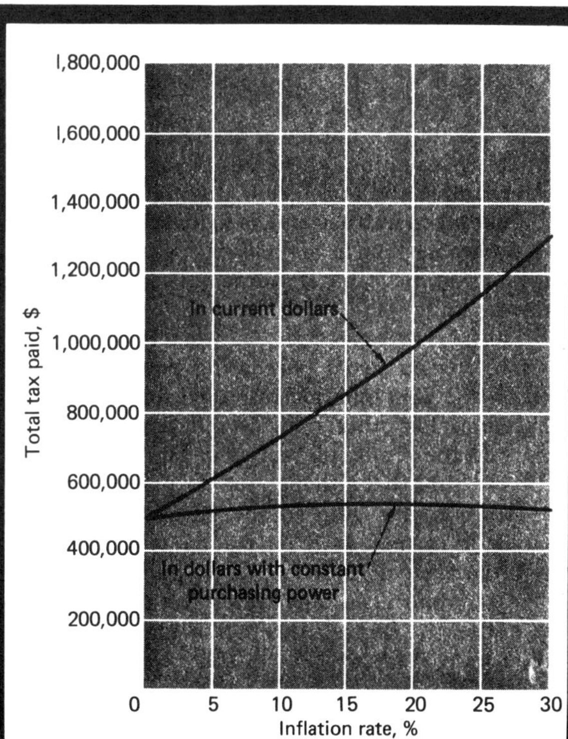

Real value of taxes paid stays almost constant Fig. 2

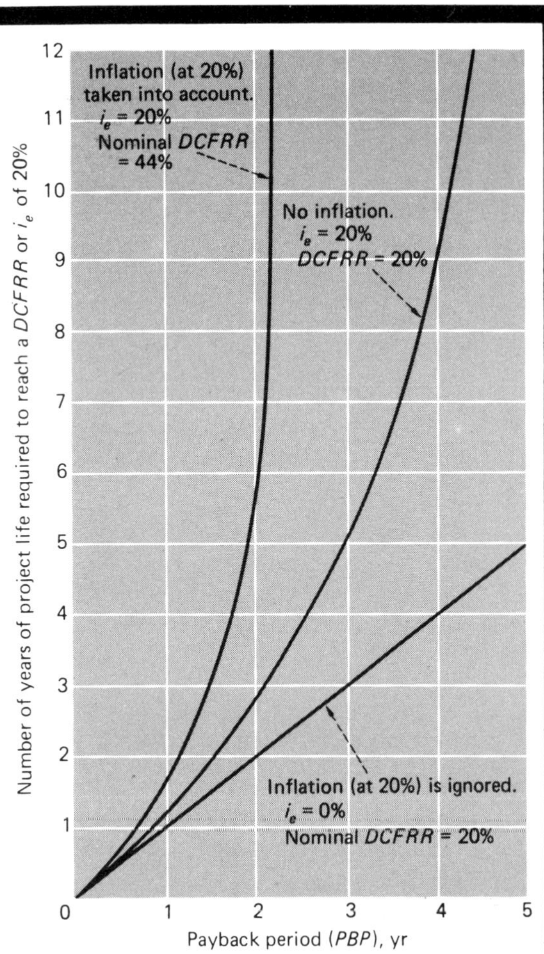

Inflation affects the relationship between payback period and discounted-cash-flow rate of return Fig. 3

NPV, of −$208,733. The lower line in Fig. 1 extends the example by assuming other rates of inflation.

The example and Fig. 1 show that the effect of inflation, if not taken into account, is to make a project seem more profitable than it actually is.

As for taxation, Table II shows that the total amount of tax actually paid over the five-year period was $988,320. This becomes $534,272 when the tax for each year is corrected to its purchasing power in year 0, based on the 20% inflation rate used for the example. Calculations for other rates of inflation can also be made, and the results plotted as in Fig. 2.

This confirms that although the tax paid will increase with inflation, the gain to the government is more apparent than real. It is interesting to note that although the tax paid corrected to its purchasing power in year 0 is almost constant irrespective of the inflation rate, it does in fact go through a maximum at an inflation rate of about 17% (in this example).

Effect of inflation on *DCFRR*

A net annual cash flow A_{CF} will have a cash value of $A_{CF}(1 + i)$ one year later if invested at an interest rate i (expressed as a fraction). If the interest rate is chosen to make *NPV* equal to zero in Eq. 3, the rate is known as the discounted-cash-flow rate of return, *DCFRR*.

If there is inflation at an annual rate i_i, then an *effective* rate of return or interest rate, i_e, can be defined by the equation:

$$A_{CF}(1 + i_e) = \frac{A_{CF}(1 + i)}{(1 + i_i)} \quad (5)$$

Eq. 5 can be simplified and rewritten to give:

$$i_e = \frac{(1 + i)}{(1 + i_i)} - 1 \quad (6)$$

and

$$i_e = i - i_i - i_e i_i \quad (7)$$

In the context of discounted-cash-flow rate of return, Eq. 7 becomes:

$$i_e = DCFRR - i_i - i_e i_i \quad (8)$$

In this equation, *DCFRR* can be viewed as the *nominal* discounted-cash-flow rate of return (i.e., the value uncorrected for inflation) and i_e can be thought of as the *true* or *real* discounted-cash-flow rate of return.

Instead of using Eq. 8, it is unfortunately common practice to try to obtain the true or effective rate of return by calculating the nominal *DCFRR* (based on actual net annual cash flows uncorrected for inflation) and then subtracting the inflation rate from it:

$$i_e = DCFRR - i_i \quad (9)$$

Eq. 8 shows that Eq. 9 is only approximately true, and should be used, if at all, solely for low interest rates and inflation rates. Consider the case of a nominal *DCFRR* of 5% and an inflation rate of 3%. Eq. 9 yields an approximate effective return rate of 2%, compared with the real effective rate of 1.94% given by Eq. 6; i.e., an error of 3.1%. Now consider the case of a nominal *DCFRR* of 25% and an inflation rate of 23%. Eq. 9 yields an approximate effective return rate of 2%, compared with 1.63% from Eq. 6; in this case, the error that results is 22.7%.

Inflation, *DCFRR*, and payback period

More insight into the effect of inflation on *DCFRR* calculations can be gained by considering the payback period (*PBP*), which is defined as the elapsed time necessary for the project's cash flows to recoup the original fixed capital expenditure. In this definition, the cash flows are not discounted to allow for the market rate of interest (or for the inflation rate), so a project with a given *PBP* could show various values for its *DCFRR*, and a given *DCFRR* could pertain to projects with various *PBP*s.

Consider the simple case of (1) a single capital expenditure made immediately before the start of production, and (2) equal positive net annual cash flows A_{CF} in all of the productive years of the project. For this case, Eq. 3 can be rewritten [1,2] in terms of the payback period and the *DCFRR* as follows:

$$PBP = \sum_{1}^{n} \frac{1}{(1 + DCFRR)^n} \quad (10)$$

The relationship set out in Eq. 10 can also be viewed

via a different chain of causality: with *DCFRR* as a given parameter, *PBP* as the independent variable, and n as the variable whose value is being sought. That approach is the basis for the lines in Fig. 3, each of which shows the number of years of project life required to achieve a *DCFRR* or an effective interest rate of 20%. by projects having various payback periods. The three lines differ from each other with respect to the matter of inflation.

If there is no inflation, then the middle line pertains. Because there is no inflation, the nominal *DCFRR* is equal to or identical with i_e, the real discounted-cash-flow rate of return, as can be seen from the relationship expressed in Eq. 8.

When inflation does exist, the relevant parameter is i_e, and it is different from the nominal *DCFRR*. Eq. 8 manipulated into the equivalent form

$$DCFRR = (1 + i_e)(1 + i_i) - 1 \tag{11}$$

shows that in order to achieve an i_e of 20% when the general inflation rate is likewise 20%, a project must generate a nominal *DCFRR* of 44%. That is the basis for the uppermost line in Fig. 3. Other lines that pertain to other rates of inflation could also be plotted in the same way.

Now assume again that 20% inflation prevails, but that the analyst ignores it and mistakenly takes a project's *DCFRR* at its nominal value instead of converting it to an i_e. Eq. 11 rearranged into the form

$$i_e = \frac{(1 + DCFRR)}{(1 + i_i)} - 1 \tag{12}$$

shows that with a nominal *DCFRR* of 20% and a general inflation rate of 20%, the true or effective rate of interest is zero. That is the basis for the lowest line in Fig. 3. Points for lines corresponding to other rates of inflation could be plotted onto that figure. And plots similar to Fig. 3 could be drawn for other *DCFRR* values.

Fig. 3 shows that the elapsed time necessary to reach a *nominal DCFRR* for a given project decreases sharply with inflation. This figure, like Fig. 1, thus shows that the effect of inflation is to make a project seem more profitable than it actually is.

That effect comes through even more clearly in Fig. 4, a plot of the time needed to reach a nominal *DCFRR* of 20% against the inflation rate, for various values of *PBP*. This plot also shows that the longer the payback period, the greater the increase in apparent profitability of the project.

The true rates of return, i_e, can be calculated from Eq. 12 to be 20, 9.09, 0 and −7.69%, respectively, for general inflation rates of 0, 10, 20 and 30%. Thus, although the time required for a project with a payback period of 4 years to reach a nominal *DCFRR* of 20% is reduced from almost 9 years under conditions of no inflation to less than 3½ years for 30% inflation, the true rate of return that prevails for the latter condition is −7.69%.

It is interesting to note that in order to reach a real *DCFRR*, or i_e, of 20% within a reasonable project lifetime when the general inflation rate is 20%, it follows

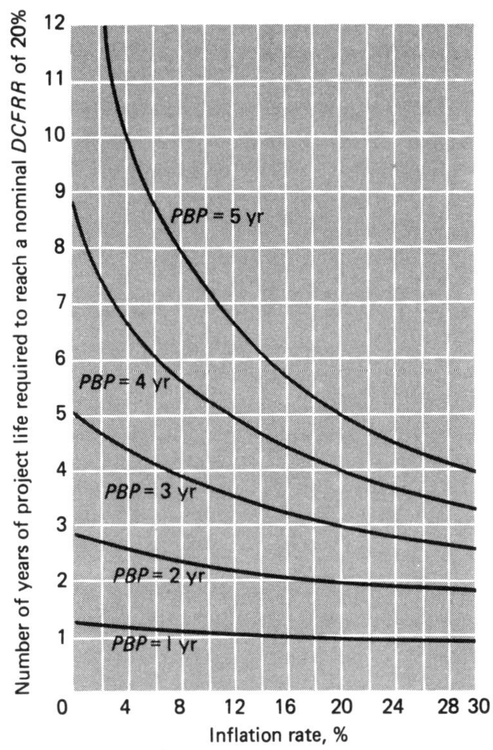

The misleading effect of inflation is more pronounced for projects having higher payback periods. Fig. 4

from Fig. 3 that the payback period for the project must not be much in excess of 2 years.

Summary and recommendations

Although it is difficult to carry out economic-feasibility studies on projects in a time of high inflation, it is important to try to predict inflation rates and allow for them in the studies.

When different people talk about inflation, they often adopt different concepts without realizing it. The area of conceptual uncertainty can be said to lie somewhere between the upper and lower lines shown on Fig. 3, in most cases.

If the analyst can approximate all the cash-flow changes at steady production rate in terms of a general rate of inflation, then the *NPV* and the rate of return for the project should be calculated in real terms, using Eq. 4 and 12. On the other hand, Eq. 9 is too great an approximation to be used in the case of high rates of inflation.

In order to reach even modest rates of return in real terms within a reasonable number of years during periods of high inflation, it is essential to choose projects having low payback periods.

References

1. Bates, A. G., Managing Capital Expenditures, Part. 3, *Hydrocarbon Process.*, Mar. 1966, p. 181.
2. Estrup, C., Evaluation of the Pay Back Methods by Means of the DCF Technique, *Brit. Chem. Eng.*, Feb./Mar. 1971, p. 171.

Opportunities to stabilize costs require closer control of expenditures for plant projects. Here are useful guidelines to prevent budget overruns, to minimize escalation, to limit prices of purchased equipment, and to forecast the plant cost index.

How To Assess Inflation of

Cost Control for Process Plants From

FORREST D. CLARK, Exxon Research and Engineering Co.

Since the cost of engineering and constructing chemical process plants has undergone unprecedented escalation in the past year, lump sum bids from contractors for sizable projects are almost impossible to obtain; and if obtained, are so studded with escalation and escape clauses that the owner cannot determine what the total cost will be.

In these circumstances, owners are turning to fully reimbursable type contracts with specified fee arrangements and, in some cases, bonus clauses to provide contractor incentive.

The owner wants to be sure that he will get the most for his money out of a reimbursable engineering or construction contract. He also feels that a declared deep dedication to cost control on the part of the contractor is not quite enough. Even the contractor's written cost-control procedures are no guarantee of effective cost control. To assure himself that the contractor is exercising effective cost control, an owner must verify that the following conditions exist:

■ The contractor's cost-control procedures must be detailed and complete, but more important, they must be followed in the everyday execution of the project.

■ The contractor's first line of supervision must be cost conscious. Engineering group and squad leaders must make their designs with cost as well as quality in mind.

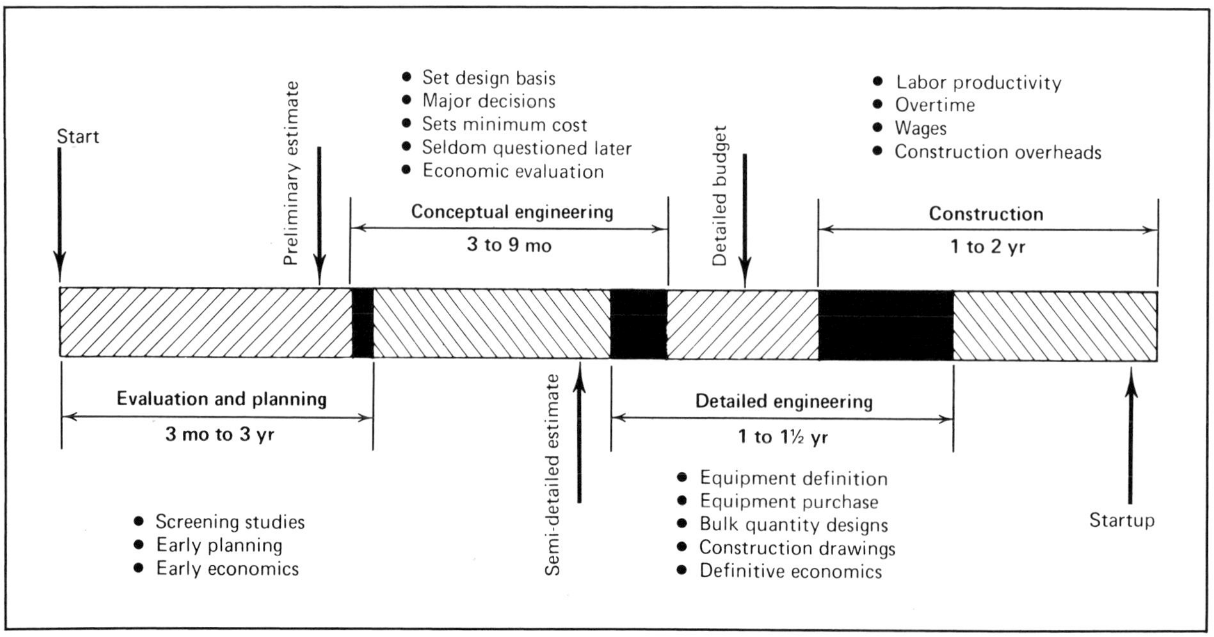

DEVELOPMENT of any project for a chemical process plant occurs in fairly distinct and overlapping periods—Fig. 1

Originally published July 1975.

Plant Costs

The Owner's View

▪ The mechanism for triggering corrective cost action must exist. The procedures should clearly spell out when and how this mechanism functions.

▪ There must be a followup procedure to verify that corrective action was taken, as well as a measure of its effectiveness.

All of these actions take time and money. There is a tendency to spend too much money on gathering cost data, and too little on analysis and control. The owner must assure himself that the contractor is gathering only the data necessary for cost control, and that these data are analyzed and used. The contractor and the owner, working together, should strike a balance between what is needed for good cost control and overemphasis on cost reports.

The Meaning of Cost Control

The purpose of cost control is to focus the attention of project management on problems areas. A good cost-control system has a device that focuses the project manager's attention to areas where cost problems may develop, and this must occur before money or labor have been committed. Potential cost problems are called to the manager's attention in time for him to take corrective action aimed at eliminating or minimizing a budget overrun. Good cost control means: spotting cost trouble areas, taking corrective action, and measuring the effectiveness of the action taken. If these things are being done, there is a good cost control; and if they are not done, there is no cost control regardless of the thickness of the monthly cost reports.

Historically, process plant projects can be divided into fairly distinct phases as shown in Fig. 1. The time interval of these phases varies significantly from project to project, and the phases themselves frequently overlap. For example, construction is normally started long before completion of detailed engineering. Nevertheless, these distinct phases nearly always exist, and generally different people are involved in each phase. It is convenient to think of cost control as being divided into similar phases, i.e.:

- Cost control during conceptual engineering.
- Cost control during detailed engineering.
- Cost control during construction.

Now, let us examine each phase in detail and see how cost control works and how the owner's cost engineer functions as a part of the project-management staff.

Cost Control During Conceptual Engineering

Cost control begins in the owner's office during the planning and conceptual-engineering phases. It is important to realize that the decisions made during the conceptual stage set the theoretical minimum cost of the project. For example, once the number of days of product storage is set, tankage capacity and a corresponding minimum cost become fixed. Similarly, once it is decided that two ship-loading berths are required and that a tanker must be loaded in ten hours, the minimum cost of the marine facilities is established.

The options open during the conceptual phase are greater than at any other time during the project. Hence, this phase offers the greatest opportunity for cost savings. Once the design basis has been set, the minimum investment is established.

Generally, the design basis, set during the conceptual phase, is not questioned throughout the remainder of the project. The pressures of a tight budget and schedule cause those engaged on the project to forge ahead without returning to an evaluation of the basis. During detailed engineering, it might occur to contractor's designer that a 12-h loading period for a tanker would save considerable money in pumps and piping. However, the contractor's designer will probably not pursue this idea if the owner has set the limit at 10 h.

Owners should be very careful in setting arbitrary limits as fixed requirements. Where flexibility is possible, the engineer or engineering contractor should be so advised. As detailed engineering progresses, the owner's staff should be reevaluating the so-called 'fixed requirements' in terms of cost.

Checklist for Rating Contractor's Cost Control

When selecting a contractor for a reimbursable contract, rate him on:

- ☐ Status of contractor's cost-control organization.
- ☐ Quality and completeness of cost-control procedures.
- ☐ Existence of the mechanism for triggering corrective action.
- ☐ Existence of followup procedures to verify effectiveness of corrective action.
- ☐ Adequacy of accounting and cost report function.
- ☐ Attitude of contractor's management toward cost control.
- ☐ Evaluation of personnel proposed for your project.

Since the basic minimum project cost is established after approval of the fixed requirements (design basis), cost control is essential while these requirements are being set. Cost control begins with a properly documented preliminary estimate.

Preliminary economics (the reason for the project) are generally based on the preliminary cost estimate made at the end of the evaluation and planning phases. In order to develop the estimate, a design basis must be established. Too often this basis is very loose and the estimator is forced to make important assumptions that the designer/planner may change. In my experience, neither the process planner nor the estimator working individually can make an adequate preliminary estimate. The two persons (or groups) should work together as a team, and be jointly responsible for the estimate and the documentation. This documentation is extremely important, because it forms the basis of cost control through the conceptual and early detailed-design phases.

If economics based on the preliminary estimate look favorable, the project moves into the conceptual phase. This work may take place in the owner's office or in that of an engineering contractor. In either case, the owner's management and top technical staff will be deeply involved. As conceptual engineering progresses, many changes are made to the original planning. As soon as this happens, the owner enters into a cost-blackout period. He no longer has a handle on the cost of the project. Often this cost-blackout period will last for months until a detailed estimate can be prepared. And, the cost may increase significantly without the owner having an opportunity to take cost reduction steps or review his economics. The objectives of cost control during the conceptual phase are to:

- Eliminate cost-blackout periods.
- Provide a continuous evaluation of total project cost.
- Take corrective action to minimize cost increases.
- Provide an economic yardstick for measuring changes to the project basis.

These objectives can be met by a program called cost trending or monitoring. The program consists of identifying and pricing changes to the preliminary estimate basis as they occur. Items that deviate from the basis are questioned, and cost increases highlighted, so that all concerned are aware of the cause and justification for each increase. This program is not costly–it works and is now being carried out by some owners and engineering contractors. This program (known under various names, such as cost trending, deviation reporting, and cost monitoring) operates as follows:

- The planner/designer, who is completely familiar with and responsible for the original estimate, identifies all changes and advises the cost engineer.
- The project manager similarly advises the cost engineer of changes in project timing or other execution phases.
- Cost engineering keeps up to date on changes in material escalation or wage rates.
- All of these changes are fed to the cost engineer, who develops the cost and schedule impact of each change and advises the cost-control committee or project manager.

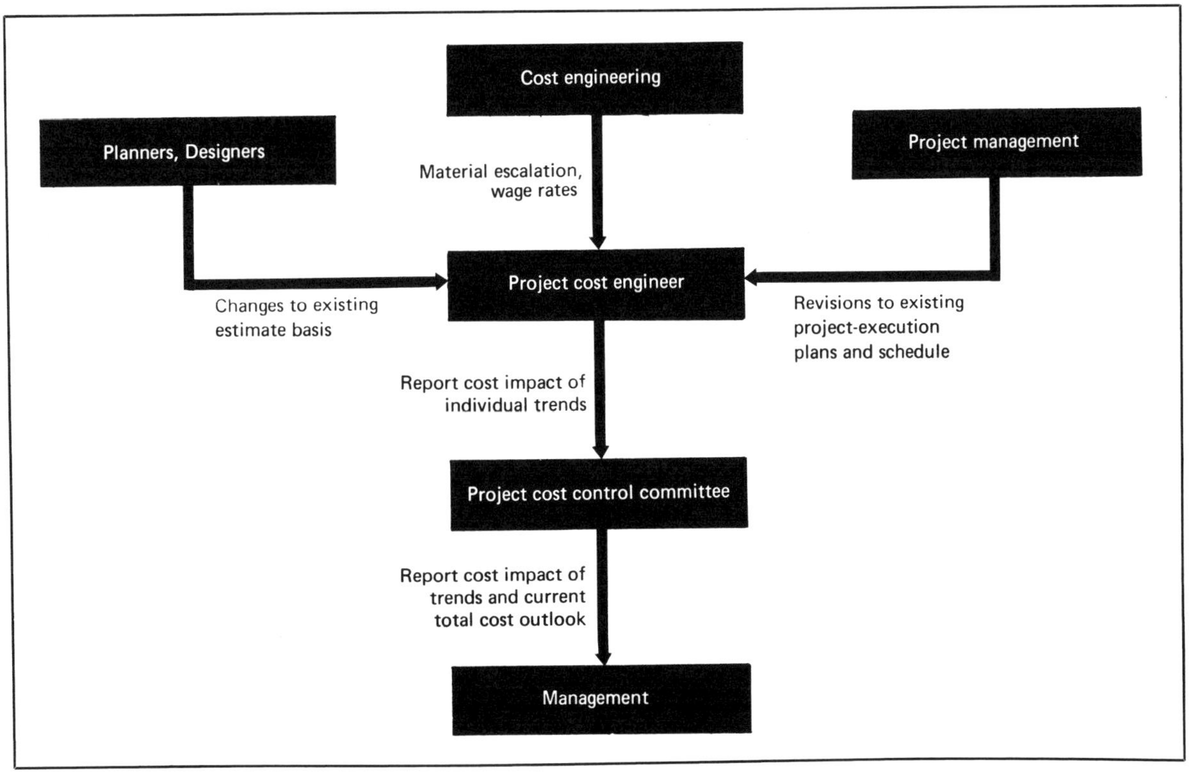

COST TRENDING requires a well-developed program and flow of information to effectively monitor plant costs—Fig. 2

A cost-control committee consisting of the lead process planner, the project manager, and the cost engineer is in position to take and follow up on corrective action. Also, the committee can develop and apply an economic yardstick for measuring the desirability of proposed changes. The committee should issue periodic (probably monthly) cost reports to management, which describe, evaluate, and justify each change to the project basis. Fig. 2 shows the flow of information in a typical cost-trending program.

In addition to meeting the stated objectives, a good cost-trending program will provide two intangible, but highly important, advantages: (1) instill a cost conscious attitude in all project personnel, and (2) keep management informed as to total project cost and permit cancellation at the earliest moment if the economics become unfavorable.

Cost Control During Detailed Engineering

Many owners turn to contractors for detailed engineering during the design and procurement phase. In today's market, the work is frequently carried out under reimbursable type contracts. Under such contracts, the owner wants to be sure that the following equally important objectives are being met by the contractor:

Quality—The plant must meet the owner's specification as visualized and prepared during conceptual engineering.

Schedule—The plant must be completed on schedule so that the owner can meet obligation to his customers.

Cost—The plant must be built within the budget so that the owner can realize his estimated profits.

In order to verify that these objectives are being met, many owners assign a project-management team to be resident in the contractor's office. The team may consist of a project manager, several project engineers, a cost engineer, and perhaps a representative from the owner's procurement department. The cost engineer operates as a key member of the team. In many respects, he is the project manager's righthand man, because he and the project manager are the only persons on the team with responsibilities that cover the entire project. Most other members of the team are confined to their specialty.

Duties of the Owner's Cost Engineer

While resident in the contractor's design office, the duties of the owner's cost engineer fall into four major categories:

First: To monitor the work of the contractor's cost-engineering group. Actual cost control can only be carried out by the contractor's personnel. The owner's cost engineer should be completely familiar with the contractor's cost-control activities. Also, he must verify that the contractor's cost-control procedure is adequate to do the job, and equally important, that it is being followed.

Second: To advise the owner's management on cost related items such as: forecasts of special owner costs, cash-flow forecasts, and cost evaluation of contractor's project-execution proposals such as overtime vs. hiring additional personnel or renting vs. purchase of construction equipment.

Third: To administrate the project change-order system. With either a reimbursable or lump-sum contract, it is advisable to set up a formal change-order system. This system will be discussed in detail later in this article.

Fourth: To prepare a monthly cost report and forecast

for owner management. Higher management of most owner companies likes to receive a monthly project-status report prepared in a standard format. The owner's cost engineer prepares the cost section of this report.

The Project Budget

The project budget, main tool of the cost engineer, is a detailed job plan realistically reflecting the project-execution scheme. As the project develops, changes to the execution plans set forth in the budget should be documented by formal change orders which, upon approval, become part of the budget. The budget is the yardstick for measuring all project cost. It must be in detail so that:

- Each piece of purchased equipment can be compared with its budget.
- Each foundation and each pipeline can be compared to the budget for both quantity and price.

Attention to detail enables the cost engineer to control cost effectively. Each project development is measured against the budget. Cost control requires that overruns be flagged before they happen so that corrective action can be considered. This means that equipment pieces must be compared to the budget at bid tabulation–not after purchase. Piping runs and structural steel layouts must be checked before being sent to the fabricator. For example, it is seldom economical to reduce the size of a platform after the steel has been fabricated, or to change the route of a pipeline after the spools have been welded.

The contractor's cost engineers must make these detailed comparisons against the budget. Hence, the budget must be available early, before equipment is purchased and bulk designs are completed.

Sophisticated estimating techniques are necessary if a meaningful budget is to be prepared in time to exercise true cost control during the detailed engineering phase. Some owners and some contractors have these techniques. Others are developing them. A contractor who provides a detailed budget after quotations are in hand for all major equipment and detailed designs are completed for 70% of the bulk materials is not providing cost control. Such a budget can only reflect the contractor's own strengths and weaknesses. It will mirror the purchasing and design expertise of personnel assigned to the project. There will be no yardstick for measuring results against a norm.

Controlled Elements of a Project

During the detailed engineering phase, certain elements of the project are to be controlled. These can be classified into three major groupings:

1. Contractor engineering manhours and other office costs. The engineering manhour budget should be prepared early–say within two months of the start of engineering. It should be broken down by functions, such as purchasing, design, drafting, scheduling, etc. As work progresses, expenditures in each category should be compared to the budget and realistic overruns and underruns forecast.

2. Equipment purchases. The estimated cost for each major piece of equipment should be listed in the budget and compared to actual cost. Since the original purchase-order value seldom corresponds to the final cost of a sophisticated piece of equipment, it is usually necessary for the cost engineer to keep a brief cost history of each equipment piece.

3. Detailed design and purchase of bulk materials. Sharpness in the design of bulks, such as concrete foundations and structural steel, can save a great deal of money. Design quantities should be compared to the budget at the earliest possible moment. Quick corrective action in this area can often minimize what would otherwise become large overruns.

Corrective Action

The cost engineer's attitude is that corrective action to reduce a forecasted budget overrun is always possible. Of course, the ultimate corrective action is cancellation of part or all of the project. In the practical world, corrective actions are often worse than budget overruns. In these cases, budget overruns are accepted and included in the cost forecast. However, available lower-cost alternatives should be considered for all budget overruns. These alternatives should be ruled out only after proven undesirable.

Corrective action is illustrated by the following examples from a recent project:

Comparison of expenditures for local field-construction staff with the budget triggered an investigation that resulted in the reduction of 18 persons and a total savings of 157 man-months.

Comparison of the budget vs. bids for surface condensers revealed that the budget basis of epoxy-lined water boxes was lower cost than the bid basis of Cu/Ni lining. Alternative bids verified that appreciable savings could be realized by going epoxy. In this case, the final decision favored Cu/Ni for maintenance reasons, but the budget comparison clearly revealed a reasonable lower first-cost route that might have been followed.

Sampling a group of computer-designed concrete footings showed that the volume of all small footings averaged between two and three times the budget estimate. Investigations revealed that all footings were being set four feet below grade regardless of location or loading. Correction of this resulted in the savings of a considerable amount of concrete, forming, and excavation.

Cost Control During Construction

If a contract is fully reimbursable with a fixed construction fee, owner's cost control during construction is a must. Even in a target-cost situation, where underrun of the target is shared between contractor and owner, the cost-control effort needs some owner direction. In many construction organizations, the role of the cost engineer has been historically limited to cost reporting and estimating change orders. In many cases, the owner must take a strong hand if the benefits of cost control are to be realized in the field. Most construction organizations are schedule-oriented and fail to consider the productivity debit that goes hand-in-hand with overtime, or the very high cost of having surplus construction equipment on hand in case it is needed to prevent a minor schedule delay. Also, construction organizations have a tendency to purchase the newest, most elaborate equipment available without really investigating lower-cost machines to see if they will do the job equally well. When this philosophy is applied to cranes and bulldozers, the cost amounts to many thousands of dollars.

Prudent use of labor and construction equipment along with well-planned temporary facilities will save money for the owner. In order to get these, an owner needs a strong, knowledgeable field team to monitor the contractor's effort, and the cost engineer is a key member.

The Owner's Field Cost Engineer

The owner's field cost engineer has the same duties as his counterpart in the design office. On many projects, he can do double duty–serving in the office during detailed engineering, and relocating to the field for the construction phase.

One special duty of the owner's field cost engineer is to verify that contractor's schedule and cost forecasts are compatible. Construction-cost forecasts are closely tied to schedule. If these functions are handled by separate groups in the contractor's organization, discrepancies are sure to occur and may result in unanticipated costs for the owner.

As in the design office, the budget is the field cost engineer's principal tool. His approach is a detailed comparison of budget cost vs. actual cost for the field components. Major items to be controlled in the field are:

- Direct labor.
- Bulk materials including field purchases.
- Subcontracts.
- Construction overheads comprising: temporary facilities, i.e., buildings, roads, fences, storage yards etc., consumables and utilities; supervision and office staff; construction equipment; and small tools.
- Miscellaneous items unique to every construction project.

Productivity and Labor Forecasts

On a direct hire, reimbursable job the cost engineer's most difficult problem is forecasting direct labor. The owner should expect the contractor to provide forecasts of total direct labor updated on a monthly basis. Formerly, labor forecasts were frequently left to the field superintendents who worked mainly by experience. Now we have some procedures that usually work better than instinct, but input from experienced field personnel can not be ignored.

Each monthly labor forecast should be a combined effort between the cost engineer and the field superintendents. In the early stages of construction, estimating procedures seem to work better than experience; however, in the final stages, experience may give better results. The following tools are available for forecasting direct field manhours:

1. The budget. This is the first forecast.
2. Material takeoffs. The budget manhour outlook is modified to reflect actual designed quantities. We can call this "recasted field budget."
3. Productivity profiles. The recasted budget is modified to reflect actual construction experience.

% Budget manhours used											
	Theoretical	14	25	34	42	50	58	65	74	85	100
	Actual	17	27	36							

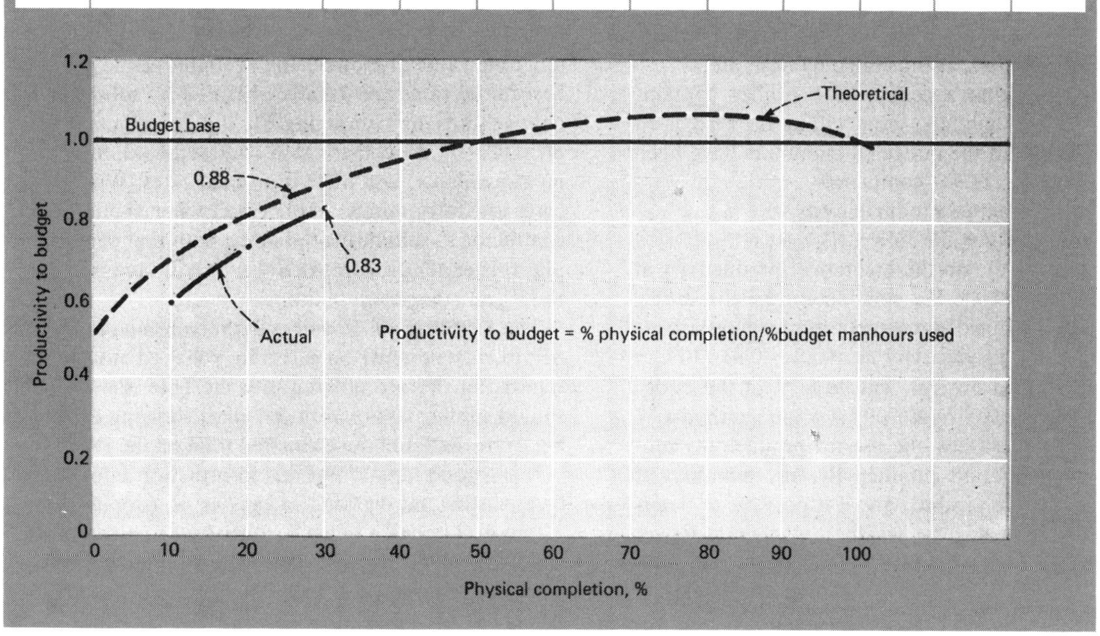

PRODUCTIVITY of aboveground piping labor compares budgeted manhours to actual manhours—Fig. 3

4. Crew sizes and layoff schedules. These become important during the final stages of construction.

A knowledge of productivity is the key to forecasting direct labor. Productivity can be measured by comparing actual production with some base or standard. For any given project, this base is the recasted manhour budget. By taking budget productivity equal to 1.0, we can define actual productivity as the budgeted manhours for any unit of work divided by the actual manhours used for that same unit of work. Since weightings used for calculation of percent complete are usually derived from budgeted manhours, it is also true that productivity for any work activity equals the percent complete divided by the percent of budgeted manhours used. For example, if the concrete activity is 80% complete and 100% of the budgeted manhours have been used, the productivity equals 0.80.

Productivity does not remain constant throughout a project but varies according to well-established patterns—sometimes referred to as learning curves, or more appropriately as productivity profiles.

The shape of productivity profiles is different for each of the various activities such as concreting, installing structural steel, and erecting pipe. These profiles are sensitive to methods used for calculating percent complete, and may vary between contractors. Typically, productivity for any activity starts out poor, reaches a maximum somewhere between 30 and 80% completion, and then tapers off to final completion. A typical productivity profile with the theoretical curve derived from an actual project is shown in Fig. 3.

The theoretical percent of budgeted manhours is derived from the theoretical profile in Fig. 3 as follows:

First: At 10% completion, we read the theoretical productivity from Fig. 3 as 0.70.

Second: We then divide the percent completion (10%) by the productivity (70%) to obtain the theoretical percent of budgeted manhours (14%).

What we have developed here is the theoretical rate of using up our budgeted manhours. By plotting the actual productivity and comparing it to the theoretical, we can forecast total manhours, as illustrated by example:

Let us suppose that a construction project contains 120,000 budgeted manhours. What will be the forecasted overrun when 36% of the budgeted manhours have been used, and the project is 30% complete?

We calculate the actual productivity as:

Actual productivity = 30/36 = 0.83

From Fig. 3, we obtain the theoretical productivity at 30% completion as 0.88. We now determine the difference between theoretical and actual productivities as:

% Difference in Productivity = (0.88 - 0.83)/100 = 5.7%. The forecasted overrun will be 5.7% of the budgeted manhours, i.e.: 0.057(120,000) = 6,800 manhours.

This example illustrates one use for productivity profiles. By plotting period productivity, say monthly, as well as accumulative productivity, it is possible to determine the impact of adverse weather, work slowdowns, overtime, changes of supervision, etc. Productivity profiles plotted for each of the major activities of crafts are the most useful tools of the cost engineer in the field. A contractor who does not use these profiles will be severely handicapped in forecasting labor on a large, complicated petrochemical or refining project.

Required Accuracy for Cost Estimates—Table I

Type of Estimate	Probability*
Preliminary For use during early conceptual engineering and preliminary economics	2 out of 5
Semi-detailed For use during final conceptual and very early detailed engineering	1 out of 5
Detailed The project budget for use during detailed engineering and construction	1 out of 10

*Probability of a cost overrun or underrun greater than 10%, number of chances out of total.

Control of Other Field Costs

A number of other field items require careful control. First among these are temporary facilities. When a construction contractor moves into the field, he should furnish the owner with sketches, layouts, and cost estimates for temporary facilities. After careful review, these estimates are accepted as the control budget. Thereafter, the estimates are used to forecast overruns or underruns. If the contractor later proposes items not in the budget, he should submit complete justification to the owner.

Construction equipment also requires careful field control. Generally, this equipment is rented; however on very large projects, purchase may be economical. Before making a decision to buy or rent, the owner should consider purchase price, rental rate, estimated time on the job, time value of money, and probable resale value. The best rental rates are usually obtained by soliciting competitive bids for supplying all the construction equipment. Often, the construction contractor will have equipment available, and offer it at fixed rates. When project costs are fully reimbursable, the owner should ask the contractor to submit his bid along with that of other suppliers. Frequently, competitive bids will save the owner 5 to 15% on equipment rental.

The numbers of persons on the contractor's field-supervisory staff must be carefully watched from the cost standpoint. Before moving into the field, the contractor should submit an organization chart showing each member of his staff and the estimated time on the job.

It is a good idea to put this information into bar-chart form, listing the various categories of people and estimated man-months in each category. As personnel move onto the job, the cost engineer compares the actual persons in each category with that estimated. Potential overruns can be quickly spotted on such a bar chart. When a potential overrun occurs, the work load of the individuals in that category should be examined to see if a possible savings can be made.

Unit-price subcontracts can also be trouble spots. In these cases, the cost engineer should estimate the number of units and forecast the total cost of the contract. Also, the forecast should include a reasonable allowance for extras. Subcontracts should be written with a clause stating that extras must be claimed when they occur and that retroactive claims will not be considered.

Other field costs that the cost engineer must watch are: computer costs, welder qualification, rework, backcharges, etc. The best way to control these miscellaneous overhead accounts is to follow them with tracking curves. From reviewing past projects, expenditure patterns can be developed. Each tracking curve should show the budget and the anticipated time/expenditure pattern. When actual expenditures are plotted on such curves, trouble spots can be located and investigated for possible corrective action.

Project Changes

For cost-control purposes, changes on a reimbursable project should be treated in the same manner as on a lump-sum project. Any revision in design or project-execution basis that differs from the budget basis should be covered by a formal change order. As soon as it is approved, the order is added to the budget. A formal documented change-order system keeps the budget alive and up-to-date with what is happening. Without continuous updating, the budget becomes obsolete and cost control as discussed here cannot be carried out.

The owner has direct control over changes, and should try to hold them to a minimum. When changes become excessive, they upset a contractor's work schedule and tend to demoralize his technical and supervisory personnel. This adds a real but intangible cost to any project.

Change orders should be governed by a formal written procedure approved by both the owner and contractor. The formal change order is normally a one-page document including:

a. Brief description of the change.
b. Reason and justification for the change.
c. Estimated cost impact.
d. Estimated schedule impact.
e. Approval of owner's project manager.

Good control for change orders will be obtained if the following procedure is followed:

Step 1: As soon as a change is contemplated, the owner's cost engineer initiates a change request and assigns it a change-order number. The contractor may recommend a change, but he should do no work on the change until he receives the official request.

Step 2: Contractor prepares a rough magnitude estimate of the cost and schedule impact of the change and advises the owner within two days.

Step 3: With the magnitude cost and schedule impact in hand, the owner gives a go-ahead or stop signal.

Step 4: Contractor prepares a detailed estimate of the cost and schedule impact within two weeks.

Step 5: With a more accurate appraisal of the cost and schedule impact in hand, the owner issues an approved change order. Or, he disapproves the change and the contractor halts all work on the change.

Step 6: If the change is approved, the contractor adds the amount of the change to his budget and the time impact to his schedule. If a change is disapproved, the contractor adds only the cost of the work already done.

Step 7: Owner deducts the amount of the change from his project contingency, or appropriates the funds from some other source.

Accuracy of Estimate

The question is often asked: How accurate must an estimate be for cost-control purposes? This question gives rise to further questions: How is the accuracy of an estimate measured,? If the final cost is greater than the estimate, was the estimate too low or was the project overspent?

One way to answer these questions is to state the accuracy of an estimate as the probability of an overrun or underrun of the more than 10%. To classify estimates on this basis requires a statistical history dating back several years. Most estimating organizations have historical data available covering estimates made at various stages of project development. For example, analyzing recent history might tell an organization that preliminary estimates made at the end of the evaluation and planning stage have three chances in five of overrunning by more than 10%. With a little further arithmetic, it can be determined what contingency should be added to reduce the overrun probability to one chance in five.

For cost-control purposes, the estimate must reflect the true project scope and the cost must represent achievable goals. Within this context, it is my experience that the accuracy of an estimate for cost control should achieve the probabilities shown in Table I.

In summary, a good cost-control program includes:

- A detailed budget reflecting actual project-execution plans.
- An accurate and up-to-date commitment register.
- Detailed and timely forecasts of total project costs.
- Provisions for comparing actual costs to budgeted costs and determining reasons for deviations.
- Provisions for taking prompt corrective action to minimize forecasted overruns.

If the contractor's procedures contain these points, and if his project-management and cost personnel carry them out, the owner can feel sure of getting cost control. #

Meet the Author

Forrest D. Clark is an Engineering Associate in the Project Management Dept. of Exxon Research and Engineering Co., P.O. Box 101, Florham Park, NJ 07932, where he coordinates the owner's cost-engineering activities on a major refinery expansion. He is a graduate of Purdue University and has a Master of Engineering degree from Yale University. His experience includes assignments in project and cost engineering in the Middle East, South America and the U.S. He is a professional engineer registered in N.J., and a member of the American Assn. of Cost Engineers.

Effects of Inflation and Escalation on Plant Costs

ALBERT C. SAVAY, Jacobs Engineering Co.

The unprecedented course of inflation and the quotation practices of suppliers over the past two years have created some extra problems for estimators of process and power plants. Though conditions have cleared, they are still confusing enough to warrant a good look at our experience, both past and present, as engineering-constructors.

We at Jacobs Engineering Co. have a plant cost index (JEC Index) that has a component weighting similar to the CHEMICAL ENGINEERING plant cost index (CE Index). As an engineering-constructor working on a wide variety of projects, we have the opportunity to routinely check the average distribution of components in the index.

The distribution of the JEC Index was 34% major equipment (called fabricated equipment, process machinery, pumps and compressors in the CE Index), 34% bulk materials (including pipe, valves and fittings, process instruments, electrical equipment, structural supports and miscellaneous, and buildings in the CE Index), 22% field labor and related costs, and 10% engineering during the volatile period from the third quarter 1973 to the third quarter 1974. Comparative values for the CE Index are 36%, 32%, 22% and 10%, respectively.

The chief difference between our own and CE's Index from 1958 to 1973 was that we did not observe an improvement in labor productivity of 2½% during that period. We acknowledge that new equipment and techniques have contributed to about 1% annual productivity improvement, and allow for that in our index.

Prior to mid-1973, the inflation rate for chemical process plants varied within a range of 3 to 5% annually. In

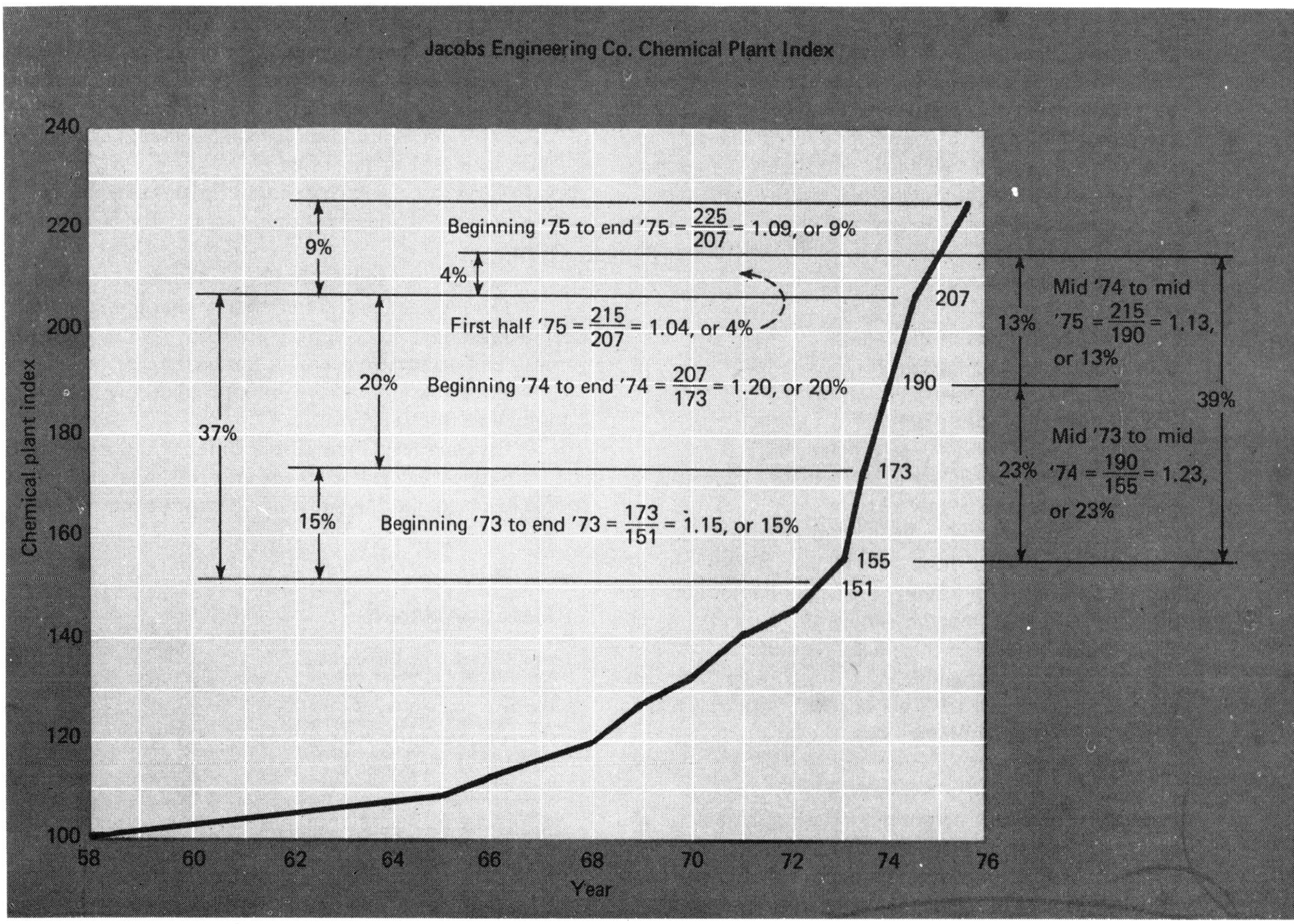

Originally published July 1975.

1973, however, a dramatic change in inflation rate occurred. The exact timing of the increase might vary from one company's index to another's. It is now obvious that a drastic increase in various plant-component costs started in the latter half of 1973. Demand surged, prices rose, and delivery times lengthened for both finished equipment and the materials and energy for making it.

According to the JEC Index (see accompanying chart), the cost of plants rose 15% from the beginning to the end of 1973. The rise continued on through early 1974, and took an upward turn in May 1974 after all wage-price controls were lifted. The index continued to gain momentum until the fall of 1974. The rise in the JEC Index from mid-1973 to mid-1974 is a startling 23%. Overall from the beginning of 1973 to the end of 1974, the JEC Index increased by 37%.

Background Causes for the Rise

Although many list prices remained relatively stable in 1973, discounts that normally applied were decreasing. Thus, government indexes, generally based on list prices, did not show the total increases. For example, a pipe fabricator no longer got an allocated shipment as large as before. Thereby, he lost a quantity discount for materials and the cheaper freight rate he normally paid for a barge load instead of a carload. The same pipe fabricator was decreasing the discount he offered from his shop labor list-price book to compensate for increased labor cost. The engineer-constructor was often surprised with incredible increases for fabricated steel, piping, etc., in rapid spurts of 10%, 20%, and more, over his expectations because fabricators had to pass on the new price structure immediately.

Apparently, the Bureau of Labor Statistics Wholesale Price Index has not provided an accurate measure of inflation for components in a chemical process plant. That index and some others cannot indicate the sudden jumps that occurred for some items of 10 to 100% in a month or two. These jumps had drastic consequences for many contractors unfortunate enough to hold fixed-price obligations.

On the other hand, producers in many industries who had experienced unprofitable histories for several years were given the opportunity to sell at more comfortable price positions. Through the first half of 1974, the heavy demand for expansion of power and chemical plants was far in excess of plant capacities for carbon and stainless steels, forgings, castings, fabricated steel products, motors, and others. Most suppliers now look upon gains of that period as "hard-earned price margins" and will be reluctant to reduce them.

Inflation vs. Escalation

Much of the current commentary on cost inflation of process plants should be read carefully for its true meaning. For example, prior to mid-1973, whether an index indicated year-end points or midyear points was relatively inconsequential because the annual change in the index was small. But rapid surges, over short periods, now make the calendar measurement important. An inflation index for "1973 plant cost" could easily be 10% lower if it represents a mid-1972 to mid-1973 change, instead of the change from the beginning of 1973 to year end.

An inflation index should not be confused with cost escalation of a particular process plant. The extension of delivery times and the now familiar "price in effect at time of shipment" vendor clauses have added a complicating dimension to cost forecasting. When we are told that "this plant now costs 50% more than it would a year ago," it does not mean that the annual index increased by 50%. The longer delivery times have stretched the center of gravity of the plant's cost to a later date. The bidders' escalation clauses, with labor-contract increases, etc., keep adding up during this extended period. Thus, a company that decided to postpone a project start from August 1973 to August 1974 might easily be faced with a plant cost increase of 45% during a period when the inflation index went up 26%.

Time Lag Causes Uncertainty to Index Data

The published data can be misleading to experts, also. At a December 1974 Workshop of the American Assn. of Cost Engineers, a speaker warned that one item in the BLS Wholesale Index was based on information from only two manufacturers. Due to this kind of background, many industry spokesmen were predicting an annual inflation of about 15% in the spring of 1974. Yet, the actual rate of inflation was well in excess of this, and destined to intensify.

Index readings are not acquired instantaneously. It takes months for the total impact of a highly complex market to be translated to a finite point on the curve. For example, the November 1974 issues of CHEMICAL ENGINEERING report the index reading for July, a four-month lapse. This time lag helped to contribute to the inaccuracy of early 1974 predictions.

Dampers on Inflation

Even as the plant-cost inflation was reaching its peak rate, leveling factors were in motion. Inflation, high interest, and decreasing rate of growth for power caused postponement and cancellation of many power-plant projects. Following closely behind, the needs for process plants, planned and already on the boards, were being reevaluated—many were delayed or cancelled. Thus, even as the full impact of equipment and materials inflation was being felt, the rate of change was diminishing. The last quarter of 1974 contained a slowdown in price rises nearly as dramatic as the previous year's increases. Firm bids are now commonplace. The inflation rate, though excessive by standards of the 1960s, is now running 1%/mo, or less.

Past practice was to include escalation directly in the various cost components of a definitive estimate for a chemical process plant (virtually all price quotations were firm through the shipping date). A double-digit inflation rate and prices that are generally good only for the month quoted, make it necessary now to calculate individual escalations for all items in an estimate. Present

Changes in Delivery Times for Selected Components

Component	Sept. 1974	Apr. 1975
Heavy-wall vessels	14 to 22 mo	9 to 14 mo
Packaged boilers	9 to 14 mo	9 to 17 mo
Centrifugal compressors	14 to 16 mo	10 to 16 mo
Switchgear	8 to 10 mo	6 to 7 mo
Fabricated stainless-steel tanks	14 to 20 wk	8 to 10 wk
Electrical instruments	40 wk	24 wk
Shell-and-tube exchangers	12 to 16 mo	10 to 13 mo
Kilns	11 to 13 mo	11 to 13 mo
Centrifugal pumps	6 to 12 mo	6 to 12 mo
Belt conveyors	6 to 7 mo	7 to 9 mo
Ball mills	18 mo	18 to 26 mo
Common-steel piping	10 to 50 wk	10 to 20 wk
Valves	30 to 70 wk	30 to 50 wk

practice for an engineer-constructor is to indicate a line item for escalation near the end of an estimate. Needless to say, this is now a major factor to be considered in the reviews of all estimates. It is likely to be the item of greatest controversy during reviews, especially on the big jobs—the ones with a long lead time to center of gravity of cost.

Shorter Delivery Time

The dramatic reduction in demand is reflected in some shorter delivery times, as indicated by our April 1975 survey (see accompanying table). Shortages of some machinery components such as gears for ball mills account for the longer delivery times for selected items, as shown in the table.

Many suppliers are willing once again to hold prices firm, some for a year or more. The year 1975 will apparently be one in which we will see a return to single-digit inflation, but still about twice the rate existing during the 1960s.

The rising cost of fuels has caused the Japanese to revert from the domestic-intensive marketing of the past decade to an export-minded attitude. At the December 1974 Workshop of the American Assn. of Cost Engineers, purchasing spokesmen for two major engineer-constructors noted that the Japanese offer good deliveries and firm prices on high-pressure vessels, turbines, special pipe, heat exchangers, and other items. They predicted that the Japanese will be reestablished competitively in our market similar to the 1960s. Our own purchasing department reports similar sales contact from Japanese firms. The infusion of this competition will tend to retain present price levels, if not reduce them, and help buyers negotiate shorter deliveries.

Despite the general downturn in demand, any prediction of a return to the pre-1973 inflation rates, or even single-digit inflation, on a long-term basis should be viewed with caution. Numerous spokesmen have stated that the U.S. can anticipate a shortfall of 25 million tons, or more, of finished-steel mill capacity by 1980. Some shortages of specific items still exist in 1975. These include steel bars, plates, stainless steel, forgings, castings, and electric motors.

Future Impact of Inflation

Significantly, many companies close to the problem have begun using 1973 as a base year for plant cost indexes. Of course, the 1973 base year emphasizes that we are in a new era of inflation rates. Since the base of 100 is chosen at 1973, an increase of say 25% inflation is indicated by only 25 points on the index. A chart with a starting date of 1958 such as the JEC Index or CE Index requires nearly 40 points to indicate the same 25% increase. This compounding effect of needing more index points each year for the same percentage rise makes the graphs of the older indexes tall and narrow.

If we are indeed in a continuous era of long-range high-inflation rates for chemical plant components, it is likely that most index makers will find 1973 a good reference point. We may well find it common practice to adjust the reference starting point ahead every decade if high inflation becomes the rule rather than the exception. Since potential shortages of critical items still exist, double-digit inflation cannot be ruled out as a strong possibility in the years ahead.

Minimizing Effects of Rising Costs

To lessen the impact of rising costs on process plants, we can suggest:

- Continuing emphasis on close client involvement in engineering and design of chemical process and power plants to speed up decisions.
- Using more standardized design—much less customizing of plant design.
- Using more identical multiples of plant units.
- Maximizing prefabrication.
- Specifying off-the-shelf equipment more often.
- Continuing efforts in the direction of closer integration of estimating, cost control and scheduling of new plants and plant expansions.

Getting early client approval is one of the most critical factors affecting the overall schedule and the ultimate project cost. We recognize, however, that client management cannot approve projects and make the proper decisions until they have an accurate cost picture, which must be provided by the engineer-constructor. #

Meet the Author

Albert C. Savay is the Chief Estimator for Jacobs Engineering Co., Pasadena, CA 91105, where he is responsible for estimating and cost control for its wide range of process plants. After receiving a civil engineering degree from the University of Southern California, he acquired project estimating and control experience with several large international engineer-constructors. He is an active member of the American Assn. of Cost Engineers, the American Soc. of Civil Engineers, and the Institute for the Advancement of Engineering.

Cost Escalation: Its Impact On Purchased Equipment

DON R. BONANO, Bechtel Corp.

Our purchasing department has established procedures to cope with rapidly changing prices. These also include special provisions in bidding documents, purchase orders and payment terms.

Although escalation provisions have evolved from rapidly increasing prices, we are now in a period where there has been some adjustment of prices downward. It is difficult to predict what will happen to prices over the next few years. For this reason, provisions for price adjustment in purchase orders both upward and downward are appropriate.

Let us examine the necessary procedures at various steps from the bidding stage to final receipt of materials or equipment.

At the Bidding Stage

Historically, we have always asked for firm-price bids. However, with rapidly increasing prices, our bid requests read:

"Firm quotations are preferred. If escalation must be quoted, the conditions shall be clearly stated.", or:

"Firm prices are desired. However, if not offered: (a) What is maximum percent of escalation?, (b) What is duration of escalation exposure?, (c) What is basis of escalation?"

On purchases from stock or for deliveries within approximately six months, we believe suppliers can quote firm prices. But for long lead-time items, an escalation clause to determine price adjustment may be the best basis.

We insist on establishment of a maximum percent of escalation so that purchase orders will provide some indication of what the final price will be. Most escalation formulas start at the time of bid, or earlier if bidding prices are based on costs from an earlier period. Escalation will normally end with shipment or at completion of manufacture. Material escalation may end on receipt of material at supplier's plant. This is an exception rather than the rule.

We use a number of documents as guides towards measuring the amount of price escalation. One of these may be the regular government indexes as published by the U.S. Bureau of Labor Statistics (BLS), or may be based on suppliers' own cost formulas for labor and material. In addition, current material prices for copper, steel, aluminum and other metal products are sometimes tied to prices published in trade publications, e.g., *Iron Age, Industry Week, American Metal Market, etc.*

This article is based on the paper, Procedures and Administration for Handling Escalation, presented at the Western Winter Workshop of the American Assn. of Cost Engineers, Carmel, Calif., Dec. 1974.

Bid Evaluation

On the basis of the foregoing procedures, one of the following is usually quoted:

- Firm prices.
- Price in effect at time of shipment.
- Prices subject to escalation based on BLS indexes.
- Prices subject to escalation based upon bidders' own cost formulas.
- Prices based on industrial publications.
- Prices based on any combination of the foregoing.

Firm prices on items with a long lead time—say over six months—have contingencies built into the price to protect the manufacturer from future unknown escalation. If all prices bid are firm, we must analyze the lowest acceptable bid to determine if the price is realistic.

Many supplier firms still quote only "price in effect at time of shipment" (no ceiling on price). This frequently is done by steel mills, other raw-material processors, and jobbers of these products. This type of bid is the least desirable, because there is no way to determine eventual cost. Policing the final result is difficult.

Although escalator clauses tied to BLS indexes are not new to U.S. industry, they were not used that much by the manufacturers we buy from except by some electrical manufacturers, wire and cable producers, etc. Now, escalator clauses are becoming more acceptable to manufacturers and buyers. Principal advantage for using BLS escalation clauses in purchase orders/contracts is that the eventual cost is determined from a set of impartial index figures that are published monthly. All such order/contract clauses will reflect both upward and downward movement of labor and material costs. The problem is: do the BLS indexes properly reflect supplier costs?

Many manufacturers and fabricators will not accept BLS escalation, and have devised their own "inhouse" formulas to average their costs, and to track the movement of these costs. This type of formula may be acceptable but generally takes more time and study to properly administer and audit.

When bids are received (reflecting all or part of the various methods of escalation previously discussed), a detailed examination and evaluation of the bids is required to determine the most attractive. When escalation is quoted with a maximum ceiling (usually expressed in percent of bid price per month through the duration of escalation), we then are reasonably assured of a maximum price. This type of escalation will gain preference over an unlimited escalation formula, e.g., price in effect at time of shipment. Either of these require that we estimate or project probable escalation in evaluating the bids.

Content of Purchase Orders

Once we have selected a supplier, the purchase order or agreement must be prepared to reflect the conditions of pricing agreed upon. A price-policy clause in purchase orders contains three basic parts:

Part 1. Prices are either shown as firm, or the method of price adjustment is shown in the body of the purchase order.

Part 2. The type of price adjustment clause agreed upon is then spelled out in detail. Depending on the basis of the bid, the different clauses reflect those based on (a) BLS Indexes, (b) prices in effect at time of shipment, and (c) manufacturers' standards.

Part 3. In all cases, a standard statement is included: "If escalation occurs, submit written justification for our approval. Do not invoice escalation portion prior to our approval."

This clause permits our buyers to thoroughly examine any increase requested by a supplier, and to determine if actual increases are within the estimates previously made as to the extent of the anticipated escalation. Then when invoices are submitted, there generally is no delay in our making prompt payment.

Clauses Based on BLS Indexes

A typical price-adjustment clause, based on BLS indexes, for purchase orders includes:

- Base price definition (usually unit price of each item).
- Labor index (selected from BLS publication, "Hours and Earnings").
- Material index (selected from BLS Wholesale Price Index).
- Base period (month for start of price adjustment, usually the month of bid).
- Final period (end date for price adjustment, usually month of shipment or period prior to shipment).
- The percent of base price subject to price adjustment for labor and for material. Typically, this will be around a 50-50 split.
- The adjustment formula and procedure must be stated: "Determine the percentage change between the base and final periods for the Labor and Material Indexes."
- Maximum adjustment, where agreed upon, will be stated (usually in percentage, or fraction of a percent, per month from base to final periods).
- A statement that price-adjustment amounts are to be invoiced separately, upon publication of BLS indexes for the final month of price adjustment.

Clauses Based on Shipping Price

A price-adjustment clause for purchase orders, which state: "price in effect at time of shipment" should include:

- Reference to the applicable price sheet, with date and applicable discount; catalog number and date; or other published reference to establish the basis for purchase-order prices.
- Statement that all requests for price adjustments be supported by revised price sheets, catalog prices and/or discounts.
- Statement that all requests for price adjustment are subject to buyers' approval prior to payment.

It may be necessary to negotiate the following provision: "Buyer reserves the right to audit seller's books to verify requested price increase."

These statements will go a long way in eliminating the wide-open implications of a price-in-effect type order. They give the purchaser better control over exorbitant requests, and alert the supplier that he is obligated to substantiate his requests for increases, and preferably prior to shipment.

We have experienced some problems with suppliers on getting justification and have, therefore, had to hold up payments. We also recognize that it is very difficult for distributors and supply houses to substantiate and justify increases that their sources pass on to them. This type of order is very undesirable because of increased administration and exposure to unlimited increases.

Approving Escalation

The final step in the process of price adjustment comes after shipment or at the final period. The supplier requests approval of escalation by letter or invoice, giving

Typical Price-Adjustment Formula

Price adjustments (PA) may be determined according to:

$$PA = \left(\frac{FL - BL}{BL}\right)L + \left(\frac{FM - BM}{BM}\right)M$$

where FL = labor index for final period; BL = labor index for base period; FM = material index for final period; BM = material index for base period; L = % of base price for labor; and M = % of base price for material.

An example will illustrate the use of this formula. Let us suppose that a purchase order in Jan. 1973 for an equipment item has a base price of $10,000 for delivery in Jan. 1975. Data for these conditions are:

BL = BLS Code 34, Jan. 1973	$4.14
FL = BLS Code 34, Jan. 1975	$4.78
BM = WPI Code 1013, Jan. 1973	132.6
FM = WPI Code 1013, Jan. 1975	195.9
L = % base price for labor	35%
M = % base price for material	55%

$$PA = \left(\frac{4.78 - 4.14}{4.14}\right)35 + \left(\frac{195.9 - 132.6}{132.6}\right)55$$

$$PA = 31.6\%$$

Since the base price is $10,000, the price adjustment becomes (10,000)(0.316) = $3,160. The revised price will therefore be:

$10,000 + $3,160 = $13,160

the amount involved and supported by index figures or other data called for in the purchase order.

The buyer examines the supporting data; checks the BLS index figures with our own internal recording of these figures, and if in agreement, notifies the supplier of approval to invoice. Many projects, for accounting and cost-control purposes, require a change to the original order to establish the revised prices.

On purchase orders having price-adjustment clauses based on suppliers' own cost formulas or on price-in-effect orders, a more detailed analysis is needed. We often check requested increases against BLS Indexes, our own price trend charts, industrial publications, our cost engineers' data, etc. We publish monthly (for internal use) price-trend charts on 30 key commodities showing our history of prices paid, and projecting futures trends through a year ahead. We also collect data monthly on 200 other commodities for our cost engineers.

New Procedures

On most projects we document the actual increased escalation costs by issuing purchase-order revisions. We are now considering the elimination of formal purchase-order revisions by using a "Commitment Authorization Form" developed by our finance and accounting department. This form is used internally to obtain approvals for total commitment amounts, beyond the dollar amount originally stated. This includes authorization for projected amounts added by escalation clauses. This form may speed up the approvals and allow accounting to pay invoices without a purchase-order revision and without undue delay.

In addition, a standard price-adjustment clause using BLS indexes has been developed. This is completed by our buyer and sent out with bid requests. Depending on the commodity to be purchased, applicable labor and material indexes are filled in, together with the percent of each to apply. A maximum monthly percentage of escalation is also entered. The purpose of these standards is to get our bidders to quote on the same basis, reducing the need for detailed analysis of dissimilar price-adjustment clauses.

One typical formula for determining price adjustment is outlined in the accompanying box. #

Meet the Author

Don R. Bonano is manager of purchasing for Bechtel Corp., P.O. Box 3965, San Francisco, CA 94119. In his position, Mr. Bonano directs the activities of purchasing agents and buyers in San Francisco for worldwide projects of the Bechtel group of companies involving oil refineries, chemical plants, power plants, pipelines, steel plants, and a variety of other industrial projects.

A Method for Forecasting The CE Plant Cost Index

DAVID W. CALDWELL and J. HARVEL ORTEGO, Monsanto Co.

The CE Plant Cost Index was introduced in 1963 by CHEMICAL ENGINEERING as a measure of the cost for a typical chemical plant. It is computed as the weighted average of 70 Bureau of Labor Statistics (BLS) price indexes for commodities and materials, five BLS indexes of wage rates, and an index of engineering costs. The index includes the costs of equipment, material, buildings, erection and installation labor, and engineering. The index does not include the cost of land, site preparation, insurance and taxes during construction, and contractor's profit. The index appears monthly on a historical basis. No forecasts are published.

Accepting the CHEMICAL ENGINEERING Plant Cost Index (PCI) as a representative measurement of escalation, our problem is to find a better way of forecasting its movement.

This article is based on the paper, Forecasting Capital Project Escalation in the Chemical Industry, presented at the first annual conference of the American Assn. of Cost Engineers/American Institute of Industrial Engineers in Houston, Apr. 10–11, 1975; and also presented at the 19th annual meeting of the American Assn. of Cost Engineers, Orlando, Fla., June 29–July 2, 1975.

Originally published July 1975.

One useful way of gaining insight into the movements of the PCI is by comparing its history to broad price indexes (macro-economic approach) for the entire economy such as:

- Gross National Product Deflator
- Plant and Equipment Deflator
- Wholesale Price Index (Overall)
- Wholesale Price Index (Industrial Goods)
- Consumer Price Index

Comparisons of the PCI with these broad indexes is best made over a long period of time to allow for historical interpretation. The PCI might be compared with these indexes by plotting the annual average values for each on the same scale.

Correlation between two time series can also be observed graphically by plotting the values of one series vs. the values of the other series, e.g., the PCI and the gross national product (GNP) deflator. (The GNP deflator is the index used to convert the nominal GNP in current dollars to the real GNP in constant dollars.) Such a crossplot indicates that the PCI rose more rapidly than the GNP deflator in the 1950s and in 1974. A crossplot of the PCI with the Plant and Equipment deflator exhibits almost a 1 : 1 relationship. Once again, 1974 is the exception.

Numerical methods may also be used to model the relationship of a price index to one of the broad economic price indexes. A good fit of the historical data was obtained when the PCI was related to economic indexes by using the simple linear model:

$$(PCI) = bI$$

where I = an economic index, and b = a constant. The results for five comparisons based on 1948–1974 annual average values are:

Economic Index, I	b	Correlation Coefficient
GNP deflator	1.05	0.98
Plant and equipment deflator	1.01	0.99
Wholesale price index (WPI)	1.00	0.81
WPI (industrial goods)	0.99	0.87
Consumer price index	1.05	0.97

In all cases, the fitted value of b was close to 1.00, suggesting that escalation permeates all of these indexes.

Although the long-run relationship is approximately 1 : 1, there is an apparent pattern to deviations from this relationship as suggested from the results of a regression analysis for the PCI vs. GNP deflator.

This pattern may be analyzed by calculating the percentage changes in each index for each year and the ratios of these percentages. The ratios express the relative rate of escalation of the two indexes over time. The ratios may be plotted, and a 10-year moving average drawn through the values. This analysis indicates that the relative rate of escalation, while averaging 1.0 over the entire time span, has cycled about that value and is now trending toward values greater than 1.0.

The primary benefit of establishing a relationship between a specific index of escalation such as the PCI and a broad aggregate index such as the GNP deflator is that:

The broad aggregate indexes are regularly forecasted by sophisticated econometric models. These forecasts are made by academic and commercial organizations such as the Wharton Business School and Data Resources Inc. Therefore, if a 7% rate of inflation is forecast for the overall economy for 1975–77 by the econometric models, the simple relative rate correlation would indicate a 9–10% rate in the PCI.

Forecast PCI Through Component Analysis

Experience has shown that it is not advisable to rely solely on aggregate econometric forecasts of escalation. A good alternative approach is to analyze the movements of significant components (micro-economic approach) of the index to be forecast.

The CHEMICAL ENGINEERING Plant Cost Index is subdivided into a number of major components. In turn, these components are based on the 70 Bureau of Labor Statistics price indexes (WPIs), wage rate indexes, etc. (See *Chem. Eng.*, Feb. 18, 1963, pp. 143–152 for complete details.)

Many of the WPIs are used in more than one PCI component, and many of the WPIs are similar in nature. Therefore, it is reasonable to expect that most of the movement of the PCI can be explained or forecast by the movement of a relatively small number of indicators such as those in Table I.

A number of possible combinations of these indicators have been tried in order to obtain a regression equation for the PCI, using quarterly data for the years 1963–74.

PCI Components by WPI Code Weightings—Table I

PCI COMPONENT	Wholesale Price Index Code*							Total
	10-13-XX	10-72-XX	11-4X-XX	11-7X-XX	XX-XX-XX	Labor	Engineering	
Fabricated equipment	8.42	8.24			0.90	5.01		22.57
Process machinery	1.24		0.53	0.30	4.55	1.92		8.54
Pipe, valves, and fittings	4.81		4.44		2.95			12.20
Process instruments and controls	0.32			2.28	0.58	1.09		4.27
Pumps and compressors			4.27					4.27
Electrical equipment and material				2.84	0.21			3.05
Structural supports, insulation and paint	2.50				3.60			6.10
Erection and installation labor						22.00		22.00
Building, materials, and labor					3.71	3.29		7.00
Engineering and supervision							10.00	10.00
Total	17.29	8.24	9.24	5.42	16.50	33.31	10.00	100.00

*WPI Code	Description
10-13-XX	Steel-mill products
10-72-XX	Metal tanks
11-4X-XX	General-purpose materials and equipment
11-7X-XX	Electrical materials and equipment
XX-XX-XX	Construction materials

Escalation Calculation for XYZ Project—Table II

Year—Quarter	PCI	Inflation, Annual Rate	Base Cost, $M	Escalated Cost, $M
73-4	147.5	—	10,000	—
74-4	177.0	+20%	—	—
75-4	204	+15%	—	—
76-1	208	+10%	100	141
76-2	214	+10%	100	145
76-3	219	+10%	800	1,188
76-4	224	+10%	1,010	1,534
77-1	228	+10%	1,040	1,608
77-2	233	+10%	1,220	1,927
77-3	237	+10%	1,570	2,523
77-4	242	+8%	1,500	2,461
78-1	245	+8%	1,210	2,010
78-2	249	+8%	730	1,232
78-3	253	+8%	540	926
78-4	256	+6%	180	312
			10,000	16,007
			Forecast escalation . . .	**6,007**

An almost perfect fit was obtained with only five independent indexes in the simple model:

$$PCI = 0.22\,WPI_{10\text{-}72} + 0.22\,WPI_{11\text{-}4} + 0.11\,WPI_{11\text{-}7} + 0.20\,WPI_{SOP} + 0.25L$$

where $WPI_{10\text{-}72}$ = wholesale price index: metal tanks, $WPI_{11\text{-}4}$ = wholesale price index: general purpose machinery and equipment; $WPI_{11\text{-}7}$ = wholesale price index: electrical machinery and equipment; WPI_{SOP} = wholesale price index for stage of processing: materials and components for construction (aggregated by BLS); and L = Chemical Engineering construction labor index. Note: the year 1967 = 100 for all indexes. The correlation coefficient for this equation is 0.999, and the standard error = 0.396. Of the 48 quarterly values, the model fit 40 points within 0.3% and all within 0.6%.

This model may be used in several ways for forecasting. The WPIs for the model are easily related to company procurement data. Buyers can be asked to project movements in these components based on their familiarity with supply contracts, market conditions, and suppliers' positions. If more detailed backup is desired, the survey can be extended to all of the 70 WPIs making up the PCI.

In early 1975, forecasts of specific wholesale price indexes such as those in the five-component PCI model became available from economic forecasting services. The accuracy of these forecasts is not yet established. Optimistically, these specific forecasts will be more useful than the broad aggregated models of the entire economy.

Alternatively, the forecaster can develop forecasts of the desired specific WPIs. This can be done by relating price movements in the desired WPI to cost of input materials and measurements of the demand for the WPI category, e.g., production indexes and order backlogs of industries using the item.

In order to derive benefits of systematic forecasting, it is necessary to have an equally systematic procedure for using the forecast.

A cost engineer is assigned to develop an estimate for a new XYZ plant with design beginning in January 1976. Using cost data for the fourth quarter of 1973, the base cost is estimated as $10 million.

The cost engineer now makes a forecast of the PCI, as shown in Table II. In order to use the forecast systematically, the base cost of $10 million is spread over the three-year project life by quarterly periods. The distribution is based on past experience with similar XYZ projects. To calculate the escalated cost, the cost engineer multiplies the expenditure for each quarter by the ratio of the forecasted PCI for that quarter to the PCI for the base period, as shown in Table II.

Since estimates of escalation are likely to be less exact than the base-cost estimate, a similar calculation should be made using the potential maximum rate of escalation. This calculation defines an additional amount termed the escalation contingency, which is added to the escalated cost of Table II. Choice of a potential maximum annual rate of escalation will depend on the expectations of each company. #

Meet the Authors

◀ **David W. Caldwell** is a senior engineer for financial and operations analysis in the Corporate Engineering Dept., Monsanto Co., 800 N. Lindbergh Blvd., St. Louis, MO 63166. Mr. Caldwell has a B.S. and M.S. in chemical engineering from Washington University and the University of Wisconsin, respectively, and an M.B.A. from St. Louis University. He is a member of AIChE, American Assn. of Cost Engineers and the Engineers Club of St. Louis.

J. Harvel Ortego is superintendent for financial and operations analysis in the Corporate Engineering Dept. of Monsanto Co. Mr. Ortego has a B.S. in chemical engineering from Louisiana State University, an M.B.A. from St. Louis University, and is working toward a doctorate in economics at St. Louis University. He is a member of the American Assn. of Cost Engineers and the St. Louis Economics Club. ▶